Human Physiology
Foundations & Frontiers

In the Möbius strip, the continuity of the loop is maintained
despite the continuous shifting of its plane. We chose the Möbius
strip as a physiological motif because in a homeostatic state,
the internal environment is not kept absolutely constant,
but rather is held within limits of fluctuations that the body
can tolerate without disruption of function. The Möbius strip
is constructed from a rectangular strip by holding one end fixed,
rotating the opposite end through 180° and attaching it
to the first end.

Human Physiology

Foundations & Frontiers

CHARLES L. SCHAUF

Professor and Chairman
Department of Biology
Indiana University -
Purdue University at Indianapolis

DAVID F. MOFFETT

Associate Professor
Program in Zoophysiology
Department of Zoology
Washington State University

STACIA B. MOFFETT

Associate Professor
Program in Zoophysiology
Department of Zoology
Washington State University

TIMES MIRROR/MOSBY COLLEGE PUBLISHING

St. Louis • Boston • Los Altos • Toronto   1990

Publisher: Edward F. Murphy
Editor: Deborah Allen
Project Manager: Patricia Gayle May
Production Editor: Donna L. Walls
Book Designer: Susan E. Lane
Illustrators: Barbara DesNoyer Stackhouse
　　　　　　　Gail Morey Hudson
　　　　　　　Graphic Works Inc.
Cover Photograph: William Leslie/Double Vision Photography

Credits for all materials used by permission appear after the glossary.

Library of Congress Cataloging-in-Publication Data

Schauf, Charles L.
　　Human physiology: foundations and frontiers / Charles L. Schauf,
David F. Moffett, Stacia B. Moffett.
　　　　p.　　cm.
　　Includes bibliographical references.
　　ISBN 0-8016-4321-X
　　1. Human physiology.　I. Moffett, David F.　II. Moffett, Stacia
B.　III. Title.
QP36.S32　1990　　　　　　　　　　　　89-39023
612—dc20　　　　　　　　　　　　　　　CIP

Cl/VH/VH　9　8　7　6　5　4　3　2　1

Dr. Charles L. Schauf received a B.S. degree with honors from the University of Chicago in 1965 and a Ph.D. degree in physics in 1969. Following a 2-year postdoctoral fellowship at the University of Maryland, Dr. Schauf was appointed Assistant Professor of Physiology at Rush University, Chicago, in 1972. He was promoted to Associate Professor in 1975 and to Full Professor in 1979. At Rush, Dr. Schauf directed the program in physiology, organized a human physiology course for nursing students, and eventually assumed administrative responsibility as Associate Chairman of the Physiology Department. Dr. Schauf assumed the chairmanship of the Biology Department at Indiana University—Purdue University at Indianapolis in 1986, where he currently teaches a 2-semester sequence in human physiology and endocrinology to a broad range of undergraduates. Dr. Schauf's research is in the general area of neurophysiology and membrane biophysics and is supported by the National Science Foundation and the National Multiple Sclerosis Society. Dr. Schauf has published over 80 research papers and has served on several national advisory review groups. In his leisure time, Dr. Schauf enjoys backpacking, cycling, and tennis.

Stacia Moffett and **David Moffett** received their Ph.D. degrees in Biology from the University of Miami in Coral Gables. They did postdoctoral training in the Departments of Physiology, Pharmacology, and Biomedical Engineering at Duke University. Currently at Washington State University, they teach human, mammalian and comparative physiology in the Department of Zoology and teach in the Graduate Program in Zoophysiology. Their research, funded by the National Science Foundation and the National Institutes of Health, emphasizes the use of invertebrates for learning about basic physiological processes. Stacia studies neurobiology and endocrinology; especially regeneration in the central nervous system and muscle and, also, the control of reproduction by neuropeptides in molluscs. David, a membrane physiologist, studies the cellular mechanisms of transepithelial ion and nutrient transport in both vertebrate and invertebrate animals. In addition to teaching, research, and writing, they enjoy hiking, sailing, and music with their three children.

Preface

*A*s the frontier of our scientific understanding expands daily, teachers face a stimulating challenge of presenting information within a conceptual framework that will allow students to understand basic physiological principles as well as provide a sound foundation for future learning. Discoveries in both pure and applied research areas are providing information that improves health care and at the same time opens new fields of inquiry. Biochemistry, immunology, and cell and molecular biology all originated from physiology in this manner.

The goal of human physiology is to understand the mechanisms that underlie the normal function of the cells, tissues, organs, and organ systems of the body. Physiological questions are asked at all levels of structural complexity. At the microscopic level, physiologists investigate how materials move across cell membranes and examine how proteins change their shape and catalyze biochemical reactions that are the molecular basis of life. It is now known that integral proteins form ion-selective, water-filled channels across cell membranes. Opening or closing these channels by voltage, physical stimuli, or chemicals such as hormones and neurotransmitters, governs nearly everything cells do, including cell division, transmission of nerve impulses, memory, glandular secretion, immune cell activation, and antibody production. Neuroscientists are finding out how the brain processes sensory input and controls movement and are making progress in understanding how brain biochemistry contributes to the neural basis of normal and abnormal human behavior. Neuroimmunologists have discovered lines of communication between the nervous and immune systems that may explain why some illnesses are associated with stress.

We wrote *Human Physiology: Foundations & Frontiers,* for today's students of human physiology. Many of those students are preparing for a career in health-care delivery or laboratory research. For them, an excellent foundation in human physiology is essential. Many other students simply are curious about how their own bodies work. For all our students, our goal is to instill the sense of excitement inherent in the study of human physiology and to provide a strong conceptual framework that will form the foundation of their ever expanding appreciation of body function through the 1990s and beyond.

DEVELOPMENT OF THIS TEXT

Many textbooks of human physiology have been written for undergraduates over the years. Why another? The answer is simply that a better job can be done.

The first step in the development of *Human Physiology: Foundations & Frontiers,* was an extensive market survey that identified the actual needs of undergraduates in various programs. We then talked with instructors of human physiology across the country and studied course outlines from various schools and programs. We used this information to guide the organization of our text. For example, most professors cover nervous system physiology very early in their course, while reproductive physiology is usually covered toward the end.

To meet the challenge of preparing a text that is thoroughly up-to-date, we drew on all possible resources. We combined our own experiences in teaching undergraduate physiology. We drew heavily on our own research backgrounds and interests and on information available in state-of-the-art reviews. Finally, we benefitted greatly from scientific feedback and updates from our colleagues.

Preliminary drafts of the text were examined by panels of reviewers selected to reflect a spectrum of users. The reviewers were asked to note any inaccuracies and to make suggestions for readability and clarity. This input not only served to ensure the accuracy of our presentation, but also helped to shape *Human Physiology: Foundations & Frontiers* to best suit its teaching goals.

In the last stages of development, we asked yet another panel of reviewers to read through the final manuscript to ensure that no errors had been inadvertently introduced. Finally, special illustration reviewers looked at each piece of labeled art to guarantee its accuracy, its quality, and its integration with textual material.

This meticulous developmental process has produced a quality text that provides the background undergraduates need as they face various educational goals.

WHAT IS DIFFERENT ABOUT THIS TEXT

National surveys show that many different types of physiology texts are used in courses with similar descriptions in college catalogs. Many of these provide outdated explanations of basic physiological processes, and few texts integrate recent advances with the core principles needed to understand the normal functions of the body and how these can be altered by disease.

Human Physiology: Foundations & Frontiers has been written with the idea that teaching physiology in the 1990s requires a readable textbook that presents basic principles and recent discoveries within a framework that facilitates students' ability to learn outside the classroom and then to retain what they have learned long after final exams are over.

Firm Foundations

One challenge in writing an undergraduate physiology text is to foster in students an understanding of basic mechanisms and encourage them to use this knowledge to solve complex physiological problems involving the interaction of several organ systems. Many students taking physiology for the first time have had limited exposure to the physical and chemical principles underlying physiology, or find that their courses presented such concepts too abstractly for them to be applied to physiological problems. *Human Physiology: Foundations & Frontiers* is self-contained because basic cell biology and chemistry as well as mathematical and physical concepts are presented in its initial chapters using physiological examples. This unique *Foundations Unit,* encompassing Chapters 2, 3, and 4, serves both as a useful review for students of physiology and also as a highlighted reference source throughout the course.

Conceptual Approach

Throughout *Human Physiology: Foundations & Frontiers,* concepts are presented systematically to encourage students to think about how organs systems interact with each other. New material is introduced without clinical jargon, and descriptive terminology is substituted for complex mathematical notation. In addition to the incorporation of up-to-date explanations of basic physiological processes, each chapter includes carefully chosen essays that describe recent scientific discoveries. We have chosen this approach because it indicates to beginning students the dynamic nature of physiology. Indeed, because many of these topics treat current events, they convey to the students the exciting field of modern physiology as it is unfolding in their own world.

A second challenge in developing an undergraduate human physiology text is to present a level of sophistication appropriate for the undergraduate. Full-length medical physiology texts assume students have had a standard premedical curriculum and are too imposing for an initial undergraduate course. Condensations of medical texts are not a solution for the undergraduate because the basic approach remains unchanged. Other undergraduate human physiology texts err in the other direction by stressing facts, not concepts. No real understanding can come from memorization of facts outside of a conceptual framework, and the field of physiology is changing so rapidly that such training quickly becomes obsolete. It is our belief that the study of physiology should be based on an understanding of principles so that the student can continue to build on this background long after the course is completed.

Although undergraduate texts should concentrate on normal physiology, many nonmedical texts neglect the many examples of physiological principles that nature has provided through the existence of disease. The intent of this text is not to catalog specific illnesses or to introduce clinical jargon, but to introduce abnormal conditions when their presentation contributes to the understanding of basic physiological processes. This challenges students to synthesize information and begin applying the course content to everyday experiences.

Up-to-Date Coverage

It is surprising that many texts fail to discuss up-to-date models of cellular and organism function. This is a disservice to students because studying from such texts does not foster an ability to think physiologically. Students may remember facts, but they remain unable to solve problems in which organ systems interact or to approach new scientific challenges later in their careers. For example, the traditional view on the kidneys was that the sodium-chlorine transport system was solely responsible for the ability to produce a urine more concentrated than the plasma, thus conserving water. This is the explanation found in many current texts. However, it has been shown that a differential permeability to urea, the primary breakdown product of proteins, plays an equally important role (Chapter 19). The patch-clamp technique was discovered nearly a decade ago and has led scientists to believe that transmembrane ion channels are intimately involved in nearly all physiological processes (Chapter 8). It is still not mentioned in other undergraduate textbooks. Likewise, there is no mention of caldesmon, which is believed to be responsible for the ability of smooth muscle to maintain tension for long periods with little expenditure of energy (Chapter 12). Many texts discuss the cardiovascular and ventilatory changes during exercise, but few discuss the recent studies suggesting that these responses are driven by the central nervous system (Chapter 18). Further integrative information about exercise physiology is provided in the *Focus Unit* following Chapter 18.

Student Usability

For maximum utility, a text must not only present up-to-date information; it must also engage student interest and be easy to read. *Human Physiology: Foundations & Frontiers* has a lively style, with analogies used to assist in presenting difficult material. As each organ system is covered, key concepts are carefully related to those systems previously studied so that students acquire an ability to think in physiological terms about how organ systems interact. Most cross-references provide page numbers to direct students to specific related portions of text. New advances, often selected from recent stories in the news media, are incorporated into the body of the text or are presented as *Frontiers* essays and used as a further means of exciting students about their study of physiology.

The level of difficulty in the text has been carefully controlled through extensive reviewer feedback to correspond to the range of abilities of undergraduates studying physiology for the first time. The text avoids the difficulties created by unfamiliar terms by using **boldface** and by fully defining all terms when they are first used. Further important words and phrases (keywords) have been included in an extensive alphabetical glossary with page references to the primary text discussion. *Human Physiology: Foundations & Frontiers* substitutes, wherever possible, descriptive terminology for complex mathematical notations. A more mathematical treatment may be desired in the area of acid-base balance, and it is provided in the form of a *Focus Unit* following Chapter 20. This allows the individual instructor flexibility of coverage.

In summary, with input from many other instructors of human physiology, we have tried to produce a lively and up-to-date text that can be used effectively by today's undergraduates, regardless of their scientific background, to develop an understanding and appreciation of the science of human physiology.

ORGANIZATION

Human Physiology: Foundations & Frontiers is divided into seven sections. Section I presents the Foundations of Physiology. The first chapter discusses the origins of the science of physiology and introduces the encompassing concept of homeostasis in physiological systems. It also introduces the tissue and organ systems of the body to provide a basis for the discussion of structure-function relationships found later in the text. Chapters 2, 3, and 4 encompass the highlighted *Foundations Unit*, which presents introductory concepts in cell biology, chemistry, physics, and mathematics with appropriate physiological examples. Chapter 5 discusses feedback mechanisms and control system loops involved in the maintenance of homeostasis. Local, neural, reflex, and hormonal control are introduced at this point in the text because they are fundamental to the integration and regulation of all body systems. Finally, Chapter 6 discusses transport across cell membranes and the establishment of an electrical resting potential in all living cells.

Section II begins with a discussion of excitable membranes in Chapter 7, followed by coverage of neurophysiological principles of transmission and integration in Chapter 8. This information is then used to show how stimulus-sensitive ion channels allow sensory receptors to convert external events to an internal neural code, and how the central nervous system processes sensory signals (Chapters 9 and 10). The body usually reacts to the environment by movements, changes in visceral muscle function, or secretion from glands, each mediated by components of the somatic and autonomic motor systems (Chapter 11). Chapter 12 presents the physiology of skeletal, cardiac, and smooth muscle.

Section III examines the functional interconnection of organ systems by the cardiovascular system. The properties of individual elements of the cardiovascular system—heart, arteries, capillaries, and veins—are covered first, followed by a description of neural and hormonal control of the cardiovascular system. Because the cardiovascular and respiratory systems combine to deliver oxygen to cells and remove carbon dioxide, the treatment of the respiratory system follows the cardiovascular system in Section IV. Renal physiology is covered in Section V because the kidney regulates salt and water balance and helps adjust blood pH. These functions relate directly to cardiovascular and respiratory functions.

Sections III through V of *Human Physiology: Foundations & Frontiers*, covering cardiovascular, respiratory, and renal physiology, compose almost a third of the text and, we believe, fittingly so. Not only do these systems include the "heart" and "breath" of modern physiology, but abnormalities in these three systems are responsible for the majority of human illnesses. Undergraduates who are planning careers as health professionals will benefit from the care we have taken to present this material at a level of detail that is both comfortable and complete.

Section VI covers gastrointestinal physiology. It begins with a description of the motility and secretion characteristics of each region, along with the neural and hormonal control of these processes. Next, students learn how the basic food groups are broken down by the gastrointestinal tract into nutrients that can be carried in the blood. The section concludes with the mechanisms and implications of nutrient absorption.

The first chapter of Section VII follows absorbed nutrients as they enter pathways of synthesis and storage during the absorptive state. It then shows how nutrients are released from storage during the postabsorptive state. The next chapter shows how the immune system protects the body from various potentially destructive invaders. The last chapters of this section discuss reproductive physiology as well as development, birth, and the physiology of the newborn.

Endocrine Coverage

Most physiology texts cover the subject of endocrinology on a gland-by-gland basis. In contrast, we have chosen to describe general endocrine mechanisms early in the text and to reserve discussion of individual hormones until they are encountered in the context of other systems. Throughout Sections III through VII of *Human Physiology: Foundations & Frontiers* there is an emphasis on how organ systems are regulated and how they interact. Neural and hormonal mechanisms are integral to this coverage. For this reason, we discuss each of the hormones and their effects as they are encountered in our presentation of systems physiology. We believe that this synthesis better enables students to understand the interactions among systems and the role each plays in the overall homeostasis of the body.

A supplement to this approach is provided by the *Focus Unit* on endocrine glands located at the end of Chapter 5, which presents, in a more traditional form, the location and function of endocrine glands. This overview allows students to grasp the entirety of the endocrine system and is also a valuable reference throughout the course.

TEACHING AND LEARNING AIDS

Presenting accurate and up-to-date information was not our only goal in the development of *Human Physiology: Foundations & Frontiers*. To be an effective teaching tool, a text must also present the information in a way that can be clearly taught and easily understood. To that end, we have included various teaching and learning aids.

Focus Units

We have mentioned several **focus units,** which are intended to provide students with integrative or in-depth coverage of specific topics. These include information on endocrine glands and their functions (after Chapter 5), exercise physiology (after Chapter 18), and acid-base balance (after Chapter 20).

Appendices

We have provided reference information at the end of the text on such topics as measurements, conver-

sions and physiological reference values. In addition to these **reference appendices,** we have also provided an **appendix of answers** to the study questions in the textbook. By providing the locations of some answers and detailed explanations of others, we encourage students to learn by example as well as to practice thinking physiologically.

Illustrations

All illustrations were chosen to complement and enhance the textual coverage of topics. All are presented in **full color** to engage students' interest and help them to visualize important physiological processes. Every opportunity is taken to use color effectively to clarify functional relationships. Approximately a third of the illustrations were prepared by Barbara Stackhouse working in close conjunction with the authors. Color was employed to demonstrate the levels of interaction of physiological variables in our many **flow charts** and **graphs.** In addition to the artwork presented in the text, we also present close to 100 **micrographs** and **photographs,** most of which were taken specifically for inclusion in *Human Physiology: Foundations & Frontiers*. These original photos are intended to enhance the students' enjoyment and appreciation of important structure-function relationships.

Vocabulary

In many ways, learning physiology is like learning a new language. New terms appear in **boldface** throughout the text and are defined on their first use. Key terms can also be found in the comprehensive alphabetical **glossary** at the end of the text. The glossary defines terms used in the text itself as well as some terms that may be encountered in other outside reading sources. Following the glossary definition of a term from the text, students will find a page reference to the area of the text where that term is defined or discussed.

Individual Chapter Elements
Chapter Outline

Each chapter is introduced with an **outline** of chapter contents. The outline lists each chapter heading and subheading along with a page reference to where each topic is covered. This outline enables students to preview the chapter contents as a whole before they begin to study, and it also serves as an easy reference guide for later review of specific topics.

Chapter Objectives

Each chapter is prefaced with a list of learning **objectives.** These objectives are not a detailed account of chapter contents, but rather a framework to which additional information can be added as the chapter unfolds. The content of these objectives is later reiterated in the chapter summary.

Chapter Overview

The chapter **overview** is a brief narrative that introduces the significance of the chapter contents in a way that is intended to engage students' interest. It is not meant to be a chapter summary; rather, it allows students to place the material in each chapter in perspective relative to their own experience as well as to what they have already studied.

Concept Checks

The material covered in the text is periodically called to the attention of the reader by timely **concept checks,** which allow students to test their recall and understanding of the material and also point out the most important concepts in the preceding section.

Tables

The extensive use of **tables** helps students to organize new material in logical order so that it is more easily understood and retained. Tables also summarize chapter content conveniently and provide an easy means of review.

Essays on Physiology

We have included over 50 essays on topics of special interest to beginning physiology students. These are intended to engage students' interest and make learning more exciting and enjoyable. **Focus essays** provide historical anecdotes or clinical applications or in-depth coverage of topics mentioned in the text. **Frontiers essays** provide information about recent scientific advances and are often related to stories in the popular media. These essays provide students with a sense of how the science of physiology has developed and instill a sense of the excitement about modern physiology.

Chapter Summary

At the end of each chapter, an enumerated **summary** provides a quick review of key concepts in the chapter. The concepts can be easily correlated with the chapter objectives, and key terms used in the chapter appear in boldface.

Study Questions

Students can realistically assess whether they have grasped the information presented in the chapter by reading through the **study questions.** These questions challenge the students to synthesize what they have learned in the chapter and to integrate it with related concepts in previous chapters. Answers to the study questions are provided at the end of the text so that students may check their performance and see concrete examples of how to analyze material in physiological terms.

Suggested Reading

Annotated sources for **additional reading** are provided for each chapter. For more information, interested students are referred to books and articles appropriate to their level of expertise.

SUPPLEMENTS

Human Physiology: Foundations & Frontiers is accompanied by a complete set of supplements to aid both students and instructors in dealing with what sometimes seems to be an immense amount of material. As with the text itself, we were concerned about producing supplements of extraordinary quality and utility. All supplements that are intended for use by students were thoroughly reviewed by panels of human physiology instructors. Often these were the same instructors who reviewed the text.

Study Guide

Written by Kevin Patton, instructor of human physiology at St. Louis University and an experienced community college instructor at St. Charles County Community College, in conjunction with Charles Schauf, the *Study Guide* far surpasses study guides that accompany other texts of this type. A direct companion to the text, the student workbook provides the student with valuable study aids such as a brief focus on key material, word part and key term exercises, a series of varied activities such as matching and labeling, quick recall, a set of concept checks for review, and a multiple choice practice test. Answers to all questions are also provided. A unique feature of the study guide is the chapter on survival, which identifies and explains the study skills necessary for success in the course, including some special techniques for reading science texts effectively.

Laboratory Manual

The *Laboratory Manual* was written by B. J. Weddell and David Moffett, both instructors in human physiology at Washington State University. Each of the exercises focuses on a key physiological concept and has been successfully performed by beginning students. Selected exercises are available to qualified adopters as videotaped demonstrations. These can be used in place of actual laboratory sessions that require equipment or preparations that may not always be available. They can also be used as a preview for more complicated techniques or as a review of data collection and analysis.

In addition to the standard physiology exercises, the manual includes special **computer exercises** designed to accompany commercially available computer software. The *Preparator's Manual,* which accompanies the lab manual, provides a list of available software for using the computer laboratory exercises, instructions for preparing solutions and setting up equipment, and the answers to the questions in the laboratory manual.

Instructor's Resource Guide
Written by Charles Schauf, the *Instructor's Resource Guide* provides useful material for both the experienced and the novice instructor. Each chapter includes a comparison of the coverage in *Human Physiology: Foundations & Frontiers* with corresponding coverage in other leading textbooks. Information is presented in a tabular form designed to allow easy conversion of lecture notes. For each chapter, there is also a summary of chapter material, a detailed outline, key terms, study questions for use in class discussion or as essay items on tests, the answers to these study questions, and up-to-date lists of related readings, audiovisual materials, and computer software.

Test Bank
The **test bank,** also written by Charles Schauf, is another feature found in the *Instructor's Resource Guide.* The test bank includes about 1200 multiple choice or short-answer items that have been class-tested and carefully reviewed and edited. Each question is coded for content and level of difficulty. Text page references where answers may be found are also provided.

Computerized Test Bank
The test bank is also available in a **computerized version** for IBM PC, Apple II, or Macintosh computers. The computerized form of the test bank includes (1) *Exam* software, which allows users to edit, delete, or add test items and to create and print tests and answer keys, (2) *Gradebook* software, which allows automated record-keeping, including display and printing of graphs that summarize information on student, class, or test performance, and (3) *Proctor* software, which allows users to provide computer-aided study materials for students.

Tutorial Software
Mosby's interactive tutorial software on human physiology is available to qualified adopters. This consists of separate modules covering topics such as regulatory mechanisms, the cardiovascular system, and special topics such as exercise physiology.

Transparency Acetates and Masters
Key physiological processes as depicted in the text are emphasized by 72 color transparency acetates and 102 transparency masters. Graphs and flow charts are provided as transparency masters to supply a common vehicle for communication between the lecturer and the student. The transparency acetates are also available as slides to qualified adopters.

ACKNOWLEDGMENTS

We are particularly grateful to colleagues who have provided original micrographs and photographs for this text, including Dr. Brenda Eisenberg at the University of Illinois; Drs. Andrew Evan, William De Meyer, Ralph Jersild, Omar Markand, Glen Lehman, Ken Julian, and Nyla Heerma of the Indiana University School of Medicine; and Drs. Carolyn Coulan and John McIntyre of the Center for Reproduction and Transplantation Immunology of Methodist Hospital. We also wish to thank Ms. Patricia Kane and the Department of Radiology at the Indiana University School of Medicine for allowing us to use the original radiographs appearing in this text, and Ms. Peggy Watson for her invaluable secretarial assistance.

This book would not have been possible without the helpful feedback provided by the reviewer panels and from the many students we have taught over the years.

CHARLES L. SCHAUF
DAVID F. MOFFETT
STACIA B. MOFFETT

REVIEWERS

Thomas Adams
Michigan State University

Patricia Berger
University of Utah

Elijah Christian
California State University at Sacramento

Benjamin Cole
University of South Carolina

Neil Ford
University of Texas at Tyler

Claude Gaebelein
St. Louis University

Frederick Goetz
University of Notre Dame

Howard Haines
University of Oklahoma

Daniel Kimbrough
Virginia Commonwealth University

David Klarberg
Queensborough Community College

Charlotte MacLeod
University of Minnesota—Duluth

Kevin Patton
St. Charles County Community College

Robert Slechta
Boston University

Greg Snyder
University of Colorado

William Stickle
Louisiana State University

Don Stratton
Drake University

Deborah Taylor
Kansas University

Carl Thurman
University of Missouri at St. Louis

Winton Tong
University of Pittsburgh

Carol Vleck
University of Arizona

Marcia Watson-Whitmyre
University of Delaware

Lloyd Yonce
University of North Carolina Medical Center

Contents in Brief

Contents

SECTION V

KIDNEY AND BODY FLUID HOMEOSTASIS

CHAPTER 1

Human Physiology in Perspective

*P*hysiology is the science that describes how the bodies of living organisms work. The name is descended ultimately from a school of Greek philosophers of the sixth and fifth centuries BC, the *physiologoi*. These philosophers, including Thales, Heraclitus, and Democritus, applied themselves to the study of all aspects of nature—mathematics, astronomy, physics, and medicine. They rejected supernatural explanations for the nature of things, arguing that the universe and all its parts are understandable and that nothing happens without cause. In modern terms they should be called scientists rather than philosophers. The inscription on their temple at Delphi was "Know yourself."

The modern forms of physiology and most other sciences had their rebirth during the revolution in thinking called the Renaissance. The first steps in the direction of modern physiology were taken by physicians of the Renaissance who attempted to understand the working of the human body by systematic study of its anatomy. However, a knowledge of the body's anatomy is not enough to explain how it functions. It is necessary to combine knowledge of the structure of body parts with observations and experiments on living structures while they are functioning.

HUMAN PHYSIOLOGY — AN EXPERIMENTAL SCIENCE

The heart and other major internal organs were well described by anatomists by the middle of the sixteenth century. In spite of the fact that one-way valves were clearly visible in the heart and veins, most physicians of that time believed that contractions of the heart simply caused blood to wash back and forth in the blood vessels. It was not until 1628 that William Harvey correctly described the pumping action of the heart and the circulation of the blood. For this achievement he utilized not only his knowledge of the structure of the heart and blood vessels, gained by human dissection, but also observations and experiments on living subjects, both human and animal.

An even longer time elapsed between description of the anatomy of the stomach and the first understanding of its function. The first direct observations of digestion of food in the stomach were not made until the beginning of the nineteenth century, most notably by William Beaumont, an American physician. Beaumont conducted his observations and experiments on a Canadian trapper named Alexis St. Martin, who had received a gunshot wound that created a permanent opening (called a fistula) between his stomach and the side of his body. Beaumont's observations started in 1835 and lasted for a number of years. Beaumont observed that the stomach of his patient responded to emotions. He removed some of the stomach's secretion and showed by chemical analysis that it contained hydrochloric acid. Beaumont's studies were the first major contribution to physiology by an American. Such studies laid the foundation of human physiology as an experimental science.

St. Martin's case is an example of how human misfortune sometimes provides an opportunity to learn about normal body function. However, human patients are not the main source of information about human physiology. Fortunately, the basic mechanisms of the human body are usually very similar to those of other animals. For example, in the muscles of all animal species studied thus far, there are two types of protein molecules, called actin and myosin, which interact to cause contraction. Furthermore, as explained in Chapter 12, the control of this interaction always seems to involve calcium. The physiological similarities are closest between humans and other mammals, so human physiology is to a very large extent equivalent to mammalian physiology.

Of all animals, mammals are the animals that most people feel they know best. The human relationship with fellow mammals is complex. Most people have known them both as treasured companions and as food. They are the characters in nursery stories and the soft animal toys that comfort children. They are also the main source of information about human physiology. An early example of how animal studies relate directly to human studies is the work of the Russian physiologist Ivan Pavlov on control of the stomach by the nervous system. During the middle decades of the nineteenth century, Pavlov created stomach fistulas in dogs, similar to that of Alexis St. Martin. He was able to demonstrate that acid secretion by the stomach occurred in response to feeding and other stimuli associated with food. He surgically disconnected the stomach from its connections with the rest of the digestive system and showed that the stomach's activity was responsive to inputs from the nervous system.

Such surgical techniques are still in use for studies of digestion and the physiological mechanisms that control hunger in experimental animals. These experiments, together with studies of human patients, have provided a basis for understanding how stress can cause stomach ulcers and why some people respond to stress by overeating. They also gave physicians knowledge and confidence needed to conduct successful surgeries on the stomachs of human patients. Had it not been possible to use nonhuman mammals as experimental subjects, today's medicine would be based to a far greater extent on speculation, trial and error, and anecdote.

The study of mammalian physiology has practical application in all fields related to human health and medicine—it forms an important part of the basis for medicine, dentistry, pharmacology, nursing, psychology, physical education, and public health. But even those who do not plan to become health professionals experience fatigue, hunger, thirst, sexual arousal, pain, fear, and anger, which are normal states of the human machine. Understanding the connections between one's life in the world and the inner life of one's body is an essential part of self-knowledge.

HOW AND WHY — EXPLANATIONS IN PHYSIOLOGY

Until recently, the state of being alive seemed so remarkable that it was possible to imagine that living organisms possessed some mysterious vital `life force. This way of thinking was called **vitalism.** Vitalism dominated physiological thinking until the middle of the nineteenth century, and only within this century has it become clear that living organisms are subject to the same physical and chemical laws as nonliving matter. Physiologists now view organisms as machines that, although incredibly complex, are nevertheless understandable by study of their working parts. This way of thinking about organisms is called the **mechanistic** approach.

Almost every aspect of human physiology is traceable to information encoded in thousands of small inherited units of DNA called genes. The mechanisms by which the genetic blueprint is passed on from parents to children are explained in

Chapter 25. Each individual inherits a mixture of the genes carried by his or her parents. The mechanisms by which the blueprint determines the structures and activities of cells are explained in Chapter 4. The large number of possible combinations of genes is largely responsible for the range of variation seen in individuals from the same population. The potential for genetic diversity is provided by small spontaneous changes in individual genes, called **mutations,** that occur from time to time.

Those individuals whose inherited characteristics are most favorable for survival and reproduction in their particular environment have a better chance of passing on to succeeding generations the genes that specify those characteristics. Over time, this **natural selection** of the fittest individuals, a concept introduced by Charles Darwin, leads to predominance in the population of the genes that favor survival and reproduction in that particular environment. For most of the physiological mechanisms that take place on the cellular level of the human body, the effect of natural selection is a continuous fine-tuning over time.

Early in evolutionary time, some cells evolved the ability to react oxygen with carbon compounds and capture the energy released by these reactions. In the beginning, these reactions, collectively called oxidative phosphorylation (Chapter 4), probably took place at the cell surface, and they still occur there in bacterial cells. As more complex cells evolved, the responsibility for energy production was transferred to structures called mitochondria, located in the interior of cells.

Many aspects of human physiology were determined long before there were human beings. Well before the evolution of mammals, natural selection had almost maximized the efficiency of oxidative phosphorylation. Subsequently, mutations that affected the proteins involved in oxidative phosphorylation continued to occur, but the mutations that survived were predominantly those that did not affect the efficiency of energy capture from carbon-based reactions. Consequently, the mitochondria of all mammalian species are almost indistinguishable in appearance and function. However, some aspects of human physiology may still be changing very rapidly on the evolutionary time scale. The fossil record suggests that there was very rapid evolution of the body form and brain of hominids, the immediate ancestors of the present-day human species *Homo sapiens.* Recognizably human creatures have walked the earth for perhaps 1.5 million years, but the human brain reached its present size only about 35,000 years ago. The brain's capabilities are probably still rapidly evolving as new selective pressures are placed on it by human culture.

Because they contribute to survival, physiological mechanisms seem to have an underlying purpose. For example, a painful injury to a finger or toe results in rapid withdrawal of the injured limb. In most cases, this withdrawal would protect the limb from further injury. However, it would be a mistake to conclude that the body makes the withdrawal response because it somehow "knows" the response is needed. It would be equally wrong to conclude that the withdrawal response evolved because it was needed. Natural selection operates only on inherited characteristics that already exist (or are produced by mutations)—it cannot call new ones into existence. Justification of a process in terms of the value of the result to an organism is referred to as a **teleological explanation.** Teleological explanations are the kind frequently given in response to the question "Why?"

In contrast, a **mechanistic explanation** of "how" the withdrawal response occurs would trace the step-by-step sequence of events that starts with the injury and ends with contraction of the muscle groups that flex the injured limb. Such an explanation could be outlined in the following way (Figure 1-1): the injury excites specific nerve endings in the skin, which relay a message by way of electrical impulses to nerve cells in the spinal cord. There are connections between the nerve cells that respond to the injury and other nerve cells in the spinal cord. These connections form a circuit, carrying electrical impulses that activate the population of nerve cells that controls contraction of the flexor muscles. Electrical impulses pass from the spinal cord to the flexor muscles, causing them to contract. The circuit that generates this response is genetically determined and is formed during early development of the nervous system.

In a mechanistic explanation, each event causes the next; a break in the chain of cause and effect would make the explanation incomplete. A complete explanation of the mechanism of the withdrawal response would involve description of the events that take place in nerve and muscle cells at each stage in the sequence, including ultimately some processes that occur on the molecular scale. A more complete description of the withdrawal response and other such responses of the nervous system, which are called **reflexes,** is presented in Chapters 5 and 11.

> 1 What is physiology? What is the relationship between human physiology and mammalian physiology?
> 2 What is vitalism? How does it contrast with the mechanistic view of life?
> 3 What is a teleological explanation?

FROM SINGLE CELLS TO ORGAN SYSTEMS

Cells are the fundamental units of organisms. Most animals that are visible without a microscope are multicellular animals. The number of cells in complex animals is typically very large—for example,

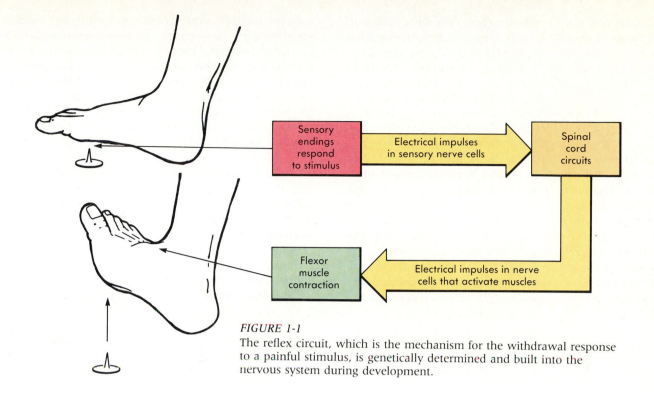

FIGURE 1-1

The reflex circuit, which is the mechanism for the withdrawal response to a painful stimulus, is genetically determined and built into the nervous system during development.

the body of an adult person contains roughly 100 trillion (100,000,000,000,000) cells. The first organisms were probably single cells, similar to some groups of modern bacteria. Free-living cells must possess all of the structures for activities such as feeding, locomotion, and reproduction. In evolutionary terms, unicellular organisms have been very successful, since they are abundant in every one of the earth's environments. However, one advantage of a multicellular organism is that individual cells can be highly specialized to carry out specific functions. The division of labor between specialized cells confers greater efficiency.

Differentiation is the developmental process by which relatively unspecialized cells become specialized to perform specific functions in a multicellular organism. Functional specialization of cells is reflected in their structures, and sometimes the cellular commitment to a single function is quite obvious. For example, as shown in Chapter 12, muscle cells are filled with a highly organized array of the contractile proteins actin and myosin, together with other structures involved in regulating contraction. The mucus-secreting goblet cells of the lining of the digestive system (Chapter 21) possess a central cavity in which mucus made by the cell is stored. The cavity typically makes up most of the total volume of the cell.

Some differentiated cells may lose the ability to reproduce themselves by cell division. In some cases, such as skeletal muscle cells and nerve cells of the central nervous system, all differentiation takes place from embryonic cells during development, and cells that die are not replaced during the life of the mature organism. In other cases, such as the skin and blood, cells that die are replaced by differentiation of some members of a population of **stem cells.**

In **tissues,** cells are often united to perform a common basic function. There are four general tissue types. **Muscle** tissue is specialized for contraction and generation of force. The different types of muscle tissue are functional adaptations of the basic contractile apparatus of actin and myosin. **Skeletal muscle** is responsible for movement of the elements of the skeleton; **cardiac muscle** is responsible for the contractions of the heart that cause blood circulation; **smooth muscle** is responsible for movement of material within soft internal organs, such as the stomach, intestines, and blood vessels.

Nervous tissue contains nerve cells, or **neurons**, which are specialized for generation and conduction of electrical impulses, and **glial cells**, which are believed to support neurons and maintain a favorable environment for their function.

Epithelial tissue is functionally more diverse than muscle or nerve tissue. Epithelial tissues include the membranes that cover body surfaces and line hollow internal organs, forming barriers between the interior of the body and the environment. Epithelial cells may be modified to serve as sensory receptors that detect stimuli from the environment, such as in the retina of the eye and the taste buds of the tongue. Epithelial cells also form the **endocrine**

glands (pituitary, parathyroid, thyroid, adrenal, ovary, and testis), which secrete substances called **hormones** directly into the blood, and the **exocrine glands,** which secrete substances via ducts (for example, the pancreas and liver secrete into the gut, whose interior is effectively outside the body).

Connective tissue is diverse and typically has large amounts of extracellular material, which contributes to the structural integrity of surrounding tissues. Examples include **bone** and the elastic elements of the skeleton and other organs. **Adipose tissue** (fat) and **blood** are usually grouped with connective tissue.

Organs are structures composed of two or more tissue types. Organs that perform related functions are grouped into **organ systems,** all of which are shown in Figure 1-2. The major organ systems are the **nervous system,** the **skeletal system,** the **muscular system,** the **integumentary system,** the **cardiovascular system,** the **respiratory system,** the **digestive system,** the **urinary system,** the **endocrine system,** the **reproductive system,** the **lymphatic system,** and the **immune system** (Table 1-1). These functional designations include some overlap. For example, hormones are secreted by several organ systems besides the endocrine system, and many components of the immune system are located within the vessels of the lymphatic system.

The organ systems of the body show a much higher degree of interdependence than is typical of subsystems in most machines. For example, if the battery of a car ceases to function, the rest of the car typically is not damaged as a direct consequence. In contrast, the continued health of all body systems depends critically on the performance of each of them. Failure of the heart, kidney, or lungs results in rapid loss of function of other organs, and with-

---

TABLE 1-1   *Organ Systems: Components and Functions*

System	Components	Functions
Nervous	Brain, spinal cord, peripheral nerves, ganglia	Controls activities of other systems; receives information from the environment; stores memories; initiates and controls behavior
Skeletal	Bone, connective tissue	Support; mineral storage; production of red blood cells
Muscular	Skeletal muscle	Movement
Integumentary	Skin, hair, associated glands, blood vessels	Protection against drying and infection; temperature regulation
Cardiovascular	Heart, blood vessels, red blood cells	Transports nutrients, gases, metabolic end products, and hormones between organ systems
Respiratory	Nose and throat, trachea, bronchi, bronchioles, lungs	Takes up oxygen and releases carbon dioxide to atmosphere; produces sounds; partly responsible for regulating blood acidity
Digestive	Mouth, esophagus, stomach, small and large intestines, salivary glands, liver, gall bladder, pancreas	Digestion, food storage, absorption of nutrients; protects against infection
Urinary	Kidneys, ureters, bladder, urethra	Homeostasis of extracellular fluid volume and composition; excretion of waste products
Endocrine	Pituitary, adrenals, thyroid, parathyroids, gonads, pancreas; many other organs secrete hormones in addition to their other functions	Regulation of reproduction, growth, metabolism, energy balance, extracellular fluid composition
Reproductive	Male: testes associated glands and ducts, penis Female: ovaries, fallopian tubes, uterus, vagina, clitoris, breasts	Reproduction; sexual gratification
Lymphatic	Lymph vessels, lymph nodes	Fluid balance; transport of digested fat; cells of the immune system are also located within it
Immune	Lymphoid tissues, bone marrow, white blood cells, thymus	Resists infection, parasitization, and cancer

Nervous System

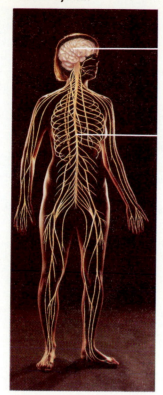

Brain

Spinal cord

Skeletal System

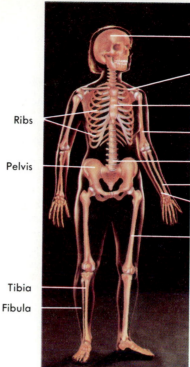

Skull

Clavicle

Sternum

Humerus

Ribs

Vertebral column

Pelvis

Radius

Ulna

Femur

Tibia

Fibula

Muscular System

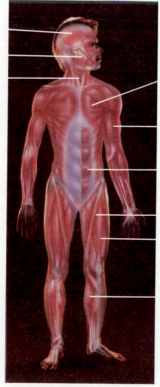

Temporalis

Masseter

Sternocleidomastoid

Pectoralis major

Biceps

Rectus abdominus

Sartorius

Quadriceps

Gastrocnemius

Cardiovascular System

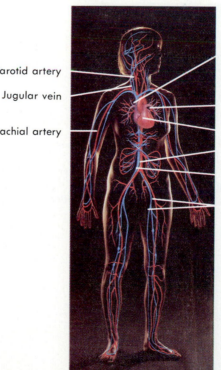

Carotid artery

Jugular vein

Brachial artery

Superior vena cava

Pulmonary artery

Heart

Aorta

Inferior vena cava

Femoral artery and vein

FIGURE 1-2
Locations of the major organ systems.

Human Physiology in Perspective

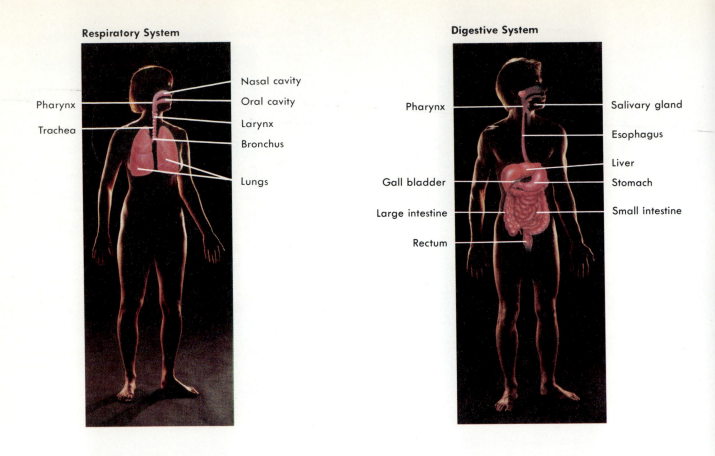

Respiratory System

Pharynx
Trachea

Nasal cavity
Oral cavity
Larynx
Bronchus
Lungs

Digestive System

Pharynx
Gall bladder
Large intestine
Rectum

Salivary gland
Esophagus
Liver
Stomach
Small intestine

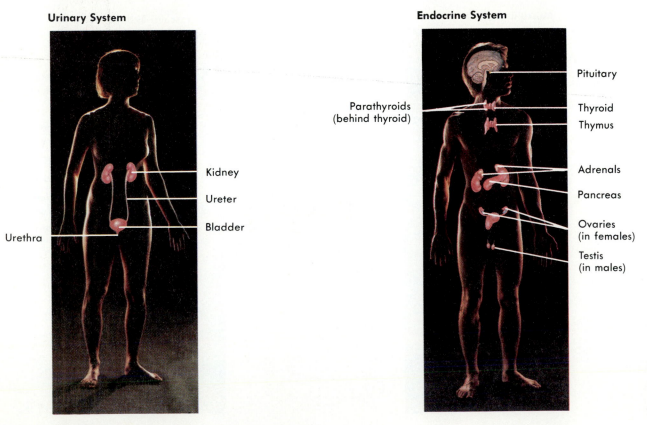

Urinary System

Urethra

Kidney
Ureter
Bladder

Endocrine System

Parathyroids
(behind thyroid)

Pituitary
Thyroid
Thymus
Adrenals
Pancreas
Ovaries
(in females)
Testis
(in males)

FIGURE 1-2, cont'd.
Locations of the major organ systems.

THE FOUNDATIONS OF PHYSIOLOGY

Reproductive System—Male

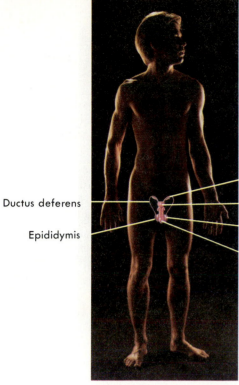

Ductus deferens

Epididymis

Seminal vesicle

Prostate gland

Testis

Glans penis

Reproductive System—Female

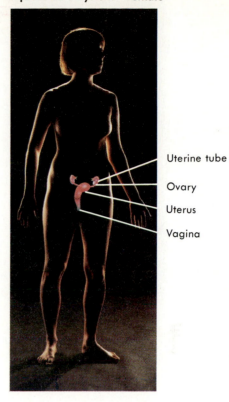

Uterine tube

Ovary

Uterus

Vagina

Immune System

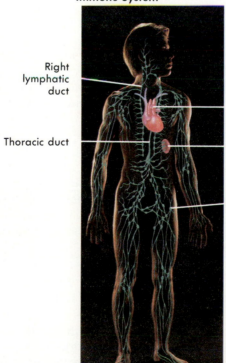

Right lymphatic duct

Thoracic duct

Thymus gland

Spleen

Lymph node

FIGURE 1-2, cont'd.
Locations of the major organ systems.

Human Physiology in Perspective

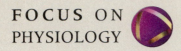

Metaphors in the History of Physiology—from the Hearth to the Computer

From the beginning of physiology as a science, about the fifth century BC, to the present day, scientists have compared the workings of the body to other natural and artificial processes that were characteristic of their time. During the pretechnological era fire was so dynamic, mysterious, and useful that it was regarded with a religious awe. Attempts to understand the equally mysterious chemistry of body metabolism led to the idea of the "fire of life." As people began to construct large dwellings and provide them with plumbing, heating, and ventilation systems, they saw that the body could be compared with a house, in which the lungs provided ventilation, the brain served to cool the blood, and the arteries provided a route for a spirit, the breath of life, or *pneuma*, to reach all of the body.

The most sophisticated machines of the eighteenth century were clocks, and almost every other natural process was understood as a form of clockwork. It was natural to see the bodies of animals as little clocks. However, even the best clocks of the time were imperfect machines that had to be wound up regularly and adjusted frequently. Vitalists could still argue for a kind of biological clock-winder in each living organism. The nineteenth century brought the beginning of the era of physiology as an experimental science. The French scientist Claude Bernard (1813-1878) showed that the liver is a source of glucose in fasting animals and also that glucose could be released by the liver after the animal itself was dead. Such experiments showed that animals, like machines, could be studied by being taken apart. If kept under favorable conditions, the individual parts could continue to function.

The physiological metaphors of the twentieth century have been chemical, electrical, and mechanical. The name of E.H. Starling is associated with so many different discoveries that one wonders if there were perhaps more than one of him. Starling demonstrated the physical forces that govern movement of fluid between the interstitial fluid and the plasma. With W.M. Bayliss he discovered a hormone, **secretin,** secreted by the duodenum of the intestine, that is responsible for stimulating pancreatic secretion in response to food in the intestine. In 1914 Starling utilized an isolated heart preparation to demonstrate the increase in its force of contraction in response to increased filling with

out medical assistance the damage to the other systems is frequently irreversible.

> 1 What is the role of differentiation in the development of multicellular animals?
> 2 What is a tissue? What are the four general categories of tissues?
> 3 What is an organ? What are the major organ systems?

HOMEOSTASIS—A DELICATELY BALANCED STATE

The English physiologist Lawrence Henderson defined an organism as "a stable structure through which flow matter and energy." The structures and chemical reactions of living organisms are highly sensitive to the chemical and physical conditions within and around cells. First of all, cells must be wet. Some small organisms can survive drying, but

blood—this was the basis of what is now called Starling's law of the heart.

The chemical connection between motor nerves and their effector tissues was most clearly demonstrated by Otto Loewi in 1921, in an experiment that is the basis for one of the classic anecdotes of physiology. The idea for the experiment came to Loewi in a dream, which he forgot on awakening. Fortunately, the dream was repeated the next night, and Loewi was able to carry out the experiment. He isolated the heart of a frog together with one of its autonomic motor nerves, the vagus, in a dish of physiological saline. Stimulating the vagus had been known to slow the heart. Loewi stimulated the vagus and collected the solution in which the heart was bathed while the vagus was being stimulated. Afterward when the heart, beating at its normal rate, was bathed with the collected saline, it slowed. Saline that bathed the heart while the vagus was not being stimulated had no effect. The substance released by the vagus was later identified as acetylcholine, a simple organic substance.

A further revolution in physiology came with the availability of radioisotopes of common elements. With radioactive labels it was possible to trace the movements and fate of atoms and molecules in the body, and it quickly became clear that all structures of the body, even bones, are subject to a continuous turnover of their molecular constituents and are thus in dynamic steady state. It was no longer possible to think of biological aging as the wearing-out of a machine, and it was clear that, in the absence of disease, aging and death were the result of a failure of homeostasis.

The newest physiological metaphor is the computer. Computer jargon has crept into physiology as it has other disciplines. Sensory signals and motor signals are the biological equivalents of input and output. The capabilities of the brain are frequently compared to those of the computer. Contemporary neurophysiologists have viewed the electrical circuits of the cerebellum and hippocampus of the brain as repetitive units, like the circuits of a microchip. The metaphor works both ways: experimenters at the Bell Laboratories are now using simple models of neural function derived from animals such as snails and slugs to improve the flexibility of information processing in computers.

while they are dry they can be said to be only potentially alive. Cells also must be surrounded by a solution compatible with life. In the case of single-celled organisms and some simple invertebrates, the surrounding solution is fresh or salt water. For complex, multicellular organisms, the surrounding fluid is referred to as the **extracellular (interstitial) fluid,** and its composition is controlled by the organism.

The **intracellular fluid,** the water inside cells, must contain particular solutes in particular concentrations. Virtually all cells contain a high concentration of potassium and low concentrations of sodium, chloride, magnesium, and calcium. In addition, cells must have a ready supply of **nutrients,** substances that serve as molecular building blocks and sources of chemical energy.

Temperature has a powerful effect on physio-

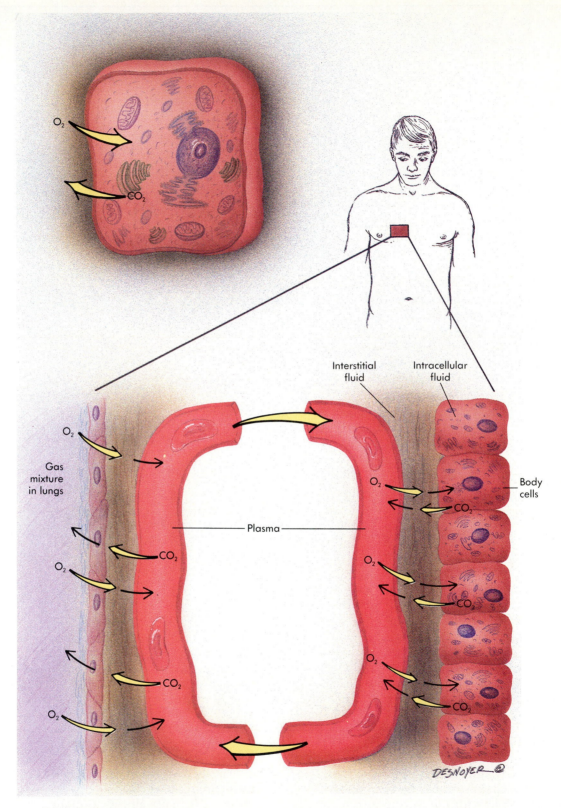

FIGURE 1-3

A   Cells exchange materials such as oxygen and carbon dioxide with their immediate environment by diffusion across the cell membrane. The path for diffusion of materials between cells and the extracellular fluid is very short.

B   In the body, cells do not have direct access to the external environment. The circulatory system carries dissolved gases and other materials over the long distance between the lungs and the cells much more rapidly than would be possible by diffusion.

logical processes. The range of temperatures in which life is possible is a tiny fraction of the known range of temperatures. Within the range of temperatures compatible with life, cooler temperatures favor preservation of cellular structure but slow the rates of the chemical reactions carried out by cells. The higher temperatures accelerate chemical reactions but also may disrupt the structure of the proteins and other large molecules within cells.

Needless to say, the temperature, water content, and chemical composition of the external environment are usually not ideal for the molecular processes carried out in the interior of cells. Unicellular organisms are highly vulnerable to environmental stress because their surfaces are in direct contact with the external environment. The larger an organism is, the less vulnerable its interior is to disruption by uncontrolled movement of materials and energy between it and the environment. On the other hand, some substances must be exchanged with the environment. In particular, the production of chemical energy in most cell types requires that oxygen and nutrients reach the interior of cells and that carbon dioxide and other chemical end products be transferred to the environment (Figure 1-3). Much exchange between cells and their immediate surroundings occurs by **diffusion** (Chapter 6), a process in which the random movements of individual molecules result in net movement of diffusing substance if there is a **concentration gradient**—a difference in the concentration of the substance over distance.

Net movement of dissolved nutrients, gases, and end products by diffusion is adequate to meet the needs of active cells if the distance is short, but much less so as the distance is increased. For example, substances dissolved in water require about 5/100 of a second to diffuse 1/100,000 of a meter, about the distance between the center and the surface of a typical cell. If the distance is increased to 1 cm, the time increases to 13 hours, and for 10 cm the time is about 53 days. Diffusion of substances over distances of meters, the scale of the human body, would require decades. Because cells take up oxygen and nutrients and give off carbon dioxide and other end products by diffusion, the diameter of individual cells is usually no greater than a few tenths of a millimeter. Thus unicellular organisms could not simply evolve into larger and larger unicellular organisms.

As large, complex multicellular animals evolved, their body plans included an internal fluid environment for the cells (Figure 1-4), called extracellular fluid. The extracellular fluid includes both the interstitial fluid immediately surrounding cells and the **plasma,** the fluid component of blood. In the circulatory system, blood is rapidly moved between the respiratory system, where gases are exchanged; the kidney, where nongaseous wastes and excess fluid

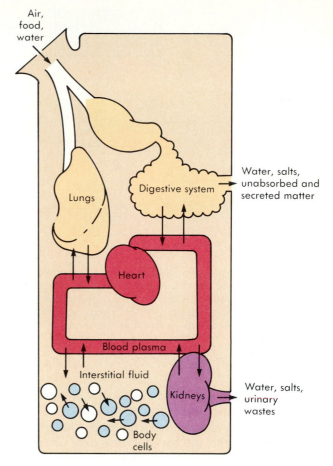

FIGURE 1-4
In the human body plan, the immediate environment of the cells is interstitial fluid. All materials exchanged between cells and the environment pass through this fluid. The circulatory system links the interstitial fluid with the sources of oxygen (the lungs) and nutrients (the digestive system) and permits carbon dioxide and waste products to be transferred to the environment by the lungs, kidneys, and digestive system.

and solutes are removed; and the digestive system, where nutrients are absorbed. Rapid transport of the substances within the internal environment by blood flow overcomes the diffusional limit on large body size. The volume of extracellular fluid in the human body is about one third the volume of intracellular fluid. The possession of a circulatory system makes it possible for this relatively small extracellular fluid volume to serve as an effective substitute for the relatively large volume of extracellular fluid that surrounds unicellular organisms.

The importance of the internal environment, or *milieu interieur,* formed by the extracellular fluid was emphasized in lectures of the French physiologist Claude Bernard as early as 1859. By maintaining a relatively stable internal environment, complex multicellular animals are able to live freely in changing external environments. The American

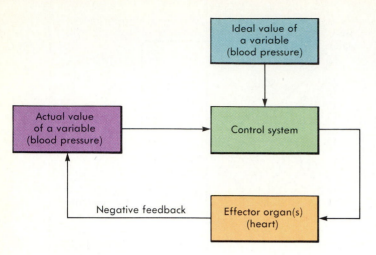

FIGURE 1-5
The principle of negative feedback control. The control system compares the actual value of a physiological variable such as blood pressure with an ideal value (set point). The control system then regulates effector organs such as the heart (and blood vessels) to correct any deviation that has occurred.

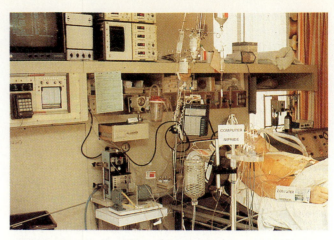

FIGURE 1-6
Photograph of a patient whose vital functions are being monitored in an intensive care unit. This patient's blood pressure is being regulated by a computer that monitors the pressure and injects a heart-stimulant drug when the pressure falls below a predetermined value.

physiologist Walter Cannon called this stable state of the internal environment **homeostasis**, from the Greek words *homeo* ("same") and *stasis* ("staying"). Homeostasis of the internal environment involves control of factors such as the chemical composition and volume of extracellular fluid, blood pressure, and body temperature. Most control systems in the body utilize the principle of **negative feedback** (Figure 1-5), described in more detail in Chapter 5. In negative feedback, the control system continuously compares a controlled variable (such as body temperature or blood pressure) with a **setpoint** value. Changes in the controlled variable initiate responses that tend to oppose the change and restore the variable to its setpoint value.

Homeostasis is a dynamic steady state—a delicate balance maintained by the many separate regulatory processes carried out by all organ systems (see Table 1-1). This balance can be overwhelmed when the external environment stresses homeostatic processes beyond their limits. For example, prolonged exposure to cold may lead to an intolerable reduction in the temperature of the body core; exercise in a hot environment may result in depletion of body fluid and an increase in the core body temperature, resulting in heat stroke.

The cells of most mammals have adapted so thoroughly to a regulated temperature that even a few degrees of variation in body temperature may have fatal effects. *Homo sapiens* is a tropical species

and has relatively weak ability to regulate its temperature. Unclothed and unprotected human beings can tolerate only a few tens of degrees Centigrade of difference between body temperature and environmental temperature. Nevertheless, *Homo sapiens* has been able to live in hostile external environments by creating engineered microenvironments (clothes, houses, space vehicles, diving suits, etc.) with appropriate conditions of temperature, atmosphere, and availability of food and water.

Many diseases impair homeostasis. The deterioration of the body in aging could also be thought of as homeostatic failure. One responsibility of physicians and other health professionals is to help maintain homeostasis. This responsibility is most obvious in critical care facilities for acutely ill patients (Figure 1-6). In such facilities a number of indicators of homeostasis are monitored, including heart rate and blood pressure, respiration, body temperature, blood chemistry, and gain and loss of body fluid. The mission of the critical care facility is to take over the responsibility for those aspects of homeostasis that the injured or diseased organ systems of the patient's own body are unable to perform.

> 1 What is diffusion? Why does it limit the size of cells?
> 2 What is homeostasis? What is the importance of the extracellular fluid in homeostasis?

SUMMARY

1. Physiology is the science that explains how the bodies of living organisms work.
2. The development of human physiology as a science depended on integration of anatomy with knowledge gained from experiments on living animals and parts of animals.
3. Vitalism attempts to explain life processes as consequences of a mysterious life principle or force. The mechanisms of body function are now believed to obey the basic principles of physics and chemistry.
4. In the course of natural selection, the stress placed on organisms by their environment causes those inherited characteristics that favor survival to predominate in succeeding generations.
5. A teleological explanation equates the cause of a process with its value to the organism. Mechanistic explanations trace the sequence of cause and effect that underlies the process.
6. In multicellular organisms, cells become specialized for performance of specific functions by differentiation. Cells are classified into four basic functional categories: muscle, nervous tissue, epithelial tissue, and connective tissue. Organs are composed of two or more tissue types; major organ systems include the nervous, muscular and skeletal, integumentary, cardiovascular, respiratory, digestive, urinary, endocrine, reproductive, and immune systems.
7. Diffusion is a process in which the random movements of individual molecules result in net movement down a concentration gradient. The slowness of diffusion over long distances limits the diameter of cells. Large complex organisms are composed of many cells bathed by an internal fluid environment. Extracellular fluid consists of the interstitial fluid, immediately surrounding the cells, and plasma, the fluid component of blood. Movement of materials between cells and the external environment is assisted by blood circulation.
8. Homeostasis is the dynamic steady state of the internal environment. Departures from the steady state are opposed by negative feedback regulation.

● **STUDY QUESTIONS**

1. Give some examples of teleological explanations. Why would vitalism favor such explanations over mechanistic ones?
2. What are two important evolutionary advantages of being multicellular?
3. Describe the relationship between distance and the rate of transport of dissolved substances by diffusion.
4. What is environmental stress? Give some examples of environmental stress.
5. What is natural selection? Do you think that natural selection is still operating on human populations?
6. What is negative feedback? Name some aspects of the internal environment that are subject to homeostatic control.

● **SUGGESTED READING**

BUTTERFIELD, H.: The Study of the Heart Down to William Harvey, Chapter 3 in *The Origins of Modern Science 1300-1800*, pp. 28-41, MacMillan, New York, 1951. Places Harvey's work in the context of the development of other sciences during the Renaissance.

CANNON, W.B.: Organization for physiological homeostatics, *Physiological Review*, volume 9, pp. 399-431, 1929. Reprinted in *Homeostasis—Origins of the Concept*, Benchmark Papers In Human Physiology, Langley, L.L. (editor), Dowden, Hutchinson & Ross, Stroudsburg, Pennsylvania, 1973. This historic paper introduces the concept of homeostasis and outlines what was known at the time about regulation of blood sugar, extracellular fluid volume, blood calcium, and body temperature.

SCHOENHEIMER, R.: *The Dynamic State of Body Constituents*, Harvard University Press, Cambridge, Mass., 1946. A pioneer report of the use of radioisotopes in physiology.

SMITH, H.W.: The philosophic limitations of physiology, pp. 31-43 in *Perspectives in Physiology*, Veith, I. (editor), American Physiological Society, Washington, D.C., 1954. An overview of the history of physiological thought by one of the foremost American kidney physiologists.

WILSON, A.C.: The molecular basis of evolution, *Scientific American*, October 1985, pp. 164-173. A short explanation of the genetic basis of evolution.

WOLF, S.: *The Stomach*, Oxford University Press, New York, 1965. This book, now somewhat dated scientifically, gives a wonderfully readable account of the life and death of a patient with a gastric fistula whose stomach was studied by Dr. Wolf over many years.

CHAPTER 2

The Chemical and Physical Foundations of Physiology

On completing this chapter you will be able to:

- Describe the structure of atoms and explain the principles of chemical bonds as they relate to the structure and behavior of biologically important compounds.

- Understand why water is the universal solvent for biological systems, and define hydrophilic and hydrophobic.

- Describe solutions in terms of their molarity and osmolarity.

- Explain the basic chemistry of acids and bases and calculate the pH of solutions.

- Express numbers in scientific notation.

- Understand how physiological variables may relate to one another as mathematical functions, and interpret graphs of functions.

- Understand the difference between equilibrium and steady state.

- Describe the relationship between force, friction, and movement that will be applied in subsequent chapters to such processes as blood flow, muscle movement, and the behavior of respiratory structures.

- Define compliance.

- Understand the application of thermodynamics to biological systems, and define entropy and free energy.

*T*he science of physiology developed as the applicability of physical and chemical laws to biological systems became apparent. Physical aspects of human body function such as blood flow, ventilation of gases, and the transmission of electrical signals in the nervous system can best be described in mathematical expressions. The ability of physiologists to predict the workings of the body by using these equations helped to demystify the human body and paved the way to development of medicine as a science.

The human body is a living system in dynamic interaction with the organic and inorganic elements in its environment. Energy from the sun enters the earth's ecosystem and is trapped by photosynthesis. This provides the chemical energy that supports animal life. The chemical environment of both plants and animals has been shaped by the properties of water molecules, because water is the solvent in which biological reactions occur. The water-based solutions of the extracellular and intracellular fluids are separated by the water-repelling properties of lipid membranes. Flow of material across these membranes is determined by defined chemical and physical forces and the physical properties of the membranes themselves.

Physical laws also govern the work performed by cells in maintaining an internal environment different from the extracellular solution and synthesizing complex molecules. Living systems do not create order out of disorder without paying the price. Cells pay by sacrificing high-energy molecules that they have synthesized for this purpose. A by-product of work is heat production, which is wasted in most man-made machines, but which can be viewed as a useful by-product in mammals because it provides the basis for maintaining a body temperature above environmental temperature. In fact, the more physiologists understand how the body works, the more they appreciate how efficiently it functions.

THE CHEMICAL BASIS OF PHYSIOLOGY
Basic Units of Matter

Atoms consist of a positive **nucleus** surrounded by negatively charged **electrons.** The positively charged particles in the nucleus are **protons,** but there are also a variable number of neutral particles called **neutrons.** The number of protons is called the **atomic number;** this number is different for each element (Figure 2-1, *A*). The sum of the protons and neutrons in the nucleus is referred to as the **atomic weight.** Changes in the number of neutrons produce **isotopes,** or variants of elements (Figure 2-1, *B*). (These isotopes are in some cases radioactive, and these can provide the basis of specific ion tracing experiments, as described in the boxed essay, p. 19.) The atomic weights of isotopes of the same element are different, but the number of protons and electrons and most chemical properties are the same. There are slightly more than 100 elements, but only 4 of them—hydrogen, oxygen, carbon, and nitrogen—make up over 95% of the substance of living organisms (Table 2-1).

In a neutral atom the number of electrons equals the number of protons in the nucleus. Electrons behave as if they orbit the nucleus in a series of planetlike orbits termed **electron shells.** Each electron shell contains a certain fixed number of electrons. The shell nearest the nucleus is filled by one pair of electrons, the next shell by four pairs, and so on. The electron shell concept is important because it explains the chemical reactivity of atoms. If the outermost shell of an atom is filled, the atom is unable to associate with other atoms to form chemical compounds and is referred to as **inert.** Examples of inert gases are neon and helium. Most atoms do not have outer electron shells that are completely filled and tend to form chemical bonds with other atoms that will result in an increase in stability for both participants.

Atoms with nearly full outer shells tend to acquire the additional electrons they need and are therefore termed **electron acceptors.** Atoms with only a few electrons in their outer shells give them up so that the next inner, complete shell becomes its outer one. They are called **electron donors.** The number of electrons an atom can donate or accept is termed its **valence.** Valence is considered positive if electrons are donated and negative if electrons are accepted (see Table 2-1). An example of an electron acceptor is oxygen, which lacks two electrons and has a valence of −2. An example of an electron donor is sodium, which has a valence of +1. Carbon has a valence of +4 and can act as either an electron donor or acceptor.

The larger the number of electron shells, the less attraction the positive nucleus exerts on the electrons in the outermost ring. Atoms with a smaller number of electron shells tend to donate their electrons less readily, but when they do enter into the formation of a chemical bond, it is a stronger bond

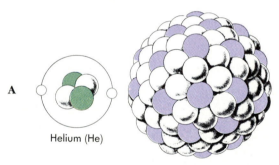

Helium (He)

Nucleus of uranium 238 (^{238}U)

Size range of atoms

B

Hydrogen (H)

Deuterium (^{2}H)

Tritium (^{3}H)

Isotopes of hydrogen

FIGURE 2-1
A The helium nucleus contains two protons and has an atomic weight of 2. The largest naturally occurring atom is uranium (atomic weight 238), whose nucleus contains 92 protons and 146 neutrons.
B The smallest atom is hydrogen (atomic weight 1), whose nucleus consists of a single proton. Hydrogen has two naturally occurring isotopes, deuterium and tritium, with one and two neutrons, respectively.

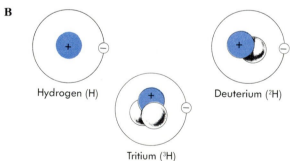

		Atomic		Body
Element	Symbol	number	Valence	weight (%)
Hydrogen	H	1	1	9.5
Carbon	C	6	+4 or −4	18.5
Nitrogen	N	7	−3	3.3
Oxygen	O	8	−2	65.0
Sodium	Na	11	+1	0.2
Magnesium	Mg	12	+2	0.1
Phosphorus	P	15	+3	1.0
Sulfur	S	16	+4 or −4	0.3
Chlorine	Cl	17	−1	0.2
Potassium	K	19	+1	0.4
Calcium	Ca	20	+2	1.5

TABLE 2-1   Elements in the Human Body

(Hydrogen, Carbon, Nitrogen, Oxygen bracketed: 96.3)

FOCUS ON PHYSIOLOGY *The Use of Isotopes in Physiology*

The *isotopes* of an element are forms of that element that possess different numbers of neutrons. Almost every element of physiological importance exists in two or more isotopic forms. Thus for hydrogen, which normally has only a proton as its nucleus, isotopes are **deuterium** with one neutron and **tritium** with two neutrons (see Figure 2-1). Some isotopes are relatively stable; others can decay to more stable atomic forms by releasing energy in the form of alpha particles, beta particles, or gamma rays and are thus called radioisotopes. Isotopes are typically indistinguishable to organisms. Stable isotopes can be separated by **mass spectrometry**; radioisotopes can be detected by their radioactivity.

Isotopes exist in nature in varying abundances; they became available as scientific and medical tools with the development of nuclear reactors, in which they can be produced. In physiology isotopes are sometimes called **tracers** or **labels** because they can be used to follow the fate of substances in the body or in isolated cells, tissues, or cytoplasmic components from one another. Tracer experiments are particularly useful for systems that are in steady state, since it might not be possible to measure the movements of substances in such systems by chemical methods.

A classic experiment provides a measure of the rate at which the thyroid gland takes up iodine from the blood. The iodine content of the gland is in steady state because the secretion of iodine-containing hormone from the gland is matched to its rate of iodine uptake from the blood; no change in its iodine content over time can be detected chemically. Radioactive iodine is injected into the blood, and the animal is placed with its thyroid over a scintillation detector. The detector converts the emissions of the isotope to small flashes of light, which are counted by a sensitive phototube. In a short time the detector begins to register an accumulation of isotope in the thyroid. At first the rate of uptake of tracer is much greater than the rate of return of tracer to the blood since the blood contains much iodine isotope and the gland initially contains none. While this condition is satisfied, the rate of increase of isotope in the gland is an accurate measure of its normal rate of uptake of nonradioactive iodine.

than is formed by the larger elements. An atom with a smaller atomic number is a stronger electron acceptor than a larger atom with the same valence because its outer shell is closer to the nucleus. Of the ten elements that have one or two electron shells, two (helium and neon) are inert. Of the remaining eight only hydrogen, carbon, nitrogen, and oxygen are abundant on earth. Thus it is apparent that the body is largely composed of those elements that are both readily available and able to form the strongest chemical bonds (see Table 2-1).

Covalent Bonds

Covalent bonds are formed by the sharing of electrons between two or more atoms. Two hydrogen atoms can share electrons with one another to form molecular hydrogen (H_2) (Figure 2-2, *A*), while two oxygen atoms can each share two electrons to form oxygen gas (Figure 2-2, *B*). Four hydrogen atoms can each share their single electron with carbon to form the gas methane (Figure 2-3). To form water, two hydrogen atoms each share an electron with oxygen. In all these compounds, the shared electrons

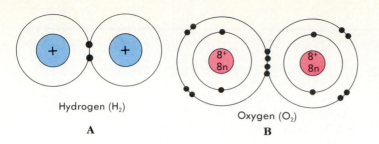

Hydrogen (H₂)

A

Oxygen (O₂)

B

FIGURE 2-2

A Hydrogen gas is a diatomic molecule composed of two hydrogen atoms, each sharing its electron with the other.

B Oxygen gas is a diatomic molecule composed of two oxygen atoms, each sharing two electrons with the other.

FIGURE 2-3

Methane is a gas composed of a carbon atom sharing a single electron with each of four hydrogen atoms. In methane each shared electron spends an equal amount of time in the vicinity of the hydrogen and carbon atoms.

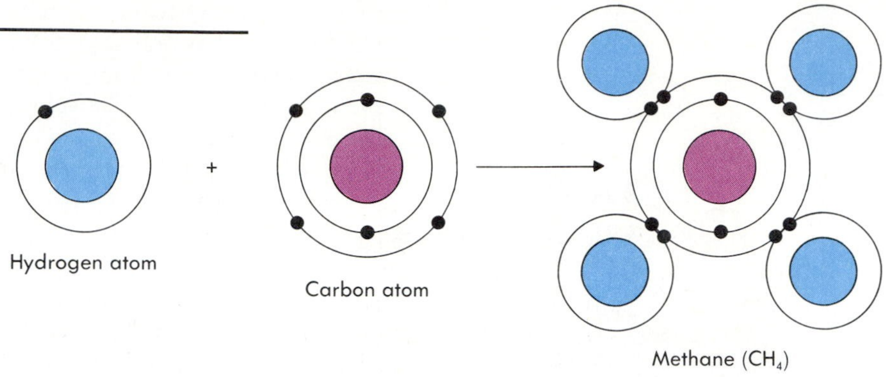

Hydrogen atom

Carbon atom

Methane (CH₄)

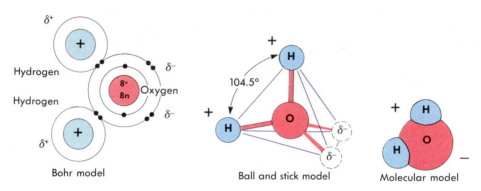

Hydrogen

Hydrogen

Bohr model

104.5°

Ball and stick model

Molecular model

FIGURE 2-4

Water is composed of one oxygen atom and two hydrogen atoms. The oxygen atom shares one electron with each hydrogen atom. Two hydrogen atoms can share electrons with the two unpaired electrons of one oxygen to form water. In water the hydrogen atoms have a slight excess positive charge, while the oxygen atom has a slight excess negative charge, indicated as δ^+ and δ^-, respectively. These charges form a dipole, and such compounds are therefore referred to as polar.

orbit all the constituent atoms so that, from the viewpoint of each nucleus, the outer electron shell appears complete. Covalent bonds are the strongest chemical bonds.

There are two classes of covalent bonds. If the shared electrons spend equal amounts of time orbiting each constituent atom, the result is referred to as a **nonpolar** covalent bond (see Figure 2-3). However, electrons are not always shared equally. One of the best examples occurs in water, where electrons donated by the hydrogen atoms spend a slightly larger fraction of their time near the oxygen nucleus (Figure 2-4). Covalent bonds of this type are termed **polar**. The oxygen side of the water molecule acts as if it has a slight negative charge, while each hydrogen molecule has a slight positive charge. Unequal average charge distributions are termed **dipoles** because such molecules have two oppositely charged regions, or "poles."

Carbon Compounds

Because carbon has four electrons in its outer shell it can form covalent bonds with four other atoms. These could be of different elements, but in many cases one or more is another carbon atom. Large molecules can thus be assembled on a carbon skeleton. No other element has a comparable ability to generate structural diversity. Biochemistry is first and foremost a study of the ability of cells to generate and rearrange the structures of molecules with carbon skeletons. Organisms probably could not have evolved if there were no element with the reactive properties of carbon. Carbon, nitrogen, and oxygen can form double or triple bonds by sharing more than one electron, increasing both the varieties of molecules and their stability. The difference between single and multiple bonds is that atoms joined by single bonds are free to rotate relative to each other, while those joined by double and triple bonds cannot.

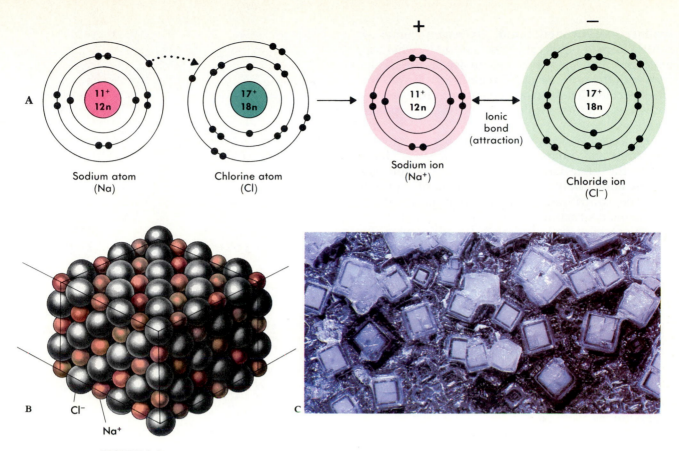

FIGURE 2-5
Ionic bonds involve outright donation and acceptance of electrons to form a crystalline array.
A When a sodium atom donates an electron to a chlorine atom, the sodium atom becomes a positively charged sodium ion, while the chlorine atom becomes a negatively charged chloride ion.
B Solid salt consists of highly ordered crystals of ionically bonded NaCl.
C The highly ordered packing of ionically bonded atoms is repeated in this example of salt crystals.

Complex carbon-containing compounds are referred to as **organic,** while all other compounds are **inorganic.** Organic compounds are subdivided into four classes: carbohydrates, lipids, proteins, and nucleic acids (see Chapter 3).

Ionic Bonds

Ionic bonds are weaker bonds formed by sharing of electrons. The sharing is so unequal in ionic bonds that the electron is effectively transferred between the atoms. In this process both participating atoms acquire opposite electrical charges which bond them electrically to one another (Figure 2-5, *A*). The ionic bonds that hold the atoms together are more easily broken than covalent bonds, so under appropriate conditions the electrically charged atoms may separate, each with a complete outer shell that resulted from the taking away or releasing of electrons. The charged atoms are thus transformed into **ions** (see Figure 2-5, *A*). Positively charged ions are **cations** and negatively charged ones are **anions.** In

table salt (NaCl) the sodium atom donates its single valence electron to chlorine, and in the process empties its outer shell and becomes a cation. The chlorine atom completes its unfilled outer shell and becomes an anion. Solid salt is a cube-shaped molecular array held together by the electrical attractions between neighboring, oppositely charged sodium and chloride ions (Figure 2-5, *B, C*). When salt is placed in water, it dissolves due to the solvent properties of the polar water molecules, as described below.

Weak Interactions

There are two types of chemical bonds that, while 10 to 20 times weaker than covalent and ionic bonds, are nevertheless very important for physiology. One such interaction involves the weak bonds formed as a result of the slight excess positive charge present on hydrogen ions that are combined in polar covalent molecules with oxygen and nitrogen. The slight negativity allows the hydrogen to produce a **hydrogen bond.** Hydrogen bonds are readily formed between water molecules, where the more positive regions of one molecule are associated with the more negative regions of adjacent water molecules. These arrangements produce what is known as a lattice throughout the solution (Figure 2-6).

Hydrogen bonds can also occur between hydrogen atoms and atoms with a slight negative charge within the same large molecule. For example, hydrogen bonds between different parts of a protein molecule contribute to stabilization of its three-dimensional structure.

The second type of weak force arises when the electrons around the atoms of one molecule interact with those of another molecule. A polar molecule can slightly shift the charge distribution in a nonpolar molecule. The nonpolar molecule develops a dipole-like character and the two molecules are attracted to one another. A similar interaction occurs between two nonpolar molecules because of random fluctuations in the position of the electrons around them. Each nonpolar molecule induces a redistribution of charge in the other. The resulting forces between the molecules are known as **van der Waals forces.**

Both hydrogen bonds and van der Waals forces are referred to as weak interactions. However, if there are many such bonds present and the molecules can fit together closely, the total energy can be significant. In large molecules such as proteins and DNA, weak interactions between parts of the large molecule play a major role in determining their three-dimensional structure.

> *1 List the types of chemical bonds in order of strength.*
> *2 Why are some chemical bonds stronger than other types?*
> *3 Why are weak bonds important in physiology?*

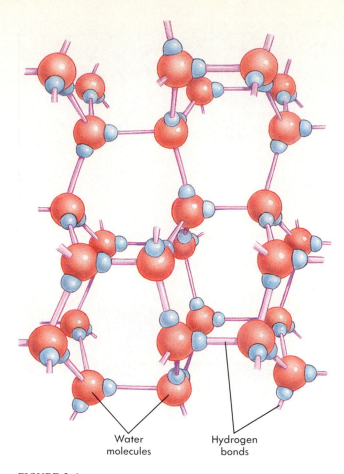

Water molecules   Hydrogen bonds

FIGURE 2-6

An illustration of the hydrogen bonds that form between water molecules.

Properties of Water

The unique properties of water (Table 2-2) deserve special attention. Water makes many compounds soluble, and most chemical reactions carried out by cells therefore take place in an environment of water molecules. Substances are rendered water-

TABLE 2-2	The Properties of Water	
Property	**Explanation**	**Physiological effect**
High polarity	Water molecules orient around ions and polar compounds	Makes a variety of compounds soluble in cells so they can react chemically
High specific heat	Hydrogen bonds absorb heat when they break and release heat when they form	Water stabilizes body temperature
High heat of vaporization	Many hydrogen bonds must be broken for water to evaporate	Evaporation of water cools the body and is an important way of regulating body temperature
High thermal conductivity	The high rate of molecular interaction in water allows heat energy to be transferred rapidly	Tissues composed largely of water are poor insulators compared with hair and fat
High surface tension	Hydrogen bonds are especially common near water surface	Layer of water on lung surfaces increases the work needed to breathe

soluble whenever the polar water molecules can overcome the forces holding the solid form of a compound together. When a compound is dissolved in water, the water is the **solvent** and the added compound the **solute.** When a salt crystal (such as NaCl) is placed in water, water molecules orient around the positively and negatively charged anions and cations in the salt and reduce the electrical force between the ions (Figure 2-7). Salts thus dissociate into their component ions and mix with water molecules. The result is termed an **electrolyte** solution.

Ions in solution are surrounded by a **hydration shell** of oriented water molecules because of the electrical forces between the ions and polar water molecules (see Figure 2-7, red circles). Like the short-range ordering between water molecules, ion hydration shells extend for only a few molecular diameters.

Polar molecules also dissolve in water. Sugar molecules have projecting polar hydroxyl (OH^-) groups. Adjacent molecules in a sugar crystal orient with oppositely charged regions attracting one another. Water molecules can get inside the sugar crystal, orient around the polar groups of each sugar molecule, and thus disrupt the attractive forces holding the compound together (Figure 2-8). The solubility of a polar compound depends on its interaction energy with water compared with the strength of its interaction with other molecules of the same compound.

A major distinction can be made between molecules that are polar enough to mix with water and those that have too few polar sites for the water molecules to interact with them. The former group is known as **hydrophilic**, or water-loving, and the type that cannot dissolve in water is **hydrophobic**, or water-hating. Most oils and fats fall into the second category. When water is present, **hydrophobic interactions** can form between hydrophobic molecules or between hydrophobic regions of large molecules. Such hydrophobic interactions hold cell membranes together and play a role in determining the three-dimensional structure of proteins (see Chapter 3, p. 48).

The physical properties of water (see Table 2-2) are important for the thermal relations of the body with the environment. The *specific heat* of a substance is the amount of energy required to increase the temperature of 1 gram of the substance by 1 degree centigrade. The specific heat of water is very high, providing protection against changes in body temperature. The *heat of vaporization* of a substance is the amount of energy absorbed by the substance as it passes from the liquid state to the gaseous state. Since water has a high heat of vaporization, evaporation of a small amount of sweat from the body surface provides effective cooling. Water conducts heat rapidly (i.e., has a high *thermal conductivity*). When one's body is immersed in cool water, heat is lost much more rapidly than in air of the same temperature.

The formation of hydrogen bonds between water molecules at the surface of a water droplet causes it to assume a spherical shape. This *surface tension* also makes a thin layer of water surprisingly resistant to stretching by forces acting along the plane of the layer. The inner surfaces of the lungs are covered by a thin layer of fluid whose surface tension

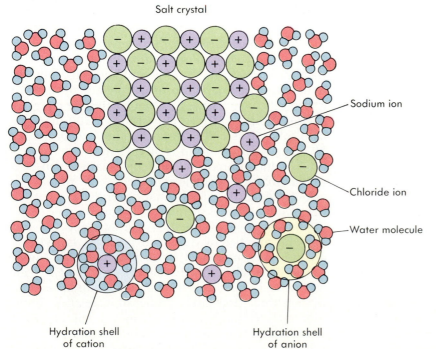

FIGURE 2-7
The molecular properties of water and solutions. When a crystal of table salt dissolves in water, individual Na^+ and Cl^- ions break away from the salt lattice and become surrounded by water molecules in hydration shells. Water molecules orient around Cl^- ions so that their positive ends face in toward the negative Cl^- ions; water molecules surrounding $Na+$ ions orient in the opposite way.

The Chemical and Physical Foundations of Physiology **23**

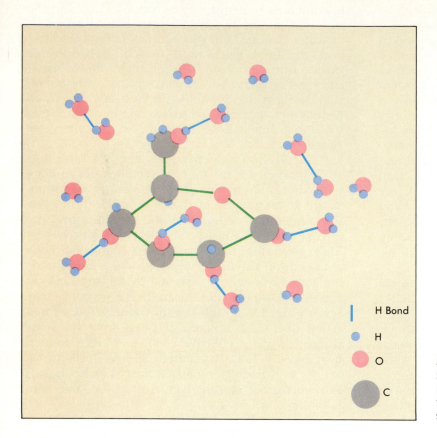

	H Bond
	H
	O
	C

FIGURE 2-8
Solvation of polar compounds, like glucose (shown here), depends on interaction of polar water molecules with polar groups on the solute.

has important consequences for the work needed to inflate the lungs during breathing.

Molarity and Osmolarity

A solution is described in terms of both the amount of solute in it and the total number of particles it contains. The **molecular weight** of a compound is defined as the sum of the atomic weights of its constituent atoms; this number, in grams, is the gram molecular weight, or a **mole**. A mole of any compound always contains 6.023×10^{23} molecules **(Avogadro's number)**. The number of moles of solute in a liter of solution is termed the **molarity** (Figure 2-9, *A*). If the molecular weight is 100, a 1.0 molar or 1000 millimolar (mM) solution is prepared by adding 100 grams of the compound to slightly less than a liter of water, and then adding water to give a final volume of exactly 1 liter. More generally, the molarity is equal to the number of grams of solute contained in a liter of solution divided by the solute's molecular weight.

Water has a molecular weight of 18 and a density of 1.00 gram/ml, so that a liter of water contains about 55 moles (1000 grams divided by 18). In biological solutions solute concentrations are seldom more than a few tenths of a mole, so there are always many more water molecules than solute molecules.

Because of its importance in situations involving the movement of water, the number of dissolved particles in a solution is described using a separate terminology. **Osmolarity** (Figure 2-9, *B*) is the number of dissolved particles in a liter of solution. For a given compound the osmolarity is the product of the molarity and the number of particles into which each molecule dissociates:

Osmolarity = (Molarity) × (Number of dissolved particles per molecule)

Polar nonelectrolytes dissolve in water but do not dissociate. The number of particles in a solution therefore equals the number of molecules in the nonelectrolyte compound originally added to it. As a result the osmolarities and molarities are equal for such substances. The osmolarity of a 100 mM glucose solution is 100 mOsm, and the osmolarity of a 50 mM glycine solution is 50 mOsm.

The total number of dissolved particles in electrolyte solutions, however, exceeds the number of salt molecules by a factor that depends on the number of ions in the salt molecule. For example, NaCl dissociates to give one Na^+ ion and one Cl^- ion, and the number of dissolved particles is twice as great as the number of NaCl molecules (see Figure 2-9, *B*). The salt $CaCl_2$ dissociates into one Ca^{++} ion and two Cl^- ions, and the number of dissolved particles is three times the number of molecules. Solute ions interact with each other to a small extent, as well as with the water molecules. To the extent that this occurs, a small correcting factor has to be added

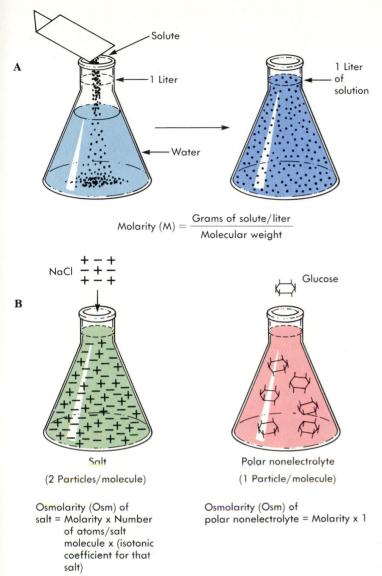

$$\text{Molarity (M)} = \frac{\text{Grams of solute/liter}}{\text{Molecular weight}}$$

Osmolarity (Osm) of
salt = Molarity x Number
of atoms/salt
molecule x (isotonic
coefficient for that
salt)

Osmolarity (Osm) of
polar nonelectrolyte = Molarity x 1

FIGURE 2-9
An illustration of the definitions of molarity (**A**) and osmolarity (**B**).

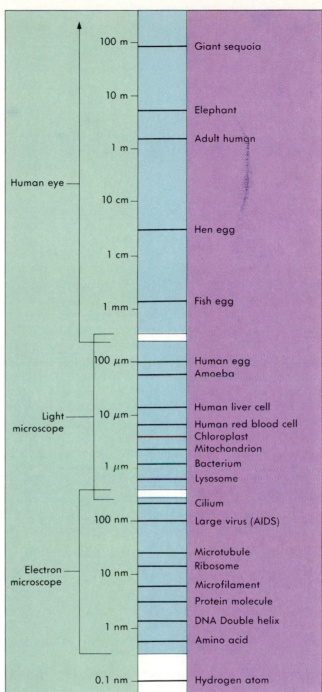

FIGURE 2-10
The size range of biologically important objects.

to the equations for osmolarity and osmotic pressure. The magnitude of this factor, called the isotonic coefficient, depends on the particular ions involved and on their concentration. This correction is small for most ions in body fluids. If this correction is neglected, a 1 molar (1.0 M) solution of NaCl is 2 osmolar (2.0 Osm); a 100 mM solution of NaCl is 200 mOsm; and a 100 mM solution of $CaCl_2$ is 300 mOsm.

Scientific Notation

In physiology it is necessary to deal with both very small and very large quantities. For example, objects of biological significance (Figure 2-10) range in size from the atomic scale (less than a billionth of a meter) to the scale of the largest organisms (one-

tenth meter to 100 meters). The ears and eyes detect sound and light intensities that vary a millionfold. The central nervous system is estimated to contain 10 billion cells. The method of scientific notation uses powers of ten to make expression of such large and small numbers easier (Table 2-3). A tenfold change in the magnitude of a variable is a change of one order of magnitude.

The logarithm to the base 10 of a number A is

TABLE 2-3   Scientific Notation

Power of ten	Prefix and symbol
10^6	mega- (M)
10^3	kilo- (k)
10^{-1}	deci- (d)
10^{-2}	centi- (c)
10^{-3}	milli- (m)
10^{-6}	micro- (μ)
10^{-9}	nano- (n)
10^{-12}	pico- (p)

$$128 = 1.28 \times 10^2$$
$$11{,}000{,}000 = 1.10 \times 10^7$$
$$0.00043 = 4.30 \times 10^{-4}$$

The logarithms of numbers in scientific notation are easy to obtain because $\log B = \log [A \times 10^y] = \log A + \log_{10} y = \log A + y$. To find the logarithm of B one looks up the logarithm of A and then adds or subtracts the whole number equal to y. Thus the logarithm of 128 equals the logarithm of 1.28 (= 0.107) plus 2, or 2.107.

Acids and Bases

An electrolyte that dissociates to yield a hydrogen ion (H^+) is called an **acid** (Figure 2-11). Because it can vary by a factor of 10,000 trillion, hydrogen ion concentration (and thus acidity) is frequently expressed in powers of 10 called **pH units,** rather than molarity. The pH of a solution is defined as the negative logarithm of the hydrogen ion concentration (pH $= -\log_{10} [H^+]$). The pH scale runs from 0 to 14 (Figure 2-12). Each change of one pH unit on the scale represents a tenfold step of H^+ concentration: a solution with an H^+ ion concentration of 1.0 molar will have a pH of zero (the log of 1 is zero); a 1 mM (1×10^{-3} M) H^+ solution will have a pH of 3.0; and a 1 μM (1×10^{-6} M) H^+ solution will have a pH of 6.0. Biological solutions usually have a pH around 7.0, so that they are in the middle of the range of H^+ ion concentrations.

A compound that combines with a hydrogen ion is a **base.** Since one product of an acid dissociation is the hydrogen ion, and chemical reactions occur in both directions, by definition the other product of dissociation must be a base; it is called the **conjugate base.** Acid-base reactions always result

written as $\log_{10} A$ (or often just as log A) and is defined as the power to which 10 must be raised to yield the number A. Logarithms to the base 10 are called common logarithms to distinguish them from cases in which a base other than 10 is used. The logarithm of 1 is zero ($1 = 10^0$); the logarithm of 10 is 1 ($10 = 10^1$); the log of 1000 is 3 ($1000 = 10^3$); and the log of 1/1000th is -3 ($0.001 = 10^{-3}$). Most logarithms are not whole numbers, and one evaluates them using a table or calculator. No power of 10 can result in a negative number, so that logarithms exist only for numbers greater than zero.

The use of logarithms is the basis of scientific notation. To write a number (B) in scientific notation one expresses it in the form $B = A \times 10^y$, where A is a number between 1 and 10, and y is the number of places the decimal point must be moved right (positive values of y) or left (y negative). In scientific notation:

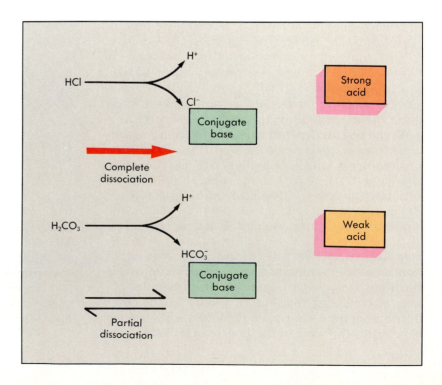

FIGURE 2-11
Strong acids dissociate completely in solution. Weak acids partly dissociate. For acids, one product is the H^+ ion; the other is called the conjugate base.

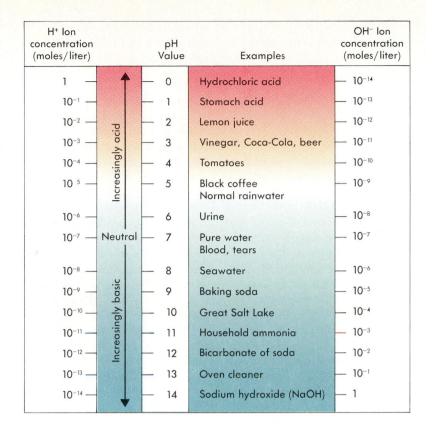

H⁺ Ion concentration (moles/liter)	pH Value	Examples	OH⁻ Ion concentration (moles/liter)
1	0	Hydrochloric acid	10^{-14}
10^{-1}	1	Stomach acid	10^{-13}
10^{-2}	2	Lemon juice	10^{-12}
10^{-3}	3	Vinegar, Coca-Cola, beer	10^{-11}
10^{-4}	4	Tomatoes	10^{-10}
10^{-5}	5	Black coffee / Normal rainwater	10^{-9}
10^{-6}	6	Urine	10^{-8}
10^{-7}	7	Pure water / Blood, tears	10^{-7}
10^{-8}	8	Seawater	10^{-6}
10^{-9}	9	Baking soda	10^{-5}
10^{-10}	10	Great Salt Lake	10^{-4}
10^{-11}	11	Household ammonia	10^{-3}
10^{-12}	12	Bicarbonate of soda	10^{-2}
10^{-13}	13	Oven cleaner	10^{-1}
10^{-14}	14	Sodium hydroxide (NaOH)	1

(Increasingly acid — Neutral — Increasingly basic)

FIGURE 2-12
The pH scale. At pH = 7.0, the concentration of H^+ and OH^- are equal; a solution with a pH of 7.0 is said to be neutral. Solutions with pH below 7.0 are acidic, reflecting H^+ concentrations greater than OH^- concentrations. Solutions with pH greater than 7.0 are basic or alkaline; at these values OH^- is more abundant than H^+. Some commonly encountered example solutions and their approximate pH values are shown below the scale. Blood and most other body fluids are slightly alkaline. A notable exception is the fluid secreted by the epithelium of the stomach.

in the formation of an acid-conjugate base pair (see Figure 2-11), so that Cl^- is the conjugate base of HCl and HCO_3^- (bicarbonate) is the conjugate base of H_2CO_3 (carbonic acid).

Acids and bases can be strong or weak. Strong acids and bases act like salts and completely dissociate, so the H^+ ion concentration in a solution of a strong acid equals the acid concentration. Hydrochloric acid, sulfuric acid, and sodium hydroxide are some examples of strong acids and bases. Weak acids and bases dissociate incompletely. Water itself is a weak acid, and only one water molecule in 10 million (1×10^{-7}) dissociates to yield H^+ and OH^- (a hydroxide ion), so that the pH of pure water is 7.0. Biologically important weak acids include many of the amino acids and carbonic acid (H_2CO_3), a compound formed when carbon dioxide reacts with water.

When weak acids or bases exist in a solution together with their salts, the result is a **buffer**. Buffers are important because they limit the changes in hydrogen ion concentration that occur when acids or bases are added to body fluids. There are two major buffer systems in the body: the **bicarbonate system** is based on the interactions of water and carbon dioxide; the **phosphate system** is based on the ability of phosphate to exist in two forms— a form that buffers strong base and a form that buffers strong acid (Table 2-4). In addition to these systems, the proteins in blood and intracellular solu-

tion act as effective buffers by absorbing or releasing hydrogen ions.

> 1 What is the difference between a mole and an osmole?
> 2 What is the difference between weak and strong acids and bases? What quality makes a compound a buffer?
> 3 If a variable is to increase by two orders of magnitude, how much should the first value be multiplied by to obtain the second value?

TABLE 2-4	Common Buffer Systems in the Body	
Buffer system	**Constituents**	**Physiological role**
Bicarbonate	$NaHCO_3$ H_2CO_3	Main buffer system that maintains the normal plasma pH
Phosphate	Na_2HPO_4 NaH_2PO_4	Main buffer system in the urine
Proteins	$-COO^-$ groups $-NH_3^+$ groups	Important buffer system regulating the internal pH of cells; accessory system for plasma pH regulation

THE DESCRIPTION OF PHYSIOLOGICAL VARIABLES
Units in Physiology

Different physical variables can be measured and expressed in different units. Mass is described in grams, time in seconds, and length in meters (Table 2-5). The exponents are abbreviated similarly, so time and length can be stated in milliseconds (msec) and millimeters (mm), microseconds (μsec) and micrometers (μm), or nanoseconds (nsec) and nanometers (nm). The nanometer has almost replaced an older unit, the Angstrom (10^{-10} meters), as the standard measure of very small lengths in physiology.

Relationships Between Variables

A **function** describes the relationship between a set of variables. In a function, the values of certain variables are specified or measured. These are called **independent variables.** Other variables must be calculated and are referred to as **dependent variables,** because their values change when the values of the independent variables are altered. Functions can include one or more constants in addition to variables. The science of physiology is to a great extent a science of relationships. It is essential to understand what is meant when variables are described as functions of one another. For example, the function that describes the relationship between the volume of a sphere (V) and its radius (r) is:

$$V = 4/3 \, \pi \, r^3$$

In this function π is a constant and V and r are variables.

A consistent set of units must be used in assigning values to variables in a function. If the radius of a sphere is expressed in centimeters, the volume will be expressed in cubic centimeters. One liter equals 1000 cm^3, so that if the answer is to be expressed in liters, the calculated volume must be divided by 1000. To avoid errors units must be "carried through" in any calculation. For example:

$$V \text{ (liters)} = V(cm^3) \times 10^{-3} \text{ liters/cm}^3$$

The cm^3 in the numerator of the right-hand side of the equation is canceled by the one in the denominator, leaving the answer in liters. As long as the units for the independent variables and constants in a function are included, the units of the dependent variable are easily determined.

Representation of Functions as Graphs

A useful way of visualizing the behavior of the dependent variables of functions is to construct graphs. For functions with two variables the independent variable is plotted on the horizontal axis, termed the x-axis, or abscissa. The dependent variable is plotted vertically on the y-axis, or ordinate. The graph for the volume of a sphere as a function of its radius is shown in Figure 2-13. The volume is determined for any radius r_o by constructing a vertical line at r_o and noting the height at which it intersects the curve that relates volume to radius. If the volume is known, the radius is found by using a horizontal line.

A more physiological example is given in Figure 2-14. Here the flow through three different tubes is plotted as a function of the pressure difference between their ends. For a rigid glass tube the flow is linearly proportional to the pressure. Rubber tubing expands when the pressure increases, so that at any given pressure difference between the ends of the tube the flow is greater in the rubber tube than in the rigid glass tube (an example of a nonlinear relationship). Finally, some blood vessels react to increased pressure by decreasing their diameter (a case of autoregulation), and the flow is less than that in the rigid glass tube at the same pressure. The flow through each tube at any selected pressure can be determined by constructing a vertical line from the pressure axis at that point and noting the level on the flow axis at which it intersects each of the curves.

TABLE 2-5	Units of Measurement in Physiology

Parameter measured	Units
Length	meter (m)
Mass	gram (g)
Volume	liter (L)
Time	second (s)
Pressure	mm Hg or torr (=atm)
Electrical resistance	ohm (Ω)
Voltage	volt (V)
Current	ampere (A)
Viscosity	poise (P)
Concentration	mole/L (M)
Osmolarity	osmole/L (OsM)

---

1 The prefixes milli- and micro- are frequently encountered in measurements of physiological variables. What are the powers of ten that correspond to these prefixes?
2 What is the difference between a dependent and an independent variable? What is a function?

PHYSIOLOGICAL FORCES AND FLOWS
Frictional Forces

Newton's second law of mechanics states that a mass subjected to an applied force will accelerate in the direction of the force with an acceleration that is proportional to the force. **Acceleration** is the rate of change of velocity with time, so that if there are no

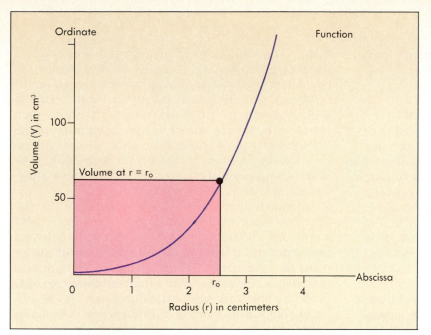

FIGURE 2-13
The volume of a sphere as a function of its radius. For any chosen radius, *r*, the corresponding volume, *v*, can be read from the graph, as is indicated for one value by the dot.

FIGURE 2-14
A graph of the rate of flow through three types of vessels as a function of the driving force, the pressure difference between their ends. The glass tube is rigid, and the flow increases linearly with respect to pressure. The rubber tube is stretched as the pressure increases, increasing the diameter of the tube, so that the flow at any given pressure difference is greater than for the rigid glass tube. Many blood vessels constrict as the pressure within them increases, reducing the flow through them below that expected for a rigid tube.

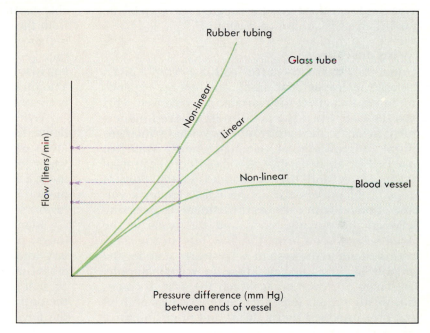

opposing forces, positive acceleration means the velocity of an object will continuously increase. Two examples of Newton's law are the gravitational acceleration experienced by an apple falling from a tree and the acceleration felt when the gas pedal of a car is pressed.

According to Newton's law, if there is no applied force there can be no acceleration, and objects should move with the same velocity forever. Although constant velocity can occur in frictionless situations, as for a body in motion in space, it is not an everyday experience. A book given a slight push along a table will come to a stop, even though no further force is exerted on it. A man jumping from a plane with a parachute will float relatively slowly to the ground at constant velocity, although he is subject to gravitational acceleration.

It is necessary to know all the forces acting on an object to apply Newton's law. In addition to gravitational and mechanical forces, there are **frictional forces** that vary with the velocity of an object and **elastic restoring forces** that arise when objects are displaced from their equilibrium positions.

Friction is a consequence of the interactions between atoms and molecules in matter. It is absent in space because there is no air. However, there is friction between the book and table, between the air and parachute, and in the flow of blood in arteries and veins. Friction occurs whenever mechanical work is performed, including muscular contraction. Whatever their origin, all frictional forces share two common characteristics: (1) the magnitude of the force is proportional to the velocity of an object; and (2) frictional forces always act in a direction exactly opposite to the direction of the object's motion.

Concept of Steady States

There are many physiological systems in which acceleration is applied to some mass of material to make it move. For example, the heart accelerates blood in the course of a heartbeat, skeletal muscles accelerate body parts in locomotion, and the volume changes of the thorax accelerate air in airways of the respiratory system. If the driving force—the acceleration—is constant, the rate of movement of the material—its **velocity**—will become constant. This is a consequence of the fact that frictional forces increase with velocity. The velocity comes into steady state when the rate of energy expenditure on the mass matches the rate of energy loss to the environment due to friction. The magnitude of the steady state velocity depends on the driving force and the **frictional coefficient**—the degree to which the moving mass is subject to friction.

Forces and Flows

A system is in **steady state** with respect to a given variable if energy-consuming and energy-expending forces that affect that variable are balanced. This was the case when the parachutist reached his terminal velocity; the acceleration provided by expenditure of gravitational potential energy was just matched by air resistance. A similar equalization of driving force and friction applies to the flow of blood in blood vessels; the heart accelerates a volume of blood with each beat; frictional forces acting between the moving blood and the walls of the blood vessels soon bring the flow into steady state. In this example the system is in steady state for the variable of flow rate, measured in volume/minute. The value of flow rate at which the system comes into steady state is dependent on the values of driving force provided by the heart and opposing force provided by the resistance of blood vessels to flow. The relationship between flow, driving force, and resistance is:

Flow rate = Driving force / Resistance

In fluids, the driving force is a pressure difference and is expressed in units of millimeters of mercury (mm Hg). Flow can be measured in liters per minute. If the pressure difference is 100 mm Hg and the flow is 1 L/minute, the flow resistance will be 100 mm Hg/L/minute. The same flow rate could be obtained with a pressure difference of only 10 mm Hg if the flow resistance fell to 10 mm Hg/L/minute (Figure 2-15).

Pressure differences are forms of potential energy. There are pressure differences of two types in blood vessels. The first type, a **hydrostatic pressure** difference (Figure 2-16), is the consequence of the mechanical force that the heart applies to the blood. The difference of hydrostatic pressure between arteries and veins drives blood through the circulatory system. Hydrostatic pressure is familiar—most children have inflated a rubber balloon with water under pressure from the tap and then watched the energy stored in the elastic walls of the balloon press the water out the nozzle of the balloon in a stream, perhaps directed at a little brother or sister. In this case the water is flowing down a difference of hydrostatic pressure.

Osmotic pressure is a less familiar driving force but is equally important in physiology. For example, the difference in osmotic pressure between interstitial fluid and blood determines the movement of fluid between the circulatory system and the tissues. Osmolarity was defined earlier (p. 24) as the number of moles of dissolved particles in a liter of solution. If there is a difference in osmolarity between two solutions separated by a water-permeable partition, water molecules will tend to move from the side of lower solute concentration (with a higher water concentration) to the more concentrated solution (a lower water concentration) to equalize osmolarity. Osmotic pressure refers to the amount of force needed to oppose water movement between a solution and pure water.

Hydrostatic pressure is a repelling force: fluid flows from higher hydrostatic pressures toward lower ones. Osmotic pressure is an attractive force: water moves from lower osmotic pressures (where the water concentration is high) toward higher osmotic pressures (where the water concentration is lower; see Figure 2-16).

Blood flows within an artery; air enters the lungs; plasma is filtered across the wall of a capillary into the kidney or interstitial space; and ions cross membranes. Different driving forces are present in each of these examples, and the nature of the opposing frictional forces also varies. Therefore the flow resistance refers to different physical processes; but the relationship between steady-state flow, driving force, and flow resistance still applies (Figure 2-17). Steady flows in such systems require expenditure of energy to overcome the frictional forces.

There are many physiological variables for which maintenance of a steady state involves a much more complex net balance of forces. For example, the total volume of water in the body is maintained in steady state by balancing losses (evaporation from body surfaces and loss in excretory products) with gain by drinking or eating. In this example, in which a steady state is an aspect of homeostasis, different forces operate on the water in different parts of the system; the effect of balancing them makes the net flow of water zero. In order to maintain this balance the body must regulate both the rate of uptake and the rate of loss.

THE FOUNDATIONS OF PHYSIOLOGY

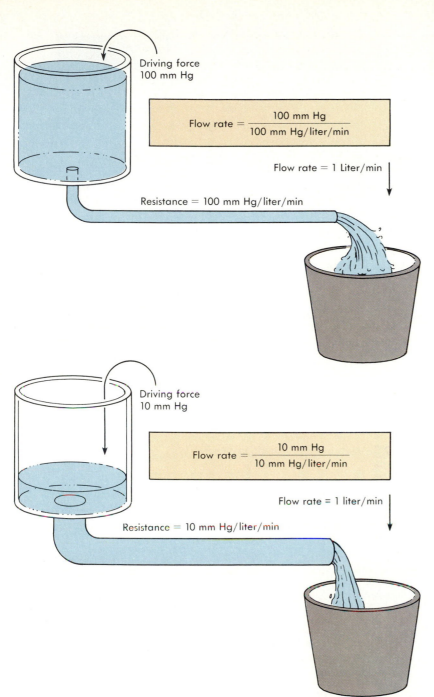

FIGURE 2-15
The flow rate is determined by dividing the driving force (in this case a gradient of hydrostatic pressure) by the resistance. If the resistance is reduced by tenfold, the driving force need be only one-tenth as large to achieve the same flow rate.

Driving force 100 mm Hg

$$\text{Flow rate} = \frac{100 \text{ mm Hg}}{100 \text{ mm Hg/liter/min}}$$

Flow rate = 1 Liter/min

Resistance = 100 mm Hg/liter/min

Driving force 10 mm Hg

$$\text{Flow rate} = \frac{10 \text{ mm Hg}}{10 \text{ mm Hg/liter/min}}$$

Flow rate = 1 liter/min

Resistance = 10 mm Hg/liter/min

FIGURE 2-16
An illustration of hydrostatic and osmotic forces in a capillary. The net force on fluid within the capillary at any point along its length is the algebraic sum of the outwardly directed hydrostatic pressure and the inwardly directed osmotic pressure.

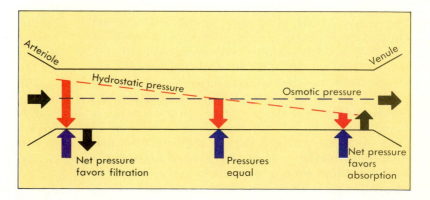

Arteriole

Hydrostatic pressure

Osmotic pressure

Venule

Net pressure favors filtration

Pressures equal

Net pressure favors absorption

The Chemical and Physical Foundations of Physiology

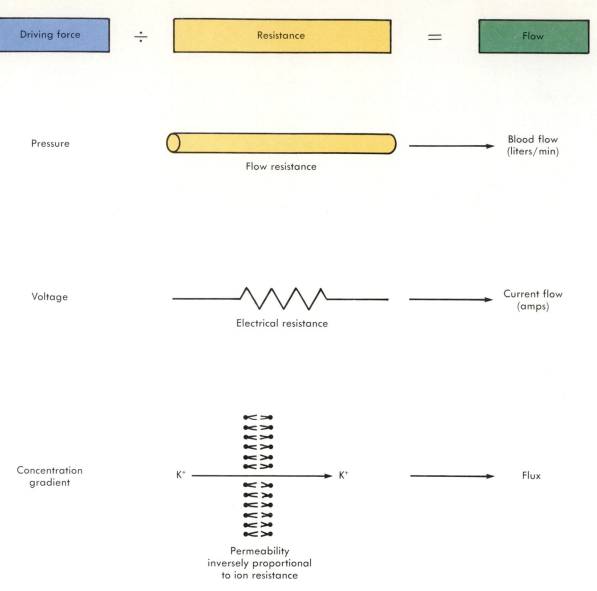

FIGURE 2-17
The steady-state/force-flow relationship can be applied to many physical situations in which there is friction, including the flow of fluids such as blood (upper diagram), the flow of electrical current (middle diagram), and the flux of ions across membranes (lower diagram).

Systems in Equilibrium

The terms steady state and equilibrium are not equivalent. A system is in **equilibrium** with respect to one of its variables only if no energy is being expended to keep the variable stable. In a house the temperature may be maintained in a steady state, given a sensitive enough heating system. However, this is not a state of thermal equilibrium—energy is being consumed to maintain the temperature at some set value which is usually different from that of the environment. If energy expenditure by the heating system were to stop, the house would begin to equilibrate with its environment, and when this

process was complete the temperature inside would be the same as that outside. On the other hand, a book resting on a table is in equilibrium; the table is not spending any energy to support the weight of the book.

The term steady state can be applied to one or more variables of a system not at equilibrium. For example, the parachutist was in a steady state with respect to his velocity, even though he was falling. A homeostatically regulated variable comes into steady state because the body's control systems direct energy-expending effector organs to balance the forces that affect that variable.

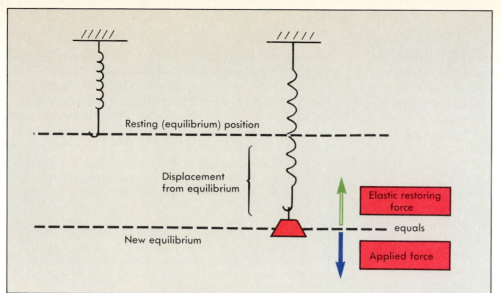

FIGURE 2-18
When objects are displaced from their equilibrium positions in elastic systems, restoring forces oppose the applied force. Thus elastic restoring forces determine the equilibrium positions or conformations of objects. The effects of such restoring forces are important in several systems, including the chest wall/lung, skeletal muscles, and arteries.

1 What is acceleration?
2 What is the general form of the function that relates steady-state flow to the driving force?
3 What is the fundamental difference between the steady-state condition and the equilibrium condition of a system?

Elastic Restoring Forces

In some of the body's organ systems, physical distortions are countered by elastic forces whose effect tends to restore the systems to mechanical equilibrium. Elastic restoring forces arise when objects are displaced or distorted, as in a spring (Figure 2-18). If a spring extends down from its resting position because a mass is hung on it, the spring will exert a force tending to pull the mass back to its original location. At the more extended position the spring and mass are in equilibrium. The externally applied force and elastic restoring force of the spring are equal in magnitude, but opposite in direction, so that no movement occurs. The greater the applied force, the larger the displacement. Another example is the balloon filled with water in the description of hydrostatic pressure. The balloon's diameter is increased by the entry of water under pressure. If the nozzle is open, the elastic restoring force returns the balloon to its original size.

The relationship between the degree of distortion and the force applied is the **compliance.** Compliance is defined as the ratio of the change in length, diameter, or volume of an object to the applied force. Compliance therefore is a measure of stiffness. The compliance of a skeletal muscle is the ratio of its change in length to the force applied to its ends; the compliance of a blood vessel is the ratio between its diameter and the pressure difference across its walls; and lung compliance is the ratio between lung volume and the pressure within it. How compliant are body structures? Compared with steel, bone is about 10 times more compliant (less

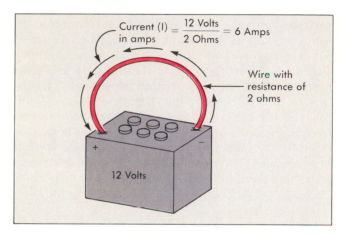

FIGURE 2-19
Ohm's Law for current flow in a conductor. The battery gives a voltage of 12 volts. If the wire has a resistance of 2 ohms, the current will be 6 amps.

stiff), a tendon 10,000 times more compliant, and a blood vessel about a million times more compliant.

Electrical Forces

Both the cytoplasm and extracellular fluid contain electrolytes. An electrolyte solution must be neutral as a whole because the total number of positive and negative ions produced by dissociation of an electrolyte are equal. However, in some situations there can be small local variations in the distribution of positive and negative charge. The most important charge variation occurs when ions move across cell membranes, a topic discussed in detail in Chapter 6. Such charge differences produce electrical forces.

Voltage (V or E) is an electrical driving force that results from charge imbalance. The relationship is best seen by considering what happens in a wire when a battery is connected to it (Figure 2-19). Electrons will move away from the negatively charged pole of the battery and toward the positive pole. The voltage is a measure of the electrical driving force of the battery. The magnitude of the resulting **current (I)** of electrons obeys the general relationship between driving force and flow developed earlier. The relationship between electrical current, driving force, and electrical resistance **(R)** is referred to as **Ohm's law** and is written:

$$I = (V2 - V1)/R$$

where V2 and V1 refer to the voltages at two different points. The unit of electrical resistance is the **ohm,** voltages are measured in **volts,** and current flow is measured in **amperes.** The reciprocal of resistance is the **conductance (G),** measured in Siemens (1/ohm). If 1 volt exists across a resistance of 1 ohm, there will be a current of 1 ampere. One ampere is equal to 1.6×10^{19} charges (one coulomb) flowing past a point per second.

All body fluids are relatively good conductors of electricity because they contain a high concentration of ionized solute. As a result, the body behaves as a **volume conductor** of electricity. Some organs, such as the heart, brain, and skeletal muscles, generate electrical currents within themselves when they are active. The flowing charges within such an organ in turn exert fields of electromagnetic forces on charges in the surrounding parts of the body. As a result, currents flow around as well as within these organs. The electromagnetic forces diminish with the square of the distance from the organ, but the resulting currents can be detected at some distance from the organ. With sensitive equipment they can even be detected at the body surface. This is the basis of recording the electrical activity of the heart, known as **electrocardiograms** (Figure 2-20, *A*), and of the brain, known as **electroencephalograms** (Figure 2-20, *B* and *C*). The electromagnetic fields generated around large skeletal muscles when they contract can also be easily detected at the body surface (the records are **electromyograms**) and can be used for diagnosis of muscle disorders.

> 1 What is meant by the compliance of a material? For what body structures is compliance a relevant value?
> 2 What is the relationship between voltage, current, and resistance? How could conductance be substituted for resistance in the relationship?

THERMODYNAMICS IN PHYSIOLOGY
First Law—Energy Conservation
Thermodynamics is the science that deals with energy transformations. In thermodynamics, **heat flow** is an energy exchange that results from a difference in temperature between two systems, while **work** refers to energy transfers that occur by other means. Heat flowing into a system increases the system's internal energy. Work done by a system on its surroundings results in a decrease in the system's internal energy. A change in the state of a system is called a **process.**

The **First Law of Thermodynamics** states that

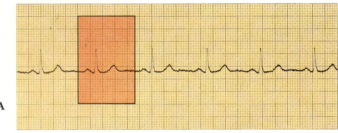

Electrical activity
during a single
heart beat

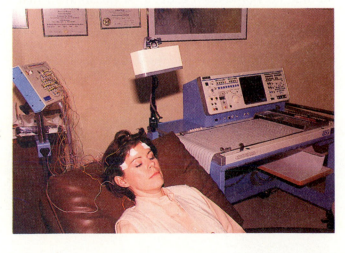

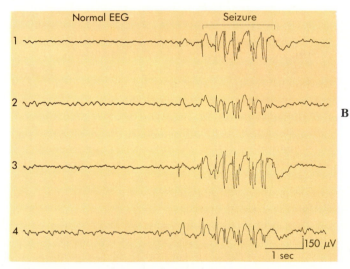

Normal EEG | Seizure

1

2

3

4

150 µV
1 sec

FIGURE 2-20

A A normal electrocardiogram (ECG). Wires are connected to the chest. The deflections seen in the electrocardiogram reflect current flow between different parts of the heart.

B An electroencephalogram (EEG). Here electrical currents in the brain are measured by electrodes placed at four different points on the scalp. Seizures are associated with a dramatic increase in the amplitude and frequency of the electroencephalogram at all four areas.

C A photograph of a patient with EEG electrodes connected to her skull.

the total quantity of energy must be conserved in any process. If there is flow of heat into a system or a system does work, the increase in internal energy equals the difference between the heat input and work output. A system can go from one state to another by doing work, by losing heat, or by a combination of both.

The basic unit of heat energy is the calorie—the amount of heat required to raise the temperature of 1 gram of water by 1 degree centigrade. The human body produces about 2500 kilocalories (kcal) of energy per day. Some of this energy is used to do work, some is used for internal repair and growth, and a portion appears as heat. Heat is not just a by-product of cellular metabolism that one would rather do without. In warm-blooded animals heat production is one essential factor in the maintenance of body temperature.

Second Law—Entropy

The bodies of living organisms are systems with a high degree of order. When an organism grows, repairs itself, or maintains homeostasis, it spends energy to maintain its internal order. The First Law said that in such processes all the energy is accounted for—none mysteriously appears or disappears.

In part, the internal energy of a system is contained in the random motion of its molecules; the rest is internal elastic energy, molecular excitations, and chemical bond energy. The transfer of thermal energy to a system (i.e., heating it) increases its random molecular motion. Work represents ordered motion. The **Second Law of Thermodynamics** says that any process results in an increase in the total amount of disorder (**entropy**) in the universe. Thus it is impossible to completely convert random motion (heat) into ordered motion (work).

Heat engines, such as steam engines and automobile motors, can convert only a portion of the heat they produce into work. The ratio of the work performed by a system to the heat transfer required to perform the work is known as **efficiency**. The concept of efficiency has immediate physiological consequences. The maximum efficiency that can be obtained in an ideal system decreases as the difference between the temperature of two systems decreases. When muscles contract they do not generate an enormous amount of heat, but suppose a difference of 10 degrees were involved. If muscles behaved as ideal heat engines, the laws of thermodynamics predict an efficiency of at most a few percent. But muscles actually convert 30% of their chemical energy into work! The reason muscles have an unexpectedly high efficiency is that muscles directly couple chemical energy and mechanical work rather than first producing heat.

1 State the First Law of Thermodynamics.
2 What is meant by the terms work and efficiency?

The Second Law dictates that for a process to occur spontaneously there must be an overall increase in randomness or entropy as a consequence of the process. To make a system more ordered requires an input of energy from somewhere. The cost of a decrease in local entropy is a net increase in the total entropy of the universe. An example occurs in protein synthesis, where making a specific protein requires 10 to 20 times as much energy as would be needed to simply form the necessary chemical bonds. The extra energy cost arises because a cell cannot put amino acids together randomly; it must do so in a very ordered way. Thus order is bought at the cost of energy. Order in the body takes many forms. Not only is order incorporated in structures, such as protein molecules, but also in differences of chemical concentration such as those between the body and the environment and between the cytoplasm of cells and extracellular fluid. The body can be thought of as a chemical engine that feeds on chemical energy from the environment to create internal order. In doing so, it counteracts for itself the inevitable increase in disorder that the Second Law prescribes for the universe as a whole.

Direction of Chemical Reactions

The direction of chemical reactions depends on the **free energy** change in the process; free energy is defined as the portion of a system's total energy that is available to perform work. The Second Law can be restated in terms of free energies: any spontaneous process, including biochemical reactions, will go in whatever direction involves a decrease in free energy. Free energies of chemical compounds can be looked up in a table. The direction of a chemical reaction can be predicted by comparing the free energies of the reactants and products.

In many reactions that take place in the body, the free energy of the products is greater than that of the reactants. These reactions cannot occur spontaneously; they must be driven by being coupled to other reactions so that the net free energy change of the driving and driven reactions is negative.

1 What is entropy?
2 State the Second Law of Thermodynamics. Do living organisms violate the Second Law when they synthesize "orderly" proteins from "disorderly" amino acids? Why not?
3 How can you predict if a chemical reaction will occur spontaneously?

SUMMARY

1. Strong chemical bonds between atoms are stable interactions in which electron sharing (for **covalent bonds**) or electron donation (for **ionic bonds**) stabilizes the outer **electron shells** of the participating atoms. **Hydrogen bonds** and **van der Waals bonds** are less stable interactions between atoms; these weak bonds nevertheless are important because, for example, they stabilize the two-dimensional structure of proteins and give water its unique properties.

2. The ability of water molecules to form hydrogen bonds accounts for many of its solvent properties, including its high specific heat and its liquidity at physiological temperatures. Water is an effective solvent for molecules with electrically charged regions (**polar** molecules) because water molecules are also polar. Nonpolar molecules are much less soluble in water and may form **hydrophobic interactions** with one another in the presence of water.

3. Solutions can be characterized by their **molarity** (the number of gram formula weights per liter) and by their **osmolarity** (the number of moles of particles per liter).

4. **Acids** and **bases** are solutes whose dissociation in solution results in hydrogen ion donation or acceptance. Strong acids and bases dissociate almost completely in solution; weak acids or bases exist in solution in an equilibrium between the dissociated and undissociated forms.

5. **Scientific notation** of quantities uses exponents of the base 10. When variables have a wide range it is useful to convert them to the logarithmic form.

6. **Functions** are mathematical statements about relationships between two or more variables. They can be visualized as graphs in which the variables are represented as scales on two axes.

7. All moving masses in the body experience **friction** that resists their movement. The general relation between forces and flows is: **velocity = driving force/frictional resistance.** If the driving force is steady, the velocity will be steady; **steady states** of variables result from steady expenditures of energy. In **equilibrium states,** forces are balanced and no energy is being expended.

8. In some body systems, physical distortions are countered by elastic forces whose effect tends to restore the systems to mechanical equilibrium. The relationship between the degree of distortion and the force applied is the **compliance.**

9. The **First Law of Thermodynamics** asserts that **systems** (the body is a system) change their states by transfer of **work, heat,** or both. In such changes of state, all of the energy transferred between the system and its environment can be accounted for.

10. The **Second Law of Thermodynamics** shows that biological order is bought at the cost of energy captured from the environment. The capture of this energy results in a net increase in entropy. Following the Second Law, chemical reactions proceed spontaneously in the direction that results in a decrease in the **free energy** of their system (a net increase in disorder).

● *STUDY QUESTIONS*

1. Define the following terms:
   valence
   electrolyte
   molecular weight
   acid

2. Molecules may interact with one another by forming covalent, hydrogen, or ionic bonds, or by van der Waals forces. Give an example of each type of bonding and describe its relative stability.

3. What factor dictates that an oxygen atom can form two covalent bonds while a carbon atom can form four?

4. What is a polar molecule? What important chemical and physical properties of water result from its being polar?

5. Under what circumstances is there a difference between one mole of a solute and one osmole of that solute?

6. Why are hydrogen bonds important in determining the structure of proteins?

7. Some physiologically important substances are present in the body in concentrations as low as 0.0000001 M. Convert this concentration into scientific notation. How would it be expressed in units of μM?

8. Imagine you are inflating a new balloon. You would probably have to blow quite hard to begin to inflate it. Once started, it would get easier to blow up. However, when the balloon is almost completely inflated, large forces would again be needed. Draw a graph that would describe the relationship between the volume of the balloon and the applied pressure.

9. State the Second Law of Thermodynamics. Why does it require energy to assemble elements into complex molecules? How does this relate to the concept of entropy?

10. Define the term resistance. Resistance enters into descriptions of blood flow, electric currents across membranes, and the movement of air in and out of the lung. These processes seem physically different. What is the relationship between force and flow that they all obey?

● *SUGGESTED READING*

BENNETT, C.H.: Demons, engines, and the second law, *Scientific American*, volume 257, November 1987, p. 108. Discusses attempts to design perpetual motion machines in violation of the second law of thermodynamics.

CAMERON, J.N.: *Principles of Physiological Measurement*, Academic Press, New York, 1986. Discusses the metric system, gas and solution concepts, pH, partial pressures, electrical measurements, and osmotic pressure.

HALLET, G., H. STINSON, and D. SPEIGHT: *Physics for the Biological Sciences*, Chapman and Hall, Agincourt, Ontario, Canada, 1982. A very readable physical science textbook that could be used to pick up needed background material.

CHAPTER *3*

The Chemistry of Cells

On completing this chapter you will be able to:

- Describe the structure of lipids, including triglycerides (triacylglycerols), fatty acids, steroids, and phospholipids, and state the major storage form of lipids in humans.
- Describe the structure of carbohydrates, including monosaccharides, disaccharides, and the starches. State the major storage form of carbohydrates in humans.
- Understand peptide bond formation.
- Compare primary, secondary, tertiary, and quaternary structure of proteins.
- State the way the sequence of amino acids in a protein is coded for by a sequence of nucleotide codons.
- Understand how enzymes catalyze chemical reactions.
- Define the terms active site, saturation, competition, and allosteric site.
- Describe DNA and how DNA replicates by complementary base pairing.
- Understand the process of transcription of the nucleotide sequence of a gene into messenger RNA.
- Describe how messenger RNA is translated into protein.

*T*he internal processes of cells involve transformations of carbon compounds, carried out and controlled by other carbon compounds. Organic chemistry, the name for the branch of chemistry that concerns itself with carbon compounds, reflects a historical belief that such compounds could only be made in living cells. Cells contain perhaps a thousand different organic compounds with molecular weights of less than 1000. Most of these fall into one of four main families: sugars, fatty acids, amino acids, and nucleotides. All of these may be assembled into larger molecules. Fatty acids are precursors of lipids; sugars are the basic subunits of carbohydrates; and nucleotides are subunits of nucleic acids such as RNA and DNA.

Amino acids are assembled into proteins. Some proteins form the basic structural elements of cells, while others act as catalysts for biochemical processes in the cell, including building other types of organic compounds. The blueprint for the cell is its DNA, a unique molecule that contains all the information necessary to manufacture all the proteins in a cell. Information contained in DNA is passed from one cell to another during cell division, and from parents to their offspring. The organization of information flow is from DNA, to RNA, to all those enzymes that catalyze intracellular chemical reactions, as well as to proteins and lipids that form intracellular structures and organelles.

LIPIDS
Triglycerides

Compounds consisting only of carbon and hydrogen are **hydrocarbons**. Hydrocarbon chains are important constituents of **lipids**. There are three classes of lipids: (1) **triglycerides**, also referred to as **triacylglycerols** or **neutral fats**; (2) **phospholipids**; and (3) **steroids** (Table 3-1).

Triglycerides have three **fatty acid** chains attached to a glycerol backbone (Figure 3-1, A). Fatty acids are hydrocarbons that have a carboxyl (−COOH) group at one end. In humans, most fatty acids contain an even number of carbons (usually 16 or more). Either single or double bonds can exist between adjacent carbons. When only single bonds are present the fatty acid is **saturated**; **unsaturated** fatty acids contain some double bonds (Figure 3-1, B).

In the triglycerides, different fatty acids may be found at each glycerol carbon. Animal fats such as butter tend to have higher proportions of saturated fatty acids than do vegetable fats such as linseed oil and corn oil. Triglycerides are not water-soluble, but exist in cells as large fat droplets.

Phospholipids

At first glance, **phospholipids** appear similar to triglycerides because they also have a glycerol backbone (Figure 3-2, A). However, the phospholipids have different properties from triglycerides because only two of the three glycerol carbons are linked to fatty acids. The third carbon has a phosphoric acid group linked to a polar molecule. The result is that one end of a phospholipid is **hydrophobic**, having properties similar to those of hydrocarbons, while the other end is **hydrophilic** because of the polar portion (Figure 3-2, B). Phospholipids are grouped by the type of polar molecule attached to the terminal carbon of glycerol. Lecithins, for example, have a choline on the terminal carbon but may have various fatty acids on the other glycerol carbons.

Because phospholipid molecules have polar and nonpolar regions, they tend to orient themselves so that their polar ends interact with water (Figure 3-3, A), while their hydrocarbon "tails" associate with other phospolipids or with nonpolar regions of other molecules. Molecules with polar and nonpolar regions are **amphipathic**—the name means "feeling both." In water, phospholipids aggregate to form either spherical **micelles** (Figure 3-3, B) or **bilayers** (Figure 3-3, C). Phospholipids are major components of cell membranes.

Steroids

The **steroids** (Figure 3-4) have a basic structure composed of four interconnected carbon rings. The side groups at various positions on these carbon rings define specific steroid classes. Physiologically important steroids include cholesterol, many vitamins, all the major sex hormones (estrogens, pro-

---

TABLE 3-1   *Major Categories of Organic Molecules in the Human Body*

Category	Subclass	Subunits	Function
Lipids	Triglycerides	3 fatty acids + glycerol	Insoluble in water; major store of reserve fuel in the body
	Phospholipids	2 fatty acids + glycerol + phosphate + charged nitrogen molecule	Molecule has polar and nonpolar ends; major constituent of membranes
	Steroids		Cholesterol and steroid hormones
	Prostaglandins	Modified phospholipid	Important regulatory agent
Carbohydrates	Monosaccharides (sugars)		Major fuel used by cells for energy
	Polysaccharides	Monosaccharides	Storage form of carbohydrate
Proteins		Amino acids	Catalysis (enzymes)
			Cell structures
			Protein hormones
			Recognition (antibodies and receptors)
			Cell membrane transport and permeability
Nucleic acids	DNA	Nucleotides containing adenine, cytosine, guanine, and thymine	Storage of genetic information
	RNA	Nucleotides containing adenine, cytosine, guanine, and uracil	Translate genetic information into protein synthesis

gesterone, and testosterone), and the adrenal hormones cortisol and aldosterone. Cholesterol is the precursor for biological synthesis of steroids.

Lipids are able to combine with other organic compounds. If a lipid and a carbohydrate are joined together, a **glycolipid** results. Lipids can also be combined with proteins to form a group of compounds called **lipoproteins**. Many of the proteins in biological membranes are lipoproteins (plasma membrane proteins). The lipid component of the molecule associates with the other lipids and attaches the protein component to the membrane. Ability to be complexed with protein also enables lipids to be transported in the blood in the form of lipoproteins. The lipoprotein complexes are polar enough to be soluble in plasma, while the lipids by themselves are not very soluble in plasma.

Prostaglandins

Prostaglandins are 20-carbon fatty acids that contain a 5-carbon ring (see Table 3-1). All are derived from arachidonic acid, a membrane phospholipid. Different prostaglandins are denoted by an abbreviation in which PG stands for prostaglandin, followed by a letter and subscript that identify the particular prostaglandin (for example, PGA_2). Prostaglandins have been implicated in a large number of regulatory functions, including inflammation and blood clotting; ovulation, menstruation, and labor; and secretion of acid by the stomach.

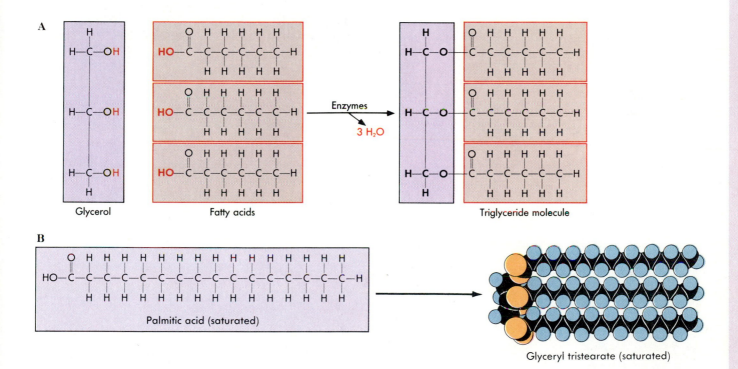

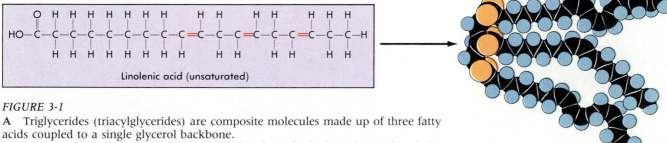

FIGURE 3-1

A Triglycerides (triacylglycerides) are composite molecules made up of three fatty acids coupled to a single glycerol backbone.

B Fatty acids can be saturated (no double bonds in the hydrocarbon tail; palmitic acid) or unsaturated (one or more double bonds; linolenic acid). Many of the animal triglycerides are saturated. Because their fatty acid chains can fit closely together, these triglycerides form immobile arrays called hard fat. Vegetable oils, such as linseed oil, by contrast, are typically unsaturated, and the multiple double bonds prevent close association of the triglycerides.

FIGURE 3-2

A Phospholipids, composite molecules similar to a triglyceride, are formed by replacing one of the fatty acids of a triglyceride with a polar phosphate compound (here choline addition forms lecithin or phosphatidyl choline).

B Phospholipids are usually oriented so that the polar portion extends from one end of the molecule and the two fatty acid chains from the other, so they are often represented diagramatically as a polar ball with a nonpolar tail.

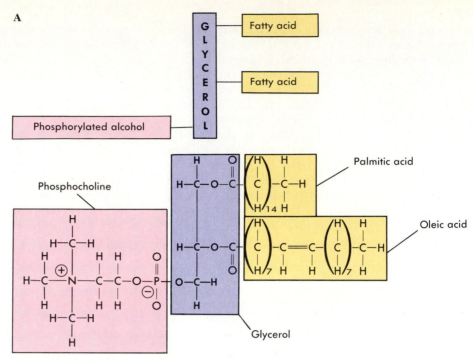

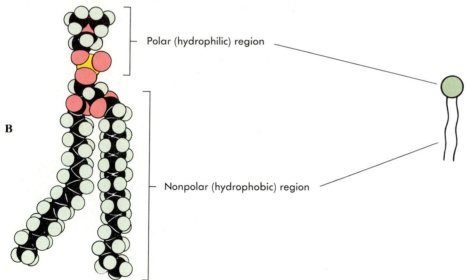

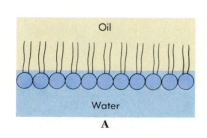

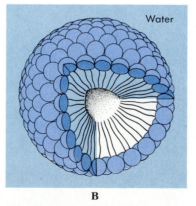

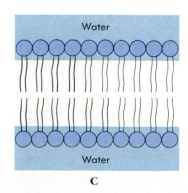

FIGURE 3-3

A At an oil-water interface, phospholipid molecules orient so that their polar heads are in water while the nonpolar tails extend into the oil phase. In an aqueous environment, phospholipids arrange themselves to form either spherical micelles (**B**) or extended bilayers (**C**).

      THE FOUNDATIONS OF PHYSIOLOGY

Human diet has changed profoundly since prehistoric times. The fat intake of prehistoric peoples living in temperate climates was probably about 20% of their total caloric intake, as compared with about 40% in the modern United States. The consumption of complex carbohydrates, such as plant starches, was much greater than that of simple carbohydrates, such as sucrose and glucose. Epidemiological evidence suggests that there may be a relationship between specific dietary elements and the incidence of some types of cancer, although it is difficult to obtain positive proof because long periods of time may elapse between the initiating event that transforms a normal cell into a potential cancer cell and the actual development of a tumor. The best evidence comes from following the dietary habits of a large study population for many years and

comparing populations of subjects with different diets. For example, the Japanese population might be compared with the U.S. population, or Japanese immigrants in the United States could be compared with Japanese who remained at home.

When such comparisons are made, the incidence of breast cancer is found to be high in natives of the United States and in Japanese immigrants who adopted typical American diets, but low in natives of Japan. Similar comparisons have been made for 39 countries and confirm a strong relationship between fat intake and mortality due to breast cancer. The rate is highest in countries where the average consumption of fat is 140 to 160 grams/day. It approaches zero in populations that consume less than 40 grams of fat per day. In contrast to breast cancer, stomach cancer is common

in Japan and rare in the United States. The incidence of stomach cancer may be related to the consumption of smoked or pickled food. Some studies suggest that it is not the calories contained in fat that matters, but rather which fatty acids a diet includes. In experimental animals certain fatty acids act as cancer-promoting substances (carcinogens).

These laboratory and epidemiological studies cannot answer the question of whether an individual's decision to change her diet would affect her risk of cancer. But even a small beneficial effect of diet modification would have important implications for public health. For every nine women who live a normal lifespan, one will get breast cancer. Even a small reduction in the risk factor for the population would mean that a large number of women would not develop the disease.

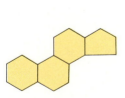

Steroid ring structure

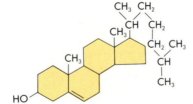

HO

Cholesterol

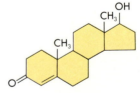

Testosterone

Cortisol

FIGURE 3-4
Steroids have a backbone of four carbon rings. Physiologically important steroids are synthesized from cholesterol; shown as examples are the sex hormone testosterone and the adrenal steroid cortisol.

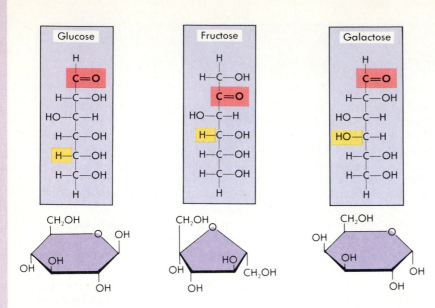

FIGURE 3-5
Structure of three monosaccharides: glucose, fructose, and galactose. In the solid form these are linear 6-carbon molecules, but in solution they form rings. Fructose is a structural isomer of glucose with identical chemical groups bonded to different carbon atoms. Galactose is a stereoisomer of glucose with identical chemical groups bonded to the same carbon atoms, but in a different orientation.

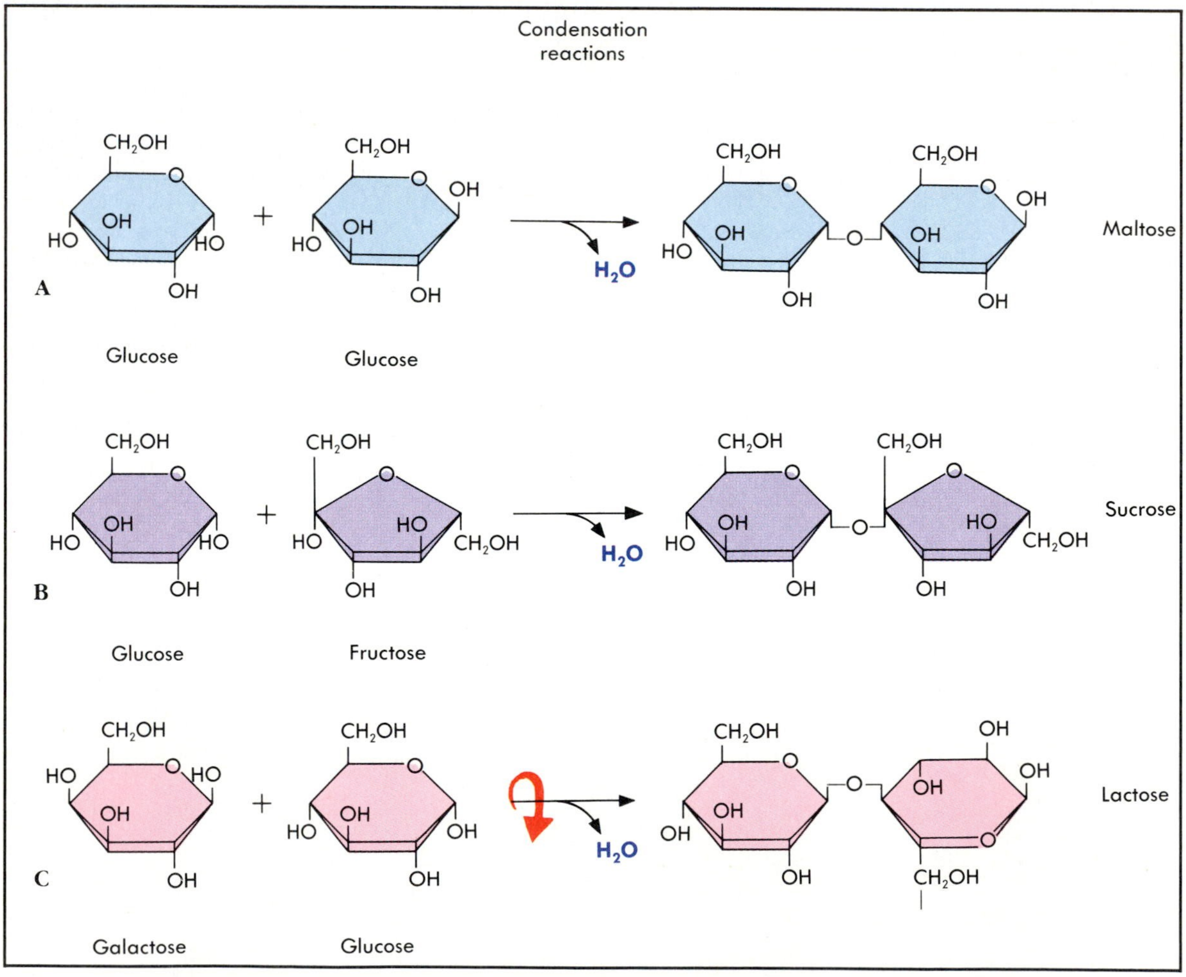

Condensation reactions

A   Glucose   +   Glucose → H_2O Maltose

B   Glucose   +   Fructose → H_2O Sucrose

C   Galactose   +   Glucose → H_2O Lactose

FIGURE 3-6
Some examples of disaccharides.
A Apples are rich in maltose, composed of two glucose molecules.
B In table sugar, glucose and fructose are joined to form sucrose.
C Milk contains a disaccharide of glucose and galactose termed lactose or milk sugar. Note that lactose contains different bonds than maltose and sucrose, a property that can make lactose difficult to digest for individuals who lack a specific enzyme (lactase).

CARBOHYDRATES
Components of Carbohydrates

Carbohydrates (see Table 3-1) are composed of carbon, hydrogen, and oxygen in the ratio 1:2:1, respectively. Among the simplest carbohydrates are the **monosaccharides** (Figure 3-5). The 6-carbon monosaccharides include glucose, its stereoisomer galactose, and its structural isomer fructose. Five of the six carbon atoms in a sugar molecule have a polar hydroxyl (⁻OH) group attached to them, so that the carbohydrates are freely water-soluble.

Disaccharides are formed by the covalent linkage of two monosaccharides (Figure 3-6). Covalent bond formation between two monosaccharides involves the removal of a hydroxyl group from one molecule and a hydrogen from the other to yield water, a type of chemical reaction called **condensation** (or **dehydration**). The reverse reaction, breaking a covalent bond by the addition of water, is **hydrolysis**. Maltose is a disaccharide of two glucose molecules; sucrose (table sugar) is the disaccharide of glucose and fructose; and lactose (milk sugar) is the disaccharide of glucose and galactose.

Polysaccharides

When many monosaccharides are joined together, the result is a **polysaccharide**. Varying numbers of monosaccharides can be linked via bonds between the different carbons. Such molecules made by joining many similar subunit molecules (**monomers**) are called **polymers**. Polysaccharides can have a wide range of molecular weights, and their physical properties depend on the nature of the internal chemical bonds and the degree of branching.

Glycogen, animal starch (Figure 3-7), is the most highly branched polymer of glucose. Glycogen is water soluble, but its size prevents it from leaving cells. These qualities suit it for its role as the major storage form for carbohydrates in animals. Plant starches are less branched, while cellulose, the rigid structural component of plants, is insoluble in water. Cellulose also possesses different chemical bonds between the component glucose molecules, so that different enzymes are needed to digest cellulose. Many body cell types possess enzymes for synthesizing glycogen from glucose and for releasing the glucose units from glycogen. The two processes are regulated so that glycogen synthesis dominates when glucose is plentiful, and glycogen breakdown takes over when glucose is scarce.

> 1 What is the chemical difference between saturated and unsaturated fatty acids?
> 2 What are the characteristics of an amphipathic molecule?
> 3 What are the different forms of carbohydrates? Which of these are polymers?

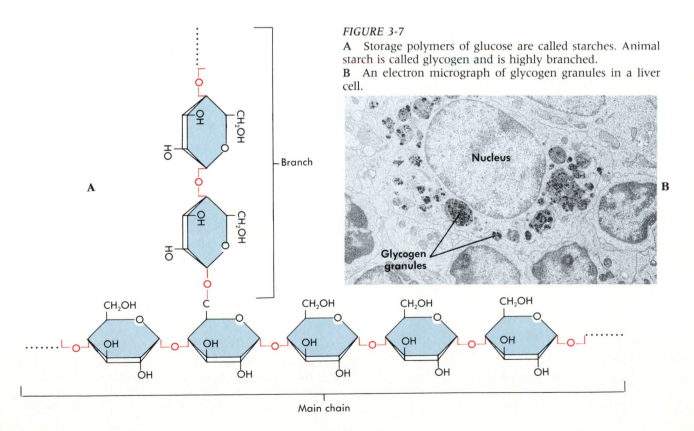

FIGURE 3-7
A Storage polymers of glucose are called starches. Animal starch is called glycogen and is highly branched.
B An electron micrograph of glycogen granules in a liver cell.

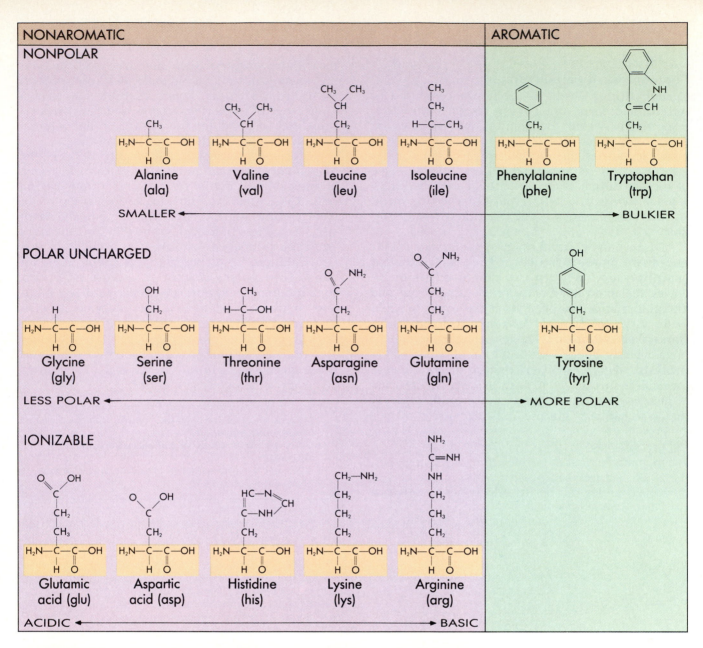

NONAROMATIC		AROMATIC

NONPOLAR

Alanine (ala) — Valine (val) — Leucine (leu) — Isoleucine (ile) — Phenylalanine (phe) — Tryptophan (trp)

SMALLER ← → BULKIER

POLAR UNCHARGED

Glycine (gly) — Serine (ser) — Threonine (thr) — Asparagine (asn) — Glutamine (gln) — Tyrosine (tyr)

LESS POLAR ← → MORE POLAR

IONIZABLE

Glutamic acid (glu) — Aspartic acid (asp) — Histidine (his) — Lysine (lys) — Arginine (arg)

ACIDIC ← → BASIC

SPECIAL STRUCTURAL PROPERTY

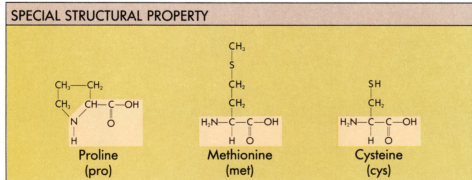

Proline (pro) — Methionine (met) — Cysteine (cys)

FIGURE 3-8
The 20 common amino acids. Amino acids consist of an amino group, a carboxylic acid group, and a side chain, which is different for each amino acid. Different side chains can be nonpolar, polar, or ionic. Some amino acids have special roles in forming links between proteins and sugars, between different parts of the same protein, or between different amino acid chains of protein complexes.

PROTEINS
Amino Acids

Approximately half the organic material of the body is **protein** (see Table 3-1). Proteins comprise many of the structural elements of cells, allow muscles to contract, form pathways by which ions and polar substances can enter and leave cells, act as catalysts for intracellular chemical reactions, regulate gene action, transport oxygen, act as carriers of electrons, and serve as markers that allow cells to identify one another. Proteins are assembled from 20 small nitrogen-containing **amino acids** (Figure 3-8). **Essential** amino acids are those for which human cells cannot synthesize the carbon backbone (Table 3-2), including histidine, which is essential in infants (but not adults). There are amino acids that are not themselves regarded as essential but which are derived from an essential amino acid, methionine, so the concept of essentiality is a relative rather than an absolute distinction.

By definition, amino acids include one carbon atom that is bound to (1) a carboxylic acid ($^-$COOH) group, (2) an amino ($-NH_2$) group, (3) a side chain of variable length and complexity, and (4) a hydrogen atom. The side chain may be nonpolar, as in alanine and leucine; polar, as in serine and glutamine; acidic, as in glutamic and aspartic acids; basic, as in lysine and arginine; or aromatic, as in tryptophan and tyrosine (see Figure 3-8).

Since the side chains of ionizable amino acids are weak acids or bases, amino acids and proteins function as buffers. The histidine groups are good buffers in the near-neutral part of the pH scale and have an important role in buffering the cytoplasm against changes in pH.

Some amino acids have special properties that are important for the structure and function of proteins. For example, prolines form attachment sites for the sugar side chains in glycoproteins. Cysteine can form disulfide bonds with other cysteine molecules in that protein and contribute to the tertiary structure of the protein (Figure 3-9).

Peptide Bonds

Proteins are assembled by the formation of **peptide bonds** between the carboxylic acid group of one amino acid and the amino group of a second

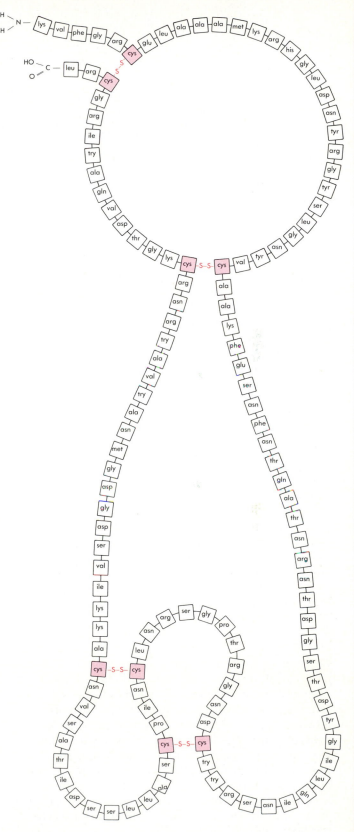

FIGURE 3-9
The amino acid sequence of the enzyme lysozyme. The four disulfide bridges are important in determining the three-dimensional shape of the enzyme.

TABLE 3-2	Essential and Nonessential Amino Acids

Essential	Nonessential
Histidine*	Alanine
Isoleucine	Arginine
Leucine	Asparagine
Lysine	Aspartic acid
Methionine	Cysteine
Phenylalanine	Glutamic acid
Threonine	Glutamine
Tryptophan	Glycine
Tyrosine†	Proline
Valine	Serine

*Histidine is essential only in infants.
†Tyrosine can be synthesized from phenylalanine.

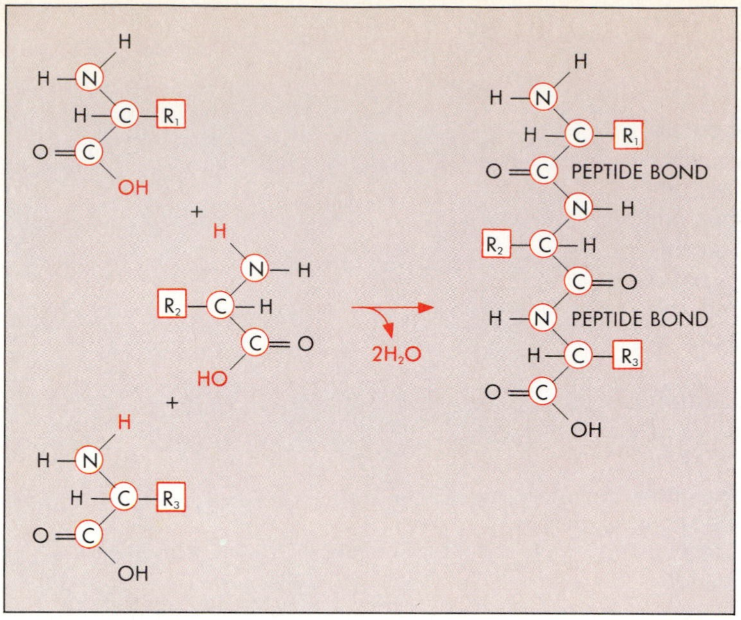

FIGURE 3-10
Amino acids combine by way of peptide bonds to form polypeptide chains. Peptide bonds are formed by the removal of one water molecule per peptide bond, referred to as a condensation reaction.

(Figure 3-10). As in disaccharide formation, this occurs through the removal of water. The dipeptide that results from the union of two amino acids has a free carboxyl group at one end, called its **C-terminal**. The opposite, **N-terminal** end has a free amino group. Peptides can therefore combine with additional amino acids to form **polypeptide** chains.

Protein Structure and Function

Life is based on the ability to create an infinite number of different protein molecules through combinations of the 20 amino acids. The linear sequence of amino acids that make up a protein is referred to as its **primary structure** (Figure 3-11, *A*). Hydrogen bonds can form between the carboxyl portion of a peptide bond and the amino group of a peptide bond further down the chain. Such interactions result in the **secondary structure** of proteins (Figure 3-11, *B*). Secondary structures include a corkscrew-shaped **alpha helix** and an association between adjacent amino acid chains called a **pleated sheet**.

The complete three-dimensional conformation that a protein assumes is its **tertiary structure** (Figure 3-11, *C*). The tertiary structure is stabilized by hydrophobic interactions and the formation of covalent disulfide bonds between cysteine subunits. The tertiary structure assumed by a particular protein represents the lowest free energy state for that protein, and the laws of thermodynamics dictate that conformational changes, like all reactions, take place in a direction that results in a decrease in free energy. In the final, three-dimensional form of a protein, single amino acids located some distance apart in terms of the primary (linear) amino acid sequence may be spatially adjacent.

Two or more polypeptides may be assembled to form a functional protein complex; this is the case for some enzymes and for hemoglobin, the oxygen-transporting protein of red blood cells. In this case the relationship of the constituent polypeptides to one another composes the **quartenary structure** of the protein (Figure 3-11, *D*).

The primary amino acid sequence determines all the intramolecular forces in a protein and therefore its tertiary structure. Changing only a few amino acids in a polypeptide can profoundly alter its tertiary structure. An example of such an effect is provided by the genetic disease **sickle cell anemia**, which is the result of a substitution of valine for glutamic acid in position 6 in the beta chain of oxygen-carrying protein hemoglobin. The result of this substitution is that the hemoglobin can aggregate, forming strands. Red blood cells in which some of the hemoglobin has aggregated are distorted into a half-moon shape, referred to as **sickled** (Figure 3-12). Such cells do not pass readily into capillaries.

The tertiary structure of a protein is highly dependent on the formation of internal disulfide bonds, hydrogen bonds, hydrophobic interactions, and on van der Waals forces. Weak bonds confer flexibility on the tertiary structure, a fact that has important consequences. Changes in the tertiary structure are sometimes involved in regulating the function of proteins. By the same token, factors that change the interactions between amino acids affect the tertiary structure of a protein and may impair its function. Physiological examples of such factors include temperature and pH. Temperatures in excess of about 45° C cause the parts of protein molecules to vibrate more strongly relative to one another, breaking some of the weak bonds and causing changes in tertiary structure called **denaturation**.

> 1 Distinguish between the primary, secondary, tertiary, and quaternary structures of a protein. What specifies the primary structure?
> 2 What are the chemical interactions that determine the secondary, tertiary, and quaternary structures of proteins?

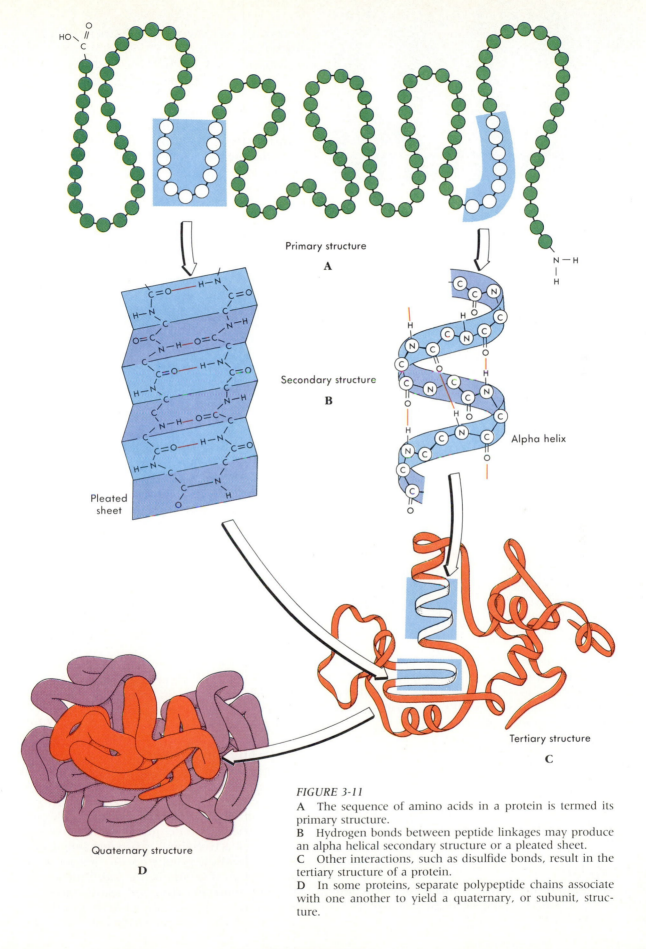

Primary structure

A

Secondary structure

B

Pleated sheet

Alpha helix

Tertiary structure

C

Quaternary structure

D

FIGURE 3-11

A The sequence of amino acids in a protein is termed its primary structure.

B Hydrogen bonds between peptide linkages may produce an alpha helical secondary structure or a pleated sheet.

C Other interactions, such as disulfide bonds, result in the tertiary structure of a protein.

D In some proteins, separate polypeptide chains associate with one another to yield a quaternary, or subunit, structure.

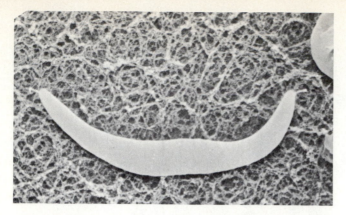

FIGURE 3-12
Sickle cell hemoglobin is produced by a mutated form of the gene that codes for the beta-chain of hemoglobin. Sickling results from aggregation of the hemoglobin into long strands, producing a change in the shape of the red blood cell, as seen in this scanning electron micrograph.

ENZYME CATALYSIS
Law of Mass Action

Suppose A, B, C, and D represent four chemical compounds reacting as follows:

$$A + B \rightleftharpoons C + D$$

For a reaction between A and B (or C and D) to occur, the molecules must collide with one another and then react. The rate of the forward reaction from left to right will depend on the product of the concentrations of the reactants A and B because this determines the number of collisions per unit time, and on a **rate constant** (f) that indicates the probability that a reaction between A and B will take place:

$$\text{Forward rate} = f\,[A]\,[B]$$

where [A] and [B] are the standard ways of designating the concentrations of A and B, respectively. A similar argument applies to the reverse reaction between C and D so that:

$$\text{Reverse rate} = r\,[C]\,[D]$$

where r is the rate constant for the reverse reaction.

At equilibrium there can be no changes in the concentrations of either the reactants or products with time. This can only occur if the forward and reverse reaction rates are exactly equal:

$$\text{Forward rate} = \text{Reverse rate}$$

or

$$f\,[A]\,[B] = r\,[C]\,[D]$$

Dividing both sides of this equation by r [A] [B] gives the **Law of Mass Action**:

$$\frac{f}{r} = \frac{[C]\,[D]}{[A]\,[B]}$$

where the ratio of the forward and reverse reaction rates (f/r) is defined as the **equilibrium constant (Keq)** of the reaction.

Enzymes and the Rates of Biochemical Reactions

Chemical reactions involving decreases in free energy can occur spontaneously. However, at normal body temperature the rate may be very slow. For example, even though it produces heat, the oxidation of sugars is so slow that there is no danger of the sugar bowl on the dining-room table spontaneously catching fire. The pace of life would be very slow if sugars were oxidized in the body by spontaneous reactions.

Chemical reactions involve **transition states**, whose free energy is higher than the total free energy of either the reactants or products (Figure 3-13, A). The difference between the free energy of the reactants and the free energy of the transition state is called the **activation energy** of a reaction. The probability that a reaction will occur in either direction depends on the magnitude of the activation energy compared with the inherent thermal energy of the reactants and products.

The difference between the activation energies of the forward and reverse reactions determines the equilibrium constant of the reaction, while the absolute rates of the forward and reverse reactions are determined by the absolute magnitude of the activation energies. That is, an increase in the energy of the transition state will decrease both the forward and reverse reaction rates, but will leave the equilibrium constant unchanged. A decrease in transition state energy accelerates both reactions, but again does not affect the final equilibrium.

Enzymes act by reducing the activation energy of a reaction, and enzymes are therefore referred to as **catalysts** (Figure 3-13, B). For example, the invertase enzyme that mediates sucrose hydrolysis reduces the activation energy from 26 kcal/mole to about 9 kcal/mole and increases the rate of the forward reaction by many orders of magnitude. Another example of physiological importance is the reaction of carbon dioxide with water to form bicarbonate and the hydrogen ion. The reaction is greatly accelerated by the enzyme **carbonic anhydrase**. This enzyme is present at various sites in the body where CO_2 must be rapidly converted to bicarbonate.

Active Sites on Enzymes

The first step in enzymatic catalysis is binding of a substrate molecule to an **active site** on the surface of the enzyme (Figure 3-14). The active site is a pocket in the enzyme into which the appropriate part of the substrate just fits. When a substrate binds to an active site it is brought into contact with specific chemical groups on the enzyme that destabilize chemical bonds in the substrate, making the substrate more reactive. Because of the large number of ways in which the 20 amino acids can combine to form enzymes, specific tertiary structures

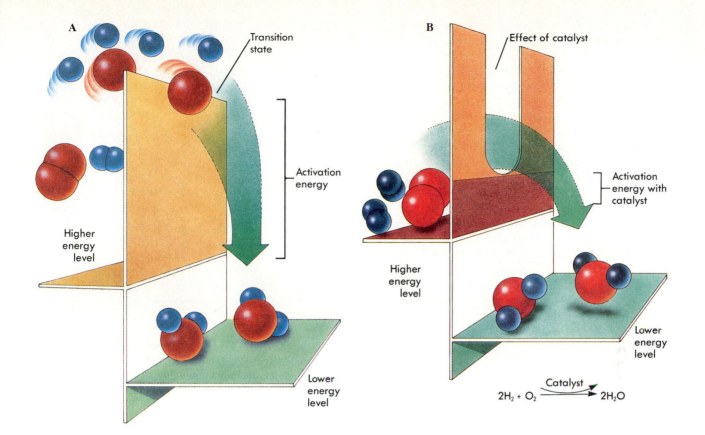

$$2H_2 + O_2 \longrightarrow 2H_2O$$

$$2H_2 + O_2 \xrightarrow{\text{Catalyst}} 2H_2O$$

FIGURE 3-13

A Chemical reactions involve intermediate transitional states that can be viewed as energy barriers somewhat higher than the mean thermal energy of reactants and products, illustrated as shelves of unequal height. Even though energy is given up in the reaction, the barrier must be surmounted for the reaction to occur. The height of the barrier is termed the activation energy, and it affects both the forward and reverse reaction rates. The difference in the energies of the reactants and products determines the ratio of reactants to products at equilibrium.

B Enzymes reduce the activation energy for chemical reactions and thus increase the probability that a reactant or product will have sufficient energy to surmount the barrier. As a result, enzymatic catalysis increases the rate of both the forward and the reverse reactions, and thus the overall reaction rate. Catalysis does not change the energy of reactants and products, so it does not change the equilibrium constant.

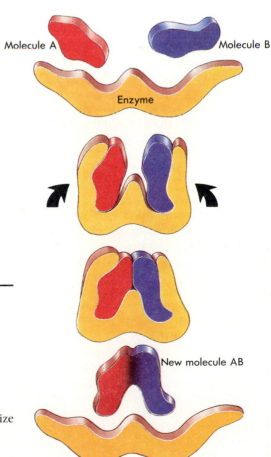

FIGURE 3-14

Enzymes have active sites that bind substrates. The active sites must recognize some aspect of the molecular structure of their substrates. While they participate in the reaction, enzymes are not permanently altered and can catalyze further reactions.

can be created that will fit almost any substrate. Usually only one or at most a limited number of substrates can fit the active site, so that enzymes are typically highly specific for particular substrates.

The number of fundamental biochemical reactions is limited. Compounds can be formed or degraded by the removal or addition of water, previously defined as condensation and hydrolysis, respectively. Compounds can be produced by the removal or addition of an electron, termed **oxidation** and **reduction**. If one molecule gives up an electron (it is oxidized), another must accept the electron (it is reduced). It is therefore usual to refer to such processes as paired **oxidation-reduction reactions**. Oxidation-reduction reactions are the basis for energy production in cells (see Chapter 4, p. 75). Other physiologically important reactions include the removal or addition of a hydrogen (**dehydrogenation** or **hydrogenation**) and the addition or removal of phosphate groups (**phosphorylation** or **dephosphorylation**).

---

1 What is the relationship between the activation energies of forward and reverse reactions and the concentrations of reactants and products at equilibrium?
2 What effect does an enzyme have on the activation energy of the reaction it catalyzes?
3 What is the role of the active site of an enzyme?

Enzyme-Catalyzed Reaction Rates

In a mixture of enzyme and substrate the rate of formation of product molecules is directly proportional to the number of active sites that are occupied by substrate molecules. If the concentration of substrate is relatively dilute compared with that of the enzyme, at any instant some, but not all, of the active sites will be occupied. If the concentration of substrate molecules is increased, the number of sites in use increases, and the rate of product formation increases. This increase of product formation can be thought of as a kind of self-control of the proportions of substrate and product in the cell. Ultimately, a substrate concentration is reached at which all the sites are busy. At this point the enzyme is said to be **saturated**; further increases in substrate concentration do not increase the rate of formation of product. Figure 3-15, A shows the rate of product formation as a function of substrate concentration for an enzyme-catalyzed reaction. If enzymatic reactions in cells are to be self-controlled, the concentrations of enzymes and their substrates must be such that the enzymes are not saturated.

If two substrates are structurally similar to one another, the active site of an enzyme may be able to bind both of them. The tenacity with which an enzyme binds a substrate is referred to as its **affinity** for the substrate. Enzymes typically have different affinities for different substrates. If the difference in

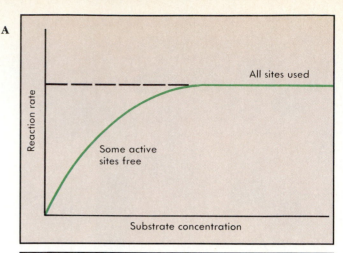

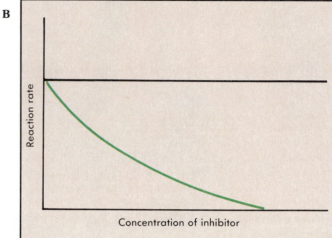

FIGURE 3-15

A The function that describes the relationship between the reaction rate and the substrate concentration for a given solution of enzyme. When all active sites of a population of enzyme molecules are occupied, the enzyme is said to be saturated.

B Substances whose molecular structure is similar to that of the natural substrates can sometimes bind to the active site and act as competitive inhibitors. The graph illustrates the relationship between the reaction rate and the concentration of a competitive inhibitor. The substrate concentration is held constant, so that if the inhibitor were not present the rate would correspond to the horizontal line.

affinities is small, substrates can **compete** with one another for the enzyme's active sites. If one of the competing substrates is bound more tightly than the normal substrate but cannot take part in the enzyme-mediated reaction, it may block the activity of the enzyme. This is called **competitive inhibition** (Figure 3-15, B).

Many enzymes have binding sites for molecules other than the substrates of the reaction they mediate. Noncatalytic binding areas are called **allosteric sites**, and molecules that affect the reaction rate when they bind to allosteric sites are called **effectors** or **regulators** (Figure 3-16). An **allosteric activator** increases the activity of the active site; an **allosteric inhibitor** decreases it. The latter effect is **noncompetitive inhibition**. Allosteric effectors

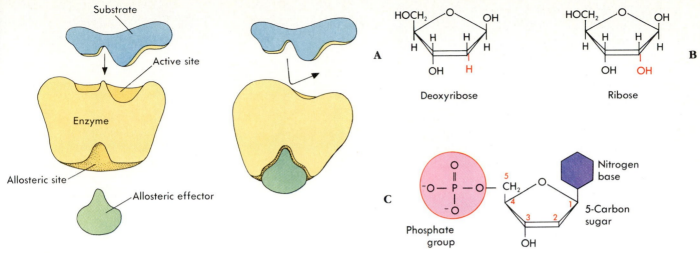

FIGURE 3-16
Many enzymes have allosteric sites that can influence enzyme activity.

FIGURE 3-17
The structures of the nucleic acids. Nucleotides are composed of a sugar, deoxyribose (A in DNA) or ribose (B in RNA), linked to a phosphate group and a base (C).

may also modify the relative affinity of an active site for different substrates.

Allosteric effects are the mechanism by which the activities of enzymes are regulated. Many of the reactants and products present in cells can act as allosteric effectors and thus influence the reactions that generate or consume them. Enzymes are also allosterically regulated by intracellular second messenger systems (Chapter 5), and allosteric effects are the major basis for intracellular negative feedback control systems.

Enzyme activity or substrate affinity can also be altered by covalent modifications of the structure of an enzyme. In most cases, either a phosphate group is added by an enzyme called a **protein kinase**, or one is removed by a **phosphatase**. Covalent modifications can occur both at the active site or elsewhere on the enzyme molecule and can either increase or decrease the enzyme's activity. Enzymes controlled by allosteric or covalent modifications are usually those at control points in intracellular metabolic pathways.

> *1 What effect would the presence of a competitive inhibitor have on the amount of substrate needed to saturate an enzyme?*
> *2 What is an allosteric effector? Why can allosteric inhibition be described as noncompetitive?*

NUCLEIC ACIDS
Structure of Nucleic Acids
The genetic material present in the nucleus of a cell is in the form of **chromosomes**. Each chromosome consists of two interconnected strands of **deoxyribonucleic acid (DNA**; see Table 3-1) about 50 nm long, with some associated smaller molecules. Human cells have 46 chromosomes in 2 sets of 23 ho-

mologous pairs. **Genes** are portions of the chromosomes that direct the synthesis of proteins. Intermediate steps in the production of proteins involve **ribonucleic acids (RNAs**; see Table 3-1), whose structures are specified by DNA. There are three types of RNA: **ribosomal RNA, messenger RNA** (usually seen abbreviated as mRNA), and **transfer RNA** (see p. 58). **Transcription** is the process by which DNA determines the composition of mRNA. After several intermediate steps, the information carried from DNA by messenger RNA is **translated** in the cytoplasm into the amino acid sequence of a particular protein.

Nucleic acids are long polymers of subunit molecules called **nucleotides** joined by phosphodiester bonds (Figure 3-17). Each nucleotide is made up of a sugar, a phosphate group, and either a **purine** or **pyrimidine** base (Figure 3-18). In the ribonucleic acids the sugar is **ribose**, while deoxyribonucleic acids contain **deoxyribose**. In DNA the base can be **adenine** (abbreviated A), **guanine** (G), **cytosine** (C), or **thymine** (T). In RNA **uracil** (U) replaces thymine as one of the pyrimidine bases.

Hydrogen bonds are formed between pairs of nucleotide bases. As a result, nucleotide chains have a particular three-dimensional form. In RNA, intramolecular hydrogen bonds cause the single nucleotide chain to coil into a helix. In DNA, two helices intertwine with one another to form a **double helix** as a result of hydrogen bonds between bases on the two different nucleotide chains (Figure 3-19).

DNA Replication
Base pairing between nucleotides follows certain rules. Adenine pairs only with thymine (uracil in the case of RNA) and guanine only with cytosine.

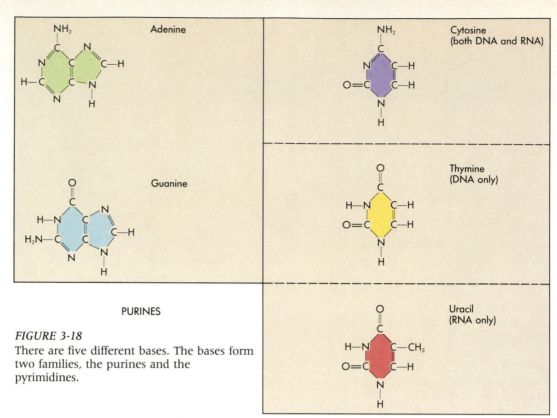

FIGURE 3-18
There are five different bases. The bases form two families, the purines and the pyrimidines.

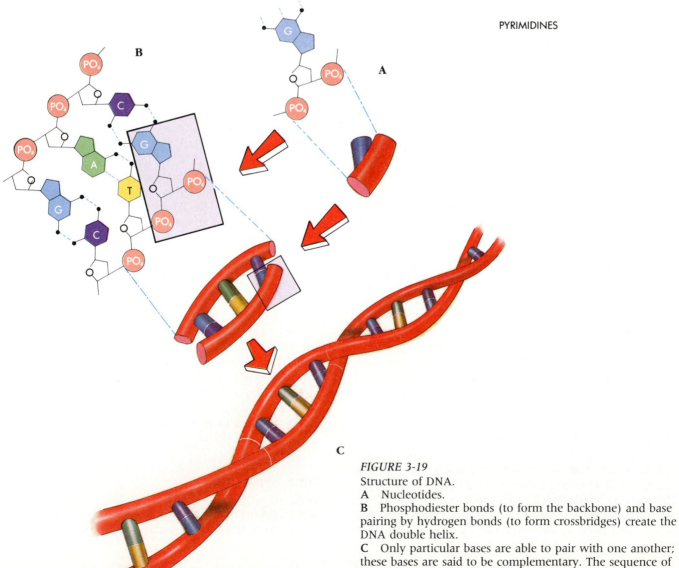

FIGURE 3-19
Structure of DNA.
A Nucleotides.
B Phosphodiester bonds (to form the backbone) and base pairing by hydrogen bonds (to form crossbridges) create the DNA double helix.
C Only particular bases are able to pair with one another; these bases are said to be complementary. The sequence of bases on one strand of DNA is matched by a sequence of their complementary bases on the other strand.

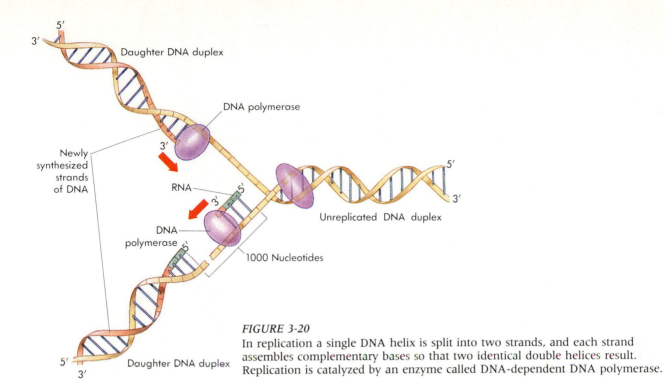

FIGURE 3-20
In replication a single DNA helix is split into two strands, and each strand assembles complementary bases so that two identical double helices result. Replication is catalyzed by an enzyme called DNA-dependent DNA polymerase.

Complementary nucleotide base-pairing enables nucleic acids to reproduce themselves, so that dividing cells can pass a copy of the blueprint to each of the daughter cells. The sequence of nucleotides in DNA is the basis of the genetic code for all of the proteins that cells contain. Because DNA molecules copy themselves, organisms can pass on genetically determined characteristics to their offspring.

The process in which a single DNA molecule makes an exact duplicate of itself is called **replication**. Replication requires the presence of several enzymes that cause unwinding of the double helix and separation of the two strands. A complementary strand develops along each single strand by the action of **DNA polymerases**, which catalyze base pairing (Figure 3-20). The result is the production of two identical double-stranded DNA molecules, each containing one of the original nucleotide strands. Replication takes place in the nucleus of every cell before **mitosis**, the process of nuclear division that provides each daughter cell with an identical copy of the genetic information present in the parent nucleus.

Tissue Growth and Cell Replacement

The life of a cell can be divided into several periods (Figures 3-21 and 3-22). During **interphase**, the chromosomes are extended and not visible by light microscopy. During the first part of interphase, the **G_1 period**, the DNA directs synthesis of proteins. During the **S period**, the chromosomes replicate in preparation for the next mitosis. Each chromosome now consists of two identical **sister chromatids**.

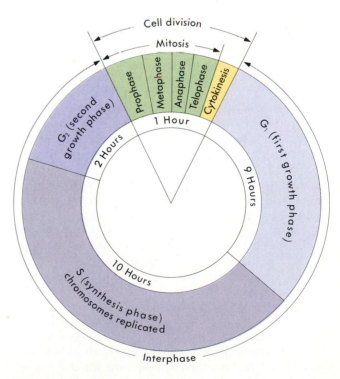

FIGURE 3-21
The cell cycle of liver cells grown in culture. During the S phase the chromosomes replicate.

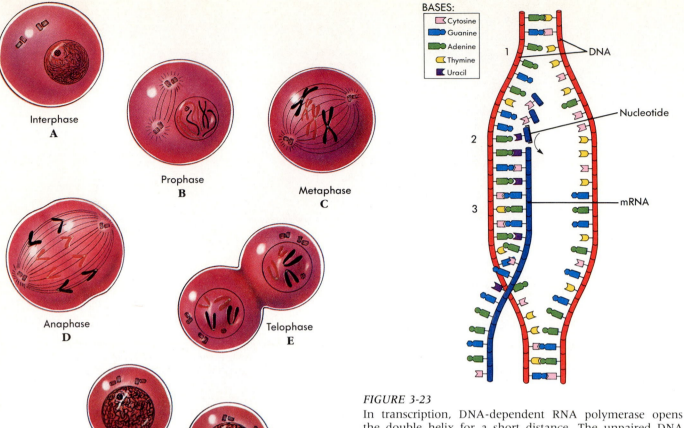

FIGURE 3-22
A diagram of the process of mitosis.

BASES:
- Cytosine
- Guanine
- Adenine
- Thymine
- Uracil

Interphase
A

Prophase
B

Metaphase
C

Anaphase
D

Telophase
E

Interphase
F

DNA

Nucleotide

mRNA

FIGURE 3-23
In transcription, DNA-dependent RNA polymerase opens the double helix for a short distance. The unpaired DNA bases of one strand of the open part of the helix are used as a template to assemble a single-stranded RNA molecule whose base sequence is complementary to that of the DNA strand. The polymerase closes the helix after the base sequence has been transcribed, at the same time opening a space for transcription of the next few bases in the sequence. The polymerase continues in this fashion down the DNA helix, trailing a lengthening strand of RNA behind it.

The S period is followed by the **G₂ period.** The first visible sign of the onset of mitosis is condensation of the chromosomes into visible entities (**prophase**). The nuclear membrane breaks down. The chromosomes become attached to the **mitotic spindle** (**metaphase**) and line up prior to separating as the spindle fibers draw sister chromatids in opposite directions (**anaphase**). The result is division of the replicated chromosomes into two equal groups. New nuclear membranes form (**telophase**), and the cytoplasm divides (termed cytokinesis) to generate two daughter cells, each with a single nucleus containing chromosomes identical to those of the parent cell. The chromosomes uncoil, and the daughter cells enter interphase.

The rate of mitosis is much higher during growth and development than during adulthood and is characteristically higher in some tissues than in others. The cells of some tissues, such as nervous tissue, have the potential to survive for the life of the individual and are not normally replaced if they die. Other tissues, such as epithelial tissue (including both the skin and the lining of the gas-

trointestinal tract), undergo continuous loss and replacement of cells during one's lifetime.

The four basic tissues of the body (epithelial, connective, nervous, and muscle) include at least 200 different cell types. What determines whether a cell eventually functions as a muscle cell, brain cell, or liver cell? The answer is that although each cell has the same genetic information, only a part of it is involved in determining the specific characteristics of each cell type. In the course of development of specific cell types, many genes become permanently "turned off."

PROTEIN SYNTHESIS
DNA Transcription into RNA
In the nucleus of a cell, specific portions of one strand of double-stranded DNA are **transcribed** into single-stranded molecules of RNA by an enzyme referred to as **DNA-dependent RNA polymerase.** The resulting mRNA has a nucleotide sequence complementary to that of the DNA and contains uracil rather than thymine (Figure 3-23).

TABLE 3-3   Amino Acid Codons

Amino acid	Nucleotide code			
Alanine	GCU	GCC	GCA	GCG
Arginine	CGU	CGC	CGA	CGG
Asparagine	AAU	AAC		
Aspartic acid	GAU	GAC		
Cysteine	UGU	UGC		
Glutamic acid			GAA	GAG
Glutamine			CAA	CAG
Glycine	GGU	GGC	GGA	GGG
Histidine	CAU	CAC		
Isoleucine	AUU	AUC	AUA	
Leucine	CUU	CUC	CUA	CUG
Lysine			AAA	AAG
Methionine				AUG
Phenylalanine	UUU	UUC		
Proline	CCU	CCC	CCA	CCG
Serine	UCU	UCC	UCA	UCG
Threonine	ACU	ACC	ACA	ACG
Tryptophan	UAU	UAC		
Tyrosine	UAU	AUC		
Valine	GUU	GUC	GUA	GUG
Stop codons			UAA	UAG
			UGA	

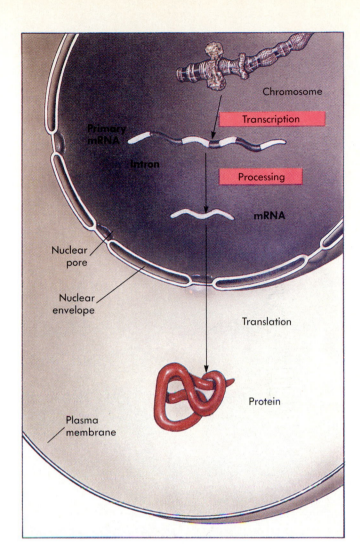

FIGURE 3-24

Messenger RNA synthesized from DNA, the primary transcript, may contain coding regions known as exons, and noncoding regions known as introns. In posttranscriptional processing, introns are removed from the primary transcript; the exons form the final messenger RNA (mRNA) that is exported to the cytoplasm to serve as a template for protein synthesis.

The functional units of nucleic acid molecules are sequences of three nucleotides called **codons**, each of which specifies a particular amino acid. For example, the base sequences UUU, UAA, and UGC represent different RNA codons. Since there are four nucleotides, there are 64 (4 × 4 × 4) different codons. Since there are only 20 amino acids, each amino acid may be specified by more than one codon (termed **redundancy**). The first two nucleotides of a codon are the most critical for amino acid specification. Thus CCA, CCC, CCG, and CCU all code for the amino acid proline, while UUx codes for either leucine or phenylalanine, depending on the third nucleotide (Table 3-3). One codon (AUG, which also codes for the amino acid methionine) serves as a **start codon** signaling the beginning of a sequence to be transcribed. Three codons (UAA, UAG, and UGA) signal the end of a sequence to be read and are therefore known as **termination sequences** or "stop" codons.

The production of mRNA involves two steps. Initially, the DNA-dependent RNA polymerase binds to a promoter region located on one strand of a DNA molecule. The two DNA strands separate locally, and the strand that will act as a template for the assembly of RNA is read to produce a complementary single-stranded mRNA molecule until a termination sequence is reached. Base pairing occurs spontaneously, but formation of bonds linking the nucleotides in linear sequence requires DNA-dependent RNA polymerase. This first portion of the process results in formation of the **primary transcript** (primary mRNA in Figure 3-24).

The primary transcript is subsequently modified in the nucleus of the cell, a procedure called **posttranscriptional processing**. As much as 75% to 90% of the primary transcript of mRNA does not specify for amino acids present in the protein; such noncoding sections are **introns**. The remaining regions of the primary transcript are coding sections called **exons**. Specific nucleotide sequences designate the beginning and end of introns. Introns are removed before the final exon-containing messenger RNA leaves the nucleus. The exons in a particular primary transcript can sometimes be spliced together in different ways so that several distinct mRNAs (and therefore proteins) can be produced

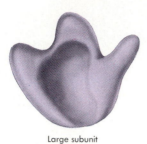

Small subunit

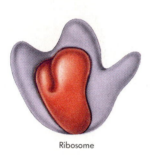

Large subunit

Ribosome

FIGURE 3-25
A ribosome is composed of a large and a small subunit, each containing both RNA and protein. The small subunit has the mRNA binding site.

from the same primary transcript. This versatility may explain how different antibodies are generated (discussed in Chapter 24).

mRNA Translation into Proteins

Eventually mRNA molecules enter the cytoplasm where they direct the synthesis of proteins in a process called **translation**. Messenger RNAs combine with a **ribosome** (Figure 3-25), a catalytic complex composed of proteins and several different RNAs called **ribosomal RNA** (rRNA), which are distinct from other types of RNA. The cytoplasm contains a third, smaller class of RNA molecules called **transfer RNA** (tRNA, Figure 3-26). There are different transfer RNA molecules for each of the 20 amino acids. At one end of the tRNA an amino acid can be bound, and at the other end there is a sequence of three nucleotides called an **anticodon** that is complementary to the codons of the mRNA molecules. The ribosome accepts particular tRNAs on the basis of what 3-nucleotide sequence appears next on the mRNA, and the associated amino acid is thus incorporated into a growing polypeptide chain. The second tRNA binding site on the ribosome is occupied by the tRNA carrying the next amino acid to be added (Figure 3-27). Translation stops when a particular stop codon is encountered.

There are several types of posttranslational modifications of proteins. Some proteins are synthesized as inactive **proenzymes** and only become active enzymes when they are later modified, usually by the removal of small peptides. The digestive enzymes of the digestive tract are generally secreted as proenzymes. Proteins destined to be incorporated

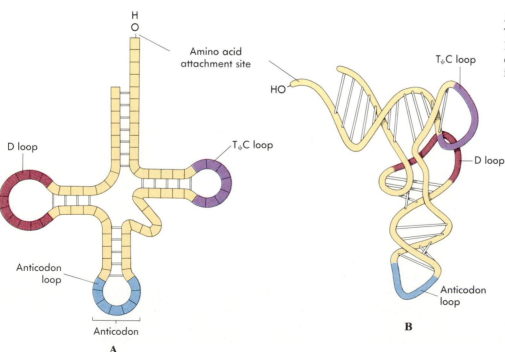

FIGURE 3-26
The structure of a transfer RNA (tRNA) molecule shown diagramatically (left) and in its three-dimensional form.

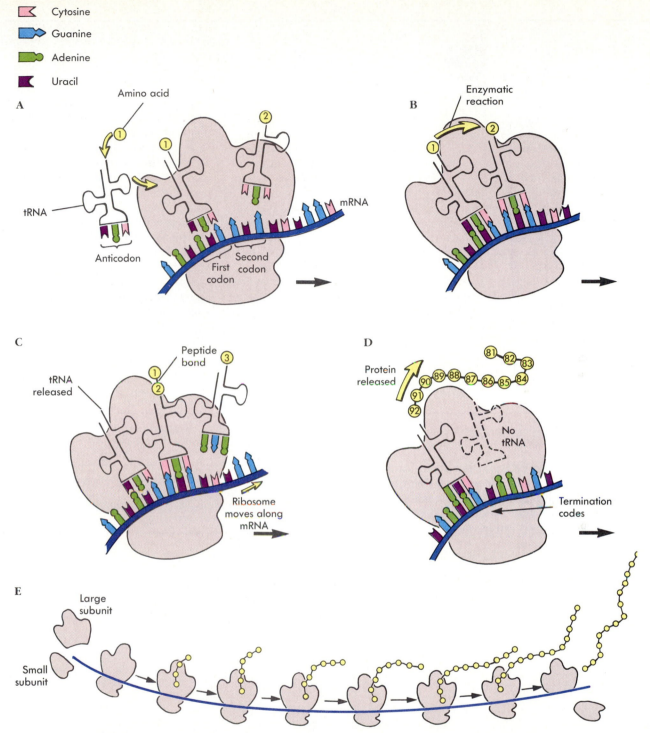

Cytosine

Guanine

Adenine

Uracil

A

Amino acid

tRNA

Anticodon

First codon

Second codon

mRNA

B

Enzymatic reaction

C

tRNA released

Peptide bond

Ribosome moves along mRNA

D

Protein released

No tRNA

Termination codes

E

Large subunit

Small subunit

FIGURE 3-27

Protein synthesis from messenger RNA by translation.

A The mRNA strand becomes associated with the ribosome. Amino acids attach to specific transfer RNA (tRNA) molecules that recognize (through their complementary anticodons) triplets of three bases on mRNA called codons. The other end of each transfer RNA binds the corresponding amino acid.

B An enzyme within the ribosome catalyzes the formation of peptide bonds between the amino acids that are delivered by tRNA.

C The formation of a dipeptide removes the amino acid from one tRNA, which is released from the ribosome as another tRNA enters the ribosome and its anticodon combines with the adjacent codon on the mRNA.

D The peptide chain is formed as amino acids from tRNAs are joined by peptide bonds until a termination codon on the mRNA stops chain elongation and the protein chain then separates from the ribosome. The ribosome then separates into its two subunits and separates from the mRNA.

E A schematic overview of the entire process.

Genes Can Be Manipulated

Genetic engineering has many goals. One is to treat or cure victims of genetic diseases. Another is to speed up the process of selection of desirable traits in domestic animals and plants. Another is to utilize the synthetic machinery of cells in culture to make large amounts of proteins that are important as drugs.

In many cases the genetic engineer's first step toward manipulating the gene that produces a particular protein is to determine the amino acid sequence of the protein. Once this has been done, a corresponding mRNA can be synthesized. The next step may be to synthesize the corresponding DNA from the mRNA, utilizing an enzyme called **reverse transcriptase**. The engineer then possesses the gene coding for the protein of interest. As an alternative to the above procedure, a cell's own DNA can serve as the starting point. There are a variety of **restriction endonucleases** that will cut DNA wherever certain sequences of 4 to 6 base pairs occur. Resulting fragments of DNA are incorporated into viruslike agents, termed **vectors**. The vectors transfer the fragments into bacteria in which the "new" DNA can be made to fuse with the bacterial DNA. Bacterial cells that receive a fragment containing the gene of interest will synthesize the corresponding protein and can be selected for large-scale culture. Foreign DNA that is incorporated into the genetic material of a cell is called **recombinant DNA**.

The hormone **somatostatin** (14 amino acids), synthesized in 1977, was the first peptide to be synthesized artificially. Since then, **growth hormone, insulin** (a hormone essential for metabolism of glucose), **interferon** (a protein important in the immune response against viruses and cancer), and **relaxin** (a hormone used to relax the birth canal in delivery) have been prepared commercially. Clearly the potential benefits of recombinant DNA technology are just beginning to be realized.

into membranes often have sugar residues added to them, and in other cases the side chains of proteins are phosphorylated after their initial translation. Large proteins may also be broken down into a series of smaller, related polypeptides, each of which may have a different function.

1 *Distinguish between the processes of replication, transcription, and translation.*
2 *What is the difference between intron and exon sequences of a primary transcript?*

SUMMARY

1. Carbon forms bonds to other carbon atoms and to hydrogen and oxygen to form **lipids** and **carbohydrates**. Lipids and carbohydrates serve as elements in cellular structure and as sources of chemical energy. In animals, the major storage form of lipid is **triglyceride**, and the major storage form of carbohydrate is **glycogen**, a polymer of **glucose**.
2. **Proteins** are polymers of **amino acids** joined by **peptide bonds**. The sequence of amino acids in a protein is coded for by a sequence of nucleotide **codons** in DNA. The three-dimensional structure of a protein is the result of the formation of internal bonds between amino acids in different parts of the molecule—the structure that is formed is the most stable of the many structures possible.
3. **Enzymes** are proteins that catalyze specific chemical reactions. Catalysis involves a decrease in the activation energy that reactants must surmount to become products.
4. The double-stranded DNA molecule provides the sequence code for all proteins made in the body. In the process of **mitosis**, the DNA molecules separate into single nucleotide strands; each strand then **replicates** a **complementary** strand by base pairing.
5. The first step in synthesizing a specific protein is **transcription** of the nucleotide sequence of a gene into messenger RNA.
6. In the second step of protein synthesis, messenger RNA passes through nuclear membrane pores into the cytoplasm, where ribosomes **translate** its message into protein. In this process, the ribosome moves along the messenger RNA, attaching molecules of transfer RNA that carry the amino acids specified by the messenger RNA.

● **STUDY QUESTIONS**

1. What are the four classes of organic compounds found in living cells? What are the unique characteristics of each class? How are their chemical structures related to their physiological function?

2. Define "enzyme." How do such molecules affect the rate of chemical reactions? What features do all enzyme-mediated reactions have in common? What is meant by the term "allosteric regulation"?

3. At what stage in the cycle of cell growth and mitosis do the following events occur:
   DNA replication
   Chromosome condensation
   Separation of sister chromatids

4. Describe the process by which a chromosome forms a replica of itself.

5. How is the base sequence of a strand of DNA transcribed into a strand of RNA? What are the differences between the DNA strand and its transcript RNA?

6. Describe the process by which a primary transcript is shortened to form messenger RNA. What is the name of this process?

7. What is a codon? The genetic code has been called redundant. Why?

● **SUGGESTED READING**

BYLINSKY, G.: **Coming: Star Wars medicine,** *Fortune,* volume 115, April 27, 1987, p. 153. Describes how monoclonal antibodies from the body's own immune system can be used to speed medical diagnosis and precisely target drugs to kill cancer cells.

HOOD, L.: **Biotechnology and the medicine of the future,** *Journal of the American Medical Association,* volume 259, March 25, 1988, p. 1837. Discusses how recombinant DNA technology can be applied to clinical medicine.

LERNER, R.A., and A. TRAMONTANO: **Catalytic antibodies,** *Scientific American,* volume 258, March 1988, p. 58. Nature has designed a limited number of enzymes. This article explores the possibility of designing antibodies to serve as catalysts in biochemical reactions.

MURRAY, A.W., and J.W. SZOSTAK: **Artificial chromosomes,** *Scientific American,* volume 257, November 1987, p. 62. Artificial chromosomes, custom-designed sequences of DNA, are being used to examine the accurate transmission of genetic information and to provide a tool for the genetic engineering of proteins.

PRESCOTT, D.M.: *Cells: principles of molecular structure and function,* Jones and Bartlett, Boston, 1988. A readable text covering the content of this chapter.

STEITZ, J.A.: **Snurps,** *Scientific American,* volume 259, June 1988, p. 56. Shows how small nuclear ribonucleoproteins (snurps) delete introns from primary messenger RNA to produce the mRNA that actually directs protein synthesis.

CHAPTER 4

The Structure and Energy Metabolism of Cells

On completing this chapter you will be able to:

- Identify the major organelles of cells and provide a brief description of their physiological role.
- Describe the structure and function of desmosomes, tight junctions, and gap junctions.
- Outline the biochemical pathways by which carbohydrates are metabolized, distinguishing between the processes that occur under anaerobic and aerobic conditions.
- Distinguish between substrate-level phosphorylation and oxidative phosphorylation.
- Describe the ways amino acids and fatty acids can enter the metabolic pathways of the cell and thus be used as energy sources.
- State the chemiosmotic hypothesis.
- Summarize the net yields of ATP obtained under anaerobic and aerobic conditions.

The structural organization of cells can be seen with microscopes of various kinds. Looking at pictures of the structures of cells is like looking into a large factory building during a business holiday. Machines, offices, raw materials, and finished goods are visible, but the processes of supply, repair, administration, shipping, and finance that underlie the operation and carry out its function can only be imagined. Life is, in the final analysis, a sum of processes. Cells may retain their structures if carefully frozen and begin to live again when thawed, but while they are frozen they cannot be said to be alive.

Living organisms are thermodynamic machines that use materials from the environment—foods—to produce energy and increase order within themselves. The organic compounds that the body metabolizes for energy are all characterized by the fact that they are able to give up electrons. Ultimately these electrons are transferred to oxygen in the powerhouses of the cell, its mitochondria. In the process of electron transfer, some of the energy contained in foods is captured in the form of high-energy phosphate bonds in ATP, the rest being released as heat. The captured energy is then used to fuel energy-requiring processes, such as synthesis of proteins and other cellular constituents, contraction of muscles, and transport of materials across cell membranes.

BASIC CELL STRUCTURE
Cell and Cell Membrane

Figure 4-1 shows an idealized cell and the major organelles found within cells. Although any particular cell may not have all the organelles illustrated, most contain a large number of them (Figure 4-2). Table 4-1 summarizes the components of cells and briefly outlines their physiological function.

All cells are surrounded by a **plasma membrane (cell membrane)** (Figure 4-3), a thin layer largely composed of phospholipids and proteins. It is important to understand the nature of this barrier. Lipids can exist as solids or liquids, depending on the temperature of their environment relative to the melting points of the lipids. Lipids containing unsaturated fatty acids have lower melting points than those containing saturated ones. In order to function, cell membranes must be neither entirely solid nor entirely liquid, but must be poised between the two states.

This combination of flexibility and rigidity is maintained by incorporation of a mixture of unsaturated and saturated phospholipids into the cell membrane. The effect is a membrane that is both ordered and disordered. Membranes are orderly in that they consist of two layers (a **bilayer**) of phospholipid molecules, all of which are standing at right angles to the surface of the membrane. Membranes are disorderly in that, within the bilayer, lipid molecules are free to jostle about like people in a crowd. Such a structure is called a **liquid crystal.**

All material transport must occur across the plasma membrane. Pure phospholipids form bilayers that are impermeable to electrolytes and large polar compounds; only nonpolar lipid-soluble compounds and small polar molecules such as water can

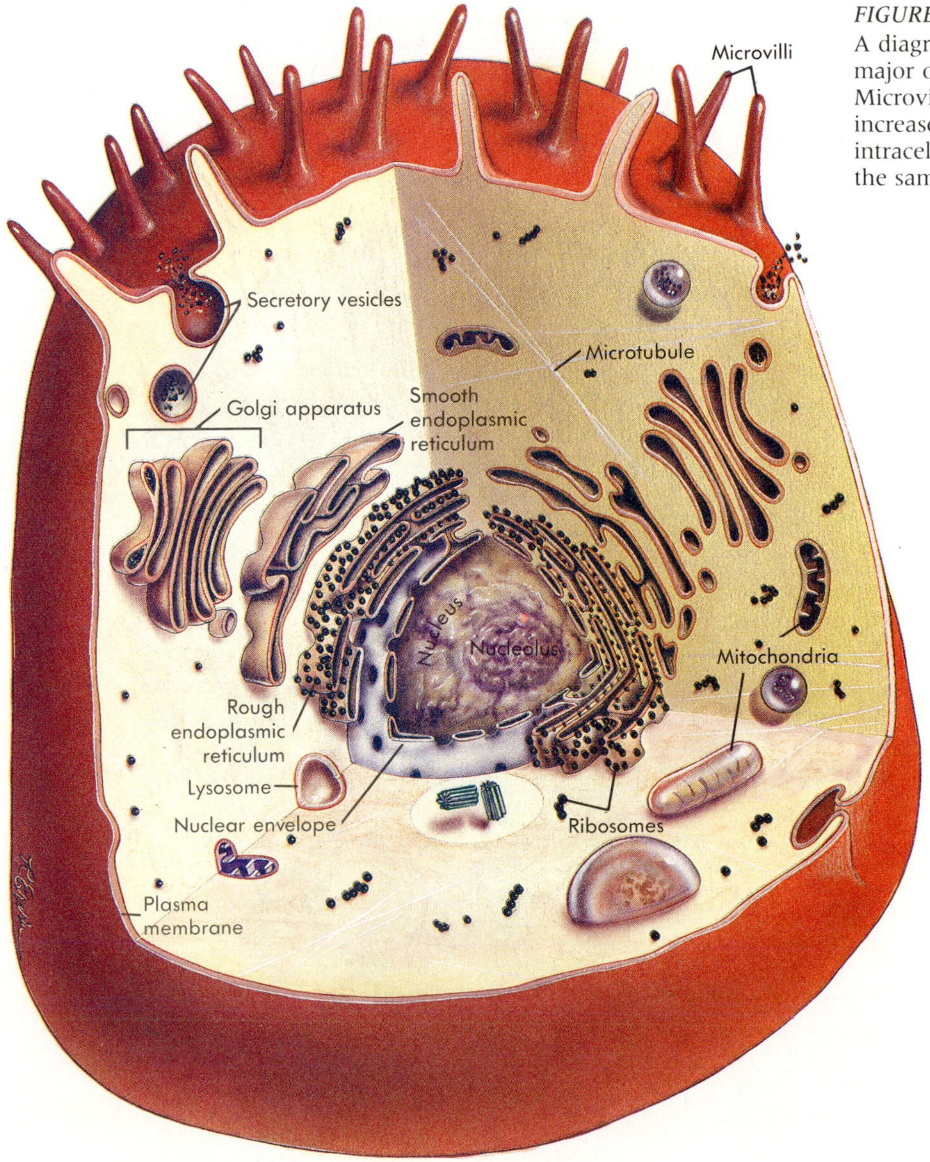

FIGURE 4-1

A diagram of an idealized cell showing the major organelles common to most cells. Microvilli are fingerlike projections that increase the cell surface area. Note that the intracellular organelles are not all drawn to the same scale.

Labels on figure: Microvilli · Secretory vesicles · Microtubule · Golgi apparatus · Smooth endoplasmic reticulum · Nucleus · Nucleolus · Mitochondria · Rough endoplasmic reticulum · Lysosome · Nuclear envelope · Ribosomes · Plasma membrane

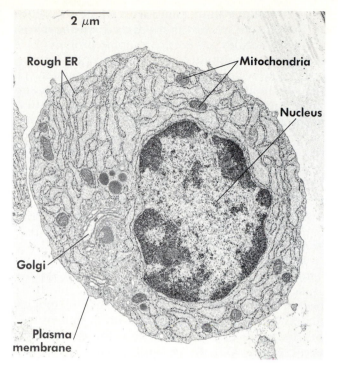

FIGURE 4-2

An electron micrograph of a plasma cell. The plasma membrane surrounds the cell; rough endoplasmic reticulum is studded with ribosomes that are synthesizing proteins for transport to the Golgi apparatus and eventual export by exocytosis, while mitochondria provide most of the energy needed by cells. The nucleus contains the diffuse form of DNA most characteristic of nondividing cells.

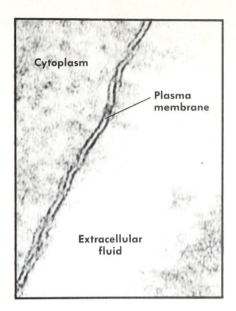

FIGURE 4-3

Electron micrograph of a human kidney cell showing the plasma membrane. The membrane has a less dense central layer surrounded by two dense outer layers.

TABLE 4-1   Intracellular Organelles

Organelle	Characteristics	Functional role
Plasma membrane	Lipid bilayer	Defines limit of cell
	Liquid crystal	Solute transport
Vesicle	Membrane-bound	Pinocytosis
	Sphere	Phagocytosis
		Exocytosis (secretion)
Nucleus	Bounded by double membrane	Controls all cell processes
	Contains DNA	
Rough endoplasmic reticulum	Sheets of membrane	Synthesis of protein to be secreted
	Ribosomes	
Smooth endoplasmic reticulum	Sheets of membrane	Membrane synthesis
	Lacks ribosomes	Fatty acid and steroid synthesis
Golgi apparatus	Series of flattened membranous sacs	Incorporation of materials into secretory vesicles
Lysosome	Specialized vesicles containing digestive enzymes	Intracellular digestive apparatus
Mitochondria	Double membrane	Oxidative phosphorylation
	Cristae	
Cytoskeleton	Microfilaments	Cell shape
	Microtubules	Cell movement
	Intermediate filaments	Anchoring of some membrane proteins

The Structure and Energy Metabolism of Cells

penetrate them. To allow the entry of water-soluble materials, cell membranes must contain specific **transport proteins** in addition to lipids. In cells frozen and then fractured, such proteins may appear as small particles (Figure 4-4). Chapter 6 discusses the aspects of membrane structure that have to do with transport of small molecules across cell membranes.

Endocytosis and Exocytosis

In addition to transport of small molecules through specialized pathways in plasma membranes, cells also carry out bulk exchange of materials with the extracellular space. This process is called either **endocytosis** or **exocytosis.** In endocytosis the membrane folds inward and pinches off to form a tiny sphere of membrane, termed a **vesicle,** whose interior contains extracellular material. There are several forms of endocytosis.

Pinocytosis (Figure 4-5) involves the uptake of bulk extracellular fluid, along with whatever solutes are present. In **receptor-mediated endocytosis,** specific molecules adhere to receptor proteins in a region of the membrane that subsequently forms a vesicle. Endocytotic vesicles may pass through a cell relatively unaltered, as is the case in capillaries and the intestines, or may combine with several different intracellular organelles. Bacteria, cell debris, and other foreign materials are engulfed by specialized cells of the immune system in a process called **phagocytosis** (discussed further in Chapter 24, p. 619). In phagocytosis, the endocytotic vesicles usually combine with digestive organelles, termed **lysosomes,** where digestive enzymes act to break down their contents in an isolated compartment of the cell (Figure 4-6).

Exocytosis is the reverse process of endocytosis. Compounds made inside cells (in the Golgi apparatus, see p. 70) are packaged into vesicles, and these vesicles are transported to the cell membrane (Figure 4-7, *A*). Here they fuse with the membrane and discharge their contents into the extracellular space (Figure 4-7, *B*). In addition to providing a

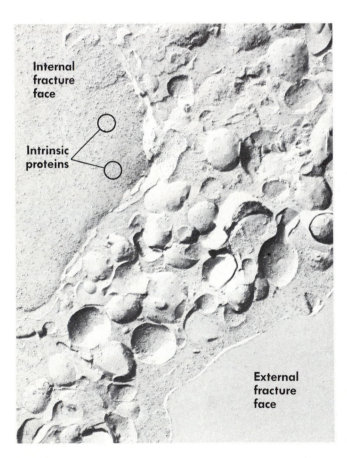

FIGURE 4-4

This scanning electron micrograph shows a white blood cell that has been frozen and then fractured. In the fracturing process, the outer leaflet of the lipid bilayer split away from the inner leaflet, revealing a population of intrinsic membrane proteins — seen as the particles that project from the fracture face at the upper left. In contrast, the external face of the membrane, shown at the lower right, appears smooth. The area between these two layers consists of several fractures through different cell processes.

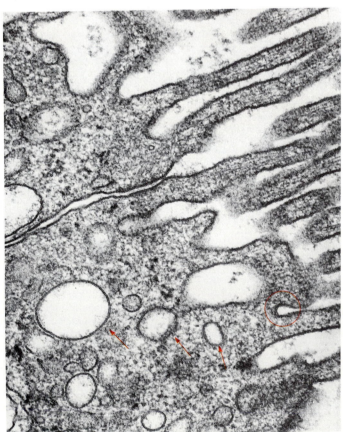

FIGURE 4-5

Pinocytosis is the ingestion by cells of water and dissolved solutes. The vesicle circled is in the process of budding off from the cell membrane. The arrows show pinocytotic vesicles within the cell.

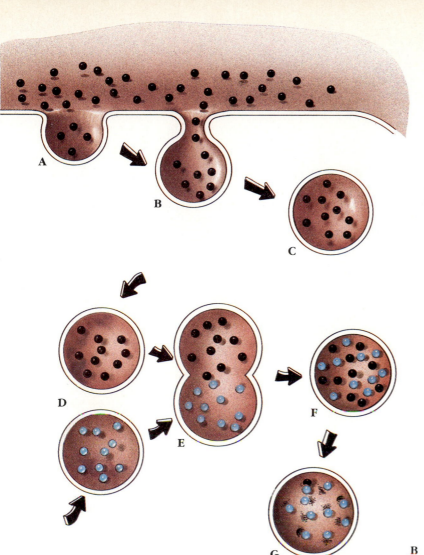

FIGURE 4-6
An illustration of phagocytosis and the action of lysosomes.
A   Material is taken into the cell by endocytosis.
B-C   The vesicle pinches off from the cell membrane.
D   A lysosomal vesicle approaches the endocytotic vesicle.
E-G   The vesicles merge, allowing digestive enzymes to degrade the contents of the endocytotic vesicle.

FIGURE 4-7
A   Exocytosis involves the fusion of cytoplasmic vesicles with the cell membrane, resulting in the release of materials into the extracellular space.
B   A transmission electron micrograph of exocytosis.

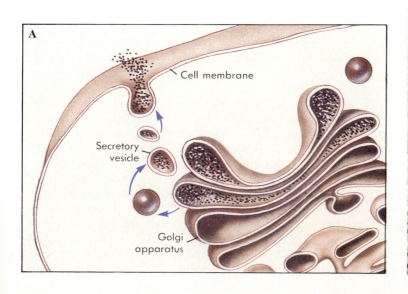

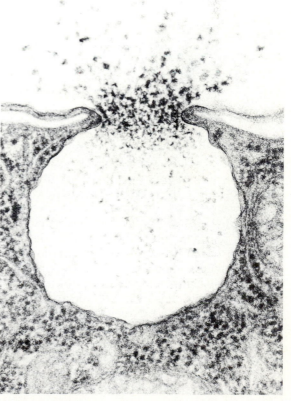

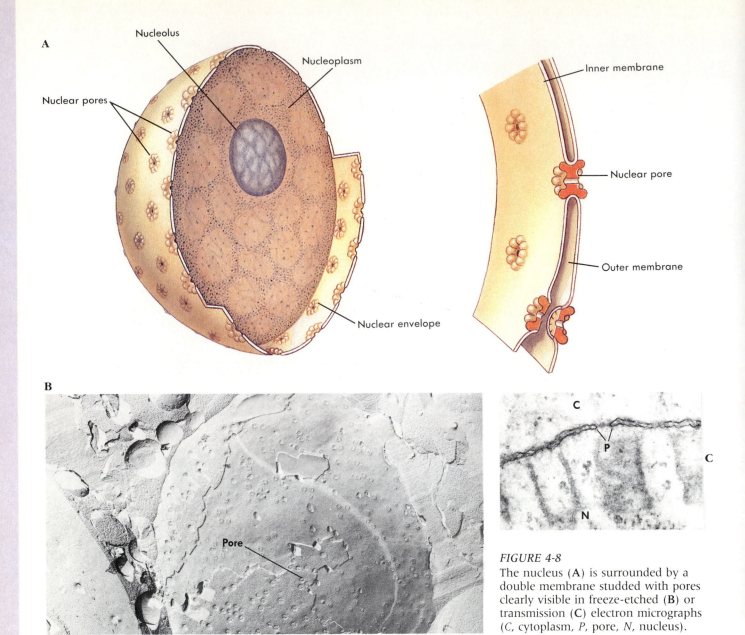

A

Nucleolus

Nucleoplasm

Nuclear pores

Nuclear envelope

Inner membrane

Nuclear pore

Outer membrane

B

Pore

C

P

C

N

FIGURE 4-8
The nucleus (**A**) is surrounded by a double membrane studded with pores clearly visible in freeze-etched (**B**) or transmission (**C**) electron micrographs (*C*, cytoplasm, *P*, pore, *N*, nucleus).

means for eliminating substances from a cell, proteins in the membranes of vesicles can be incorporated into the plasma membrane or into the membranes of other intracellular organelles. Exocytosis is a way a cell can add to its surface area or replace membrane lost in the process of endocytosis.

The plasma membrane has a variety of nontransport functions. Cells must communicate with and recognize one another, and these processes generally involve protein receptor molecules located in cell membranes. For example, electrically excitable cells communicate by exocytosis of a chemical, called a **neurotransmitter,** from a neuron; this chemical then interacts with a receptor on another neuron or muscle cell. Many hormones interact with specific membrane receptors, as do components of the immune system called **antibodies** (Chapter 24).

During development, cells that must remain together to form tissues identify one another by virtue of markers in the cell membrane.

Nucleus

Except for mature red blood cells, all cells possess a **nucleus** (see Table 4-1). The nucleus contains the DNA molecules that determine the structure and function of the cell, and it is bounded by a double **nuclear membrane.** The inner and outer nuclear membranes fuse in places to form **nuclear pores** 80 to 100 nm in diameter. In the nucleus DNA is combined with certain specific protein molecules to form chromosomes, which are extended in a lace-like network except during cell division (Figure 4-8). Within the nucleus of many cells there is a dense **nucleolus,** composed of ribosomal RNA (rRNA)

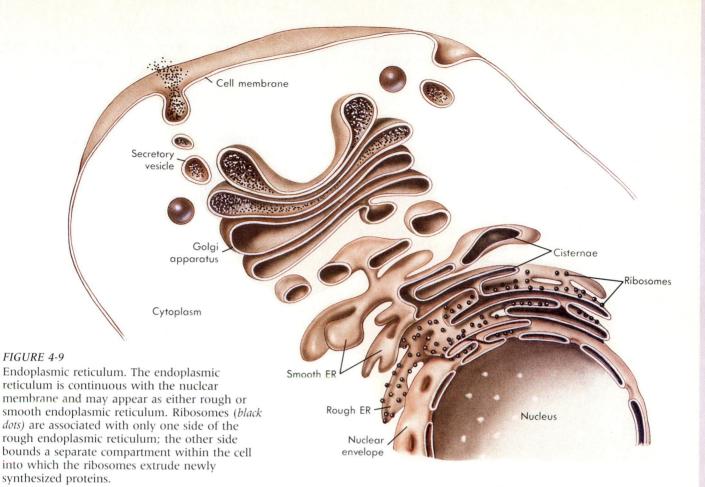

FIGURE 4-9
Endoplasmic reticulum. The endoplasmic reticulum is continuous with the nuclear membrane and may appear as either rough or smooth endoplasmic reticulum. Ribosomes (*black dots*) are associated with only one side of the rough endoplasmic reticulum; the other side bounds a separate compartment within the cell into which the ribosomes extrude newly synthesized proteins.

and associated ribosomal proteins. It is in the nucleolus that the basic structure of ribosomes is assembled.

Nuclear pores are selective, although the precise mechanism is unknown. Transfer RNA (tRNA) and rRNA can pass through them, as can a variety of small molecules and ions. Nuclear pores prevent DNA from leaving the nucleus and only allow messenger RNA (mRNA) to enter the cytoplasm when it has been processed and is ready to guide protein synthesis (see Chapter 3, p. 58).

> 1 What is the structure of the cell membrane? Why are membranes sometimes referred to as liquid crystals?
> 2 What are the functional differences between endocytosis, pinocytosis, and phagocytosis?
> 3 What is the role of nuclear pores?

CYTOPLASMIC ORGANELLES
Endoplasmic Reticulum

The **endoplasmic reticulum** looks like a pile of flattened sheets of membrane (Figure 4-9). These membranes form an extensive intracellular membrane system. Two types of endoplasmic reticulum are found in most cells. **Rough endoplasmic reticulum** appears granular because it has ribosomes attached

to its surface, while **smooth endoplasmic reticulum** lacks ribosomes. The smooth endoplasmic reticulum is the site of synthesis of new membrane, fatty acid, and steroids, but not protein synthesis (see Table 4-1). In muscle, the smooth endoplasmic reticulum stores and releases calcium (Ca^{++}) ions and is called the **sarcoplasmic reticulum.**

Ribosomes (see Figure 4-9) attached to the rough endoplasmic reticulum are the site of intracellular synthesis of proteins destined to be secreted by cells or to be inserted into their plasma membranes or the membranes of intracellular organelles (see Table 4-1). Ribosomes attached to the rough endoplasmic reticulum release completed proteins into its lumen, so that these proteins are immediately segregated from the cytoplasm. Parts of the lumen together with some of the rough endoplasmic reticulum membrane pinch off to form vesicles that eventually fuse with the Golgi apparatus for further processing.

Proteins used internally in cells are not synthesized on ribosomes attached to the endoplasmic reticulum but on ribosomes that are free in the cytoplasm. Free ribosomes do not appear to differ from those on the rough endoplasmic reticulum. Rather, ribosomes are directed to the endoplasmic reticulum only when they are translating mRNA for a protein that is destined for secretion. In particular, if an mRNA has an appropriate initial **signal sequence,** it will attach to ribosomes bound to the endoplasmic reticulum.

Golgi Apparatus

The **Golgi apparatus** is a series of flattened membranous sacs (Figure 4-10). Proteins from the rough endoplasmic reticulum migrate to the Golgi apparatus, where carbohydrate residues are removed or replaced. These modified proteins are then repackaged into vesicles. Vesicles from the Golgi apparatus can be transported to the periphery of the cell and undergo exocytosis (see Figure 4-9), or they can fuse with other organelles. In general, the Golgi apparatus seems to direct intracellular trafficking of proteins, probably by means of carbohydrate "destination labels" attached to the protein molecules (see Table 4-1).

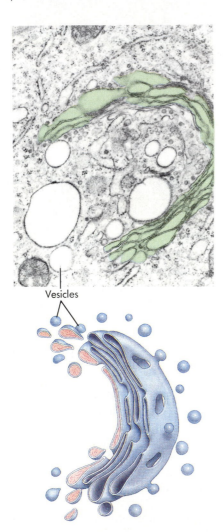

Vesicles

FIGURE 4-10
A Golgi body in cross section. Within the Golgi complex the membrane is smooth and pinches together at its terminus to produce membrane-bound vesicles containing whatever enzymes or other substances have been transported from the endoplasmic reticulum.

Lysosomes—Digestive Organelles

Lysosomes (Figure 4-11) are vesicles manufactured in the Golgi apparatus that contain digestive (hydrolytic) enzymes capable of destroying a variety of substances. Lysosomes keep potentially harmful reactions isolated from the remainder of the cell interior. Within the cell, lysosomes encapsulate and break down damaged organelles and other structural components. When foreign substances are taken up by endocytosis or phagocytosis, the resulting vesicle often fuses with a lysosome that destroys the engulfed material.

Mitochondria and Oxidative Energy Metabolism

Mitochondria (Figure 4-12) are surrounded by a double membrane. The outer membrane is smooth, while the inner is folded; the folds are called **cristae.** Mitochondria contain all of the enzymes involved in **oxidative metabolism,** a metabolic pathway that converts carbohydrate substrates to CO_2 and water, while capturing the released energy in the form of high-energy phosphate bonds in the compound **adenosine triphosphate (ATP)** (Figure 4-13), often referred to as the energy "currency" of cells. The enzymes and coenzymes needed for oxi-

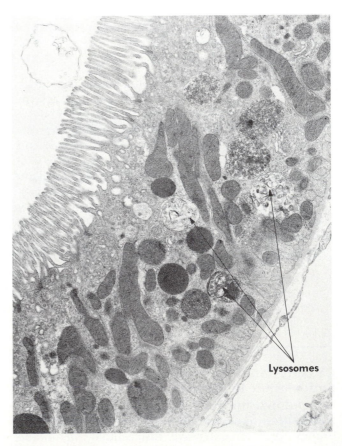

Lysosomes

FIGURE 4-11
An electron micrograph showing lysosomes (*arrows*) at various stages of intracellular digestion.

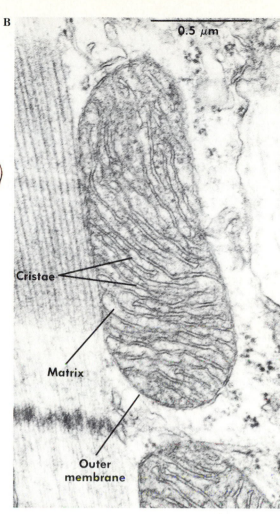

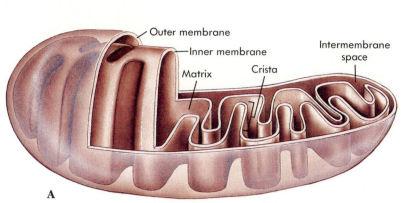

FIGURE 4-12

A schematic illustration (**A**) and electron micrograph (**B**) of a mitochondrion (the latter in a muscle cell). The mitochondrial outer membrane surrounds the inner membrane, which is folded into cristae. The enzymes of oxidative phosphorylation are located on the inner membrane, while the other enzymes for oxidative metabolism are located in the matrix.

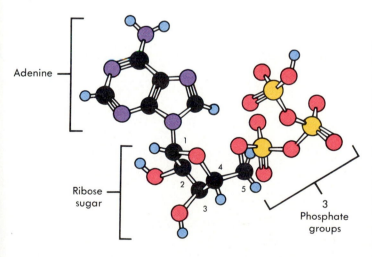

= Oxygen

= Carbon

= Nitrogen

= Phosphorus

= Hydrogen

FIGURE 4-13

The structure of adenosine triphosphate (ATP).

dative metabolism are located within the **inner mitochondrial membrane** and in the innermost space, called the **mitochondrial matrix**. The region between the outer and inner mitochondrial membranes is where H^+ ions are translocated by the electron transport chain (discussed later, p. 78).

Internal Cytoskeleton

All cells have a network of interconnecting and interlacing proteins collectively referred to as the **cytoskeleton** (Figure 4-14, *A*). The cytoskeleton provides the framework that determines the shapes of cells, controls the distribution of the cell's internal organelles, and allows some cells to move. The cytoskeleton is capable of holding some membrane proteins of a cell in a fixed location. There are three types of cytoskeletal elements. **Microfilaments** are largely composed of **actin**, the same protein found in muscle (Figure 4-14, *B*). **Microtubules** are hollow tubes composed of a protein known as **tubulin**. **Intermediate filaments** are made up of another type of protein and are the most stable component of the cytoskeleton. In general, intermediate fila-

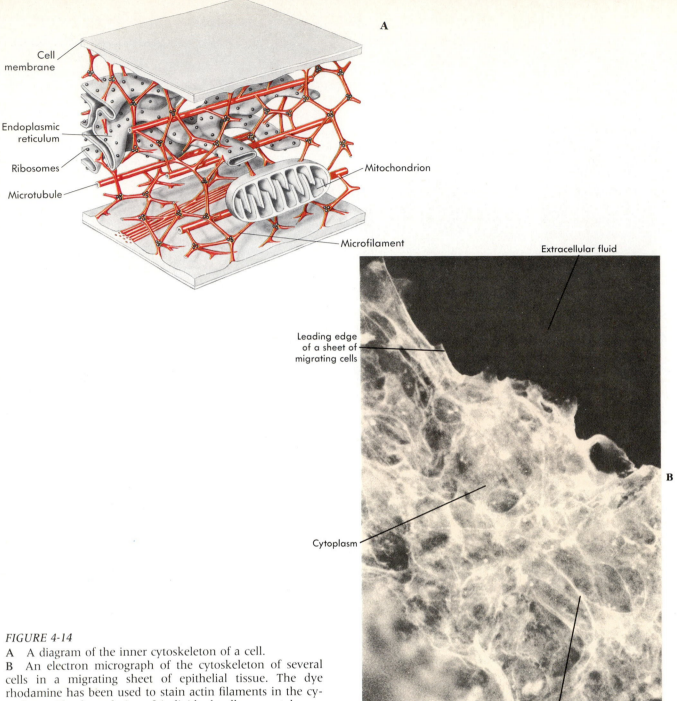

FIGURE 4-14

A   A diagram of the inner cytoskeleton of a cell.

B   An electron micrograph of the cytoskeleton of several cells in a migrating sheet of epithelial tissue. The dye rhodamine has been used to stain actin filaments in the cytoplasm. The boundaries of individual cells cannot be resolved.

ments probably maintain the overall structures of cells, while microfilaments and microtubules cooperate to allow intracellular and cellular movements (see Table 4-1).

> 1 *What is the difference between the rough endoplasmic reticulum and the smooth endoplasmic reticulum?*
> 2 *What is the role of the Golgi apparatus in the function of cells?*
> 3 *What is the functional role of mitochondria?*

JUNCTIONS BETWEEN CELLS
Desmosomes

The structures that bind the cell membranes of adjacent cells to one another may have three roles, depending on the function and structure of the tissue in which they are found (Table 4-2). Besides maintaining the structure of tissues, **junctions** may serve as lines of electrical and chemical communication between adjacent cells and are important in determining the tightness of barriers between different body compartments.

TABLE 4-2   Junctions between Cells

Junction	Characteristics	Functional role
Desmosomes	Filaments attached to cytoskeleton	Binds cells together Impermeable
Gap junction	Connexons	Couples cytoplasm of adjacent cells
Tight junction	Fusing of cell membranes	Joins cells in epithelial sheets

The job of binding cells together to maintain tissue integrity is accomplished by structures called **desmosomes** (shown diagramatically in Figure 4-15). Desmosomal filaments extend between the membranes of the joined cells and attach to the cytoskeleton on each side of the junction. There are several types of desmosomes, including **zonula adherens** and **macula adherens** (Figure 4-16).

Gap Junctions

Gap junctions (Figure 4-17) consist of paired subunits called **connexons.** Each pair of connexons has a central pore large enough to allow small molecules to pass from one cell to another, coupling the cytoplasm of one cell to that of its neighbors. Electrical currents, carried by small ions, can also flow through gap junctions. Electrical coupling is seen in

heart muscle and in some types of smooth muscle. In both cases the net effect is to ensure that electrical excitation spreads to all of the coupled muscle cells. All epithelial cells are also connected by gap junctions (see Figure 4-16).

When cells are coupled, it becomes possible for damage to one cell to affect the cytoplasmic composition of neighboring cells. Spread of injury is prevented by the fact that the permeability of gap junctions is only high when the intracellular Ca^{++} concentration is very low. When a cell is damaged, Ca^{++} ions enter the cytoplasm and the increased cytoplasmic Ca^{++} levels close the connexons, rendering gap junctions impermeable to solutes and electric currents and uncoupling the damaged cell from its neighbors. Tissues in which cell-cell coupling is important thus have a built-in "damage control" mechanism.

Tight Junctions

Epithelial tissues that separate fluid compartments of the body, such as the linings of the stomach and intestine, the gallbladder, the nephrons of the kidney, and the walls of exocrine glands, must be selective in their permeability to solutes and water. Substances that might diffuse through epithelia

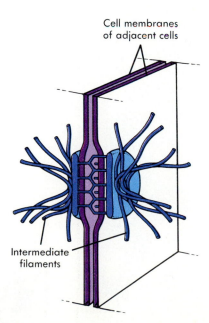

FIGURE 4-15
Desmosomes hold cells together, as shown in this schematic diagram.

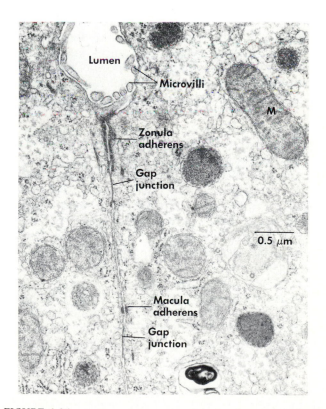

FIGURE 4-16
An electron micrograph of intestinal epithelial cells. Two types of desmosomes can be seen connecting intestinal epithelial cells—the zonula adherens and the macula adherens. In addition to desmosomes, these epithelial cells are coupled via gap junctions.

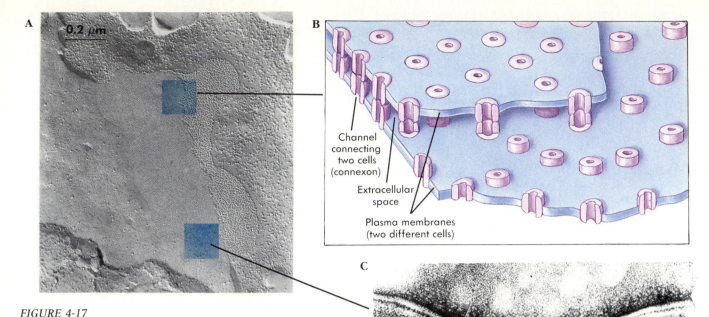

FIGURE 4-17

Gap junctions provide electrical and chemical communication between adjacent cells.

A   A freeze-etch micrograph of a gap junction.
B   A schematic illustration of a gap junction.
C   A gap junction in cross section.

take one or both of two possible routes: through the cells (the **transcellular route**) or between them (the **paracellular route**). The transcellular route is possible for lipid-soluble substances and for substances for which there are specific membrane-transport proteins. The paracellular route is guarded by structures that hold the epithelial cells together. Traditionally, these have been called **tight junctions** (Figure 4-18). In tight junctions the membranes of the two cells appear to fuse.

In spite of their name, the tight junctions can be arranged so that the paracellular pathway is either tight or leaky to small polar solutes and water. **Tight epithelia** are found where an osmotic gradient between fluid compartments must be maintained. An example is the urinary bladder, which stores urine that must remain at a different concentration from intercellular fluid. **Leaky epithelia** are found where osmotic flow of water is important. An example is the small intestine, in which absorption of nutrients is accompanied by a proportional absorption of water. Even leaky epithelia usually do not permit the passage of large molecules such as proteins.

> 1 What are the structural differences between desmosomes, tight junctions, and gap junctions?
> 2 Which types of junctions permit materials to pass from one cell to another?
> 3 What is the difference between tight and leaky epithelia?

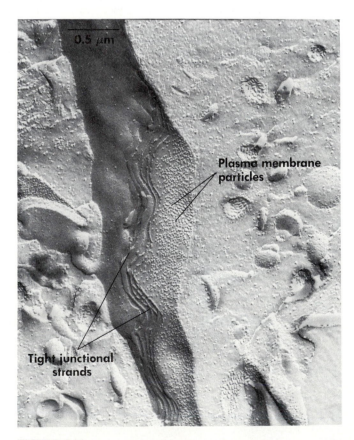

FIGURE 4-18

A freeze-fracture electron micrograph of a tight junction showing the strands that connect adjacent cells.

ENERGY METABOLISM
ATP—The Energy Currency of Cells

The breakdown of organic compounds to yield energy is referred to as **catabolism.** Catabolism releases chemical potential energy, some of which takes the form of heat. An adult whose weight is in a steady state uses about 2500 kcal per day. Energy is required for muscular contraction, for the transport of substrates and ions across cell membranes, and in the formation of complex biochemical compounds such as proteins and nucleic acids. Synthesis of materials from simpler substrates is called **anabolism,** the predominant process in tissue replacement, growth, and development. For example, peptide bond formation requires 0.5 to 4.0 kcal of energy, depending on the specific amino acids to be linked. In anabolic processes, chemical energy that originated outside the body is used to increase order and structure.

A major source of energy is oxidation of glucose to carbon dioxide and water. If glucose were burned in a furnace, its complete oxidation would yield 686 kcal of heat energy per mole. The energy contained in glucose and other substrates can be packaged, because in cells glucose is not broken down into carbon dioxide and water all at once, but rather in a series of biochemical reactions involving numerous intermediate compounds. The energy is ultimately transferred to phosphorylated nucleotides: **guanosine triphosphate (GTP), cytosine triphosphate (CTP), uracil triphosphate (UTP),** and **adenosine triphosphate (ATP).** Of these the most important is ATP (see Figure 4-16); the other phosphorylated nucleotides are interconvertible with ATP.

Hydrolysis of the terminal phosphate group of ATP to **adenosine diphosphate (ADP)** and inorganic phosphate produces a free energy change of about 7.3 kcal/mole, a convenient amount for cells to work with. The energy in ATP is made available to the cell by coupling ATP hydrolysis to chemical reactions that require energy. There is a relatively small reserve of ATP in the cytoplasm, so that as ATP is used it must constantly be replaced by energy-yielding reactions that phosphorylate ADP back to ATP.

There are two mechanisms for regeneration of ATP. The first is referred to as **substrate-level phosphorylation** because a phosphate group is transferred to ADP from an intermediate compound to form ATP. Some substrate-level phosphorylation can occur in the absence of oxygen, a condition termed **anaerobic metabolism.** One step in the pathway of oxidative metabolism also involves substrate-level phosphorylation. The second means of ATP production, **oxidative phosphorylation,** occurs in the inner mitochondrial membranes, requires oxygen, and produces most of the ATP used by the body. Oxidative phosphorylation and the re-

action sequences that serve it are referred to collectively as **aerobic metabolism.**

Electron Carriers—NADH and FADH

If one molecule gives up an electron and becomes oxidized, another must accept the electron and be reduced, so that it is helpful to think of such processes as paired **oxidation-reduction reactions.** Most of the potential energy carried by glucose is captured by oxidation-reduction reactions. A key element in such reactions is the participation of **coenzymes.** The most important electron-carrying coenzymes are **nicotinamide adenine dinucleotide** and **flavine adenine dinucleotide** (Figure 4-19, *A*). The oxidized forms of these coenzymes are abbreviated **NAD^+** and **FAD^+,** respectively; the reduced forms are **NADH** and **FADH.** Coenzyme A (Figure 4-19, *B*) transfers acetyl (2-carbon) groups and is discussed later.

The electron-carrying coenzymes are present in the cell only in very small amounts. At some steps in the pathways of energy metabolism, the oxidized coenzymes are converted to NADH and FADH. If

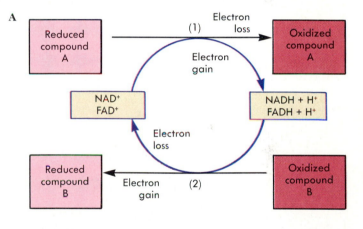

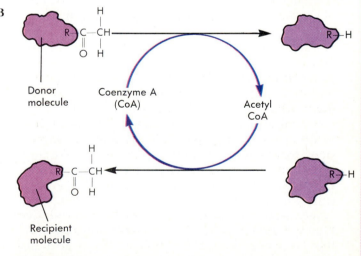

FIGURE 4-19

The role of the coenzymes NAD^+ and FAD^+ as electron carriers (A), and the role of coenzyme A as a carrier of acetyl groups (B).

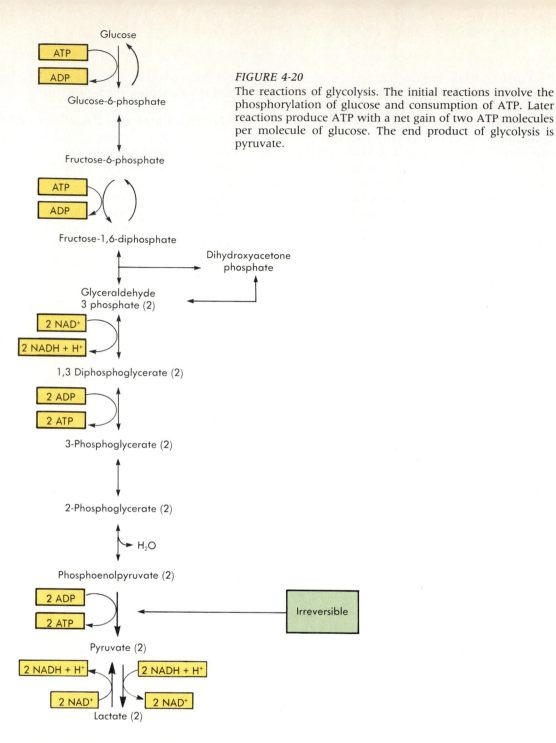

FIGURE 4-20
The reactions of glycolysis. The initial reactions involve the phosphorylation of glucose and consumption of ATP. Later reactions produce ATP with a net gain of two ATP molecules per molecule of glucose. The end product of glycolysis is pyruvate.

there were no means to regenerate NAD^+ and FAD^+, all of the oxidized coenzyme would quickly be converted to the reduced forms, and the reaction steps that require the oxidized forms would stop. The electron-carrying coenzymes can be reoxidized in two different ways, depending on whether oxygen is available. If oxygen is not available, some other metabolic intermediate can be reduced (see Figure 4-19). Reoxidation of NADH and FADH by this means has two disadvantages: (1) it does not yield any energy; and (2) it must involve the build-up of the reduced compound, either locally or in the body as a whole.

Under aerobic conditions a more effective solu-

tion is to reduce oxygen to regenerate NAD^+. Reduced NADH and FADH molecules are converted into NAD^+ and FAD^+, while at the same time electrons are transferred to molecular oxygen. This reaction takes place in the mitochondria during oxidative phosphorylation. Any NAD^+ or FAD^+ generated in the cytoplasm must be transported into the mitochondria for reoxidation, at an energy cost equivalent to one molecule of ATP per molecule of reduced coenzyme.

Glycolysis

The first major pathway in energy metabolism of glucose is **glycolysis** (Figure 4-20). Glycolysis does not require oxygen and can occur in its absence. Glycolysis yields 2 ATP molecules by substrate-level phosphorylation for each glucose molecule consumed. Cells that obtain energy through glycolysis are not limited by their own internal glucose stores but can also use glucose supplied by the blood.

In most cells the concentration of glucose in cytoplasm is much lower than in the extracellular fluid. This difference in concentration between the exterior and interior of cells is maintained by two processes. The extracellular glucose concentration is under homeostatic regulation, so that it normally does not drop below about 5 mM. Glucose molecules entering the cytoplasm react almost immediately to enter one of the metabolic pathways of a cell: glucose oxidation, glucose storage as glycogen, or a tissue-building pathway. The end result is that the intracellular glucose concentration remains much less than 5 mM, and a concentration gradient favorable for glucose to diffuse into cells is always maintained.

Glucose is a highly polar molecule, so even when a large concentration gradient exists, it does not easily cross cell membranes. A specific route of entry must be provided. Glucose transport in most tissues requires the presence of the pancreatic hormone **insulin,** which opens specific routes for diffusion of glucose into the cells.

In the first step of glycolysis (see Figure 4-20), glucose is converted to **glucose-6-phosphate,** a reaction driven by ATP. The second reaction of glycolysis is conversion of glucose-6-phosphate to **fructose-6-phosphate** (with the use of another ATP molecule). In the next several steps in glycolysis, the 6-carbon ring of glucose is broken into two 3-carbon pieces, yielding two molecules of **glyceraldehyde-3-phosphate,** and ultimately two molecules of **pyruvate** that enter the citric acid, or Krebs, cycle (Figure 4-21, *A*). Conversion of two glyceraldehyde-3-phosphate molecules to two pyruvate molecules generates two ATP molecules by substrate-level phosphorylation, two molecules of NADH, and 56 kcal of heat. Since two ATP molecules were used in the initial steps of glycolysis, the net yield is two ATP molecules per glucose molecule.

Conversion of 1 mole of glucose to 2 moles of pyruvate results in production of 2 moles of NADH. The NADH must be reoxidized if glycolysis is to continue. If the pyruvate enters the pathway of oxidative metabolism, the NAD^+ will be regenerated by that pathway. If a cell does not have the enzymes of the oxidative pathway or if oxygen is not available, it must regenerate the NAD^+ by converting pyruvate to lactic acid (Figure 4-21, *B*).

The interconversion of lactate and pyruvate is catalyzed by an enzyme called **lactate dehydrogenase (LDH).** Different cell types make different forms of the enzyme. One form favors production of lactate from pyruvate; this form predominates in

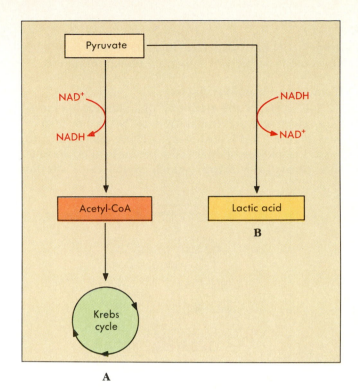

FIGURE 4-21
Alternative fates of pyruvate. Pyruvate can enter the citric acid (Krebs) cycle via acetyl-CoA **(A)** or be converted to lactic acid to regenerate NAD^+ **(B)**.

skeletal muscle cells that depend exclusively on anaerobic glycolysis. Another form favors the conversion of lactate to pyruvate; this is the form that predominates in heart muscle cells that depend on oxidative metabolism.

> 1 What are the differences between substrate-level phosphorylation and oxidative phosphorylation?
> 2 What is the role of NADH and FADH in conserving the chemical energy of substrate?
> 3 Why is it necessary for cells metabolizing anaerobically to convert pyruvate to lactate?

Krebs Cycle

To enter the Krebs cycle, pyruvate must first be **decarboxylated,** releasing CO_2 and forming NADH (Figure 4-22). The two remaining carbon molecules (in the form of an acetyl group) are accepted by **coenzyme A** (CoA; see Figure 4-19, *B*). Acetyl groups derived from fats also enter the cycle in the form of acetyl-CoA. Some amino acids derived from protein catabolism can be converted to intermediates of the Krebs cycle and thus enter it at several levels.

The Krebs cycle (Figure 4-23) is a circular sequence of eight reactions that occurs in the interior (matrix) of the mitochondria. In the Krebs cycle 2-carbon acetyl groups are degraded into two molecules of carbon dioxide, while at the same time four electrons are transferred to electron-carrier coenzymes (three to NAD$^+$ and one to FAD$^+$) and one ATP molecule is formed by substrate phosphorylation.

Oxidative Phosphorylation

At five steps in the pathway of oxidative metabolism, electrons are transferred to NAD$^+$ or FAD$^+$ (see Figures 4-20 and 4-23). These electrons are passed to oxygen as part of a process called **oxidative phosphorylation,** which yields most of the

ATP molecules produced by cells under aerobic conditions. A series of iron-containing enzymes called **cytochromes** (Figure 4-24) associated with the inner mitochondrial membrane first transfer electrons from NADH and FADH among themselves and finally give the electrons up to oxygen, reducing it to water (Figure 4-25, *A*).

In the process of electron transfer, energy is released. Some of this energy is inevitably lost as heat, but some of it is used to transport H$^+$ across the inner mitochondrial membrane from the matrix to the intermembrane space. As a result of this process, the energy is stored in the gradient of H$^+$. The gradient is a difference of chemical concentration (a pH gradient); it is also an electrical gradient because each of the transported H$^+$ carries a positive charge. Both of these aspects of this **electrochemical gradient** favor movement of H$^+$ from intermembrane space to matrix of the mitochondrion (Figure 4-25, *B*).

The inner mitochondrial membrane has protein complexes that form routes by which the hydrogen ions, driven by their electrochemical gradient, can return to the matrix of the mitochondrion from the space between the inner and outer membranes, giving up stored energy. Each protein complex is analogous to a turnstile. A current of H$^+$ flows through

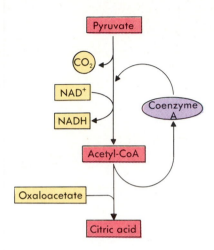

FIGURE 4-22

The decarboxylation of pyruvate yields acetyl-CoA. Acetyl-CoA donates its acetyl group to oxaloacetate to form citric acid, the initial stage of the Krebs cycle.

FIGURE 4-23

In the Krebs cycle, pyruvate, amino acids, and fatty acids transfer 2-carbon segments to coenzyme A to form acetyl-CoA. The 2-carbon segments enter a pathway in which they become part of a 6-carbon backbone. Successive reactions split off 2-carbon segments from the backbone to form CO$_2$, restoring a 4-carbon molecule that can accept another 2-carbon segment and start the cycle again. In the process, energy is captured in the form of NADH and FADH. There is also one substrate-level phosphorylation step.

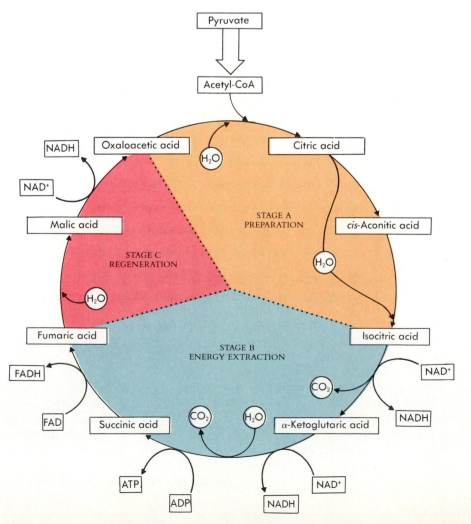

FIGURE 4-24
The structure of cytochrome C, one of the components of the electron transport chain.

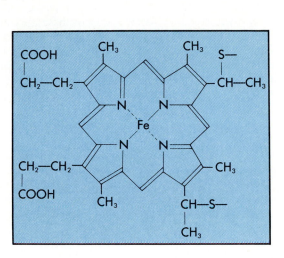

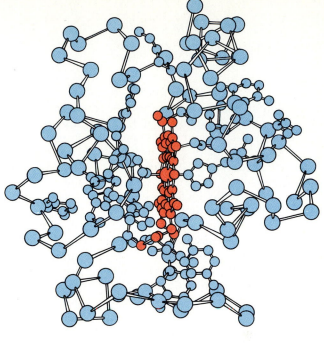

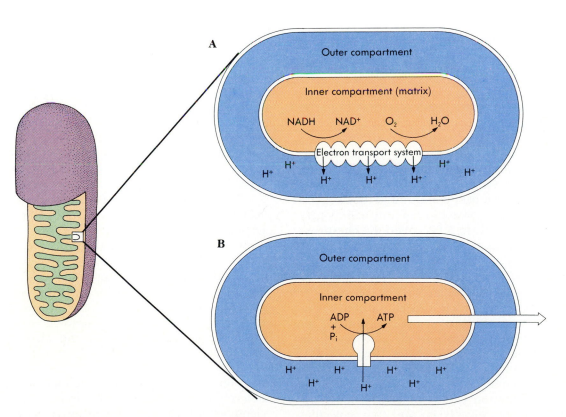

FIGURE 4-25
A The electron transport chain in the mitochondrial inner membrane. Electrons are transferred between a series of cytochromes and eventually combined with molecular oxygen to form water.
B Proton movement in mitochondria. Energy released in electron transport is used to establish a gradient of H^+ ions across the mitochondrial inner membrane. The energy stored in this gradient is used to phosphorylate ADP.

The Actual Yield of ATP in Oxidative Phosphorylation

Before the acceptance of the chemiosmotic hypothesis, biochemists thought they knew exactly how many ATP molecules were produced by complete oxidation of 1 molecule of glucose. There is still no doubt about the yield of ATP from the substrate-level phosphorylation steps, but, paradoxically, the yield of ATP from oxidation of reduced coenzyme has become less certain.

The previously accepted yield was based on the now-outdated hypothesis of oxidative phosphorylation in which exchange of electrons between the respiratory enzymes resulted directly in phosphorylation. The inability to determine how many ATP molecules each molecule of reduced coenzyme is worth stems in part from the fact that it is not clear exactly how many H^+ molecules are pumped out across the mitochondrial inner membrane when electrons are transferred from NADH and FADH to oxygen. The number of H^+ molecules that must return through the mitochondrial inner membrane to bring about the phosphorylation of one ADP molecule is also unknown. Of course, if more than two H^+ molecules are required to phosphorylate each ADP molecule, the efficiency drops.

Furthermore, there are hidden costs in the process that have only recently been appreciated. These are costs of getting the reactants ADP and phosphate into the mitochondria and the product ATP out, and they are paid for by the use of some energy from the H^+ electrochemical gradient. If these costs are considered, each molecule of NADH is worth 2 ATP molecules, and each molecule of FADH is worth 1.3 ATP molecules. In this case the net yield of ATP for the complete oxidation of glucose would be a little more than 25 per glucose molecule, rather than 36. These discoveries are reassuring because the yields of ATP actually observed in experiments with isolated mitochondria very seldom approached the theoretical value.

---

it like a stream of subway commuters at rush hour. Every turn of the turnstile transfers energy from the H^+ gradient to the protein complex, which applies it to phosphorylate ADP. Current experimental evidence suggests that at least two H^+ molecules are needed to phosphorylate one ADP molecule. In the analogy to the subway turnstile, this means that at least two commuters must push through the turnstile simultaneously. This explanation of how energy gets from reduced coenzyme to ATP, called the **chemiosmotic hypothesis,** is now generally accepted.

ATP Production in Oxidative Phosphorylation

Assuming that two H^+ molecules are needed per ATP molecule and that each molecule of NADH or FADH results in transport of six H^+ or four H^+, respectively, complete oxidation of glucose might yield as many as 36 ATP molecules (Figure 4-26). Two ATP molecules come from glycolysis and two more from the Krebs cycle. The remainder are the result of the oxidation of 10 molecules of NADH (including two from glycolysis) and two molecules of FADH. One mole of NADH yields 6 moles of H^+, and 1 mole of FADH yields 4 moles of H^+. By themselves, coenzyme reoxidations produce 34 molecules of ATP, and the total ATP yield per glucose molecule should be 38 molecules of ATP. The reason it is not is that the two NADH molecules produced by glycolysis have to cross from the cytoplasm into the mitochondria. This transport requires the equivalent of two ATP molecules. One molecule of glucose contains a total of 686 kcal of energy; if cells packaged 36×7.3 kcal = 263 kcal as ATP, the energy efficiency for complete carbohydrate metabolism would be about 40%.

This balance sheet for the theoretical yield of ATP from complete glucose oxidation is the one generally accepted until very recently. The current view is that oxidative metabolism yields about 25 moles of ATP per mole of glucose, giving an efficiency of about 27%. Although the net yield of ATP for the complete aerobic oxidation of glucose is probably less than the number accepted until recently, it is clear that the complete process yields much more energy as ATP, and also much more heat, than incomplete metabolism of glucose to lactate.

Figure 4-27 summarizes the metabolic pathways of a cell. Nucleic acids can be broken down into nucleotides, which enter the Krebs cycle. Proteins are degraded into their component amino acids, while fat breakdown yields fatty acids and glycerol. These products can be converted to acetyl-CoA and enter the Krebs cycle. The same sort of arithmetic applied to carbohydrate metabolism can be used to understand the metabolism of proteins and fats. Amino

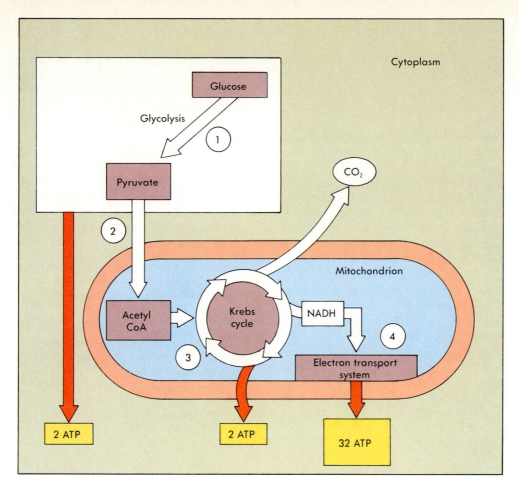

FIGURE 4-26
An illustration of ATP synthesis by carbohydrate metabolism. The figure of 32 ATP molecules obtained by oxidation of the NADH and FADH represents a conventional assumption about the yield of ATP from this process. See the essay (p. 80) for discussion of the uncertainty about this yield.

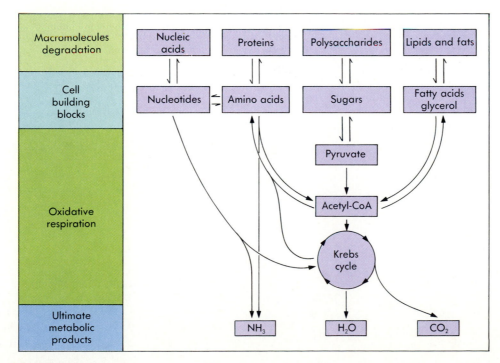

FIGURE 4-27
An illustration of the ways cells extract chemical energy from nucleic acids, proteins, carbohydrates, and fats.

acids yield at most the 15 ATP molecules that would result from the metabolism of 1 molecule of pyruvate. The efficiency and energy content per gram are about the same for protein metabolism as for carbohydrate metabolism.

In the catabolism of fats, glycerol is metabolized by entering the glycolytic pathway, so that the fatty acid chains end up being the main source of ATP. Each 2-carbon segment from a fatty acid generates one molecule of acetyl-CoA, which then enters the Krebs cycle. Although one ATP molecule is consumed when acetyl-CoA is formed, two molecules of NADH are produced so that there is a net gain of five ATP molecules prior to the entry of fatty acid fragments into the Krebs cycle. In the Krebs cycle, acetyl-CoA provides one ATP molecule, three NADH molecules, and one FADH molecule. These NADH and FADH then generate (with the traditional assumptions) 11 additional ATP molecules via oxidative phosphorylation, bringing the total to 17 molecules of ATP for each 2-carbon fatty acid fragment. Overall the maximum possible yield for complete breakdown of an 18-carbon fatty acid is 147 ATP molecules. The energy content of fats per gram is 2.3 times that of glucose. The high concentration of energy that fat represents makes it appropriate for its role as the major form of energy storage in the body.

Energy metabolism involves the participation of three major categories of nutrients that are oxidized to carbon dioxide and water, with formation of ATP by substrate-level phosphorylation. Reducing power transferred to reduced coenzyme is converted to phosphate bond energy of ATP by oxidative phosphorylation.

1 In the chemiosmotic hypothesis, what are the forms that the energy passes through before it is finally captured as ATP?

2 Why are the mitochondrial cytochromes called the "electron transport chain"?

3 What is the difference in terms of net ATP production between anaerobic glycolysis and oxidative phosphorylation?

SUMMARY

1. The major organelles of cells are:
   a. the **cell membrane** or **plasma membrane**, which serves as a selective barrier between cytoplasm and extracellular fluid.
   b. the **nucleus**, which contains the DNA.
   c. the **endoplasmic reticulum** and **Golgi apparatus**, which serve as the site of synthesis of cell membrane elements and of proteins destined for secretion by exocytosis.
   d. **lysosomes**, which are the source of enzymes for intracellular digestive processes.
   e. **mitochondria**, the sites of the **Krebs cycle** of substrate metabolism and of **oxidative phosphorylation** of ADP.
   f. the **cytoskeleton**, which determines the forms of cells and is involved in cellular movement.
2. Three types of junctions couple cells together: **desmosomes**, **tight junctions**, and **gap junctions**. Gap junctions serve as routes of direct communication between adjacent cells.
3. Carbohydrates can be metabolized to lactate in the absence of oxygen, with the capture by **substrate-level phosphorylation** of a small fraction of the potential energy of the carbohydrate.
4. In the presence of oxygen, **pyruvate** is converted to **acetyl-CoA**, which enters the Krebs cycle. Amino acids and fatty acids can contribute to the Krebs cycle by conversion to pyruvate, acetyl-CoA, or a Krebs cycle intermediate.
5. Complete metabolism of substrates to CO_2 and water (**oxidative metabolism**) involves the use of oxygen as an electron receptor. In the **chemiosmotic hypothesis** of **oxidative phosphorylation**, reducing power captured by **NADH** and **FADH** is converted to phosphorylation power by way of a transmembrane H^+ gradient, yielding ATP with high efficiency.

● **STUDY QUESTIONS**

1. Describe the function of each of the following organelles:
   nucleus
   cell membrane
   Golgi apparatus
   mitochondria
   ribosome
   endoplasmic reticulum
   lysosomes
2. What are the differences between gap junctions, tight junctions, and desmosomes?
3. What is meant by the terms "tight" and "leaky" epithelia?
4. Red blood cells lack nuclei and mitochondria. How does this affect the functions they can perform?
5. What are the major metabolic pathways that contribute to phosphorylation of ADP? What is the functional role of each pathway in the body?
6. What is the advantage of aerobic metabolism over anaerobic metabolism in energy production in living systems?
7. How is energy liberated in the electron transport chain? Where do the electrons come from and what is their final destination?
8. What are the routes by which proteins and fats can be metabolized? How does the energy yield compare with carbohydrate metabolism?

● **SUGGESTED READING**

BRETSCHER, M.S. How Animal Cells Move, *Scientific American*, volume 257, December 1987, p. 72. Discusses the role of the internal cytoskeleton in the movements of white blood cells and fibroblasts.

DULBECCO, R.: *The Design of Life*, Yale University Press, New Haven, Conn., 1987. A noted cell biologist gives his view of cellular function.

HAROLD, F.M.: *The Vital Force: A Study of Bioenergetics*, W.H. Freeman and Co., New York, 1986. Describes the mechanisms of cellular energy production and the problem of caloric yield from oxidative metabolism.

HOCHACHKA P.W., and G.N. SOMERO: *Biochemical Adaptation*, Princeton University Press, Princeton, NJ, 1984. A classic discussion of the evolution of enzymes for their function.

PRESCOTT, D.M.: *Cells: Principles of Molecular Structure and Function*, Jones and Bartlett, Boston, 1988. A readable text covering the content of this chapter.

CHAPTER 5

Homeostatic Control: Neural and Endocrine Mechanisms

On completing this chapter you will be able to:

- Describe the elements of a negative feedback control system and show how such systems stabilize physiological variables.
- Understand how positive feedback leads to instability.
- Provide examples of intracellular, local, and extrinsic control processes in the body.
- Distinguish between somatic, autonomic, and hormonal reflexes.
- State the characteristics of hormones and describe the three major chemical classes of hormones.
- Understand what is meant by down-regulation and up-regulation.
- Describe the factors controlling hormone release from the posterior pituitary.
- Describe the role of releasing and release-inhibiting hormones in control of anterior pituitary hormone secretion.
- Provide examples of several second messenger systems.
- Compare the intracellular mechanisms of action of steroid and peptide/protein hormones.

*T*he human body is a self-controlling (homeostatic) unit. The principles of homeostatic control that are so important to physiology were first worked out by engineers and physical scientists. Biological machines can be shown to follow the same control-system principles that were put to use in the thermostat, the governor that regulated the speed of primitive industrial engines, and the electronic amplifier. What is most striking about biological control systems is their formidable complexity and the enormous range of size and time scales over which they operate.

It is difficult to separate homeostatic control mechanisms from the physiology of the various organ systems of the body. The most effective way to conceptualize the complex interactions that take place is to integrate coverage of specific neural and hormonal influences with the discussion of the physiology of each organ system. The purpose of this chapter is to describe the basic principles of negative-feedback control (reflexes), briefly introduce the differences between the somatic (voluntary) and autonomic (involutary) nervous systems, and provide a focused introduction to the endocrine system.

Intracellular control systems frequently involve a many-step cascade of chemical reactions that amplifies a single initiating event thousands of times. A similar complexity is seen in the central nervous system, in which millions of neurons may be involved in as simple an act as walking up stairs. Some control systems in the nervous system operate on the time scale of milliseconds; intracellular regulatory processes operate at the size scale of individual molecules or ions. On the other end of the time and size scales, the developmental plan of the human body, orchestrated by the endocrine system and involving many billions of cells, is fulfilled on a time scale of decades.

HOMEOSTATIC CONTROL THEORY
General Properties of Negative Feedback

Homeostasis demands that important physiological parameters such as body temperature, blood composition, and blood pressure be maintained within appropriate limits (Figure 5-1). Negative feedback provides the mechanism for opposing the departure of a **controlled variable** from its appropriate value.

The ideal level of a controlled variable is defined as its **setpoint**. The controlled variable is monitored by **sensors** or **receptors** that pass information to an **integrator**, which compares the sensor's input with the setpoint. Any deviations from the acceptable range give rise to an **error signal** when the deviation is sensed by the integrator. An **error signal** arises when there is a difference between the setpoint and the value indicated by the sensor. An error signal results in activation of **effectors** that oppose the departure from the setpoint. The term *negative* is used because the effector's response opposes the departure from the setpoint. The effector's response completes a **feedback loop** that runs from the controlled variable through the sensor to the integrator and back to the controlled variable by way of the effector (Figure 5-2, *A*). Such systems are sometimes called **closed loop systems.**

A Practical Example—Driving

A car and driver form a feedback control system (Figure 5-2, *B*). The controlled variable is the position of the car in its lane. The sensors are the eyes of the driver; the driver's brain (the integrator) constantly compares the car's position with the center of the lane. Deviations from the setpoint that occur as a result of **perturbing factors,** such as bumps or curves in the road, are opposed by a system of effectors that includes the muscles of the driver and the car's steering wheel (see Figure 5-2, *B*).

The setpoint in physiological systems may be changed from time to time. A change in the setpoint corresponds to a lane change in the analogy of the

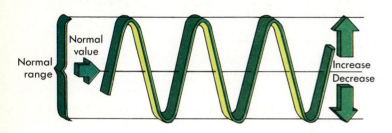

FIGURE 5-1

Negative feedback. Shown is the range over which a given homeostatic condition is maintained. Values for the condition fluctuate above and below a normal value within a normal range.

FIGURE 5-2

A   A generalized diagram of a negative feedback loop.
B   Application of the negative feedback loop to the analogy of a car and driver.

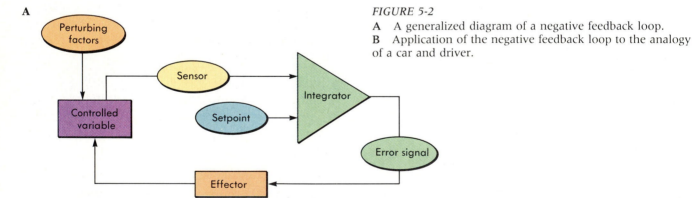

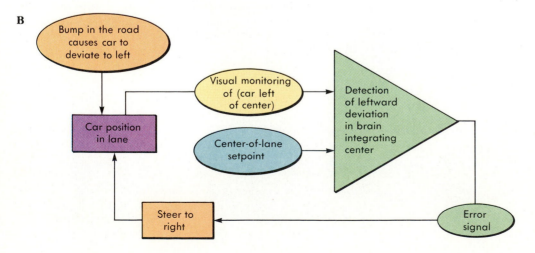

car and driver. For example, body temperature is regulated at a lower value during sleep and at a higher value during fever. In women, body temperature also varies predictably during the menstrual cycle.

The error signal is proportional to the difference between the setpoint and the value of the controlled variable. Thus the body's effectors are usually capable of making larger or smaller efforts, depending on the magnitude of the error signal. In the driving analogy, the driver makes a larger steering correction the farther the car is from the center of the lane.

Open Loop Systems

Open loop systems are those with no negative feedback (Figure 5-3). In the analogy of car and driver, the feedback loop is opened if the driver shuts his eyes or takes his hands off the steering wheel. Deviations from the setpoint cannot be corrected, and the car will soon leave the road. Open loops can result from disease or damage to some part of the feedback loop. For example, damage to parts of the motor control system of the brain may result in uncontrolled body movements, as in Parkinson's disease (see Chapter 11, p. 272).

In some physiological systems, open loops are part of normal function. Body movements that must occur very rapidly, such as the movements that the eyes make to follow an object when the head moves, the pitcher's pitch or the boxer's punch, must be carried out according to a learned pattern because they must be completed before feedback could be effective. The skill attained through learned modification of such open loop behaviors is called the **feed-forward** component of the effector command.

Gain of Negative Feedback Systems

Two factors determine how closely the controlled variable will hover around the setpoint. The first

factor is its **gain,** equivalent to sensitivity. The gain of a feedback system is defined as the degree to which the output of an effector is altered for a given change in the controlled variable. Gain is a kind of leverage built into the control system. In the analogy of driving a car, power steering would increase the gain of the feedback system over manual steering.

More generally, if the gain is low, changes in the error signal will have a relatively small effect on the output of an effector and the controlled variable. Low gain means that larger error signals will be needed to change the output of the effector by a certain amount. Because larger error signals are needed, the result of low gain is that steady perturbing factors can force the controlled variable further from the setpoint than would be the case for larger gains.

Time Lags in Negative Feedback

Physiological feedback loops usually have a finite **time lag** between departure of the controlled variable from the setpoint and the response of effectors. Time lag is the second factor that determines how closely the controlled variable will hover around its setpoint. Time lags may have several origins: the time needed to send sensory information to the integrator, the time for the signal to be processed in the integrator, the time needed for the command to travel to the effector, and the delay in the response of the effector itself.

In the body, messages carried by nerve impulses may involve time lags of only a fraction of a second. However, if hormones are used to carry information, the delay due to the transit time of the messages may be much longer. Time lags may cause the system to oscillate around the setpoint if the controlled variable is rapidly altered by perturbing factors.

Physiological Examples

Larger gains in negative feedback systems combined with shorter time lags give better control because the average deviations of the controlled variable about the setpoint are smaller. Physiological control systems usually have gains and time lags that match the response characteristics of the control systems to the rate and magnitude of change of the physiological variables they regulate. For example, the system that controls blood pressure (Figure 5-4) has a large gain and little time lag. This is appropriate, since changes in the orientation of the body relative to the pull of gravity could cause rapid and substantial changes in arterial blood pressure. The sensors for this control system are in the aorta and in the carotid arteries that serve the head. When a person stands up, the control system usually responds rapidly enough to keep the blood pressure in these arteries very close to the setpoint. If disease

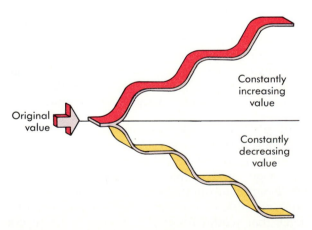

FIGURE 5-3
Open loop systems continue to change in one direction once perturbed from their original value.

Original value

Constantly increasing value

Constantly decreasing value

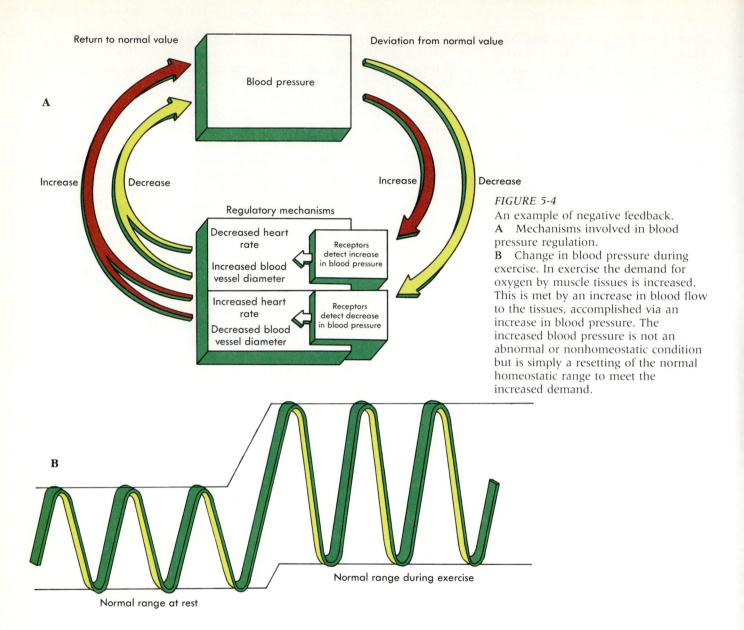

Return to normal value

Deviation from normal value

Blood pressure

A

Increase   Decrease

Increase   Decrease

Regulatory mechanisms

Decreased heart rate

Increased blood vessel diameter

Receptors detect increase in blood pressure

Increased heart rate

Decreased blood vessel diameter

Receptors detect decrease in blood pressure

B

Normal range during exercise

Normal range at rest

FIGURE 5-4
An example of negative feedback.
A Mechanisms involved in blood pressure regulation.
B Change in blood pressure during exercise. In exercise the demand for oxygen by muscle tissues is increased. This is met by an increase in blood flow to the tissues, accomplished via an increase in blood pressure. The increased blood pressure is not an abnormal or nonhomeostatic condition but is simply a resetting of the normal homeostatic range to meet the increased demand.

weakens the gain of the control system, blood pressure in the arteries serving the head cannot be maintained close to the setpoint when the body is erect; the condition sometimes results in fainting when the person stands up.

Rate of Change and Negative Feedback

Inexperienced drivers tend to oversteer, so that the car wanders around its setpoint in the lane. This is in part the result of the driver's response to the magnitude of the error rather than the rate at which the car's position is changing relative to the setpoint. Negative feedback control can be dramatically improved if the integrator takes the rate of change of the controlled variable into account. This allows the integrator to predict the future state of the system. If effectors are rapidly returning the controlled variable to the setpoint, the signal to effectors can be diminished before the setpoint is reached. Thus

the tendency for the controlled variable to overshoot the setpoint is decreased. The result is a series of oscillations that decrease with time until they become quite small. Such anticipation is seen in most physiological control systems. Some physiological sensors are adapted for measurement of rates of change of stimuli; the presence of such receptors allows integrators to take rates of change into account.

Matching of Feedback Response to Physiological Role

The refinements in feedback systems discussed above usually ensure that the error is small. These small deviations from the setpoint can be thought of as an acceptable cost of avoiding the much larger changes that would occur if the variable were not feedback regulated. However, in some cases even the small error can be significant for health. If one begins to eat a lot of salt, the kidney will react

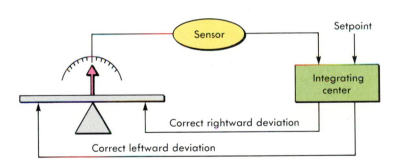

FIGURE 5-5
A control system with two sets of effectors; one increases the controlled variable and the other decreases the controlled variable. Opposing effectors are typical in physiological systems.

within a few days and excrete the excess salt to maintain the body's salt content near its ideal setpoint. However, the gain of the renal control system is not infinite, and the consequences of a continued high salt intake will be an elevation of the salt content of the body by 1% to 2%. This small elevation constitutes the residual error, and it may be a contributing factor in certain types of high blood pressure (hypertension).

Most physiological control systems need to monitor several variables because of the complex interactions between different organ systems. The cardiovascular control center receives information on arterial blood pressure, blood volume, and oxygen and carbon dioxide content. In muscle control, both the absolute position and direction of movement of the limbs are constantly assessed. In both of these

examples the relative importance of each controlled variable must be evaluated by the integrator to determine an optimum response.

In physiological systems, feedback may be applied by two sets of effectors that have opposite effects on the controlled variable (Figure 5-5). The most obvious examples are in the skeletal muscle system, in which the position of a limb is controlled by muscles that work to move it in opposite directions. Another example is the heart, which is inhibited by input from the parasympathetic nervous system and stimulated by input from the sympathetic system. Both branches are under the control of an integrating center in the brainstem; a reflexive increase in heart rate involves both an increase in sympathetic input and a decrease in parasympathetic input.

Positive Feedback

Negative feedback systems stabilize variables near their setpoints because the response of the effector minimizes the error signal. In **positive feedback,** a change in the controlled variable causes the effector to drive it even further away from the initial value. Systems in which there is positive feedback are highly unstable. The effect of the initial perturbation is analogous to that of a spark that ignites an explosion.

Because it causes instability, positive feedback is usually an undesirable trait for physiological systems. Destructive positive feedback commonly occurs in disease, where it results in very rapid deterioration of homeostasis. For example, in certain types of heart disease, the heart becomes overloaded and cannot pump out all the blood returning to it. The volume of the heart increases, and this volume increase further diminishes the ability of the heart to pump blood. The end result is the conversion of a stabilizing negative feedback loop into an unstable positive feedback loop, with dangerous consequences for the patient.

Positive feedback is not always abnormal. It may sometimes be put to use to accomplish a specific purpose. One example occurs in the female reproductive cycle, where release of a pituitary hormone and its detection by the pituitary gland increases its level of production and release. The result is a rapid rise of blood levels of the hormone that triggers ovulation (Chapter 25). Another example is a nerve impulse, in which the opening of a few sodium channels causes the remainder to open (Chapter 6). The normal response of the immune system to bacteria and viruses is a third example of positive feedback (Chapter 24). Positive feedback is useful in expulsive processes such as the contrac-

tions of the uterus in childbirth, in which a positive feedback loop increases uterine contractions in response to the pressure of the baby's head on the cervix (Figure 5-6).

1 What is the definition of gain of a feedback system?
2 What is the value to the integrator of information about the rate of change of a controlled variable?
3 In what situations is positive feedback useful?

LEVELS OF PHYSIOLOGICAL REGULATION

Three levels of regulation can be distinguished (Figure 5-7):
1. Regulation at the level of single cells (**intracellular regulation**)
2. **Local regulation** at the level of regions within organs, typically involving many cells of different types (also called **intrinsic regulation** or **autoregulation**)
3. **Extrinsic regulation,** in which the activity of tissues and organs is controlled by hormones or nerve impulses that originate outside the structures being controlled.

Intracellular Regulation

Intracellular control almost always involves changes in the rates of enzyme-catalyzed reactions. One way to control enzymes is by allosteric effects. As discussed in Chapter 3, allosteric modification involves changes in the activity of the active or catalytic site of an enzyme by the binding of a regulator molecule at a separate site on the same enzyme (Figure 5-8, A). In this case, the molecules of enzyme that are on hand in the cell are made to be more or less active. An enzyme regulated in this way is like a faucet that senses how full the washbasin is.

FIGURE 5-6

An example of positive feedback in childbirth.

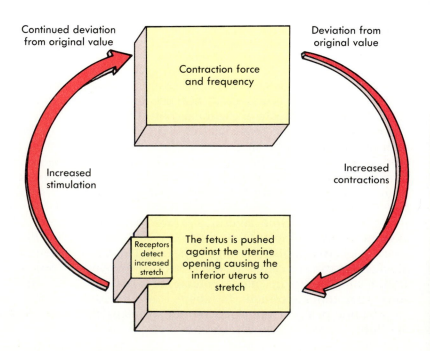

Continued deviation from original value

Deviation from original value

Contraction force and frequency

Increased stimulation

Increased contractions

Receptors detect increased stretch

The fetus is pushed against the uterine opening causing the inferior uterus to stretch

THE FOUNDATIONS OF PHYSIOLOGY

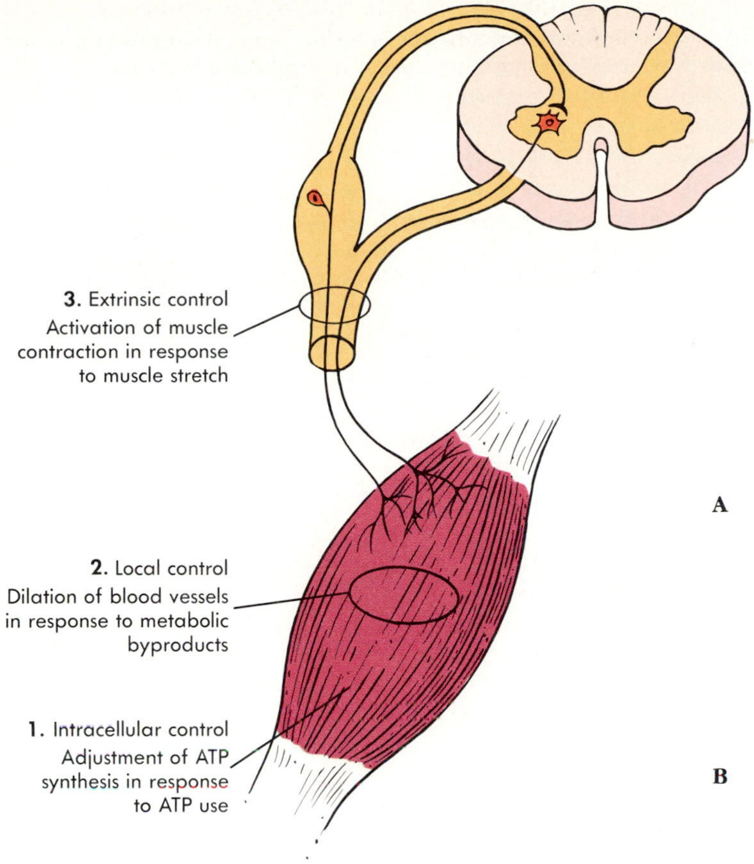

3. Extrinsic control
Activation of muscle contraction in response to muscle stretch

2. Local control
Dilation of blood vessels in response to metabolic byproducts

1. Intracellular control
Adjustment of ATP synthesis in response to ATP use

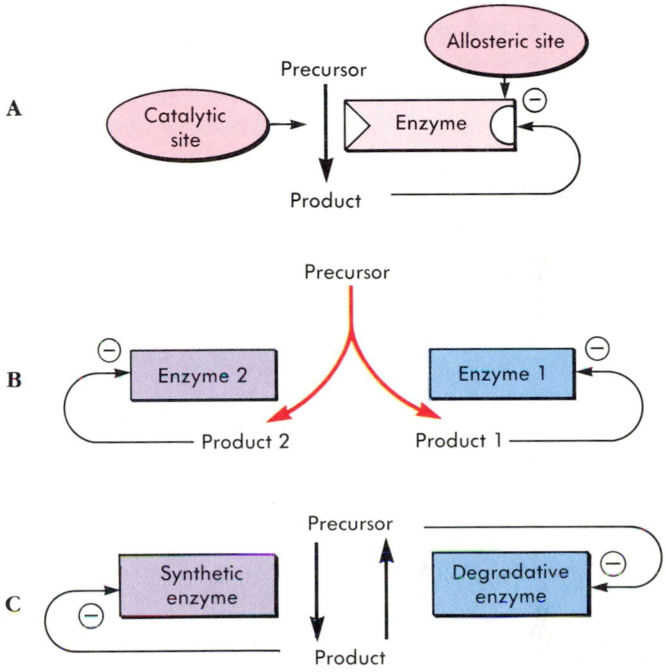

FIGURE 5-8
A   Allosteric regulation of an enzyme by the product of the reaction it catalyzes.
B   Control of a metabolic branch-point by feedback from both products.
C   Reversible reaction in which the forward reaction and the backward reaction are catalyzed by different enzymes, each of which is inhibited by its product.

Intracellular reactions located at critical branch points in metabolic pathways are frequently catalyzed by two different enzymes. Each of these enzymes is subject to allosteric regulation, so that the flow of reactants into one branch or the other is adjusted to meet the need for the end products of each branch (Figure 5-8, B). This case resembles two faucets connected to the same pipe but serving two different washbasins. Frequently one of the pair of enzymes serves a synthetic pathway while the other serves a degradative pathway. This kind of regulation affects the balance between net synthesis and degradation within cells, and the relative flow through the two branches is controlled by both substrate and end-product levels (Figure 5-8, C). For example, after a person eats a meal the blood glucose, amino acid, and fatty acid levels increase. Nutrient availability is detected by receptors in the hypothalamus of the brain and in the gastrointestinal tract. These receptors cause the release of hormones, including insulin, which activate enzymes that promote glucose uptake into the body tissues and thus net energy storage and protein synthesis (Figure 5-9, A). The same hormones tend to inhibit those enzymes that catalyze the corresponding catabolic pathways.

An increase in energy expenditure in muscle decreases intracellular ATP. Decreasing the ATP stores results in increased ATP synthesis via glycolysis (see Chapter 4, p. 76), decreasing blood glucose. When blood sugar levels drop, insulin secretion is inhibited (see Figure 5-9, A). In addition, low blood sugar and stressful situations result in the release of the hormone epinephrine from the adrenal medulla. Epinephrine activates enzymes that promote the breakdown of glycogen to glucose, increasing blood sugar levels to help the body maintain its energy expenditure (Figure 5-9, B).

Sometimes intracellular control involves inhibition or activation of mRNA synthesis or translation of mRNA into protein. This is a means of regulating the synthesis of specific proteins. In this way, the

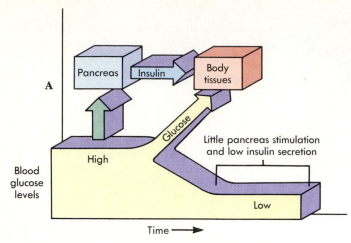

FIGURE 5-9
Control of hormone secretion by blood glucose is an example of an intracellular negative feedback system.

A The concentration of glucose in the blood affects the rate of insulin secretion. After a meal, elevated blood glucose stimulates insulin secretion. Insulin acts on the tissues to increase the rate of glucose uptake. As glucose levels decline, so does the release of insulin. When blood glucose levels are low, the resulting decrease in insulin secretion prevents the tissues from taking up too much glucose.

B Epinephrine is released from the adrenal medulla in response to sympathetic stimulation induced by stress or low blood glucose. Epinephrine helps the body respond to stressful conditions by enabling it to mobilize its reserves of glycogen and fat.

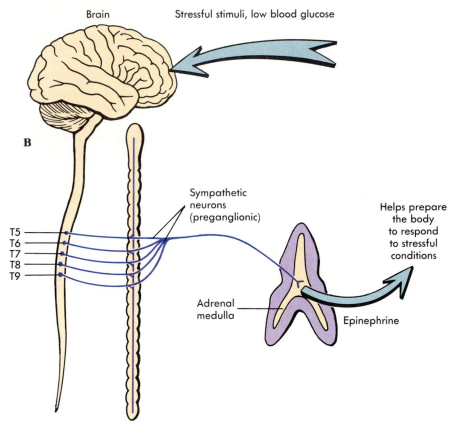

appropriate mix of structural proteins and enzymes is synthesized as cells grow and divide.

Control by Local Chemical Factors

In some cases local control is valuable because it gives tissues or organs the ability to respond to changing conditions on a size scale too small to be dealt with effectively by the external nervous or hormonal control mechanisms. For example, the flow of blood through each individual capillary bed is adjusted according to the changing needs of the small population of cells served by that bed. Increased metabolic activity, such as occurs in a muscle that is beginning to contract, causes carbon dioxide to accumulate in the active tissues, while oxygen is depleted. Active tissues release a number of substances, termed **local factors,** that accumulate in the interstitial fluid if not flushed out by blood flow. These factors include potassium, end products of metabolism such as carbon dioxide and lactate, and also members of a family of regulatory molecules called **prostaglandins.** The effect of an increase in the concentration of local factors is to dilate arterioles serving the active tissue.

The negative feedback loop is closed when the increased blood flow increases oxygen delivery to

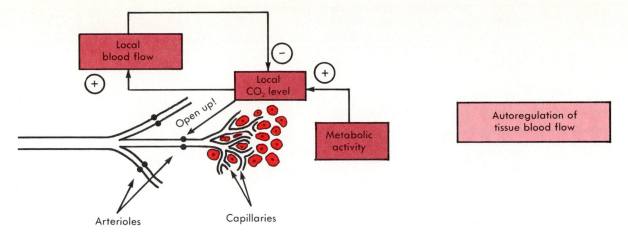

Local blood flow

–

Local CO₂ level

+

+

Open up!

Metabolic activity

Autoregulation of tissue blood flow

Arterioles

Capillaries

FIGURE 5-10

Local flow autoregulation is an example of negative feedback control at the local tissue level. An increase in metabolic activity causes accumulation of local factors, which dilate the arterioles and increase blood flow. The increase in blood flow minimizes the increase in concentration of the local factors.

the active tissue and increases the rate at which local factors are flushed out. For any steady level of muscle activity, there is a corresponding setpoint for blood flow, a process called **flow autoregulation.** In flow autoregulation the error signals are chemical signals such as CO_2, one product of metabolic activity (Figure 5-10), and the effector is the arteriolar muscle.

Regulation by Prostaglandins

Prostaglandins are 20-carbon fatty acids that contain a 5-carbon ring (see Chapter 3, p. 41). All are derivatives of arachidonic acid, a membrane phospholipid. Different prostaglandins are denoted by an abbreviation in which PG stands for prostaglandin and is followed by a letter and subscripts that identify the particular prostaglandin (Table 5-1). This class of molecules has been implicated in a large number of local regulatory functions, including inflammation and blood clotting (Chapter 14), ovulation, menstruation, and labor (Chapters 25 and 26), and secretion of acid by the stomach (Chapter 21).

The initial step in prostaglandin production is the splitting off of a compound known as arachidonic acid by a membrane-bound phospholipase enzyme. In the next step the 5-carbon ring is formed. This reaction is inhibited by the drugs aspirin and acetaminophen, which is the reason that these drugs are effective against inflammation. At the same time inhibition of PGE_2 promotes increased acid secretion in the stomach, so large doses of aspirin can irritate the stomach lining. Prostaglandins are rapidly converted to inactive forms, so their effects are usually confined to the local level. Chemical signals that operate at the local level are called **paracrine** agents to distinguish them from the endocrine agents (hormones) that travel

TABLE 5-1	Effects of Prostaglandins
Prostaglandin	*Physiological effects*
Tx$_{A2}$ (Thromboxane)	Constricts blood vessels
	Promotes blood clotting
PGI$_2$ (Prostacyclin)	Inhibits blood clotting
	Dilates blood vessels
PGE$_2$	Stimulates inflammation
	Dilates blood vessels
	Increases urine formation
	Contracts uterus
	Inhibits acid secretion in the stomach
PGF$_2$	Dilates airways
	Constricts blood vessels
	Contracts uterus

with the blood to all body organs and other types of chemical messengers (Table 5-2).

Intrinsic Autoregulation

In some cases, regulation at the tissue or organ level is a built-in phenomenon. Two examples are provided by the heart and the kidney. The contractile elements of striated heart muscle cells are arranged so that moderate stretching of the cells increases the strength of contraction, while excessive stretching reduces the strength of contraction. In the healthy heart the moderate stretching caused by an increased return of blood to the heart results in increased strength of contraction. This is an example of intrinsic homeostasis.

TABLE 5-2   Chemical Messengers in the Body

Messenger	Examples	Description
Hormone	Insulin Estrogen	Secreted by specialized cells Travels to target cells Influences specific activities
Parahormone	Prostaglandins Histamine	Secreted into the tissue spaces Secreted by a variety of cell types Usually has local effect
Neurohormone	Oxytocin Antidiuretic hormone	Produced by neurons Acts like hormone
Neurotransmitter	Acetylcholine Norepinephrine	Produced by neurons Immediate influences on postsynaptic neurons
Neuromodulator	Endorphins Enkephalins	Produced by neurons Alters the sensitivity of postsynaptic neurons to neurotransmitters

In the kidney, the initial step in urine formation is filtration of fluid from renal capillaries into nephrons of the kidney. The driving force for this is blood pressure. If the total volume of blood plasma increases, the pressure tends also to increase, and the increased pressure causes an increased rate of filtration, which helps to reduce the plasma volume.

By itself, intrinsic regulation is not sufficient to provide all the control necessary for either the heart or the kidney, and in almost all cases intrinsic regulation is supplemented by extrinsic homeostatic processes.

> 1 How can allosteric effects stabilize levels of metabolic intermediates within a cell?
> 2 How does the rate of production of mRNA affect the rate of synthesis of specific proteins in a cell?

EXTRINSIC REGULATION: REFLEX CATEGORIES

Reflex arcs or loops are circuits that link a detection system to a response system (Figure 5-11). A reflex circuit must contain at least three elements: (1) an **afferent,** or sensory, component that detects variations in external or internal variables and relays information about the variable using neural or chemical signals; (2) an **integrator,** often called an **integrating center** or a collection of **association neurons** when it exists within the central nervous system, that determines the magnitude of the response that is appropriate; and (3) an **efferent** or **motor** component that sends neural or hormonal signals from the integrator to the effector organ.

Reflexive control can be used to stabilize physiological variables if the circuit is arranged in such a way as to provide negative feedback. Some exam-

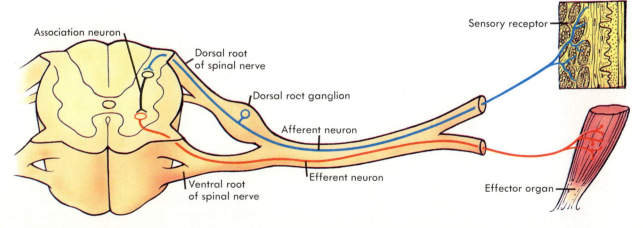

FIGURE 5-11
Basic diagram of a spinal reflex arc. Information from sensory receptors enters the spinal cord via the dorsal root. After synapsing with one or more association neurons, the efferent fiber transmits nerve impulses to effector organs.

ples of homeostatic reflexes are the **baroreceptor reflex,** which regulates arterial blood pressure, and the **respiratory reflexes,** which regulate ventilation of the lungs.

Neural and Endocrine Reflexes

In some reflex loops, integrator cells that are neurons (anatomically speaking) synthesize and release into the blood substances that act as hormones. These are **neurosecretory** or **neuroendocrine** cells. Endocrine glands that are not of neural origin may serve as lines of communication between the nervous system and effectors if their hormonal secretions are controlled by nervous input.

In some cases, endocrine glands combine the functions of sensor and integrator and respond to changes in the controlled variable by increasing or decreasing their rates of secretion. Although the term reflex has traditionally stood for a process involving the central nervous system, it is possible to refer to such a feedback loop as a **hormonal reflex** or **endocrine reflex.** Several organs not readily recognizable as glands serve an endocrine function. This is the case for the kidney and the heart, both of which secrete hormones important in regulating the salt and water economy of the body.

Reflexes can be divided into three classes:

1. **Somatic motor reflexes** that control skeletal muscle. A familiar example is the **withdrawal reflex,** in which a painful stimulus to an arm or leg results in rapid flexing of the appendage, removing it from threat of further injury.
2. **Autonomic reflexes** that modulate the activities of smooth muscle, exocrine glands, and the heart.
3. **Endocrine reflexes** in which the feedback loops may or may not involve the nervous system.

The basic principles of reflex control are the same for somatic motor, autonomic, and endocrine reflexes.

Somatic Motor Control

One function of somatic motor reflexes is to preserve constancy of body position with respect to the surroundings. Another function is to protect the body from dangerous stimuli delivered to the body surface. "Voluntary" movements of the body have a substantial reflexive component.

Afferent input to the central nervous system arrives in the spinal cord by way of nerve fibers from the special sense organs, muscles, and joints. Skeletal muscles are controlled by fibers from nerve cells called **motor neurons.** The pathway of information from sensory neurons to motor neurons almost always includes **interneurons.** These are cells that do not project out of the central nervous system. The specific connections that interneurons make between sensory neurons and motor neurons determine the reflex response that occurs. These connections are established during development so that sensory information is relayed to effectors that can make an appropriate response. The interneurons in a reflex pathway typically increase the possibility for control and modification of the response. Many somatic motor reflexes are so constant in their response pattern that they are referred to as **stereotyped.** If a muscle's tendon is lightly tapped, as by a physician testing the knee jerk response, only the muscle that was stretched by tapping on the tendon will contract. The contraction results from a **stretch reflex** (sometimes called a **tendon reflex** or **myotactic reflex**) that resists or opposes stretching of the muscle (Figure 5-12, *A*). Such reflex circuits are important in maintenance of posture because their negative feedback loop tends to return limbs to their original position.

The skeletal muscles of the human body are, for the most part, arranged as opposing muscle groups, or **antagonists.** Activity in the antagonistic muscles at a joint sets the position of the skeleton relative to

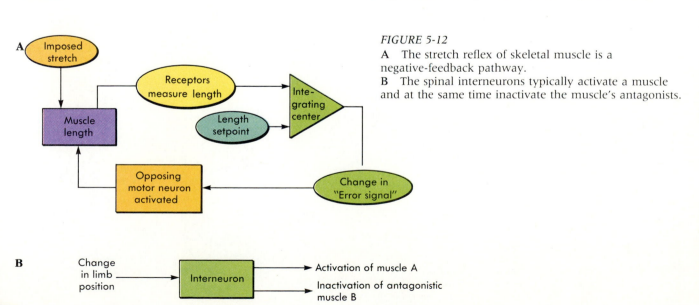

FIGURE 5-12
A   The stretch reflex of skeletal muscle is a negative-feedback pathway.
B   The spinal interneurons typically activate a muscle and at the same time inactivate the muscle's antagonists.

a joint. If one muscle is to alter the limb position, it is necessary for the antagonistic muscle or muscle group to not oppose that action by contracting simultaneously. Typically interneurons in the spinal cord connect the motor neurons of antagonists in such a way that activation of a muscle is automatically accompanied by deactivation of its antagonists (Figure 5-12, B). Stereotyped spinal reflexes are the building blocks of far more complicated motor activities, as discussed in Chapter 11.

Autonomic Reflexes

The internal variables of the body are monitored by receptors that have functional names, such as baroreceptors (blood pressure detection), chemoreceptors (sensitivity to chemicals such as oxygen and carbon dioxide), osmoreceptors (measurement of osmolarity), and thermoreceptors (which monitor the temperature of the skin and body core). As is the case in the somatic motor system, the afferent limb for visceral reflexes is neural. Sensory information is evaluated by integrating centers within the central nervous system. Commands are sent out over efferent nerves and may activate or relax visceral smooth muscle, cause glandular secretion, or alter intracellular metabolism. These neural signals correct deviations from the setpoints programmed within the central nervous system.

There is no difference between the reflex mechanisms for visceral reflexes and those that control skeletal muscles. However, there are anatomical differences in the pathway between the central nervous system and visceral and somatic (skeletal) effectors, and it was formerly believed that voluntary control of visceral functions was not possible. This misconception led to the convention of classifying the neurons that constitute the efferent limb of visceral reflexes as the **autonomic**, or **involuntary**, **division** of the central nervous system.

ENDOCRINE REFLEXES
Hormones as Chemical Messengers

Hormones are one of the many types of chemical messengers in the body (see Table 5-2). Some of these chemical signals have been known for a long time, while others have only recently been discovered. One example of what may be a new category of chemical signals, **nerve growth factor,** is discussed in the essay on p. 98. This chapter uses specific examples to illustrate the principles of endocrine control, but most endocrine systems will be considered in detail when the organ systems they regulate are discussed. A comprehensive summary of endocrine systems is provided as an Appendix to this chapter.

There are two important facts about the mechanism of hormonal information transfer. First, although the message travels throughout the body, it is received only by **target cells**, which are those cells that possess **receptors** for the hormone (Figure 5-13). Hormonal receptors are proteins that recognize and bind to the hormone in question.

FIGURE 5-13
Hormones act only on target cells having specific receptors for a particular hormone.

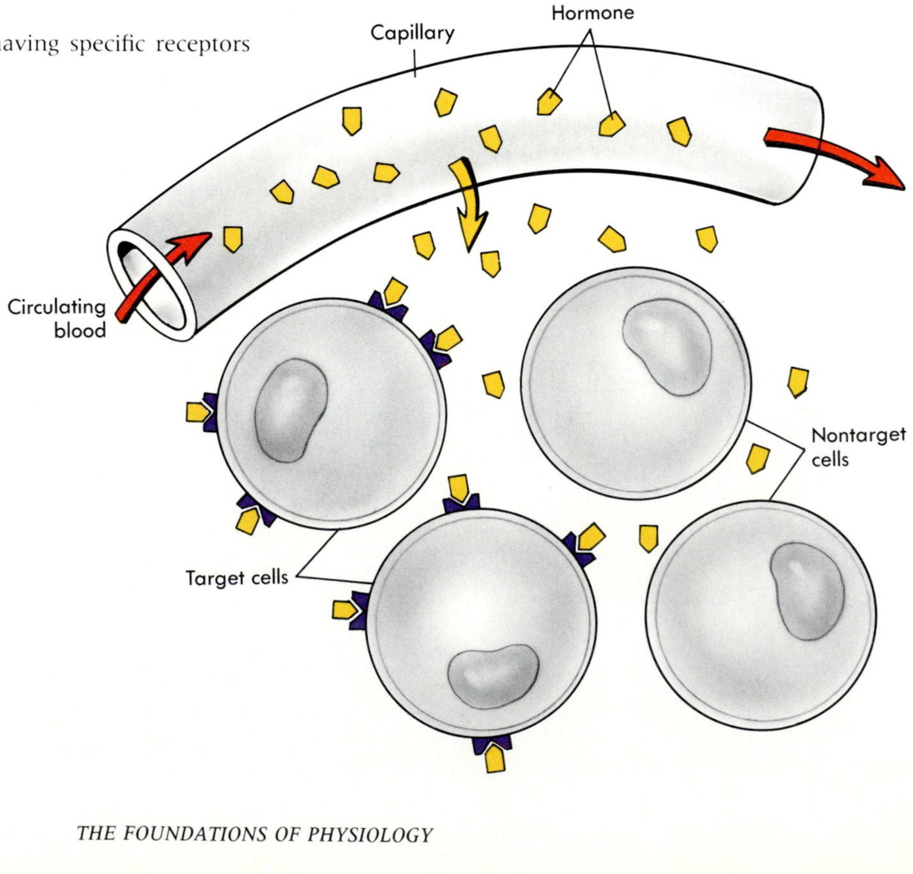

THE FOUNDATIONS OF PHYSIOLOGY

This occurs by a process similar to the interaction between enzymes and substrates. In this case, the complex of hormone and receptor causes changes in specific activities of the target cells, without a change in the chemical structure of the hormone.

The second important fact is that the response of the target cells to the message depends on the capabilities of those cells. A hormone binding to receptors on a gland cell might cause secretion; the same hormone binding to identical receptors in a smooth muscle cell might cause contraction rather than secretion. Thus the meaning of the message depends on the response capabilities of the target cells.

Duration of Hormone Effects

Many endocrine glands synthesize and release their hormones only in response to other hormones, various metabolites, or external signals. Other glands constantly secrete hormones at a fixed (basal) rate, which is then increased or decreased by the control systems that affect the particular gland. Basal hormone release may be steady or periodic; in the latter case there is often a **diurnal** (24-hour) pattern. For example, in the case of **growth hormone** (also called **somatotropin**) secreted by the pituitary, a large fraction of the daily secretion of the hormone occurs in a pulse during the hour or two following the onset of deep sleep (Figure 5-14). Only during this pulse do the levels of hormone in the plasma rise high enough to have a significant stimulatory effect on growth.

Different hormone molecules have different life spans in the blood. The level of hormone in the blood is determined by the secretion rate and the rate at which the hormone is removed by metabolism or excretion. Usually the **half-life**, the time required for half of a given quantity of hormone in the plasma to be metabolized or excreted, is used as a measure of how rapidly a particular hormone is turned over by the body. The shorter the half-life, the higher the secretion rate must be to maintain a given level of hormone in the plasma. The rate of hormone secretion, hormone half-life in the plasma, and the duration of the hormone's effect on target cells all determine the duration of the response to a pulse of hormone. For example, epinephrine has a plasma half-life of a few minutes and does not produce a prolonged response because the response of the target cells is also brief. Growth hormone has a plasma half-life of 20 minutes, but its effects persist because growth hormone initiates long-lasting changes in its target cells.

Hormone Functional Classes

Hormones can have either **direct** or **permissive** actions. In a direct effect, the binding of hormone with receptor initiates a change in the activity of the target tissue. In some cases a second hormone must also be present in small quantity in order to permit the primary hormone to exert its full effect; the second hormone is called a **permissive hormone**. For example, many of the direct metabolic effects of the hormone epinephrine, secreted by the adrenal medulla, are dependent on the presence of an adequate concentration of the hormone cortisol, secreted by the adrenal cortex. The cortisol cannot substitute for epinephrine, but epinephrine is much less effective

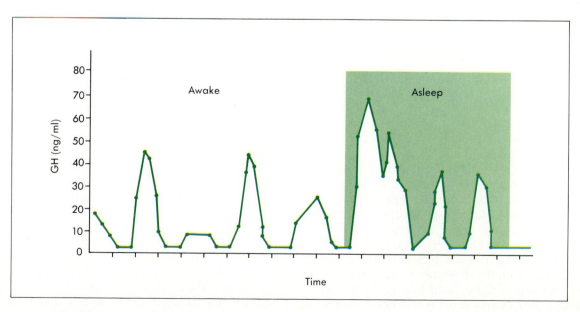

FIGURE 5-14

Secretion of growth hormone in a normal male adolescent occurs in small pulses during waking, with a large pulse just after the onset of sleep.

Nerve Growth Factor

Occasionally, a single discovery opens an unknown realm of scientific inquiry. Scientists are supposed to be open to the unexpected, but it takes an exceptional person to grasp the significance of an observation that does not fit into the accepted framework. The discovery of **nerve growth factor** has such broad implications for cellular communication and control that the field of research it spawned has not yet found its place in the framework of science.

The observation that started this new field was made by Rita Levi-Montalcini and for it she and her biochemist collaborator, Stanley Cohen, were awarded the Nobel Prize in 1986. Dr. Levi-Montalcini was trained as a physician in Turin, Italy, and completed her degree just as Fascist control of Italy closed the door to medical practice for Jewish women. Nevertheless, she had received training in research, and she decided to pursue questions of interest to her in a tiny laboratory that she set up in her bedroom. Working with a crude microscope

and performing operations with a sharpened sewing needle, she repeated experiments on chick embryos published earlier by an American scientist, Victor Hamburger. She published her interpretations in a Belgian journal, because Italy was still not a place where a Jewish writer could publish. This led to an invitation from Dr. Hamburger to join him at Washington University in St. Louis. Dr. Levi-Montalcini's interpretations proved correct, and with the encouragement of Dr. Hamburger, she remained to collaborate with him and enjoy the more open atmosphere for scientific inquiry in the United States.

The line of observations that led to the unexpected discovery began in 1949 as the result of culturing a chick embryo together with a mouse tumor. What happened was that growth of several classes of nerve cells was stimulated enormously in response to the presence of the tumor. The nerve cells penetrated the tumor and grew with such wild abandon that they overran it. This experiment

had been performed before, but its significance had not been grasped. Levi-Montalcini proposed an explanation that had no foundation in the theoretical framework of the day: she thought that the sprouting was due to the release of a chemical factor, a trophic factor, from the tumor. To test this hypothesis, she cultured tumor cells with isolated sympathetic nerve cells. Within hours the nerve cells grew outward in a thick halo, indicating the presence of a powerful stimulant of nerve growth released by the tumor. After years of patient effort and Cohen's discovery that normal salivary glands contain large quantities of the same factor, they determined the chemical composition of nerve growth factor. Since then, many other growth factors have been isolated and studied, including **epidermal growth factor** discovered by Cohen. Nerve growth factor is believed to be released normally during development by target cells that receive connections from sympathetic nerves.

---

in its absence. The nature of the permissive effect is not known in all cases, but one mechanism by which permissive hormones can exert their effects is by stimulating the target cells to synthesize receptors for the other hormone.

In some cases, two or more chemically distinct hormones may have similar effects on the target tissue. If the effect of both hormones applied together is greater than the sum of their individual effects, the hormones are said to be **synergists.** If, when two hormones are present, their effects on the target tissue are merely additive, such hormones are called **agonists. Antagonistic** hormones are ones whose effects on the target tissue tend to cancel each other out. Antagonistic hormones are another example of the principle that homeostatic regulation is often achieved by opposing forces. In general, the sensitivity and versatility of control of a target tissue are

enhanced by multiple lines of hormonal communication.

The term **tropic (or trophic) hormone** is potentially a source of confusion because it can be used in two distinct (but related) senses. In one sense, it applies to hormones that are necessary for the long-term survival or growth of their target tissues, as well as causing immediate changes in their activity. An example of this sense of the term is the hormone **gastrin,** which is necessary for adequate growth and repair of the stomach lining and also stimulates secretion of acid by the stomach. In its second sense, the term tropic hormone is applied to hormones that control the secretion of other hormones. The regulatory pathways that run from the brain to endocrine systems by way of the anterior pituitary are excellent examples of control of endocrine systems by tropic hormones. The common

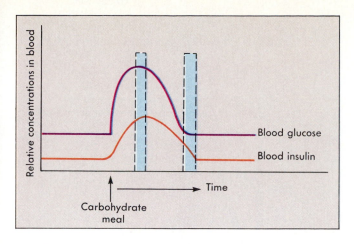

FIGURE 5-15
Blood levels of glucose and insulin after a carbohydrate meal. Note that the peak of blood insulin is reached shortly after the peak of blood glucose, and that insulin levels remain elevated for some time after the glucose has returned to within the normal range. These are indications of the time lag in this feedback system.

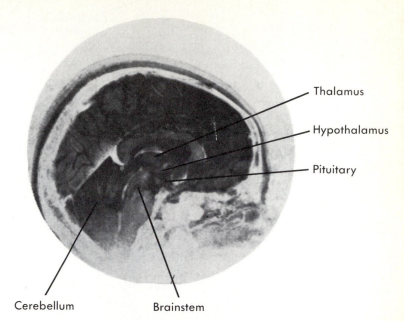

Thalamus

Hypothalamus

Pituitary

Cerebellum   Brainstem

FIGURE 5-16
A technique known as nuclear magnetic resonance imaging (MRI) can be used to examine the brain without using X-ray irradiation. This particular MRI scan represents a section of the brain through the midline and shows the location of the pituitary.

thread of meaning between the two senses is that tropic hormones always have an effect on growth and maintenance. For example, **thyroid-stimulating hormone** (or **thyrotropin**) secreted by the anterior pituitary gland affects both the size of the thyroid gland and the rate of secretion of thyroid hormones.

Hormonal Regulation

Insulin was mentioned earlier in the chapter as an excellent example of negative feedback (see Figure 5-9). Insulin is secreted by β cells of the **islets of Langerhans,** which compose the endocrine portion of the pancreas. The rate of secretion of insulin is controlled primarily by blood glucose levels. An increase in blood glucose is a consequence of consuming a meal containing carbohydrates. One important effect of insulin is to stimulate glucose uptake across cell membranes.

Pancreatic β cells are sensitive to the level of glucose in the blood flowing through nearby capillaries. Insulin is secreted at a low rate during fasting, but an increase in blood glucose above the setpoint level stimulates insulin secretion. Circulating insulin binds with its receptors, which are present on almost every cell type of the body. An increase in the rate of transfer of glucose from blood into the tissues results. The glucose is used for energy or it enters anabolic pathways, leading to storage as fat and glycogen. As a result of the insulin-stimulated movement of glucose into cells, blood levels of glucose fall back in the direction of the setpoint (Figure 5-15). As the setpoint is approached, insulin secretion diminishes. The negative feedback circuit employs the sensitivity of insulin-releasing cells as the sensor and the release of insulin as the effector that returns the variable (blood glucose level) to its setpoint.

Posterior Pituitary

The pituitary gland (hypophysis) is a small but complex structure that is attached to the ventral side of the hypothalamus of the brain (Figure 5-16). From an embryological and functional view, its anterior and posterior parts are distinct. The posterior part of the pituitary gland is, embryologically, a part of the brain. Branches of neurons run from the hypothalamus into the posterior pituitary. Electrical activity in the neurons results in the release of two hormones: **antidiuretic hormone** (abbreviated **ADH,** also called **vasopressin**) and **oxytocin.** The hormones are the outputs of neural reflexes whose inputs come from sensory projections that run to the brain, or from sensors within the brain itself (Figure 5-17, *A*). ADH is involved in fluid and electrolyte balance (Section V), and oxytocin is important in the processes of labor and lactation (Chapter 25).

Anterior Pituitary

A more complex process of endocrine communication between the brain and the body is seen in the anterior pituitary. Unlike the posterior pituitary, the anterior pituitary does not arise from the nervous system in development, but from a cleft of epithelial tissue (Rathke's pouch) that folds and separates from the pharynx (throat) in early development.

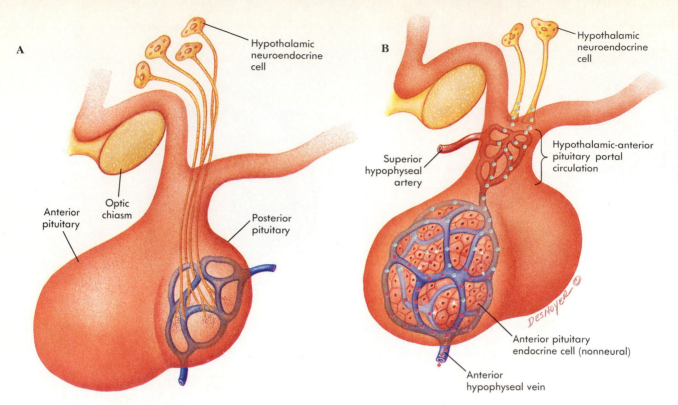

FIGURE 5-17

A The control of hormone secretion by the posterior pituitary. Some hypothalamic neurons send processes down through the stalk of the pituitary into the posterior pituitary. The peptide hormones oxytocin and ADH are synthesized in the neuron cell bodies and transported down the processes into the pituitary, where they are stored. Electrical activity in the neurons results from sensory inputs and leads to release of the hormone.

B The control of hormone secretion by the anterior pituitary. In this case, hypothalamic neurons secrete releasing hormones that travel along the portal circulation to non-neuronal endocrine cells in the anterior pituitary. Binding of the releasing hormone *(aqua)* by the target cell stimulates release of the anterior pituitary hormone *(red)*.

The anterior pituitary contains five recognizable populations of cells that, together, secrete a total of at least seven hormones (see Table 5-3). Six of these hormones have well-characterized roles in metabolism, growth, and reproduction: adrenocorticotropic hormone (ACTH), thyroid-stimulating hormone (TSH), growth hormone (GH), luteinizing hormone (LH), follicle-stimulating hormone (FSH), and pro-

lactin (Pr). The role of melanocyte-stimulating hormone in humans is less well understood.

Of these six, four are tropic hormones, by means of which the brain controls the thyroid gland (TSH), the adrenal cortex (ACTH), and the gonads (FSH and LH). The other two hormones are released into the blood from storage sites in the anterior pituitary and have direct effects on growth (GH) and lactation (Pr). To control secretion of one of the anterior pituitary hormones, the brain secretes the appropriate **releasing hormone** (or in some cases a **release-inhibiting hormone**) into a branch of the circulatory system that carries blood from the brain to the anterior pituitary (Figure 5-17, *B*). Such circulatory systems linking the capillary networks of two organ systems are called **portal circulations;** this one is called the **hypothalamic–anterior pituitary portal circulation.** With one exception, all of the known releasing and release-inhibiting hormones are small peptides (see Table 5-3).

Feedback control of hormones controlled by the anterior pituitary (Figure 5-18) is potentially quite complex because feedback can be applied at several levels, often referred to as **short-loop** or **long-loop** feedback. The controlled hormone may inhibit its

TABLE 5-3	Anterior Pituitary Hormones

Hormone	Common abbreviation
Adrenocorticotropic hormone	ACTH
Thyroid-stimulating hormone	TSH
Growth hormone	GH
Luteinizing hormone	LH
Follicle-stimulating hormone	FSH
Melanocyte-stimulating hormone	MSH
Prolactin	Pr

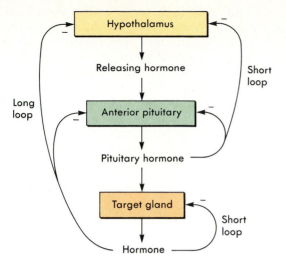

FIGURE 5-18
Feedback may act at several levels in endocrine systems controlled by the hypothalamus and anterior pituitary.

own secretion, the secretion of the tropic hormone, or the secretion of the releasing hormone. The tropic hormone may also inhibit secretion of the releasing hormone. Recent evidence suggests that in addition to the portal circulation from hypothalamus to pituitary, there is a second portal circulation running from anterior pituitary to hypothalamus, which could mediate feedback of anterior pituitary hormones on the hypothalamus.

Hormone Receptor Regulation

Hormone receptors, like all other cellular elements, are subject to turnover. The rate of synthesis of the receptor proteins is influenced by the circulating levels of the hormone. In some cases, continued exposure of cells to a high level of hormone results in a decrease in the number of receptors, called **down-regulation** (Figure 5-19, *A*). Down-regulation can be seen as a process of negative feedback. In one variety of the metabolic disease **diabetes mellitus**,

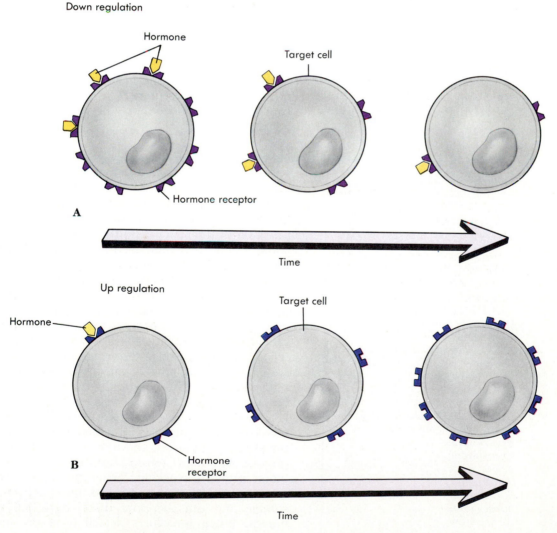

FIGURE 5-19
Down-regulation (**A**) and up-regulation (**B**) of hormone receptors.

chronic elevation of the hormone **insulin** leads to down-regulation of insulin receptors (Chapter 23). This impairs the feedback loop that controls blood glucose.

The opposite process, **up-regulation,** can also occur (Figure 5-19, *B*). This process results in positive feedback. A good example is found in the hormonal control of maturation of follicles in the female ovary (Chapter 25). Maturing follicles secrete the female sex hormone estrogen; as a follicle matures, it augments its secretion of estrogen by increasing the number of binding sites for a pituitary hormone that stimulates estrogen production. Sometimes this is referred to as a **priming effect.** This positive feedback cycle is brought to an end by ovulation (release) of the follicle.

---

1 *What is a local factor? What is a paracrine agent? Are all local factors properly termed paracrine agents?*
2 *What are the differences between direct effects, tropic effects, and permissive effects of hormones?*
3 *What two factors determine the level of a hormone in the blood?*

Chemical Classes of Hormones

The chemical classes of hormones include proteins, glycoproteins, polypeptides, hormones derived from the amino acids tyrosine and tryptophan, and steroids (Table 5-4). There are many protein and poly-peptide hormones. These range in size from thyrotropin-releasing hormone, which has only 3 amino acids, to growth hormone, containing 191 amino acids. Hormones derived from the amino acid tyrosine include epinephrine, secreted by the adrenal medulla, and the thyroid hormones (T_4, sometimes called thyroxine, and T_3). Steroid hormones are synthesized from cholesterol and include the sex hormones estrogen, progesterone, and testosterone produced by the gonads, as well as cortisol and aldosterone produced by the adrenal cortex. Vitamin D, also derived from cholesterol, acts by mechanisms similar to the steroid hormones and is now considered a hormone by most endocrinologists.

Hormonal Signal Amplification

Hormones are typically present in endocrine glands in amounts so small that purification of an amount large enough for chemical analysis may require starting with hundreds of kilograms of tissue. The typical blood concentrations of hormones are very low, and only a few molecules of hormone delivered to each target cell may be enough to cause a response. The target cells must greatly amplify this small message. For example, each molecule of insulin that arrives at a target cell is capable of causing the cell to take up many molecules of glucose. The details of the amplification process are different for different types of hormone, but in all cases the effect is due to multiplication of the hormonal message within the cell.

Mechanism of Action of Steroid Hormones

The effects of steroid hormones on their target tissues are largely due to the ability of these hormones to induce the production of specific proteins in the target cells. This is the result of "turning on" transcription of the specific portions of the target cells' chromosomes that code for the proteins. As a result, the structure of the target tissue may be changed, as happens when sex hormones induce the development of sex-specific body structures during early development and at puberty (Chapter 25). If the new proteins are enzymes, the tissue function served by those enzymes will be stimulated; this is what happens when aldosterone, the hormone secreted by the adrenal cortex, stimulates sodium transport in the distal portion of the nephrons of the kidney (Chapter 20).

Steroids are not sufficiently water soluble to travel as free molecules in blood plasma, so they travel in the blood in the form of complexes with protein. Although steroids can bind to a variety of blood proteins, specific carrier proteins exist for many steroid hormones. Steroid hormones are lipid soluble and can therefore dissolve in cell membranes. When molecules detach from the protein carrier and dissolve in the cell membrane, they can enter a cell. If the cell is not a target cell for that hor-

TABLE 5-4	Structural Categories of Hormones

Structural category	Examples
Proteins	Growth hormone
	Prolactin
	Insulin
	Parathyroid hormone
Glycoproteins	Follicle-stimulating hormone
	Luteinizing hormone
	Thyroid-stimulating hormone
Polypeptides	Thyrotropin-releasing hormone
	Oxytocin
	Antidiuretic hormone
	Calcitonin
	Melanocyte-stimulating hormone
	Adrenocorticotropic hormone
	Hypothalamic hormones
	Somatostatin
Amino acid derivatives	Epinephrine
	Thyroxine
	Melatonin
Steroids	Estrogen
	Progesterone
	Testosterone
	Cortisol

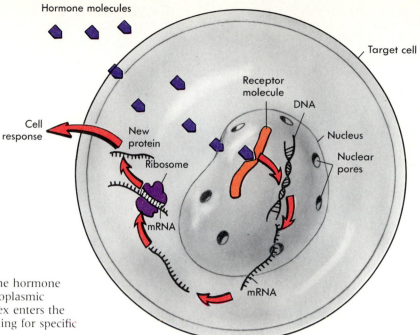

Hormone molecules

Cell response

New protein

Ribosome

mRNA

Receptor molecule

DNA

Nucleus

Nuclear pores

Target cell

mRNA

FIGURE 5-20

The mechanism of effect of steroid hormones. The hormone penetrates the cell membrane and binds to a cytoplasmic (mobile) receptor. The hormone-receptor complex enters the nucleus and affects the transcription of DNA coding for specific proteins.

mone, the process of information transfer goes no further, and the hormone is ultimately metabolized as if it were any other membrane lipid.

Target tissues possess steroid hormone receptors in their cytoplasm (Figure 5-20). These **cytoplasmic receptors** are probably proteins, but not all of them have been chemically characterized. In some cases, the steroid hormone must undergo enzymatic conversion to a different steroid before it can bind effectively with its receptor. For example, many tissues convert the male sex hormone testosterone to dihydrotestosterone (Chapter 25). This is particularly important in the case of embryonic external genitalia, which follow the female pattern of development unless testosterone is present. Genetic males who lack the DNA coding for the enzyme that catalyzes the conversion (called 5-reductase) are born with feminized genitalia.

Once formed, the complex of steroid hormone and cytoplasmic receptor migrates into the nucleus. The necessity of this step led to the name **mobile receptor hypothesis** for this description of events. Once in the nucleus, the complex reacts with a second **nuclear receptor** that is associated with the DNA. The nuclear receptor splits, with part staying at the original site on the chromosome. The other part attaches to an adjacent region of DNA, initiating transcription of mRNAs carrying a copy of the code for the protein specified by that part of the chromosome. The mRNAs enter the cytoplasm, where their message is translated into protein by ribosomes.

Each steroid-receptor complex can initiate the production of a large number of mRNAs, and each mRNA can result in the synthesis of a large number of copies of the protein. Thus a single hormone mol-

ecule might easily result in production of several thousand protein molecules, giving the feedback loop served by the hormone a gain of several thousandfold.

Mechanism of Action of Peptide and Protein Hormones

Many protein- or peptide-type hormones act by stimulating or inhibiting specific enzymatic pathways; for example, the hormone epinephrine causes the liver to increase glycogen breakdown and decrease glycogen synthesis. It does this by activating and inhibiting enzymes present in the target cell at the time the hormonal message arrives. The responses to protein and peptide hormones are initiated by binding with receptors present on the surface of target cells; these are thus **membrane-bound or fixed receptors** (Figure 5-21). Although some of the effects that hormones of this type produce are mediated by movement of the hormone into the cell (possibly by endocytosis of both hormone and receptor) with subsequent effects on mRNA synthesis, most of the obvious effects are mediated by **second messengers.** Interaction with receptors at the cell surface releases second messengers, which carry the message to enzymes in the cell interior. Several such second messengers are known.

One common second messenger system (Figure 5-22, A) involves a compound called **cyclic AMP** (a form of adenosine monophosphate, abbreviated **cAMP**). Binding of the hormone with its receptor results in activation of a class of membrane protein now referred to as **G-proteins.** Different types of G-proteins in turn activate (G_s) or inhibit (G_i) an enzyme called **adenylate cyclase** located on the cyto-

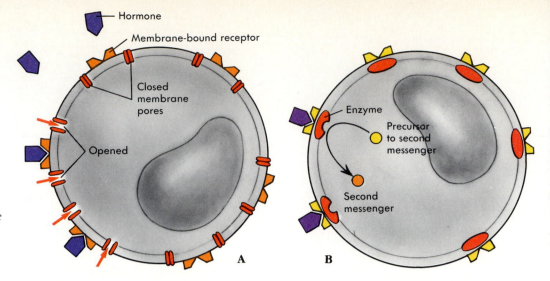

FIGURE 5-21

The mechanism of action of most protein and peptide hormones. The hormone binds with receptors on the exterior of the cell membrane, setting in motion second messengers.

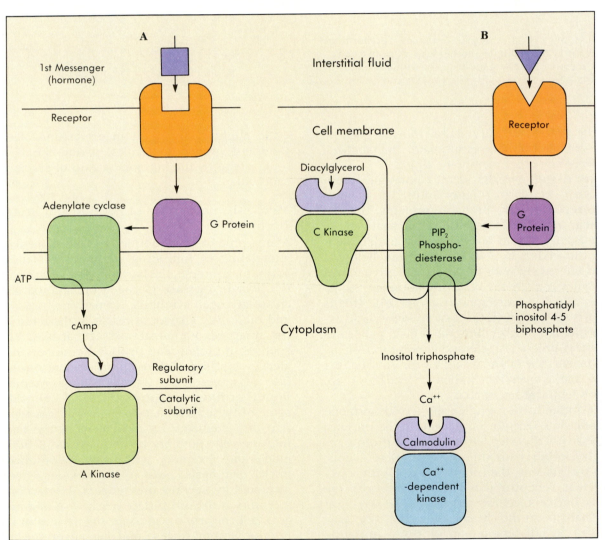

FIGURE 5-22

Two specific second messenger systems.

A The sequence of events that leads to the generation of the cAMP second messenger. Hormone binding can either increase or decrease adenylate cyclase activity via intermediate, membrane-bound G-proteins. The effects of cAMP within cells are mediated by protein kinases termed *A kinases*. The cAMP acts by binding to an allosteric site on a regulatory subunit of the enzyme.

B The sequence of events collectively termed the *phosphoinositol system*. Hormone binding increases phosphodiesterase activity via an intermediate G-protein. Activation of the phosphodiesterase results in the formation of two messengers, diacylglycerol and inositol triphosphate. Diacylglycerol can bind to the regulatory subunit of a family of C kinases. Inositol triphosphate can bring about entry of Ca^{++} into the cell. The Ca^{++} messenger binds to calmodulin, the regulatory subunit of Ca^{++}-dependent protein kinases.

FIGURE 5-23

Calcium-calmodulin as an intracellular second messenger. Certain hormones bind to receptors on the cell membrane, often causing an increase in the membrane permeability to calcium. Calcium entering the cell binds to calmodulin, and the complex regulates the activity of a variety of intracellular enzymes.

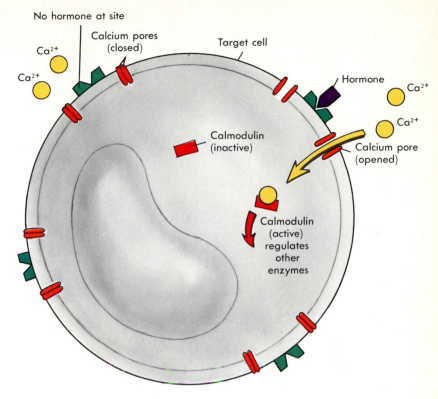

plasm-facing surface of the cell membrane. Adenylate cyclase catalyzes the conversion of a small amount of the cell's stock of ATP into cAMP, the second messenger. The duration of the hormone effect is determined by the activity of a second enzyme called **phosphodiesterase** that converts cAMP into plain AMP, which is not active as a messenger. A similar mechanism utilizes cyclic guanosine monophosphate (cGMP) as a second messenger.

An equally important but more recently discovered second messenger system (Figure 5-22, *B*) involves phospholipid compounds called **phosphoinositides.** In this case the hormone receptor is coupled by a membrane-bound G-protein (structurally similar to G_i) to an enzyme that breaks down polyphosphoinositides to **inositol triphosphate (ITP)** and **1-2 diacylglycerol** (DAG, a diglyceride). Hormone binding activates this enzyme and both ITP and DAG act as internal second messengers at different sites.

Intracellular Ca^{++} is involved in control of many cellular functions. In some cases, it makes sense to call it a second messenger. It is important to realize that the use of Ca^{++} to control cell activities depends on maintaining a very low concentration of Ca^{++} in the cytoplasm (approximately 10^{-6} M). Hormonal signals can then be relayed by relatively few Ca^{++} ions. Depending on the particular cell type, messenger Ca^{++} may be allowed to enter the cell from extracellular fluid as a result of hormone binding, or it may be released from intracellu-

lar stores located in such organelles as mitochondria and the endoplasmic reticulum.

The most important general mechanism for translating a Ca^{++} second message into effects on cellular function involves an intracellular Ca^{++} receptor protein called **calmodulin** (Figure 5-23). Calmodulin is frequently attached to enzymes, where it serves as a **regulatory subunit** (see Figure 5-22, *B*). This is true for the enzyme glycogen phosphorylase kinase, in which attachment of Ca^{++} to the regulatory subunit results in activation of the **catalytic subunit** of the enzyme that actually catalyzes the breakdown of substrate (Figure 5-24).

Second Messenger Regulation of Protein Kinases

A great many enzymes that catalyze or control key steps in cell metabolism, or important cell functions such as muscle contraction, can be activated or inhibited by attachment of phosphate groups to allosteric control sites. This process is referred to as **enzyme phosphorylation.** Phosphorylation of specific enzymes is catalyzed by a family of control enzymes called **protein kinases** (see Figure 5-22). Protein kinases are, in turn, under the control of second messengers. In the case of glycogen breakdown (Figure 5-24), the key step in conversion of glycogen to glucose is catalyzed by glycogen phosphorylase. This enzyme is activated by glycogen phosphorylase kinase, which is activated by another protein kinase, which is activated by cAMP.

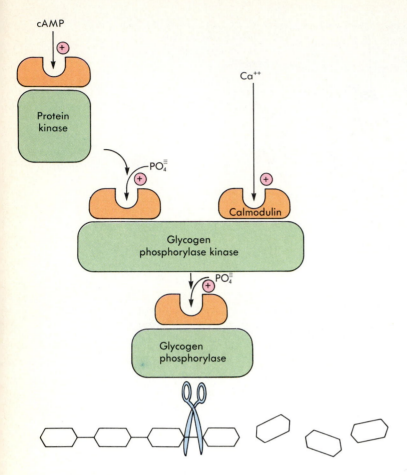

cAMP

Protein kinase

Ca⁺⁺

PO$_4^{\equiv}$

Calmodulin

Glycogen phosphorylase kinase

PO$_4^{\equiv}$

Glycogen phosphorylase

FIGURE 5-24

An example of the effects of cAMP and Ca^{++} second messengers. Both stimulate glycogen breakdown in liver cells. The protein kinase activated by cAMP is an A kinase. Calmodulin is a regulatory subunit of glycogen phosphorylase kinase, so this enzyme is an example of a Ca^{++}-dependent kinase. The actual conversion of glycogen to glucose-6-phosphate is catalyzed by glycogen phosphorylase, the final step in the regulatory cascade. The + symbols indicate an allosteric activating effect.

Why are there so many control steps? Just as in the case of steroid hormones, each step gives amplification. A few molecules of hormone arrive at the cell, each resulting in the formation of a few molecules of cAMP. Each cAMP molecule activates many protein kinase molecules. Each protein kinase molecule activates many phosphorylase kinase molecules. Finally, each active molecule of phosphorylase releases many glucose units from glycogen.

Note that both cAMP and Ca^{++} can activate glycogen phosphorylase kinase at different allosteric sites; for this enzyme they are agonistic second messengers. There are other sites on the enzyme that can be phosphorylated by other protein kinases; some of these yield inhibitory effects. Why are there so many control sites on this molecule? The answer is that the key step of glycogen breakdown controlled by glycogen phosphorylase kinase must be responsive to many different homeostatic signals from different hormones and from other elements of the cell's energy metabolism. The enzyme molecule serves as an integrator, summing the inputs to arrive at a level of activity appropriate to serve the needs of the liver cell and the body's need for glucose.

1 Why must steroids be complexed with protein in order to travel in the blood?
2 Why does the low lipid solubility of peptides and tyrosine derivatives dictate that they act by way of second messengers?
3 How is amplification of the hormonal message obtained in the case of steroid hormones? How about protein hormones?
4 Why do key enzymes frequently have multiple allosteric sites?

SUMMARY

1. In a **negative feedback loop**, information flows from the **sensors** to the **integrator**, where it is compared with a **setpoint**. A difference between the sensor's input and the setpoint results in generation of an **error signal** that is passed to effectors. The effect of the loop is to minimize the difference.
2. **Positive feedback loops** result in dramatic increases or decreases in effector activity, driving the controlled variable to extreme values.
3. In the body, homeostatic processes occur on three levels: **intracellular control**, which frequently involves enzyme regulation; **local control** on the tissue level, which frequently involves the feedback effects of local chemical factors; and **extrinsic control**, in which the feedback loops are neural or hormonal reflexes.
4. Extrinsic feedback loops can be divided into three categories of reflexes: **somatic, autonomic,** and **hormonal.**
5. Hormone effects can be categorized as **direct, permissive,** or **tropic.**
6. The number of hormone receptors is under cellular control and may demonstrate either negative feedback (**down-regulation**) or positive feedback (**up-regulation**).

7. The brain releases two neurohormones, **antidiuretic hormone** and **oxytocin,** by way of the **posterior pituitary.** Using **releasing** and **release-inhibiting hormones,** it controls the release of at least seven **anterior pituitary** hormones. Four of the anterior pituitary hormones (ACTH, GH, FSH, and LH) are tropic hormones for endocrine glands.
8. There are three major chemical classes of hormones: **steroids, peptides/proteins,** and **amino acid derivatives.**
9. Steroid hormones are recognized by **mobile cytoplasmic receptors;** typically their action involves increases in the synthesis of specific proteins.
10. Peptide/protein hormones and amino acid derivatives are recognized by **fixed receptors** on the cell membrane; binding of hormone to receptor results in activation of an intracellular **second messenger.** Common second messengers are **cyclic adenosine monophosphate (cAMP), Ca^{++}, inositol triphosphate,** and **1-2 diacylglycerol.** The effects of second messengers typically involve **phosphorylation** of an **allosteric control site** on key enzymes of the target cells.

● STUDY QUESTIONS

1. What are the basic elements of a feedback loop?
2. Why is the gain of a feedback loop important?
3. What happens when there is a delay between the origin of an error signal and the response of the effector?
4. At what three levels is negative feedback exerted?
5. Name some reflexes that act as negative feedback loops.
6. What are the major chemical classes of hormones?
7. What determines whether a given hormone will affect a particular target organ?
8. What are the two main intracellular mechanisms of hormone action? For each, what are the steps that lead from arrival of the hormone at the target cell to the response of the target cell?
9. What membrane constituents are the starting material for prostaglandin synthesis? At what level do prostaglandins operate?
10. What are the two definitions of tropic hormone? Name some examples and state the target tissue that each controls.
11. What are the functional differences between the two lobes of the pituitary?
12. Why might it be a mistake to attempt to evaluate the rate of secretion of a hormone on the basis of a single blood sample?

● SUGGESTED READING

DeCHERNEY, A.H., and F. NAFTOLIN: Recombinant DNA-derived human luteinizing hormone, *Journal of the American Medical Association,* volume 259, June 10, 1988, p. 3313. Discusses one case where recombinant DNA technology has been rapidly applied to an important problem in clinical medicine.

Growing up with the endocrine system, *Current Health,* volume 1, February 1987, p. 3. Discussion of the structure and function of the endocrine system, including neuroendocrine mechanisms.

HOUK, J.C.: Control strategies in physiological systems, *FASEB Journal,* volume 2, February 1988, p. 97. An article written for a very general audience that discusses electrical and chemical messages and control systems from both a cellular and integrative perspective.

THOMPSON, C.C., C. WEINBERGER, R. LEBO, and R.M. EVANS: Identification of a novel thyroid hormone receptor in the central nervous system. *Science,* volume 237, September 25, 1987, p. 1610. Shows how thyroid hormone influences the development of the central nervous system.

Endocrine Glands and Their Hormones

The classical definition of endocrinology was that it comprised the area of physiology concerned with the "internal secretions" of the body. The original concept that a chemical substance called a hormone, liberated by one type of cell, was carried by the bloodstream to act on a distant target cell represented a major advance in physiological understanding. Over time, this original definition of a hormone and an endocrine gland has been expanded to include locally acting paracrine agents, neurohormones, and various releasing and release-inhibiting factors. Target cells of hormone action now include other hormone-producing cells, and some hormones are known to require chemical modification at intermediate sites between the gland of origin and the target cell(s) before their mission can be accomplished. Hormones once thought to be confined to the gastrointestinal tract have been demonstrated to be present in the brain, and receptors for some central nervous system neurotransmitters have been identified on cells in the immune system. The latter finding has even resulted in the definition of a new research area—"neuroendocrine immunology."

Advances in our understanding of the mechanisms of action of hormones, hormonal interactions, and physiological processes have made it increasingly difficult to discuss endocrinology as an isolated topic. If an endocrine section occurs early in a textbook, students have not been exposed to the physiology of the target tissues and organs and thus cannot easily grasp the role of hormones in homeostasis. On the other hand, if the hormonal influences on a particular organ system are to be understood, they must be carefully integrated with the coverage of basic physiology. Thus endocrinology cannot be deferred to the end of a textbook.

Many hormones regulate more than one physiological process and, conversely, an individual organ system may be regulated by several hormones. For example, antidiuretic hormone, the renin-angiotensin-aldosterone system, and atrial natriuretic hormone are all released from different glands and interact to affect cardiovascular performance and fluid and electrolyte balance. The cardiovascular response to exercise and stress involves the autonomic nervous system, cortisol, and epinephrine. Such considerations preclude the use of a gland-by-gland approach to endocrinology.

This text integrates endocrinology and autonomic control with the basic physiology of each of the organ systems. However, it is useful to have an idea of how the endocrine system is organized, what the most important glands are, and what they do. The following set of figures and tables describes the endocrine glands of the body, lists the hormones secreted, and provides page references to the physiological processes they regulate.

Figure 5-A shows the location of the major endocrine organs of the body: pituitary, thyroid, parathyroid, thymus, adrenal gland, pancreas, ovaries (in females), and testes (in males). As noted in the main part of the chapter, some of the peripheral endocrine glands (thyroid, adrenal cortex, ovaries, and testes) are controlled by the anterior pituitary hormones, whose secretion, in turn, depends on a set of hypothalamic releasing and release-inhibiting factors. The posterior pituitary, an extension of the central nervous system, secretes antidiuretic hormone and oxytocin (Figure 5-B). The role of antidiuretic hormone is discussed within the context of body fluid control (Chapter 20), and oxytocin is detailed in the reproductive section (Chapter 26).

Pancreatic hormones are discussed in the section on metabolism (Chapter 23), and the adrenal hormones are described in Chapter 15 (cardiovascular regulation) and Chapter 23. Detailed coverage of FSH, LH, prolactin, and the sex steroids is found in Chapters 25 and 26.

To these easily identified endocrine glands must be added collections of cells within the gastrointestinal tract that secrete substances such as gastrin, motilin, and gastric inhibitory peptide (Chapter 22). There are specialized cells in the kidney that release the hormone renin (Chapter 20), and stretch-sensitive cells in the wall of the right atrium secrete atrial natriuretic factor (Chapters 15 and 20). During pregnancy, even the placenta takes on a role as an endocrine organ (Chapter 26). Prostaglandins have become increasingly important as hormones and are discussed at several points in the text. This focus unit provides a list of hormones produced by the body, identifying their target tissue(s) and summarizing their physiological actions. The tables can be used as a reference when each hormone is discussed in detail within the context of the tissues and organ systems it regulates.

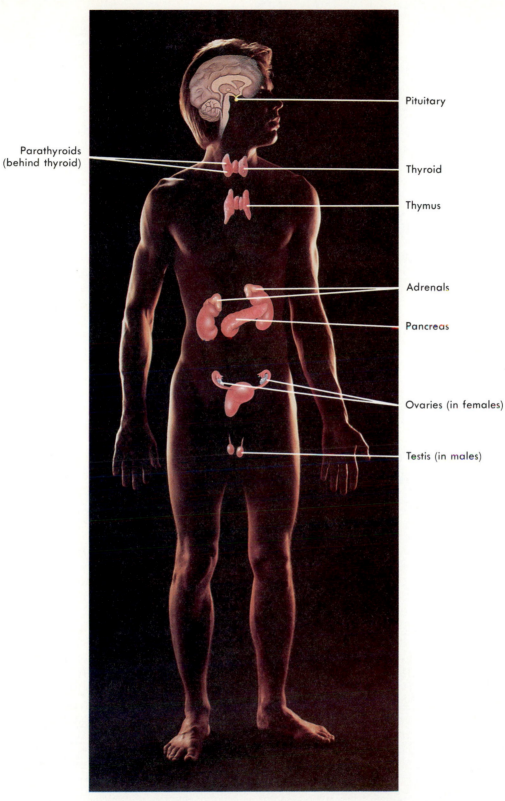

Pituitary

Parathyroids
(behind thyroid)

Thyroid

Thymus

Adrenals

Pancreas

Ovaries (in females)

Testis (in males)

FIGURE 5-A
Location of the major endocrine glands of the body.

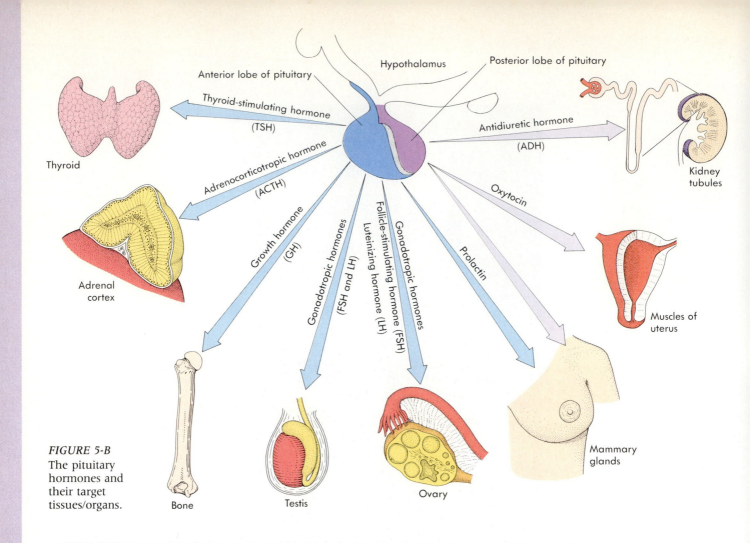

FIGURE 5-B
The pituitary hormones and their target tissues/organs.

Hypothalamus

Hormone	Source	Target	Action/Reference
Corticotropin-releasing hormone	Hypothalamus	Anterior pituitary	Increased ACTH secretion p. 612
Gonadotropin-releasing hormone	Hypothalamus	Anterior pituitary	Increased FSH, LH secretion p. 655
Growth-hormone–releasing hormone	Hypothalamus	Anterior pituitary	Increased GH secretion p. 602
Prolactin-inhibiting hormone	Hypothalamus	Anterior pituitary	Decreased Pr secretion p. 699
Prolactin-releasing hormone	Hypothalamus	Anterior pituitary	Increased Pr secretion p. 699
Somatostatin	Hypothalamus	Anterior pituitary	Decreased GH secretion p. 602
Thyrotropin-releasing hormone	Hypothalamus	Anterior pituitary	Increased TSH secretion p. 608

Pituitary

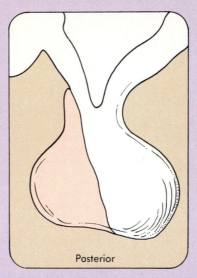

Posterior

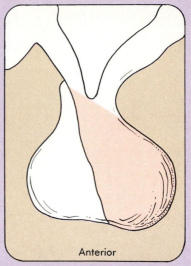

Anterior

Hormone	Source	Target	Action/Reference
Adrenocorticotropic hormone	Anterior lobe	Adrenal cortex	Secretion of adrenal cortex hormones p. 612
Follicle-stimulating hormone	Anterior lobe	Ovaries, testes	Oogenesis p. 653 Spermatogenesis p. 655 Menstrual cycle p. 661
Growth hormone	Anterior lobe	Most tissues	Anabolism, growth and development p. 602
Leutinizing hormone	Anterior lobe	Ovaries, testes	Oogenesis p. 653 Spermatogenesis p. 655 Menstrual cycle p. 661
Melanocyte-stimulating hormone	Anterior lobe	Varied	Not fully understood in humans
Prolactin	Anterior lobe	Mammary glands	Milk production p. 699
Thyroid-stimulating hormone	Anterior lobe	Thyroid	Production and secretion of thyroid hormone p. 608
Antidiuretic hormone p. 608	Posterior lobe	Kidney	Water reabsorption p. 511
Oxytocin	Posterior lobe	Mammary glands, uterus	Milk letdown p. 699 Uterine contraction p. 696

Gastrointestinal Tract

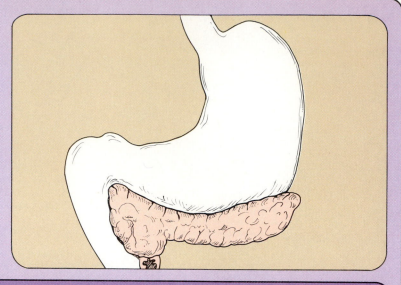

Hormone	Source	Target	Action/Reference
Cholecystokinin (CCK)	Duodenal mucosa	Gallbladder, pancreas	Contraction of gallbladder p. 583 Pancreatic enzyme secretion p. 580
Gastric inhibitory peptide (GIP)	Duodenal mucosa	Stomach, pancreas	Inhibits stomach motility p. 582 Increases insulin and glucagon secretion p. 592 At higher concentrations, motility and secretion p. 582
Gastrin	Stomach mucosa	Stomach	Acid secretion p. 578 Motility p. 580
Motilin	Stomach mucosa	Stomach, intestine	Motility p. 580
Secretin	Duodenal mucosa	Stomach, pancreas	Inhibits acid secretion and gastrin release p. 583 Pancreatic enzyme and fluid secretion p. 583
Glucagon	Pancreas (alpha cells)	Liver	Increases breakdown of glycogen p. 596
Insulin	Pancreas (beta cells)	Liver, muscle, fat	Increases uptake and use of glucose and amino acids p. 590
Somatostatin	Pancreas (delta cells)	Alpha and beta cells	Inhibits insulin and glucagon secretion p. 595

Thyroid and Parathyroid

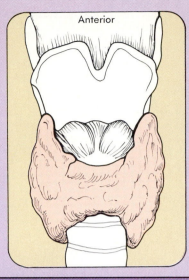

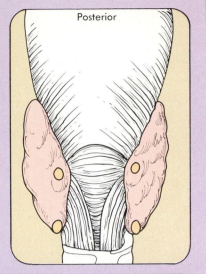

Hormone	Source	Target	Action/Reference
Calcitonin	Thyroid	Bone	Bone deposition p. 517
			Plasma calcium concentration p. 518
Thyroxin	Thyroid	Most tissues	Growth and development p. 604
			Response to stress and cold p. 608
Parathormone	Parathyroid	Bone, kidneys	Calcium and phosphate homeostasis p. 518

Adrenals and Kidney

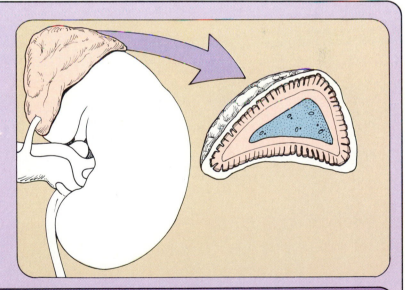

Hormone	Source	Target	Action/Reference
Aldosterone	Adrenal cortex	Kidney, most tissues	Na^+ retention, K^+ excretion p. 514
Androgens	Adrenal cortex	Most tissues	In females, sexual drive, hair growth p. 658
Cortisol	Adrenal cortex	Most tissues	Catabolism p. 599
			Inhibition of inflammation p. 634
Epinephrine	Adrenal medulla	Most tissues	Response to stress p. 612
Cholecalciferol (DOHCC)	Kidney	Intestine	Calcium and phosphate homeostasis p. 517
Erythropoietin	Kidney	Bone marrow	Erythrocyte synthesis p. 349

Reproductive Hormones

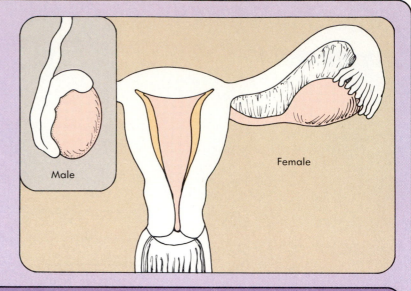

Male

Female

Hormone	Source	Target	Action/Reference
Testosterone	Testes	Most tissues	Spermatogenesis p. 648 Secondary sex characteristics p. 655
Inhibin	Testes	Pituitary	Feedback from testes to the pituitary p. 660
Estrogen	Ovaries	Most tissues	Menstrual cycle p. 661 Development of mammary glands and uterus p. 656 Secondary sex characteristics p. 656
Progesterone	Ovaries	Most tissues	Menstrual cycle p. 661 Development of mammary glands and uterus p. 656 Secondary sex characteristics p. 656
Human chorionic gonadotropin (HCG)	Placenta	Corpus luteum	Maintain corpus luteum in pregnancy p. 683
Human placental lactogen	Placenta	Mammary glands, fetus	Preparation for lactation p. 687 Fetal bone growth p. 687
Relaxin	Placenta	Bones and ligaments of pelvis, cervix	Preparation for delivery p. 687
Estrogen	Placenta	Breasts, uterus	Maintains uterine lining p. 670 Stimulates breasts and menstrual cycle p. 698
Progesterone	Placenta	Breasts, uterus	Inhibits uterine contraction p. 697 Stimulates breasts and menstrual cycle p. 698

Other Structures

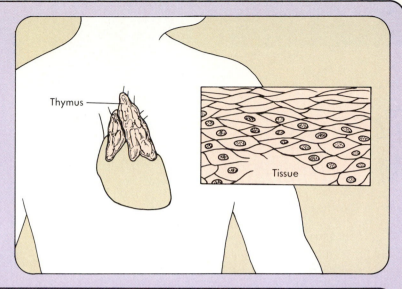

Thymus

Tissue

Hormone	Source	Target	Action/Reference
Atrial natriuretic hormone (ANH)	Atrium	Kidney	Regulates solute and water loss p. 513
Thymosin	Thymus	Immune tissues	Development and function of the immune system p. 627
Prostaglandins	Most tissues	Most tissues	Inflammation p. 631 Uterine contraction p. 697

CHAPTER 6

Transport across Cell Membranes

On completing this chapter you will be able to:

- Describe the bilayer structure of cell membranes, identifying intrinsic and extrinsic proteins.
- Understand why pure lipid bilayers are impermeable to polar solutes and ions.
- Describe the characteristics of mediated transport systems.
- Understand the factors that determine the rate of diffusion.
- Define the terms isotonic, hypotonic, and hypertonic and be able to apply them to determine whether cells will change their volume.
- Understand why cell volume regulation requires that Na^+ be effectively impermeant.
- Describe the origin of the membrane potential and be able to determine the magnitude and sign of the resting potential, knowing the permeabilities and concentration gradients of Na^+ and K^+.
- Distinguish between facilitated diffusion, active transport, and gradient-driven countertransport or cotransport.

The cell membrane is a thin film of lipid studded with proteins that encloses each cell. It maintains an appropriate chemical environment for the metabolic processes of the cell, regulates the volume of cytoplasm, and mediates information transfer in the form of chemical and electrical signals.

The membrane is a dynamic barrier. There is a steady traffic of nutrients, such as glucose and amino acids, across the membrane. Some solutes that are not lipid soluble can penetrate the membrane with the aid of protein molecules that span the membrane and serve as channels or carriers. Concentration gradients of Na^+ and K^+ across the membrane are maintained by ATP-driven active transport of these ions. One effect of these ionic concentration gradients, together with the differential permeability of the cell membrane to Na^+ and K^+, is an imbalance of electrical charge across the membrane, called the membrane potential. The molecular constituents of the cell membrane itself are constantly being exchanged with intracellular membrane systems such as the endoplasmic reticulum.

THE MOLECULAR STRUCTURE OF CELL MEMBRANES

Properties of Phospholipids in Membranes

When they are surrounded by water molecules, the phospholipid molecules behave in a fashion reminiscent of a herd of bison surrounded by wolves. The hydrophilic ends of the phospholipids point outward toward the water, while the hydrophobic hydrocarbon chains face away from the water and interact with one another. This is, from an energetic point of view, the most favored state for the phospholipids. The resulting phospholipid structure can be a spherical **micelle** (see Figure 3-3, *B*), as when fat digestion occurs in the gastrointestinal tract (Chapter 22). In the case of the cell membrane, as well as the membranes of intracellular organelles, the favored state consists of two monolayers of phospholipids and is referred to as a lipid **bilayer** (see Figure 3-3, *C*). The hydrocarbon interior of the membrane constitutes a barrier to **permeation**—passage through the membrane—of polar solutes, such as amino acids, sugars, and ions.

Membrane Proteins

The structure of a cell membrane is shown in Figure 6-1. This representation is referred to as the **fluid mosaic model** because the membrane is a mosaic of phospholipids and proteins, and, in many cases, both these constituents are free to move in the two-dimensional plane of the membrane.

Membrane proteins can be divided into two classes based on how tightly they interact with membrane phospholipids. **Extrinsic** proteins are loosely bound to the external or internal surfaces by electrostatic forces and can be removed by mild chemical procedures, such as changes in salt concentration, that disrupt weak electrostatic forces (see Chapter 3).

Extrinsic proteins are typically associated with the second category of membrane proteins, **intrinsic,** or **integral**, membrane proteins. Intrinsic membrane proteins associate more closely with membrane lipids than extrinsic proteins because intrinsic proteins contain a large number of amino acids with hydrophobic side groups. These side groups interact strongly with the hydrophobic tails of the membrane phospholipids, so that intrinsic membrane proteins can only be dissociated from membranes by drastic treatments, such as addition of strong detergents or organic solvents. Some intrinsic membrane proteins extend completely across the membrane, while others only partially penetrate it (Figure 6-1). When membrane proteins extend across the membrane, the membrane-spanning seg-

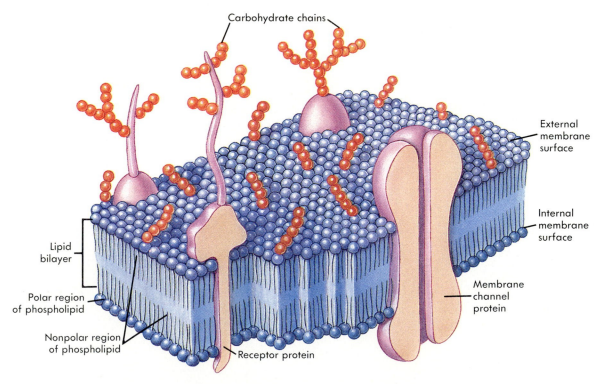

Carbohydrate chains

External membrane surface

Internal membrane surface

Membrane channel protein

Lipid bilayer

Polar region of phospholipid

Nonpolar region of phospholipid

Receptor protein

FIGURE 6-1

Fluid mosaic model of a cell membrane composed of a lipid bilayer with proteins "floating" in the membrane. The nonpolar hydrophobic portion of each phospholipid molecule is directed toward the center of the membrane, and the polar hydrophilic portion is directed toward the water environment either outside or inside the cell. Intrinsic proteins partially or fully penetrate the membrane and may serve as transport or receptor molecules.

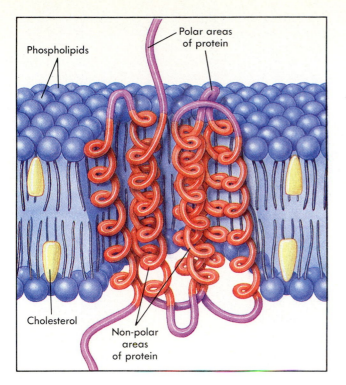

FIGURE 6-2
Membrane proteins are anchored within the lipid bilayer by nonpolar segments having a helical secondary structure. Many membrane proteins consist of several hydrophobic segments within the membrane connected by hydrophilic segments projecting into the aqueous environment on either side of the membrane.

Labels in figure:
Phospholipids
Polar areas of protein
Cholesterol
Non-polar areas of protein

ments are generally helical in their secondary structure (Figure 6-2). Several membrane-spanning segments may be present in such proteins. In many cases, portions of a membrane protein that project from the membrane have carbohydrate chains attached to them (see Figure 6-1). The result is referred to as a **glycoprotein.**

Membrane proteins serve a variety of functions (Table 6-1). Some are **transport proteins** and serve as **channels** or **carriers** for solutes that cross the cell membrane, while others are **receptor proteins** (Figure 6-3). Several classes of these membrane proteins are involved in converting chemical messages arriving at membrane surface receptors from other parts of the body into intracellular responses. Some receptor proteins are also transport proteins because the conversion process involves changes in ion movements across the cell membrane. Other receptor proteins can be termed **recognition proteins** because they allow transplanted cells or cancer cells to be recognized by the immune system (see Chapter 24, p. 626).

Enzymes may be attached to the interior or exterior surface of membranes. Enzymes that face the interior may be involved in metabolic activities of the cell. A good example of this is adenylate cyclase (see Chapter 5, p. 103). Those that face the exterior may function as hormone receptors (see Chapter 5, p. 104). They may also catalyze reactions that take place near the cell surface. For example, intestinal cells carry some of the enzymes involved in carbohydrate and protein digestion on their exterior surfaces. The components of junctions between cells (see Figure 4-17, p. 74) are also composed of membrane proteins.

Two-Dimensional Mobility of Membrane Proteins
One piece of evidence for fluidity of cell membranes was provided by experiments in which certain intrinsic proteins were labelled and shown to be capable of moving freely within the plane of the membrane. This freedom of movement does not include rotation; the protein-phospholipid interactions are strong, and therefore the energy required to turn an intrinsic protein around so that its external and internal ends are interchanged would be very large.

TABLE 6-1	Functional Classes of Cell Membrane Proteins

Class	Function	Examples
1. Receptors	Bind molecules with specific properties	Acetylcholine receptor (Chapter 8)
2. Recognition proteins	Allow the immune system to distinguish cancer cells and invading organisms from normal body cells	Histocompatibility antigens (Chapter 24)
3. Transport proteins	Permit specific polar solutes and ions to cross the cell membrane; in some cases these proteins are also enzymes that split ATP	Na$^+$-K$^+$ ATPase (This chapter)
4. Junctional proteins	Form links between adjacent cells	Gap junctional protein (Chapter 4)
5. Enzymes	Catalyze specific reactions of substrates in intracellular or extracellular fluid	Acetylcholine esterase (Chapter 8)

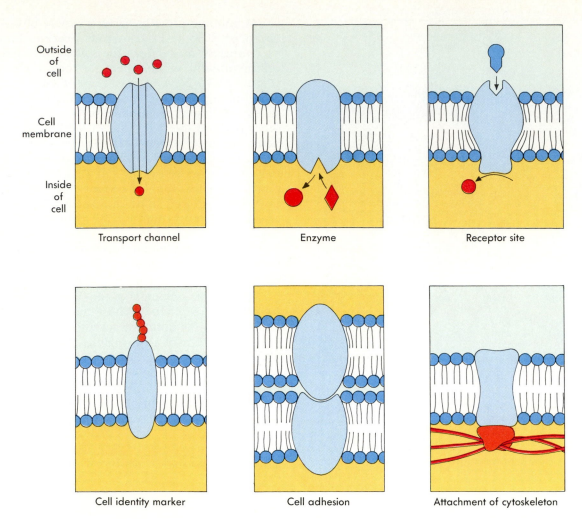

Outside of cell

Cell membrane

Inside of cell

Transport channel

Enzyme

Receptor site

Cell identity marker

Cell adhesion

Attachment of cytoskeleton

FIGURE 6-3
An illustration of the various functions membrane proteins may serve.

Intrinsic membrane proteins thus maintain a specific orientation to the membrane and cannot reorient themselves in any way that would result in an interchange of their internal and external ends. Intrinsic membrane proteins are thus like giants in a crowd; they can shuffle along, but they cannot turn handsprings. It is this property that led to the use of the term **fluid mosaic.**

Anchoring of Membrane Proteins
Some regions of cell membrane are specialized for particular functions that require that the membrane proteins in it be anchored in place. For example, membrane proteins are segregated in myelinated nerve axons (Chapter 7), in which the channels that allow the ion movements responsible for nerve impulses are confined to regions, called nodes, that are free of insulating myelin. Spatial segregation also occurs in muscle fibers (Chapter 12), where the receptors that must be activated for contraction to occur are limited to regions called endplates. Specific transport processes are often confined to only one

side of a cell in epithelial cells of the gastrointestinal tract (Chapter 21) and renal tubules (Chapter 19). Lack of lateral mobility is due in part to the interaction of membrane proteins with the cytoskeleton (see Figure 4-14). The question of how different proteins get to the right membrane sites and stay there is a major issue in current research.

> 1 What are the general functional classes of membrane proteins?
> 2 What is the difference between extrinsic and intrinsic membrane proteins?
> 3 What chemical feature of intrinsic proteins accounts for their close association with the membrane lipids?

DIFFUSION
Random Motion of Molecules
Molecules in solution are in constant motion, and their pathways consist of frequent, random changes in direction (Figure 6-4). This random movement is a manifestation of the thermal energy all molecules

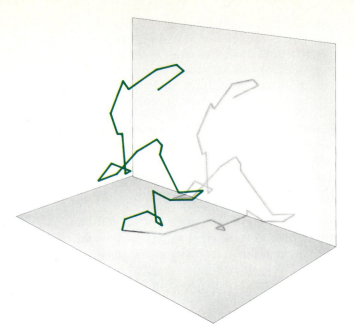

An illustration of the random path a molecule may take due to its inherent thermal energy. This random movement is the basis of diffusion.

possess. If a **concentration gradient** is present, the random motion of solute molecules causes net movement from areas of high concentration to those of low concentration, lessening the concentration gradient. This form of solute movement is called **diffusion.**

The dissipation of concentration gradients by diffusion can be illustrated with an example of some crystals of glucose that are poured into an unstirred vessel containing water (Figure 6-5, *A*). In the first stage of equilibration, the water that is in the immediate vicinity of the glucose dissolves it, but all of the glucose is near the bottom of the vessel (Figure 6-5, *B*). The result is a gradient of glucose concentration in the vessel.

At any instant, some glucose molecules are moving toward the top of the vessel, others are moving toward the bottom, and some are moving sideways. Since the glucose concentration is initially high near the bottom, there is a large unidirectional **flux,** or flow of molecules toward the top. Toward the top of the vessel, the concentration is low, so the unidirectional flux from the top toward the bottom is low. The **net flux** is the algebraic sum of these two unidirectional fluxes. As more molecules approach the top, the concentration gradient diminishes (Figure 6-5, *C*), and ultimately the glucose will be uniformly distributed throughout the water and the system will have come into **equilibrium,** with equal concentrations at top and bottom and equal fluxes of glucose molecules moving in all directions (Figure 6-5, *D*).

Concentration gradients represent stored energy because work must have been done to concentrate the solute. The stored energy can provide a driving force for net solute movement. Diffusion releases the stored energy and increases disorder. When the glucose molecules are randomly scattered throughout the system, disorder (entropy) is at a maximum, the free energy of the system is at a minimum, and the system is in equilibrium.

Determinants of the Rate of Diffusion
The magnitude of the net flux of a substance moving by diffusion depends on properties of the system involved and properties of the solute (Table 6-2). It is apparent from the discussion above that the magnitude of the concentration gradient is an important determinant. The dimensions of the path available to the diffusing molecules are also important. The greater the cross-sectional area through which the molecules are diffusing and the shorter the distance, the greater the net flux. Another factor is the rate of thermal motion of the molecules. The average velocity of thermal motion depends on the temperature (motion is faster at higher temperatures) and on the size of the molecule (small, com-

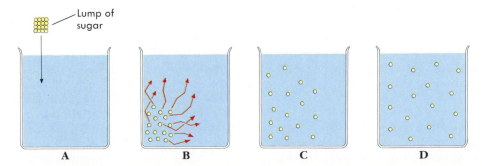

Lump of sugar

A   B   C   D

FIGURE 6-5
Diffusion. If a lump of glucose is dropped into a beaker of water, its molecules dissolve (**A**) and diffuse (**B,C**). At equilibrium the concentration of glucose is the same throughout the vessel (**D**), and the average number of glucose molecules moving in any given direction is approximately the same at any instant.

TABLE 6-2 *Factors that Affect the Rate of Solute Movement by Diffusion*

Factor	Rationale
The magnitude of the concentration gradient	Provides the driving force for net movement
The diffusion coefficient	Incorporates the factors of temperature and the size of the solute molecule. The diffusion coefficient is higher if the molecules are small and the temperature is high.
The geometry of the path	The greater the area available to the diffusing solute and the shorter the distance that must be covered, the greater the rate of diffusion.

pact molecules move more rapidly than large ones). The **diffusion coefficient** of a solute is a measure of its rate of thermal motion at a standard temperature and has units of square centimeters per second. The **diffusion equation** relates net flux of an uncharged solute by diffusion to the determining factors of gradient, area, distance, and diffusion coefficient:

$$\text{Net flux} = DA(C_1 - C_2)/L$$

where D is the diffusion coefficient of the solute, A is the area available for diffusion, $C_1 - C_2$ is the concentration gradient, and L is the distance the diffusing particles must travel. The net flux has units of molecules per second.

The rate at which molecules can move by diffusion sets important limits on the scale of biological structures across which diffusion must occur, such as the walls of capillaries, the lining of alveoli of the lung, and the diameter of cells. A molecule with a diffusion coefficient of 1×10^{-5} cm^2/sec can diffuse a distance of 10 μm (roughly the radius of a cell) in 50 msec; for the same molecule to diffuse 1 cm requires 13 hours.

Permeability of Membranes

The diffusion coefficient describes the random thermal movement of molecules in free solution. Along with the distance to be traveled and the concentration gradient, it is one factor determining the net rate of diffusion of a solute. Cell membranes are for all practical purposes infinitely thin, so for cell membranes the element of distance becomes unim-

portant. Cell membranes can be visualized simply as barriers between two solutions. The **permeability coefficient** describes the ease with which substances cross membranes. Permeability coefficients differ for various solutes and resemble diffusion coefficients in free solution because they are one of the factors that determine the net rate at which substances move across membranes. The units of permeability coefficients are cm/sec, and the units of diffusion coefficients are cm^2/sec. The different units reflect the fact that for membrane permeation the distance factor has been eliminated.

For polar solutes, permeability coefficients are typically three to five orders of magnitude less than the diffusion coefficient for the same solutes in water. Permeability coefficients for polar solutes vary greatly from one solute to another because polar solutes require the presence of specific pathways to cross the membrane. Thus the quality of being permeable or impermeable to particular solutes is a property of individual membranes; the corresponding solutes are termed **permeant** or **impermeant**.

The same principles apply to permeation through membranes as to diffusion in solution. The net flux is greater if the permeability coefficient is greater, the area of the cell membrane occupied by solute-specific transport pathways is larger, or the concentration difference across the membrane (often still referred to as a concentration gradient) is greater. At equilibrium, the concentration throughout the system is equal and the unidirectional fluxes are equal. For convenience, the term coefficient is often omitted and one simply refers to the **permeability** of a membrane to a particular solute.

If biological membranes were composed of pure lipid, the permeability of the membrane to a given solute would be determined solely by the solute's molecular weight and lipid solubility. In this case, membrane permeability would be high to lipid-soluble hydrocarbons, such as fatty acids, and very small polar solutes, such as water. Larger polar solutes, such as glucose, would have lower membrane permeabilities. For example, the permeability of a pure lipid membrane to glucose is five orders of magnitude less than that to water. Passage of ions directly through pure lipid membranes is particularly difficult because ions are surrounded by a **hydration shell** of water molecules oriented toward the charge of the ion (see Chapter 2, p. 22). As a result, for pure lipid membranes, ionic solutes such as Na$^+$ and K$^+$ have membrane permeabilities some 10 orders of magnitude less than expected, given their diffusion coefficients in water.

Since cell membranes may be permeable to glucose, Na$^+$, K$^+$, and other solutes that would cross pure lipid membranes with great difficulty, there must be special routes through the cell membrane for such solutes. Certain intrinsic membrane proteins called **transport proteins** determine the per-

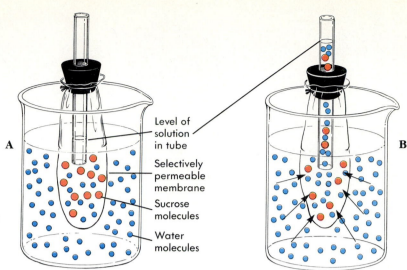

FIGURE 6-6
Osmotic pressure is the result of a difference in the water concentration between pure water and a solution. The osmotic pressure of a solution is defined as the driving force developed by the difference in water concentration between a particular solution and pure water. Osmotic pressure can be measured by placing the solution into a bag made of a membrane permeant to water but not to the solute **(A)**, connecting a vertical tube to the bag, and measuring the height of the solution in the tube **(B)**. The osmotic pressure is equal, but opposite in direction, to the hydrostatic pressure exerted by the column of solution in the tube. The more concentrated the solute, the less concentrated the water and thus the greater the osmotic pressure of the solution.

Level of solution in tube

Selectively permeable membrane

Sucrose molecules

Water molecules

meability of cell membranes to specific polar solutes and ions.

1 How do the magnitudes of the concentration gradient and the cross-sectional area available for diffusion affect the rate of diffusion?
2 How do the variables of temperature and molecular size affect the diffusion coefficient of a solute?
3 Describe the relationship between the two unidirectional fluxes of a solute in diffusional equilibrium (it can be done in three words).

Osmosis

The rules of diffusion apply to water as well as solutes. Water diffuses from a solution of higher water concentration to a solution with a lower concentration of water. Differences in water concentration arise because solute particles dilute the water (solvent) in which they are dissolved. The diffusion of water across a membrane that is permeable to water, but not all solutes, is referred to as **osmosis**. Osmosis is important to cells because volume changes caused by water movements disrupt normal cell function.

If a concentration gradient for water exists between two solutions that are separated by a water-permeable membrane, there will be a net water flux due to osmosis. The **osmotic pressure** of a solution is defined as the driving force developed by the difference in water concentration between a particular solution in question and pure water. The osmotic pressure of any solution can be measured experimentally by placing the solution into a bag made of a membrane permeable to water (but not to the solute), connecting a vertical tube to the bag, and measuring the height of the solution in the tube (Figure 6-6). The osmotic pressure is equal, but opposite in direction, to the hydrostatic pressure (see Chapter 2, p. 30) exerted by the column of solution in the tube. Osmotic pressure can be expressed in centimeters of water (cm H_2O) but is usually given in millimeters

of mercury (mm Hg) because mercury is much denser than water and the resulting numbers are smaller.

The greater the concentration of solute particles in a solution, the greater the difference between the water concentration of the solution and pure water, and thus the greater the osmotic pressure of the solution. At body temperature a solution with an osmolarity of 1 milliosmole per liter (1 mOsm) produces an osmotic pressure difference of 19.3 mm Hg compared with pure water. Two solutions differing by 1 mOsm (for example, 300 mOsm and 301 mOsm) would also have an osmotic pressure of 19.3 mm Hg between them. By way of comparison, extracellular and intracellular fluids have an osmolarity of about 300 mOsm.

Although water molecules are polar, their small size allows them to move quite rapidly through lipid bilayers. As a result of their high water permeability, lipid membranes are unable to oppose significant osmotic forces. At even a small osmotic pressure difference between the cytoplasm and the extracellular fluid there would be net water movement across the cell membrane, causing cells to swell or shrink.

Three terms are frequently used to compare the osmotic pressure of solutions. Solutions with the same concentration of solute particles will have the same osmotic pressure, even though the solute particles may differ, and are therefore termed **isosmotic**. A solution having a greater concentration of solute particles than another is referred to as **hyperosmotic** with respect to the more dilute solution. A more dilute solution, having a lower osmotic pressure, is termed **hyposmotic**.

Determinants of Cell Volume

Three additional terms describe the tendency of cells to shrink or swell when placed in a solution. A solution is defined as **isotonic** if it produces no change in the volume of a cell placed in it, **hyper-**

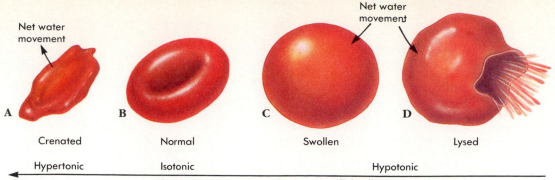

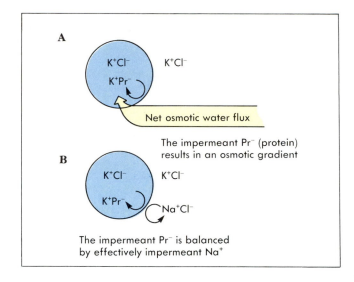

FIGURE 6-7

Effects of hypertonic, isotonic, and hypotonic solutions on red blood cells.
A   Hypertonic solutions result in crenation (shrinkage) of the cell.
B   Isotonic solutions result in normal-shaped cells.
C   Hypotonic solutions result in swelling and, eventually, lysis of cells (**D**).

tonic if cells shrink, and **hypotonic** if cells swell (Figure 6-7). A distinction between **tonicity** and osmolarity is necessary because the effect of a given solute on osmotic water movement across a cell membrane depends on the membrane permeability to the solute. This can be illustrated by two contrasting examples. If a cell is bathed in a solution that is initially isotonic, addition of some solute to the solution outside the cell would have opposite effects on the cell's volume, depending on whether the solute could enter the cytoplasm. If the cell is permeable to the added solute, some of the solute will enter the cytoplasm, increasing the intracellular concentration of solute particles. Osmosis will then cause water to enter the cell in proportion to the amount of solute that entered, and the cell will swell.

If the cell is impermeable to the added solute, the effect will be opposite. The increase in the concentration of solute particles outside the cell creates a gradient of osmotic pressure that will cause water to move from the cytoplasm to the extracellular solution, shrinking the cell (see Figure 6-7, *A*). Since even a small osmotic pressure gradient can cause cells to experience dangerous volume changes, the extracellular fluid must be isotonic.

To be isotonic, the extracellular fluid must contain a concentration of impermeant solute particles outside the cell equal to whatever impermeant solute particles are present in the cytoplasm. Protein molecules, important intracellular solutes, are large and normally membrane impermeant. To be isotonic, the extracellular fluid must therefore contain a concentration of impermeant solute particles equal to the protein concentration of the cytoplasm. Most smaller solutes (for example, sugars, amino acids, and ions) can cross cell membranes, although they may do so quite slowly. Since there is relatively little protein in the extracellular fluid, osmosis should drive the movement of water from extracellular fluid into the cytoplasm.

FIGURE 6-8

A   A cell bathed in isosmotic KCl swells as water moves into the cell along the osmotic pressure gradient caused by the impermeant intracellular protein anions (Pr$^-$).
B   Swelling does not occur in normal extracellular fluid because the osmotic pressure of the internal impermeant solutes is matched by the osmotic pressure of the effectively impermeant extracellular Na$^+$.

What keeps a cell from swelling? To answer this question it is necessary to understand that three requirements must be satisfied simultaneously. The numbers of solute particles inside and outside the cell must be equal to satisfy the conditions for an osmotic equilibrium. In the absence of a voltage difference, diffusional equilibrium requires that there also be no concentration gradients for any permeant solutes. Finally, solutions must be electrically neutral; otherwise enormous electrical forces would exist within them (see Chapter 2, p. 33).

These three conditions can be applied to a cell permeant to both K$^+$ and Cl$^-$ and bathed in an extracellular fluid containing only KCl (Figure 6-8).

TABLE 6-3   *Classes of Mediated Transport Systems*

Class	Characteristics	Example
1. Channel	A pore through the membrane that mediates a high rate of solute transfer (up to 10^8 ions/sec); driving force is the solute gradient	Na^+ channels of nerve and muscle (Chapter 7)
2. Carrier	An intrinsic protein binds the transported solute(s) and undergoes a conformational change that moves the solute(s) to the other side of the membrane; rates of transfer slower than channels (up to 10^4 molecules/sec)	Transport of glucose into cells (This chapter)
a. "Facilitated diffusion"	Driving force is the transported solute's gradient	Glucose uptake in most cells (Chapter 23)
b. Active transport	Driven directly by ATP	Na^+-K^+ pump (This chapter)
c. Gradient-driven cotransport	Gradient of one transported substance drives transport of other(s) in same direction	Na^+-coupled glucose and amino acid absorption in intestine and kidney (Chapters 20 and 22)
d. Gradient-driven countertransport	Gradient of one transported substance drives other(s) in opposite direction	HCO_3^--Cl^- exchange in intestinal cells (Chapter 22)

The total number of positive charges in the cytoplasm and extracellular solution must equal the numbers of negative charges in each. Outside the cell the numbers of K^+ and Cl^- ions will be equal. Inside the cell there will be more K^+ ions than Cl^- ions because the intracellular negatively charged proteins (*Pr*$^-$ in Figure 6-8, *B*) are confined to the cytoplasm, and excess K^+ ions are needed to balance the negative charges on these protein anions. Given these conditions, cells cannot come into equilibrium. The osmotic imbalance would drive an osmotic movement of water from extracellular fluid into the cytoplasm that would burst the cell.

Living cells do not burst because they expend energy to make the Na^+ in the extracellular fluid act as if it were impermeant, thus balancing the osmotic effect of negatively charged cytoplasmic proteins (see Figure 6-8, *B*). The **Na^+-K^+ pump**, discussed in the next section, is an intrinsic protein enzyme in the membrane that returns Na^+ to the extracellular fluid in exchange for K^+ from the extracellular fluid. In this process, ATP is split to provide the energy needed to move both the Na^+ and the K^+ against their concentration gradients. Consequently, the enzyme is sometimes called the **Na^+-K^+ ATPase.** Although cell membranes are actually slightly permeant to Na^+, from an osmotic point of view Na^+ is rendered impermeant because it is continuously being removed from the cytoplasm as rapidly as it leaks in.

> 1 What is meant by the osmotic pressure of a solution?
> 2 What is the effect of a hypotonic solution on cell volume?

MEMBRANE TRANSPORT PROTEINS
Mediated Transport
Polar solutes must take specific avenues through lipid membranes. The processes by which they cross the cell membrane are called **mediated transport** (Table 6-3). Mediated transport processes can be saturated in a manner similar to enzymes (see Figure 3-15, *A*, p. 52), and, like enzymes, they are specific for particular chemical structures. In some cases, solutes with similar structures can compete for the active site of the transport process. Transport processes that are directly energized by ATP are referred to as **active transport** processes. Mediated processes driven by a gradient of transported substance are called **passive** because cells spend no energy to drive such processes directly; historically such processes have been called **facilitated diffusion** because the transport process allows the transported substance to pass through the membrane much more rapidly than would be expected from their lipid solubility.

Membrane Channels
The simplest way for polar solute to cross the cell membrane is by way of pores, or **channels** (Figure 6-9, *A*). A channel is a tube through the membrane formed by one or more intrinsic membrane proteins. On the outside of the tube facing the phospholipid are many hydrophobic groups. The inside of the tube is lined with hydrophilic groups, allowing the channel to be filled with water. Thus polar solutes can diffuse through the membrane in aqueous solution. Channels typically contain a region called the **selectivity filter** that determines which solutes may enter the channel. For example, the per-

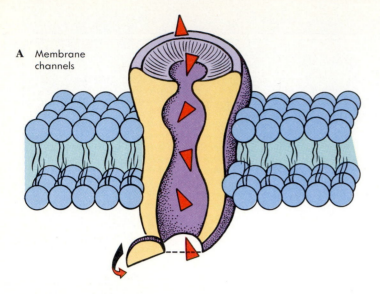

A Membrane channels

FIGURE 6-9
Comparison of the operations of membrane channels and membrane carriers.
A A channel is a tube through the membrane formed by one or more intrinsic membrane proteins.
B In contrast to channels, *carriers* must undergo a cycle of binding and conformational change.

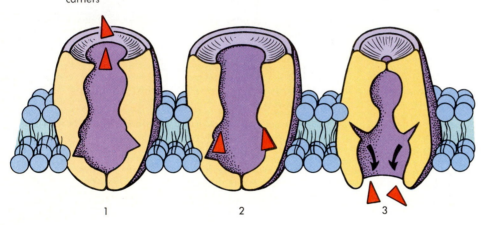

B Membrane carriers

1   2   3

meability of the cell membrane to Na^+ and K^+ is due to the presence of structurally different Na^+ and K^+ channels. Channels typically open and close spontaneously; some are **gated channels,** whose opening is regulated by an external factor.

Carrier-Mediated Transport

In contrast to channels, **carriers** must undergo a cycle of binding and conformational change (Figure 6-9, *B*). The driving force for net solute movement mediated by a carrier may be provided by the concentration gradient or by ATP, depending on the particular carrier. The entry of glucose into most cells is mediated by a carrier; the driving force is provided by a gradient of glucose concentration across the membrane, so that this transport process is energetically passive.

The Na^+-K^+ pump (Na^+-K^+ATPase) is an example of a carrier-mediated active process. The pump consists of a protein that spans the membrane more than once. The elements of the protein that are within the membrane can assume several different conformations (Figure 6-10). In the conformation labelled *A*, the pump is open to the cytoplasm, where it can bind Na^+, and closed to the external solution. The pump hydrolyzes ATP to phosphorylate a specific site. As a result of this phosphorylation, the pump undergoes a conformational change (*B*) that has three effects: (1) the affinity for Na^+ is decreased; (2) binding sites for K^+ are uncovered; and (3) the pump cavity becomes open to the external solution and closed to the cytoplasm (*C*). When external K^+ ion is bound, the phosphate dissociates and reverses the conformational change, returning the pump to its initial state (*D*). The result is that Na^+ is bound on the inside and released into the external solution, while at the same time K^+ is taken up at the external surface and discharged into the cytoplasm. The pump is believed to exchange K^+ for Na^+ in the ratio of 3 Na^+/2 K^+/ATP. The rate of cycling of the pump is regulated by the intracellular Na^+ concentration. Accumulated Na^+ is removed before its concentration gradient can change significantly (Figure 6-11).

FIGURE 6-10

Schematic diagram of the Na^+-K^+
pump cycle.

A Three Na^+ ions bind from the
cytoplasmic side.

B The pump protein is phosphorylated
by ATP.

C Phosphorylation causes a
conformational change of the protein
that also involves a decrease in the
affinity of Na^+ binding sites and an
increase in affinity of K^+ binding sites.
The 3 Na^+ ions are then released to the
extracellular side.

D Two K^+ ions occupy the K^+
binding sites, and the pump protein
releases the phosphate and returns to
its original conformation. The affinity of
the K^+ binding sites decreases, and that
of the Na^+ increases. The K^+ ions are
released to the cytoplasmic side of the
membrane, the pump is ready to bind 3
Na^+, and the cycle starts again.

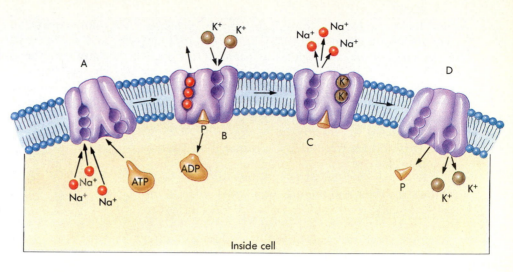

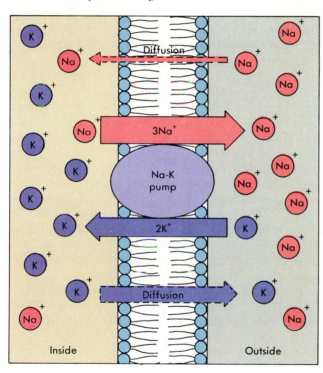

FIGURE 6-11
The Na^+-K^+ pump maintains Na^+ and K^+ gradients, coun-
tering diffusional entry of Na^+ and exit of K^+.

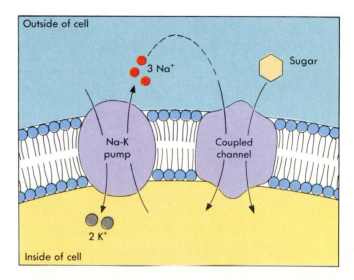

FIGURE 6-12
The Na^+ gradient produced by the Na^+-K^+ pump can be
used to drive absorption of glucose against its concentration
gradient by coupling glucose absorption to the diffusion of
Na^+ down its concentration gradient.

The Na^+-K^+ pump is specifically inhibited by a
plant toxin called **ouabain,** which binds to the external part of the pump. Ouabain is one of a family
of drugs long known to have effects on the heart
and kidney. The discovery of its mechanism was a
key step in experimental characterization of the
pump protein. Inhibitors have been identified for
many important transport proteins; such inhibitors
are useful as drugs and in the experimental study of
membrane transport physiology. In the kidney, for
example, pump inhibitors act as diuretics (see
Chapter 19, p. 498).

Na$^+$ Gradient-Driven Cotransport

The Na^+ concentration gradient across the cell
membrane results from the continuous activity of
the Na^+-K^+ ATPase. The energy stored in this gradient is released uselessly if Na^+ leaks back into the
cell, but in some cells the energy stored in the Na^+
gradient can be used to transport other substances
against their concentration gradients. Such processes are examples of **gradient-driven cotransport** (see Table 6-3).

Sodium-linked cotransport requires a carrier
whose affinity for the cotransported substance depends on whether Na^+ is bound at the same time.
An example is glucose uptake into intestinal cells
(Figure 6-12). Glucose binds to its carrier only
when Na^+ is also bound. Since the Na^+ concentration is high outside and is maintained low inside

the cells by the Na^+-K^+ pump, the carrier cycles between binding glucose and Na^+ at the external face of the membrane and releasing them at the cytoplasmic face. Similar cotransport systems exist for many small molecules, including amino acids. One interesting example is the Na^+, K^+, 2 Cl^- cotransporter, which is important for solute reabsorption in renal tubules (Chapter 19). In this case the Na^+ gradient drives both K^+ and Cl^- against their electrochemical gradients.

There are gradient-driven transport systems in which the driver and the driven solutes move in opposite directions: these are referred to as **countertransport** systems (see Table 6-3). An example is exchange of Cl^- for HCO_3^- in the kidney (Chapter 20) and the gastrointestinal tract (Chapter 21).

Two ionic solutes may appear to be countertransported or cotransported even if they do not share the same carrier, due to electrical coupling. In such cases, primary active transport of one charged solute creates an electrical gradient that provides the driving force for the movement of a solute of like charge in the opposite direction or an oppositely charged solute in the same direction. An example of this is seen in **H^+-Na^+ exchange** (Figure 6-13). In this case, the electrical energy provided by the H^+ pump can drive Na^+ absorption from a lower concentration outside the cell to a higher one inside the cell.

---

1 What are the sources of energy for:
a. glucose absorption by most cells?
b. active transport of Na^+?
c. Na^+-linked glucose transport in intestinal cells?
2 What characteristics do all of these forms of mediated transport have in common?
3 What are the differences between the way channels and carriers transport solutes?

THE MEMBRANE POTENTIAL
Electrical Gradients across Cell Membranes

All living cells maintain an imbalance of electrical charge across their cell membranes. Gradients of electrical potential energy are generally called **membrane potentials.** The term **resting potential** is a specific designation used to refer to the membrane potential of quiescent cells. Membrane potentials can be measured by first inserting a fine glass capillary, or micropipette, filled with conductive solution into the cell and then connecting a voltmeter between the fluid-filled capillary, which functions as a **microelectrode,** and the extracellular solution (Figure 6-14). The resting membrane potential is typically between 40 and 80 mV, depending on the cell type, with the inside of the cell negative to the outside.

The amount of charge imbalance needed to produce resting potentials of the size typical of those across biological membranes is very small in com-

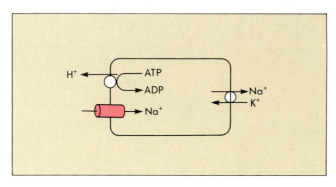

FIGURE 6-13
H^+ secretion in epithelial cells is driven by an H^+ pump, and Na^+ entry is driven by the Na^+ gradient. The coupling between the two processes is electrical: a positive charge entering the cell tends to force out one that is inside, and vice-versa.

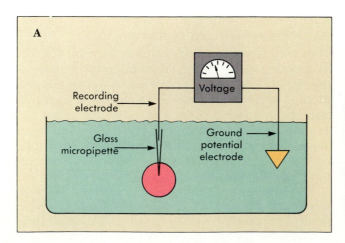

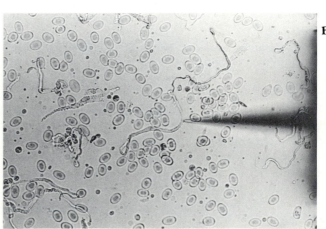

FIGURE 6-14
A   Cell membrane potentials can be recorded using glass capillary microelectrodes.
B   A microelectrode is shown penetrating a tissue-cultured muscle cell.

FRONTIERS IN PHYSIOLOGY

Artificial Membranes as Valuable Experimental Tools

Cell membranes are complex structures that carry out many functions simultaneously. Simpler membrane systems can be created in the laboratory for study of the basic principles of transport and permeation. An artificial hydrocarbon membrane can be made by placing a drop of hydrocarbon on a small hole drilled in a partition that separates two reservoirs of solution. The hydrocarbon forms a film 2 molecules thick across the hole (Figure 6-A).

Some substances that previously had been known for their antibiotic activity were the first models for the behavior of transport proteins. These substances, called **ionophores** (the name means "ion carriers"), are molecules with a highly polar interior that can dissolve in artificial lipid membranes and confer permeability to specific ions One example is **valinomycin**, a small doughnut-shaped peptide with 6 polar oxygen atoms facing the interior of the molecule. A K^+ ion just fits in the hole of the doughnut. The polar oxygen molecules replace its hydration shell. Valinomycin can pass from one side of the membrane to the other, shuttling K^+ ions. Biological carrier proteins were once thought to function like this, but intrinsic proteins are now known to be much too bulky to shuttle back and forth within the membrane.

Some ionophores form relatively stable pores in lipid membranes. One example is a helical peptide antibiotic called gramicidin. Polar groups within the helix allow ions with the correct diameter to discard their hydration shells and move through the pore. Gramicidin is a useful model of the behavior of biological channels.

Recently it has been proved possible to extract transport proteins from biological membranes and reconstitute them in artificial membranes. In some cases the mRNA coding for transport proteins has been injected into frog oocytes, where the oocyte's translational machinery is able to synthesize the transport proteins and insert them into its cell membrane. Several important channels and pumps have been studied in reconstituted form with electrical measurement techniques similar to those employed for any living cell (for example, the patch-clamp described in the boxed essay on stretch-activated channels in Chapter 8, p. 178).

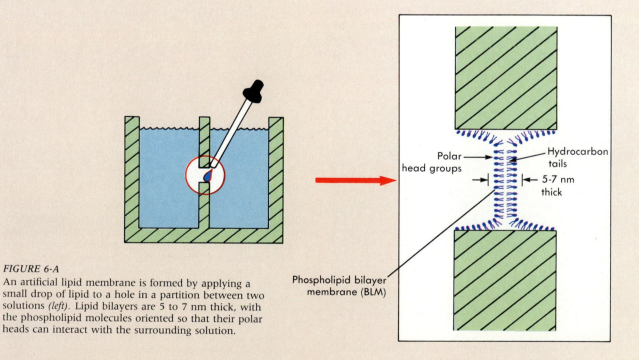

FIGURE 6-A

An artificial lipid membrane is formed by applying a small drop of lipid to a hole in a partition between two solutions *(left)*. Lipid bilayers are 5 to 7 nm thick, with the phospholipid molecules oriented so that their polar heads can interact with the surrounding solution.

parison with the total number of ions in the intracellular or extracellular solutions (Figure 6-15). That is, even though there will be a slight excess of negative charge near the interior of the cell membrane and the same amount of excess positive charge near the exterior, there will be no measurable change in the overall concentrations of any ion. For example, for a membrane 8 nm thick, a voltage of 100 mV (inside negative) could be produced if only those K^+ ions located within one membrane thickness were to move from the inside to the outside.

Gradients of ionic solute represent stored energy that has the potential to do electrical work as well as chemical work. To understand the nature of the membrane potential and to be prepared to understand how changes in membrane potential are used in communication in the nervous system and muscles, it is necessary to understand the relationship between the chemical gradients and the electrical gradient across the cell membrane.

The relationship between the magnitude of the concentration gradient and the electrical driving force that can be generated by the gradient can be demonstrated by a simple system in which two reservoirs of solution, one containing 1 M KCl and the other 0.1 M KCl, are separated by a membrane permeant to K^+ but not to Cl^- (Figure 6-16). Initially in each solution the number of K^+ equals the number of Cl^-, so that the solutions themselves are electrically neutral (see Figure 6-16, A). After this system is set up, some K^+ will cross the membrane driven by the tenfold concentration gradient. Each K^+ that crosses carries a positive charge to the right side and leaves a negative charge in the form of an unpaired Cl^- on the left side (see Figure 6-16, B). Before very many K^+ have crossed, the electrical force exerted by the separated charges becomes sufficient to cause a right-to-left unidirectional flux equal to the left-to-right flux (see Figure 6-16, C). At

this point, the system has come into electrochemical equilibrium. At electrochemical equilibrium the electrical gradient, or **equilibrium potential**, exactly balances the force applied to K^+ by the concentration gradient. The equilibrium potential for an ion is generally abbreviated using the letter E with the ion identified by a subscript. For example, E_K stands for the K^+ equilibrium potential.

Nernst Equation
The relationship between chemical potential energy and electrical potential energy for K^+ in the situa-

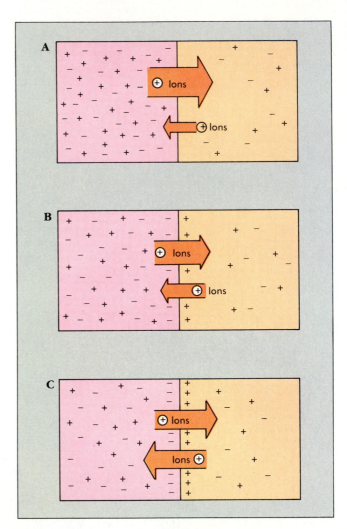

FIGURE 6-16
Development of electrochemical equilibrium in a system in which the membrane is permeant only to the positive ion (K^+, indicated simply as + signs) of the salt in solution.
A   Initially, the concentration gradient and lack of opposing charge result in a large net flux across the membrane.
B   As ion movement results in charge separation, the size of the flux driven by the concentration gradient declines, and the opposite flux increases, driven by the electrical attraction of the charge on the membrane.
C   At equilibrium, the electrical and concentration forces are equal and opposite, and there is no net flux across the membrane. So few ions have moved relative to the total number in solution that the change in the concentrations of the two compartments has been negligible.

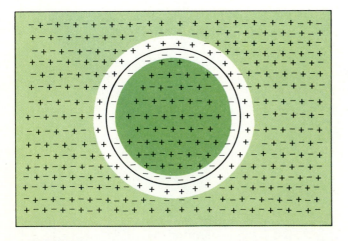

FIGURE 6-15
This diagram emphasizes that membrane potentials involve a very small number of the ions present in the intracellular and extracellular solutions.

tion shown in Figure 6-16 is described by the **Nernst equation** for K^+:

$$E_K = C \log_{10} ([K^+]_{right} / [K^+]_{left})$$

The constant C in the Nernst equation incorporates the universal gas constant, the temperature, an electrical constant called Faraday's number, and the valence of the ion (in this case +1). At body temperature, C = 60 mV for a monovalent (single unit of charge) cation. In this case where the K^+ concentration on the left is 10 times that on the right, $E_K = 60$ mV $\log_{10} (0.1) = -60$ mV with the left side neg-

ative. Increasing the K^+ concentration ratio by an order of magnitude (to one hundredfold greater on the left) would double the equilibrium potential to −120 mV. If the gradient were reversed (with the K^+ concentration on the right side 10 times the left, for example), E_K would have the same value but the opposite sign.

With adjustments, the Nernst equation can be written for any ion that is in equilibrium across a membrane. The sign of the constant becomes minus if the equation is written for an anion. In the case shown in Figure 6-16, the Nernst equation for Cl^- would be:

$$E_{Cl} = (-60 \text{ mV}) \log_{10} ([Cl^-]_{right} / [Cl^-]_{left})$$

For a tenfold concentration gradient for Cl^- (greater on the left), $E_{Cl} = (-60$ mV$) \log_{10} (0.1) = 60$ mV. The value of the constant C in the Nernst equation is halved if the equation is written for a divalent ion having two units of charge.

The concentration ratio for K^+ between cytoplasm and extracellular fluid is approximately thirtyfold for nerve and muscle cells (Table 6-4). There is an outwardly directed concentration gradient for K^+ (Figure 6-17), and the K^+ equilibrium potential (abbreviated E_K) that corresponds to this thirtyfold

TABLE 6-4	The Ionic Composition of Cytoplasm and Extracellular Fluid (Concentrations in mMoles)			
	Cytoplasm	Extracellular fluid	Ratio	Equilibrium potential
Na^+	15	150	10:1	+60 mV
K^+	150	5	1:30	−90 mV
Cl^-	7	110	15:1	−70 mV

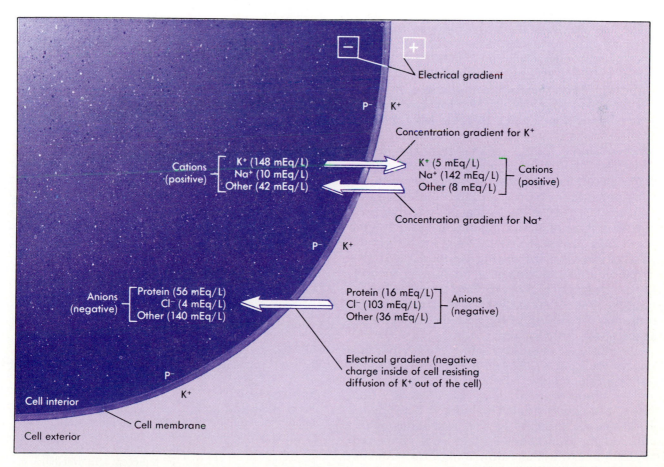

FIGURE 6-17
A diagram of a cell showing the intracellular and extracellular concentrations of ions and molecules and the gradients for Na^+, K^+, and voltage.

gradient is approximately 90 mV, cytoplasm negative to extracellular fluid. The Na^+ concentration is low inside cells and high outside, so there is an inward-directed concentration gradient for Na^+ (see Figure 6-17), with a concentration ratio of about tenfold for typical nerve and muscle cells; thus the corresponding Na^+ equilibrium potential (E_{Na}) will be 60 mV (see Table 6-4). Because the concentration gradient is opposite to that for K^+, E_{Na} is of opposite sign (that is, the cytoplasm is positive relative to the extracellular fluid).

Na^+ and K^+ Ion Distribution across Cell Membranes

In order for charges to move, there must be a driving force acting on them. The equilibrium potential for an ion was defined as that voltage where the two unidirectional fluxes exactly balance. At the equilibrium potential for an ion there is no net movement of charge or, equivalently, no net driving force (see Chapter 2, p. 30, for a general discussion of forces and flows). When the membrane potential is displaced from the equilibrium potential of an ion, a net driving force exists. The magnitude of the electrical driving force is equal to the difference between the membrane potential and the equilibrium potential for that particular ion.

The magnitudes of the ionic gradients across the membrane can be measured with chemical techniques or with microelectrodes modified to be sensitive to only one type of ion (ion-selective microelectrodes). Such measurements show that the membrane potential (about -70 mV for the cells whose composition is given in Table 6-4), although inside negative, is not equal to the calculated equilibrium potential for K^+ and far from the inside-positive Na^+ equilibrium potential. The electrical driving force for K^+ is the difference between the resting potential (-70 mV) and the equilibrium potential for K^+ (-90 mV), so that K^+ experiences a net driving force of 20 mV, tending to make it leave the cell. The electrical driving force for Na^+ is the difference between the resting potential (-70 mV) and the equilibrium potential for Na^+ ($+60$ mV), and Na^+ experiences a net driving force of 130 mV, tending to make it enter the cell. As a result there is a continual net leakage of K^+ from the cell and a net leakage of Na^+ into the cell.

The existence of steady-state concentration gradients of Na^+ and K^+ that do not correspond to the equilibrium potentials of these ions implies that cells are doing work to keep Na^+ and K^+ from coming into equilibrium. The work is done by the Na^+-K^+ pump proteins in the membrane, which counteract the continuous leakage of K^+ ions from cytoplasm to extracellular fluid and of Na^+ from extracellular fluid to the cytoplasm. In contrast to K^+ and Na^+, Cl^- in the cell, described in Table 6-4, is in electrochemical equilibrium, indicating that it is

not actively transported. Chloride is not in electrochemical equilibrium for all cell types, however.

Determinants of the Resting Potential of a Cell

The Na^+ gradient represents a force that could drive the resting potential in the direction of E_{Na}; the K^+ gradient could drive the resting potential in the opposite direction, toward E_K. The fact that the resting potential is much nearer to E_K than E_{Na} (Figure 6-18, line E) represents a partial victory for the K^+ gradient over the Na^+ gradient. The victory is due to two factors: (1) the K^+ gradient is about three times the Na^+ gradient, and (2) the membranes of most cells are from 10 to 75 times more permeant to K^+ than Na^+.

The resting potential of a cell can be altered in two ways: the permeability to a particular ion may change, or the gradient for an ion may change. For example, an increase in the permeability of the membrane to Na^+ relative to that for K^+ would tend to decrease the margin of victory of K^+ over Na^+ and thus make the resting potential less inside-negative (Figure 6-18, lines B-D). An increase in the K^+ permeability would further increase the margin of victory of K^+ over Na^+, making the resting potential more inside-negative (Figure 6-18, line A).

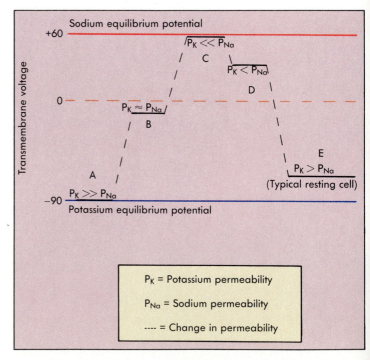

FIGURE 6-18

Effect of different Na^+-K^+ permeability ratios on the membrane potential. At A the potassium permeability (P_K) is much greater than the sodium permeability (P_{Na}). At position B, P_K and P_{Na} are nearly equal, while at C, P_K is much less than P_{Na}. D and E represent intermediate cases. Changes in the relative magnitude of P_{Na} and P_K cause shifts in the membrane potential, as illustrated by the dashed lines.

THE FOUNDATIONS OF PHYSIOLOGY

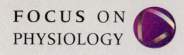
The Goldman Equation Describes the Membrane Potential

If the Na^+ and K^+ concentrations and the relative permeabilities of the cell membrane to Na^+ and K^+ are known, the membrane potential can be calculated using the **Goldman equation:**

$$E = C \log_{10} \frac{(P_K[K^+]_{out} + P_{Na}[Na^+]_{out}}{P_K[K^+]_{in} + P_{Na}[Na^+]_{in})}$$

where C = 60 mV because only univalent cations are involved. The equation can be expanded to include other monovalent ions such as Cl^-. The Goldman equation describes the steady-state membrane potential, which is neither a K^+ nor an Na^+ equilibrium potential. When it is solved for the ion distributions given in Table 6-4 using measured values for the membrane permeabilities to K^+ and Na^+, one can calculate the normal resting potential of -70 mV, discussed in the text. If either permeability is very much larger than the other, the terms for the less permeant ion can be neglected, and the algebraic form of the Goldman equation becomes the same as the Nernst equation for the more permeant ion.

Changes in the intracellular or extracellular concentrations of Na^+ or K^+ change the magnitude of their concentration gradients. Since the relative concentration gradients are one factor determining the margin of victory of K^+ over Na^+, changes in these gradients would be expected to change the resting potential. Increasing the extracellular concentration of K^+ would decrease the gradient for K^+ (E_K would be less internally negative), allowing a greater margin of victory for Na^+ and making the resting potential less inside-negative. An increase in extracellular Na^+ would increase the magnitude of the Na^+ gradient and thus make E_{Na} more inside-positive. The increase in the net driving force for Na^+ would increase its margin of victory, again causing the resting potential to become less inside-negative.

Role of Na^+-K^+ Pumps and Homeostatic Systems

Since neither Na^+ nor K^+ is in equilibrium and the membrane is permeant to both, Na^+ leaking into cells and K^+ leaking out would cause the system to come to equilibrium with no concentration gradients for Na^+ and K^+ and no membrane potential. In living cells the low Na^+ and high K^+ concentrations of the cytoplasm are maintained by the Na^+-K^+ pump. The Na^+ and K^+ concentrations of the extracellular fluid are regulated by the kidney (Chapter 20). As long as the concentration gradients and permeabilities remain the same, the membrane potential will be in steady state.

1 How can the balance between the effect of an ion's concentration gradient and the electrical driving force on it be described algebraically?

2 What are the net driving forces for Na^+ and K^+ in a cell where the resting potential is not equal to either E_K or E_{Na}?

3 What factors determine the value and sign of the membrane potential?

SUMMARY

1. Cell membranes consist of phospholipid molecules arranged as a **bilayer** in which **intrinsic proteins** are embedded. Extrinsic proteins are attached to the surface.

2. Pure lipid membranes are impermeant to polar solutes and ions; specialized **carriers** or **channels** are needed for such substances to cross cell membranes and are the basis of **mediated transport.** Mediated transport systems are characterized by specificity and saturability.

3. **Diffusion** is a process in which net solute movement down concentration gradients is the result of random thermal motion of solute molecules. The rate of solute diffusion is affected by the magnitude of the concentration gradient, the cross-sectional area and length of the diffusion path, the temperature, and the size of the diffusing solute.

4. Lipid membranes are highly permeant to water. The **osmotic pressure** of a solution is defined as the driving force developed by the difference in water concentration between a particular solution in question and pure water. Gradients of **osmotic pressure** across cell membranes can result when there is an imbalance of impermeant solutes across the membrane.

5. **Isotonic** solutions are solutions that contain impermeant solutes sufficient to balance the osmotic pressure of impermeant solutes in the cytoplasm, so that cells bathed in isotonic solution neither swell nor shrink. **Hypotonic** solutions contain insufficient impermeant solute, and cells bathed in such solutions swell. **Hypertonic** solutions contain excessive impermeant solute, resulting in cellular shrinkage.

6. Cell volume stability requires there to be a high concentration of Na^+ in the extracellular fluid and for Na^+ to be impermeant. Even though it slowly leaks into the cell, Na^+ is made to be effectively impermeant by the **Na^+-K^+ pump** of the cell membrane.

7. The **membrane potential,** an inside-negative gradient of electrical charge across the membrane, results from the chemical energy of Na^+ and K^+ concentration gradients and the differential permeability of the membrane to the two ions. If only one of the two ions were permeant, the resulting potential would be an **equilibrium potential** whose magnitude would be predicted by the **Nernst equation.** Since membranes are typically permeant to both Na^+ and K^+, the membrane potential is not an equilibrium potential for either ion.

8. The magnitude and sign of the membrane potential is determined by the relative magnitudes of the permeabilities and concentration gradients of Na^+ and K^+. Since typical membranes are more permeant to K^+ than to Na^+ and the K^+ concentration gradient is larger than the Na^+ gradient, the membrane potential is close to, but not equal to, the K^+ equilibrium potential and far from the Na^+ equilibrium potential.

9. There are three basic mechanisms of mediated transport of solutes across cell membranes: **facilitated diffusion,** in which the solute's own concentration gradient drives the **net flux; active transport,** in which the solute is driven against its concentration gradient by energy released from ATP; and **gradient-driven countertransport** or **cotransport,** in which a net flux of one solute against its concentration gradient is driven by a net flux of another (usually Na^+) down its concentration gradient.

10. Cotransported and countertransported solutes may share the same carrier molecule. Alternatively, cotransport or countertransport of ions may involve electrical coupling in which the routes of the two solutes are physically distinct, but the electrical potential created by transport of one ion drives the movement of the other.

1. Describe the fluid mosaic model of membrane structure. What features of this model would allow the internal surface of a membrane to have properties different from its external face?

2. What features of membrane and solute would determine the permeability coefficient of a particular solute through a particular membrane?

3. Specify the concentration in millimoles of three solutions with an osmolarity of 300 mOsm. One of these solutions should contain only a nonelectrolyte such as glucose, another should contain only univalent ions, and the third should contain divalent cations and univalent anions.

4. The plasma has an osmolarity of 300 mOsm. What must the intracellular osmolarity be? Why?

5. What is the role of the Na^+-K^+ pump in regulating cell volume?

6. What factors make the membrane potential inside-negative?

7. If a cell were bathed in a solution like normal extracellular fluid except containing 15 mM K^+ instead of 5 mM, what effect would this have on the membrane potential?

8. What are the two common features of all mediated transport systems?

● *SUGGESTED READING*

BRETSCHER, M.S.: The molecules of the cell membrane, *Scientific American*, October 1985, p. 100. Covers the fluid mosaic model and the transport functions of integral membrane proteins.

DE DUVE, C.: *A Guided Tour of the Living Cell*, Scientific American Library, W.H. Freeman & Co., New York, 1984. One of the best surveys of the structure and function of the major intracellular organelles.

UNWIN, N., and R. HENDERSON: The structure of proteins in biological membranes, *Scientific American*, February 1984, p. 78. Describes the three-dimensional configuration of membrane transport proteins and receptor molecules.

YEAGLE, P.: *The Membranes of Cells*, Academic Press, Orlando, Fla., 1987. A more advanced, but still readable, summary of transport across membranes. Covers the fluid mosaic model, lipid-protein interactions in membranes, the various types of mediated diffusion, and the intracellular synthesis of cell membrane components.

CHAPTER 7

Excitable Membranes

On completing this chapter you will be able to:

- Describe the differences and similarities between action potentials, receptor potentials, and synaptic potentials.
- Describe the behavior of voltage-gated Na^+ and K^+ channels in excitable membranes as a function of voltage and time.
- Understand how an action potential is initiated, defining the concept of threshold.
- Understand how inactivation of Na^+ channels and opening of K^+ channels combine to terminate the action potential.
- Distinguish between the absolute and relative refractory periods.
- State the factors governing the velocity of conduction of action potentials in an axon or muscle cell.
- Understand the role of myelin as an insulator and characterize saltatory conduction.
- Understand the origin of the compound action potential of a nerve trunk and account for the origin of its various components.

The most beautiful thoughts, the greatest human performances, and the simplest everyday acts depend on a rapid, complex flow of information within the nervous system and between the nervous system and the rest of the body. The information is encoded in electrical signals. The ability to send information rapidly over long distances makes high-speed control of the body possible. In the nervous system and muscles, signals are carried over long distances by rapid changes in membrane potential called action potentials.

The ability to generate and transmit action potentials is a cellular specialization; cells that can produce action potentials are said to be excitable. Excitability is a property of the cell membrane and is due to the presence of transport proteins called ion channels that span the membrane. The membrane potential, which is a characteristic of all living cells, provides the stored energy that allows changes in membrane potential to serve as the basis of information transfer in excitable cells.

Channels that can open and close rapidly in response to environmental influences respond to the stimuli that give sensations such as taste, sight, hearing, or touch. An open channel permits a flow of ions across the membrane—an electrical current. These tiny individual currents, multiplied manyfold, provide the information flow that links the 10 billion brain cells with one another. Currents flowing in a group of motor neurons selected by the brain excite the muscles in the surgeon's hand or carry the gymnast through her routine.

ELECTRICAL EVENTS IN MEMBRANES
Categories of Electrical Events

The plasma membranes of some cells are able to undergo rapid, reversible changes in ion permeability that result in a displacement of the membrane potential from the resting potential. These permeability changes are the basis for transfer of information in the nervous system and are also important in initiating contraction in muscles and secretion from glands.

Electrical events in cells fall into two categories. **Action potentials,** the subject of this chapter, are the mechanism used for rapid long-distance signalling in the nervous system and muscle. They take place in nerve axons, the long tubular extensions of nerve cells (Figure 7-1), and also in muscle cells. **Receptor potentials** and **synaptic potentials** (see Chapter 8) are important in the acquisition and analysis of information and in the communication between nerves and muscles. It is important to ap-

preciate the differences between the basic types of electrical events, as summarized in Table 7-1.

Receptor potentials and synaptic potentials have three common features: (1) their amplitudes are **graded,** increasing with the magnitude of the chemical or physical **stimulus** that activates them; (2) they cannot be transmitted over long distances because the voltage change diminishes rapidly with distance from the stimulus; and (3) they can add together, or **summate.**

Action potentials differ from receptor potentials and synaptic potentials in two important respects (see Table 7-1): (1) the amplitude of an action potential is constant and not related to the size of the stimulus, so that action potentials are **all-or-none** events; and (2) action potentials do not fade away with distance. The process by which action potentials spread along cell membranes is called **propagation,** to distinguish it from the **decremental** process in which receptor potentials and synap-

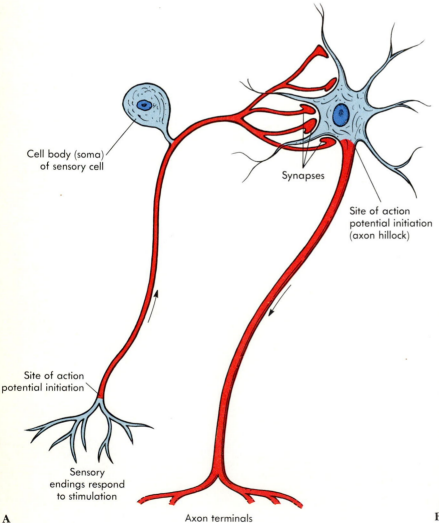

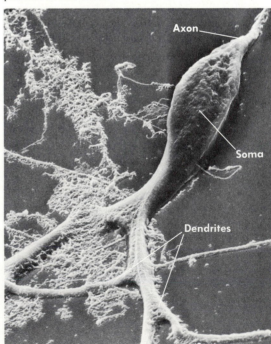

FIGURE 7-1

A   A schematic drawing of a sensory and a motor nerve cell. Electrically excitable regions *(indicated in red)* are present in the membranes of the axons of these two cells, at the synaptic nerve terminals of the sensory cell, and at the synaptic nerve endings of the motor cell at the muscle. The sensory endings responding to stimulation, dendrites of the motor cell, and the cell bodies (soma) are electrically inexcitable. The region of the cell body of the motor cell nearest the point at which the axon exits is electrically excitable and is referred to as the axon hillock.

B   A scanning electron micrograph of a neuron showing the cell body, dendrites, and a small portion of the axon.

TABLE 7-1   Comparison of Electrical Events in Membranes

Characteristic	Receptor and synaptic potentials	Action potentials
Location	Sensory endings, synapses, junction between nerve and muscle	Axons of nerve cells, muscle
Kind of stimulus-initiating event	Physical or chemical	Electrical
Nature of event	Graded	All-or-none
Size	Variable but usually less than 10 mV	Usually greater than 100 mV
Propagated	No	Yes
Mechanism of spread of current	Decremental conduction	Nondecremental conduction
Threshold	No	Yes
Summation	Yes	No
Direction of potential change	Depolarizing or hyperpolarizing	Depolarizing
Membrane type	Electrically inexcitable	Electrically excitable
Type of ion channel present	Stimulus-gated	Voltage-gated

tic potentials fade away with distance. Cells that can experience action potentials are referred to as **electrically excitable,** to distinguish them from **electrically inexcitable** cells which only exhibit graded receptor or synaptic potentials.

Cells may possess some regions that are electrically inexcitable and some that are electrically excitable. For example, the terminal portion of sensory nerve endings is electrically inexcitable, while the axon is electrically excitable (electrically excitable regions are indicated in red in Figure 7-1). This distinction is important in understanding the initiation of action potentials in sensory receptors and the transmission of activity from one cell to another.

Resting Potential

The previous chapter showed how the membrane potential is determined by the relative magnitudes of the membrane permeabilities to Na^+ and K^+ (see Figure 6-18). In cells that can undergo rapid changes in permeabilities that produce shifts in membrane potential, these shifts carry information. For this information to be meaningful, it must be compared with a baseline, or stable state. This relatively stable state, the **resting potential,** is the membrane potential that the cell exhibits when it is not responding to stimulation of some kind.

The point at which the membrane potential equals 0 mV lies between the physiological extremes set by the K^+ equilibrium potential (E_K, about -90 mV) and the Na^+ equilibrium potential (E_{Na}, around $+60$ mV). The resting potential is inside-negative, typically around -70 mV (see Chapter 6, p. 132). Another way of saying that a membrane is charged is to say that it is **polarized,** with a membrane potential not equal to zero. If the resting potential is the baseline against which membrane potential changes are compared, then a shift that makes the membrane potential more polarized

(internally more negatively charged and moving further away from 0 mV) is referred to as a **hyperpolarization** (Figure 7-2). Changes that make the membrane potential less inside-negative (moving closer to 0 mV) are termed **depolarizations** (see Figure 7-2). Action potentials are characterized by a large depolarization of the membrane, receptor potentials involve small depolarizations, and synaptic potentials (again much smaller than action potentials) may involve depolarizations or hyperpolarizations (see Table 7-1).

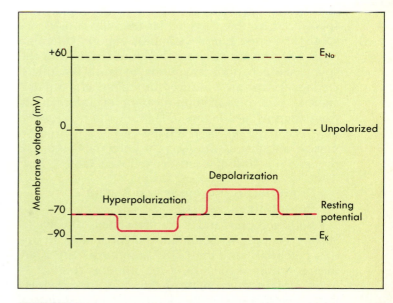

FIGURE 7-2

Definition of hyperpolarizing and depolarizing voltage shifts. The Na^+ and K^+ equilibrium potentials ($E_{Na} = +60$mV; $E_K = -90$ mV) are shown in relationship to the resting potential (-70 mV) and the unpolarized state of the membrane (0 mV).

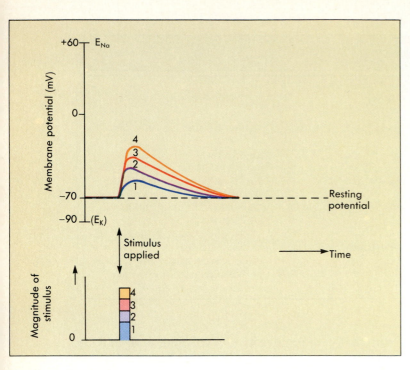

FIGURE 7-3
Graded responses. The lower portion of the figure shows four stimuli of increasing size; the upper portion shows the corresponding receptor potentials on the same type of graph. These responses would typically last several milliseconds.

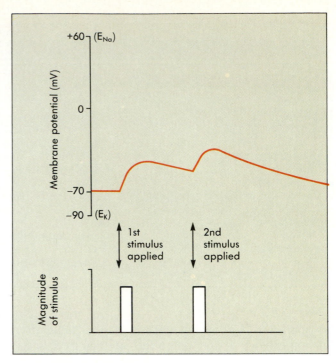

FIGURE 7-4
Summation. The lower portion of the figure shows two stimuli occurring close together in time. The upper portion shows how the two receptor potentials would add together when a second stimulus occurs before the first receptor potential has faded away.

Receptor and Synaptic Potentials

Receptor potentials are shifts in potential brought about by sensory stimulation of sensory nerve terminals (see Figure 7-1). These nerve terminals are specialized, so that a particular terminal is most responsive to one type of stimulus energy. Different classes of sensory cells are therefore required to respond to touch, odor, or sounds. Synaptic potentials are signals that are produced in regions of a cell designed to receive communication of information from neurons (see Figure 7-1). They can be thought of as occurring on the receiving end of a neuron or muscle cell, just as the receptor potentials occur on the receiving end of a sensory cell (see Table 7-1).

Properties of Receptor and Synaptic Potentials

Both receptor and synaptic signals involve a smaller change in membrane potential than in action potentials, and their size (amplitude) is graded in proportion to the size of the stimulus (Figure 7-3). A large displacement of the skin, for example, will result in a greater membrane potential shift in a pressure-sensitive sensory receptor in the skin than will a small displacement. Synaptic potentials and receptor potentials immediately begin to decrease in size after the stimulus that caused channels to open is no longer present (see Figure 7-3).

Synaptic and receptor potentials decrease in size with the distance from the stimulated patch of membrane. They are therefore said to spread along the membrane to other regions of a cell by **decremental conduction.** While neither receptor potentials nor synaptic potentials can travel very far along the membrane before they become insignificantly small, individual signals typically affect the activity of the cell by adding to other similar signals received by the cell in an overlapping time frame. This is possible because the change in the membrane potential that results from receptor and synaptic potentials persists long enough (4 msec or much longer, up to many seconds and even minutes) so that other signals can add to it, a process called **summation** (Figure 7-4). The ability of neurons to combine information arriving from several sources or arriving from the same repeatedly activated source within a short time is the basis of **integration,** the decision-making capability of the nervous system. This topic is treated in Chapter 8.

Action Potentials

An action potential either occurs or it does not, depending on the magnitude of the stimulus that is applied. **Threshold** is defined as the minimum value of the membrane potential at which an action potential will occur 50% of the time (Figure 7-5).

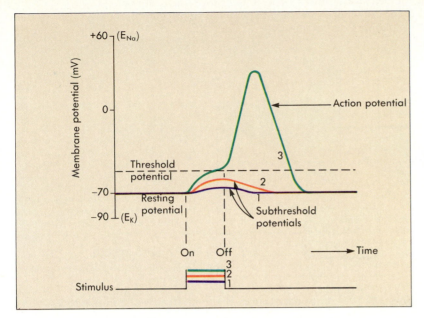

FIGURE 7-5

An illustration of the all-or-none property of action potentials. The membrane potential is shown in the upper part of the figure, while the lower part shows when the stimulus is applied *(ON)* and removed *(OFF)*. When a stimulus is applied, the membrane potential shifts to a depolarized value; the magnitude of the depolarization increases with larger stimuli. The responses to subthreshold stimuli fade away when the stimuli are removed. At threshold, an all-or-none action potential is generated.

Levels of membrane potential below the value needed to produce an action potential are referred to as **subthreshold.** If a stimulus fails to shift the membrane potential to threshold, the partial subthreshold response fades away altogether.

Action potentials are relatively large compared with receptor and synaptic potentials. Except for heart muscle and some smooth muscles (see Chapter 12, p. 306), action potentials are very short-lived shifts in membrane potential that last between 1 and 5 msec. During an action potential the membrane potential goes from its resting, inside-negative value to an inside-positive value before returning to the resting potential.

Action potentials represent a response of standard size that moves along the length of the axon, rather like the burning of a fuse. The size of the change in membrane potential does not diminish as the action potential moves along, but unlike the burnt region left behind as a fuse burns, the region behind the moving action potential recovers its ability to transmit an action potential. Transmission without loss of strength of the signal is called **nondecremental conduction** (see Table 7-1).

> 1 What are the differences between action potentials and receptor or synaptic potentials?
> 2 What is meant by the terms decremental and nondecremental conduction?
> 3 Define the threshold for an action potential.

MOLECULAR BASIS OF ELECTRICAL EVENTS IN MEMBRANES
Ion Channels in Membranes
The ability to undergo rapid changes in permeability is due to the presence of transport proteins in membranes (see Figure 6-9, *A*) that act as **gated ion channels.** Gated channels are water-filled pores that are usually shut (by a regulatory part of the protein structure, called the "gate") unless an appropriate signal causes the gated channel to temporarily open. Gated channels responsible for receptor and synaptic potentials open and close in response to physical or chemical stimuli, such as pressure or odor molecules, and are referred to as **stimulus-gated** channels (see Table 7-1). Other gated channels, including those in nerve axons, open and close in response to changes in the electrical potential across the membrane and are referred to as **voltage-gated** channels.

Gated channels are further identified by specifying the ions that they allow to pass through them. This chapter discusses three types of channels: (1) voltage-gated Na^+ **channels,** (2) voltage-gated K^+ **channels,** and (3) **leak** channels that let most ions pass through them and are not sensitive to membrane potential. It has recently become possible to biochemically isolate various types of gated ion channels, determine their amino-acid sequence and three-dimensional structure, and even to clone them using the recombinant DNA techniques described in Chapter 3 (see essay, p. 60).

Determinants of the Membrane Potential
The membrane potential of all cells is determined by: (1) the relative permeability of the membrane to specific ions, and (2) the concentration gradients of those ions across the membrane. Ion concentration gradients rarely change very much because the body's homeostatic systems maintain the composition of the extracellular fluid, while the Na^+-K^+ ATPase (see Chapter 6, p. 126) maintains constant intracellular concentrations.

Changes in membrane potential can be brought

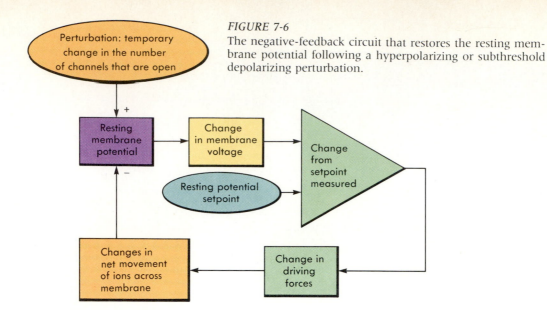

FIGURE 7-6
The negative-feedback circuit that restores the resting membrane potential following a hyperpolarizing or subthreshold depolarizing perturbation.

about in excitable cells by temporary alteration of the permeability to specific ions. In these cells, the membrane potential undergoes a shift from the resting potential due to changes in ion permeabilities. Such shifts in membrane potential are the foundation for the way information is coded and then processed within the nervous system, the main topic discussed in the following chapter.

A flow of ions across the membrane, which occurs when the permeability for an ion changes, results in an electrical current. The current carried by any particular ion is a function of the **driving force** for that ion and the **conductance** of the membrane for an ion. The driving force for any ion is the difference (in millivolts) between that ion's equilibrium potential and the membrane potential (see Chapter 6, p. 131). For example, at the normal resting potential of -70 mV, the driving force for K^+ is 20 mV (the difference between -90 mV and -70 mV), while the driving force for Na^+ is 130 mV (the difference between $+60$ mV and -70 mV). In most cases, the terms permeability and conductance can be used interchangeably because both depend on the proportion of open channels that will admit a particular type of ion.

Membrane Repolarization

To appreciate what occurs in the generation of an action potential, it is important to first understand how the membrane potential is restored to the resting level following a single graded event, such as a synaptic or receptor potential. These events can be viewed as perturbations in a system regulated by a negative-feedback loop (Figure 7-6; also see Chapter 5, p. 86). When an event such as binding of a chemical with a membrane protein or mechanical distortion of the membrane elements associated with a gated ion channel causes additional channels to open, ions will flow down their concentration gradients, locally changing the charge distribution along the membrane.

For instance, if the channels are of a size to permit K^+ to move through, stimuli will open more gates on these K^+ channels. The additional flow of K^+ down its concentration gradient (out of the cell) through the newly opened K^+ channels will make the charge on the membrane more negative (closer to the K^+ equilibrium potential, E_K). When these K^+ channels close, the condition of membrane permeability that existed formerly is restored, but the driving force for Na^+ and K^+ will have changed.

Figure 7-7, A relates the resting potential and the equilibrium potentials for K^+ and Na^+ ions to the driving forces on each ion. At rest, the membrane potential is much closer to the equilibrium potential for K^+ than to that for Na^+, which makes the driving force for K^+ relatively small. The resting membrane potential is closer to the K^+ equilibrium potential because the membrane is much more permeable to K^+ than to Na^+ (Figure 7-7, B). Under resting conditions the negative internal potential attracts positive K^+ ions, thus opposing the concentration gradient forcing K^+ out of the cell. The result is that few K^+ ions flow out at the resting potential, despite the presence of many open K^+ channels.

The situation is different for Na^+, which is far from its equilibrium potential (see Figure 7-7, A), so that the driving force for Na^+ is large. The negative internal charge tends to draw Na^+ ions into the cell, but in a resting cell there are few Na^+ channels with open gates, and the net entry of Na^+ ions into the cell at rest is not much different from the net exit of K^+ ions. For simplicity, the contribution of chloride and other ions will be ignored, but these also can move across the membrane and form part of the negative-feedback response.

If the membrane potential is shifted away from the resting potential, the driving forces for Na^+ and K^+ will both change. Take, for example, a situation in which the cell is hyperpolarized (Figure 7-8, A). The driving force for K^+ will decrease, while the driving force for Na^+ increases. The amount of Na^+

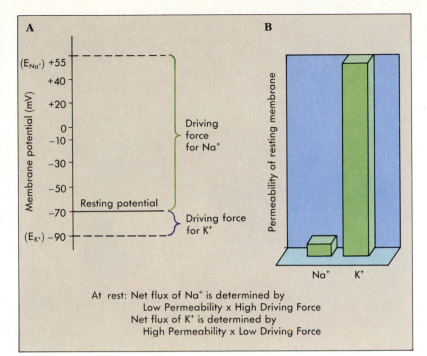

FIGURE 7-7
Comparison of the driving forces (**A**) and permeabilities (**B**) for Na⁺ and K⁺ at the resting membrane potential.

FIGURE 7-8

A The effect of a hyperpolarizing stimulus is to increase the driving force for Na⁺ and decrease the driving force for K⁺. When the stimulus is removed, the combination of an increased net entry of Na⁺ and a decreased net exit of K⁺ returns the membrane potential to the resting potential.

B The effect of a depolarizing stimulus is to decrease the driving force for Na⁺ and increase the driving force for K⁺. When the stimulus is removed, the combination of a decreased net entry of Na⁺ and an increased net exit of K⁺ returns the membrane potential to the resting potential.

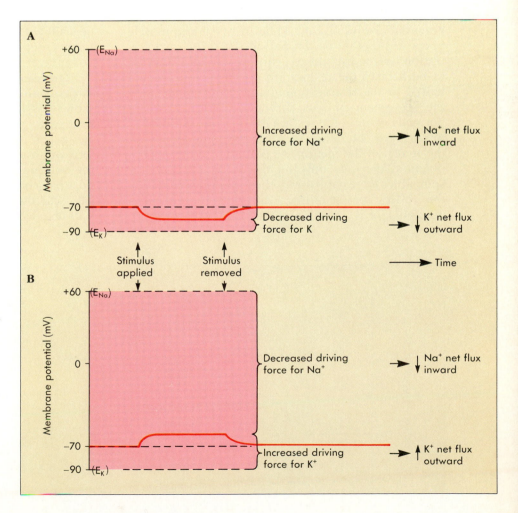

Excitable Membranes

145

entering the cell will increase even though there are relatively few open Na^+ channels in the resting membrane because the more negative interior of the cell is now even more attractive than it was at the resting potential. Likewise, the K^+ outward movement that normally roughly balanced the Na^+ inward movement is depressed, because the extra inside-negative charge tends to keep the positively charged K^+ ions inside the cell.

If the initial shift in membrane potential is in a depolarizing direction, the opposite response will occur (Figure 7-8, *B*). There will be an increased net flow of K^+ out of the cell and a decreased net flow of Na^+ into the cell. The resulting loss of positive charge will restore the membrane potential to the resting potential. In both cases, as a result of the altered driving forces for K^+ and Na^+, the change in membrane potential is wiped out, and the membrane potential is restored to the resting potential.

Concept of Threshold

In sensory receptors and at synapses, the cause of the initial displacement of the membrane potential from the resting potential is a physical or chemical stimulus (see Table 7-1), whereas in nerve and muscle there is electrical activation of voltage-gated channels. In sensory receptors and at synapses, restoration of the membrane potential to the resting potential always occurs, while in nerve and muscle fibers, restoration occurs only if the initial depolarization fails to initiate an action potential. However, in both cases the restoration of the membrane potential to the resting potential is accomplished by the movement of ions through channels that are normally open in the resting membrane. That is, following any alteration of the membrane potential, the return of the membrane potential to the resting potential depends only on the forces that are responsible for establishing the resting potential— the resting permeabilities and the concentration gradients for K^+ and Na^+.

The initiation of an action potential can be viewed as a combination of: (1) failure of the negative-feedback system to restore the membrane potential, and (2) the contribution of voltage-gated Na^+ channels that exhibit positive-feedback properties. The voltage-gated K^+ and Na^+ channels that are responsible for an action potential are different from ion channels in membrane regions that are nonexcitable (such as acetylcholine-activated channels at the junction between a nerve and a muscle, see Chapter 8, p. 175) because voltage-gated K^+ and Na^+ channels have gates that are triggered to open by depolarizing changes in the membrane potential.

To understand what is happening around the threshold membrane potential, it is helpful to picture the competition between opposing forces at this point (Figure 7-9). As the membrane potential becomes depolarized, the driving force for K^+ increases and there is an increased movement of K^+ ions out of the cell. This restorative, negative-feedback force tends to repolarize the membrane and gets stronger as the cell becomes more depolarized.

At the same time that there is an increased exit of K^+ from the cell, some of the gates on voltage-gated Na^+ channels are opening in response to the depolarization. The driving force for Na^+ entry is very large, so the opening of even a few Na^+ channels allows a large increase in Na^+ entry, which will further depolarize the cell and cause more of the voltage-gated Na^+ channels to open.

Threshold is the point at which either of the forces could win control of membrane potential. It represents the minimum depolarizing voltage shift needed to produce an action potential. If the voltage-gated Na^+ channels let just a few more Na^+ ions into the cell than the number of K^+ ions flowing out due to the change in driving force, the extra positive charge will open still more voltage-gated Na^+ channels, and the positive-feedback system will take over.

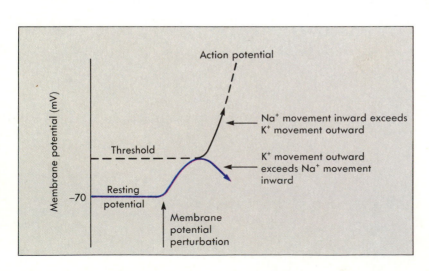

FIGURE 7-9
The positive-feedback and negative-feedback forces that determine membrane responses at threshold. At threshold, the net entry of Na^+ exceeds the net exit of K^+, resulting in an additional depolarization and the opening of still more voltage-gated Na^+ channels.

NERVE AND MUSCLE

1 What is the definition of driving force?
2 How is the membrane potential restored to its resting state following a hyperpolarizing or a subthreshold depolarizing shift in membrane potential?
3 What opposing forces are balanced at the threshold potential?

VOLTAGE CLAMP STUDIES
Voltage Clamp of Giant Axons

The study of the mechanism underlying action potentials was aided by the development of an experimental technique called the **voltage clamp.** The electronic circuitry of the voltage clamp makes it possible to displace the membrane potential of a neuron or muscle cell from its resting value to some other value and hold it there while the current flow resulting from movement of ions across the membrane is measured. The principle of the voltage clamp circuit (Figure 7-10) is that of negative feedback: when the flow of current across the membrane starts to alter the membrane voltage, the electronic circuitry forces current of opposite polarity into the cell to keep the membrane potential from changing. Thus the currents injected by the voltage clamp negative-feedback circuit (the "bucking currents") are equal in magnitude and opposite in charge to the currents flowing across the membrane.

Squid Giant Axon

In order to apply the voltage clamp to the nervous system, it was necessary to find an animal with nerve cells that had axons sufficiently large that a pair of fine wires could be inserted along the length of the inside of the axon. No mammal's axons approach this size, but members of several groups of invertebrates possess large axons. The axons are typically found in neural pathways that activate reflex responses important for the survival of the animal, and the large size of these axons increases speed of action potential propagation. These **giant axons** may be as much as 1 mm in diameter, two to three orders of magnitude larger than typical vertebrate axons. The most useful ones are from the squid, a marine mollusk (Figure 7-11).

It is possible to insert internal electrodes for voltage clamp studies and even to remove the cytoplasm from a squid giant axon and perfuse the interior of the cell with solutions of desired compositions. It is safe to say that without the squid much less would be known today about how the nervous system functions.

Voltage clamp experiments were first performed in the 1930s and 1940s by the American scientists Kenneth Cole and Howard Curtis and the British scientists Alan Hodgkin, Andrew Huxley, and Bernard Katz. At that time, very little was known about the structure of cell membranes and the functional role of proteins in them. A major observation made in the initial recordings from squid giant axons was that the membrane potential did not just go to the point of zero potential but briefly became internally positive at the peak height of the action potential. Another general observation was that the overall flux of ions across the membrane increased dramatically during the action potential.

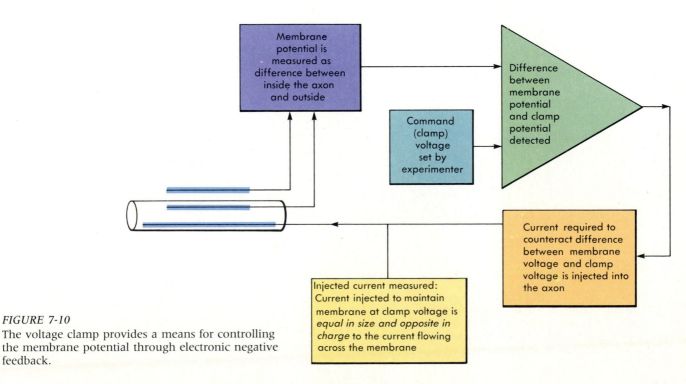

FIGURE 7-10
The voltage clamp provides a means for controlling the membrane potential through electronic negative feedback.

A

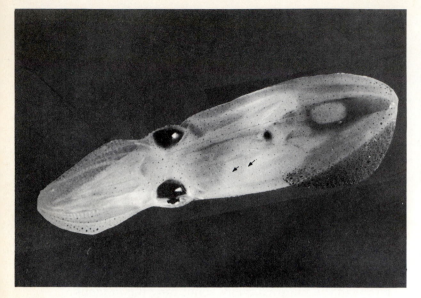

B

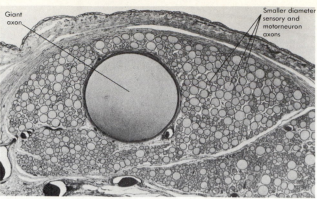

FIGURE 7-11
A Photograph of a giant squid. The arrows show the locations of giant axons in the mantle.
B Cross section of the mantle of a squid showing one of the giant axons.

Measuring Membrane Conductances

The problem for the investigators became one of determining which ions were flowing across the membrane during the early and late periods of the brief action potential. They tackled this problem by rapidly clamping the membrane potential at values

above the resting potential and holding the membrane potential at that command level for a period of time longer than the duration of an action potential. The analysis of which ions were flowing across the membrane was simplified by conducting the experiments with the axon maintained in two conditions: in normal seawater, which is very similar to the blood of a squid, and in an artificial solution in which a large impermeant positive ion was substituted for the Na^+ ion. Thus the contribution of K^+ ions alone was apparent from the responses of the membrane when no Na^+ ions were present, and the pure contribution of Na^+ could be calculated by subtracting the response in Na^+-free seawater from the response recorded from the combined action of both ions in normal seawater. Luckily, only these two ions contribute to the action potential in the squid axon.

Characteristics of Na^+ and K^+ Conductances

The results revealed the flow of ions at each clamp potential over time. The Na^+ and K^+ were found to differ in how soon they responded to a membrane potential change and in their rate of change in permeability (Figure 7-12). At a membrane potential just above threshold (-40 mV), there is an elevation of Na^+ conductance; but this is not maintained, even under constant voltage clamp conditions, and declines within a few milliseconds (see Figure 7-12, *A*). The response of K^+ ions to a clamping current just above threshold is a much slower increase in K^+ ion flux that is maintained at a constant level as long as the clamp current is on (see Figure 7-12, *B*).

The measurement of responses to clamp currents around zero membrane potential and near the peak of the action potential shows a similar pattern that differs in the magnitude of the response. In all cases, the Na^+ response involves a relatively rapid onset (less than 1 msec) and a self-determined turn-

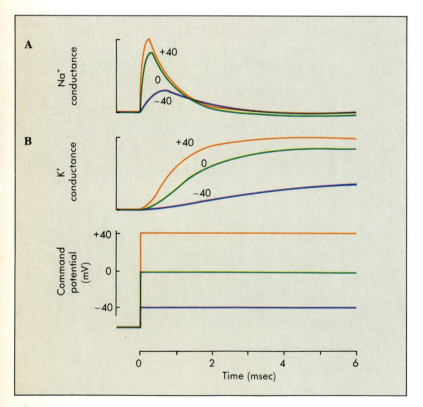

FIGURE 7-12
Responses under voltage clamp for three command potentials show that the Na^+ conductance (**A**) is low around threshold (-40 mV), intermediate at 0 mV, and very high at $+40$ mV. The Na^+ conductance is initiated rapidly (activation) and is subsequently inactivated. The K^+ conductance (**B**) has a slower onset and does not inactivate as long as the command potential is held.

off, whereas the K⁺ response has a slower onset (1 to 2 msec) and is maintained. Note that these measurements are representations of absolute current flow and that the direction of ion flow is determined by the concentration gradients, which drive Na⁺ into the cell and K⁺ out of the cell. These data allowed Hodgkin and Huxley to describe the magnitude of the conductance of Na⁺ and K⁺ at any time after the action potential is triggered.

The voltage clamp recordings revealed that the rapid depolarization that occurs at the beginning of an action potential is due to a dramatic increase in Na⁺ permeability, which quickly takes the membrane potential close to the Na⁺ equilibrium potential. Repolarization results from: (1) the time-limited nature of the increase in Na⁺ permeability, and (2) the increase in K⁺ permeability due to opening of voltage-gated K⁺ channels. These changes rapidly repolarize the membrane and then carry its potential even further in the direction of the K⁺ equilibrium potential for a short time before the resting condition is restored (referred to as an **after-hyperpolarization**).

Figure 7-13 summarizes the voltage changes during the action potential on the same time scale as the conductances for Na⁺ and K⁺. When the membrane potential is allowed to change, rather than being clamped, the rising phase of the action potential is driven by the Na⁺ influx, and the falling phase can occur because the Na⁺ current turns off as the K⁺ efflux turns on. Even though K⁺ conductance is maintained under voltage clamp conditions, it is turned off by the repolarization of the membrane that results from the K⁺ flow in the unclamped condition.

Note that there is actually a time delay between the onset of the voltage changes of the action potential and the conductance changes in Na⁺ and K⁺. This seeming contradiction can be resolved if it is understood that the charge on the membrane (which is what the voltage is) changes first, before

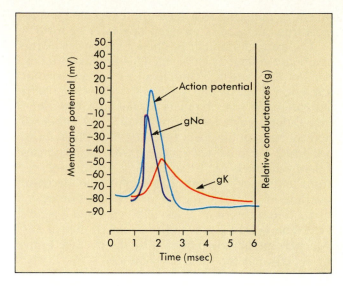

FIGURE 7-13
An illustration showing how voltage changes during the action potential are related to the time course of the Na⁺ and K⁺ conductances (abbreviated gNa and gK).

there is a net flow of ions across the membrane. In other words, the first flow of charges is onto the membrane rather than across the membrane.

VOLTAGE-GATED CHANNELS
Electrically Excitable Membrane Proteins

It is now possible to interpret the voltage clamp observations in the light of what is known about the membranes of excitable cells. Ion channels are presently thought to possess a region that acts as a selectivity filter and a second region that acts as a gate (Figure 7-14). The gate is probably a portion of the channel structure that bears a net charge and can swing to block the pore through which ions permeate. Changes in the membrane voltage will alter the electrical forces that these gates experience and can therefore trigger conformational rearrangements of

FIGURE 7-14
A hypothetical model of the Na⁺ channel illustrating the three states: resting, activated, and inactivated.

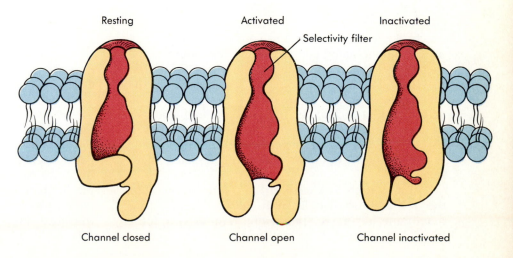

the entire molecule that have the effect of converting it from a closed to an open (activated) pore.

The average number of open gates varies as the membrane voltage is changed. This is why different levels of Na⁺ and K⁺ conductances are recorded at different voltage clamp potentials. Activation of a population of ion channels is not all-or-nothing, even though each individual channel is either all the way open or all the way closed. The behavior of individual Na⁺ channels and the average responses of many channels can be seen using a technique, referred to as patch-clamp, that permits the electrical isolation of a small region of cell membrane (a patch) containing only a few Na⁺ channels (Figure 7-15). The overall permeability is determined at any particular membrane potential by the average number of activated Na⁺ and K⁺ channels.

Properties of Na⁺ and K⁺ Channels

The gates of resting Na⁺ channels can be thought of as closed, but not securely latched. A depolarization can open the gates. Opening occurs rapidly (within a few tenths of a millisecond) and is called **Na⁺ activation.** Once a Na⁺ channel has opened, it closes spontaneously after a time, even though the membrane potential is still depolarized. Closing of the opened channels, called **Na⁺ inactivation,** takes about 10 times longer than Na⁺ activation. An inactivated channel is not simply a closed channel; it is a closed and latched channel (see Figure 7-14). The inactivated state lasts as long as the membrane remains depolarized. When the membrane repolarizes, the channels revert to the initial resting state and can be reopened again by depolarization. On the average, potassium channels open approximately 10 times more slowly than Na⁺ channels. Potassium channels do not inactivate spontaneously; they can close only if the membrane potential returns to its original resting level.

Changes in Permeability During Action Potentials

The movement of ions across the membrane at various stages during the generation of an action potential can be illustrated by events as they occur in a patch of membrane. Resting conductances, as well as the voltage-gated conductances, contribute to the membrane potential changes (Figure 7-16). The driving forces for ions can be determined for the membrane potential changes that occur during an action potential just as for subthreshold changes in membrane potential, but the ease with which the Na⁺ and K⁺ ions move through the membrane is altered by the opening of the voltage-gated channels.

At rest, there is a weak driving force for K⁺, but the K⁺ conductance is high because the resting membrane is very K⁺-permeable. The driving force for Na⁺ is large, but the resting Na⁺ conductance is

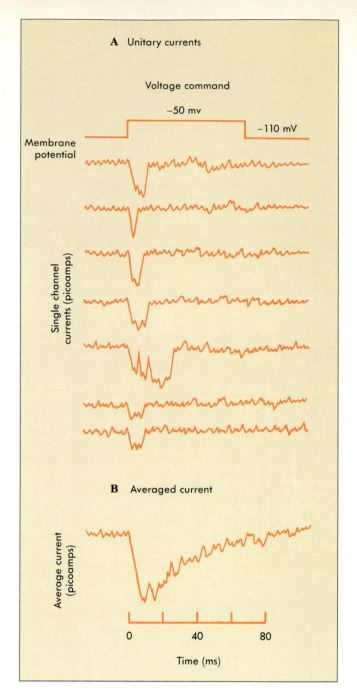

FIGURE 7-15
A Currents associated with individual Na⁺ channels in response to a depolarizing voltage step show that individual channels open and close independently. Currents associated with individual channels are on the order of 1 picoampere. **B** The averaged currents for individual channel openings resemble the activation-inactivation time-course characteristic of the whole cell.

nevertheless small because the Na⁺ permeability of the membrane is relatively low. When some Na⁺ gates open in response to depolarization, the conductance to Na⁺ increases by the positive-feedback mechanism of depolarization of the voltage-gated gates, which quickly produce a very high Na⁺ conductance (Figure 7-17). The limits to the increase in this conductance are (1) the decrease in driving

FIGURE 7-16

Changes in membrane permeability at selected points during the action potential. At the resting potential some K⁺ channels are open, and the exit of K⁺ is balanced by an entry of Na⁺ through "leak" channels. At threshold, enough Na⁺ channels have opened so that the influx of Na⁺ exceeds the exit of K⁺, causing more Na⁺ channels to open until all have opened. During the falling phase of the action potential, voltage-gated K⁺ channels are opening while Na⁺ channels are becoming inactivated. During the after-hyperpolarization, the number of open K⁺ channels is temporarily increased, so that the membrane potential is closer to the K⁺ equilibrium potential. When these K⁺ channels close, the membrane potential returns to the resting potential.

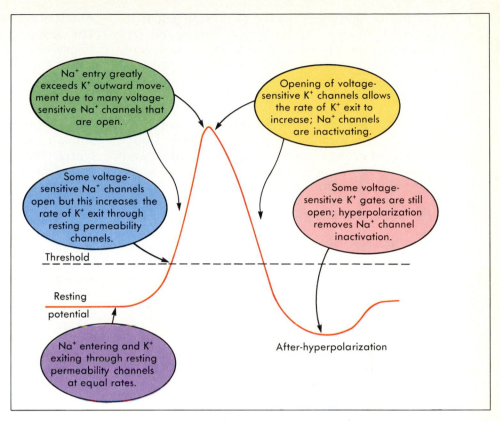

Na⁺ entry greatly exceeds K⁺ outward movement due to many voltage-sensitive Na⁺ channels that are open.

Opening of voltage-sensitive K⁺ channels allows the rate of K⁺ exit to increase; Na⁺ channels are inactivating.

Some voltage-sensitive Na⁺ channels open but this increases the rate of K⁺ exit through resting permeability channels.

Some voltage-sensitive K⁺ gates are still open; hyperpolarization removes Na⁺ channel inactivation.

Threshold

Resting potential

Na⁺ entering and K⁺ exiting through resting permeability channels at equal rates.

After-hyperpolarization

FIGURE 7-17

The positive-feedback contribution to the rising phase of the action potential is made through the effect of depolarization on voltage-gated Na⁺ channels. The depolarization also activates voltage-sensitive K⁺ channels after a time delay. The contribution of current flow through the K⁺ channels, together with coincident inactivation of Na⁺ channels, repolarizes the membrane. An additional effect of membrane repolarization is to remove the inactivation of the Na⁺ gates, so that they return to a closed but voltage-sensitive state.

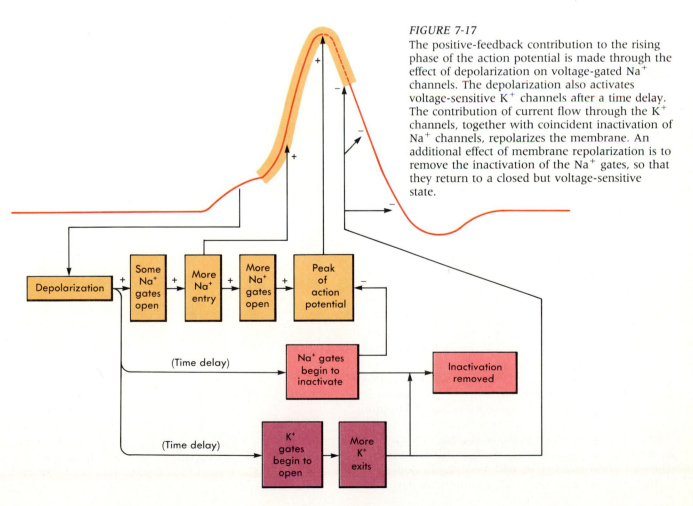

Depolarization

Some Na⁺ gates open

More Na⁺ entry

More Na⁺ gates open

Peak of action potential

(Time delay)

Na⁺ gates begin to inactivate

Inactivation removed

(Time delay)

K⁺ gates begin to open

More K⁺ exits

Excitable Membranes

151

force for Na$^+$ movement as the membrane potential approaches the Na$^+$ equilibrium potential, and (2) the spontaneous closing (inactivation) of Na$^+$ gates within a set time after their opening.

Inactivation of Na$^+$ channels causes the Na$^+$ permeability to return to the low value characteristic of the resting membrane. Inactivation of Na$^+$ channels, by itself, would result in membrane repolarization, but repolarization would be relatively slow. K$^+$ channels activate at about the same rate that Na$^+$ channels inactivate (see Figure 7-17). Activation of K$^+$ channels accelerates repolarization and limits the duration of the peak of the action potential to 1 to 2 msec. When the membrane repolarizes, the voltage-gated K$^+$ channels begin to close. Closing takes time, so the K$^+$ conductance does not return to its resting level immediately. During this part of the action potential, after the Na$^+$ channels have inactivated and while the voltage-gated K$^+$ channels remain open, the K$^+$ permeability of the membrane is even greater than during rest, and the membrane potential hyperpolarizes past the resting value and approaches the K$^+$ equilibrium potential (see Figure 7-16). This has been called either a **hyperpolarizing afterpotential** or an after-hyperpolarization potential.

> 1 What differences in the Na$^+$ and K$^+$ conductances were revealed by voltage clamping?
> 2 What is meant by Na$^+$ inactivation?
> 3 What are the positive- and negative-feedback elements that determine the action potential?

REFRACTORY PERIOD AND FIRING FREQUENCY
Absolute Refractory Period
The inactivated state of the Na$^+$ gates prevents them from responding to any degree of depolarization. As long as the vast majority of the channels are inactivated, no amount of stimulation can open them. The **absolute refractory period** is the period of time from the triggering of an action potential to the time that most of the Na$^+$ gates have their inactivation removed by membrane repolarization. This term is defined by the experimental observation that it becomes virtually impossible to alter the membrane potential enough to initiate another action potential during the falling phase of the action potential (Figure 7-18). Because inactivation does not occur synchronously for all the channels, but is a population phenomenon, the membrane gradually becomes more and more susceptible to reactivation. Another factor that makes it difficult to activate an action potential when some of the Na$^+$ channels are inactivated is the fact that during this same period the K$^+$ conductance is still very high due to the number of voltage-gated K$^+$ channels that are open.

Relative Refractory Period
The absolute refractory period is followed by a somewhat longer **relative refractory period** in which a second action potential can be initiated, but the membrane potential that corresponds to the threshold voltage is higher (more internally positive) than is true when the cell has not recently fired an action potential (see Figure 7-18). Therefore the stimulus must be stronger to initiate an action potential. The duration of the relative refractory period is determined by how long it takes for the K$^+$ conductance to return to normal.

Repetitive Firing During Prolonged Depolarization
When a long-lasting depolarization results from many synaptic or receptor potentials on the receiving end of the cell, a series of action potentials can often be produced (Figure 7-19). This behavior is known as **repetitive firing**. Repetitive firing is important because it allows the all-or-none action potentials to give information on the strength of the stimulus in terms of a relationship between the number of action potentials and the intensity of the stimulus (see Chapter 8).

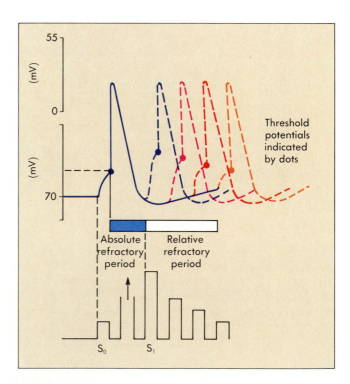

FIGURE 7-18
No stimulus, however great, will produce a second action potential immediately following a stimulus (S_o) that initiates an action potential; this is the absolute refractory period. The magnitude of a second stimulus (S') needed to elicit a second action potential during the relative refractory period is greater (that is, the threshold is higher) than for the first action potential, but decreases with time until the end of the relative refractory period when it has returned to the initial value.

NERVE AND MUSCLE

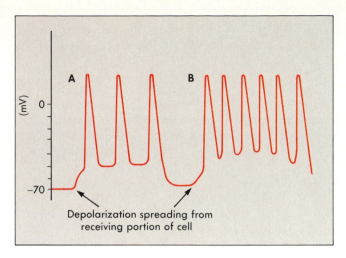

FIGURE 7-19
The rate of repetitive firing is a function of the membrane depolarization level.
A The maintained depolarization was slightly above threshold, and three action potentials were generated.
B The depolarization was much higher, and action potentials were generated during the relative refractory period, producing a higher frequency of firing.

To understand the basis of repetitive firing, consider what is happening as the membrane repolarizes following an action potential. With a maintained depolarization of the membrane, at some point during the relative refractory period the inward current flowing from the receptor or synaptic potentials would become equal to the outward current due to the K^+ conductance. Another action potential would then occur. If the membrane depolarization is larger, the second action potential will occur sooner, equivalent to an increased frequency. The result is that the magnitude of a steady depolarizing current is encoded into the frequency of action potentials that arise in the nerve.

Extracellular Ca⁺⁺ and Excitability

The polar heads of membrane phospholipids and many side groups of membrane proteins are ionized and carry a negative charge. To the extent that these **fixed negative charges** are unpaired, they cancel out some of the unpaired positive charges that are associated with the extracellular surface of the membrane of a resting excitable cell. The effect of the fixed negative charges is similar to that of depolarizing current. Calcium ions, because they have two positive charges, can interact with the fixed negative charges and neutralize or "screen" them. The changes in extracellular Ca^{++} concentration can thus affect membrane excitability by altering the threshold for action potential generation (Figure 7-20).

In low-Ca^{++} solutions threshold decreases, and the excitable membranes of nerve cells and muscle cells become **hyperexcitable.** They may even dis-

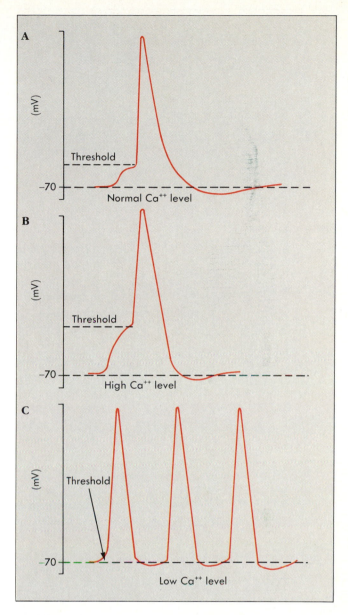

FIGURE 7-20
The effect of extracellular Ca^{++} on electrically excitable membranes. The normal condition (**A**) is contrasted with the elevated threshold that results from high Ca^{++} levels (**B**) and the decreased threshold that results from low Ca^{++} levels (**C**). In the latter case, spontaneous generation of action potentials can occur if the threshold falls to the resting membrane potential.

charge action potentials continuously, just as if they were being stimulated. In high-Ca^{++} solutions threshold is increased. Since a larger stimulus is needed to produce an action potential, the excitable membranes of nerves and muscles become refractory to stimulation in high-Ca^{++} solutions.

Disorders of calcium homeostasis produce large fluctuations in heart rate and are thought to be a causative factor in some cases of hypertension (elevated blood pressure). In **hypocalcemia** (low plas-

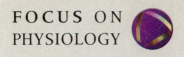

Toxins that Block Channels: The History of TTX and STX

Until the mid-1960s biophysicists had no clear idea of how ions actually moved across cell membranes. Were there transient water-filled pores? Did ions dissolve in a homogenous lipid membrane? Were lipid-soluble protein carriers involved? Answers to these questions were provided in part by the discovery of chemical "magic bullets" that aimed themselves unerringly at membrane channel proteins. Tetrodotoxin (TTX) is a poison found in the sex organs and liver of marine puffer fish and other species of the order Tetraodontiformes, as well as in the blue-ringed octopus and some salamanders. Saxitoxin (STX) is a related molecule found in single-celled marine protozoa called dinoflagellates. Both these toxins are small, water-soluble compounds with a positively charged guanidine group at one end. They both block voltage-gated Na^+ channels in nerve and muscle fibers at nanomolar concentrations, making the conduction of action potentials in these cells difficult or impossible. Their high affinity for the Na^+ channels makes them rank among the most potent known poisons.

The earliest reference to TTX appeared in the first Chinese pharmacopoeia, written approximately 200 BC, where it was recommended as an effective means of arresting convulsions. A more detailed account appeared in a pharmacopoeia published in 1600 AD, in which various descriptions of its effect can be found, such as: "In the mouth they (the liver and eggs of puffer fish) rot the tongue; internally they rot the guts"; "the poisoning no remedy can relieve"; and "to throw away life, eat puffer fish." In Japan the puffer fish, called *fugu*, is still regarded as both a cult object and a delicacy. If the flesh is to be eaten safely, the fish must be prepared without puncturing either the liver or the reproductive organs. Restaurants that serve fugu are strictly licensed, but fascination with the consumption of fugu accounts for some 200 to 400 cases of fugu poisoning annually. Most poisonings

---

ma Ca^{++}), motor nerves and skeletal muscle fibers discharge spontaneously, causing muscle spasms that are called **hypocalcemic tetany.**

Role of the Na^+-K^+ ATPase in Excitability

The membrane potential is maintained in the face of continual leakage of Na^+ and K^+ across the membrane by the actions of the Na^+-K^+ ATPase (Na^+-K^+ pump). The Na^+-K^+ ATPase operates steadily to prevent a loss of excitability that would result from loss of the concentration gradients. During an action potential, some additional K^+ moves from the cytoplasm to the exterior; an equal amount of Na^+ moves into the cytoplasm. The amounts of Na^+ and K^+ that move in the course of a single action potential are typically quite small compared with the amounts of these ions that produce the transmembrane concentration gradients. In an extreme exam-

ple, a single action potential in a squid giant axon results in loss of only about one ten-millionth of the cytoplasmic K^+. The Na^+-K^+ ATPase system of an axon can be blocked pharmacologically without interfering with the steady generation of action potentials for many hours in cells with such large ion stores. This shows that the production of action potentials is not an energy-dependent event, but rather is driven by the stored energy of the concentration gradient. However, no matter whether the cells in question are big or small, losses of K^+ and gains of Na^+ during action potentials are added to the steady leakage of Na^+ in and K^+ out that occurs continuously in resting cells.

Problems related to the effect of action potential conductances on concentration gradients develop sooner in smaller cells because their surface areas are larger relative to their cytoplasmic volumes. A

occur when the fish is prepared by amateurs, but there is reason to believe that risk-taking behavior, in the form of deliberate addition of some of the toxin-containing parts of the animal to the flesh that is eaten, plays at least as large a role as carelessness or ignorance.

TTX was a new encounter for Europeans who began to visit the Orient in the seventeenth century. A detailed account of the symptoms of fugu poisoning appears in the log of Captain Cook's second circumnavigation of the globe in 1670. Mild cases involve limb numbness, flushing of the skin, muscle weakness, and a tingling sensation in the mouth and tongue. This tingling is regarded by some Japanese as part of the sought-after experience of consuming fugu. Larger doses produce severe disturbances of the heart rhythm. Lethal doses paralyze the diaphragm, leading to asphyxiation.

STX-containing dinoflagellates "bloom" at certain times of the year in certain regions of the ocean. The concentration of organisms sometimes reaches 20 million per liter, giving the water a reddish tint referred to as a "red tide." The dinoflagellates are consumed by clams, mussels, and scallops that thus become contaminated, causing economic losses despite the fact that these waters are constantly monitored for the presence of the dinoflagellates, and several accidental deaths from shellfish poisoning occur each year.

These toxins have been used by neurophysiologists to show that sodium channels are physically distinct from other membrane channels, such as the K^+ channels, and have helped in characterizing the "entryway" of the Na^+ channel and in separating and counting the channel molecules. They are not the only natural toxins that have proved useful in studies of membrane channels: other useful toxins have been extracted from plants, the skin of amphibians, and the poison glands of scorpions, spiders, and sea anemones.

measurable stimulation of the Na^+-K^+ ATPase activity occurs after a high level of repetitive firing of small axons. Without the restoring action of the Na^+-K^+ ATPase, leaks would sooner or later discharge the transmembrane concentration gradients and lead to a loss of excitability; but ion pumps operate on a much slower time scale than the events of the individual action potential and thus have no direct role in the conductance changes that occur during action potentials.

> 1 What two factors are responsible for the refractory period?
> 2 How is level of depolarization reflected in the information carried by action potentials?
> 3 How does extracellular Ca^{++} affect membrane excitability?

PROPAGATION OF ACTION POTENTIALS
Spread of Excitation

Under normal conditions, initiation of an action potential occurs in membrane regions close to the electrically inexcitable portion of the cell where stimuli are received. This region in nerve cells is called the **axon hillock** or **initial segment** (see Figure 7-1). The current flow affects the gates of voltage-gated Na^+ channels, and if the depolarization is above threshold, an action potential will be initiated. Once an action potential is initiated, it will flow along the membrane surface as adjacent voltage-gated Na^+ channels are affected by the charge distribution caused by entry of Na^+ (Figure 7-21).

The Na^+ ions that enter across the membrane produce a positive cloud of charge locally, and this upsets the balance between positive and negative charges in the cell interior, so that positive charges

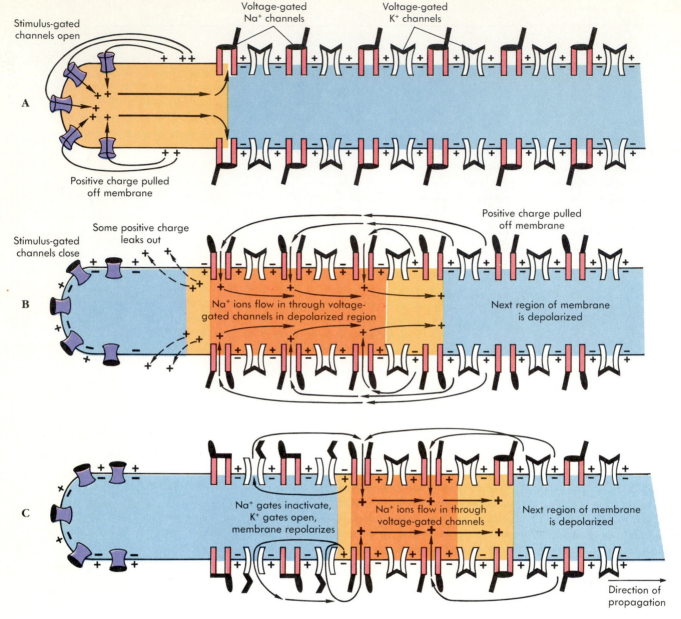

FIGURE 7-21

A The decremental spread of current entering receptor channels in the receiving portion of an excitable cell may be great enough to cause opening of voltage-gated Na$^+$ channels in the electrically excitable portion of the cell.

B Current flow through the Na$^+$ channels spreads in one direction, down the axon, because voltage-gated channels are not present in the electrically inexcitable, receiving portion of the cell.

C The path of current flow along an axon when an action potential is propagating down it includes depolarizing, repolarizing, and as yet unexcited regions of the membrane.

are repelled into adjacent regions of the membrane, where they partially neutralize the negative charges on the membrane. The effect is greatest close to the site of ion entry and diminishes with distance along the membrane. However, if an action potential has been initiated, the quantity of ions flowing across the membrane will be more than sufficient to open Na$^+$ gates nearby, and so the action potential will not fail to be propagated along the membrane. After a time delay, the K$^+$ gates will open and the Na$^+$ gates will close, so the path of current flow will re-

store the resting membrane potential in the membrane region behind the moving wave of depolarization (see Figure 7-21).

In this way an action potential, once initiated, can spread along the axon. The necessity of the non-decremental conduction that characterizes action potentials is apparent when the relatively large distances over which axons project are taken into consideration. The distance between a sensory nerve ending in the foot and its cell body near the spinal cord is about 1 meter. Without some way of re-

A

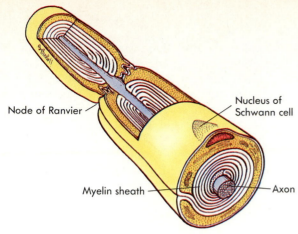

Node of Ranvier

Nucleus of
Schwann cell

Myelin sheath — Axon

B

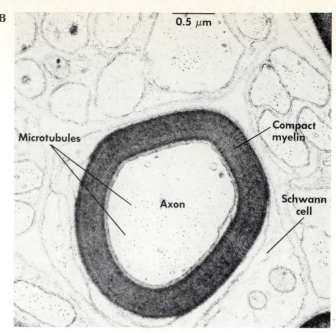

0.5 µm

Microtubules

Compact
myelin

Axon

Schwann
cell

C

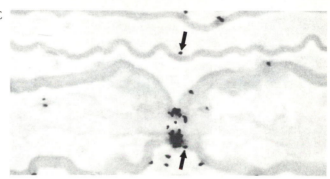

FIGURE 7-22

A   A longitudinal section of a myelinated axon from the central nervous system, showing a node of Ranvier and the terminal loops of the myelin sheath.
B   A cross section of a myelinated axon showing the layers of membrane forming compact myelin and their relationship to the surrounding Schwann cell.
C   An autoradiograph of a node of Ranvier. The black dots indicated by the arrows represent monoclonal antibodies directed at sodium channels and are thus concentrated at the nodes.

charging the signal as it spreads along the membrane, an electrical signal would travel only a few millimeters before it died out. The cycle by which Na^+ channels open, conduct a current driven by the transmembrane gradient, and inactivate after the current has opened adjacent Na^+ channels provides continuous regeneration of the signal so that it does not die away with distance.

Differences in Propagation Velocity

Some responses made by the body require rapid conduction, whereas others can be transmitted much more slowly without affecting the ultimate effectiveness of the response. Rapid conduction of information relating to the development of the internal state associated with hunger is unnecessary, but rapid reflex withdrawal from a hot object or to avoid the path of a moving projectile is crucial. Some pathways of the nervous system have been subjected to intense evolutionary pressure to increase propagation velocity, and this is apparent in many animal groups. The giant axons found in the nervous systems of many types of invertebrate animals are an adaptation for speed and are associated with pathways that mediate escape.

Effect of Diameter and Myelin on Propagation Velocity

The velocity with which an action potential can propagate in an excitable membrane is determined by how much adjacent membrane the action potential can bring to threshold, or to put it another way,

how far ahead of themselves the open Na^+ channels can open more Na^+ channels. Two factors determine this distance. One is the diameter of the fiber. The larger the diameter, the better a conductor of electricity it will be and the further current will spread. In vertebrate animals, however, another factor contributes much more to the increase in conduction velocity than diameter.

The current that flows away from the activated channels down the length of the nerve fiber is diminished as it travels by leakage of current between the cell and the exterior. In many axons of the vertebrate nervous system, insulation against current leakage is provided by supporting cells of the nervous system called **glial cells.** The glial cells that provide insulation are called **Schwann cells** or **oligodendrocytes,** depending on where in the nervous system they are located.

In the course of development, these glial cells send out tonguelike processes of lipid-rich membranes that wrap themselves around nearby axons, forming **myelin sheaths.** Each myelin sheath is composed of many layers of phospholipid cell membrane. The myelin sheaths are interrupted periodically along the length of the axons to form small gaps 1 to 2 µm wide called **nodes of Ranvier.** Figure 7-22, *A* shows a longitudinal section of a myelinated fiber including a node of Ranvier, while Figure 7-22, *B* shows the appearance of the myelin sheath in cross section. In a fiber with a diameter of 10 to 20 μm, the nodes are about 1 mm apart. The axonal cell membrane does not have voltage-gated channels in the cell membrane under the myelin; these are present only in the membrane at the nodes (the black dots in Figure 7-22, *C*).

Propagation in Myelinated Nerve Fibers

The effect of the myelin sheaths on the pattern of current flow is shown in Figure 7-23. Node 1 is producing an action potential. During the action poten-

tial, Na$^+$ ions enter the fiber, carrying an inward current. Because of the insulation, most of the current flows down the axon, with only a small leakage out through the myelin. As a result of the insulation, the current is conserved and is adequate to depolarize node 2 to threshold, even though it is relatively far away.

When node 2 is depolarized to threshold, an action potential will be produced. The inward current spreads forward to node 3, but also backward to node 1. An action potential will result at node 3, but not at node 1 because it is in its refractory period (see Figure 7-23, *B*). Thus the action potential is

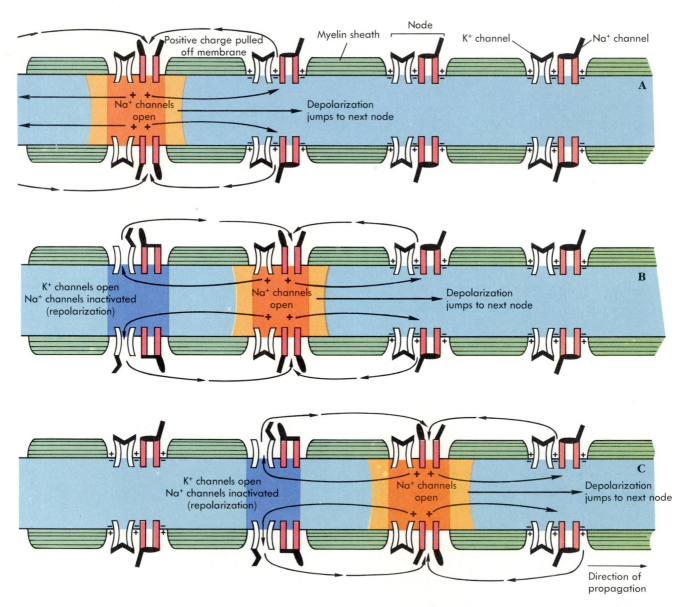

FIGURE 7-23
A   Saltatory conduction along a myelinated axon involves spread of depolarization along the core of the axon and regeneration of the signal at the nodes.
B and **C**   Current flow is unidirectional because the channels behind the moving wave of depolarization are inactivated. (Note that the distance between nodes is not drawn to scale.)

transmitted in only one direction. As this process is repeated at each node, the action potential appears to jump from one node to another. This is called **saltatory conduction.** The speed is gained because the electrical repelling of charge along the length of the axon to the next node is fast compared with the flow of ions through open channels. The jumps between nodes are almost instantaneous, and then the progress of the action potential is slowed at each node as the recharging takes place.

The spacing of nodes is an important feature of the development of the nervous system. The distance cannot be so long that the current becomes too weak in its travel to depolarize the next node. The developing nervous system actually seems to err on the safe side—the distances between nodes are such that current flow between them is several times the minimum needed to bring a node to threshold. The proportion of extra current over the minimum needed to bring the node to threshold is referred to as the **safety factor.**

Comparison of Nonmyelinated and Myelinated Nerves

The relative advantages of size and myelination can be seen in the following examples. An unmyelinated nerve 10 μm in diameter would conduct action potentials at a velocity of about 0.5 meter/sec. If axons like these were present in the reflex pathway that withdraws an appendage from a painful stimulus, the response would have a lag time of about 4 seconds. If the diameter of these axons were increased to the size of a squid giant axon, about 500 μm, the conduction velocity would increase to approximately 25 meter/sec (Table 7-2), and the lag time would drop to about 80 msec. If, on the other hand, the 10-μm axon were myelinated, the conduction velocity would rise to about 50 meter/sec and the lag time would be about 40 msec. Clearly myelination results in a saving of space compared with the strategy of increasing the axon diameter.

Myelination is used selectively in the nervous system for those pathways where speed is essential. This includes sensory nerve fibers from pressure receptors in the skin and motor nerves that run to skeletal muscles in the arms and legs, where conduction velocities can exceed 100 meter/sec. Sensory fibers that carry information from temperature receptors are smaller but still myelinated, whereas many motor nerves leading to internal organs (where the response time is not so critical) are not myelinated and conduct at less than 1 meter/sec (see Table 7-2).

Compound Action Potential in Nerve Trunks

The structures referred to in gross anatomy as "nerves" are actually bundles containing many axons surrounded and supported by glia (Figure 7-24). Electrodes placed on or near these nerve trunks can record electrical signals that represent the summation of a large number of individual action potentials. If all of the axons are stimulated simultaneously by application of electrical stimulation at some point on the nerve, the composite response of the axons is a **compound action potential.** The axons in nerves are typically distributed into several size groups (Figures 7-24 and 7-25), each having a different conduction velocity. If a pair of recording electrodes is placed some distance away from the

Nerve	Diameter	Myelin	Conduction velocity
Squid giant axon	500 μm	No	25 m/sec
Large motor nerve to leg muscle	20 μm	Yes	120 m/sec
Nerve from skin pressure receptor	10 μm	Yes	50 m/sec
Nerve from temperature receptor in skin	5 μm	Yes	20 m/sec
Nerve from pain receptor	1 μm	No	2 m/sec

TABLE 7-2 *Conduction Velocities of Nerve Fibers*

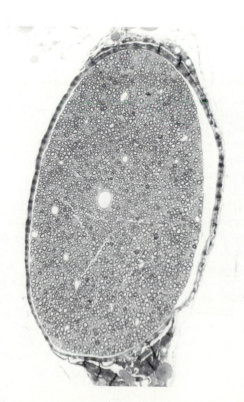

FIGURE 7-24
An electron micrograph of a cross section of a spinal nerve. Each small circle is a myelinated axon. Note that the axons fall into several size classes.

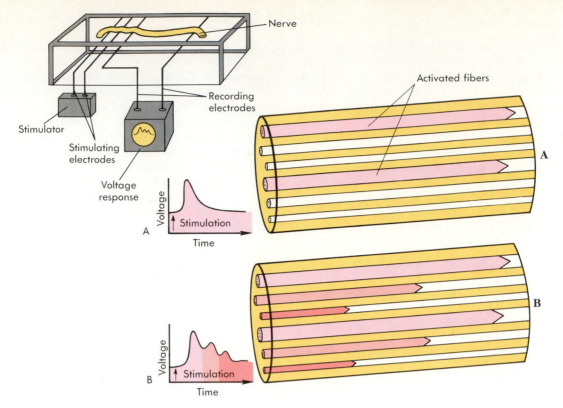

FIGURE 7-25

Stimulation of a whole nerve causes a compound action potential.

A   Small stimulating currents activate only the largest class of axons. These have similar conduction velocities and produce a single peak at the recording electrode.

B   Larger stimulating currents can activate all classes of axons in the nerve. The different conduction velocity of the large, intermediate, and small axons results in multiple peaks in the compound action potential as the action potential of the different axon classes arrive at the recording electrode at different times after the stimulus.

point of stimulation, action potentials traveling in the different size groups will arrive at the recording electrodes at different times after the stimulus has been delivered. The compound action potential thus consists of several bumps occurring over time, each bump corresponding to a family of axons of about the same diameter (see Figure 7-25).

Fiber groups in a nerve trunk have different apparent thresholds; larger fibers are depolarized to threshold by smaller currents. This does not represent a difference in their thresholds in response to natural stimulation, but is simply a result of the way the stimulating current spreads in the nerve trunk. As a result, the compound action potential is not all-or-none, even though the action potentials of the individual fibers are such. As the stimulus current is increased, more and more fibers are brought to threshold. Each contributes its individual increment to the composite response.

The compound action potential often seen in laboratory exercises is produced when an isolated nerve is stimulated. When a nerve is stimulated in this manner, the action potential is initiated in all the axons that reach threshold at the point of stimulation. The action potentials will propagate in both

directions from the point of stimulation. This is an unphysiological situation because the normal initiation of an action potential results in unidirectional flow of the action potential. Thus there is a normal direction for action potential conduction that is determined by the source of stimulation—for example, in the nerve endings of sensory receptors.

In the body, all fibers in a nerve trunk are never activated simultaneously. Some fibers are carrying sensory information into the nervous system, while others are delivering commands to the muscles and glands. Each axon constitutes a separate line of communication, and different axons carry different types of information in a code based on the frequency of action potentials. The way this information is sorted out by the nervous system is the subject of the next chapter.

> 1 What two factors determine the velocity with which a neuron or a muscle cell conducts an action potential?
> 2 What is the immediate source of the energy that is used to regenerate the electrical signal in an action potential?
> 3 What is a compound action potential? Under what circumstances might one be recorded?

The Consequences of Nerve Demyelination

Nerve fibers conduct reliably even when the firing rate is as high as several hundred action potentials per second. In a myelinated nerve axon the amount of current generated during an impulse at one node is five to seven times the amount needed to excite the next node; this is the safety factor.

The situation is quite different in diseases that result in loss of myelin. Myelin loss, or demyelination, occurs in diabetes and can also be a consequence of alcoholism. However the most common demyelinating disease is multiple sclerosis. This disease is presently understood to be an autoimmune disorder in which an immune attack is mounted against myelinated parts of the central nervous system. Depending on the site of the demyelination, symptoms may involve muscle weakness, paralysis, or abnormal sensations in any part of the body; the symptoms may come and go. The more extensive the myelin loss and the more nerve tracts that are affected, the worse the symptoms become.

In myelinated axons, the voltage-dependent Na^+ channels are confined to the membrane of the nodes of Ranvier, held in place by the cytoskeleton. Myelin loss increases the amount of current that leaks out of the fiber (Figure 7-A). The first effect is an increase in the time needed to bring the next node to its threshold, so that conduction slows. As demyelination progresses, there may not be enough current to bring the next node to threshold, and conduction is blocked. Conduction block in a group of motor axons results in paralysis of the muscles they control; in sensory nerves the effect is a loss of sensation. Even slowing of conduction has profound effects, because information processing within the central nervous system often depends on the timing of incoming signals.

The effects of demyelination reveal the crucial effects of transmission success and timing of signal flow within the nervous system. There is hope that a better understanding of factors that normally prevent the immune system from attacking normal body tissues can be applied to produce a therapy for those suffering from demyelinating diseases.

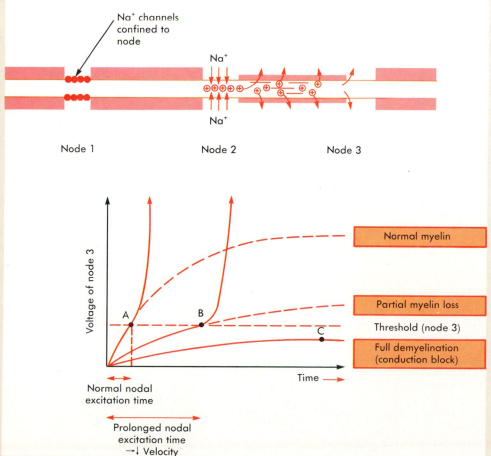

FIGURE 7-A
The upper part of the figure shows the Na^+ channels localized at the nodes of Ranvier. Activating channels at one node normally results in sufficient current flow down the core of the axon to bring the next node rapidly to threshold. Demyelination is occurring between nodes 2 and 3. The lower part of the figure shows the consequences for node 3 of progressing demyelination. Before demyelination starts *(A)*, the node is rapidly brought to threshold by currents that reach it from node 2. As demyelination progresses, more and more current is lost through the myelin, and node 3 reaches threshold more slowly; the conduction velocity is reduced. At last, the current leakage is so great that node 3 is not brought to threshold at all; the demyelination has blocked conduction, and the axon no longer functions.

SUMMARY

1. **Action potentials** differ from receptor potentials and synaptic potentials in that they have a **threshold,** are all-or-none, cannot summate, are propagated regeneratively across excitable membranes, and leave the membrane temporarily refractory to a second stimulus.

2. Action potentials are explained on the basis of the behavior of **voltage-gated ion channels** in the excitable membranes: Na^+ channels are activated rapidly by depolarization and inactivate spontaneously; K^+ channels are activated more slowly by depolarization and do not inactivate, but rather close in response to repolarization.

3. An action potential is initiated when a depolarizing stimulus opens enough voltage-gated Na^+ channels that the Na^+ conductance exceeds the K^+ conductance, setting up a positive-feedback cycle in which depolarization rapidly opens the remaining Na^+ channels.

4. Repolarization is the result of two factors:(1) spontaneous inactivation of Na^+ channels, and (2) the opening of K^+ channels. These two factors make the membrane **absolutely refractory** to a second stimulus for a few milliseconds after an action po-

tential is initiated. When a sufficient number of inactivated Na^+ channels have reverted to the closed but potentially responsive state, the membrane is still **relatively refractory** until the K^+ channels have closed.

5. The velocity of conduction of an action potential down an axon or muscle cell is determined by the diameter of the fiber and is also greatly increased by myelination in the case of axons. In myelinated axons, conduction is **saltatory:** a regenerative action potential occurs at intervals along the fiber at nodes of Ranvier, and the signal covers the intervening distance rapidly in the form of a current between the depolarized node and the next polarized node.

6. In a nerve trunk containing many axons, stimulation evokes many simultaneous action potentials that propagate in both directions from the site of stimulation and whose currents can be recorded as a **compound action potential.** Since different fibers conduct at different rates, the compound action potential separates with distance into several peaks that correspond to the action potentials of axons of different size populations in the nerve trunk.

1. The venoms of most scorpions and many insecticides act by slowing the inactivation of Na$^+$ channels. What effects would such venoms have on the duration of an action potential? Would its amplitude be changed significantly? What would happen to the refractory period?

2. How does decremental conduction differ from nondecremental conduction?

3. Relate the membrane potential changes that occur during an action potential to the ion permeability changes that produce them.

4. What determines the height of the action potential? Why is it that a small stimulus does not produce a small action potential and a large one produce a large action potential?

5. What are the negative feedback forces that return the membrane potential to the resting level after a hyperpolarizing shift?

6. What is Na$^+$ inactivation? What removes the inactivation and what state of the Na$^+$ channel results?

7. What effect does myelin have? What are nodes of Ranvier and what function is associated with them?

8. Why does a compound action potential appear to violate the all-or-none law?

9. Local anesthetics used by dentists to deaden pain are blockers of Na$^+$ channels. If half of the Na$^+$ channels in a sensory nerve were blocked, how would this affect the threshold? Could it affect the ability of the nerve to propagate an action potential?

● SUGGESTED READING

COREY, D.P.: Patch clamp: current excitement in membrane physiology, *Neuroscience Commentaries*, volume 1, p. 99, 1983. A brief, lucid description of a technique that allows the examination of the electrical properties of previously inaccessible cells and allows one to look at the opening and closing of single-ion transport pathways in cell membranes.

EDWARDS, D.D.: Still stalking MS, *Science News*, volume 132, October 10, 1987, p. 234. Article summarizes progress in understanding the cause of multiple sclerosis and reviews the efficacy of several current therapies.

HILLE, B.: *Ionic Channels of Excitable Membranes*, Sinauer Associates, Sunderland, Mass., 1984. The only comprehensive textbook in the field. While there are many parts that are quantitative, several chapters provide a descriptive discussion of the varieties of ion channels present in cell membranes and their physiological function.

HILLE, B., and D.M. FAMBROUGH, editors: *Proteins of Excitable Membranes*, John Wiley and Sons, Somerset, N.J., 1987. Relatively difficult monograph focusing on the isolation, sequencing, and three-dimensional structure of membrane proteins. Provides a number of examples of how molecular genetics can be applied to this field.

JUNGE, D.: *Nerve and Muscle Excitation*, ed. 2, Sinauer Associates, Sunderland, Mass., 1981. Readable text that presents a complementary discussion of the origin of nerve and muscle action potentials, propagation, and excitation-contraction coupling.

KUFFLER, S.W., J.G. NICHOLLS, and A.R. MARTIN: *From Neuron to Brain*, ed. 2, Sinauer Associates, Sunderland, Mass., 1984. A medium-level general neurobiology text.

MORELL, P., and W.T. NORTON: Myelin, *Scientific American*, May 1980, p. 88. Still current; describes the composition of myelin, its synthesis, and how myelination increases the conduction velocity.

CHAPTER 8

Initiation, Transmission, and Integration of Neural Signals

On completing this chapter you will be able to:

- Understand how sensory information reaches the central nervous system along labelled lines.
- Explain how stimulus intensity is coded.
- Understand how the activation of stimulus-sensitive ion channels produces generator (receptor) potentials and be able to compare their properties with those of action potentials.
- Summarize the mechanisms of receptor adaptation.
- Understand the sequence of events occurring in synaptic transmission.
- Distinguish between excitatory postsynaptic potentials and inhibitory postsynaptic potentials.
- Understand how presynaptic inhibition can modulate transmitter release.
- Use your knowledge of synaptic transmission to design simple logical circuits.
- Understand how neuromodulators can produce long-term changes in synaptic efficacy.

*T*he nervous system detects external and internal physical and chemical stimuli and initiates muscular and glandular responses. Since action potentials are the only signals that nerve fibers are able to conduct over long distances, all of the many types of information relayed within the nervous system must be translated into a neural code based on action potentials. In the case of sensory receptors, this translation is called sensory transduction.

Coded messages arriving in the central nervous system initiate appropriate responses in pathways leading to effector organs, so sensory transduction is the initial event in many homeostatic processes. How the central nervous system uses sensory information to determine appropriate responses is one of the most interesting problems in modern neurobiology. Two processes, both involving communication between cells, must occur: (1) action potentials in axons must be able to be transmitted to other neurons and to effector organs; and (2) neurons must be able to combine information from more than one source, so that their output depends on the weights of many different inputs. This second process is called neural integration and is thought to be the basis for information processing in the nervous system. For a single neuron, outputs take the form of either initiating or not initiating an action potential at any given instant. Many complex neural circuits in the central nervous system are involved in homeostatic regulatory processes that go on at a subconscious level. Others function in conscious choice, such as how to respond to a threatening stimulus or whether to order pancakes or waffles for breakfast.

Processes of communication occur at specialized junctions between cells called synapses. At a synapse, all-or-none action potentials in one neuron are transduced into graded electrical events, called postsynaptic potentials, in a second neuron. In most cases in the central nervous system, the activity of a neuron depends on the sum of postsynaptic potentials coming from a large number of synaptic connections.

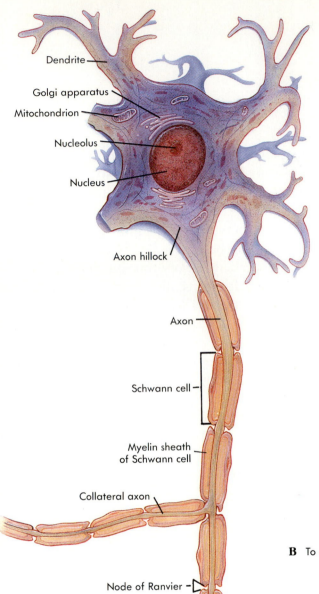

A

Dendrite

Golgi apparatus

Mitochondrion

Nucleolus

Nucleus

Axon hillock

Axon

Schwann cell

Myelin sheath
of Schwann cell

Collateral axon

Node of Ranvier

Synaptic terminals

NEURON STRUCTURE AND INFORMATION TRANSFER

Input and Output Segments of Neurons

All neurons must be able to accept information from other nerve cells or from a form of energy in the environment and communicate this information to other nerve cells or effector cells. These processes take place at separate specialized input and output segments of the neurons. In neurons that relay information over considerable distance, an axon must be part of the output segment.

The anatomy of neurons differs with anatomical location and function, but each neuron has a cell body, or soma, that provides metabolic support for other parts of the neuron. The neuron shown in Figure 8-1, *A* is typical of the general structure of neurons that process information within the central nervous system, including **motor neurons** and some types of **interneurons**. In such cells the **cell body (soma, perikaryon)** is covered with fine branching processes called **dendrites.** The dendrites and cell body receive many synapses of other neurons, and thus these parts comprise the **input segment** in these cells. The **output segment** takes the form of a single cablelike process, the axon, which arises from an **axon hillock** on the surface of the cell body and carries action potentials to axon terminals, where activity can be communicated to other cells by **synaptic transmission.**

Not all neurons have the same morphology. For example, in afferent **sensory neurons,** the input segment does not include the cell body (Figure 8-1, *B*). Cell bodies of afferent neurons are located either within the central nervous system or, in the case of

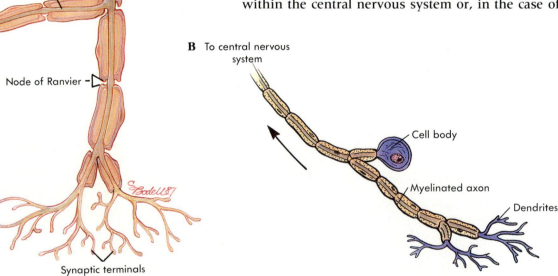

B To central nervous system

Cell body

Myelinated axon

Dendrites

FIGURE 8-1

A A generalized neuron. The input segment of the neuron consists of the dendrites and cell body. The output segment is composed of the axon and synaptic terminals. Action potentials are initiated at the axon hillock, where the axon sprouts from the cell body, and are conducted to the synaptic terminals. The cell body contains the usual intracellular organelles.

B Most sensory neurons are unipolar with a single axon extending from the cell body and then branching in two directions.

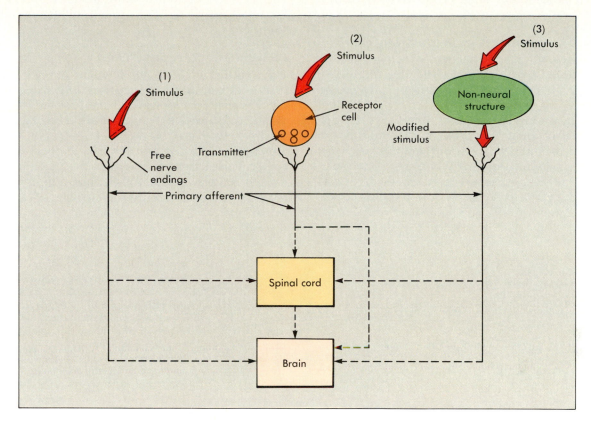

FIGURE 8-2

A diagram illustrating the three ways that primary afferent nerve terminals can be activated by an environmental stimulus *(red arrows)*. In some sensory systems, stimuli directly activate the free nerve endings of primary afferents *(1)*. In the most complex sensory systems, receptor cells release a chemical transmitter to activate the primary afferent nerve endings *(2)*. In others, stimuli are modified by surrounding, generally nonneural tissues, and the free nerve endings sense this modified stimulus *(3)*. Depending on the specific stimulus, afferent fibers may synapse in the spinal cord or go directly to the brain.

spinal afferents, in the **dorsal root ganglia** immediately adjacent to the spinal cord. Cell bodies in the dorsal root ganglia are electrically inexcitable and are usually separated from the axon by a short process, so they play no direct part in information transfer. Instead, the input segment is connected to an axon, which passes into the central nervous system. Once within the central nervous system, the axon branches to form synapses that communicate with other neurons.

Even though there are morphological differences, some general statements about the properties of input and output segments can be made. The input segment typically possesses many ion channels that are able to respond to the type of stimulation that the cell is designed to receive (for instance, odor molecules or pressure) but possesses few or no voltage-sensitive channels. It therefore has quite a high threshold and usually does not produce an action potential. Instead, signals in the input segment take the form of **local currents** (see Chapter 7, p. 155). In this form, the signals are conducted to a

site of low threshold near the beginning of the output segment. In neurons the axon hillock (see Figure 8-1, *A*) has the lowest threshold. Synaptic currents summate at the axon hillock and, if the currents result in sufficient depolarization, bring the membrane to threshold and generate one or more action potentials.

Sensory Neurons

Figure 8-2 shows several ways afferent neurons can carry sensory information into the central nervous system. In some sensory systems, such as touch and pressure sensors, the afferent neuron is directly responsive to the stimulus (shown as pathway 1 in Figure 8-2). This is the simplest arrangement. In the visual and auditory pathways, in the taste or gustatory system, and in the vestibular system of the ear (which mediates balance and equilibrium), specialized epithelial cells are the primary receptors for the stimulus (Figure 8-2, pathway 2). The sensory epithelial cells do not generate action potentials, but their permeability response results in release of a chemical message that changes the activity of the afferent neurons.

Nonneural Components in Sensory Systems

In some cases, the responsiveness of sensory receptors is modified by nonneural structures (Figure 8-2, pathway 3). For example, **Pacinian corpuscles** (Figure 8-3) are vibration-sensitive receptors found in the deeper skin layers and other sites in the body. In the Pacinian corpuscle, nerve endings are surrounded by multiple layers of connective tissue that can slip over one another when steady pressure is applied, relieving the pressure on the receptor endings. The effect is to render the receptor sensitive only to pressure changes, not steady pressures.

This is one example of how associated structures can tune sensory receptors to selected features of the stimulus.

Concept of a Receptive Field

The location of the input segment of each afferent neuron determines its **receptive field,** which is defined as that area in which an appropriate stimulus of adequate intensity can result in a change in permeability in that neuron (Figure 8-4). For instance, the endings of a receptor for skin temperature have a specific distribution in the skin that corresponds to their receptive field, and the position of a photoreceptor cell in the retina determines its receptive field.

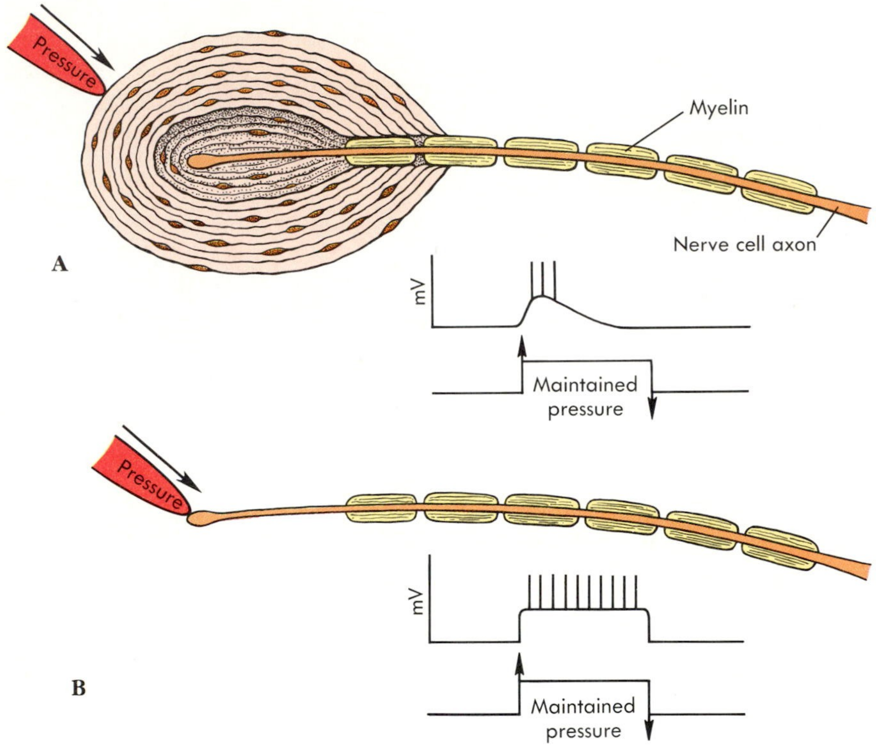

FIGURE 8-3

A Illustration of a Pacinian corpuscle. Primary afferent nerve endings are surrounded by a capsule consisting of onionlike layers of tissue. In the intact corpuscle, maintained pressure produces a transient response consisting of a burst of action potentials.

B Removal of the capsule alters the response properties of the cell so that maintained pressure is reflected by a constant level of action potential firing.

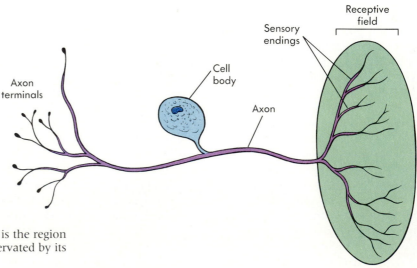

FIGURE 8-4

The receptive field of a primary sensory neuron is the region of a sensory surface (for example, the skin) innervated by its receptive endings.

Afferent Neurons as Labelled Lines

Three types of information about sensory signals must be supplied to the central nervous system: What kind? Where? and How much? The answers to the first two questions are provided by the structure of the nervous system itself. During development of the nervous system, each afferent neuron sends an axon out to establish a receptive field in an appropriate place in the body or on its surface. The peripheral end of the sensory neuron develops in such a way that it is selectively responsive to a single type, or **modality,** of stimuli, such as pressure, vibration, sound, or taste. At the same time its central processes establish axon terminals on neurons in a specific brain or spinal region that is devoted to processing information about that modality (Figure 8-5). As a consequence of these connections, **labelled lines** are established between central regions specialized for processing different modalities and the sensory inputs from receptors specific for that modality. Thus light does not produce a sensation of sweetness in the mouth, and a light directed into the ear does not give either a visual or an auditory perception. The labelled lines between particular categories of receptors and the brain regions that analyze that type of sensory input provide the answer to the question "What kind?"

Afferent neurons form connections within each specialized region of the central nervous system. Afferent neurons form orderly connections as determined by the location of their receptive fields, so that the central regions literally become living maps of the location of receptive fields. Not only is the information about body surface and in-

terior sites mapped, but visual and auditory spaces are also mapped within the sensory processing centers of the brain. The nature of the maps differs for the different sensory systems (see Figure 9-17), but each provides the answer to the question "Where?" The answer to the question "How much?" is provided by response characteristics of the receptors, as described in the following paragraphs.

PHYSIOLOGICAL PROPERTIES OF SENSORY RECEPTORS
Sensory Receptors as Transducers

Sensory receptors transform external forms of energy into a neural code and are thus one example of **transducers** (see Chapter 5, p. 86). Stimuli fall into categories of sensory perception, the modalities, according to their physical nature (see Figure 8-5). Cutaneous sensation has several submodalities, including pressure, vibration, pain, and temperature. Internal receptors fall into two catagories. Proprioceptors provide information about the position of body segments relative to one another (joint angles, for example) and the position of the body in space. Interoceptive systems measure internal bodily events such as oxygen and carbon dioxide levels, osmolarity and acidity of body fluids, limb position, and muscle tension.

An important property of sensory receptors is an exquisite sensitivity to one particular stimulus modality. The rods and cones of the visual system can detect single photons of light; the auditory system can detect sounds that displace the cilia of auditory hair cells by a distance equivalent to the di-

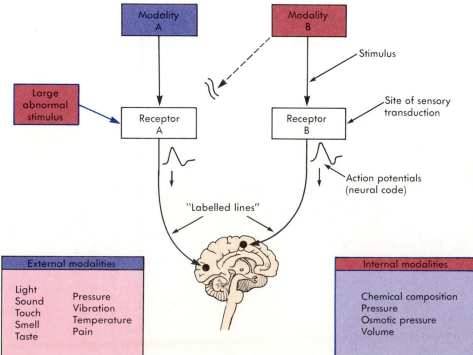

FIGURE 8-5
The concepts of stimulus modality and labelled lines. Sensory receptors have low thresholds for specific modalities and are connected to the central nervous system in fixed ways. Abnormally large stimuli of other modalities can activate sensory receptors, and in such cases, the sensation experienced is that which normally would be appropriate for the given receptor. For example, pressing on the lids of closed eyes gives rise to sensations of light, but the mechanical energy that must be applied to the receptors to excite them is far greater than the light energy detected by visual receptors in normal function.

ameter of a single atom; and the olfactory system can detect single odor molecules. Despite their high specificity, sensory receptors can sometimes be excited by inappropriate but intense stimuli (see Figure 8-5).

The remarkable sensitivity to some stimuli is obtained by limiting the range of effective stimuli. For example, the human eye is sensitive to only a tiny fraction of the spectrum of electromagnetic energy that includes, in addition to visible light, infrared and ultraviolet light, microwaves, radio waves, x-rays, and gamma rays. The human ear is quite insensitive to sound energy of frequencies below about 10 cycles/sec and above about 20,000 cycles/sec, although other mammals are able to respond to other portions of the sound spectrum. The selectivity of our sensory receptors limits to a large extent our experience of the world.

Filtering Sensory Signals

Only a small part of the information passing along afferent pathways is destined to form part of conscious experience. For example, information from receptors that sense blood pressure is simply not analyzed at the conscious level. More importantly, one of the functions of the central nervous system is to filter out information that is distracting, unimportant, or lacking in novelty. The parts of the brain involved in conscious sensation attend selectively to some sensations, especially those that are novel or those that experience has shown to be important. This ability to focus attention is important for higher brain functions such as learning.

> 1 What is a stimulus modality? What are some examples of stimulus modalities?
> 2 What is the relationship between receptor specificity and stimulus modality?
> 3 What is a labelled line?

GENERATION OF THE NEURAL CODE
Receptor Potential Production

Sensory receptors encode information through electrical events that occur in sensory receptor endings. Cells respond to stimuli because they possess **stimulus-sensitive receptor proteins** in their cell membranes. The receptor proteins are either ion

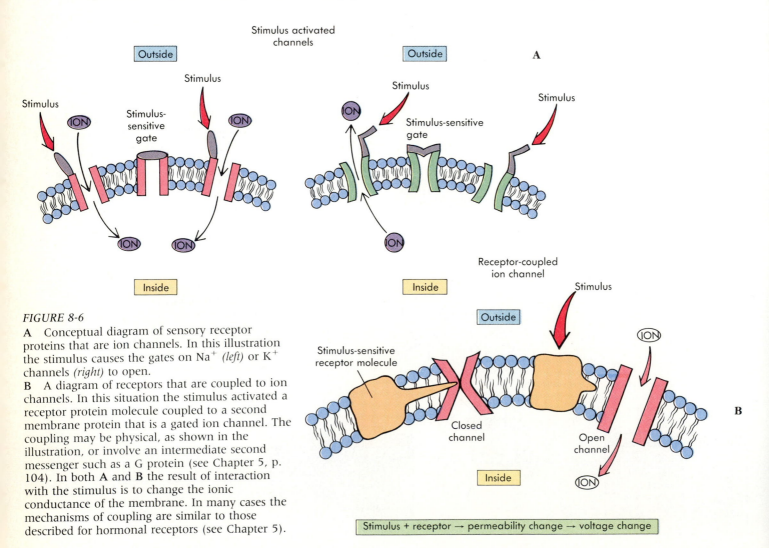

FIGURE 8-6

A Conceptual diagram of sensory receptor proteins that are ion channels. In this illustration the stimulus causes the gates on Na$^+$ *(left)* or K$^+$ channels *(right)* to open.

B A diagram of receptors that are coupled to ion channels. In this situation the stimulus activated a receptor protein molecule coupled to a second membrane protein that is a gated ion channel. The coupling may be physical, as shown in the illustration, or involve an intermediate second messenger such as a G protein (see Chapter 5, p. 104). In both **A** and **B** the result of interaction with the stimulus is to change the ionic conductance of the membrane. In many cases the mechanisms of coupling are similar to those described for hormonal receptors (see Chapter 5).

Stimulus + receptor → permeability change → voltage change

NERVE AND MUSCLE

channels themselves (selective for Na^+ or K^+) (Figure 8-6, *A*) or are coupled to ion channels (as in Figure 8-6, *B*). Note that the term "receptor" is now used in a different context than in preceding paragraphs, where it referred to an entire cell. Applying the appropriate stimulus to the receptor proteins can result in either opening or closing of channels, depending on the sensory system involved.

Opening or closing ion channels results in permeability changes that can cause membrane voltage changes (Figure 8-7). The voltage changes that occur in sensory receptors upon stimulation are referred to as **generator potentials** or **receptor potentials**. In most sensory systems, the stimulus-gated channels in sensory receptors allow both Na^+ and K^+ to pass through them. Since the resting potential of the endings is normally fairly close to the K^+ equilibrium potential, the driving force for Na^+ is much larger than that for K^+ (Figure 7-7, *A*). The dominant effect will be an influx of Na^+, resulting in depolarization. In the visual system, receptor potentials are hyperpolarizations that result from a decrease in Na^+ influx (see Chapter 10, p. 273).

Initiation of Action Potentials in Receptors

The nerve endings themselves are electrically inexcitable, but a difference between the resting potential of the nerve axon and the depolarized or hyperpolarized nerve endings results in a flow of current between endings and axon (Figure 8-8, recording site *A*). Chapter 7 (see p. 152) described how a steady electrical current produced a series of action potentials. The behavior of sensory neurons is similar, except that a receptor with stimulus-sensitive ion channels has been substituted for the electrical stimulator. If the permeability change is depolarizing, it may bring the axon hillock to threshold and initiate a single action potential or a train of action potentials (Figure 8-8, *B*).

Encoding Stimuli

The receptor potential gives rise to a decremental current that fades rapidly with distance and is at best adequate to depolarize only the most distal parts of the axon. However, the fact that it is graded to the stimulus intensity makes it possible for sensory receptors to encode information about intensity into their signals. Intensity is reflected by the level of the membrane potential change. For example, larger depolarizations can result from two types of summation: (1) **spatial summation** and (2) **temporal summation**.

Spatial summation involves the addition of inputs arising from more than one site on the receptor

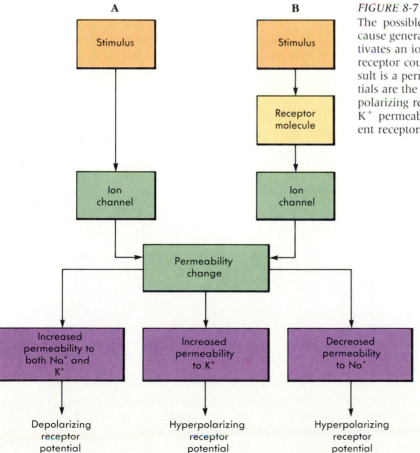

FIGURE 8-7
The possible sequences of events by which stimuli may cause generator (receptor) potentials. In **A** the stimulus activates an ion channel, while in **B** the stimulus activates a receptor coupled to an ion channel. In both cases the result is a permeability change. Depolarizing receptor potentials are the result of increases in Na^+ permeability; hyperpolarizing receptor potentials are the result of increases in K^+ permeability or decreases in Na^+ permeability. Different receptors utilize different pathways in this scheme.

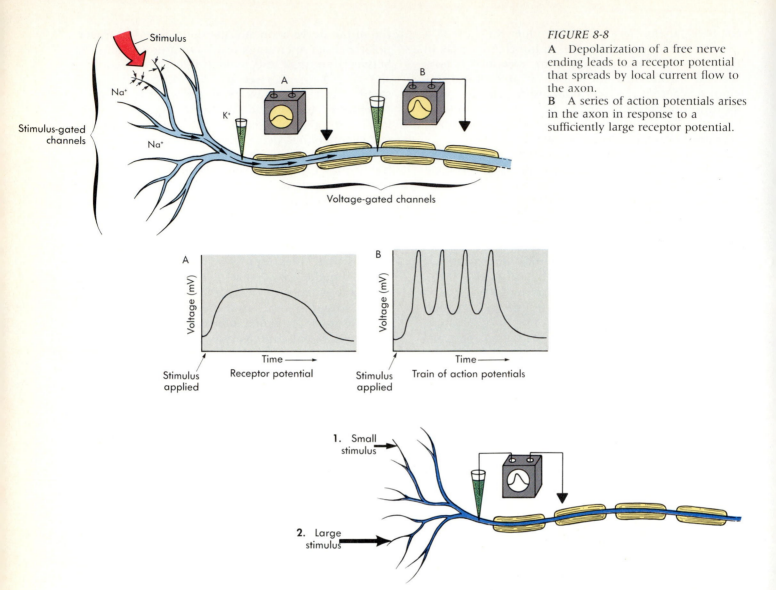

FIGURE 8-8
A Depolarization of a free nerve ending leads to a receptor potential that spreads by local current flow to the axon.
B A series of action potentials arises in the axon in response to a sufficiently large receptor potential.

FIGURE 8-9
Spatial summation occurs in the input segment of a sensory receptor. Two simultaneous stimuli to adjacent regions of the receptor result in a larger receptor potential than either one by itself.

ending. Only sensory cells with branching or distributed endings are able to display this type of summation. The typical example of a receptor that exhibits spatial summation is a cutaneous (skin) touch receptor. Inputs can add in "space" on the surface of the receptor membrane and together determine the overall receptor potential. Stimulation of a larger portion of the receptor's receptive field (Figure 8-9) results in correspondingly higher frequencies of action potentials in the axon.

Temporal summation is the result of rapidly repeated stimulation of the same input region (Figure 8-10). As with spatial summation, the effect is an increase in the magnitude of the receptor potential.

Linear and Logarithmic Receptors
The frequency of action potentials is always proportional to the amplitude of the receptor potential. The relationship between the stimulus intensity and the amplitude of the resulting receptor potential

FIGURE 8-10
Temporal summation in the input segment of a sensory receptor. The receptor potential generated at a single input site stimulated once is compared with the receptor potential resulting from repeated stimulation.

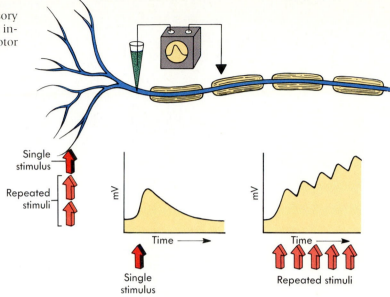

Temporal summation

varies with the nature of the stimulus that a sensory receptor detects. There are two basic patterns. In one pattern (Figure 8-11, *A*), the receptor potential is directly proportional to the stimulus intensity (a **linear response pattern**). In the other pattern, the receptor is quite sensitive to differences in stimulus intensity when the stimuli are weak (compare stimuli 1 and 2 in Figure 8-11, *B*) but becomes progressively less sensitive to stimulus intensity as the stimulus intensity is increased (a **logarithmic response pattern**).

The logarithmic response pattern is typically seen in receptors that must register the intensity of stimuli that vary over many orders of magnitude, such as light and sound. The refractory period limits the frequency of action potentials to no more than several hundred per second. A linear response would take the output of the receptor to its maximum value if the stimulus intensity were increased by about two orders of magnitude. The logarithmic response pattern of visual and auditory receptors results in action potential frequencies that change by only sevenfold in response to a 10 millionfold change in stimulus intensity (the difference between a barely audible whisper and a rock concert).

Tonic Sensory Receptors

In sensory physiology, adaptation refers to the way a particular receptor responds to a long-lasting stimulus. Sensory receptors differ in their adaptive responses, falling into three general classes. The first class continues to discharge at the same rate no matter how long the stimulus lasts — these are called **tonic** or **nonadapting.** Tonic receptors monitor such things as pain, temperature, limb position,

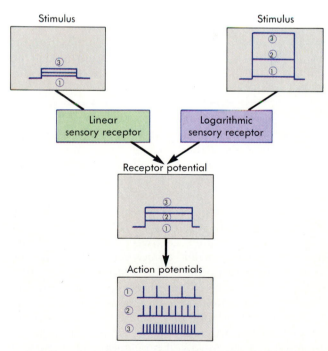

FIGURE 8-11

Intensity coding by sensory receptors. In systems where the intensity does not vary over a wide range, the relationship between stimulus intensity and receptor potential is linear. For systems that must represent a wide range of stimulus intensity, a logarithmic relationship is seen. In both cases, action potential frequency is directly proportional to receptor potential amplitude.

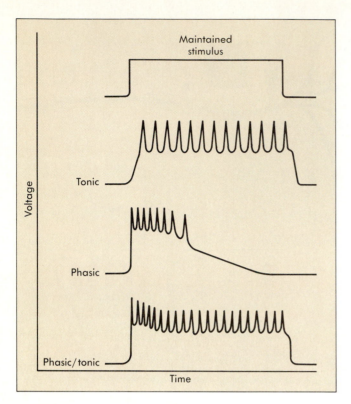

FIGURE 8-12
Response of a tonic receptor, a phasic receptor, and a mixed (phasic/tonic) receptor to a maintained stimulus.

blood pressure, and blood oxygen content. In these receptors a receptor potential continues as long as the stimulus is present, and the magnitude of the receptor potential is proportional to the stimulus intensity (Figure 8-12).

Phasic Sensory Receptors

The second category of receptors exhibits a burst of action potentials at the onset of stimulation, but as the stimulus continues, the rate of action potentials diminishes, and ultimately no more action potentials occur. These are termed **phasic** or **rapidly adapting.** There are several possible mechanisms of adaptation, and these generally differ from one sensory system to another. In the example of Pacinian corpuscles given earlier, an accessory structure, the capsule, was responsible for adaptation. In other systems the nature of the processes that open and close stimulus-sensitive channels seems to be responsible for adaptation. Clearly, the ability to adapt is so important to sensory systems that mechanisms for it have evolved repeatedly.

The response of phasic receptors generally reflects the rate of change of a stimulus rather than its final magnitude (see Figure 8-12). In these receptors the receptor potential disappears after a time, even if the stimulus continues at a steady level. Phasic receptors are specialized to register novel stimuli or changes in stimulus intensity and are found in

systems where change is especially significant. Examples are phasic receptors in skin, which respond only to initial contacts, and muscle receptors, which detect the speed of muscle length changes.

Mixed Receptors

The third category incorporates features of both phasic and tonic receptors. The receptor potential in these phasic/tonic receptors, or **mixed receptors,** displays an initial transient response, which is proportional to the rate of change of the stimulus (see Figure 8-12), followed by a steady-state response of smaller magnitude, which is proportional to the intensity of the stimulus. There is therefore an initial burst of action potentials at a rate proportional to the rate of onset of the stimulus, followed by a decline in the frequency of action potentials to a steady rate, proportional to the final magnitude of the stimulus.

Modified Epithelial Cells as Receptors

The sequence of events in sensory transduction is somewhat more complicated in those sensory systems in which the receptors are not neurons but are instead modified epithelial cells. This is the case in the auditory, vestibular, visual, and gustatory systems. However, these systems all obey the general principles of sensory transduction that have been discussed.

In neurons that serve as both receptors and afferents, such as the Pacinian corpuscle, the initial event of sensory transduction, the receptor potential, occurs in an electrically inexcitable part of the cell and is graded to stimulus intensity. In sensory epithelia the primary receptors are not nerve cells and cannot experience action potentials. They do undergo permeability changes as a consequence of stimulation. The receptor cells communicate with sensory afferents by releasing chemicals onto sensory afferents (Figure 8-13) or in a few cases by being directly coupled to sensory afferent neurons by gap junctions.

The epithelial receptors can be thought of as whole-cell input segments for their sensory pathways. The permeability changes that cause receptor potentials in epithelial cells vary from one sensory system to another, and in some cases the receptor potential may be hyperpolarizing rather than depolarizing. For example, in rod and cone photoreceptors, photons cause a decrease in Na^+ permeability that results in hyperpolarization.

1 In what two senses is the term "receptor" used?
2 What is the effect of receptor adaptation on transmission of information about stimulus intensity?
3 What stimulus modalities require a logarithmic response of receptors? Why?

NERVE AND MUSCLE

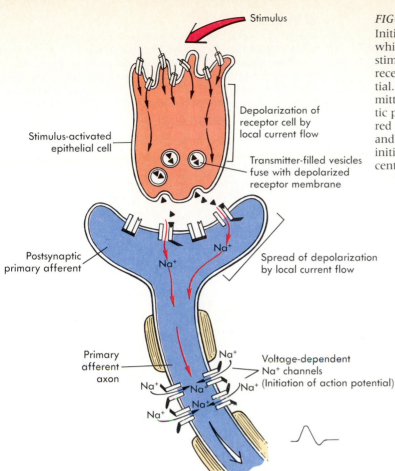

Stimulus

Stimulus-activated
epithelial cell

Depolarization of
receptor cell by
local current flow

Transmitter-filled vesicles
fuse with depolarized
receptor membrane

Postsynaptic
primary afferent

Na⁺
Na⁺

Spread of depolarization
by local current flow

Primary
afferent
axon

Na⁺

Na⁺
Na⁺

Voltage-dependent
Na⁺ channels
(Initiation of action potential)

Na⁺

Na⁺

Na⁺

FIGURE 8-13

Initiation of action potentials in a sensory system in which the input segment is a modified epithelial cell. A stimulus activates the epithelial cell, which experiences a receptor potential but cannot generate an action potential. The receptor potential modulates the rate of transmitter release. The transmitter depolarizes the postsynaptic primary afferent ending. A local current, shown by the red arrows, flows between the depolarized nerve ending and the axon (containing voltage-gated Na⁺ channels), initiating action potentials in the axon that leads to the central nervous system.

THE PHYSIOLOGY OF SYNAPTIC TRANSMISSION
Electrical Versus Chemical Synapses

Communication between cells of the nervous system and between neurons and effectors occurs at cell-to-cell junctions called **synapses**. **Electrical synapses** (Figure 8-14) are sites at which gap junctions (see Figure 4-17) form direct electrical connections between cells, so that the arrival of an action potential at the synapse results in flow of depolarizing current between the cells. This type of synapse allows information to be transferred in both directions across the synapse. Electrical synapses are fairly common in invertebrates, but their potential importance has received relatively little attention because although they are present in some areas of the mammalian central nervous system, they are relatively difficult to isolate and study.

In the **chemical synapse** (Figure 8-15), there is a narrow intervening **synaptic cleft** about 30 to 50 nM wide between the membranes of the two cells. This cleft does not allow electrical currents to flow from cell to cell; instead, chemical messengers are released as a result of activity in one of the cells. They diffuse across the synaptic cleft and affect the electrical activity of the other cell. This mechanism of information transfer is called **chemical synaptic transmission** (Figure 8-16). Information crosses chemical synapses in one direction only, from the

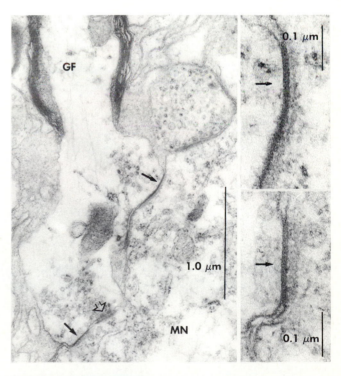

FIGURE 8-14

An electron micrograph of an electrical synapse made by a giant fiber on a motor neuron in a hatchetfish. *Left*, A giant fiber *(GF)* loses its myelin sheath near the top of the figure and forms gap junctions, the substrate of electrical transmission, with a motor neuron *(MN)* near the center and at the lower left *(solid arrows)*. *Right*, Enlargement of two gap junctions.

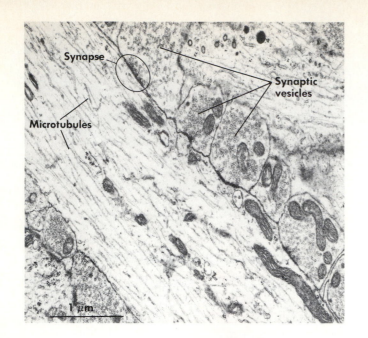

Synapse

Synaptic vesicles

Microtubules

1 μm

FIGURE 8-15
Example of several chemical synapses onto a single postsynaptic cell. Note that the synaptic terminals are full of vesicles containing transmitter chemical, while no vesicles are observed on the postsynaptic side of the membrane. Densely staining material is present on the interior surface of both presynaptic and postsynaptic membranes. The postsynaptic cell contains a number of microtubules.

FIGURE 8-16
The process of synaptic transmission. A presynaptic action potential *(1)* opens Ca^{++} channels in the nerve terminal *(2)*, facilitating the fusion of transmitter-containing vesicles with the nerve terminal membrane *(3)*. Exocytotic release of the transmitter *(4)* allows the neurotransmitter to diffuse across the synaptic cleft and bind to receptors *(5)*, activating ion channels in the postsynaptic membrane *(6)*. The resulting current flow spreads to adjacent areas of the postsynaptic membrane *(7)*. Some of the released transmitter molecules may be returned to the nerve terminal *(8)* or escape by diffusion *(9)*.

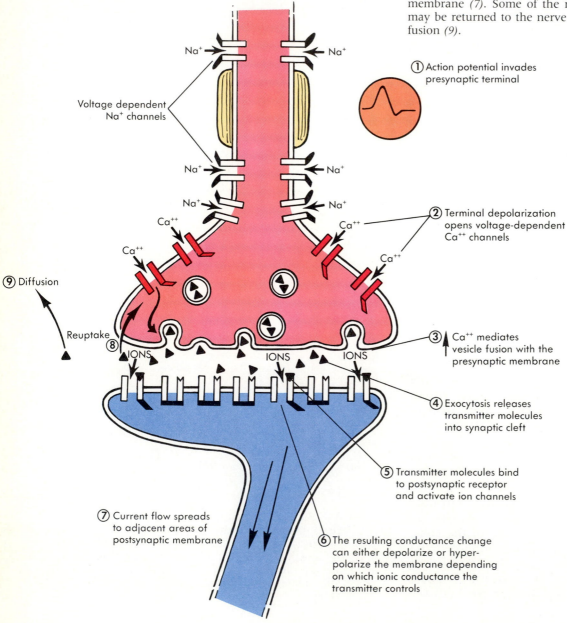

Voltage dependent Na^+ channels

Na^+ Na^+

Na^+ Na^+

Na^+ Na^+

Ca^{++} Ca^{++} Ca^{++} Ca^{++}

(9) Diffusion

Reuptake

(8)

IONS   IONS   IONS

(1) Action potential invades presynaptic terminal

(2) Terminal depolarization opens voltage-dependent Ca^{++} channels

(3) Ca^{++} mediates vesicle fusion with the presynaptic membrane

(4) Exocytosis releases transmitter molecules into synaptic cleft

(5) Transmitter molecules bind to postsynaptic receptor and activate ion channels

(7) Current flow spreads to adjacent areas of postsynaptic membrane

(6) The resulting conductance change can either depolarize or hyperpolarize the membrane depending on which ionic conductance the transmitter controls

NERVE AND MUSCLE

presynaptic cell, which sends the chemical message, to the **postsynaptic cell**, which receives it. Thus the terminal of the presynaptic cell is part of its output segment, while it functions as part of the input segment of the postsynaptic cell.

Differences in Chemical Synapses

The details of synaptic structure are different for the central nervous system, the junctions between neurons and skeletal muscle, and the junctions between neurons of the autonomic nervous system and their effectors. In the central nervous system, the amount of cell surface taken up by each synapse is relatively small, and transmitter chemical easily diffuses away from the synaptic cleft. In neuromuscular synapses on skeletal muscle (Figure 8-17), the axon terminal is relatively large and complex and is buried within folds of the postsynaptic membrane. The postsynaptic membrane underneath the synapse is also thrown into elaborate folds. The role of this structure is to prevent the loss of transmitter by diffusion and provide plenty of postsynaptic membrane for the transmitter to act upon.

In the autonomic nervous system, efferent nerve fibers penetrate the organs they innervate and release transmitter molecules from a series of enlargements, or **axon varicosities** (Figure 8-18). The distance between the axon varicosities and the cell membranes of surrounding effector cells is quite large compared with that of the narrow synaptic clefts of the central nervous system. The effect is that the transmitter released from each varicosity is distributed over a relatively wide area.

Synaptic Vesicles and Quantal Transmission

The presynaptic nerve terminal possesses a supply of intracellular **synaptic vesicles**, which are small spheres of membrane containing transmitter chemical (see Figures 8-15 and 8-16). The appearance of the vesicles and their diameter vary characteristically between synapses of different types; the vesicles may be dense or light when viewed with the electron microscope, and their diameter may range from 30 to 50 nM. For any given type of synapse, each vesicle contains about the same amount of transmitter. Since the smallest quantity of transmitter that can be released is that contained in a single vesicle, this amount is termed a **quantum** of transmitter.

Axoplasmic Transport

Nerve terminals may be located at a considerable distance from the cell body. For a sensory neuron whose receptive field is on a toe, this distance may be as much as a meter. Neurotransmitters, synaptic vesicles, ion channels, and enzymes must be synthesized in the nerve terminals, taken up by the terminals from extracellular fluid, or somehow carried down the axon. The latter is the case for many materials. The processes involved are known collectively as **axoplasmic transport.** The structural basis for at least one form of axoplasmic transport is provided by microtubules arranged lengthwise

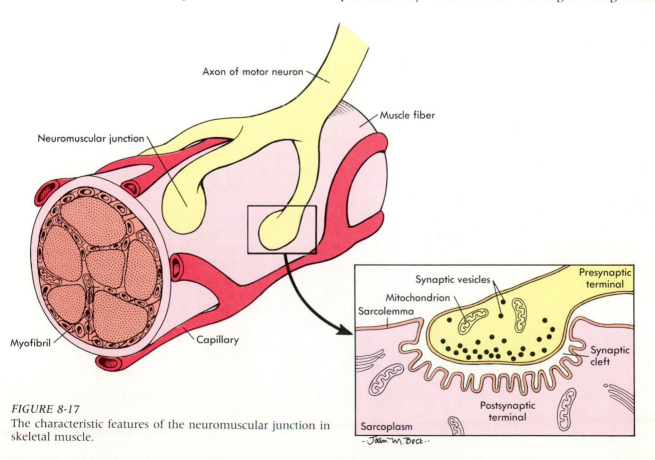

FIGURE 8-17
The characteristic features of the neuromuscular junction in skeletal muscle.

Stretch-Activated Ion Channels

A technique called patch-clamping, in which a patch of cell membrane small enough to contain a single channel is isolated, has shown that the number of different types of channels that neurons possess is much larger than previously suspected. The patch-clamp technique uses blunt glass pipettes with tip diameters of about 1 μm. The tip of the pipette is pressed against the membrane of a cell, and

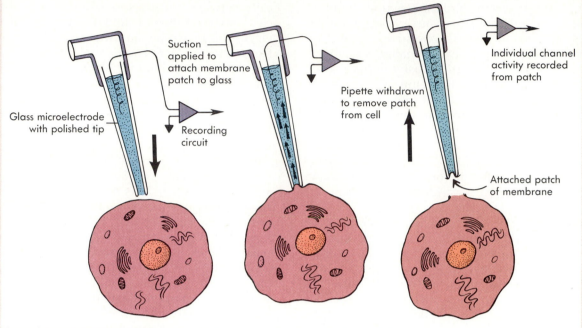

FIGURE 8-A

Steps in the formation of an inside-out membrane patch. Contact with the cell membrane *(left)*. Application of suction to attach the membrane to the electrode *(middle)*. The withdrawal of the electrode to remove the membrane patch from the cell. The size of the patch is about 1 μm² — the relationship of a membrane ion channel to this patch would be about the size of a thumbtack in a football field *(right)*.

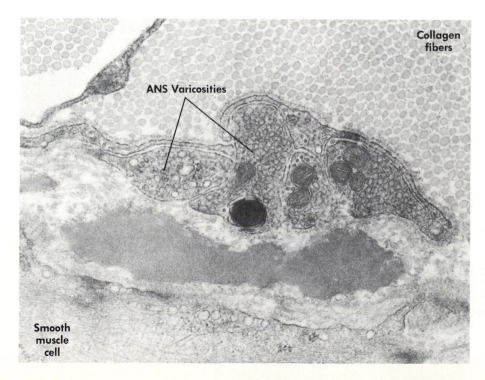

FIGURE 8-18

A photomicrograph showing synapses in the autonomic nervous system. Synaptic vesicles are present in swellings, referred to as axon varicosities in the vicinity of a smooth muscle cell, but there is not direct contact between them.

a small amount of suction is applied. An electrically tight seal forms between the cell membrane and the glass tip (Figure 8-A). If an ion channel is in this patch, its opening and closing can be seen as a steplike change in the resistance of the membrane (Figure 8-B). If desired, the patch pipette can be pulled away from the cell, taking with it the attached membrane (Figure 8-A). Both sides of these excised patches then can be exposed to solutions of controlled composition. For example, it is possible to add cAMP, the catalytic subunit of adenyl cyclase, Ca^{++}-calmodulin, or protein kinase C to determine the effect of these second messengers on ion channels of various types.

Stretch-activated channels are present in sensory cells that are able to detect distortion of their membrane. These include hair cells, touch cells, muscle stretch receptors, and receptors of blood pressure and osmolarity. The sensitivity of these cells to stretch can now be tested by applying controlled suction to a patch of membrane. An increase in membrane area of less than 2% can

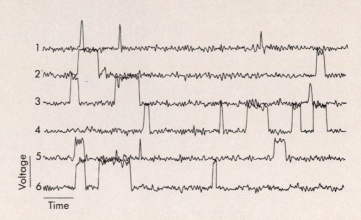

FIGURE 8-B
Sample of currents recorded from a single ion channel under patch clamp. Each steplike change in the level of the trace represents an opening or closing of the ion channel.

maximally activate stretch-sensitive channels. The ions that flow through the stretch-activated channels are typically cations, with the relative selectivity to K^+, Na^+, and Ca^{++} varying in the cell types that have been examined.

The mechanism by which stretch opens channels has not yet been elucidated, but investigators propose that the cytoskeletal fibers may be involved in focusing membrane distortion from a relatively large area onto the channel site. Osmoregulation is such an essential biological function that receptor mechanisms that can activate appropriate responses probably evolved very early. Stretch-activated channels are present in bacterial membranes and the membranes of protozoa, as well as in many cell types in multicellular organisms. It is possible that stretch-activated channels may represent the type of channel from which voltage and stimulus-gated channels evolved.

within the axon (Figure 8-19). In video micrographs of axons, mitochondria and synaptic vesicles can be seen riding along these tubules as if on a railway. Remarkably, the railway runs in both directions: full synaptic vesicles run in the direction of axon terminals, while the membrane remnants of empty vesicles, collected into bundles called **multivesicular bodies,** are carried toward the cell body for degradation.

Control of Neurotransmitter Release by Calcium Channels

The synaptic vesicles are located within a few µm of the presynaptic membrane, but as long as the concentration of Ca^{++} in the cytoplasm remains at its resting value of less than 0.1 micromolar, fusion of a vesicle to the membrane and release of its contents is a relatively rare event. The axon terminals of resting neurons release transmitter at such a low rate that its effect on the postsynaptic cells is negligible.

Axon terminals contain voltage-dependent Na^+ and K^+ channels like those in axons and also **voltage-dependent Ca^{++} channels.** When an action potential propagates into a presynaptic terminal and depolarizes it (see event 1 in Figure 8-16), these Ca^{++} channels open and Ca^{++} ions enter from the external solution (event 2). When the cytoplasmic Ca^{++} level rises above the resting level, various intracellular second messengers are activated, and the probability that vesicles will fuse with the presynaptic membrane is greatly increased. Many vesicles therefore fuse with the membrane (events 3 and 4).

Quantal Content and Synaptic Efficacy

There is a **synaptic delay** of about 0.5 msec between the arrival of the action potential at the axon terminal and the beginning of the response of the postsynaptic cell. Much of this delay reflects the time needed for activation of second messengers in the axon terminal; other events such as diffusion of

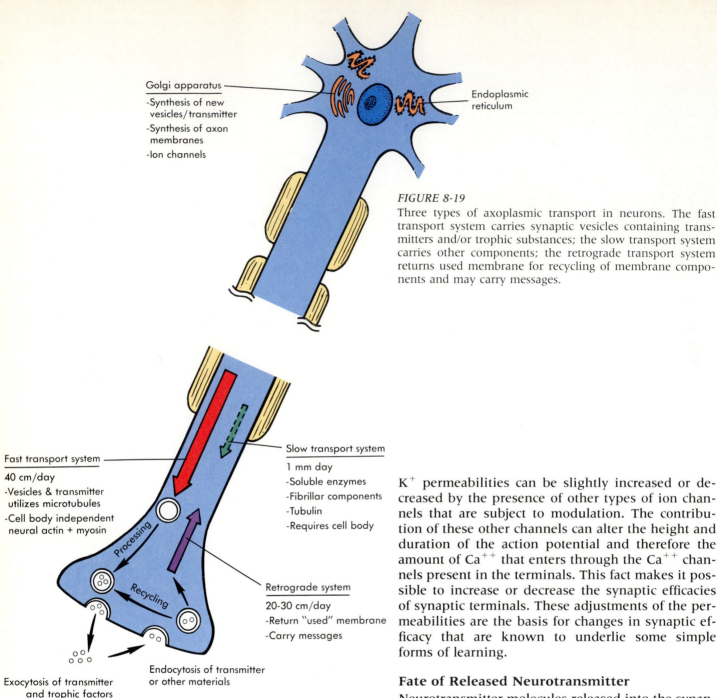

Golgi apparatus
-Synthesis of new vesicles/transmitter
-Synthesis of axon membranes
-Ion channels

Endoplasmic reticulum

FIGURE 8-19

Three types of axoplasmic transport in neurons. The fast transport system carries synaptic vesicles containing transmitters and/or trophic substances; the slow transport system carries other components; the retrograde transport system returns used membrane for recycling of membrane components and may carry messages.

Fast transport system
40 cm/day
-Vesicles & transmitter utilizes microtubules
-Cell body independent neural actin + myosin

Processing

Recycling

Slow transport system
1 mm day
-Soluble enzymes
-Fibrillar components
-Tubulin
-Requires cell body

Retrograde system
20-30 cm/day
-Return "used" membrane
-Carry messages

Exocytosis of transmitter and trophic factors

Endocytosis of transmitter or other materials

the transmitter across the synaptic cleft and activation of postsynaptic channels appear to account for only a small fraction of the delay. The **quantal content,** the number of quanta of transmitter released as the result of a single action potential, is determined by the Ca^{++} concentration attained by the axon terminal. The quantal content is one of the factors that determines the **synaptic efficacy.** The synaptic efficacy is the degree to which any single synapse can affect the activity of the postsynaptic cell.

Action potentials are all-or-none events, and they have constant amplitudes in axons because the Na^+ and K^+ permeability changes are always the same. In nerve terminals, however, overall Na^+ and

K^+ permeabilities can be slightly increased or decreased by the presence of other types of ion channels that are subject to modulation. The contribution of these other channels can alter the height and duration of the action potential and therefore the amount of Ca^{++} that enters through the Ca^{++} channels present in the terminals. This fact makes it possible to increase or decrease the synaptic efficacies of synaptic terminals. These adjustments of the permeabilities are the basis for changes in synaptic efficacy that are known to underlie some simple forms of learning.

Fate of Released Neurotransmitter

Neurotransmitter molecules released into the synaptic cleft are free to diffuse. Random movement leads each molecule into one of several possible fates. Some reach sites on the postsynaptic membrane (see event 5 in Figure 8-16). Others disappear into the extracellular fluid without ever reaching the postsynaptic cell (event 9). Some presynaptic terminals can actively recover their transmitters for recycling (event 8). At the neuromuscular junction, transmitter molecules may be degraded to an inactive form by specific enzymes on the postsynaptic membrane. The effect of these processes is to ensure that the neurotransmitter concentration in the vicinity of the postsynaptic membrane remains high for only a few milliseconds. During this brief period transmitter molecules are available to react with receptor molecules on the postsynaptic membrane

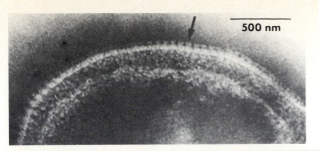

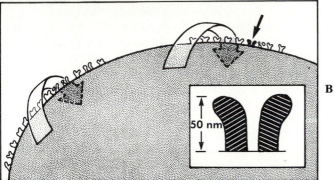

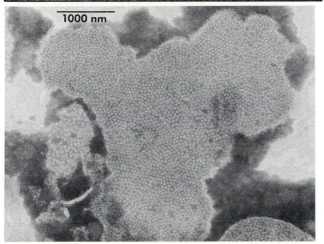

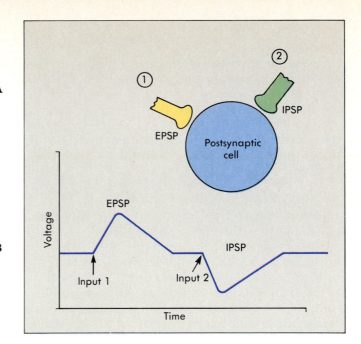

FIGURE 8-21
Illustration of an EPSP and IPSP resulting from activation of two synaptic inputs to a postsynaptic cell.

FIGURE 8-20
A   An edge view of a membrane with acetylcholine receptors *(arrow)*.
B   A diagram of the projecting part of the receptor in cross section. The binding site for acetylcholine *(arrow)* is located on the part of the receptor projecting from the outside of the membrane.
C   Acetylcholine receptors are seen end-on; the pore in the center of each receptor is the ion channel.

(event 6). The binding of transmitter with receptor is a reversible reaction, so that even those transmitter molecules that succeed in binding with a receptor are gone from the synaptic cleft within a few milliseconds after transmitter release. The brevity of synaptic events makes high rates of information transfer possible.

Neurotransmitter-Induced Permeability Changes

Just as for sensory receptor proteins, **neurotransmitter-sensitive receptors** may themselves be channels or they may be coupled to channels. Fig-

ure 8-20 shows a high-magnification electron micrograph of purified receptors for the transmitter acetylcholine. Each receptor consists of a transmitter-binding site that projects from the membrane surface and controls a transmembrane protein. This protein is made up of five subunits that form a gated channel. The interaction of a neurotransmitter with its receptor results in opening or closing of channels, causing a transient change in the potential of the postsynaptic membrane called a **postsynaptic potential.**

The polarity of the postsynaptic potential depends on the type of channel or channels affected by the binding of transmitter with receptor, so that the same transmitter binding to the same receptor type can exert different effects in different cells. Conversely, more than one receptor type may be coupled to a single type of channel. In cases in which the channel itself does not possess the receptor site, the coupling between receptor protein and channel protein is brought about by a **G protein,** one of a family of intrinsic membrane proteins that act as second messengers within membranes. This mechanism of signaling to the ion channels using a chemical messenger system within the membrane is similar to that described for the initiation of hormonal second messengers (Chapter 5).

Excitatory Postsynaptic Potentials
Postsynaptic potentials are divided into two categories according to their effects on the activity of the postsynaptic cell. **Excitatory postsynaptic potentials (EPSPs)** are depolarizing potentials that increase the probability that the postsynaptic cell will initiate an action potential (Figure 8-21). They re-

sult from an increase in the permeability of the postsynaptic membrane to both Na^+ and K^+. Just as for the sensory receptors, such a change results in depolarization because the driving force for Na^+ is so much larger than that for K^+. The equivalent event occurring at the neuromuscular junction of skeletal muscle is called an **endplate potential (EPP)**.

Inhibitory Postsynaptic Potentials

Inhibitory postsynaptic potentials (IPSPs) reduce the chance that the postsynaptic cell will initiate an action potential (see Figure 8-21). This effect can be accomplished in three ways: by increasing K^+ permeability, by increasing Cl^- permeability, or both. Simply increasing K^+ permeability results in a hyperpolarizing postsynaptic potential because the equilibrium potential for K^+ is usually more internally negative than the resting potential (see Chapter 7, p. 143). If a cell is hyperpolarized, threshold is increased, inhibiting action potentials. To understand the effect of increasing Cl^- permeability, it only is necessary to remember that Cl^- is typically in electrochemical equilibrium across the neuronal cell membrane (Chapter 6), and that to prevent action potentials it is only necessary to prevent depolarization of the axon hillock to threshold. An increase in Cl^- permeability stabilizes the membrane potential at the Cl^- equilibrium potential, which is also the resting potential, and thus opposes depolarization. This is termed **silent inhibition**.

Neurotransmitters in the Nervous System

Table 8-1 lists some of the different neurotransmitters in the nervous system and the processes they mediate. Acetylcholine was the first neurotransmitter to be discovered. In addition to being a transmitter at the neuromuscular junction and in the autonomic nervous system, a deficiency of acetylcholine has been suggested as a cause of the type of dementia referred to as Alzheimer's disease. Norepinephrine, again a transmitter in the sympathetic nervous system, appears to have a role in regulating arousal and emotional responses in the brain.

Norepinephrine is an example of a class of compounds known as **monoamines.** Other monoamine transmitters include **epinephrine, dopamine**, and **serotonin.** Epinephrine is involved in responses to stress (see Chapter 23). Dopamine is involved in emotional behavior, with excesses of dopamine causing certain types of schizophrenia. Dopamine is also important in motor control; the lack of dopamine leads to the shaking and uncontrolled movements seen in Parkinson's disease. Overproduction of dopamine may cause some types of depression. Serotonin appears to be important in the control of the sleep cycle, and the depletion of serotonin also seems to lead to depression.

Another important class of transmitters is the amino acids: **glutamate, glycine,** and **gamma aminobutyric acid (GABA).** Glutamate is probably the main excitatory transmitter in the central nervous system, while glycine is one of the major inhibitory transmitters. GABA is also inhibitory, and a deficit of GABA appears to be responsible for certain types of anxiety because many of the anti-anxiety drugs (for example, Valium) increase GABA levels or the effectiveness of this transmitter. In all these examples, the specification of a function for a particular neurotransmitter usually is the result of correlating changes in behavior with the administration of drugs known to affect the action of specific neurotransmitters.

TABLE 8-1   The Classical Neurotransmitters

Transmitter	Location	Function(s)
Acetylcholine	Neuromuscular junction	Transmission
	Autonomic nervous system	Transmission
	Hippocampus and cortex	Memory
Norepinephrine	Sympathetic nervous sytem	Transmission
	Brain	Regulation of emotional responses to the environment
Dopamine	Corpus striatum	Parkinson's disease; schizophrenia
Serotonin	Raphe nucleus	Involved in mood and sleep; depletion results in depression
Histamine	Many brain pathways	Involved in thirst, body temperature, emotions
Glycine	Wide distribution	Major inhibitory neurotransmitter present at 30%-40% of synapses of brainstem and spinal cord; also retina
Glutamate	Wide distribution	Major excitatory neurotransmitter released by afferent neurons in spinal cord and brainstem
GABA (gamma aminobutyric acid)	Wide distribution	Inhibitory transmitter at 25%-40% of all synapses in the brain; also present in retina

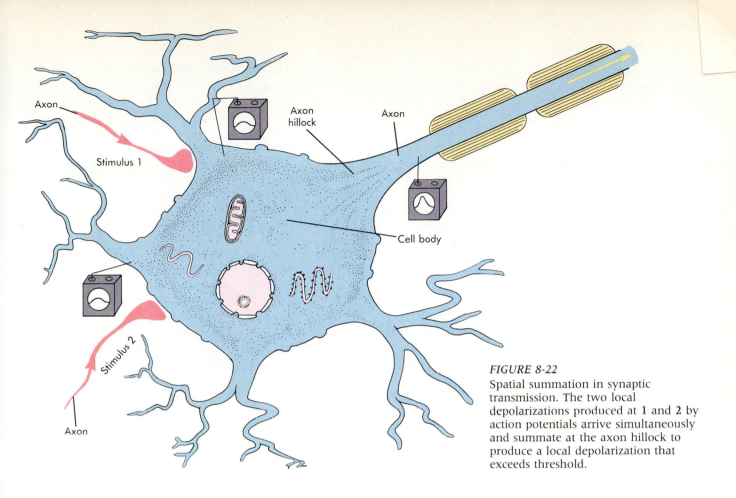

FIGURE 8-22
Spatial summation in synaptic transmission. The two local depolarizations produced at **1** and **2** by action potentials arrive simultaneously and summate at the axon hillock to produce a local depolarization that exceeds threshold.

Labels in figure: Axon; Stimulus 1; Axon hillock; Axon; Cell body; Stimulus 2; Axon

1 What are the characteristic features of chemical synapses?
2 What is a quantum of transmitter? What is meant by the quantal content of a synaptic discharge?
3 What conductance changes cause an EPSP? Changes in the conductance of which two ions can cause an IPSP?

PRINCIPLES OF INFORMATION PROCESSING IN THE NERVOUS SYSTEM
Graded Response Properties of Postsynaptic Potentials

Like sensory receptor potentials, postsynaptic potentials are graded. This is possible because increasing the amount of transmitter released activates more and more receptors, thus increasing the magnitude of the permeability change and the resulting postsynaptic potential. Just as with sensory receptor potentials, postsynaptic potentials are not propagated. This is because the input segment of excitable cells is characteristically lacking in voltage-sensitive channels and is thus difficult or impossible to bring to threshold. Instead, a current flows away from the synapse into all surrounding regions of the cell interior. This current fades rapidly with distance. The extent to which activity at a certain synapse can affect the electrical activity of the postsynaptic cell depends on the distance between the synapse and the electrically excitable membrane of the cell body. Synapses that are located near the axon hillock are much more influential than those located far out on a dendrite. Synapses that release a large amount of transmitter are, all other things being equal, more influential than synapses that release a small amount, because a larger amount of transmitter can activate more channels and result in a larger initial current.

At the neuromuscular junction of skeletal muscle, the quantal content is about 100 vesicles per presynaptic impulse. The resulting EPSP is about 50 mV, far greater than the minimum needed to bring the surrounding excitable muscle cell membrane to threshold. As a result, a single impulse at a muscle endplate always results in an action potential in the muscle cell, and thus no summation of EPSPs is necessary. For neurons of the central nervous system, the quantal content is much lower, and the distance from the synapses to electrically excitable membrane is usually longer. As a result it is typically impossible for a single EPSP to trigger an action potential. Therefore activity in these cells depends on summation of PSPs.

Spatial and Temporal Summation
Spatial summation results if two or more PSPs are generated simultaneously at different synapses on the cell's input segment (Figure 8-22). The currents

flow together through the cell. If they are of the same polarity (all EPSPs or IPSPs), they summate constructively, and the effect felt at the axon hillock is greater. If they are of opposite polarity (mix of EPSPs and IPSPs), the IPSPs tend to cancel the EPSPs.

Temporal summation of postsynaptic potentials occurs if two or more action potentials arrive at a presynaptic terminal in rapid succession. In such cases an additional batch of transmitter can be released before the first PSP is over, and the resulting PSPs will summate (Figure 8-23).

Each central neuron receives many synaptic inputs, a property referred to as **convergence** (Figure 8-24). Because of convergence there is combined spatial and temporal summation of both excitatory and inhibitory postsynaptic potentials (Figure 8-25, *A*). The postsynaptic potentials summate over time in the input segment in proportion to their quantal contents and their respective distances from the axon hillock. The resulting voltage fluctuation of the postsynaptic cell is the weighted average of the many excitatory and inhibitory synaptic inputs (Figure 8-25, *B*). Whenever the voltage of the postsynaptic cell exceeds threshold, an action potential will be produced (Figure 8-25, *C*).

Presynaptic Inhibition and Facilitation

There are many instances in the central nervous system in which an axon terminal that is presynaptic to its target neuron is itself postsynaptic to another pathway, as shown in Figure 8-26. This is an **axo-axonic synapse**: one axon (terminal 2) ends on another axon (terminal 1) rather than on a cell body or dendrite. An axo-axonic synapse does not itself initiate an action potential, but if the axo-axonic synapse on an axon terminal is activated at the same time that an action potential arrives at the nerve terminal, the effect of the axo-axonic synapse is seen in a reduction or an increase in the quantal content of transmitter released. These effects, called **presynaptic inhibition** or **presynaptic facilitation,** respectively, are two examples of the modulation of synaptic efficacy discussed earlier (see p. 180).

How close the membrane potential at the peak of an action potential comes to the equilibrium potential for Na^+ depends on the ratio of the Na^+ conductance to the K^+ conductance at that instant (Chapter 7). The all-or-none rule of action potential propagation is modified by current understanding of how neuromodulation occurs. The Ca^{++} influx into a synaptic terminal is a sensitive function of the peak amplitude and duration of the presynaptic ac-

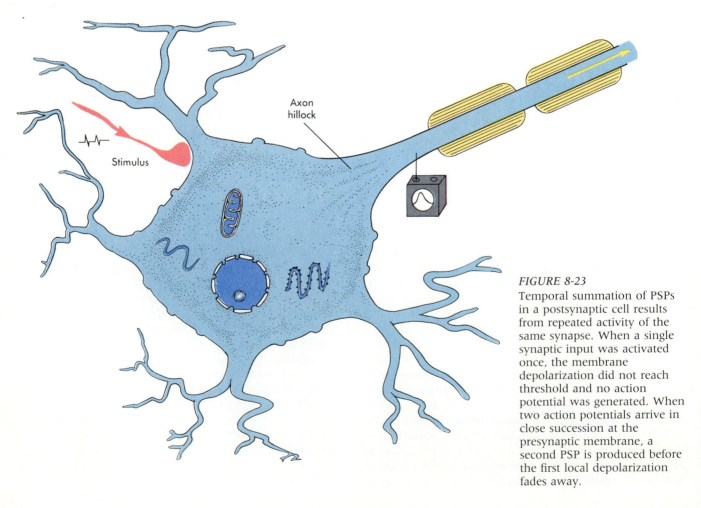

Axon hillock

Stimulus

FIGURE 8-23
Temporal summation of PSPs in a postsynaptic cell results from repeated activity of the same synapse. When a single synaptic input was activated once, the membrane depolarization did not reach threshold and no action potential was generated. When two action potentials arrive in close succession at the presynaptic membrane, a second PSP is produced before the first local depolarization fades away.

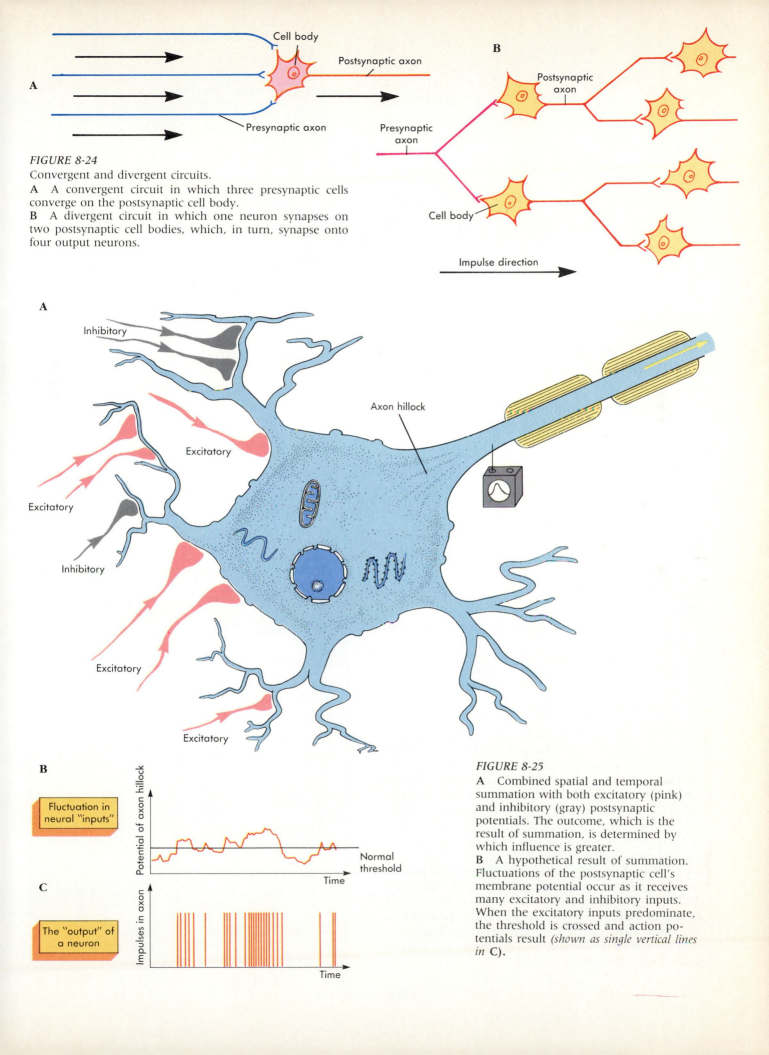

A

Cell body

Postsynaptic axon

Presynaptic axon

B

Postsynaptic axon

Presynaptic axon

Cell body

Impulse direction

FIGURE 8-24
Convergent and divergent circuits.
A A convergent circuit in which three presynaptic cells converge on the postsynaptic cell body.
B A divergent circuit in which one neuron synapses on two postsynaptic cell bodies, which, in turn, synapse onto four output neurons.

A

Inhibitory

Excitatory

Excitatory

Inhibitory

Excitatory

Excitatory

Axon hillock

B

Fluctuation in neural "inputs"

Potential of axon hillock

Normal threshold

Time

C

The "output" of a neuron

Impulses in axon

Time

FIGURE 8-25
A Combined spatial and temporal summation with both excitatory (pink) and inhibitory (gray) postsynaptic potentials. The outcome, which is the result of summation, is determined by which influence is greater.
B A hypothetical result of summation. Fluctuations of the postsynaptic cell's membrane potential occur as it receives many excitatory and inhibitory inputs. When the excitatory inputs predominate, the threshold is crossed and action potentials result *(shown as single vertical lines in* **C**).

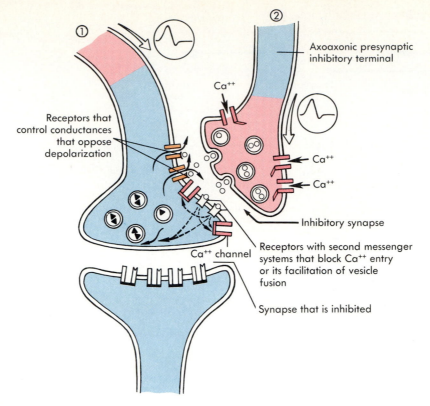

Axoaxonic presynaptic
inhibitory terminal

Receptors that
control conductances
that oppose
depolarization

Ca++

Ca++

Ca++

Inhibitory synapse

Ca++ channel

Receptors with second messenger
systems that block Ca++ entry
or its facilitation of vesicle
fusion

Synapse that is inhibited

FIGURE 8-26

In presynaptic inhibition a particular presynaptic terminal *(labelled 1)* receives synaptic input from another neuron *(labelled terminal 2, called an axo-axonic synapse).* Presynaptic inhibition can change the K^+ conductance of the terminal and therefore alter the amplitude of the invading potential and the quantal content. In some cases presynaptic inhibition involves modulation of voltage-sensitive Ca^{++} channels by intracellular second messenger systems.

tion potential. Alteration of the K^+ channels that are responsible for repolarization can increase or decrease the amplitude and duration of the action potential and therefore the amount of Ca^{++} that enters. One possible mechanism for presynaptic inhibition is as follows: if the K^+ permeability of terminal 1 has just been increased by transmitter released by terminal 2, the ratio of the Na^+ and the K^+ permeabilities at the peak of an arriving action potential will be decreased and the action potential will be briefer. The action potential also will be reduced slightly in amplitude. Both effects decrease the amount of Ca^{++} entering terminal 1 and thus the quantal content. As a result the PSP in the postsynaptic cell will be smaller. This is presynaptic inhibition. If the transmitter released by terminal 2 caused a decrease in K^+ permeability of terminal 1, the amplitude and duration of the action potential would be increased, and an increase in the quantal content would result in presynaptic facilitation. The Ca^{++} channels of the presynaptic terminal may themselves be modified by the neuromodulator (Table 8-2). While this is one probable mechanism whereby axo-axonic synapses may work, others may well remain to be discovered.

TABLE 8-2   *Some Substances Believed to Serve as Neuromodulators and Examples of Their Effects*

Neuromodulator	Effects
Angiotensin	Affects voltage-gated Ca^{++} channels of heart cells
Enkephalins	Voltage-gated Ca^{++} channels in dorsal root ganglion; modulation of pain perception
Substance P	Afferents that transmit pain sensations; inhibits K^+ channel activation
Histamine	Stimulates adenyl cyclase system in cardiac muscle
Norepinephrine	Alters the AMP system that controls voltage-gated channels in dorsal root ganglion
GABA	Decreases transmitter output from sensory afferents
Somatostatin	Decreases the Ca^{++} current in cultured nerve cells and dorsal root ganglion
Dopamine	Inhibits Ca^{++} channels activated by receptors for norepinephrine

Processing of Synaptic Inputs in Central Neurons

Summation of PSPs is the basis of neural integration: processing of synaptic inputs for an appropriate response. One kind of neural integration involves determination of whether a sensory input meets a predetermined criterion; for example, in nervous regulation of arterial blood pressure, the information provided by the baroreceptors that monitor blood pressure is compared with a setpoint determined by the cardiovascular center of the brain. If the message from the receptors indicates that the blood pressure has deviated from the setpoint value, neurons in the cardiovascular center initiate motor activity that restores the blood pressure to normal. This is an example of neural integration as part of a homeostatic reflex.

In a second kind of neural integration, the nervous system monitors sensory inputs for external stimuli. A threatening stimulus activates neurons that initiate escape behavior, such as the reflexive withdrawal of an injured limb or the startle response. The nervous system may prioritize other kinds of sensory input; for example, sleeping parents may ignore many sounds but will awaken in response to relatively faint sounds made by their children. In this example the nervous system discards sensory information that it has determined to be irrelevant. In the nervous system, such decisions are made by many thousands or millions of neurons acting in concert but can be understood in principle on the basis of the possible responses of single neurons.

Logical Circuits

Consider an idealized neuron with two synapses (Figure 8-27). Suppose each synapse produces an EPSP that reaches the axon hillock with a value of 10 mV, but to reach threshold the axon hillock must be depolarized by 15 mV. A single impulse at either synapse would not bring the postsynaptic cell to threshold, but if impulses arrive within a few milliseconds of each other at both synapses, spatial and temporal summation will depolarize the axon hillock by 20 mV and action potentials will be initiated. This neuronal circuit answers the question: A and B? Suppose that the threshold was decreased to 10 mV. Now the circuit would answer the question: A or B? Suppose additionally that B is an inhibitory synapse. If this were the case, the circuit would answer the question: A and not B? If A and B carry information from sensory inputs, such simple circuits could be combined into more complex circuits whose outputs would represent quite sophisticated analyses of the sensory information provided by the inputs.

It is a challenging task to trace such complex circuits in the brain, but many neurophysiologists now believe that some parts of the nervous system are arrays of repeated "simple" units. "Simple" is a

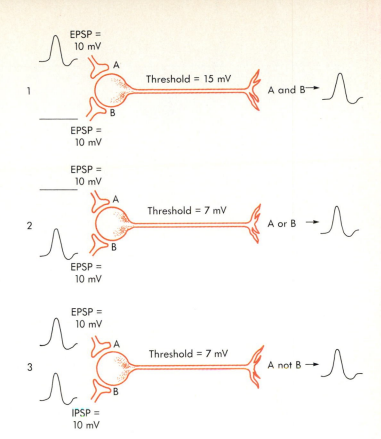

FIGURE 8-27
An illustration of how neurons can be constructed to form logical circuits.
1 A logical AND circuit. Inputs *A* and *B* are both excitatory and produce EPSPs of 10 mV, while the threshold for the postsynaptic cell is 15 mV.
2 A logical OR circuit. Inputs *A* and *B* are the same, but the threshold for the postsynaptic cell is 10 mV.
3 A logical NOT circuit. Input *A* is excitatory with an EPSP of 10 mV, but now input *B* is inhibitory with an IPSP of −10 mV. The threshold for the postsynaptic cell is 10 mV.

relative term. Each individual neuron is capable of complex response patterns in response to its many inputs. In most cases, a single neuron makes synaptic contacts with many different postsynaptic neurons, an organizational feature termed **divergence** (Figure 8-24, *B*), which further complicates the neural circuitry of the brain.

It would also be a misconception to regard the circuitry of the brain as made up, like a computer, of elements whose properties never change. At many synapses the responsiveness of the postsynaptic membrane's receptors to transmitter decreases as the synapse is used repeatedly. This process, termed **desensitization**, is analogous to adaptation of sensory receptor cells. It is one way in which the use of nerve pathways can cause changes in their responsiveness.

Long-Term Modulation of Synaptic Transmission

Up to this point the brevity of events in the information transfer process has been emphasized. Yet

Learning and Memory Can Depend on Calcium Entry

Distinction is commonly drawn between two kinds of gated channels: those that are gated by membrane voltage changes and those at sensory endings and postsynaptic membranes that are gated by the incoming stimuli. One category of receptors for the transmitter glutamate represents an exception to this rule. These receptors are referred to as NMDA receptors because this compound strongly mimics glutamate at this class of glutamate receptors. This type of glutamate-controlled receptor has been identified at synapses in the hippocampus, a brain region that has long served as a model for learning in the mammalian nervous system. The synapses function only when two conditions are met: (1) the transmitter molecules must bind to their receptors; and (2) the postsynaptic cell must already be depolarized for the excitatory effect of the synapse to be seen. This is just the type of contingency that is required for the type of learning known as classical conditioning: two stimuli arising from different sources must converge on the response system and they must overlap in time.

The mechanism that has served as a model for learning in the hippocampus is posttetanic potentiation, a long-lasting (days to weeks) increase in the effectiveness of stimulation after a pathway in the hippocampus is stimulated strongly only once. In slices of hippocampal tissue, the pairing of stimuli at two inputs to a single cell will produce the same effect, and the effect is blocked by an antagonist of the NMDA class of glutamate receptors.

The way in which the contingency of the glutamate channels are determined is now understood. The channel associated with the glutamate receptor is normally not only closed but blocked by Mg^{++}. The depolarization of the membrane by another source drives the Mg^{++} out of the channel and makes the opening of the channel by glutamate binding effective. When the channel is open, it admits Ca^{++} into the cytoplasm. The Ca^{++} then acts as a messenger that in some way strengthens the synaptic effectiveness of the cell on a long-term basis. There is evidence that this might be accomplished through reshaping the synaptic contacts made by the postsynaptic cell. For example, Ca^{++} might activate enzymes that alter the cytoskeleton. Anatomical changes that could increase the action potential current flow into the synaptic endings could increase the amount of transmitter released. Changes of this nature are observed in cultures of hippocampal cells stimulated by neurotransmitters.

everyday experience shows that sensory inputs can lead to changes in the activity of the nervous system that can persist for minutes, hours, or days. Some of these long-term changes appear to be mediated by a family of transmitter chemicals, sometimes termed **neuromodulators** (see Table 8-2), whose effects on neuronal activity are much more subtle than the simple, brief changes in Na^+, K^+, and Cl^- permeability that result in postsynaptic potentials. The long-term effects of neuromodulators represent a second level of information processing.

How can a single synapse exercise both short-term and long-term effects on the activity of a postsynaptic cell? One possibility is that when activated the synapse releases one transmitter that mediates the rapid effects and at the same time a second transmitter that mediates the long-term effects. Evidence for this is provided by many findings of two or more transmitters stored in the same nerve ending. In many cases the additional transmitters are members of a family of small proteins called **neuropeptides.** Some members of this family were first known to physiologists as hormones secreted by the nervous system into the blood. Examples of neuropeptides secreted by the brain as hormones are **vasopressin** (antidiuretic hormone), **somatostatin**, and **oxytocin.** In their roles as hormones, these peptides control the kidney, the secretion of growth hormone, and contractions of the uterus and breasts, respectively. All of these are now believed to be co-released with other transmitters and to function in the brain as neuromodulators. The distinction between neuromodulators and transmitters, if indeed one can be made, is not a matter of chemical identity, since the same substances are thought to carry out both functions in different pathways. Rather the distinction lies in the time scale and nature of the effect of the substance on information processing. Transmitters mediate effects on the postsynaptic cell that are immediate and quickly over; neuromodulators mediate effects that are slower, more lasting, and typically involve second messengers within the cell. The effects of neuromodulators may be either postsynaptic or presynaptic.

Neuromodulation of Pain

Excellent examples of the importance of neuromodulators are found in the pathway that mediates information about painful stimuli. An injury to the body that is severe enough to damage tissue results in the release of several neuromodulators at the injury site. Among these are **prostaglandins** and **bradykinin.** These substances enhance the activity of nerve endings that initiate pain messages. The sensory impulses pass along pain afferents into the spinal cord, where synaptic connections with ascending interneurons allow the information to be passed on to the brain. The transmitter substance believed to be released by pain-sensitive sensory neurons is a peptide called **substance P.**

The intensity with which this information is ultimately perceived depends very strongly on the nature of other sensory inputs taking place at the same time and on the emotional state of the pain sufferer. Variations in pain perception are attributable in large part to the effects of a family of peptide neuromodulators called **enkephalins.** These substances are believed to be released by interneurons descending from the brain and making presynaptic inhibitory connections with neurons in the spinal cord. Enkephalin release appears to reduce or even shut off the passage of pain information into the nervous system. In this way the nervous system regulates its own responsiveness to pain. Opium derivatives, such as morphine or heroin, have an analgesic, or pain-reducing, effect because they are similar enough in chemical structure to bind to the receptors normally utilized by enkephalins.

Second-Messenger Neuromodulation

A second general mechanism of long-term effects involves the generation of multiple "second messengers" by the transmitter-receptor complex. Because of G protein coupling, a single transmitter can activate a number of different channel types, as the G protein message circulates within the postsynaptic cell's membrane. Some of these channels may open and close rapidly; others may remain open for longer periods and continue to influence the subsequent electrical activity of the cell. Neuromodulators that activate Ca^{++} channels or K^+ channels in nerve terminals can change synaptic efficacy. Activated G proteins can also set in motion within the cytoplasm of the neuron other second messengers that affect its metabolism, the efficacy of its synapses, and the behavior of its ion channels. Two such intracellular second messengers that are now believed to operate in neurons as well as other cell types (see Chapter 5) are **cyclic adenosine monophosphate (cAMP)** and **phosphatidyl inositol 4,5 biphosphate.** Second messengers such as these have multiple, long-lasting effects within neurons that may ultimately be found to have a part in the changes in behavior that result from life experiences.

> 1 *What is synaptic efficacy? What factors determine it?*
> 2 *What is the difference between temporal and spatial summation of postsynaptic potentials?*
> 3 *Describe the general relationship between synaptic inputs to a neuron and the frequency of action potentials.*

SUMMARY

1. Sensory information reaches the central nervous system along **labelled lines** from **receptors** specific for single **sensory modalities.**

2. The intensity of stimuli is encoded by the frequency of action potentials in afferent nerves. The initial event in intensity coding is a graded potential change (usually a depolarization), termed a **generator** or **receptor potential**, that results from activation of **stimulus-sensitive ion channels**. The magnitude of the local current that results from this event determines the frequency of action potentials that arise in adjacent excitable parts of the receptor.

3. **Adaptation**, a decrease in receptor responsiveness with stimulation, makes receptors more sensitive to changes in stimulus intensity. Receptors that must encode a large range of stimulus intensities display a logarithmic rather than a linear relationship between stimulus intensity and action potential frequency.

4. Action potentials in a neuron can affect the electrical activity of another neuron or an effector cell if the two are connected by a **synapse.** The sequence of events is as follows. Depolarization of the synaptic terminal leads to an influx of Ca^{++}, stimulating fusion of transmitter vesicles with the presynaptic membrane. Some of the transmitter chemical diffuses to the postsynaptic cell and is bound by receptor molecules on its surface. Receptor binding activates ion channels directly or by way of a **G protein**, leading to a change in the membrane conductance and a **postsynaptic potential.**

5. The postsynaptic potential may **summate** with other synaptic inputs to increase or decrease the probability of an action potential in the postsynaptic cell. **Excitatory** (depolarizing) **postsynaptic potentials** result from an increase in conductance to all small cations; **inhibitory postsynaptic potentials** may be hyperpolarizing (resulting from an increase in K^+ conductance) or silent (resulting from an increase in Cl^- conductance that stabilizes the membrane potential near its resting value).

6. If the synapse is on another synaptic terminal, it can influence the synaptic efficacy of the second synapse, either increasing it (**presynaptic facilitation**) or decreasing it (**presynaptic inhibition**).

7. Arrangement of neurons into circuits with appropriate synaptic connections makes the activity of postsynaptic cells depend on the inputs they get from presynaptic cells; this is the basis of neural integration in the nervous system. The long-term behavior of such circuits can be tuned by **neuromodulator** chemicals that affect the efficacy of synapses in the circuits.

STUDY QUESTIONS

1. Describe the distinctive features of the input and output regions of a neuron.

2. Is your telephone a labelled line?

3. How could the receptive field of a skin touch receptor be mapped experimentally? What stimuli could cause spatial summation in such a receptor?

4. Both spatial and temporal summation require that the stimulation arrive at the input region of a cell within a brief period. What special feature of temporal summation distinguishes it from spatial summation?

5. What ions flow through channels that produce depolarizing receptor potentials? Hyperpolarizing receptor potentials?

6. How does the neural code give information about stimulus intensity? How does the fact that the initial response is in an electrically inexcitable part of the cell make this possible?

7. Compare the response properties of tonic and phasic receptors. How might the two types of information they provide be used in integrating information about a single stimulus modality into an appropriate response?

8. What are the key differences between neurotransmitters, neuromodulators, and neurohormones? Could the same chemical compound serve in all three roles?

SUGGESTED READING

BLOOM, F.: Brain drugs, *Science 85*, November 1985, p. 100. Discusses how synthetic drugs that mimic the action of CNS neurotransmitters are being used to treat Alzheimer's disease and clinical depression.

COTMAN, C.W., and L.L. IVERSEN: Excitatory amino acids in the brain—focus on NMDA receptors, *Trends in Neurosciences*, volume 10, p. 263, 1987. Discusses the relationship between dopamine receptors and drugs used to treat psychological disorders.

EDELMAN, G.M., W.E. GALL, and W.M. COWAN, editors: *Synaptic Function*, John Wiley & Sons, Somerset, N.J., 1987. Collection of papers on synapses, some of which consider the evidence for and against the classical quantal hypothesis of synaptic transmission.

FIELDS, R.D., and M.H. ELLISMAN: Synaptic morphology and differences in sensitivity, *Science*, volume 228, p. 197, 1985. Attempts to correlate the structure and location of synaptic connections with their physiological function.

HILLE, B.: *Ionic Channels of Excitable Membranes*, Sinauer Associates, Sunderland, Mass., 1984. The only comprehensive textbook in the field. It contains a good description of the acetylcholine channel and other types of postsynaptic, neurotransmitter-activated ion channels.

KUFFLER, S.W., J.G. NICHOLLS, and A.R. MARTIN: *From Neuron to Brain*, ed. 2, Sinauer Associates, Sunderland, Mass., 1984. A comprehensive textbook in the neurosciences.

Neurotransmitters and their actions, Special issue of *Trends in Neurosciences*, volume 6, number 8, 1983. Brief summary of compounds thought to serve as CNS neurotransmitters and their physiological role.

SHEPPARD, G.M.: *Neurobiology*, ed. 2, Oxford University Press, New York, 1988. Available in paperback, it is one of the most readable texts in the field. Contains a particularly good discussion of the higher functions of the nervous system such as perception, learning, and memory.

TERMAN, G.W., et al.: Intrinsic mechanisms of pain inhibition: activation by stress, *Science*, volume 226, p. 1270, 1984. Summary of the pain pathways and the role of endorphins and opiates in pain perception.

VECA, A., and J.H. DREISBACH: Classical neurotransmitters and their significance within the nervous system, *Journal of Chemical Education*, volume 65, p. 108, 1988. Written for non-neuroscientists, provides an up-to-date survey of the role of acetylcholine, norepinephrine, and amino acid transmitters.

CHAPTER 9

Sensory Systems and Brain Function

On completing this chapter you will be able to:

- Distinguish between the peripheral and central nervous system, and describe the location of the major subdivisions of the brain: the brainstem, cerebellum, thalamus, hypothalamus, and cerebral hemispheres.
- Understand what is meant by a topographic representation and why such representations are important in the central processing of sensory information.
- Outline the elements of the somatosensory afferent pathways, identifying the different routes taken by the submodalities of touch, vibration, temperature, and pain.
- State the function of the reticular activating system.
- Distinguish between habituation and sensitization and associative learning.
- Understand the difference between short-term memory and long-term memory.
- Appreciate the significance of the term dominant cerebral hemisphere and identify the location of the language areas of the brain.
- Identify the characteristics of the stages of sleep.

*T*he brain and the spinal cord make up the central nervous system. The nervous system collects and integrates sensory information for appropriate response. The nervous system mediates reflexes; controls many endocrine systems, including those that regulate growth, metabolism, and reproduction; and generates behavior, the complex patterns of activity in the body's effector systems. Finally, the nervous system learns from experience; that is, it alters behavior patterns on the basis of stored information.

Acquisition of sensory information is a key element in most of these functions. Sensory information is collected and analyzed in discrete regions of the brain. Information provided by different sensory pathways is integrated with information stored in memory. For example, when we hear a friend's voice, we turn, recognize his face, and remember his name. The voice, the face, and the name are felt to be part of one reality. In principle, everything that the central nervous system does can be understood as arising from the capabilities of individual neurons. Nevertheless, the ability of the nervous system to sort through billions of bits of information and recognize a kindred face is one of the most remarkable feats of living systems and requires a vast underpinning of neuronal activity.

ANATOMY OF THE NERVOUS SYSTEM
Segregation of Axons and Cell Bodies

The nervous system has many distinct components (Figure 9-1), but each is made up of neurons and their processes (axons and dendrites), along with several types of supporting glial cells (see Chapter 8, p. 166). In the nervous system the input segments of neurons, dendrites and cell bodies, are often segregated from the output segments, axons. Cell bodies and their dendrites form the **gray mat-**ter of the nervous system, while the **white matter** is composed of axons. Within the gray matter, neuronal cell bodies are often densely packed together in groups.

In the spinal cord and brainstem, the gray matter is in the center, surrounded by an outer layer of white matter (Figure 9-2). Spinal cord white matter consists of nerve fibers that often are segregated into anatomically distinct groups, referred to as ascending **sensory tracts** and descending **motor tracts.**

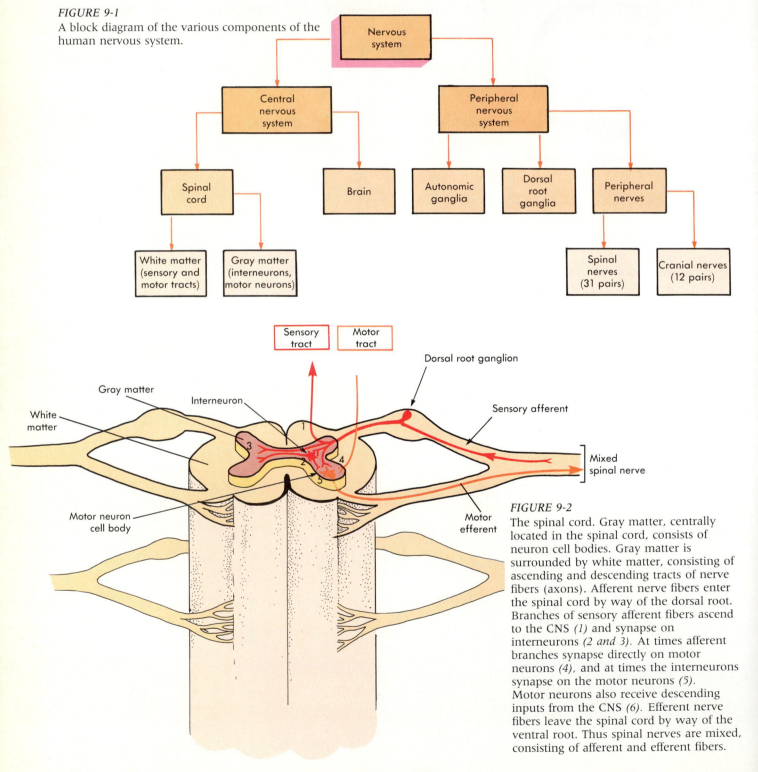

FIGURE 9-1
A block diagram of the various components of the human nervous system.

FIGURE 9-2
The spinal cord. Gray matter, centrally located in the spinal cord, consists of neuron cell bodies. Gray matter is surrounded by white matter, consisting of ascending and descending tracts of nerve fibers (axons). Afferent nerve fibers enter the spinal cord by way of the dorsal root. Branches of sensory afferent fibers ascend to the CNS *(1)* and synapse on interneurons *(2 and 3)*. At times afferent branches synapse directly on motor neurons *(4)*, and at times the interneurons synapse on the motor neurons *(5)*. Motor neurons also receive descending inputs from the CNS *(6)*. Efferent nerve fibers leave the spinal cord by way of the ventral root. Thus spinal nerves are mixed, consisting of afferent and efferent fibers.

NERVE AND MUSCLE

Sensory neurons carry action potentials from sensory receptors to the dorsal roots of the spinal cord and the brain (see Figure 9-2). The axons of **motor neurons** carry action potentials from the brain or the ventral root of the spinal cord to skeletal muscles, smooth muscles, glands, and visceral organs. The term **afferent** is used for fibers that carry sensory information from the periphery to the central nervous system; **efferent** fibers carry information toward the periphery. Most efferent fibers are motor neurons that control or modulate the activities of effectors. However, some specialized sensory organs receive efferents that allow the central nervous system to increase or decrease the sensitivity of sensory receptors.

Central and Peripheral Components of the Nervous System

The nervous system can be divided into two components. The **central nervous system** consists of the spinal cord and the many subdivisions of the brain (Figure 9-3). The **peripheral nervous system** refers to neural elements outside the brain and spinal cord. The **autonomic nervous system** (see Chapter

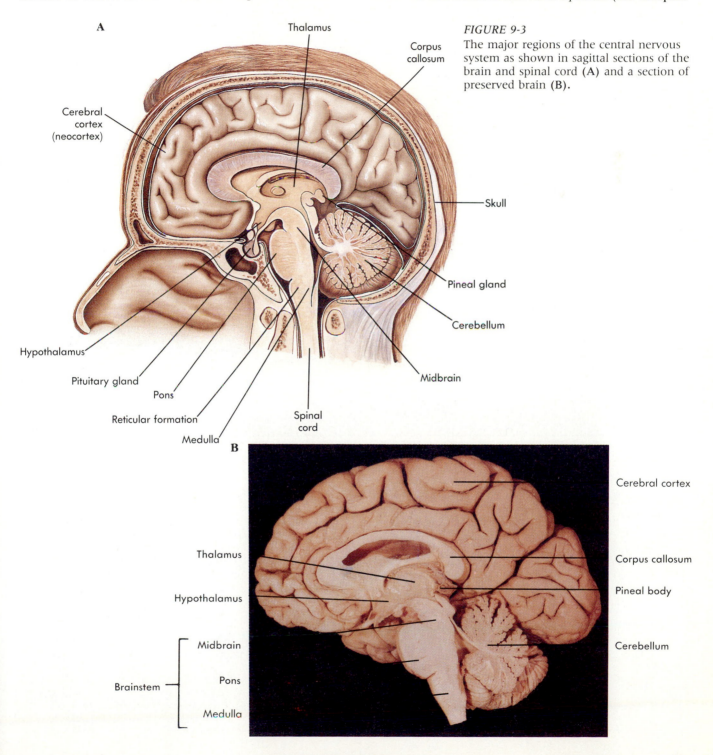

FIGURE 9-3
The major regions of the central nervous system as shown in sagittal sections of the brain and spinal cord (**A**) and a section of preserved brain (**B**).

Sensory Systems and Brain Function

5, p. 96 and Chapter 11, p. 275) refers to the collection of motor neurons in the central and peripheral nervous systems that innervate visceral smooth muscle and glands. The vast majority of neuronal cell bodies are found within the central nervous system. Estimates of the total number of neurons in the central nervous system vary somewhat, but about 10^{10} cells is a currently accepted figure.

By far the largest number of neurons in the nervous system are neither sensory neurons nor motor neurons and are called **association neurons** or **interneurons**. All parts of these neurons are within the central nervous system. As their names suggest, these neurons are involved in linking sensory inputs to appropriate motor outputs. The large number of possible connections provided by the many

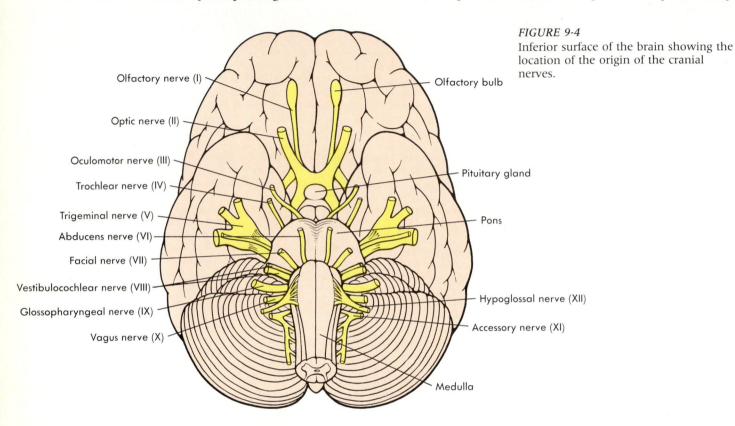

FIGURE 9-4
Inferior surface of the brain showing the location of the origin of the cranial nerves.

Labels (left side): Olfactory nerve (I); Optic nerve (II); Oculomotor nerve (III); Trochlear nerve (IV); Trigeminal nerve (V); Abducens nerve (VI); Facial nerve (VII); Vestibulocochlear nerve (VIII); Glossopharyngeal nerve (IX); Vagus nerve (X).

Labels (right side): Olfactory bulb; Pituitary gland; Pons; Hypoglossal nerve (XII); Accessory nerve (XI); Medulla.

TABLE 9-1 The Cranial Nerves

Nerve	Pathway and function
I Olfactory	Sensory—from olfactory epithelium to olfactory bulb
II Optic	Sensory—from retina to lateral geniculate nucleus of thalamus
III Oculomotor	Carries motor and muscle stretch receptor pathways between midbrain and eye that mediate movement of eyeball and eyelid, focusing, and control of pupillary diameter
IV Trochlear	Carries motor and muscle stretch receptor pathways that mediate eye movement
V Trigeminal	Carries motor pathways from pons to muscles involved in chewing; carries sensory pathways from cornea, face, lips, tongue, and teeth to pons
VI Abducens	Carries motor and muscle stretch receptor pathways that mediate eye movement
VII Facial	Carries pathways that mediate facial expression, secretion of saliva and tears, and taste sensation
VIII Vestibulocochlear	Sensory—carries auditory and vestibular afferents
IX Glossopharyngeal	Carries taste afferents and pathways involved in swallowing and salivary secretion
X Vagus	Carries sensory and autonomic pathways running between medulla and thoracic and upper abdominal viscera
XI Accessory	Carries pathways between medulla and throat and neck muscles involved in swallowing and head movements
XII Hypoglossal	Carries pathways between medulla and tongue involved in speech and swallowing

interneurons provides enormous integrating power.

The peripheral nervous system consists of peripheral nerves and neuronal cell bodies. There are two categories of peripheral nerves: **cranial nerves** extend from the brain (Figure 9-4), while **spinal nerves** extend from the spinal cord. Twelve pairs of cranial nerves (Table 9-1) arise from the undersurface of the brain, in most cases from the brainstem. Of these, three pairs (I, II, and VIII) contain only sensory afferent fibers. Five pairs (III, IV, VI, XI, and XII) contain efferent motor fibers, along with muscle sensory afferents from the relevant muscles, and are thus primarily motor in function. Four pairs of cranial nerves (V, VII, IX, and X) are **mixed nerves,** containing both motor pathways and non-muscle sensory afferents.

Spinal nerves are formed by the unification of dorsal (sensory) and ventral (motor) roots on each side of each spinal segment (see Figure 9-2) and are thus mixed nerves. There are 31 pairs of spinal nerves (Figure 9-5). In order from the most anterior, there are: 8 cervical, 12 thoracic, 5 lumbar, 5 sacral, and 1 coccygeal.

Neuronal cell bodies outside the central nervous system include the cell bodies of sensory neurons, located in the dorsal root ganglia immediately outside the spinal cord (see Figure 9-2), and the cell bodies of postganglionic cells of the autonomic nervous system, located in **autonomic ganglia. A ganglion** is defined as any association of neuronal cell bodies within the peripheral nervous system, while groups of cell bodies within the central nervous system are referred to as **nuclei.**

The axons of **somatic** motor neurons, those that control skeletal muscles, run from the spinal cord to the muscle fibers in spinal nerves without interruption. Motor axons to visceral effectors in the autonomic motor system run from the spinal cord to peripheral autonomic ganglia (see Chapter 11, p. 258). Here the axons of spinal motor neurons synapse on postganglionic cells, which in turn innervate the effectors.

Regions of the Central Nervous System

The major anatomical subdivisions of the central nervous system (Table 9-2; Figure 9-3) are:

1. **Spinal cord**
2. **Medulla**
3. **Pons**
4. **Midbrain** or **mesencephalon**
5. **Cerebellum**
6. **Diencephalon**
7. **Telencephalon**

The medulla, pons, and midbrain are referred to collectively as the **brainstem.** The brainstem links the rest of the brain with the spinal cord and contains the motor neurons of cranial nerves. Many nuclei in the brainstem control the internal organs (referred to as cardiovascular control centers, re-

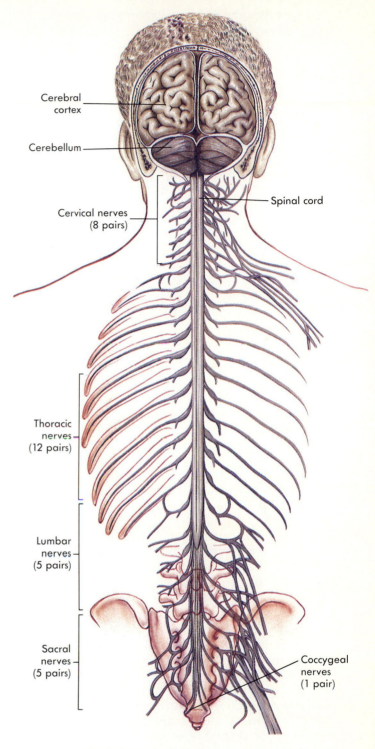

FIGURE 9-5
The spinal cord and spinal nerves.

spiratory control centers, etc.). The brainstem also contains a less densely packed network of neurons, referred to as the **reticular formation.** Most sensory pathways give off branches (collaterals) that lead to the reticular formation. The function of this nonspecific sensory information is to maintain a level of conscious awareness, a topic discussed

TABLE 9-2 *Subdivisions of the Central Nervous System*

Major subdivision	Components	Function
Spinal cord		Spinal reflexes; relay sensory information
Brainstem	Medulla	Sensory afferent nuclei; reticular activating system; visceral control centers
	Pons	Reticular activating system; visceral control centers
	Midbrain	Similar to pons
Cerebellum		Coordination of movements; balance
Diencephalon	Thalamus	Relay station for ascending sensory and descending motor neurons; control of visceral function
	Hypothalamus	Visceral function; neuroendocrine control
Telencephalon	Basal ganglia	Motor control
	Red nucleus	Motor control
	Corpus callosum	Connects the two hemispheres
	Hippocampus (limbic system)	Memory; emotion
	Olfactory lobes	Smell
	Neocortex	Higher functions

later in the chapter (see p. 212). The reticular formation is also one of the areas where autonomic activity is initiated and emotions are generated. Another important structure in the midbrain is the **red nucleus,** named for its characteristic appearance when the brainstem is sectioned and stained.

The two hemispheres of the cerebellum, quite different in appearance from the pons and medulla, are located dorsal to the brainstem. In the cerebellum the outermost layer is gray matter, with the white matter on the interior. The two most important components of the diencephalon are the **thalamus** and the **hypothalamus** (Figure 9-6). The **pituitary gland** (see Chapter 5, p. 99) is located just beneath the hypothalamus.

The telencephalon consists of the **cerebral hemispheres.** The telencephalon arises from the most anterior part of the embryonic nervous system. It has become so well-developed in humans and other primates that it almost covers the other parts of the brain. Deep within each cerebral hemisphere there is a complex structure called the **corpus striatum.** The most important elements of the corpus striatum are several large groups of neurons in nuclei referred to collectively as the **basal ganglia** (the globus pallidus, caudate nucleus, and putamen in Figure 9-7). This designation is an exception to the general convention that refers to collections of neurons within the central nervous system as nuclei. The basal ganglia, together with the cerebellum and the red nucleus of the midbrain, are involved in generating the patterns of activity in motor neurons, which drive complex behavior (see Chapter 11).

Communication between neurons in different parts of a cerebral hemisphere is provided by fibers

that course from one part of the hemisphere to the other by way of the corpus striatum. The two hemispheres communicate with one another by way of the **corpus callosum** (see Figures 9-3 and 9-7), a thick bundle of fibers (white matter) that connects the two hemispheres.

The Cerebral Cortex

The **cerebral cortex,** a cap of gray matter about 3 mm thick (see Figure 9-3), overlies the inner, evolutionarily older structures of the brain ("cortex" means rind or bark). The cortex contains perhaps two thirds of all the neurons in the central nervous system. It has three subdivisions: the **neocortex,** the **hippocampus,** and the **olfactory lobes.** The neocortex is the largest in area and is what is seen when one looks at the surface of the brain. The neocortex is believed to be the newest part of the brain in an evolutionary sense. The presence of large amounts of neocortex is a mammalian characteristic, and it is common to simply use the term cortex to refer to the neocortex. In primates the neocortex has a predominant role in sensory processing (except for smell) and in generating behavior.

The cerebral cortex of some mammals has fissures, called **sulci** (singular, sulcus), and ridges, termed **gyri** (singular, gyrus), that increase its surface area. The folding is most pronounced in the human brain, where it increases the surface area of the cortex by about three times. The folding creates important anatomical landmarks of the surface of the cortex. The longitudinal fissure divides the cerebrum into its two hemispheres (see Figure 9-7). The fissure of Rolando (central sulcus) separates the **frontal** and **parietal lobes;** the fissure of Sylvius divides the **temporal lobe** from the frontal and pa-

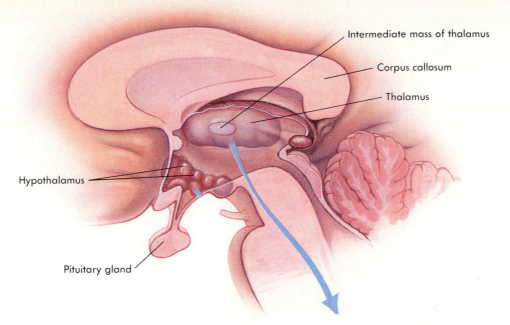

Intermediate mass of thalamus

Corpus callosum

Thalamus

Hypothalamus

Pituitary gland

FIGURE 9-6
The diencephalon showing the thalamus, hypothalamus, and the location of the pituitary gland.

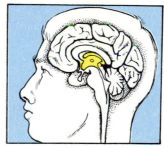

FIGURE 9-7
A section through the middle of the brain showing the corpus striatum and basal ganglia. White myelinated nerve fiber tracts run between the areas of gray matter of the basal ganglia, producing the striped appearance of the corpus striatum.

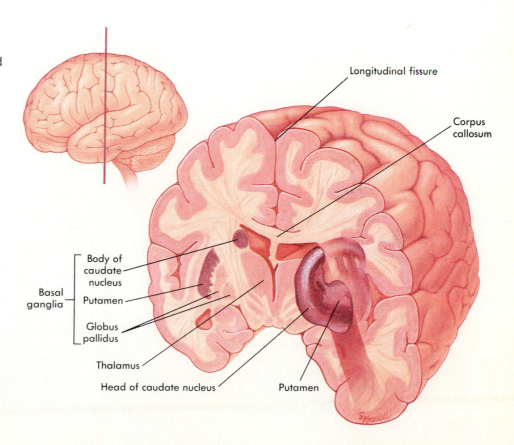

Longitudinal fissure

Corpus callosum

Basal ganglia
- Body of caudate nucleus
- Putamen
- Globus pallidus

Thalamus

Head of caudate nucleus

Putamen

Sensory Systems and Brain Function

rietal lobes; while the parieto-occipital fissure (not as easily identified) separates the **occipital lobe** from the parietal lobe (Figure 9-8).

The hippocampus and olfactory lobes are two of the components of the **limbic system** (Figure 9-9), an evolutionarily old group of linked structures deep within the telencephalon and diencephalon that are responsible for emotional responses. The

hippocampus is folded under the neocortex and is believed to be important in the formation and recall of memories.

Cortical Organization

The cortex is composed of six layers (Figure 9-10, *A* and *B*), each of which has a characteristic pattern of synaptic connections with other parts of the cortex

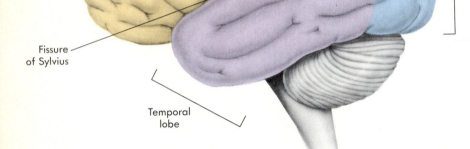

FIGURE 9-8
The cerebral cortex can be divided into four anatomical regions, or lobes: frontal, parietal, temporal, and occipital.

Fissure of Rolando

Frontal lobe

Parietal lobe

Occipital lobe

Fissure of Sylvius

Temporal lobe

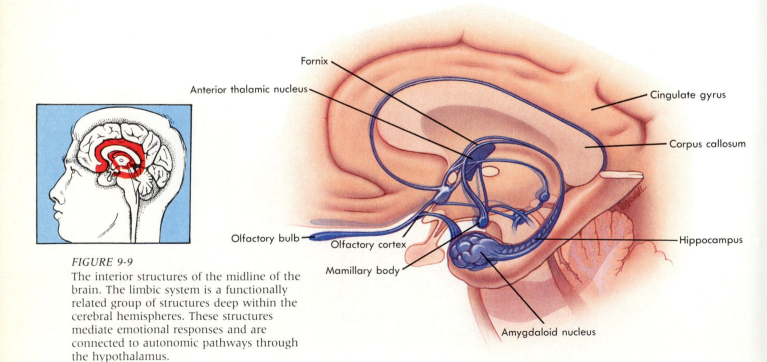

FIGURE 9-9
The interior structures of the midline of the brain. The limbic system is a functionally related group of structures deep within the cerebral hemispheres. These structures mediate emotional responses and are connected to autonomic pathways through the hypothalamus.

Fornix

Anterior thalamic nucleus

Cingulate gyrus

Corpus callosum

Olfactory bulb

Olfactory cortex

Mamillary body

Hippocampus

Amygdaloid nucleus

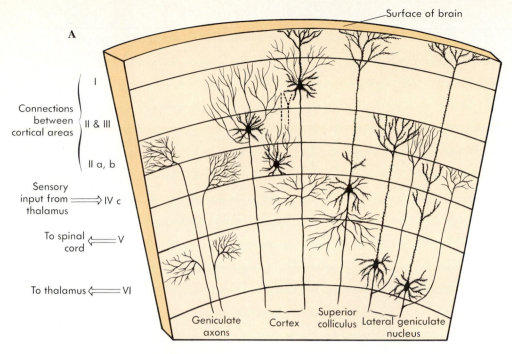

A

Surface of brain

I

Connections between cortical areas

II & III

II a, b

Sensory input from thalamus ⟹ IV c

To spinal cord ⟸ V

To thalamus ⟸ VI

Geniculate axons   Cortex   Superior colliculus   Lateral geniculate nucleus

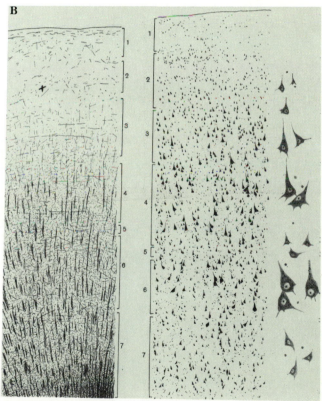

B

FIGURE 9-10

A   A diagram of the six-layer structure of the cortex with representative types of neurons.

B   Micrographs of the cortex illustrating its six-layer structure. The left micrograph has been stained to show the location of nerve fibers, while the right side has been stained to illustrate the locations of cell bodies. The structure of the cell bodies in each layer is shown enlarged on the far right.

and other parts of the nervous system. The two innermost layers (V and VI) contain axons of cortical cells that pass to the spinal cord and thalamus, while layer IV receives sensory input via the thalamus. Some regions of the cortex, particularly the primary sensory and motor areas, have a columnar organization. A **cortical column** is a small (about 1 mm diameter) cylinder of cortex containing neurons that are involved in processing information about a single discrete part of the body. For example, in the somatosensory system, a single column may represent information arising from a small area on the fingertip. In the motor system, a single column may concern itself with movement of a single joint.

Imaging the Brain

In order for a new diagnostic tool to become useful to the medical community, it must satisfy certain requirements. First, the risks of the procedure must be determined in animal experiments. If the risks and side effects are acceptably low, then tests are conducted on healthy adult volunteers. Extension of a technique to all segments of the population, including young children, pregnant women, and other high-risk groups requires additional careful assessment.

Another criterion that must be met in order for a new tool to be accepted is that the information it provides must be interpretable. It is unacceptable to subject a patient to a test that provides a wealth of data that cannot be used by health care professionals in treating the patient. One way to establish the utility of a new diagnostic tool is by conducting a parallel study on volunteer patients. In such a study a familiar technique and a new

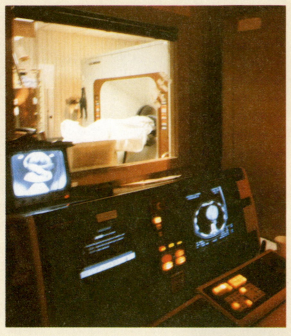

FIGURE 9-A
Apparatus for computerized axial tomography (CAT scan).

FIGURE 9-B
A sagittal magnetic resonance image (MRI scan) of a normal brain.

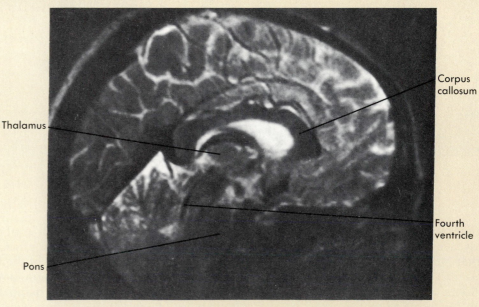

Thalamus

Corpus callosum

Fourth ventricle

Pons

technique are both used, so that the information provided by the new technique can be evaluated in the light of past experience with the established technique. An example of this method of evaluating a new diagnostic tool is provided in studies of two noninvasive methods of imaging the soft tissues of the body, including the brain: (1) computerized axial tomography (CAT scans), and (2) a newer technique, magnetic resonance imaging (MRI). In a CAT scan (Figure 9-A) differences in the absorbance of tissues to X-rays are used to construct an image. In MRI, differences in the way hydrogen nuclei (protons) vibrate in the body's tissues are detected by an externally applied magnetic field. A common feature of these techniques is computer-assisted reconstruction of an image of the brain (for example, Figure 9-B shows an MRI scan equivalent to the anatomical drawing in Figure 9-3). Neither CAT scans nor MRIs would be feasible without the advances in computer speed and memory capacity that have developed over the past two decades.

Both CAT and MRI scans can provide information about the location of tumors, sites of hemorrhage following injury, and other abnormalities. In Figure 9-C, bleeding following a concussion is revealed by both the MRI and CAT scans, but more information about the site and extent of intracranial bleeding is available in the MRI. The difference is due to the fact that the MRI technique has greater contrast resolution than the CAT scan, and bony structures do not obscure the MRI images.

Both CAT and MRI scans have been used to noninvasively locate tumors and other abnormalities in other parts of the body. For many applications, MRI may be the diagnostic tool of choice in the future because it does not require that a patient receive X-rays. However, if information about the presence of bony structures in relation to the site of injury or abnormality is needed, then other diagnostic tools may be more effective than the MRI.

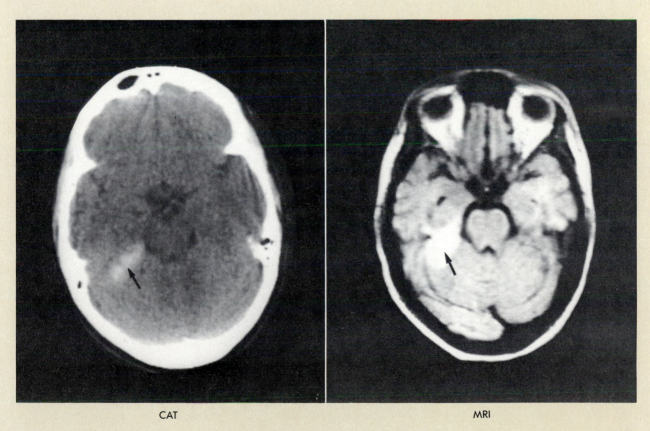

CAT   MRI

FIGURE 9-C
Comparison of an MRI of a hemorrhage in the brain to that obtained via CAT scan. In both photographs the site of bleeding is indicated by the arrow.

1 *What are the major divisions of the central nervous system?*

2 *What is the function of the limbic system? What is the relationship between olfactory pathways and the limbic system?*

3 *What is a cortical column? What is the functional relationship between neurons that make up a cortical column?*

SENSORY PROCESSING IN THE BRAIN
Sensory Receptive Fields

One characteristic of sensory receptors is that they respond to stimulation delivered to specific areas on the body surface. These areas were defined in the previous chapter as **receptive fields** (see Figure 8-4). For example, the receptive field of an individual cutaneous receptor is the area of skin within which its dendrites lie. A stimulus delivered to this area excites the cutaneous receptor. The receptor is indifferent to stimuli delivered to other areas. In the visual system, a photoreceptor responds only to photons that impinge directly on it.

Projection of Sensory Inputs

Conscious sensations arise from activity in sensory pathways that project to the cerebral cortex. The cortex has highly organized regions dedicated to detailed analysis of features from a given sensory modality. Chapter 7 showed how the questions of "What modality?", "Where?", and "How much?" are answered in the nervous system. Information arising from a single population of receptors can follow **divergent pathways** (Figure 9-11) and be mapped in several regions of the brain, so that different relevant features can be extracted and integrated for appropriate responses. This is the principle of **parallel processing**.

Exaggeration of Differences in Input Signals

We tend to ignore parts of objects that are uniform but can easily detect even slight differences in texture (for example, we can easily distinguish between a tennis ball and baseball by touch alone). We also readily identify the boundaries of an object. How is this possible when adjacent receptors in any sensory system often are subjected to stimuli that differ only slightly in intensity?

The ability of the central nervous system to detect small spatial differences in stimulus intensity is enhanced by a process called **lateral inhibition** (Figure 9-12). In this process, the weaker of two signals is diminished by inhibitory inputs arising from the stronger signal. Thus the difference in the

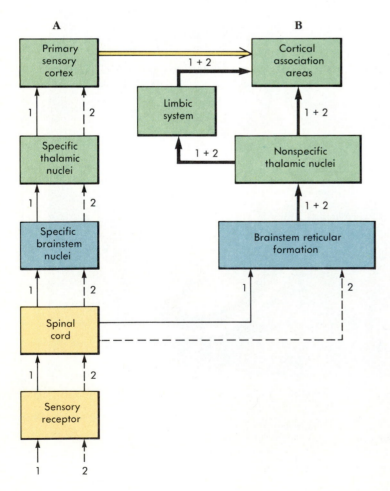

FIGURE 9-11

Divergent pathways of somatosensory inputs.

A   One route by which information from two sensory modalities *(1 and 2)* ascends to the primary cortex is via specific brainstem and thalamic nuclei.

B   In addition to the specific route, the same sensory information also diverges from the spinal cord and projects to the brainstem reticular system. The reticular formation then exerts a nonspecific effect of arousal on a variety of areas of the brain, including the limbic system and cortical association areas. The effect of reticular system activity is to make these areas more sensitive to the specific pathways that project to the association areas from the modality-specific primary cortical regions.

FIGURE 9-12

A A stimulus affects the central one of three sensory receptors most intensely and the lateral ones less intensely. The rate of action potentials of each of the three receptors is shown on the right. Localization of the stimulus may not be precise because the receptors lateral to the stimulus are also activated.
B Lateral inhibitory connections have been added. The most intensely stimulated receptor is still just as active, but it also suppresses the signals from the receptors lateral to it before these signals can reach central processing areas. Thus localization of the stimulus is more precise.

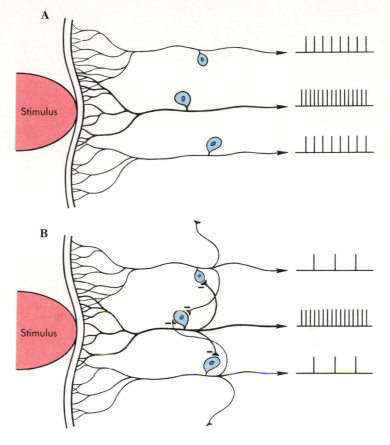

inputs reaching higher centers is magnified. Detection of the boundaries of an object is one example of the general process of **feature extraction,** in which important aspects of a particular modality are emphasized at the expense of others. Another example of feature extraction is the fact that moving objects in the visual field are much more readily detected than stationary ones. This is because many elements of the visual processing system are dedicated to extracting the feature of motion.

Physical Reality and Sensory Perception

The important features of stimuli are those that have survival value. Therefore the sensory receptors themselves, and the sensory processing systems, are tuned to those aspects of the environment that must elicit appropriate responses. Receptors deliver a relative rather than an absolute representation of stimuli. Processing involves detecting differences and recognizing that different stimuli may come from the same source under a variety of conditions. For example, a basin of water may be perceived as warm or cool, depending on the previous experience of temperature receptors. The retina's response to color depends on the overall intensity of light in the surroundings, so that a red object is correctly identified as red even when the spectrum of wavelengths arising from the object is altered in dim lighting. Sensory systems further alter the representation of reality by extracting important features and discarding others. The central nervous system is deluged with unimportant sensory information. Some less-important information must be prevented from reaching the conscious level or it would be impossible to devote selective attention to important features. Different sensory modalities compete for conscious attention, and the level of attention depends on the overall state of arousal of the central nervous system. This is the **filtration principle** of sensory processing.

> 1 What is a receptive field?
> 2 What is the function of lateral inhibition? How does it aid in feature extraction?
> 3 What function is served by divergent pathways for sensory information?

THE SOMATOSENSORY SYSTEM
Dermatome Organization of Somatosensory Receptive Fields

Each spinal nerve innervates a restricted region of the body surface, known as a **dermatome** (Figure 9-13). Sensory nerve trunks contain many fibers, but all sensory fibers fall into one of four fiber size classes, three of which are myelinated and one unmyelinated (Figure 9-14 and Table 9-3). Two classification schemes can be used. In the fiber Group

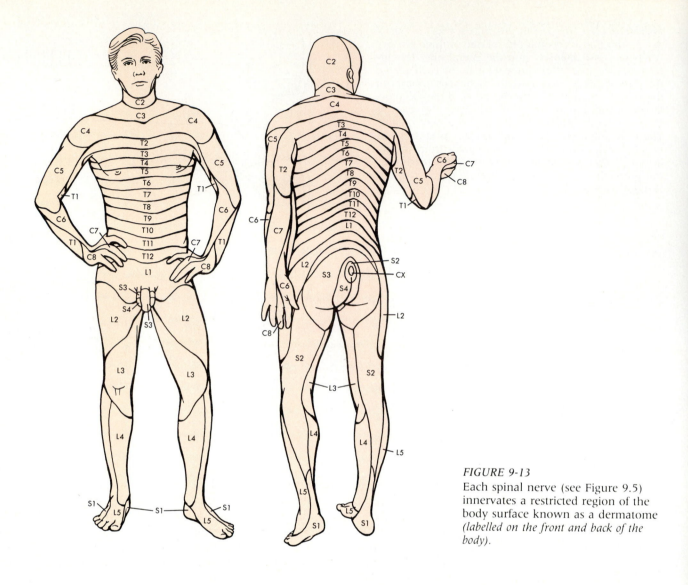

FIGURE 9-13
Each spinal nerve (see Figure 9.5) innervates a restricted region of the body surface known as a dermatome *(labelled on the front and back of the body).*

FIGURE 9-14
Fiber groups are separated by their rates of axonal conduction in the compound action potential of a spinal nerve trunk. Sensory nerve fibers can be divided into four fiber size classes.

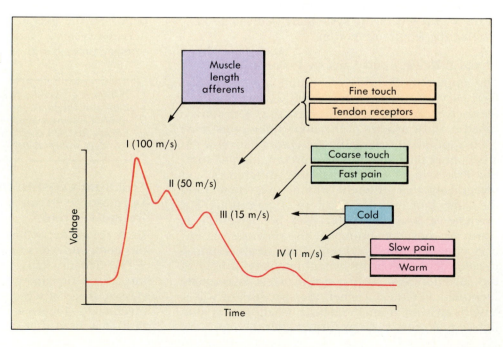

TABLE 9-3   Classes of Sensory Fibers

Class	Diameter	Conduction velocity	Myelination	Function
Group I	13-20 μm	70-110 m/s	Yes	Muscle length
Group II	6-12 μm	25-70 m/s	Yes	Tendon receptor; rapidly adapting touch receptors; Pacinian corpuscle
Group III	1-5 μm	3.5-20 m/s	Yes	Touch; fast pain; cold
Group IV	1 μm or less	Less than 1 m/s	No	Slow pain; temperature; itch; tickle

classification, **Group I** consists of the largest myelinated fibers, 13 to 20 μm in diameter. Myelinated fibers of 6 to 12 μm in diameter are called **Group II**, while myelinated fibers with diameters of 1 to 5 μm are termed **Group III**. Unmyelinated fibers are referred to as **Group IV** and are about 1 μm in diameter.

The second classification scheme refers to fiber types. Type A_{alpha} myelinated fibers have conduction velocities similar to Group I fibers, Type A_{beta} velocities are similar to Group II fibers, and Type A_{delta} velocities are similar to Group III fibers. Unmyelinated fibers (Group IV) are referred to as Type C fibers. Sensory fibers from muscles have all four fiber classes and are often referred to using the Groups I, II, III, and IV designation. Sensory fibers from the skin lack the largest size class of myelinated fibers (Group I or A_{alpha}) and are frequently described using the Type classification. The Group classification scheme will be used in the remainder of this text.

The somatosensory system has deep receptors and surface, or cutaneous, receptors. Deep receptors include proprioceptors signalling joint position and muscle length. These proprioceptors are innervated by Group I fibers. Other deep receptors are sensitive to pressure and are innervated by Group II fibers. Cutaneous receptors detect simple contact; complex mechanical stimuli, such as vibration; temperature (hot and cold are separate sensations); and surface pain. They are innervated by Groups II, III, and IV fibers.

Touch Receptors

Several types of mechanically sensitive receptors are present in the dermis and subcutaneous tissue (Figure 9-15). Morphologically specialized receptors mediate fine touch and vibration and are most dense on areas such as the fingertips and face. They are innervated by Group II myelinated fibers and are used to localize cutaneous stimuli precisely. Morphologically specialized receptors can be phasic or tonic. Phasic receptors include **hair follicle receptors** and **Meissner's corpuscles** in hairless

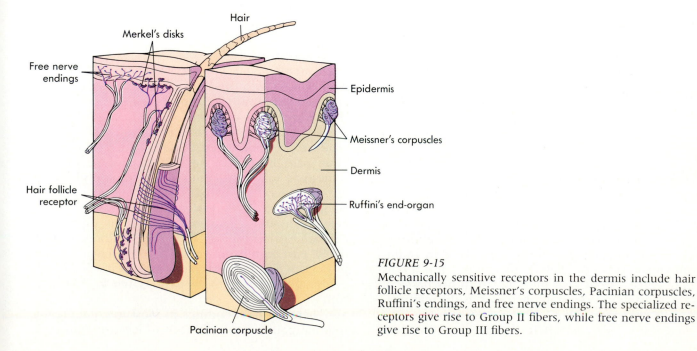

FIGURE 9-15

Mechanically sensitive receptors in the dermis include hair follicle receptors, Meissner's corpuscles, Pacinian corpuscles, Ruffini's endings, and free nerve endings. The specialized receptors give rise to Group II fibers, while free nerve endings give rise to Group III fibers.

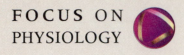
Selective Stimulation and Anesthetic Block Demonstrate Fiber Specificity

Sensory fiber groups differ in size and thus conduction velocity. Peaks in the compound action potential of a nerve trunk correspond to conduction velocity, so that fiber groups will appear as distinct bumps. Somatosensory modalities will be segregated to different components of the compound action potential because modalities are confined to specific fiber groups. This can be demonstrated in several ways. When a nerve is stimulated electrically, the largest fibers are activated first. If electrodes are attached to a subject's finger and the subject is asked to describe what she feels as the current is increased, she will first report the illusion of mechanical contacts as Group II fibers are activated. When Group III fibers are stimulated, the subject will feel fast pain, but when Group IV (Type C) fibers are activated the subjective experience is slow pain. At the lowest concentration a local anesthetic blocks smaller nerve fibers, so that slow pain is lost. Increasing the concentration eliminates first temperature, then fast pain, and lastly touch. Pressure blocks large fibers first, so that touch is lost before pain (think about what happens when your arm "falls asleep").

areas. The most effective stimulus for both is a periodic displacement at a frequency of 30 to 40 Hz (cycles per second) and results in the sensation of flutter. The **Pacinian corpuscle** is also phasic but is found deeper in the subcutaneous tissue, and its most effective stimulus is a high-frequency periodic displacement (300 Hz) sensed as vibration. There are two types of tonic receptors. **Touch dome endings (Merkel's discs)** are located near the surface of the skin, while **Ruffini's endings** are in the dermis.

In most areas of the body surface, mechanical contact causes activity in the smaller myelinated Group III fibers. The ability to localize stimuli delivered to these receptors is less acute. The resulting sensation is referred to as "coarse" touch; it arises from diffusely distributed **free nerve endings** present throughout the dermis and subcutaneous tissue.

Temperature Receptors

There is no such thing as absolute cold, only varying degrees of warmth. However, separate **warm** and **cold receptors** are identified by the way they respond to changes in temperature. Temperature receptors are free nerve endings (specialized in an unknown way) that can detect changes of a few tenths of a degree. "Cold" receptors begin to discharge at 35° C, and their firing rate increases as the temperature decreases to 20° C. Warm receptors are activated above 30° C, and their discharge rate increases

with increasing temperature to 45° C. The combined response of warm and cold receptors can be compared to the shape of an inverted bell, with minimum activity at temperatures slightly lower than normal body temperature. Cold receptors are 5 to 10 times more abundant than warm receptors. Both are most dense on the face and hands. Axons of warm receptors are Group IV (Type C) fibers, while axons of cold receptors may be either Group III or Group IV (Type C) fibers.

Pain Receptors

There are two types of pain. **Fast pain** is the type of pain evoked by a needle jab, for example. It is well localized but rapidly disappears without residual effects. **Slow pain** describes the unpleasant burning sensation that follows tissue injury such as might be produced by prolonged contact with a hot object. Slow pain is difficult to localize and persists for some time following removal of the painful stimulus. Slow pain also evokes autonomic reflexes and emotional reactions. Fast and slow pain are both detected by free nerve endings, but these pain signals are carried by Group III and IV fibers, respectively. Fast pain is not sensitive to the narcotics that physicians use to block pain, and its mechanism is poorly understood. Slow pain is blocked by narcotics.

There are pain endings in most internal organs. Internal pain is more difficult to localize than cutaneous pain because visceral pain fibers enter the spinal cord along with pain fibers from the skin, and the nervous system seems unable to distinguish the exact origin of these stimuli. As a result, visceral pain is often perceived as if it had come from a cutaneous site. For example, individuals experiencing a heart attack feel pain in the chest and left arm. This is called **referred pain** because the nervous system maps the sensation to a site different from its actual origin.

For a time after amputation, the amputee experiences **phantom limb** sensations—an illusion that the amputated limb is still present. This effect is due to the fact that action potentials arising at the severed sensory nerve axons are interpreted by the brain as sensory inputs to the receptors of the absent limb.

The Dorsal Column Pathway

The spinal cord mediates **segmental reflexes** in which the reflex loop is completed within one spinal segment. These are especially important in control of posture and movement (see Chapter 11, p. 259) and involve branches of sensory afferents that remain within a given segment of the spinal cord. For coordinated activities of the body, sensory information must also be passed to higher centers of the central nervous system.

Group I and II fibers mediating proprioception and fine touch branch after entering the spinal cord and ascend to higher centers by a specific, anatomically ordered route that is called the **dorsal column pathway** (or **lemniscal system**) (Figure 9-16). The dorsal column pathway is the most direct way somatosensory signals can reach the cortex because there are no synapses until the axons reach the dorsal column nuclei of the medulla. Postsynaptic fibers **decussate** (cross to the opposite side of the nervous system) in the upper brainstem and then ascend to synapse in a specific sensory area of the thalamus. From the thalamus, axons go to the primary somatosensory region of the cortex.

Somatosensory Maps in the Dorsal Column Pathway

The dorsal column pathway is the only spinal somatosensory tract that contains an internal map of the body surface. Any such map can be referred to as a **somatotopic representation**. An internal neural map is generated because afferent fibers enter the dorsal columns of the spinal cord and ascend to higher centers in an orderly fashion. Fibers entering a particular segment of the spinal cord (corresponding to a single dermatome) push the fibers already present from lower segments toward the center of the cord, so that fibers from the lower portions of the body (the legs, for example) are more medial within the spinal cord and brainstem.

The afferent fibers from individual dermatomes

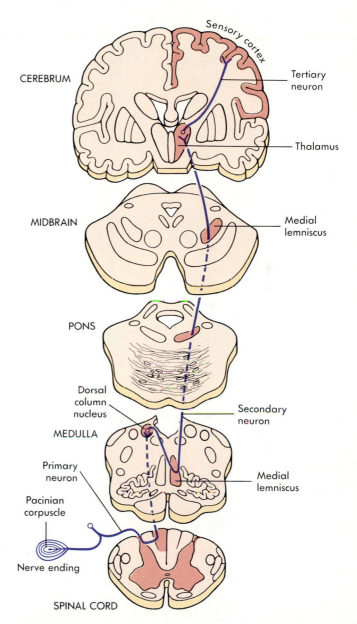

FIGURE 9-16

The dorsal column (medial lemniscal) pathway is the most direct route for somatosensory information carried by Group I and II fibers to reach the cortex. Branches of sensory fibers ascend the spinal cord to the dorsal column nucleus (nucleus gracilis) in the medulla. The postsynaptic fibers of secondary neurons decussate (cross) in the medulla and ascend through the midbrain into the somatosensory nuclei of the thalamus, where they synapse on tertiary neurons whose fibers pass to the sensory cortex. The dorsal column pathway is the only tract that contains a map of the body surface (referred to as somatotopic). The primary somatosensory cortex gets information only from the opposite side of the body.

FIGURE 9-17

The somatosensory homunculus of the left cortex. The medial surface of the somatosensory part of the parietal cortex receives information from the legs and feet. More lateral regions project from the anterior parts of the body, with the hand and face regions occupying a large part of the map. These are the body regions with the highest density of receptors.

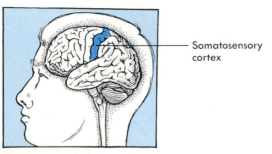

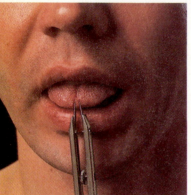

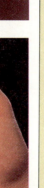

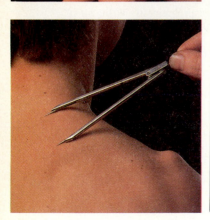

A
Somatosensory cortex

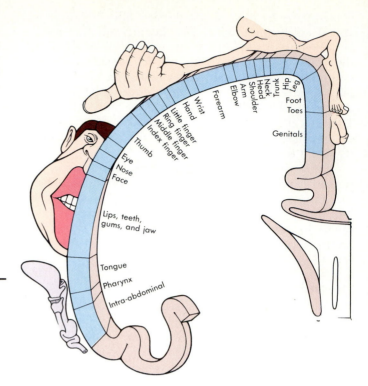

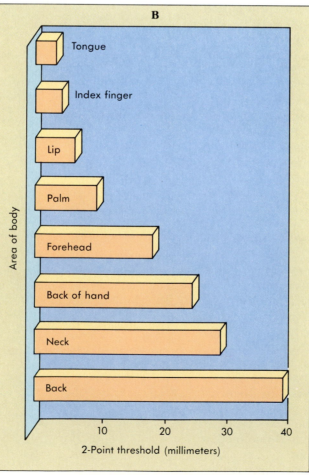

B

- Tongue
- Index finger
- Lip
- Palm
- Forehead
- Back of hand
- Neck
- Back

Area of body

10   20   30   40

2-Point threshold (millimeters)

FIGURE 9-18

A Two-point discrimination can be demonstrated by touching a person's skin with the two points of a compass. When the two points are close together, the individual perceives only one point. When the points of the compass are opened wider, the person becomes aware of two points.

B A bar graph showing typical two-point discrimination thresholds for different parts of the body. Below the threshold, two points are perceived as one. Above the threshold, two separate points can be discriminated. Compare the size of the thresholds with the relative size of the sensory cortex for each region as shown in Figure 9-17.

and specific submodalities also remain segregated, so the somatosensory system contains a three-dimensional map of position and modality by the time impulses reach the dorsal column nuclei. A similar three-dimensional map is maintained when fibers from the dorsal column nuclei project to the thalamus. The result is that a particular stimulus on some point on the body surface activates small groups of central neurons corresponding to that spot on the brain's maps. Fibers from the thalamus produce several maps of the body surface on the cortex. For example, in the primary somatosensory cortex, receptive fields from more distal regions are located at the top of a cerebral hemisphere and also on the medial surface. Figure 9-17 shows the somatotopic representation of the body surface on the primary somatosensory cortex. This cortical map, called the **somatosensory homunculus,** is exaggerated for those areas of the body surface that have the greatest density of receptors. Because the ascending nerve tracts that pass to the sensory cortex decussate, the left cortex represents the right side of the body and the right cortex represents the left side of the body.

Two-Point Discrimination

One measure of sensitivity of the somatosensory system is the **two-point discrimination test.** This is performed by stimulating the skin with a divider and measuring the minimum separation that can be perceived by the subject as two separate points. One determinant of the two-point resolution is the relative density of receptors. More afferents come from the finger tips than from regions where tactile sensation is poor, so the ability to resolve nearby stimuli varies from 1 to 2 mm on the face and fingers to as much as 40 mm on the back (Figure 9-18). Another determinant of sensitivity is the degree of **convergence** of afferent pathways. Convergence in the somatosensory system is the extent to which a single central neuron receives inputs derived from different points on the skin surface. If the level of convergence is high for a given body region, two-point resolution will be correspondingly poor.

In contrast to the complete decussation seen in the pathways leading to the primary somatosensory cortex, the other cortical somatosensory areas receive fibers from both sides of the spinal cord and thus receive bilateral signals. The internal maps in these areas are superimposed. For example, a particular column of cortical neurons might be activated (or inhibited) by stimuli to the left and right forefingers. The usefulness of bilateral input is reflected in the ability to compare the weights of hand-held objects.

Nonspecific Somatosensory Pathways

The pathway used by Group III and IV fibers (Figure 9-19) is different from that of Group I and II fibers. Group III and IV fibers enter the spinal cord, giving off branches that cross the spinal cord in the same spinal segment and ascend in a different tract of the spinal cord (compare this route with that shown for the dorsal column projections in Figure 9-16). All Group IV fibers (warmth, cold, and slow pain) project either to nonspecific thalamic nu-

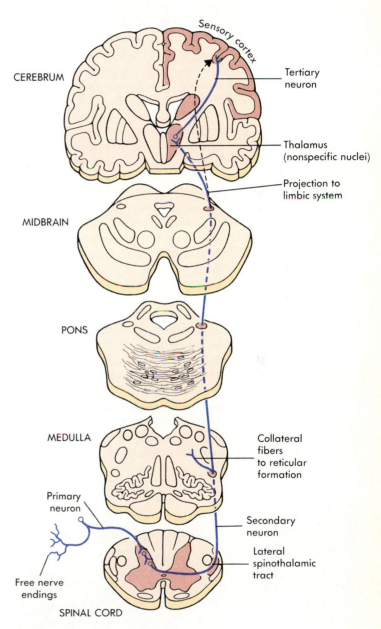

FIGURE 9-19

The nonspecific ascending pathways. Group III and IV (Type C) fibers synapse on interneurons that cross the cord and ascend in different tracts. All Group IV fibers project to nonspecific thalamic nuclei or to the reticular formation. Their projections influence many cortical regions and the limbic system. Group III fibers project to the thalamic nuclei that receive dorsal column fibers by way of the spinothalamic tract (or anterolateral system) but do not reach the cortex directly.

clei or to the brainstem reticular formation. These tracts include the slow pain pathways. Since these pathways do not reach the cortex directly, the brainstem and limbic system must mediate perception of slow pain, with the result that slow pain cannot be precisely localized.

Group III fibers (crude touch, cold, and fast pain) go to the same thalamic nuclei that receive dorsal column fibers. This pathway is the **spinothalamic tract** (or **anterolateral system**). The fast pain pathway does not reach the cortex directly either, but since fast pain can be localized relatively well, the pathway must project to a cortical map.

The Reticular Activating System

Nonspecific inputs from all sensory modalities and feedback from the cortex converge on a network of neurons, the **reticular activating system,** located in the reticular formation of the brainstem and thalamus (Figure 9-20). The activator portion of the system maintains consciousness and alertness and lies in the medial portion of the mesencephalon and upper pons. It has two major outputs: (1) to higher levels of the brain, and (2) through the reticulospinal tracts into the spinal cord. The major neurotransmitter in this part of the reticular activating system is acetylcholine. The inhibitory portion of the system is called the **substantia nigra.** The dopamine secreted at axon terminals of these neurons inhibits cells in the corpus striatum, hypothalamus, cerebral cortex, and limbic system.

The **locus ceruleus** is a portion of the reticular activating system that utilizes norepinephrine as its transmitter. Projections from the locus ceruleus go to many areas in the brain, exciting some areas and inhibiting others. The locus ceruleus is important for one form of sleep, **REM (rapid eye movement) sleep.** The **raphe nuclei,** also located in the brainstem, project to the diencephalon and spinal cord. The predominant transmitter for this region of the reticular activating system is serotonin, and the release of serotonin is associated with other forms of sleep. The raphe nuclei have an additional, important role in the perception of pain.

Inputs to the reticular activating system from cutaneous sensory receptors or from muscle and joint receptors increase alertness through these projections to almost all areas of the diencephalon and cortex. Pain and proprioceptive inputs signifying a change in body position are signals that typically require immediate action by the brain and are quite effective in altering the output of the reticular activating system. Such sensory signals can immediately arouse a sleeping person, which is why shaking is an effective means of awaking someone who is sleeping.

Projections from the brainstem reticular formation to the cortex and other brain regions are depressed by anesthetics and altered by specific neurotransmitters during sleep states. In addition to its role in regulating the wakeful and sleep states, the reticular activating system plays an important role in cardiovascular, pulmonary, and digestive system regulation.

1 What is meant by the term somatotopic representation? Why are the areas occupied by the hands and face so large in proportion to those occupied by the legs and torso?
2 What is the functional difference between specific sensory pathways and nonspecific pathways?
3 What is the general role of the reticular activating system? What inputs affect it?

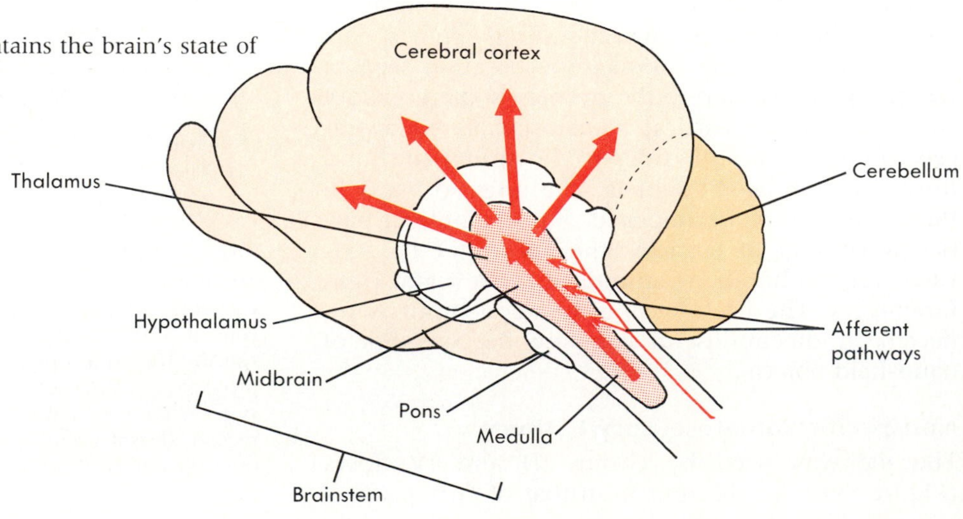

FIGURE 9-20
The reticular activating system maintains the brain's state of awareness.

Cerebral cortex

Cerebellum

Thalamus

Afferent pathways

Hypothalamus

Midbrain

Pons

Medulla

Brainstem

LEARNING AND MEMORY
Distinctions Between Learning and Memory

Learning is the process whereby experience modifies centrally controlled behavior. **Memory** is the storage and retrieval of experience. The two concepts are related in that any relatively long-lasting change in the response to a particular set of conditions requires storage and retrieval of sensory events or ideas. Memories can be subdivided into those that are verbal (declarative, or "what") and nonverbal (procedural, or "how").

Short-Term and Long-Term Memory Stores

Memory operates in two different time scales: (1) short-term memory, and (2) long-term memory. Short-term memory stores information over a time scale of seconds to minutes. For example, this is the memory mode used when one looks up a telephone number and remembers it long enough to dial correctly. This information resides in short-term memory as long as it is the focus of immediate attention. In contrast, a telephone number that has been transferred to the long-term memory store can sometimes be retrieved hours, days, or months later. Long-term memory involves some relatively permanent change in the brain. Shifting events and ideas (or portions of them) from short-term to long-term memory is referred to as **memory consolidation.**

Since long-term memory is not disrupted by procedures that affect the normal conduction of nerve impulses, there must be some reasonably permanent alteration in the brain. Neuroscientists now believe that changes in synaptic efficacy are the basis of long-term memory. A structural change could take place at particular synapses, the response of postsynaptic receptors to a pattern of inputs could be altered, or new receptors could be synthesized. Some experiments suggest that consolidation of short-term into long-term memory can be blocked by agents that inhibit protein synthesis, and, more recently, neuroscientists have found that the enkephalins and endorphins (see Chapter 7) also interfere with memory formation. While long-term memory clearly exists, specific cortical sites cannot be identified for particular memories because relatively extensive cortical damage does not selectively remove memories. Many memories persist in spite of the damage, and ability to access them gradually recovers with time. Either there are multiple copies of memories throughout the brain or long-term memory involves the storage of individual bits of information from which the original experience can be reconstructed, even though the data may be incomplete.

Cerebral Structures Contributing to Memory

Regions of the temporal lobes, the hippocampus, and a structure known as the **amygdala** (a part of the limbic system) are involved in both short-term memory and memory consolidation. These areas are thought to contribute to memory consolidation because lesions in them selectively affect the ability to process recent events into long-term memories. Such lesions do not affect previously stored long-term memories but prevent subsequent formation of long-term memories.

Models of Learning and Memory

Many organisms can modify their behavior, and some of these changes can persist for extended periods of time. While the human brain has billions of neurons, certain invertebrates have only a few thousand neurons. Some neurons with known functions in behavior can be identified from one animal to another. This makes it possible to record behavioral changes controlled by a particular cell. Two simple forms of learning that occur in a snail called *Aplysia* are **habituation** and **sensitization.** In habituation a stimulus that elicits a reflex is presented on a regular basis, and the animal gradually stops responding. If a train comes by every night at 2 AM, for example, an individual will eventually no longer wake up. Sensitization involves an increased reflex response to some harmless event when the animal has been aroused by a noxious stimulus. Both of these types of learning are called **nonassociative learning,** because there is no association formed between a stimulus that normally evokes a response and some other neutral stimulus.

Habituation and sensitization in *Aplysia* demonstrate both short-term and long-term patterns of memory formation. Short periods of stimulation produce habituation lasting for only a few minutes, but repeated training sessions modify the animal's response for weeks. Behavioral modifications lasting for days or weeks involve intracellular biochemical processes. In both cases there are changes in the amount of Ca^{++} that enters the presynaptic terminal that alter the release of neurotransmitter from one synapse in the motor pathway. In particular, the activation of an intracellular second messenger produces a long-lasting structural modification of membrane ion channels, causing them to respond differently to a particular synaptic input. Are these behavioral changes related to human memory? The mechanisms revealed in the experiments in *Aplysia* have also been found in the vertebrate nervous system and are likely to be included among the types of neuronal change that contribute to learning and memory in human beings.

Associative Learning

Associative learning causes an animal to make a response to a neutral stimulus after that stimulus has been paired with reward or punishment. This was first described experimentally by the Russian scientist Ivan Pavlov, who taught dogs to salivate at the sound of a bell by pairing the neutral bell stimulus with a food reward. The same type of learning is present in simple animals such as snails and slugs. The mechanism both in invertebrates and in mammals may also involve a cascade of events that changes membrane channel properties.

One category of associative learning that is exceptionally effective is **aversion learning** that involves taste pathways. A lasting modification of behavior can result from a single pairing of a novel taste stimulus with a visceral response such as nausea, even though the two stimuli may occur hours apart. This helps animals learn to avoid poisonous foods. The pathway for taste information meets the pathway for sensory information from the intestines in a nucleus in the brainstem that is also involved in arousal. This observation points to one feature that is common to everyday experience with learning and memory: that memory consolidation operates against a background of emotion. Very vivid and detailed memories can be consolidated at times of emotional stress. Much research effort is being concentrated on factors that operate in the brain to modulate the process of memory consolidation.

SPECIALIZATIONS OF THE HUMAN BRAIN
Hemispheric Specializations

The two hemispheres of the human brain are roughly symmetrical, but different functions can be identified with the right and left hemispheres. The best studied hemispheric specialization is language. The hemisphere in which language ability resides has been called the **dominant hemisphere.** Cerebral dominance is not the same as handedness because the left side of the cortex is dominant in about 95% of all right-handed people and 70% of all left-handed people. There are two language areas in the dominant hemisphere: (1) **Wernicke's area,** and (2) **Broca's area** (Figure 9-21). Wernicke's area, located in the parietal lobe between the primary auditory and visual areas, is important for interpretation of language and for formulation of thoughts into speech. Broca's area, found near that part of the motor cortex controlling the face, is connected with Wernicke's area by a thick bundle of nerve fibers. Broca's area is responsible for generation of the patterns of motor output that result in meaningful speech.

Evidence for Language Specialization

One type of evidence of hemispheric specialization is provided by studies of individuals who have suffered strokes that caused defects in their ability to communicate. Damage to Broca's area leads to a condition termed **expressive aphasia** in which concepts are communicated accurately, but speech does not flow smoothly and consists of a series of phrases with numerous omissions. However, the phrases are logically related and contain informa-

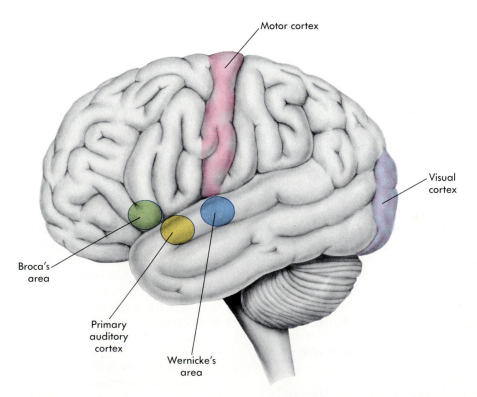

Motor cortex

Visual cortex

Broca's area

Primary auditory cortex

Wernicke's area

FIGURE 9-21

The location of cortical areas specialized for language in relation to the primary motor cortex and primary sensory cortical areas.

The Ear Can Learn from the Eye

In several places in the brain, topographic maps of space are formed by overlapping projections from two or more sensory systems. For instance, pit vipers, snakes that can locate prey on the basis of infrared radiation from warm bodies, have maps of the world in which inputs from visual and pit organ receptors provide information from the same point in space. Thus a mouse can be recognized on the basis of both its appearance and its pattern of radiation above background level.

Sounds are located in space by comparing the inputs from the two ears. This information is mapped in a separate place in the brain from the tonotopic map that functions in recognition and comparison of sounds of different frequencies. The formation of this topographic map has not been extensively studied in human beings, but auditory maps of space have been studied in other species. The best understanding of

how an auditory map of space is organized has come from studies of owls. These birds, which are nocturnal predators, are able to catch prey on the basis of a single rustling sound, and blind owls perform this task as efficiently as ones with normal vision.

Studies in which young owls had one ear plugged for a period of time during development have revealed that visual and auditory information are used together in the formation of the map. The auditory inputs from the owls' plugged ears were greatly reduced, but the owls still learned to locate prey by comparing the sounds arriving at their two ears. If the plug was then removed, tests performed in the dark revealed that the owls misguessed the location of the prey in a predictable way: they assumed that a sound originating close to the midline was closer to the side with the formerly plugged

ear, because they were used to receiving little information from that ear unless the sound was very close to that side of the head.

With experience, the young owls could adjust to the removal of the plug and again learn to accurately locate auditory stimuli in space. The type of information that is crucial to the adjustment is visual experience. Only through visually locating the source of a sound in space can the owls learn to properly interpret the information arriving at the two ears. Thus comparison of sights with sounds is necessary for the formation of an auditory map of space. While experience influences the strength of synaptic connections in young owls, this adaptability is lost after a critical period, and owls that did not have the ear plug removed until they were adults never learned to accurately locate prey in the dark.

---

tion. Similar defects occur in writing. Individuals with expressive aphasia are aware of the errors they are making and are able to perform tasks of comparable complexity, such as singing complex melodies or drawing. They also comprehend verbal messages normally. The ability to comprehend language suggests that incoming auditory information can proceed directly to Wernicke's area for decoding without having to be first processed by Broca's area.

Damage to Wernicke's area is more serious because it impairs the ability to express and comprehend language. Patients with damage to Wernicke's area exhibit **receptive aphasia.** The speech of receptive aphasics has been described as "word salad." They can speak fluently and produce grammatically correct sentences, but the sentences are illogical and unrelated to actual events or intentions. Persons with receptive aphasia cannot comprehend either verbal or written commands.

Two other forms of aphasia further illustrate the role of the language areas. In **conduction aphasia** the neural connection between Broca's area and

Wernicke's area is disrupted and individuals can understand words, but they cannot repeat what has just been said. Lesions in the pathways connecting Wernicke's area with the primary auditory cortex lead to a condition called **word deafness** in which comprehension is impaired only for spoken language, so that speaking, reading, and writing remain unaffected.

In addition to Wernicke's area, the parietal lobe contains the primary somatosensory processing area. Lesions in the primary somatosensory cortex cause sensory deficits. Adjacent areas are the integrating sites for the sensory modalities. One such area is believed to help synthesize sensory experiences into an overall body image. Lesions at that site in one parietal lobe cause individuals to neglect the opposite side of the body, almost as if it did not exist. The parietal lobe of the dominant hemisphere has a major role in controlling movement of the whole body. One consequence of this is that most left-dominant individuals are right-handed and right-footed.

Split-Brain Individuals

An important source of information about hemispheric specialization was provided accidentally by some surgeries in which the corpus callosum was severed. The corpus callosum is a large bundle of axons that connect the two hemispheres. The surgeries were performed in hopes of limiting the spread of epileptic seizures from one hemisphere to the other, but resulted in what have been termed **split-brain** patients. In right-handed patients, the left cortex receives sensory input from the right half of the body, including the right eye and ear. However, the two hemispheres cannot communicate with one another. The left cortex can initiate voluntary movements of the right arm and leg so that the patient can write and speak. The right hemisphere lacks the language areas, so that split-brain patients cannot identify objects held in the left hand or present in the left visual field. However, such patients could use their left hands to solve nonverbal test problems or to draw pictures.

Specializations of the Right Hemisphere

A well-studied specialization of the right cortex confers the ability to recognize individuals by their facial features alone. Trauma to the ventral portion of the occipital lobe anterior to the primary visual cortex in the nondominant hemisphere eliminates the capacity to recall faces. Reading, writing, and oral comprehension remain normal, and patients with this disability can still recognize their acquaintances by their voices.

Trauma to other parts of the right hemisphere may lead to an inability to appreciate spatial relationships and may impair musical activities such as singing, but does not usually result in aphasia. Consolidation of verbal memories requires the help of the left hemisphere. The right hemisphere is important for consolidation of memories of nonverbal experiences.

All of these observations suggest that the two hemispheres handle information differently. The dominant hemisphere is adept at sequential reasoning, such as that needed to formulate a sentence. The nondominant hemisphere is adept at spatial reasoning, such as that needed to assemble a puzzle or draw a picture.

Cortical Regions Involved in Memory, Emotion, and Self-Image

Each lobe of the cortex is associated with specific functions that have important implications for personality. Lesions in the frontal lobe affect emotions by eliminating rage and decreasing anxiety. Damage to the frontal lobe also impairs the ability to elaborate thoughts, to set priorities, and to anticipate the consequences of actions.

During the late 1950s it became possible to perform brain surgery to relieve epilepsy. During such surgeries the patient can remain conscious under local anesthesia because the brain itself is insensitive to pain. In fact, the cooperation of the patient is important, because frequently it is necessary to map the cortex to determine exactly what functional areas will be disrupted by the surgery. In the course of many such surgeries, the American scientist Wilder Penfield had the opportunity to explore most of the cortical surface with stimulating electrodes. He found that stimulation of the primary visual, auditory, and somatosensory areas did not mimic well-defined sensory experiences, but only rudimentary hallucinations such as flashes of color, simple sounds, or vague contacts. On the other hand, temporal lobe stimulation often evoked coherent, detailed memories or fantasies of sensory experiences along with associated emotions. This accounts for the observation that the patients often had a dreamlike appearance. Initially, these experiments were interpreted as indicating that the temporal lobe had access to a large store of memories of past experiences, even those that had been "forgotten." More recent evidence suggests that in such experiments the temporal lobe stimulation spreads to the limbic system, a portion of the brain known to be involved with both memory and emotion. However, patients with damage to the right temporal lobe do have deficits in processing spatial and visual information and in remembering such information.

SLEEP AND THE WAKEFUL STATE
Sleep as an Active Function

During the waking hours of the daily cycle, activity reaching the cortex by way of the reticular formation stimulates the cortex and increases the ability of the brain to attend to information provided by specific sensory pathways.

An early hypothesis about sleep held that falling asleep involved a decrease in the activation provided to the cortex by the reticular formation. Sleep would then involve a transient loss of consciousness resembling the comatose state that follows severe head injury. It is now known that the reticular activating system controls both sleep and the waking state. Sleep is an active multistate process that involves output from the reticular activating system and input to it from many regions of the brain. Although the regional pattern of brain activity differs in sleep from that during waking, overall the brain consumes as much energy during sleep as during the waking state. In contrast, energy use by the brain is depressed in comatose persons.

Stages of Sleep

Three criteria define the stages of sleep: (1) the electrical activity of the cortex as measured in an **electroencephalogram (EEG),** (2) the ease with which

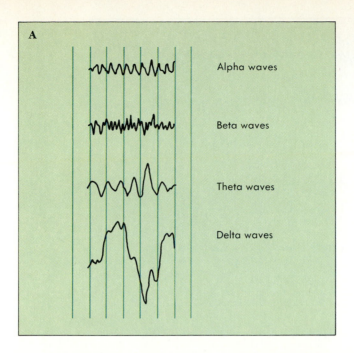

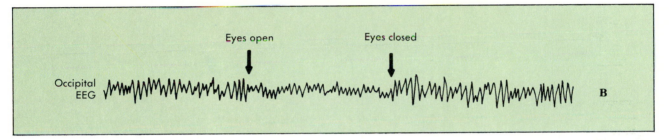

FIGURE 9-22
A   The EEG is a measure of the activity level of the cortex. There are four basic rhythms of the brain: alpha, beta, theta, and delta waves.
B   Recordings from the occipital cortex change when a relaxed subject with eyes closed is asked to open his or her eyes.

the sleeping individual can be aroused, and (3) muscular tone. An EEG represents the external current generated by neuronal activity in the cortex. The more synchronized this activity is, the larger the extracellularly recorded signal.

In a relaxed individual whose eyes are shut, the pattern of activity in the EEG consists of large slow waves that occur at a frequency of 8 to 13 Hz; these are referred to as **alpha waves** (Figure 9-22). Alpha waves are most easily recorded from the occipital lobe. In an alert subject whose eyes are open, electrical activity of the cortex is more desynchronized and thus has a lower amplitude. These waves are faster (18 to 20 Hz) and are designated as **beta waves.** The EEG of an alert individual is less synchronized because multiple sensory inputs are being received, processed, and translated into motor activities.

The first change seen in the EEG with the onset of drowsiness is slowing and a reduction in the overall amplitude of the waves (Figure 9-23, A). In the first stage of sleep, **theta waves** make their appearance. In the second stage, bursts of high-frequency activity known as sleep spindles appear, along with occasional large-amplitude slow **delta waves.** Delta waves become more frequent in the third stage of sleep. In the fourth stage, **deep sleep,** the EEG is dominated by delta waves. Slow-wave sleep also involves a decrease in arousability, skeletal muscle tone, heart rate, blood pressure, and respiratory frequency.

Rapid Eye Movement Sleep

Although the stages of slow-wave sleep are relatively orderly, another phase of the sleep cycle appears paradoxical. During the rapid eye movement (REM) phase of sleep, the EEG resembles that seen in a relaxed awake individual, and the heart rate, blood pressure, and respirations are all increased. Although eye movements occur, skeletal muscle tone is decreased. Individuals in REM sleep are difficult to arouse and are more likely to awaken spontaneously than those in the deeper phases of slow-wave sleep. Experimental studies in which subjects are purposely awakened during various stages have demonstrated that dreaming occurs during REM

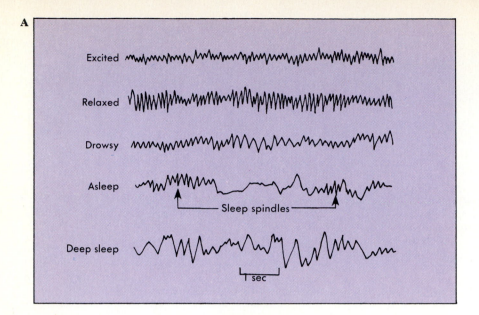

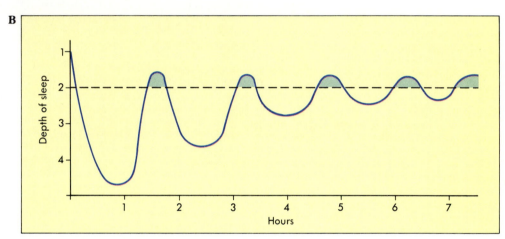

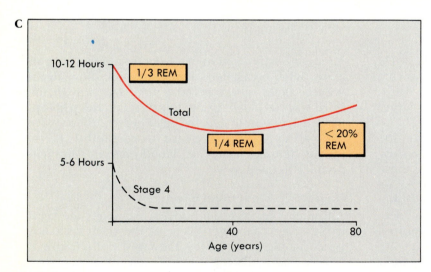

FIGURE 9-23

A Changes in the EEG in a relaxed individual, someone who is drowsy, and one in successively deeper stages of sleep.

B The sequence of sleep cycles in a typical night. Each cycle is terminated by an episode of dreaming in REM sleep *(shading)*. Spontaneous waking can occur most readily from the REM sleep stage.

C The overall amount of sleeping time *(solid curve)*, the proportion of this that is REM sleep *(boxes)*, and the amount of time spent in slow-wave sleep *(broken curve)* change with age.

stage sleep. The rapid eye movements that occur during this stage are similar to the tracking movements made by the eyes during waking, suggesting that dreamers "watch" their dreams.

The Sleep Cycle

Figure 9-23, *B* shows that when an individual starts to fall asleep there is an initial progression through successively deeper levels of slow-wave sleep (stages 1 to 4) that typically requires a half-hour or so. After a relatively brief period in stage 4 sleep, the process begins to reverse, but then is usually interrupted by an abrupt transition to REM sleep. After a variable amount of time the REM phase ends, and there is a return to progressively deeper levels of slow-wave sleep. This cycle is repeated several times and is terminated by spontaneous awakening, usually during the final period of REM sleep. There are changes in the sleep cycle as the night progresses. Slow-wave sleep is reached very early, with more time being spent in stages 2 and 3 as morning approaches. In contrast, REM duration increases from only a few minutes initially to 30 to 45 minutes as successive REM cycles are entered throughout the night.

The human sleep pattern is also age dependent (Figure 9-23, *C*). The total amount of sleep time and the amount of stage 4 sleep decrease with age, reaching a minimum in middle age. The relative amount of REM sleep decreases from 50% in infants, to 25% in middle age, to less than 20% in the elderly. Thus as age advances, more time is spent in the transitional stages of sleep and less in REM sleep and slow-wave sleep.

1 *What is the difference between learning and memory?*
2 *What is the difference between associative and nonassociative learning?*
3 *What functions does the dominant hemisphere dominate?*
4 *What are the features that distinguish slow-wave sleep from REM sleep?*

SUMMARY

1. The nervous system is divisible into a **central** component, consisting of the brain and spinal cord, and a **peripheral** component, consisting of nerve axons, dorsal root ganglia, and autonomic ganglia. The brain is divisible into: the **brainstem** (made up of the pons, medulla, and midbrain), the cerebellum, the **diencephalon** (made up of the thalamus and hypothalamus), and **telencephalon** (cerebral hemispheres). In the central nervous system, regions containing cell bodies are **gray matter**; those consisting of axons are **white matter**.

2. Sensory pathways commonly **diverge** to carry signals to more than one brain region for **parallel processing**. Central processing areas **filter** the incoming signals, **extracting** novel or important **features** for further analysis. Central processing areas are usually organized **topically** (somatotopic, tonotopic, etc.), so that receptors from a spot on the body surface project to a corresponding spot in the relevant brain area.

3. Somatosensory receptors are morphologically specialized for different submodalities of touch, vibration, temperature, and pain. Ruffini endings and touch dome endings mediate touch, and Pacinian corpuscles sense vibration. Temperature and pain are sensed by free nerve endings.

4. The **dorsal columns** of the spinal cord are labelled lines that carry sensory pathways to specific cortical processing regions. The nonspecific pathways exercise a general arousal effect on the brain by way of the reticular activating system.

5. **Learning** occurs when sensory inputs result in altered behavior. The behavior alteration may involve responses to repeated stimuli **(habituation and sensitization)** or induction of a response to a neutral stimulus **(associative learning)**. Information may be stored in **verbal** and **nonverbal** forms, and it is not possible to identify a single discrete brain region in which memory stores are located. **Short-term memory** depends on electrical activity in neuronal circuits. In simple nervous systems, long-term storage appears to involve a change in synaptic responsiveness that requires protein synthesis.

6. In the course of development, the **dominant hemisphere** differentiates a specialization for control of voluntary motor activities and language. Usually the left hemisphere becomes dominant. The **nondominant hemisphere** differentiates a specialization for nonverbal intellectual functions. Damage to language regions of the dominant hemisphere results in **aphasias** (language disabilities) and impairments of verbal memory. Damage to corresponding regions of the nondominant hemisphere results in impairment of spatial and nonverbal functions.

7. Sleep is an activity of the brain. It can be divided into **slow-wave sleep** and **REM sleep** on the basis of differences in the EEG, muscle tone, and arousability.

STUDY QUESTIONS

1. Students found the two-point threshold to be 1 mm on the lip and 5 mm on the shoulder. What predictions can they make about each of the following:
   a. The relative size of the lip and shoulder representations in the somatosensory cortex
   b. The relative sizes of the receptive fields on the lip and shoulder
   c. The density of receptors on the lip and shoulder

2. What is the role of the reticular activating system during wakefulness? During sleep? What evidence indicates that sleep is an active state of the brain?

3. What experimental evidence suggests that memories are distributed over wide areas in the brain?

4. A patient speaks very slowly and with long pauses between words, but can sing quite well. What neurological deficit might he have?

5. What is a mixed nerve? Give some specific examples.

6. Which kinds of sensory information are carried by myelinated fibers and which kinds by unmyelinated fibers?

7. What structures are present in white matter? How are gray matter and white matter distributed in the brain? How does the brain differ from the spinal cord in this regard?

8. What are the anatomical names of the two main ascending pathways for sensory information in the spinal cord? What types of information does each characteristically carry?

9. Name the types of cutaneous receptors. Which of these are phasic and which are tonic receptors?

SUGGESTED READING

BAHILL, A.T., and T. LARITZ: Why can't Batters Keep Their Eyes on the Ball? *American Scientist*, volume 72, p. 249, 1984. Describes the reflexes that control eye movement and the tracking of moving objects.

BLACK, I.B., et al.: Biochemistry of Information Storage in the Nervous System, *Science,* volume 236, June 5, 1987, p. 1263. Discusses short- and long-term changes in synaptic efficacy and the appearance of specific intracellular proteins correlated with learning and memory in invertebrate model systems.

BURKHOLTZ, H.: Pain: Solving the Mystery, *New York Times Magazine*, volume 137, September 27, 1987, p. 16. General discussion of pain perception and the new drugs used to combat pain.

DICHTER, M.A.: Cellular Mechanisms of Epilepsy: a Status Report, *Science,* volume 237, July 10, 1987, p. 137. Good explanation of the factors that may act to increase the excitability of neurons. Includes a description of the mechanisms of action of antiepileptic drugs.

EDWARDS, D.D.: A Common Medical Denominator, *Science News*, volume 131, January 25, 1986, p. 60. Describes how specific neuropathological changes in Alzheimer's disease and Down's syndrome are helping researchers understand both disorders.

LENT, C.M., and M.H. DICKINSON: The neurobiology of feeding in leeches, *Scientific American*, volume 259, June 1988, p. 98. Shows how simple nervous systems can mediate complex behavior. In this case a single neurotransmitter, serotonin, mediates the entire feeding pattern.

Magnetic Resonance Imaging of the Central Nervous System, *Journal of the American Medical Association*, volume 259, February 28, 1988, p. 1211. Illustrated review of how MRI scans are performed and what they can tell us about central nervous system diseases.

MISHKIN, M., and T. APPENZELLER: The Anatomy of Memory, *Scientific American*, June 1987, p. 80. Discusses how memories may be stored in the central nervous system and the role of different regions of the brain in memory storage and recall.

The Nightmare of Sleepless Nights, *Science News*, volume 1, November 1, 1986, p. 280. Describes the role of the thalamus in sleep.

Primates with a Preference, *Discover*, volume 9, July 1988. Explores the hemisperic specializations that cause primates to prefer one hand over the other.

SELIGMAN, J.: The Fear of Forgetting, *Newsweek*, volume 108, September 29, 1986, p. 51. Discusses Alzheimer's disease and other conditions resulting in amnesia, their psychological effects, and possible explanation.

SPRINGER, S.P., and G. DEUTSCH: *Left Brain, Right Brain*, ed. 2, W.H. Freeman & Co., San Francisco, 1985. Excellent discussion of hemispheric specialization and language. Demonstrates how the brain can be tricked by limiting sensory input to one side or the other.

SQUIRE, L.R.: *Memory and Brain*, Oxford University Press, New York, 1987. Available in paperback, provides an up-to-date summary of the cellular mechanisms for learning and memory seen in invertebrate model systems. It also discusses a variety of psychological approaches to the study of memory.

TANK, D.W., and J.J. HOPFIELD: Collective Computation in Neuronlike Circuits, *Scientific American*, volume 257, December 1987, p. 104. Explores how electronic circuits designed along the lines of biological nervous systems are being used to recognize objects and what this may tell us about information processing in the nervous system.

CHAPTER 10

The Special Senses

On completing this chapter you will be able to:

- Describe how the tympanic membrane and auditory ossicles transmit sound energy to the cochlea.
- Understand how the resonance properties of the cochlea translate sound energy into displacements of hair cells on the basilar membrane.
- Understand how sounds are localized in space.
- Appreciate how the structure of the vestibular system translates gravitational or inertial forces into hair cell displacement.
- Understand the process of image formation in the eye and describe accommodation.
- Appreciate how lateral inhibition makes retinal ganglion cells respond to spots on contrasting backgrounds. Distinguish between ON center/OFF surround and OFF center/ON surround cells in terms of the synaptic connections in the retina.
- Summarize the principles of feature extraction in the visual system.
- Describe the organization of cortical columns in the visual cortex.
- Describe the modalities of taste and olfaction.

*T*he senses of hearing, equilibrium, vision, taste, and smell are mediated by specialized multicellular receptor systems. The inputs from these receptors are rich in information and are subjected to extensive analysis by the nervous system. In a multicellular sense organ such as the eye, each sensory cell provides a piece of the puzzle that can be reassembled by the brain into a representation of the external world. The eye projects an image on photoreceptors, each of which provides a small portion of the picture to the visual cortex. The same effect could not be obtained if photoreceptors were distributed like touch receptors over the entire body surface. Similarly, the ear transforms different sound frequencies to distinct areas of an array of receptors in the inner ear. The brain's comparison of the activity of different cells in the array provides frequency discrimination. In the vestibular system, a group of sensory structures associated with the inner ear, the structure of the system causes different receptors to be stimulated by different orientations or movements of the head.

Taste and olfaction (smell) are complicated sensory systems to understand because there appear to be only a few basic tastes and no fundamental odors. Instead, a large variety of stimulus molecules interact with single olfactory or taste receptors, somehow giving rise to fairly specific sensations.

The entire range of information provided by the inputs of receptor cells in a multicellular receptor system is much greater than the information provided by individual receptors. In the brain's reconstruction of the sensory world, the whole is greater than the sum of the individual parts.

THE AUDITORY SYSTEM
Physics of Sound

The auditory system transduces (converts) sound energy into electrical signals that are relayed to the central nervous system for interpretation. Sound waves consist of periodic increases and decreases in density (**compression** and **rarefaction**) of the air (Figure 10-1). A **pure tone** is a sound with only one frequency, expressed in cycles per second or **Hertz** (Hz). There are laboratory sound sources capable of generating pure tones, but musical instruments, voices, and other sound sources usually generate sounds containing a mixture of frequencies. A sound is said to have **pitch** if there is a dominant frequency in the mixture. The human auditory system is normally able to discriminate pitches in the range of frequencies between 16 and 20,000 Hz.

The **intensity** of a sound is a reflection of the amplitude of the sound pressure wave (see Figure 10-1). Audible sounds may vary in intensity (loudness) by many orders of magnitude, so for convenience a logarithmic scale is used. A change in intensity of 10 **decibels** (dB) represents a change in sound pressure amplitude of a factor of 10. The threshold for detection of sound is defined as 0 dB. Intensity differences of as little as 0.1 dB can be detected. Normal conversation generates about 50 to 60 dB, a city street has a sound level of around 75 dB, and an auto horn at close range is about 90 dB. Sounds become painful at 130 dB, a level approached in some rock music concerts. There is a 10 millionfold difference in amplitude between conversation (60 dB) and painful sounds (130 dB).

The Outer Ear and Tympanic Membrane

The ear has three components: (1) the **outer ear,** (2) the **middle ear,** and (3) the **inner ear,** or **cochlea** (Figure 10-2). The outer ear consists of the **pinna, auditory canal,** and **tympanic membrane** (eardrum). The pinna is the visible outer ear. The tympanic membrane, a drumlike structure with an area of 50 mm^2, spans the auditory canal and separates the outer and middle ear. Sound waves traveling into the auditory canal strike the tympanic membrane and cause it to vibrate. The tendency of the eardrum to vibrate best over a particular range of frequencies (its resonance properties, see p. 228), combined with the sound-gathering properties of the outer ear, approximately doubles the pressure of sound waves in the middle range of audible frequencies.

Ossicles of the Middle Ear

The middle ear has three small bones, or **ossicles:** (1) the **malleus,** (2) the **incus,** and (3) the **stapes** (see Figure 10-2). These bones mechanically relay vibrations from the tympanic membrane to the 3 mm^2 membrane of the **oval window.** The middle ear bones also form a lever system which produces a mechanical advantage and thus amplifies sounds

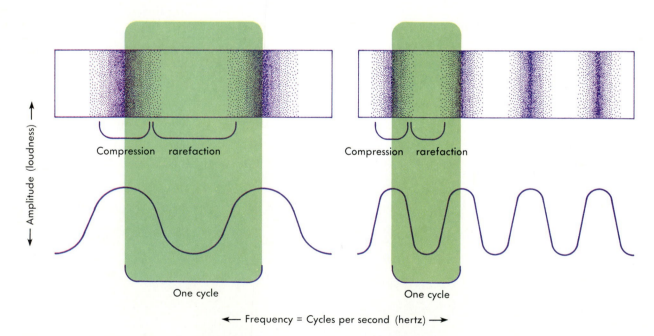

FIGURE 10-1
Sound waves consist of alternating regions of compression and rarefaction of air molecules. Two pure tones are shown; most sounds are mixtures of pure tones and have a much more complex waveform. The two properties of a pure tone are its frequency (the number of wave peaks per second) and its amplitude (the difference between the pressure in the peaks of compression and in the valleys of rarefaction). The frequency is experienced as pitch (Hertz) and the amplitude as loudness (decibels).

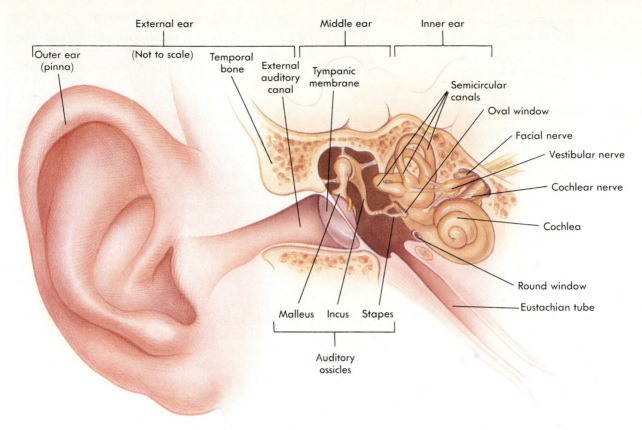

External ear — (Not to scale) — Middle ear — Inner ear

Outer ear (pinna)

Temporal bone

External auditory canal

Tympanic membrane

Semicircular canals

Oval window

Facial nerve

Vestibular nerve

Cochlear nerve

Cochlea

Round window

Eustachian tube

Malleus   Incus   Stapes

Auditory ossicles

FIGURE 10-2

View of the outer ear and auditory canal, leading to the middle ear and inner ear. Note the location of the tympanic membrane (eardrum) and the relationship between the bones of the middle ear and the cochlea. The middle ear is connected to the pharynx by means of the Eustachian tube. The structures shown are not drawn to scale.

by about a factor of two. The difference in area between the tympanic membrane and the smaller oval window concentrates the sound energy by a factor of about 20. In all, three factors contribute to an approximately eightyfold increase in sound wave pressure: tympanic resonance, ossicle mechanical advantage, and concentration of sound at the small oval window. These factors constitute an **impedance matching system** that increases the wave pressure so that it is effectively transmitted from the low-resistance medium of air into the higher-resistance medium of the fluid in the inner ear.

Sound Transmission in the Middle Ear

The ossicles in the middle ear transmit sound properly only if the pressures are equal on both sides of the tympanic membrane. The **Eustachian tube** (see Figure 10-2) connects the middle ear with the atmosphere by way of the throat and normally prevents the development of pressure differences between the middle ear and the atmosphere. However, in colds or illnesses such as **otitis media** (an infection of the middle ear), inflammation can close the Eustachian tube or otherwise increase the pressure in the middle ear, damping vibrations of the

tympanic membrane and the oval membrane and causing temporary hearing impairment. In severe cases, middle ear infections can cause permanent hearing loss by altering the properties of the ossicle chain.

Two muscles, the **tensor tympani,** attached to the tympanic membrane, and the **stapedius,** attached to the stapes, control the effectiveness of sound transmission through the ossicles (see Figure 10-2). Loud sounds initiate a reflex that activates these muscles, stiffening the chain of ossicles, and decreasing sound transmission. This reflex has a latency of about 30 milliseconds, so it cannot protect the ears from sudden loud sounds such as gunshots. It also only partly protects the receptor cells from prolonged intense sounds such as those associated with jackhammers or rock music, and overexposure to sounds such as these is responsible for hearing loss in an ever-increasing number of individuals in industrialized societies. The cumulative damaging effect of loud sounds on hearing acuity is not seen in societies isolated from artificial sources of sounds, and individuals in such societies show much less hearing loss with advancing age.

Structure of the Cochlea

The ability of the ear to discriminate frequencies is dependent on the arrangement of sensory cells and other structures within the **cochlea.** The cochlea is a 3.5 cm long, spiral-shaped chamber that looks like a snail's shell (Figure 10-3, *A*). The end nearest the middle ear is called the **base** and the opposite end the **apex.** Figure 10-3, *B* shows a cross-sectional view of the cochlea, with the **organ of Corti** enlarged in Figure 10-3, *C*. The organ of Corti contains the auditory receptor cells called **hair cells** (Figures 10-4 and 10-5).

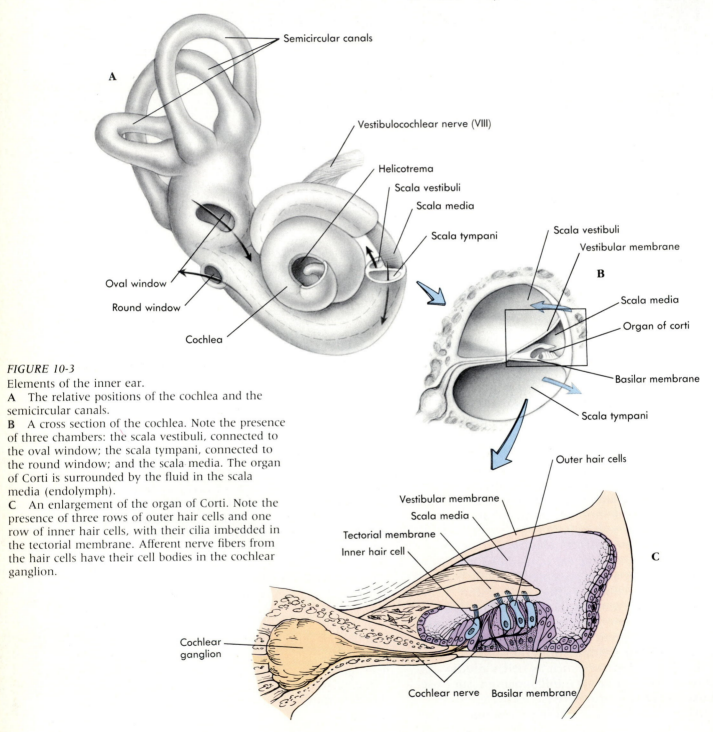

FIGURE 10-3
Elements of the inner ear.
A The relative positions of the cochlea and the semicircular canals.
B A cross section of the cochlea. Note the presence of three chambers: the scala vestibuli, connected to the oval window; the scala tympani, connected to the round window; and the scala media. The organ of Corti is surrounded by the fluid in the scala media (endolymph).
C An enlargement of the organ of Corti. Note the presence of three rows of outer hair cells and one row of inner hair cells, with their cilia imbedded in the tectorial membrane. Afferent nerve fibers from the hair cells have their cell bodies in the cochlear ganglion.

NERVE AND MUSCLE

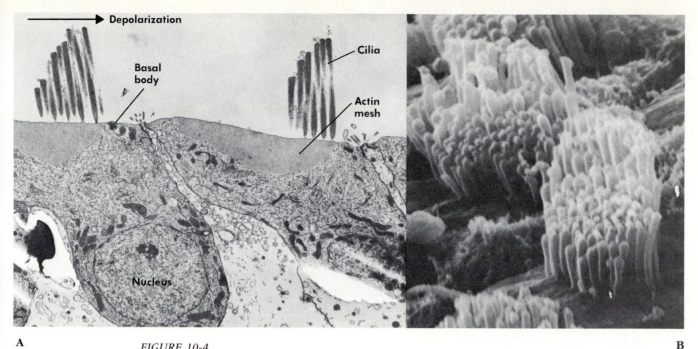

A

B

FIGURE 10-4

Transmission (A) and scanning (B) electron micrographs of cochlear hair cells.

The cochlea is divided into an upper chamber, the **scala vestibuli,** and a lower chamber, the **scala tympani,** by a middle chamber, the **scala media,** or **cochlear duct** (see Figure 10-3, *B*). The stapes acts on the oval window to set up waves in the fluid within the scala vestibuli (Figure 10-6). At the apical end of the cochlea, the scala vestibuli is connected by a narrow passageway (the helicotrema) to the scala tympani. The scala tympani is separated from the middle ear by the round window (Figure 10-3, *A*).

The fluid in the scala vestibuli and the scala tympani is called **perilymph.** Between the scala vestibuli and the scala tympani there is a closed chamber, called the **scala media,** bounded by the **vestibular membrane** and the **basilar membrane** (see Figure 10-3, *C*). The **endolymph** present within the scala media is a fluid that differs in composition from perilymph, and the difference in ionic composition contributes to the magnitude of the membrane potential changes that occur in the auditory receptor cells.

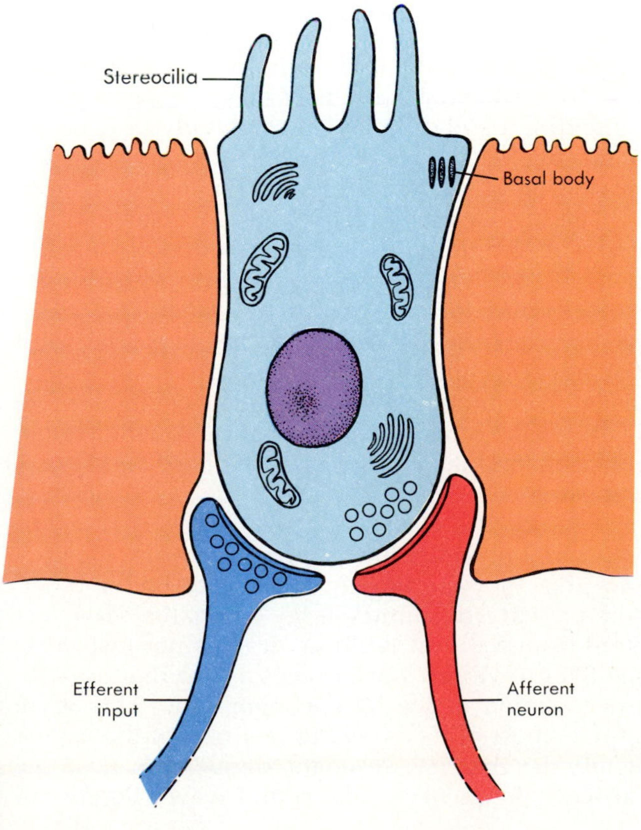

FIGURE 10-5

Diagrammatic representation of an auditory hair cell. Auditory hairs cells in the adult possess stereocilia and are morphologically polarized. They are depolarized when the stereocilia are deflected toward the basal body. The hair cell communicates with the afferent neuron by release of transmitter and also receives efferent synaptic inputs that modulate its sensitivity.

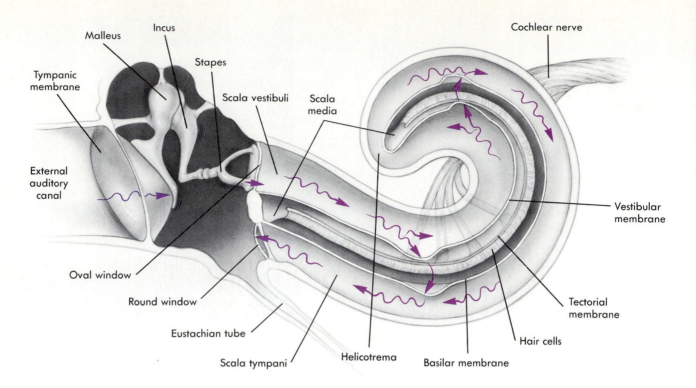

FIGURE 10-6
Effect of sound waves on the cochlear structures. Sound waves *(arrows)* strike the tympanic membrane and cause it to vibrate. This vibration is transmitted to the oval window by the ossicles of the middle ear. Vibration of the oval window causes the perilymph in the scala vestibuli to vibrate. As a consequence the basilar membrane is deflected.

Frequency Localization in the Cochlea

The ability of the ear to analyze the frequency components of sounds depends on the elastic properties of the basilar membrane and is based on the principle of **resonance.** When struck, a tuning fork vibrates at a characteristic **resonant frequency,** as do the strings of musical instruments. A swing also shows resonance. It tends to oscillate at a resonant frequency that depends on its length and total weight of swing and swinger. When the swinger pumps with her legs at the resonant frequency of the swing, the amplitude increases, because the force always occurs at the same point in time relative to the natural oscillation of the swing. This phenomenon is a type of positive feedback and is referred to as **sympathetic resonance.** While it is possible to cause a structure with a resonant frequency to move at other frequencies by applying a force to it (someone pushing the swing, for example), positive feedback will not occur and therefore the amplitude of the oscillation will be smaller.

Sympathetic resonance also occurs when a violin note is played near a piano. Thus, a middle C (512 Hz) on a violin causes the corresponding piano string to vibrate. In stringed musical instruments, the resonant frequency of a string depends on the length of the string and its tautness (corresponding to the length and total weight of a swing). Thus a string on a cello or bass is normally of lower frequency than a string on a violin, but a musician can raise the frequency of the note played by a string by tightening it and can play notes of higher frequency by placing her fingers on the fingerboard to shorten the portion of the string free to vibrate.

The basilar membrane consists of elastic fibers of varying length and stiffness imbedded in a gelatinous matrix. It resembles a piano into which someone has poured a large quantity of Jello. At its base, the fibers of the basilar membrane are short and stiff, but at its apex the fibers are 5 times longer (the basilar membrane is wider) and 100 times more flexible. Like a piano, the resonant frequency of the basilar membrane is highest at the base and lowest at the apex.

Because the elastic fibers in the basilar membrane are imbedded in a gelatinous matrix, they behave like they are loosely coupled to one another. When a wave of sound energy enters the scala vestibuli from the oval window, it does not just cause one fiber to vibrate but initiates a traveling up-and-down motion of the basilar membrane. This wave imparts most of its energy to that part of the basilar membrane that has resonant frequencies near the particular frequency of the sound wave (Figure 10-

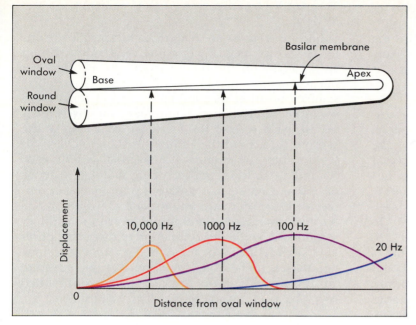

FIGURE 10-7

Frequency localization in the cochlea. The cochlea is shown unwound at the top, while the resonance of the basilar membrane in response to different sound frequencies (tones) is illustrated below. The parts of the basilar membrane nearest the oval window resonate preferentially to sounds of high frequency; at increasing distances from the oval window progressively lower resonant frequencies are encountered.

7), resulting in a maximum deflection of the basilar membrane at that point. Other areas of the basilar membrane do not experience the same degree of positive feedback because their resonant frequencies differ from that of the sound. The coding of sound frequency is therefore based on selective response properties **(place coding)** along the length of the basilar membrane.

The upper range of the frequencies that the auditory system can detect (about 20,000 Hertz) are much higher than the maximum rate at which action potentials can be generated in an axon, so frequency coding in this portion of the range could not be accomplished by one-for-one following of sound frequency by action potentials of the individual neurons. However, there is a tendency for the overall activity of the auditory nerve to follow the frequency of the sound wave in the lower range of frequencies. Intensity of sounds, on the other hand, is reflected in the degree of displacement of the basilar membrane and is therefore coded for by the rate of firing of neurons and the number of cells that are affected.

Hair Cells—The Auditory Receptors

The organ of Corti includes about 20,000 hair cells attached to the basilar membrane in three outer rows and a single inner row (see Figure 10-3, C). Hair cells in the adult auditory system have a surface covered with several rows of **stereocilia** and also a cytoplasmic structure, termed the **basal body,** which provides polarity to the hair cells (see Figures 10-4, A and 10-5).

Deflections of the stereocilia toward the basal body depolarize hair cells, while movements away from the basal body hyperpolarize them. Thus the receptor potential can be either depolarizing or hyperpolarizing, and during the movements of the basilar membrane caused by passage of a sound wave, both changes typically occur in succession. Like presynaptic nerve terminals (Chapter 8), the bases of hair cells contain transmitter-filled vesicles (see Figure 10-5). Resting (unstimulated) hair cells release some transmitter continuously, causing a steady low level of action potentials in auditory afferent neurons. Depolarization increases transmitter release and elevates afferent activity above this basal level, whereas hyperpolarization inhibits release.

There are 5 to 6 times as many outer hair cells (about 20,000) than inner hair cells (3500). In addition, the cilia of the outer hair cells are arranged in three V-shaped rows, whereas the inner hair cells are arranged in a single row. The stereocilia of outer hair cells are embedded in the **tectorial membrane,** which forms a roof over the basilar membrane. In this position, they can be excited in proportion to the degree of displacement of the basilar membrane. The stereocilia of the inner hair cells extend into the gelatinous matrix (see Figure 10-3, C). They also respond to the movement of the basilar membrane, but the pattern of their response differs from that of the outer hair cells. Another important feature of frequency coding in the cochlea, not fully understood, is that the inner hair cells are more densely innervated than are the outer hair cells.

Hair cells are also innervated by efferent fibers from the central nervous system (see Figure 10-5). Increases in activity in the efferent fibers can make hair cells less sensitive and less sharply tuned in particular portions of the audible frequency range. This example of central control of receptor sensitivity has been demonstrated to increase the ability of individuals to concentrate on one part of the auditory range (for example, a conversation) in the midst of background noise, which is effectively "tuned out" by the efferent control.

Auditory Afferent Pathways

Primary auditory afferents join fibers from the vestibular system to form the eighth cranial nerve, which passes into the brainstem (Figure 10-8). The auditory fibers synapse on interneurons in cochlear nuclei of the medulla. Some neurons in another nucleus of the medulla, known as the **superior olive**, receive **binaural inputs** (inputs from both ears). Most fibers in the auditory pathways decussate in the brainstem and ascend into the contralateral (opposite side of the body) thalamus, but some fibers project to the ipsilateral (same side of the body) thalamus (see Figure 10-8). From the thalamus, pathways project to the auditory cortex in the temporal lobe. Since many higher auditory neurons receive inputs from both ears, unilateral lesions in the auditory pathways usually do not greatly impair hearing ability.

Tonotopic Organization of the Auditory System

Sound frequencies are mapped onto specific regions of the basilar membrane, while mechanical stimuli are localized on the body surface. As a result, frequency discrimination in the auditory system resembles position discrimination in the somatosensory system. The first requirement for discrimination in each case is the existence of a map of the sensory surface. In the auditory system, cochlear afferents enter the nervous system in an orderly way. Ascending auditory pathways are mapped in terms of frequencies, so that different pure tones are represented in a sequence on the cortical surface, called a **tonotopic** map. Lateral inhibition in the auditory system aids in the discrimination of tones that differ only slightly in frequency. When a tone excites one population of neurons, neurons that project from adjacent regions of the organ of Corti are inhibited by pathways from the more excited neurons.

Localization of Sounds in Space

In the lower frequency range, **phase differences** (time of arrival of a particular point on the sinusoi-

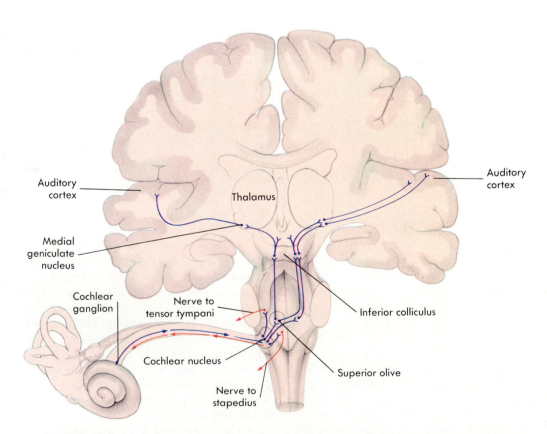

FIGURE 10-8
Central nervous system pathways for hearing. Afferent fibers *(blue arrows)* from the cochlear (spiral) ganglion terminate in the cochlear nucleus in the brainstem. Axons from neurons in the cochlear nucleus project to the superior olive, inferior colliculus, medial geniculate nucleus, and the auditory cortex. Note that auditory input projects to both sides of the cortex and that neurons in the superior olive send efferent axons back to the cochlea and to the muscles of the middle ear *(red arrows)*.

NERVE AND MUSCLE

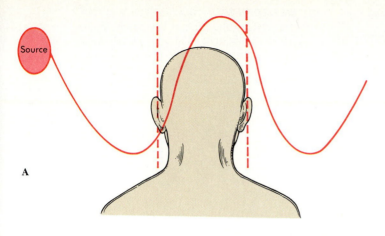

A

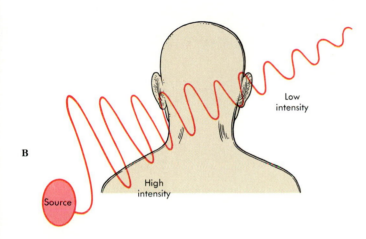

B

Low
intensity

High
intensity

Source

FIGURE 10-9

A Sound localization clues are provided by phase differences for sounds of low frequency, in which each pressure wave strikes the right ear and the left ear at slightly different times.

B For sounds of high frequency, phase differences would give ambiguous information because there is time for more than one wave cycle while the sound travels the width of the head. Thus the intensity difference must be used as a directional clue. The head absorbs some of the sound, emphasizing the intensity difference.

dal sound wave) in the sounds that strike the two ears are most important for sound localization. At higher frequencies (above 3000 Hz), there is very little difference in the time of arrival, and **intensity differences** of the sound seem to be most important for sound localization (Figure 10-9). Separate populations of superior olivary neurons sensitive to small differences in phase and intensity seem to be most instrumental in localizing sounds in space.

> *1 What is meant by resonant frequency?*
> *2 How does the gradient of resonant frequencies along the basilar membrane help resolve sounds into pure tones?*
> *3 Why is the tonotopic organization of the auditory cortex like the somatotopic organization of the somatosensory cortex?*

THE VESTIBULAR SYSTEM
Vestibular System Functions

The vestibular system is a part of the inner ear adjacent to the cochlea. Both the cochlea and the canals that house the vestibular system are located in a bony labyrinth within the temporal bones on the right and left sides of the skull (Figure 10-2). In this

location, they are protected from irrelevant mechanical distortion such as would be caused by muscle contraction or the pulsation of large blood vessels.

The information provided by the vestibular system is integrated with information from the visual system, inputs from receptors signaling the angle of joints, and signals from muscle length detectors to give a sense of position or movement in space. The vestibular inputs provide information that is critical to those somatic reflexes that maintain the position of the head and body with respect to gravity and stabilize images on the retinas of the eye during head movements.

Vestibular Receptors

Hair cells are the primary sensory receptors of the vestibular system, as well as the auditory system. In the vestibular system, the same basic receptor type responds to displacement due to changes in the orientation and acceleration of the head. The fact that the auditory system responds to sound whereas the vestibular system detects head movements is accounted for by the physical coupling of the two sets of receptors to different types of mechanical stimulation.

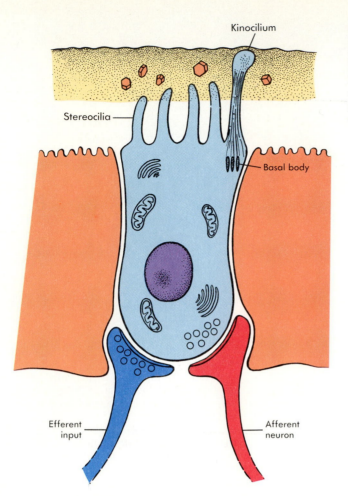

FIGURE 10-10
Hair cells of the human vestibular system possess both stereocilia and a knoblike kinocilium projecting from a basal body. Like auditory hair cells, vestibular hair cells transmit to an afferent neuron and receive modulation from vestibular efferents.

Kinocilium

Stereocilia

Basal body

Efferent input

Afferent neuron

FIGURE 10-11
A The position of the saccular macula and utricular macula in relation to the cochlea and semicircular canals.
B Enlargement of a section of the utricle showing the otoliths embedded in the gelatinous matrix that covers the hair cell receptors.

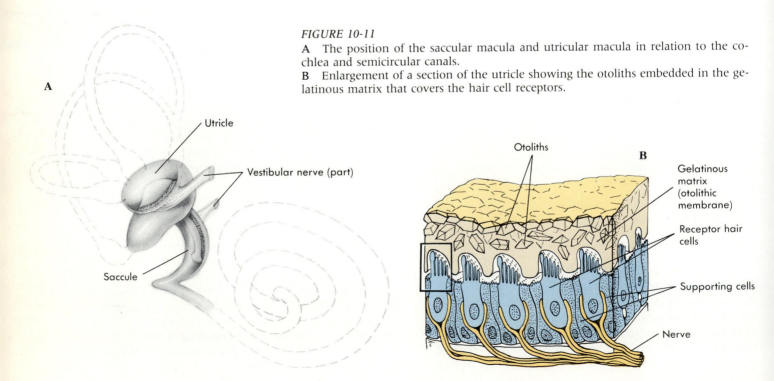

A

Utricle

Vestibular nerve (part)

Saccule

Otoliths

B

Gelatinous matrix (otolithic membrane)

Receptor hair cells

Supporting cells

Nerve

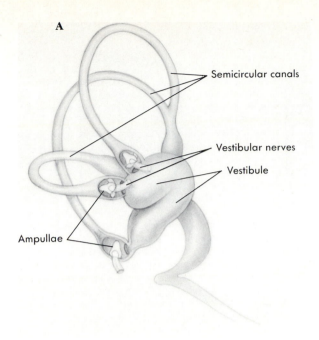

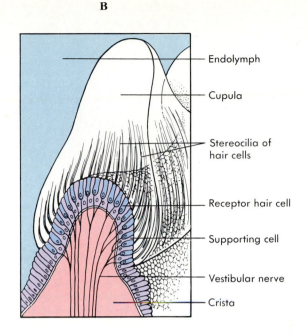

FIGURE 10-12

A The position of the semicircular canals in relation to the rest of the inner ear showing the location of the ampulla.
B Enlargement of a section of the ampulla showing how hair cell stereocilia insert into the cupula.

Hair cells in the vestibular system have a single knoblike process known as a **kinocilium** in addition to the rows of smaller stereocilia (Figure 10-10). In the vestibular system, the hair cells are positioned so that they can respond to the position of the head with respect to gravity and to acceleration of the head. Head position and linear acceleration are detected by two structures in each inner ear, the **utricles** and **saccules** (Figure 10-11, *A*). Rotation is sensed by the **semicircular canals,** one set in each inner ear (Figure 10-12, *A*).

The structures of the vestibular system are filled with endolymph. The utricle and saccule are sacs containing gelatinous **otolithic membranes** weighted with small crystals of calcium carbonate called **otoliths** (Figure 10-11, *B*). The stereocilia of hair cells are embedded in the otolithic membrane. Each vestibular apparatus has three semicircular canals positioned at right angles to one another (see Figure 10-12, *A*). Each semicircular canal has an enlargement, called the **ampulla,** at the end closest to the utricle. The ampulla contains a mound of tissue, the **crista,** that supports a tuft of vestibular hair cells. The vestibular stereocilia extend into a ball of gelatinous material, called the **cupula,** that occludes the semicircular canal (Figure 10-12, *B*).

Detection of Head Position and Linear Acceleration

The utricle is oriented so that when the head is vertical, gravity pulls the utricular otolithic membrane straight down on the stereocilia (Figure 10-13, *A*). When the head is tilted sideward, gravity shifts the utricular otolith relative to the hair cells, bending the stereocilia. The orientations of hair cells (defined by the location of their kinocilia) vary over the utricular surface. Any particular head tilt excites one population of hair cells and inhibits another, generating a unique pattern of afferent activity (Figure 10-13, *B*).

Forward **linear accelerations** (such as are experienced in an automobile when the driver steps on the gas) are like backward head tilts because inertia causes the dense utricular otolithic membrane to be left behind. The hair cells in the saccule extend horizontally when the head is in the upright position. Some hair cells are excited more by gravity than others when the head is in this normal position. Head tilts will alter the normal discharge pattern from the saccule, with an opposite pattern of excitation generated for the right and left saccule sensory cells. When the head is upside down, gravity pulls the otolithic membrane away from the hair cells. Because they are oriented at right angles, the saccule and utricle of each ear provide information on all head positions and linear accelerations.

Detection of Head Rotations

Angular acceleration is a change in the rate of rotation of the head or whole body. Detection of angular acceleration is the result of movement of the

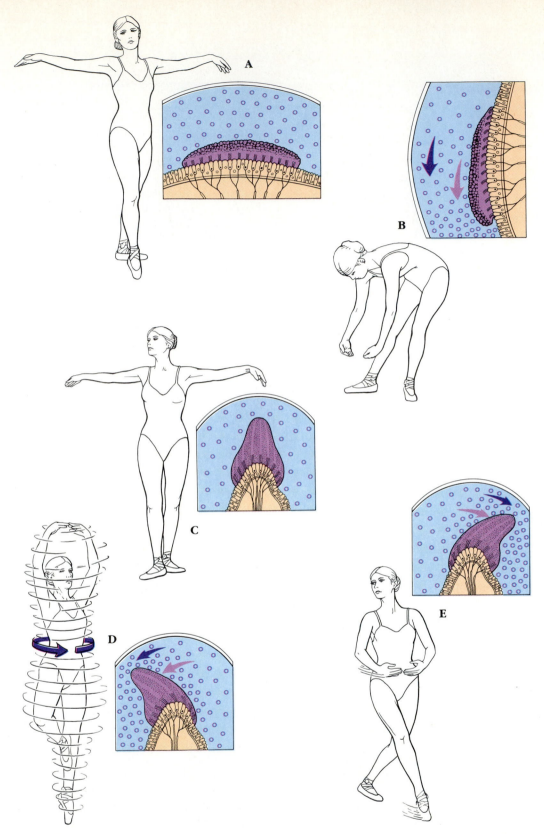

FIGURE 10-13
Function of the otoliths and the semicircular canals in maintaining balance.

A   The utricle responds to changes in the position of the head relative to the direction of gravity. When the head is upright the utricular macula exerts a uniform downward force on the vestibular hair cells.

B   Tilting the head forward causes a shearing force. Similar deflections can also be caused by linear accelerations, such as are experienced in a car or elevator.

C-E   The semicircular canals respond to rotations. As a person begins to spin, the cupula is displaced by the endolymph in a direction opposite to the direction of spin. This displacement exerts a shearing force on the vestibular hair cells. When the person stops spinning, the inertia of the endolymph causes the cupula to be displaced in the same direction as the original spin.

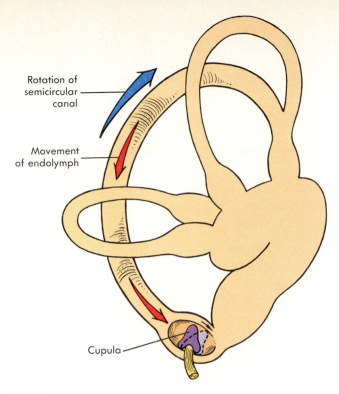

Rotation of
semicircular
canal

Movement
of endolymph

Cupula

FIGURE 10-14
The effect of inertia on the movement of endolymph in an idealized representation of a semicircular canal. Rotation results in acceleration of the semicircular canal relative to the endolymph. The inertia of the endolymph causes it to lag behind, displacing the cupula and stimulating the hair cells.

endolymph relative to the cupula (Figure 10-14). For example, if the head begins to rotate in the horizontal plane, the inertia of the endolymph causes it to lag behind the canal rotation. From the point of view of the canal, the endolymph is flowing in a direction opposite to the head rotation. When the head rotates to the right, the relative endolymph flow in the right horizontal canal will be toward the left, while that in the left canal will be to the right (see Figure 10-13, *C* and *D*). This flow deflects the cupula.

All hair cells in the ampulla are oriented in the same way, so that deflections of the cupula either excite or inhibit all of them, depending on whether the stereocilia are bent toward or away from the kinocilium. In the horizontal canals, hair cells are oriented so that their kinocilia face the utricle, and movements of endolymph toward the utricle therefore excite them. Head rotations to the right excite hair cells in the right horizontal canals and inhibit those in the left canals. The reverse occurs when rotations stop because the inertia of the endolymph causes it to continue to flow, deflecting the cupula in a direction opposite to that occurring when the rotation began.

Rotational Nystagmus
If a subject is rotated for some time in a swivel chair, the motion of endolymph catches up with

that of the walls of the horizontal semicircular canals. When rotation is stopped, the endolymph continues to move in the direction of the rotation for a few seconds (see Figure 10-13, *E*). During this time, the subject finds it difficult to maintain his balance if asked to walk, and his eyes rotate slowly in the direction of endolymph movement, snap back rapidly, and move slowly in the original direction again, as if to keep up with the perceived head rotation. This eye movement is called **rotational nystagmus**. In the superior and posterior canals, the same physical responses occur, and the only difference is the plane of rotation that maximally stimulates each canal. Each semicircular canal is excited (or inhibited) to the extent that there is a component of rotation in its plane of orientation. Together, the canals can provide information about rotation in any plane.

In the absence of rotation, the inputs from the left and right semicircular canals are the same and nystagmus does not occur. However, some diseases (multiple sclerosis, strokes) cause these inputs to become unbalanced. The result is a resting nystagmus and is used by neurologists as an important diagnostic tool.

> 1 What is the importance of inertia for the functions of the vestibular system?
> 2 What is the difference between angular acceleration and linear acceleration?
> 3 What aspects of the construction of the macular organs and the semicircular canals result in specialization of the two systems for different types of acceleration?

THE VISUAL SYSTEM
Structure of the Eye
Vision requires that light from the large area that constitutes the visual field be concentrated into a small image projected on light-sensitive receptors. The gross structure of the eye reflects its specialization for image formation.

The eyeball is covered with a protective fibrous coat called the **sclera** (Figure 10-15). Light passes through a transparent anterior portion of the sclera called the **cornea**. Interior to the sclera is the **choroid** layer, which contains blood vessels in the posterior part of the eye. In the anterior region of the eye the choroid layer is elaborated into the lens and associated structures responsible for regulating light entry and for image focusing. The **iris** is a pigmented disc of smooth muscle with a central aperture called the **pupil**. Reflexive changes in pupillary diameter regulate the amount of light that reaches the lens. The lens is attached to a structure called the **ciliary body** via a series of elastic **zonular fibers** (Figure 10-16) that are continuous with similar fibers in the lens. The lens is normally transparent, but cataracts (regions of the lens that become

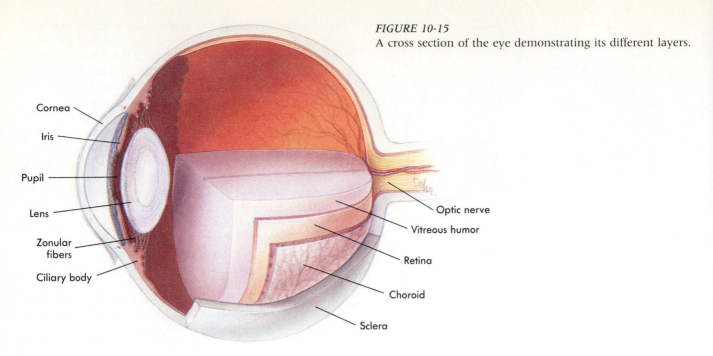

Cornea

Iris

Pupil

Lens

Zonular fibers

Ciliary body

Optic nerve

Vitreous humor

Retina

Choroid

Sclera

FIGURE 10-16
The detailed structure of the anterior portion of the eye illustrating the attachment of the lens to the ciliary body.

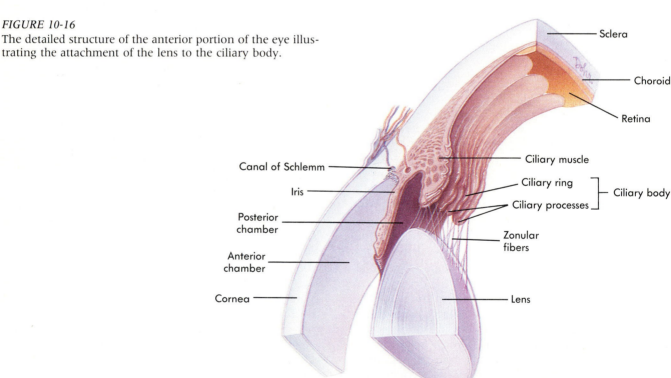

Sclera

Choroid

Retina

Canal of Schlemm

Iris

Posterior chamber

Anterior chamber

Cornea

Ciliary muscle

Ciliary ring

Ciliary processes

Ciliary body

Zonular fibers

Lens

opaque) (Figure 10-17) can impair vision or even cause blindness.

The **anterior cavity,** the part of the eye between the lens and cornea, is divided into two compartments—the **anterior** and **posterior chambers** (see Figure 10-16). The anterior cavity is filled with a watery fluid, the **aqueous humor.** The **posterior cavity** behind the lens contains a much thicker fluid, the **vitreous humor** (see Figure 10-15).

Aqueous humor is secreted from the ciliary body into the posterior chamber and drains into the venous blood from the anterior chamber via the **canal of Schlemm.** This drainage is important because if the canal of Schlemm is plugged, the pressure in the eye increases, resulting in a condition called **glaucoma.**

The retina, the innermost layer of the posterior cavity, consists of photoreceptors and several

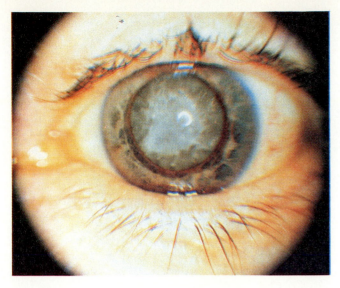

FIGURE 10-17

Photograph of a cataract. The cataract is the white spot approximately in the center of the cornea.

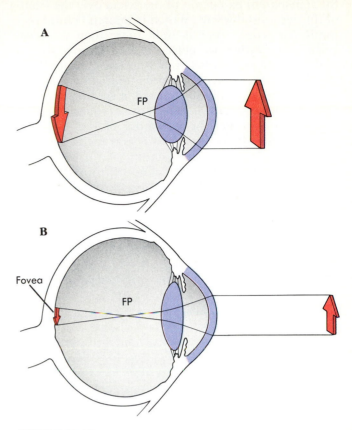

FIGURE 10-18

Image formation on the retina. Note the location of the fovea, the area with the greatest photoreceptor density. *FP* indicates the focal point.

A For near vision the ciliary muscles contract, decreasing the force exerted by the suspensory ligaments and causing the lens to become less flattened (more convex). A more convex lens bends the entering light rays more.

B Distant objects are focused by a flattened (less convex) lens. The lens is flattened because the ciliary muscle is relaxed, and the suspensory ligaments are exerting their greatest force on the lens.

types of interneurons. The photoreceptors are specialized epithelial cells. The interneurons are derived from the brain during embryological development; the retinal interneurons are thus a part of the brain.

The Eye as an Image-Generating Device

In a camera the lens forms an inverted image of an object on film by bending incoming rays of light. The total amount of light entering a camera is controlled by adjusting its **aperture,** the diameter of the opening through which light reaches the lens. The inside of a camera is painted black to keep light from being reflected inside.

In many ways the eye resembles a camera. For example, both the eye and the camera form an inverted image (Figure 10-18). The retina is analogous to the film in a camera. The choroid layer behind the retina contains pigment that absorbs the light not captured by photoreceptors. The amount of light entering the eye depends on the diameter of the pupil, and pupillary diameter is adjusted by the iris.

Dynamic Range of Vision

In a camera, the aperture can typically be varied by a factor of about a thousand. Further control of the total amount of light that enters is provided by shutter speed. The human eye is sensitive to changes in light intensities of about 10^{10}-fold, but only a small part of this range can be accounted for by changes in the diameter of the pupil, which can be changed by only about sixfold, resulting in a thirtyfold difference in light entry. The large range of light intensities over which the eyes can operate is therefore largely a property of the photoreceptors

and associated neural pathways of the retina. By analogy with the camera, the photoreceptors correspond most nearly to the film, although they respond much more effectively to changing light conditions because they have the adaptability of a living system.

Accommodation of the Eye

In a camera the amount of bending of light depends on the difference between the refractive index of the air and the glass of a lens and on the curvature of the lens. In a camera both of these are constant for any particular lens, so that the camera is focused by changing the distance between the lens and film, not by altering the properties of the lens.

When the eye focuses light, the distance between the lens and retina does not change significantly. The ability of the eye to focus light depends on the combined optical properties of the cornea, the transparent extension of the sclera over the front of the eye, and the lens, which is located behind the pupil. The cornea bends light more than the lens because the difference between the refractive index of the cornea and air is about five times greater than the difference between the lens and cornea. While the cornea is optically stronger than the lens, adjustment of its curvature is not very significant for focusing. In contrast, the curvature of the lens depends on fibers that attach to it and the activity of small ciliary muscles (see Figure 10-16). Focusing on objects, a process termed **accommodation,** is accomplished by control over the ciliary muscles.

The lens has a natural elasticity. If it is removed from the eye, the lens assumes a rounded shape that is appropriate for focusing at close range (see Figure 10-18, *A*). In the released eye, the lens is subject to forces exerted by the zonular fibers of the suspensory ligament attached to the lens. These fibers act against the intrinsic elasticity of the lens to pull it more nearly flat, a condition that is appropriate for focusing on distant objects (see Figure 10-18, *B*). The zonular fibers suspend the lens in the center of a doughnut-shaped structure called the **ciliary body.** When the ciliary body thickens due to contraction of the ciliary muscles, the doughnut-ring becomes smaller and the tension in the zonular fibers diminishes, allowing the elastic lens to assume the more rounded shape it would have if removed from the eye. The fact that active contraction of the ciliary muscles is required for focusing on close objects, whereas distance vision is performed with a relaxed eye, explains why focusing on near objects can result in eyestrain.

Disorders of Image Formation

The natural elastic properties of the lens permit the eye to focus on objects as close as a few inches. The closest point at which a clear image can be formed is termed the **near point** for vision. In older individuals the lens loses some of its elasticity, increasing the near point so that it becomes difficult to focus on objects close to the eye. This condition is called **presbyopia,** or restriction of accommodation.

Although the lens and cornea of a normal eye will focus a distant object on the retina, in some individuals the eyeball is too long or too short relative to the power of the lens and cornea. If the eyeball is too long (Figure 10-19, *A*), the image will be focused in front of the retina; this situation is **myopia,** or **nearsightedness.** If the eyeball is too short, the condition is called **hyperopia,** or **farsightedness** (Figure 10-19, *C*). While myopia often occurs with a normal lens because the eyeball is too long, normal individuals tend to become farsighted as they get older because as the lens ages it loses some of its elasticity and cannot bend incoming light rays enough. Thus the changes of aging tend to correct myopia. Myopia is also corrected by glasses (or contact lenses) that cause the incoming light rays to di-

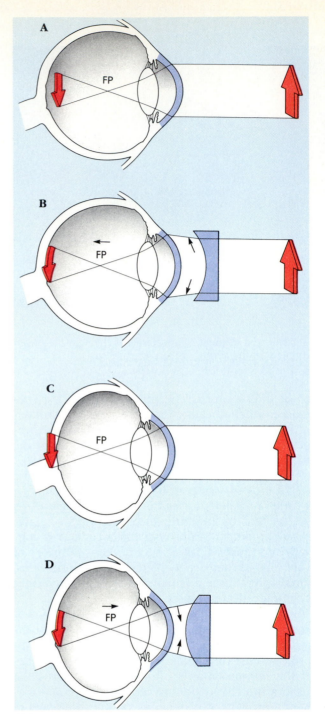

FIGURE 10-19

Visual disorders and their correction by various lenses. *FP* is the focal point.

A Myopia (nearsightedness) is a defect in which the cornea and lens are too powerful, or the eyeball is too long, causing the focal point to be too near the lens. The image *(red arrow)* is formed in front of the retina.

B The effect of placing a concave lens in front of the myopic eyeball is to spread out the light rays *(black arrows),* moving the focal point away from the lens so that there is a focused image on the retina.

C Hyperopia (farsightedness) is a disorder in which the cornea and lens are too weak or the eyeball is too short. The image is formed behind the retina.

D The effect of placing a convex lens in front of the hyperopic eyeball is to bend the incoming light more, moving the focal point toward the lens.

FIGURE 10-20

A A diagram of the retina. The afferent nerve fibers from the ganglion cells are closest to the light source. The bipolar cell layer is between the ganglion cell layer and the photoreceptors. Synaptic contacts are made in the inner and outer plexiform layers.

B A photograph of a normal retina. The blood vessels enter and leave the retina at the optic disk.

C The retina in an individual with glaucoma. In glaucoma, the blood supply to the retina is reduced by the pressure in the eye, causing the retina to appear pale.

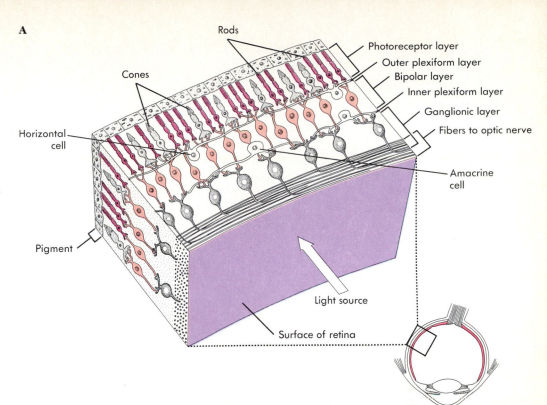

A

Rods — Photoreceptor layer — Outer plexiform layer — Bipolar layer — Inner plexiform layer — Ganglionic layer — Fibers to optic nerve — Cones — Horizontal cell — Amacrine cell — Pigment — Light source — Surface of retina

verge slightly (Figure 10-19, *B*). Hyperopia is treated with glasses that help converge parallel rays of light (Figure 10-19, *D*).

> 1 How do the roles of the cornea and the lens in image formation differ from one another?
> 2 What physical factors limit the range of accommodation of the eye?
> 3 What changes in the properties of the eye give rise to myopia and hyperopia?

The Retina

The cells of the retina are organized into layers (Figure 10-20, *A*). The retinal cells can be divided into **photoreceptors,** which are modified epithelial cells, and interneurons. Two types of retinal interneurons are **ganglion cells,** the axons of which form the optic nerve, and **bipolar cells,** which couple photoreceptors and ganglion cells. Other interneurons, **horizontal cells** and **amacrine cells,** couple adjacent photoreceptors and bipolar cells to one another.

The formation of the retina in embryonic development results in an inside-out design, so that the photoreceptors are nearest the sclera at the back of the eye, and light must pass through the retinal interneurons to reach the photoreceptors. The blood vessels serving the retina, on the other hand, are nearest the interior of the eye and can be easily seen when the retina is viewed with an ophthalmoscope (Figure 10-20, *B*). In glaucoma, referred to during the discussion of fluid drainage, the increased intraocular pressure closes off the blood vessels that supply the retina, making the retina appear paler

B

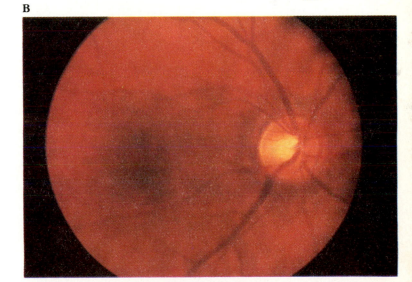

C

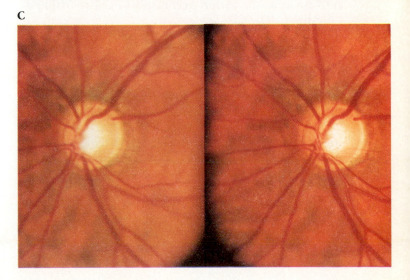

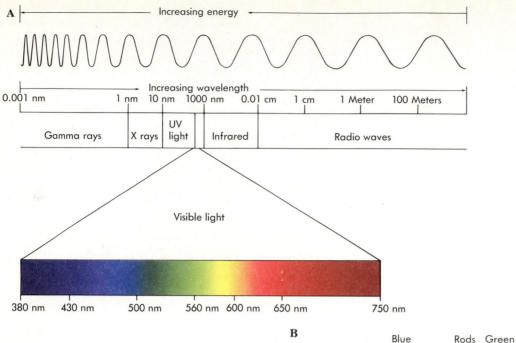

FIGURE 10-21

A The electromagnetic spectrum and the small segment of it that is visible, with the corresponding color spectrum.

B The sensitivity of rods and the three categories of cones to light of different wavelengths. Each receptor responds most vigorously to an optimum wavelength and less vigorously to higher and lower wavelengths.

and causing starvation and death of the retinal cells (Figure 10-20, *C*).

Rod and Cone Photoreceptors

Visible light is electromagnetic energy with wavelengths between 400 (the color violet) and 700 nanometers (nm) (red) (Figure 10-21, *A*). The portion of the electromagnetic energy spectrum that is visible to the human eye is only a small part of the total electromagnetic spectrum, which includes wavelengths longer than visible light (infrared radiation, microwaves, and radio waves) and shorter than visible light (ultraviolet rays, X-rays, and gamma rays).

The two types of photoreceptors, **rods** and **cones,** have some common structural features (Figure 10-22). The **outer segment** (nearest to the sclera) consists of stacked **membrane disks** that contain a high concentration of the **visual photo-**

pigment. The **inner segment** contains mitochondria, the photoreceptor nucleus, and transmitter-filled vesicles. The outer and inner segments are joined by a thin connecting zone containing a cilium, showing that photoreceptors evolved from ciliated epithelial cells.

There are four basic types of retinal photopigments; any photoreceptor possesses only one type. The absorption spectrum of each photopigment determines what portion of the visible range that photoreceptor can respond to. The information provided by the cone receptors is used to mediate color vision. There are three types of cones, each having a different wavelength-sensitivity. **Blue-sensitive cones** absorb maximally at 420 nm, **green-sensitive cones** at 530 nm, and **red-sensitive cones** at 660 nm (see Figure 10-21, *B*). The cones mediate **photopic,** or daytime color, vision because

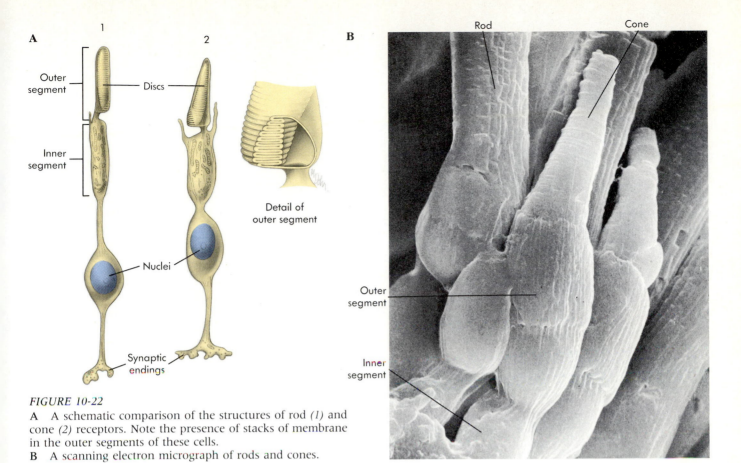

FIGURE 10-22
A A schematic comparison of the structures of rod *(1)* and cone *(2)* receptors. Note the presence of stacks of membrane in the outer segments of these cells.
B A scanning electron micrograph of rods and cones.

the information pathway to the visual cortex is organized in labelled lines that preserve color information provided by each type of cone. Rods absorb best at about 500 nm (see Figure 10-21, *B*) and are responsible for visual sensation of light intensity from black, through shades of gray, to white; they mediate **scotopic,** or low-light, vision. Although rods are wavelength selective, their inputs are analyzed by the visual system for intensity rather than color.

Photoreceptor Density

The distribution of photoreceptors on the retina is not uniform. The **fovea** (see Figure 10-18), a region of the retina specialized to provide information about color and detail, has the highest photoreceptor density. Each fovea contains about 4000 receptor cells, all of which are cones. The cone density of the fovea (about $10^5/mm^2$) gives the human visual system a resolution 100,000 times greater than that of a color TV set. There are about 100 million rods in the eye and only 6 million cones. The relative density of cones diminishes, and that of rods increases, as the distance from the fovea increases. Nerve fibers leave the eye via the **optic nerve** (see Figure 10-15) at a location called the **optic disk.** The optic disk has no photoreceptors and therefore is a "blind spot" in the retina.

Photopigment Bleaching — The First Step in Photoreception

The process of photoreception begins when light activates visual pigments. Visual pigment molecules are made up of two parts: a **chromophore** that absorbs light, coupled to one of four different proteins termed **opsins.** For both rods and cones, the chromophore is **retinal,** a modified form of vitamin A. The opsins confer wavelength specificity to the photopigment. Rods contain one type of opsin, and the resulting visual pigment is called **rhodopsin,** or **visual purple.** Each of the three types of cones possesses a different opsin.

Rhodopsin was the first photopigment to be studied and will serve as an example of photopigment reactions in general. In the dark, retinal is in a conformation called **11-cis-retinal** (Figure 10-23, *A*). When a photon of the appropriate wavelength is absorbed by a rhodopsin molecule, 11-**cis**-retinal undergoes a conformational change to **all-trans-retinal;** this process is called **bleaching,** and it initiates a sequence of events leading to a change in the rod cell membrane potential (Figure 10-23, *B* and *C*).

Bleached rhodopsin cannot respond to light, so that all-**trans**-retinal must be restored to the 11-**cis**-retinal conformation if the photosensitivity of the retina is to be maintained (Figure 10-23, *D*). This

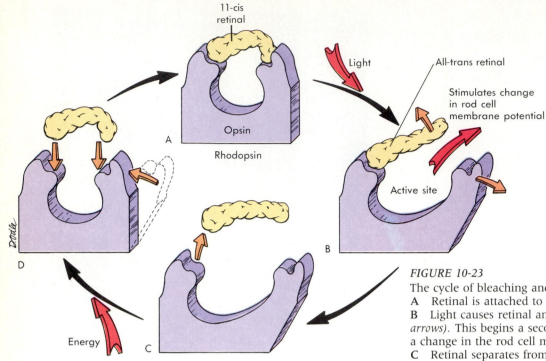

FIGURE 10-23

The cycle of bleaching and restoration of rhodopsin.
A   Retinal is attached to opsin.
B   Light causes retinal and opsin to change shape *(red arrows)*. This begins a second messenger cascade that causes a change in the rod cell membrane potential.
C   Retinal separates from opsin (bleaching).
D   Energy is required to bring opsin back to its original form and to attach retinal to it. Until this is done, the photopigment is unable to respond to light.

FIGURE 10-24

The molecular events of transduction of light energy to membrane permeability changes in photoreceptors.
A   In the dark, relatively high levels of cGMP keep Na^+ channels in the cell membrane of the outer photoreceptor segment open.
B   When photons interact with retinal in photoreceptor disk membrane, formation of all-trans rhodopsin activates G proteins, which activate phosphodiesterase. Phosphodiesterase degrades cGMP, closing Na^+ channels. The decrease in Na^+ conductance hyperpolarizes the receptor and reduces transmitter release.

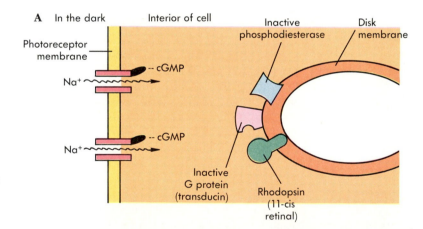

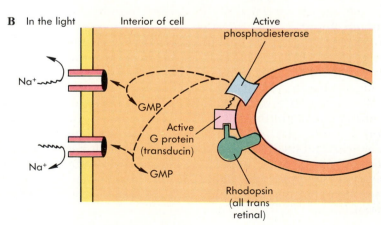

Color Sensation Is Genetically Determined

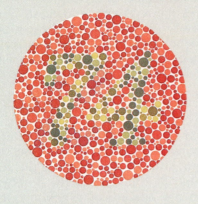

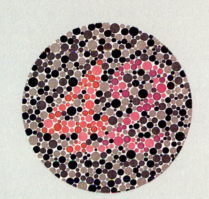

The ability to distinguish subtle variations in color is the result of integration of signals from photoreceptor classes that are selective for overlapping ranges of wavelengths. For example, the sensation of yellow is obtained when the retina is stimulated with **monochromatic** (single wavelength) light of 580 nm, which stimulates red cones about twice as effectively as green cones. The same sensation can be obtained if two beams of light of 560 nm (yellow-green) and of 600 nm (orange) are mixed. The net effect of the mixed light is to stimulate red and green cones in the same proportion as did the monochromatic light. Thus relative responses of different cone classes are integrated by the brain to confer the sensation of color.

Normal individuals with three cone pigments are referred to as **trichromats. Color blindness** is the absence of one or more of the cone pigments. It may be complete, involving the total absence of a particular pigment, or partial, with reduced levels of a pigment. **Dichromats** have only two cone pigments. If the red pigment is lacking **(protanopia)**, discrimination of red and green is impossible and the visual system is insensitive to deep red colors. If the green pigment is lacking **(deuteranopia)**, it is still impossible to distinguish red and green, but the visual system is nevertheless sensitive to light in the range normally served by the green pigment because the responsiveness of the red and blue pigments overlaps into this range. **Tritanopes** lack the blue pigment and cannot discriminate shades of blue and green. **Monochromats** lack all three cone pigments and cannot distinguish colors at all. Because monochromats must use the scotopic system for seeing in bright as well as dim light, they find bright light very unpleasant.

In humans, most color blindness is a recessive X-linked inherited trait (the gene is on the X chromosome). In Western Europe, approximately 8% of all males have some form of color blindness, but only about 1% of females are color blind. Color blindness is often assessed using charts in which normal individuals can see one number, while color-blind individuals see a different number (Figure 10-A).

---

process is ATP-dependent and occurs at a rate much lower than the rate of bleaching in bright light. After the retina has been exposed to dim light for some minutes, almost all of the rhodopsin is in the unbleached form and the retina is **dark adapted.** When the retina is exposed to bright light of appropriate wavelengths, almost all of the rhodopsin is converted to the bleached form, and the scotopic pathway becomes much less sensitive. This is one way in which the retina adjusts its sensitivity to the intensity of the stimulus. Since rods are not responsive to red light (see Figure 10-21, *B*), red-lit instruments are used in airplane cockpits and other situations where it is important to preserve the high sensitivity of the dark-adapted retina.

Origin of Photoreceptor Potentials

Photoreceptors have Na^+ channels in the membranes of their outer segments. In the dark these Na^+ channels are open (Figure 10-24). The resulting inward flow of current passes through the connecting segment of the photoreceptor and depolarizes the inner segment. When photons bleach photopigment in the outer segment membrane, the resulting conformational change of rhodopsin begins a second-messenger cascade that closes the Na^+ channels that were open in the dark, decreasing current flow and allowing the inner segment to hyperpolarize.

Rhodopsin is chemically very similar to membrane receptors for neurochemical transmitters described in Chapter 7. The second messenger cascade set in motion by rhodopsin bleaching involves a G protein called **transducin** that activates phosphodiesterase. Phosphodiesterase degrades **cyclic guanosine monophosphate (cGMP)**, an intracellular second messenger. cGMP keeps the Na$^+$ channels open in the dark, and the effect of light acting on rhodopsin is to reduce the levels of cGMP, closing the channels (see Figure 10-24). This second messenger system is much like those used by protein-type hormones (Chapter 5). However, cGMP seems to act directly to gate ion channel proteins, rather than acting by phosphorylation. Intracellular Ca^{++} does not seem to have a direct role in phototransduction but may be involved in visual adaptation.

Pathways from Photoreceptors to Ganglion Cells

Ganglion cells are the retinal cells whose axons form the optic nerve; their output constitutes the final product of information processing in the retina. In the dark, ganglion cells produce a low, steady rate of action potentials. The activity of ganglion cells can be increased or decreased in response to illumination, depending on the inputs they receive from bipolar cells. The **ON** and **OFF pathways** between receptor cells and ganglion cells result from two types of bipolar cells. These differ in their response to the transmitter that is continuously released by photoreceptors in the dark. The transmitter causes hyperpolarization in one type of bipolar cell and depolarization in the other (Figure 10-25). Bipolar cells that are hyperpolarized by the photoreceptor transmitter become relatively depolarized when light excites the receptors, and these constitute the ON pathway (Figure 10-25). Those bipolar cells that are depolarized by the photoreceptor transmitter become relatively hyperpolarized when light excites the receptors and constitute the OFF pathway.

Ganglion Cell Receptive Fields

The receptive field of a ganglion cell is that area on the retina, consisting of a population of photoreceptors, in which light stimulation can affect the activity of the ganglion cell. The ratio of the number of photoreceptors to the number of ganglion cells is approximately 100 million to 1 million, so that in regions outside the fovea there is a large degree of convergence of photoreceptor inputs on ganglion cells. The scotopic system in particular is characterized by a high degree of convergence of rod inputs on ganglion cells; this increases the probability that a single photon will excite a ganglion cell and improves the sensitivity of dim light vision.

Visual acuity refers to the ability to distinguish objects (such as the letters on an eye chart) at a dis-

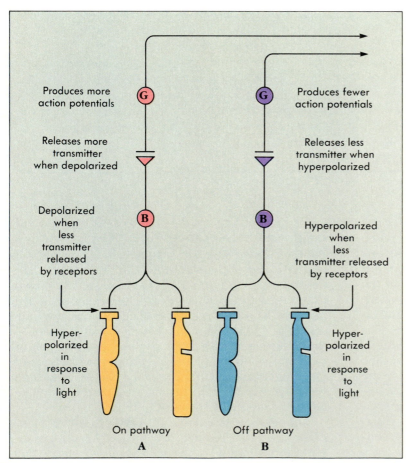

FIGURE 10-25
The response of the photoreceptors (both rods and cones) to light is always hyperpolarization.
A   The bipolar cell (B) in this pathway is excited by the decrease in transmitter release from the photoreceptors, that is, the transmitter had an inhibitory effect that was turned off. The bipolar excitation increases the number of action potentials in the ganglion cell (G) and therefore this is known as the ON pathway.
B   Hyperpolarization of the photoreceptors decreases transmitter release, which inhibits the bipolar cell. This effect indicates that receptor cell transmitter has an excitatory effect on this class of bipolar cells. The inhibited bipolar cell releases less transmitter at its terminals on the ganglion cell, and this results in a decreased number of action potentials from ganglion cells in this pathway. This is therefore the OFF pathway.

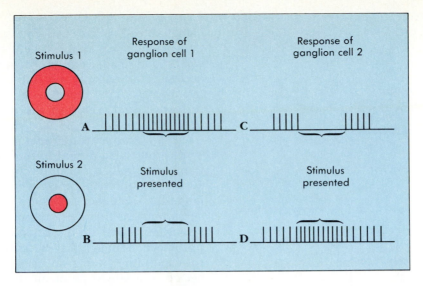

FIGURE 10-26
Electrical recordings from two categories of retinal ganglion cells that are responding to two stimuli. The traces show the rate of action potentials *(indicated by vertical spikes)* over time.
A   Ganglion cell 1 is turned on (the frequency of action potentials increases) by a stimulus *(1)* consisting of a bright spot (the "center") surrounded by a dark background (the "surround").
B   Ganglion cell 1 is turned off by a stimulus *(2)* consisting of a dark center surrounded by a bright background. It is therefore called an "OFF center/ ON surround" cell.
C   Ganglion cell 2 is turned off by stimulus 1 and turned on *(D)* by stimulus 2. It is therefore called an "ON center/ OFF surround" cell.

tance. It is greatest in the fovea, where only a few adjacent photoreceptors converge onto a ganglion cell. The high degree of convergence of rod inputs onto ganglion cells increases light sensitivity, but does so at the expense of acuity.

The receptive field of a ganglion cell can be mapped experimentally by stimulating the retina with a small spot of light while recording action potentials from the ganglion cell's axon in the optic nerve. Some ganglion cells respond to the general level of illumination in their receptive fields, but typically ganglion cells have receptive fields with central regions in which a stimulus results either in a burst of action potentials or a decrease in the steady, low rate of action potentials. These are called **ON center** and **OFF center** ganglion cells, respectively.

Most receptive fields combine excitation and inhibition, so that if light in the center of the receptive field excites a ganglion cell, the same cell will be inhibited by light in a circular area around the center (the **surround**); these are **ON center/OFF surround** cells (ganglion cell 1 in Figure 10-26). In other ganglion cells, a spot of light falling on the center is inhibitory, while in the surrounding region it is excitatory; these are **OFF center/ON surround** cells (ganglion cell 2 in Figure 10-26). The optimum stimulus for an ON center/OFF surround cell is a spot of light of the right size on a dark background. The optimum stimulus for an OFF center/ON surround cell is a dark spot on a white background.

In some cases, the basic receptive field organization incorporates selectivity for colors. For example, a ganglion cell may be excited by green light in its center and red light in its surround but inhibited by red in the center and green in the surround.

Each ganglion cell may send three different messages to the brain. A burst of action potentials constitutes a signal to the brain that a ganglion cell is receiving its optimum stimulus. A decrease in the rate of action potentials means that it is receiving an opposite stimulus pattern. No change in the rate of action potentials means that light, if present, does not vary in intensity over its receptive field. The effect of the antagonistic center/surround organization of ganglion cell receptive fields is to favor response to contrast in the visual field and to suppress response to uniformity.

Retinal Interneurons

There are two types of interneuron that make the retina sensitive to contrast and movement (Figures 10-20 and 10-27). **Horizontal cells** make connections in the region where the photoreceptors and bipolar cells synapse with one another (the outer plexiform layer of the retina). Horizontal cells interconnect adjacent photoreceptors and bipolar cells

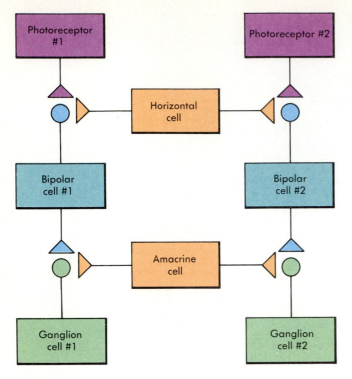

FIGURE 10-27

Synaptic interactions within the retina. Horizontal cells couple receptor cells, including those some distance away. Their interaction with the direct bipolar-to-ganglion cell pathway contributes to the lateral inhibition that produces the center/surround antagonism of the previous figure. This is the basis of spatial analysis of objects. Amacrine cells also mediate lateral transfer of information in the retina, but, in contrast to horizontal cells, amacrine cells act to make certain types of ganglion cells especially sensitive to movement and are therefore thought to mediate temporal analysis.

and are generally associated with **spatial analysis;** that is, they enhance contrast through lateral inhibition.

Amacrine cells make connections in the region where bipolar cells and ganglion cells synapse (the inner plexiform layer). Amacrine cells also couple adjacent in-line pathways in the retina but, in contrast to horizontal cells, seem to function in **temporal analysis**—detection of moving elements in the visual picture. The synaptic connections and the neurotransmitter chemicals used differ among amacrine cells, so that this category of retinal cells probably carries out a number of different functions.

> 1 What factors determine the ability of different parts of the retina to resolve the details of an image?
> 2 How does bleaching affect the photosensitivity of the visual receptors?
> 3 How does the spontaneous rate of action potentials in ganglion cells facilitate contrast detection in the retina?

Central Visual Pathways

Ganglion cell axons enter the optic nerve in an orderly fashion, so that adjacent axons in the nerve correspond to adjacent receptive fields on the retinal surface. The pathway ascends to the lateral geniculate nucleus of the thalamus and then projects to the primary visual cortex in the occipital lobes of the brain (Figure 10-28, A).

The visual fields of human eyes overlap. Objects in the right visual field form images on the lateral (temporal) half of the retina of the left eye (Figure 10-28, B) and on the medial (nasal) half of the retina of the right eye. To avoid double vision, the images of objects that lie in the overlapping part of the visual field must be projected onto **corresponding points** on both retinas. Receptors located at these corresponding points on each retina must be connected to a single point on the brain's map of the visual field.

This is accomplished by a controlled violation of the rule that sensory inputs from the body surface project to the opposite side of the brain. Ganglion cell axons from the temporal half of the left eye project to the lateral geniculate nucleus without crossing (decussating). Afferent fibers from the nasal half of the right eye decussate at the optic chiasm (shown in an MRI of the brain in Figure 10-28, C) and go to the opposite lateral geniculate. The left lateral geniculate receives binocular input from the right visual field, while the right lateral geniculate receives input from the left visual field. Fibers from the lateral geniculate go only to the corresponding primary visual areas (see Figure 10-28, A). This binocular input is essential for depth perception.

As in the cortical maps of other special senses, the visual map is not uniform. The number of visual cortex neurons surveying a particular region of the visual field is inversely related to the degree of convergence in the corresponding area of the retina. Even though the fovea is a small portion of the retinal surface, it dominates the visual map.

Detection of Shape and Movement

In contrast to ganglion cells, visual cortical cells are most responsive to lines or edges rather than spots. One type of cortical neuron, the **simple cell,** is most sensitive to lines oriented in a particular angle to the vertical. Simple cells probably receive a convergence of inputs from lateral geniculate cells whose receptive fields partly overlap (Figure 10-29), so that the most effective stimulus would be a line covering all the excitatory areas. Other cortical neurons, designated **complex cells,** differ from simple cells in being sensitive to both orientation and movement. For example, a complex cell might be stimulated by a line or edge with a particular tilt only if it was also moving in the right direction. Complex cells are less sensitive to the exact position of lines in the visual field than simple cells. **Hypercomplex**

A

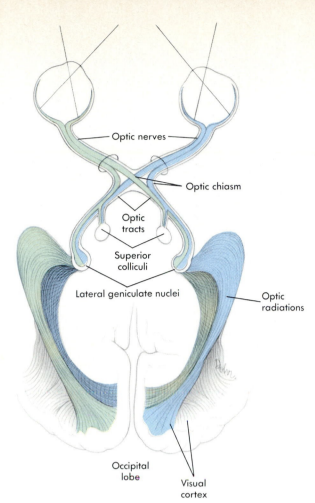

Optic nerves

Optic chiasm

Optic tracts

Superior colliculi

Lateral geniculate nuclei

Optic radiations

Occipital lobe

Visual cortex

FIGURE 10-28

A Central visual pathways. The axons of ganglion cells of the retina synapse with cells of the lateral geniculate nucleus of the thalamus. They also send collaterals to the superior colliculi where visual information is used to control eye movements. Axons from the lateral geniculate project to the visual cortex (the optic radiations). Note the partial crossing (decussation) of optic nerve fibers at the optic chiasm.

B The axons of ganglion cells that project from both eyes to the left visual cortex have receptive fields that collect information from the right visual field *(blue circle)*. This includes ganglion cells on the temporal (lateral) side of the left retina and on the nasal (medial) side of the right retina. The axons of ganglion cells that project from both eyes to the right visual cortex have receptive fields that collect information from the left visual field *(green circle)*. This includes ganglion cells on the nasal side of the left retina and on the temporal side of the right retina. The nasal part of each visual field overlaps to produce a central area of binocular vision.

C An MRI through the optic chiasm.

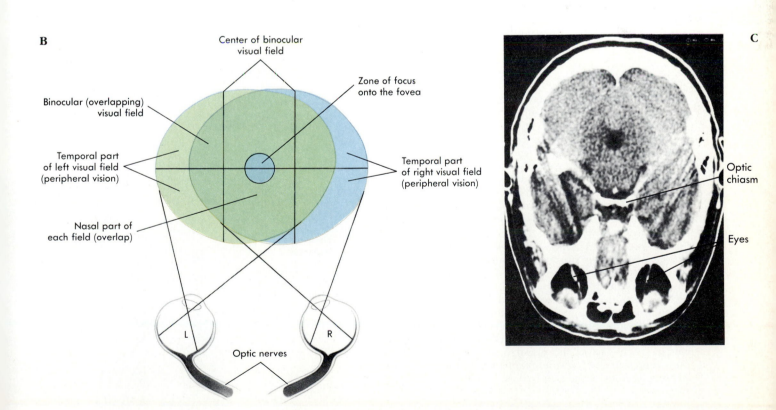

B

Center of binocular visual field

Zone of focus onto the fovea

Binocular (overlapping) visual field

Temporal part of left visual field (peripheral vision)

Temporal part of right visual field (peripheral vision)

Nasal part of each field (overlap)

L

R

Optic nerves

C

Optic chiasm

Eyes

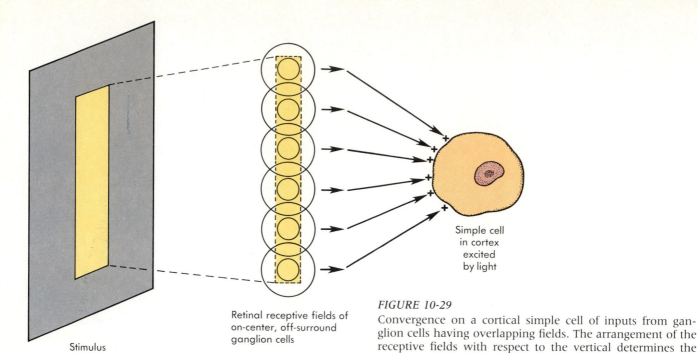

Stimulus
presented to eye

Retinal receptive fields of
on-center, off-surround
ganglion cells

Simple cell
in cortex
excited
by light

FIGURE 10-29
Convergence on a cortical simple cell of inputs from ganglion cells having overlapping fields. The arrangement of the receptive fields with respect to the vertical determines the orientation to which the cortical cell will respond.

cells require not only that the line or edge be moving but also that it be of a certain length or have a corner.

The columnar organization of the cortex is the basis for analysis of the visual image for the feature of orientation. The visual cortex consists of **macrocolumns,** three of which are shown schematically in Figure 10-30. Each macrocolumn takes up about 1 mm² of the cortical surface. All the cells in the various layers of a macrocolumn have receptive fields in the same area of the visual field, and adjacent macrocolumns have adjacent receptive fields. Macrocolumns consist of **orientation microcolumns,** each of which contains simple, complex, and hypercomplex cells that all respond maximally to a single orientation.

FIGURE 10-30
Projections from the lateral geniculate nucleus to the cortex relay information in a very orderly fashion. The path of an electrode that penetrates the LGN perpendicular to the surface penetrates a series of neurons that process information from one area of the visual field. As the electrode passes through the successive layers, it encounters cells driven first by one eye and then the other. The projections from each eye are segregated as the information is relayed to the visual cortex, producing macrocolumns consisting of the two adjacent regions dominated by corresponding right and left eye monocular responses.

The monocular dominance macrocolumns (stippled for the left eye projections) run at right angles to the orientation columns. In the more superficial layers of the cortex, interneuronal connections have crossed the boundary between the left and right eye projections and provided neurons with binocular inputs. The projections shown here relate to only one of the many orientation columns detected in that part of the visual field, and such projections are repeated over and over for each part of the visual field. Adjacent macrocolumns represent the bar orientations for all the other corresponding regions of the two retinas.

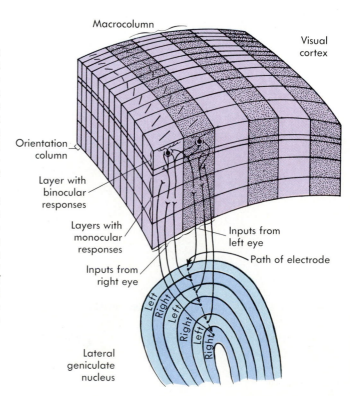

Macrocolumn

Visual cortex

Orientation column

Layer with binocular responses

Layers with monocular responses

Inputs from left eye

Path of electrode

Inputs from right eye

Left Right Left Right Left Right

Lateral geniculate nucleus

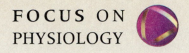

Stimulation is Critical for the Developing Visual System

The responsiveness of the nervous system is easily influenced during its development by changes in the environment; this quality is referred to as **plasticity.** Plasticity of the visual system has been demonstrated by experiments in which experimental animals were deprived of normal visual stimulation. These experiments have shown that there are times during development — **critical periods** — during which even brief periods of deprivation have profound effects on the visual system. In the most extreme case, if one eye is prevented from receiving any light during the critical period, binocular vision will be permanently impaired.

The visual cortex contains bands of microcolumns that are dominated by input from one eye, alternating with bands that are dominated by input from the other eye. This can be demonstrated experimentally by injecting the eye of an animal with a radioactively labelled amino acid. The cells of the retina pass the amino acid to successive cells in the visual pathway, and ultimately some of the labelled molecules are taken up by visual cortical cells. Autoradiograms of the visual cortex of animals that have been visually deprived in one eye show that the portion of the visual cortex that is dominated by input from the undeprived eye increases. The synapses from the stimulated eye literally take over the cortical neurons that would normally have been dominated by inputs from the unstimulated eye.

In other experiments, visual stimulation during the critical period was restricted to lines of a single orientation. In this case, most of the animal's behavior after the critical period seems normal, but behavioral tests show that the animal is blind to visual stimuli of all orientations except the one that it was exposed to during its critical period. When the visual cortex of such an animal is explored with microelectrodes, all of the orientation microcolumns are found to have become sensitive to the single orientation that the animal had been exposed to. The effect was not reversed by exposure to a rich visual environment after the critical period was over.

These experiments have important implications for humans, whose critical period for visual development may continue until about 6 years of age. First, they suggest that medical treatments that restrict visual input—for example, surgery that would necessitate bandaging the eye for a long period—be undertaken with care to minimize developmental effects if the patient is a child. More broadly, these experiments emphasize the importance to children's mental development of providing an environment rich in all kinds of sensory stimulation.

Within a macrocolumn, the preferred orientations of adjacent microcolumns shift systematically in increments of about 10° of arc (shown as the lines oriented in different directions on the surface of the microcolumns in Figure 10-30). Within a single macrocolumn is a band of microcolumns dominated by input from the left eye and a band dominated by input from the right eye. Each cortical macrocolumn seems to contain all of the circuitry needed to analyze a small portion of the visual world.

The columnar organization of the primary visual cortex converts information about the shape of an object in the visual field into a pattern of activity in many discrete columns. There is no such thing as a picture of the visual world on the cortex. Instead, the identification of objects in the visual field requires that the visual cortex acquire, store, and compare many small bits of data from a wide area.

THE CHEMICAL SENSES: TASTE AND SMELL
Submodalities of Taste

There are about 10,000 taste buds located in the crevices of small, domelike papillae that cover the tongue, as well as portions of the palate, pharynx, larynx, and upper third of the esophagus. Each taste bud resembles an orange in which individual receptor cells are arranged like the segments (Figure 10-31). The primary receptor cells, which are modified epithelial cells, extend short, microvillus-like **taste (gustatory) hairs** through a pore to the surface of the tongue. Stimulation of taste receptor cells by a large variety of solute molecules can give rise to relatively few sensations: the ones classically recognized as primary are sour, sweet, salty, and bitter (Figure 10-32). Current research suggests that there might be additional primary taste sensations, such as metallic, but in any case the number of primary taste sensations is very small in comparison with the variety of different stimulus molecules.

The molecular structure of stimulus molecules does not closely correlate with the quality and intensity of the resulting taste sensation. Some artificial sweeteners available commercially, for example, are quite different in structure from sugars and many times more effective than glucose on a mole-for-mole basis. The inorganic salts of lead are sweet enough to appeal to animals and children, who may consume paint and other products that contain them and ultimately suffer lead poisoning.

Afferent information from the taste buds is relayed directly to specific locations near the mouth region of the somatosensory cortex by way of the brainstem and the thalamus. In addition, many collaterals project to areas of the brain concerned with feeding behavior, including the hypothalamus and several brainstem structures. The subjective sensations of taste are strongly influenced by input from the olfactory receptors.

Olfactory Receptors

Olfactory receptors are located in a small patch of **olfactory mucosa** on the dorsal surface of the nasal cavity (Figure 10-33). Each receptor is a neuron with a long, thin dendrite that terminates in a small knob bearing several cilia (Figure 10-34). Supporting cells of the olfactory epithelium secrete mucus; in order to be smelled, odor molecules must first dissolve in the layer of mucus that covers all surfaces of the nasal cavity. Dissolved substances may then bind to receptor molecules on the surface of the olfactory receptors and open ion channels, initiating receptor potentials.

Olfactory neurons are not specialized to detect single fundamental odors. A given receptor may respond vigorously to some stimulus molecules,

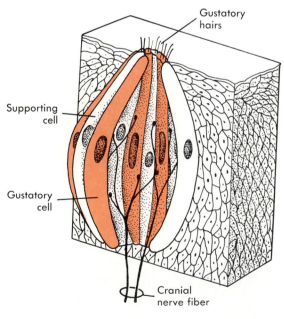

FIGURE 10-31
An enlarged illustration of a taste bud showing the sensory (gustatory) cell and the taste pore with its gustatory hairs.

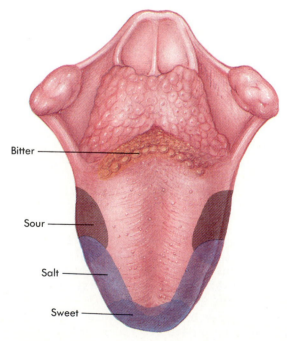

FIGURE 10-32
Regions of the tongue sensitive to particular tastes.

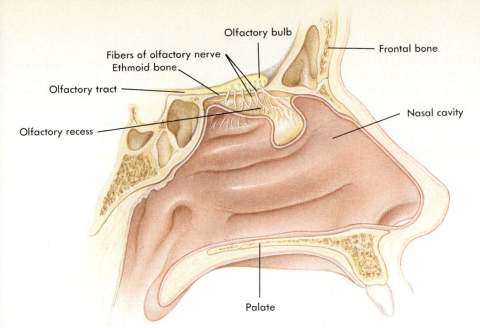

FIGURE 10-33
Lateral wall of the nasal cavity showing the location of the olfactory bulb and the fibers of the olfactory nerve.

Olfactory bulb

Fibers of olfactory nerve
Ethmoid bone

Olfactory tract

Olfactory recess

Frontal bone

Nasal cavity

Palate

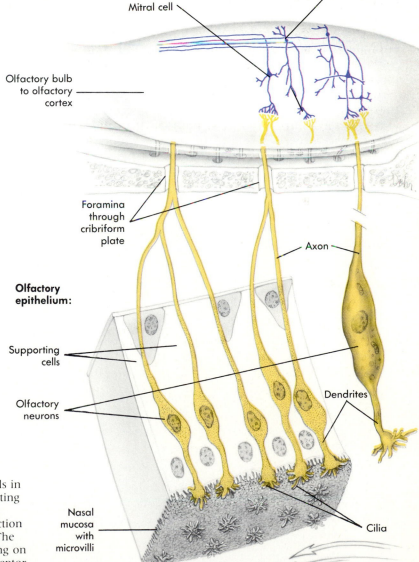

Mitral cell

Interneuron

Olfactory bulb to olfactory cortex

Foramina through cribriform plate

Axon

Olfactory epithelium:

Supporting cells

Olfactory neurons

Dendrites

Nasal mucosa with microvilli

Cilia

FIGURE 10-34
Schematic diagram of receptors and supporting cells in the olfactory mucosa. The cytoplasm of the supporting cells is filled with material that forms mucus when secreted. The olfactory receptors send a long projection tipped with cilia to the surface of the epithelium. The cilia of the olfactory receptors form a shaggy coating on the surface of the epithelium. The axons of the receptor cells pass through the cribriform plate and enter the olfactory bulb of the brain, where they synapse with interneurons.

weakly to several others, and not at all to others, so that the concept of a few primary submodalities is not useful for understanding the olfactory system. Sensations of odor must therefore arise in the brain from the integration of many different receptor responses. Consequently, the olfactory system can mediate a large number of different odor sensations. The number may be very large indeed. Each person possesses a unique odor signature that trained dogs can distinguish from those of a practically unlimited number of other odors.

Central Olfactory Pathways

Axons from olfactory receptors pass through perforations in the cribriform plate of the ethmoid bone (see Figure 10-34) and synapse with mitral cells in the olfactory bulb of the brain. The subsequent pathways followed in the processing of olfactory information do not parallel those described for the other special senses. A major pathway leads into the limbic system, the seat of emotion and memory. This route can affect visceral responses and influence the activities of the hypothalamus and pituitary. The importance of olfaction in mammalian evolution is reflected in the direct line that the olfactory receptors have to relatively primitive parts of the brain responsible for sex, aggression, feeding, and visceral homeostasis. It is possible that olfactory signals processed along this route influence moods and behavior without entering conscious awareness. The second route leads to the olfactory cortex. Connections between the olfactory cortex and other sensory regions of the cortex allow integration of olfactory sensations with those arising from other sensory modalities. Signals that are processed along this route are more likely to lead to conscious sensations.

The sensitivity of the olfactory receptors for odor stimulants is several orders of magnitude greater than the sensitivity of taste receptors. The sensitivity of the olfactory system is modified by the rapid rate of adaptation exhibited by the central neural pathways processing the odor inputs. Adaptation is responsible for the loss of perception of an odor soon after it is first encountered and serves to filter non-novel stimuli from the stream of olfactory information presented to higher processing areas. The central nervous system also alters odor perceptions through efferent control of sensory synapses in the olfactory bulb. This control reflects the overall state of the individual and is altered, for example, by hunger and reproductive status. As with other sensory systems, the efferent control modulates the selectivity of the filtration process.

> 1 What functional differences are there between taste and olfactory receptors?
> 2 What differences are there between the central processing of taste and smell information?

SUMMARY

1. The **tympanic membrane** and auditory **ossicles** transmit sound energy to the **cochlea.** The sensitivity of this transmission is regulated by small muscles in the middle ear.

2. The resonance properties of the cochlea translate sound energy into displacement in such a way that specific hair cells are stimulated according to the pure tone components of the sound.

3. Sound localization in space depends on the detection of phase and intensity differences between the two ears.

4. The structure of the vestibular system translates gravitational or inertial forces into hair cell displacement. Each **semicircular canal** is oriented in a different plane and is preferentially stimulated by rotations of the head in that plane. The **saccule** and **utricle** are oriented at right angles to one another.

5. The saccule and utricle are responsive to orientation and linear acceleration of the head; the semicircular canals are responsive to angular acceleration.

6. The **cornea** and **lens** of the eye focus images on the **retina**, which contains **photoreceptors** and several types of **interneurons.**

7. In the retina, lateral inhibition filters the image to increase contrast. Retinal **ganglion cells**, whose axons project to the thalamus, are tuned to respond to spots on contrasting backgrounds.

8. The central processing of visual images involves feature extraction at increasing levels of analysis. Each **orientation microcolumn** of the visual cortex contains cells that respond to lines of a single orientation, with additional conditions of motion and length that depend on the cell type. A complete analysis of the part of the image that falls on any two corresponding points of the retina is carried out by a **macrocolumn** containing orientation microcolumns, which analyze the input from each retina for all possible orientations.

9. The taste modality is characterized by a small number of primary taste perceptions; in contrast, the olfactory modality gives rise to a wide variety of olfactory perceptions. The taste receptors are epithelial cells; the olfactory receptors are neurons. Taste information is processed in a specific region of the neocortex.

10. The olfactory system is unusual in that processing of olfactory information goes on in the limbic system in parallel with processing in the olfactory cortex.

● STUDY QUESTIONS

1. Under what conditions might the mechanisms that operate to decrease the effect of loud noises on the ear fail to provide adequate protection?

2. What are the principles that determine how pure tones are resolved in the organ of Corti?

3. How is it that excitation of a relatively large fraction of the hair cells can result in activation of only a narrow band of cells in the auditory cortex?

4. Describe factors that determine whether the hair cells in the ampulla will depolarize or hyperpolarize in response to a motion of the head.

5. Why is the nystagmus that occurs following rotation opposite in direction to the nystagmus occurring at the start of a rotation?

6. Trace the pathways of light detection from receptor cells to ganglion cells in the retina. What factor distinguishes the ON pathway from the OFF pathway?

7. What is the sequence of events that leads from photopigment bleaching to a membrane potential change in a photoreceptor?

8. What pattern of connections in the projections from the lateral geniculate nucleus could account for the response pattern of a simple cell? A complex cell?

9. Compare the developmental origin of the sensory cells that are responsible for the two chemical senses. Describe the ambiguous nature of the information provided by any single receptor's responses to stimulation. How does the brain process this type of information?

10. Why is the concept of primary odors relatively useless in explaining the responses of the olfactory system?

● SUGGESTED READING

FINGER, T.E., and W.L. SILVER: *Neurobiology of Taste and Smell,* John Wiley & Sons, Inc., Somerset, N.J., 1987. A current summary of what is and is not known about these special senses.

KUFFLER, S.W., J.G. NICHOLLS, and R.A. MARTIN: *From Neuron to Brain,* edition 2, Sinauer Associates, Inc., Sunderland, Mass., 1984. A comprehensive textbook in the neurosciences.

PIERCE, J.R.: *The Science of Musical Sound,* W.H. Freeman & Co., Publishers, New York, 1983. Nice description of the principles of sound localization in the cochlea.

POGGIO, T., and C. KOCH: Synapses that Compute Motion, *Scientific American,* May 1987, p. 46. Describes how certain cells in the visual system become differentially sensitive to moving objects.

SCHNAPF, J.L., and D.A. BAYLOR: How Photoreceptor Cells Respond to Light, *Scientific American,* April 1987, p. 40. Up-to-date discussion of the role of second messengers in visual transduction.

TROTTER, D.M.: How the Human Eye Focuses, *Scientific American,* July 1988, p. 92. Explores the biochemical and geometrical factors that cause the eye to gradually lose its ability to focus on nearby objects.

CHAPTER 11

Somatic and Autonomic Motor Systems

On completing this chapter you will be able to:

- Compare innervation of skeletal muscle fibers by somatic motor neurons with autonomic innervation of visceral smooth muscle and glands.
- State the anatomical differences between the pathways followed by parasympathetic and sympathetic efferents to their effectors.
- Understand the role of dual innervation in the autonomic nervous system.
- Describe the neurotransmitters used in the autonomic nervous system.
- Outline the spinal circuitry and functional role of the stretch reflex, withdrawal reflex, and the Golgi tendon reflex.
- Appreciate the role of subcortical structures in the initiation and control of movements.
- Distinguish between the lateral and proximal motor control systems.
- Understand the pharmacological differences between postsynaptic receptors in the autonomic nervous system.

*T*his chapter is about the output side of the nervous system: the motor pathways that control effectors of the body. Historically, the motor nerves were subdivided into a somatic motor system, which controls skeletal muscle, and an autonomic motor system, which controls visceral tissues such as the heart, exocrine glands (those that secrete via ducts in contrast to ductless endocrine glands), blood vessels, and the gastrointestinal system. The somatic motor system has also been called the voluntary nervous system because it is involved in consciously initiated activities; the autonomic nervous system was so named because visceral activities such as blood circulation, secretion of saliva, and digestion seemed to go on by themselves.

The two motor systems operate according to the principles of negative feedback control explained in Chapter 5. With some slight differences, the same principles of action potential propagation and synaptic transmission apply to both systems. They differ most importantly in the anatomy of their peripheral components and in the nature of the effectors they serve. From a functional point of view, the role of the somatic nervous system generally is to command, while the role of the autonomic system generally is to modulate or influence. Skeletal muscles do not contract unless stimulated. Many of the effectors served by the autonomic nervous system, such as the heart and intestines, are active in the absence of any nervous input, and their activity can be increased or decreased by the autonomic nervous system.

The distinction between purely voluntary and purely involuntary activities became less sharp with advances in understanding of both systems. For instance, although the somatic motor neurons respond to voluntary commands that originate in the brain cortex, they also are responding continually to signals from muscle receptors whose inputs are ordinarily not consciously perceived. On the other hand, autonomic regulation of functions normally not consciously controlled, such as blood pressure, can be brought under voluntary control if appropriate feedback is provided.

MOTOR PATHWAYS OF THE PERIPHERAL NERVOUS SYSTEM
Motor Units

Each skeletal muscle fiber is innervated by only one motor neuron. However, most motor nerve axons innervate, and therefore control, more than one muscle fiber. The set of muscle fibers innervated by all branches of the axon of a single motor neuron is called a **motor unit** (Figure 11-1, *A*; Figure 11-2). Every time the motor neuron produces an action potential, all muscle fibers in the motor unit contract together. Thus a motor unit, rather than a muscle fiber, is the smallest functional element of a skeletal muscle.

Motor units differ in size within any given muscle, and the average number varies greatly from one muscle to another. Muscles that can perform delicate movements, such as those of the fingers or those that position the eyes, have motor units con- sisting of as few as 2 or 3 muscle fibers. At the other extreme, muscles that produce the force for power- ful, rapid movements such as jumping may have motor units consisting of several hundred muscle fi- bers. Individual fibers in a motor unit are not neces- sarily adjacent to one another, but may be distrib- uted throughout the entire cross-sectional area of the muscle (see Figure 11-1, *A*).

Control of Contractile Force

The force of contraction of a muscle depends on two factors: (1) the number and size of the motor units activated, or **recruited,** by excitatory synaptic in- puts, and (2) the frequency of action potentials in axons serving each active motor unit. Not all motor units need be active at the same time, and the total tension of a whole muscle is an average over all the motor units in it (see Figure 11-1, *B*). At the level of the motor neurons there is a high degree of conver- gence; each spinal motor neuron may receive as many as 15,000 synaptic inputs. Most of these in- puts come from spinal interneurons, but a small fraction come directly from brain motor centers. The integration of all these inputs at the level of each motor neuron is what causes particular muscle units to be activated at any given instant. The pioneer neurophysiologist C.S. Sherrington referred to so-

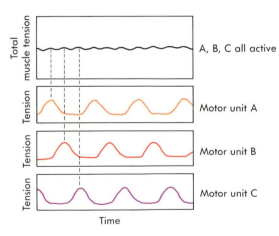

A

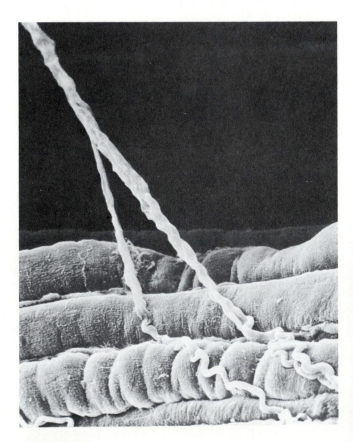

FIGURE 11-2

A scanning electron micrograph of a nerve fiber innervating several skeletal muscle fibers.

FIGURE 11-1

A motor unit is a group of skeletal muscle fibers controlled by a single motor neuron.
A A schematic diagram of three different motor units in a single muscle, consisting of 3 to 5 individual muscle fibers.
B The total tension in a muscle is given by summating the activity of all the active motor units.

NERVE AND MUSCLE

matic motor neurons as the "final common path" from neurons in motor areas of the brain to muscle cells.

Innervation of Skeletal Muscle

The cell bodies of somatic motor neurons that control the skeletal muscles of all parts of the body except the head are located in the ventral horns of spinal cord segments (Figure 11-3). The corresponding motor neurons for the face and head are in brainstem nuclei and project through cerebral nerves. In the spinal cord, motor neurons innervating a single muscle or group of muscles are typically found within 2 to 4 adjacent spinal segments. The axons of somatic motor nerves leave the ventral horns and pass into spinal nerves.

Spinal nerves innervate the muscles of a single muscle segment of the body trunk, or **myotome.** Myotomes are functionally similar to the dermatomes of the somatosensory system. There is an approximate anatomical correspondence as well, with each dermatome and its underlying myotome being served by the same spinal nerve (see Figure 9-13, p. 209). The muscles of the head and limbs have a slightly different embryonic origin from trunk muscles, being derived from embryonic muscle tissue that does not originate in segmental divisions, but innervation is nevertheless traceable to particular spinal nerves.

Four major **plexuses,** or networks of axons, result from the intermingling of axons from more than one spinal nerve. These are the **cervical, brachial, lumbar,** and **sacral plexuses**; the nerve branches that emerge from these plexuses bear the names of the regions that they innervate. The largest nerve in the body, the **sciatic nerve,** emerges from the lumbar plexus and innervates the posterior thigh, leg, and foot muscles.

Pathways to the Visceral Effectors

Autonomic motor pathways to visceral organs are composed of two neurons. **Preganglionic neurons** arise in the central nervous system and project axons to **postganglionic neurons,** whose cell bodies are in the autonomic ganglia. The preganglionic neurons of the **sympathetic branch** are in the in-

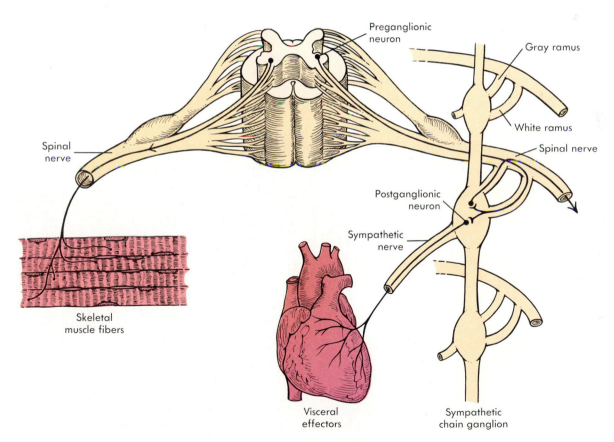

FIGURE 11-3

The somatic pathway *(left)* and autonomic sympathetic pathway *(right).* For clarity, the two pathways are shown on opposite sides of the spinal segment; sympathetic and somatic motor neurons are actually present on both sides of the cord. Most sympathetic postganglionic axons run to effectors along sympathetic nerves, but some are found in spinal nerves along with the axons of somatic motor neurons.

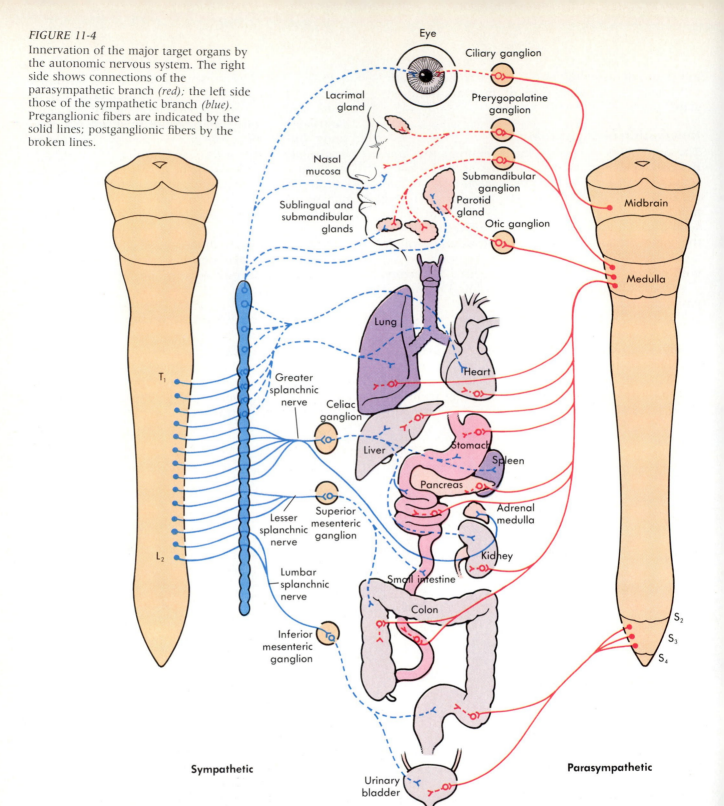

FIGURE 11-4

Innervation of the major target organs by the autonomic nervous system. The right side shows connections of the parasympathetic branch *(red);* the left side those of the sympathetic branch *(blue).* Preganglionic fibers are indicated by the solid lines; postganglionic fibers by the broken lines.

Eye

Ciliary ganglion

Lacrimal gland

Pterygopalatine ganglion

Nasal mucosa

Submandibular ganglion

Sublingual and submandibular glands

Parotid gland

Otic ganglion

Midbrain

Medulla

T_1

Greater splanchnic nerve

Celiac ganglion

Lung

Heart

Liver

Stomach

Spleen

Pancreas

Lesser splanchnic nerve

Superior mesenteric ganglion

Adrenal medulla

L_2

Kidney

Lumbar splanchnic nerve

Small intestine

Colon

Inferior mesenteric ganglion

S_2

S_3

S_4

Sympathetic

Parasympathetic

Urinary bladder

termediate, lateral gray matter (intermediolateral horn) of the thoracic and lumbar spinal cord, and are indistinguishable in microscopic appearance from the somatic motor neurons.

The **paravertebral sympathetic ganglia** form a chain on each side of the spinal cord (see Figure 11-3); sympathetic **prevertebral ganglia** are found in the abdomen adjacent to the aorta (the main ar-

tery of the body). Preganglionic fibers of the sympathetic branch run out of the ventral horn and pass into sympathetic ganglia through short connectives called **white rami** (solid blue lines in Figure 11-4). Within the ganglia, preganglionic axons synapse on postganglionic cells. The axons of most sympathetic postganglionic cells pass into sympathetic nerves that run to blood vessels and internal organs

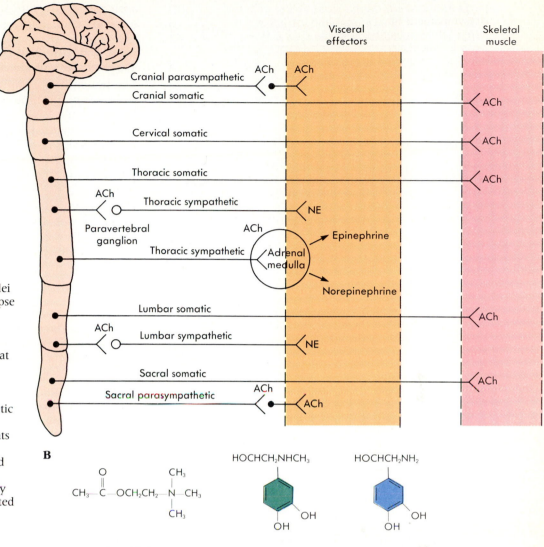

FIGURE 11-5

A A comparison of the central origins of somatic and autonomic motor pathways. Somatic motor neurons are present in cranial nuclei and all spinal segments; they synapse directly on their effectors after leaving the spinal cord and release acetylcholine. Parasympathetic preganglionic neurons are present at cranial and sacral levels only and acetylcholine is the transmitter at both the ganglionic and the postganglionic synapses. Sympathetic preganglionic neurons are present in thoracic and lumbar segments of the spinal cord; acetylcholine is the transmitter in paravertebral and prevertebral ganglia, but the postganglionic transmitter is usually norepinephrine (exceptions are noted in the text).

B The chemical structures of acetylcholine (ACh), epinephrine, and norepinephrine (NE).

(shown as broken blue lines in Figure 11-4). The axons of some sympathetic preganglionic neurons pass through the **gray rami** into spinal nerves, mingle with the axons of somatic nerves, and run to ganglia located near the effectors themselves (for example, the inferior mesenteric ganglion in Figure 11-4).

The **adrenal medulla** is derived during embryonic development from a sympathetic ganglion and is directly innervated by preganglionic fibers. It is a neurosecretory organ because the postganglionic cells, called **chromaffin cells,** are modified neurons that do not possess axons. Instead, they release sympathetic hormones directly into the blood.

The preganglionic neurons of the **parasympathetic branch** are located in two regions of the CNS: the motor nuclei of some cranial nerves and sacral segments of the spinal cord (see Figure 11-4). An important difference between the sympathetic

and parasympathetic branches of the autonomic nervous system is in the location of their ganglia. In the parasympathetic system preganglionic fibers are long (solid red lines in Figure 11-4), leading to ganglia located in or on the organs they serve (except for those that serve the eye, tear glands, and salivary glands). Conversely, in the sympathetic system the ganglia are either next to the spinal cord or in the abdomen, but relatively remote from the organs they serve. Figure 11-5, *A* and Table 11-1 summarize the similarities and differences between the somatic nervous system and the two branches of the autonomic system.

Dual Autonomic Innervation of Visceral Effectors

Effectors innervated by the autonomic system typically receive inputs from both the parasympathetic and sympathetic branches; this is referred to as

TABLE 11-1	Comparison of Somatic and Autonomic Motor Systems	

Characteristic	Somatic	Autonomic
Effectors	Skeletal muscle	Cardiac muscle Heart Smooth muscle Gastrointestinal tract Blood vessels Airways Exocrine glands
Effect of motor nerves	Excitation	Excitation or inhibition
Innervation of effector cells	Always single	Typically dual
Number of neurons in path to effector	One	Two
Peripheral ganglia	No	Yes
Transmitters	Acetylcholine	Acetylcholine Norepinephrine
Receptor types	Nicotinic	1. Cholinergic Nicotinic (ganglia) Muscarinic (effectors) 2. Adrenergic α or β (depending on the effector)

dual innervation (see Figure 11-4). Unlike the somatic nervous system, which can only excite its effectors, the autonomic nervous system generally can mediate either excitation or inhibition. This is possible because visceral effectors typically have some degree of intrinsic activity; the sum of autonomic inputs may either increase the activity or suppress it. Organs that receive dual autonomic innervation are generally excited by one of the branches and inhibited by the other; which branch is excitatory and which inhibitory varies from one organ system to another.

The morphology of synapses, described in Chapter 8, differs between the somatic and autonomic systems. Somatic motor neurons have elaborate motor end plates on each of the muscle fibers of their motor units (see Figure 8-17, p. 177). The receptors for their neurotransmitter, acetylcholine, are confined to the immediate vicinity of the motor end plate. Autonomic synapses are less elaborate, and may take the form of enlargements in autonomic axons, termed axon **varicosities**, which spread transmitter over a large area and do not have a one-to-one relationship with effector cells (see Figure 8-18,

p. 178). In such cases there is no discrete postsynaptic membrane and the effector cells possess receptors over their entire surfaces.

Acetylcholine and Norepinephrine as Transmitters in the Motor Pathways

The transmitter chemical released at the synapse between somatic motor neurons and skeletal muscle fibers is acetylcholine, which is also the transmitter used between preganglionic and postganglionic cells in ganglia of both branches of the autonomic system (Figure 11-5, *B*). Thus the rule for motor pathways is that the first synapse in the periphery (which is also the only synapse in the somatic system) is cholinergic. The postganglionic transmitters differ for the two autonomic branches: as a rule, acetylcholine is released on effectors by postganglionic axons of the parasympathetic nervous system, while **norepinephrine** is released by most postganglionic axons of the sympathetic system (see Figure 11-5).

The discovery that **peptide neuromodulators** (see Chapter 8, p. 188) are coreleased with acetylcholine or norepinephrine has greatly expanded appreciation of the integrative functions of autonomic peripheral synapses. For instance, in sympathetic pathways, enkephalin is often found in the preganglionic axons and somatostatin is often found in the postganglionic neurons. Peptide neuromodulators are also believed to be particularly important in the parasympathetic plexuses of the gastrointestinal tract.

The chromaffin cells of the adrenal medulla are an exception to the rule about postganglionic transmitters because chromaffin cells release primarily epinephrine, along with a small amount of norepinephrine. Epinephrine differs from norepinephrine only by the presence of a methyl ($-CH_3$) group; both are **catecholamines** (see Figure 11-5, *B*). Both epinephrine and norepinephrine are regarded as neurohormones as well as neurotransmitters.

Sympathetic postganglionic fibers that release acetylcholine, termed **sympathetic cholinergic fibers,** are another exception to the rule that sympathetic postganglionic fibers usually release norepinephrine. These sympathetic postganglionic fibers include the pathways to sweat glands and some of the fibers that serve blood vessels of skeletal muscle. In those sympathetic cholinergic fibers which serve blood vessels of skeletal muscle, acetylcholine mediates dilation of the blood vessels. Those which serve sweat glands are responsible for increased sweating during exercise and emotional stress.

Somatic motor neurons resemble autonomic preganglionic neurons in two ways: both originate embryonically in similar spinal regions, and both release acetylcholine. Skeletal muscle fibers, while different in their embryonic origin, functionally resemble autonomic postganglionic neurons in that both receive input from central motor neurons and in both cases the transmitter is acetylcholine.

FIGURE 11-6
The size principle of motor neuron recruitment: the same level of synaptic input more easily recruits a small motor neuron than a large one. In this simplified illustration, synaptic inputs are symbolized by a single synapse that depolarizes equivalent membrane areas on the three motor neurons, but only brings the smallest one to threshold.

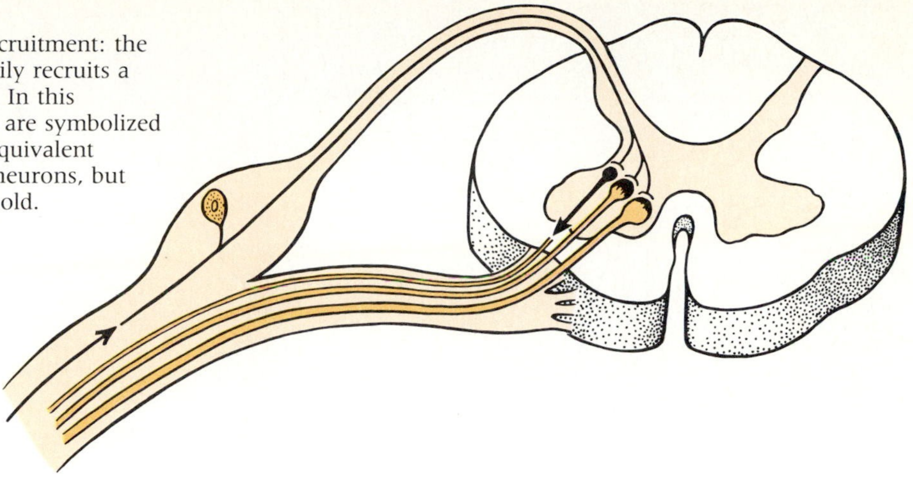

---

1 Describe the paths from central nervous system to effectors in the autonomic nervous system. How do they differ from the path of the somatic nervous system?

2 What is the definition of a motor unit in the somatic nervous system? Why is the name "autonomic nervous system" a misnomer?

3 What rule describes the locations of synapses that use acetylcholine as the transmitter? Which pathways don't follow the rule?

SOMATIC MOTOR CONTROL AT THE SPINAL LEVEL

The Size Principle of Motor Unit Recruitment

One way in which the total force exerted by a skeletal muscle can be regulated is through selective activation of its motor units. The motor units activated at the lowest levels of effort are those with the smallest number of muscle fibers. The initial increments to the total force generated by a muscle are therefore relatively small. If greater force is required, larger and larger motor units are recruited and the force increments become larger. Recruitment of motor units in the order of increasing size results in smooth increases in force.

The basis for the size principle of recruitment is that spinal motor neurons have cell bodies of different sizes. Small motor units are innervated by spinal motor neurons whose cell bodies are small; larger motor units have motor neurons with correspondingly larger cell bodies. Synaptic inputs more readily depolarize a small neuron to threshold than a large one. This is largely because of the greater surface area present in the integrating region of the bigger cells; a given amount of depolarization provided by one excitatory synaptic event will more effectively depolarize the axon hillock region of a small neuron than a big one (Figure 11-6). Recruitment according to size is seen in both reflexive and voluntary contractions of a muscle. One exception to

this rule occurs in learned motor patterns that can recruit the large motor units earlier, thereby favoring very rapid shortening velocity.

When a muscle acts to lift a weight, individual motor units are recruited asynchronously so that an individual fiber is not active continuously, but rather cycles on and off repeatedly. At any instant during the contraction a constant number of fibers may be active, but the composition of this population changes as different individual units turn on and off. Continuously changing the population of active motor units helps prevent any particular motor unit from becoming fatigued.

Spinal Reflexes and Postural Stability

Spinal reflexes are negative feedback systems (see Chapter 5, p. 95). The most familiar example of a spinal reflex is the knee jerk a physician elicits in a physical examination; it is referred to as the **stretch,** or **myotatic,** reflex. When the physician taps the tendon of a muscle, the muscle is stretched momentarily, activating receptors, referred to as **muscle spindles** (described in the following section), in the muscle. The reflex response is a rapid contraction of the particular muscle that was stretched (Figure 11-7, *A*). The stretch reflex is a feedback loop that maintains muscle length and opposes passive increases in length.

Muscles that cooperate to move a joint in one direction are called **synergists;** opposing muscles are referred to as **antagonists.** Muscles that decrease the angle of a joint are called **flexors,** while those increasing the angle of a joint are termed **extensors.** In the knee-jerk reflex, while the stretched extensor muscle is contracting, its antagonist flexors are inhibited (Figure 11-7, *B*). This is the result of **reciprocal inhibition** in the pathways that activate the spinal neurons of antagonistic muscles. Reciprocal inhibition is a common feature of motor pathways and is advantageous for two reasons: (1) energy is wasted if a muscle has to overcome the force

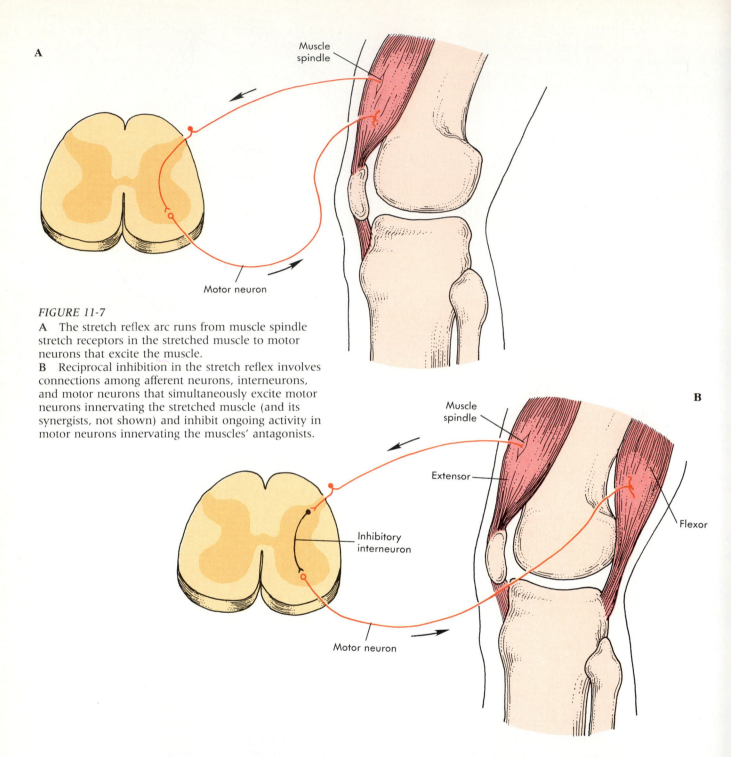

A

Muscle
spindle

Motor neuron

FIGURE 11-7

A The stretch reflex arc runs from muscle spindle stretch receptors in the stretched muscle to motor neurons that excite the muscle.

B Reciprocal inhibition in the stretch reflex involves connections among afferent neurons, interneurons, and motor neurons that simultaneously excite motor neurons innervating the stretched muscle (and its synergists, not shown) and inhibit ongoing activity in motor neurons innervating the muscles' antagonists.

B

Muscle
spindle

Extensor

Flexor

Inhibitory
interneuron

Motor neuron

of its antagonists, and (2) movements are smoother when opposing muscles do not interfere with one another.

Three examples illustrate the importance of the stretch reflex in both extensors and flexors. The muscle involved in the knee jerk extends the lower leg. In a standing person, gravity tends to buckle the knee, stretching the knee joint extensor. Gravitational stretch activates the stretch reflex, maintaining sufficient steady contraction in extensor muscles to prevent collapse of the knee joint. This steady

low level of contraction in skeletal muscles is referred to as **muscle tone.**

The second example illustrates the ability of the stretch reflex to increase muscle force in order to compensate for an unexpected additional load that tends to lengthen an actively contracting muscle. Suppose several individuals are working together to lift a heavy object. If one of them suddenly slips, the others experience an increased load, and this load stretches their flexors. Stretch of the flexors produces a feedback signal from muscle receptors that increases the activity level of motor units already in use and recruits additional, larger motor units that had been inactive. The result is a rapid increase in muscle force that compensates for the increased load.

The third example involves an abrupt decrease in the load on a muscle, such as would occur if the rope broke during a game of tug-of-war. This is a case of rapid unloading of active muscles, the opposite of the previous example. The muscles that had been stretched would shorten because the force that had been opposing their contraction is removed. As the stretched muscle suddenly shortens, positive feedback activity in the stretch reflex pathway rapidly diminishes and limits the degree of excitatory drive on the motor neurons. In this way the spinal reflex, which works faster than conscious adjustment of muscle activity by the brain, reduces the amount of shortening the pulling muscles exhibit when the rope breaks.

Structure of the Muscle Spindle

Scattered throughout the mass of a muscle are tiny stretch receptors, the **muscle spindles,** that provide the central nervous system with information about muscle length (Figure 11-8; Table 11-2). A muscle spindle consists of a small bundle of modified muscle fibers, called **intrafusal fibers,** to which the endings of several sensory nerves are attached. The nerve endings and the intrafusal muscle fibers are enclosed within a capsule. The muscle fibers that make up by far the greater part of the mass of the muscle and are responsible for generating all of its force are called **extrafusal fibers.** Intrafusal muscle fibers are too weak to develop enough force to move body parts; their contractions regulate the sensitivity of the muscle spindles. Extrafusal fibers are innervated by **α-motor neurons;** intrafusal fibers are controlled by **γ-motor neurons.**

There are actually two different types of intrafusal fibers (see Figure 11-8; Table 11-2): (1) a **nuclear bag fiber** has a swollen central region, and (2) a **nuclear chain fiber** is smaller and uniform in diameter throughout its length. The central regions of bag and chain fibers contain no myofibrils and are the most elastic parts of the fibers, so stretching the receptor stretches the central regions of the fibers preferentially. These elastic central regions are wrapped by endings of **group I** afferent neurons (sometimes termed annulospiral endings; see Chap-

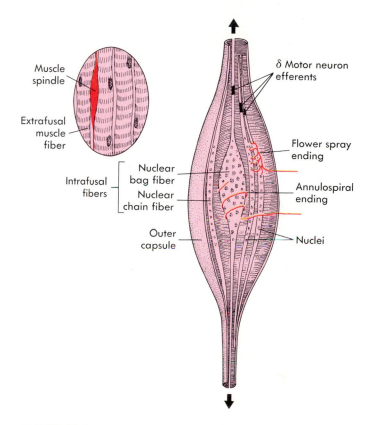

FIGURE 11-8
A diagram of the muscle spindle. The spindle consists of modified intrafusal muscle fibers. There are two types of intrafusal fibers: nuclear bag fibers are innervated by annulospiral endings, and nuclear chain fibers are innervated by flower-spray endings. The ends of the muscle spindles are also innervated by efferent fibers from γ-motor neurons. The muscle spindle is in parallel with the normal (extrafusal) muscle fibers.

TABLE 11-2	Muscles and Their Receptors	

Component	Innervation	Functional role
Extrafusal fibers	α-Motor neurons	Generation of muscle force
Intrafusal nuclear bag fibers	Group I (annulospiral) afferents	Phasic stretch receptors
	γ-Motor neurons	Regulate spindle sensitivity
Intrafusal nuclear chain fibers	Group II (flower spray) afferents	Tonic length detectors
	γ-Motor neurons	Regulate spindle sensitivity
Golgi tendon organ	Group II afferents	Monitor muscle tension

ter 9, p. 207 for a discussion of nerve fiber groups). The lateral, contractile regions of chain fibers are where the endings of smaller **group II** afferent neurons are located (flower spray endings). There are no endings of group II neurons on nuclear bag fibers.

Muscle Spindles as Length and Length Change Detectors

When a muscle is stretched, the responses of its spindles include a phasic component in which action potential frequency is a measure of the rate of stretching, and a tonic component in which action potential frequency is a measure of relative muscle length. Bag fibers are affected primarily by stretch, and chain fibers primarily by length. Both bag and chain fibers have group I afferent endings in their central regions, so the group I response contains both a phasic and a tonic component. Group II afferents have endings only on chain fibers, so their activity only relays information on muscle length (see Table 11-2).

The density and the response characteristics of the stretch receptors present in a particular muscle are related to its function. Muscles typically involved in fine, highly controlled movements, such as finger muscles, have a high density of stretch re-

ceptors containing both bag and chain fibers. Muscles that are not involved in delicate movements, but rather support the body and move large body parts, contain relatively few spindles and those spindles consist in large part of chain fibers.

Group IA afferents from bag and chain fibers branch within the spinal cord. Some branches ascend to higher centers; others synapse with interneurons or with the α-motor neurons. The branches that synapse on α-motor neurons are responsible for the reflexive contraction of a stretched muscle. The effect of spindle afferents on α-motor neurons is excitatory. The efficacy of these synapses must be high, because impulses resulting from spindle stretch are sufficient to recruit motor neurons, even though the number of synapses coming from the spindles is a small fraction of the total number of synapses on these neurons. The time interval, or **latency**, between activation of spindle afferents and the reflexive response is short because there is only one synapse in the reflex arc.

The importance of the stretch reflex for maintenance of muscle tone can be demonstrated by **denervation**: cutting the dorsal roots of a spinal segment, which interrupts afferent signals from the muscle spindles of that segment. When this is done, the tone of the muscles innervated by the segment

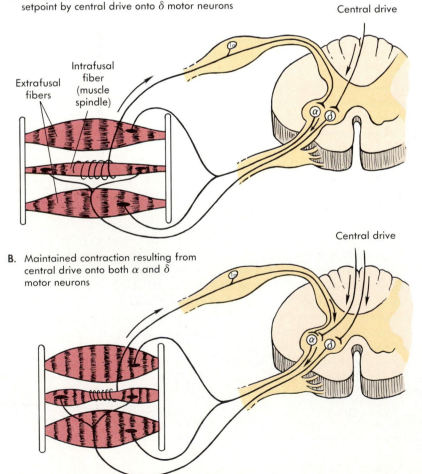

A. Muscle tone resulting from maintenance of setpoint by central drive onto δ motor neurons

Central drive

Extrafusal fibers

Intrafusal fiber (muscle spindle)

B. Maintained contraction resulting from central drive onto both α and δ motor neurons

Central drive

FIGURE 11-9

A In a relaxed muscle, some tone is maintained by central drive onto γ-motor neurons, which sets the activity level of the spindle afferents and maintains a low level of input onto α-motor neurons.

B During a maintained contraction, central drive onto α-motor neurons summates with the feedback from IA afferents.

decreases dramatically, even though the ventral roots containing the motor pathways to those muscles are still intact.

The exact role of the group II endings on chain fibers is not fully understood. Because the group II afferents respond only to maintained length changes, sudden tendon taps have little effect on them, so they do not contribute to reflexes such as the knee jerk. Activity of the group II neurons probably is more important in the slower reflex responses associated with maintenance of posture.

Control of the Stretch Reflex by γ-Motor Neurons

The stretch reflex opposes departures from a set point of muscle length. This setpoint is achieved by the central excitatory drive onto the γ-efferents, which determines how much the ends of the spindle fiber are contracted and therefore how much the center region that contains the sensory endings is stretched. Muscle length is at its setpoint when spindles are "loaded," that is, taut enough to provide a small amount of stimulation to the receptor endings. Under these circumstances, feedback from spindle afferents contributes to muscle tension (Figure 11-9, A).

If the muscle is exhibiting the tone characteristic of a relaxed muscle, the spindle afferent may be almost the only drive on the motor neurons; if the muscle is maintaining tension during a voluntary contraction, spindle afferent feedback adds significantly to the excitatory drive from higher motor centers in determining which motor units are recruited (Figure 11-9, B).

Coactivation of α- and γ-Motor Neurons

Figure 11-10, A illustrates the situation in a muscle prior to the initiation of a voluntary contraction. When voluntary changes in muscle length are initiated by motor control regions of the brain, the motor commands include changes of the set point of the muscle spindle system. These are initiated by a volley of impulses arriving at the α- and γ-motor neurons by way of interneurons that descend from higher motor centers (Figure 11-10, B). If the extrafusal fibers shortened without a compensatory shortening of intrafusal fibers, the spindles would become unloaded, that is, so slack that they would no longer function as length sensors (Figure 11-10, C). Adjustment of the sensitivity of spindles over the entire range of muscle lengths allows the sensory system to be functional at all times (see Figure 11-10, D). The simultaneous activation of extrafusal fibers (by way of α-motor neurons) and intrafusal fibers (by way of γ-motor neurons) is called **α-γ coactivation.**

The important benefit that results from incorporation of sensitivity to length into voluntary activities is illustrated by the following example. When a suitcase must be lifted, the motor centers of the brain estimate on the basis of experience approximately how many motor units must be activated to lift it. If the weight is assessed correctly, the amount of force called for will be adequate to the job. Extrafusal fibers will be able to shorten as rapidly as intrafusal fibers, so the activity of spindle afferents will not change much as the muscles shorten.

But if the suitcase is full of books instead of clothes, the amount of force initially called for by the motor centers may not be adequate to lift the load, and certainly the rate of shortening will be slower than expected. In this case, extrafusal fibers will not shorten much, but intrafusal fibers, which bear none of the load, will shorten. The central region of the spindles will be stretched and a burst of excitatory postsynaptic potentials (EPSPs) from the spindle afferents will summate with those already reaching the α-motor neurons from the descending pathways. Additional motor units will be recruited by this reflexive input, and the force of the muscle will increase until the load is lifted. In this example, the stretch reflex is activated by a difference between the desired muscle length and its actual length. Thus there are few "voluntary" motor actions that do not incorporate a substantial unconscious reflexive component.

> 1 What features of the muscle spindle cause it to act as a detector of muscle length?
> 2 What is the significance of α-γ motor neuron coactivation?

Golgi Tendon Organs: Monitors Of Muscle Tension

Tendons that connect muscles with the skeleton contain receptors called **Golgi tendon organs** (Figure 11-11). Golgi tendon organs detect changes in the tension in a tendon and therefore monitor the force exerted by a muscle. Golgi tendon organs are sensitive to small changes in tension and discharge during muscle contractions, as well as during imposed stretch. Unlike primary spindle afferents, fibers from Golgi tendon afferents do not synapse directly on α-motor neurons, but instead on spinal interneurons projecting to the motor neurons that control the relevant muscle, its synergists, and its antagonists. The effect of activity in Golgi tendon afferents is to inhibit the motor neurons of the muscle whose tendon was stretched and its synergists, while exciting the motor neurons of the antagonists (Figure 11-12). One possible function of Golgi tendon organs is to provide information to motor centers about the force of muscle contraction. Golgi tendon organs may also provide overload protection to relax muscles that are experiencing a dangerous level of tension.

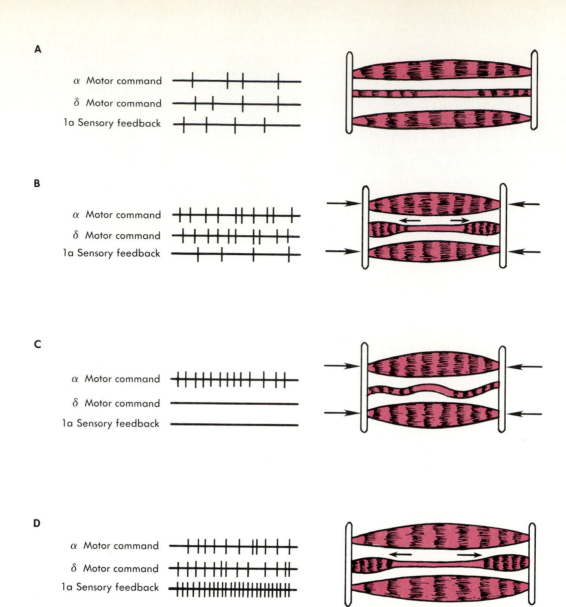

FIGURE 11-10

A   Condition of the muscle and sensory system prior to initiation of a voluntary contraction.

B   Result of α-γ coactivation is shortening of the extrafusal fibers and shortening of the contractile ends of the intrafusal fibers to maintain sensitivity of the spindle afferents. Note that the rate of action potentials in the spindle afferent is similar in **A** and **B**.

C   The loss of sensory feedback information that would result if α-motor neurons were activated in the absence of γ coactivation.

D   α-γ coactivation in a situation in which the muscle is contracting against an excessive load produces elevated levels of action potentials in IA afferent neurons. The effect of this feedback is to supplement the excitatory drive on α-motor neurons.

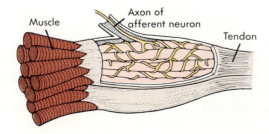

FIGURE 11-11
The Golgi tendon organ is located in tendons and is thus in series with the muscle.

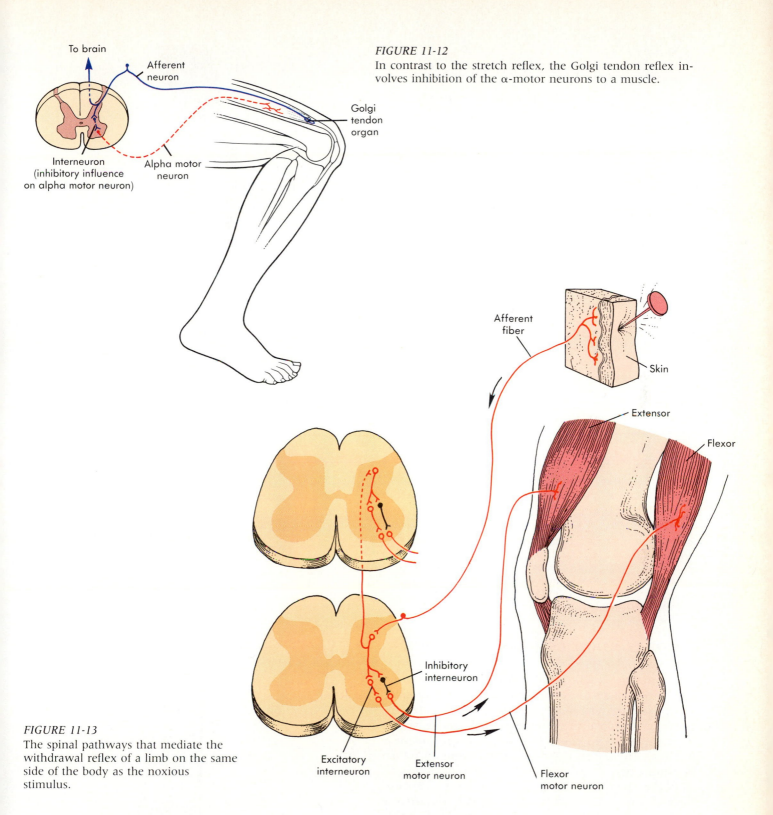

FIGURE 11-12
In contrast to the stretch reflex, the Golgi tendon reflex involves inhibition of the α-motor neurons to a muscle.

To brain

Afferent neuron

Golgi tendon organ

Interneuron (inhibitory influence on alpha motor neuron)

Alpha motor neuron

Afferent fiber

Skin

Extensor

Flexor

Inhibitory interneuron

FIGURE 11-13
The spinal pathways that mediate the withdrawal reflex of a limb on the same side of the body as the noxious stimulus.

Excitatory interneuron

Extensor motor neuron

Flexor motor neuron

The Withdrawal Reflex and Crossed Extension

Nociceptive cutaneous afferents synapse on spinal interneurons, some of which eventually relay information, which is interpreted as pain to the brain. On the ipsilateral (same side) of the spinal cord they immediately excite the flexor muscles of the stimulated limb and inhibit extensors (Figure 11-13). Injury to an extremity of the body results in re-

flexive flexing of the injured arm or leg, withdrawing it from the source of the injury.

Branches of cutaneous nociceptor afferents also cross the spinal cord to excite extensors and inhibit flexors on the opposite, or contralateral side of the body (Figure 11-14). The effect is **crossed extension,** that is, bracing or extension of the opposite limb. The significance of crossed extension lies in

FIGURE 11-14
The spinal pathways that mediate crossed extension in the withdrawal reflex.

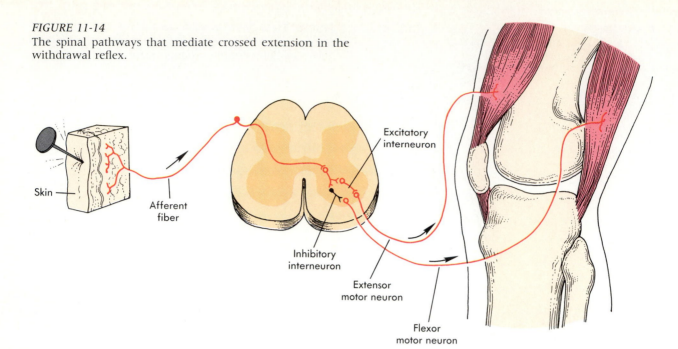

the fact that when a limb that is supporting the body must suddenly be flexed, the weight previously carried by the flexed limb is transferred to the opposite limb. The spinal reflex connections thus simultaneously bring about flexion of the injured limb and preparation of the muscles of the opposite limb to support more weight, all without the participation of the brain.

> 1 The reflex arc of the stretch reflex is a negative feedback loop. What structures correspond to the sensors, integrating center, and effectors?
> 2 Why do Golgi tendon organs provide information on muscle tension?

MOTOR CONTROL CENTERS OF THE BRAIN
Brainstem Motor Areas

The degree to which spinal reflexes contribute to actual movements can be regulated by the motor areas of the brain by way of descending pathways that increase or decrease the overall excitability of spinal interneurons. The control of reflex responsiveness by the brain was revealed by experiments in which the motor pathways were cut at different points. A cut below the brainstem (*cut A* in Figure 11-15) suppressed, for a time, all spinal reflexes beyond the point at which the spinal cord was sectioned (Table 11-3). Before the cut was made, spinal interneurons and motor neurons must have been receiving a steady stream of excitatory postsynaptic potentials from areas within or above the brainstem. Steady excitation that moves the spinal neurons closer to threshold and makes them more likely to respond to other excitatory inputs, such as those from stretch receptors, is termed **facilitation.**

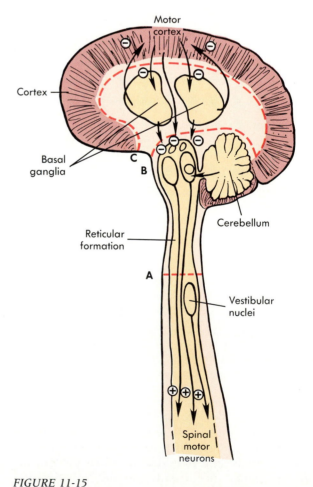

FIGURE 11-15

Brainstem areas that modulate the sensitivity of stretch reflexes. Cut A results in a spinal animal. Cut B results in a decerebrate animal. Cut C results in a decorticate animal.

TABLE 11-3 Results of Interruption of Motor Pathways

Level of section	Terminology	Functional consequences
Below brainstem	Spinal animal	Temporary loss of spinal reflexes below level of section
Above vestibular nuclei in brainstem	—	Enhanced spinal stretch reflexes in antigravity muscles
Above brainstem reticular formation	Decerebrate	Severe rigidity in all limbs
Removal of cortex	Decorticate	Severe handicaps in primates, but relatively minor deficits in nonprimates

The reticular formation of the brainstem includes several nuclei that are sources of descending pathways that facilitate spinal motor neurons (see Figure 11-15). Of particular importance are the **vestibular nuclei**, which coordinate antigravity reflexes. Sudden accelerations or changes of orientation elicit vestibular reflexes that help maintain equilibrium. For example, a sudden tilt to one side, such as might occur from catching one's toe in a crack in the pavement, results in a powerful extension of the limb on the side toward which the body is tilted. This is called a **righting reflex.**

Decerebrate Rigidity

The general effect of the vestibular nuclei on spinal motor neurons can be demonstrated by severing the brainstem above the vestibular nuclei. This enhances stretch reflexes, particularly in those muscles that oppose the pull of gravity. The result is that the limbs are rigidly extended. If the brainstem cut is made still higher (*cut B* in Figure 11-15) so as to leave all of the reticular nuclei attached to the spinal cord **(decerebration)**, and the cerebellum is also removed, the rigidity is even greater (see Table 11-3).

The fact that decerebrate rigidity also involves sensory input from the muscle spindles can be demonstrated by cutting the dorsal roots of a decerebrate animal. The effect is an immediate loss of rigidity in the myotomes served by the cut roots. Decerebrate rigidity can also be reduced by destruction of the vestibular nucleus (or the nerve fibers that descend from it to the spinal cord) or by stimulation of the cerebellum. These experiments show that the major role of the brainstem is continuous facilitation of spinal neurons, particularly of those that serve "antigravity" muscles, which must operate against the pull of gravity. Higher motor centers in the cerebrum and the cerebellum suppress the output of the reticular formation to a greater or lesser degree, regulating it so that the responsiveness of stretch reflexes is sufficient to maintain posture. By turning off the brainstem's facilitation of specific spinal neurons, the higher centers can "unlock" for use in voluntary motion those muscles that normally are involved in resisting the pull of gravity.

Role of the Basal Ganglia in Stereotyped Movements

The **basal ganglia** (see Figure 11-15; also Chapter 9, p. 198) are subcortical forebrain nuclei including the **caudate nucleus, putamen,** and **globus pallidus.** Neurons in the basal ganglia receive input from wide areas of the cortex and send output to the reticular formation and the cortex. Damage to the basal ganglia or to connections between them and other motor centers results in a family of movement disorders called **dyskinesias.** Typically these disorders involve spontaneous, repeated, inappropriate movements of entire muscle groups. Patients may fling their limbs around or exhibit writhing movements or tremors.

The basal ganglia are believed to be organizing centers for motor programs. A **motor program** is a set of commands that is assembled in the brain before a movement begins and is sent to the motor units with the proper timing and sequence to result in the desired movement. A motor program thus is analogous to a script that specifies the parts that individual motor units will play in a complex movement. It still is not possible to describe the physical form in which motor programs are stored and how they are translated into a pattern of activity in the appropriate neurons of the motor system. In diseases of the basal ganglia, the inappropriate movements are thought to be the result of inappropriate execution of motor programs that are activated spontaneously by the injured nuclei.

The Cerebellum and Coordination of Rapid Movements

The cerebellum is an important destination for sensory information from muscle spindles, Golgi tendon organs, the vestibular system, and the visual system. All of the output of the cerebellum is inhibitory (see Figure 11-15); it is one source of the inhibition of the reticular formation that prevents decer-

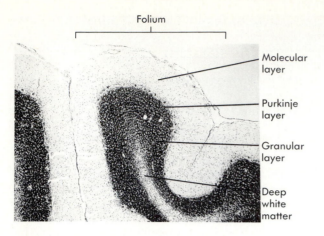

FIGURE 11-16

A micrograph showing the precisely layered structure of the cerebellum. This highly ordered structure is thought to be a critical feature of the cerebellum's ability to coordinate movements.

Folium

Molecular layer

Purkinje layer

Granular layer

Deep white matter

ebrate rigidity in intact animals. The structure of the cerebellum is highly ordered and this organization is thought to be a critical feature of its ability to coordinate movements (Figure 11-16).

Damage to the cerebellum results in motor activity that is rough, jerky, and uncoordinated, especially for rapid movements. Movements tend to be **decomposed**; for example, an individual trying to move a limb might do so by activating muscles around each joint separately, and the result is that the limb does not move smoothly. There are obvious disturbances in walking and postural equilib-

rium, and starting or terminating movements is difficult. Unlike damage to the basal ganglia, cerebellar damage does not result in inappropriate movements; therefore the cerebellum does not seem to be an organizing center for motor programs. Rather, the role of the cerebellum is to coordinate and smooth the execution of programs generated by the basal ganglia. Another important aspect of the role of the cerebellum in motor control is prediction of the future position of moving body parts. When athletes and musicians practice, the effect is to train the cerebellum to achieve perfect timing and coordination of particular motor programs.

Regions of the Motor Cortex

There are at least three motor regions in the cortex: the **primary motor cortex** is located in the frontal lobe of the brain anterior to the primary somatosensory region of the parietal lobe, the **premotor cortex** is in front of the primary motor cortex, and a **supplementary motor area** is located on the medial surface of the cerebral hemispheres. Like the somatosensory cortex, the primary and premotor cortex are topographically organized. In the resulting motor cortical map or **motor homunculus** (Figure 11-17), the representation of the fingers, lips, tongue, and vocal cords is large in comparison to that of the trunk, legs, and upper arms. This is because many cortical neurons are needed to control the large number of small motor units involved in finely controlled movements of these body regions. Like sensory areas of the cortex, the motor areas are layered and contain distinctive neurons referred to as Betz cells (Figure 11-18).

FIGURE 11-17

The somatotopic organization, or motor homunculus, of the motor cortex.

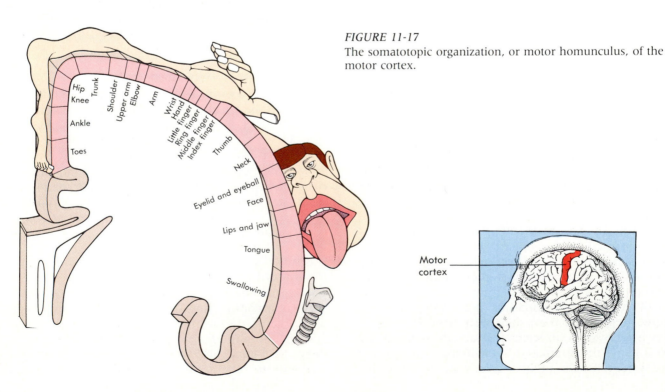

Hip
Trunk
Knee
Shoulder
Upper arm
Elbow
Arm
Wrist
Hand
Little finger
Ring finger
Middle finger
Index finger
Thumb
Ankle
Toes
Neck
Eyelid and eyeball
Face
Lips and jaw
Tongue
Swallowing

Motor cortex

NERVE AND MUSCLE

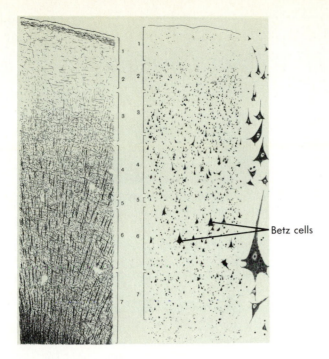

FIGURE 11-18

Sections through the motor cortex. The section on the left has been stained to show nerve fibers, while that on the right shows nerve cell bodies (further enlarged on far right). A characteristic feature of the motor cortex is the presence of large Betz cells in layer 6.

Neurons in the motor cortex that activate particular muscles receive three types of sensory information from the somatosensory cortex: (1) muscle spindles signal static and dynamic muscle length and tension, (2) joint capsule receptors provide information on the position and movement of all of the joints affected by a muscle or group of muscles, and (3) cutaneous receptors provide data on stimuli on the body surface in those areas affected by activation of a muscle or muscle group. Thus the motor cortex is informed about all of the important factors needed to determine the rate, range, and force of movement.

The supplementary motor area differs from the primary motor and premotor areas in two ways: (1) its efferents do not reach the spinal cord by the same pathway, and (2) neurons in the supplementary motor area activate functionally-related groups of postural muscles.

Motor Programs in the Central Nervous System

The most thoroughly studied motor programs are those involved in limb movements. Several levels of programing can be identified by lesion experiments. For example, if the spinal cord segments that innervate an animal's leg are isolated by cutting the spinal cord above and below the region innervating the leg, the animal's leg will exhibit movements that resemble walking. These movements are crude

compared with those seen in a normal animal, but the experiment demonstrates that some aspects of locomotion are programed in the spinal segments. In decerebrate animals the brainstem is intact, and relatively normal movements of all four limbs are seen. When the **mesencephalic locomotor area** is stimulated, animals walk normally even if inputs from the muscle spindles and skin are blocked. Increasing the intensity of the stimuli eventually results in a transition from walking to running. Because coactivation of α- and γ-motor neurons occurs in these experiments, there is reason to think the movements are typical of those seen during voluntary locomotion.

If the cortex is surgically separated from the subcortical structures, a **decorticate** animal results (see Table 11-3). The cuts that are made in decortication are shown diagramatically by line *C* in Figure 11-15. In primates, the cortex is so important for all behavior that decortication results in severe handicap, but in nonprimate animals such as cats or rats, decortication does not abolish all goal-directed behavior. For example, decorticate nonprimate animals can make exploratory movements directed toward sources of food and can care for their newborn offspring. This shows that in these animals the cortex is neither necessary for initiation of motor programs, nor is it solely responsible for generating them.

Readiness Potentials

How primate brain centers interact to initiate motor programs is not entirely clear, but some insight into the process has been gained by recording electrical signals called **evoked potentials** from different regions of the cortex. An evoked potential represents a change in the activity of a population of brain neurons in response to stimulation of an afferent pathway. In human subjects who are carrying out a task on command, there is a burst of neuronal activity (a form of evoked potential called a **readiness potential**) in several areas of association cortex and in the premotor area nearly a second before actual movements occur. Activity in the association cortex is followed by a nearly simultaneous increase in activity in both the basal ganglia and cerebellum; this takes place about 200 to 300 msec prior to movement. Activity in the primary motor cortex follows the onset of activity in the basal ganglia.

These experiments suggest that motor activity associated with voluntary tasks is planned in regions of the cortex other than the primary motor area. This planning involves the integration of sensory information with experience, selection of appropriate motor programs, and transmission of preliminary commands to the basal ganglia and the cerebellum, which organize an appropriate motor program. The results are communicated to the primary motor cortex, which in turn activates descend-

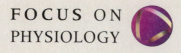

Parkinsonism: Without Dopamine We Would All Freeze Up

One of the inevitable consequences of aging is the loss of neurons through a normal process of cell death. Parkinson's disease is the direct result of loss of neurons in the substantia nigra, a brainstem nucleus whose fibers project to the caudate nucleus. The incidence of Parkinson's disease is approaching one million in America, and the probability of developing clinical symptoms is over one in forty for those living a normal life span. The functional disorders of Parkinson's disease appear very gradually, with slight tremors in extremities. These are followed by stiffness of the trunk and slowed movements. The characteristic tremors at rest, involuntary movements and ultimately debilitating rigidity, result from loss of the transmitter dopamine, which is released from the nerve endings from the substantia nigra which project to cells of the caudate nucleus.

Parkinson's patients have been treated with L-dopa, a precursor of dopamine that can enter the brain and be converted to dopamine. The problem with this therapy is that most patients progressively fail to respond, and at higher doses L-dopa causes hallucinations and convulsions.

Because Parkinson's disease does not appear to develop with age in laboratory animals, there have been no natural animal models in which to develop treatments of the disease. However, the tragic story of drug-induced Parkinson's disease has revealed a potent chemical toxin that specifically targets the substantia nigra neurons. Overnight transformation of young, healthy individuals into patients with incapacitating paralysis resulted from abuse of a synthetic narcotic that differed slightly from the intended chemical structure. Synthesis of the parkinsonism-inducing drug occurred in the course of attempts of street drug chemists to synthesize meperidine (MPPP), a "designer drug" similar to heroin. The drug that actually resulted, 1-methyl-4-phenyl-1,2,3,6-tetrahydropyridine (MPTP), is converted into a toxic form by a naturally occurring enzyme, monoamine oxidase. The resulting compound is identical to the herbicide cyperquat, which is chemically similar to the more commonly used herbicide paraquat. Once this fact was recognized, the incidence of Parkinson's disease was studied in relation to herbicide use, and a positive correlation was found. Additional evidence strongly supports the hypothesis that Parkinson's disease is an environmentally provoked disease of the industrial age, with exposure to a variety of chemicals probably contributing to the accelerated rate of neuronal death in the substantia nigra.

Researchers have tried to use this information to produce animal models in which to improve on the treatment of Parkinson's disease. The common laboratory rat does not develop Parkinson-like symptoms when given MPTP, but monkeys do. Thus primate relatives of humans provide an experimental model of the disease in which new therapies can be tested. One such treatment is the grafting of cells from the patient's own adrenal medulla into the brain. These cells release some dopamine because they produce it as an intermediate in synthesis of norepinephrine and epinephrine.

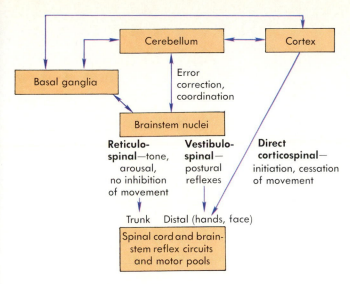

FIGURE 11-19
Interactions of motor areas in the brain.

ing motor fibers running directly to the spinal motor neurons about 20 msec before muscles contract. If the subject is carrying out a program that requires postural changes, the pathways to postural muscle will also be active. In addition, the basal ganglia use a route to spinal motor neurons that does not involve the motor cortex. During movements, continuous sensory feedback to the cerebellum and subcortical motor areas allows for the moment-to-moment coordination of motor activity. Figure 11-19 summarizes current ideas concerning the interactions between the motor control regions of the brain.

> 1 What are the major sources of the synaptic inputs integrated by spinal motor neurons? How does the nervous system normally prevent decerebrate rigidity?
> 2 What is a motor program? What is the role of the motor cortex in the execution of motor programs?
> 3 The cerebellum has no direct connections to spinal neurons. How can it exert its effects on the execution of motor programs?

CONNECTIONS BETWEEN THE BRAIN AND SPINAL MOTOR NEURONS
Comparison of the Medial and Lateral Motor Systems
Skeletal muscles can be divided into two rough categories on the basis of their location in the body and their functions: (1) **proximal muscles** are the muscles of the trunk, involved mainly in gross body movements and the maintenance of postural stability; (2) **distal muscles** are the muscles of the extremities, important for controlled **manipulatory** movements of the arms, legs, and fingers. Of course, control of the two categories must be integrated, because fine movements require postural stability. For

example, it is hard to perform delicate tasks with one's fingers unless the arm is steady.

In most cases, proximal muscles are innervated by motor neurons in the central, or medial, region of the ventral horn, whereas motor neurons that control distal muscles are located laterally (Figure 11-20, *A*). Similarly, spinal interneurons in the intermediate zone of the spinal cord can be subdivided into those located medially and those located laterally.

Descending motor tracts that control refined movements project to both lateral motor neurons and lateral interneurons (Figure 11-20, *B*). Lateral interneurons project to lateral motor neurons, but only on one side of the spinal cord. Medial interneurons synapse onto medial motor neurons on both sides of the spinal cord and thus can control muscles on both sides of the body. Descending motor tracts that regulate postural equilibrium synapse on medial spinal interneurons. They do not terminate directly on medial motor neurons.

Some spinal interneurons send their axons up and down the spinal cord where they synapse in other segments. These are **propriospinal interneurons,** and they also fall into two classes. Axons from lateral propriospinal interneurons extend for only a few segments and terminate on lateral motor neurons or lateral interneurons. Their function is to interconnect motor neurons that innervate small groups of distal muscles, such as those of a single joint. Medial propriospinal interneurons affect segments located at long distances up or down the spinal cord and coordinate the variety of proximal muscles needed to maintain postural stability. Motor control centers coordinate medial and lateral propriospinal interneurons in those cases in which manipulation and postural stability are simultaneously important.

Descending Motor Pathways from the Brain
A pathway, the **corticospinal tract** (Figure 11-21), running directly from the motor cortex to spinal motor neurons, has evolved in parallel with coordinated use of the fingers and reaches its highest evolutionary development in primates. Sometimes this pathway is called the **pyramidal tract** because it forms pyramid-shaped structures as it passes through the medulla.

Several fiber tracts connect the basal ganglia and brainstem motor areas with the spinal cord. These pathways are sometimes called the **extrapyramidal tracts** (Figure 11-22). They are all slower pathways because they are polysynaptic, that is, they include several synapses. The **rubrospinal tract** arises in the reticular formation, crosses to the opposite side of the body in the midbrain, and runs parallel to the lateral corticospinal tract. Like the lateral corticospinal tract, the rubrospinal tract synapses on lateral motor neurons and serves the contralateral side of the body.

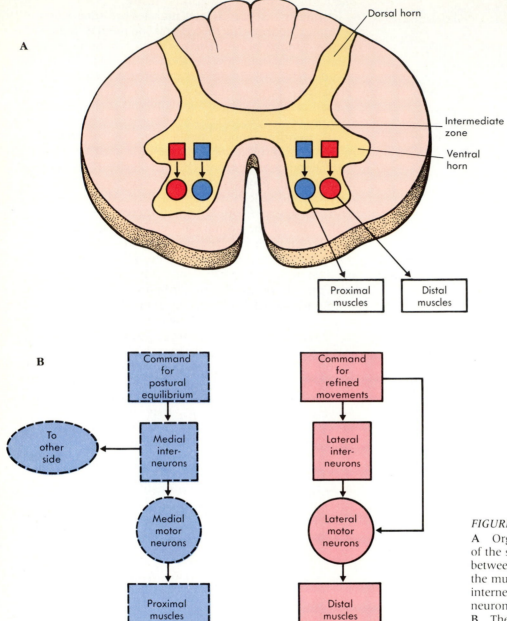

A   Dorsal horn

Intermediate
zone

Ventral
horn

Proximal
muscles

Distal
muscles

B

Command
for
postural
equilibrium

Command
for
refined
movements

To
other
side

Medial
inter-
neurons

Lateral
inter-
neurons

Medial
motor
neurons

Lateral
motor
neurons

Proximal
muscles

Distal
muscles

FIGURE 11-20

A   Organization of the motor components
of the spinal cord. Note the distinction
between lateral and medial elements and
the muscles they control. Squares represent
interneurons; circles represent motor
neurons.
B   The pathways for proximal and distal
muscles are contrasted.

The other polysynaptic routes (reticulospinal
tract, tectospinal tract, tectobulbar tract in Figure
11-22) serve medial motor neurons that control
proximal musculature and are the ones by which
the brainstem facilitates postural reflexes. These
routes include some fibers that decussate and others
that do not, so that these pathways project to both
sides of the body.

When the corticospinal tracts are cut, fine movement control is lost. Individuals with corticospinal
lesions can walk normally because the medial motor
pathways are unaffected. In contrast, if the medial
motor pathways are interrupted, control of overall
postural stability is lost and individuals wobble

when trying to walk and cannot use their fingers for
complex manipulatory tasks. However, if external
support is provided, for example by holding an arm,
the individual's ability to carry out fine movements
is actually fairly normal.

1 *What is the general difference between the functions
of distal and proximal muscles? In what parts of the
spinal cord are the motor neurons of proximal and
distal muscles located?*
2 *Contrast the kinds of activities that demand the use of
the corticospinal pathway with those that are
controlled by the polysynaptic pathways.*

NERVE AND MUSCLE

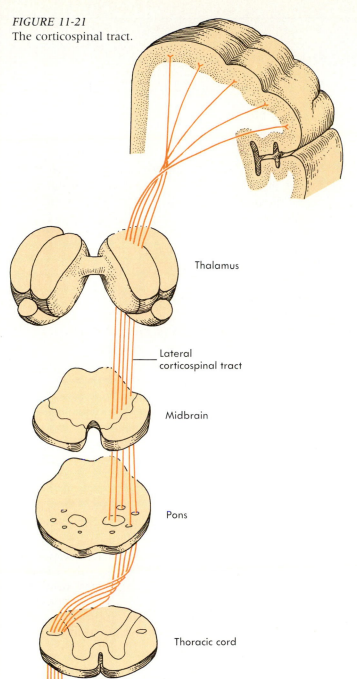

FIGURE 11-21
The corticospinal tract.

Thalamus

Lateral
corticospinal tract

Midbrain

Pons

Thoracic cord

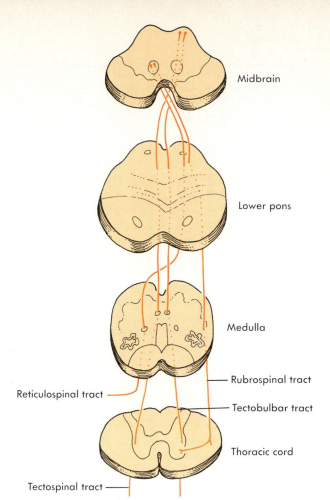

Midbrain

Lower pons

Medulla

Reticulospinal tract

Rubrospinal tract

Tectobulbar tract

Thoracic cord

Tectospinal tract

FIGURE 11-22
Extrapyramidal motor pathways.

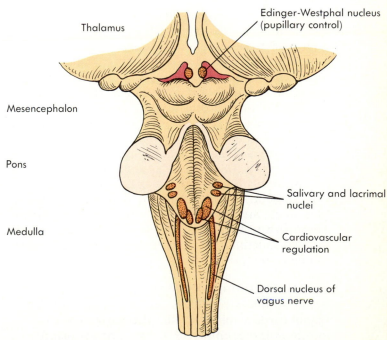

Thalamus

Edinger-Westphal nucleus
(pupillary control)

Mesencephalon

Pons

Salivary and lacrimal
nuclei

Medulla

Cardiovascular
regulation

Dorsal nucleus of
vagus nerve

FIGURE 11-23
Locations of brainstem nuclei of the autonomic motor system.

THE AUTONOMIC NERVOUS SYSTEM
Brainstem Autonomic Control Centers

Afferents from visceral receptors enter the spinal cord along with axons from muscle spindles and cutaneous sensory endings. This internal sensory information is integrated within the central nervous system, either at the level of spinal segments or within the brain. The reticular formation of the pons, medulla, and midbrain (Figure 11-23) contains several nuclei associated with the autonomic nervous system.

The **cardiovascular centers** of the pons are involved in regulation of blood pressure. The **dorsal vagal nuclei** contain the parasympathetic preganglionic cell bodies of the vagus nerves, which are important for control of all visceral organs of the thorax and upper abdomen. The **Edinger-Westphal nuclei** of the midbrain mediate reflexive changes in pupillary diameter and accommodation by way of sympathetic and parasympathetic fibers to the eye. The **lacrimal** and **salivary nuclei** control secretion of tears and saliva. These nuclei receive inputs from other parts of the reticular system that are involved in control of the overall state of arousal of the nervous system, and also from areas of the hypothalamus that are involved in emotional responses. These connections are responsible for changes in visceral function that accompany changes in arousal or emotion.

Spinal Autonomic Reflexes

Some autonomic reflexes are mediated within the spinal cord. A relatively simple example of an autonomic reflex is emptying of the urinary bladder in infants. When the bladder is empty, stretch receptors in the bladder wall are inactive, but when it fills their activity increases. These receptors synapse on interneurons in the sacral part of the cord. The interneurons synapse on parasympathetic motor neurons whose effect is to increase the activity of the smooth muscle of the bladder. As the bladder fills, parasympathetic activity increases until the pressure in the bladder is sufficient to overcome the resistance of the internal urethral sphincter, a ring of smooth muscle that guards the junction between the bladder and urethra. At this point reflexive urination occurs.

Voluntary control of urination, learned in childhood, is largely the result of learning to maintain contraction of the skeletal muscles of the pelvic diaphragm, which closes the neck of the bladder. Relaxation of the pelvic diaphragm permits urine to flow through the urethra, but this process is assisted by the tone of the smooth muscle of the bladder. When voluntary pathways from the cortex to the sacral cord are disrupted by injury in adulthood, bladder emptying reverts to being a local spinal reflex as it was in infancy. A similar autonomic reflex initiates reflexive defecation. Children learn to override reflexive defecation by voluntary control of the external anal sphincter, a skeletal muscle. These are examples of functional interaction of autonomic and somatic nervous systems.

The Role of Dual Autonomic Inputs

The importance of spontaneous activity of effectors and the regulatory role of tonic inputs from the autonomic nervous system can be illustrated by several examples. Smooth muscles around small airways in the lungs are activated by parasympathetic stimulation, causing the airways to **constrict**, or become smaller, while sympathetic stimulation causes **dilation** of airways.

TABLE 11-4	Autonomic Innervation of Target Tissues	
Organ	**Sympathetic stimulation**	**Parasympathetic stimulation**
Eye		
Ciliary muscle	Relaxed	Contracted
Pupil	Dilated	Constricted
Glands		
Salivary	Vasoconstriction Slight secretion	Vasodilation Copious secretion
Gastric	Inhibition of secretion	Stimulation of secretion
Pancreas	Inhibition of secretion	Stimulation of secretion
Lacrimal	None	Secretion
Sweat	Sweating	None
GI tract		
Sphincters	Increased tone	Decreased tone
Wall	Decreased tone	Increased motility
Liver	Glucose released	None
Gallbladder	Relaxed	Contracted
Bladder		
Muscle	Relaxed	Contracted
Sphincter	Contracted	Relaxed
Heart		
Muscle	Increased rate and force	Slowed rate
Coronaries	Dilated	Dilated
Lungs	Bronchi dilated	Bronchi constricted
Blood vessels		
Abdominal	Constricted	None
Skin	Constricted	None
Muscle	Constricted (α receptors)	None
	Dilated (β receptors)	None
Sex organs	Ejaculation	Erection
Metabolism	Increased	None

The effect of parasympathetic input to the heart is to reduce the heart rate; one of the effects of sympathetic input is to increase it. The heart is spontaneously active, and in the absence of any input it would beat at about 100 beats per /minute. The resting heart rate is approximately 70 beats per /minute, so that at mild to moderate activity levels the heart is continuously inhibited by the parasympathetic nervous system. This steady input is referred to as parasympathetic **tone.** At high levels of physical activity the heart is dominated by its sympathetic input, which increases the force of contraction of the heart and dilates the coronary arterioles, as well as increasing the heart rate.

Some organs are dominated by a single branch of the autonomic nervous system, either because they receive no other innervation, or because inputs from one of the two are more abundant. For example, the blood vessels of most organs are innervated only by the sympathetic nervous system. The normal functions of the gastrointestinal tract are largely controlled by the parasympathetic nervous system, which (1) stimulates smooth muscle in the esophagus, stomach, and intestines, (2) increases secretion, and (3) relaxes sphincters to allow food to pass from one part of the tract to another. The gastrointestinal tract also receives sympathetic fibers, but their influence is less important for normal function.

Membrane Receptors for Acetylcholine

Table 11-4 summarizes the effects of autonomic activation on each of the major organ systems. Subsequent chapters will describe some of these autonomic effects in more detail. These effects are mediated through specific receptors. The receptors can be classified according to their responses to drugs and toxins, so the distinctions are based on pharmacology. A drug or toxin that activates a particular receptor type is termed an **agonist**; agents that interfere with the function of a receptor type are **antagonists** (Figure 11-24).

The two types of **cholinergic** receptors (for acetylcholine) are **nicotinic**, named for the agonist nicotine, and **muscarinic**, named for the agonist muscarine (Figure 11-25; Table 11-5; see also Table 11-1). Nicotinic receptors are found on the postganglionic cells of both branches of the autonomic motor system and also on the postsynaptic membranes of skeletal muscle cells. Muscarinic receptors are found on effectors innervated by the parasympathetic branch. Thus the rule for cholinergic receptor types is that nicotinic receptors are found on the postsynaptic cells of the first synapse outside the central nervous system; muscarinic receptors at the second synapse if it is cholinergic.

Membrane Receptors for Norepinephrine

The situation with **adrenergic** receptors (for epinephrine and norepinephrine) is more complex.

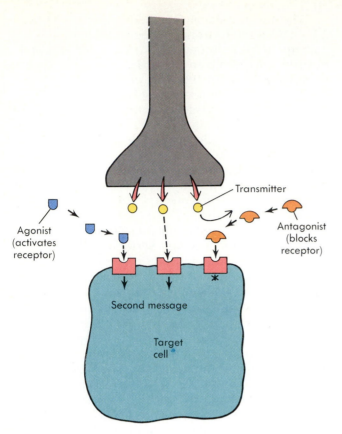

FIGURE 11-24
Agonists bind to postsynaptic receptors and have an effect similar to that of neurotransmitters. Antagonists block the receptor, interfering with transmission.

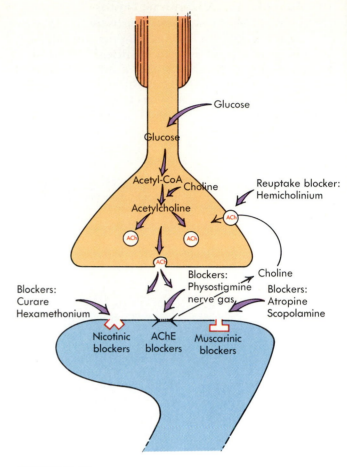

FIGURE 11-25
Neurotransmission at cholinergic synapses. Both nicotinic and muscarinic receptor types are shown together for simplicity, although actually both are not found together at the same synapse. Cholinergic transmission can be blocked by receptor antagonists and by inhibitors of choline absorption. Transmission can be enhanced by inhibitors of acetylcholinesterase.

TABLE 11-5	Location of Cholinergic Receptor Classes

Nicotinic	Muscarinic
Neuromuscular junction of skeletal muscle	Visceral effectors: heart pacemakers, gastrointestinal tract, arterioles of genitalia, iris of eye, salivary and sweat glands, lung airways, urinary bladder
Parasympathetic and sympathetic postganglionic neurons	
Cholinergic synapses of central nervous system	

Historically two different types, α and β, were identified. As new antagonist and agonist drugs were identified, each of these types was subdivided into two subtypes designated by number subscripts (Figure 11-26; Table 11-6; see also Table 11-1).

Each receptor type initiates characteristic responses in effector organs. Each has a unique set of agonists and antagonists that form the basis of medical therapies that result in selective activation or differential block of visceral responses. For any given receptor, the effects of transmitter and agonists are determined by the nature of the second messenger systems set in motion by receptor binding in that particular tissue. Thus, for example, acetylcholine binding to muscarinic receptors results in inhibition of heart pacemaker fibers and activation of intestinal smooth muscle.

Autonomic Receptor Antagonists
Transmission across autonomic ganglionic synapses (and across the neuromuscular junction) is specifically blocked by the nicotinic antagonists **curare** and **hexamethonium** (see Figure 11-25). Curare-like compounds are useful in surgery because an anesthetist can achieve complete skeletal and

smooth muscle relaxation without resorting to high levels of general anesthetic that would inhibit brainstem visceral control systems.

Muscarinic antagonists such as **atropine** and **scopolamine** (see Figure 11-25) decrease gastrointestinal smooth muscle activity and gastrointestinal secretion. Such drugs are valuable in the treatment of gastrointestinal spasms, as well as disorders of hypersecretion such as ulcers. Atropine-like drugs are common ingredients of cold tablets and pills for motion sickness because they inhibit respiratory secretions and suppress the parasympathetically-mediated vomiting reflex.

Distribution of Autonomic Receptor Types

Different subclasses of adrenergic receptors are found in different parts of the circulatory system (see Figure 11-26 and Table 11-6). Heart muscle fibers, for example, possess β_1 receptors. A β_1 agonist such as **isoproterenol** would be appropriate therapy in situations of impaired pumping ability of the heart. Adrenergic β_2 receptors are present on the smooth muscle cells of arterioles that control the size of coronary arterioles. Activation of β_2 receptors relaxes these cells, causing dilation and an increase in blood flow to the heart. Thus β_2 agonists are important in the treatment of diseases in which coronary blood flow is inadequate.

In most tissues, the smooth muscle that surrounds arterioles and veins contracts when norepinephrine activates α_1 receptors. This contraction can be inhibited by α_1 antagonists such as **phentolamine.** Drugs that selectively stimulate α_1 receptors cause constriction of arterioles and veins. An example is **phenylephrine,** a common ingredient of cold medications and nasal sprays. The vasoconstriction it produces relieves nasal congestion. Similar drugs are useful in the treatment of low blood pressure because they increase the total flow resistance of the systemic vessels.

In the blood vessels of skeletal muscles, both α_1 and β_2 receptors are present. The vessels of most visceral organs contain predominantly α_1 receptors. Activation of the α_1 receptors leads to vasoconstriction; activation of the β_2 receptors leads to vasodilation. This differential distribution of adrenergic receptors is important in the redistribution of blood flow into skeletal muscles during exercise.

The β_2 receptors differ from the other adrenergic receptors in being less sensitive to activation by

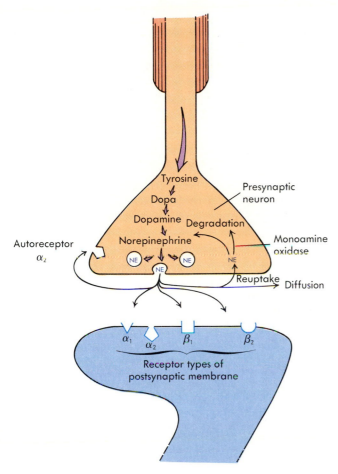

FIGURE 11-26

Neurotransmission at adrenergic synapses. Some antagonists and inhibitors are shown. Adrenergic transmission is blocked by antagonists specific for receptor types, such as propranolol for β receptors and phentolamine for α receptors. Norepinephrine is reabsorbed and subsequently degraded by monoamine oxidase.

TABLE 11-6	Location of Adrenergic Receptor Classes		
α_1	α_2	β_1	β_2
Sweat glands	Gastrointestinal tract	Heart muscle	Blood vessels in heart and gastrointestinal tract
Blood vessels in skeletal muscle and kidney	Adrenergic axon terminals		Lung airways
Iris of eye			

norepinephrine than epinephrine. In tissues, such as the skeletal muscle vasculature, that possess both α_1 and β_2 receptors, a low dose of epinephrine can activate predominantly β_2 receptors, while either norepinephrine or a high dose of epinephrine elicits predominantly an α_1 effect. Thus the sympathetic nervous system can mediate either vasodilation or vasoconstriction in such tissues, depending on the level of input and whether the input is primarily neural (norepinephrine) or hormonal (epinephrine).

In the respiratory system, airways dilate in response to stimulation of β_2 receptors, so β_2 agonists are useful in asthma, a disease in which respiratory distress is caused by an abnormal constriction of airways. These examples emphasize the importance of autonomic receptor specificity.

Transmitter Synthesis in Adrenergic and Cholinergic Synapses

Synaptic transmission depends on a sequence of events that starts with transmitter synthesis and packaging; continues with transmitter release and inactivation, recycling, or loss by diffusion; and ends with a postsynaptic potential. The steps of this process differ between adrenergic and cholinergic synapses, and the differences are important for physiology and medicine.

Cholinergic transmission was illustrated in Figure 11-25. The precursors of acetylcholine are acetyl-CoA (derived from glucose in glycolysis; see Chapter 4) and choline. Choline, which results when acetylcholine is inactivated by postsynaptic acetylcholinesterase (AChE), is recycled into cholinergic terminals (see Chapter 8, p. 180). Blockers of AChE such as **physostigmine** can be used to increase the synaptic efficacy of cholinergic synapses; some nerve gases and pesticides also block this enzyme. Recycling of acetylcholine can be blocked by the choline analog **hemicholinium**; the result is a decrease in the quantal content of the cholinergic synapses.

Figure 11-26 shows the processes that support adrenergic transmission. The immediate precursor of norepinephrine is dopamine. Dopamine may be synthesized from tyrosine that has been taken up by the terminal. Norepinephrine is reabsorbed from the synaptic cleft and then degraded within the presynaptic terminal by the enzyme **monoamine oxidase.**

Adrenergic terminals possess α_2 receptors for their own transmitter. The α_2 receptors are believed to be part of a negative feedback loop (see Figure 11-26) that regulates the excitability of the terminal. Some effectors possess α_2 receptors also.

1 *What are the different sympathetic and parasympathetic receptor types? Define receptor agonists and antagonists.*
2 *Nicotinic antagonists can block both sympathetic and parasympathetic actions. Why?*
3 *Many visceral effectors are spontaneously active. How does innervation by both branches of the autonomic motor system help in the control of such organs?*

SUMMARY

1. Somatic **motor neurons** project directly to skeletal muscle fibers, innervating each muscle fiber with a single excitatory synapse. Autonomic **preganglionic** neurons whose cell bodies are in the central nervous system synapse in peripheral ganglia with postganglionic fibers that innervate visceral effectors.

2. The **parasympathetic** and **sympathetic** branches of the autonomic motor system differ anatomically in that the ganglia in the sympathetic branch are anatomically remote from the effector organs, while the ganglia of the parasympathetic branch are located in or on the effector organs.

3. Visceral effectors typically have **dual innervation** by autonomic fibers from both the sympathetic and parasympathetic branches and may possess more than one motor synapse on each effector cell.

4. As a general rule of organization of the motor systems, the transmitter chemical at the first synapse outside the nervous system is **acetylcholine.**

5. Spinal segmental reflexes that contribute to posture, protection, and voluntary movement include (1) the **stretch reflex,** which maintains muscle length constant; (2) the **withdrawal reflex,** which mediates rapid flexion of an injured limb; (3) the **crossed extension reflex,** which extends the limb contralateral to a flexed limb; and (4) the **Golgi tendon reflex,** which may protect muscles and their tendons from extremes of tension.

6. Motor programs may be generated at several levels in the central nervous system. Simple stepping can be carried out by the spinal cord. More complicated motor programs require the participation of the **basal ganglia, motor cortex,** and **cerebellum.**

7. The cell bodies of motor neurons in spinal cord segments are arranged so that **distal muscles** are served by lateral motor neurons, and medial motor neurons control **proximal muscles.**

8. Pathways from the motor centers of the brain to spinal motor neurons can be divided roughly into two categories: one class that serves primarily proximal muscles and mainly mediates adjustments of posture, and one that serves distal muscles and is important for fine movements. Control of distal muscles is exerted mainly by the direct **corticospinal** (pyramidal) tract; control of proximal muscles involves polysynaptic (extrapyramidal) tracts that pass through brainstem nuclei.

9. The responses of visceral effectors to autonomic inputs are mediated by pharmacologically distinguishable types of receptor molecules. **Nicotinic** cholinergic receptors are located on autonomic postganglionic cells of both branches, but **muscarinic** receptors are found on effectors innervated by cholinergic postganglionic fibers. Adrenergic receptors are divided into α and β types, and into α_1, α_2, β_1, and β_2 subtypes. The subtypes are typically associated with particular effectors, although some effectors have more than one subtype.

10. For any given tissue, the effect of autonomic inputs depends entirely on the second messenger systems activated by transmitter-receptor binding; the same receptor subtypes can frequently mediate opposite responses in different tissues.

1. Movements can be broadly categorized as finely coordinated or postural. Where are the respective motor neurons located in the spinal cord? Are any motor neurons activated directly by the motor cortex? Which ones?

2. Describe the structure of a muscle spindle. What does the muscle spindle measure? How can the sensitivity of the spindle be changed?

3. The stretch reflex has been described as the "power steering" of voluntary movements. How does this reflex contribute to the ability to carry out voluntary movements? Why must the α- and γ-motor neurons be coactivated if this "power steering" is to occur?

4. What are the Golgi tendon organs? What information do they provide?

5. What is meant by the term reciprocal innervation? In which spinal reflexes is this organization most important? Why?

6. What is meant by the term facilitation? What is the physiological role of connections between the vestibular system and the spinal motor neurons?

7. What is the physiological role of the cerebellum? What are the general consequences of cerebellar damage?

8. State the major nuclei that make up the basal ganglia. Can the basal ganglia initiate movements by themselves?

9. What is the evidence for the existence of central motor programs for locomotion?

10. State the two different classes of receptors for acetylcholine and the four classes for norepinephrine and epinephrine. Where is each found? Give two examples of the physiological importance of differential distribution of receptor types.

11. What differences are there between the anatomy of the sympathetic and parasympathetic pathways? In what ways are both distinct from the somatic system?

12. What organs receive single innervation from the autonomic nervous system? Give three examples of organs that receive dual innervation. What characteristics of the target cells determine whether the effect of innervation by a branch of the autonomic system will be inhibitory or excitatory?

● SUGGESTED READING

CARMICHAEL, S.W., and H. WINKLER: The Adrenal Chromaffin Cell, *Scientific American*, August 1985, p. 40. General explanation of the mechanisms of excitation-secretion coupling (exocytosis) and how they resemble synaptic transmission.

CLARKE, E., and B. JACYNA: *Nineteenth Century Origins of Neuroscientific Concepts*, University of California Press, Berkeley, 1987. Survey of the history of neurophysiology and the mind-body problem.

FINE, A.: Transplantation in the Central Nervous System, *Scientific American*, August 1986, p. 52. Discusses the use of adrenal transplants to treat Parkinson's disease and possible extensions of this approach to other neurological disorders.

GEVINS, A.S., et al: Human Neuroelectric Patterns Predict Performance Accuracy, *Science*, volume 235, 1987, p. 580. Discusses cortical evoked potentials associated with activity in the motor system.

GRILLNER, S: Neurological Bases of Rhythmic Motor Acts in Vertebrates, *Science*, volume 228, 1985, p. 143. Describes how the properties of individual neurons and neural networks can generate motor programs.

MAUGH, T.H.: Medicine to Change Your Mind, *Discover*, January 1988, p. 39. Description of how implants of adrenal tissue into the brain have been used to treat Parkinson's disease.

Ultramarathon man, *Discover*, February 1988, p. 16. Describes changes in pain perception and in the muscle composition of the legs of an individual who ran 2200 miles.

CHAPTER 12

Skeletal, Cardiac, and Smooth Muscle

On completing this chapter you will be able to:

- Identify the components of the contractile machinery of muscle.
- Understand the origin of force generation in muscle.
- Describe the structure of a sarcomere.
- State the sliding filament model of muscle contraction.
- Describe the sequence of events mediating excitation-contraction coupling in muscle.
- Distinguish between isotonic and isometric contractions.
- State the functional differences between cardiac and skeletal muscle.
- Understand the initiation and control of contraction in smooth muscle.
- Appreciate the different roles of calmodulin and caldesmon in smooth muscle contraction.
- Distinguish between multiunit and unitary smooth muscle.

Muscle accounts for approximately 40% to 50% of the body weight of an adult. Muscles consume 25% of the total oxygen used at rest; during strenuous exercise, the oxygen consumption of muscle can increase 10 to 20 times. Thus muscle metabolism accounts for a major part of the body's energy budget. There are three basic muscle types: skeletal, cardiac (heart), and smooth muscle. Both skeletal and cardiac muscle have alternating light and dark bands called striations, which result from the ordered way in which contractile proteins (actin and myosin) are arranged within the muscle cells. Thus skeletal and cardiac muscle are referred to collectively as striated muscle.

Skeletal muscles are connected to the skeleton by tendons and cause movement about joints. Each skeletal muscle contains many muscle cells, or muscle fibers, so called because of their elongated shape. Motor neurons innervate small groups of fibers, the motor units, which are the smallest functional elements of skeletal muscle. Skeletal muscles are subject to fatigue with continued use, although the rate of fatigue development varies greatly from one muscle fiber population to another.

Cardiac muscle is functionally different from skeletal muscle because individual cardiac muscle fibers are in electrical communication with each other. Also, cardiac muscle has built-in protection against the fatigue seen in skeletal muscles.

Smooth muscle contains the same elements as striated muscle. However, because the contractile proteins are not organized in a regular array, smooth muscle lacks striations. Smooth muscle is a major constituent of the muscular internal organs of the body, including the arteries and veins, the urinary bladder, and the gastrointestinal tract. In many cases, smooth muscle exhibits spontaneous activity, which can be modified by neural and hormonal factors.

THE STRUCTURE OF SKELETAL MUSCLE
Connective Tissue Boundaries of Muscle Cells

Most skeletal muscles are attached to the skeleton by **tendons** composed of **collagen** (Figure 12-1). Tendons are of various lengths. For example, the muscles that control finger movements are located in the arm, and their tendons are quite long, whereas the tendons of many postural muscles are fairly short. The two places a muscle attaches to bone, the **origin** and **insertion,** are defined by the result of shortening the muscle. The origin is the less movable of the two points, and contraction pulls the insertion toward the origin.

Limb muscles that cause the angle of a joint to decrease are classified as **flexors; extensors** have the opposite effect. Typically, the angle of a joint is affected by several flexors and extensors. Muscles that cooperate to produce movement in the same direction are **synergistic** muscles; those which oppose each other are **antagonistic.**

In skeletal muscle, the cells, or fibers, frequently extend the entire length of the muscle. A muscle may contain from a few hundred to several thousand fibers. The diameter of an individual fiber ranges from 10 to 100 μm. Skeletal muscle fibers are formed by the fusion of cells in development, so each fiber has many nuclei.

Each muscle is surrounded by a connective tissue sheath (the **epimysium**) that is continuous with the muscle's tendons (Figure 12-2, A). Within the muscle, bundles of fibers are bound together by sheaths of **perimysium** to form **fascicles.** Fascicles contain many muscle fibers, each enclosed in its own sheath of **endomysium.** These connective tissue elements transfer the force generated by muscle contraction to the tendons.

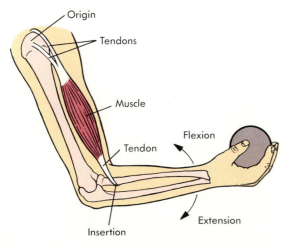

FIGURE 12-1
The gross anatomy of a skeletal muscle, showing its attachment to the skeleton by tendons.

The Structure of Sarcomeres

The cell membrane of a skeletal muscle fiber is called the **sarcolemma** (Figure 12-2, B). Muscle fibers contain long, parallel **myofibrils** 1 to 2 μm in diameter, which make up 75% of a muscle's total volume. Myofibrils are made up of repeating subunits called **sarcomeres.** Adjacent sarcomeres are stacked "in register" within the muscle cells, presenting patterns of repeating bands and lines; these are the characteristic striations seen in muscle viewed with a light microscope.

A sarcomere is a cylindrical structure formed from the interdigitation of **thin** and **thick filaments,** which contain the contractile proteins **actin** and **myosin,** respectively (Figure 12-2, C). The thin filaments appear smooth, whereas the thick filaments have numerous protrusions referred to as **crossbridges.** A relaxed sarcomere is 1.5 to 2.0 μm long. Seen from the side, it has a central **H zone** in which the thick filaments are not overlapped by thin filaments. On either side of the H zone is a region of overlapping of thick and thin filaments.

The entire region of the sarcomere occupied by thick myosin filaments is the **A band.** Toward the ends of the sarcomere the thin filaments are not overlapped by thick filaments; these two regions of nonoverlap in each sarcomere are the **I bands,** which continue into the adjacent sarcomeres. The ends of the sarcomere and beginnings of adjacent sarcomeres are marked by disklike **Z structures,** to which the thin filaments are attached. The Z structures appear as lines in longitudinally sectioned muscle. The sarcomeres of all the myofibrils within a muscle fiber are typically lined up so that all the Z lines form continuous, thin bands (Figure 12-3).

Sliding Filament Model of Muscle Contraction

There is a long history of theories about the mechanism of muscle contraction, beginning with the Greek physiologist Galen, who held that muscles contracted when they were inflated with a fluid delivered to them by nerves. A later theory explained contraction as a form of shrinkage. This theory was disproved by the finding that muscles do not change their volume as they contract.

A key piece of evidence about the nature of muscle contraction was provided by the discovery that the degree of overlap of thick and thin filaments was greater and sarcomere length was shorter in muscles that were contracting when they were fixed for microscopy (Figure 12-4). On the basis of such evidence, the British scientists H. E. Huxley and J. Hanson proposed the **sliding filament theory** of contraction in 1955.

If thick and thin filaments slide across one another, physical links that generate force must form between the two filament types during contraction. Evidence for such links was provided by electron micrographs of contracting muscle, which revealed

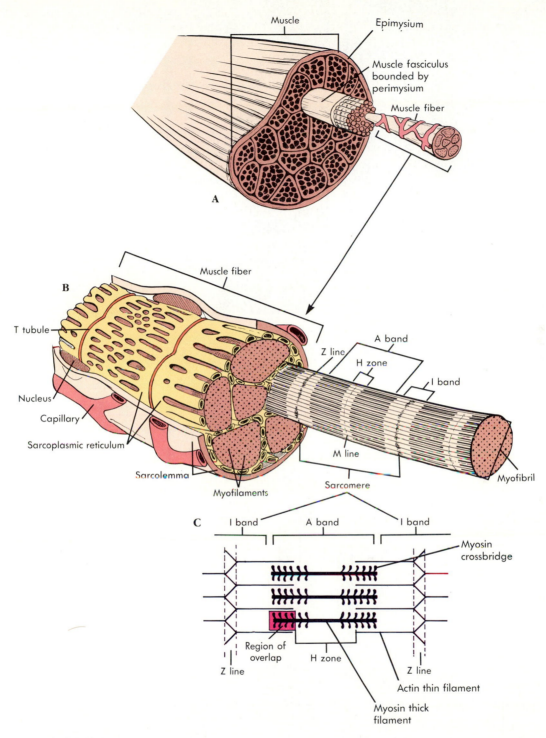

FIGURE 12-2
A The outer covering of a muscle is composed of layers of connective tissue and an inner coat of epimysium. A muscle fascicle consists of a bundle of muscle fibers (cells), each surrounded by perimysium. Within a fascicle, each individual fiber is bounded by its sarcolemma and is separated from adjacent fibers by a wrapping of endomysium (not illustrated).
B The levels of organization of a striated muscle fiber. The contractile elements of a muscle fiber are the myofibrils, each composed of many repeating subunits called sarcomeres.
C Within a sarcomere thousands of individual actin and myosin molecules form a regular array of thick and thin filaments.

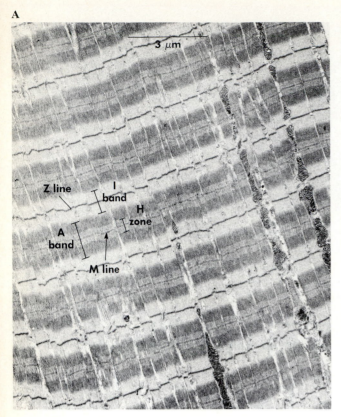

A

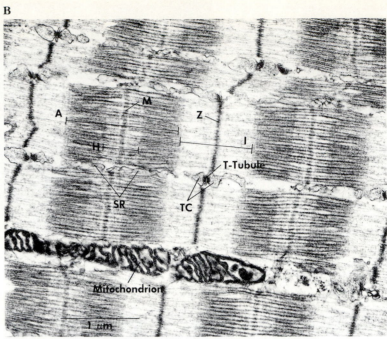

B

FIGURE 12-3

A Longitudinal view of a portion of a muscle fiber including many myofibrils. The sarcomeres of the myofibrils are lined up, giving the whole muscle fiber a banded appearance. Mitochondria are scattered among the myofibrils.

B Several sarcomeres from the muscle shown in **A,** viewed at higher magnification. *TC* is terminal cisternae, *SR* is sarcoplasmic reticulum, and *M* is mitochondria.

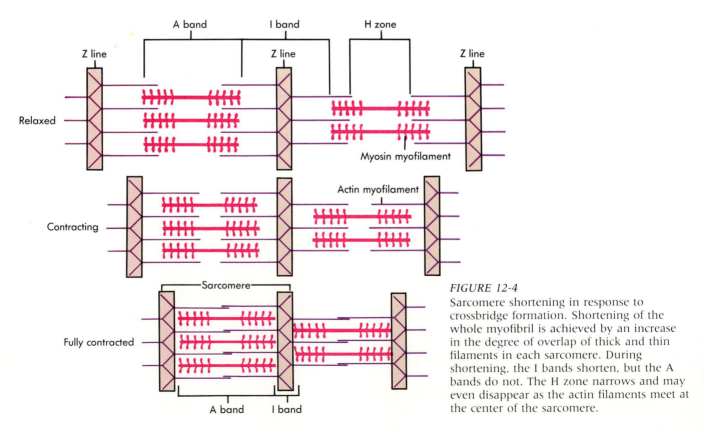

FIGURE 12-4

Sarcomere shortening in response to crossbridge formation. Shortening of the whole myofibril is achieved by an increase in the degree of overlap of thick and thin filaments in each sarcomere. During shortening, the I bands shorten, but the A bands do not. The H zone narrows and may even disappear as the actin filaments meet at the center of the sarcomere.

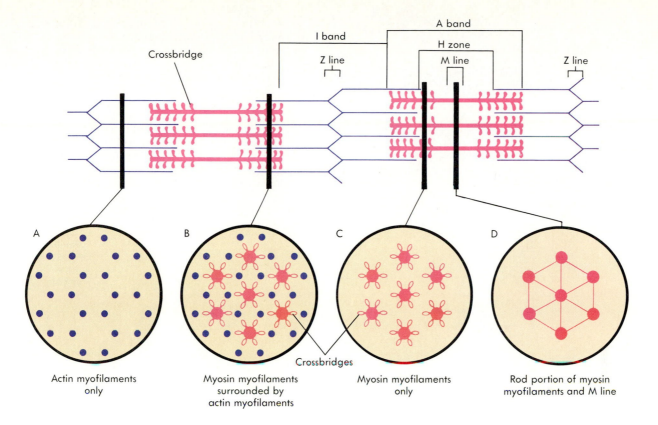

FIGURE 12-5
Cross sections at various points in a sarcomere.
A   In the I band there are only actin filaments.
B   Cross sections in the region of overlap of thin and thick filaments show that each thick
filament is surrounded by a hexagonal array of thin filaments.
C   In the region of nonoverlap only myosin molecules are present.
D   In the middle of a sarcomere, the myosin molecules lack crossbridges.

numerous crossbridges—connections between thick and thin filaments—in muscles fixed during activity. These crossbridges are best seen in cross section (Figure 12-5). Sections through the I band (see Figure 12-5, A) show only thin filaments. In the region of filament overlap (see Figure 12-5, B), there is a characteristic pattern of thick filaments, with crossbridges located 60 degrees apart and surrounded by hexagonal arrays of thin filaments. Each thick filament interacts with six thin filaments, whereas each thin filament interacts with three thick filaments.

Hungarian scientist A. Szent-Gyorgyi extracted actin and myosin from muscle and showed that when the actin and myosin were mixed together, they formed actomyosin. This interaction is the basis of crossbridge formation. Two questions are posed by this discovery. How do crossbridges generate force? What regulates formation of crossbridges? These questions have largely been answered by studies of the protein molecules that compose the thick filaments and thin filaments.

MUSCLE CONTRACTION AT THE MOLECULAR LEVEL
The Structure of Thin Filaments

Thin filaments are composed of a double helical strand of identical globular actin molecules that resembles two strands of pearls twisted around each other (Figure 12-6, A). The single globular molecules of actin are called G-actin (4 nm in diameter; molecular weight about 50,000 daltons). Each strand of G-actin subunits is about 1.0 μm long and is called **F-actin.** A single turn of the F-actin helix is 70 nm long and contains 13.5 G-actin subunits. Each G-actin subunit has a single site where a myosin head can bind to it. Because the thin filament is formed by the twisting of two F-actin strands, the binding sites of some of the G-actin subunits face the inside of the helix and cannot be reached by myosin heads. G-actin subunits that have myosin binding sites projecting outward and available for myosin binding occur every 2.7 nm along the thin filament.

In addition to actin, thin filaments contain two proteins that have a regulatory function (see Figure

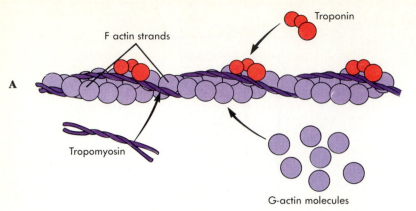

F actin strands

Troponin

Tropomyosin

G-actin molecules

FIGURE 12-6

A   The components of thin filaments. G-actin is assembled into single-stranded F-actin. F-actin strands are twisted together to form the backbone of a thin filament. Double helixes of tropomyosin lie on the F-actin strands. Each tropomyosin has a molecule of troponin positioned at one end.

B   A view of a thin filament showing the relationship of troponin, tropomyosin (TM), and F-actin. Troponin consists of three subunits. Binding of myosin heads (M) is impossible when tropomyosin covers the binding sites (left), but is possible when Ca^{++} binding to troponin causes the tropomyosin to move away from the binding sites (right).

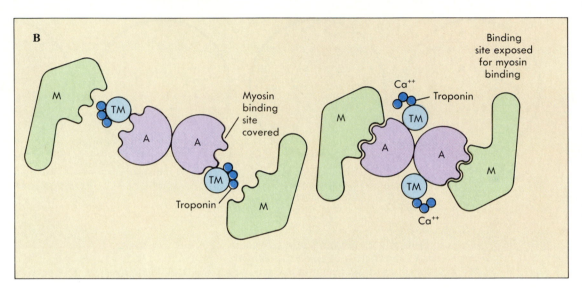

12-6, A). Strands of **tropomyosin** lie on the actin filament. Each tropomyosin strand extends along about 7 G-actin subunits, and the strands are joined end to end. The second regulatory protein of thin filaments, **troponin,** has three globular subunits. One subunit binds to a strand of tropomyosin, the second attaches to actin, and the third has a binding site for Ca^{++}.

When Ca^{++} is not present, troponin is bound to both actin and tropomyosin in such a way that the tropomyosin strands block the myosin binding sites present on each G-actin subunit of F-actin (Figure 12-6, B). The presence of Ca^{++} and its binding to troponin causes a conformational change in the troponin molecule that shifts the tropomyosin strand and exposes the myosin binding sites on the actin, allowing actin and myosin to interact (see Figure 12-6, B, right). Thus part of the question of how contraction is controlled in striated muscle is answered by an understanding of the roles of troponin, tropomyosin, and Ca^{++}.

THE MYOSIN SUBUNIT STRUCTURE OF THICK FILAMENTS

Each thick filament is 1.5 μm long and 15 nm in diameter and is composed of several hundred myosin molecules, each with a molecular weight of about 500,000 daltons. The backbone of the thick filament is a bundle of the long chains of myosin molecules, the heads of which extend out to the sides (Figure 12-7). The myosin molecules are oriented with their heads extending away from the center of the filament, so that the two ends of the filament have heads oriented oppositely and the center region has no heads (see Figure 12-5).

Myosin molecules have a polarity that affects how they associate with each other to form the thick filaments. Each myosin unit is a double structure composed of two heavy polypeptide chains that form a rodlike tail and a double head (see Figures 12-7 and 12-8). The heads, which bind actin and engage in ATPase activity, are joined to the rest of the myosin molecule by peptide regions that act as hinges. Three or four smaller (light) polypeptide chains are associated with the head and play a regulatory role.

Myosin molecules in solution cannot generate force when they interact with actin, but when the myosin molecules associate to form the thick filaments of the sarcomeres, binding to actin allows the energy to produce a force directed toward the center of the thick filament.

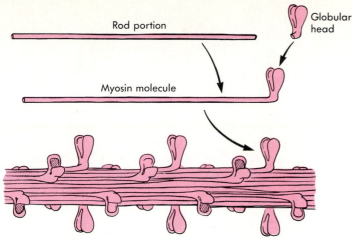

Rod portion

Globular head

Myosin molecule

FIGURE 12-7
Single myosin units have a double globular head at the end of a long rod. The long rod is composed of two heavy polypeptide chains; the ends of the heavy chains, together with the light polypeptide chains of myosin, form the heads. If myosin units are mixed together, they self-assemble into thick filaments. The myosin heads project from the thick filament at the proper angles to interact with the hexagonal array of thin filaments.

> 1 What structural feature gives striated muscle its banded appearance?
> 2 What connective tissue structures give muscle its structural integrity?
> 3 Distinguish between the roles of actin, myosin, troponin, and tropomyosin.

Powerstroke of Muscle Contraction

In a muscle at resting length there is some overlap between the thick and thin filaments. During contraction, the thick filaments pull the populations of thin filaments on each side of the sarcomere toward the center of the sarcomere (see Figure 12-4). This force is caused by the combined action of all the crossbridges formed when myosin heads are able to contact actin binding sites. In a contracting muscle, each myosin head undergoes a cycle of (1) attachment to an adjacent thin filament, followed by (2) a **powerstroke** that moves the head about 10 nm relative to the site of attachment, followed by (3) detachment from the thin filament, followed by (4) the beginning of another cycle. An individual myosin head can perform this cycle several times a second.

A description of the myosin cycle should start at the point just before the head attaches to the thin filament. It is at this time that the myosin is in the state that it would be in at the instant of activation of a previously resting muscle (Fig. 12-9, *A*). A myosin head that is ready to bind to actin has hydrolyzed a molecule of ATP. ADP and inorganic phosphate remain bound to the myosin, forming a **myosin:ADP complex.** Myosin:ADP represents a "high-energy" form of myosin that can bind to exposed actin sites in the presence of Ca^{++}.

Myosin heads are thought to have several sites that can interact with several sites on actin. After the first site on the myosin head has bound to the first actin site, the energy of the myosin:ADP complex is gradually transformed into mechanical energy by sequential binding with sites of greater and greater affinity. Sequential binding rotates the myo-

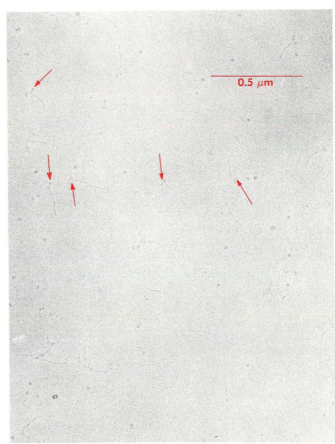

0.5 μm

FIGURE 12-8
An electron micrograph of free myosin units. The arrows indicate the myosin heads.

sin head, generating tension between the thick and thin filaments (Figure 12-9, *B*). This is the powerstroke of the crossbridge cycle. During the powerstroke, ADP and phosphate are released from the head.

At the end of the powerstroke, the myosin head is tightly bound to the thin filament in a low-energy form of actomyosin. This state is called **rigor complex** because the thick and thin filaments cannot slide past each other and the muscle is stiff. In liv-

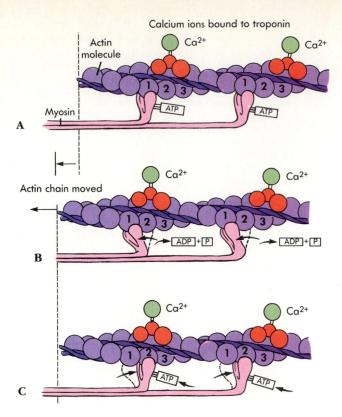

FIGURE 12-9
The crossbridge cycle. In the relaxed state (Ca^{++} concentration less than 10^{-7} M), the myosin binding site on the actin filament is blocked by tropomyosin, and myosin and actin are unable to interact.
A A molecule of ATP attaches to the globular head of a myosin molecule. In the presence of elevated Ca^{++}, troponin complexes form, moving tropomyosin away from actin binding sites. Myosin then binds to exposed active sites *(1)* on the actin molecules.
B ATP is hydrolyzed to ADP and phosphate *(P)*. The released energy causes the myosin head to rotate, causing the thin filament to move relative to the thick filament. This is the powerstroke.
C After the powerstroke, ATP binding releases the head from the rigor complex; the myosin head then recoils to its original position. The cycle is repeated with myosin binding to the next site *(2)* on the actin filament.

ing muscle, the rigor complex is quickly broken by the binding of a molecule of ATP to the head (Figure 12-9, *C*). When blood circulation and breathing stop at death, the energy metabolism of muscle halts. The small supply of ATP is quickly used up, and actin and myosin remain locked together in the rigor complex. This complex is responsible for the rigor mortis of muscle that occurs after death.

In biochemical terms (Figure 12-10), when the rigor complex is broken by the binding of a molecule of ATP to the head, the low-energy actin:myosin complex dissociates to yield actin and a **myosin: ATP complex.** As soon as ATP can bind to myosin, it is hydrolyzed to ADP and inorganic phosphate to regenerate the high-energy state of myosin, which can begin another crossbridge cycle.

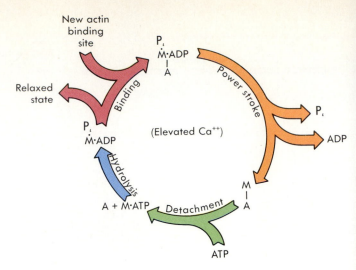

FIGURE 12-10
Summary of the biochemical reactions associated with the crossbridge cycle. *A* stands for actin; *M* for myosin; *P* for phosphate.

The rate at which ATP can be hydrolyzed is a measure of the cycling rate and is termed the **ATPase activity** of the myosin.

Muscle Contractility

The **contractility** of a muscle cell, the extent to which it is able to generate force and to shorten rapidly, is determined by the molecular properties of the thick and thin filaments. For example, different forms of myosin, or **myosin isozymes,** have different ATPase activities. If the rate of ATP hydrolysis is rapid, force development and shortening occur rapidly and the force developed is relatively great. Each myosin head repeats the crossbridge cycle five times per second in fibers that have a myosin isozyme with high ATPase activity. Such muscles are called **fast-twitch muscles.** A slower rate of ATP hydrolysis and smaller contractile force are characteristic of **slow-twitch muscles,** which possess a different isozyme. The contractility of skeletal muscle can also be regulated by phosphorylation of allosteric sites on the myosin head.

EXCITATION-CONTRACTION COUPLING
T-Tubules of Striated Muscle

Every few micrometers along the length of a striated muscle fiber, the sarcolemma invaginates to form **transverse tubules,** fingers of membrane that penetrate the muscle cell interior (Figure 12-11). Transverse tubules branch extensively so as to come into close proximity to each myofibril. The lumina of the transverse tubules are near the junctions between sarcomeres. Because transverse tubules are continuous with the muscle membrane, the interior of the tubule is connected to the extracellular space.

The membrane of the transverse tubule system is part of the plasma membrane, and when the membrane is depolarized during a muscle action

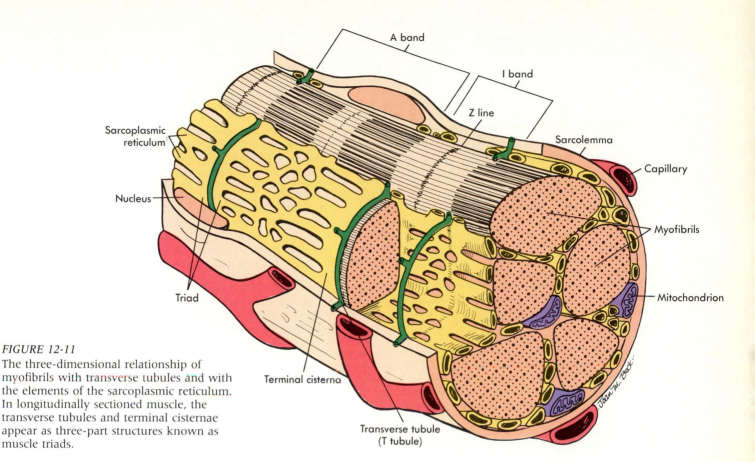

FIGURE 12-11

The three-dimensional relationship of myofibrils with transverse tubules and with the elements of the sarcoplasmic reticulum. In longitudinally sectioned muscle, the transverse tubules and terminal cisternae appear as three-part structures known as muscle triads.

potential, the action potential is brought close to the cell interior. The transverse tubule system is an adaptation for rapid activation of all parts of a muscle fiber. Each fiber is several hundred micrometers in diameter. The passage of an action potential across the sarcolemma, an event termed **excitation,** must be communicated rapidly to the innermost parts of the muscle if all parts of the fiber are to contract together. The transverse tubules carry the electrical signal into the cellular interior, minimizing the distance that must be covered by a diffusing intracellular second messenger.

Sarcoplasmic Reticulum

The second tubule system, which is greatly elaborated in skeletal muscle cells and less elaborated in cardiac muscle cells, is the smooth endoplasmic reticulum. It forms an intracellular system of interconnected tubules known in these cells as the **sarcoplasmic reticulum** (see Figure 12-11). The interior of the sarcoplasmic reticulum surrounds all the myofibrils, but, although it comes close, it never unites with the transverse tubules. The sarcoplasmic reticulum enlarges into structures known as **terminal cisternae (lateral sacs)** located next to the transverse tubules. The association of a transverse tubule with two adjacent terminal cisternae is termed a **muscle triad** (Figure 12-12). The triad is the site

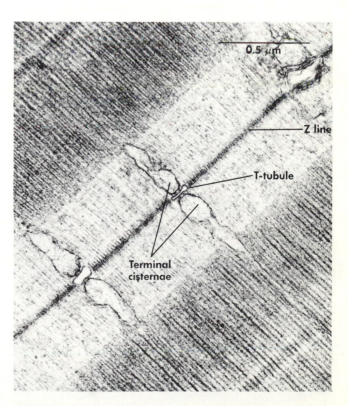

FIGURE 12-12

A electron micrograph illustrating the structure of a muscle triad.

where muscle action potentials are coupled with the internal mechanisms responsible for force generation, a process known as **excitation-contraction coupling.**

Calcium and Excitation-Contraction Coupling

When a muscle fiber is at rest, an active transport system in the sarcoplasmic reticulum pumps Ca^{++} out of the cytoplasm into the sarcoplasmic reticulum and maintains the intracellular Ca^{++} concentration at less than 0.1 μmol/L, well below the threshold for significant binding to troponin. In a resting muscle, the amount of Ca^{++} that is **sequestered** (taken up as a result of Ca^{++} pumping) inside the sarcoplasmic reticulum is so great that the pumps must work against a concentration gradient of about six orders of magnitude. This gradient would be even greater (about eight orders of magnitude), except that inside the sarcoplasmic reticulum are large quantities of a protein called **calsequestrin.** Calsequestrin binds Ca^{++} reversibly, taking some of it out of solution.

The events of excitation-contraction coupling are shown in Figure 12-13 and summarized in Table 12-1. The first event is the production of an action potential at the neuromuscular junction (see Figure 8-17), which rapidly propagates over the sarcolemma and also spreads into the transverse tubules. The electrical signal is communicated to the terminal cisternae of the sarcoplasmic reticulum across a significant distance of muscle cell cytoplasm. Recent evidence suggests that depolarization of the transverse tubule activates a membrane-bound phosphodiesterase, which hydrolyzes phosphatidylinositol to inositol triphosphate and diacylglycerol. Inositol triphosphate diffuses from the inner surface of the transverse tubule to the sarcoplasmic reticulum, where it binds to receptors that control the Ca^{++} channels.

The end result is that Ca^{++} channels in the sarcoplasmic reticulum are opened, allowing Ca^{++} to leave the sarcoplasmic reticulum and enter the cytoplasm. As the Ca^{++} levels in the sarcoplasmic reticulum drop, calsequestrin gives up its bound Ca^{++}, making additional Ca^{++} available. The cytoplasmic Ca^{++} level increases rapidly from less than 0.1 μM

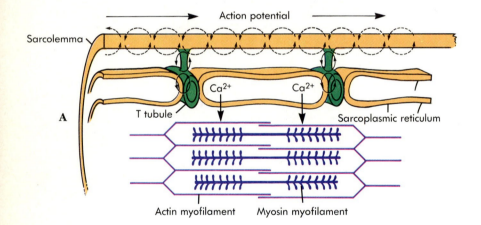

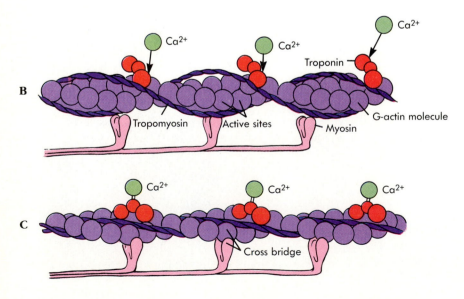

FIGURE 12-13
The sequence of events in excitation-contraction coupling.
A   The initial event is an action potential that propagates across the sarcolemma and into the transverse tubule system. The action potential causes the formation of an inositol triphosphate second message *(not shown)*, which opens Ca^{++} channels of the terminal cisternae. Ca^{++} rushes out into the cytoplasm.
B   The released Ca^{++} binds to troponin.
C   This binding causes the tropomyosin molecule to move deeper into the groove along the actin myofilament, exposing active sites on actin and initiating the crossbridge cycle.

<table>
<tr><td colspan="2">TABLE 12-1   The Events of Muscle Contraction</td></tr>
</table>

Sequence	Event
1	Muscle action potential
2	Spread of depolarization into the transverse tubule system
3	Communication of transverse tubule depolarization to the terminal cisternae of the sarcoplasmic reticulum
4	Release of Ca^{++} from the terminal cisternae
5	Binding of Ca^{++} to troponin
6	Exposure of a myosin binding site on actin
7	The powerstroke of muscle contraction, a process repeated so long as Ca^{++} and ATP are present (crossbridge cycling)
8	Uptake of Ca^{++} into the sarcoplasmic reticulum, resulting in the actin binding sites for myosin becoming covered again by troponin

---

TABLE 12-2   *The States of the Contractile Machinery of Striated Muscle*

	Resting	Contracting
Cytoplasmic Ca^{++} concentration	Less than $10^{-7}M$	About $10^{-5}M$
Troponin	Ca^{++} binding sites free	Ca^{++} binding sites occupied
Tropomyosin	Blocks actin binding sites	Exposes actin binding sites
State of myosin heads	High-energy myosin:ADP	Cycling

to more than 10 μM, a concentration sufficient to saturate binding sites on troponin. Binding of Ca^{++} to troponin shifts the tropomyosin molecules, exposing the myosin binding sites on the G-actin subunits so that myosin:ADP and actin interact with each other (see Figure 12-13, *B* and *C*). The combination of troponin and tropomyosin functions as a **Ca^{++}-activated switch** that turns on the contractile machinery.

The contraction that results from a single action potential is a **twitch**. After an action potential, the contractile machinery will remain active as long as cytoplasmic Ca^{++} levels remain elevated (Figure 12-14). The period of activity depends on the rapidity with which the released Ca^{++} can be returned to the sarcoplasmic reticulum. The active transport system of the sarcoplasmic reticulum is stimulated by an increase in the intracellular free Ca^{++} concentration. In fast-twitch muscles, the Ca^{++} pump is so powerful that after only 10 to 20 msec, all the released Ca^{++} is pumped into the sarcoplasmic reticulum and the intracellular Ca^{++} concentration drops below the threshold for significant crossbridge activation. In slow-twitch fibers, the sarcoplasmic reticulum contains fewer Ca^{++} pump proteins and the duration of twitch contractions may be as long as 40 to 50 msec.

The states of the contractile machinery of striated muscle during relaxation and activity are summarized in Table 12-2.

FIGURE 12-14
The time relationships of the muscle action potential, rise and fall of cytoplasmic Ca^{++}, and force development by sarcomeres during a twitch.

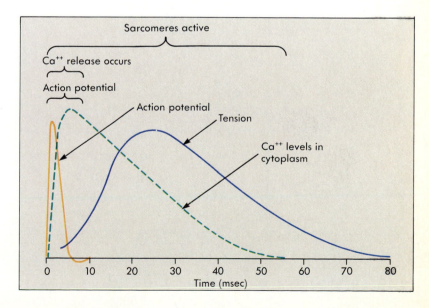

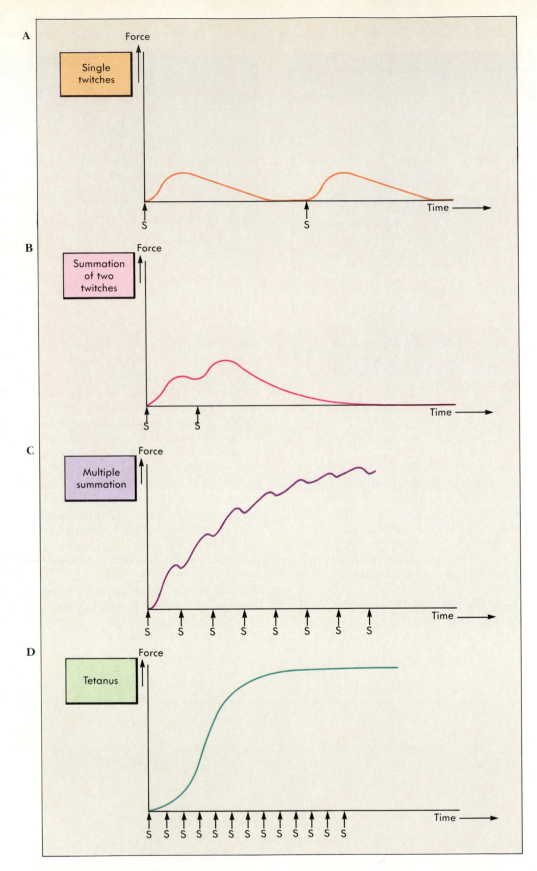

FIGURE 12-15
A   Single twitches in a muscle fiber initiated by stimuli *(arrows labelled S)* well separated in time.
B and **C**   Individual twitches in a skeletal muscle can summate with each another when stimuli are closely spaced.
D   At high rates of stimulation, the contractile force rises smoothly in a tetanus.

Summation of Contractions in Single Muscle Fibers

The total amount of force developed by skeletal muscles depends on two factors: (1) **summation** of twitches in individual muscle fibers, and (2) increases in the number of fibers active in a given muscle, a process referred to as **recruitment** (see Chapter 11, p. 261).

Repeated activation of the motor neuron innervating a given muscle fiber results in summation. If the muscle action potentials occur very infrequently, the Ca^{++} concentration returns to rest levels before the arrival of the next action potential, and the muscle's sarcomeres return to rest length (Figure 12-15, A). When two or more muscle action potentials occur in rapid succession, not all of the Ca^{++} released in response to the first activation can be resequestered before more Ca^{++} is released. As a result, the contractions summate (Figure 12-15, B).

What summation means in terms of the crossbridge activity is that some of the sliding that resulted from the first action potential is retained, and the second surge of Ca^{++} allows additional shortening of the sarcomere. As the stimulation rate is increased (multiple summation), the Ca^{++} concentration increases and the total force rises (Figure 12-15, C). At high stimulation frequencies, **tetanus** occurs, and force development becomes smooth (Figure 12-15, D).

The ability of a muscle to undergo tetanic contraction depends on two factors: (1) the contractile machinery must be able to respond to maintained high levels of internal Ca^{++} with continual crossbridge activity, and (2) the muscle action potential must not last as long as the resulting twitch contraction. This second requirement is met in skeletal muscle, where tension development lasts much longer than the action potential, but the action potential in cardiac muscle is so long that the sarcomeres return to precontraction length before the action potential is over. This protects the heart from tetanic contractions, which would prevent filling of the heart.

---

1 What is meant by excitation-contraction coupling?
2 At what stage in crossbridge cycling is energy transferred from ATP to myosin?
3 What is a twitch? What process determines the duration of a twitch? What factors determine the ability of a muscle to undergo tetanic contraction?

THE MECHANICS OF MUSCLE CONTRACTION
Properties of Isometric Contractions

An **isometric** (constant length) contraction occurs when an activated muscle is unable to shorten. In this kind of contraction, the effect of crossbridge formation is to increase the **tension,** or tautness, of the muscle. Force generation without shortening is a normal activity for postural muscles, and this also occurs when a person tries to lift an immovable object. In laboratory experiments, isometric contractions can be studied by fixing the tendons of an isolated muscle between rigid supports.

If a muscle is loaded by attachment of a weight that stretches it, the tension of the muscle is proportional to the weight applied, regardless of whether crossbridges are active. An **isotonic** (constant tension) contraction occurs if the loaded muscle contracts and lifts the load. In practice, a contraction can change from isometric to isotonic and vice versa. For instance, when a bucket of water is lifted from the ground, the contraction is isotonic as long as the load is moving. If the load is held at arm's length, the force exerted by the muscles is equal to the bucket's weight, and muscle length is not changing, so the contraction becomes isometric.

The relationship between the length of a resting muscle and its tension can be measured by attaching a muscle to two rigid supports that can be set at different distances apart. One of the supports has a **force transducer** —a device that measures the tension transmitted to the support from the muscle. The muscle assumes its **rest length** when it is removed from the body and has no load. Passive tension is zero at rest length (Figure 12-16, A—note the transducer shows zero force). Stimulation of the motor nerve to the muscle results in an active force on the lower support (Figure 12-16, B). If an unstimulated muscle is stretched beyond its rest length (Figure 12-16, C), the increase in tension becomes greater with each increase in length. Stimulation will then produce a total force that is the sum of the active and passive forces (Figure 12-16, D).

Relation between Maximum Force and Muscle Length

The **passive-tension** curve of a muscle (Figure 12-17) is a measure of the muscle's **elasticity,** or tendency to return to rest length when stretched. The **series elastic elements** include all elastic structures that lie between the attachments of the crossbridges and the ends of the tendons. When a muscle performs an isometric contraction, the active tension generated by the crossbridges is added to the passive tension to get the total tension (see Figure 12-17). Conversely, the length-tension relationship for the sarcomeres can be determined by subtracting the passive length-tension curve from the length-tension curve of the active muscle.

Active tension is maximal near rest length; at this length the overlap between thick and thin filaments is such that all myosin heads may form crossbridges. When the muscle is stretched by more than a few percent of rest length, the overlap is decreased and some myosin heads can no longer form cross-

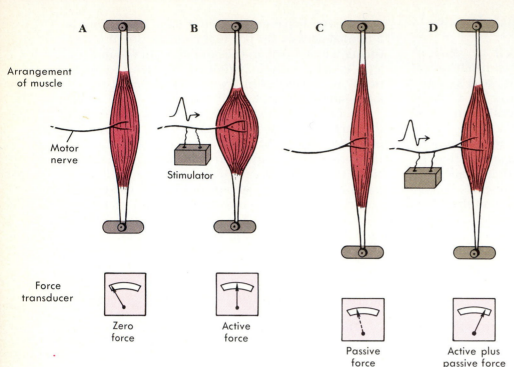

FIGURE 12-16
The method of stimulating a muscle and recording the resulting force during an isometric contraction.
A   An unstimulated muscle is fixed between two supports at its resting length.
B   The muscle is stimulated, producing an active force that can be measured by a transducer.
C   The muscle has been stretched slightly, resulting in an increase in passive force.
D   The stretched muscle is stimulated. Note that the total force in an isometric contraction is the sum of the active force produced by actin-myosin interaction and the passive force that is continuously present if the muscle is stretched beyond its rest length.

A Arrangement of muscle — Motor nerve

B Stimulator

Force transducer — Zero force / Active force / Passive force / Active plus passive force

FIGURE 12-17
The curves of passive tension as a function of length, active tension as a function of length, and total (active plus passive) tension as a function of length. The curve of active tension as a function of length is obtained by subtracting the passive curve from the active plus passive curve. The operating range of skeletal muscle length in the body is within a few percent of rest length.

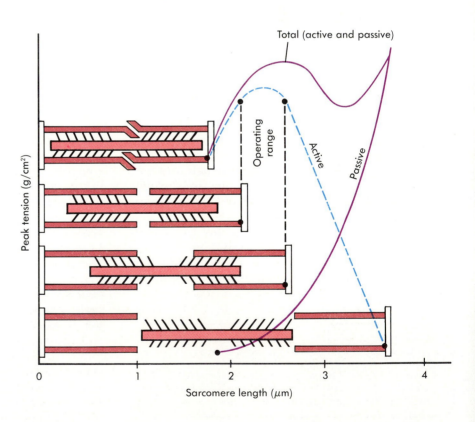

bridges (see Figure 12-17). The consequence is a sharp decrease in active tension.

Active tension also decreases if the muscle is allowed to shorten by more than a few percent of rest length. This decrease is probably caused by the hindrance to crossbridge formation that results from overlap of thin filaments from opposite ends of the sarcomere. With further shortening, active tension declines to zero because both ends of the thick filaments bump into the Z structures. In summary, skeletal muscle delivers its maximum force at rest length and is adapted to give optimum force development over a very narrow length range. This range of optimum performance corresponds closely to the actual **operating range** of muscles in the body (see Figure 12-17).

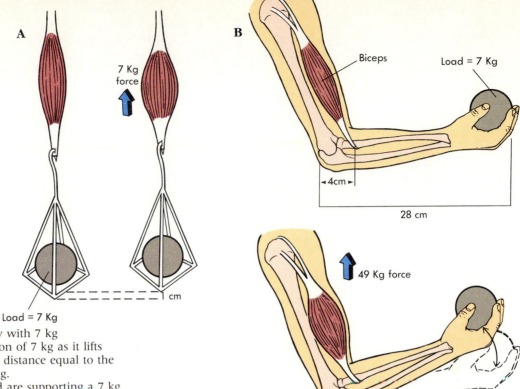

FIGURE 12-18
A A muscle loaded directly with 7 kg experiences a constant tension of 7 kg as it lifts the load. The load is lifted a distance equal to the amount of muscle shortening.
B When the arm and hand are supporting a 7 kg load, the tension in the biceps is multiplied by the leverage of the arm. If the distance between elbow and biceps insertion is 1/7 of the total length of the arm, the tension in the biceps will be 7 × 7 kg, or 49 kg. If the biceps shortens 1 cm, the load will be moved 7 cm.

Skeletal Muscle and Skeletal Lever Coadaptation

When a muscle lifts a load directly (Figure 12-18, A), the muscle tension must equal the pull of gravity on the load. In many cases, muscles are attached to parts of the skeleton that act as levers. Muscle insertions are usually close to joints, placing skeletal muscles at a mechanical disadvantage. To lift a loaded appendage, they must exert a force many times greater than that of the load. The **leverage factor** is the factor by which the muscle force must exceed the load. It is determined by dividing the total length of the lever by the fraction of its length that lies between the joint and the muscle insertion. For example, to lift a weight held in the hand, the biceps muscle must exert a force about seven times that of the load (Figure 12-18, B). However, because the insertion is close to the elbow, the biceps can lift the loaded hand through a wide range of movement with a relatively small amount of shortening. The construction of sarcomeres, which maximizes force generation over a narrow operating range, allows skeletal muscles to function well within the lever system of the skeleton.

Latent Period in Afterloaded Contractions

A delay, or **latent period,** occurs between excitation of a muscle and the beginning of shortening (Figure 12-19, A). Part of this delay involves excitation-contraction coupling. Another phase of the latent period is attributable to the necessary stretching of the series elastic elements in proportion to the load. A **preloaded muscle** supports the load before it is stimulated to lift it. The latent period of a preloaded muscle does not change much with increased load (Figure 12-19, B).

An isolated muscle is **afterloaded** if it does not support the load until it begins to shorten. An afterloaded muscle is under no tension until its sarcomeres begin to contract. Before the muscle can begin to shorten and lift the load, its series elastic elements must be stretched in proportion to the weight to be lifted. If the load is light, very little stretching is necessary. If the load is greater, more stretching is necessary, just as a rubber band will stretch more if it is supporting a heavier load. The latent period of an afterloaded muscle is greater (Figure 12-19, C) than that in a preloaded muscle, and it increases still more with greater loads (Figure 12-19, D). The reason is that the series elastic elements of the afterloaded muscle must stretch in proportion to the load. This stretching requires time. In contrast, the latent period of a preloaded muscle increases only slightly with increased load. The series elastic elements of the preloaded muscle are already stretched

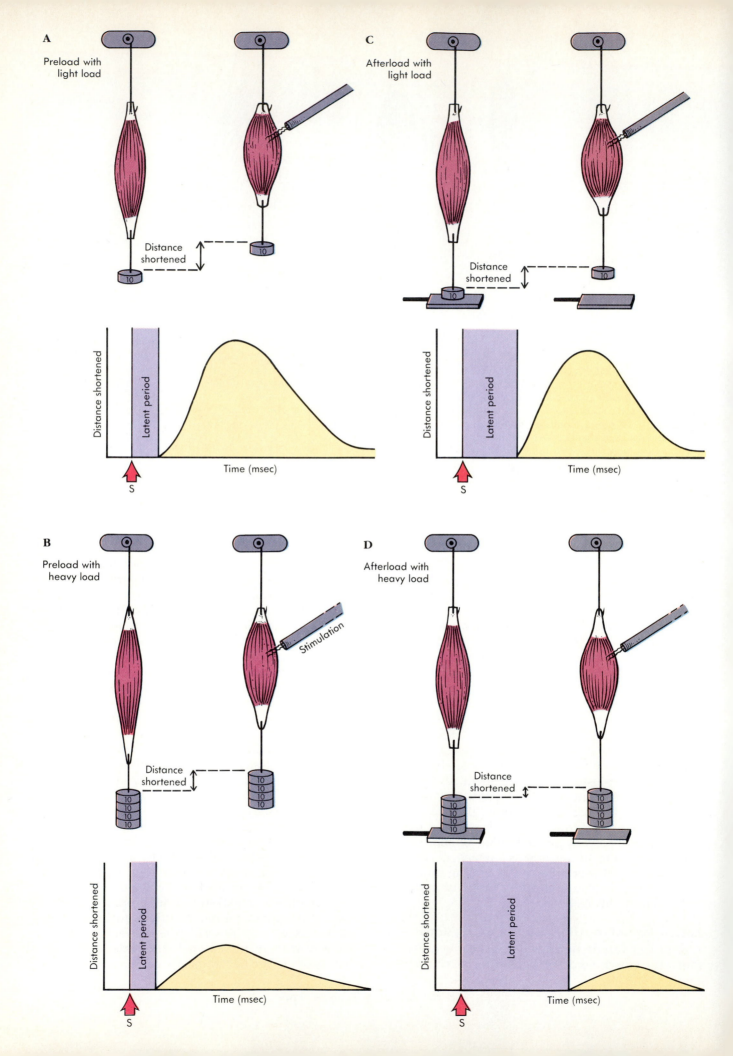

A

Preload with
light load

Distance
shortened

Distance shortened

Latent period

S

Time (msec)

B

Preload with
heavy load

Stimulation

Distance
shortened

Distance shortened

Latent period

S

Time (msec)

C

Afterload with
light load

Distance
shortened

Distance shortened

Latent period

S

Time (msec)

D

Afterload with
heavy load

Distance
shortened

Distance shortened

Latent period

S

Time (msec)

FIGURE 12-19

The effect of preloading and afterloading on the latent period and the shortening attained in a twitch. The effect on a preloaded muscle of an increased preload (compare **B** with **A**) is a slightly decreased amount of shortening without a substantial decrease in the latent period. **C** and **D** show a comparable experiment performed on an afterloaded muscle. In the afterloaded muscle, the latent period is longer than that of the preloaded muscle (compare **C** with **A** and **D** with **B**). Increasing the load in an afterloaded muscle causes a much larger increase in the latent period and a much larger decrease in the shortening attained in a twitch than is seen in the preloaded muscle (compare **C** and **D** with **A** and **B**).

---

in proportion to the load; the slight increase in the latent period reflects the greater inertia the active force must overcome.

Because twitches are brief, the longer the latent period, the less time remaining for muscle shortening. The period during which Ca^{++} levels are elevated is the same for all four examples in Figure 12-19 because each response is the result of a single action potential. The afterloaded muscle is prevented from shortening as much as does the preloaded muscle because the Ca^{++} levels decline before a similar degree of shortening can occur.

> 1 What is meant by the term operating range? What is the general effect of the leverage factor of skeletal levers on the operating range of skeletal muscles?
> 2 What is the difference between isometric and isotonic contraction?
> 3 What makes a preloaded contraction different from an afterloaded contraction? Which has the shorter latent period?

MUSCLE ENERGETICS AND METABOLISM
Muscle Energy Consumption and Activity

In a resting muscle, most of the ATP is used by the sodium-potassium (Na^+-K^+) pump, which maintains the concentration gradients across the sarcolemma. In a contracting muscle, the Na^+-K^+ pump uses ATP at a higher rate because it must counteract the additional ion movement resulting from the action potentials. Active muscle also uses more ATP for resequestration of Ca^{++} than does resting muscle. However, the active crossbridges account for most of the extra ATP use in contracting muscles. Because each crossbridge cycle requires the hydrolysis of one ATP molecule, the energy requirements of active muscle are greatly influenced by the rate of crossbridge cycling. The rate of energy use is higher in isotonic than in isometric contractions because crossbridges cycle more rapidly when filaments are in motion.

The concentration of ATP in skeletal muscles is about 5 mM. This ATP concentration is the immediate source of energy for contraction, and by itself can support only a few seconds of activity. Muscles contain a second immediate reserve of energy in the form of **creatine phosphate,** which can donate phosphate to ADP, becoming **creatine** (Figure 12-20). The creatine phosphate reserve prevents a large decrease in ATP levels during the few seconds needed for energy metabolism to respond to the increased rate of ATP use that occurs when contraction begins. The reaction is reversed to recharge the creatine phosphate reserve in resting muscle. Both forward and reverse reactions are catalyzed by the same enzyme, **creatine phosphokinase.** The reaction of creatine phosphate with ADP to yield creatine and ATP is favored by a decrease in ATP concentration and an increase in ADP concentration. When the muscle is not contracting the abundance of ATP drives the reaction toward creatine phosphate synthesis.

Skeletal Muscle Fiber Types and Energy Metabolism

Different muscle groups serve different functions and display corresponding contractile properties. For example, finger and eye muscles contract rapidly and become fatigued with continued use, whereas postural muscles contract slowly and do not become fatigued as easily. Functional specialization of muscle is possible because three types of muscle fibers exist, and the relative numbers of each fiber type are different in different muscles. **Fiber types** are distinguished by several factors, including the metabolic pathways used for energy and resistance to fatigue.

The three muscle fiber types are (1) **slow oxidative,** or **Type I,** (2) **fast oxidative,** or **Type IIA,** and, (3) **fast glycolytic,** or **Type IIB.** The characteristics of these three types are summarized in Table 12-3. The speed of contraction is related to the type of myosin a fiber contains. Fast fibers contain myosin with a high ATPase activity, whereas slow fibers have myosin with low ATPase activity. The higher the rate of ATP hydrolysis, the faster the crossbridges can cycle and the more rapidly a muscle can shorten. The price of faster shortening is a higher rate of ATP use.

Type I muscle fibers are found in the greatest numbers in **postural muscles** (for example, the long muscles in the back) and generate some 10 grams of force per square centimeter of muscle cross-sectional area. Oxidative metabolism is appro-

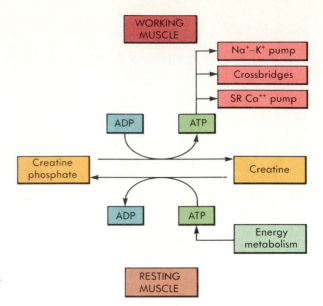

FIGURE 12-20
The reserve of creatine phosphate is drawn on in working muscle and is recharged by energy metabolism after the muscle returns to rest.

TABLE 12-3   Comparison of Skeletal Muscle Fiber Types

Characteristics	Type I	Type IIA	Type IIB
Contractile			
Contraction velocity	Slow	Fast	Fast
Myosin ATPase	Slow	Fast	Fast
Twitch duration	Long	Short	Short
Ca^{++} sequestration	Slow	Rapid	Rapid
Metabolic			
Capillaries	Abundant	Intermediate	Sparse
Glycolytic capacity	Low	Intermediate	High
Oxidative capacity	High	High	Low
Myoglobin content	High	Intermediate	Low
Glycogen content	Low	Intermediate	High
Fiber diameter	Small	Intermediate	Large
Motor unit size	Small	Intermediate	Large
Recruitment order	Early	Intermediate	Late

priate for muscles that must maintain contraction for long periods because it produces ATP with high efficiency. The commitment of Type I fibers to oxidative energy metabolism is reflected in their enzyme profiles, high mitochondrial density, and rich capillary blood supply. Their relatively small diameter facilitates gas exchange with the circulatory system. Type I fibers are red in color because they contain an oxygen carrier known as **myoglobin.** Myoglobin facilitates oxygen diffusion within the cell and provides a reserve oxygen supply when dissolved oxygen levels in the cytoplasm fall very low.

The two types of fast muscle (Types IIA and IIB) produce more than 100 grams of force per square centimeter of muscle cross-sectional area. They are distinguished from each other by their

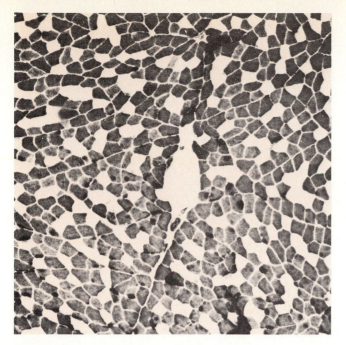

FIGURE 12-21
A micrograph of a muscle containing both fast and slow fiber types. Myosin ATPase reacts with ATP in a procedure that stains the reaction product. The darkly stained cells have high levels of myosin ATPase and thus are Type II fibers. The unstained cells are Type I fibers.

susceptibility to fatigue. Type IIA fibers (found, for example, in the soleus muscle of the calf) primarily use oxidative processes and are moderately resistant to fatigue, but not as resistant as are Type I fibers. They have many mitochondria, considerable myoglobin, and a moderate amount of **glycogen,** which can be broken down to provide glucose and thus ATP via glycolysis (see Chapter 4, p. 76). Like Type I fibers, Type IIA fibers are red.

Type IIB fibers rely on anaerobic glycolysis almost exclusively and thus become fatigued quite easily. They have an extremely high glycolytic capacity, large stores of glycogen, and few mitochondria. They possess little myoglobin and thus are sometimes called white fibers. They are mixed with the other muscle fiber types in the limb muscles, particularly in the flexors, and the eye and finger muscles have a high proportion of this fiber type.

Anaerobic glycolysis is advantageous for muscle fibers that are specialized for brief, powerful contraction because the maximum force can be much greater if it is not limited by the rate at which glucose and oxygen can be delivered by the circulatory system. The duration of use of Type IIB fibers is limited by their stores of glycogen. Although Type IIB fibers contain substantial stores of glycogen, anaerobic glycolysis is relatively inefficient in converting the stored energy to ATP. The rapid fatiguing of these fibers during sustained, submaximal exercise is highly correlated with the depletion of their glycogen stores.

Type I fibers are innervated by small motor neurons that activate only a small number of fibers; the size principle of recruitment (see Chapter 11, p. 261) causes them to be recruited when small forces are required. Type IIB fibers are supplied by large motor neurons that activate large numbers of fibers; these fibers are recruited for heavier loads. Type IIA fibers have intermediate properties. Thus the size principle provides for smooth increases in force and also activates first and most frequently those fibers (Type I) which are most resistant to fatigue.

The types of individual fibers can be identified under the microscope in tissue stained for particular metabolic enzymes. For example, the fast fibers of the muscle shown in Figure 12-21 are darkened by a stain specific for myosin ATPase.

The Cori Cycle

After strenuous exercise, oxygen consumption does not fall to its resting level immediately. Full return to the resting level of oxygen consumption may require up to several hours. The extra oxygen consumption occurring after exercise is referred to as **oxygen debt.** Some of the oxygen debt is the result of accumulated lactic acid, which must be metabolized to CO_2, and of a corresponding decrease in the glycogen stores of Type IIB muscle fibers. Repayment of the debt occurs when the lactate produced during exercise is returned to the Type IIB fibers in the form of glucose. The Cori cycle is a process by which some of the lactic acid can be converted to glucose in the liver and returned to muscle (Figure 12-22).

During heavy exercise and for a time after exercise is stopped, lactic acid is released by muscle cells into the bloodstream. Some lactic acid is taken up by the liver and resynthesized to glucose by an energy-requiring pathway (described in Chapter 4, p. 77) supported by the oxidative metabolism of the liver. The resulting glucose is released into the bloodstream and can be taken up by any cells that need it, including muscles that have depleted their glycogen reserves. In principle, the extra oxygen used by the liver for converting lactic acid to glucose could account for the oxygen debt.

In fact, however, the oxygen debt cannot be accounted for by the Cori cycle alone. First, lactate metabolism goes on during, as well as after, exercise. Also, much of the lactate produced during exercise does not enter the Cori cycle, but is metabolized to CO_2 by the oxidative metabolism of the heart, liver, brain, and Type I fibers of skeletal muscle (see Figure 12-22). Finally, much of the increased oxygen consumption occurring during recovery apparently does not result from lactic acid metabolism. A large part of the postexercise metabolic rate can be explained by physiological changes that occurred during exercise persisting into the recovery phase. These include hormonal responses,

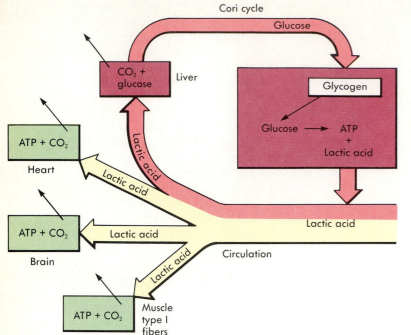

FIGURE 12-22
Metabolic fate of lactic acid produced during exercise. The Cori cycle *(in pink)* represents one possible fate; other tissues that respire oxidatively during exercise, such as the heart, kidney, and Type I fibers, can convert lactic acid to pyruvate and metabolize it to CO_2 by way of the Krebs cycle.

circulatory changes, and elevated body core temperature.

Effects of Training on Specific Fiber Types

Muscle fibers that are frequently active respond by increasing their performance capabilities. The changes that occur differ for different fiber types. Oxidative fibers respond with increases in the myoglobin content, the number of mitochondria, and the number of capillaries serving the muscle (see Table 12-3). These changes involve little change in the mass of the muscle. Glycolytic fibers respond by increasing the number of myofibrils and the size of the glycogen stores; these changes necessitate an increase in fiber diameter and thus an increase in muscle mass (see Table 12-3). Tissue growth resulting from an increase in cell size, rather than in the number of cells, is called **hypertrophy.**

Hypertrophy of skeletal muscle in human athletes is not believed to involve an increase in the number of fibers, although some animal experiments suggest that such increases are possible. In any individual, the proportion of the three fiber types in a particular muscle is not altered by training and appears to be determined largely by heredity. However, selective use of fibers of a single type results in hypertrophy of those fibers. Exercise that calls for sustained effort, such as distance swimming, running, or skiing, results in development of endurance. This result causes an increase in the circulatory supply, mitochondria, and oxidative enzymes and myoglobin in all fibers. Athletes trained for endurance increase oxygen uptake capacity by up to 20% and produce lower levels of lactic acid during exercise than do untrained individuals.

The situation is quite different for brief, high-intensity training such as weight lifting. The increase in muscle mass experienced by bodybuilders is the result of two changes: (1) increase in the diameters of glycolytic fibers, and (2) addition of collagen and other connective tissues required to sustain the passive tension of heavy loads. The rate and amount of tension developed during training exercises are the most important factors in increasing contractile protein incorporation into muscle. The short-term demands of maximum force development do not cause the circulatory and metabolic changes associated with endurance training. The effects of both types of training are rapidly reversed if the regimen is not maintained; this process is called **disuse atrophy.**

Muscle Fatigue

Fatigue is a use-dependent decrease in the ability of muscle to generate force. The **rapid fatigue** seen in high-intensity exercise apparently results from operating muscles under anaerobic conditions. High-intensity exercise causes an increase in the lactic acid in muscle and blood. There is evidence that the resulting acidic conditions alter several aspects of muscle physiology, including the activity of key glycolytic enzymes and possibly the contractile machinery. Anaerobiosis also depletes high-energy phosphate compounds and results in an alteration of the ionic gradients across the cell membrane and transverse tubules. Changes in the energy status of

What Determines Muscle Fiber Type?

The differences between the three types of muscle fiber arise during development. Mammalian muscle cells are all slow-contracting at birth, and those fibers destined to be fast (Types IIA and IIB) differentiate by two processes: (1) the "switching off" of the synthesis of myosin with low ATPase activity in some fibers, and (2) the dying off of some fibers to produce the proportion of Types I, IIA, and IIB that is appropriate for each particular muscle. The destiny of muscle cells can be altered by changing the signals the cells receive from the nervous system. A pair of muscles, one fast and the other slow, can have their innervation cut and the cut ends of the nerves redirected so that the nerves regenerate connections with the wrong muscles.

The result of this **cross innervation** is that the properties of the two muscles are dramatically altered: the former slow muscle will contract faster, and the fast muscle will contract slower (Figure 12-A). The biochemical and structural properties of the two muscles alter accordingly. Thus the motor neurons exercise a **trophic effect**—they determine the specific functional properties of their own motor units. The influence is exerted both by chemical messengers that travel between the muscle and the nervous system via axoplasmic flow and by the frequency and pattern of recruitment of the muscles by the motor neurons. Although the results of cross innervation are more dramatic when they are performed on very young animals, similar effects can be demonstrated in adult animals. These studies demonstrate the capacity of muscle fibers to adjust their properties in response to the signals they receive from the nervous system.

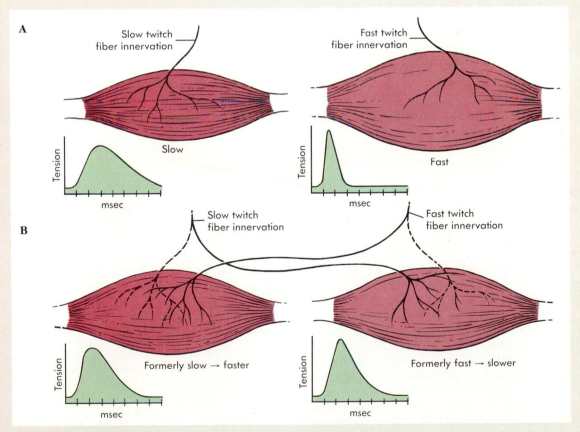

FIGURE 12-A

The effect of changes in innervation on the contractile properties of muscle. The nerve running to a muscle composed predominantly of slow-twitch fibers is surgically switched with a nerve innervating a muscle composed predominantly of fast-twitch fibers. The original twitch properties of the muscles are shown in **A**; the effects of the crossed innervation on twitch properties are shown in **B**.

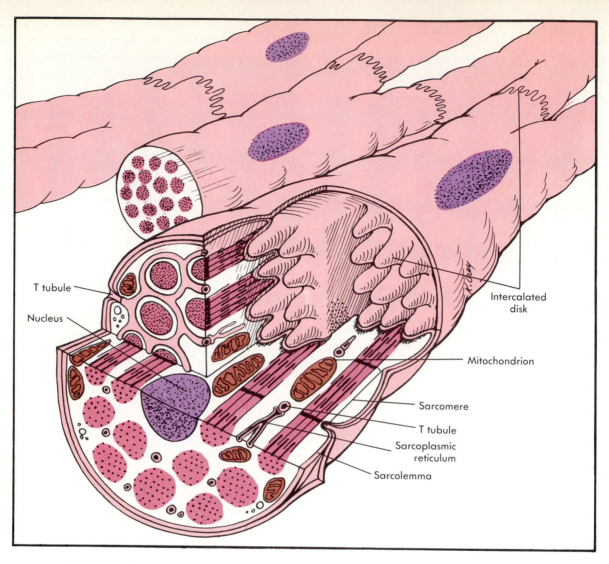

FIGURE 12-23
The major structural features of cardiac muscle fibers. Note the presence of intercalated disks connecting successive sarcomeres.

the fibers may impair the ability of the sarcoplasmic reticulum to control intracellular Ca^{++} levels.

The **slow fatigue** that results from prolonged submaximal exercise coincides with depletion of the glycogen stores of the liver and working muscle cells. When this occurs, the body must use fat as the sole energy source. Energy production from fat occurs at about half the rate of energy production from glucose, so the depletion of glycogen stores is marked by a substantial decrease in muscle performance. Distance runners refer to the onset of slow fatigue as "hitting the wall." With adequate amounts of oxygen, fatigue of glycolytic motor units is initially partly compensated for by increased reliance on oxidative units. However, repeated activation of oxidative units results in failure of neuromuscular transmission and a further decline in performance.

Endurance in athletic events requiring sustained exercise can be improved by **glycogen loading,** a dietary regimen in which consumption of a low-carbohydrate diet is followed by a switch to a high-carbohydrate diet just before an important event. Muscle cells adapt to the low-carbohydrate diet by increasing their ability to take up glucose from the blood; the switch to the high-carbohydrate diet results in a temporary but substantial increase in muscle glycogen levels that confers greater endurance.

1 *What processes might contribute to elevated oxygen consumption for a time after exercise (oxygen debt)?*
2 *What are the differences between hypertrophy in oxidative muscle and in glycolytic muscle? Which muscle type is trained selectively in bodybuilding?*
3 *What is the general definition of muscle fatigue?*

THE PHYSIOLOGY OF CARDIAC MUSCLE
The Structure of Cardiac Muscle

Cardiac muscle fibers (Figure 12-23), like skeletal muscle fibers, contain myofibrils and a network of transverse tubules and sarcoplasmic reticulum. The molecular mechanism of force generation and its

NERVE AND MUSCLE

control by Ca^{++} are similar in cardiac and skeletal muscle. Cardiac muscle cells resemble Type I skeletal muscle cells in having abundant myoglobin, many mitochondria, and a rich supply of capillaries. These characteristics make heart muscle very fatigue resistant but also very dependent on a continuous oxygen supply. Cardiac muscle fibers are much shorter than skeletal fibers. The fibers are linked together by strong connections between their ends that form a mesh; the connections are called **intercalated disks** (see Figure 12-23).

The Heart as an Electrical Unit

One distinction between cardiac and skeletal muscle is that the cells of cardiac muscle are electrically coupled to each other. When one cell is excited, the action potential spreads throughout the entire muscle, allowing the cardiac muscle to contract as a unit. In this respect, cardiac muscle resembles single-unit smooth muscle, described later in this chapter. The electrical communication is mediated by gap junctions (see Chapter 4, p. 73; Figure 12-24) in the intercalated disks.

Lack of Summation in Cardiac Muscle

The characteristics of action potentials in cardiac cells are described in Chapter 13, but the important points to understand now are that (1) some cardiac cells initiate action potentials spontaneously in a regular rhythm, and (2) action potentials last much longer in the heart than in nerve axons and skeletal muscle fibers. A typical nerve or skeletal muscle action potential lasts several milliseconds; some cardiac cells remain depolarized for several hundred milliseconds (Figure 12-25). Because of the prolonged depolarization, the membrane of cardiac muscle cells does not recover from its refractory period until the active state of the myofibrils is almost over; thus individual contractions of cardiac muscle cannot summate. The advantage of the prolonged action potential is that it prevents tetanic contractions, which would interfere with the heart's pumping cycle of contraction and relaxation.

Just as in skeletal muscle, the force delivered by cardiac muscle is affected by stretch. In skeletal muscle, the leverage factors of the skeleton ensure that the muscle is always near the sarcomere length optimal for force production. The heart is not constrained to operate within a narrow range of lengths by origins and insertions. In its general shape, the length-tension curve of cardiac muscles (Figure 12-26, *A*) resembles that of skeletal muscle. The difference is that the operating range is in a region of the length-tension curve that is considerably shorter than the length at which maximal force is generated (Figure 12-26, *B*). Thus, in the healthy heart, additional stretch results in additional force.

The contractility of cardiac cells may be altered by nervous or hormonal inputs. There are several mechanisms involved; one is based on a positive re-

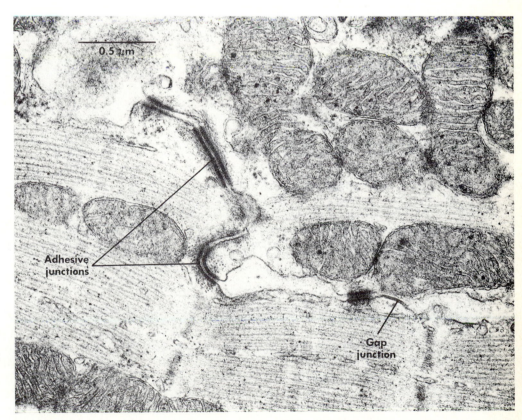

FIGURE 12-24
A longitudinal section of an intercalated disk between two cardiac muscle cells. Desmosomes (adhesive junctions) within the intercalated disk weld the two cells together. Gap junctions line the disk, forming electrical connections between the cells.

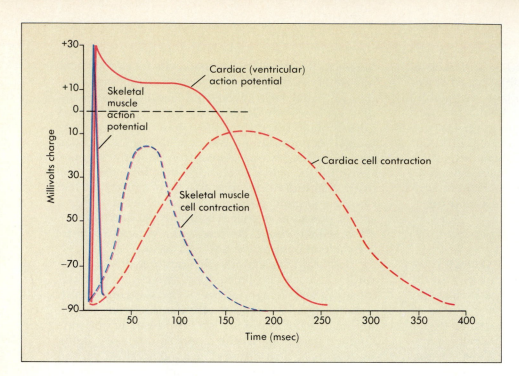

FIGURE 12-25
Comparison between the action potentials of a skeletal muscle fiber and a fiber of the heart ventricle. Note that the contractions of the two cells are similar, but the heart action potential lasts much longer.

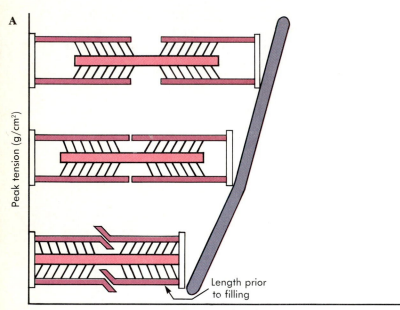

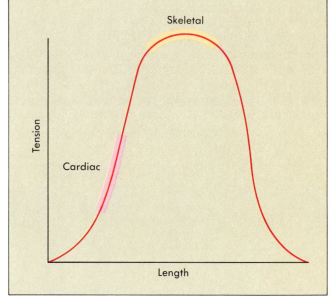

FIGURE 12-26
A The relationship between sarcomere length and active tension in a cardiac muscle cell. If the curve were continued to greater lengths, the tension would decrease, as it does in the length-tension curve of skeletal muscle shown in Figure 12-17.
B The general relationship between length and tension in muscle. The normal range of lengths for skeletal muscle is near the peak of the curve; that of cardiac muscle is on the rising part of the curve.

lationship between the intracellular Ca^{++} concentration achieved in activation and the force developed. In cardiac muscle, the plasma membrane has both voltage-dependent Ca^{++} channels and a Ca^{++} ion pump. The intracellular Ca^{++} level, and thus contractility, depends on the balance between two factors: (1) Ca^{++} influx and active extrusion across the surface membrane, and (2) the release and re-uptake of Ca^{++} by the sarcoplasmic reticulum.

A second way of modulating contractility is by effects on the cardiac contractile machinery. Both the thick and thin filaments of cardiac muscle have regulatory sites for second messengers. The ATPase activity of cardiac-muscle myosin can be modified by allosteric effects to alter the crossbridge cycling rate; the Ca^{++} binding properties of troponin may also be affected by intracellular messengers.

1 What is meant by the contractility of muscle? What intracellular processes affect contractility?
2 What process normally prevents tetanic contractions of cardiac muscle?
3 How does the length-tension relationship of cardiac muscle compare with that of skeletal muscle? In what part of the length-tension curve is the cardiac operating range?

THE PHYSIOLOGY OF SMOOTH MUSCLE
Classes of Smooth Muscle

Smooth muscle surrounds hollow internal organs including the gastrointestinal tract and all blood vessels except capillaries. Many aspects of smooth muscle contraction are particularly relevant to the functioning of organ systems and are described in later chapters, for instance, in Chapter 15 (discusses smooth muscle of blood vessels) and Chapter 23 (discusses gastrointestinal smooth muscle).

Two types of smooth muscle are distinguishable on the basis of features associated with electrical

A Relaxed

B Fully contracted

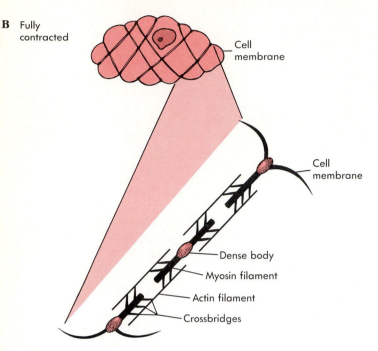

Cell membrane

Cell membrane

Dense body

Myosin filament

Actin filament

Crossbridges

FIGURE 12-27
A Contractile units span the diameter of a relaxed smooth muscle cell and are connected to the cell membrane by dense bodies.
B When the cell contracts, its surface becomes rounded and irregular.
Inset The relationship between thick filaments, thin filaments, and dense bodies in a single active contractile unit.

coupling between cells. The cells of **single-unit (unitary)** smooth muscle are electrically and mechanically coupled to each other. This type of smooth muscle is typical of visceral organs. Gastrointestinal smooth muscle is largely single-unit, as is the muscle around small blood vessels. **Multiunit** smooth muscle contains relatively few gap junctions, so that at best only small numbers of adjacent fibers can act as a unit. Multiunit smooth muscle is found in the iris of the eye, the walls of larger blood vessels, the airways of the lung, and the skin, where smooth muscle cells around hair follicles cause **piloerection** (goose bumps).

Length-Tension Relationships in Smooth Muscle

Smooth muscle contains actin and myosin, but these contractile proteins are not organized into sarcomeres. Instead, contractile units consist of parallel arrangements of thick and thin filaments crossing diagonally from one side of the cell to the other. Within these units, myosin molecules are attached either to structures called **dense bodies** (the func-

tional equivalents of Z structures) or to the sarcolemma. Actin filaments have no constant relationship to myosin filaments (Figures 12-27 and 12-28). The ratio of actin filaments to myosin filaments is much larger in smooth muscle than in striated muscle; most smooth muscles contain 10 to 15 actin filaments per myosin filament, as compared with the 3 thin filaments per thick filament of striated muscle.

In smooth muscle there is no practical limit to the distance that individual thick and thin filaments can slide relative to each other. In this respect smooth muscle differs radically from striated muscle, in which the thick and thin filaments can slide only a short distance before interfering with each other. The organization of smooth muscle is an adaptation for force development over a broad range of operating lengths. This adaptation is especially appropriate for muscle cells in the walls of organs that experience large volume changes, such as the stomach and intestine.

Excitation-Contraction Coupling in Smooth Muscle

Excitation of smooth muscle cells occurs by means of intrinsic and extrinsic mechanisms. In single-unit smooth muscles, spontaneous depolarizations of the muscle membrane occur at regular intervals, resulting in a **basic electrical rhythm,** or **BER** (Figure 12-29). The depolarizing waves of the BER may or may not result in action potentials, depending on the smooth muscle involved. This intrinsic **pacemaker activity** can be modified by extrinsic inputs from the sympathetic and parasympathetic divisions of the autonomic nervous system. The effects of autonomic input differ in different smooth muscle systems (see Table 11-4). Hormones may also activate or inhibit contraction. Single-unit smooth muscle resembles cardiac muscle in its intrinsic excitability and possession of gap junctions.

In multiunit smooth muscle, changes in contractile activity usually result from extrinsic inputs from the autonomic nervous system rather than from an intrinsic BER. Because there are few gap junctions, each fiber must typically receive direct motor input. In these respects, multiunit smooth muscle is like skeletal muscle.

Smooth muscle cells do not have the well-developed sarcoplasmic reticulum characteristic of striated muscle cells; instead, the sarcoplasmic reticulum takes the form of small vesicles near the cell membrane. They are much smaller than skeletal muscle cells and contract much more slowly, so they can depend primarily on Ca^{++} influx from the extracellular space for the activation of contraction. Electrical events are difficult to study in cells that are electrically coupled to each other. The solution has been to dissociate them. The membrane properties of single muscle cells can be studied using the

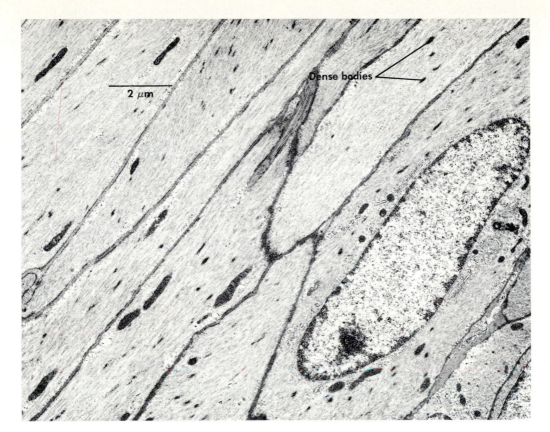

FIGURE 12-28
Micrograph of longitudinally sectioned smooth muscle fibers. The cytoplasm is filled with thick and thin filaments. Note the dense bodies *(DB)* scattered throughout the cytoplasm. The sarcoplasmic reticulum is represented by the small vesicles just next to the sarcolemma in each cell.

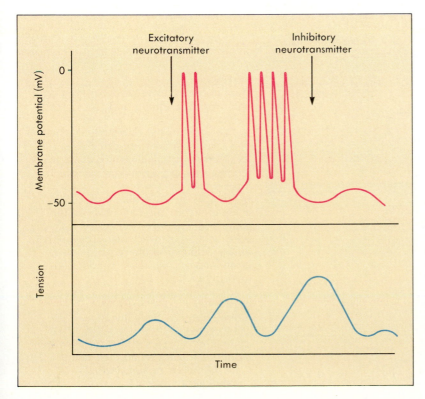

FIGURE 12-29
The basic electrical rhythm of a smooth muscle cell consists of slow waves of depolarization and repolarization. Depending on the specific smooth muscle, the waves themselves may or may not cause changes in force production. Application of excitatory transmitter or hormone increases the probability that each depolarizing wave will trigger action potentials. Action potentials cause larger force increases than does slow depolarization alone. Inhibitory neurotransmitter decreases the probability of action potentials.

patch clamp technique (see the boxed essay, Chapter 8, p. 178). Smooth muscle cells isolated from the stomach have spontaneous action potentials and can also be activated by neural inputs. The depolarization results from Ca^{++} rather than Na^+ entry, and the repolarization depends on the opening of K^+ channels, which are sensitive to the rise in the internal Ca^{++} level.

As in skeletal muscle, Ca^{++} is the link between excitation and contraction. Inside the smooth muscle cell the Ca^{++} combines with at least two target proteins: the light chain of myosin and a protein called **calmodulin**. Both of these molecules are structurally similar to one of the subunits of troponin. Initiation of contraction by Ca^{++} is the result of control steps involving the thick filament. In this aspect, smooth muscle differs from striated muscle, in which the control of contraction primarily involves the thin filament.

Excitation-contraction coupling in smooth muscle occurs by the following sequence of events: (1) the Ca^{++}-calmodulin complex activates the enzyme **myosin light-chain kinase** (MLCK), then (2) MLCK phosphorylates one of the light chains of the myosin heads, and the **myosin phosphorylation** exposes a binding site for actin. The myosin ATPase can now be activated by myosin binding to actin, and the interaction generates force. The actomyosin complex formed is broken down by a second enzyme called **myosin phosphatase**.

Control of Smooth Muscle Contractility

A second level of contractility control involves the thin filament and a molecule called **caldesmon**

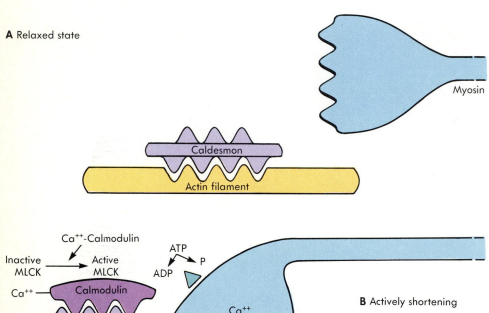

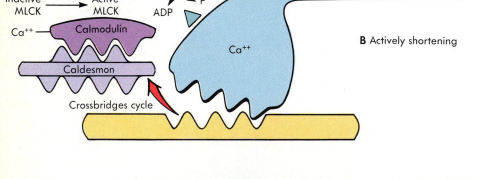

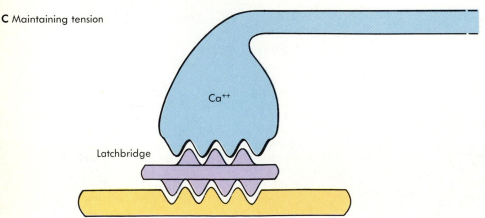

FIGURE 12-30
The molecular interactions between actin and myosin in smooth muscle contraction.

A The state of the contractile machinery in a relaxed smooth muscle. Calcium levels are low. Caldesmon blocks actin binding sites, preventing crossbridge formation.

B In actively shortening muscle, Ca^{++} levels are high. The Ca^{++}-calmodulin complex *(1)* removes caldesmon from actin, and *(2)* activates myosin light-chain kinase, which phosphorylates myosin, promoting cycling. Ca^{++} is bound to myosin heads *(3)*.

C Tension is maintained in smooth muscle when Ca^{++} levels are intermediate. At these levels, the calmodulin system is not activated, so caldesmon occupies actin sites. However, the myosin heads retain their bound Ca^{++} *(3)*, allowing noncycling crossbridges to form.

(Figure 12-30). This control system is believed to be responsible for the ability of smooth muscle to maintain tension for long periods with relatively little expenditure of energy. When cytoplasmic Ca^{++} levels are insufficient to maintain myosin phosphorylation, but are still high enough to allow myosin to bind Ca^{++}, crossbridges that are formed do not cycle but are maintained. These are called **latchbridges**. Caldesmon is believed to form a link between actin and myosin heads that keeps the latchbridges from cycling. At higher Ca^{++} levels, latchbridges do not form because caldesmon is bound by the Ca^{++}-calmodulin complex and is not available to link actin and myosin. The ability to form latchbridges that do not hydrolyze ATP makes tension maintenance in smooth muscle considerably more economical than in skeletal muscle, in which force maintenance requires continuous crossbridge cycling and ATP hydrolysis.

Table 12-4 summarizes the features of the contractile machinery of smooth muscle in three basic states of activity: relaxed, shortening, and maintaining tension.

Similarities and Differences between the Three Muscle Types

Muscle cells of all three types are highly specialized cells in which almost every cellular process and organelle is specifically adapted to provide particular contractile properties. Some of the important functional and structural similarities and differences between the three basic types of muscle are summarized in Table 12-5 and Figure 12-31.

1 What are the differences between single-unit and multiunit smooth muscle?
2 What are the two major systems by which contractility of smooth muscle is regulated? What is the role of calmodulin in each of the control systems?
3 What conditions favor the formation of latchbridges?

TABLE 12-4	Contractile States of Smooth Muscle		
	Relaxed	*Actively shortening*	*Maintaining tension*
Cytoplasmic Ca^{++} concentration	$10^{-7}M$	3 to $5 \times 10^{-6}M$	$10^{-7}M$ to ($3 \times 10^{-6}M$)
Calmodulin	Binding sites open	Binding sites occupied	Binding sites open
MLCK	Inactive	Active	Inactive
Myosin	Dephosphorylated, no Ca^{++} bound	Phosphorylated, Ca^{++} bound	Dephosphorylated, Ca^{++} bound
Myosin ATPase activity	Very low	High	Low
Caldesmon	Bound to actin	Free	Bound to actin
Crossbridge state	No crossbridges	Crossbridges cycling	Latchbridges

TABLE 12-5	Differences between Skeletal, Cardiac, and Smooth Muscle		
Feature	*Skeletal*	*Cardiac*	*Smooth*
Calcium receptor in excitation-contraction coupling	Troponin	Troponin	Calmodulin, myosin
Ca^{++} sources	Sarcoplasmic reticulum	Sarcoplasmic reticulum and extracellular space	Primarily extracellular space
Force regulation	Summation, recruitment	Stretch, contractility modulation	Contractility modulation
Organization	Many motor units	Single-unit	Single-unit or multiunit

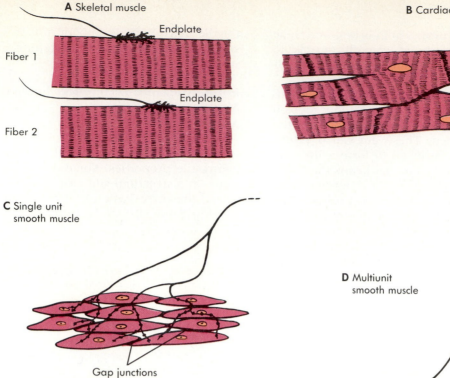

A Skeletal muscle

Endplate

Fiber 1

Endplate

Fiber 2

B Cardiac muscle

Electrical coupling via intercalated disc

C Single unit smooth muscle

Gap junctions site of electrical coupling

D Multiunit smooth muscle

Autonomic nerve

FIGURE 12-31
The characteristic structural features of skeletal muscle and cardiac muscle and the two basic forms of smooth muscle.

SUMMARY

1. The contractile machinery of muscle consists of thick filaments of **myosin** and thin filaments containing **actin.**
2. Force generation occurs when **crossbridges** on myosin interact with actin binding sites.
3. **Striated** (skeletal and cardiac) muscle contains **myofibrils** in which thick and thin **filaments** are organized into **sarcomeres,** an adaptation for rapid force development.
4. The thin filaments of striated muscle contain the regulatory proteins, **troponin,** which binds Ca^{++}, and **tropomyosin,** which controls myosin binding sites.
5. Contraction of striated muscle is triggered when a muscle action potential causes the release of Ca^{++} from the **terminal cisternae** of the sarcoplasmic reticulum in a process refered to as excitation-contraction coupling. Calcium binds to troponin and causes tropomyosin to shift away from myosin binding sites.
6. In **isotonic** contractions, muscle tension is constant, and shortening results from force development. In isometric contractions, muscle length is constant, and sarcomere shortening is accompanied by stretching of series elastic elements and an increase in muscle tension.

7. Cardiac muscle differs from skeletal muscle in that the spontaneously generated cardiac action potential outlasts contraction so that contractions cannot summate. Cardiac muscle fibers are connected by intercalated disks containing **gap junctions** so that the heart contracts as a **single unit.**
8. The thick and thin filaments of smooth muscle are not organized into sarcomeres but form a network that is able to develop force over a broad operating range.
9. Contraction in smooth muscle is controlled by a second messenger system in which the primary step is Ca^{++} entry. Ca^{++} binds to **calmodulin** and also to myosin. The Ca^{++}-calmodulin complex activates myosin light-chain kinase, which phosphorylates myosin, causing it to form crossbridges that cycle. In smooth muscle, tension can be maintained with little expenditure of ATP. A second level of control of smooth muscle contractility occurs when **caldesmon** links actin and myosin leads to produce crossbridges that do not cycle.
10. Each fiber is innervated in **multiunit** smooth muscle. The fibers of **single-unit** smooth muscle are electrically coupled, and the tissue is usually spontaneously active as the result of a basic electrical rhythm.

1. An action potential in a motor nerve fiber produces a single muscle contraction. Describe the sequence of events that occurs between the time that a nerve impulse arrives at a muscle and the onset of the muscle twitch.

2. What is the role of each of the following in excitation-contraction coupling in skeletal muscle?

Transverse tubules	Sarcoplasmic reticulum
Inositol triphosphate	Troponin
Tropomyosin	

3. What is the physical state of the crossbridges in a skeletal muscle when its internal ATP stores are exhausted? What is their state when ATP is present, but there is no free Ca^{++}?

4. What is the difference between an isometric and an isotonic contraction? Give a real-life example of each.

5. If you were holding a 1-kilogram weight in your right hand and a 100-gram weight in your left, how would the pattern of motor neuron activity differ for the muscles in your right and left arms?

6. Large amounts of stored glycogen are present in which type of skeletal muscle fiber? What is the functional role of this stored glycogen?

7. What specific processes contribute to rapid fatigue and slow fatigue?

8. How does the process of contraction in smooth muscle differ from that in skeletal muscle? What other features distinguish these two types of muscle?

9. Why are latchbridges particularly useful in the typical functions of smooth muscle?

10. Define single-unit and multiunit smooth muscle. How can the heart be viewed as a single-unit muscle?

● SUGGESTED READING

ALEN M., M. REINILA, and R. VIKKO: Response of Serum Hormones to Androgen Administration in Power Athletes, *Medicine and Science in Sports and Exercise,* volume 17, June, 1985, p. 354. Explores the long-term changes in the endocrine system in athletes taking large doses of anabolic steroids to improve performance.

FRANKLIN, D.: Steroids Heft Heart Risks in Iron Pumpers, *Science News,* Volume 126, July 21, 1984, p. 38. Describes evidence that anabolic steroid use is associated with an increased risk of acute myocardial infarcts.

GOLLNICK, P.D.: Metabolism of Substrates: Energy Substrate Metabolism During Exercise and as Modified by Training, *Federation Proceedings,* volume 44, February 1985, p. 353.

HOYLE, G.: *Muscles and Their Neural Control,* John Wiley and Sons, Inc., New York, 1983. General review of motor control from spinal reflexes to cortical motor areas.

JUNGE, D.: *Nerve and Muscle Excitation,* edition 2, Sinauer Associates, Inc., Sunderland, Massachusetts, 1981. Readable text that presents a complementary discussion of the origin of nerve and muscle action potentials, propagation, and excitation-contraction coupling.

MCARDLE, W.D., F.I. KATCH, and V.L. KATCH: *Exercise Physiology,* edition 2, Lea & Febiger, Philadelphia, 1986. A textbook on the subject written for courses in sports medicine.

RASMUSSEN, H., Y. TAKUWA, and S. PARK: Protein Kinase C in the Regulation of Smooth Muscle Contraction, *FASEB Journal,* volume 1, September 1987, p. 177. Comprehensive discussion of the mechanisms of smooth muscle contraction and excitation and of contraction coupling in smooth muscle.

WARSHAW D.M., W.J. MCBRIDE, and S.S. WORK: Corkscrew-Like Shortening in Single Smooth Muscle Cells, *Science,* volume 236, June 12, 1987, p. 1457. Describes the use of digital video microscopy to track the movement of marker beads on the surface of smooth muscle cells and show how these cells contract.

THE CARDIOVASCULAR SYSTEM

CHAPTER 13

The Heart

On completing this chapter you will be able to:

- Describe the components of the circulatory system: the heart, arteries, capillaries, and veins.
- Understand the role of the valves in the heart.
- Distinguish between the three types of cardiac muscle cells: pacemakers, conducting fibers, and contracting fibers, which generate the force of systole.
- Understand how pacemaker discharge rate is affected by sympathetic and parasympathetic input.
- Describe the ionic basis of the plateau phase of the cardiac action potential.
- Identify the five phases of the cardiac cycle: early diastole, atrial contraction, isovolumetric ventricular contraction, ejection, and isovolumetric ventricular relaxation.
- Understand how the Frank-Starling law of the heart can be derived from the length-tension relationship of cardiac muscle fibers.
- Identify the major factors that determine cardiac output.

*E*ach minute, the heart of a resting adult pumps about 5 liters of blood, or approximately the person's total blood volume. This works out to at least 7200 liters/day; this volume of blood weighs about 100 times more than the body. Moreover, the rate of pumping may increase to 20 to 30 liters/minute during strenuous exercise. The relentless work of the heart uses 5% to 10% of the body's total energy budget.

Regulation of the heart to meet the demands of the body is exceedingly complex; thus it has been difficult to construct mechanical hearts with anything approaching the flexibility of the human heart. The heart has a large measure of intrinsic regulation—its basic rhythm is established by its own pacemaker cells, and the force of its contractions is largely self-adjusting. Inputs to the heart from both branches of the autonomic nervous system cause the changes in heart activity that accompany excitement and emotion.

The impression that the heart has a life of its own may have led French philosopher Blaise Pascal to coin the aphorism "The heart has its reasons that reason does not know." On a more mundane level, pulse rate is one of the responses that is measured in the so-called "lie detector," or polygraph, test, which is designed to evaluate the state of activity of the autonomic nervous system.

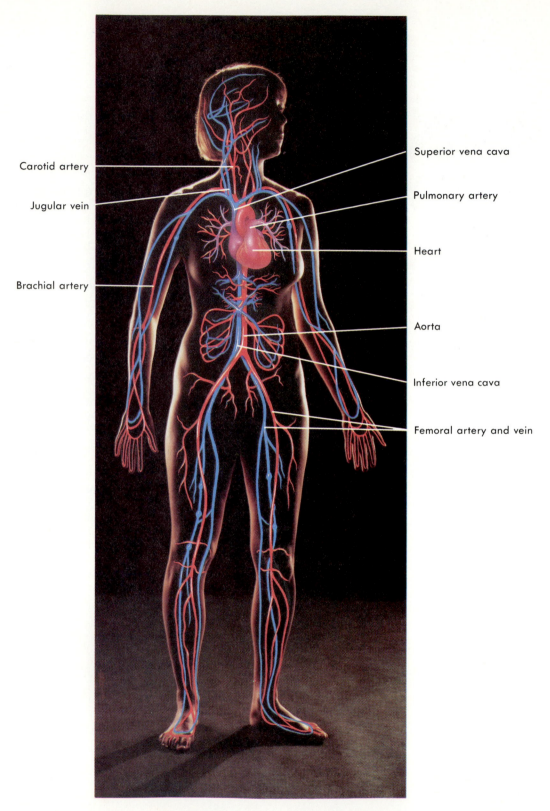

Carotid artery

Jugular vein

Brachial artery

Superior vena cava

Pulmonary artery

Heart

Aorta

Inferior vena cava

Femoral artery and vein

FIGURE 13-1
Overview of the circulatory system, including the pulmonary and systemic systems.

ANATOMY OF THE HEART AND CIRCULATORY SYSTEM

Relationship of the Heart to the Circulation

Figure 13-1 provides an anatomical overview of the circulatory system and shows the location of some of the major blood vessels of the body. From a physiological point of view, the circulatory system can be represented as two loops of blood vessels, the **systemic loop** and the **pulmonary loop,** both connected to the heart (Figure 13-2). The two loops function in a cycle of gas exchange between metabolizing tissues of the body and the atmosphere.

The systemic loop begins with the **aorta,** which branches to give rise to the systemic arteries, highly elastic structures that supply all the body's tissues (see Figure 13-1). The systemic loop includes the **systemic capillaries,** where the blood gives up part of its oxygen (O_2) and takes on additional carbon dioxide (CO_2). Blood leaving the systemic capillaries enters the **systemic veins,** which return blood to the heart, completing the systemic loop. The **superior** and **inferior venae cavae** return venous blood to the heart from the upper and lower parts of the body, respectively.

The systemic venous blood is pumped by the heart into the **pulmonary loop,** including the **pulmonary arteries** and **pulmonary capillaries,** and returns to the heart through **pulmonary veins.** During its passage through the pulmonary capillaries, the blood unloads some of its CO_2 and picks up an amount of O_2 equal to that delivered to tissues in the systemic loop. Blood returning to the heart from the lungs is pumped into the systemic loop to begin a new cycle.

The two loops are joined at the heart, which transfers blood from one loop to the other. When removed from the body in preparation for transplant or when exposed during open-heart surgery (Figure 13-3), the heart appears beautifully simple, but it is a highly complex organ. The right and left halves of the heart constitute separate pumps for each of the two loops. Thus it is sometimes convenient to refer to the **right heart** and **left heart** as if they were separate units (see Figure 13-2), although both sides of the heart act simultaneously in a heartbeat.

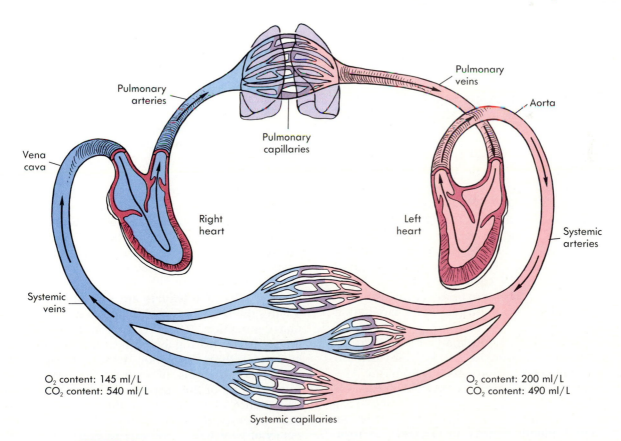

Pulmonary arteries

Pulmonary veins

Pulmonary capillaries

Aorta

Vena cava

Right heart

Left heart

Systemic arteries

Systemic veins

O_2 content: 145 ml/L
CO_2 content: 540 ml/L

O_2 content: 200 ml/L
CO_2 content: 490 ml/L

Systemic capillaries

FIGURE 13-2

The circulatory system can be divided into the pulmonary loop and systemic loop, which together form a complete circuit. All of the blood pumped by the left heart passes through the systemic loop, returning to the right heart as systemic venous blood, which is enriched in CO_2 and partly depleted of O_2. The right heart pumps the blood into the pulmonary loop, where some CO_2 is lost and O_2 is gained by gas exchange with the atmosphere. This blood returns to the left heart for another trip through the circuit.

The Heart

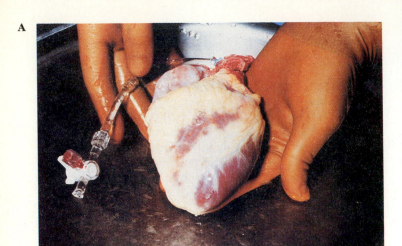

A

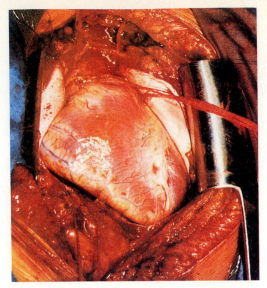

B

C

FIGURE 13-3
A A photograph of a human heart prepared for transplantation into a patient.
B A photograph of a heart during open-heart surgery. The tubes shown lead to the heart-lung machine shown in C.

Chambers and Valves of the Heart

Figure 13-4 shows two views of the exterior of the heart, displaying (1) the origin of the aorta and pulmonary artery and (2) the points at which the superior and inferior vena cavae and the pulmonary veins enter the heart. The aorta not only supplies the systemic circulation with blood, but also gives rise to the left and right **coronary arteries,** which serve the heart itself. The **coronary sinus** then returns venous blood from the coronary circulation (see Figure 13-4).

The heart consists of four chambers surrounded by cardiac muscle (Figure 13-5). The **right atrium** receives low-oxygen venous blood from the systemic circulation. The **right ventricle** alternately fills with blood from the right atrium and contracts to pump blood into the pulmonary circulation. Reoxygenated blood returns to the **left atrium,** whereas the **left ventricle** propels blood into the systemic circulation.

The right and left atria are separated by the **interatrial septum;** the **interventricular septum** separates the two ventricles. The muscular wall of the heart, or **myocardium,** is lined with an inner layer, the **endocardium,** and is covered with **epicardium** (see Figure 13-5). The entire heart is enclosed in a fibrous sac, the **pericardium,** which is continuous with the epicardium. A small quantity of **pericardial fluid** is present in the space between the epicardium and pericardium.

The sheet of connective tissue separating atria and ventricles is penetrated by **atrioventricular (AV) valves,** which allow one-way flow of blood from atrium to ventricle. The **right AV valve** has three **cusps,** or leaflets, and is also called the **tricuspid valve** (Figure 13-6, *A*). The **left AV valve,** or **bicuspid valve,** has two leaflets that resemble a bishop's mitre, so it is also known as the **mitral valve** (Figure 13-6, *B*). The AV valves allow the ventricles to fill from the atria between heartbeats but prevent backflow of blood when the ventricles contract.

The AV-valve leaflets are thin and delicate, allowing them to close rapidly at the onset of ventricular contraction. The high pressure developed by the ventricles during a heartbeat might evert the AV leaflets if they were not held back by **chordae tendineae,** threads of connective tissue running from the tips of the cusps to the **papillary muscles** (see Figure 13-6, *B*). The papillary muscles contract during ventricular contraction; this opposes the tendency of ventricular pressure to evert the valves.

When the ventricles are not contracting, the **pulmonary valve** prevents backflow of blood from the pulmonary arteries into the right ventricle, and the **aortic valve** (Figure 13-6, *C*) prevents backflow from the aorta into the left ventricle. Each of these valves consists of three leaflets shaped like half-

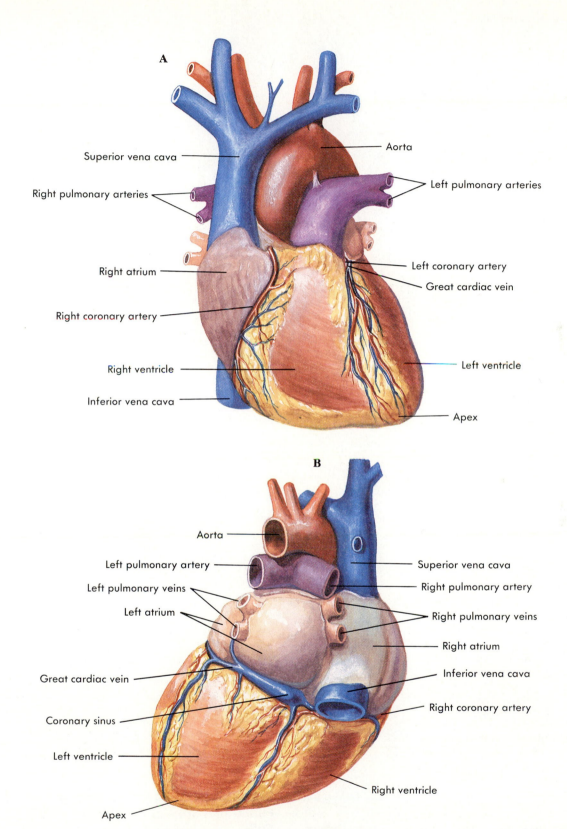

A

Superior vena cava

Right pulmonary arteries

Right atrium

Right coronary artery

Right ventricle

Inferior vena cava

Aorta

Left pulmonary arteries

Left coronary artery

Great cardiac vein

Left ventricle

Apex

B

Aorta

Left pulmonary artery

Left pulmonary veins

Left atrium

Great cardiac vein

Coronary sinus

Left ventricle

Apex

Superior vena cava

Right pulmonary artery

Right pulmonary veins

Right atrium

Inferior vena cava

Right coronary artery

Right ventricle

FIGURE 13-4
Two diagrammatic external views of the heart and its major vessels.

FIGURE 13-5
An internal view of the heart.

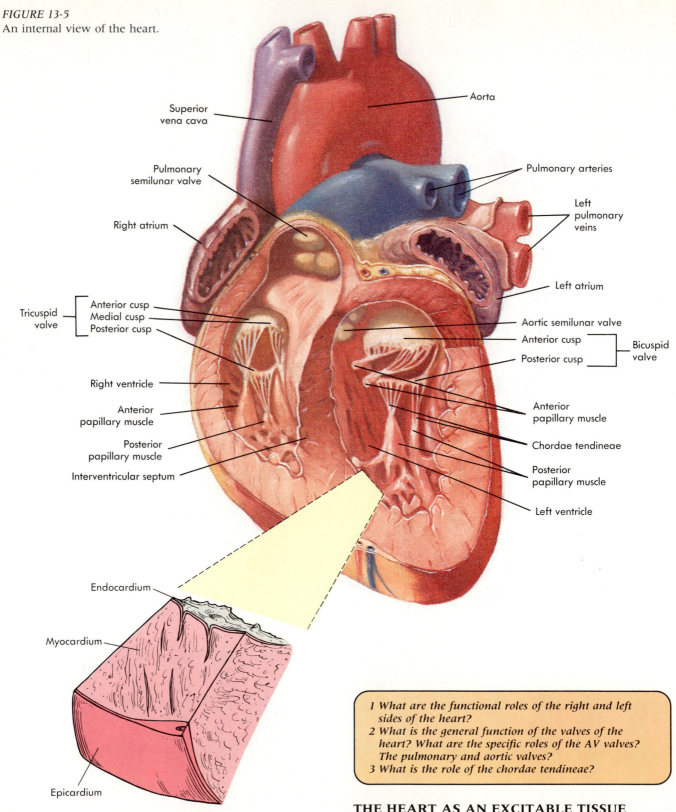

Superior vena cava

Aorta

Pulmonary semilunar valve

Pulmonary arteries

Right atrium

Left pulmonary veins

Left atrium

Tricuspid valve
- Anterior cusp
- Medial cusp
- Posterior cusp

Aortic semilunar valve

Anterior cusp
Posterior cusp
} Bicuspid valve

Right ventricle

Anterior papillary muscle

Posterior papillary muscle

Interventricular septum

Anterior papillary muscle

Chordae tendineae

Posterior papillary muscle

Left ventricle

Endocardium

Myocardium

Epicardium

1 What are the functional roles of the right and left sides of the heart?
2 What is the general function of the valves of the heart? What are the specific roles of the AV valves? The pulmonary and aortic valves?
3 What is the role of the chordae tendineae?

moons, so they are also known as **semilunar valves.** Because the pulmonary and aortic valves must withstand high pressure during much of the heart cycle, their leaflets are much more heavily constructed than those of the AV valves.

THE HEART AS AN EXCITABLE TISSUE
Functional Specializations of Cardiac Muscle Cells

Several types of cells are found in the heart (Table 13-1). The bulk of the heart is made up of **myocardial,** or **contractile,** cardiac muscle cells, whose role is to generate force. The contractions of these

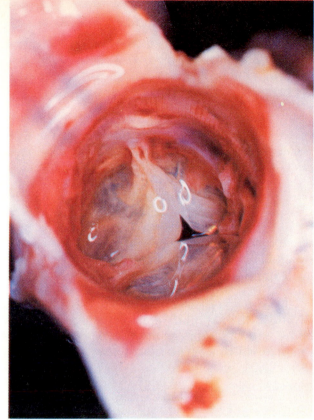

A

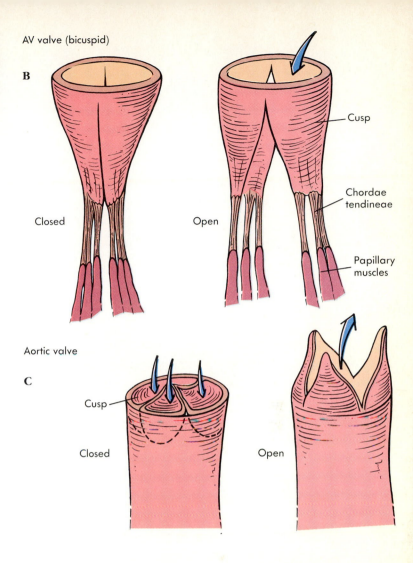

AV valve (bicuspid)

B

Cusp

Chordae tendineae

Closed   Open

Papillary muscles

Aortic valve

C

Cusp

Closed   Open

FIGURE 13-6
A   A photograph of the tricuspid valve during open-heart surgery.
B   The structure of the AV valves and papillary muscles.
C   The structure of the pulmonary and aortic valves.

myocardial cells are not spontaneous under normal conditions (but they may be in disease—see the discussion of fibrillation on p. 329), nor are they directly initiated by nervous commands. Instead, each heartbeat is initiated by action potentials originating in the **pacemaker fibers** of the heart.

TABLE 13-1	*Types of Myocardial Cells*	
Cell type	**Location**	**Physiological function**
Myocardial (ordinary) muscle	Atria and ventricles	Generate force
Pacemaker	SA node and AV node	Initiate heartbeat; control heart rate
Conducting	Bundle of His and branches; Purkinje fibers; some in atria	Coordinate contraction

Pacemaker cells are present in several regions of the heart (see Table 13-1). Normally heart rate is controlled by pacemakers in the **sinoatrial (SA) node** (Figure 13-7), which is located where the superior vena cava and right atrium join. Cells of the SA node undergo regular spontaneous action potentials. Other spontaneously active cells are present in the **atrioventricular (AV) node.** This node is located in the inferior portion of the interatrial septum.

Conducting fibers (see Figure 13-7 and Table 13-1) form a system that rapidly relays action potentials from the SA node to the AV node (the internodal pathways) and then to all parts of the ventricle. **Purkinje fibers** compose the **bundle of His,** which arises from the AV node and divides into one **right-** and two **left-bundle branches.** The bundle branches extend down the interventricular septum to the tip, or **apex,** of the ventricle and then project upward toward the **base,** or most superior part, of the ventricle, branching extensively to form a dense network of Purkinje fibers beneath the endocardium.

Electrical Activation of the Heart
Cardiac cells are electrically coupled to each other by intercellular gap junctions located in intercalated

The Heart   **325**

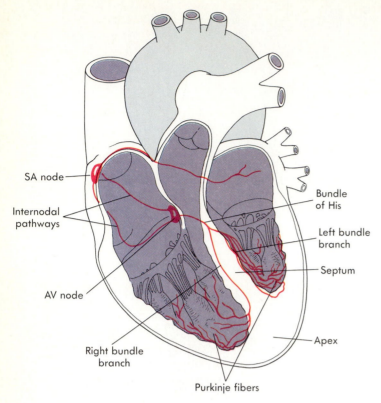

FIGURE 13-7
The conducting fiber system of the heart.

Labels on figure:
SA node
Internodal pathways
AV node
Right bundle branch
Purkinje fibers
Bundle of His
Left bundle branch
Septum
Apex

disks, structures that link adjacent cardiac muscle fibers (see Chapter 12, p. 307). By way of gap junctions, an action potential initiated in one cardiac muscle cell can spread throughout the entire heart. The sequence of activation of different parts of the heart in a heartbeat is determined by the pathways taken by impulses as they spread through the heart.

Systole is defined as the period during which first the atria and then the ventricles contract. The term **diastole** refers to the period during which the ventricles are relaxed and both the atria and ventricles are filling with blood. The initial event of systole is the production of an action potential in the cells of the SA node (see Figure 13-7; Table 13-2).

This action potential spreads among the myocardial fibers of the atria, causing both atria to contract. The effect of atrial contraction (atrial systole) is to add to the volume of blood that has already entered the ventricles during the previous diastole.

While it is spreading throughout the atrial myocardium, the electrical excitation enters the internodal bundles, reaching the AV node about 40 msec after its origin in the SA node. At this point the atria have not completed their contraction. Spreading of the action potential through the AV node into the conducting fiber system of the ventricle is delayed by the slowly conducting fibers surrounding the AV node (see Table 13-2), resulting in an **AV delay** of about 110 msec. During the AV delay, the atrial fibers complete their contraction and enter the refractory period of their action potentials.

Once it has passed into the bundle of His, electrical activity is conducted along the branch bundles and into the network of Purkinje fibers that supply the ventricular myocardium within about 30 msec (see Figure 13-7 and Table 13-2). Because the Purkinje fibers branch from apex to base within the innermost part of the myocardium, excitation of the ventricles proceeds from the apex of the ventricles towards the base and from endocardium to epicardium. Purkinje fibers are the largest in diameter of all heart fibers. Because conduction velocity increases with diameter, the Purkinje fibers are the most rapidly conducting fibers (see Table 13-2) and are able to activate all parts of the ventricles much more rapidly than if the activation were spread by only contractile fibers. The result is a forceful, coordinated ventricular contraction (ventricular systole).

> 1 What are the three types of cardiac muscle cells? How do they differ functionally?
> 2 Define the terms systole and diastole.
> 3 What is the location of the cells responsible for initiating a heartbeat?

Cellular Basis of the Cardiac Rhythm

A heart that has been removed from the body can continue to beat if maintained under favorable

TABLE 13-2 Electrical Characteristics and Innervation of Myocardial Cells

Type	Conduction velocity (m/sec)	Pacemaker activity	Autonomic innervation
SA node	0.05	Normal (60 to 80/min)	Dual
Atrium	0.8 to 1.0	None	Minor
AV node	0.05	Reserve	Dual
Purkinje fibers	2.0 to 4.0	Latent	Sympathetic
Ventricle	0.4 to 0.8	None	Sympathetic

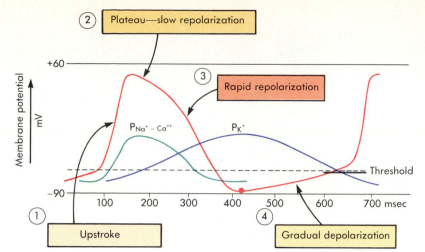

conditions. Thus the cardiac rhythm is established because action potentials are discharged at regular intervals, even in the absence of any nervous or hormonal input. All three types of heart cells (contractile, pacemaker, and conducting) are somewhat capable of spontaneous activity, but this capability is highly developed in the pacemaker cells of the SA and AV nodes. The rate of rhythmic discharge is highest in the cells of the SA node (see Table 13-2), so the rest of the heart normally proceeds at their rhythm.

Disorders of the heart rhythm, or **arrhythmias,** can arise in several ways. One cause of an arrhythmia is an **AV block,** in which action potentials from the SA node fail to reach the AV node. The block may be partial or complete. During complete, or third-degree, AV block the atria continue to beat at the rate determined by the SA node, but the ventricles beat at the slower rate of the AV node. An AV block discoordinates the activities of the atria and ventricles and slows the heart rate, but it is not necessarily life threatening because the ventricles can still fill adequately. The AV nodal cells can be thought of as reserve pacemakers.

The initial rise of depolarization, or **upstroke,** in a nodal or pacemaker fiber is relatively slow as compared with the action potentials of nerve and skeletal muscle (Figure 13-8). The action potential pauses before it repolarizes; this pause, or **plateau phase,** is found to some extent in the action potentials of all cardiac cells. In pacemaker cells the membrane potential does not return to a steady resting value after each action potential; rather, it drifts off in the direction of depolarization until threshold is reached and another action potential ensues. All these features of pacemaker action potentials are traceable to the behavior of ion channels in the cell membrane.

The membrane channels of cardiac pacemaker cells are characterized by relatively slow activation and inactivation as compared with those of nerve axons and skeletal muscle. The upstroke of the action potential is caused by activation of **slow Na^+-Ca^{++} channels** (see Figure 13-8). During the plateau phase, the membrane potential moves slowly in the direction of repolarization as the Na^+-Ca^{++} channels slowly inactivate. The K^+ channels of cardiac cells require about 400 msec to become fully activated. With the closing of the Na^+-Ca^{++} channels and the completion of opening of the K^+ channels, the plateau comes to an end, and the membrane rapidly completes its repolarization.

The K^+ channels remain open for some time after the Na^+-Ca^{++} channels have closed. As the K^+ channels slowly close, the membrane potential moves in the direction of the Na^+ equilibrium potential. In a nerve cell axon or skeletal muscle fiber, the membrane potential returns to a steady resting value more inside-negative than the threshold. In pacemaker cells, the membrane is much more permeable to Na^+ than that in nerve or skeletal muscle cells. As the K^+ channels close, the membrane potential crosses its threshold before it can arrive at a steady resting potential. Another action potential results, and the cycle continues.

The rate at which the K^+ channels of pacemaker fibers close sets the time interval between action potentials, which determines the **heart rate,** or number of beats per minute. If the K^+ channels closed more rapidly, the heart rate would increase, whereas a slower closing rate would decrease the heart rate. The pacemaker fibers receive synaptic input from the sympathetic and parasympathetic branches of the autonomic nervous system (Table 13-3). Autonomic transmitters or drugs that affect the heart rate are said to exert a **chronotrophic effect.** An increase in heart rate represents a positive chronotrophic effect; the rapid heart rate is termed **tachycardia.** A decrease in heart rate, or **bradycardia,** represents a negative chronotrophic effect.

The effect of sympathetic input to the SA node, in the forms of norepinephrine from sympathetic

TABLE 13-3   The Effects of Autonomic Transmitters on the Heart

Transmitter	Receptors	Cell type	Membrane effect	Heart effect
Acetylcholine	Muscarinic	Pacemaker	Slower closing of K^+ channels	Decreased heart rate
		Myocardial	Not innervated	Not innervated
Norepinephrine	Beta$_1$ adrenergic	Pacemaker	Faster closing of K^+ channels	Increased heart rate
Epinephrine		Myocardial	Increased Ca^{++} entry	Increased force of contraction
			Increased Ca^{++} pumping	Decreased duration of systole

nerves to the heart and epinephrine from the adrenal medulla, is to accelerate the closure of the slow K^+ channels. The faster the channels close, the more rapidly the pacemaker fibers return to threshold, and the higher the heart rate (Figure 13-9, *A*). Heart rates as high as 200 beats/minute can be attained as a result of sympathetic input to the heart during exercise or excitement.

The effect of the parasympathetic transmitter, acetylcholine (Figure 13-9, *B*), is to retard the closing of the slow K^+ channels. In the absence of any nervous or hormonal input to the SA node, the heart rate might be as high as 100 beats/minute. The heart rate of about 70 beats/minute, typical of resting adults (see Table 13-2), reflects a low level of steady parasympathetic input, or **parasympathetic tone,** which reduces the rate of SA nodal discharge.

Irregularities of the Cardiac Rhythm

Arrhythmic beats, called **extrasystoles** or **premature ventricular contractions** (PVCs), can result from various conditions. Most extrasystoles are caused by **ectopic pacemakers** (ectopic means "out of place") that occasionally discharge spontaneous action potentials. An action potential originating at an ectopic site can spread throughout the ventricle, causing an extrasystole if it happens to occur during diastole when the heart is not refractory. Frequently, the extrasystole is followed by a prolonged interbeat interval, or "skipped beat," because the extrasystole has made the ventricular myocardium refractory just at the time that the normal signal to contract enters the bundle of His. Anything that increases the excitability of the myocardium can cause an extrasystole in an otherwise nor-

FIGURE 13-9
Factors influencing the rate of depolarization in pacemaker cells and thus the heart rate. The time between successive action potentials is determined by two factors: the magnitude of depolarization needed to reach threshold, and the rate at which depolarization occurs. The rate of depolarization is increased by sympathetic stimulation (**A**), which hastens closing of K^+ channels after an action potential, and is decreased by parasympathetic stimulation (**B**), which retards K^+-channel closing.

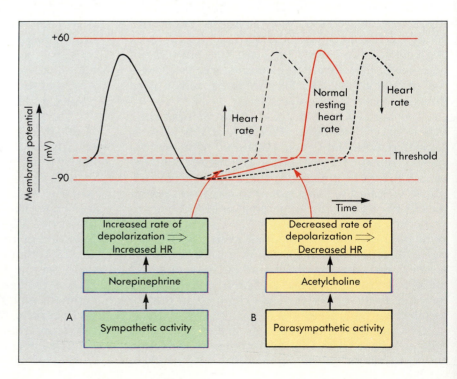

A
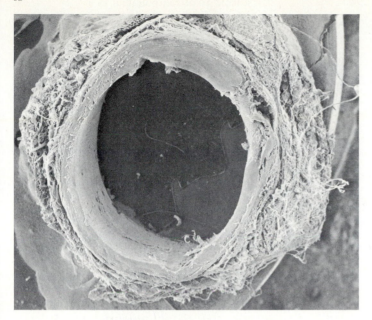

B
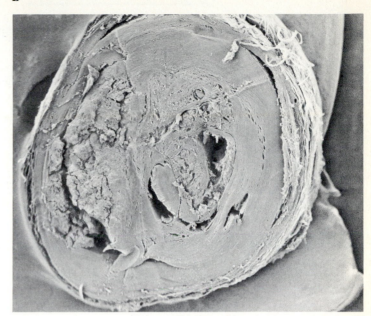

FIGURE 13-10
Scanning electron micrographs of A, a normal coronary artery and B, an artery occluded by a thrombus.

mal heart: caffeine and sympathetic input can have this effect.

When a coronary artery is blocked by a blood clot, or **thrombus** (Figure 13-10), the result is damage to a region of heart muscle. This damage is referred to as a **myocardial infarct,** or simply a **heart attack.** Extrasystoles often occur following heart attacks because, provided the patient survives the immediate damage, cells in the region of the damaged area can become hyperexcitable. The chance of clots forming is increased when coronary arteries are narrowed by the fatty deposits seen in **atherosclerosis** (discussed further in Chapter 15, p. 394).

The almost simultaneous contraction and relaxation of all fibers of the ventricles can be disrupted by damage or disease that slows the flow of excitation through some part of the conducting system of the myocardium. If the decrease in conduction velocity is sufficiently large, normal areas of the ventricles may recover from their refractory state before the injured, slowly conducting regions have been fully activated. The injured regions will then excite the normal areas of the affected ventricle, and the entire process of cardiac depolarization may be repeated before the next normal action potential arrives from the SA node. Continual self-excitation of the ventricle is known as **reentry;** this condition often follows heart attacks.

Reentry excitation may lead to a state called **fibrillation,** in which the contraction of the myocardium is so uncoordinated that pumping stops. If untreated, ventricular fibrillation leads to death within minutes, but frequently fibrillation can be

stopped in clinical situations by applying a strong electrical shock to the heart. This procedure, called **defibrillation,** resynchronizes the electrical activity of the myocardial fibers.

> 1 What process causes action potentials to be discharged at regular intervals in pacemaker cells?
> 2 How do the action potentials of cardiac pacemakers differ from those of nerve cell axons and skeletal muscle fibers?
> 3 What is an arrhythmia? What is an ectopic pacemaker?

Excitation-Contraction Coupling in Myocardial Fibers

Figure 13-11 shows an action potential typical of myocardial cells (small differences between atrial and ventricular cells can be neglected for most purposes). In contrast to the slow upstroke of the action potential of pacemaker cells (see Figure 13-9), the upstroke of the action potential of myocardial fibers occurs as quickly as in nerve cell axons and skeletal muscle fibers. As a result, the conduction velocity is relatively high (see Table 13-2). The plateau phase is longer, and the membrane potential is stable during the interval between successive action potentials. The rapidity of the upstroke results from the fact that, unlike pacemaker cells, myocardial fibers have fast Na^+ channels in addition to slow Na^+-Ca^{++} channels similar to those of pacemaker cells.

Myocardial cells have slow K^+ channels that open during the plateau phase. The summation of

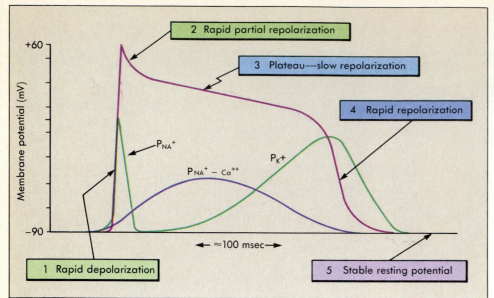

FIGURE 13-11
The permeability changes that occur during an action potential in a myocardial fiber. The pattern is similar to that in Purkinje cells. Rapid depolarization results from the activation of fast Na⁺ channels, whereas rapid partial repolarization is caused by Na⁺ inactivation. The plateau is produced by activation of slow Na⁺-Ca⁺⁺ channels, whereas repolarization is the combined result of K⁺ activation and Na⁺-Ca⁺⁺ inactivation. During diastole, the membrane potential is stable at the resting potential.

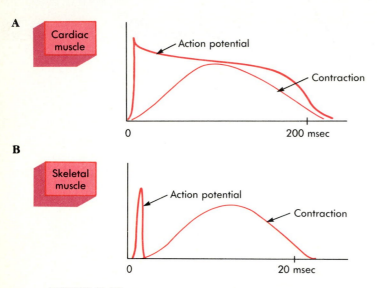

FIGURE 13-12
The time relationship between the cardiac action potential and mechanical activity in the heart, A, compared with that in, B, skeletal muscle. Note the tenfold difference in the time scale.

the outward K⁺ current and the inward Na⁺ and Ca⁺⁺ currents holds the membrane potential relatively depolarized during the plateau when both types of channels are open. The force a myocardial cell develops in response to an action potential peaks shortly before the end of the plateau phase and declines considerably by the time the phase of rapid repolarization begins (Figure 13-12, A).

An important effect of the plateau is to prevent tetanic contractions of the heart. During each heartbeat, each fiber undergoes one action potential and enters an active state that lasts several hundred milliseconds. If the active state were to be prolonged by a second action potential, the heart would remain in systole and its pumping would be disrupted. Until the plateau phase is over, cardiac fibers are refractory to the initiation of another action

potential. Thus the plateau prevents summation of contraction in heart muscle when it is beating rapidly. In contrast, the action potential in skeletal muscle is so brief that the refractory period is over even before peak force development is achieved (Figure 13-12, B); repeated stimuli can therefore summate and lead to a sustained contraction.

Excitation-contraction coupling in heart muscle (Figure 13-13) resembles that in skeletal muscle in that the key event is an increase in intracellular Ca⁺⁺, which enables actin and myosin to interact and generate force. However, unlike contraction in skeletal muscle, contraction in heart muscle depends highly on the presence of Ca⁺⁺ in the extracellular fluid. The small amount of Ca⁺⁺ that enters the cardiac fiber from the extracellular fluid during the plateau phase of the action potential is believed to trigger release of a larger amount of Ca⁺⁺ from the sarcoplasmic reticulum (SR). This process is an example of positive feedback and is termed **Ca⁺⁺-mediated Ca⁺⁺ release.** The Ca⁺⁺ released into the cytoplasm binds to troponin and causes crossbridge activity.

Relaxation occurs as some of the Ca⁺⁺ that was released is pumped back into the SR (see Figure 13-13) while the rest is pumped out across the cell membrane by a gradient-mediated countertransport mechanism energized by the Na⁺ gradient. The Na⁺ that enters during action potentials in exchange for Ca⁺⁺ is removed from the cytoplasm by the Na⁺-K⁺ pump.

Normally the contractile machinery of the heart is less than maximally activated during a systole because the amount of Ca⁺⁺ released into the cytoplasm is less than enough to saturate all the troponin molecules. Adjusting the ability of the heart's contractile machinery to deliver force (usually referred to as the inotrophic state) is one important means by which the sympathetic nervous system affects heart function (see Table 13-3). Activation of β₁-adrenergic receptors by norepinephrine or epi-

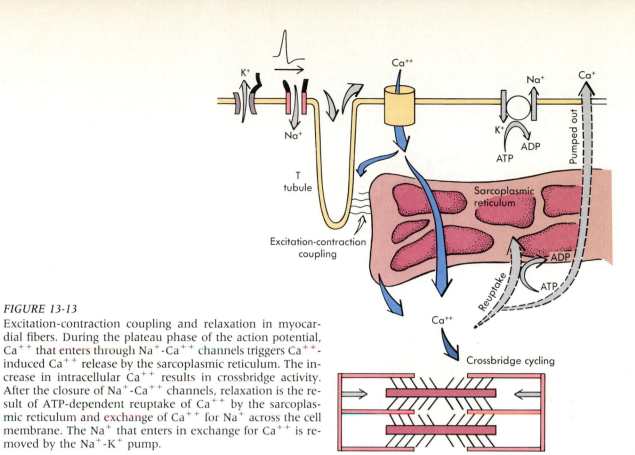

FIGURE 13-13
Excitation-contraction coupling and relaxation in myocardial fibers. During the plateau phase of the action potential, Ca^{++} that enters through Na^+-Ca^{++} channels triggers Ca^{++}-induced Ca^{++} release by the sarcoplasmic reticulum. The increase in intracellular Ca^{++} results in crossbridge activity. After the closure of Na^+-Ca^{++} channels, relaxation is the result of ATP-dependent reuptake of Ca^{++} by the sarcoplasmic reticulum and exchange of Ca^{++} for Na^+ across the cell membrane. The Na^+ that enters in exchange for Ca^{++} is removed by the Na^+-K^+ pump.

nephrine initiates a second messenger, adenosine 3':5'-cyclic monophosphate (cAMP). One effect of cAMP is to increase the number of Na^+-Ca^{++} channels that open during each action potential, thereby increasing the peak concentration of Ca^{++} attained in response to an action potential (Figure 13-14).

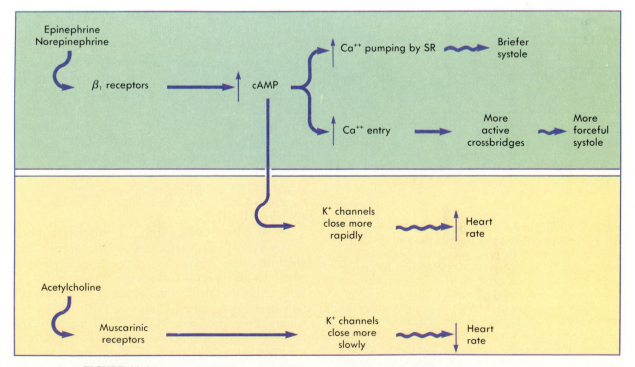

FIGURE 13-14
Summary of direct effects of autonomic transmitters on the heart.
A Adrenergic transmitters act on β_1-receptors to increase cAMP. This second messenger enhances both Ca^{++} release and Ca^{++} entry, resulting in a greater number of active crossbridges. cAMP also causes a more rapid uptake of Ca^{++} into the sarcoplasmic reticulum (SR) and an increase in heart rate.
B Acetylcholine acts on muscarinic receptors to decrease the heart rate.

The result increases the number of crossbridges active during each systole, with consequent increases in the force applied to pump blood and in the velocity of sarcomere shortening. A second effect of the cAMP message is to make each systole briefer by increasing the rate at which released Ca^{++} is cleared from the cytoplasm (see Figure 13-14).

In addition to the direct effects of sympathetic inputs on the contractility of myocardial fibers, the sympathetic effect on heart rate indirectly increases contractility. This effect occurs because when the heart is beating rapidly, there is less time to clear Ca^{++} from the cytoplasm between systoles. Thus sympathetic tachycardia adds to the peak amount of

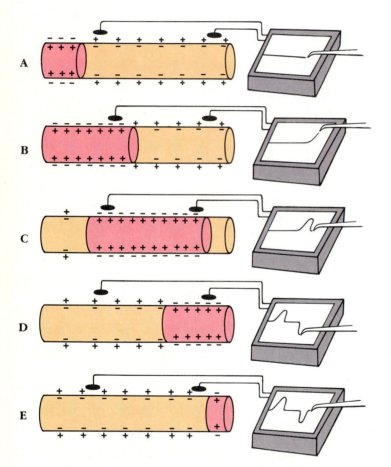

FIGURE 13-15
Extracellular recording of an action potential in a single cardiac fiber. The electrocardiograph records the voltage difference between the two electrodes. An upward deflection of the pen corresponds to an excess of negative charge at the first electrode. In A the action potential has not yet reached the first electrode, so the voltage difference is zero.
In B the first electrode senses an excess of negative charge as compared with the second electrode, so the pen is deflected. In C the action potential occupies all of the fiber between the two electrodes, so no voltage difference exists between them and the pen returns to baseline. A deflection in the reverse direction is recorded as the tail of the action potential passes under the second electrode (D). The trace returns to baseline after the action potential has passed under the second electrode (E).

Ca^{++} available to bind to troponin and increases contractility (see Figure 13-14).

The parasympathetic nervous system makes relatively few synapses on ventricular myocardial fibers, and, in the absence of simultaneous sympathetic input to the heart, parasympathetic nerves have little effect on the contractility of the ventricles. However, parasympathetic fibers do make presynaptic inhibitory connections on postganglionic sympathetic fibers innervating the ventricles, so parasympathetic inputs can exert an indirect effect by turning off simultaneous sympathetic inputs.

The effects of autonomic inputs to the heart can be summarized as follows (see Table 13-3). Activation of sympathetic nerves serving the heart results in (1) increasing the heart rate, and in (2) making each systole both shorter and more forceful. Epinephrine released by the adrenal medulla has the same effects. Activation of parasympathetic nerves to the heart slows the rate of discharge of the SA node, but, in contrast with sympathetic stimulation, parasympathetic activity has only a weak directly negative effect on heart contractility. However, parasympathetic stimulation can reduce the effect of sympathetic inputs.

> 1 What is the importance of the slow Na^+-Ca^{++} channels in the action potential of myocardial fibers?
> 2 What is the role of the slow Na^+-Ca^{++} channels in excitation-contraction coupling?
> 3 What factors normally determine the contractility of the heart?

Electrocardiogram

Electric currents flow around any excitable cell that is conducting an action potential. The extracellular current is driven by the voltage difference between the polarized and depolarized regions of the cell. The flow of the current through the surrounding tissues results, in turn, in voltage gradients that can be recorded by extracellular electrodes. The electrical events that occur when an action potential travels along a single cardiac muscle fiber can be recorded by electrodes placed at two points along the fiber (Figure 13-15, A).

When the leading edge of the action potential (the wave of depolarization) passes under the first electrode, the electrode becomes negative with respect to the second electrode (Figure 13-15, B). When the membrane under both of the electrodes becomes depolarized (Figure 13-15, C), both electrodes sense the same voltage, and the trace returns to zero. When the cardiac action potential has left the first electrode and is passing under the second electrode, a voltage difference between the two electrodes is again recorded by the amplifier. This time polarity of the charge differences between the two electrodes is reversed (Figure 13-15, D), and the pen deflects downward. The final record of voltage changes occurring over time is shown in Figure 13-15, *E*.

In the heart, many fibers are activated simultaneously, generating large extracellular currents that can be detected by electrodes placed on the body surface, most commonly on the right and left arms and left leg. As in nerve trunks (see Chapter 7, p. 159), the recorded currents are an example of a compound action potential. These records are referred to as **electrocardiograms (ECGs or EKGs)** (Figure 13-16).

The initial event of a heartbeat, depolarization of the SA node, cannot be seen in the ECG because the amount of tissue involved is too small. The wave of depolarization that spreads over the atria following SA-node depolarization results in the **P wave** (see Figure 13-16, *A*). When the atrial fibers are all in the plateau phase of their action potentials, the ECG trace returns to its baseline. In the meantime, the wave of activation has entered the bundle of His. As with the SA node, the signal generated by depolarization of the conducting fibers is too small to be recorded at any distance from the heart.

The next event visible in the ECG is the **QRS complex,** the result of ventricular depolarization (see Figure 13-16, *A*). At the same time that the ventricle is depolarizing, the atria are repolarizing, but the small signal of atrial repolarization is masked by the much larger signal from the ventricle. When all parts of the ventricle have entered the plateau phase of the cardiac action potential, the ECG trace returns to its baseline and remains there until ventricular repolarization begins.

The ventricular plateau is followed by a wave of repolarization, giving rise to the **T wave.** The T wave characteristically is smaller and lasts longer than the QRS complex because (1) repolarization is slower than depolarization, and (2) the duration of the plateau phase differs slightly in different parts of the myocardium. Contrary to what might be expected, the T wave has the same polarity as the R wave, the major component of the QRS complex. The reason is that the first cells to be activated in the ventricle are also the last ones to repolarize. Thus the wave of repolarization follows a route through the heart that is opposite to that of the wave of depolarization, and the net current flow through the ventricle during repolarization has a polarity similar to that of depolarization.

The ECG gives information about the timing of electrical events in the heart (see Figure 13-16, *A*). The **P-R interval** corresponds to the time between the onset of atrial depolarization and the beginning of ventricular depolarization. Most of this time reflects the AV delay. Normally the P-R interval lasts about 0.2 second; a lengthening of this interval occurs in the mildest form of AV block. Complete ventricular activation normally requires about 0.1 second; if the QRS complex lasts significantly longer than this, some part of the ventricular conducting system has failed.

Activation progresses through the heart in a complex, three-dimensional pattern and the electrocardiograph is most sensitive to those currents which flow along the axis formed by its recording electrodes. A picture of the spread of activation in the heart can be obtained by recording the ECG from several different electrode locations, or **leads.**

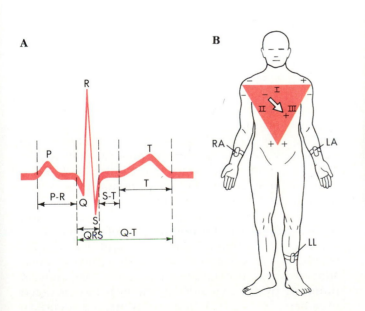

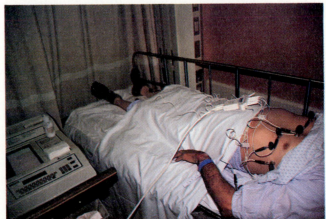

FIGURE 13-16

A An idealized ECG of a single heartbeat. The P-R interval is normally measured at the beginning of the QRS complex. **B** The arrangement of the standard limb leads (I, II, and III). The axes of the three leads form Einthoven's triangle (shown in red). The white arrow illustrates the mean electrical axis of the heart. **C** A person undergoing electrocardiography.

The three **standard limb leads** are illustrated in Figure 13-16, *B*. In lead I, the negative input of the amplifier is connected to the right arm, the positive input to the left arm, and the ground to the left leg. In lead II, the positive input trades positions with the ground. In lead III, the negative input is connected to the left arm, the positive input to the left leg, and the ground to the right arm.

In these arrangements, the two arms and the left leg are simply more convenient places to attach the electrodes than is the chest, and the limbs could be thought of as wires connected to the chest. The axes of the three limb leads form **Einthoven's triangle** (colored red in Figure 13-16, *B*). The limb leads are sensitive to currents that flow in right-to-left and superior-to-inferior directions. Currents that flow along front-to-back axes are not detected by the standard limb leads. At present, records from 12 different leads are made in a typical ECG evaluation.

Electrical Axis of the Heart

The sum of currents flowing within the plane of Einthoven's triangle is referred to as the **mean electrical axis** of the heart. For example, if the relative magnitudes of the R wave in the three leads are plotted on Einthoven's triangle, the resulting current vector typically points downward and to the left (arrow in Figure 13-16, *B*).

The flow of excitation through the bundles of His causes ventricular depolarization to begin in the septum and apex of the heart. Thus, during the peak of the R wave, the septum and apex of the ventricle are depolarized. Extracellular currents flow between these areas and the surrounding parts of the ventricle that are not yet depolarized (Figure 13-17); the bulk of these currents flow from the base of the ventricle toward its apex. The left ventricle's mass is greater, so more current flows around that side of the heart, and the mean electrical axis is rotated toward it.

> 1 What are the three major waves of the ECG? What events correspond to each of them?
> 2 What are the standard limb leads in electrocardiography?
> 3 What is the mean electrical axis of the heart?

THE HEART AS A PUMP
Events of the Pump Cycle

The ventricles alternately relax and fill with blood from the vena cava, pulmonary veins, and coronary sinus and contract to eject blood into the aorta, pulmonary arteries, and coronary arteries. The pump cycle involves changes in ventricular and arterial pressure that can be measured by inserting a small probe into a peripheral vein and then guiding it

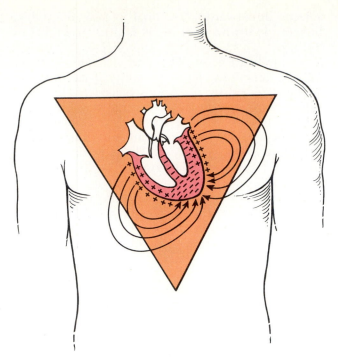

FIGURE 13-17

The flow of current around the heart at a time when the ventricles are partly depolarized (about the time of the peak of the R wave). Current flows around the heart from the excess of positive charge in the undepolarized regions towards the excess of negative charge in the cells of the depolarized regions, as indicated by the arrows. As measured in the axis of lead I, the current surrounding the left ventricle is partly canceled by that flowing in the opposite direction around the right ventricle. The largest part of the total current is flowing parallel to the axis of lead II and at a slight angle to that of lead III. Thus the largest deflection will be seen in lead II, and the smallest in lead I.

into the heart in a diagnostic procedure known as **cardiac catheterization** (Figure 13-18).

The pressure of blood in the veins entering the heart is only a few mm Hg. At the instant of ventricular contraction, the diastolic pressure in the aorta is typically 60 to 80 mm Hg and in the pulmonary arteries is about 8 mm Hg. The ventricular systole briefly raises the pressure in the aorta to about 120 mm Hg and in the pulmonary arteries to about 25 mm Hg. The large pressure difference between the pulmonary venous blood entering the left atrium and that of the aorta means that the left heart must fill with blood under low pressure and eject it into the aorta under high pressure.

For simplicity, the following paragraphs and figures emphasize the pump cycle of the left heart only. The function of the right and left cardiac pumps is similar, except for the much lower pressure of blood in pulmonary arteries. The rates of pumping by the two sides of the heart, as averaged over periods longer than a few beats, are identical. Blood flows more easily through the pulmonary

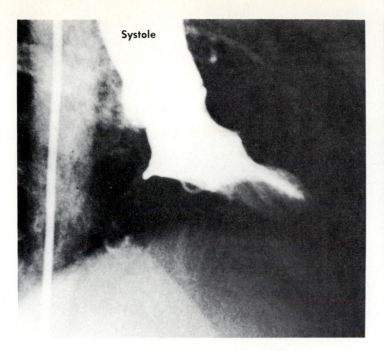

Systole

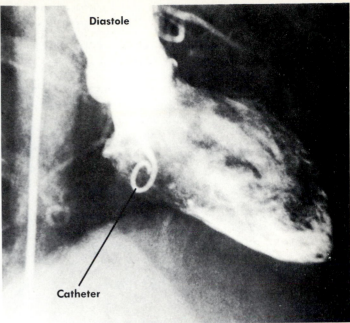

Diastole

Catheter

FIGURE 13-18
Two chest radiographs taken at the end of systole (**left**) and diastole (**right**) during cardiac catheterization. The probe used to measure pressures in the heart, seen clearly during diastole, was inserted into a vein and carefully guided into the heart.

loop than through the systemic loop because the pulmonary loop has a lower flow resistance. Therefore the right ventricle does not have to generate as much pressure, and its wall is markedly thinner and less muscular than that of the left ventricle.

The time relationships of the pressure and volume changes to the electrical and mechanical events within the heart cycle can be followed in a simultaneous recording of the arterial pressure, ventricular pressure, atrial pressure, ECG, heart sounds, and ventricular volume (Figure 13-19). The cycle can be divided into five phases: (1) **passive filling,** (2) **atrial contraction,** (3) **isovolumetric ventricular contraction,** (4) **ejection,** and (5) **isovolumetric ventricular relaxation** (Table 13-4). Despite the fact that both the atria and ventricles contract and relax during a heartbeat, the term, systole, is generally reserved for the period in which the ventricles are contracted (phases 3 to 4) and the term, diastole, for the period during which the ventricles are relaxed (phases 1, 2, and 5).

Most ventricular filling occurs during phase 1 (Figure 13-20, *A*), when all parts of the myocardium are relaxed. During this phase the pressure of blood in the left atrium is very slightly higher than that in the ventricles. Therefore the AV valve is open, and blood flows from the vena cava into the atrium and on into the ventricle. Ventricular pressure rises as filling expands the ventricular volume. The arterial blood pressure is much higher than the ventricular pressure, so the aortic valve remains

closed. The arterial pressure falls steadily during diastole as blood ejected into the arteries by the previous heartbeat runs off into the systemic vascular tree.

The phase of passive filling concludes with atrial contraction (Figure 13-20, *B*), which delivers only about 20% of the final ventricular volume under resting conditions. The total volume of blood in the ventricle at the end of atrial contraction is termed the **ventricular end-diastolic volume.**

The onset of ventricular contraction raises ventricular pressure above atrial pressure and closes the AV valves. The closing of these valves causes the **first heart sound,** which is frequently described as sounding like "lubb." At this point, the AV and aortic valves are closed, and the ventricle contracts against a constant volume of blood —thus the term, isovolumetric contraction (Figure 13-20, *C*; see Table 13-4). While isovolumetric contraction is rapidly increasing the ventricular pressure, the arterial pressure continues its slow decrease.

When ventricular pressure exceeds aortic pressure, the aortic valves are forced open, and a volume of blood—the **stroke volume**—is expelled from each ventricle into the respective artery. This is the ejection phase of the heart cycle (Figure 13-20, *D*). As the stroke volume is squeezed into the aorta, the aortic pressure rises rapidly from its minimum value immediately before ejection to a maximum that is normally about 120 mm Hg.

In clinical terminology, "blood pressure" refers

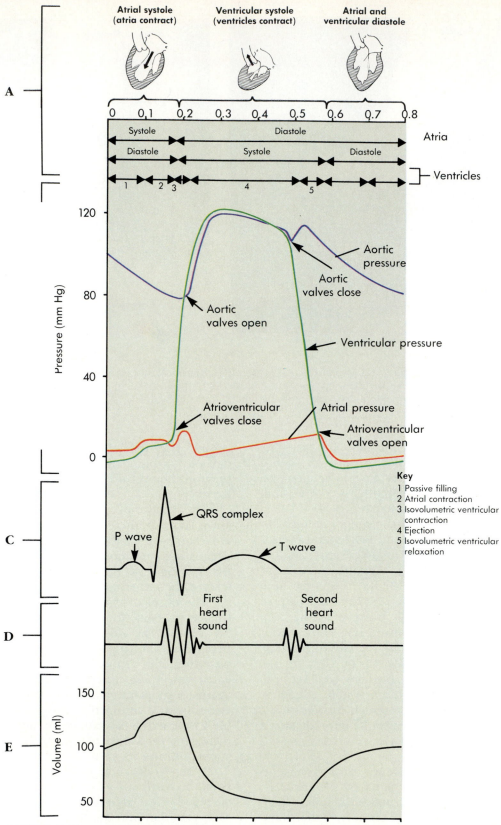

FIGURE 13-19

The cardiac cycle. Phases and approximate durations are indicated at the top of the figure. The blue, green, and red curves show the pressures in the aorta, left ventricle, and atrium, respectively. Below this are shown the ECG, the heart sounds, and the ventricular volume. The phases of the cardiac cycle occur in the following order: (1) passive filling (early diastole), (2) atrial contraction (late diastole), (3) isovolumetric ventricular contraction, (4) ejection, and (5) isovolumetric ventricular relaxation.

TABLE 13-4   Phases of the Cardiac Cycle

Phase	Approximate duration	Atrial state	Ventricular state	Pulmonary and AV valves	Aortic valves
1	250 msec	Relaxed	Relaxed: filling	Open	Closed
2	150 msec	Contracting	Relaxed: filling	Open	Closed
				First heart sound	
3	50 msec	Relaxed	Isovolumetric contraction	Closed	Closed
4	400 msec	Relaxed	Contraction: ejection	Closed	Open
				Second heart sound	
5	50 msec	Relaxed	Isovolumetric relaxation	Closed	Closed

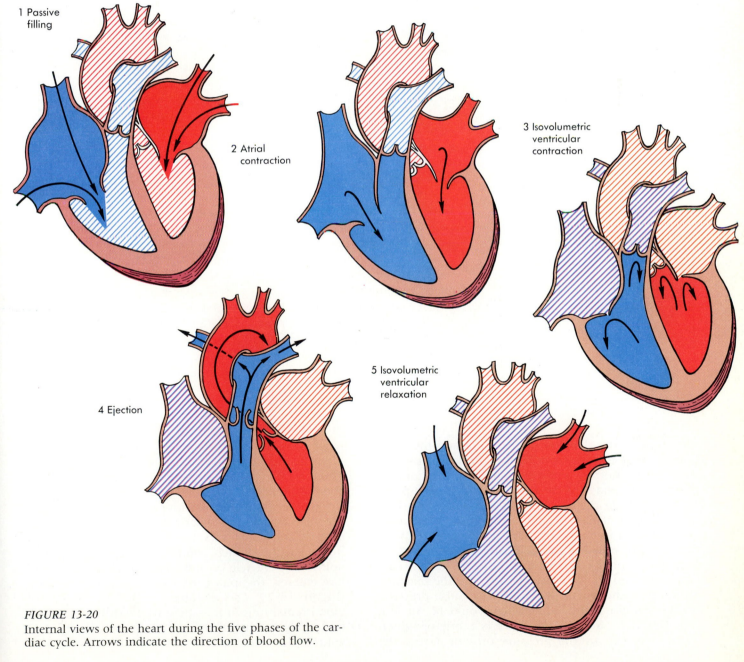

1 Passive filling

2 Atrial contraction

3 Isovolumetric ventricular contraction

4 Ejection

5 Isovolumetric ventricular relaxation

FIGURE 13-20
Internal views of the heart during the five phases of the cardiac cycle. Arrows indicate the direction of blood flow.

The Heart

Consequences of Valvular Heart Disease

There are two types of heart valve defects. In one type, **stenosis,** a valve is narrowed, restricting blood flow through it. In the second type, **insufficiency,** the valves are leaky and permit backflow of blood.

In mitral stenosis, blood has difficulty entering the left ventricle from the atrium. Left atrial pressure is increased, whereas the left ventricular end-diastolic volume and cardiac output are decreased. During atrial contraction, the flow of blood through the narrowed mitral valve produces a distinctive blowing sound referred to as a **presystolic murmur.**

In mitral insufficiency, backflow of blood from the left ventricle into the atrium during systole increases the left atrial pressure. It also decreases effective cardiac output because some of the stroke volume does not enter the aorta. The backflow of blood to the left atrium is added to the normal venous return, increasing left ventricular end-diastolic volume and pressure and causing the ventricle to work harder to pump the same amount of blood than it would in a normal heart. Such **volume overloads** cause the left ventricle to hypertrophy. Backflow through the mitral valve produces a **systolic murmur.**

Aortic valve stenosis results in an increased pressure gradient across the aortic valve during ventricular ejection. The left ventricular pressure is increased, and ejection velocity is decreased. The aortic pressure rises more slowly during ejection, reducing the pulse pressure. A systolic murmur results from turbulent flow through the narrow aortic valve. Aortic stenosis generates a **pressure overload** on the left ventricle, causing it to hypertrophy.

Ventricular hypertrophy is a natural adjustment of the heart muscle mass to the increased workload imposed on it. As with skeletal muscle (see Chapter 12), the myofibrillar content of the muscle cells increases without an increase in the number of myocardial cells. Hypertrophy is initially beneficial because it enables the ventricle to generate higher pressures. The problem is in providing an adequate oxygen supply to the enlarged myocardium because the diffusion distances become greater. In long-term overloads of the heart, some muscle cells die and are replaced by less compliant connective tissue. Furthermore, the internal work of the heart—the work it must do on itself to pump a given volume of blood—increases with the increasing thickness of the ventricle. Ultimately, overloads may result in heart failure.

to the pressure of blood in the aorta and the large arteries that branch from it. It is indicated by two values: (1) the maximum pressure recorded during the ejection phase of the heart cycle **(systolic pressure),** and (2) the minimum value reached at the end of ventricular diastole **(diastolic pressure).** The blood pressure of a patient with a systolic pressure of 120 mm Hg and a diastolic pressure of 70 mm Hg would be recorded as "120/70." The difference between the systolic pressure and the diastolic pressure is the **pulse pressure.** The pulse pressure of the patient mentioned above would be 50 mm Hg. The **mean arterial pressure** is a time-weighted average of the arterial pressure present during the heart cycle. It is approximately equal to the sum of diastolic pressure plus a third of the pulse pressure. Thus mean arterial pressure in this example is about 87 mm Hg.

As ventricular relaxation begins, ventricular pressure falls below arterial pressure and the pulmonary and aortic valves close, making the **second heart sound,** which sounds like "dupp." At this point the ventricle is once again a closed box and has entered the phase of isovolumetric relaxation (Figure 13-20, *E;* see Table 13-4). The amount of blood remaining in it—the **end-systolic volume**—is roughly a third of the end-diastolic volume. Isovolumetric relaxation of the left ventricle begins with the closing of the aortic valve and continues until ventricular pressure falls below atrial pressure and the AV valves open (see Table 13-4). At this point the cycle returns to the phase of early diastole.

> 1 Why is atrial contraction relatively unimportant in the performance of the heart?
> 2 What events are responsible for the first and second heart sounds?
> 3 What is meant by each of the following terms: systolic pressure, diastolic pressure, and pulse pressure?

Mechanisms of Ventricular Filling in Diastole

The ventricle is said to fill passively before atrial contraction because all parts of the heart are relaxed in this phase. The small pressure gradient between the blood in central veins and that in the heart itself has been regarded as the driving force for ventricular filling during this phase. Recent studies suggest that the rapid ventricular filling that occurs immediately after the AV valves open is aided by suction generated by the rebound of the ventricle from the distortion of its structure that occurred during systole (see Fig. 13-19). As the heart contracts, energy is stored in collagen strands that lace the myocardial cells together. As the ventricle begins to relax, this stored energy is released. The relaxing ventricle acts like the rubber bulb of a medicine dropper in that squeezing the bulb to empty it corresponds to systole and releasing the bulb allows the energy stored in the rubber to refill the dropper. This effect may be especially important when the heart is contracting vigorously.

Later in diastole the ventricular muscle has completed its rebound and begins to be distended by the pressure of venous blood, so that both ventricular volume and pressure are increasing. The amount of blood that enters the heart during the later stages of diastole mainly depends on the driving force of central venous pressure, which, with the assistance of atrial contraction, stretches the fully relaxed ventricle.

Autoregulation of Stroke Volume

Early studies of the length-tension relationship of cardiac muscle were performed on the amphibian heart by the German physiologist Otto Frank in the late 1800s. During the early 1900s the English physiologist Ernest H. Starling and his associates developed a **heart-lung preparation** for studies of the mammalian heart. In this preparation the heart of a dog was disconnected from the systemic loop but remained connected to the pulmonary loop. Reservoirs were attached to the right atrium and the aorta so that the right atrial pressure and the aortic pressure could be manipulated as experimental variables.

It was natural for Frank and Starling to apply the concepts developed in studies of skeletal muscle

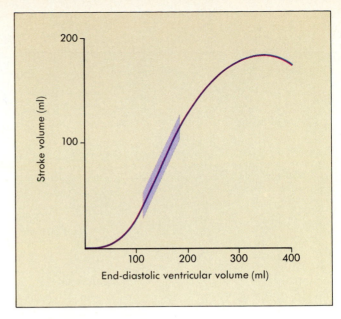

FIGURE 13-21

The relationship between end-diastolic volume and stroke volume in the heart-lung preparation (Frank-Starling curve). The range of values typical of the normal heart is shown by the shaded portion of the curve.

to the isolated heart. Isotonic twitches of isolated skeletal muscle are affected by the preload applied to the muscle before it contracts as well as by the afterload experienced by the contracting muscle. During the last part of diastole, the ventricle is stretched by the pressure of venous blood, so the central venous pressure is sometimes called the **preload** of the ventricle. After it begins to contract, the ventricle opens the valves that stand between it and the arterial blood; at this point, it must work against the arterial pressure, which is sometimes referred to as the **afterload** of the heart. The descriptive terms "central venous pressure," "filling pressure," and "right atrial pressure" are used instead of preload in this text, and "mean arterial pressure" and "aortic pressure" are used instead of afterload.

Skeletal muscle is specialized to deliver large amounts of force over a relatively narrow range of its length-tension relationship. Within this range, tension development is optimal. Changes in the overlap of actin and myosin filaments can change the ability of skeletal muscles to develop force, but this effect is small in the physiological length range. In contrast, Starling found that the force developed by the heart during systole is very responsive to changes in central venous pressure over the physiological range. An increase in central venous pressure increases the volume of blood in the ventricle at the end of diastole; the increased stretch of cardiac muscle fibers increases both the velocity and the force of the contraction during the subsequent systole. The result is that, over the normal range of end-diastolic volumes, an increase in end-diastolic ventricular volume automatically causes a proportionate increase in stroke volume (Figure 13-21).

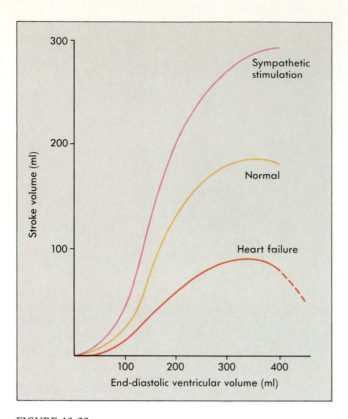

FIGURE 13-22
Frank-Starling curves from both a heart strongly activated by sympathetic input and a failing heart, as compared with the normal curve.

The phenomenon of the heart autoregulating its stroke volume according to its end-diastolic volume is called the **Frank-Starling Law of the Heart.** The Frank-Starling Law arises from the fact that the normal range of lengths of myocardial fibers in the functioning heart is shorter than the optimum for force generation. Recent studies suggest that this is a direct response of the contractile machinery to stretch rather than changes in the degree of overlap of thick and thin filaments.

The stroke volume pumped by the heart at any particular end-diastolic volume is affected by the intrinsic myocardial contractility. An increase in contractility shifts the Frank-Starling curve upward (Figure 13-22), increasing the stroke volume pumped for any given end-diastolic volume. Because the stroke volume is equal to the end-diastolic volume minus the end-systolic volume, an increase in the contractility causes the ventricle to pump a larger fraction of its end-diastolic volume if that volume remains constant.

Damage to the myocardium or reduction of its blood supply lowers the Frank-Starling curve (see Figure 13-22). Disease processes that progressively weaken the myocardium result in a steady increase in the end-diastolic and end-systolic volumes, termed **congestive heart failure.** In the beginning

of the disease process, the increased ventricular volume enables the weakened ventricle to continue pumping the volume of blood that returns to it. However, as the end-diastolic volume continues to increase, the weakened heart enters a range of the Frank-Starling curve in which further increases in end-diastolic volume do not allow it to maintain a normal stroke volume, and ultimately into a range in which further stretch decreases the stroke volume rather than increasing it. This leads to a condition of positive feedback that rapidly abolishes the pumping ability of the heart and leads to death.

Several drugs that increase contractility can be used to combat congestive heart failure. Typically, these act by increasing the cytoplasmic Ca^{++} levels of myocardial cells. Digitalis, a drug extracted from the foxglove plant, was one of the first drugs used for this purpose.

> 1 *What processes are responsible for ventricular filling?*
> 2 *What is the Frank-Starling Law? What property of heart muscle is responsible for it?*
> 3 *What is congestive heart failure?*

Determinants of Cardiac Output

The volume of blood pumped by a single ventricle per minute is the **cardiac output (CO).** It is determined by multiplying the stroke volume (SV) of that ventricle by the heart rate (HR):

$$CO = SV \times HR$$

The customary units of cardiac output are liters per minute (L/min). The cardiac output of a resting adult is roughly 5 L/min; the value varies considerably depending on the individual and especially on his or her physiological state. During exercise, cardiac output may increase four to six times; the increases are especially large in trained athletes. Obviously, any factor that alters the stroke volume, the heart rate, or both will affect cardiac output.

The heart rate is determined by the pacemaker cells' integration of the inputs they receive from parasympathetic and sympathetic postganglionic neurons (see Table 13-3). The increases in heart rate that accompany exercise or arousal are the result of a simultaneous decrease in parasympathetic input and increase in sympathetic input, together with an increase in the level of circulating epinephrine. Thus the heart rate is controlled by the autonomic nervous system.

The stroke volume has two major determinants: (1) sympathetic inputs to the heart, and (2) the end-diastolic ventricular volume (Figure 13-23). The end-diastolic volume is determined in turn by two variables: (1) the time available for filling, and (2) the central venous pressure. The positive effect of increased heart rate on cardiac output might seem to be offset by a decrease in the time available for

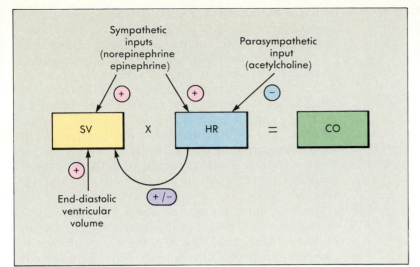

FIGURE 13-23
Major factors that determine cardiac output. The effect of heart rate on stroke volume is positive at moderate-to-high heart rates but becomes negative at the highest rates.

FIGURE 13-24

Changes in heart rate and stroke volume (right vertical scale) and their product, cardiac output (left vertical scale), as work load (horizontal axis) is increased during exercise. Note that heart rate increases continuously as the work load is increased, but stroke volume first becomes stable, then decreases as the heart rate becomes very high. At the highest work loads, cardiac output is limited by the decreasing stroke volume.

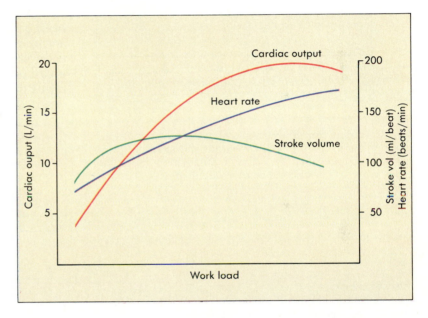

ventricular filling, and very high heart rates actually do decrease stroke volume. However, one of the effects of sympathetic activation of the myocardium is to make each systole shorter. This effect protects diastolic filling time when the heart rate is increased by sympathetic input. When combined with the increased force of contraction of the rapidly beating heart, this effect allows the stroke volume to remain almost independent of heart rate until rates as high as 160 to 180 beats/min are reached (Figure 13-24).

Central venous pressure, the second variable important for end-diastolic volume, is affected by the **venous return**—the rate at which blood returns to the atria from the vasculature. The importance of the Frank-Starling Law is that, by its autoregulation of stroke volume, the heart automatically adjusts the cardiac output to equal the venous return. The same mechanism also ensures that output

of the right and left ventricles is closely matched. The Frank-Starling Law could be rephrased as, "The heart pumps out the volume delivered to it by the veins." The relationship between end-diastolic volume and stroke volume can be translated into terms of cardiac output and central venous pressure. Central venous pressure is proportional to end-diastolic volume; cardiac output is proportional to stroke volume, assuming that the heart rate is constant. The resulting plot is called the **cardiac function curve** (Figure 13-25). The axes differ, but the shape of the curve, like that of the Frank-Starling curve, reflects the length-tension relationship of cardiac muscle fibers. Just as for the Frank-Starling curve (see Figure 13-22), sympathetic stimulation raises the cardiac function curve; weakening of the heart by disease lowers it.

Because the left ventricle must work against the

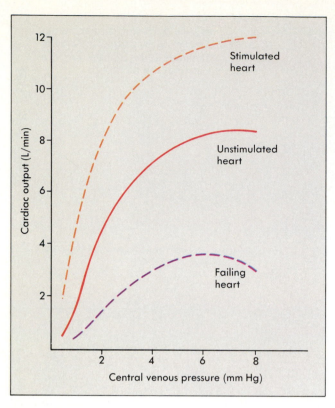

FIGURE 13-25

A The cardiac function curve, a statement of the Frank-Starling Law in terms of cardiac output and left atrial (central venous) pressure. Sympathetic stimulation raises the curve, whereas decreases in contractility depress it.

pressure of blood in the aorta to eject blood, arterial blood pressure might be expected to be a major determinant of stroke volume and thus of cardiac output. However, data from the heart-lung preparation show that arterial pressure has a relatively minor effect on cardiac output. In experiments in which the mean arterial pressure is progressively increased while the central venous pressure and heart rate are constant, the cardiac output of the isolated heart-lung preparation declines relatively slowly as the mean arterial pressure is increased until values considerably higher than the normal range are reached (Figure 13-26).

In the normal range of arterial pressures, the heart obeys the Frank-Starling Law almost independently of how much work it must perform to accomplish the task. Consequently, arterial pressure is a relatively minor determinant of stroke volume in the healthy heart. Pumping fails suddenly when the arterial pressure approaches a value so great that the heart can no longer muster enough force to open the aortic valve. Changes in the contractility of the ventricle shift the relationship between mean arterial pressure and cardiac output: sympathetic input raises the cardiac function curve; weakening of the ventricle by disease lowers it (see Figure 13-26).

1 State the equation that defines cardiac output. What are the major factors that affect the stroke volume?
2 In the cardiac function curve, what parameter corresponds to end-diastolic volume? What corresponds to stroke volume?
3 What is the effect of arterial pressure on cardiac output?

FIGURE 13-26

The relationship between cardiac output and mean arterial pressure for three states of contractility of an isolated heart: an unstimulated normal heart, a heart made hypereffective by sympathetic stimulation, and a heart made hypoeffective by disease. An important feature of these curves is that cardiac output is stable over a range which includes the normal values of mean arterial pressure (shaded area).

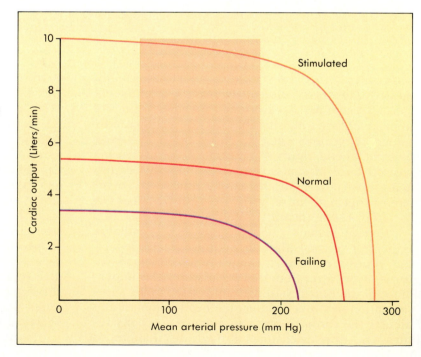

THE CARDIOVASCULAR SYSTEM

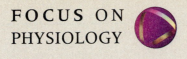
Coronary Blood Flow and Cardiac Metabolism

Blood flow through the coronary arteries (Figure 13-A) varies greatly during the heart cycle, particularly in the left ventricle. When isovolumetric contraction begins, the tension in the walls of the ventricle is very high, compressing the coronary blood vessels and preventing blood flow. When ejection begins, the coronary blood flow increases and decreases with the rise and fall of the aortic pressure. Early in diastole, the combination of a high aortic pressure with the relaxation of the ventricle allows coronary blood flow to reach its maximum for the cycle, decreasing as the aortic pressure approaches the diastolic value. The flow pattern in the vessels of the right ventricle is similar, but the changes are less exaggerated because the changes in wall tension of the right ventricle are less dramatic.

A normal heart consumes 8 to 10 ml of oxygen/min/100 grams of cardiac tissue under resting conditions. The oxygen consumption increases with increases in myocardial work. Myocardial work is affected by aortic blood pressure, ventricular diastolic filling, and changes in the heart rate and contractility. The resting heart extracts about 75% of the oxygen carried by the coronary arterial blood, so the heart cannot accommodate increased metabolic need by extracting more oxygen. Instead, the coronary blood flow must increase when the heart's workload increases. How does this occur?

The oxygen content of the blood in the capillaries of the heart is determined by the relative rates of oxygen transport in the blood and oxygen consumption by the heart. When oxygen content decreases, the local concentration of local metabolic factors increases. One of these factors is **adenosine.** Adenosine dilates coronary arterioles, increasing coronary blood flow. Under some conditions, such as exercise, the internal autoregulation of coronary blood flow is augmented by sympathetic stimulation. In the coronary arterioles, the dominant sympathetic receptors are β_2-receptors, which mediate dilation.

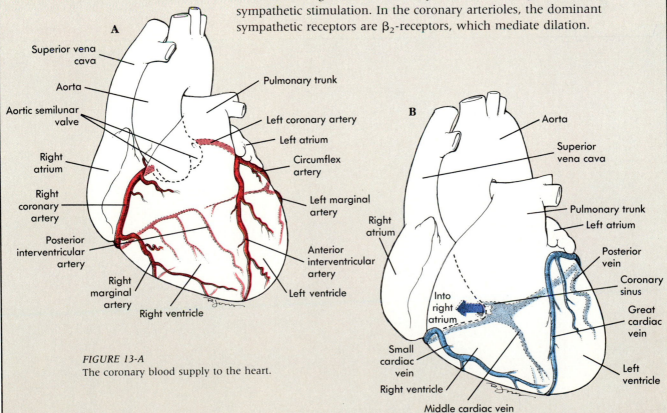

FIGURE 13-A
The coronary blood supply to the heart.

SUMMARY

1. In a complete circuit of the circulatory system, blood passes alternately through the **pulmonary loop** and the **systemic loop.** Blood pumped into the pulmonary loop by the right heart returns from the lungs to the left heart, which pumps it into the systemic loop.

2. The heart consists of right and left **atria** and right and left **ventricles.** The **atrioventricular valves** prevent backflow of blood from ventricles to atria; the **pulmonary** and **aortic** valves prevent backflow of arterial blood into the ventricles.

3. The three types of cardiac fibers are:
   a. **Contracting,** or myocardial, fibers, which generate the force of systole.
   b. **Pacemakers** located in the SA and AV nodes, which discharge action potentials at regular intervals.
   c. **Conducting** fibers specialized for impulse conduction, which carry excitation between nodes and from the AV node to the ventricle.

4. The rate at which pacemakers discharge action potentials is increased by sympathetic (adrenergic) input and decreased by parasympathetic (cholinergic) input. These effects are caused by increase and decrease, respectively, in the rate of the closing of K^+ channels following each action potential.

5. Cardiac fibers possess "slow" channels for inward current of Na^+ and Ca^{++}, which are responsible for the **plateau phase** of the cardiac action potential. Ca^{++} that enters during the plateau phase triggers release of Ca^{++} from the sarcoplasmic reticulum. In contractile fibers, the Ca^{++} concentration attained during the plateau phase determines the amount of force the contractile machinery can develop (contractility). Adrenergic (sympathetic) inputs increase the contractility.

6. The pump cycle of the heart consists of five phases:
   a. **Passive filling.**
   b. **Atrial contraction** (late diastole).
   c. **Isovolumetric ventricular contraction.**
   d. **Ejection.**
   e. **Isovolumetric ventricular relaxation.**

7. The length-tension relation of heart muscle is such that increased stretch by increased filling (increased end-diastolic volume) increases the force of systole and thus the stroke volume. This mechanism allows the heart to autoregulate its stroke volume to make cardiac output equal to venous return over a wide range of end-diastolic volumes (the **Frank-Starling Law** of the Heart).

8. The **cardiac output** is equal to the **stroke volume** multiplied by the **heart rate.** The major determinants of cardiac output are:
   a. The effects of adrenergic and cholinergic inputs to pacemakers.
   b. The **contractility** of the heart, determined largely by sympathetic input to contractile fibers.
   c. The **end-diastolic volume,** determined by the central venous pressure and the filling time between beats.

1. How is the force of contraction controlled in the heart? Increases in plasma Ca^{++} concentration increase the strength of contraction of cardiac muscle but not of skeletal muscle. Why is there a difference?

2. Describe the changes in aortic pressure, ventricular pressure, and atrial pressure over a single heart cycle. How do these pressure changes relate to the opening and closing of the cardiac valves?

3. Arrange the following events in the order in which they occur (some may overlap).

   QRS complex   Isovolumetric contraction
   Atrial contraction   First heart sound
   Atrial repolarization   Ejection
   T wave   P wave
   Isovolumetric   Second heart sound
   relaxation

4. What is happening in the atria during the phase of isovolumetric ventricular contraction? Is ventricular pressure greater than or less than atrial pressure during this phase? Is it greater than or less than aortic pressure during this phase?

5. What properties of cardiac muscle membrane give rise to the plateau of the cardiac action potential? What is the functional significance of the plateau?

6. What events or processes in the electrical activation of the heart correspond to each of the following events of the ECG?

   P wave   S-T interval
   QRS complex   P-R interval
   T wave

7. What determines which population of spontaneously active cells will serve as the actual pacemakers for the heart cycle?

8. Define each of the following terms:

   Arrhythmia   Bradycardia
   AV block   Ectopic pacemaker
   Tachycardia

9. Diagram the pathway of electrical conduction of the heart. What is the role of gap junctions in spreading excitation? What is reentry? What property of heart cells normally prevents reentry?

10. What sequence of events occurs within myocardial cells as a result of activation of the β_1 receptors? What changes in heart performance result?

11. State the Frank-Starling Law of the Heart. What property of cardiac muscle is responsible for this law?

12. What factors affect the cardiac output?

● *SUGGESTED READING*

BEAN, B.P., M.C. NOWYCKY, and R.W. TSIEN: **Beta-Adrenergic Modulation of Calcium Channels in Frog Ventricular Heart Cells,** *Nature,* 1984, vol. 307, p. 371. Shows how norepinephrine increases calcium influx in cardiac muscle, thus augmenting its inherent contractility.

BRANDT, R.: **These Pictures Could Save Your Life,** *Business Week,* July 1987, p. 112. Describes new advances in imaging the cardiovascular system.

ERON, C.: **Take Heart: Ventricular Tachycardia Cure,** *Science News,* August 1988, p. 133. Describes various drugs used to treat cardiac arrhythmias and their mechanism of action.

GRAHAM, J.: **A Guide to Stress Tests,** *New York Times Magazine,* March 1987, p. 40. Describes the use of stress tests, such as treadmill exercise, in the diagnosis and treatment of coronary artery disease.

KATONA, P.G., M. MCLEAN, D.H. DIGHTON, and A. GUZ: **Sympathetic and Parasympathetic Cardiac Control in Athletes and Nonathletes at Rest,** *Journal of Applied Physiology,* 1986, vol. 52, p. 1652. Discusses how cardiovascular performance in exercise is regulated and the effect of athletic training on the cardiovascular system.

RALOFF, J.: **Sounding Out 3-D Interiors,** *Science News,* May 1983, p. 314. Describes the technique of echocardiography and how ultrasonic imaging can improve the diagnosis of heart disease.

ROBINSON, T.F., S.M. FACTOR, and E.H. SONNENBLICK: **The Heart as a Suction Pump,** *Scientific American,* June 1986, p. 84. Readable discussion of the cellular mechanisms underlying the Frank-Starling Law of the Heart.

CHAPTER 14

Blood and the Vascular System

On completing this chapter you will be able to:

- Identify the components of blood and define hematocrit.
- Describe the process of blood clotting (hemostasis), noting the differences between the intrinsic and extrinsic pathways.
- Appreciate the role of the arteries in maintaining blood flow into the vasculature during diastole.
- Understand why veins are referred to as capacitance elements of the circulation.
- Distinguish between linear velocity of flow, average flow velocity, and volume flow rate.
- Describe how vessel length, fluid viscosity, and vessel radius affect vessel flow resistance.
- Understand the determinants of total peripheral resistance.
- Describe the effects of extrinsic autonomic control on vascular smooth muscle.
- Describe the factors that determine capillary fluid exchange.
- Understand the anatomy and circulation of lymph.

*T*he blood vessels are sometimes referred to as a "vascular tree." The metaphor is apt. Major arteries branch from the trunk of the aorta to course into each major body region—the head, arms, abdomen, and legs. The left heart pours its output into the aorta, the trunk of the arterial tree. Further branching forms successively smaller arteries. The final branches are arterioles, the smallest arteries, from which spring a profusion of capillaries, the twigs of the tree. At each successive branch the blood flow is further subdivided. Ultimately, a tiny fraction of the cardiac output passes through each capillary, bringing oxygen, nutrients, and hormones to each cell and removing carbon dioxide and metabolic end products. The veins form a second, reversed, tree as capillaries merge to form first venules and then veins, finally joining the trunk of the vena cava. Thus the vasculature can be divided into a distributing system (the arteries and arterioles), an exchange system (the capillaries), and a collecting system (the venules and veins). Altogether, there are perhaps 60,000 miles of blood vessels in an adult.

Circulation of blood is a relatively modern concept. Physicians in earlier times held ideas about the function of the cardiovascular system that seem quite bizarre now. At first, arteries were thought to contain air rather than blood because the arteries of cadavers sometimes filled with air when cut during dissection. By the time of the Greek physician Galen (200 AD), blood was thought to ebb and flow from the heart and to pass through imagined pores in the septum of the heart. The English physician William Harvey introduced the hypothesis of blood circulation in 1628 on the basis of careful observation of the heart, arteries, and veins. Harvey's ideas revolutionized physiology and were all the more remarkable because in his time the smallest blood vessels, capillaries, had not yet been seen. When the microscope was invented and used by Marcello Malpighi in 1661 to examine a frog's lung, it was finally possible for physiologists to observe the movement of blood through capillaries.

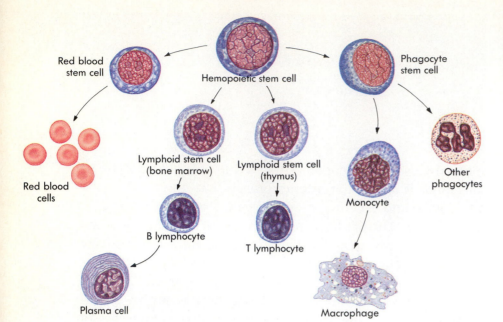

FIGURE 14-1
An illustration showing how
erythrocytes and several types of white
blood cells are all derived from a single
stem cell in the bone marrow.

Red blood
stem cell

Hemopoietic stem cell

Phagocyte
stem cell

Red blood
cells

Lymphoid stem cell
(bone marrow)

Lymphoid stem cell
(thymus)

Monocyte

Other
phagocytes

B lymphocyte

T lymphocyte

Plasma cell

Macrophage

BLOOD—A LIQUID TISSUE
Blood Composition

Blood consists of several types of cells suspended in
a liquid called **plasma.** All of the cellular elements
of blood have a common ancestor cell type, the hemopoietic stem cell, found in the bone marrow
(Figure 14-1).

If a blood-filled capillary tube is centrifuged, the
plasma and cells are separated; the cells are packed
in the bottom, whereas the less dense plasma occupies the top of the tube (Figure 14-2). A thin **buffy
coat** of **white cells (leukocytes)** and **platelets
(thrombocytes)** separates the packed **red cells
(erythrocytes)** from the clear, straw-colored plasma.
The **hematocrit,** the percentage of the total blood
volume that is red cells, can be calculated by dividing the length of the column that is filled with red
cells by the total length of the column of blood and
multiplying the resulting fraction by 100. Normal
hematocrit values range from 40% to 50% in men
and 35% to 45% in women. These values correspond to cell densities of 5.1 to 5.8 million cells/mm^3
in men and 4.3 to 5.2 million cells/mm^3 in women
(Table 14-1).

Plasma contains inorganic ions, many organic
compounds produced or consumed in metabolism,
and **plasma proteins** (see Table 14-1). Plasma proteins include **albumins,** which serve as transport
proteins for lipids and steroid hormones and are
important in body fluid balance; **immunoglobulins,** which mediate specific immunity; and
fibrinogen and several other plasma proteins involved in the formation of blood clots.

Red cells (Figure 14-3, *A*) are disks that are 8.1
μm in diameter and about 2.7 μm thick with a concave center. About 25% of their cytoplasmic volume

is taken up by hemoglobin, an iron-containing protein that accounts for the red color of blood (see
Chapter 17, p. 432). Hemoglobin is almost entirely
responsible for O_2 transport in blood and plays an
important role in CO_2 transport and regulation of
blood acidity. Red cells lack a nucleus and cannot
undergo mitosis. They have a lifetime of about 120

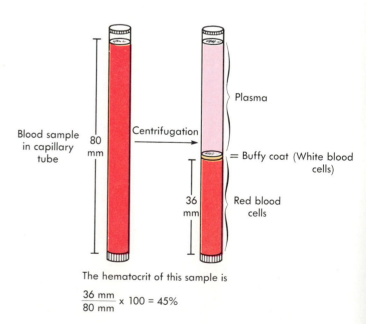

Blood sample
in capillary
tube

80
mm

Centrifugation

Plasma

= Buffy coat (White blood
cells)

36
mm

Red blood
cells

The hematocrit of this sample is

$$\frac{36 \text{ mm}}{80 \text{ mm}} \times 100 = 45\%$$

FIGURE 14-2
Separation of blood into cellular and fluid (plasma) components by centrifugation. A blood sample is placed in a glass
capillary tube treated to prevent clotting. After centrifugation, the red cells are packed at the bottom of the tube with
the white cells above them; the plasma remains at the top of
the tube. The hematocrit level of this sample is calculated as
shown.

TABLE 14-1   Elements of the Blood

Parameter	Normal range	Units
Cellular elements		
Hematocrit	40-54	
Hemoglobin	14-18	gm/dl
Red blood cells	4.6-6.2	Million/mm³
(erythrocytes)		
White blood cells	5000-10,000	Cells/mm³
(leukocytes)		
Platelets	0.2-0.4	Million/mm³
Plasma components		
Water	91.5% of plasma volume	
Proteins	7.0% of plasma volume	
Albumin	3.2-5.6	gm/dl
Globulins	2.3-3.5	gm/dl
Fibrinogen	0.2-0.4	gm/dl
Other solutes		
Bicarbonate	21-27	mEq/L
Ca⁺⁺	2.1-2.6	mEq/L
Chloride	95-103	mEq/L
Cholesterol	150-250	mg/dl
Glucose	65-100	mg/dl
Iron	60-150	μg/dl
Magnesium	1.5-2.6	mEq/L
Phosphate	1.8-2.6	mEq/L
Potassium	4.0-4.8	mEq/L
Sodium	136-142	mEq/L
Sulfate	0.2-1.3	mEq/L
Urea	8-20	mg/dl
Uric acid	2.1-7.6	mg/dl

days in the circulation. Aging red cells are removed and destroyed by macrophages in the spleen and must be continuously replaced by formation of new ones from red blood stem cells (erythroblasts) in the bone marrow, a process called **erythropoiesis** (see Figure 14-1).

About 200 billion red blood cells are produced each day in an adult, equivalent to the number of red cells in about 100 ml of whole blood, so that a pint of blood can be replaced in less than a week. Erythropoiesis is regulated by the hormone, erythropoietin, which is released by the kidney in response to decreases in arterial oxygen content.

Platelets are disks that are 2 to 5 μm in diameter formed by the fragmentation of **megakaryocytes** in bone marrow. Like red cells, they lack nuclei and cannot reproduce. Their lifetime in the circulation is perhaps as little as 10 days.

In adults, formation of red blood cells, platelets, and some types of white blood cells occurs in bone marrow (Table 14-2). Formation of the types of white blood cells involved in the specific immune response occurs in the lymph nodes. White blood cells (Figure 14-3, *B*) are agents of the immune system; their origin and function are discussed in detail in Chapter 24.

Hemostasis

After a person suffers a cut or scrape, blood may at first flow readily from damaged blood vessels, but after several minutes the flow slows or stops. The prevention of further blood loss is the result of several processes, collectively called **hemostasis**. He-

A

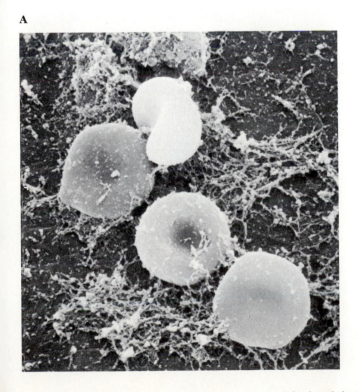

B

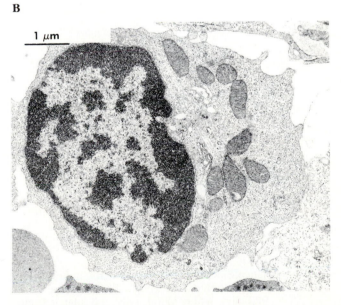

1 μm

FIGURE 14-3

A   A scanning electron micrograph of red blood cells.

B   A transmission electron micrograph of a white blood cell.

Life stage	Site of blood cell production
Fetus	Liver and spleen
Adolescent	Marrow of sternum, ribs, vertebrae, skull, pelvis, femur, and tibia
Adult	Marrow of sternum, ribs, skull, vertebrae, and pelvis

mostasis involves the action of substances produced by injured blood vessels, blood platelets, and a family of plasma proteins called clotting factors.

There are three stages of hemostasis: (1) the initial sealing of a damaged vessel by a temporary **platelet plug,** (2) the formation of a **clot**—a network of threadlike **fibrin** molecules that forms a more lasting patch over the break in the blood vessel, and (3) the dissolution of the clot after vessel repair.

Formation of a platelet plug (Figure 14-4, *A*) requires a change in the behavior of platelets so that they are able to bind to the damaged surface of the vessel and to each other. The transformed platelets are called **sticky platelets.** Platelet sticking is triggered by exposure to an abnormal surface. Undamaged vessels carry a net positive charge on their surfaces, but injured cells and the structural protein collagen have a negative surface charge, which causes platelets that contact it to bind. Factors essential for this initial step are secreted by platelets and the endothelial cells that line vessels.

Binding to the damaged surface causes the platelets to release **arachidonic acid,** a fatty acid. In the plasma, some of the arachidonic acid is converted to **thromboxane** A_2 (a prostaglandin). Thromboxane A_2 attracts additional platelets to the damage site and causes the damaged vessel to constrict. It also causes platelets to release ADP. The ADP is a platelet-to-platelet signal that causes platelets to flatten, send out processes, and expose receptors for fibrin on their cell surfaces. This process is an example of positive feedback because a few sticking platelets can cause many more to become sticky.

The next event in the formation of a blood clot is the activation of **Hageman factor (factor XII)** by contact with a damaged area of a blood vessel (Figure 14-4, *B*). Active Hageman factor triggers a series of reactions involving clotting factors (numbered I to XII) and, in many cases, Ca^{++} as an essential cofactor. Sticky platelets participate in the intrinsic pathway for blood clotting by releasing **phospholipid PF3,** a cofactor necessary for the activation of factor X. The fibrin formed in the clotting

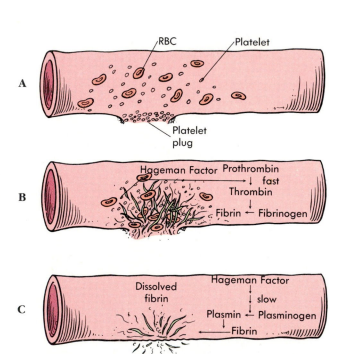

FIGURE 14-4
The sequence of events in formation and subsequent dissolution of a blood clot.
A   A platelet plug is formed.
B   Damage activates Hageman factor, which initiates a cascade of reactions *(arrows)* that ends with formation of a blood clot. Red and white blood cells and platelets are trapped among the fibrin strands of the clot.
C   The clot is eventually dissolved by plasmin which is produced by a much slower reaction sequence also triggered by Hageman factor.

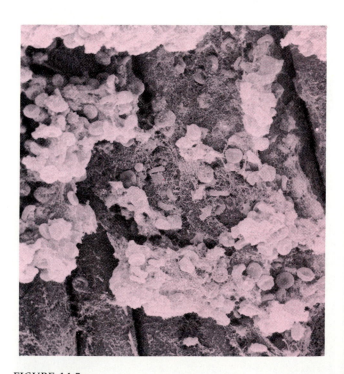

FIGURE 14-5
A scanning electron micrograph of a blood clot, showing trapped blood cells.

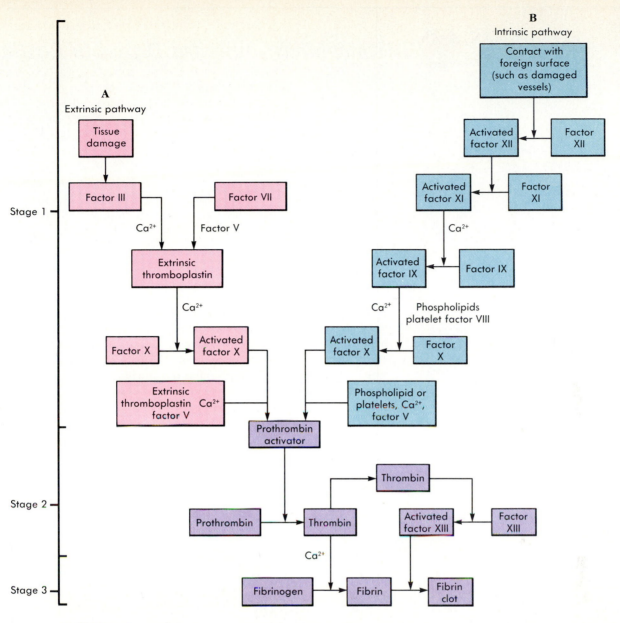

FIGURE 14-6
A   The extrinsic pathway, which is stimulated by tissue damage.
B   The intrinsic pathway for clot formation, which mediates clot formation within blood vessels. Both these pathways cause the production of prothrombin activator at the end of stage 1 of coagulation. Stage 2 involves the production of thrombin, which then causes the polymerization of fibrinogen into fibrin in stage 3.

process glues the platelets to each other and traps them within the clot. This stage of the clotting cascade ends with the formation of prothrombin activator, which causes the conversion of **prothrombin** to **thrombin** (see Figure 14-4, *B*). In the next step thrombin mediates a process in which many molecules of the soluble plasma protein, **fibrinogen**, interact (polymerize) to form long, insoluble strands of fibrin. Fibrin strands form a tight, meshlike lattice that binds the edges of the injured vessel together and traps platelets, erythrocytes, and leukocytes (Figure 14-5). Because several of the clotting factors require Ca^{++}, adequate calcium levels are required for the clotting process to be normal.

The details of the clotting cascade are illustrated in Figure 14-6. Clotting may be triggered by an **extrinsic pathway**, which is initiated by chemical factors released from damaged tissue (see Figure 14-6, *A*), or by an **intrinsic pathway** which requires only components present in blood (see Figure 14-6, *B*). Damaged tissues initiate the extrinsic pathway by release of **tissue thromboplastin**, or **factor III**, a complex mixture of lipoproteins. Factor III in the presence of Ca^{++} and activated factor V forms a complex with **factor VII** called **extrinsic thromboplastin**, which activates **factor X**. Stage 1 of the extrinsic pathway ends with the production of prothrombin activator.

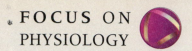
Hemophilias: Inherited Defects in Clotting Factors

There are several inherited diseases of blood clotting, all of which involve defective genes for clotting factors. Hemophilia A, the most common form, is the result of the presence of a defective factor VIII protein; a less common form, hemophilia B, results from a defective factor IX protein. In severe cases of the disease, spontaneous bleeding into the joints, skin, and soft tissues is frequent. Without effective treatment, death usually occurs at an early age from complications resulting from internal bleeding. Bleeding episodes can be treated by transfusion of plasma fractions rich in clotting factors. This treatment only lessens the bleeding, and patients ultimately develop an immune response to the proteins of foreign plasma, so its use has to be reserved for the severest episodes.

Hemophilia almost always affects males but is inherited through maternal lines; such diseases are said to be sex linked. The basis for sex linkage lies in the genetics of sex determination. Human cells possess 23 pairs of chromosomes. One member of each pair is inherited from each parent. Generally the pairs are similar in appearance, but in the case of the sex chromosomes the pair is similar only in females, who carry two X chromosomes. In males the pair consists of an X chromosome and a much smaller Y chromosome, which carries few genes. The genes for factors VIII and IX are carried on the X chromosome. Defective genes for these factors are rare in the population. Possession of only one

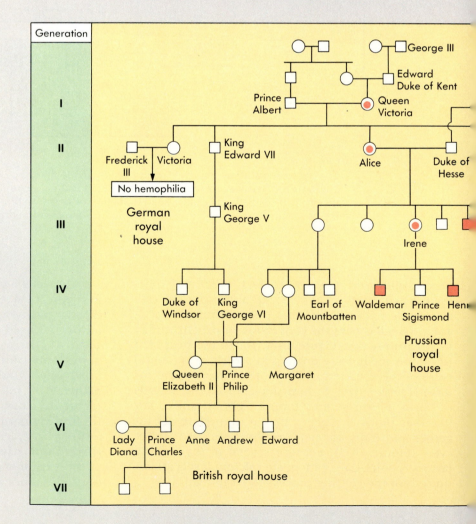

nondefective gene is enough to supply the needs of the body for normal factors, and a woman rarely inherits a defective gene on both the X chromosome from her mother and that from her father. Males have no such backup. If a male's single X chromosome, inherited from his mother, bears a defective gene, only the defective amount of factor is made. A mother who carries one X chromosome with the defective gene has a 50:50 chance of transmitting the disease to a male child and a 50:50 chance of passing her carrier status to a female child.

The descendants of Queen Victoria present an interesting example of inheritance of hemophilia A. The defective gene in this family apparently arose as a mutation in one of Queen Victoria's parents or at an early embryonic stage in the queen herself. Relatively complete pedigrees, termed family histories, have been kept for royal families, and royal medical problems receive close attention. Consequently we know that, of Queen Victoria's five female children, two were carriers of the gene. Their marriages to other European royalty transmitted the gene to several royal families (Figure 14-A). Of Victoria's four male children, one affected son survived long enough to pass the trait to descendants; an unaffected son became King Edward VII, establishing the current line of succession, into which the defective gene did not pass.

Women who have a family history of hemophilia may want to know whether they carry the defective gene. Usually this can be determined by testing the woman's plasma for the defective clotting factor. Genetic counseling can help those who carry the defective gene to decide whether to bear children. Cures for such inherited errors of protein structure are a major promise of genetic engineering.

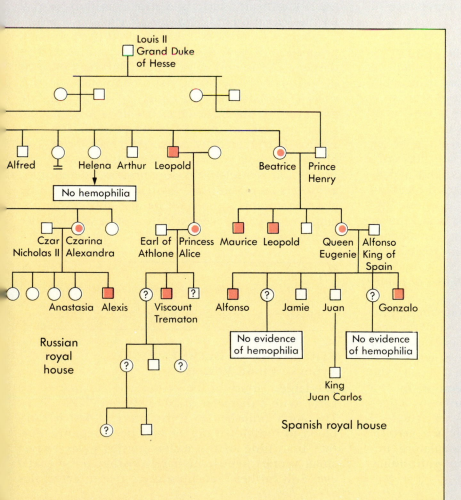

FIGURE 14-A
Family tree (pedigree) of descendants of Queen Victoria showing the sex-linked inheritance of hemophilia A. The circles represent female members of the family; squares represent males. Each branch in the family tree represents one generation of offspring. Filled squares indicate males that have hemophilia. Circles with a dot indicate women assumed to be carriers of the defective factor.

TABLE 14-3   Scale of the Circulatory System

Vessel type	Pressure* (mm Hg)	Length (mm)	Typical diameter	Number in parallel	Area (cm²)	Relative resistance (%)
Aorta	100	500	25 mm	1†	2.5	4
Arteries	98	10	2 mm	600	5	5
Arterioles	90	3	70 μm	5×10^7	40	40
Capillaries	35	1	5 μm	10^9	1700	25
Venules	15	10	30 μm	10^8	375	5
Venae cavae	5	300	13 mm	2	10	2

*The pressure is given at the start of each vessel type.
†The aorta branches after leaving the heart to give rise to the carotid arteries, subclavian arteries, and thoracic aorta.

The intrinsic coagulation pathway is so named because it is initiated by the activation and aggregation of platelets plus the activation of a plasma coagulation factor, rather than by tissue factors. In addition to damaged blood vessel walls, various artificial surfaces such as glass are able to trigger the intrinsic pathway. The intrinsic and extrinsic pathways have a common end point, the production of prothrombin activator (see Figure 14-6).

As in other regulatory system cascades (see Chapter 5, p. 102), the function of the cascade of reactions that leads to clotting is to greatly amplify the signal set in motion by sticking platelets or tissue damage. This is an example of the importance of positive feedback in hemostasis.

The clotting mechanism must be reliable because either failure of clot formation at an injury or spontaneous clot formation at an inappropriate site can result in death. The great number of reaction steps allows for much control over the clotting process, but it also provides many points at which the process could be arrested. Failure of clotting can result from a lack of or defect in any one of the factors in the cascade. For example, vitamin K is necessary for synthesis of some clotting factors, and a deficiency of this vitamin results in prolonged bleeding times after injuries. **Hemophilia** is a genetic disease in which a single clotting factor is lacking or insufficient (see essay on p. 352). Before modern methods of treating the disease were developed, most hemophiliacs died in childhood, typically from uncontrolled bleeding from minor internal or external injuries.

A clot that forms at an inappropriate site within vessels is called a **thrombus,** and can block the vessel in which it forms. Sometimes clots break loose from their sites of formation and travel within the circulatory system to block a smaller vessel. The general name for such a block is an **embolism.** If a clot forms in a coronary artery or a thrombus lodges in a coronary artery, the result is a **heart attack.** If either occurs in the brain, the result is a **stroke.** Clots that form in the systemic vasculature frequently lodge in the lungs, forming **pulmonary emboli.**

Normal vessels antagonize thromboxane A_2 by secreting **prostacyclin,** a prostaglandin that acts as an **anticoagulant.** Some tissues, such as the lungs, produce the anticoagulant protein, **heparin,** which helps prevent embolisms. Heparin is extracted from the tissues of slaughterhouse animals and used medically to prevent clotting. Aspirin exerts several effects that are related to its ability to inhibit prostaglandin synthesis. One of its effects is to inhibit clot formation; this is the result of inhibition of thromboxane formation. Some evidence suggests that continuous use of a low dose of aspirin could reduce the incidence of embolism and thus decrease the risk of heart attacks and strokes in those at risk for the formation of inappropriate clots.

Once a blood vessel has repaired itself through production of new endothelial cells and connective tissue, a clot is no longer needed and may even be dangerous because of the risk of embolism. In this stage of hemostasis (see Figure 14-4, C), clots are dissolved by **plasmin.** This enzyme is formed from **plasminogen,** an inactive blood protein. The conversion is triggered by active Hageman factor, one of the initiating factors in the clotting cascade. The conversion is a relatively slow process, so appropriate levels of plasmin are achieved some time after clot formation is complete. Clot dissolution is an example of negative feedback in a process otherwise dominated by positive feedback.

---

1 What is the hematocrit?
2 What are the respective roles of red cells? white cells? platelets?
3 What are the three stages of hemostasis?

STRUCTURE OF BLOOD VESSELS
Types of Blood Vessels

The largest artery of the systemic circulation is the aorta, with a diameter of 20 to 30 mm (Table 14-3). The aorta branches to form arteries, which carry blood to individual organs and body regions. Further branching of the major arteries gives rise to small arteries, which are barely visible to the naked eye. Still further branching gives rise to **arterioles**

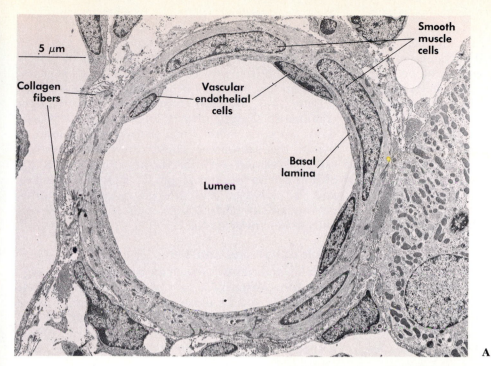

5 μm

Collagen
fibers

Vascular
endothelial
cells

Smooth
muscle
cells

Basal
lamina

Lumen

A

FIGURE 14-7

A Transmission electron micrograph of an
arteriole.
B Transmission electron micrograph of
several capillaries, two of which contain red
blood cells *(arrows).*

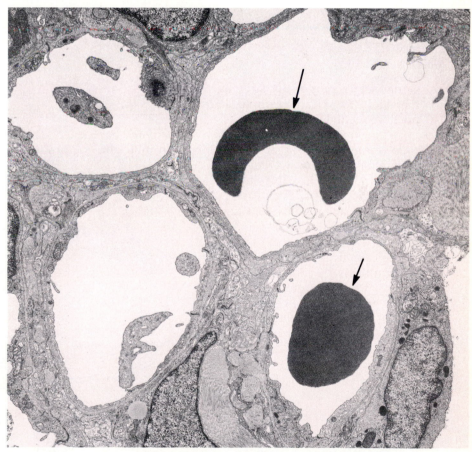

B

(Figure 14-7, *A*), whose diameter of about 70 μm
places them on the microscopic scale (see Table 14-
3). Each arteriole usually branches into several
metarterioles of 10 to 20 μm in diameter, the walls
of which are wrapped with smooth muscle at intervals along their length. Metarterioles branch to form
capillaries (Figure 14-7, *B*), thin-walled vessels
that are 5 to 10 μm in diameter, just wide enough
for red cells to pass through in single file. At the origin of each capillary is a cufflike **precapillary
sphincter** consisting of a single smooth muscle fiber. Precapillary sphincters open and close their
capillaries at intervals in response to changes in
their immediate environment.

The capillaries are the sites of exchange of nutrients and wastes between tissues and blood. The capillaries arising from a single arteriole typically form a network called a **capillary bed** that serves a discrete area of tissue. Capillary density varies greatly from one tissue to another and is usually related to the maximum metabolic needs of the tissue concerned. Capillary density is high in skeletal and cardiac muscle, glands, and the CNS, but very low in cartilage and subcutaneous tissue.

Capillaries join together to give rise to **veins.** The smallest veins are called **venules** and have a diameter of about 20 to 30 μm. The veins draining the lower thorax, abdomen, and legs join to form the inferior vena cava; those draining the upper thorax, arms and head join in the superior vena cava. The venae cavae are about 13 mm in diameter. Veins in parts of the body subject to compression by the skeletal muscles also contain one-way valves.

Generally, blood must pass through capillaries to go from arteries to veins, but two types of vessels allow blood to bypass capillary beds. In some cases, metarterioles make connections directly to venules (Figure 14-8, *A*). Such connections are particularly common in skeletal muscle. In some capillary beds there are even more direct pathways for blood to bypass the capillaries. These direct pathways are called **arteriolar-venular shunts** (Figure 14-8, *B*). Shunts can be found between arterioles and venules and also between small arteries and veins. Arteriolar-venular shunts are a prominent feature of the cutaneous microcirculation.

Walls of Arteries and Veins

Arteries and veins, viewed in cross section in Figure 14-9, consist of a coat of longitudinal smooth muscle, the **tunica externa,** surrounding an inner coat of circular smooth muscle, the **tunica media.** The inner surface is lined with a single layer of endothelial cells, the **tunica intima.** The arteries and

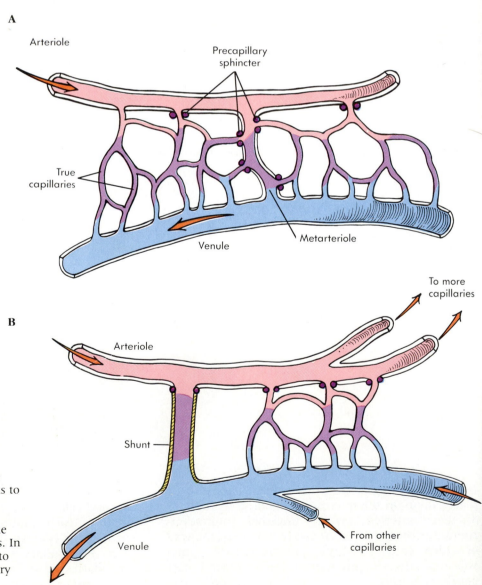

A

Arteriole

Precapillary
sphincter

True
capillaries

Venule

Metarteriole

B

Arteriole

To more
capillaries

Shunt

From other
capillaries

Venule

FIGURE 14-8

A   The arrangement of the microcirculation in muscle. Note the presence of metarteriolar connections to venules.
B   The arrangement of the microcirculation in the skin. Note the presence of arteriolar-venular shunts. In both muscle and skin, the entrance to capillaries is controlled by precapillary sphincters.

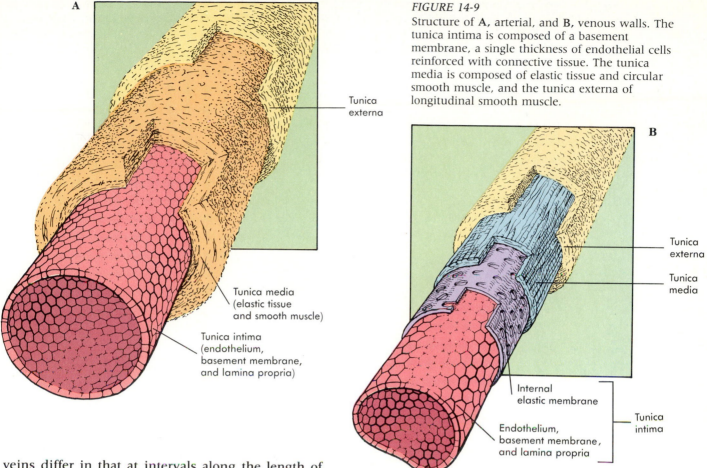

A

Tunica
externa

Tunica media
(elastic tissue
and smooth muscle)

Tunica intima
(endothelium,
basement membrane,
and lamina propria)

FIGURE 14-9
Structure of **A**, arterial, and **B**, venous walls. The
tunica intima is composed of a basement
membrane, a single thickness of endothelial cells
reinforced with connective tissue. The tunica
media is composed of elastic tissue and circular
smooth muscle, and the tunica externa of
longitudinal smooth muscle.

B

Tunica
externa

Tunica
media

Internal
elastic membrane

Endothelium,
basement membrane,
and lamina propria

Tunica
intima

veins differ in that at intervals along the length of
veins, the tunica intima forms one-way valves that
favor blood flow in the direction of the heart. Also,
the thickness of the smooth muscle of arteries is
considerably greater, and the inner diameter
smaller, than that of corresponding veins.

The connective tissue of the walls of arteries
and veins contains a mixture of the proteins, collagen and elastin. Elastin is easily stretched; collagen
is stiffer. The proportions of the two strongly influence the **compliance,** or ease of stretch, of the vessel wall. The tone, or state of contraction, of the vascular smooth muscle is a second major determinant
of a vessel's compliance. The compliance of the
vessel wall determines the rate at which pressure
inside a vessel increases as the volume of blood
contained in it is increased. A vessel with high
compliance can accept a large volume of blood with
a relatively small increase in pressure; forcing the
same volume of blood into a vessel with low compliance would produce a larger change in pressure.

Arterial walls have a relatively high collagen
content, which makes the arteries considerably less
compliant than the veins (Figure 14-10). The low
compliance of arteries has an important role in
maintaining blood flow through the circulation during diastole. During the ejection phase of the heart
cycle, blood is entering the aorta much more rapidly
than it is able to flow away into the rest of the vas-

culature. The aorta is stretched by the additional
volume, but because its compliance is low, a large
fraction of the energy applied by the heart to the
blood during systole is stored in the stretched wall
of the aorta and large arteries. During diastole, the
elastic recoil of the aorta and large arteries slowly
releases this stored energy, maintaining a relatively
high mean arterial pressure and a relatively constant
flow of blood through the capillaries. Thus the arteries tend to smooth out the pulsatile flow of blood
from the heart (Figure 14-11).

The arteries must be neither too compliant nor
too rigid. The effect of too little compliance is illustrated by the effect of the increase in collagen content of arteries that occurs with age. The arteries of a
70-year-old person have become about half as compliant as those of a young adult. As a result, the
heart of a 70-year-old person does considerably
more work to eject the same stroke volume than
that of a 20-year-old person. The decreasing compliance of arteries is at least partly responsible for the
increase in normal arterial blood pressure that occurs with age. The normal blood pressure for 20-
year-old men is 123/76; that for 70-year-old men is

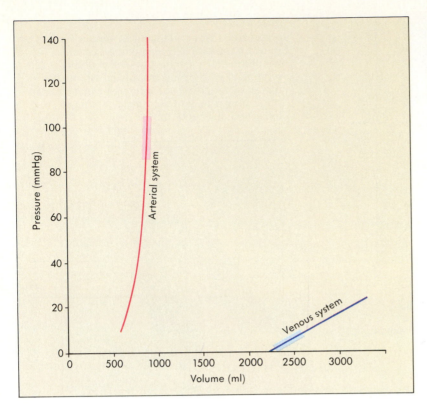

FIGURE 14-10
Pressure-volume curves for the arterial system and the venous system. The colored portions of the curves indicate approximate physiological ranges. Arterial pressure rises rapidly with increased volume, so that at the normal arterial volume of about 700 ml, the mean arterial pressure is almost 100 mm Hg. In the much more compliant venous system, the normal volume of more than 2 liters is held under a central venous pressure of only a few mm Hg.

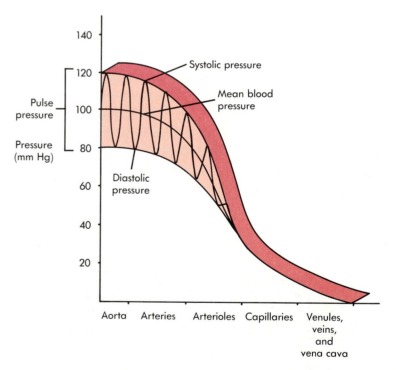

FIGURE 14-11
Blood pressure in each of the major blood vessel types.

145/82. An even larger increase over this age interval is normal for women.

The volume of blood in the veins at any instant is more than half of the total blood volume (Figure 14-12), but the venous pressure is only a few mm Hg. This is possible because the walls of veins contain a high proportion of elastin to collagen and thus are very compliant (see Figure 14-10). Because they contain so much of the blood volume, small changes in venous compliance have dramatic effects on blood pressure. The veins are often called the **capacitance elements** of the circulatory system because their capacity to hold blood is several times greater than that of the arteries.

THE CARDIOVASCULAR SYSTEM

FIGURE 14-12
The relative volumes of blood in different parts of the circulatory system at any instant (values are typical of resting adults).

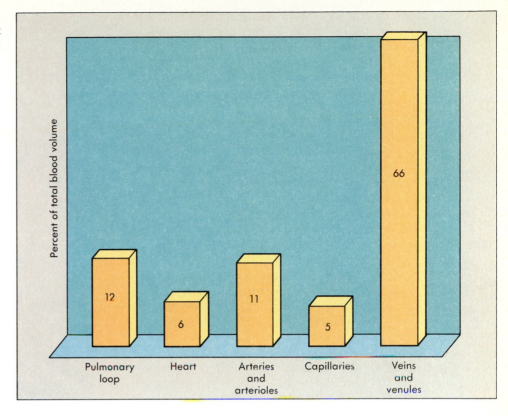

Structure of Capillaries

Capillaries (Figure 14-13; see Figure 14-7, *B*) are not just miniature arteries or veins; they lack the smooth muscle coats of other vessels. Generally speaking, capillaries are leaky to solutes and fluid. The structure of the capillary endothelium, and thus the degree of leakiness, differs significantly in different organs. The capillaries of most organs have closely joined endothelial cells whose intercellular junctions allow water and small hydrophilic solutes to pass readily between plasma and interstitial fluid. Typical capillaries have a low permeability to molecules the size of plasma proteins. The capillaries of most regions of the central nervous system have a low permeability even to small organic solutes. The capillary endothelium of these capillaries constitutes a **blood-brain barrier.** In the kidney, endocrine glands, intestine, and liver, capillaries may have large pores (**fenestrated capillaries;** see Figure 19-11) or even gaps in the endothelium (**discontinuous capillaries).**

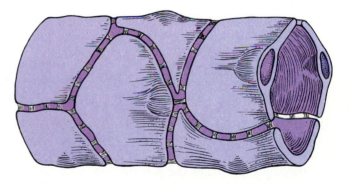

FIGURE 14-13
The structure of a generalized capillary. The spaces between endothelial cells are exaggerated in this diagram; they are normally about 4 nm wide, and the capillary wall is about 1 μm thick.

MECHANICS OF BLOOD FLOW
Definition of Blood Flow

Flow may be used in three different contexts: (1) as **volume flow;** (2) as **average flow velocity,** and (3) as the **linear velocity** of a small element of fluid. The volume flow rate is the total volume of fluid that passes a given point per time unit. In the cardiovascular system, cardiac output and the blood flow through the kidneys are examples of different volume flows. Volume flow can be measured by collecting the blood at some location (such as the heart or kidney) over a known interval of time. When the term "flow" is used by itself, volume flow is meant.

---

1 What are the distinguishing features of each the following vessel types?

Arteries	*Capillaries*
Arterioles	*Veins*
Metarterioles	

2 What is compliance? What is the significance of the difference in compliance between arteries and veins?
3 Why can the veins be termed capacitance elements?

Average flow velocity is defined as the total volume flow at a given point divided by the cross-sectional area. Average flow velocity is an indicator of how much time blood spends in particular vessels of the circulatory system. The units of average flow velocity are volume/time/area. The linear flow velocity is the rate of travel of a single drop of blood. The units of linear velocity are distance/time.

The distinctions between the three types of flow can be illustrated by the flow of water in a river. If there are no tributaries and if evaporation and rainfall are neglected, the volume flow in the river must be the same at all points along its length. Where the banks are close together, the cross-sectional area is reduced, and rapids—areas of increased average flow velocity—are the result. Where the river is broad, the cross-sectional area through which flow is occurring is great, and water motion may be unnoticeable. At any given point along the river's course, not all of the water molecules are moving with the same velocity. The current, which represents the linear velocity of flow of a small element of water in the river, is swiftest near the river's center and slowest near the bank. The difference results from the fact that frictional forces at the bank of the river hinder movement of water molecules close to the bank. These molecules in turn slow down others sightly further out, and so on. As the distance from the bank increases, the bank's influence diminishes and flow velocity increases.

Flow in blood vessels is almost always **laminar**—each element of the stream flows in a straight line without swirling or mixing across the axis of the vessel (Figure 14-14, A). A difference between the flow of blood in vessels and the flow of water in a river is that the flow is frequently turbulent in rivers. Turbulent flow in a river is evidenced by the formation of eddies. The tendency to form eddies is increased as the size of the stream and its velocity increase. This tendency is also a result of irregularities in the river's banks. Even the largest blood vessels in the body do not experience turbulent flow at normal mean volume flow rates. Turbulent flow does occur in the vicinity of the pulmonary and aortic valves of the heart, and it may be increased by irregularities of the valve surface or by stenosis (narrowing) of the valves (Figure 14-14, B). The increased turbulence that results causes the aorta to vibrate audibly and is one cause of heart murmur.

If each of the elements of the vascular system were merged into one single vessel of corresponding total volume and net cross-sectional area, the result would look something like Figure 14-15. In this figure, arteries and arterioles and veins and venules have been lumped together for simplicity. The volume flow rate (Q) must remain constant throughout the system. Volume flow in each of the three elements of the system is the product of the total cross-sectional area of each class of vessels (A_1,

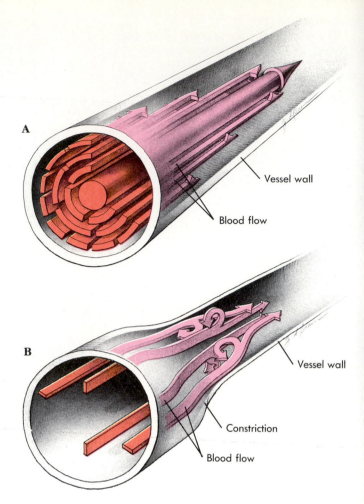

FIGURE 14-14
A Laminar flow. Fluid flows in long smooth-walled tubes as if it were composed of a number of concentric layers. Velocity is zero in the thin shell of fluid immediately next to the wall. It increases progressively in more central layers, rising to a maximum in the center.
B Turbulent flow. Turbulent flow is caused by numerous small currents flowing crosswise or oblique to the long axis of a vessel. It is more likely to occur at points where flow is abnormally constricted.

A_2, and A_3) and the average flow velocity in those vessels (V_1, V_2, and V_3). The constancy of the product of area and average flow velocity is called the **continuity equation of flow**. Continuity of flow means that a larger total cross-sectional area is associated with a lower average flow velocity.

The average flow velocity of blood in different types of vessels depends on the total cross-sectional area presented by each vessel type. Average flow velocity through arteries is rapid, so blood spends little time in passing from the heart to the capillaries. Because the overall cross-sectional area of capillaries is large, they have the lowest flow velocity of all the vessels in the circulation. A typical systemic capillary has a length of approximately 1 mm, and

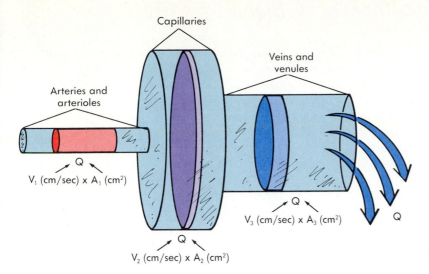

Capillaries

Veins and
venules

Arteries and
arterioles

Q
V_1 (cm/sec) x A_1 (cm²)

Q
V_2 (cm/sec) x A_2 (cm²)

Q
V_3 (cm/sec) x A_3 (cm²)

Q

FIGURE 14-15
The principle of continuity of flow. Because there is no place where a net fluid gain or loss occurs, the total volume flow *(Q)* must remain constant. Volume flow is the product of average velocity *(V)* and area *(A)*. Therefore, as the cross-sectional area increases (for example, going from arterioles to capillaries), average linear velocity of flow must decrease *(V_1 to V_2)*.

the flow velocity is about 1 mm/second, so a red cell spends about 1 second in a capillary.

Capillary walls consist of a single layer of endothelial cells (see Figure 14-13), the minimum possible barrier to diffusion. Diffusional equilibrium of the tissues with the blood in a capillary is achieved because the permeability of the capillary walls is extremely high, the blood is moving relatively slowly, and capillaries approach to within 10 to 30 μm of every cell. As blood enters venules and veins, the average flow velocity increases to a value

similar to, but less than, that of arteries. The pattern of flow velocity through the circulatory system is characterized by a rapid flow of blood to and from capillaries and with a slow flow of blood through capillaries (Figure 14-16).

Relationship of Blood Flow to Driving Force and Resistance

The laws of fluid mechanics apply to the flow of blood through the circulation and to the distribution of the cardiac output among the different or-

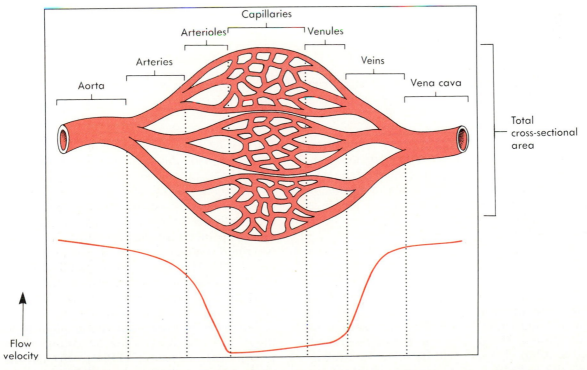

Capillaries

Arterioles

Venules

Arteries

Veins

Aorta

Vena cava

Total
cross-sectional
area

Flow
velocity

FIGURE 14-16
The relationship between net cross-sectional area and average flow velocity in different vessel types.

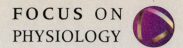
Measurement of Blood Pressure

Systemic arterial blood pressure can be measured by several techniques. In animal experiments and during cardiac catheterization or heart surgery in human patients, pressure at various points in the vasculature or even within the heart itself can be measured by inserting a tube, or cannula, into a blood vessel and connecting this tube to a blood pressure transducer, which generates an electrical signal that can be recorded using a polygraph (Figure 14-B, *1*). An advantage of the pressure transducer is that it can respond to rapid changes in pressure such as those which occur during systole.

In the classical method of **auscultation,** a pressure cuff is inflated around a major peripheral artery, usually the brachial artery in the arm, and connected to a pressure-measuring device known as a sphygmomanometer (Figure 14-B, *2*). Initially the cuff is inflated so that cuff pressure exceeds arterial systolic pressure, collapsing the brachial artery and preventing flow at all stages of the heart cycle. The pressure in the cuff is then gradually decreased. As cuff pressure reaches the arterial systolic pressure, blood can spurt through only when the arterial pressure is at its highest value. Each spurt is heard as a tapping sound, called the first Korotkoff sound, in a stethoscope placed on the arm distal to the cuff. The first sound corresponds to the arterial systolic pressure, typically 120 mm Hg. As the cuff pressure is decreased further, more blood passes through at each beat. The resulting sound first changes from a tapping to a louder thud and then becomes muted as the cuff pressure continues to decrease. When the cuff pressure falls to arterial diastolic pressure, typically 70 mm Hg, the brachial artery is not closed by the cuff at any stage of the heart cycle, and the sounds disappear.

gans. Blood flow through a blood vessel is the result of a driving force, in the form of a pressure gradient, or **perfusion pressure,** along the length of the vessel. The relationship of blood flow to perfusion pressure follows the general form of force-flow relationships described in Chapter 2:

$$\text{Blood flow} = \frac{\text{Perfusion pressure}}{\text{Flow resistance}}$$

The ratio of the pressure difference between any two points (the pressure gradient) and the resulting blood flow equals the flow resistance between the same two points. The decrease in the pressure with distance in a blood vessel will therefore be proportional to the vessel's flow resistance. Think of the resistance of a circulatory element as "using up" the driving force of the blood that travels through it.

The greater the resistance of an element, the greater the fraction of the arteriovenous pressure gradient that is used up as blood passes through it.

The relationship between blood flow, perfusion pressure, and resistance can be restated in terms that apply to the entire vascular system acting as a unit. The mean arterial pressure (MAP) is a time-weighted average of the arterial pressure over the heart cycle. The driving force for blood flow through the systemic loop is the difference between the MAP and the central venous pressure (CVP) which is the same as right atrial pressure. (Because the venous pressure does not change significantly over the heart cycle, it need not be averaged.) The **total peripheral resistance (TPR)** is the net flow resistance of the systemic loop. The relationship between flow through the systemic loop (Q), the driving force (MAP − CVP), and total peripheral resis-

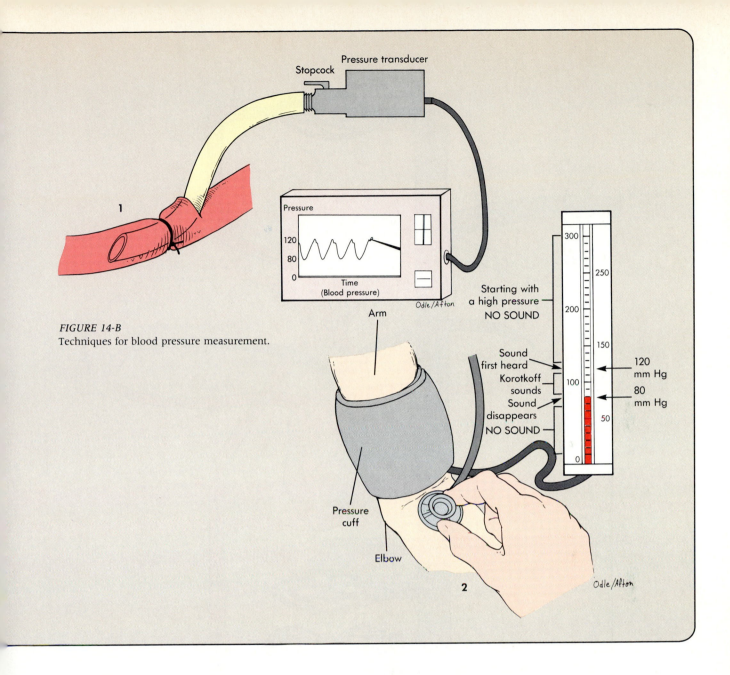

FIGURE 14-B
Techniques for blood pressure measurement.

Pressure transducer

Stopcock

1

Pressure

120
80
0
Time
(Blood pressure)

Odle/Afton

Arm

300

250

200

150

Starting with
a high pressure
NO SOUND

Sound
first heard
Korotkoff
sounds
Sound
disappears
NO SOUND

100

50

0

120
mm Hg

80
mm Hg

Pressure
cuff

Elbow

2

Odle/Afton

tance (TPR) follows the form of the general relation between force, resistance, and flow stated above:

$$Q = \frac{(MAP - CVP)}{TPR}$$

According to the Frank-Starling Law, the heart pumps out all the blood that comes to it, and the cardiac output equals the venous return as averaged over times longer than a few heartbeats. Thus in this equation, **Q** can stand equally well for the cardiac output or the venous return.

> 1 *What are the three measurements of flow?*
> 2 *State the general relationship between driving force, flow rate, and resistance.*
> 3 *For the systemic loop, what specific parameter corresponds to driving force? Flow rate? Resistance?*

Effects of Vessel Dimensions and Branching on Resistance and Flow

The flow resistance of individual segments of the vasculature depends highly on the geometry of the vessels involved. The simplest case is a vessel of uniform diameter that does not branch. The flow resistance of such a vessel is given by the equation:

$$R \propto \frac{L \times \eta}{r^4}$$

where **R** is flow resistance, **L** is the length of the vessel, η is the viscosity, or thickness, of blood, and **r** is the radius of the vessel.

Of the factors in the equation, the vessel radius is the most powerful determinant of flow resistance because it is raised to the fourth power, so small radius changes cause large resistance changes. For ex-

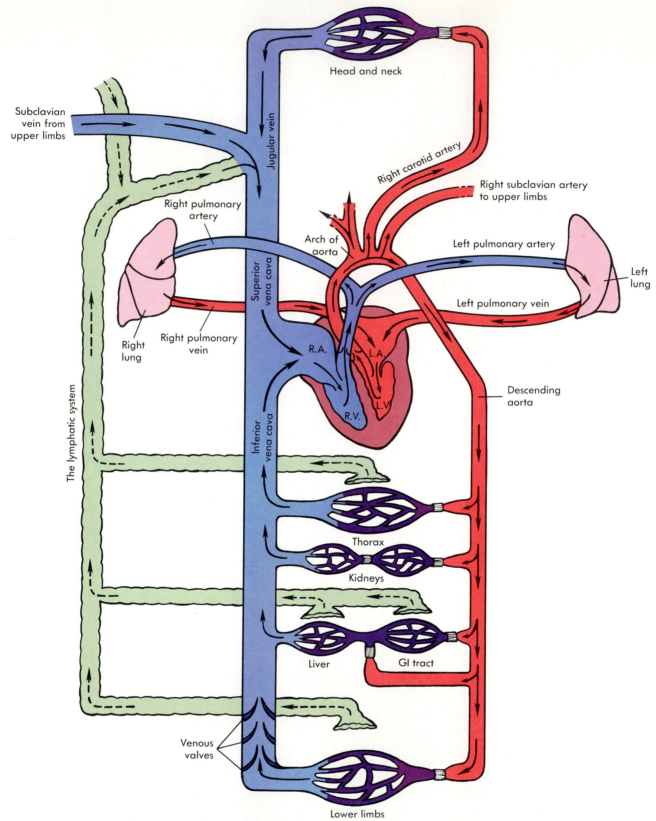

FIGURE 14-17
Diagram of the circulatory system and lymphatic system. The vessels of the lymphatic system begin as blind tubes in the tissues and return extracellular fluid to the heart via the superior vena cava and right atrium.

ample, if the radius of a vessel is halved, its flow resistance is increased by 2^4, or 16 times. The vessel length is a less powerful factor in the equation, so it is possible for major arteries and veins to be quite long and still have relatively low resistance as compared with that of shorter but narrower vessel types. Radius changes are an important aspect of the control of blood flow and pressure; vessel length changes are usually small and unimportant in control of the vasculature.

The viscosity of a fluid is a measure of the internal work necessary to make the fluid flow. A practical example of the effect of viscosity is the extra work that is performed by diners to get ketchup or salad dressing to flow from a narrow-mouthed bottle, from which less viscous fluids such as water would flow freely. The viscosity of blood is determined by the composition of blood, the nature of the vessel in which it is flowing, and the mean flow velocity. Plasma has a viscosity about twice that of water; whole blood has a viscosity approximately triple that of water. Thus red blood cells contribute significantly to blood viscosity, and an increase in the hematocrit level increases the work the heart must do to maintain the cardiac output.

When blood vessels branch, parallel pathways for blood flow result. Branching of major arteries causes the circulations of almost all organs to parallel each other (Figure 14-17). Within individual organs (with few exceptions), all arteries parallel each other, all capillaries parallel each other, and so on. The effect of branching on the net flow resistance depends on two factors: (1) the radii of the branches, and (2) the number of branches.

The key to understanding how vessels connected in parallel behave is the fact that the pressure gradient is the same across every vessel of the parallel combination. The volume flow through any single vessel (Q_1, Q_2, and Q_3 in Figure 14-18, A) is inversely proportional to its resistance. The total volume flow through the ensemble of parallel vessels is the sum of each of the individual volume flows. Each element of fluid may take one of several possible pathways and effectively experiences only the resistance of the particular vessel it enters. The total flow resistance of a set of parallel vessels is always less than the flow resistance of a single branch vessel because fluid has "somewhere else to go"; that is, to another parallel vessel.

An analogy can be made with traffic patterns: if a large volume of freeway traffic passes through parallel toll booths, each individual driver will encounter the delay (resistance to flow) only of his or her own pathway in the parallel array. Whether the flow of traffic will be slowed or not depends on two factors: (1) how many booths there are, and (2) how

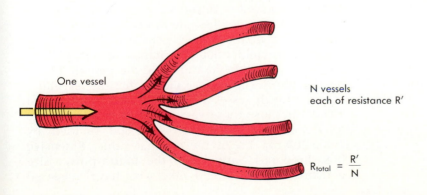

All vessels have
the same pressure gradient

Vessel 1

Q_1

Vessel 2

Q_2

Vessel 3

Q_3

Total Q = Q_1 + Q_2 + Q_3

One vessel

N vessels
each of resistance R'

$R_{total} = \dfrac{R'}{N}$

FIGURE 14-18
A Flow resistances in parallel. The total resistance of the parallel set of vessels (1 to 3) is less than the lowest of the individual flow resistances because the total flow (Q) is greater than any single flow (Q_1, Q_2, Q_3) and the pressure gradient is the same.
B When a large number of identical vessels are connected in parallel, the overall flow resistance is just the resistance of a single vessel divided by the number of vessels in parallel with each other.

Blood is Thicker than Water

As the hematocrit increases, so does the viscosity of blood. At a normal hematocrit of 45%, the viscosity is about 3 times higher than water, but blood may become 6 to 8 times as viscous as water if the hematocrit reaches 60%. High hematocrits are seen in disorders in which red cell production is elevated (polycythemia) and in severe water loss (dehydration). At the opposite extreme, hematocrits as low as 20% to 25% can be observed in disorders in which red cell production is low (anemia) and following severe blood loss. At a hematocrit of 25%, the viscosity of blood is only twice that of water.

Some athletes have used a technique called **blood doping** in an attempt to improve their performance in endurance events, such as bicycling or distance running. In this technique, some of the athlete's blood is drawn 5 to 6 weeks in advance of an important meet, and the red cells are separated and frozen for storage. In the time before the event, erythropoiesis restores the hematocrit to normal. Immediately before the meet, the stored red cells are injected, raising the hematocrit by 8% to 20%. The increased hematocrit persists for at least 2 weeks. In principle, large increases in the hematocrit could dramatically increase maximal performance during exercise because the maximum sustained effort of a trained athlete is limited by the rate at which oxygen can be delivered to the working muscles. However, the usefulness of blood doping is limited because the increased viscosity of the blood increases the flow resistance of all the organs. With increases in the hematocrit above 20%, the advantage of higher oxygen-carrying capacity is outweighed by the resulting limitation of perfusion of working tissues and the increased load on the heart. At present, it appears that increases in performance of, at most, 15% can result from this technique.

rapidly the toll can be collected from vehicles passing each booth.

To apply this analogy to blood vessels, suppose a stem vessel gives rise to a number, N, of identical branch vessels, each with a resistance R′ (Figure 14-18, *B*). A single drop of blood approaching the branch vessels may take any one of N identical paths. The total flow in the entire ensemble of branches is N times the flow through any single branch. But the pressure difference between the ends of the ensemble is the same as the pressure difference between the ends of each vessel. The flow resistance of the parallel combination of N vessels of resistance R′ must therefore be N times less (R′ divided by N) than the single vessel resistance R′. That is:

$$R_{(total)} = \frac{R'}{N}$$

As in the analogy of toll booths on the freeway, when a stem blood vessel branches into vessels of smaller diameter, the resistance of the collection of branch vessels can be either greater than or less than the resistance of the stem. The outcome depends on the number of branches (the more branches, the lower the net resistance) and the resistance of each branch (the smaller the branch vessels as compared with the stem, the higher the net resistance). This is illustrated in two examples.

For the first example, consider a vessel that branches to yield 1,000,000 branch vessels, each of which has a tenth the radius of the parent. The r^4 relationship for resistance means that the resistance of each branch vessel will be increased by a factor of 10,000. Because these vessels are all in parallel, the total collective resistance is 10,000 divided by 1,000,000, or 1% of the original value. Extensive branching has outweighed the fourth-power size dependence that would otherwise have increased the equivalent flow resistance. This example corresponds to the degree of branching occurring at the level of the capillaries and explains why, as a group, capillaries have a relatively low flow resistance, despite the small size of each individual capillary.

If branching is less extensive, the situation can be different. For the second example, suppose that the stem vessel gives rise to only 1000 branch vessels instead of 1,000,000. The resistance of each branch vessel would still be increased by 10,000. However, with only 1000 branches in parallel, the collective resistance of the branches would be 10 times greater than that of the stem. This example corresponds to the transition between arteries and arterioles. When blood flows from arteries into arterioles, the number of arterioles is not large enough to compensate for the large resistance of each single arteriole. As a result, the arterioles make the largest contribution to the total peripheral resistance of all

the vessel types and are referred to as the **resistance vessels** of the circulation.

The different net resistances of different vessel types are reflected in the pressure profile of the systemic loop (Figure 14-19). More than half of the pressure drop occurs as blood passes through arterioles, with another drop about half as great in the passage through capillaries. Flow through the large arteries and veins is accomplished with very small pressure drops.

Importance of Arterioles for Control of Blood Flow Distribution and Peripheral Resistance

The dominating resistance of arterioles has two important consequences. First, the arterioles determine how the cardiac output is distributed among different tissues and organs. Contraction of the smooth muscle of arteriolar walls results in a decrease in radius, or constriction; relaxation of the

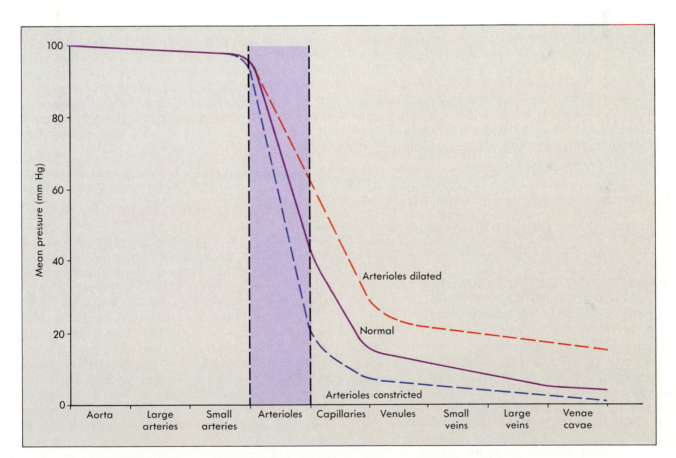

FIGURE 14-19

Pressure profile of the systemic loop. Pressure drops by only a few mm Hg between the heart and the arteriolar branch points. The pressure drop is much larger as blood flows through arterioles. A smaller pressure drop occurs in capillaries, and the last leg of the circuit through venules and veins is completed under very low pressure. Arteriolar dilation reduces the pressure drop in arterioles and proportionately increases the pressures in capillaries, venules, and veins. Arteriolar constriction increases the pressure drop in arterioles and reduces capillary and venous pressures. For simplicity, mean arterial pressure is assumed to remain constant.

muscle has the opposite effect and is termed dilation. Dilation of the arterioles of a particular tissue decreases the flow resistance of that tissue and has the effect of diverting more of the cardiac output to the capillary beds of that tissue.

The sum of the resistances of the arterioles in the various branches of the systemic loop is the major determinant of the **total peripheral resistance.** Total peripheral resistance is increased by arteriolar constriction and decreased by arteriolar dilation. These changes also affect the pressure profile. Arteriolar dilation decreases the pressure drop in arterioles and increases capillary and venous pressure; arteriolar constriction increases the pressure drop in arterioles and reduces capillary and venous pressure (see Figure 14-19). The importance of total peripheral resistance as a determinant of cardiac output and mean arterial pressure is discussed in the next chapter.

An important difference between the systemic loop and the pulmonary loop is that the pressure gradient between arteries and veins is only about 14 mm Hg in the pulmonary loop as compared with the approximately 100 mm Hg of the systemic loop. Because the output of the right and left hearts must be the same over time, the resistance of the pulmonary loop must be about 14% of that of the systemic loop. One important factor in the low resistance of the pulmonary loop is the fact that the pulmonary arterioles are substantially larger than the systemic ones. Pulmonary circulation is discussed further in Section IV, Respiration.

> 1 *What three factors determine the flow resistance of a single vessel?*
> 2 *Which vessels contribute most to the total peripheral resistance?*

TRANSFER OF FLUID AND SOLUTES BETWEEN CAPILLARIES AND TISSUES
Exchange of Nutrients and Wastes across Capillary Walls

In almost all tissues, solute transfer between plasma and interstitial fluid is driven solely by concentration gradients rather than by the expenditure of metabolic energy. For example, glucose uptake by metabolizing cells tends to reduce the concentration of glucose in the interstitial fluid, resulting in a gradient of glucose concentration between plasma and interstitial fluid. Glucose is poorly soluble in lipid, so almost all of the glucose that diffuses between the plasma and interstitial fluid must pass through the intercellular pores of the capillary endothelium. At the same time, carbon dioxide produced by oxidative metabolism diffuses down its concentration gradient from the cytoplasm into the interstitial fluid, and from there it diffuses into the plasma. Dissolved CO_2 and O_2 are quite lipid soluble, so

their route passes mainly through the endothelial cells.

The brain is an important exception to the rule that solute exchange across capillary walls occurs by diffusion. Capillaries in most regions of the brain lack open intercellular pores, and hydrophilic solutes cannot readily pass across them, although lipid-soluble substances can diffuse through brain capillary walls quite easily. This feature of brain capillaries has been termed the **blood-brain barrier.** The glucose needed for energy and the amino acids needed for synthesis of proteins and peptide neurotransmitters are actively transported across the walls of brain capillaries. The blood-brain barrier gives the brain a large measure of control over the composition of its interstitial space.

Fluid Exchange between Capillaries and Interstitial Spaces

A capillary wall can be thought of as a filter whose pores retain large plasma solutes, allowing the passage of water and small solutes. At each point along its length, a capillary has both an outward hydrostatic and an inward osmotic pressure gradient across its wall. The occurrence of a net loss or gain in the fluid volume carried by the capillary is determined by which of these forces is stronger.

The movement of fluid between the blood and interstitial-fluid compartment is described quantitatively by Starling's Hypothesis of Capillary Filtration:

$$\text{Net fluid filtration} = K \text{ (Net hydrostatic forces} - \text{Net osmotic forces)}$$

According to Starling's Hypothesis, the constant **K,** representing the leakiness of the capillaries of a given tissue, is multiplied by the difference between the net **hydrostatic forces** and the net **colloid osmotic forces** exerted by the proteins in blood and the interstitial fluid. These colloid osmotic forces are often referred to as oncotic forces.

Fluid exit from the capillary is driven by the hydrostatic pressure gradient that drives blood through the capillary; this pressure is greatest at the arteriolar side and least at the venular side (Figure 14-20). The tissue, or interstitial, hydrostatic pressure is very small and may even be a negative value, so the blood pressure that drives fluid out of the capillary is relatively unopposed by tissue hydrostatic forces.

Fluid entry into the capillary is driven by the colloid osmotic force attributable to plasma proteins that are present in the blood when the blood enters the capillary and that are not filtered out. This value does not change significantly as the blood moves through the capillary (see Figure 14-20) and is opposed along the length of the capillary by a small but measurable tissue colloid osmotic force attributable to the small amount of protein that does leak

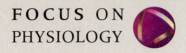

The Blood-Brain Barrier

Communication between nerve cells of the brain depends on extremely careful regulation of the composition of the brain's interstitial fluid. This necessity exists because the excitability of nerve cells depends on ionic gradients between the cytoplasm and interstitial fluid and because many small organic molecules normally found in the blood are also used for chemical communication between brain cells. The blood concentrations of ions such as K^+ and amino acids such as glycine, glutamate, and aspartate can vary significantly as a result of absorption of a meal. If the brain were exposed to these variations, overall alterations in excitability and random activation of the pathways that use amino acid transmitters would undermine communication within the brain.

An important consequence of the presence of the blood-brain barrier is that antibiotics such as penicillin cannot be delivered to the brain by injection into the systemic vasculature. In contrast, caffeine, nicotine, cocaine, and alcohol are lipid soluble and readily cross the blood-brain barrier where they affect cerebral function. An important task of pharmacology is to identify or modify drugs to increase their lipid solubility, allowing them to cross the blood-brain barrier.

The blood-brain barrier is less tight in some specialized regions of the brain, such as the hypothalamus, pituitary, and pineal glands. In these regions the leakiness allows peptide hormones and glucose to enter the blood from the brain.

Recently, a method of temporarily opening the blood-brain barrier was discovered. An injection of hypertonic solute into the blood circulation of the brain disrupts the continuous tight junctions of the capillary endothelial cells, probably by shrinking the cells themselves. The effect lasts for up to several hours. This technique permits drugs that normally do not cross the blood-brain barrier to be delivered to the brain cells.

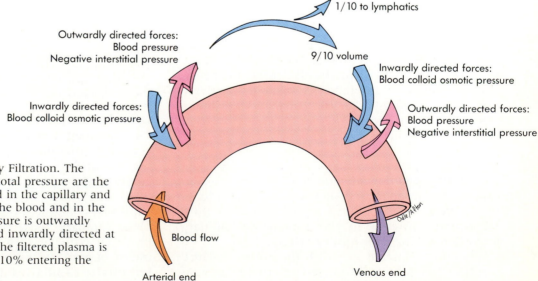

FIGURE 14-20
Starling's Hypothesis of Capillary Filtration. The major pressures composing the total pressure are the hydrostatic pressure of the blood in the capillary and the colloid osmotic pressure of the blood and in the interstitial spaces. The total pressure is outwardly directed at the arteriolar end and inwardly directed at the venous end. About 90% of the filtered plasma is reabsorbed, with the remaining 10% entering the lymphatics.

out into the interstitial fluid. The smaller organic molecules and salts tend to be at the same concentration in blood and interstitial fluid and therefore do not contribute to the net osmotic force.

In general, there is a slight excess of filtration over absorption (see Figure 14-20). However, just where this occurs on the microscopic level has been difficult to determine. Starling believed the tissue hydrostatic pressure to be a small positive value (that is, slightly unfavorable for fluid movement into the tissues). In this view, the tissues could be likened to a sponge saturated with water. Some investigators now believe that the tissue hydrostatic pressure may be negative by as much as 7 mm Hg. If so, the tissues would be like a sponge that is not saturated. If this is the case, both the capillary blood pressure and the suction of the negative tissue hydrostatic pressure favor filtration of fluid from the blood into the interstitial-fluid compartment at all points along the capillary.

Small, local variations in the capillary blood pressure, the water permeability of different parts of the capillary, or the tissue colloid osmotic and hydrostatic pressures could combine to favor filtration or reabsorption in other capillaries or in the same capillary at a different time. Some investigators now believe that for open capillaries (those whose precapillary sphincters are open), the net force favors filtration along their entire length, and that for closed capillaries, the net force favors reabsorption. In any case, the net effect for the body is that about 10% of the plasma filtered is not reabsorbed, resulting in net transfer of 2 to 4 liters of fluid per day from the systemic circulation to the interstitial fluid.

The balance between capillary filtration and reabsorption can be altered by changes in any of the elements in Starling's Hypothesis. One example of a disease process in which elevated mean capillary pressure alters the filtration-reabsorption balance is right ventricular failure. When this occurs, the venous pressure increases, increasing the mean pressure in the systemic capillaries and thus increasing the rate of filtration relative to reabsorption. Accumulation of fluid in the interstitial spaces causes swelling, referred to as **edema** (Table 14-4).

The lungs are particularly susceptible to edema. The total mass of lung tissue is small, so relatively small shifts of fluid from the plasma to the lungs can have serious effects. The normal capillary blood pressure of the lungs is only about 7 mm Hg, consistent with the low pressures throughout the pulmonary loop. These low blood pressures reduce the probability of pulmonary edema under normal conditions. Failure of the left ventricle results in a backup of blood in the pulmonary circulation, with a consequent increase in the pulmonary capillary pressure, causing **pulmonary edema.**

Another situation in which the normal balance of capillary filtration is altered is chronic protein

TABLE 14-4	Causes of Edema
Factor altered	*Possible causes*
1. Increased capillary blood (hydrostatic) pressure	Arteriolar dilation Heart failure Prolonged standing
2. Decreased plasma osmotic pressure	Protein undernutrition Failure of plasma protein synthesis (liver disease) Excessive loss of plasma proteins in urine (kidney disease)
3. Increased capillary permeability	Inflammation Allergies Burns
4. Increased tissue osmotic pressure	Release of protein by damaged cells
5. Obstruction of lymphatic vessels	Injury Parasitism of lymphatic system (filariasis)

undernutrition. In this case, the protein content of plasma is depressed. The resulting decrease in osmotic pressure reduces absorption. The resulting general edema causes the bloated appearance of seriously undernourished people.

A shift in the balance of capillary filtration may have a beneficial effect. For example, if bleeding significantly decreases blood volume, capillary blood pressure falls. As a result, there is a net shift of fluid from the interstitial spaces to the plasma. As much as 1 liter of volume can be transferred to the plasma over about an hour. The interstitial fluid can thus be regarded as a reserve that can be drawn on to maintain plasma volume.

> *1 What is the Starling Hypothesis of Capillary Filtration?*
> *2 What is the source of the pressure that favors movement of fluid from capillaries to interstitial spaces?*
> *3 What feature of the capillary wall is responsible for the colloid osmotic gradient between interstitial fluid and plasma?*

Lymphatic System

The lymphatic system (see Figure 14-17) returns excess interstitial fluid formed in capillary filtration to the circulatory system and salvages the small amounts of protein that escape through the capillary walls. The system contains blind-ended **lymphatic capillaries** (Figure 14-21, *A*) that terminate among blood capillary beds. The lymphatic capillaries are provided with large openings that permit the entrance of proteins as well as smaller solutes. The **lymphatic capillaries** drain into **lymphatic veins,**

A

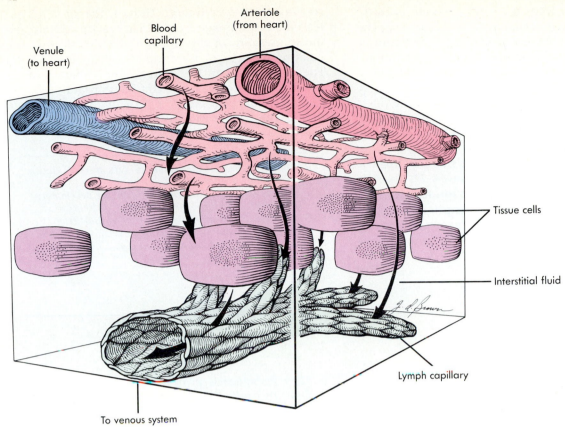

Venule (to heart)

Blood capillary

Arteriole (from heart)

Tissue cells

Interstitial fluid

Lymph capillary

To venous system

B

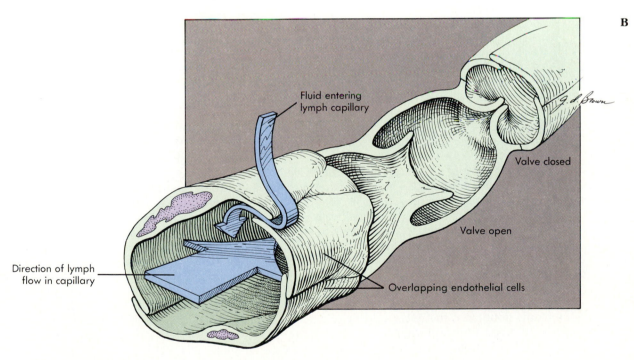

Fluid entering lymph capillary

Valve closed

Valve open

Direction of lymph flow in capillary

Overlapping endothelial cells

FIGURE 14-21

A Movement of fluid from blood capillaries into tissues and from tissues into lymph capillaries. Lymphatic capillaries are close to blood capillaries. Major features of their structure include valvelike pores between their endothelial cells and supporting fibers running from lymphatic walls into surrounding tissue to prevent collapse of the lymphatic capillaries.

B One-way flow of lymph in lymph capillaries. The overlap of the lymph capillaries' endothelial cells allows easy entry of interstitial fluid but prevents movement back into the tissue. Valves in lymphatic veins, like those of blood veins, ensure one-way flow in the direction of the heart.

Blood and the Vascular System

which ultimately join to form two **lymphatic ducts;** these join with the subclavian veins near the heart. The duct on the left side, called the **thoracic duct,** serves the entire body except the right shoulder and the right side of the head, which are served by the **right lymphatic duct.**

Lymph is propelled in the direction of the ducts by contractions of surrounding skeletal muscle and, to a lesser extent, by contractions of smooth muscle surrounding the lymphatic vessels themselves. Like veins of the blood circulatory system, lymphatic veins have one-way valves (Figure 14-21, *B*). The lymphatic vessels are an ideal location for elements of the immune system, called **lymph nodes** (see Chapter 24, p. 627), because lymph coming from an infected region of the body carries invading bacte-ria. In infected or injured tissues, the proteins released from damaged cells increase the colloid osmotic pressure of the interstitial fluid, tipping the balance in favor of filtration. Consequently a local edema develops in these areas (see Table 14-4). In the case of infection, the extra interstitial fluid accumulation flushes bacteria into the lymphatic system, where they more readily trigger an immune response.

---

1 What is the role of the lymphatic system in regulating the volume of interstitial fluid?

2 What is edema? What changes favor the development of edema?

3 What changes would favor transfer of fluid from tissues to plasma across capillary walls?

SUMMARY

1. **Blood** is a liquid tissue consisting of **red cells, white cells,** and **platelets** suspended in **plasma.** Red cells function in O_2 and CO_2 transport, white cells in immunity, and platelets in **hemostasis.** The **hematocrit** level is the ratio of red cell volume to total volume.

2. Three phases of hemostasis are (1) formation of a platelet plug at the site of vessel injury, (2) clot formation, and (3) clot dissolution. Clotting is the result of conversion of **fibrinogen** to **fibrin,** which is triggered by a factor cascade initiated either by factors present in blood (the **intrinsic pathway**) or by factors released by injured tissue (the **extrinsic pathway**). Clot dissolution is initiated by the factors that trigger clotting but becomes effective more slowly.

3. The ratio of volume to pressure in a vessel is determined by its **compliance.** Arteries are low-compliance elements containing a small volume of blood under high pressure. The arterial walls store energy during systole and recoil to maintain blood flow into the vasculature during diastole. Veins are the high-compliance, **capacitance** elements of the circulation.

4. Depending on context, "flow" indicates either the volume of blood passing a point over time (volume flow rate), the linear velocity of a single drop of blood (flow velocity), or the average velocity of all drops in a stream (average flow velocity).

5. The volume flow rate through a vessel is equal to the perfusion pressure divided by the vessel's **flow resistance.** In terms of the whole systemic loop, the **venous return,** or **cardiac output,** corresponds to volume flow rate; the **mean arterial pressure** minus the **central venous pressure** to perfusion pressure; and the **total peripheral resistance** to flow resistance.

6. Flow resistance is proportional to vessel length, fluid viscosity, and the inverse fourth power of vessel radius. Of the factors, vessel radius has the most important implications for flow through blood vessels.

7. Branching causes a decrease in net flow resistance. Decreases in radius increase net flow resistance. When both occur, there may be an increase or decrease in the net flow resistance, depending on the number of branches and the radius of each branch. In the case of arterioles, a net increase in resistance occurs; for capillaries, branching is sufficient for a decrease in resistance to occur. The net resistance of arterioles largely determines total peripheral resistance; dilation or constriction of individual arterioles controls the perfusion of individual tissues.

8. Extrinsic control of vascular smooth muscle is generally mediated by the sympathetic branch of the autonomic nervous system, which mainly causes vasoconstriction.

9. **Capillaries** are the sites of all exchange of materials between tissues and plasma. This exchange is favored by the slow average flow velocity of blood in capillaries. Solutes cross capillary walls by diffusion; brain capillaries are an exception in that they actively transport glucose and amino acids.

10. The hydrostatic pressure gradient across capillary walls tends to drive filtration in the arteriolar ends of capillaries, and the colloid osmotic gradient tends to drive absorption in the venular ends. Excess interstitial fluid is returned to the circulation by **lymphatic** vessels. Pressure imbalances favoring filtration cause **edema;** decreased blood pressure following blood loss causes the colloid osmotic force to dominate and thus favors fluid movement into the plasma.

1. What advantage is conferred by the complexity of the factor cascade that triggers clotting? Which of the stages of hemostasis involve positive feedback? Which one represents negative feedback?

2. The diameter of a capillary is smaller than that of an arteriole, yet collectively the capillaries have a lower flow resistance than the arterioles. Explain.

3. What would be the effect of an increase in tissue colloid osmotic pressure on the volume of interstitial fluid? An increase in venous pressure?

4. Why does systolic pressure normally increase with age?

5. What effect would a decrease in venous compliance have on the volume of blood in veins? On the venous pressure? On the venous return? On the total peripheral resistance? What effect would arteriolar dilation have on the total peripheral resistance? On venous return?

6. What factors determine whether branching will increase or decrease the net flow resistance of a stem vessel and its branches?

7. Describe the role of each of the following in blood clotting.
   Hageman factor   Thrombin
   von Willebrand's factor Tissue thromboplastin

8. A sample of blood centrifuged in a glass capillary tube yields the following measurements: length occupied by plasma: 29 mm; length occupied by red blood cells: 21 mm. What is the hematocrit?

9. What is the effect on vascular resistance and blood flow of activating each of the following pathways:
   a. Sympathetic adrenergic input to fibers with alpha receptors.
   b. Adrenergic fibers on vascular muscle with beta receptors.
   c. Sympathetic cholinergic fibers.
   d. Parasympathetic fibers.
   In which tissue types is each of these pathways important?

10. An artery gives rise to two branches, one of which has half the radius of the other. If the blood flow rate in the artery is 10 ml/minute, what would be the flow rate in each of the branches?

11. What would be the effect of an increase in tissue osmotic pressure on the volume of interstitial fluid? An increase in central venous pressure?

● SUGGESTED READING

DOOLITTLE, R.F.: Fibrinogen and Fibrin, *Scientific American*, December 1981, p. 126. Still reasonable current summary of clot formation.

EDWARDS, D.D.: Searching for the Better Clot-Buster, *Science News*, April 1988, p. 230. Describes recent drugs that offer improved treatment of fibrolytic blood clots.

GOLDE, D.W., and J.C. GASSON: Hormones That Stimulate the Growth of Blood Cells, *Scientific American*, July 1988, p. 62. Shows how specific hormones cause the progenitor cells in the bone marrow to differentiate into red blood cells or into one of the varieties of white blood cells. Soon it may be possible to treat diseases by calling up an extra supply of the cells needed.

LAWN, R.M., and G.A. VEHAR: The Molecular Genetics of Hemophilia, *Scientific American*, March 1986, p. 48. Describes inherited defects in clotting factors responsible for hemophilia.

TALBOT, L, and S.A. BERGER: Fluid-Mechanical Aspects of the Human Circulation, *American Scientist*, volume 62, 1974, p. 671. Explains the origins of Poiseuille's law and pressure drops within the circulatory system.

WEISS, R.: In a Similar Vein, *Science News*, September, 1987, p. 201. Discusses fluosol-DA as a possible substitute for blood.

CHAPTER 15

Regulation of the Cardiovascular System

On completing this chapter you will be able to:

- Describe the autonomic innervation of blood vessels, distinguishing special features of blood vessels in skeletal muscle, the coronaries, and the external genitalia.
- Diagram the baroreceptor reflex.
- Describe how antidiuretic hormone, the atrial natriuretic hormone system, and the renin-angiotensin-aldosterone system regulate mean arterial blood pressure.
- Appreciate the significance of the cardiovascular operating point in the regulation of central venous pressure, mean arterial pressure, and cardiac output.
- Understand how standing up results in a drop in venous return and how mean arterial pressure is maintained by the baroreceptor reflex, venous valves, and the skeletal and respiratory pumps.
- Outline the changes that follow moderate hemorrhage.
- Describe the cardiovascular responses to exercise.
- List factors that might be expected to lead to hypertension.

When the primate ancestors of man began to walk upright, they imposed new stresses on many parts of the body, including the cardiovascular system. In most animals, the major axis of the circulatory system is horizontal, and most of the blood volume is close to heart level. In standing human beings, about 70% of the blood volume is in highly compliant veins below the heart level. To respond to postural changes, the cardiovascular system requires receptors that sense relative hydrostatic pressure fluctuations and initiate appropriate reflex responses.

In addition to the problems associated with an upright posture, extra burdens are placed on the cardiovascular system by stresses such as exercise, blood loss, heat, and disease. For example, strenuous exercise causes, roughly, a twentyfold increase in blood flow to skeletal muscle and a tenfold increase in cardiac output, with very little change in mean arterial pressure. This stability requires the integration of neurally and hormonally mediated cardiovascular reflexes with the intrinsic autoregulation that occurs in the heart and blood vessels.

Some of the mechanisms involved in cardiovascular integration are explained in this chapter; others are not yet fully understood. For example, how do the cardiovascular centers of the brain "know" how much sympathetic input to give the heart and vessels to meet the challenge of rapidly changing workloads? Some evidence suggests that the cardiovascular regulatory centers of the brain receive input from motor centers that provide "feedforward" information about the demands that contracting skeletal muscle is about to place on the cardiovascular system.

CARDIOVASCULAR REFLEXES

Components of the Cardiovascular System

The conceptual organization needed to understand regulation of the cardiovascular system is shown in Figure 15-1. The properties of the heart are myocardial contractility and heart rate. Properties of the peripheral vessels include the total peripheral resistance, blood volume, and venous compliance. However, the heart and blood vessels are not isolated but form an integrated system. System variables include the cardiac output, mean arterial pressure, central venous pressure, and venous return.

Extrinsic Control of Vascular Smooth Muscle

Vascular smooth muscle is present in the walls of all blood vessels except capillaries. Changes in vascular smooth muscle tone (tension) control total peripheral resistance, the compliance of arteries and veins, and the distribution of blood flow. Two separate major control systems affect vascular smooth muscle tension: (1) extrinsic control by hormones and the autonomic nervous system (Table 15-1), and (2) intrinsic control, or autoregulation, by chemical factors produced in the immediate vicinity of the blood vessels (Table 15-2). The relative importance of these two systems depends on the type of vessel and the organ or tissue served by that vessel.

The smooth muscle of most blood vessels is innervated by the sympathetic branch of the autonomic nervous system. The sympathetic fibers to blood vessels usually exhibit some degree of **resting tone** (action potential frequency in the efferent fibers). An increase in sympathetic adrenergic input to a vessel that possesses mainly α_1-adrenergic receptors increases smooth muscle tension above the resting level; the effect of smooth muscle contraction on the vessel is a decrease in radius, or **vasoconstriction**. A decrease in the sympathetic tone allows the smooth muscle to relax; the resulting increase in the radius of the vessel is **vasodilation**.

The arterioles and veins of most vascular beds contain α_1-adrenergic receptors, but the effect of sympathetic stimulation differs. The effect on arterioles (**arteriolar vasoconstriction**), the major site of flow resistance in the body, is to redirect blood flow and increase the total peripheral resistance. The effect of sympathetic stimulation on veins (**venoconstriction**) does not affect total peripheral resistance very much because venous resistance is low compared with arteriolar resistance. Instead, the major effect is to decrease the compliance of vein walls (stiffen them), thus decreasing the volume of blood in the veins.

There are three exceptions to this pattern of response to sympathetic stimulation. The arterioles that supply blood to skeletal muscle possess a mixture of α_1- and β_2-adrenergic receptors (see Table 15-1). The result is that low levels of circulating epinephrine preferentially activate the β_2-receptors and induce dilation. Higher plasma levels of epinephrine, or norepinephrine released directly from sympathetic fibers, activate α_1-receptors, causing arteriolar constriction.

The arterioles of the skin and skeletal muscle are an exception to the rule that vascular smooth muscle is innervated only by sympathetic adrenergic inputs. These vessels are innervated by sympathetic cholinergic fibers that mediate dilation. These inputs are important for diverting blood flow to the

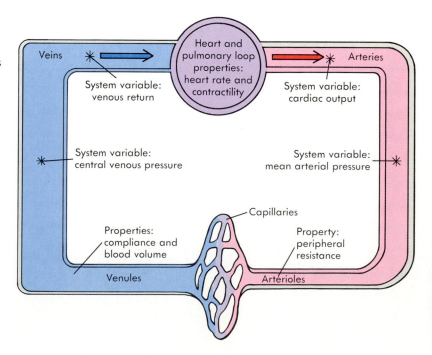

FIGURE 15-1

Schematic diagram of the heart and systemic circulation showing the properties associated with each part of the system and the variables that affect both the heart and the vasculature.

TABLE 15-1   Extrinsic Regulation of Blood Vessels

Agent	Effect	Comments
Sympathetic nerves		
α_1-Adrenergic	Constriction	The dominant effect of the sympathetic nervous system on the vasculature and the major cause of changes in the total peripheral resistance and venous compliance.
β_2-Adrenergic	Dilation	Skeletal muscle arterioles. Effect is minor compared with α_1-effect but may be important during moderate elevation of epinephrine.
Cholinergic	Dilation	Skeletal muscle arterioles. Important in preparation for activity, but effect is small compared to the autoregulatory increase in working muscle.
Parasympathetic nerves		
Cholinergic	Dilation	Important in genital sexual response. Organs that are activated by parasympathetic input, such as the salivary glands and GI tract, may vasodilate as indirect effect.
Hormones		
Antidiuretic hormone (vasopressin)	Constriction	Octapeptide released by hypothalamic neurons that project to posterior pituitary in response to signals, which include decreased atrial pressure.
Angiotensin I, II	Constriction	Results from renin secretion by kidneys in response to signals, including decreased renal perfusion, a consequence of decreased mean arterial pressure.

TABLE 15-2   Intrinsic Control of Systemic Blood Vessels (Autoregulation)

Autoregulatory agents	Comments
Metabolic vasodilators	
Decreased O_2	Due to high rate of oxidative metabolism.
Decreased pH	Reflects increased levels of CO_2 and lactic acid.
Increased K^+	Due to activity of excitable cells.
Paracrine agents	
Prostaglandins	May dilate or constrict depending on the particular prostaglandin involved. Also important in vascular responses in inflammation and hemostasis.
Bradykinin	Dilator important in inflammation and in autoregulation in sweat glands.
Histamine	Dilator important for vascular responses in inflammation and allergic response.
Adenosine	Vasodilator important in some tissues such as the heart.

skin to increase heat loss and to skeletal muscle to increase perfusion in advance of exercise.

The arterioles of the external genitalia are an exception to the rule that vessels receive only sympathetic innervation because the external genitalia are also innervated by parasympathetic fibers that mediate dilation. These fibers have an important role in the vascular changes associated with sexual excitement (see Table 15-1).

Intrinsic Control of Vascular Smooth Muscle

In **flow autoregulation**, the arterioles of individual vascular beds change their diameter in such a way as to adjust the flow to meet the metabolic demands of the tissue, despite variations in perfusion pressure. This process was presented in Chapter 5 (see p. 92) as an example of negative feedback. The feedback of tissue metabolism on arteriolar radius may be mediated by several substances referred to collectively as local factors.

Depending on the tissue involved, local factors are generally either end products of energy metabolism, such as lactic acid and CO_2, substances released in the course of activity, such as K^+, or paracrine substances, such as **prostaglandins, adenosine**, or **bradykinin** (see Table 15-2). Oxygen and other nutrients may also exert such effects when their concentrations are decreased by an increase in tissue activity. With an increase in metabolic rate, the changes in the concentrations of the local factors cause arterioles to dilate. Such changes in the local environment are particularly important for metarterioles because they receive little or no innervation. Similar factors stimulate opening and inhibit closing of precapillary sphincters.

Skeletal muscle provides an excellent example of how local factors affect flow autoregulation. The fraction of the cardiac output that enters a resting skeletal muscle is relatively low because arterioles are constricted and precapillary sphincters are frequently closed. Much of the blood flow that does enter the muscle passes through shunts because the shunt resistance is lower than the resistance of the path through the constricted arterioles and most of the precapillary sphincters, which are closed.

If the muscle contracts repeatedly, end products of metabolism accumulate and nutrients and oxygen are depleted. In response, arterioles dilate and precapillary sphincters remain open more of the time.

The decrease in the tissue flow resistance caused by arteriolar dilation delivers a larger share of the cardiac output to the working muscle. The open precapillary sphincters allow this increased flow to pass through capillaries. Thus the net effect of a mismatch between production of local factors and their removal by blood flow is an increase in the delivery of blood to the affected area.

Differences between Vascular Beds

The relative weight of extrinsic inputs and autoregulation in determining blood flow differs from tissue to tissue and even in the same tissue under different conditions, as demonstrated by the following examples. Sympathetic nervous system activation has its largest vasoconstrictive effect on the vessels of the skin and abdominal organs. The dually innervated vessels of the external genitals are very responsive to changes in their autonomic inputs occurring during the sexual-response cycle (see Chapter 26, p. 680). In contrast, autoregulation is believed to be much more important than is extrinsic control in the heart and brain.

In skeletal muscle, the level of tissue activity shifts the balance of extrinsic and intrinsic control. In resting skeletal muscle, sympathetic (norepinephrine) inputs are dominant and exert mainly a vasoconstrictive effect. The blood flow to resting skeletal muscle can be increased somewhat by autonomic vasodilation (the result of cholinergic fibers and epinephrine); this occurs when exercise is anticipated. In working skeletal muscle, autoregulation is able to increase blood flow far above the increment attributable to the autonomic input; thus autoregulation is the mechanism by which the skeletal muscles capture a large fraction of the cardiac output during exercise.

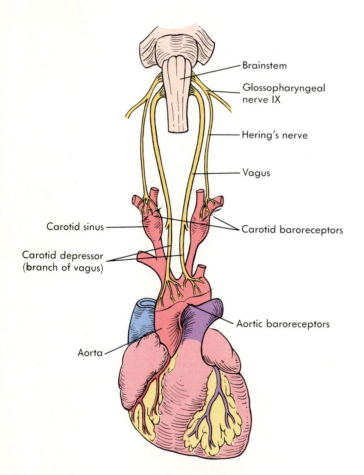

FIGURE 15-2
The location of carotid and aortic baroreceptors and their pathways into the central nervous system. The afferent axons of carotid baroreceptors course in Hering's nerve, which joins the glossopharyngeal (cranial IX) nerve. The aortic arch baroreceptor afferents course into the brainstem by way of the cardiac depressor branch of the vagus (cranial X) nerve.

Labels on figure:
Brainstem
Glossopharyngeal nerve IX
Hering's nerve
Vagus
Carotid sinus
Carotid baroreceptors
Carotid depressor (branch of vagus)
Aortic baroreceptors
Aorta

> 1 What variables describe the overall function of the cardiovascular system?
> 2 What two variables best describe the function of the heart? What three variables describe the vascular system?
> 3 What property of the vasculature is affected by arteriolar constriction or dilation? Which one is affected by changes in venous compliance?

Baroreceptor Reflex

Figure 15-2 illustrates the location of the cardiovascular **baroreceptors.** One important population of baroreceptors is located in the walls of the common carotid artery where it branches to form the internal and external carotid arteries. At this point the carotid bulges slightly, forming the **carotid sinus.** Other baroreceptors are scattered throughout the wall of the aortic arch. The axons of these baroreceptor cells enter cranial nerves IX and X and ascend to the cardiovascular center of the brainstem medulla. There they synapse on interneuron that determine the outflow of impulses in parasympathetic and sympathetic pathways to the heart, sympathetic pathways to the vessels, and sympathetic activation of the adrenal medulla (Figure 15-3).

The importance of carotid sinus receptors was discovered in the 1920s by a physician named Hering, who noticed that merely stroking the area of the neck above the carotid sinuses caused slowing of the heart and fainting in some sensitive patients. Hering then showed with animal experiments that distending the carotid sinuses results in slowing of the heart and a drop in arterial blood pressure, whereas reducing pressure at the sinuses by clamping the carotid arteries below the carotid sinus stimulates an increase in heart rate and a rise in blood pressure.

Baroreceptors discharge action potentials at a rate proportional to the degree of stretch of the arterial wall (Figure 15-4, *A*). The rate rises and falls in a pattern determined by the heart cycle. For the cardiovascular integrating center in the medulla, there is a "setpoint" firing pattern that corresponds to the normal mean arterial pressure. If mean arterial blood pressure in the carotid arteries begins to fall, the firing rates of the aortic and carotid baroreceptors decrease (Figure 15-4, *B*). The medullary cardiovascular control center reacts to a drop in the frequency of action potentials from baroreceptors by increasing sympathetic activity to the heart and blood vessels and by reducing parasympathetic tone to the heart pacemakers (see Figure 15-3). These changes in autonomic inputs to the heart increase its rate and force of contraction. Increased sympathetic input to arterioles of the systemic loop causes a general arteriolar constriction and an increase in total peripheral resistance. Together, these changes rapidly restore carotid arterial pressure to normal and maintain cardiac output.

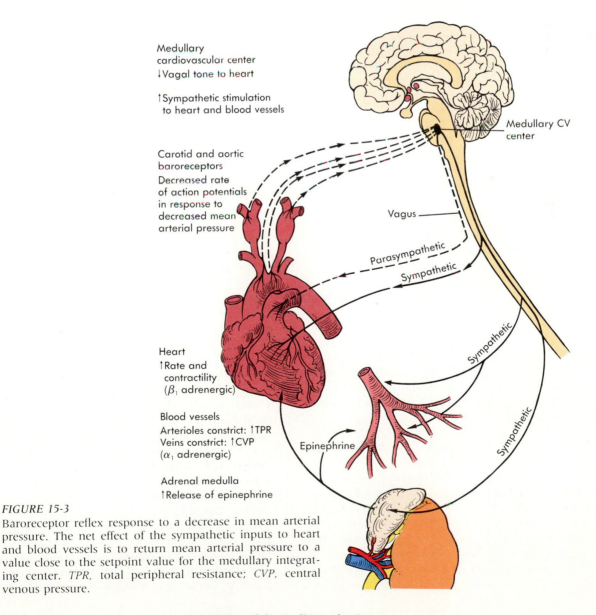

FIGURE 15-3
Baroreceptor reflex response to a decrease in mean arterial pressure. The net effect of the sympathetic inputs to heart and blood vessels is to return mean arterial pressure to a value close to the setpoint value for the medullary integrating center. *TPR,* total peripheral resistance; *CVP,* central venous pressure.

Regulation of the Cardiovascular System

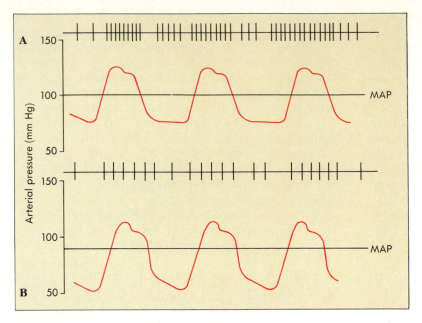

FIGURE 15-4
Idealized response of a carotid baroreceptor to
two mean arterial pressures. The upper trace in
each block shows the rate of action potentials;
the lower trace, the record of arterial pressure.
An increase in the rate of action potentials
occurs with the pressure rise of each systole.
The medullary cardiovascular center averages
the rate of baroreceptor action potentials over
time so that its response reflects the mean
arterial pressure.

The baroreceptors can also oppose an increase in blood pressure above the reflex set point. For example, if epinephrine is injected into a normal subject, the immediate effects are the same as those described for activation of the sympathetic nervous system: an increase in heart rate, arteriolar and venous constriction, and a dramatic increase in blood pressure. However, the heart slows within a few seconds. The slowing results from an increase in parasympathetic input to the heart, reflexively elicited by the pressure increase detected by the baroreceptors. At the same time, any sympathetic input to the cardiovascular system that may have been occurring at the time of the epinephrine injection is diminished. These adjustments soon restore the blood pressure to a normal value.

Carotid baroreceptors play a more significant role in the baroreceptor reflex than do the baroreceptors in the aortic arch. However, a maximum response requires participation of both sets. The response of these receptors is not linearly related to the increase or decrease in blood pressure; instead, their response becomes more intense the farther the blood pressure departs from the setpoint. This nonlinearity increases the speed and therefore the effectiveness of the baroreceptor reflex.

Role of Blood Volume Regulation in Blood Pressure Regulation

The baroreceptor reflex can respond in seconds to blood pressure changes caused by postural changes or moderate hemorrhage, but blood pressure regulation over periods of hours to days requires the cooperation of a family of neural and hormonal mechanisms that regulates both plasma volume and the baseline tone of vascular smooth muscle. The mechanisms of plasma volume regulation by the kidney

are treated more fully in Chapter 20. The volume regulatory systems are outlined here because of their importance in blood pressure homeostasis. The three main volume regulatory systems are the antidiuretic hormone system, the atrial natriuretic hormone system, and the renin-angiotensin-aldosterone system (Figure 15-5).

Antidiuretic hormone (ADH or vasopressin) has three major effects: (1) it regulates the fraction of water recycled into the body from urine in the kidney, thus affecting the rate of fluid loss from the plasma; (2) the ADH system affects the sensation of thirst, so it contributes to regulation of fluid intake as well as loss; and (3) it is a vasoconstrictor, so increases in plasma ADH tend to increase total peripheral resistance.

As described in Chapter 5 (p. 99), ADH is a small peptide neurohormone secreted by hypothalamic neurons whose axons pass into the posterior pituitary. The ADH neurons integrate two types of sensory input. One type of input is provided by baroreceptors in the wall of the right atrium, which are responsive to changes in central venous pressure. Venous pressure is a sensitive indicator of blood volume. Decreases in central venous pressure signal abnormally low plasma volume and increase the rate of secretion of ADH, causing increased thirst, reduction in the rate of fluid loss in urine, and increased vascular tone (see Figure 15-5). The effect on vascular tone requires especially high levels of ADH and may not be important under most circumstances.

The ADH neurons also respond to changes in the osmotic pressure of plasma, increasing their secretion when plasma becomes more concentrated and decreasing it when the plasma is too dilute. As a result, factors that affect plasma osmolarity also affect blood pressure; implications of this are treated further in Chapter 20.

Recently, the atrial walls have been shown to contain endocrine cells that secrete a peptide called **atrial natriuretic hormone (ANH).** Like ADH, ANH secretion is increased in proportion to atrial stretching, which reflects central venous pressure (see Figure 15-5). The possible effects of ANH are still being explored, but at present it appears that a major effect of ANH is to regulate the rate of loss of plasma solutes and water in the urine.

Secretion of the steroid hormone, **aldosterone,** by the adrenal cortex is controlled in part by a regulatory cascade that originates in the kidney. Endocrine cells of the kidney increase their secretion of **renin** in response to either decreased renal perfusion or decreased plasma Na^+ concentration. Decreased renal perfusion is a signal of decreased mean arterial pressure because, when arterial pressure falls, the baroreceptor reflex constricts the arterioles of the kidney as well as those of most of the other abdominal organs. Thus renin secretion is part of a feedback loop that regulates mean arterial pressure (see Figure 15-5).

In the plasma, renin catalyzes the conversion of a plasma protein, **angiotensinogen,** into **angiotensin I.** Angiotensin I is converted to **angiotensin II** by a converting enzyme present in organs such as the lung and liver. Angiotensin II is converted to **angiotensin III** in the adrenal cortex; this form of angiotensin stimulates aldosterone secretion. Like ADH, angiotensin II is believed to stimulate thirst. Both angiotensin I and II stimulate contraction of vascular smooth muscle, increasing peripheral resistance.

Aldosterone decreases the rate of Na^+ loss in urine. Because Na^+ is ingested continually with food, increased aldosterone secretion increases the body's total content of Na^+. Because almost all Na^+ remains in the extracellular fluid, the ultimate effect of an increase in total Na^+ is an increase in the volume of tissue interstitial fluid and plasma, elevating the blood pressure. Excessive secretion of renin is one cause of chronic elevation of arterial blood pressure **(hypertension).** The hypertension is caused both by the increased plasma volume **(hypervolemia)** that results from excess aldosterone secretion and by the increased total peripheral resistance resulting from the vasoconstriction caused by the angiotensins.

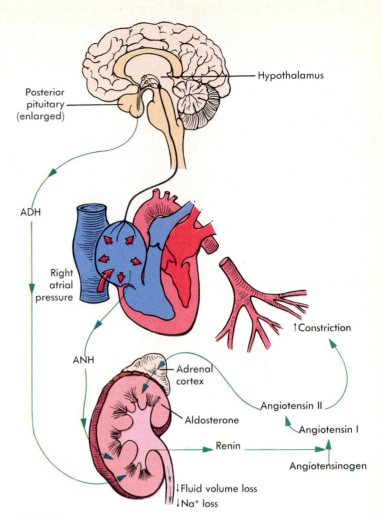

FIGURE 15-5
Summary of three hormone systems that regulate blood volume in response to changes in atrial pressure (ADH and ANH) or renal perfusion (renin-angiotensin-aldosterone). A significant loss of extracellular fluid results in increased secretion of ADH, ANH, and aldosterone; their net effect is to decrease renal excretion of Na^+ and water and to increase thirst.

INTERACTION OF THE HEART AND VESSELS
Effect of Cardiac Output on Central Venous Pressure

Central venous pressure both affects and is affected by the pumping action of the heart. Consider what would happen if the heart were stopped for a time. After heart pumping ceased, blood flow through the circulation would continue until pressure was equal throughout the system. Because the veins are the most compliant vessels, the blood pressure throughout the system after flow had ceased would be determined by the compliance of the veins and the volume of blood in them. A typical value for this pressure would be about 7 mm Hg. If the heart began pumping again, it would move some blood from the venous side into the arterial side of the circulation. Because the arteries are far less compliant

> 1 What is the functional role of the baroreceptor reflex?
> 2 Both ADH and ANH release are affected by stretch of atrial walls. What variable is directly regulated by this input?
> 3 Both ADH and angiotensins cause arteriolar constriction. What property of the vascular system is regulated by these hormones?

Measurement Of Cardiac Output

Cardiac output can be measured using a method based on a dilution principle, first introduced by the physiologist A. V. Fick. An indicator substance is added to the volume of venous blood returning to the heart, and its concentration is measured at an arterial site with the assumption that it has been completely mixed with the blood that passed with it through the heart.

The original version of the Fick Method used oxygen taken up by the lungs as the indicator. Oxygen is added to the blood in a steady stream at a known rate by respiration, and its subsequent dilution can be measured by collecting data on the rate of oxygen uptake and the difference between the oxygen concentration in systemic venous blood and in systemic arterial blood (the A-V O_2 difference). It then becomes possible to calculate cardiac output using the equation:

$$\text{Cardiac output} = \frac{O_2 \text{ consumption}}{\text{A-V}O_2 \text{ difference}}$$

For example, suppose that venous blood oxygen is 14 ml/dl blood and arterial blood oxygen is 19 ml/dl blood. This means that each deciliter of blood that passes through the heart receives 5 milliliter of oxygen, the A-V O_2 difference. A liter (1000 milliliters of blood) would therefore receive 50 milliliters of oxygen. If the rate of oxygen consumption is 250 ml/min, the cardiac output is:

$$\text{Cardiac output} = \frac{250 \text{ ml/min}}{50 \text{ ml } O_2/L}$$
$$= 5 \text{ L/min}$$

This method sounds simple, but it is complicated by the fact that venous bloods draining from various organs differ in oxygen content. Differing venous bloods become well mixed only after they pass through the right ventricle. Therefore the venous blood sample must be collected from the pulmonary artery. This requires cardiac catheterization—passing a cannula into a vein in the arm or leg and from there to the vena cava, through the right atrium, right ventricle, and into the pulmonary artery—entailing some risk to a cardiac patient.

Various dyes can also be used as indicators. The dye is injected rapidly into a vein, and the blood is repeatedly sampled from an artery.

than the veins, the effect of shifting a small volume of blood from the venous side to the arterial side of the circuit would be a large increase in arterial pressure and a small decrease in venous pressure. The greater the cardiac output, the greater the volume shifted would be and the lower the venous pressure would drop. To restate the relationship in terms of venous return, the greater the difference between mean arterial pressure and central venous pressure generated by heart activity, the greater the venous return.

The graph of the relationship between cardiac output (=venous return) and central venous pressure is the **vascular function curve** (Figure 15-6). The vascular function curve describes the behavior of the peripheral vessels, while the cardiac function curve (see Figure 13-25) describes the behavior of the heart as a pump. Both curves relate cardiac output and central venous pressure. Because the vascular function curve will later be superimposed on the same axes as the cardiac function curve, the central venous pressure has been plotted on the horizontal axis, and cardiac output on the vertical axis. The upper limit of venous pressure is the value when

In its course through the heart, the dye becomes well mixed with the blood. Some seconds after the dye is injected, it starts to appear at the sampling site (Figure 15-A). Subsequently it passes through the rest of the systemic loop, returning to the sampling site more and more diluted as it mixes with the total blood volume and is taken into the tissues. The final dilution of the dye is an approximate measure of the total blood volume. A problem with dyes is that the second pass of dye starts to appear at the sampling point just as the first pass is ending; this pileup makes computing the mean concentration of the first pass more difficult. More recently, a pulse of heat has been used; the heat dissipates into the body after passing the sampling site and does not return for a second pass.

FIGURE 15-A
The indicator dilution method for determining cardiac output. The indicator is injected into a major vein as a pulse. As it mixes with the rest of the venous blood in the heart and pulmonary loop, the indicator is diluted and the pulse spreads out because of the variations in linear flow velocity in different parts of the vessels. The pulse arrives at the sampling site in a major artery, where its mean concentration is measured by rapid sampling. The spreading effect allows the arrival of the second passage of the dye to catch up with the tail of the first pass. After two passes the indicator has mixed almost completely with the total blood volume.

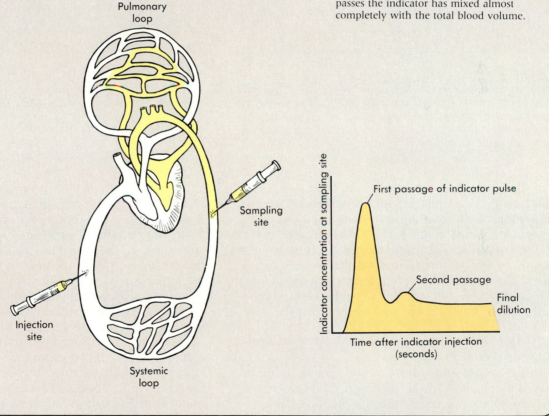

flow is zero; the lower limit is set by the fact that veins begin to collapse and restrict flow at pressures less than the atmospheric pressure (indicated by the plateau in the curve). The vascular function curve can be measured in animal experiments in which the heart is replaced by a mechanical pump and vascular reflexes are blocked.

Cardiovascular Operating Point

Depending on its autonomic inputs and state of health, the heart can maintain a cardiac output of as little as 2 to 3 L/min or of more than 20 L/min. The vasculature could, if maximally dilated, permit a venous return even greater than the heart could pump. The actual flow through the circuit at any instant depends both on the state of the vasculature, as specified by the properties of plasma volume, total peripheral resistance, and venous compliance, and on the performance of the heart, as specified by the heart rate and myocardial contractility. The vascular function curve describes the performance range of the vasculature; the cardiac function curve (see Figure 13-25) describes the performance range of the heart.

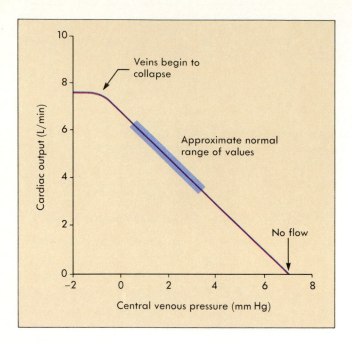

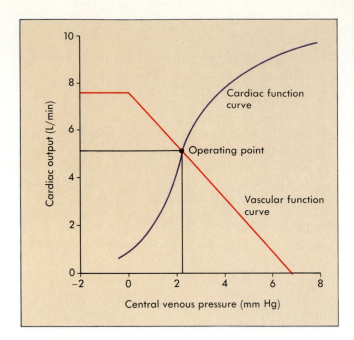

FIGURE 15-6
Vascular function curve. When flow through the vasculature is stopped, venous pressure is maximal; as cardiac output increases, venous pressure decreases. When venous pressures become lower than atmospheric pressure, veins begin to collapse, allowing no further increase in cardiac output.

FIGURE 15-7
When the vascular function curve and the cardiac function curve are plotted on the same axes, they intersect at an operating point, whose coordinates give the cardiac output and central venous pressure determined by the interaction of the two systems.

If the vascular function curve and the cardiac function curve are plotted on the same axes, the intersection of the two curves is the **cardiovascular operating point** (Figure 15-7). This intersection is the one point on the graph at which cardiac output equals venous return. Thus, of the wide range of blood flows possible for the heart and vasculature, the operating point specifies the actual flow that results when the heart and vasculature interact.

Vascular Determinants of Cardiac Output and Mean Arterial Pressure

Cardiac output and central venous pressure are affected by changes in the total peripheral resistance, the compliance of veins, and the blood volume. For simplicity, the effects of vascular changes on cardiac output and venous pressure are first described with the assumption that the heart rate and contractility do not change. In reality, changes in the status of the vessels are almost always accompanied by changes in autonomic inputs to the heart.

An increase in blood volume, such as from a blood transfusion, appears primarily in the veins. Thus the effect of increased blood volume is to increase the central venous pressure (Figure 15-8, A). Consequently, the cardiac output increases as the heart obeys the Frank-Starling Law. An opposite shift of the vascular function curve occurs when there is a decrease in blood volume, such as by hemorrhage (see Figure 15-8, A).

When arterioles dilate, the flow from the arterial side to the venous side increases, filling the veins and increasing the central venous pressure (Figure 15-8, B). As shown in Figure 15-8, the increased stretching of the ventricles during diastole causes the cardiac output to increase to the value (the new operating point) represented by the intersection of the new vascular function curve and the cardiac function curve. Arteriolar constriction has the opposite effect (see Figure 15-8, B).

As metabolically active tissues increase their metabolic rate, the resulting decrease in total peripheral resistance causes the heart to increase the cardiac output. This is a very effective means of regulating cardiac output because it does not require any direct nervous or hormonal communication between the active tissues and the heart. It is the major means by which cardiac output is regulated in heart transplant recipients because a transplanted heart is not neurally innervated. However, greater increases in cardiac output are possible in normal individuals because in them the heart is also stimulated by adrenergic input.

The difference between the actual cardiac output that matches venous return at any instant and the maximum output that could be maintained by the heart is called the **cardiac reserve** (Figure 15-9). The cardiac reserve of a healthy resting person is about 3 L/min, so that without sympathetic stimulation the cardiac output could rise no higher than

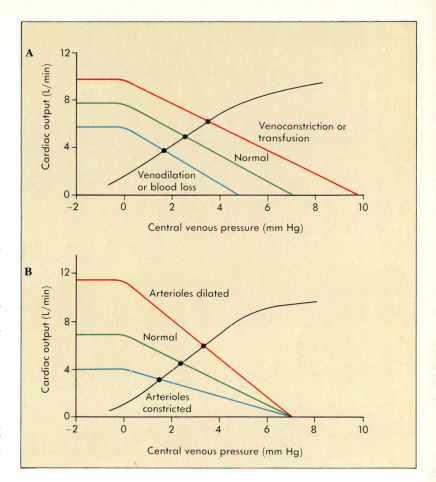

FIGURE 15-8

A   The effect of changes in blood volume or venous tone on the vascular function curve and the cardiovascular operating point. Venoconstriction and blood transfusion increase the cardiac output and venous pressure; venodilation or blood loss decreases cardiac output and venous pressure.

B   The effect of arteriolar dilation or constriction on the vascular function curve and the operating point. Arteriolar dilation (that is, a decrease in total peripheral resistance) increases cardiac output and venous pressure; arteriolar constriction (that is, an increase in total peripheral resistance) decreases cardiac output and central venous pressure. The common feature of the venous tone, volume, and arteriolar diameter changes is that they affect central venous pressure and thus increase or decrease cardiac output according to the Frank-Starling Law.

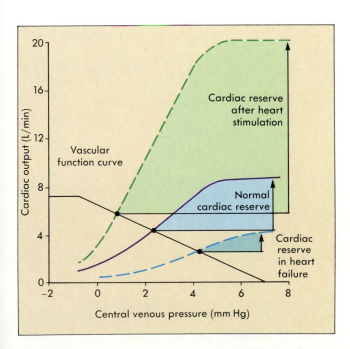

FIGURE 15-9

Cardiac reserve is the difference between the actual cardiac output that matches venous return at any instant and the maximum output that could be maintained by the heart.

about 8 L/min, assuming a resting cardiac output of about 5 L/min. The cardiac reserve sinks toward zero in those with progressive heart failure. Sympathetic stimulation can more than double the cardiac reserve; thus one advantage of extrinsic control of the heart is that it makes possible much greater increases in cardiac output than would be possible from the effect of the Frank-Starling Law alone.

Central control of the heart and blood vessels is also necessitated by the fact that flow autoregulation by individual tissues is a selfish process that can compromise maintenance of mean arterial pressure. For example, consider blood loss, whose immediate effects on cardiac output and central venous pressure are illustrated in the bottom curve in Figure 15-8, *A*. The drop in cardiac output would cause all tissues to become underperfused. In the absence of any overriding central control, a general arteriolar dilation would occur because local autoregulation in the underperfused tissues acts to increase their blood flow. The mean arterial pressure would therefore fall further. In this case autoregulation, if not countered by central control mechanisms such as the baroreceptor reflex, would result in a dangerous lack of control over the driving force for perfusion of essential organs such as the brain and the heart itself.

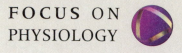

Causes and Consequences of Heart Failure

Acute damage to the myocardium can result from a sudden occlusion of coronary blood flow (a myocardial infarct, or heart attack) or from the effects of hemorrhagic shock. Coronary artery disease may slowly reduce coronary blood flow over a period of time, resulting in progressive damage to the heart muscle. Decreases in myocardial contractility that result from disease or damage are generally referred to as **heart failure.**

In heart failure, right atrial pressure rises as unpumped blood piles up in the vena cava. The increase in right atrial pressure stretches the ventricles, until cardiac output equals venous return. Thus the ventricles compensate for their decreased contractility at the expense of an increased end-diastolic volume. Some decrease in arterial blood pressure also occurs. As the ventricles continue to fail, right atrial pressure becomes higher and higher, so that the volume of blood left in the ventricles at the end of a systole may become several times greater than the stroke volume. Ultimately the cardiac muscle is stretched so much that further stretching does not elicit an increase in cardiac output. In fact, there is a decrease in the strength of contraction and a marked drop in systemic arterial blood pressure.

If heart failure takes place slowly, the decrease in arterial pressure that occurs leads to an increased release of aldosterone. Aldosterone acts to retain Na^+, so that during the course of the failure, plasma volume actually increases. This extra blood volume further increases the load on the failing heart. Some of the extra volume is transferred to the interstitial fluid because the increased venous pressure "backs up" into the capillaries. Edema in the lower parts of the body occurs when the patient is erect, and pulmonary edema, causing difficulty in breathing, may occur when the patient is reclining. These are the symptoms of congestive heart failure, congestive being a term used to describe the accumulation of fluid.

One frequent procedure used in the diagnosis of heart failure, the stress test, involves monitoring cardiac function in an individual during exercise on a treadmill (Figure 15-B).

The treatment of heart failure depends on the cause, and interventions appropriate for one cause may well result in additional damage for another. If the failure results from inadequate blood flow (cardiac ischemia), treatment is aimed at increasing the oxygen supply to the heart. In such cases, stimulating cardiac contractility with drugs such as epinephrine or digitalis may be a two-edged sword. These drugs

---

1 In the vascular function curve, which system variable is taken as the dependent variable? Which is the independent variable?
2 How is the cardiovascular operating point determined graphically?
3 What is the cardiac reserve? What factors affect the magnitude of the cardiac reserve?

THE CARDIOVASCULAR SYSTEM IN EXERCISE AND DISEASE
Standing Upright: the Baroreceptor Reflex and Venous Pumps

In a person who is lying down, the mean arterial blood pressure of the major arteries of the head, trunk, and legs is similar; the same is true for the

THE CARDIOVASCULAR SYSTEM

increase the contractility of the undamaged portion of the heart, but they also increase the oxygen consumption of the heart, usually causing further damage to a region that is inadequately perfused. The internal work of the heart depends on the systolic pressure it must generate and on its end-diastolic volume. Any intervention that serves to decrease right atrial pressure or mean arterial pressure should be beneficial to a failing heart. Consequently, diuretics (drugs that increase urine production) are given to reduce right atrial pressure, and vasodilators to reduce peripheral resistance.

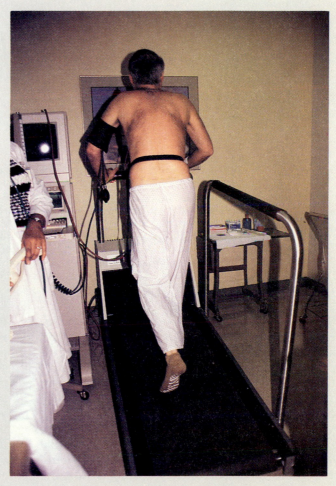

FIGURE 15-B
A patient undergoing a stress test on a treadmill. Cardiac function is monitored as the exercise level is progressively increased.

pressure in great veins (Figure 15-10, *A*). Standing up adds a gravitational component to the arterial and venous pressures of the lower body and reduces the pressures in the upper body above the heart (Figure 15-10, *B*). As a consequence, arterial pressure in the lowermost parts of the body increases, whereas in the head it decreases. The venous pressure in the feet is increased from a few mm Hg to as much as 150 mm Hg, whereas in the head it becomes significantly negative. Prolonged standing or chronic elevated central venous pressure may distend the veins of the legs enough to weaken their walls, resulting in permanently distended **varicose veins.**

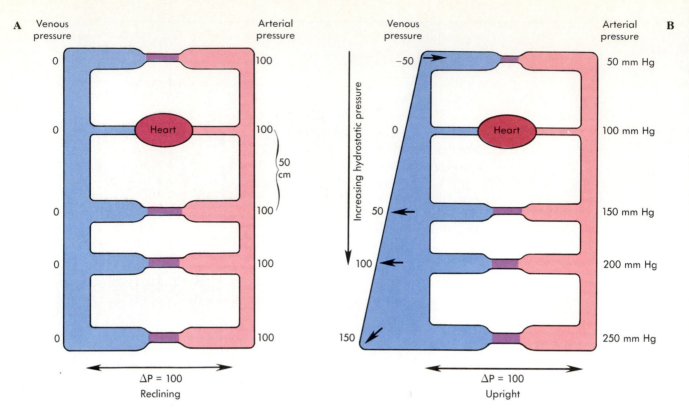

A | Venous pressure | | Arterial pressure | B
Venous pressure | | Arterial pressure

FIGURE 15-10

Pressures in the cardiovascular system for a person who is **A**, lying down and **B**, standing. The pressures in mm Hg are given at various levels in both the arteries and veins. In a recumbent person, the mean arterial and venous pressures are uniform because they depend only on the pumping action of the heart. In a standing individual, the hydrostatic pressure component must be added (below the heart) or subtracted (above the heart). However, the hydrostatic pressure affects only the transmural pressures across the blood vessels *(small solid arrows)*. The pressure difference between the arteries and veins (perfusion pressure) remains unchanged. As a result of the increase in transmural pressure, blood tends to pool in the veins of lower extremities (indicated by the changes in size).

The pressure changes that occur on standing up have implications for the movement of blood through the circulation. The increased venous pressure in the lower body increases the volume of blood in lower veins by as much as 500 to 700 ml, reducing the central venous pressure that is responsible for returning blood to the heart. If these changes were unopposed, the drop in venous pressure and venous return would result in a substantial decrease in cardiac output. Arterial pressure would soon become insufficient to drive blood into the parts of the body above the heart, and fainting would result from inadequate perfusion of the brain. In its effect on arterial pressure, standing up is equivalent to immediately losing more than 500 ml of blood. Standing erect without fainting is possible because of the response of the baroreceptor reflex and because venous return is assisted by the contraction of limb muscles and the respiratory system musculature.

The baroreceptor reflex has three effects: (1) a slight elevation of the heart rate, (2) an increase in the total peripheral resistance, and (3) a decrease in venous compliance. All of these changes are completed within a minute or so after standing up and serve to maintain arterial blood pressure at its normal value at the level of the baroreceptors. Blood pressure measured using a manometer and arm cuff will actually rise slightly after a person stands up because the arm cuff is positioned lower than the baroreceptors and includes a hydrostatic component.

In some people, the baroreceptor reflex is weak or absent, and standing frequently results in fainting. This condition is called **orthostatic hypotension.** Even if the baroreceptor reflex is functioning normally, people standing quietly still experience some venous pooling, causing the stroke volume to drop by about 40%, a drop that may decrease arterial blood pressure enough to cause fainting. However, when an erect person is physically active, pooling is reduced by the operation of two secondary pumps, the **skeletal muscle pump** and the **respiratory pump.**

The veins that pass between and next to the skeletal muscles of the limbs are squeezed and partly emptied with every contraction of the muscles (Figure 15-11). The one-way valves ensure that blood squeezed out of veins moves in the direction of the heart. When the skeletal muscles relax, the veins refill from their arterial ends. Soldiers standing for prolonged periods at attention are taught to rhythmically contract their leg muscles to reduce the pooling of blood and prevent fainting.

During the inspiratory phase of respiration the pressure within the chest cavity falls below atmospheric pressure; during expiration it rises above atmospheric pressure. These pressure changes are transmitted to the portion of the inferior vena cava that is located within the chest cavity (Figure 15-12). The result is that some blood is sucked into the vena cava from the abdominal veins during inspiration and then driven into the right atrium during expiration. The respiratory pump is more effective during exercise because the thoracic pressure changes are greater. The skeletal and respiratory pumps are usually sufficient to overcome the gravitational effect on central venous pressure and maintain stroke volumes equal to or higher than those of people who are reclining quietly.

Hemorrhage and Hypovolemic Shock

The decreased blood volume (**hypovolemia**) caused by hemorrhage may reduce venous pressure to the point that end-diastolic volume is insufficient to maintain cardiac output and arterial blood pressure. Life is threatened when the decreased arterial pressure (**hypotension**) causes critical organs such as the heart and brain to become underperfused

(Figure 15-13). The baroreceptor reflex is the first line of defense against decreasing arterial blood pressure. The reflexive increase in heart rate and peripheral resistance partly compensates for the decreased stroke volume and cardiac output. The gen-

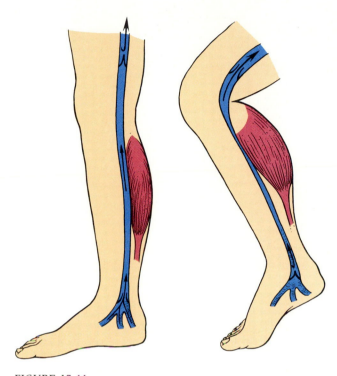

FIGURE 15-11
The skeletal muscle pump depends on the presence of valves in the veins and the fact that muscles adjacent to the veins squeeze these thin-walled vessels when the muscles contract.

FIGURE 15-12
The respiratory pump draws venous blood toward the heart as a result of the changes in pressure in the thoracic cavity during the respiratory cycle. In inspiration, the thoracic cavity enlarges, and, as it does so, the subatmospheric pressure pulls blood from the portion of the inferior vena cava in the abdominal cavity into the thoracic portion of that vein. When deep breaths are taken, the diaphragm moves downward during inspiration, pressing on the abdominal organs and further increasing the pressure difference experienced by the blood in the abdominal and thoracic portions of the inferior vena cava.

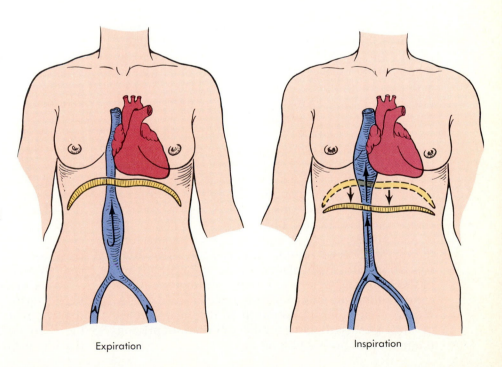

Expiration

Inspiration

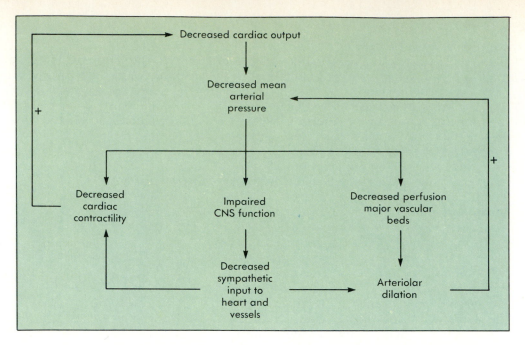

FIGURE 15-13
The positive feedback that can lead to death from hypovolemic shock.

eral arteriolar constriction caused by the reflex does not affect the arterioles of the heart and brain because autoregulation is the major determinant of flow resistance in these organs. In effect, perfusion of the skin, skeletal muscle, and abdominal organs is sacrificed to maintain cardiac and cerebral blood flow.

The drop in mean arterial pressure and the reflexive increase in arteriolar constriction after hemorrhage cause a general decrease in capillary hydrostatic pressure. As noted in Chapter 14, decreases in capillary hydrostatic pressure favor transfer of fluid from interstitial spaces into the blood. Within a few hours after blood loss, as much as several hundred milliliters of plasma may be replaced by this process and by a shift of volume from the lymphatic circulation to the blood circulation. Hours to days may be required to replace the water, ionic constituents, and plasma proteins lost in severe bleeding. Replacement of lost blood cells may require several weeks.

Sweating, diarrhea, vomiting, or kidney diseases that cause excessive urine production can result in large losses of extracellular fluid volume without loss of blood cells. The threat of these fluid losses to circulatory function is similar to that seen in hemorrhage, and the reflexive and hormonal compensations are also similar.

Hemorrhage that results in rapid loss of more than 20% of blood volume may result in **hypovolemic shock.** In hypovolemic shock, underperfusion of the heart and nervous system compromises heart contractility and blocks cardiovascular reflexes. A vicious cycle is set in motion in which the arterioles of underperfused tissues dilate, further depriving the heart and brain of blood.

> *1 What processes make it possible to stand upright without fainting?*
> *2 What are the immediate reflexive responses to hemorrhage?*
> *3 What role does capillary filtration play in homeostasis of plasma volume after hemorrhage?*

Performance in Sustained Exercise: the Cardiovascular Limit

Sustained exercise relies heavily on slowly fatiguing, high-oxidative skeletal muscle fibers. In principle, the workload these fibers can sustain could be limited by their own metabolism or contractility, by the ability of the lungs to exchange CO_2 for O_2, or by the rate at which the circulation can move blood from the pulmonary loop to working muscles. When the cardiac output of an exercising human reaches a maximum, no further increases in O_2 uptake or in workload can be sustained. Thus cardiovascular adjustments are crucially important for maximum performance in endurance exercise. In well-trained athletes, cardiovascular performance may be the limiting factor that separates champions from also-rans. World-class endurance athletes typically can generate maximum cardiac outputs in excess of 40 L/min, so much greater than those of average well-trained athletes as to suggest that truly superior performers in sports such as the marathon, cross-country ski racing, and bicycle racing owe something to genetics as well as to training.

Maximization of blood flow to working muscles necessitates that the cardiac output be increased and that the working muscles receive the largest possible share of the increased output. In nonathletes, the blood flow to skeletal muscle may increase by

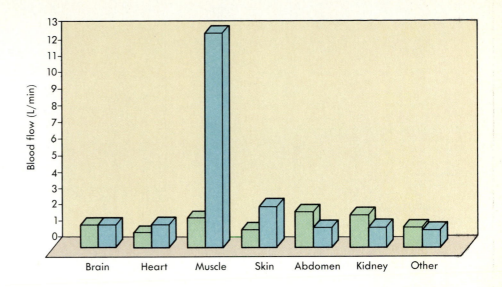

FIGURE 15-14
Typical changes in organ perfusion with exercise. The left bar in each pair shows the resting blood flow; the right bar shows the flow during exercise.

twentyfold; in elite endurance athletes, the increase can be as much as fortyfold. These increases are achieved by a combination of four processes: (1) a large vasodilation in the working muscle, (2) autoregulation of cardiac output according to the Frank-Starling Law, (3) an increase in heart rate and contractility resulting from increased adrenergic input, and (4) sympathetically mediated vasoconstriction in many vascular beds of the visceral circulation (Figure 15-14 and Table 15-3).

In resting muscle, arterioles are highly constricted and much of the blood flow passes through metarteriolar connections to venules; only about 1% of the capillaries are open at any given instant. When skeletal muscle fibers begin to contract, local factors dilate the muscle's arterioles, delivering a greatly increased share of the cardiac output to the muscles (see Table 15-3). Without any other cardiovascular changes, dilation of muscle arterioles would decrease the peripheral resistance. Venous return would increase, causing the cardiac output to increase as a consequence of the Frank-Starling Law.

The activation of sympathetic efferents to arterioles causes a general arteriolar constriction, especially in the abdominal viscera, which diverts some blood flow from the viscera into working muscles (see Figure 15-14). However, the increase in resistance of the abdominal organ circulations is far overbalanced by the decrease in resistance of the vascular beds of working muscles. In sum, total peripheral resistance typically falls during moderate-to-heavy exercise (see Table 15-3), and the greater the mass of working muscle the greater the decrease in total peripheral resistance.

The large increases in cardiac output typical of exercise could not be achieved without changes in the properties of both the heart and vasculature. An increase in cardiac performance alone would cause only a fractional increase in cardiac output. Vascular changes by themselves are slightly more effective but would increase cardiac output only by the

---

TABLE 15-3   Typical Circulatory Changes with Submaximal Exercise

Parameter	Resting	Exercise
Cardiac output	5 L/min	20 L/min
Mean arterial pressure	90 mm Hg	105 mm Hg
Systolic pressure	120 mm Hg	160 mm Hg
Diastolic pressure	70 mm Hg	60 mm Hg
Stroke volume	70 ml	130 ml
Total peripheral resistance	18 mm Hg/L/min	5.25 mm Hg/L/min
Muscle blood flow	1 L/min	10 L/min
Flow to abdominal viscera	1.2 L/min	0.5 L/min
Cutaneous blood flow	0.5 L/min	1.5 L/min

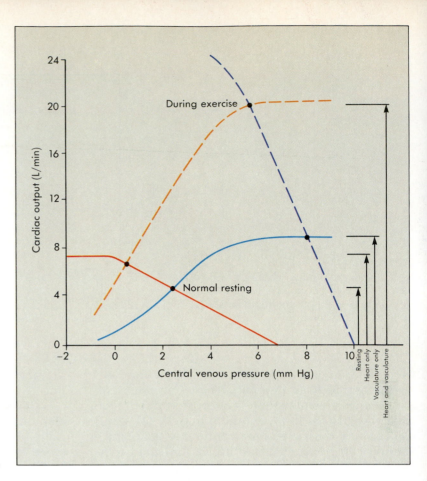

FIGURE 15-15
The importance of both cardiac and vascular responses in attaining high cardiac output in exercise. The two solid curves are the resting cardiac and vascular function curves. The broken curves are curves typical of submaximum exercise. Extrinsic stimulation of the heart alone would raise the cardiac output only slightly. The vascular responses (arteriolar dilation and venoconstriction) can increase cardiac output only to the extent of the cardiac reserve. Interaction of the vascular responses with the cardiac response raises cardiac output by a large factor: fourfold in this example. Considerably larger increases are possible for trained athletes engaging in maximum exercise.

amount of the cardiac reserve (see Figure 15-9). To achieve even the severalfold increase in cardiac output typical of submaximal exercise (see Table 15-3), cardiac performance and vascular changes must occur together (Figure 15-15).

During moderate exercise, increased contractility of the heart and increased stroke volume (see Figure 13-24 and Table 15-3) usually cause pulse pressure to increase (Figure 15-16). At the same time, the decrease in total peripheral resistance speeds blood runoff from the aorta into the arterial tree, allowing aortic pressure to fall rapidly between heartbeats. In well-muscled people the diastolic pressure may even decrease with increasing workload. The net effect of these changes is to leave mean arterial pressure almost unchanged. Thus changes in the driving force for blood flow do not play an important part in increasing performance of the cardiovascular system during exercise. Blood pressure is regulated by the baroreceptor reflex during exercise as during rest. In this process the medullary cardiovascular center must balance the demand of working muscle for blood flow against the necessity of maintaining a mean arterial blood pressure adequate to perfuse the brain.

Vascular Changes in Regulation of Core Body Temperature

In cold environments, the arterioles and veins nearest the surface of the skin are highly constricted, di-

verting most of the blood flow through arteriovenular shunts (see Chapter 14) and reducing the rate of heat loss to the environment from the body core. This vasoconstriction causes the paleness seen in chilled hands and feet.

As environmental temperature increases, heat loss from the body core is increased by reflexive dilation of the cutaneous vessels. A shift of blood volume into the cutaneous circulation occurs (Figure 15-17). If the heat stress is great, the volume shift and the decrease in total peripheral resistance may be only partly compensated for by vasoconstriction in other vascular beds. As a result, venous return and arterial blood pressure may decrease. Standing quietly in an erect position is therefore more likely to cause fainting if the environment is warm than if it is cold.

A hot environment makes maintenance of arterial blood pressure during exercise a difficult balancing act for the cardiovascular center. The increased metabolism of exercising heart and muscles increases the heat load; perfusion of the skin for heat loss conflicts with diversion of cardiac output to muscles. Also, the fluid loss from the plasma caused by sweating further reduces venous return. This combination of stresses can be life-threatening for athletes performing heavy exercise in hot environments.

The effect of exercise on cutaneous perfusion varies depending on whether the environment is

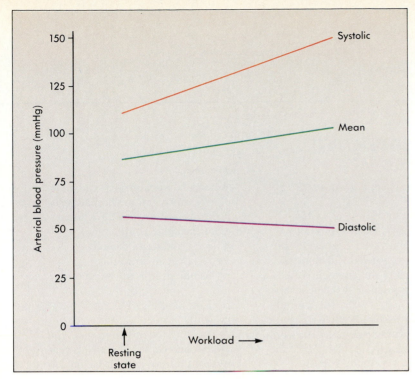

FIGURE 15-16
The effect of increasing workload on systolic pressure, diastolic pressure, and mean arterial pressure.

hot or cold. In cold environments, cutaneous blood flow is low at rest and increases during exercise; in hot environments, cutaneous blood flow is high at rest and decreases with increasing workload.

Hypertension

Abnormally high mean arterial pressure (**hypertension**) may result from elevated plasma volume, elevated sympathetic activity, or other disorders that increase the total peripheral resistance. In some cases, excessive secretion of renin is the cause. Hypertension may be worsened by excessive consumption of NaCl beause salt retention causes increases in plasma volume. In more than 95% of individuals with hypertension, the underlying cause of the disease is unknown. Hypertension usually causes no immediate symptoms but poses a long-term threat to health because of the stress it places on the heart and vessels. Whatever the cause of hypertension, therapy typically involves some combination of **diuretics** (drugs that increase urinary loss of fluid and solutes) and adrenergic antagonists (drugs that decrease cardiac contractility and dilate arterioles by blocking the effect of sympathetic stimulation).

1 *What mechanisms are responsible for the large increase in cardiac output possible during exercise?*
2 *What adjustments enable mean arterial pressure to remain essentially constant during exercise?*
3 *What is hypertension? In what ways can it be reversed by therapy?*

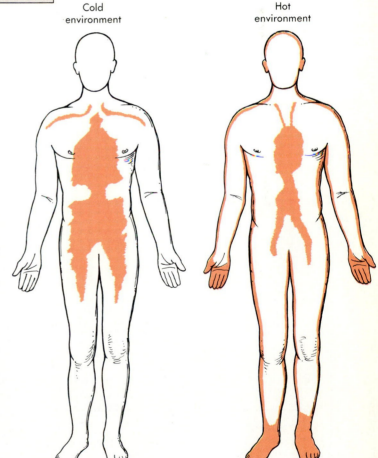

FIGURE 15-17
Major locations of venous volume in a cold environment *(left)* and in a hot environment. In the cold, cutaneous arterioles are highly constricted, and most flow takes the deeper arteriovenous shunts. In nonexercising people in hot environments, cutaneous arterioles are greatly dilated, and a significant fraction of the blood volume is near the skin surface.

Atherosclerosis

About 50% of all deaths from heart disease in developed nations are the result of atherosclerosis in the arteries of the heart or brain. Atherosclerosis is a disease process that results in the formation of abnormally thickened regions of the vascular wall called plaques (Figure 15-C). Plaques are characterized by an abnormal proliferation of modified smooth muscle cells and deposits of large quantities of cholesterol. Damage to the heart or brain can result from progressive blood deprivation as plaque development narrows arteries. Because the plaque is an abnormal surface, a clot (thrombus—see Chapter 14, p. 354) may form on it and block the artery, causing a heart attack or stroke.

A recent technique, called **angioplasty,** has been used to enlarge narrowed arteries. In this procedure, a deflated balloon is inserted into the diseased vessel and then briefly expanded several times (Figure 15-D). This procedure resembles cardiac catheterization and, because angioplasty does not require surgery, it may reduce the number of coronary bypass operations performed and thus decrease total health-care costs.

Considerable progress has occurred in understanding the development of plaques. The initial event in the formation of a plaque is believed to be an injury to the arterial endothelium that causes an inflammatory lesion. Of all aspects of the disease, the least is known about the cause of the initial lesions. Candidates include chemical agents from the environment, substances produced in the body itself, or the mechanical stress of elevated blood pressure.

Once the lesion has formed, the white blood cells and platelets that invade it release growth factors (see Chapter 14, p. 350) that cause surrounding smooth muscle cells to proliferate, migrate towards the interior of the arterial wall, and accumulate cholesterol. The progress of atherosclerosis is greatly accelerated by eating foods

FIGURE 15-C
A micrograph showing an atherosclerotic plaque in a coronary artery.

FIGURE 15-D
In angioplasty, a catheter with a deflated balloon on the end is inserted into a peripheral vein and carefully manipulated into a narrowed coronary artery. It is then inflated several times to restore the artery to its normal diameter.

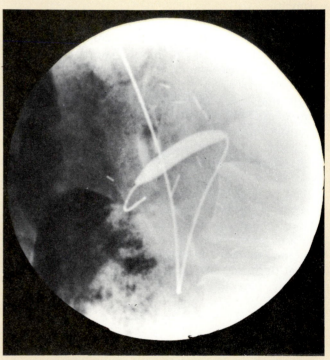

Angioplasty

excessively high in cholesterol content and by having high plasma levels of low-density lipoprotein, one form in which cholesterol is transported in the blood. High-density lipoproteins, on the other hand, tend to remove cholesterol from the circulation, and high levels of these lipoproteins are associated with a decreased incidence of atherosclerosis.

Although a large fraction of the population develops at least some plaques, not all people develop clinical disease as a result. The incidence of coronary artery disease, one manifestation of atherosclerosis, increases with age and is higher in men than in women in all age categories (Figure 15-E). The Framingham study was a landmark investigation of 5000 inhabitants of a small Massachusetts town who were monitored more than 20 years. This study found that hypertension, elevated blood lipid levels, cigarette smoking, obesity, low levels of physical activity, and diabetes are all risk factors associated with a higher incidence of coronary artery disease.

In addition to reducing cholesterol intake, there is evidence that other dietary factors can reduce the risk of atherosclerosis. For example, fish oils contain omega-3 fatty acids, which seem to increase high-density lipoprotein levels. Exercise seems to decrease low-density lipoprotein levels while simultaneously increasing high-density lipoprotein levels. Because the body can synthesize cholesterol from other substances, it is important to know the entire lipid profile of an individual to accurately assess his or her risk of coronary artery disease.

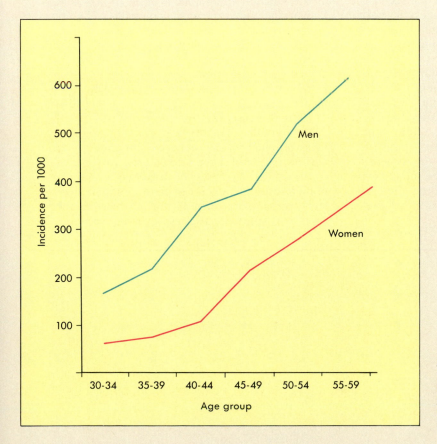

FIGURE 15-E
Incidence of coronary heart disease in the Framingham study population. Incidence is given as the cumulative number of people who developed disease over the 24 years of the study.

SUMMARY

1. Most blood vessels are innervated only by sympathetic adrenergic fibers. **Vasoconstriction**, the predominant adrenergic effect, is mediated by α_1-receptors. In some tissues such as skeletal muscle, **vasodilation** is mediated by β_2-adrenergic receptors and by cholinergic sympathetic fibers. Vessels of the external genitalia are dually innervated; parasympathetic inputs cause dilation and sympathetic inputs, constriction.

2. The **baroreceptor reflex** counters mean arterial pressure changes over a time scale of seconds to minutes. The medullary **cardiovascular center** integrates inputs from baroreceptors in the walls of the carotid sinus and aortic arch. Decreased mean arterial pressure, indicated by decreased baroreceptor firing, results in decreased vagal tone to the heart, increased sympathetic input to heart and vessels, and increased release of epinephrine from the adrenal medulla. The increased heart effectiveness and vasoconstriction return arterial blood pressure towards the setpoint value.

3. Regulation of mean arterial pressure over hours to days involves three hormonal systems that regulate extracellular fluid volume: the **antidiuretic hormone** system, the **atrial natriuretic hormone** system, and the **renin-angiotensin-aldosterone** system.

4. The properties of the cardiovascular system are **total peripheral resistance, blood volume, venous compliance, heart rate,** and **contractility.** Given a particular set of values for these properties, the **cardiovascular operating point** is the one flow value for which the cardiac output equals the venous return. A change in one or more of the system's properties changes the outcome of the interaction between heart and circulation and affects the system variables: central venous pressure, mean arterial pressure, and cardiac output.

5. Standing results in pooling of blood in compliant veins of the lower body and a drop in venous return and central venous pressure, with a consequent drop of mean arterial pressure in the arteries above the heart. Mean arterial pressure is maintained by the baroreceptor reflex, which is assisted by the pumping action applied to large veins by muscular movements of locomotion and respiration (the **skeletal** and **respiratory pumps**).

6. The responses to moderate **hemorrhage** resemble those involved in standing, except that lost fluid, plasma proteins, and blood cells must be replaced. Fluid transfer from interstitial spaces to plasma can make up for some lost volume; ultimately, fluid replacement involves responses of volume-regulatory hormonal control systems. Replacement of proteins and cells requires days. Loss of more than 20% of blood volume may cause **hypovolemic shock,** a positive feedback cycle in which underperfusion of the heart and brain reduces the ability of the system to compensate for the lost blood.

7. The cardiovascular responses to exercise include:
   a. Vasodilation and a large increase in perfusion of working muscle, mediated mainly by flow autoregulation.
   b. An increase in the **cardiac reserve**, which allows the heart to match cardiac output to increased venous return from working muscle.
   c. Sympathetically mediated vasoconstriction of abdominal viscera, partly compensating for the decreased resistance of working muscle.

   The response of cutaneous vessels to exercise reflects a compromise between **thermoregulation** and regulation of mean arterial pressure.

8. In most cases, causes of hypertension are unknown; drug therapy for hypertension may involve some combination of reduction of total peripheral resistance, heart effectiveness, or blood volume.

● STUDY QUESTIONS

1. How would venous return be altered by transfusion of blood? By arteriolar constriction? By decreases in venous compliance?

2. The cardiac output of a denervated heart still increases significantly during exercise. What mechanism makes this possible?

3. If myocardial contractility is increased by a drug, what will be the effect on central venous pressure? On cardiac output? On mean arterial pressure?

4. Define cardiac reserve. What factors increase the cardiac reserve? What factor might decrease it?

5. A healthy person lying on a tilt table is shifted into an upright position. What happens to the baroreceptor firing rate? To the action potential frequency in the nerve fibers innervating the SA node? To the mean capillary pressure in the foot? To the venous compliance? What other changes are initiated by the baroreceptor reflex?

6. A person's hematocrit is measured before and several hours after donation of a pint of blood. The second hematocrit is found to be lower than the first. Explain the changes that decreased the hematocrit during the hours following the donation.

7. A healthy person loses one liter of blood. How would the following factors change as an immediate consequence of the blood loss, before any compensation has had time to take place?

Cardiac output	Mean arterial pressure
Central venous pressure	Total peripheral resistance

8. What effect will the reflexive compensation for loss of the liter of blood in Question 7 have on:

Total peripheral resistance	Cerebral blood flow
Central venous pressure	Renal blood flow

9. Why is regulation of both mean arterial pressure and body temperature more difficult for people exercising in a hot environment?

● SUGGESTED READING

ALDER, V: Beyond Balloons, *American Health*, March 1988, p. 14. Describes how lasers and subminiature drill bits have been used to treat coronary artery disease.

DAWBER, T.R.: *The Framingham Study*, Harvard University Press, Cambridge, Mass., 1980. Landmark investigation of the epidemiology of atherosclerosis in the inhabitants of a small Massachusetts town. This study linked elevated lipid levels, cigarette smoking, obesity, and low physical activity to arteriosclerosis.

EDWARDS, D.D.: Lasers and Tips and Drills, *Science News*, December 12, 1987, p. 376. Describes the use of lasers and microminiaturized drills to open coronary arteries blocked by atherosclerotic plaques.

EDWARDS, D.D.: Repairing Blood Pressure Damage, *Science News*, May 1988, p. 292. Describes how new antihypertensive drugs can improve long-standing damage to the cardiovascular system.

EISENBERG, S.: Type A and Coronary Artery Disease, *Science News*, November 7, 1987, p. 292. An examination of data linking personality characteristics to the incidence of heart disease. It also examines whether stress-control techniques are effective in preventing coronary artery disease.

EISENBERG, S.: Smoking Raises Female Heart Attack Risk, *Science News*, November 22, 1987, p. 341. An examination of epidemiological data linking increased risk of heart attack in women with increased use of tobacco.

ERON, C.: Fatty Acids Cut Heart Artery Narrowing, *Science News*, September 24, 1988, p. 197. Discusses studies suggesting that omega-3 fatty acids can reduce the risk of atherosclerosis.

GRADY, D.: Can Heart Disease Be Reversed? *Discover*, March 1987, p. 54. Discusses the ways diet modification, exercise, and drugs affect coronary artery disease.

GUNBY, P.: Laser May Provide Better Channel, Smoother Lumen in Future Coronary Artery Occlusions, *Journal of the American Medical Association*, March 13, 1987, p. 1283. Self-explanatory, but points out that the technique needs further refinements before it can be generally applied.

MCARDLE, W.E., F.I. KATCH, and V. L. KATCH: *Exercise Physiology*, second edition, Lea & Febiger, Philadelphia, 1986. A readable, comprehensive textbook on the subject written for courses in sports medicine.

ROWELL, L.B.: *Human Circulation Regulation During Physical Stress*, Oxford University Press, New York, 1986. Comprehensive coverage of cardiovascular responses to exercise.

SCHMECK, H.M.: The Revolution in Heart Treatment, *New York Times Magazine*, October 9, 1988, p. 14. Describes a variety of new therapeutic approaches to the prevention and treatment of heart attacks.

SILBERNER, J.: No Free Lunch, *Science News*, volume 1, December 1986, p. 360. Discusses the fact that drugs cannot substitute for exercise in maintaining cardiovascular fitness.

SILBERNER, J.: Artery Clogging and APO-B, *Science News*, volume 1, February 1987, p. 90. Describes how apolipoprotein B may reduce blood cholesterol levels and the incidence of coronary artery disease.

THE RESPIRATORY SYSTEM

CHAPTER 16

Respiratory Structure and Dynamics

On completing this chapter you will be able to:

- Describe the components of the respiratory system, identifying the classes of airways, gas exchange elements, and the muscles responsible for ventilation.

- Describe the defense mechanisms of the lung.

- Understand how the terminal bronchioles are neurally and hormonally regulated.

- Appreciate the factors that affect gas exchange in the lung.

- Define the lung volumes and capacities and provide approximate normal values for them.

- Understand the factors that affect respiratory minute volume.

- Explain why intrapleural pressure is always less than alveolar pressure.

- Distinguish between the elastic and flow-resistive components of the work of breathing.

- Understand the role of pulmonary surfactant. Summarize the basic characteristics of obstructive and restrictive lung disease.

- Compare pulmonary and systemic blood flow.

- Describe the conditions under which there may be an imbalance in the ventilation-perfusion ratio of regions of the lung, and show how this affects the O_2 and CO_2 content of alveolar gas.

*T*he respiratory system links the circulatory system with the atmosphere, an infinite source of oxygen and an infinite sink for the carbon dioxide generated by cellular oxidative metabolism. In some amphibians, gas exchange by diffusion across the skin (cutaneous respiration) is sufficient to support the animal's relatively low metabolic rate. If this were the mechanism of gas exchange in human beings, the high rate of metabolism would necessitate a body surface area so great that the entire body plan would have to be unrecognizably different. The lungs constitute only a few percent of total body mass but contain within their structure the great amount of surface needed to support the required rate of diffusional gas exchange. Much of respiratory physiology is concerned with the problems posed by ventilation of such a complex structure.

In the mythology of the ancient Greeks, *pneuma*, or breath, was an invisible personal spirit that gave its possessor life. For healthy people, breathing is easily taken for granted, as it is almost effortless and goes on for the most part without conscious awareness. For those with respiratory disease, every breath may be hard won. Respiratory diseases are in large part due to breathing contaminated air. Poor air quality is sometimes regarded as an exclusively modern problem, but autopsies of mummified Egyptians and the frozen bodies of prehistoric Inuits recovered in Alaska suggest that inhalation of sand particles and smoke from cooking fires caused lung disease long before the industrial revolution.

STRUCTURE AND FUNCTION IN THE RESPIRATORY SYSTEM
Lungs and Thorax

The **thoracic cavity,** or chest cavity, is a compartment bounded by the rib cage (made up of the ribs, **intercostal muscles** connecting one rib to another, and connective tissue) and closed by a sheet of skeletal muscle called the **diaphragm** (Figure 16-1, *A*). The **pleura** is a sheet of epithelial tissue that lines the interior of the thorax and surrounds each lung (Figure 16-1, *B*). The portion of the pleura lining the interior of the thoracic cavity is the **parietal pleura;** the **visceral pleura** covers each lung. The **intrapleural space** is enclosed by the parietal and visceral pleura. Within the intrapleural space there is a small amount (several milliliters) of **intrapleural fluid** that minimizes friction between the wall of the thoracic cavity and the lungs as the lungs inflate and deflate. The left lung is smaller than the right and consists of two lobes, whereas the right lung has three.

Structures of the Airway

The lungs are ventilated with atmospheric air by way of the treelike **airway.** The airway consists of all structures that conduct air between the atmosphere and the **alveoli,** the membranous lung structures across which gas exchange takes place. The parts of the airway that lie outside the lung are the nose, **pharynx** (throat), **larynx** (vocal cords), **trachea,** and right and left **primary bronchi.**

Within the lung, each primary bronchus branches extensively (Figure 16-2; Table 16-1). The first branching gives rise to secondary bronchi; the second branching gives rise to segmental (tertiary, quaternary, etc.) bronchi that supply discrete areas within the lobes. Subsequent branchings result ultimately in about 150,000 **terminal bronchioles** (enlarged in Figure 16-2). Terminal bronchioles are surrounded by multiunit smooth muscle cells that can adjust the diameter of these passageways in response to autonomic innervation and circulating

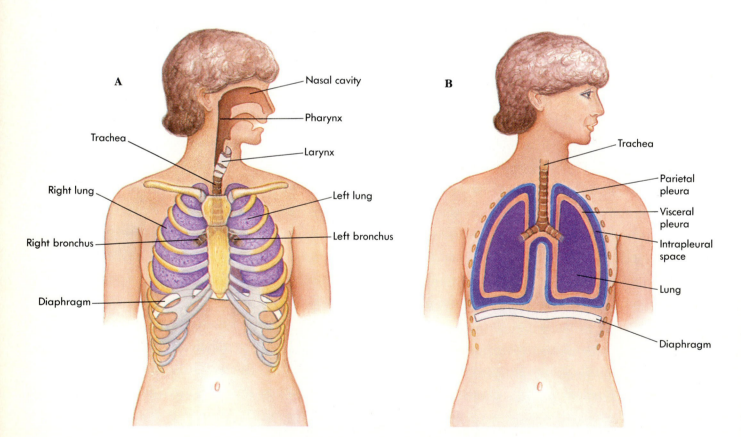

A — Nasal cavity
— Pharynx
Trachea — Larynx
Right lung — Left lung
Right bronchus — Left bronchus
Diaphragm

B — Trachea
— Parietal pleura
— Visceral pleura
— Intrapleural space
— Lung
— Diaphragm

FIGURE 16-1

A The anatomy of the thorax showing major airway components (pharynx, larynx, trachea, and left and right bronchi), the chest wall, and abdomen. The thoracic cavity is separated from the abdominal cavity by the diaphragm.
B The lungs are separated from each other and from the chest wall by the parietal and visceral pleura. The intrapleural space is exaggerated for clarity.

FIGURE 16-2

The anatomy of the respiratory airway. Air moves through the conducting zone of the lung (the trachea, left and right primary bronchi, secondary and tertiary bronchi, and bronchioles) by bulk flow. The figure shown is a simplification; there are believed to be 12 generations of branches, which cannot all be shown on the same scale. Terminal and respiratory bronchioles, alveolar ducts, aveolar sacs, and alveoli are shown in the enlargement. The alveolar ducts and sacs form the exchange zone of the lung. Here diffusion replaces bulk flow because the amount of branching is so great as to reduce the average flow velocity to nearly zero.

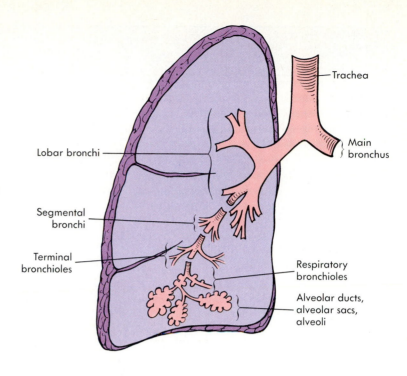

TABLE 16-1   Subdivisions of the Respiratory Tree

Name	Diameter	Number	Cartilage
Trachea	1.8 cm	1	Mucus secretion
Main bronchi	1.2 cm	2	Smooth muscle
Lobar bronchi	0.8 cm	3 on right, 2 on left	Cilia
Segmental bronchi	0.6 cm	10 on right, 8 on left	
Terminal bronchioles	1.0 mm	About 48,000	
Respiratory bronchioles	0.5 mm	About 300,000	Gas exchange
Alveolar ducts	400 μm	About 9×10^6	
Alveoli	250 μm	About 3×10^9	

epinephrine. The absence of cartilage in the terminal bronchioles makes them susceptible to collapse if enough external pressure is applied across their walls. Bronchioles with diameters smaller than 0.5 mm are referred to as **respiratory bronchioles.** Respiratory bronchioles differ slightly from terminal bronchioles in that they contain a few alveoli in their walls.

The respiratory bronchioles eventually terminate in the exchange zone of the lung, the **alveolar ducts** and **alveolar sacs** (see Figure 16-2). Alveolar ducts contain smooth muscle cells, and thus their diameter can be regulated by the contraction or relaxation of the muscle. The alveolar ducts and alveolar sacs give rise to numerous spherical alveoli, so gas exchange can occur in both. However, alveoli located in the alveolar sacs provide most of the surface area for gas exchange.

The bronchi and bronchioles are major sites of **airway resistance.** Airway resistance is affected by autonomic input and by paracrine agents (Table 16-2). Parasympathetic activity constricts bronchiolar smooth muscle. Sympathetic input is mainly in the form of norepinephrine and adrenal epinephrine. Activation of adrenergic receptors leads to relaxation of bronchiolar smooth muscle, which dilates the airways and decreases their flow resistance. The increase in sympathetic input makes increased ventilation easier in exercise. Histamine is a paracrine substance released in response to infection and, in susceptible people, to substances called **allergens** that trigger inappropriate responses of the immune system. The effects of histamine include constriction of bronchiolar smooth muscle, which increases the effort of ventilation and decreases alveolar ventilation. In severe allergic attacks, the airway resistance

TABLE 16-2	Factors that Affect Bronchiolar Smooth Muscle

Bronchiolar dilators—decrease airway resistance

Sympathetic input (epinephrine, norepinephrine)
Adrenergic agonists (for example, theophylline)
Antihistamines
Increased CO_2 in alveolar air*

Bronchiolar constrictors—increase airway resistance

Parasympathetic input (acetylcholine)
Irritants (smoke, dust, chemicals)
Histamine†
Prostaglandins D_2, F_2 α
Leukotrienes

*This factor is involved in ventilation-perfusion matching described in the last part of this chapter.
†The last three factors are produced in immune and allergic responses (see Chapter 24).

may rise by twentyfold, and the effect on ventilation may be life threatening.

The airway structures are specialized for air flow, and no exchange of O_2 or CO_2 with the blood occurs across their surfaces. During passage across the warm, moist surface of the airway, the air is warmed to body temperature and saturated with water vapor. The water vapor protects the delicate surfaces of the lungs from desiccation. The airways are surrounded by smooth muscle that can respond to activation of the autonomic nervous system, inspired particles and chemicals, and substances released locally by cells along the respiratory tract. Except for the terminal bronchioles, the conducting airways are supported by incomplete rings of cartilage that make it quite difficult to collapse them. With each subsequent branching the flow velocity of air becomes less, reaching very low values in the smallest branches and the alveoli themselves.

> 1 What is the intrapleural fluid?
> 2 What are the subdivisions of the airways between the trachea and the alveoli?

Sound Production

Sound production is one of the functions of the respiratory system (Figure 16-3, *A*). The vocal cords (Figure 16-3, *B*) of the **larynx** are abducted toward the wall of the larynx and away from the airstream in normal breathing. For sound production, the vo-

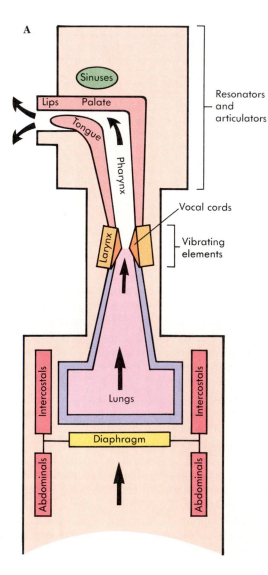

A

Sinuses

Lips Palate

Tongue

Pharynx

Resonators and articulators

Vocal cords

Larynx

Vibrating elements

Intercostals

Lungs

Intercostals

Diaphragm

Abdominals

Abdominals

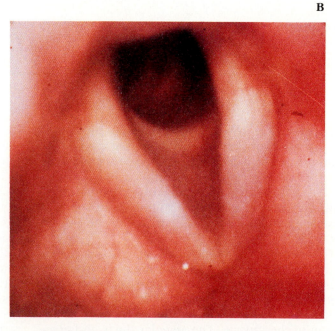

B

FIGURE 16-3

A Mechanical elements that contribute to sound production.
B The vocal cords as seen through a fiber-optic device known as an endoscope.

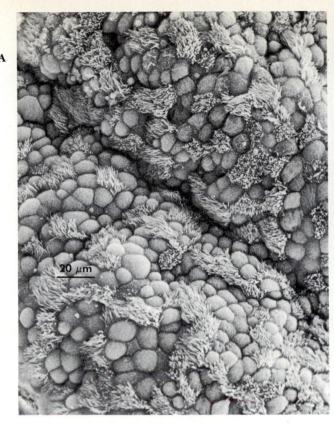

A

B

FIGURE 16-4

The epithelium of a small bronchus is shown at low (A) and high (B) magnification in scanning electron micrographs. Note that the epithelial surface is provided with tufts of cilia.

cal cords are moved into the airstream, which vibrates them as it passes. The intensity of the sound produced is controlled by the position of the vocal cords in the airstream. The pitch is controlled by muscles attached to the cords that regulate their tautness; for high-pitched sounds the muscles are contracted to increase the tautness. Increased tautness also causes the edges of the cords to become thinner. The natural resonant frequency of the cords (see Chapter 10, p. 228, for a discussion of resonance) is increased by both the tightening and the thinning.

The range of sound frequencies produced by adult men is typically lower than for women because of sex-related differences in the dimensions of the vocal cords. The sounds produced by the vocal cords are altered by the articulators and resonators that consist of the lips, tongue, mouth, nose and nasal sinuses, pharynx, and chest cavity (see Figure 16-3, A). Differences in resonances are responsible in large part for the different voice qualities of different individuals.

Speech requires the coordination of many sets of muscles that are innervated by motor neurons that are widely distributed in the spinal cord and the brain. The high level of sophistication of human vocalization is correlated with a correspondingly large representation of respiratory muscles in the motor cortex. Because sound production is based on the principle of forced air, production of speech requires exquisite control of the respiratory muscles: the diaphragm, the intercostal muscles, and the abdominal muscles. The pattern of inspiration and ex-

piration recorded from an experimental subject is altered significantly when the subject is asked to change from quiet breathing to speaking or singing. Holding a note requires that the pressure in the airway be constant so the pressure source must be operated in a different manner from the pattern of rising and falling pressure characteristic of quiet breathing (see Figure 16-3, A).

Protective Functions of the Airway

Along with inspired air comes an incredible array of environmental materials, including dust particles, pollen, bacteria, fungal spores, and viruses. The airway possesses several lines of defense against particulate contaminants. Sensory endings in the nasal passages, trachea, and bronchi elicit the sneezing and cough reflexes, which reduce the intake of foreign particles to a limited extent. Large particles (over 5 to 10 microns in diameter) settle out or are caught on the hairs and mucus in the nasal passages. Breathing through the nose rather than the mouth improves the chances that larger particles will be stopped in the first levels of the respiratory tract, because air is drawn across a longer pathway before it enters the pharynx.

The second mechanism for cleansing the air is provided by the ciliated epithelial lining of the airway (Figure 16-4). Mucus is continuously secreted by epithelial cells and glands. The cilia propel the mucus upward toward the trachea at a rate of as much as 1 cm/min. Trapped particles are thus carried from the airway into the pharynx to be swallowed or coughed up; the process has been called

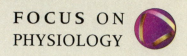
Asbestosis and Mesothelioma— Consequences of Exposure to Asbestos

Asbestos is a general term for several fibrous, heat-resistant minerals used in insulation, floor tiles, roofing, and other products. A large number of workers have been exposed to high levels of asbestos fibers; the entire population has received some level of exposure from asbestos products. The full potential of asbestos to cause disease has been recognized only recently, since in many cases decades pass between the first exposure to asbestos and the onset of illness. As asbestos fibers accumulate in the lungs, they are encapsulated, forming fibrotic lesions similar to those caused by other dusts. About 9% of asbestos insulation workers die of the consequences of fibrotic disease.

One recent survey shows that about 19% of insulation workers die of lung cancer, a rate about 7 times greater than the general population. The effect of cigarette smoking on the incidence of lung cancer in asbestos workers is an example of how risk factors may interact to multiply the incidence of a disease. Nonsmokers with asbestos exposure die of lung cancer at a rate about five times greater than nonsmokers who do not have documented asbestos exposure. Smokers die of lung cancer at a rate about 10 times greater than nonsmokers. Asbestos workers who are also smokers die of lung cancer at a rate that may be as great as 50 times that of nonsmokers without asbestos exposure. About 8% of insulation workers die of mesothelioma, a cancer of the pleura that is extremely rare in the general population. This cancer appears to be the result of migration of asbestos fibers from the lung to the pleura.

Data from sources cited in CHURG, A., and F. H. Y. GREEN, *Pathology of Occupational Lung Disease*, Chapters 7 and 8, Igaku-Shoin, New York, 1988.

the **mucociliary escalator.** The mucociliary escalator is inhibited by a variety of materials including tobacco smoke, cold air, and many drugs. Heavy smokers, for example, typically have an episode of early morning coughing elicited by the accumulation of mucus and the irritation caused by the foreign particles.

Defense against infection is provided by white blood cells that leave nearby capillaries and move about on the surface of the airway and alveoli. The specific mechanisms whereby these cells inactivate infectious organisms and remove small particles will be described in Chapter 24.

In spite of the airway's defenses, very small particles (of the order of 5 microns diameter or less) reach the alveoli; some lodge there permanently. The lungs of all citizens of industrialized nations accumulate a burden of carbon particles that is clearly visible at autopsy. The burden of foreign material in the lungs is heavier in smokers than nonsmokers.

A family of disabling lung diseases results from exposure to specific particulates. The "black lung" of coal miners, the "brown lung" of textile mill workers, the silicosis of glass workers, and the asbestosis of steamfitters and insulation workers are a few examples of such diseases. A common feature of all of these diseases is a scarring, or **fibrosis,** of the alveolar membrane that affects both the length of the path for gas diffusion and the elasticity of the lung.

Acini and Alveoli

The basic functional unit of the lung is an **acinus,** a structure that resembles a bunch of grapes (Figure 16-5). Each acinus arises from a single terminal bronchiole, the stem of the bunch, which branches into about 100 alveolar ducts. Each alveolar duct terminates in about 20 alveoli, each 100 to 300 microns in diameter. There are about 300 million alveoli in each lung, giving it a spongelike consistency. The alveoli are supported by a mesh of elastic connective tissue cells called **parenchymal cells.**

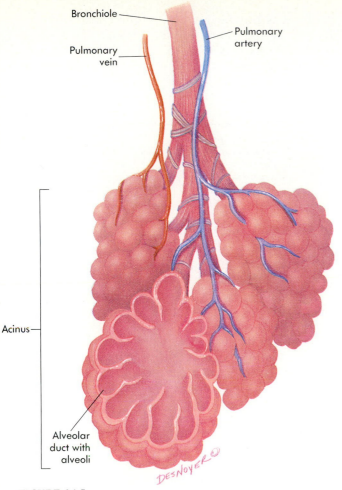

FIGURE 16-5

A pulmonary acinus consists of a terminal bronchiole, alveolar ducts and sacs, and associated circulation.

Movement of gas between the atmosphere and the blood occurs in two stages: bulk flow and diffusion. In the **conduction zone** of the lung, consisting of the network of branching bronchi and bronchioles, there is bulk flow of air (convection). Bulk flow is a much more effective mechanism than diffusion for transporting substances over long distances. Between the respiratory bronchioles and the alveoli, referred to as the **diffusion zone** of the lung, gas exchange occurs by diffusion. Gas exchange between alveoli and pulmonary capillary blood also occurs by diffusion.

Diffusion is favored by large surface area and impaired by long path length. Almost every aspect of the structure of the lung and pulmonary circulation reflects two general adaptations that maximize the effectiveness of diffusive gas exchange in the lung—minimization of the distance that gases must move by diffusion between alveolar air and capillary blood, and maximization of the surface available for diffusional gas exchange.

The barriers that lie in the diffusional path between alveolar gas and capillary blood (Figure 16-6) are (1) film of moisture on the alveolar surface, (2) the single thin layer of epithelial cells with their basement membrane, (3) the parenchyma and interstitial fluid, and (4) the capillary wall. The total path length posed by these structures is only about 0.5 to 1.5 microns (Figure 16-7). The surface area for gas exchange of alveoli in the lung is 60 to 80 m^2, about the same as a tennis court! The large surface area of alveoli is matched by dense vascular beds—in all there are about 30 billion pulmonary capillaries, or about 100 capillaries per alveolus. Alveoli can be vi-

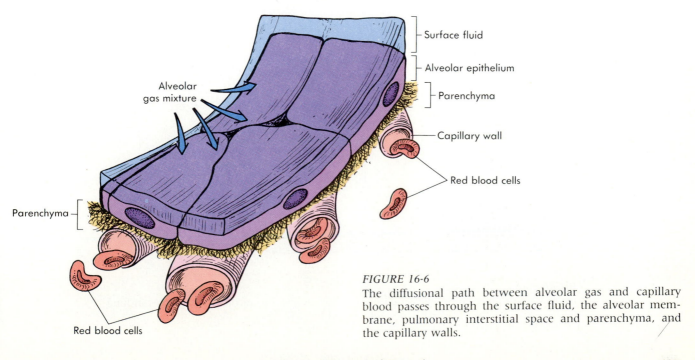

FIGURE 16-6

The diffusional path between alveolar gas and capillary blood passes through the surface fluid, the alveolar membrane, pulmonary interstitial space and parenchyma, and the capillary walls.

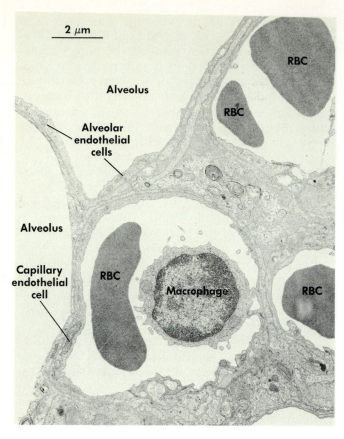

FIGURE 16-7
Pulmonary capillaries and adjacent alveoli in cross section in the electron microscope.

sualized as tiny air bubbles whose entire surface is bathed by the blood flowing through the lung. In all, the lungs may contain as much as 1500 miles (about 2400 km) of blood vessels, and each milliliter of blood may be spread out over about 70 miles (about 110 km) of capillaries.

> *1 How are sounds produced?*
> *2 What mechanisms protect the lung from air particulates?*
> *3 What structural features maximize the diffusing capacity of the lung?*

RESPIRATORY VOLUMES AND FLOW RATES
Lung Volumes and Capacities

The lung volume changes that occur during inspiration and expiration can be measured by **spirometry** (Figure 16-8). In the most common method, the subject breathes from a tube connected to an inverted drum reservoir that floats in the water of a basin and is free to rise as air enters it during expiration and sink as air is withdrawn during inspiration. The volume changes in the reservoir correspond to fluctuations of the lung volume. The change in vertical position of the reservoir is thus a measure of change in lung volume. The volume of air contained by the lungs if they are maximally filled can be divided into four nonoverlapping **volumes** (Figure 16-9, *A*; Table 16-3). The term **capacity** is used to refer to measures of lung function

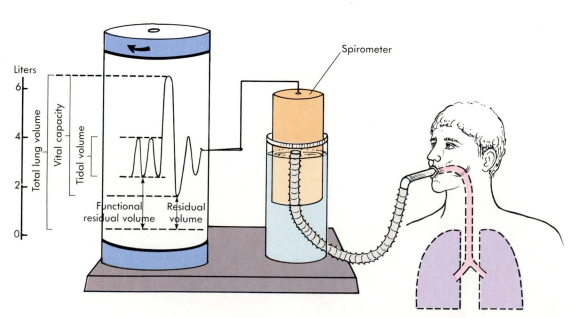

FIGURE 16-8
Static lung volumes can be measured in an intact subject by spirometry. Changes in lung volume are measured during normal quiet breathing and when a subject is asked to inhale maximally and then to exhale as much air as possible.

A

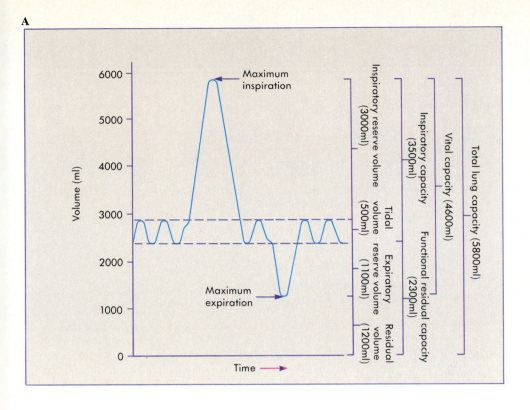

B

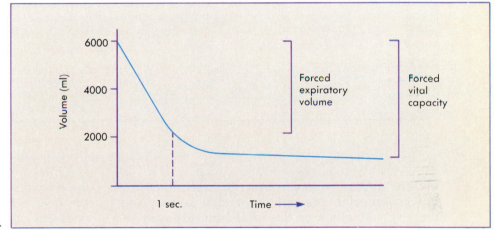

FIGURE 16-9
A The static lung volumes and capacities.
B The 1 second forced expiratory volume and the forced vital capacity, referred to as the dynamic lung volumes.

that encompass more than one volume. The volume of air alternately inspired and expired in a single breath is the **tidal volume.** A normal tidal volume is about 500 ml. The volume of air that remains in the lung after a passive expiration is the **functional residual capacity** (FRC) and is about 2300 ml.

The volume of air that can be inspired after a normal inspiration is the **inspiratory reserve volume** and may be in the range of 3000 ml. The volume that can be inspired following a normal expiration is called the **inspiratory capacity.** The inspiratory capacity is the sum of the tidal volume and the inspiratory reserve volume and is typically about 3500 ml. The volume of air that can be expelled from the lung after the end of a passive expiration is called the **expiratory reserve volume** and is about 1100 ml. The sum of the tidal volume and the inspiratory and expiratory reserve volumes equals the maximum amount of air that can be moved into and out of the lungs by muscular effort. This is called the **vital capacity.** The average vital capacity is 4.6l in young men and 3.1l in young women; the vital capacity is highly correlated with body size. The air remaining in the lung after a maximum forced expiration is the **residual volume,** typically about 1200

TABLE 16-3   *Respiratory Volumes and Capacities*

Volumes	Definition	Typical resting values (ml)
Tidal volume (TV)	The volume of gas inspired or expired in a single respiratory cycle	500
Inspiratory reserve volume (IRV)	The maximum volume of gas that can be inspired starting at the end of a normal inspiration	3000
Expiratory reserve volume (ERV)	The maximum volume of gas that can be expired starting from the lung-chest wall midpoint (that is, making a maximum expiration starting at the end of a normal expiration)	1100
Residual volume (RV)	The volume of gas that remains in the lungs after a maximum expiration	1200
Forced expiratory volume ($FEV_{1.0}$)	The maximum amount of gas that can be expelled in 1 second, following a maximum inspiration	4000
Capacities		
Total lung capacity (TLC)	The total amount of gas in the lungs at the end of a maximum inspiration (the sum of all four lung volumes)	5800
Vital capacity (VC)	The maximum volume of gas that can be inspired after a maximum expiration (the sum of the expiratory reserve volume, normal tidal volume, and inspiratory reserve volume)	4600
Inspiratory capacity (IC)	The maximum amount of gas that can be inspired starting at the lung-chest wall midpoint (the sum of the tidal and inspiratory reserve volumes)	3500
Functional residual capacity (FRC)	The amount of gas in the lungs at the end of a normal expiration (the sum of the expiratory reserve and residual volumes)	2300
Forced vital capacity (FVC)	The amount of gas that can be expelled from the lungs by expiring as forcibly as possible, after a maximum inspiration	5000

ml. The residual volume cannot be determined using a spirometer. The sum of the vital capacity and the residual volume is the **total lung capacity.** A normal value for the total lung capacity of an adult male is 5800 ml (see Figure 16-9, *A*).

The rate at which air can be forced through the airways can also be an important measure of pulmonary function. The volumes described above are called **static volumes** because they are measured in the absence of air flow. In the case of tidal volume, flow is zero at the ends of inspiration and expiration. The other volumes are measured during breath-holding at various stages of inspiration. The changes in lung volume measured during a forced expiration are called the **dynamic lung volumes.** The two most commonly measured dynamic lung volumes are the 1 second **forced expiratory volume** ($FEV_{1.0}$) and the **forced vital capacity** (FVC).

The forced expiratory volume and forced vital capacity are measured by requesting that a subject inhale to the total lung capacity and then exhale as forcibly as possible (Figure 16-9, *B*). The 1 second forced expiratory volume is the amount of air that can be expelled from the lung during the first second. The total amount of air eventually expelled from the lung is defined as the forced vital capacity. In a normal individual the 1 second forced expiratory volume is about 4 liters, and the forced vital capacity is about 5 liters (a ratio of 0.8). Normally the forced vital capacity is about the same as the vital capacity measured during a slow expiration because little or no gas is trapped in the lung. The maximum air flow is usually measured, and a normal value would be 8 to 10 liters/min.

Alveolar Ventilation and the Anatomical Dead Space

The **respiratory rate** is the number of breaths per unit time. The total amount of air entering and leaving the respiratory system per minute, or **respiratory minute volume** (Table 16-4), is the product of the tidal volume and the respiratory rate in breaths/min. A normal respiratory minute volume is around 5 liters/min, but minute volumes of as much as 130

TABLE 16-4   *Factors Involved in Ventilation*

Factor	Definition	Typical resting value
Respiratory rate	Number of breaths/min	12
Respiratory minute volume	Tidal volume × Respiratory rate	6000 ml
Anatomical dead space volume	Volume of airway from nose to terminal bronchioles	150 ml
Alveolar minute volume	$\dfrac{\text{Tidal volume} - \text{Dead space volume}}{\text{Respiratory rate}}$	4200 ml

liters/min can be attained by young adult males during strenuous exercise.

Not all the air inhaled and exhaled enters the alveoli where it can be exposed to the blood. Air in the trachea, bronchi, and nonrespiratory bronchioles does not contact the pulmonary capillaries, and these areas are referred to as the **anatomical dead space** of the lung. A fraction of the respiratory minute volume ventilates the dead space; the remainder, the **alveolar minute volume,** ventilates the alveoli and is subject to gas exchange. The amount of fresh air that reaches the alveoli with each breath is equal to the tidal volume minus the dead space volume. The anatomical dead space volume of a healthy 70 kg man is about 150 ml. Thus for a single breath of 500 ml, only 350 ml of new air reaches the alveoli. The alveolar minute volume is equal to the fraction of each tidal volume that reaches the alveoli multiplied by the number of breaths/min (see Table 16-4). For a person whose respiratory rate is 12 breaths/min, anatomical dead space volume is 150 ml, and tidal volume is 500 ml, the alveolar minute volume is 12 breaths/min times 350 ml/breath = 4200 ml/min.

The effect of anatomical dead space on the dynamics of gas movement in the respiratory tract is illustrated in Figure 16-10. The airway, which has no capillary contact area, is represented by a single tube leading to the cluster of alveoli. Just before air is exhaled from the lungs, the alveoli are full of air that has been accumulating CO_2 and giving up O_2 (see Figure 16-10, A). Although the actual amounts of these two gases in the alveoli have not changed very much, as Chapter 17 will show, the air in the alveoli is represented as "old" air, in contrast to the "fresh" air that will enter the respiratory system from the atmosphere upon inspiration. At expiration, some of the "old" air is left in the anatomical dead space (see Figure 16-10, B). Upon the next inspiration, this "old" air enters the lungs and mixes with the "fresh" air (see Figure 16-10, C).

Different combinations of tidal volume and res-

FIGURE 16-10
A conceptual illustration of the respiratory dead space. In **A** the lung and dead space are illustrated as they appear just before exhalation. In **B** exhalation forces some of the air out of the lung and dead space. During inhalation (**C**), alveolar air that entered the dead space at the end of the previous exhalation returns to the alveoli, followed by fresh air. In **D** tidal volume does not exceed dead space volume, and no fresh air enters the lungs.

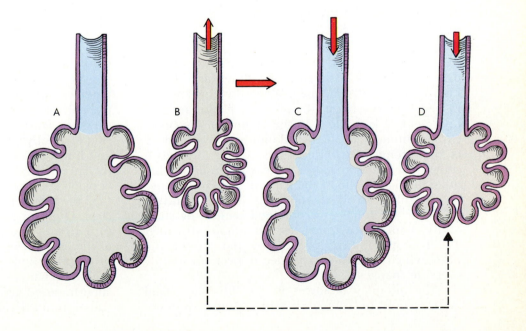

TABLE 16-5	Alveolar Ventilation Produced by Different Combinations of Respiratory Rate and Tidal Volume		
Alveolar minute volume (ml/min)	Respiratory rate	Tidal volume (ml)	Dead space volume (ml)
5250	15	500	150
150	15	250	150
0	15	150	150
5250	52.5	250	150
5250	6	1025	150

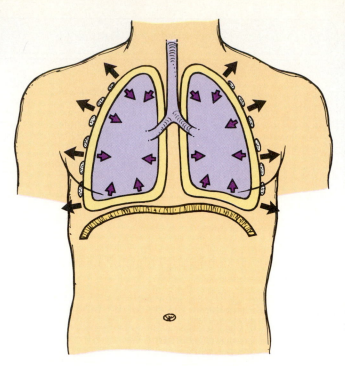

FIGURE 16-11

At the midpoint of the lung-chest wall system, the elastic recoil of the lung *(inward-pointing arrows)* is balanced by the outward spring of the chest wall. At this point, the alveolar pressure is equal to atmospheric pressure and the intrapleural pressure is negative to atmospheric by about 4 mm Hg.

piratory rate can give quite different alveolar ventilation rates (Table 16-5). As the tidal volume approaches the anatomical dead space volume, alveolar ventilation approaches zero (see Figure 16-10, *D*). Thus alveolar ventilation (and therefore gas exchange) is compromised if breathing is very shallow. Alveolar ventilation may be threatened by additions to the dead space. For example, when swimmers breathe through a snorkel tube, tidal volume must be increased by the volume of the additional dead space if alveolar ventilation is to remain at its normal value. Alveolar ventilation is regulated by respiratory control reflexes described in Chapter 18.

---

1 What is a pulmonary volume? What is a pulmonary capacity? Name and define the four static volumes and four static capacities.

2 What is the difference between static and dynamic measures of pulmonary volumes and capacities?

3 What variables determine the alveolar minute volume?

THE MECHANICS OF BREATHING
Lung-Chest Wall System

Compliance (see Chapter 2, p. 33) is the ratio of the change in volume of a container to the pressure difference across its walls. The lungs are elastic, highly compliant structures. When the lungs are removed from the chest, the alveoli collapse and the lung volume is reduced to the volume of the bronchi and bronchioles, which do not collapse because of their connective tissue rings. When the alveoli contain air, the lung exerts an elastic recoil like that of an air-filled balloon, which will tend to collapse by expelling air if its opening is unobstructed. However, the lungs are contained within the chest cavity, which is a much less compliant structure. The two are linked together by the intrapleural fluid, which is neither expanded nor compressed by pressure changes.

The interplay of forces in the lung-chest wall system could be compared with a tug of war between the chest wall and the lungs in which the intrapleural fluid is an inelastic rope. When the intercostal muscles are relaxed, the inward pull or recoil of the lung is matched exactly by the outward spring of the chest wall and diaphragm (Figure 16-11). The system is said to be at its **midpoint.** This is the state of the system between the end of an expiration and the beginning of the next inspiration in a quietly breathing person. Since the trachea is open to the atmosphere, the alveolar pressure is equal to the atmospheric pressure and no air flow is occurring.

As a result of the inward recoil of the lung and outward spring of the chest wall, the intrapleural pressure is several mm Hg less than atmospheric pressure. During the late prenatal and postnatal period of a baby's life, differential growth of the chest and lungs creates a permanent negative pressure in the intrapleural space. The chest cavity initially is very pliable, but it increases in size more rapidly than the lungs. The result is that the tendency of the chest cavity to expand to the dimensions its structure dictates is always opposed by the elastic recoil of the lungs. The expansion of the chest cavity that produces volume changes in the lungs results from pressure changes in the intrapleural fluid that are superimposed upon the negative pressure that resulted from the differential growth.

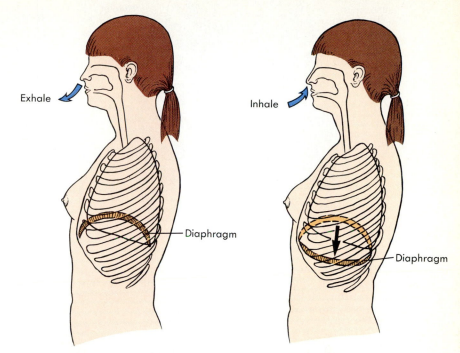

FIGURE 16-12
A The movement of the diaphragm. At rest, the negative intrathoracic pressure pulls the relaxed diaphragm upward toward the thoracic cavity. Contraction of the diaphragm pulls it downward toward the abdomen, increasing the thoracic volume.
B Chest radiographs taken at the end of a normal inspiration *(left)* and at the end of a normal expiration *(right)*.

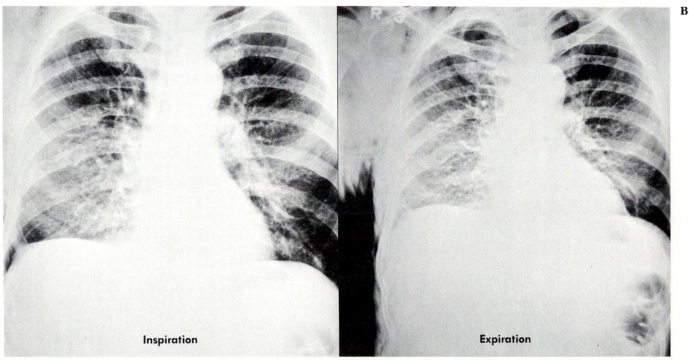

The significance of the balance of forces between the lung and chest wall is evident in what happens during thoracic surgery or penetrating injury to the chest. When the intrapleural space is punctured, air is pulled into the intrapleural space by the negative pressure. The lung and chest cavity are now uncoupled, so their volumes can change independently of one another. The lung collapses to its minimum volume, while the chest expands slightly. This condition is called a **pneumothorax.** It is treated by resealing the intrapleural space and connecting a tube that can be used to create a negative intrapleural pressure during the healing process.

Respiratory Movements

Inspiration during quiet breathing is mainly the result of contraction of the diaphragm, which moves downward in the thorax and increases the intrathoracic volume (Figure 16-12). Larger tidal volumes are achieved by a change in the balance of forces in the lung-thorax system by contraction of the exter-

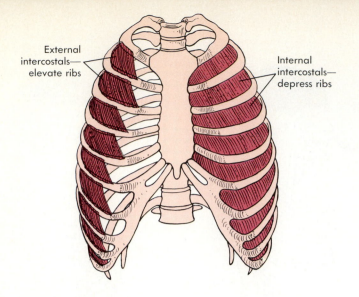

External
intercostals—
elevate ribs

Internal
intercostals—
depress ribs

FIGURE 16-13
The intercostals are the respiratory muscles that move the rib cage. External intercostals elevate the ribs, increasing thoracic volume. Internal intercostals depress the ribs during forced expiration, decreasing thoracic volume.

nal intercostal muscles (Figure 16-13). Contraction of external intercostal muscles lifts the rib cage outward and upward, while the diaphragm moves downward. This increases the volume of the thorax (compare Figure 16-14, *A* and *B*). The lung walls must follow the expansion of the thorax because in-

trapleural fluid couples them to the chest wall and diaphragm.

The intrapleural pressure becomes more negative on inspiration (Figure 16-15), because as the chest volume is increased the lungs are stretched, and the recoil force of the lungs also increases. The bronchi and bronchioles are relatively rigid structures compared with alveoli, so almost all of the increase in lung volume is attributable to increased alveolar volume. The increase in volume decreases the alveolar pressure (see Figure 16-15), creating a pressure gradient between alveoli and atmosphere that drives air through the airway and into the alveoli.

When external intercostals relax, the recoil of the lung and chest wall restores the balance of force between chest wall and lungs, and the system returns to its midpoint position (Figure 16-14, *C* and *D*). While the volume of the lung is dropping back to the end-expiratory value, the alveolar pressure rises above atmospheric pressure (see Figure 16-15), driving air from alveoli through the airway into the atmosphere. Intrapleural pressure rises toward the midpoint value. This is the mechanism of expiration in quiet breathing. It is called **passive expiration** because it is not driven by any muscle contractions. As noted earlier, the volume of the lung at the end of a passive expiration is referred to as the functional residual capacity (FRC).

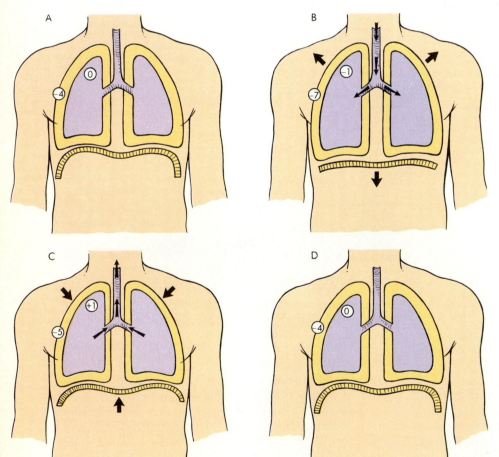

FIGURE 16-14
Figures of the chest during a cycle of inspiration (**A** to **B**) followed by expiration' (**C** to **D**). The direction of air movement, the intrapleural pressure, and the alveolar pressure are shown.

THE RESPIRATORY SYSTEM

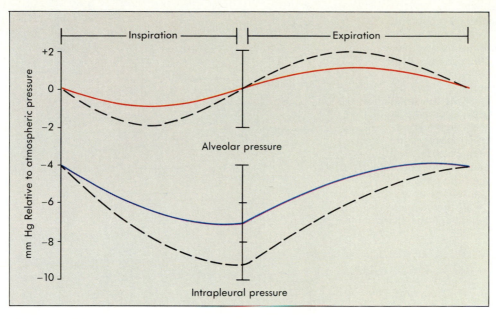

FIGURE 16-15

Changes in alveolar pressure and intrapleural pressure over the respiratory cycle. The solid traces show a cycle of 500 ml tidal volume (quiet breathing). The broken traces show a cycle in which a tidal volume of 1 liter was moved in a cycle lasting the same amount of time.

The increases in tidal volume that occur during effort or excitement are the result of larger excursions of the chest wall. The changes of intrapleural pressure and alveolar pressure during the respiratory cycle are correspondingly greater (illustrated by the broken lines in Figure 16-15). The forceful expiration of vigorous breathing is the result of contraction of internal intercostals (see Figure 16-13) along with muscles of the abdominal wall. The effect of contracting these muscles is to increase the alveolar pressures attained during expiration and to carry the volume of the thorax to a value lower than that of the midpoint. The transients of alveolar pressure are reflected in changes of intrapleural pressure, but intrapleural pressure is always less than atmospheric pressure whether breathing is quiet or forceful.

Pressure, Flow, and Resistance in the Airway

For air to flow along the airway and enter the alveoli, the intraalveolar pressure must be less than atmospheric pressure. Conversely, the intraalveolar pressure must exceed atmospheric pressure when air is leaving the lungs. **Boyle's law** says that, at constant temperature, the product of the pressure and volume of a gas is constant.

Suppose that the lungs have an initial volume, V_0, when connected to the atmosphere, and that the entrance to the lungs is controlled by a valve (Figure 16-16, A). The initial pressure, P_0, is equal to the atmospheric pressure, P_{Atm}. If the valve is shut, an increase in the lung volume from V_0 to the new volume V_1 will result in a decrease in the in-

traalveolar pressure below atmospheric pressure (P_1 in Figure 16-16, B). That is:

$$P_{Atm} \times V_0 = P_1 \times V_1$$
or
$$P_1 = P_{Atm} (V_0/V_1)$$

Since V_1 is greater than V_0, P_1 must be less than P_A. If the valve is opened (Figure 16-16, C), the pressure gradient causes air to enter the lung until the intraalveolar pressure increases to equal atmospheric pressure (Figure 16-16, D). The rate of air flow depends on the pressure gradient and on the resistance of the airway. Unless the airways are closed, pressure differences between the alveoli and the atmosphere can only exist momentarily. At the end of each inspiration and expiration there is no air flow, and the intraalveolar pressure is the same as atmospheric pressure.

The pressure gradient that drives respiration is produced by changes in lung volume. Alveolar volume changes occur because the lungs are effectively attached to the chest cavity by the fluid in the intrapleural space, and as long as the intrapleural space remains free of air, changes in the volume of the chest cavity caused by movements of the respiratory muscles are immediately transmitted to the lung. During inspiration the chest cavity expands, and this expands the lung. Expansion of the lung increases the volume of the alveoli because alveoli are components of the lung tissue. Increasing alveolar volume decreases alveolar pressure, creating the pressure gradient that drives the flow of air into the lung.

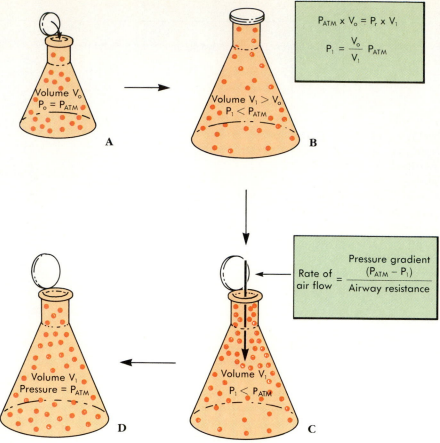

FIGURE 16-16
Boyle's law applied to the lung. The lung is illustrated as a single large alveolus whose entrance is controlled by a valve. At rest the pressure inside the lung (P_o) will be equal to the atmospheric pressure (P_{Atm}), and the volume (V_o) will equal the functional residual capacity (part **A**). Expansion of the alveoli to a larger volume (V_1 in part **B**) reduces the alveolar pressure below the atmospheric pressure (to P_1). If the valve is opened, air will flow into the lung (part **C**) until the pressure again equals the atmospheric pressure (part **D**). The rate of air flow depends on the pressure gradient and the airway resistance.

Atmospheric pressure
P_{ATM}

A — Volume V_o / $P_o = P_{ATM}$

B — Volume $V_1 > V_o$ / $P_1 < P_{ATM}$

Boyle's law

$$P_{ATM} \times V_o = P_r \times V_1$$

$$P_1 = \frac{V_o}{V_1} P_{ATM}$$

$$\text{Rate of air flow} = \frac{\text{Pressure gradient } (P_{ATM} - P_1)}{\text{Airway resistance}}$$

C — Volume V_1 / $P_1 < P_{ATM}$

D — Volume V_1 / Pressure = P_{ATM}

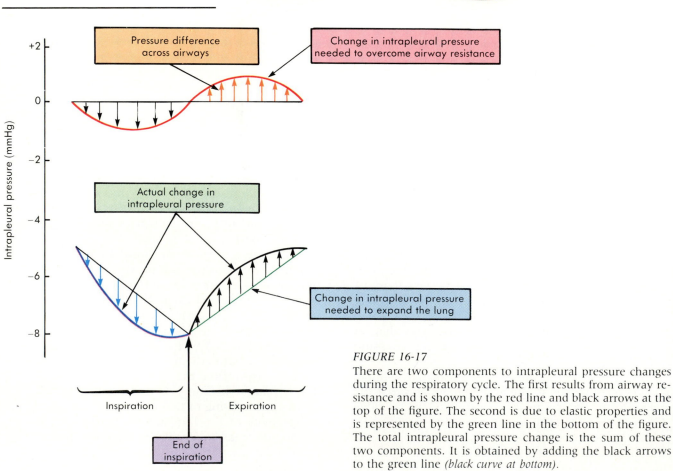

Pressure difference across airways

Change in intrapleural pressure needed to overcome airway resistance

Actual change in intrapleural pressure

Change in intrapleural pressure needed to expand the lung

Intrapleural pressure (mmHg)

Inspiration — Expiration

End of inspiration

FIGURE 16-17
There are two components to intrapleural pressure changes during the respiratory cycle. The first results from airway resistance and is shown by the red line and black arrows at the top of the figure. The second is due to elastic properties and is represented by the green line in the bottom of the figure. The total intrapleural pressure change is the sum of these two components. It is obtained by adding the black arrows to the green line (*black curve at bottom*).

THE RESPIRATORY SYSTEM

The rate of air flow at any point in the inspiratory or expiratory cycle equals the pressure gradient between the alveoli and the atmosphere divided by the airway resistance.

$$\text{Air flow} = \frac{(P_{Atm} - P_{Alv})}{\text{Airway resistance}}$$

The upper curve in Figure 16-17 shows the alveolar pressure change during one complete cycle of inspiration followed by passive expiration. The height of the arrows is equal to the flow-resistive pressure differences at different times in the respiratory cycle. For a normal tidal volume of 500 ml, the flow-resistive pressure gradient varies from about +1 mm Hg during expiration to about −1 mm Hg on inspiration. The arrows would become larger if either the flow velocity increased or airway resistance increased.

The total change in intrapleural pressure during a respiratory cycle (lower trace in Figure 16-17) is the sum of two components. Part of the intrapleural pressure change is used to move air through the airway (the flow-resistive component), but part is used to overcome the elasticity of the lung-chest wall system (the elastic component). In Figure 16-17, the black line representing the total intrapleural pressure change is the sum of the alveolar pressure change, or flow-resistive component (red line) and the lung-chest wall elastic component (green line). This is shown by adding to the elastic component the black arrows representing the magnitude of the flow-resistive component, giving the total change in intrapleural pressure (blue arrows). The magnitude of the elastic component at any instant depends on how far away the lung-chest wall system is from the midpoint, while the flow-resistive component depends on the airway resistance and the rate of air flow at that instant.

For very slow inspirations and expirations, the flow-resistive component is negligible and the elastic component accounts for the entire change in intrapleural pressure. At high rates of ventilation, air flows more rapidly through the airway, and the pressure needed to overcome airway resistance becomes a larger part of the total intrapleural pressure change.

Work—force times distance (or volume)—is a measure of energy expenditure (see Chapter 2, p. 36). The work that is done by respiratory muscles to maintain a constant level of ventilation is the sum of work done against elasticity and work done against resistance. For any constant level of ventilation, there is an optimum respiratory rate at which the work of breathing is minimum (Figure 16-18). Tidal volume and respiratory rate tend to be set by the nervous system at the levels that minimize the effort needed to maintain the necessary level of ventilation. For typical resting rates of ventilation, the

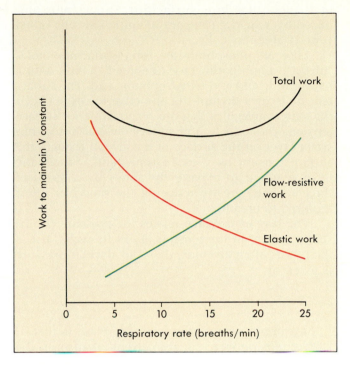

FIGURE 16-18
The elastic component and the flow-resistive component of the work needed to maintain a constant alveolar minute volume are affected by tidal volume. If tidal volume increases, the same alveolar ventilation requires fewer breaths per minute. The resistive work of breathing is low, but elastic work is high. If respiratory rate increases, the tidal volume is diminished accordingly. The elastic work decreases, but the resistive work increases. The sum of the two components is minimum at some intermediate point—about 12 breaths/min for a resting rate of alveolar ventilation. V̇—alveolar minute volume.

work is least at a value very close to the actual resting respiratory rate of about 12 breaths/min.

> 1 What is the midpoint of the lung-chest wall system? What forces are involved in passive expiration?
> 2 Which muscle groups are involved in inspiration? Which groups are involved in forced expiration?
> 3 What are the two components of respiratory work?

Roles of Pulmonary Surfactant

The **surface tension** of water is the result of attractive forces between water molecules near an air-water interface (see Chapter 2, p. 23). This is the force that makes water droplets bead on a nonabsorbent surface. The alveolar membranes themselves are so compliant that the surface tension of the thin layer of water on the air side of the alveolar membrane contributes about 70% of the elastic recoil of the alveoli. Even so, the measured surface tension in the alveolar walls is much lower than would be expected from calculations of the surface tension of a layer of pure water. The reason is that the water on the surface of the alveoli contains phospholipid molecules called **lung surfactant.** Surfactant is produced by a special category of epithelial cells called **type II alveolar cells.**

Like all phospholipids, lung surfactant molecules are amphipathic (see Chapter 3, p. 40), with a hydrophilic end and a hydrophobic end. The surfactant molecules disrupt the attractive forces between the water molecules that are responsible for surface tension. The reduction of surface tension in the thin water layer on the surface of the alveoli reduces the surface tension by 7 to 14 times and reduces the force required to expand the lungs by a factor of three to four, greatly reducing the muscular effort required to breathe.

The **law of LaPlace** relates the pressure (P) inside a sphere to the tension (T) in its walls. If r is the radius of an alveolus, then the alveolar pressure is given by

$$P = \frac{2T}{r}$$

An implication of the law of LaPlace is that, if surface tension were equal in alveoli of all sizes, the pressure within large alveoli should be less than in small ones (Figure 16-19, A). If so, pressure differences between larger and smaller alveoli would force air out of smaller alveoli into large ones, and all of the alveoli of an acinus would merge to form one large alveolus. The large alveolus thus formed would be less effective because its surface area would be smaller relative to the volume of air in it. However, surfactant makes the surface tension of small alveoli less than that of larger alveoli. This effect can be understood by visualizing the surfactant molecules as closer together in the smaller alveolus than in the larger alveolus, just as the color is more dense in a balloon that has less air in it. The more concentrated surfactant reduces the surface tension more in smaller alveoli than in larger ones. The volume-dependent effect of lung surfactant almost completely prevents the pressure differences that would otherwise exist between alveoli of different sizes (Figure 16-19, B).

Infants born prematurely may suffer from low lung compliance in a condition called **respiratory distress syndrome** or **hyaline membrane disease.** This condition is caused by inadequate quantities of lung surfactant. Lung surfactant begins to be synthesized about the 32nd week of fetal development. Surfactant synthesis requires the hormone cortisol and is completed only a few weeks before normal delivery. Infants with respiratory distress

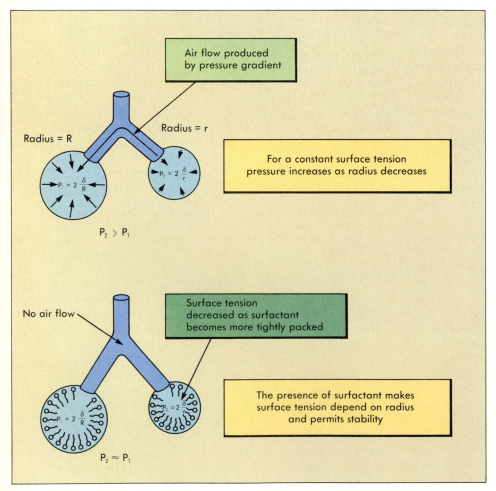

FIGURE 16-19
A The law of LaPlace states that a smaller alveolus should have a higher internal pressure than a large alveolus, and thus air should flow into the largest alveolus.
B Collapse of smaller alveoli is prevented by the presence of lung surfactant because surfactant reduces the magnitude of the surface tension in the alveoli and compensates for the differences in alveolar size. As alveoli expand, surfactant concentration decreases, so the pressure in small alveoli is very close to that in large alveoli.

In figure:

Air flow produced by pressure gradient

Radius = R   Radius = r

For a constant surface tension pressure increases as radius decreases

$P_1 = 2\frac{\delta}{R}$ $P_2 = 2\frac{\delta}{r}$

$P_2 > P_1$

No air flow

Surface tension decreased as surfactant becomes more tightly packed

The presence of surfactant makes surface tension depend on radius and permits stability

$P_1 = 2\frac{\delta}{R}$ $P_2 = 2\frac{\delta}{r}$

$P_2 \approx P_1$

A

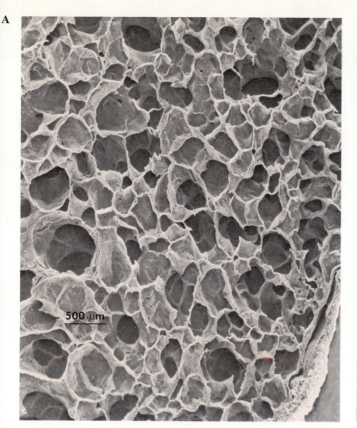

500 µm

B

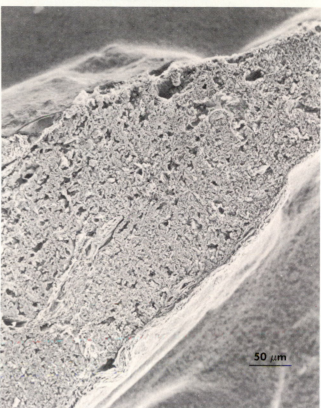

50 µm

C

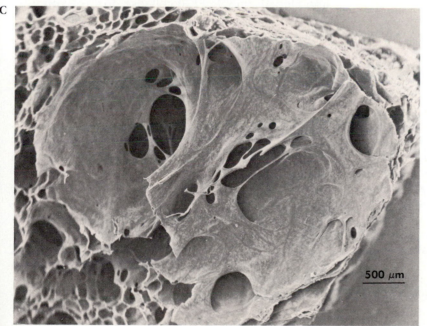

500 µm

FIGURE 16-20
Scanning electron micrographs of a normal lung (A), a surfactant-depleted lung (B), and a lung with emphysema (C). Note that the micrographs in **A** and **C** were taken at 10 times the magnification as that in **B**.

syndrome must make strenuous efforts to breathe, and in extreme cases this increased work of breathing can lead to exhaustion and death. Furthermore, the effect of the law of LaPlace may lead to the collapse of one or both lungs.

The importance of pulmonary surfactant can be appreciated by comparing surfactant-depleted lung tissue with normal lung tissue (Figure 16-20, *A* and

B). The alveolar ducts are enlarged, but the alveoli themselves tend to be collapsed. The decrease in lung compliance caused by a lack of surfactant greatly increases the work of breathing. Infants with respiratory distress syndrome can now be treated with synthetic surfactant until normal synthesis of surfactant begins. It is also possible for women who are threatened with premature delivery

Respiratory Structure and Dynamics

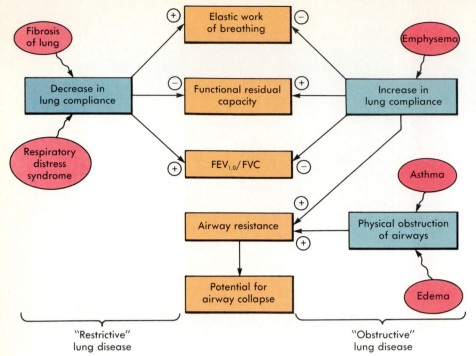

FIGURE 16-21
A diagram illustrating the changes in respiratory function in obstructive and restrictive pulmonary disease. Restrictive disease involves a decrease in lung compliance. It increases the elastic work of breathing, decreases the functional residual capacity (FRC), and increases the ratio of forced expiratory volume to the forced vital capacity ($FEV_{1.0}$/FVC). Obstructive diseases can involve an increase in lung compliance or a physical obstruction of the airways. In the first case, the elastic work of breathing is decreased, the FRC is increased, and the $FEV_{1.0}$/FVC ratio is reduced. In all obstructive diseases the small airways tend to collapse because airway resistance is increased.

to be given an injection of the adrenal hormone, cortisol, which will stimulate the early synthesis of surfactant in the fetus.

Pulmonary Resistance and Compliance in Disease

Pulmonary diseases can be divided into two general families (Figure 16-21). **Restrictive diseases,** in which lung compliance is decreased, include pulmonary fibrosis and respiratory distress syndrome. Restrictive diseases limit the volume changes of the lung and increase the elastic work of breathing. **Obstructive diseases** increase the airway resistance and thus the flow-resistive work of breathing. Obstructive diseases are subdivided into two types: (1) diseases such as emphysema, in which a loss of lung tissue (see Figure 16-20, *C*) decreases the elastic recoil of the lung; and (2) diseases such as asthma, in which bronchiolar constriction and increased mucus secretion increase airway resistance, and pulmonary edema, in which the alveoli and airways are plugged by fluid.

Increased airway resistance causes the collapse of some smaller airways during expiration. This is because a forced expiration increases the intrapleural pressure and produces a significant pressure drop between the alveoli and the atmosphere. If airway resistance is increased by disease, the pressure drop is larger and may drop below intrapleural pressure in the terminal bronchioles. They will then be compressed, trapping alveolar gas.

Cigarette smoking is a major cause of emphysema, a disease that results in loss of the parenchymal elastic elements of the lung. Loss of these components of lung tissue results in increased lung compliance. This has the same effect on alveoli as an

increase in airway resistance because the intrapleural pressure is less negative, and negative intrapleural pressure normally helps distend the small airways. Even if terminal bronchioles are otherwise unoccluded, they tend to collapse during expiration and require the opposing force of the negative intrapleural pressure to reopen them during inspiration.

Obstructive diseases are diagnosed by measuring the $FEV_{1.0}$. The $FEV_{1.0}$ is normally about 4.0 liters, compared with a forced vital capacity (FVC) of 5.0 liters. This produces a ratio of forced excitatory volume to forced vital capacity of about 0.80. In a person with obstructive disease, the functional residual capacity is increased. However, the main abnormality is a decreased expiratory flow rate and an increase in the amount of gas trapped behind collapsed airways. Both the forced expiratory volume measured in the first second and the forced vital capacity are decreased. Typical values in an emphysemic person would be 1.3 liters and 3.1 liters, respectively. The ratio is thus about 0.40. In contrast, the vital capacity measured during slow expiration may be normal.

People with obstructive disease tend to breathe slowly, maintaining a large functional residual capacity. When the FRC is large the lungs are more fully inflated at all stages of the respiratory cycle. Under these conditions the intrapleural pressure is more negative, helping to keep airways distended. Such people also obtain some relief by pursing their lips. The resulting external resistance increases the pressures all the way back along the airways and therefore increases the distending force.

The restrictive lung diseases cause a decrease in compliance, resulting in a more negative intrapleural pressure. Airway collapse during forced expira-

THE RESPIRATORY SYSTEM

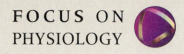

FOCUS ON PHYSIOLOGY

Cardiopulmonary Resuscitation and the Heimlich Maneuver

A failure at any point in gas transport between tissues and the atmosphere impairs the function of the heart and brain within seconds. In heart attacks, the weakened heart pumping may lead rapidly to cessation of ventilation. If ventilation is interrupted (asphyxiation), the heart soon ceases to pump. Cardiopulmonary resuscitation is a means of restoring and maintaining both ventilation and blood pumping until the victim recovers or other medical assistance can be obtained.

In cardiopulmonary resuscitation the rescuer(s) ventilate the victim with air from their own lungs (mouth-to-mouth resuscitation) and simultaneously apply rhythmic pressure to the victim's chest. When the chest is compressed, intrathoracic pressure rises, causing the equivalent of a forced expiration and at the same time expelling blood from the heart and major thoracic blood vessels. When the compression is released, thoracic pressure decreases to the midpoint value, filling the lungs and the heart. Enough ventilation and blood circulation can be maintained in this way to prevent brain and heart damage.

Cardiopulmonary resuscitation must be applied immediately, is easily learned, and cannot be used effectively by untrained people. Substantial increases in survival of heart attacks and other accidents have been shown in cities in which a significant portion of the population has been trained in the method. The American Heart Association suggests that all Americans learn cardiopulmonary resuscitation, starting in the eighth grade.

Choking on food is a major cause of accidental death. It results from inspiration of food into the trachea or lodging of food in the pharynx, blocking air flow into the trachea. Powerful reflexive responses act to oppose blockage of the trachea or pharynx, but these are less effective if blunted by alcohol, a common accompaniment of meals. Sudden inability to speak or breathe is a distinguishing characteristic of choking.

The Heimlich maneuver is an effective means of relieving choking. To carry it out on an adult victim, the rescuer embraces the victim from behind, placing both arms around the victim's waist just above the beltline. The rescuer makes a fist with one hand and presses the thumb side of the fist upward into the victim's abdomen with the other hand, repeatedly making a sharp flexion of the elbows. The effect is to press the diaphragm upward into the thorax, elevating thoracic pressure. The increased pressure in the airway ejects the food, as a cork is ejected from a bottle. If the victim is lying down, the upward thrust can be applied by placing the heel of one hand on the abdomen above the navel and just below the sternum. Once the airway is clear, cardiopulmonary resuscitation can be applied if necessary.

tion does not occur. It is easy to expel air, and the ratio of vital capacity to functional residual capacity is usually closer to 1.0 than in normal subjects. However, the absolute values of $FEV_{1.0}$ and forced vital capacity are decreased because the lung volumes are generally decreased.

1 What is the law of Laplace? How does it apply to alveoli?
2 What are two important functions of pulmonary surfactant?
3 What are the differences between obstructive and restrictive pulmonary disease?

Respiratory Structure and Dynamics **421**

ADAPTATIONS OF THE PULMONARY CIRCULATION
Pressure and Flow in the Pulmonary Loop

Because the flow resistance of the pulmonary vessels is low compared with that of the systemic circulation, the normal pulmonary arterial pressure of only about 18 mm Hg is sufficient to drive blood through the pulmonary loop at a rate equal to that of the systemic loop. Unlike the systemic loop, the flow resistance of the pulmonary circulation is almost completely independent of changes in cardiac output. There are two reasons for this independence. When the cardiac output increases and the pulmonary pressure begins to rise, blood vessels that were previously collapsed begin to open. This process is termed **recruitment.** The pulmonary arteries are also more compliant than the systemic arteries. They expand as the pressure within them increases, and thus their flow resistance falls. These features of the pulmonary loop make it possible for the large increases in output of the left heart that occur in exercise to be matched by the relatively weak right heart, without a substantial increase in pulmonary arterial pressure.

Capillary Filtration and Reabsorption in the Lung

The low pulmonary arterial pressure results in a mean pressure of about 7 mm Hg in the pulmonary capillaries, biasing the balance of filtration and absorption strongly in favor of absorption. This normally protects the lung from edema. Pulmonary edema can occur as a consequence of inflammation or left ventricular failure and significantly impairs lung function in two ways. First, expansion of the pulmonary interstitial fluid lengthens the diffusional path between capillaries and alveoli. Second,

some of the excess interstitial fluid is forced into the alveoli, impairing alveolar ventilation as well as gas diffusion.

Ventilation-Perfusion Matching

The ratio of alveolar ventilation to alveolar blood flow is the **ventilation-perfusion ratio** (abbreviated **V/Q**). For the respiratory system as a whole, the ratio is the alveolar minute volume divided by the cardiac output; about 4.0 liters/min air / 5.0 liters/min blood, or 0.8. For most efficient gas exchange, V/Q ought to be uniform throughout all regions of the lung. Regional imbalances of V/Q result in **physiological dead space** and **physiological shunt.** The meaning of these terms can be illustrated by the most extreme cases of V/Q mismatch: an alveolus that is ventilated but receives no perfusion and an alveolus that is perfused but receives no ventilation. Ventilation of an alveolus that receives no perfusion has the same effect as ventilation of an equivalent volume of the anatomical dead space. Perfusion of an alveolus that is not ventilated is the same as allowing that amount of blood flow to bypass (shunt) the lung.

Ventilation-perfusion matching is a trivial problem in a person who is lying down, because both intrapleural pressure and blood pressure are essentially identical throughout the lung. As explained in Chapter 14, the effect of gravity on hydrostatic pressure at different points within the body has implications for the vascular systems of standing humans; there are also effects on the lungs. In a standing person, the hydrostatic component of blood pressure decreases above the heart and increases below it. Thus blood pressure is greater at the base of the lung than at the apex. Pulmonary blood vessels are more compliant than those in the systemic

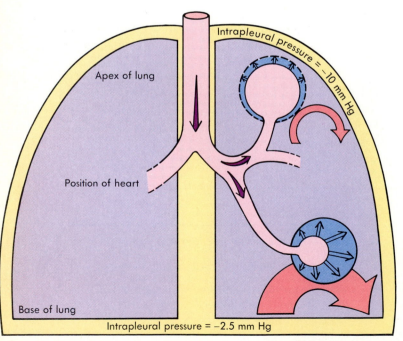

Alveoli at base
receive greater fraction
of total lung ventilation and perfusion

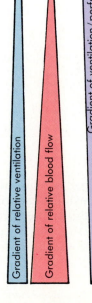

FIGURE 16-22
Ventilation-perfusion relationships in the lung. Because the intrapleural pressure is more negative at the apex than at the base, alveoli at the base of the lung are less expanded than those of the apex prior to inhalation and receive a larger fraction of the inspired air. The base of the lung also receives a larger blood flow, because the outward transmural pressure across the pulmonary vessels is greater, increasing their diameter and decreasing their flow resistance. However, these gradients do not balance, and the ratio of ventilation to perfusion (V/Q) decreases from the apex to the base. Arrows to the right of the figure indicate the magnitude of the three gradients of blood flow, ventilation, and the ventilation to perfusion ratio.

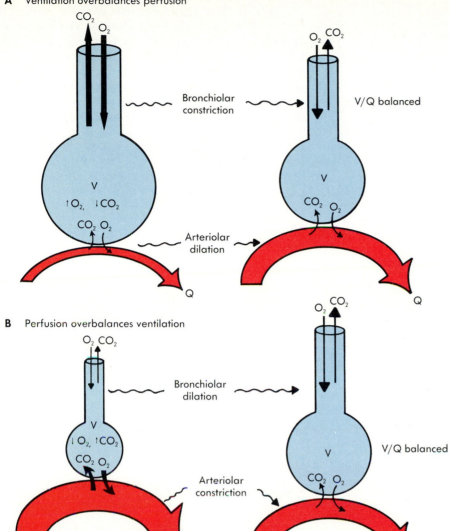

A Ventilation overbalances perfusion

CO_2 O_2

Bronchiolar
constriction

V
$\uparrow O_2$, $\downarrow CO_2$
CO_2 O_2

Arteriolar
dilation

Q

O_2 CO_2

V/Q balanced

V
CO_2 O_2

Q

B Perfusion overbalances ventilation

O_2 CO_2

Bronchiolar
dilation

V
$\downarrow O_2$, $\uparrow CO_2$
CO_2 O_2

Arteriolar
constriction

Q

O_2 CO_2

V
CO_2 O_2

V/Q balanced

Q

FIGURE 16-23

Compensation by bronchioles and pulmonary arterioles for ventilation-perfusion mismatch is driven by changes in alveolar gas composition. The arrows show relative rates of gas movement between alveoli and blood and alveoli and airway. The response of arteriolar smooth muscle is reflected in a change in local perfusion shown by the magnitude of the flow arrow. The response of the smooth muscle of terminal and respiratory bronchioles is reflected by a change in local alveolar minute volume indicated by the diameter of the bronchiole in the figure.

circulation, so the effect of the greater blood pressure at the base of the lung is to expand the vessels of the base and substantially increase perfusion at the base compared with the apex.

In a standing individual, the intrapleural pressure at the midpoint is more negative at the apex of the lung (about -10 mm Hg) than at its base (about -2.5 mm Hg; Figure 16-22). Since the alveoli experience the intrapleural pressure, alveoli at the apex of the lung tend to be more fully expanded than alveoli near the base. If alveolar compliance were constant, expansion would not matter, but it is not. As explained earlier in this chapter, larger alveoli have lower compliances. Alveoli at the base of the lung are smaller and thus expand more during inspiration than those at the apex. That is, in addition to greater perfusion, they receive a larger fraction of the inspired air than those at the apex and are thus better ventilated.

From the description just given, it might appear that the effects of gravitation on ventilation and perfusion would cancel each other out. However, the apex-to-base ventilation gradient and perfusion gradient are not identical. With no compensation by blood vessels and bronchioles, the base of the lung would receive about twice as much ventilation as the apex, but about ten times more blood flow.

The lung compensates for this variation in local ventilation and blood flow in two ways: (1) increases in the local levels of carbon dioxide and decreases in oxygen cause the nearby bronchioles to enlarge, redirecting the inspired air toward relatively underventilated alveoli at the apex of the lung; while (2) decreases in oxygen or increases in carbon dioxide cause a local vasoconstriction and shift blood away from the underventilated apical portions of the lung (Figure 16-23; Table 16-6). The response of pulmonary arterioles to local concentra-

TABLE 16-6   Ventilation-Perfusion Matching

Imbalance	Change in alveolar gas	Response of pulmonary arterioles	Response of bronchioles
Ventilation too great relative to perfusion	Increased O_2 Decreased CO_2	Dilate	Constrict
Perfusion too great relative to ventilation	Decreased O_2 Increased CO_2	Constrict	Dilate

tions of O_2 and CO_2 is exactly opposite to the response that systemic arterioles make to the same factors in systemic flow autoregulation.

As a result of these intrinsic compensations, the actual discrepancy between alveolar ventilation and perfusion is less severe than it would be in their absence. However, there is still a difference in V/Q between the apex and base of the lung in a standing person. A large measure of compensation for this difference is provided by the processes of gas exchange to be described in the next chapter.

1 *What is the function of recruitment of pulmonary blood vessels?*
2 *What are the advantages of the low flow resistance and low pressures of the pulmonary loop?*
3 *What changes in pulmonary circulation and ventilation result from standing up?*

SUMMARY

1. The respiratory system consists of the **airways**: the nasal passages, **pharynx, trachea, bronchi,** and **bronchioles;** the gas exchange elements: the **alveolar sacs** and **alveoli;** and the **diaphragm** and **intercostal muscles** that are responsible for ventilation.

2. The airway warms and humidifies inspired air. Its **mucous secretions** and **cilia** filter and remove particulate contaminants and microbes. The flow of air through the airway is the energy source for sound production in the larynx. The diameter of the terminal bronchioles of the airway is under neural and hormonal control, which affect air flow by changing the **airway resistance.**

3. The gas exchange function of the lung is maximized by the very short path length between pulmonary capillary blood and alveolar gas and by the enormous surface areas of alveoli and capillaries.

4. The volume of gas in the lungs can be divided into the tidal, inspiratory reserve, expiratory reserve, and residual volumes. The lung capacities are the sum of two or more volumes. The **vital capacity** is a measure of the largest possible tidal volume; the **functional residual capacity** is the volume of air present in the lung at the end of an expiration. These may be measured as static quantities in the absence of air flow or as dynamic quantities affected by air flow.

5. The **respiratory minute volume** is the total volume of air moved by the respiratory system per minute. Of this volume, part ventilates the anatomical dead space and the rest ventilates the alveoli.

6. Respiratory muscles cause volume changes of the thorax, which result in pressure changes in the thorax according to Boyle's law. The lungs and chest wall are mechanically coupled by the intrapleural fluid. The **intrapleural pressure** is always less than alveolar pressure because of the inward **elastic recoil** of the lung.

7. The total work done in ventilation is the sum of work done in changing the volume of the lung-chest wall (elastic component) and the work done to move air through the system (flow-resistive component). Both components are dependent on tidal volume and respiratory rate, which are adjusted to minimize the work of breathing.

8. **Pulmonary surfactant** reduces the surface tension of the air-water interface in the alveoli, with two major consequences for the lung. First, the work of expanding the alveoli is decreased. Second, the tendency of small alveoli to merge into larger ones is reduced, so that the large surface area of the alveoli is maintained. Pulmonary diseases can be divided into **obstructive disease,** in which air flow is impeded, and **restrictive disease,** in which the elasticity of the alveoli is reduced. The two types of disease can be distinguished on the basis of their effects on the static and dynamic lung volumes and capacities.

9. Large increases in pulmonary blood flow are possible with only small pressure changes because of the high compliance of pulmonary vessels and because small increases open (recruit) additional pulmonary vessels. The low resistance of the pulmonary loop favors capillary absorption over filtration and thus protects the lung from edema.

10. In a standing person, gravity tends to cause the base of the lung to be better ventilated and better perfused than the apex. The resulting inequities of the **ventilation-perfusion** ratio affect the O_2 and CO_2 content of alveolar gas and initiate compensatory responses of bronchiolar and arteriolar smooth muscle.

1. Where are the following located and what are their main functions?

Bronchi	Terminal bronchioles
Respiratory bronchioles	Alveoli

2. What structures compose a pulmonary acinus?

3. What structures correspond to the anatomical dead space? If a subject who has an anatomical dead space volume of 100 ml is breathing through a snorkel tube that has a volume of 200 ml, what must her tidal volume be to maintain an alveolar minute volume of 3 liters at 12 breaths/min?

4. Define the following lung volumes:

Residual volume	Functional residual capacity
Tidal volume	Vital capacity

5. Why is the intrapleural pressure in a normal individual negative relative to the atmospheric pressure?

6. When the lung volume is at the functional residual capacity, in which direction is the chest recoiling? The lung? How do these recoil forces compare with one another?

7. Why do intrapleural pressure changes not mirror alveolar pressure changes?

8. How does the intrapleural pressure compare with the intraalveolar pressure during inspiration and expiration?

9. How would the functional residual capacity differ between a premature infant lacking pulmonary surfactant and a normal full-term infant?

10. Subject A has a lung compliance that is lower than that of Subject B. All other physiological parameters are the same for the two subjects. How would the work of breathing differ between the two individuals?

11. Compare the apex of the lung with the base in terms of their:

Alveolar ventilation	Ventilation-perfusion ratio
Blood flow	Intrapleural pressure

12. Airway resistance in a subject is doubled. What effect would this have on the intrapleural pressure during inspiration? On the intraalveolar pressure? On the work of breathing?

● SUGGESTED READING

DAVIS, J.M., et al.: Changes in Pulmonary Mechanics After the Administration of Surfactant to Infants with Respiratory Distress Syndrome. *New England Journal of Medicine,* volume 319, August 1983, p. 476. Describes the breathing pattern of infants with respiratory distress syndrome.

EDWARDS, D.D.: Mending a Torn Screen in the Lung, *Science News,* volume 132, May 2, 1987, p. 277. Describes new treatments for alpha-antitrypsin deficiency.

Preemies Reprieve, *Discover,* volume 6, November 1987, p. 16. Describes how extracts of pulmonary surfactant from calves' lungs can be given to treat infants suffering from respiratory distress syndrome.

RALOFF, J.: New Clues to Smog's Effect on Lungs, *Science News,* volume 132, August 8, 1987, p. 86. Describes how the lung's defense mechanisms can be damaged by environmental contamination.

SEYMOUR, R.J.: *The Heart Attack Survival Manual—A Guide to Using CPR in a Crisis,* Prentice-Hall, Inc., Englewood Cliffs, N.J., 1981. Practical manual of cardiopulmonary resuscitation.

CHAPTER 17

Gas Exchange and Gas Transport in the Blood

On completing this chapter you will be able to:

- Apply the gas laws to gas mixtures in the respiratory tract and gases dissolved in the blood and tissues.
- Understand the concept of partial pressure.
- Describe how O_2 is transported in the blood by hemoglobin.
- Appreciate why the sigmoid oxyhemoglobin dissociation curve maximizes O_2 loading in the lungs and unloading in the tissues.
- Describe the factors that shift the oxyhemoglobin dissociation curve to the right.
- Compare the oxyhemoglobin dissociation curve for adult hemoglobin with those for fetal hemoglobin and myoglobin.
- Describe how CO_2 in the blood is carried as HCO_3^-.
- Be able to apply the Henderson-Hasselbalch equation to bicarbonate transport in the blood.

*R*espiration could not be understood until the birth of chemistry as a science. The English physician Joseph Black both discovered CO_2 and showed that it is produced in metabolism. One of his experiments was performed in 1754 in an air duct in the ceiling of a church in Glasgow where 1500 people had gathered for a 10 hour service. Black dripped limewater—containing $Ca(OH)_2$—over rags and collected the precipitate of $CaCO_3$ that resulted from its reaction with the CO_2 exhaled by the congregation below. The French chemist Antoine Lavoisier (1743-1794) found that, when placed in a sealed container, candles went out and animals died as soon as most of the enclosed O_2 had been converted into CO_2.

The mechanisms by which O_2 and CO_2 are transported in the blood began to be understood in the latter half of the 19th century. Christian Bohr showed the quantitative relationship between O_2 availability and O_2 binding by hemoglobin in 1886. In 1904, together with K.A. Hasselbalch and August Krogh, he showed that CO_2 can force hemoglobin to give up O_2 in systemic capillaries, a process now referred to as simply the Bohr effect.

In the last decades of the 19th century, the mechanism whereby O_2 and CO_2 are exchanged across the alveolar membrane was the subject of a scientific controversy. One hypothesis held that the lung can actively secrete CO_2 and absorb O_2 into the blood against a concentration gradient. In 1904 August Krogh and his wife Marie showed that under normal conditions diffusion can account for alveolar gas exchange. The final defeat of the secretion theory was provided by Joseph Barcroft in 1919. Barcroft spent 6 days in a glass chamber maintained at subatmospheric pressure. During this exposure to simulated high altitude, Barcroft found that gas exchange always occurred down a concentration gradient.

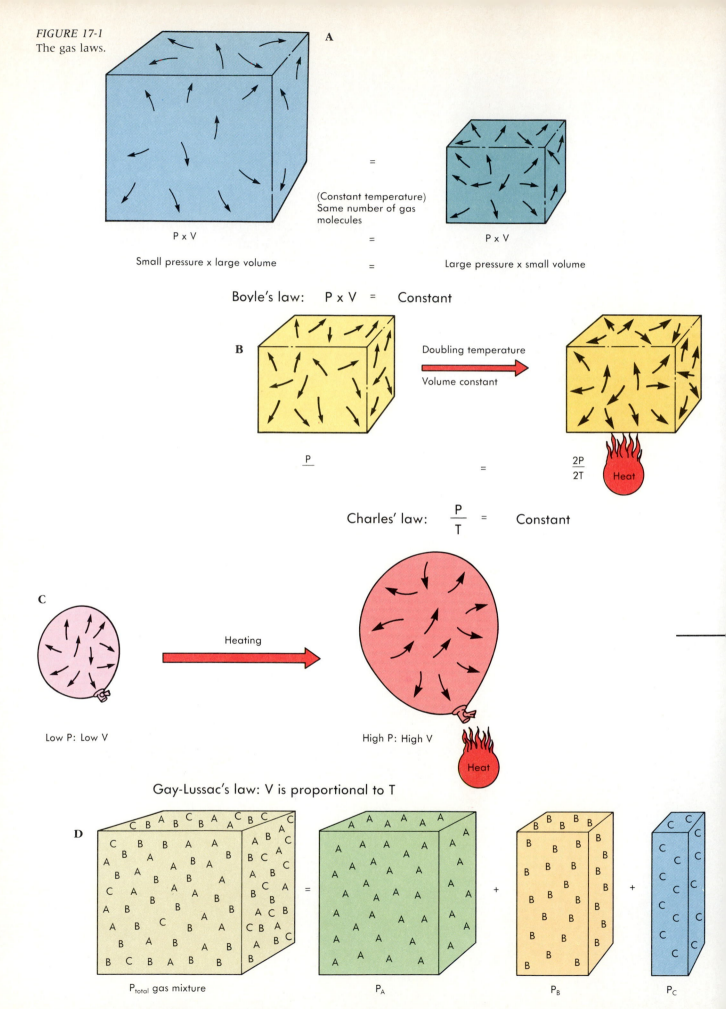

FIGURE 17-1
The gas laws.

A

= (Constant temperature) Same number of gas molecules

P x V

Small pressure x large volume

= = P x V

Large pressure x small volume

Boyle's law: P x V = Constant

B

Doubling temperature

Volume constant

$\frac{P}{}$

= $\frac{2P}{2T}$

Heat

Charles' law: $\frac{P}{T}$ = Constant

C

Heating

Low P: Low V

High P: High V

Heat

Gay-Lussac's law: V is proportional to T

D

P_{total} gas mixture

= P_A + P_B + P_C

Dalton's Law: $P_{total} = P_A + P_B + P_C$

PHYSICS AND CHEMISTRY OF GASES
Molecular Properties of Liquids and Gases

Water and many other compounds can exist as solids, liquids, or gases. Both liquids and gases are fluids, forms of matter in which individual molecules can move freely relative to one another. In some respects, however, liquids and gases behave differently. The differences have to do with the density of the constituent molecules. Liquids are dense enough that intermolecular effects have significant consequences; surface tension and viscosity are examples of properties that arise from such interactions. For most gases (or gas mixtures) at physiological pressures, the density is so low that collisions between molecules very rarely occur; thus the gas molecules do not interact. Such gases are termed **ideal gases.** A consequence of the large distances and infrequent interactions between gas molecules is that gases are **compressible.** In contrast, the forces between molecules in liquids are so great that the volume of a liquid will not change significantly if the pressure on the liquid is increased.

Gas Laws

Gas molecules continuously strike the walls of whatever container they are in and produce a **force per unit area,** or **pressure.** The pressure of a gas depends on two factors: (1) the average number of collisions per unit time; and (2) the average velocity of the molecules. Collision frequency is proportional to the number of molecules per unit volume, or concentration of the gas. The average velocity of the gas molecules is greater the higher the temperature.

The gas laws describe the relationship between the volume, pressure, and temperature of an ideal gas. If one of the three variables is held constant, the relationships between the other two are given by Boyle's law and Charles' law. In **Boyle's law** (Figure 17-1, *A*), which has already been shown to apply to ventilation (Chapter 16), temperature is not allowed to change and, under that condition, increases in the volume of a gas decrease gas molecule density, thus decreasing the pressure (and vice versa). In **Charles' law** (Figure 17-1, *B*), volume is not allowed to change, and under that condition, the average kinetic energy (pressure) of gas molecules increases or decreases with rising or falling temperature. **Gay-Lussac's law** allows volume to change with changes in temperature, and as a consequence, pressure also changes (Figure 17-1, *C*).

The three gas laws described in the previous paragraph can be combined into the **gas equation:**

$$PV = nRT$$

where **n** is the number of moles of a gas, **R** is the **universal gas constant,** and **T** is the temperature in °Kelvin. For any compound, a mole of a gas is its molecular weight in grams, which contains 6.023 x 10^{23} molecules **(Avogadro's number).** At a temperature of 0° C (273° Kelvin) and a pressure of 760 mm Hg (1 atmosphere), 1 mole of any gas occupies exactly 22.4 liters. At body temperature (37° C or 310° K) and one atmosphere of pressure, a mole of gas will occupy 25.5 liters.

Dalton's law (Figure 17-1, *D*) says that the gas equation applies to gas mixtures as well as a pure gas. Because gas molecules do not interact substantially and ideal gas of whatever type exerts the same pressure per number of molecules, Avogadro's number of gas molecules will occupy the same volume (22.4 L) regardless of the types of gases in the mixture.

Air is a mixture of gases whose major constituents are O_2 and N_2 (Table 17-1). That part of the total pressure of a gas mixture which is attributable to one of the gases is the **partial pressure** of that gas. The total pressure of a mixture of gases equals the sum of the individual partial pressures of its components. The partial pressure of one constituent in a

Gas molecules in gaseous phase

Gas molecules in liquid phase

Water in beaker

(At equilibrium, P_{gas} is equal throughout the system)

Henry's Law: Concentration of gas in solution = P_{gas} x solubility of gas

TABLE 17-1	Constituents of Clean, Dry Atmospheric Air at Sea Level	

Gas	Percent of total	$P_{partial}$ (mm Hg)*
Nitrogen (N_2)	78.08	593.40
Oxygen (O_2)	20.95	159.22
Inert gases (argon, neon, helium, krypton, xenon)	0.9654	7.34
Carbon dioxide	0.0314	2.39

*Partial pressures total is slightly higher than 760 mm Hg because of rounding.

mixture is calculated by multiplying the percentage of the total pressure which that gas exerts by the total pressure of the mixture (see Figure 17-1, D). The partial pressure of O_2 in dry atmospheric air at sea level is computed as follows: atmospheric pressure is 760 mm Hg, and the O_2 content of the air is about 21%. The partial pressure is :

$$760 \text{ mm Hg} \times 0.21 = 160 \text{ mm Hg}$$

Thus the part of the total pressure that is attributable to O_2 is 160 mm Hg. The partial pressure of O_2 is abbreviated Po_2.

Water Vapor

If a beaker of water is placed in a closed compartment containing dry air, some water molecules will leave the liquid phase and enter the gas phase (Figure 17-1, E). As this occurs, an oppositely directed flux of water molecules develops. When the system has come to equilibrium, the vapor-to-liquid and liquid-to-vapor water fluxes are equal, and the gas phase is saturated with water vapor. The partial pressure of water vapor depends on the average thermal energy of the water molecules (an application of Charles' law); the saturation point is higher, that is, more molecules of water will move into the vapor phase at a higher temperature. At the normal lung temperature of 37° C, the partial pressure of water vapor (or **vapor pressure**) is 47 mm Hg in air saturated with water.

Inspired air passes over the warm, moist surfaces of the airway and becomes saturated with water at body temperature. The addition of water vapor does not change the total pressure of the air, but

its inclusion means that the partial pressures of other component gases must decrease. During inspiration, the partial pressure of O_2 in the tracheal air is 21% of (760 − 47) mm Hg, or about 150 mm Hg (Figure 17-2).

Gases in Solution

Not only do gases become saturated with water, but water will become saturated with dissolved molecules of the gases with which it is in contact. As the gas molecules collide with the gas-water interface, some of them will dissolve in the water. The number of surface collisions per unit time made by molecules of each of the types of gas in the air is proportional to the partial pressure of each type. As the number of dissolved molecules of each gas type increases, the number of molecules that pass from water back to air also increases. At equilibrium, the partial pressure of each constituent in the gas phase is equal to the partial pressure of that constituent in solution, and there is no net movement of gas in either direction across the interface.

Henry's law says that at equilibrium the concentration of dissolved gas (number of moles of gas per unit volume of water) depends on two factors: the partial pressure of the gas and its **solubility** (see Figure 17-1, E). A liquid exposed to two gases that have the same partial pressure but different solubilities will have different numbers of molecules of the two gases dissolved per unit volume at equilibrium. Gas molecules of all ideal gases bombard the liquid with the same energy, but the more soluble a gas, the more likely that the collision will result in entry of the gas molecule into the liquid (or the less likely

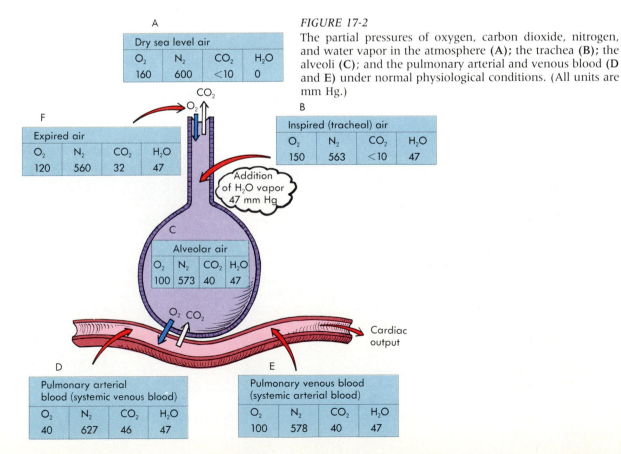

FIGURE 17-2
The partial pressures of oxygen, carbon dioxide, nitrogen, and water vapor in the atmosphere (**A**); the trachea (**B**); the alveoli (**C**); and the pulmonary arterial and venous blood (**D** and **E**) under normal physiological conditions. (All units are mm Hg.)

A

Dry sea level air

O_2	N_2	CO_2	H_2O
160	600	<10	0

F

Expired air

O_2	N_2	CO_2	H_2O
120	560	32	47

B

Inspired (tracheal) air

O_2	N_2	CO_2	H_2O
150	563	<10	47

Addition of H_2O vapor 47 mm Hg

C

Alveolar air

O_2	N_2	CO_2	H_2O
100	573	40	47

O_2 CO_2

Cardiac output

D

Pulmonary arterial blood (systemic venous blood)

O_2	N_2	CO_2	H_2O
40	627	46	47

E

Pulmonary venous blood (systemic arterial blood)

O_2	N_2	CO_2	H_2O
100	578	40	47

that a molecule colliding with the surface from the liquid phase will escape from the liquid). The solubility coefficient for CO_2 in plasma at 37° C is approximately 21 times greater than that of O_2.

The rate at which dissolved gas diffuses in solution is determined by the magnitude of the partial pressure gradient and the diffusion coefficient of the particular gas involved. The diffusion coefficient depends on temperature and the nature of the diffusion path. For living tissue at body temperature, the diffusion coefficient of CO_2 is about 20 times greater than that of O_2.

1 *What are the three variables whose relationship is described by the gas laws? State the equation that summarizes the relationship.*
2 *What is meant by the partial pressure of a gas? What two factors determine the partial pressure?*
3 *For gases in solution, what two factors determine the number of gas molecules per unit volume of water?*

GAS EXCHANGE IN ALVEOLI
Composition of Alveolar Gas
The ratio of CO_2 production to O_2 consumption is the **respiratory quotient** (RQ). The respiratory quotient varies with the specific metabolic pathways used for energy production. The overall equation for glucose metabolism is:

$$C_6H_{12}O_6 + 6O_2 \longrightarrow 6CO_2 + 6H_2O$$

If glucose were the sole fuel used by the body, the RQ would be 1.0. Fats and proteins contain more carbon atoms than oxygen molecules. Metabolism of fat to CO_2 results in an RQ of about 0.7; similarly, protein metabolism results in an RQ of slightly less than 0.8. Typically a mixture of all three fuels is being metabolized at any instant in resting people, so that the RQ is in the range of 0.8 to 0.85. At rest a normal individual consumes about 250 ml of O_2 per minute. With an RQ of 0.8, this individual would produce about 200 ml of CO_2 per minute. That is, averaged over the entire body, slightly more than 4 molecules of CO_2 are produced by the tissues for every 5 molecules of O_2 consumed.

Even though the alveoli are continually ventilated, the partial pressure of O_2 in the alveoli is not as high as in the inspired air because the inspired air mixes with the volume of "old" air retained in the anatomical dead space at the end of each expiration. The partial pressure of CO_2 in the alveoli is higher than in inspired air for the same reason. Composition of the alveolar air (for which typical values were given in Figure 17-2) depends on the ventilation of the alveoli (frequency and depth of breathing) and the rates of O_2 uptake and CO_2 release.

The relationship between alveolar P_{CO_2}, alveolar P_{O_2}, the P_{O_2} of inspired air, and RQ is described by the **alveolar gas equation**:

Alveolar P_{O_2} = Inspired air P_{O_2} −
(Alveolar P_{CO_2}/RQ)

At sea level, the inspired P_{O_2} is about 150 mm Hg (see Figure 17-2). A normal alveolar P_{CO_2} under these conditions is 40 mm Hg. For a respiratory quotient of 0.8, the alveolar gas equation predicts that alveolar P_{O_2} will be 150 mm Hg − (40 mm Hg/0.8), or 100 mm Hg.

The alveolar gas composition changes very little during each respiratory cycle. There are two reasons for this stability. First, as noted in Chapter 16, the rate of airflow in the smallest airways is so slow that diffusion is the dominant mode of gas movement. There is relatively little bulk flow of air in and out of each alveolus with each breath. Also, the approximately 350 ml of alveolar ventilation of each breath is diluted into more than 2 liters of functional residual volume.

During exhalation, the alveolar air mixes with air present in the anatomical dead space that was not altered by gas exchange, so expired air has an average P_{O_2} of 120 mm Hg and P_{CO_2} of 32 mm Hg (see Figure 17-2).

Gas Composition of Pulmonary Blood
Mixed venous blood typically has a P_{O_2} of 40 mm Hg and a P_{CO_2} of 46 mm Hg (see Figure 17-2). The typical resting values for alveolar P_{O_2} and the alveolar P_{CO_2} are 100 mm Hg and 40 mm Hg, respectively. The driving force for O_2 diffusion at the arteriolar end of the pulmonary capillaries is the difference between the P_{O_2} of alveolar air (100 mm Hg) and that of the pulmonary arterial blood (40 mm Hg), or 60 mm Hg. The driving force for CO_2 diffusion into the alveoli, calculated similarly, is only 6 mm Hg. Nevertheless, approximately equal volumes of O_2 and CO_2 are exchanged across the alveolar membrane. This is possible because CO_2 has a much higher diffusion coefficient in water than O_2.

Blood gas partial pressures come into equilibrium with the partial pressures of alveolar air in the first third of the 1 second that each drop of blood spends in a pulmonary capillary. Figure 17-3 shows how P_{O_2} varies with distance along the capillary. The mean flow rate of blood through the capillary may be twice as fast in exercise (*curve B* in Figure 17-3), but full equilibration of pulmonary capillary blood with alveolar air still occurs in healthy individuals, because the diffusion path between alveolar air and blood (Figure 17-4) is so short.

Since complete equilibration occurs between alveolar air and pulmonary blood, the partial pressures of dissolved gas in systemic arterial blood might be assumed to be identical to those of alveolar air. This is usually not exactly the case because as shown in Chapter 16, a small fraction of the pulmonary blood flow bypasses the alveoli, and because the ventilation and perfusion are not perfectly matched in all alveoli. However, the differences are normally so slight that they can be neglected in a discussion of the basic mechanisms of gas exchange.

FIGURE 17-3

The upper portion of the figure schematically illustrates a red blood cell traversing a pulmonary capillary, while the lower portion shows how the partial pressure of oxygen varies from the arterial to the venous end of the capillary. Full equilibration of oxygen between the alveoli and pulmonary capillaries occurs in the first third of a pulmonary capillary under resting conditions **(curve** A) and occurs only slightly further along when cardiac output is increased **(B).** Oxygen fails to equilibrate fully only in disease conditions in which the barrier to diffusion is increased, a condition referred to as diffusion limitation **(curve** C).

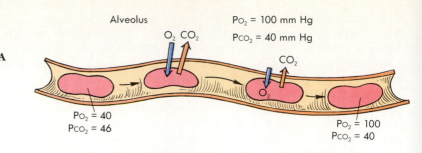

A

Alveolus

P_{O_2} = 100 mm Hg
P_{CO_2} = 40 mm Hg

O_2 CO_2

CO_2

O_2

P_{O_2} = 40
P_{CO_2} = 46

P_{O_2} = 100
P_{CO_2} = 40

B

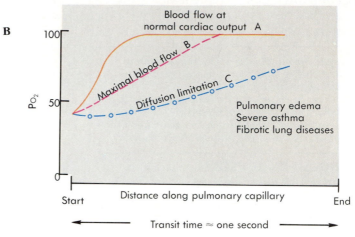

Blood flow at normal cardiac output A

Maximal blood flow B

Diffusion limitation C

Pulmonary edema
Severe asthma
Fibrotic lung diseases

P_{O_2}

100

50

0

Start

Distance along pulmonary capillary

End

Transit time ≈ one second

A

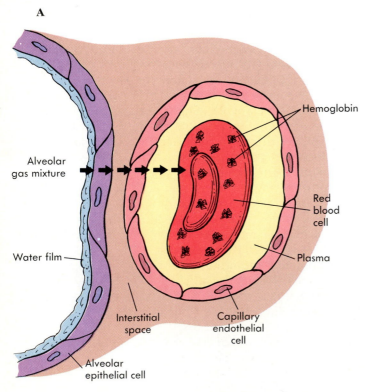

Hemoglobin

Alveolar
gas mixture

Red
blood
cell

Water film

Plasma

Interstitial
space

Capillary
endothelial
cell

Alveolar
epithelial cell

B

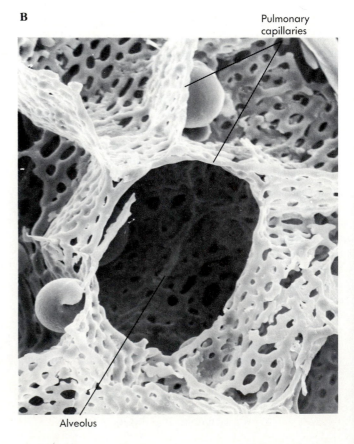

Pulmonary
capillaries

Alveolus

FIGURE 17-4

A The diffusional path between alveolar air and hemoglobin.

B A scanning electron micrograph of a vascular cast of lung tissue. The capillaries were injected with a resin that was allowed to harden. The tissue was then digested away from the hardened resin, leaving the cast. The large space is an alveolus surrounded by a dense mesh of capillaries.

THE RESPIRATORY SYSTEM

	Systemic arterial (= pulmonary venous)	Mixed systemic venous (= pulmonary arterial)	Arteriovenous difference
Oxygen			
P_{O_2}	100 mm Hg	40 mm Hg	60 mm Hg
In physical solution	0.3	0.1	
As HbO_2	19.5 ml	14.4	
TOTAL	19.8 vol%	14.5 vol%	5.3 vol%
Carbon dioxide			
P_{CO_2}	40 mm Hg	46 mm Hg	6 mm Hg
In physical solution	2.7	3.1	
As bicarbonate	43.9	47.0	
As carbaminoHb	2.4	3.9	
TOTAL	49.0 vol%	54.0 vol%	5.0 vol%

1 What is the respiratory quotient? What determines its value?

2 What physical factors determine the rates of diffusion of O_2 and CO_2 between blood and alveoli?

3 How is it possible for CO_2 to be exchanged at approximately the same rate as O_2 when the partial pressures of the two gases differ by about tenfold?

OXYGEN TRANSPORT IN THE BLOOD
Hemoglobin and the Problem of Oxygen Solubility

At body temperature and an arterial P_{O_2} of 100 mm Hg, 0.3 ml of O_2 is dissolved per 100 ml of plasma (0.3 vol%). A normal cardiac output of 5 liters/min could supply 15 ml/min of dissolved O_2 (5 liters/min × 3.0 ml O_2/liter), but this is only 6% of the body's resting oxygen consumption of 250 ml O_2/min. Therefore if dissolved O_2 were the only form in which O_2 were transported in the blood, the cardiac output would have to be 80 liters/min just to meet the needs of tissues during rest. For the heart to pump this volume of blood is clearly impossible.

The actual O_2 content of systemic arterial blood is about 20 ml/100 ml of blood, or about 20 vol% (Table 17-2). About 99% of the O_2 is carried by **hemoglobin,** an oxygen-binding protein contained within the red blood cells (Figure 17-5; see also Chapter 14, p. 346). Hemoglobin, abbreviated **Hb,**

FIGURE 17-5
The molecular structure of hemoglobin.

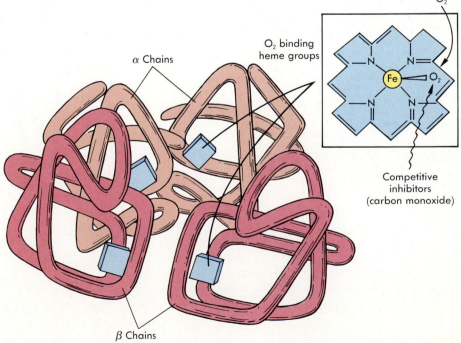

is a protein in which a **heme group** is attached to each of four subunit polypeptide chains. The two **α chains** and two **β chains** are linked to one another between the α chains. Each polypeptide chain folds to produce a cavity in which an iron-containing heme group binds O_2. Therefore four O_2 molecules can combine with one molecule of hemoglobin.

Hemoglobin fully saturated with oxygen (**oxyhemoglobin**) is a brilliant red. Hemoglobin that has lost one or more of the four O_2 molecules (**deoxyhemoglobin**) is dark red. Venous blood is darker than arterial blood because it contains more deoxyhemoglobin than oxyhemoglobin. The reaction of deoxyhemoglobin with O_2 to form oxyhemoglobin takes less than 0.01 second, so it is never the rate-limiting step in O_2 transfer.

The ease with which hemoglobin accepts an additional molecule of O_2 depends on how many of the sites are already occupied by O_2 molecules. If one of the four sites is occupied, binding of O_2 to the second and third sites becomes easier. This is an example of an allosteric effect (see Chapter 3, p. 52). The result of this cooperativity between O_2 binding sites is that the amount of oxygen bound to hemoglobin increases in an S-shaped, or sigmoid, fashion as the P_{O_2} increases. The relationship between the P_{O_2} and the extent of O_2 binding to hemoglobin is the **oxyhemoglobin dissociation curve** (Figure 17-6). The oxyhemoglobin dissociation curve can be described either in terms of the **relative saturation** of hemoglobin (left-hand vertical scale in Figure 17-6) or by the **oxygen content** of the arterial blood (right-hand vertical scale). The P_{50} for hemoglobin is the value of P_{O_2} at which the hemoglobin is half saturated with oxygen (that is, on the average, each hemoglobin molecule has two O_2 molecules bound).

The **O_2 carrying capacity** is the volume of O_2 contained in a volume of O_2-saturated blood; the carrying capacity depends on the concentration of effective hemoglobin. The carrying capacity is reduced in various forms of **anemia,** which may be caused by a reduction in the number of red blood cells, insufficient hemoglobin production, or the production of abnormal hemoglobin. An anemic condition can be produced for a period of days to weeks after significant blood loss in otherwise normal individuals. Failure of hemoglobin production occurs in dietary **iron deficiency anemia,** because iron is needed for synthesis of heme groups. Dietary deficiency or failure of absorption of vitamin B_{12} causes **pernicious anemia,** in which formation of red blood cells is impaired. In **sickle cell anemia,** substitution of a single amino acid in the β chains causes hemoglobin to aggregate into large polymers when P_{O_2} is low. These polymers distort the red blood cells into a characteristic sickle shape that prevents their passage through small blood vessels. In various forms of **thalassemia,** there is an inherited defect in the DNA that codes for either the α or β chains. Sufferers of these diseases have higher than normal hematocrit levels, but the red blood cells contain less hemoglobin, and the O_2 binding characteristics of the hemoglobin are abnormal.

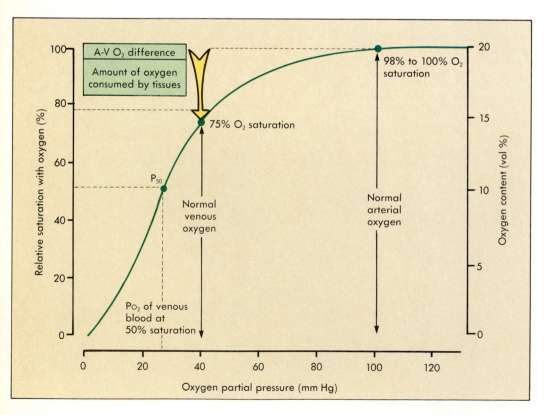

FIGURE 17-6
Oxyhemoglobin dissociation curve for blood under systemic arterial conditions. The left-hand vertical axis gives percentage saturation; the right axis gives the corresponding values in vol% for blood with a typical O_2 carrying capacity. The P_{50} is the value of P_{O_2} at which the hemoglobin is one half saturated. At the P_{O_2} of mixed venous blood the hemoglobin is about 75% saturated; in other words, about 25% of the O_2 carried by systemic arterial blood has been unloaded.

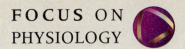

FOCUS ON PHYSIOLOGY

Hypoxia and Oxygen Therapy

Hypoxia is a condition of inadequate O_2 delivery to a body region or to the whole body. Inadequate O_2 delivery may result from inadequate blood perfusion (**ischemia**); from a disorder in the function of hemoglobin, as in anemia or carbon monoxide poisoning; or from failure of ventilation or blood oxygenation in the lungs (**asphyxiation**). Breathing pure O_2 is effective therapy if ventilation is depressed or gas exchange is impaired. It is less effective when ventilation and pulmonary gas exchange are normal, since in those cases arterial hemoglobin will already be almost saturated with O_2, and the only effect of O_2 breathing will be to increase the dissolved O_2 content of plasma.

Oxygen Loading in the Lungs

The shape of the oxyhemoglobin dissociation curve has important implications for O_2 uptake in the lungs and delivery in the tissues. The normal alveolar P_{O_2} of 100 mm Hg is on the flat portion of the oxyhemoglobin dissociation curve, so that the hemoglobin remains almost fully saturated even in the face of relatively large changes in the alveolar P_{O_2}. This means that the O_2 loading is very close to 100% even when ventilation is not optimum.

The safety factor for loading is particularly im-portant at high altitudes. For example, at an altitude of 3000 meters the atmospheric pressure is 520 mm Hg, as compared with the sea level value of 760 mm Hg, and the P_{O_2} of the inspired air is 110 mm Hg rather than 150 mm Hg. The alveolar gas equation can be applied to this situation by assuming an RQ of 0.8. The alveolar P_{CO_2} will be 40 mm Hg, and the alveolar P_{O_2} would be reduced by 50 mm Hg to 60 mm Hg (*point A* in Figure 17-7). At a P_{O_2} of 60 mm Hg, hemoglobin is still 88% saturated with O_2.

FIGURE 17-7
The effects of a decrease in the inspired P_{O_2} on the delivery of oxygen to the tissues viewed using the oxyhemoglobin dissociation curve. The normal arteriovenous O_2 difference of about 5 vol% is shown as the larger, downward-directed yellow arrow and results in a fall in venous P_{O_2} to 40 mm Hg (*yellow, left-directed arrow*). At 3000 meters the inspired P_{O_2} is about 60 mm Hg, hemoglobin is 88% saturated with oxygen, and the arterial O_2 content is 18 vol% (*point A*). The same arteriovenous O_2 difference of 5 vol% (*yellow arrow under point A*) causes the venous P_{O_2} to fall only slightly below the normal value of 40 mm Hg (*black, left-directed arrow and point B*).

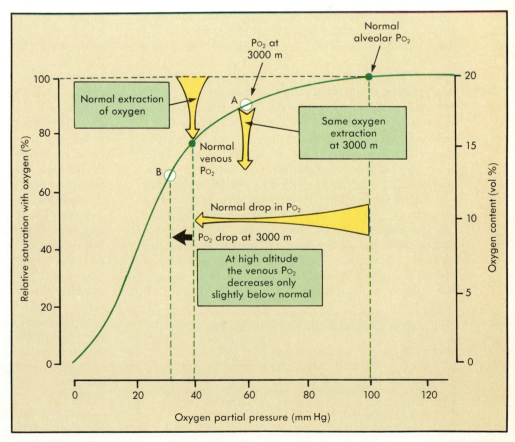

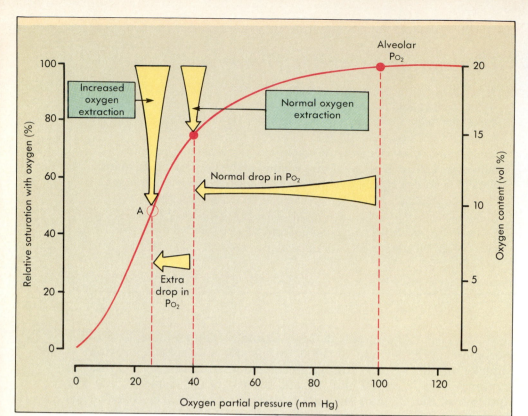

FIGURE 17-8
The effects of an increase in total body metabolism (assuming constant cardiac output) viewed using the oxyhemoglobin dissociation curve. The normal arteriovenous O_2 difference of 5 vol% is shown as the small, downward-directed arrow and results in a fall in venous P_{O_2} to 40 mm Hg *(large, left-directed arrow)*. If the arteriovenous O_2 difference is increased to 10 vol% *(large, downward-directed arrow)*, the venous P_{O_2} falls only slightly below the normal value of 40 mm Hg *(small, left-directed arrow and point A)*.

Oxygen Unloading in Systemic Capillaries

The P_{O_2} in the extracellular fluid surrounding the tissues is about 40 mm Hg. When arterial blood enters capillaries and encounters this lower P_{O_2}, it gives up, or **unloads,** some of its bound O_2 according to the dissociation curve. For values of P_{O_2} in this range, the oxyhemoglobin dissociation curve is steep, and small decreases in P_{O_2} can cause relatively large changes in the amount of bound O_2 unloaded. The significance of the steepness of this part of the curve is illustrated by the change in O_2 delivery that occurs in exercise. At rest, the amount of O_2 unloaded is about 5 vol%, which corresponds to delivering, on the average, only one of the four O_2 molecules bound by the hemoglobin molecules. During strenuous exercise the O_2 content drops, on the average, from 20 vol% to 10 vol% between arteries and veins (Figure 17-8, *right side*), representing an unloading of 10 vol%. On the average, therefore, the hemoglobin molecules have delivered two rather than one of the four O_2 molecules that they carried. An decrease in average tissue P_{O_2} of only about 16 mm Hg is able to result in this doubling of the amount of O_2 delivered to the active tissues.

> 1 What is measured by the percent saturation of hemoglobin?
> 2 What percent saturation values for hemoglobin are characteristic of the pulmonary venous blood? Of the systemic venous blood?
> 3 On average, about what percent of the O_2 content of systemic arterial blood is unloaded in the tissues in a resting individual?

Effects of CO_2, pH, Temperature, and 2,3 DPG on Oxygen Unloading

The delivery of O_2 to the tissues is affected by four factors that alter the affinity of hemoglobin for O_2 (Figure 17-9; Table 17-3). The first two factors, CO_2 and pH, are closely related. Some of the CO_2 that enters the capillaries of the systemic loop reacts with water to form carbonic acid, H_2CO_3, which dissociates to give a bicarbonate ion, HCO_3^-, and a hydrogen ion, H^+. When the H^+ concentration increases (lowering the pH), some H^+ binds to hemoglobin and reduces the affinity of hemoglobin for O_2. The oxyhemoglobin dissociation curve is shifted to the right, in the direction of increased O_2 delivery (*dashed curve* in Figure 17-9). Binding of some of the CO_2 to hemoglobin results in formation of **carbaminohemoglobin,** which has the same effect as H^+ binding on O_2 delivery (a right shift).

This pH and CO_2-dependent right shift is referred to as the **Bohr effect.** The significance of the Bohr effect is that it occurs in response to the metabolites released by active tissues, which always have a lower pH and higher CO_2 release rate than inactive tissues. The P_{O_2} does not have to change for active tissues to be preferentially supplied with O_2 because the properties of hemoglobin are altered in response to these by-products of metabolism. In human beings, a decrease in plasma pH from 7.4 to 7.2 increases the P_{50} from 26 mm Hg to 30 mm Hg. The increase in P_{50} results in a 15% greater unloading of O_2.

Active skeletal muscle may have a local temperature several degrees higher than resting muscle be-

FIGURE 17-9

Factors shifting the oxyhemoglobin dissociation curve to the right (increase the P_{50}) correspond to a decrease in the affinity of hemoglobin for O_2 and increase the amount of O_2 that can be extracted from hemoglobin at a given P_{O_2} *(black portion of the downward-directed arrow)*. Such factors include a decrease in the pH (Bohr effect), an increase in the P_{CO_2}, an increase in temperature, and an increase in the concentration of 2,3 DPG. Note that all these factors combine to favor combination of oxygen with hemoglobin in the lung while, at the same time, facilitating unloading of O_2 to metabolically active tissues.

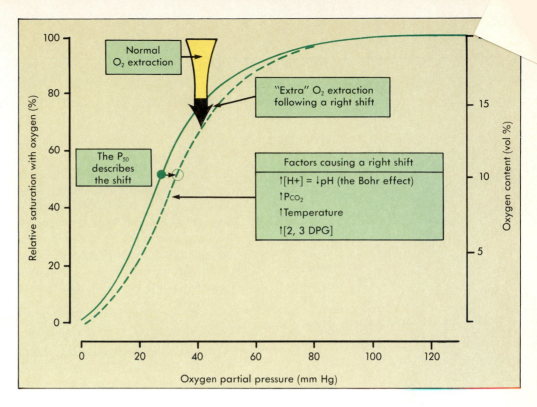

cause heat is also a by-product of oxidative energy metabolism. The increased temperature favors additional O_2 unloading. For example, at 43° C the P_{50} is increased from 26 mm Hg to 37 mm Hg. Conversely, at 30° C the P_{50} is reduced to 18 mm Hg. The temperature effect is reversed when blood passes through the pulmonary circulation, since the temperature of the lungs is about 1° C lower than core body temperature.

The last of the four factors that affects the oxy-hemoglobin dissociation curve is **2,3 diphosphoglycerate (2,3 DPG)**, a metabolite produced by red blood cells. Production of 2,3 DPG is increased at low P_{O_2}, so when red blood cells pass through inadequately oxygenated tissues they produce more 2,3 DPG. Doubling the 2,3 DPG concentration increases the P_{50} of hemoglobin from 26 mm Hg to 35 mm Hg. Thus the effects of pH, CO_2, elevated temperature, and 2,3 DPG production all complement one another in the sense that they favor oxygen unloading from hemoglobin in metabolically active tissues. The 2,3 DPG levels become depleted in stored blood, and the oxyhemoglobin dissociation curve is shifted far to the left even though the pH, P_{CO_2}, and temperature are all normal. If stored blood is given without any pretreatment (such as heating and oxygenation), the P_{O_2} in the tissues can fall to low levels.

Oxygen Transfer from Mother to Fetus

A physiological problem posed by the period of intrauterine development characteristic of mammals is delivery of adequate O_2 to the fetus. The placenta, the organ for exchange of gases and nutrients, does not normally allow maternal and fetal bloods to mix. Instead, the fetal capillaries in the placenta are bathed in maternal blood. The P_{O_2} of the maternal blood leaving the placenta is 32 mm Hg, similar to the P_{O_2} that would be seen in blood at an altitude of 35,000 feet, well above the normal limits of human habitation. If the fetus contained adult hemoglobin, the blood entering its circulation would only be 60% saturated, and O_2 extraction by the developing fetus would cause dangerously low O_2 levels in fetal

TABLE 17-3	Factors That Bring About a Rightward Shift of the Oxyhemoglobin Dissociation Curve

Factor	Source
Decreased pH	Results from increased production of CO_2 and lactic acid in active tissues
Increased P_{CO_2}	Increased in venous blood from tissues with high rate of oxidative metabolism; acts by formation of carbaminohemoglobin
Increased temperature	Higher in fever and in active tissues, especially skeletal muscle
Increased 2,3 DPG	Produced by red blood cells; increased in anemia and high-altitude adaptation

Physiology of Carbon Monoxide Poisoning

Carbon monoxide (CO), a product of incomplete combustion of hydrocarbons, competes with O_2 for the hemoglobin binding site and thus blocks its O_2-carrying capability. Binding of CO results in formation of **carboxyhemoglobin** (COHb). The affinity of the Hb binding site for CO is more than 200 times greater than that for O_2, so once CO has bound to a heme group it may prevent that heme group from transporting O_2 for as much as several hours. Binding of CO to one of the four sites of hemoglobin also abolishes the cooperativity of O_2 binding at the other sites, shifting the oxyhemoglobin dissociation curve to the left. This shift makes unloading of O_2 more difficult.

Because of its very high affinity for hemoglobin, very low levels of CO in environmental air can result in formation of significant amounts of COHb and a decrease in the blood's O_2 carrying capacity. In nonsmokers breathing clean air, the COHb level in the blood is less than 1%. Heavy smokers, nonsmokers who breathe tobacco smoke in confined areas, and freeway drivers or those who live and work on heavily travelled streets may accumulate levels of COHb as high as 7%. The problem posed by the reduced carrying capacity and the left shift are most serious for organs such as the heart that extract a large fraction of the arterial O_2. Elevated levels of COHb are regarded as a potential contributing factor in heart attacks and premature births.

---

tissue. Fetal blood is able to load O_2 from maternal blood because of synthesis of a different form of hemoglobin, **fetal hemoglobin,** during fetal life.

Fetal hemoglobin resembles adult hemoglobin in that it also binds oxygen cooperatively and therefore has an S-shaped dissociation curve (Figure 17-10). The difference is that γ polypeptide chains are substituted for the adult β chains (Table 17-4). This shifts the dissociation curve of fetal hemoglobin to the left along the P_{O_2} axis (Figure 17-10). A position to the left of the adult curve means that fetal hemoglobin binds O_2 more readily and can remove more of it from the maternal blood. Consequently, fetal hemoglobin can be fully saturated with O_2 at a lower partial pressure than adult hemoglobin. It does not get a chance to be fully saturated, however, because the maternal blood arriving in the placenta has a P_{O_2} considerably lower than that of arterial blood.

Blood passing from the placenta to the fetus in the umbilical vein is about 80% saturated with O_2. The gain in loading provided by the higher O_2 affinity of fetal hemoglobin means that the fetal tissues must fall to a lower P_{O_2} before they provide a gradient that favors O_2 delivery. Fetal tissues, therefore, operate at a lower P_{O_2} than adult tissues. Fetal hemoglobin is relatively insensitive to 2,3 DPG, so this mechanism of enhancing unloading is not im-

portant before birth. The fetal O_2 carrying capacity is higher than that of adult blood because of two additional adaptations: the hematocrit of fetal blood is higher, and the fetal red blood cells contain a higher concentration of hemoglobin.

Myoglobin—an Intracellular Oxygen Binding Protein

Myoglobin is a heme-containing protein found in energetic tissues such as red skeletal muscle and cardiac muscle. Myoglobin differs from hemoglobin in that it consists of a single α chain and can bind only a single O_2 molecule. Because each molecule has only one heme unit, binding of O_2 to myoglobin cannot involve cooperativity, and thus the dissociation curve for myoglobin is hyperbolic rather than S-shaped (see Figure 17-10). Myoglobin binds O_2 more strongly than hemoglobin (see Table 17-4). The oxymyoglobin dissociation curve lies to the left of the hemoglobin curves, so that the loading portion of the myoglobin curve lies within the range of partial pressures in which hemoglobin unloads.

Myoglobin has two roles: first, it represents an intracellular reserve of O_2. Second, when the P_{O_2} in the muscle cell has fallen very low, the presence of myoglobin increases the rate of O_2 diffusion from the cell surface to the interior. Recall that the rate of diffusion is related to the magnitude of the gradient

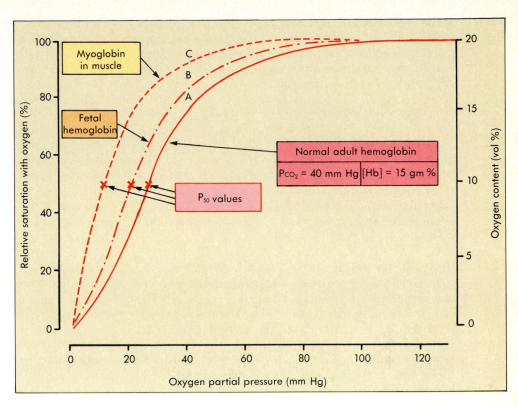

FIGURE 17-10

The dissociation curves for adult hemoglobin (**A**), fetal hemoglobin (**B**), and myoglobin (**C**). These dissociation curves depend on the P_{O_2} and can be expressed either in terms of the percent saturation of hemoglobin *(left-hand vertical axis)* or, given the hemoglobin concentration, as the O_2 content of the blood *(right-hand vertical axis)*. Note the sigmoidal (S-shaped) nature of the dissociation curves for adult and fetal hemoglobin compared with the dissociation curve for myoglobin. The position of each dissociation curve can be described by stating its P_{50}. The further to the left a particular curve is, the more tightly the corresponding molecule binds O_2.

of the diffusing substance. If myoglobin is present, O_2 can move through the cytoplasm in two forms: (1) physically dissolved, and (2) "riding" from one myoglobin to another in the direction of the lower O_2 concentration. This effect is particularly strong when the P_{O_2} has fallen very low, because many of the myoglobin molecules have given up their bound O_2 and can serve to move O_2 from the region just inside the muscle cell membrane toward the cell interior. Thus when O_2 is readily available, the cell can accept and store more O_2 due to the presence of myoglobin, and when O_2 is in short supply, myoglobin can help move it to the areas where it is most needed by providing a second route for diffusion.

TABLE 17-4	Subunit Composition and P_{50} Values for Hemoglobin, Fetal Hemoglobin, and Myoglobin	

Hemoprotein	Subunits	P_{50} (arterial conditions)
Adult hemoglobin	$\alpha, \alpha, \beta, \beta$ (tetramer)	26 mm Hg O_2
Fetal hemoglobin	$\alpha, \alpha, \gamma, \gamma$ (tetramer)	20 mm Hg O_2
Myoglobin	α (monomer)	8 mm Hg O_2

1 What portion of the oxyhemoglobin dissociation curve is most strongly affected by temperature, CO_2, pH, and 2,3 DPG?
2 What is the functional significance of the right shift caused by these factors?
3 What is the difference between the dissociation curve of myoglobin and hemoglobin? What is the functional significance of the difference?

CARBON DIOXIDE TRANSPORT AND BLOOD BUFFERS

The CO_2-H_2CO_3-HCO_3^- Equilibrium

Carbon dioxide is the major waste product of oxidative metabolism. CO_2 reacts with water to form carbonic acid (H_2CO_3). The rate of carbonic acid formation is increased one hundred-fold within red blood cells and many other tissues by an enzyme called **carbonic anhydrase.** Carbonic anhydrase is present inside the red blood cells and catalyzes both the forward and reverse reactions. Which direction is dominant depends on the relative concentrations of CO_2 and H_2CO_3. At equilibrium the rates of the forward and reverse reactions are the same. Once formed, carbonic acid rapidly and spontaneously dissociates to give a bicarbonate ion and a hydrogen ion:

$$CO_2 + H_2O \underset{\text{anhydrase}}{\overset{\text{Carbonic}}{\rightleftharpoons}} H_2CO_3 \rightleftharpoons H^+ + HCO_3^-$$

In solution, CO_2, H_2CO_3, H^+, and HCO_3^- coexist in a chemical equilibrium. The relative proportions of the reactants are constant, as predicted by the law of mass action (see Chapter 2, p. xx). The relationship is described by the **Henderson-Hasselbalch equation,** written in terms of pH, the negative logarithm of H^+ concentration:

$$pH = 6.1 + \log_{10} \frac{[HCO_3^-]}{0.03 \, P_{CO_2}}$$

In this equation, the plasma pH is proportional to the plasma concentrations of HCO_3^- and CO_2. The constant 6.1 is the effective acid dissociation constant for the overall reaction of CO_2 and H_2O. The concentration of dissolved CO_2 in the plasma is equal to the P_{CO_2} multiplied by the solubility coefficient of CO_2 in plasma (0.03).

Blood Buffers and Control of Plasma pH

A **buffer** is a substance that opposes changes in H^+ concentration by binding some H^+ when H^+ becomes more abundant and releasing some H^+ when H^+ becomes less abundant. The CO_2-H_2CO_3-HCO_3^- equilibrium is a buffer system (usually referred to as "bicarbonate buffer"). Addition of H^+ results in formation of H_2CO_3 and CO_2; a decrease in H^+ is opposed by dissociation of some CO_2 to keep the proportions of reactants constant. The CO_2-H_2CO_3-HCO_3^- system is an important buffer for extracellular fluids. Other buffer systems are also important.

For comparative purposes, Table 17-5 includes all of the major buffer systems of the body. Hemoglobin is an important blood buffer because it is the most abundant protein of the red blood cell cytoplasm, and like all proteins it has ionizable -COOH and -NH_3 groups. The phosphate buffer system is of some importance in blood but is more important as a urinary buffer (the importance of urinary buffer systems will be discussed in Section V). Proteins are relatively insignificant as buffers in plasma and interstitial fluid but are the major buffers of the intracellular fluid.

The normal pH of arterial plasma is 7.4. This pH corresponds to a plasma H^+ concentration of 4×10 M, or 40 nanomolar (nM). Plasma pH values of less than 6.9 and more than 7.6 are life threatening,

yet they correspond to a change in H^+ concentration of less than 1 μmole/L from the normal value! The normal whole-body CO_2 production of 200 ml/min would result in formation of about 15 moles of H^+ ions per day, much more than the capacity of the body's buffer systems, if it were not for the fact that the lungs eliminate CO_2 as rapidly as it is produced.

The continuous production of CO_2 by oxidative energy metabolism potentially has a very large impact on the acidity of the extracellular fluid. If the plasma pH has become either too acid or basic, small changes in P_{CO_2} can restore the plasma pH to normal. The CO_2-H_2CO_3-HCO_3^- system cannot oppose changes in pH that are caused by addition of one of the reactants of the system, such as CO_2. Precise balance of the rate of CO_2 loss by ventilation to the rate of CO_2 production by metabolism is thus essential. The respiratory system has a central role in the regulation of acid-base balance through its ability to control the plasma P_{CO_2}.

Carbon Dioxide Loading in Systemic Capilaries—the Haldane Effect

Dissolved carbon dioxide gas diffuses readily from cells into interstitial fluid (Figure 17-11). As red blood cells pass through systemic capillaries, CO_2 that diffuses into them from plasma is converted to H^+ and HCO_3^-, a reaction accelerated by the carbonic anhydrase inside red cells. Hemoglobin accepts H^+, driving the reaction to the right, and allowing the blood to accept additional CO_2. The HCO_3^- ions are returned to the plasma in exchange for Cl^- ions (Figure 17-11). This Cl^- movement to balance bicarbonate movement is called the **"chloride shift."** It prevents the build up of HCO_3^- inside the red cell that would otherwise oppose further dissociation of CO_2.

The amount of CO_2 the blood can load is greater the more the hemoglobin is deoxygenated. This phenomenon, called the **Haldane effect** (Figure 17-12), links CO_2 loading to O_2 unloading because it involves the same two factors as the Bohr effect (see Figure 17-11). These two factors are binding of H^+ and formation of carbamino complexes. First, de-

TABLE 17-5	Major Buffer Systems of the Body	

System	Reactor	Importance
Bicarbonate	$CO_2 + H_2O \overset{+}{\underset{+}{\rightleftharpoons}} H_2CO_3 \rightleftharpoons H^+ + HCO_3^-$	Most important buffer system of blood plasma
Hemoglobin	$Hb - H \rightleftharpoons Hb^- + H^+$	Buffers interior of red blood cells
Phosphate	$H_2PO_4^- \rightleftharpoons H^+ + HPO_4^=$	Most important buffer of urine
Protein	$Pr - H \rightleftharpoons Pr^- + H^+$	Most important buffer system of intracellular fluid; quantitatively the most important buffer of the whole body

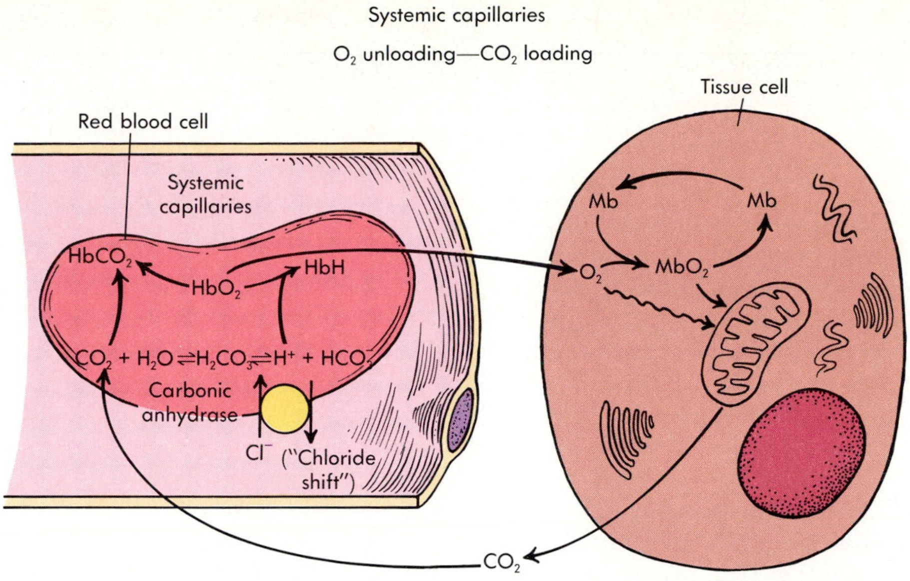

Systemic capillaries

O_2 unloading—CO_2 loading

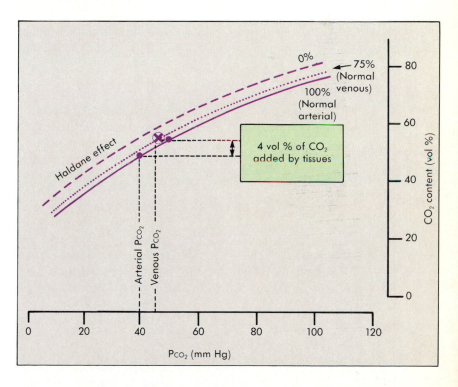

FIGURE 17-11
Cooperation between CO_2 loading and O_2 unloading in systemic capillary blood. The increase in P_{CO_2} within red blood cells drives the CO_2 dissociation reaction to the right, liberating H^+ and HCO_3^-. The reaction is accelerated by carbonic anhydrase. Some of the H^+ is accepted by hemoglobin, which becomes a better buffer as it becomes deoxygenated (the Haldane effect). The HCO_3^- is exchanged across the red cell membrane for Cl^-, allowing the rightward reaction to continue. The oxyhemoglobin more readily becomes deoxygenated in the more acidic environment (the Bohr effect). Some of the oxyhemoglobin exchanges its O_2 for CO_2, forming carbaminohemoglobin (shown as $HbCO_2$). Within the cells, O_2 diffusion to mitochondria is assisted by myoglobin.

FIGURE 17-12
CO_2 dissociation curve for systemic arterial blood (*lower curve*), systemic venous blood (*middle curve*), and completely deoxygenated blood (*upper curve*). This plot demonstrates that the more deoxygenated blood is, the greater its CO_2 content is at any value of P_{CO_2}. In contrast to the oxyhemoglobin dissociation curve, the dissociation curve for CO_2 is nearly linear over the physiological range of CO_2 partial pressures. Normally, 4 vol% of CO_2 are added to the mixed venous blood by the tissues (*vertical arrow*) and would increase the venous P_{CO_2} from point A to point B. However, when O_2 dissociates from hemoglobin there is an increased formation of reduced hemoglobin and carbamino compounds, both of which increase the CO_2 carrying capacity of the blood. This is referred to as the Haldane effect, and as a result the venous P_{CO_2} only increases to 46 mm Hg (**X** *on the 75% saturation curve*).

oxyhemoglobin is a better buffer than oxyhemoglobin. As hemoglobin becomes deoxygenated, even more CO_2 can be accepted in the blood. Second, deoxyhemoglobin forms carbamino complexes more readily than does oxyhemoglobin. Of the two factors, 70% comes from increased carbaminohemoglobin formation and 30% from increased buffering.

The total CO_2 content of the blood increases from 49 vol% in the arteries to 54 vol% in the veins (see Table 17-2). Of this total, 44 to 47 vol% is transported as bicarbonate, giving a plasma bicarbonate concentration of 24 mEq/liter. The remainder of the CO_2 carried by the blood is nearly equally divided between carbaminohemoglobin and dissolved CO_2.

Changes of the RQ over its normal range of 0.7 to 1.0 can be accommodated by changes of about 2 vol% of the CO_2 content of mixed venous blood. For example, the RQ value corresponding to the blood gas values in Table 17-2 is equal to the ratio of AV differences for O_2 and CO_2, or 5 vol%/5.3 vol% = 0.94. An increase in venous CO_2 content of 0.3 vol% would result from elevating the RQ to 1.0; a decrease of RQ to 0.7 would result in a decrease of a little more than 1 vol% from the value of venous CO_2 given in Table 17-2.

Physiology of Diving

People trained in free diving can spend a minute or more under water without breathing; some aquatic mammals such as whales, seals, and porpoises can undergo dives lasting tens of minutes. During these periods of no gas exchange the needs of energy metabolism are provided by O_2 stored in lung air, bound to hemoglobin and myoglobin, and dissolved in body fluids. The central physiological problem of prolonged diving is that of maintaining heart and brain function. Almost all mammals show some cardiovascular response to immersion of the head; this has been called the **diving reflex.** In humans the diving reflex typically results in a slowing of the heart rate by several beats/min and some inhibition of neurons involved in ventilation when the face is immersed in water. The reflex is more profound if the water is cold. It is probably of little importance in recreational and sport diving, but when combined with the peripheral vasoconstriction and reduced metabolism induced by hypothermia, it may increase chances of survival in people threatened with drowning in cold water.

In aquatic mammals, diving is usually accompanied by dramatic reflexive bradycardia and peripheral vasoconstriction. The effect is to force skeletal muscle first to draw on its store of O_2 in myoglobin and then shift to anaerobic glycolysis. The O_2 that would otherwise have been used by the muscle is spared for the heart and brain. During the dive, some of the lactic acid that is produced by the anaerobic muscle can enter the pathway of oxidative metabolism in heart muscle; some can enter the pathway of gluconeogenesis in the liver and be reconverted to glucose (see the discussion of **oxygen debt** in Chapter 12). Most of the lactic acid accumulates in the muscle and is washed out when blood flow to skeletal muscle is restored at the end of the dive.

SCUBA or helmet divers who work at depths greater than a few tens of meters breathe air at higher than atmospheric pressure and accumulate dissolved gases in blood and tissues. If the diver's return to the surface is too rapid, some of the dissolved gas comes out of solution, forming bubbles in tissues and blood vessels. The bubbles are mainly N_2, since N_2 is the major constituent of air, and is less soluble than either O_2 or CO_2. The resulting pain caused the condition to be named **the bends.** Permanent damage can result if blood flow to parts of the brain is blocked. The condition can be prevented by controlled decompression, which allows the gas partial pressures of the blood to return gradually to atmospheric values.

Diving animals typically exhale at the beginning of a dive. The increased pressure experienced by the animal if it dives to a considerable depth further reduces the volume of gas in the lungs (Boyle's law) so that most of the remaining gas is in the dead space and the lungs are collapsed. Reducing air volume in the lungs, where the gases are exposed to capillaries, minimizes the transfer of dissolved N_2 to the animal's blood so that bubble formation is not a problem when the animal returns to the surface.

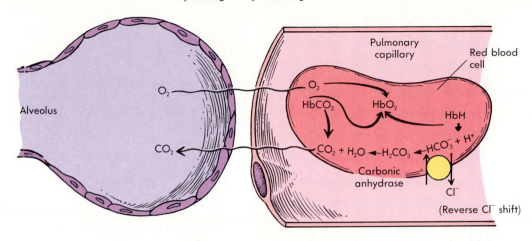

FIGURE 17-13

Cooperation between CO_2 unloading and O_2 loading in pulmonary capillary blood. The decreasing P_{CO_2} pulls the CO_2 dissociation reaction to the right; the reaction is accelerated by carbonic anhydrase. H^+ is released by deoxyhemoglobin as it becomes fully oxygenated (the Haldane effect in reverse). HCO_3^- is provided by the reverse chloride shift. The increase in pH slightly favors O_2 binding to hemoglobin (the reverse of the Bohr effect). Some CO_2 is released from carbaminohemoglobin, freeing the hemoglobin to accept O_2.

Carbon Dioxide Unloading in the Lungs

The conditions that favored O_2 unloading and CO_2 loading in the tissues are reversed when blood arrives in the lungs (Figure 17-13). The pH of the blood in the lungs is normal, P_{CO_2} is low, the temperature may be cooler than in very active systemic tissues, and the high levels of O_2 reduce 2,3 DPG formation. Thus the same factors that shift the hemoglobin dissociation curve to the right in active tissues cause a left shift in the lung and facilitate the combination of oxygen with hemoglobin.

The higher P_{O_2} in the alveoli drives O_2 into the blood, loading the hemoglobin to near 100% saturation. As the deoxyhemoglobin is converted to oxyhemoglobin, it unloads some of H^+ that it had been carrying (see Figure 17-13). The law of mass action dictates that the addition of H^+ to the red blood cell cytoplasm will drive the formation of H_2CO_3, which

is rapidly converted into CO_2 and H_2O in the presence of carbonic anhydrase. The dissociation of H^+ from the hemoglobin and its reaction with available HCO_3^- creates a gradient favorable for entry of bicarbonate ions into the red cells. The movement of HCO_3^- into the red cell is electrically balanced by exit of Cl^-; this is the **reverse Cl^- shift**. Some CO_2 carried in the blood as carbaminohemoglobin is released, adding to the amount of CO_2 unloaded and freeing additional hemoglobin for binding with O_2.

---

1 In what forms is CO_2 carried in the blood? What is the role of carbonic anhydrase in CO_2 transport?
2 What is the Haldane effect? How is it distinct from the Bohr effect?
3 What is the Cl^- shift? How does it facilitate CO_2 transport?

SUMMARY

1. Gas mixtures in the respiratory tract and gases dissolved in the blood and tissues behave according to the gas laws.
2. Gases always diffuse from an area with a higher partial pressure to an area with a lower partial pressure.
3. The majority of the O_2 transported in the blood is carried by **hemoglobin** (19.7 vol%); the remainder (0.3 vol%) is dissolved in the plasma.
4. The plateau portion of the **S**-shaped **oxyhemoglobin dissociation curve** allows hemoglobin to load O_2 to near 100% saturation even if alveolar P_{O_2} falls below normal, and the steep portion of the curve allows a large amount of O_2 to be delivered to the tissues with a relatively small drop in P_{O_2}.
5. The oxyhemoglobin dissociation curve is shifted to the right by conditions prevailing in the tissues (increases in H^+, CO_2, 2,3 DPG, and temperature). This right shift decreases the affinity of hemoglobin for O_2; the greater the shift the greater the O_2 delivery in the tissues.
6. The leftward position of the fetal hemoglobin oxyhemoglobin dissociation curve in comparison with the adult hemoglobin curve reflects a greater affinity for O_2. Thus fetal tissues can selectively take up oxygen from the maternal circulation. The affinity of **myoglobin** for O_2 is higher than that of either form of hemoglobin, representing a reserve supply in muscle.
7. The majority of the CO_2 in the blood is carried as HCO_3^-. Smaller amounts are dissolved in the plasma or associated with the hemoglobin molecule. The presence of **carbonic anhydrase** in the red blood cells catalyzes the reaction of CO_2 with H_2O that leads to H_2CO_3. Carbonic acid dissociates to form H^+ and HCO_3^-. The chemical equilibrium between the reactants and products is represented by the **Henderson-Hasselbalch equation**.

1. Why are the steady-state alveolar P_{CO_2} higher and the alveolar P_{O_2} lower than those of the inspired air? Suppose overall metabolic activity doubled without a corresponding change in alveolar ventilation. What would happen to the alveolar partial pressures of O_2 and CO_2? Why?

2. What effect does the respiratory quotient have on the composition of alveolar gas?

3. What determines the O_2 carrying capacity of blood?

4. An individual has a normal rate of O_2 consumption but a hemoglobin concentration 50% of normal. How would the magnitude of the arteriovenous O_2 difference in this individual be different from a normal individual?

5. How is the difference between the dissociation curves for myoglobin and hemoglobin related to the physiological roles of hemoglobin in gas transport in the blood and myoglobin as an oxygen reserve in muscle?

6. An individual voluntarily doubles his alveolar ventilation. What would happen to plasma pH? Why?

7. The normal arterial P_{O_2} is 95 mm Hg, slightly less than the normal alveolar P_{O_2} of 100 mm Hg. Why? The alveolar-arterial P_{O_2} difference can be very large in cases of emphysema. Why? Would there be a similar difference for CO_2?

8. Diagram (A) the steps in the pathway for diffusion of CO_2 from tissues into the blood and (B) from the blood into the alveolar gas mixture.

9. When an individual begins to breathe from an atmosphere whose total pressure is one-half that of sea level, what changes would occur in the following: arterial P_{O_2}, venous P_{O_2}, arterial O_2 content?

● SUGGESTED READING

CAMERON, J.N.: *Principles of Physiological Measurement*, Academic Press, New York, 1986. Discusses gas and solution concepts, pH, and partial pressures.

ELSNER, R., and B. GOODEN: *Diving and Asphyxia: A Comparative Study of Animals and Man*, Cambridge University Press, Cambridge, Mass., 1983. Describes the diving reflex in humans and animals and what it can tell us regarding cardiovascular and respiratory control.

PERUTZ, M.F.: Hemoglobin Structure and Respiratory Transport, *Scientific American*, volume 239, p. 92, 1978. Still current description of how oxygen binds to hemoglobin.

CHAPTER 18

Regulation of the Respiratory System

On completing this chapter you will be able to:

- Describe the movements of the diaphragm, intercostal muscles, and abdominal muscles during inspiration and expiration.
- Understand how respiratory motor movements are affected by centers in the medulla and pons.
- List the factors affecting the intensity of respiratory drive, rate of breathing, and tidal volume.
- Explain how, by monitoring the pH of CSF, the central chemoreceptors serve to regulate the arterial P_{CO_2}.
- Appreciate how inputs from the peripheral chemoreceptors become important during hypoxia and acid-base imbalance.
- Describe the ventilatory responses to high-altitude exposure.
- Understand the role of the respiratory system in responding to acute metabolic acidosis and lung disease.
- Describe the factors determining the ventilatory increase during exercise.

*B*reathing, like the heartbeat, goes on automatically. Yet the generation and control of the ventilatory cycle are in many ways different from the heart cycle. The respiratory muscles are skeletal muscles. Unlike cardiac muscle, respiratory muscles do not contract if deprived of nervous stimulation. Each intake of breath must be driven by a burst of action potentials in motor neurons of the respiratory muscles. The neuronal mechanisms by which the central pattern of respiratory activity is generated are still poorly understood.

There is a substantial amount of voluntary control over the rhythm of respiration, but the overriding importance of internal stimuli for control of respiration is evidenced by the fact that most people cannot hold their breath for much longer than 1 minute. The main internal stimuli for regulation of breathing are the P_{CO_2} and pH of arterial blood. The respiratory control system protects the extracellular fluid from the threat of acidification posed by metabolic production of CO_2, and it provides an immediate compensation for alterations in pH caused by ingestion or production of acids other than CO_2, and bases. The respiratory system is one of the two pillars of acid-base regulation described in Chapter 20, the kidneys being the other. Under most circumstances the "safety zone" of the oxyhemoglobin dissociation curve ensures that the hemoglobin of systemic arterial blood will be oxygen saturated no matter what changes in alveolar ventilation occur as part of acid-base regulation.

THE RHYTHM OF BREATHING
The Central Motor Program for the Respiratory Cycle

The breathing pattern is produced by interactions between specific brainstem neurons, stretch receptors in the lung, neural signals from higher levels of the central nervous system, and inputs from the chemoreceptors (Figure 18-1). Groups of neurons in the medulla and pons generate the **respiratory motor program**—a pattern of action potentials that drives the spinal motor neurons innervating the diaphragm, the internal and external intercostal muscles of the rib cage, and the muscles of the abdominal wall. The diaphragm is innervated by the paired **phrenic nerves** from the cervical spinal cord, and the intercostal respiratory muscles are innervated by

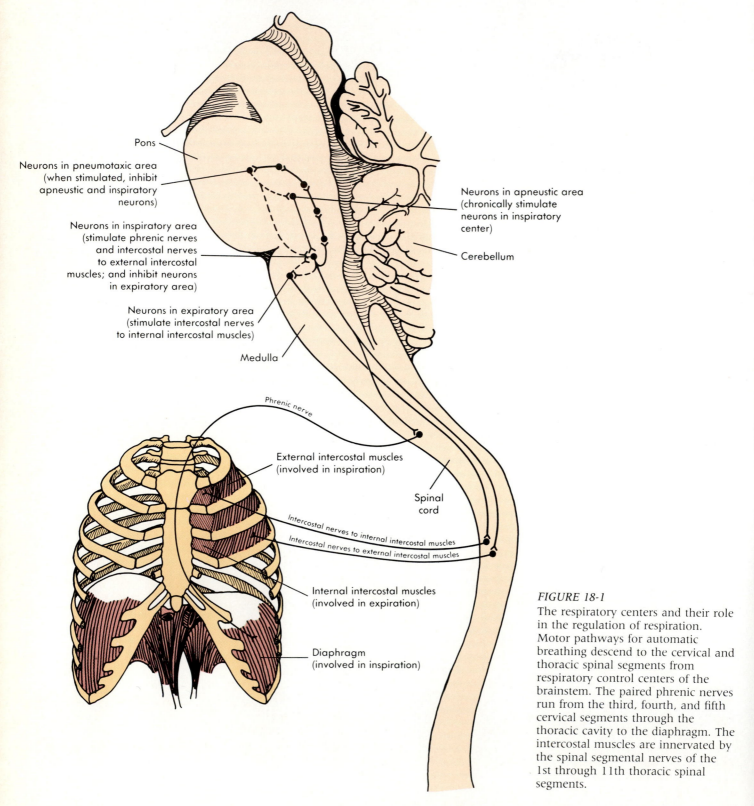

Pons

Neurons in pneumotaxic area (when stimulated, inhibit apneustic and inspiratory neurons)

Neurons in inspiratory area (stimulate phrenic nerves and intercostal nerves to external intercostal muscles; and inhibit neurons in expiratory area)

Neurons in expiratory area (stimulate intercostal nerves to internal intercostal muscles)

Medulla

Neurons in apneustic area (chronically stimulate neurons in inspiratory center)

Cerebellum

Phrenic nerve

External intercostal muscles (involved in inspiration)

Intercostal nerves to internal intercostal muscles

Intercostal nerves to external intercostal muscles

Spinal cord

Internal intercostal muscles (involved in expiration)

Diaphragm (involved in inspiration)

FIGURE 18-1
The respiratory centers and their role in the regulation of respiration. Motor pathways for automatic breathing descend to the cervical and thoracic spinal segments from respiratory control centers of the brainstem. The paired phrenic nerves run from the third, fourth, and fifth cervical segments through the thoracic cavity to the diaphragm. The intercostal muscles are innervated by the spinal segmental nerves of the 1st through 11th thoracic spinal segments.

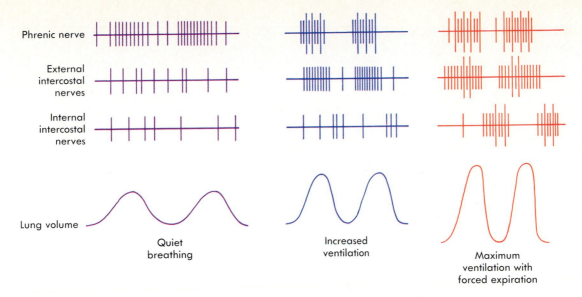

Phrenic nerve

External
intercostal
nerves

Internal
intercostal
nerves

Lung volume

Quiet
breathing

Increased
ventilation

Maximum
ventilation with
forced expiration

FIGURE 18-2

An idealized pattern of action potentials in the phrenic nerve, a nerve to external intercostals, and a nerve to internal intercostals, in three levels of pulmonary ventilation.

thoracic spinal segmental nerves (see Figure 18-1). As a result, high thoracic spinal cord injuries may paralyze the accessory respiratory muscles of the rib cage and abdominal wall but spare neural control of the diaphragm.

In quiet breathing, the main motor output of the respiratory control system is to the diaphragm, which is responsible for about two-thirds of the thoracic volume change. Figure 18-2 shows the pattern of action potentials in the phrenic nerve during quiet breathing. After a passive expiration the volume of the lung is equal to the functional residual capacity, and only a few motor axons in the phrenic nerve are active. An increase in the number of active fibers in the phrenic nerve and an increase in the action potential frequency in each active fiber signals the start of an inspiration. These changes are the result of a burst of activity in medullary respiratory neurons that synapse on the spinal motor neurons of the phrenic nerve. The burst of action potentials in the respiratory motor neurons comes to an end rather abruptly. Without excitatory input, the respiratory muscles relax, and the lung-chest system passively recoils to the functional residual capacity. After a short time the cycle is repeated.

At ventilation rates higher than the resting rate, motor neurons innervating the external intercostal muscles are recruited during inspiration (see Figure 18-2). As the total motor drive to the inspiratory muscles increases, so does the total inspiratory force. Further increases in tidal volume are achieved by activation of the internal intercostal and abdominal muscles in forced expirations. Contraction of internal intercostal and abdominal muscles forces more air out of the thoracic cavity, making the end-expiratory volume of the lung in a forced expiration

smaller than the functional residual capacity (Chapter 16, p. 414). When expiration is forced in this manner, the motor program for a respiratory cycle consists of a burst of action potentials in the motor units of inspiratory muscles, followed by a burst of action potentials in the motor units of internal intercostal nerves (see Figure 18-2).

In summary, the tidal volume of a respiratory cycle is determined by the number of motor neurons recruited and their firing rate during inspiration and expiration; the respiratory rate is determined by the time interval between inspiratory bursts. The rate and depth of cycles are adjusted unconsciously by the respiratory centers of the brain to minimize the energy cost of breathing.

The pathway for conscious control of respiratory muscles is used, for example, in speaking, singing, and voluntary breath-holding. Central control does not involve inputs to the brainstem respiratory centers. Instead, there is an anatomically separate corticospinal pathway (Chapter 11, p. 271) by which the cortex can share control of the spinal respiratory motor neurons with the brainstem. Some axons in this pathway also project to motor neurons of other somatic muscle groups. Damage to the corticospinal pathway causes a general paralysis that abolishes voluntary control of respiration but preserves automatic respiration.

Damage to the pathway between medulla and respiratory motor neurons causes a paradoxical condition called **Ondine's curse,** in which the capacity for automatic control is lost but voluntary control remains. The patient must therefore remember to breathe. The name comes from a legend in which the water nymph Ondine punished her faithless husband by taking away his automatic body func-

Receptor	Location	Stimulus
Stretch receptors	Smooth muscle layer of bronchi and bronchioles	Stretch of lungs during inflation; mediate Hering-Breuer reflex
Irritant receptors	Among airway epithelial cells	Noxious gases, allergens, lung inflammation; stimulate rapid, shallow breathing
Central chemoreceptors	Ventrolateral surface of medulla, near roots of cranial nerves IX and X	pH changes in CSF-induced changes of arterial P_{CO_2}
Carotid bodies	In crotch of carotid bifurcation on top of carotid sinus	Mainly sensitive to decreased arterial P_{O_2}; also to pH and P_{CO_2}
Aortic bodies	Scattered over surface of aorta and major thoracic arteries	

tions. In the legend, the husband died when he fell asleep. Modern sufferers of the curse must sleep in respirators.

Important feedback about the degree of inflation of the lung is provided by stretch receptors in the bronchi and bronchioles that are activated by increasing lung volume (Table 18-1). The axons of the lung receptors run to the CNS in the vagus nerves; this is the **Hering-Breuer reflex.** After the vagus nerves of an experimental animal are cut, the tidal volume increases. Stimulation of the ascending stump of one of the cut vagus nerves causes the animal to maintain an expiration as long as the stimulation continues. These results suggest that, in the intact system, the rising level of activity in lung stretch receptors as the lungs are expanded helps to terminate each inspiration.

Scattered among the epithelial cells of the airway are **irritant receptors** (see Table 18-1) responsible for the sensation of distress when noxious or irritating gases or particulates are inhaled. Irritant receptors are also responsive to chemical mediators of the immune system released upon exposure to allergens and during inflammation. These receptors tend to promote rapid, shallow breathing and to stimulate bronchiolar constriction.

How Does the CNS Generate the Basic Rhythm of Breathing?

There are two basic ways in which rhythmic activity might be generated in the nervous system. In a **connectivity mechanism,** the rhythmic output is a consequence of synaptic connections between neurons, each of which by itself would be incapable of generating such an output. The oscillation of such a system could be analogized to the rocking of a seesaw; the rocking motion is the result of give and take between the two children and could not be generated by one child alone. In an **intrinsic mech-**

anism, the driving cells are rhythmic by virtue of oscillations in their membrane potentials. Cardiac pacemaker cells are examples of intrinsically rhythmic cells, and other examples are known from the central nervous systems of some invertebrates.

Some early recordings of neuronal activity suggested that the medulla contained separate inspiratory and expiratory centers (Table 18-2), each of which drove the appropriate population of spinal motor neurons. It was tempting to hypothesize that mutual inhibitory synaptic connections between the two kept the system oscillating, just as in the seesaw analogy. More recently, the organization of the respiratory areas in the medulla has been shown to be complex, with some cells in both areas active during inspiration and some active during expira-

TABLE 18-2	**Respiratory Control Centers**

Center	Location	Effect
Pneumotaxic	Upper pons	Tends to terminate inspiration
Apneustic	Lower pons	Promotes inspiration
Inspiratory	Dorsal medulla	Active during inspiration; stimulation promotes inspiration; generates basic respiratory rhythm
Expiratory	Ventral medulla	Some neurons active during all phases of the cycle, especially during forceful expiration

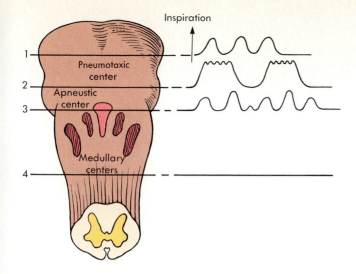

Inspiration

1 —

Pneumotaxic
center

2 —

Apneustic
center

3 —

Medullary
centers

4 —

FIGURE 18-3
Identification of respiratory centers in the pons-medulla by cutting it at the levels indicated by the numbered lines. The resulting respiratory patterns are shown to the right of each cut. See the text for further explanation.

tion. In fact, some inspiratory cells are active only early in the cycle and others later in the cycle. These findings do not rule out a "connectivity" mechanism, but it has been difficult to incorporate them into a satisfactory working model.

It now appears likely that some of the respiratory neurons in the medulla possess intrinsic rhythmicity, analogous to heart pacemakers. Unlike heart pacemakers, the respiratory pacemakers probably generate bursts of action potentials rather than single action potentials. These neurons might set a basic respiratory rhythm, or their rhythmicity may help to stabilize a rhythm that is determined by synaptic connections.

A crude test of the connectivity hypothesis was made in experiments in which cuts were made at different points in the brainstems of experimental animals. These demonstrated that the entire basic respiratory control system lies within the pons-medulla (between cuts 1 and 4 in Figure 18-3) because a cut above the pons has no effect on the cycle and a cut below the medulla abolishes ventilation. Separation of the upper pons from the lower pons and the rest of the brainstem (a cut at level 2 in Figure 18-3) produces a breathing pattern characterized by deep and very long inspirations, followed by rapid expiration. This pattern is particularly evident when the vagus nerves are also cut. Thus loss of the part of the brainstem between cut 1 and cut 2 biases the system in the direction of inspiration, and more so if the remaining part of the control system does not receive the inhibitory stretch receptor input. The area located somewhere between cut 1 and cut 2 has been termed the **pneumotaxic center** (see Table 18-2; Figure 18-1). Its role is to terminate each inspiration.

If the medulla is separated from the pons (a cut at level 3 in Figure 18-3), the respiratory rate is increased and the system spends most of its time in expiration. The area between cut 2 and cut 3 was named the **apneustic center** (see Table 18-2; Figure 18-1). The activity of this center can prolong inspiration. The pneumotopic and apneustic centers can be thought of as providing a modulatory influence on the basic rhythm generated by the medulla (Figure 18-4). Neither the pneumotaxic center nor the apneustic center nor the connections that run between them are necessary for the generation of a basic rhythm because breathing after cut 3 is still rhythmic although the tidal volumes are irregular.

Whatever the origin of the rhythm, the basic respiratory rhythm is strengthened and modified by tonic inputs from chemoreceptors that monitor arterial P_{CO_2}, P_{O_2}, and pH, and inputs from the reticular activating system that sets the brain's level of arousal. Other less well-defined inputs are important in driving respiration at a high level during exercise. Somatosensory inputs can affect the respiratory centers. For example, people gasp reflexively when slapped or plunged into cold water. Emotion has an effect on respiration; anxiety increases ventilation, and depression decreases it. These inputs are integrated at various levels in the control system (see Figure 18-4), providing a continuous **respiratory drive** whose magnitude is reflected in the intensity and frequency of respiratory cycles driven by the medullary center. Under resting conditions much of the respiratory drive is attributable to chemoreceptor inputs, which are discussed in the following sections.

1 *Describe the motor program for respiration during quiet breathing.*
2 *What is the difference between the intrinsic oscillator and connectivity oscillator hypotheses for the generation of the respiratory rhythm?*
3 *What is respiratory drive?*

THE CHEMICAL STIMULI FOR BREATHING
Arterial P_{CO_2}—the Central Chemoreceptors

At normal or near normal arterial P_{O_2}, changes in the arterial P_{CO_2} have a much greater effect on respiratory minute volume than do variations in the arte-

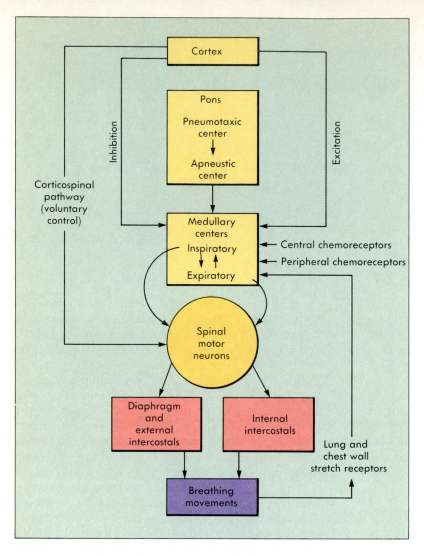

FIGURE 18-4
A schematic diagram of information flow in the respiratory control system, including some of the inputs that establish respiratory drive. The feedback loops based on the effect of ventilation on blood gas composition are not shown.

rial P_{O_2}. For example, an increase in the arterial P_{CO_2} **(hypercapnia)** of 2 to 3 mm Hg will approximately double the respiratory minute volume. In contrast, the arterial P_{O_2} would have to decrease by 30 mm Hg to stimulate respiration to the same degree. The greater sensitivity to CO_2 rather than O_2 provides effective regulation of breathing for two reasons: (1) the S-shape of the oxyhemoglobin dissociation curve protects against decreases in the O_2 content of blood that might occur as a result of small decreases in alveolar P_{O_2}; and (2) small changes in the arterial P_{CO_2} produce large shifts in the plasma pH, so that CO_2 production is a constant threat to total body acid-base balance.

The respiratory response to CO_2 originates within the central nervous system. The central respiratory chemoreceptors are thought to be located near the ventrolateral surface of the medulla; the particular cells in this area that serve as chemoreceptors have not been pinpointed. Like all areas of the central nervous system, the central chemoreceptors are surrounded by extracellular fluid. The interstitial spaces of the brain are continuous with the **ventricles,** large cavities in the interior of the brain. **Cerebrospinal fluid (CSF)** circulates throughout

the ventricles and serves as a large reservoir of extracellular fluid for the brain cells. CSF is formed by the **choroid plexuses,** highly vascularized structures on the walls of the third and fourth ventricles. For convenience, the total extracellular fluid volume of the brain, consisting of the CSF and a smaller volume of interstitial fluid, will be referred to as CSF in subsequent discussion.

Because the walls of brain capillaries constitute a blood-brain barrier (see Chapter 14, p. 367) not freely permeable to ions such as H^+ and HCO_3^-, they effectively isolate the brain interstitial fluid and CSF from the arterial blood. Ionic solutes such as H^+ and HCO_3^- cannot readily cross the blood-brain barrier in the absence of specific transport mechanisms, but dissolved CO_2 can cross readily without help (Figure 18-5). The central respiratory chemoreceptors do not detect plasma P_{CO_2} directly but respond to the pH of their immediate environment, the CSF. However, any increase in the arterial P_{CO_2} results in an immediate and equivalent increase in the P_{CO_2} of the CSF. Once in the CSF, the CO_2 reacts with water to form carbonic acid (H_2CO_3), which dissociates to produce H^+ and HCO_3^-:

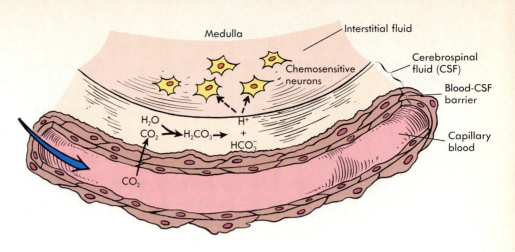

FIGURE 18-5

Activation of central chemoreceptors by changes in arterial P_{CO_2}. Although the receptors are sensitive to small changes in the P_{CO_2}, they are not directly stimulated by CO_2, but instead by the pH of the cerebrospinal fluid. CO_2 readily diffuses across the blood-brain barrier and forms carbonic acid, which dissociates in the CSF and interstitial fluid of the brain. By this mechanism, an increase in arterial P_{CO_2} acidifies the CSF and interstitial fluid.

$$CO_2 + H_2O \longrightarrow H_2CO_3 \longrightarrow HCO_3^- + H^+$$

According to the law of mass action (see Chapter 3, p. 50), elevating the P_{CO_2} of CSF drives the reaction to the right and thus increases both the H^+ and HCO_3^- concentrations. Hydrogen and HCO_3^- ions formed within the CSF are trapped there because they cannot cross the blood-brain barrier into the plasma (see Figure 18-5).

Compared with the blood, CSF is weakly buffered because, like interstitial fluid, it contains little protein. The low buffer capacity of the CSF makes the pH of CSF sensitive to changes in plasma P_{CO_2}. When the arterial P_{CO_2} increases, the P_{CO_2} of the CSF increases proportionately. The subsequent reaction of CO_2 with H_2O in the CSF decreases its pH. The response of the central chemoreceptors to this change in pH increases the respiratory drive and thus the alveolar minute volume (see Figure 18-4). The **ventilatory response** is the change in alveolar minute volume that results from any change in the inspired gas or the plasma chemistry. In the case of CO_2, the ventilatory response is nearly linearly related to the change in arterial P_{CO_2} (Figure 18-6).

FIGURE 18-6

The relationship between alveolar P_{CO_2} and ventilation. The subjects breathed either air or air containing 2%, 4%, or 6% CO_2. The resulting values of alveolar P_{CO_2} can be expected to be closely paralleled by the P_{CO_2} of arterial blood.

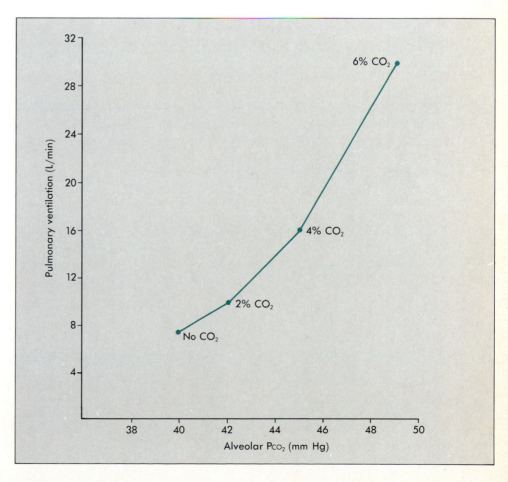

Regulation of the Respiratory System

The Choroid Plexuses and the Ventilatory Setpoint

Although, chemically speaking, the CSF is not well buffered, the changes in CSF pH that result from changes of arterial P_{CO_2} are smaller and less lasting than the corresponding changes in arterial pH (Figure 18-7). This is because the choroid plexuses oppose changes in CSF pH by increasing or decreasing their transport of HCO_3^- from CSF to blood. As the pH of the CSF is restored toward its normal value, the ventilatory drive of central chemoreceptors diminishes. In effect, in the process of protecting the brain from changes in blood pH, the choroid plexuses reset the central chemoreceptors, a process which serves the same function in the respiratory system as receptor adaptation does in sensory systems. If the central chemoreceptors were not reset, they would become unresponsive to variations in the metabolic load of CO_2. Such changes in the setpoint of the ventilatory response are important in situations in which there is a chronic change in arterial P_{CO_2}, such as lung disease and adaptation to high altitude, discussed later.

Arterial P_{O_2} and pH—the Carotid and Aortic Chemoreceptors

Although arterial P_{CO_2}, as monitored by the central chemoreceptors, is responsible for most of the respiratory drive, the respiratory centers are also responsive to changes in arterial P_{O_2} and arterial pH detected by **peripheral chemoreceptors.** The major peripheral chemoreceptors that are well described are located in the **carotid** and **aortic bodies** (Figure 18-8). Unlike the central chemoreceptors, the peripheral chemoreceptors are in direct contact with the arterial blood. The peripheral chemoreceptors are most responsive to changes in arterial P_{O_2}. As arterial P_{O_2} falls, chemoreceptor activity increases slowly at first but rapidly as lower values are reached (Figure 18-9). In addition to the P_{O_2}, chemoreceptors are also responsive to arterial pH, especially in the case of the aortic chemoreceptors. Being responsive to pH makes the receptors respond indirectly to P_{CO_2} because the reaction of CO_2 with water produces HCO_3^- and H+.

Distinguishing the effects of peripheral chemoreceptors from those of the central receptors is

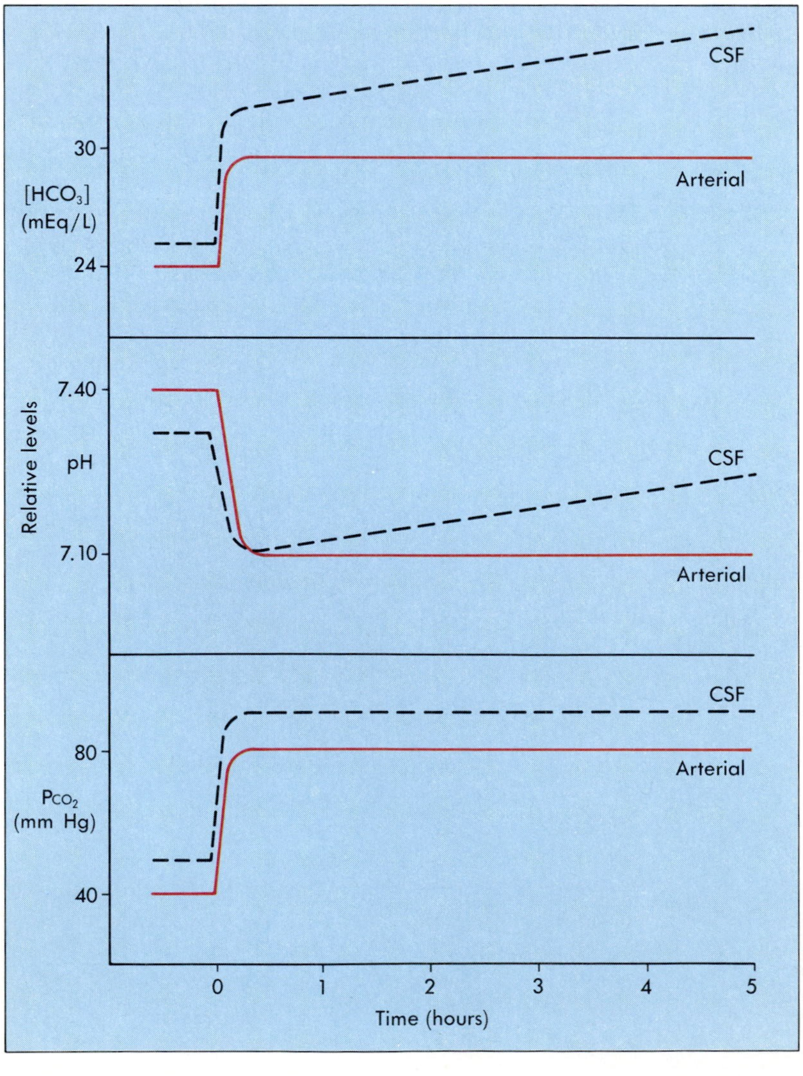

FIGURE 18-7
Relative changes in P_{CO_2}, pH, and HCO_3^- concentrations in CSF and arterial plasma after addition of CO_2 to inspired air at time 0. The P_{CO_2} of CSF tracks that of arterial blood, but addition of bicarbonate to the CSF by the choroid plexuses causes the CSF pH to become less acid over the subsequent hours, a process that is continuing at the end of the time scale shown. In the case of a decrease in plasma P_{CO_2}, the pattern would be similar except that the restoration of CSF pH toward normal would involve net loss of HCO_3^- from the CSF.

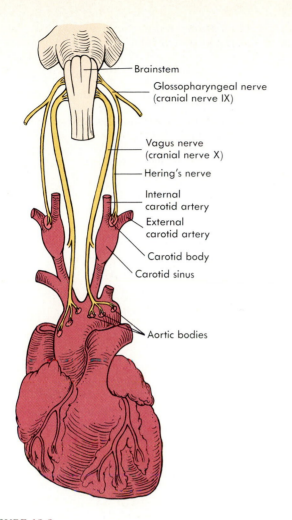

FIGURE 18-8

The locations of carotid and aortic arch chemoreceptors. The carotid bodies are near the carotid sinus baroreceptors discussed in Chapter 15, but are anatomically and functionally distinct from them.

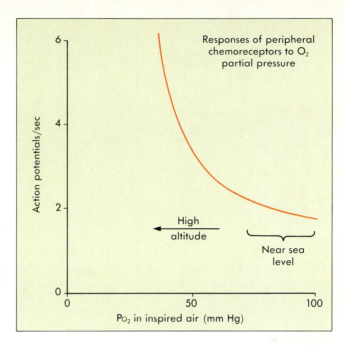

FIGURE 18-9

The rate of discharge of action potentials by a single carotid chemoreceptor as a function of ambient P_{O_2}.

difficult because it is difficult to change just one of the important variables. In a normal individual, O_2 and CO_2 levels change in opposite directions when alveolar ventilation changes, and changes in CO_2 also affect the plasma pH. Experimenters had to develop ways to vary O_2, CO_2, and pH independently in order to study the function of particular chemoreceptors. For instance, the central effects of peripheral chemoreceptor activity can best be studied when the cerebrospinal fluid is artificially perfused so that its values do not change when arterial values of CO_2, O_2, and pH are experimentally altered.

When arterial P_{O_2} is decreased, the frequency of action potentials in the afferent nerves from both the aortic and carotid bodies increases. This increased neural activity is transmitted to the brainstem respiratory control centers along the glossopharyngeal and vagus nerves. The respiratory

centers respond by increasing respiratory frequency and tidal volume, therefore increasing the respiratory minute volume and alveolar ventilation. A similar effect is seen when the plasma P_{CO_2} is increased and when the plasma pH is decreased.

When arterial P_{O_2} is decreased and arterial P_{CO_2} and pH are held at normal values (40 mm Hg and 7.40, respectively), there is little increase in ventilation until the arterial P_{O_2} falls below about 60 to 70 mm Hg (Figure 18-10). At a P_{O_2} of 60 to 70 mm Hg, hemoglobin is still 90% saturated with O_2. The peripheral chemoreceptors start to become very effective at generating respiratory drive at arterial P_{O_2} valves at which hemoglobin is significantly less than fully saturated with O_2. The respiratory minute volume rapidly increases as the arterial P_{O_2} decreases below 60 mm Hg. Such a condition might be encountered at an altitude of 3000 meters. At this altitude, arterial P_{O_2} is about 60 mm Hg, and the respiratory minute volume is about three times its normal resting value. Central chemoreceptor responses to high altitude would tend to result in the cessation of breathing due to hyperventilation and subsequent loss of CO_2, but the peripheral response to lowered P_{O_2} provides for an adaptive response.

Peripheral chemoreceptors are responsive to CO_2 as well as to O_2 (Figure 18-11). If P_{CO_2} is elevated, ventilation is increased at all partial pressures of O_2. For instance, if the arterial P_{CO_2} were elevated to 50 mm Hg (upper, dotted curve), the increase sums with the increase in lung ventilation that was attributable to decreased P_{O_2}. The lower,

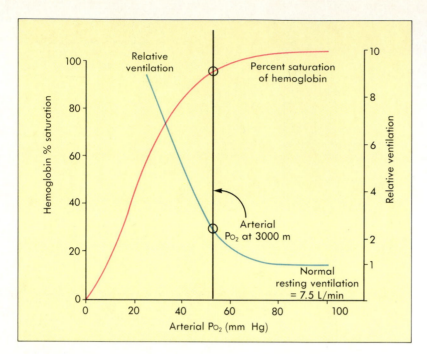

FIGURE 18-10
Regulation of ventilation by oxygen. The increase in respiratory minute volume (relative to the resting minute volume of 7.5 liters/min) produced by activation of the peripheral chemoreceptors is shown by the green curve; it does not become significant until the arterial P_{O_2} drops below 80 mm Hg. This corresponds to the shoulder of the oxyhemoglobin dissociation curve *(shown as the red curve)*. The values of hemoglobin saturation at 3000 meters and the corresponding value of relative ventilation are indicated by the vertical line.

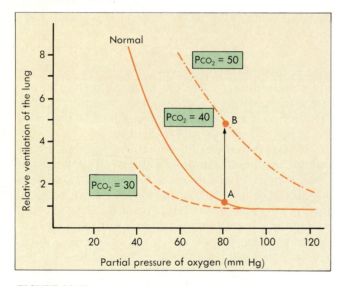

FIGURE 18-11
The effect of the arterial P_{CO_2} on the ventilatory response to O_2 driven by peripheral chemoreceptors. When the CO_2 level increases, the response to decreasing arterial P_{O_2} is augmented at all P_{O_2} levels. Conversely, respiratory minute volume is increased by increases in the P_{CO_2} at any given P_{O_2} *(point A to point B)*.

dashed curve shows that the responsiveness to O_2 is decreased when the P_{CO_2} is lowered to 30 mm Hg. If arterial P_{O_2} is held constant while arterial P_{CO_2} is increased, the relative ventilation of the lung increases (vertical arrow from point A to point B in Figure 18-11).

These changes in the sensitivity of the peripheral chemoreceptors to oxygen are not produced directly by dissolved CO_2, but instead by the effect that P_{CO_2} has on plasma pH. The evidence that H^+ rather than CO_2 is the actual stimulus for the peripheral chemoreceptors was obtained by experimentally maintaining the pH constant while increasing the arterial P_{CO_2}. When pH is held constant, P_{CO_2} has no effect on the activity of peripheral chemoreceptors. The peripheral chemoreceptors are responsible for at most about 15% to 20% of the overall respiratory response to CO_2, so the response of central chemoreceptors to CO_2 is far more important for regulation of arterial P_{CO_2}.

The respiratory control centers are sensitive to very small changes in arterial blood pH detected by peripheral chemoreceptors. If P_{O_2} and P_{CO_2} are normal, a decrease of 0.2 pH units increases ventilation by a factor of five (Figure 18-12). The peripheral chemoreceptors are the only receptors that are capable of responding to acids or bases that cannot cross the blood-brain barrier.

> 1 Why does a change in arterial P_{CO_2} have a greater effect on CSF pH than on plasma pH? What is the significance of the sensitivity of CSF pH to arterial P_{CO_2}.
> 2 What is meant by ventilatory response?
> 3 What are the functional roles of the central chemoreceptors? The peripheral chemoreceptors?

INTEGRATED RESPONSES OF THE RESPIRATORY CONTROL SYSTEM
Interactions between Peripheral and Central Chemoreceptors

The effects of central and peripheral chemoreceptors on ventilation are usually synergistic, as shown in Figure 18-13. Four situations in which the effects of

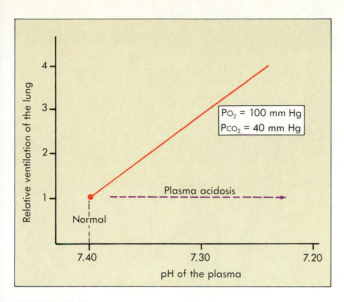

FIGURE 18-12
The ventilatory response to plasma pH driven by the peripheral chemoreceptors. In contrast to the O_2 response, the ventilatory response driven by the peripheral chemoreceptors is highly sensitive to small increases in plasma H^+ ion concentration (acidosis). The peripheral chemoreceptors are thus a major element in the control of total body acid-base balance.

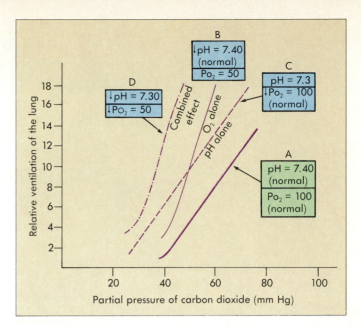

FIGURE 18-13
The overall ventilatory response of an intact individual to oxygen, carbon dioxide, and plasma pH. **Curve A** shows that under normal conditions (arterial pH = 7.40; arterial P_{O_2} = 100 mm Hg), the arterial P_{CO_2} represents the major factor controlling the respiratory minute volume. However, this response is augmented by decreases in the arterial P_{O_2} and pH. **Curve B** shows the effect of the P_{CO_2} when the P_{O_2} is reduced to 50 mm Hg. Ventilation is increased and so is the sensitivity to additional changes in the P_{CO_2}. **Curve C** illustrates the effect of a decrease in arterial pH to 7.3. Ventilation is increased, but the CO_2 sensitivity is about the same as normal. **Curve D** shows the large increases in ventilation produced by CO_2 when both the P_{O_2} and pH are changed.

central and peripheral chemoreceptors are *not* synergistic are breath-holding after hyperventilation, high-altitude acclimatization, chronic acidosis, and respiratory disease. These are discussed in following sections.

Breath-Holding

During breath-holding, plasma P_{CO_2} rises and P_{O_2} falls. At some point the increase in ventilatory drive, resulting primarily from the increased P_{CO_2}, will overcome any psychological resistance, and breathing will begin again. It is possible to hold one's breath longer if breath-holding is preceded by inspiration rather than expiration, because the additional lung air provides an additional O_2 reserve and CO_2 sink. Even though the drop in P_{O_2} has less of an effect in overriding the central control and initiating involuntary ventilation, the hypoxia that can result from voluntary breath-holding rarely threatens the heart and brain. However, when breath-holding is preceded by a period of hyperventilation, the duration of breath-holding increases, because the hyperventilation does not significantly increase P_{O_2}, but it does produce a lower arterial P_{CO_2} before breath-holding begins. Consequently, the P_{CO_2} may not rise high enough to restart ventilation before arterial P_{O_2} has fallen so low that brain hypoxia causes unconsciousness. In a number of reported instances, swimmers who hyperventilated to increase breath-holding time became unconscious underwater and died.

Acclimatization to Altitude

The first example of opposing chemoreceptor effects is seen when a person ascends rapidly to a high altitude. Many people live and vacation at altitudes sufficient to cause significant effects on blood oxygenation. For example, at an altitude of 3000 meters (about the altitude of many mountain resort areas in Colorado and California, for example), the inspired P_{O_2} is decreased to about 100 mm Hg, causing arterial P_{O_2} to fall from its normal value of 95 mm Hg to about 50 mm Hg. At this arterial P_{O_2} the systemic hemoglobin is about 87% saturated. When people acclimated to sea level travel to such altitudes, the lower arterial P_{O_2} stimulates the peripheral chemoreceptors and increases alveolar ventilation (Figure 18-14). The increased ventilation decreases the arterial P_{CO_2} and inhibits the central chemoreceptors. Given these contradictory inputs, the output of the respiratory center may oscillate between hyperventilation and short periods of apnea; generally it regulates ventilation at a level below that needed to maintain tissue P_{O_2} at normal levels. The immediate sensations experienced by the transition to high altitude are fatigue, dizziness, headache, and nausea; when they are severe enough to cause distress, these symptoms of generalized hypoxia are called **acute mountain sickness.** The

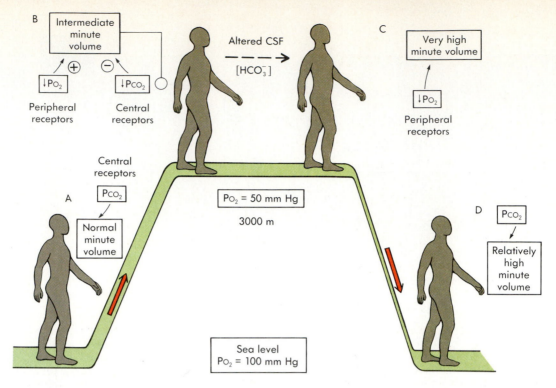

FIGURE 18-14

Changes in respiratory regulation during and following the process of altitude acclimatization. Initially ventilation is less than it should be because the central chemoreceptors counteract the peripheral anoxic drive *(box B)*. With time, the pH of the cerebrospinal fluid returns toward its normal value because the rate of HCO_3^- transport across the blood-brain barrier changes and ventilation increases *(box C)*. After an acclimatized individual returns to sea level, respiration remains elevated until the central chemoreceptors can be reset *(box D)*.

symptoms tend to be worse the more rapid the ascent; travel by air from sea level to a region that is several thousand feet above sea level is more likely to provoke these symptoms than hiking to a high altitude.

The choroid plexuses respond to the abnormally high CSF pH by reducing their transport of HCO_3^-. After a few days the CSF pH returns almost to normal, even though the arterial P_{CO_2} is still low. As a result, the central chemoreceptors no longer prevent the respiratory control center from responding to the reduced arterial P_{O_2}, and respiratory minute vol-

ume gradually increases. At this point the respiratory center is said to have adapted to the altitude. Adaptation is not something that needs to occur only at extreme altitude; some degree of adaptation of the respiratory center occurs in people who live at any altitude above sea level.

The respiratory changes described in preceding paragraphs are part of a general process of adaptation to altitude that also involves an increase in the hematocrit and a right shift of the hemoglobin dissociation curve (Table 18-3). The increase in the hematocrit is brought about by increased secretion of

TABLE 18-3	*Physiological Adaptation to High Altitude*		
Variable	**Change**	**Time scale**	**Comments**
Ventilation	Increased	7 to 10 days	Respiratory control shifts from CO_2 to O_2
Hematocrit	Increased	7 days	Results from decrease in plasma volume (early) and increased secretion of erythropoietin (later)
Blood viscosity	Increased	2 to 7 days	Due to increased hematocrit
Cardiac output	Increased	Immediate onset	Rises in proportion to decreased blood O_2
Hemoglobin P_{50}	Increased	24 hr	Due to increased 2, 3 DPG; opposes left shift due to lowered plasma pH

Sleep Apnea, Sudden Infant Death Syndrome, and Respiratory Depression

Some adults and children stop breathing for a few minutes periodically during sleep, a condition called **sleep apnea.** In sleep, the major alteration is an increase in the arterial P_{CO_2}, the consequence of a decrease in the sensitivity to CO_2. In sufferers from sleep apnea, this decrease in sensitivity to the major source of respiratory drive is apparently sufficient to stop breathing. After breathing stops hypoxia develops, and the distress that this causes is generally sufficient to awaken the patient and temporarily correct the problem. In infants such an episode of apnea may end in death.

The infant sleep apnea fatalities are believed to be at least one major cause of sudden, unexplained infant death during sleep, or **sudden infant death syndrome** (SIDS). In adults hypoxia enhances the response of the respiratory centers to CO_2. In newborn preterm infants the reverse is true: hypoxia produces respiratory depression, especially in a cool environment. SIDS victims are outwardly healthy, but some postmortem evidence is consistent with the hypothesis that these infants have subtle metabolic or respiratory problems. Many SIDS victims have enlarged carotid bodies, and in some cases there is also an abnormal persistence of deposits of brown fat, a heat-generating tissue that is important for thermoregulation in young infants but normally disappears in older infants. Studies of other children in the families of victims suggest a familial decreased responsiveness to hypoxia and CO_2. This evidence suggests that in SIDS victims both the normal regulation of P_{CO_2} and the "backup" response to hypoxia may fail, allowing the infant to fall into a vicious cycle in which hypoxia and CO_2 acidosis depress the respiratory centers into unresponsiveness.

Drugs whose action depresses the central nervous system, including anesthetics, opiate drugs such as heroin and morphine, barbiturates, and alcohol, also decrease responsiveness to CO_2. Respiratory depression is the most common cause of death in cases of overdose of such drugs. Physicians should use such drugs with extra care on patients whose respiratory drive may be diminished already, such as those with long-standing lung disease. The susceptibility of the respiratory centers to the action of opiate drugs has suggested that endogenous opiates produced by the brain (see Chapter 8, p. 187) may be involved in setting the sensitivity of the respiratory system. If so, some cases of sleep apnea might be successfully treated with drugs that antagonize the binding of opiates to their receptors.

---

the hormone erythropoietin (Chapter 14) by the kidneys. The normal hemoglobin concentration at sea level is 15 g/100 ml, and in the Peruvian Andes (at about 5000 meters) it may be elevated above 19 g/100 ml. With the resulting increase in O_2 capacity, the blood may actually carry a normal load of O_2 even though the hemoglobin is not O_2 saturated. The right shift is the result of increased formation of 2,3 DPG by red blood cells (Chapter 17). Although the right shift does not benefit O_2 loading at the reduced P_{O_2}, it does increase the ability of the tissues to extract O_2 from the hemoglobin. It also opposes a left shift of the curve that might be expected as the plasma P_{CO_2} decreases **(hypocapnia).** The threshold for such adaptations is about 1600 meters; a number of major cities (for example, Mexico City, Guadalajara, Denver, and Albuquerque) are near or above the threshold.

When an adapted person descends to sea level, the process of adaptation is reversed. The arterial P_{O_2} increases to its sea-level value of 100 mm Hg, and the component of respiratory drive due to peripheral chemoreceptors is greatly diminished. As a result, the arterial P_{CO_2} rises above normal until the choroid plexuses have had time to adjust the CSF pH and reset the central chemoreceptors (see Figure 18-14).

Acute Acidosis

Acids other than CO_2 are called **fixed acids** since, unlike volatile CO_2, they cannot be removed from the body by the respiratory system. Fixed acids dissociate in plasma to yield H^+ and an anion called the **conjugate base.** Any change in plasma pH due to fixed acids is conventionally referred to as **metabolic acidosis,** even if the acid is not of metabolic origin. Generally, conjugate base ions do not readily cross the blood-brain barrier. Nevertheless, ventilation is increased in metabolic acidosis, because a decrease in plasma pH stimulates the peripheral receptors. The resulting decrease in plasma P_{CO_2} opposes the change in pH caused by the fixed acid.

In acute metabolic acidosis, just as with acute exposure to high altitude, a conflict develops between the peripheral and central chemoreceptors

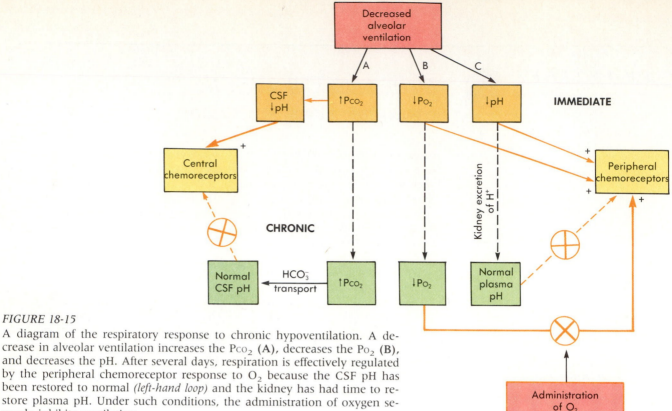

FIGURE 18-15

A diagram of the respiratory response to chronic hypoventilation. A decrease in alveolar ventilation increases the P_{CO_2} (A), decreases the P_{O_2} (B), and decreases the pH. After several days, respiration is effectively regulated by the peripheral chemoreceptor response to O_2 because the CSF pH has been restored to normal *(left-hand loop)* and the kidney has had time to restore plasma pH. Under such conditions, the administration of oxygen severely inhibits ventilation.

that prevents the respiratory compensation for acute acidosis from being completely effective. With the onset of the acidosis, the peripheral chemoreceptors stimulate increased ventilation. The resulting decrease in P_{CO_2} starts to restore the arterial pH to normal. But at the same time, the pH of the CSF begins to become more alkaline. The conjugate base and H^+ could not penetrate the blood-brain barrier to make the CSF more acid, but the decrease in arterial P_{CO_2} caused the CSF P_{CO_2} to decrease and the pH of the CSF to increase. As this happens, the response of central chemoreceptors begins to oppose the excitatory input from the peripheral receptors, preventing full compensation for the acidosis.

If acidosis continues for days or longer (that is, becomes chronic), there is time for the change in CSF pH to be adjusted by the choroid plexuses. Then the peripheral chemoreceptors regain full control, and a more complete compensation can occur. There are a number of disease conditions in which acid end products of metabolism accumulate in the plasma; the respiratory response is an important factor in protecting the plasma pH in such conditions. The cooperation of the respiratory control system and the renal system in regulation of acid-base balance will be described in Chapter 20.

Respiratory Disease

Arterial P_{CO_2} may be chronically elevated in lung disease, injury to the brainstem, or drug-induced depression of the central nervous system. If alveolar ventilation decreases, the arterial P_{CO_2} is increased (Figure 18-15), and the arterial pH must thus decrease. The arterial P_{O_2} also decreases, and all of these changes stimulate respiration.

After a time, the central chemoreceptors adapt, and the peripheral chemoreceptors become the major factor that is driving respiration. During the same time, the kidneys secrete H^+ ions and restore the plasma pH to normal. At this point the person's hypoxia is the only factor driving respiration. If the person is now given 100% O_2 to relieve his hypoxia, respiratory drive may be insufficient to maintain ventilation and he may well stop breathing and die. Such patients are also much more likely to stop breathing under anesthesia, since even moderate depression of the CNS by anesthetics can abolish their weakened respiratory drive.

The Ventilatory Response to Exercise

Within a few seconds of the beginning of exercise, the ventilation rate rises. It may increase from its resting level of about 7.5 liters/min to 70 liters/min in heavy exercise. This increase in ventilation supports the oxidative component of energy metabolism of skeletal muscle during exercise. The degree of the increase in respiration must be appropriate for the amount of effort that is being expended. Given what is known about regulation of ventilation during rest, it was natural to hypothesize that working muscles send a signal about the change in respiratory demand in the form of an increase in ar-

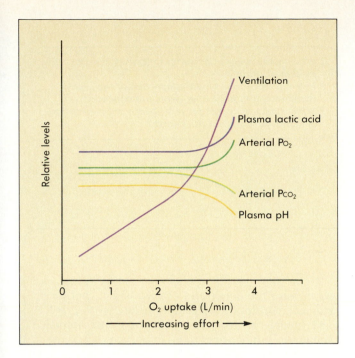

FIGURE 18-16
Changes in blood gas composition, plasma pH, and plasma lactic acid levels with increasing effort, as measured by the rate of O_2 uptake. The beginning of oxygen debt is indicated by the inflections in the curves.

terial P_{CO_2}, a decrease in arterial P_{O_2}, or a decrease in arterial plasma pH. In fact, arterial P_{CO_2}, P_{O_2}, and pH change slightly or not at all in exercise involving moderate effort (Figure 18-16). None of these changes are large enough to drive the respiratory system to ventilation rates several times higher than rest by the known feedback control mechanisms.

During exercise the amount of O_2 extracted by muscles increases, so the venous P_{O_2} decreases. One hypothesis suggested that this decrease was the signal for increased ventilation. However, there are no known chemoreceptors on the venous side of the circulation. Furthermore, if venous blood is prevented from returning to the heart of an experimental animal, the animal's ventilation still increases at the start of exercise. Such experiments rule out the release of a chemical messenger from exercising muscle to the respiratory centers. If there is no way for a signal from exercising muscles to reach the respiratory centers, there can be no negative feedback loop.

Presently, the increase in ventilation at the start of exercise is regarded as an example of "feed-forward" regulation, in which the control system is able to anticipate the needs of the body for gas exchange. Possibly the parts of the brain that are concerned with carrying out motor programs learn how much stimulation to give the respiratory centers in the preparatory ("on your mark—get set") phase of exercise. This would account for the fact that the onset of elevated respiration occurs too soon to be accounted for by any of the changes in blood chemistry associated with exercise (Figure 18-17). After muscular exercise has begun, sensory input from muscle and joint receptors could augment the respiratory drive. Also, the respiratory centers have been shown to be sensitive to temperature. A small increase in brain temperature due to exercise could be one of the signals that stimulates the respiratory centers.

At the highest levels of effort, the body's ability to deliver O_2 is not sufficient to support the work

FIGURE 18-17
Changes in ventilation during and after a brief period of exercise at a steady level of effort. There is an initial, very rapid increase in ventilation and a slower secondary increase to a new higher steady-state level. These changes are almost mirrored during recovery, except that respiration may continue at a slightly higher rate for some time after exercise is over.

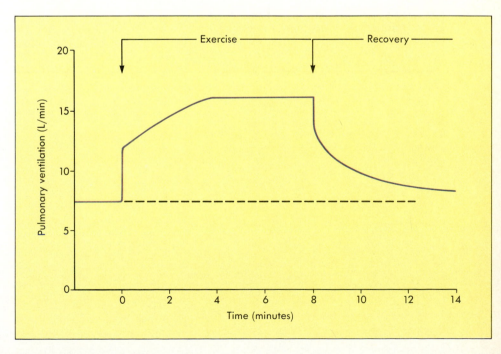

performed. At this point oxidative metabolism is replaced by anaerobic metabolism in some tissues, and an oxygen debt develops (Chapter 12, p. 301). At these levels of effort, ventilation increases substantially as peripheral chemoreceptors are stimulated by increasing levels of lactic acid in the plasma (see Figure 18-16), and ventilation continues at a high level even though arterial P_{CO_2} is decreasing and arterial P_{O_2} is increasing.

1 What is the role of the choroid plexuses in adaptation to high altitudes?
2 In what situations might adaptation of the central chemoreceptors have dangerous consequences?
3 What evidence argues for a feed-forward mechanism of respiratory control in exercise?

SUMMARY

1. During quiet breathing, inspiration is driven by bursts of action potentials passing along axons in the **phrenic nerve** to the **diaphragm.** Increases in ventilation are obtained by recruitment of the spinal motor neurons of **internal intercostal** muscles during inspiration and of **external intercostal** and abdominal muscles during expiration.

2. The respiratory motor program is initiated by centers in the medulla of the brainstem and modulated by inputs from the **pneumotaxic** and **apneustic** centers of the pons.

3. The intensity of respiratory drive determines the rate of cycling and the **tidal volume** of each cycle. A large number of inputs are integrated to determine the **respiratory drive.** These include peripheral and **central chemoreceptors,** somatosensory input, and input from parts of the brain involved in arousal and parts of the brain concerned with voluntary movement.

4. Under normal conditions, the input from the **central chemoreceptors** provides the major component of respiratory drive. These receptors monitor the pH of CSF, which is determined over the time scale of minutes to hours by the arterial P_{CO_2}.

5. Input from the **aortic** and **carotid** bodies, **peripheral chemoreceptors** sensitive to arterial P_{O_2}, pH, and (indirectly) P_{CO_2}, is less important in normal conditions but becomes more important during **hypoxia** and **acid-base imbalance.**

6. The peripheral and central chemoreceptors have a synergistic effect on the respiratory control system: input from either enhances the sensitivity of the system to input from the other. This enhancement is appropriate in most but not all circumstances.

7. In circumstances such as high-altitude exposure, acute **metabolic acidosis,** and lung disease, the peripheral and central chemoreceptors may give contradictory signals to the respiratory control system initially, preventing a fully effective ventilatory response. After some days, the contradictory signals from the central chemoreceptors diminish as the **choroid plexuses** correct the CSF pH. This process resets the sensitivity of the central chemoreceptors.

8. Presently a feed-forward mechanism is hypothesized for the ventilatory increase during exercise, since it cannot be accounted for by changes in any of the parameters monitored by the chemoreceptors, and it occurs too rapidly to be driven by metabolic signals.

1. Describe the pattern of activity in the phrenic nerve during the inspiration and expiration of quiet breathing. How does the pattern in this nerve and the innervation of the internal and external intercostals change with increases in respiratory drive?

2. Where are the peripheral chemoreceptors located? To what factors do they respond? What is the magnitude of their contribution to respiratory drive under normal conditions?

3. What is known about the location of the central chemoreceptors? To what factor do they respond? What is the magnitude of their contribution to respiratory drive under normal conditions?

4. The ventilatory response to small changes in P_{CO_2} and to P_{O_2} is different. How does this difference relate to the way in which these two gases are carried by the blood?

5. How do the ventilatory responses to changes of CSF pH and arterial P_{O_2} differ?

6. On climbing a 14,000 foot peak in Colorado most individuals will experience fatigue. Account for this in terms of the blood gas composition at 14,000 feet and the responses of central and peripheral chemoreceptors.

7. A normal individual has an arterial P_{O_2} of 50 mm Hg, an arterial P_{CO_2} of 40 mm Hg, and an arterial pH of 7.4. Describe her ventilation. Which chemoreceptors are most important for her ventilatory response?

8. A normal individual has an arterial P_{O_2} of 100 mm Hg, a P_{CO_2} of 40 mm Hg, and an arterial pH of 7.2. Describe his ventilation. Which chemoreceptors are most important for his ventilatory response?

9. A normal individual has an arterial P_{O_2} of 100 mm Hg, an arterial P_{CO_2} of 45 mm Hg, and an arterial pH of 7.4. Describe her ventilation. Which chemoreceptors are most important for her ventilatory response?

10. In an experimental animal whose respiratory centers are intact, both vagus nerves are cut. What will be the immediate effect on tidal volume? A few minutes afterward, alveolar ventilation is measured. Do you expect it to be different from the alveolar ventilation measured just before the vagus nerves were cut? Why or why not? What respiratory feedback loop is interrupted by the vagotomy? Which ones remain intact after both vagus nerves are cut?

● *SUGGESTED READING*

DAVSON, H.K., et al.: *The Physiology and Pathophysiology of the Cerebrospinal Fluid,* Chapter 9, Acid Base Characteristics of the CSF, p. 453, Churchill-Livingstone, Edinburgh, 1987. Describes the role of the blood-brain barrier in the regulation of the partial pressure of carbon dioxide in the arterial blood.

LOPEZ, B.J., J.R. LOPEZ, J. URINA, and C. GONZALEZ: Chemotransduction in the Carotid Body, *Science,* volume 241, July 29, 1988, p. 580. Describes patch-clamp (see Chapter 7) studies showing that oxygen tension may be sensed by K^+ channels in type I chemoreceptor cells.

WARRILL, E.G.: Where are the Real Respiratory Neurons? *Federation Proceedings,* volume 40, p. 2389, 1974. The respiratory control centers of the brain are still hypothetical concepts. This reviews evidence for and against specific centers controlling rate and depth of breathing.

PALLOT, O.J. (editor): *Control of Respiration,* Croon Helm, London, 1983. Comprehensive textbook on this subject.

RIGHTER, D.W., D. BALLANTYNE, and J.H. REMMERS: How is the Respiratory Rhythm Generated? A model, *News in Physiological Sciences 1,* p. 109, 1986. Talks about how some fairly simple neuronal interactions could generate various patterns of respiration.

ROLBEIN, B.: For No Reason, *Boston Magazine,* volume 78, May 1986, p. 109. Discusses the incidence of sudden infant death syndrome and speculates on possible causes and cures.

SCHMIDT, C.F., and J.H. COMROE, JR: Functions of the Carotid and Aortic Bodies, *Physiological Reviews 20,* p. 115, 1940. Detailed discussion of the peripheral chemoreceptors and their effects on the respiratory and cardiovascular systems.

Physiological Aspects of Exercise

WHAT IS EXERCISE PHYSIOLOGY?

Exercise is usually thought of as structured physical activity that increases fitness or improves ability in sports. It is important to realize that, from a physiological point of view, any voluntarily undertaken activity that uses skeletal muscles is exercise—lifting a barbell is not fundamentally different from lifting a bag of groceries, nor is running to catch a bus different from competitive sprinting. From this point of view, light-to-moderate exercise is part of almost everyone's life. More extreme exercise tests the capabilities of almost all of the body's homeostatic systems.

Exercise physiology is the branch of physiology concerned with how homeostatic mechanisms meet the short-term stresses of exercise and with how the long-term adaptive changes occur in the structure and biochemistry of the musculoskeletal, cardiovascular, and respiratory system during training. One purpose of this focus unit is to show how the musculoskeletal, cardiovascular, and respiratory systems (described separately in Sections II, III, and IV) work together during exercise. Also, it is convenient to note some implications of exercise for the digestive, endocrine, and renal systems; these are treated in subsequent chapters. Table 18-A summarizes much of the following discussion.

ENERGY SOURCES FOR EXERCISING SKELETAL MUSCLE

As described in Chapter 4, p. 76, ATP may be produced rapidly but inefficiently by the substrate-level phosphorylation steps of glycolysis, or it may be produced more efficiently by the oxidative reactions coupled with the Krebs cycle. As described more fully in Chapter 12, (see p. 302) skeletal muscle fibers have evolved into fiber types that show varying degrees of specialization for either brief, powerful contractions (Type II) or sustained, less powerful contractions (Type I). The highest degree of specialization for brief contractions is seen in fast-twitch glycolytic fibers (Type IIB), which have few mitochondria and are committed to anaerobic metabolism. At the other extreme are Type I slow-twitch

oxidative fibers, which deliver much slower, weaker contractions but are also resistant to fatigue. Fast oxidative fibers (Type IIA) are regarded as an intermediate between these two extremes.

Individual muscles are mixtures of glycolytic and oxidative fibers, but for each muscle the mix matches the functions normally assumed by that muscle. For example, the muscles of the neck and those attached to the spine are predominantly composed of oxidative fibers, whereas flexors of the arms are predominantly composed of glycolytic fibers. In each individual, the numbers of fast and slow fibers in each skeletal muscle are thought to be essentially constant after maturity, having been established by the muscles' innervations (see Chapter 12, p. 303). Thus both the total number of fibers and the relative numbers of fast and slow fibers are probably more strongly affected by heredity than by sex or training. Some evidence provided by muscle samples taken from individual athletes suggests that exceptional performance in events requiring either strength or endurance is correlated with disproportionate numbers of either fast or slow fibers. The contractile capabilities and, to some extent, the metabolic properties of skeletal muscle fibers are affected by training, as discussed below.

WHOLE-BODY METABOLISM IN EXERCISE

The immediate energy supply for muscle contraction is the intracellular store of high-energy phosphates—ATP and creatine phosphate. By themselves, these stores could support only very brief efforts—a 6-second sprint, a 1-minute walk, or lifting a moderate weight. Additional work requires continuous regeneration of the stores of high-energy phosphate by anaerobic glycolysis or by oxidative phosphorylation.

During brief, heavy exercise, metabolism of glycolytic fibers is almost independent of that of the rest of the body because glycolytic fibers do not require oxygen and possess large internal stores of glycogen. In contrast to glycolytic fibers, oxidative fibers depend continuously on the circulatory system to supply glucose, other metabolic intermedi-

TABLE 18-A *Physiological Changes in Exercise*

System	Factors involved	Consequences
Cardiovascular		
Increased heart rate and contractility	Increased sympathetic input, decreased parasympathetic input	Increases cardiac output
Vasodilation in active muscle	Local vasodilatory substances and β-adrenergic effect of epinephrine	Increases blood flow to skeletal muscle, decreases peripheral resistance
Vasoconstriction in abdominal and renal beds	Increased sympathetic input to blood vessels (α-adrenergic)	Helps maintain arterial pressure, diverts blood to active muscles
Increased venous return	Decreased peripheral resistance, sympathetic venoconstriction, action of respiratory and skeletal muscle pumps	Supports high cardiac output
Respiratory		
Increased tidal volume and respiratory rate	Increased central respiratory drive	Increases gas exchange
Increased diffusing capacity	Effect of increased lung perfusion	Increases gas exchange
Endocrine		
Increased levels of circulating epinephrine and norepinephrine	Increased sympathetic activity	Stimulates glycogen and fat breakdown
Increased plasma levels of cortisol	Increased release of ACTH-releasing hormone by hypothalamus	Stimulates glycogen and fat breakdown
Decreased insulin levels	Effect of sympathetic input to pancreas	Inhibits conversion of glucose to glycogen
Increased glucagon levels	Effect of sympathetic input to pancreas	Increases breakdown of glycogen and fat
Thermoregulatory		
Elevated core body temperature	Elevated setpoint of central thermostat	Increases rates of metabolic reactions, favors O_2 unloading
Increased cutaneous blood flow	Effect of cutaneous sympathetic afferents	Increases rate of heat loss to environment
Increased sweating	Effect of cutaneous sympathetic afferents	Increases rate of heat loss to environment

ates, and oxygen. This difference in dependence on the circulatory system is reflected in the numbers of capillaries that serve each fiber type (see Table 12-2).

The magnitude of the glycogen stores of muscle poses an important limitation on the duration of maximum exercise because the circulatory system cannot deliver glucose to glycolytic fibers rapidly enough to sustain maximum force development. Depletion of muscle glycogen stores during maximum exercise is one of several factors highly correlated with both a sharp decline in performance and a subjective feeling of exhaustion. Marathon runners sometimes refer to this subjective effect as "hitting the wall."

Glycogen loading (carbohydrate loading) is a training regimen that results in a temporary but significant increase in the amount of glycogen stored in both fast and slow fibers. The regimen is based on the observation that carbohydrate deprivation increases the ability of muscles to take up and store glucose. Accordingly, athletes who are preparing for a major endurance event switch to a high-carbohydrate diet 2 or 3 days before the event and stop intensive training. Glycogen levels in muscle are temporarily driven to levels as much as two times higher than those obtained with muscle hypertrophy alone. Doubling of glycogen stores has been shown to increase the maximum duration of endurance exercise by as much as 50%.

Production of lactic acid rises with the intensity of exercise, but the rise becomes much steeper as the power developed by exercising muscles approaches about half the workload that could be sup-

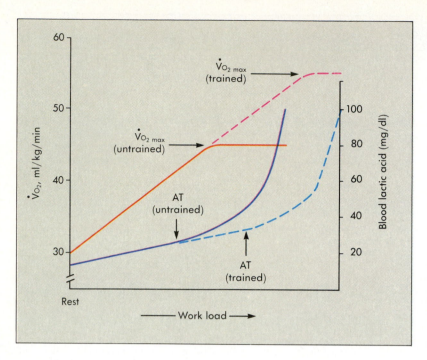

FIGURE 18-A

The relationship between oxygen uptake *(red curves)* and blood lactic acid levels *(blue curves)* as the intensity of exercise increases. At about 50% of the maximimum oxygen uptake, the blood lactic acid levels begin to rise exponentially—this point is called the anaerobic transition *(AT)*. In endurance-trained people *(dashed curves)*, the maximum oxygen uptake ($Vo_{2\ max}$) is increased and the AT occurs at a higher workload as compared with those of untrained people *(solid curves)*.

ported by oxidative metabolism (Figure 18-A). The level of effort at which the the steep rise begins is called the **anaerobic transition**. The maximum blood level of lactic acid that can be developed is a measure of an individual's potential for brief, intense muscle work. Athletes involved in sports that require such work (for example, a half-mile sprint) can achieve very high blood lactic acid levels, whereas untrained individuals and endurance athletes have low or moderate maximum lactic acid levels.

Some of the lactic acid produced by glycolytic fibers may be reconverted to glucose by way of the Cori cycle (see Chapter 12, p. 301); the rest is reconverted to pyruvate and further metabolized to CO_2 and water by oxidative skeletal muscle, the heart, and other organs that continuously rely on oxidative metabolism. Thus, from the body's point of view, lactic acid is a metabolic intermediate rather than an end product of muscle contraction, and all exercise is supported by oxidative metabolism. The energy cost of a particular exercise can be estimated by measuring the total amount of oxygen consumed in the process, including both that consumed during the exercise and that consumed after the exercise is over (the "oxygen debt"; see p. 301). The total energy expended by a person weighing 70 kg in performing various activities is shown in Table 18-B. Note that the overall energy utilization can increase by about 13-fold during sustained, heavy exercise (marathon running) and by as much as 50-fold during brief periods of very intense effort (sprinting).

Carbohydrates and fats are the main metabolic fuels of exercise; protein metabolism accounts for only about 10% of the total energy expended. The carbohydrate stores are within the muscle cells and in the liver. Carbohydrate metabolism accounts for about 40% of the total energy used during rest. During maximum exercise, the contribution of carbohydrate metabolism to total energy use increases to 75% to 80%. As exercise duration increases, fat metabolism accounts for a larger part of the total energy. Fat is stored in deposits adjacent to muscle cells within muscles and at other sites in the body. The protein that may be consumed during sustained heavy exercise is in the muscle cells. Replacement of all the energy sources occurs during a recovery period following the exercise. When glucose is the substrate, the uptake of 1 liter of O_2 results in the

TABLE 18-B	Total Energy Expended in Various Activities By A Person Weighing 70 Kg

Activity	Kcal/min
Sleeping	1.21
Sitting	1.67
Standing	1.83
Light housework, office work	2.5
Bicycle riding (5.5 mph)	3.17
Walking with a 43-pound load	4.83
Running (5.7 mph)	12.00
Marathon running	16.5
Sprinting (15.8 mph)	65.17

release of about 5.1 kcal of energy. When fat is the substrate, the amount of energy released is 4.69 kcal/L O_2. For mixed substrates (the normal situation), an approximate value of 5.00 kcal/L O_2 can be used.

ENDOCRINE RESPONSES

The adrenal glands secrete epinephrine and cortisol, two hormones that stimulate breakdown of fat and glycogen. At the beginning of exercise, an immediate increase occurs in plasma catecholamine levels. This is the result of increased sympathetic activity, which includes elevated adrenal secretion. Both the increased levels of circulating catecholamine (primarily epinephrine) and the increased input of norepinephrine delivered by the sympathetic innervation of liver and adipose tissue promote mobilization of stored lipid and glycogen. Epinephrine also stimulates glycogen breakdown in skeletal muscle and inhibits glycogen formation (see Chapter 23, p. 599). The rise in sympathetic activity is followed within a few minutes by an increase in plasma levels of adrenocorticotropic hormone (ACTH). One effect of the increased levels of ACTH is an increase in cortisol secretion by the adrenal cortex. Cortisol has several effects that contribute to increased resistance to physical stress. It is important during exercise because of its permissive effect—it must be present for epinephrine and norepinephrine to be fully effective.

The pancreas secretes insulin and glucagon, two hormones that have opposing effects on fat and glycogen metabolism (see Chapter 23). Exercise inhibits insulin secretion and increases glucagon secretion, probably as results of increased sympathetic input to the pancreas. Thus muscles must increase their glucose uptake during exercise despite a decrease in plasma insulin levels. Because the number of insulin receptors in muscle is stimulated by exercise, regular exercise can reduce the amount of insulin needed by diabetics to control their plasma glucose levels. The increased glucagon secretion adds to the effects of epinephrine and cortisol in releasing energy substrates from storage.

OXYGEN UPTAKE AND THE RESPIRATORY SYSTEM

Metabolic needs at rest correspond to an oxygen uptake of about 250 ml/min. Oxygen uptake increases in proportion to the mass of muscle used; the highest rates of oxygen uptake are attained only when both arms and both legs are active. During maximum exercise with an energy consumption of 10 to 25 kcal/min, oxygen uptake rises to 2 to 5 L/min, an increase of 8 to 20 times. This increase is reflected in a similar increase in pulmonary minute ventilation, which rises from 5.6 L/min to as much as 120 to 140 L/min. The ventilatory increase is attained by an increase in both tidal volume and respiratory rate (Figure 18-B). The increase in tidal volume demands about 50% of the vital capacity and is taken mostly from the inspiratory reserve, requiring only a minor contribution from the expiratory reserve. The factors involved in the rapid increase in ventilation are poorly understood (see Chapter 18, p. 460). However, it has been concluded that feedback processes from receptors that monitor blood gas composition are not involved because plasma P_{CO_2}, P_{O_2}, and pH are essentially unchanged in moderate exercise (see Figure 18-16).

The increased ventilation increases the elastic and flow-resistive work of breathing, thereby in-

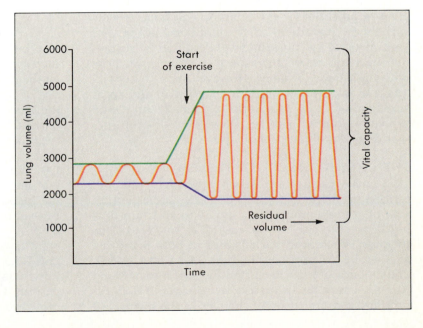

FIGURE 18-B
Typical changes in the pulmonary volumes with the onset of exercise.

creasing the total energy requirement for breathing (see Figure 16-8). At rest, O_2 consumption of the respiratory muscles is approximately 2 ml/min; it may increase to 100 ml/min during heavy exercise.

Exercise increases the **diffusing capacity** of the lungs; this is a measure of the amount of lung surface effectively exposed to gas exchange. The increase results from the greater expansion of the lungs and also from improvement of the ventilation/perfusion ratio, which results from increased blood flow through the lungs (see Chapter 16, p. 422).

Exceptional athletes can consume O_2 at rates as high as 7 L/min. When scaled to body weight, this may translate into a value as high as 90 ml/kg/min. Maximum O_2 consumption ($Vo_{2\ max}$) is relatively constant for each individual. It declines with inactivity and increases with training, but the change that otherwise healthy individuals can effect with training is, at most, perhaps 25% to 35%. In both men and women, $Vo_{2\ max}$ is maximum at about the time growth is completed and declines by about 20% by middle age (Figure 18-C). There is a difference in the mean $Vo_{2\ max}$ of men and women as expressed on a total-body-weight basis (see Figure 18-C, *A*), but this difference almost disappears when the values are scaled to lean body weight (see Figure 18-C, *B*).

The variability of $Vo_{2\ max}$ in the healthy population is greater than can be readily accounted for by the effects of training, and the $Vo_{2\ max}$ values of pairs of identical twins are much more similar than those of pairs of nonidentical twins. These findings suggest that $Vo_{2\ max}$ is primarily determined by heredity. The performance of athletes in events requiring sustained high levels of exertion is almost certainly limited by the rate of O_2 delivery to the heart and working skeletal muscle because glycolytic fibers cannot support sustained effort. These facts have led to the suggestion that athletes with the greatest potential for excellence in endurance events such as the marathon, cross-country ski racing, and long-distance cycling could be identified by screening the population of prospective athletes for those with high $Vo_{2\ max}$ values. However, many other factors are involved in athletic excellence, and measurement of $Vo_{2\ max}$ can, at best, identify those who probably do not have the potential to develop into world-class endurance athletes.

THE CARDIOVASCULAR SYSTEM

The cardiovascular responses to exercise are described in Chapter 15 (pp. 390-392). Cardiac output may be 5 to 6 L/min at rest and 30 to 40 L/min during heavy exercise, an increase of 6- to 7-fold. This increase is achieved by increases of both heart rate and stroke volume (see Figures 13-24 and 15-15). The amount of oxygen carried by each liter of blood (the **O_2 carrying capacity**; see Chapter 17) is essentially constant if small changes in the hematocrit

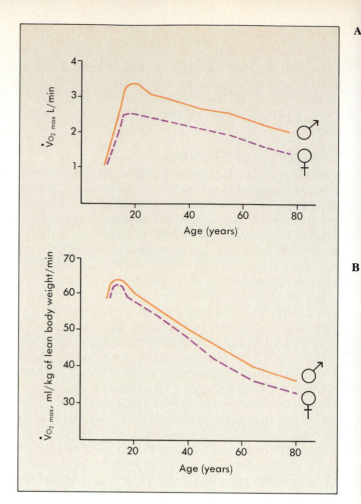

FIGURE 18-C

A Maximum O_2 uptake of men *(solid curve)* and women *(dashed curve)* reaches a maximum in late teens and declines thereafter. These values are on a per person basis and not scaled to body weight. There is some evidence suggesting the decline can be minimized by training.

B Maximum O_2 uptake of men *(solid curve)* and women *(dashed curve)* scaled on the basis of lean body weight. Note the much smaller difference between the male and female curves in this plot as compared with the unscaled curves shown in **A**.

(see below) are neglected. If oxygen delivery is to increase by as much as 20-fold, there must also be an increase in the **oxygen extraction**—the percentage of bound oxygen released from hemoglobin as the blood passes through the capillaries of working tissues. Similarly, each liter of blood that travels to the lungs must deliver a larger fraction of its burden of CO_2. Factors that affect oxygen extraction and CO_2 transport include the pH, Pco_2, and Po_2 of the working tissues (see Chapter 17) and the fraction of the blood flow that passes through true capillaries instead of through arteriovenous shunts (see Chapter 15).

The hematocrit increases slightly during exercise because fluid is redistributed from plasma to interstitial space. This change adds at most a few percent to the oxygen carrying capacity. Various measures adopted by serious athletes can tempo-

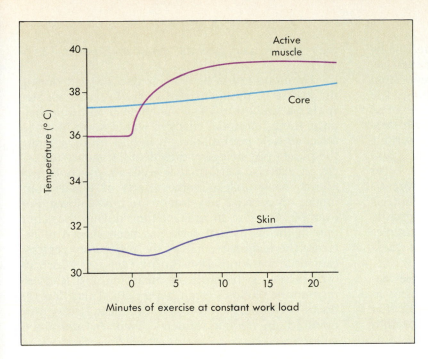

FIGURE 18-D
Changes in core body temperature, muscle temperature, and skin temperature during exercise.

rarily increase the oxygen carrying capacity more significantly. One of these measures is a period of training at a high altitude, which increases the rate of red cell synthesis (see Chapter 18, p. 459). Another method, blood doping (see Chapter 14, p. 364) involves increasing the hematocrit by transfusing a person with his or her own concentrated red blood cells that had been removed and stored. It is not clear that increasing the hematocrit improves performance at sea level.

Exercise changes the partitioning of cardiac output between different tissues (see Figure 15-14 and Table 15-3). These changes can be summarized as:

1. Increase in blood flow to exercising muscle, primarily determined by local vasodilator substances.
2. Decrease in blood flow to nonexercising muscle and abdominal and renal vascular beds, caused by central sympathetic activity.
3. Increase in cutaneous perfusion, primarily determined by central thermoregulatory centers.

The cardiovascular changes are affected by the nature of the exercise. **Dynamic exercise** is characterized by movement of body parts so that muscle contractions are accompanied by shortening (that is, isotonic contractions). The total peripheral resistance typically decreases during dynamic exercise; the extent of the decrease is determined by the fraction of the total muscle mass that is involved. **Static exercise** is characterized by tension development without length change (that is, isometric contractions). Some bodybuilding exercises are purely static, but weightlifting and wrestling also have substantial static components. In static exercise at more than about 20% of maximum effort, blood flow through working muscles is impaired as the contracting fibers collapse blood vessels. There are two important consequences of flow impairment:

(1) most of the energy used in static exercise must come from anaerobic glycolysis, and (2) the net effect of static exercise is a large increase in total peripheral resistance. This results in much larger increases in arterial blood pressure than are seen in dynamic exercise, making static exercise potentially harmful for people with heart or vascular disease.

THERMOREGULATION AND FLUID BALANCE

Because most of the energy released from substrate in energy metabolism appears as heat, the 10 to 20-fold increases in metabolism that occur in exercise (Table 18-B) represent a significant change in the body's heat budget. In the first several minutes of exercise, muscle temperature rises rapidly while core body temperature rises more slowly (Figure 18-D). Skin temperature decreases initially because of sympathetic vasoconstriction, but, as core body temperature rises, the central thermoregulatory centers cause cutaneous vasodilation and sweating, which allow more rapid heat transfer to the environment. After some minutes of exercise, core body temperature attains a new steady state at a value that may be as high as 40° C to 41° C. This higher core body temperature reflects the resetting of the central thermostat to a higher setpoint. Probable beneficial effects of increased temperature include an increase in the velocity of all chemical reactions in the active tissues and an increase in O_2 delivery because of the right shift of the hemoglobin-oxygen dissociation curve (see Figure 17-9).

If heat loss to the environment is restricted by high environmental temperatures, high humidity, or heavy clothing, the heat produced by muscular activity is a serious threat to thermal homeostasis that may result in collapse (heat stroke) or even death.

At rest, about 15 ml of fluid per/hr are lost by evaporation from the body surface. During exercise, an increase in fluid loss occurs through evaporation from respiratory surfaces and from sweating. The net loss may be as much as 2 to 5 L/hr. Loss of as little as 3% of body weight begins to degrade performance, and loss of 8% to 10% may have serious consequences. Plasma electrolytes (Na^+, K^+, and Cl^-) are lost in sweat, but their rate of loss is lower than the rate of fluid loss because sweat is more dilute than plasma. Maintaining electrolyte balance is of much less concern for exercising people than is maintaining fluid balance, and using salt supplements or isotonic fluids during training is of questionable value and may be dangerous.

The American College of Sports Medicine recommends that athletes training and competing in hot weather should have free access to fluid that contains not more than 25 grams of glucose, 10 mEq of Na^+, and 5 mEq of K^+ per liter. Such dilute solutions are physiologically appropriate and also rapidly emptied from the stomach into the intestine. In contrast, most commercial soft drinks, including some that are advertised as appropriate for use by athletes, contain 80 to 100 mg of sucrose/L, concentrations high enough to retard stomach emptying (see Chapter 22) and thus slow their absorption. However, such beverages may be appropriate for use in cold weather when rapid absorption is not as critical.

GENDER-RELATED DIFFERENCES IN ATHLETIC PERFORMANCE

There are general differences in the construction, composition, and capabilities of male and female bodies after puberty (see Chapter 25). In particular, the musculoskeletal system makes up a larger percentage of the average total body mass in men than in women, and the weight-adjusted cardiac output, vital capacity, hematocrit, and blood volume of men are greater than those of women. Some of these differences are attributable to the fact that, on average, fat makes up a larger fraction of total body weight in women than in men; adjustment of the values for lean body weight reduces many of the differences between men and women (see Figure 18-C). In general, athletic training minimizes the gender-related differences in performance that are seen in the general population.

The gender-related differences in the mass of muscle, skeleton, and fat are partly consequences of the different metabolic and developmental effects of male and female sex hormones, and partly consequences of differences in male and female activity patterns (see Chapter 25). In particular, the male sex hormone, testosterone and its metabolites stimulate the growth and functional state of skeletal muscle and bone. These effects are also obtained from synthetic derivatives of testosterone (anabolic steroids), which are used by many athletes, both

male and female (see the essay in Chapter 12, p. 307).

Hormone levels in women rise and fall during the course of the menstrual cycle. Because sex steroids are known to affect metabolism and fluid balance, these hormonal changes would seem likely to affect athletic performance. It is difficult to separate the purely physiological effects of the menstrual cycle on exercise performance from its psychological effects. Nevertheless, some effects of female hormones on exercise physiology are now known. For example, the rise in progesterone during the latter part of the menstrual cycle has been shown to increase the responsiveness of the respiratory centers, causing hyperventilation during exercise, which in turn increases the work of breathing and leads to respiratory alkalosis (see Chapter 20). This effect reduces performance, although the effect is less pronounced in female athletes than in untrained women.

Some evidence suggests that to have normal menstrual cycles, women must possess a certain amount of body fat (see Chapter 25, p. 655). Some dedicated female athletes have very small amounts of body fat, and also have irregular menstrual cycles or even fail to have them (a condition called **amenorrhea**). Amenorrhea seems to have little effect, either good or bad, on athletic performance. It may favor development of a higher hematocrit because it reduces the rate of iron loss from the body. However, if amenorrhea continues for long periods, the lack of estrogen normally provided by active ovaries might contribute to loss of bone calcium and to the development of osteoporosis (see the essay, Chapter 23, p. 605).

Although it has been taken for granted that androgens (male sex hormones) favor athletic performance, very respectable performances have been given by women who were in the early stages of pregnancy. Hormones released by the placenta, the organ that nourishes the fetus before birth, cause changes in the mother's physiology that contribute to the success of the pregnancy (see Chapter 26). These changes may also favor athletic performance. Some concern has arisen that female athletes might exploit this effect to improve their performance in important events, such as Olympic competition.

EXERCISE AND THE IMMUNE SYSTEM

Moderate exercise may reduce the incidence of minor infectious disease as a result of its beneficial effect on overall health. However, exhaustive exercise does not promote disease resistance and may even reduce immune responses. The B and T lymphocytes are the cells of the immune system responsible for recognition of infectious organisms that have previously been encountered by the body (see Chapter 24). Intensive exercise has been found to reduce both the effectiveness of T lymphocytes and the levels of antibody secretion by B lymphocytes.

The reasons for these effects are not clear. However, intensive exercise induces some effects which are similar to those of an immune response to infection. These effects include an increase in the core body temperature and an increase in the number of circulating white blood cells. As described in Chapter 24, the immune system, like other major body systems, includes some negative feedback controls. Possibly, the infection-like effect of severe exercise is followed by increased negative feedback, which blunts the response to a real infection. Also, intense exercise makes such severe demands on the body's metabolic stores and repair processes that it constitutes a form of physical stress.

ADAPTIVE EFFECTS OF ACTIVITY AND TRAINING

The capabilities of skeletal muscles are responsive to levels and types of activity; this responsiveness maintains a level of muscle development which corresponds to the demands placed on that muscle over time. A period of bed rest or immobilization of a limb in a cast results in **disuse atrophy,** a decrease in muscle mass and contractility. Training can bring about increases in muscle capability, or **hypertrophy.** Hypertrophy and atrophy are relative terms that refer to the adaptive responses of muscle to use. As shown in Table 18-C, the effects of training depend on which aspects of muscle performance the training regimen emphasizes.

Strength training involves contracting muscles against heavy resistance. It can lead to increased peak isometric tension and increased shortening speed. One of the first effects of a strength training regimen is an increased ability to activate motor units voluntarily. Trained weight lifters appear to be able to cause simultaneous tetanic contraction in almost all of their motor units. When all motor units can be voluntarily activated, additional increases in strength result from increases in contractility. Strength training selectively affects the fast fibers of a muscle. The major effects of fast fiber hypertrophy are increases in:

1. The number of myofibrils.
2. The amount of stored glycogen.
3. The fiber diameter, reflecting changes 1 and 2. On the whole muscle level, this is seen as an increase in muscle mass.

Note that this type of training regimen does not increase capillary density, but rather the increase in muscle mass actually decreases the number of capillaries per gram of muscle. The same is true for the number of mitochondria per gram of muscle. Strength training is not currently believed to result in formation of additional skeletal muscle fibers in human athletes. Some animal experiments suggest that additional fibers may be added in muscle hypertrophy, but these studies used training regimens more severe than would be appropriate even for elite human athletes.

TABLE 18-C	Effects of Training	
Factor	**Strength**	**Endurance**
Muscle mass	Increased	Little effect
Fiber diameter	Increased diameter of fast fibers	Little effect on either fiber type
Glycogen content	Increased	Increased
Number of myofibrils	Increased in fast fibers	Little effect
Glycolytic enzymes	Increased in fast fibers	Little effect
Oxidative enzymes	Little effect	Increased in oxidative fibers
Capillary density	Decreased	Increased
Mitochondrial density	Decreased	Increased
Cardiovascular system	Little effect	Maximum CO and $V_{O_{2max}}$ increased

Endurance training (sometimes called aerobic training) can increase the ability to sustain submaximum exercise and the threshold level of effort at which an oxygen debt begins to be incurred (see Chapter 12, p. 301). Endurance training selectively affects oxidative fibers. The changes seen include increased:

1. Size and number of mitochondria.
2. Capillary/muscle fiber ratio.
3. Capacity to oxidize fat, resulting in a greater reliance on fat as a fuel.
4. Levels of myoglobin.

As noted above, endurance training increases $V_{O_{2\ max}}$. This is the result both of adaptive changes in the respiratory and cardiovascular systems that increase oxygen delivery and of the increases in the oxidative capacity of skeletal muscle described above. The cardiovascular effects include:

1. Increased mass and contractility of the myocardium.
2. A lower resting heart rate, which increases the range of the cardiac output, given the increased stroke volume that results from effect 1.
3. Increased capillary density in the myocardium.
4. Decreased peripheral resistance during rest.

All of these changes contribute to increased maximum cardiac output. In untrained but otherwise healthy people, such changes can typically be attained by an exercise regimen that involves 15 to 60 minutes of exercise 3 to 5 days a week in which the heart rate rises to 60% to 90% of the maximum value. The maximum heart rate can be estimated by subtracting the person's age from 220.

KIDNEY AND BODY FLUID HOMEOSTASIS

CHAPTER 19

The Kidney

On completing this chapter you will be able to:

- Describe the anatomy of the excretory system.
- Outline the reflex and voluntary pathways controlling urination.
- Understand the structure of the nephrons and distinguish between cortical and juxtaglomerular nephrons.
- Describe the process of ultrafiltration across glomerular capillaries and state the composition of primary urine.
- Describe the mechanism for reabsorption of most of the filtered Na^+ and water in the proximal tubule and understand how this leads to the passive reabsorption of other small solutes.
- Understand the mechanisms of active reabsorption and secretion, defining the transport maximum.
- Appreciate how the loop of Henle acts to create an osmotic gradient in the kidney and understand the role of urea in urine concentration.
- Describe the transport processes occurring in the distal tubule and how these are affected by aldosterone.
- Understand how ADH affects the water permeability of the collecting duct.
- Appreciate the role of the vasa recta in preserving the medullary osmotic gradient.
- Define clearance and be able to solve problems involving clearance.

*T*he kidneys receive about a fifth of the resting cardiac output, or about 1.1 L/min of blood flow. This is a larger fraction of the cardiac output than the brain receives and is about the same as that received by resting skeletal muscle. A tiny fraction (1 to 2 ml/min) of the plasma volume that perfuses the kidneys is transformed into urine. This small amount of urine is the end product of a process in which almost 200 liters of a protein-free filtrate of plasma generated within the kidney in 1 day passes through tiny tubular structures that modify its composition and greatly reduce its volume. Typically, about 99% of the filtered plasma volume is reabsorbed before the stage of final urine is reached.

As filtered plasma is modified in the kidney, the changes in its composition reflect three major functions of the kidney. The kidney is the primary route by which metabolic end products such as urea, uric acid, creatinine, ammonia, phosphate, and sulfate leave the body. Another major function of the kidney is homeostasis of body fluid volume and solute composition. Dietary intake of water and mineral ions such as Cl^-, Na^+, K^+ and Ca^{++} must be closely balanced with urinary excretion. A third function is control of the plasma pH. The kidney excretes acidic end products of metabolism and regulates the plasma bicarbonate concentration.

The importance of functioning kidneys is seen in the profound consequences of end-stage kidney disease: people whose kidneys have failed completely can live comfortably for only a few days while their bodies accumulate solutes, fluid, and acidic metabolic end products. To survive, these people must rely on periodic cleansing of the blood by a dialysis unit, or artificial kidney, or must receive a kidney transplant.

STRUCTURE OF THE EXCRETORY SYSTEM
Anatomy of the Kidney, Ureters, and Bladder

Each human kidney weighs about 200 to 300 grams and is about the size and shape of a fist (Figure 19-1, *A*). The bulk of the kidney is divided into an outer **cortex** and an inner **medulla.** The initial steps of urine formation and modification occur in the cortex. In the last step in the process of urine production, the urine flows into tubes, called **collecting ducts,** whose parallel arrangement gives a striped appearance to the innermost parts of the medulla, the **renal pyramids.** Urine from the collecting ducts of the renal pyramids flows into the **renal calyces** (singular, **calyx**), the **renal pelvis,** and eventually the **bladder** by way of the **ureter.** Figure 19-1, *B* shows an X-ray image of kidneys in which the calyces and pelvis of each kidney have been made visible by filling them with a contrast medium that absorbs X-rays.

The Urinary Bladder: Urine Storage and Urination

Urine passes from the renal pelvis through the ureters into the urinary bladder (see Figure 19-1), a hollow sac of smooth muscle lined with epithelial cells that stretches as it is filled with urine. Urine is stored in the bladder until it is allowed to flow out through the **urethra** in the process of **urination (micturition).**

The smooth muscle of the bladder forms the **internal sphincter** at the junction of the urethra with the bladder. A second, **external, sphincter** located at the base of the bladder is composed of skeletal muscle (Figure 19-2, *A*). Stretch receptors are present in the bladder and also in the muscle of the internal sphincter. Filling of the bladder is detected by the stretch receptors of the bladder. The excitation of these receptors initiates a reflex contraction of the smooth muscle of the bladder, and each contraction leads to another contraction because the stretch receptors are strongly excited each time the bladder contracts but does not empty (Figure 19-2, *B*). After several bladder contractions, the reflex pathway becomes refractory and the stretch receptors do not cause bladder contraction for a period of

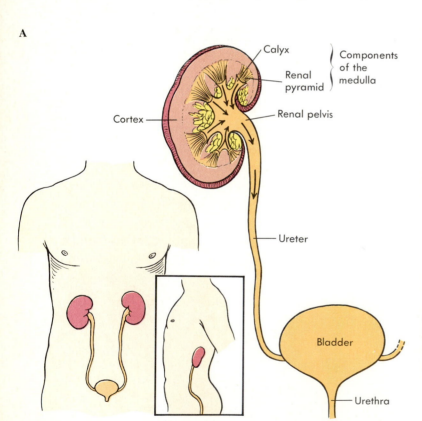

A

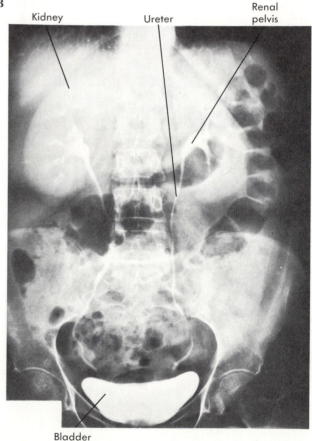

B

FIGURE 19-1

A The anatomy of the kidneys. Each kidney is composed of an outer cortex and an inner medulla. Urine flows into the renal pyramids, to the calyces and pelvis, and finally through the ureter to the bladder. The bladder empties to the exterior by way of the urethra.

B An X-ray pyelogram made by passing contrast medium from the bladder up the ureters to the pelvis. The contrast medium absorbs X-rays better than does tissue, so it outlines the hollow, fluid-filled parts of the excretory tract. The calyces are surrounded by contrast medium, giving them a cuplike appearance.

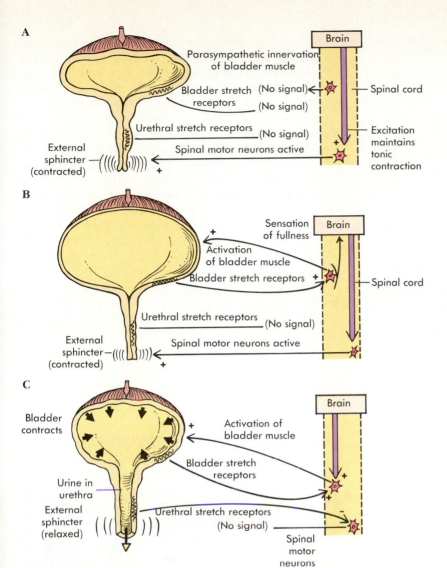

A

Parasympathetic innervation of bladder muscle

Brain

Bladder stretch receptors (No signal)

(No signal)

Spinal cord

Urethral stretch receptors (No signal)

External sphincter (contracted)

Spinal motor neurons active

Excitation maintains tonic contraction

B

Sensation of fullness

Brain

Activation of bladder muscle

Bladder stretch receptors +

Spinal cord

Urethral stretch receptors (No signal)

External sphincter (contracted)

Spinal motor neurons active

C

Brain

Bladder contracts

Activation of bladder muscle

Bladder stretch receptors

Urine in urethra

External sphincter (relaxed)

Urethral stretch receptors (No signal)

Spinal motor neurons

FIGURE 19-2
Micturition.

A   The bladder contains little urine and therefore no stretch reflex is activated. The brain maintains the tonic drive on the motor neurons that innervate the skeletal muscle of the external sphincter.

B   The volume of urine in the bladder has increased enough to activate the stretch receptors in the bladder wall. The stretch receptors excite the parasympathetic innervation of the bladder, and each contraction activates a positive feedback loop. The external sphincter continues to be closed in response to signals from the brain, but the sensation of the need to void has been felt.

C   Voluntary control of bladder emptying occurs when the signals from the brain to the external sphincter motor neurons cease and stimulation of the bladder parasympathetic innervation occurs. This is assisted by the stretch reflex and the inhibition of external sphincter contraction while urine is in the urethra.

from several minutes to an hour, despite the increasing distension of the bladder. After this period, however, the cycle is repeated.

The greater the volume of urine in the bladder, the stronger the bladder contractions become. At some point, the contractions are sufficient to stretch open the internal sphincter and force some urine into the urethra. This initiates a second stretch reflex that inhibits the spinal motor neurons that maintain tonic contraction of the external sphincter muscles. At this point, urination will occur if the urethral stretch receptor input is able to prevail over the brain's control of the external sphincter. Once urination has begun, positive feedback continues as long as urine is flowing through the urethra, allowing the bladder to be completely emptied (Figure 19-2, C).

This description of control of the bladder is a simplified one, and many other reflex pathways also contribute. For instance, sudden, frightening events interrupt the brain's maintenance of external sphincter control and also activate bladder contrac-

tion by increased autonomic activity. Involuntary urination may result.

In infants, urination is purely reflexive. Voluntary control of urination develops about 2 to 3 years after birth. The series of bladder contractions that are initiated by bladder stretch receptors probably results in a signal to the brain that provides a sense of the need to void the bladder. Voluntary initiation of urination is accomplished by descending pathways that inhibit the spinal motor neurons of the external sphincter and surrounding muscles of the pelvic floor. Once these pathways are inhibited, urine enters the urethra, initiating the positive feedback that promotes urination. The reflexive emptying of the bladder is assisted by contraction of muscles of the lower abdomen, which increases the pressure of the bladder and promotes reflexive contraction. Damage to pelvic nerves may abolish or weaken the reflex, causing the bladder to be chronically overfilled (**urinary retention**). Damage to nerves serving the external sphincter and skeletal muscles of the pelvic floor results in a loss of sphincter tone and a periodic leakage of urine (urinary incontinence).

1 What is the role of each of the following structures in transport and storage of urine?
   Bladder
   Ureter
   Urethra
2 What muscular structures are involved in urination? Which of these are skeletal muscle, and which are smooth muscle?
3 How does voluntary initiation of urination differ from the reflexive urination of infants?

Nephrons: The Functional Units of the Kidney

Each kidney contains about 1.2 million **nephrons,** or **renal tubules** (Figure 19-3). At the head of each nephron is a **glomerulus,** a tuftlike capillary bed enclosed in a funnel-shaped structure called **Bowman's capsule.** The glomerulus and its capsule form a **renal corpuscle.** All of the renal corpuscles are in the cortex of the kidney.

Primary urine is filtered from blood in the renal corpuscles. The steps in processing primary urine to produce final urine occur as urine flows from renal corpuscles through the intermediate regions of the tubule to the **collecting duct.** Descriptions of the tubule employ the terms proximal for the region closer to the corpuscle and distal for the region farther away from the corpuscle. The **proximal convoluted tubule** is a twisted region that lies entirely within the cortex of the kidney. There is a hairpin-like loop in the tubule called the **loop of Henle,** which connects the proximal convoluted tubule with the distal tubule. Because the loop of Henle dips downward onto the medulla and then bends back upward to the cortex to join the distal tubule, it is described as having **descending** and **ascending limbs** (see Figure 19-3, *A*).

The distal tubule has specialized regions. The initial region of the distal tubule immediately following the ascending limb of the loop of Henle is twisted and is referred to as the **convoluted** region of the distal tubule. The next region of the distal tubule is relatively straight and is called the **connecting** region of the distal tubule. The connecting region delivers urine into the **collecting** region of the distal tubule (see Figure 19-3, *A*).

A

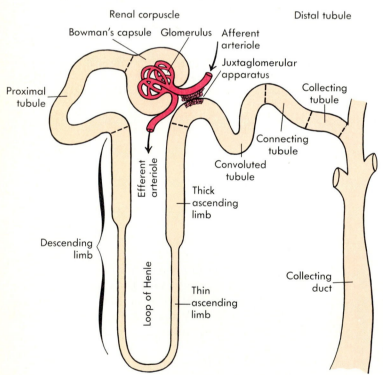

B

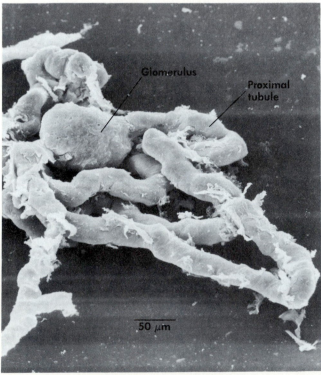

FIGURE 19-3

A The functional anatomy of a nephron. All nephrons begin with the glomerulus, the site of plasma filtration. The hollow portion of the nephron surrounding the glomerulus is Bowman's capsule. From Bowman's capsule, the tubular fluid passes through the proximal tubule, loop of Henle, distal tubule, and collecting duct. The afferent arteriole contacts a spot on the initial portion of the distal tubule; the juxtaglomerular apparatus is located at this junction.

B A scanning electron micrograph of the renal corpuscle and proximal tubule of a single nephron dissected from the kidney.

The collecting regions of the distal tubules of 5 to 10 nephrons drain into each cortical collecting duct. Hundreds of cortical collecting ducts converge to form each of the final collecting ducts that descend into the medulla. Each of these ducts carries the urine from approximately 3000 nephrons into the hollow renal pelvis, from which it drains via the ureters into the bladder (see Figure 19-1).

The nephrons can be divided into two structural classes: about 85% are **cortical nephrons,** and the rest are **juxtamedullary nephrons** (Figure 19-4). The glomeruli of cortical nephrons are located in the outer regions of the cortex, and almost all of the length of cortical nephrons is within the renal cortex, with only a small portion of the loop of Henle descending into the outer zone of the medulla. The glomeruli of juxtamedullary nephrons are located deeper within the cortex, and their loops of Henle have long, **thin descending limbs** that plunge deeply into the renal medulla. The ascending limbs have thin walls within the medulla, but the wall thickness is greater in the cortical portion. Thus the ascending part of the loops of Henle of the juxtamedullary nephrons can be divided into **thin** and **thick ascending limbs.** In contrast, the loops of Henle of cortical nephrons have short, thin descending segments and remain thick throughout the rest of their length. Although the juxtamedullary neph-

rons are fewer in number, their total bulk is greater than that of the cortical nephrons because their loops of Henle are so much longer (the loops of nephrons in Figure 19-4 are not drawn to scale).

The epithelial cells of the nephron differ in structural details in different parts of the nephron, but some common features are recognizable. The cells are knit together by bands of tight junctions (Figure 19-5, *A* and *B*; see also Chapter 4, p.73) in much the same way that beverage cans are linked by flexible plastic to form a six-pack. The **apical** surfaces of the cells face the urine on the inside of the nephron and correspond to the tops of the cans.

In most parts of the nephron, the apical surfaces are covered with microvilli, projections of membrane that increase the surface area available for transport. Microvilli are particularly evident in the proximal tubule (Figure 19-5, *C*). The **basolateral** surfaces of tubular cells face away from the interior of the tubule and correspond to the sides and bottoms of the cans. The transport proteins of apical and basolateral membranes differ, with the result that the cells can move a substance in across one surface and out the other side. For example, almost all parts of the nephron reabsorb Na^+ from urine. The first step in this process is entry of Na^+ down its concentration gradient across the apical membrane. The entry of Na^+ is frequently coupled to en-

FIGURE 19-4

The structures and locations of the two classes of nephrons. The renal corpuscles of cortical nephrons are near the surface of the cortex, and their short loops of Henle penetrate only the outermost layer of the medulla. The renal corpuscles of juxtaglomerular nephrons are typically deeper in the cortex, and their long loops of Henle penetrate into the innermost parts of the medulla, paralleling the collecting ducts.

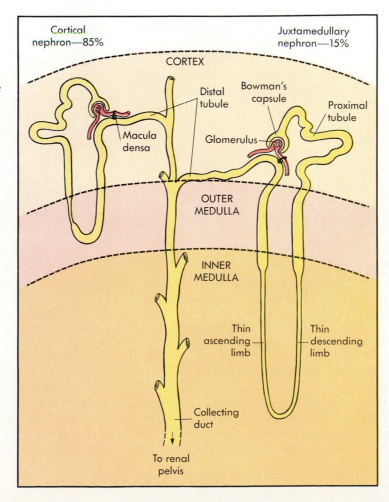

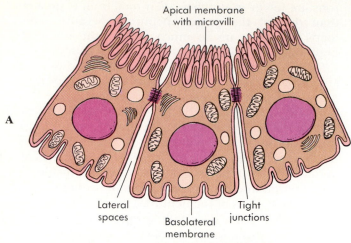

Apical membrane
with microvilli

A

Lateral
spaces

Basolateral
membrane

Tight
junctions

FIGURE 19-5

A General features of renal tubular epithelial cells.

B Section through a portion of the wall of a distal tubule, showing parts of two tubular epithelial cells and a peritubular capillary. The tight junction connecting the adjacent cells is visible. Between the two cells, a lateral space separates the basolateral membranes of the two cells.

C Photomicrograph of section through a proximal tubule. Proximal tubular cells have extensive apical microvilli and contain many mitochondria.

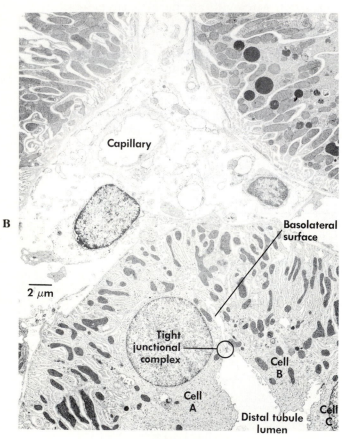

B

Capillary

Basolateral
surface

2 μm

Tight
junctional
complex

Cell
B

Cell
A

Cell
C

Distal tubule
lumen

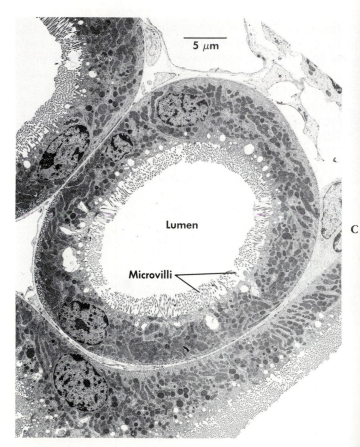

5 μm

Lumen

Microvilli

C

try or exit of other substances, a phenomenon called gradient-mediated cotransport or countertransport (see Chapter 6, p. 127). Exit of Na^+ from the cytoplasm is driven by the Na^+-K^+ pump, which is confined to the basolateral membrane.

Renal Blood Vessels and the Juxtaglomerular Apparatus

The structural relationship between nephrons (Figure 19-6, *A*) and their vasculature (Figure 19-6, *B*) has important functional implications. Each kidney is served by a renal artery that branches within the cortex to form **afferent arterioles** that serve individual glomeruli. Glomerular capillaries join to form **efferent arterioles.** The efferent arterioles branch into diffuse **peritubular capillary beds** (Figure 19-6, *C*). In the case of cortical nephrons, the peritubular capillaries surround the short loops of Henle and the proximal and distal tubules. In the case of juxtamedullary nephrons, the peritubular capillary beds include capillaries that serve the cortical portions of the nephrons and also the **vasa recta,** which are capillaries that accompany the loops of Henle into the medulla (see Figure 19-6, *C*).

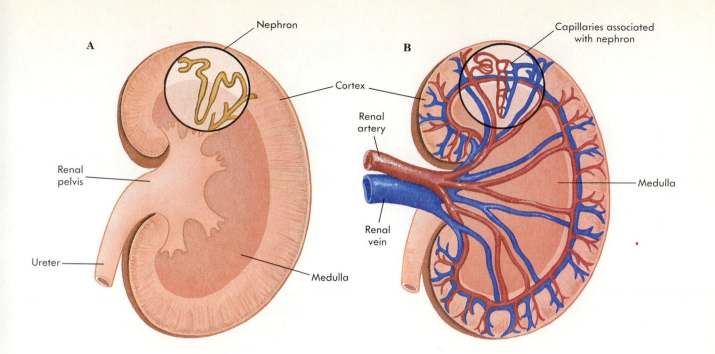

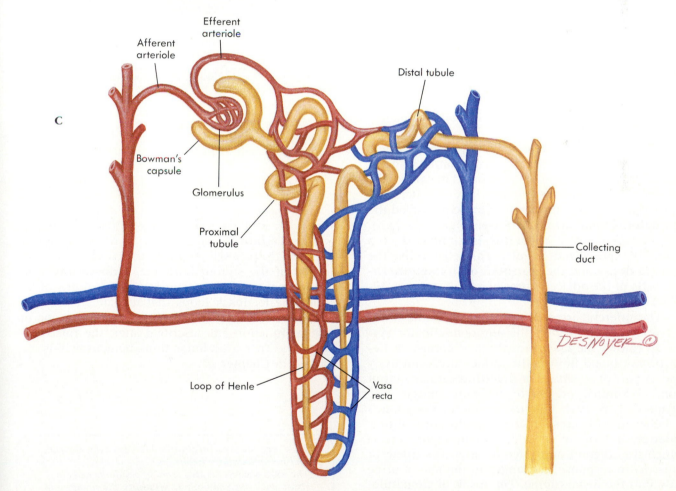

FIGURE 19-6

A Location of a juxtaglomerular nephron in the kidney.
B Relationship of its glomerular and peritubular capillaries to the renal blood vessels.
C An enlarged view of the nephron with its afferent and efferent arterioles, capillary beds, and venules. Branches of the peritubular capillaries, the vasa recta, follow the loop of Henle of juxtaglomerular nephrons in their plunge into the medulla.

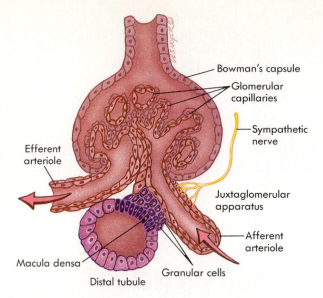

FIGURE 19-7

The structure of the juxtaglomerular apparatus in relation to the glomerulus and afferent and efferent arterioles.

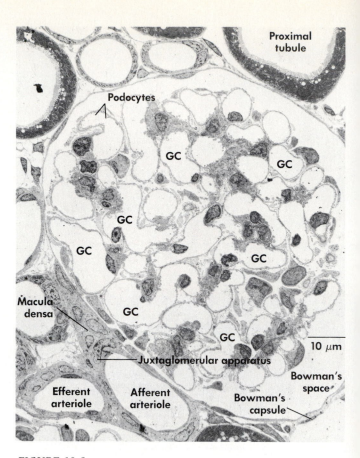

FIGURE 19-8

A photomicrograph of a section through a glomerulus, showing the juxtaglomerular apparatus *(lower left)* adjacent to the distal tubule and afferent and efferent arterioles.

The vasculature of the kidney is unusual in that it consists of two sets of capillary beds connected in series with each other: the glomerular capillaries and the peritubular capillaries. Peritubular capillaries join to form venules, small veins, and ultimately the renal vein.

The cortical portions of each nephron are intertwined like a tangled telephone cord, and the initial part of the distal tubule bends around to contact the afferent arteriole at the place where it enters the glomerulus. At this junction is a structure called the **juxtaglomerular apparatus** (see Figure 19-3, *A*). The juxtaglomerular apparatus is composed of a group of specialized tubular epithelial cells, the **macula densa** and the **granular,** or **juxtaglomerular, cells** of the adjacent arteriolar wall (Figures 19-7 and 19-8).

The afferent arterioles receive sympathetic innervation, which allows the autonomic motor system to control the rate of renal blood perfusion. Reduction of blood flow to the kidney occurs in exercise as part of a pattern of diverting cardiac output from abdominal organs to working muscle (see Chapter 15, p. 391). The juxtaglomerular apparatus is believed to be responsible for **glomerotubular balance,** a form of blood flow autoregulation in which the afferent and efferent arterioles dilate or constrict in response to changes in the rate of urine flow into the distal tubule. The result of glomerulotubular balance is a constant rate of glomerular filtration despite changes in renal perfusion pressure.

The granular cells of the juxtaglomerular apparatus secrete an enzyme called **renin,** which initiates a regulatory cascade that produces angioten-

sins I, II, and III (the effects of the angiotensins on the cardiovascular system were discussed in Chapter 15, p. 381). Angiotensin III stimulates secretion of the steroid hormone, **aldosterone** from the adrenal cortex. Aldosterone is part of one of the endocrine feedback loops that regulates extracellular fluid homeostasis. The effects of aldosterone on tubular function are described later; the role of aldosterone in extracellular fluid homeostasis is described in Chapter 20.

1 *What are the structural differences between cortical and juxtamedullary nephrons?*
2 *The kidney contains two types of capillary beds in series with one each other. What are the names and locations of the two beds? How do the capillary beds of cortical and juxtamedullary nephrons differ?*
3 *What structures compose the juxtaglomerular apparatus? Where in the nephron is this apparatus located? What are its two major functions?*

When Kidneys Stop Working: Dialysis and Kidney Transplant

There are many causes of chronic kidney failure. The most common causes are infection, diabetes mellitus, inherited renal degeneration, and damage by the body's own immune system. If the failure becomes complete, two treatment options exist: dialysis and kidney transplant.

Dialysis is a method of removing toxic wastes from the blood. This may be done in two ways. In hemodialysis, the original procedure, the patient receives surgical connections (usually located on the lower arm) that allow arterial blood to pass from the patient's circulation into the dialysis unit and back into a vein. Inside the dialysis unit (Figure 19-A, *top*), the blood passes through hollow fibers of semipermeable membrane that allow small solutes to pass down their concentration gradients into an isotonic dialyzing fluid. In this procedure the patient is connected to the dialysis unit three times a week. The patient must carefully manage salt and water intake because the dialysis machine, unlike the kidney, does not regulate blood volume and total body Na^+.

In the newer approach, the peritoneal membrane lining the abdominal cavity is used as the dialysis membrane. For this procedure, called continuous ambulatory peritoneal dialysis (CAPD), the patient receives an abdominal catheter that allows the abdominal cavity to be filled with dialysis fluid (see Figure 19-A, *bottom*). The dialysis fluid is changed several times daily. This procedure has the advantages of making work and travel easier. Diet and fluid restrictions are eased, but great care is needed to prevent peritoneal infections. In some cases a combination of hemodialysis and peritoneal dialysis is used.

Dialysis is a short-term solution to renal failure. Not only is dialysis expensive, but it severely restricts the activity of an individual. Because a single healthy kidney can meet the excretory needs of the body, a more permanent solution is the transplantation of a kidney from a healthy donor. A transplanted kidney from a living donor genetically related to the recipient has the best chance of success, but donors are frequently unrelated victims with healthy kidneys, who have died in accidents. A successful transplant means that the recipient is cured of renal disease, but there are some drawbacks. Even organs from closely related donors may be rejected by the recipient's immune system (the immune basis of transplant rejection is described in Chapter 24). To reduce chances of rejection, the recipient must be treated with immunosuppressive drugs, and such drugs not only increase susceptibility to infections, but may be directly toxic to the liver and bone marrow.

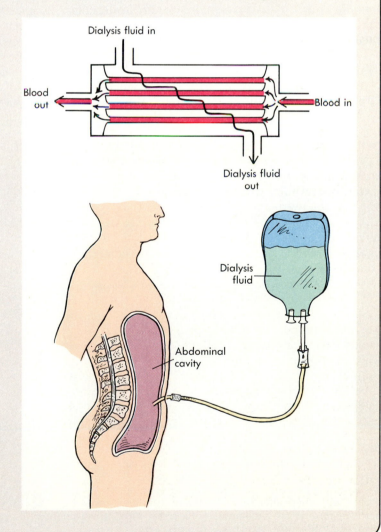

FIGURE 19-A

Top The interior of a hemodialysis unit. The blood passes through hollow fibers of dialysis membrane. The small solutes of blood are exchanged into the isotonic dialysis fluid, which circulates around the fibers.
Bottom Continuous ambulatory peritoneal dialysis. The abdominal cavity is being filled with isotonic dialysis fluid, which will accumulate small solutes, including waste products, for several hours before being replaced with fresh fluid.

MECHANISMS OF URINE FORMATION AND MODIFICATION
Ultrafiltration and the Composition of Primary Urine

The structure of Bowman's capsule (Figure 19-9, *A*) and the glomerular capillaries are structurally adapted for a high rate of capillary filtration. The surface area of glomerular capillaries is extensive (Figure 19-9, *B*). The glomerular capillary wall (Figure 19-10) contains three structural layers. Of these, the **fenestrated capillary endothelial layer** is a very coarse screen that probably serves only to keep blood cells from blocking the glomerular filter (Figure 19-11). The **basement membrane,** or **basal lamina,** appears to serve as an intermediate filter that restricts only larger plasma proteins. The most selective filtration takes place at the diaphragms of the **slit pores,** which are formed by footlike projections of supporting cells (**podocytes,** see Figure 19-10). The glomerular filter is freely permeable to water, mineral ions such as Na^+, K^+, Ca^{++}, and Cl^-, and to small organic molecules such as glucose. A substantial fraction of small plasma proteins (those up to molecular weights of about 20,000 daltons) is filtered, but plasma proteins with molecular weights above 40,000 daltons are filtered only in trace amounts.

Per unit of surface area, glomerular capillaries are about 100 times more permeable than capillaries in most parts of the circulation. Thus a much larger volume of capillary filtrate is formed in the glomer-

uli than in all of the rest of the body's capillary beds put together. The glomerular filtrate passes into Bowman's capsule, where it is referred to as the **primary urine.** Ordinarily about 20% of the plasma volume that flows through the renal artery is filtered; this is called the **filtered fraction.** The kidneys receive about 1580 L/day of whole blood. Assuming a hematocrit of 45%, the total renal plasma flow is about 870 L/day. If the filtered fraction is 20%, the **glomerular filtration rate,** or **GFR,** is 174 L/day, or about 60 times the plasma volume (about 3 liters). In 1 day, the kidney thus processes a volume equal to the total blood volume in less than 30 minutes.

The hydrostatic and osmotic forces involved in glomerular filtration are summarized below.

Hydrostatic forces: the mean hydrostatic pressure in the glomerular capillaries is 45 mm Hg (Figure 19-12). Normally, the opposing hydrostatic pressure within Bowman's capsule is 10 mm Hg or less, so the net hydrostatic driving force for glomerular filtration is about 35 mm Hg.

Osmotic forces: plasma proteins are negatively charged; thus the effective osmotic pressure that plasma proteins produce is slightly greater than expected on the basis of plasma protein concentration. This is sometimes referred to as the oncotic pressure (see Chapter 14, p. 368) and is about 25 mm Hg.

Because very few proteins are filtered, the effective osmotic pressure of the filtrate is nearly zero; thus the net osmotic pressure gradient across the af-

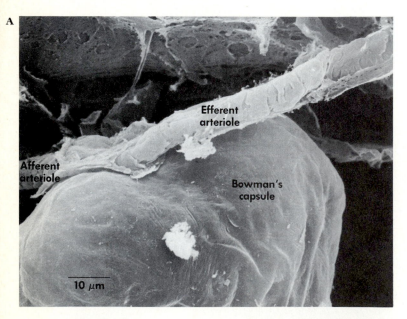

A

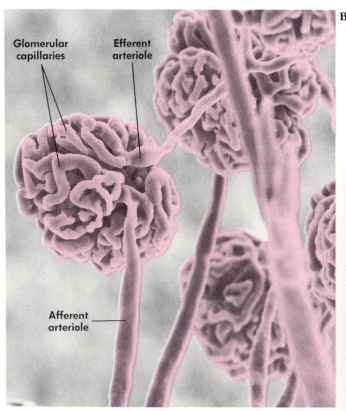

B

FIGURE 19-9
A Scanning electron micrograph of Bowman's capsule showing afferent and efferent arterioles.
B Scanning electron micrograph of casts of glomerular capillaries with their afferent and efferent arterioles, made by perfusing the renal circulation with resin. After the resin hardened, the kidney tissue was digested away.

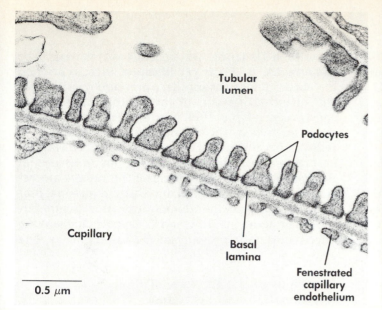

FIGURE 19-10
Transmission electron micrograph of a portion of glomerular capillary wall showing structures involved in glomerular filtration.

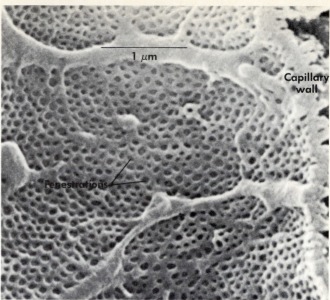

FIGURE 19-11
The surface of a glomerular capillary at much higher magnification. The first step in urine formation is filtration of water and solutes through numerous small openings termed fenestrations.

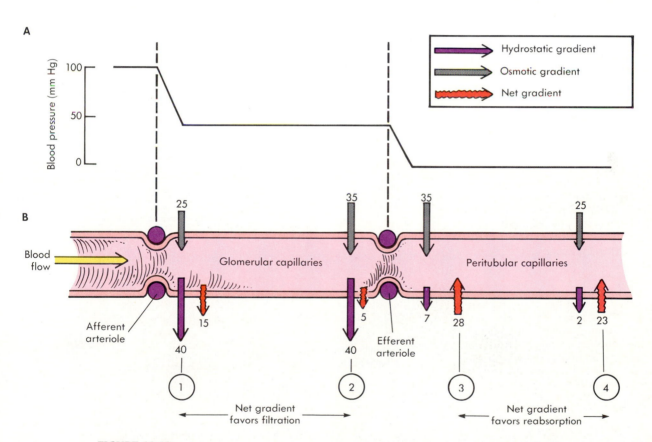

FIGURE 19-12
The graph shows the capillary hydrostatic pressure in the kidney. Note that because the afferent and efferent arterioles are connected in series, the pressure in the glomerular capillaries is 45 mm Hg, much higher than in the systemic capillaries. The hydrostatic pressure in the peritubular capillaries is about 10 mm Hg, only slightly higher than the venous pressure. The diagram shows the balance of hydrostatic and osmotic forces in the two renal capillary beds. The purple arrows (and the associated values) show the net hydrostatic pressure, the gray arrows show the net osmotic pressure, and the red arrows indicate the net force for filtration or reabsorption at various points along the nephron (**A** to **D**). Note that fluid is always filtered in the glomerular capillaries and reabsorbed in the peritubular capillaries.

ferent arteriolar end of glomerular capillaries is about 25 mm Hg in the direction of reabsorption (open, inwardly directed arrow in Figure 19-12, *A*). The net filtration force at the afferent arteriolar end of a glomerular capillary is the difference between the net hydrostatic force and the net osmotic force, or:

$$35 \text{ mm Hg} - 25 \text{ mm Hg} = 10 \text{ mm Hg}$$

Although most plasma proteins filtered are reabsorbed by specific transport pathways, there is normally a small but significant protein loss in urine, about 100 mg/day. In healthy individuals this loss is readily compensated by protein intake. In some kidney diseases, the normal recovery of filtered proteins is impeded. In other kidney diseases, damage to the glomeruli causes an increase in the amount of plasma protein filtered. In both situations not all of the protein can be recovered, and abnormally large quantities of protein appear in the final urine **(proteinuria).** In severe kidney disease the loss of protein can be so large as to cause muscle wasting unless dietary protein is supplemented. In the aftermath of strenuous exercise, hemoglobin and myoglobin released into the plasma from stressed muscle may also appear in urine **(hemoglobinuria),** giving it a brownish tinge.

The glomerular filtration rate is determined by the same four factors that determine the filtration-absorption balance in systemic capillaries (see Chapter 14, p. 370): (1) filtration increases when the mean glomerular capillary pressure rises as a result of either dilation of the afferent arterioles or constriction of the efferent arterioles; (2) filtration increases when the plasma osmotic pressure is reduced because this reduces the force favoring reabsorption; (3) filtration falls if the pressure in Bowman's capsule rises because the ureters are occluded; and (4) filtration increases if plasma proteins escape into the glomerulus—this occurs because protein in Bowman's capsule makes the effective osmotic pressure of the urine greater than zero.

Functions of the Proximal Tubule: Reabsorption and Secretion

Of the 174 L/day of primary urine, only about 60 L/day leaves the proximal tubules and enters the loops of Henle. The **standing gradient hypothesis** explains how fluid reabsorption occurs across the proximal tubules without the existence of a substantial osmotic gradient between urine and blood. According to this hypothesis (Figure 19-13), the cells of the tubular wall actively remove Na$^+$ and some other filtered solutes from the urine (Table 19-1). Relatively permeable tight junctions (see Chapter 4, p. 73) between the cells of the tubular wall of this part of the nephron allow water to flow across

FIGURE 19-13

The cellular mechanisms of Na$^+$ and water reabsorption in the proximal tubule. Active transport of Na$^+$ from the renal epithelial cells into the interstitial fluid occurs on the basolateral surface of the epithelial cells, reducing the intracellular Na$^+$ concentration. Sodium enters the epithelial cells from the renal tubule by facilitated diffusion down its electrochemical gradient, driving the absorption of numerous other solutes by Na$^+$ gradient-driven cotransport. Electroneutrality is maintained in part by secretion of H$^+$ and in part by facilitated diffusion of Cl$^-$. The result is salt reabsorption, creating an osmotic gradient for water flow. Small, nonpolar solutes such as urea are reabsorbed passively as their concentrations in tubular urine rise above plasma concentrations.

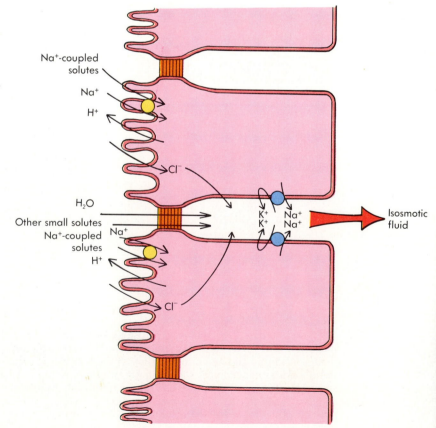

KIDNEY AND BODY FLUID HOMEOSTASIS

Reabsorbed substances (coupled to Na⁺ gradient)	Actively secreted substances
Sugars	*Substances generated within the body*
D-Glucose	H⁺ ion (coupled to Na⁺ absorption)
D-Galactose	Hydroxybenzoates
	Hippurates (for example, paraaminohippuric acid)
Amino acids	Neurotransmitters (for example, dopamine, acetylcholine,
Neutral (for example, L-alanine, L-glutamine)	epinephrine)
Acidic (for example, L-glutamate, L-aspartate)	Bile pigments (responsible for yellow color of urine)
Basic (for example, L-arginine, L-ornithine)	Uric acid
Inorganic ions	*Drugs and toxins*
Phosphate	Antibiotics (for example, penicillin G, cephalothin)
Sulfate	Atropine
	Morphine
Metabolites	Saccharin
L-Lactate	Paraquat (an herbicide)
Succinate	
Citrate	

the wall in response to an osmotic gradient. The transported solute enters the narrow intercellular spaces between the basolateral surfaces of the tubular cells, drawing an osmotic flow of water. The gradient of solute concentration that provides the osmotic driving force for fluid movement is confined to these lateral intercellular spaces. As a result of equilibration of water with actively transported solute within the lateral spaces, the reabsorbed fluid is isosmotic with the tubular urine and with the interstitial fluid of the kidney by the time it exits the lateral spaces. In summary, both water and solute are reabsorbed in the proximal tubule without affecting the overall ratio of solute to water in the urine (Figure 19-14).

For many substances, the energy for active reabsorption is provided by the Na⁺ concentration gradient across the apical membrane of the renal tubule cells, a process termed **Na⁺ gradient-coupled cotransport.** These substances include monosaccharides, amino acids, and sulfate and phosphate ions (see Table 19-1). Active reabsorption driven directly by ATP or secondarily by the Na⁺ gradient is necessary for the kidney to approach complete reabsorption of a filtered solute.

Water reabsorption and active solute reabsorption tend to increase the concentrations of solutes that are not actively reabsorbed. Some of these solutes are reabsorbed by diffusing passively down their developing concentration gradients. Urea, a nitrogen-containing product of protein catabolism, is an example of such a **passively reabsorbed** solute.

Most renal transport pathways specifically transport a small family of solutes with similar chemical structures. For example, there are five or six separate transport systems for the different amino acids. Substances using the same carrier may compete with or block each other, and carrier-mediated transport is often allosterically regulated (see Chapter 4). Because there is a finite number of transport sites for any particular substance along the renal tubule, the tubular reabsorption rate has an upper limit—the **transport maximum,** or **Tm.**

Reabsorption of filtered glucose by the proximal tubule provides an example of the general principles of renal reabsorption. For any substance X, the **filtered load** L_x or the amount filtered per unit of time, is computed as:

$$L_x = P_x \times GFR$$

where P_x is the plasma concentration of the substance usually expressed per deciliter as a percent. For example, the normal plasma glucose concentration is about 100 mg% (100 mg/dl). The filtered load of glucose is 1.00 gm/L × 174 L/day = 174 gm/day, or about 121 mg/min.

Curves *1, 2,* and *3* in Figure 19-15 show how the amounts of glucose filtered, reabsorbed, and excreted vary when plasma glucose concentration increases. The transport maximum for glucose is 375 mg/min. Because the normal filtered load of glucose is well below the Tm, glucose reabsorption is normally complete (see Figure 19-15, *curve 2*), and only trace amounts of glucose remain in the urine. However, the filtered load of glucose increases with increasing plasma concentration (Figure 19-15, *curve 1*). If the plasma glucose concentration increases above 3 mg/ml, the filtered load exceeds the trans-

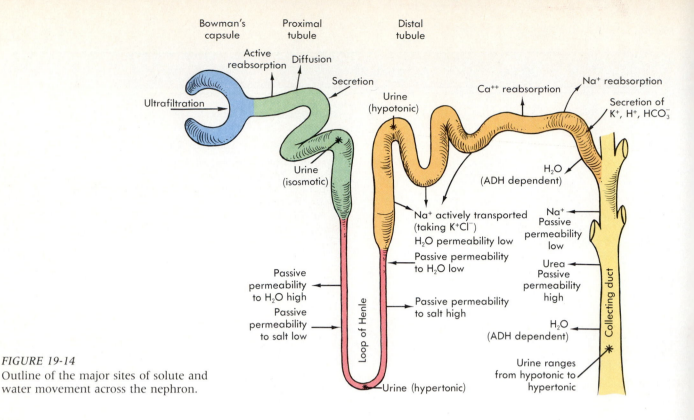

FIGURE 19-14
Outline of the major sites of solute and water movement across the nephron.

port maximum of 375 mg/min (see Figure 19-15, *point X*). When this occurs, glucose begins to be excreted in the final urine (see Figure 19-15, *curve 3*), and the amount of glucose excreted increases as the plasma glucose concentration increases. Such abnormally high levels of plasma glucose are characteristic of diseases that affect carbohydrate metabolism such as diabetes mellitus. High levels of

plasma glucose may also occur briefly in healthy individuals under conditions of stress or after a meal high in carbohydrates.

In addition to its reabsorptive processes, the proximal tubule is the site at which secretion of numerous substances occurs (see Table 19-1 and Figure 19-14). Some secreted substances are generated by normal body metabolism; other secreted sub-

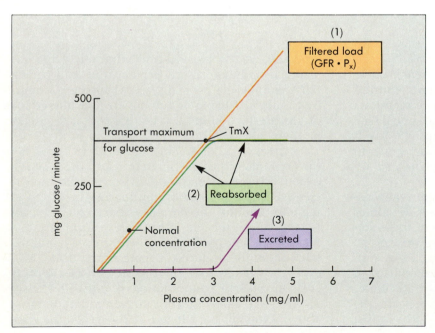

FIGURE 19-15
Curve 1 The filtered load of glucose as a function of plasma glucose concentration. The filtered load varies in direct proportion to the plasma concentration.
Curve 2 The rate of glucose reabsorption as a function of plasma glucose concentration. Reabsorption equals the filtration rate until the T_m is reached. At glucose filtration rates above the T_m, glucose reabsorption does not increase further.
Curve 3 The rate of glucose excretion as a function of plasma glucose concentration. The rate of glucose excretion equals the filtration rate (**curve 1**) minus the rate of reabsorption (**curve 2**). The rate remains negligible until the T_m is reached. The vertical axis of this plot is in units of mg glucose/min filtered (**curve 1**), reabsorbed (**curve 2**), and excreted (**curve 3**).

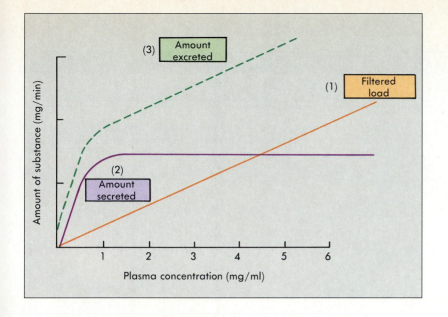

FIGURE 19-16
Curve 1 The filtered load of a filtered and secreted substance as a function of concentration of the substance in plasma.
Curve 2 The rate of tubular secretion as a function of plasma concentration.
Curve 3 The rate of excretion of the substance as a function of plasma concentration. The rate of excretion is the sum of the rate of filtration (curve 1) and the rate of secretion (curve 2). The vertical axis of this plot is in units of mg substance/min filtered (curve 1), secreted (curve 2), and excreted (curve 3).

stances are drugs or toxins. The tubular cells actively transport these substances from the unfiltered blood within peritubular capillaries. Active secretion by the proximal tubule is particularly important in the case of metabolites or toxins that are able to bind to plasma proteins. Excretion of these substances would be very slow if it depended on filtration alone. Secretion greatly reduces the lifetime of some drugs in the body, necessitating larger and more frequent doses than would otherwise be necessary. Penicillin is an example of such a rapidly secreted drug.

Secretion via specific transport pathways involves the same general principles as does absorption. The filtered load of a secreted substance increases linearly with its plasma concentration (Figure 19-16). The amount secreted increases rapidly until the T_m for secretion is reached, and then the amount secreted remains constant. The amount excreted is the sum of the filtered load and the amount secreted. Once the plasma concentration of the substance exceeds the T_m for secretion, the contribution of secretion is maximal, and further increases in the plasma concentration of the substance increase only the filtration rate.

Small proteins (of molecular weights of 20,000 daltons or less) are able to pass through the glomerular filter in significant amounts. However, as noted previously, a large fraction of the filtered proteins is reabsorbed in the proximal tubule. Reabsorption occurs by endocytosis (see Chapter 4, p. 66), and subsequently these proteins are metabolized in the proximal tubule cells. Almost all protein-type hormones (see Chapter 5, p. 102) are of a size that allows them to be filtered. Filtration and subsequent metabolism of protein hormones by the kidney significantly shortens their lifetime in the blood. For some protein hormones, a percentage of the filtered molecules escape reabsorption and can be detected in the urine, making it possible to assess a person's hormonal status from an analysis of the urine. An example of such a test is the at home pregnancy test, which detects human chorionic gonadotropin, a specific protein hormone produced by an embryo after it implants in the uterus (discussed in Chapter 26).

1 Define the following terms.
   Glomerular filtration rate (GFR)
   Filtered fraction
   Tubular maximum (T_m)
2 What is the standing gradient hypothesis? In this hypothesis, what is the role of lateral spaces between the cells of the proximal tubular epithelium?
3 How does the protein content of the urine compare with that of the plasma? Why is there a difference?

Distal Tubule: Production of Dilute Urine

Although the proximal tubule greatly reduces the volume of the tubular fluid, it has no effect on the overall ratio of water to solute, so urine entering the loop of Henle is isosmotic to plasma. The thick ascending limb of the loop of Henle and the segments of the distal tubule play an important role in producing urine with a lower osmotic pressure than that of plasma. Thus the thick ascending loop and distal tubule cause reabsorption of solute without proportionate reabsorption of water (see Figure 19-14). This is possible because these segments, unlike the proximal tubule, are almost impermeable to water (Table 19-2).

The major transport process of the thick ascending loop and the initial part of the distal tubule (convoluted region) is reabsorption of Na^+ and Cl^-. In the first step of salt reabsorption, an Na^+ ion and a K^+ ion are taken across the apical membrane of the tubule cells in company with two Cl^- ions

TABLE 19-2 *Transport and Permeability Properties of Nephron Segments Related to Urine Concentration and Dilution*

Segment	Active salt transport	Permeability			Osmolarity	
		H₂O	NaCl	Urea	Beginning	End
Proximal tubule	+	++++	+++	+++	Iso	Iso
Loop of Henle						
Descending limb	0	++++	0	+	Iso	Hyper
Thin ascending limb	0	0	+++	+	Hyper	Hyper
Thick ascending limb	++++	0	0	0	Hyper	Hypo
Distal tubule						
Convoluted tubule	+	0	0	0	Hypo	Hypo
Collecting tubule	+ to >+++*	0 to >+++*	0	0	Hypo	Hypo/iso*
Collecting duct	+	0 to +++*	0	+++	Hypo/iso*	Hypo/hyper

*Varies in response to hormonal regulation.

(Figure 19-17). In this process, called **Na⁺,K⁺,2Cl⁻ cotransport,** the Na^+ gradient provides the energy for uptake of both the K^+ and the Cl^- ions. Subsequently, the Na^+ that enters the cell by this route is actively pumped across the basolateral membrane, the Cl^- diffuses passively into the interstitial fluid, and the K^+ transported in at both membranes replaces K^+ that is constantly leaking out of the cells.

The middle part of the distal tubule (the connecting region) is believed to be responsible for reabsorption of filtered Ca^{++}. This segment of the nephron plays a key role in homeostasis of plasma Ca^{++}, described in the next chapter. The dominant transport process of the final part of the distal tubule (the collecting region) is reabsorption of NaCl, but it is also able to secrete H^+ and to both reabsorb and secrete K^+.

Loops of Henle and the Collecting Duct: Formation of Concentrated Urine

Formation of a final urine more concentrated than plasma allows the body to rid itself of toxic metabolic end products and excess ions with a minimum loss of water. The kidneys concentrate urine by maintaining the medullary interstitial fluid at a concentration typically as much as 4 to 5 times that of the plasma (that is, about 1400 mOsm versus the 300 mOsm of plasma). The osmotic gradient between the concentrated medullary interstitial fluid and the dilute urine entering the collecting duct provides a driving force for water to move from the urine in the collecting duct to the medullary interstitial fluid.

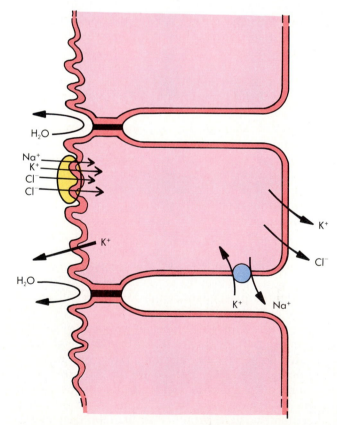

FIGURE 19-17

Mechanisms of salt transport in the thick ascending limb and distal convoluted tubule. Entry of Na^+, Cl^-, and K^+ is coupled, and all are driven by the Na^+ gradient. Na^+ and Cl^- leave the tubular cells for the interstitial fluid. The tight junctions are not permeable to water in these segments of the nephron, so relatively little water movement occurs and the urine becomes hypotonic.

Physiologists began to understand the urine concentrating mechanism when they realized that only those animals with loops of Henle that penetrate the renal medulla can form concentrated urine and that maximum urine tonicity depends on the overall length of these loops. Four factors are required for urine concentration: (1) the loop of Henle must plunge deep into the renal medulla and then sharply bend upward so that the ascending limb stays near the descending limb, (2) the descending and ascending limbs must have different water permeabilities, (3) NaCl transport must occur only in the ascending limb, and (4) the capillaries of the vasa recta must accompany the loops of Henle.

The permeability properties and solute transport activities (and thus energy requirements) of the loop of Henle and the collecting duct (see Table 19-2) are reflected in the structure of their cells. The descending limb has relatively flat cells with few mitochondria. Consequently, the tubule itself is relatively thin. These cells are not believed to transport Na^+. The changes in urine composition that occur as the urine passes through the descending limb are the result of passive exchange with the interstitial fluid (Figure 19-18). The descending limb is highly permeable to water but much less permeable to salt. The thin part of the ascending limb of the loop also has relatively flat cells. Like the descending limb, this portion of the tubule does not actively transport ions. In contrast to the descending limb, the thin ascending limb has a low water permeability but a high salt permeability (see Figure 19-14).

The cells of the thick ascending limb have many mitochondria and other intracellular structural features associated with active ion transport. This part of the nephron actively transports NaCl from urine to interstitial fluid, but its passive permeability to both salt and water is low (see Table 19-2 and Figure 19-14). Historically there was controversy about whether Na^+ or Cl^- was the ion actively transported. It is now known that both Na^+ and Cl^- move from the urine in the thick ascending limb to the interstitial fluid against electrochemical gradients. In this process, ATP is spent to move Na^+ across the basolateral membrane. The uptake of Cl^- and K^+ across the apical membrane is driven by the Na^+ gradient across the apical membrane, which results from the action of the basolateral Na^+-K^+ pump.

The urine entering the loop of Henle from the proximal tubule has a NaCl concentration and osmolarity similar to that of plasma, and, as in plasma, NaCl is its most abundant solute. As urine passes into the descending limb, it passes through regions of the kidney with increasingly concentrated interstitial fluid. The wall of the descending limb is permeable to water, so as water leaves the descending limb, the urine comes into osmotic equilibrium with more and more concentrated in-

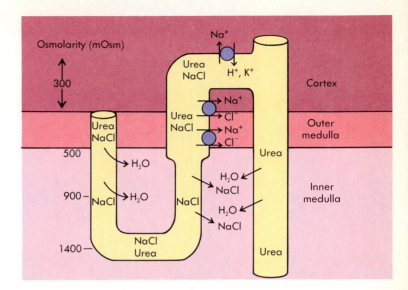

FIGURE 19-18
Mechanism of formation of concentrated urine according to the two-solute hypothesis. The osmolarity of the interstitial fluid at different levels of the medulla is shown on the scale at the left. The urine leaving the proximal tubule is isotonic. As the urine travels in the descending limb of the loop of Henle, water leaves the descending limb, drawn by the increasing osmotic pressure of interstitial fluid in the medulla. As a result, the urine in the descending limb becomes progressively more concentrated. As the urine passes through the thin ascending limb, NaCl diffuses out as the osmotic pressure of interstitial fluid decreases. In the thick ascending limb, more salt is removed by active reabsorption. The urine entering the distal tubule is more dilute than plasma, and urea has been concentrated by the reabsorption of water. Urea and water move down their concentration gradients as urine passes through the collecting duct. The remaining solutes in the urine are concentrated further by the water reabsorption, and a urine as concentrated as the interstitial fluid at the innermost part of the medulla may be formed if ADH levels are high. If ADH levels are low, a final urine similar to the dilute urine in the distal tubule is excreted.

terstitial fluid. Osmotic withdrawal of water from the descending limb concentrates the solutes in the urine (see Figure 19-14).

As the urine rounds the bend of the loop and moves along the thin ascending limb, it passes through regions with more and more dilute interstitial fluid. The wall of the thin ascending limb is permeable to NaCl, but not permeable to water, so as the urine ascends, osmotic equilibrium is reached by movement of NaCl out of the urine.

In the thick ascending limb, NaCl is actively transported from the urine and water cannot follow, so a concentration gradient develops between the urine and the interstitial fluid. The urine leaving the loop has a lower NaCl concentration than the plasma. As the urine passes through the distal tubule, more NaCl is actively transported out of the

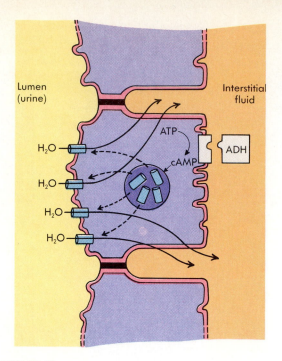

FIGURE 19-19
Current hypothesis explaining the action of ADH on water permeability of the collecting duct. The binding of hormone to receptor increases intracellular levels of cAMP, causing preformed water pores to be incorporated in the apical cell membrane.

urine. Active transport of NaCl out across the thick ascending limb and distal tubule makes those solutes remaining behind more important in determining the osmotic pressure of the urine. One such solute is urea. By the time urine reaches the collecting duct, urea has become one of its major solutes.

The wall of the collecting duct actively absorbs

NaCl at a low rate, but its passive permeability to NaCl is low. As the urine descends the collecting duct, it passes once again through regions of increasing interstitial fluid osmotic pressure. Permeability of the duct wall to water is a key factor in regulation of urine concentration because it determines the extent to which water can follow the osmotic gradient from urine in the collecting duct to the medullary interstitial fluid. The water permeability of the final segment of the distal tubule and the collecting duct is regulated by **antidiuretic hormone (ADH),** a peptide hormone released by the posterior pituitary (see Chapter 5 p. 99). ADH increases the water permeability by opening pores in the epithelial cells (Figure 19-19).

When ADH is present, water can equilibrate across the walls of the final distal tubule and collecting duct, and the descending urine becomes concentrated as water is reabsorbed (Figure 19-20). Because the wall of the collecting duct is very permeable to urea, some urea is reabsorbed, but relatively little reabsorption of NaCl occurs. Thus the final urine that leaves the collecting duct is reduced in volume, and the urine may be as concentrated as the extracellular fluid in the innermost, highly concentrated regions of the renal medulla (see Figure 19-14).

How the kidney generates the high concentration of solute in the interstitial fluid of the medulla is still not fully understood. There is no doubt that an essential factor is the parallel, opposite flows of urine in the two limbs of the loop of Henle and in the collecting duct. This arrangement has been called a **countercurrent multiplier.** As the name suggests, the parallel flow of descending and as-

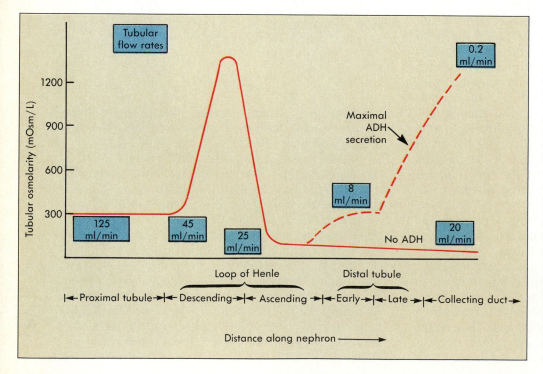

FIGURE 19-20
An overview of the osmotic pressure and volume flow changes along the length of the nephron in the presence and absence of ADH. The solid line on the graph indicates the osmolarity of the tubular fluid at various points along the nephron. The tubular fluid is initially isotonic, increases in osmolarity in the descending loop of Henle, and decreases in osmolarity in the ascending loop of Henle. The tubular fluid enters the distal tubule hypoosmotic to plasma. It then can become either more hypoosmotic (if ADH is absent—*solid line*) or hypertonic (if ADH is present—*dashed line*). The numbers in the boxes are the tubular flow rates; they indicate the degree of water reabsorption that has occurred. Specific transport properties of the different regions are summarized in Table 19-2.

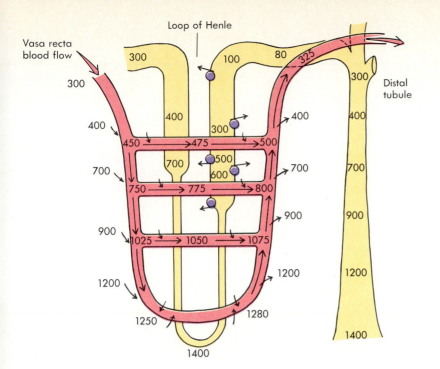

Loop of Henle

Vasa recta blood flow

Distal tubule

FIGURE 19-21

Overview of the concentrations of solute in the renal tubule and the vasa recta. A countercurrent system is one in which fluid flows in parallel tubes in opposite directions. Heat, solutes, and water diffuse from one tube to the other so that, in the absence of any energy-driven process, the fluid in both tubes has nearly the same composition. The renal tubule acts as a countercurrent multiplier system because salt is actively transported. The vasa recta accompany the loop of Henle and act as a countercurrent exchange system to prevent dissipation of the medullary concentration gradient by blood flow. The numbers give the osmolarity of the urine, plasma, and interstitial fluid in the renal tubule and vasa recta at several levels.

cending urine multiplies the power of ion transport processes of the tubular wall to create a concentration gradient.

Historically, several attempts have been made to explain how the countercurrent multiplier generates the osmotic gradient between the outermost and innermost parts of the medulla. In the **classical hypothesis,** a high concentration of solute at the tip of the loop of Henle was maintained by active transport of ions out of the ascending loop and water diffusion out of the descending loop. In this hypothesis, the loop of Henle could be likened to a pair of escalators running in parallel, one going up and the other going down. If riders wanting to go to the top were transferred from the ascending escalator to the descending escalator at various points (countercurrent exchange), there would be a crowd of frustrated people repeatedly getting on the up escalator only to be forced to ride back down.

Unfortunately, when the permeability and transport properties of the loop were measured in pieces of tubule dissected from the kidney, the results (see Table 19-2) did not entirely support this hypothesis. A point against the classical hypothesis is the fact that the descending loop is impermeable to ions. In the escalator analogy, this impermeability to ions would prevent ions that are ejected from the up escalator (the thick ascending limb) from getting on the down escalator (the descending limb).

Currently the most plausible hypothesis is the **two-solute hypothesis** (see Figure 19-18). This hypothesis builds on the finding that, along with NaCl, urea makes up a large fraction of the total solute of the medullary interstitial fluid. The high concentrations of NaCl and urea in the medullary interstitial fluid result because (1) NaCl is the major sol-

ute of urine in the thin ascending limb, and urea is a major solute in the urine of the medullary collecting duct; and (2) the thin ascending limb is more permeable to NaCl than to urea, and the collecting duct is more permeable to urea than to NaCl.

Two driving forces are at work in the two-solute hypothesis: the NaCl gradient across the thin ascending limb and the urea gradient across the collecting duct. Both of these gradients are created by the active reabsorption of NaCl by the thick ascending limb. Both gradients drive solute into the medullary interstitial fluid, concentrating both NaCl and urea in the interstitial fluid.

Role of the Vasa Recta in Urine Concentration

Only about 10% of the filtered water is still present in the urine entering the collecting ducts. When additional reabsorption of water occurs as the collecting ducts travel through the medulla, the amount that moves into the interstitial fluid of the medulla is not great enough to overwhelm the ability of the medullary structures to form a concentrated interstitial fluid. However, if the composition of plasma flowing through the medulla were identical to the composition of the plasma in other parts of the circulation, the medullary interstitial fluid would rapidly equilibrate with the more dilute plasma. This does not happen because the structure of the vasa recta forms a **countercurrent exchanger.**

Countercurrent exchange of solute and water between ascending and descending blood across the walls of the vasa recta is a passive process driven by the medullary solute gradient. The vasa recta capillaries follow the hairpin course of the loops of Henle, and these capillaries are permeable to both NaCl and water (Figure 19-21). The plasma within the capillaries becomes as concentrated as the adja-

Experimental Methods of Renal Physiology

Several experimental techniques have been used to characterize the function of different parts of the nephron. These include the **micropuncture method,** the **perfused renal tubule** preparation, and methods in which pieces of renal tubules are suspended in an appropriate medium. Much of what is known about the human kidney is the result of experiments using the kidneys of other mammals or even of amphibians, reptiles, and fish on the assumption that many of the fundamental mechanisms of urine formation and modification are common to all vertebrates.

In the micropuncture method, a micropipette is introduced into nephrons at various points to remove a sample of the urine (Figure 19-B, 1). The rates at which fluid is reabsorbed by different parts of the nephron can be determined by measuring the concentration of a nonreabsorbed marker substance (usually inulin, see the text) at different points along the nephron. Similarly, the ratio between the inulin concentration and that of other solutes can reveal solute absorption or secretion. The success of this method depended on accurate methods for handling, and

chemical assay of, very small fluid samples. Also, micropuncture can be applied only to those parts of cortical nephrons which are accessible from the kidney surface. Thus it is not possible to apply this method to the loop of Henle or to the juxtamedullary nephrons.

Portions of nephrons can be dissected from the kidney and mounted between two perfusion pipettes (see Figure 19-B, 2). A very slow flow of fluid through the tubule is set up, and the effect of the mounted piece of tubule on the composition of the fluid can be measured. This method permits study of the parts of the nephrons that are deeper in the kidney, but it has the disadvantage of disrupting the normal structural and functional relationships between the different parts of the nephron and between the nephron and its circulation.

If the kidney cortex is gently minced, pieces of proximal and distal tubules are released. The ends of these segments tend to seal themselves off. If a substance that is actively secreted by these parts of the nephron is dissolved in the medium, it becomes concentrated in the tubules. This is especially

evident if the substance is colored because an optical assay of the tubular fluid is possible. Chemical or radioisotope methods may also be used to follow renal handling of a particular substance. In a further refinement of this basic approach, sonication of the tubule pieces with a generator of high-frequency sound waves breaks up the cells of the tubules, causing the formation of membrane vesicles derived from either apical or basolateral surfaces of the cell membrane. The vesicles from the different regions can be separated for measurement of their ability to transport specific solutes.

Most recently, the composition of urine in the different parts of nephrons has been measured using the **X-ray microanalysis** method. In this method, living kidney tissue is quickly frozen and sliced. A beam of electrons scans across the slice. Different chemical elements scatter the electrons in characteristic ways; thus, with appropriate calibration, the elemental composition of the urine and extracellular fluid of different kidney regions can be determined from the energy spectrum of the scattered electrons.

---

cent interstitial fluid as the vasa recta follows the descending loop, developing a very high osmolarity deep within the medulla. As the blood reascends to the cortex, the extra solute is lost and water is absorbed. In the end, the plasma leaves the vasa recta with an osmotic pressure only slightly greater than it had when it entered; thus the vasa recta removes solute from the medulla at a low rate. However, the rate of solute removal by the vasa recta depends on the rate of flow in these capillaries, so blood flow in the vasa recta must be low to preserve the medullary osmotic gradient.

Endocrine Control of Urine Concentration

In the absence of ADH, the collecting portion of the distal tubule and the collecting duct are not water

permeable, and water recovery from urine in the collecting duct is minimal (see Figure 19-20). The 20% of the filtered load of water that remains unreabsorbed in the absence of ADH corresponds to a urine output of about 20 ml/min, or a little less than 30 L/day. This flow rate represents the upper limit of the kidney's ability to handle a water load. Such a high urine flow rate might be maintained for a brief period by a normal individual after rapidly drinking several liters of water. Damage to the pituitary or hypothalamus can result in a disease called **diabetes insipidus,** in which little or no ADH is secreted. A chronic high rate of production of dilute urine is characteristic of this disease.

Normal levels of ADH result in the production of about 1 L/day of slightly hyperosmotic urine. If

Because some other epithelial tissues share some of the transport mechanisms of the nephron, it is frequently possible to study these transport processes in a more convenient tissue. For example, the skin epithelium of the frog is similar in many ways to the distal tubule. This skin has been a valuable experimental preparation in which to study the mechanisms of diuretic drugs and the ways in which the distal tubule responds to the hormones, aldosterone and antidiuretic hormone.

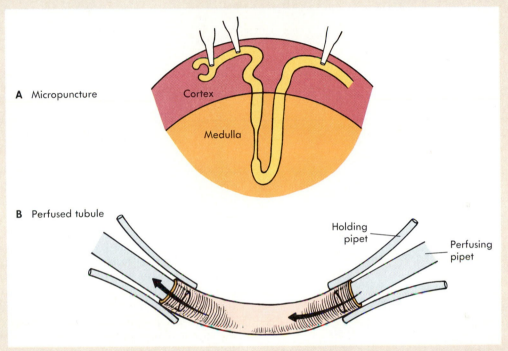

A Micropuncture

Cortex

Medulla

B Perfused tubule

Holding pipet

Perfusing pipet

FIGURE 19-B
1   Micropuncture of cortical nephrons with fine glass pipettes can provide nanoliter samples of fluid from Bowman's capsule, the proximal tubule, and the distal convoluted tubule.
2   Pieces of tubule can be dissected from any region of the kidney and mounted between fine glass pipettes. The outer pipette holds the tubule in place; the inner pipette is for perfusion of the tubule. The isolated tubule is bathed in physiological saline solution. Changes in the composition of the perfusate as it passes through the tubule can be measured. The permeability of the tubule to water and salt are measured using chemical techniques or radioisotope markers.

water intake is restricted or if sweating is profuse, plasma levels of ADH rise. As a result, urine osmolarity may increase to 1200 to 1400 mOsm/L, four to five times plasma osmolarity (see Figure 19-20), and the urine flow rate can be as low as 0.2 ml/min, or 300 ml/day. The factors involved in control of ADH secretion are discussed in detail in the next chapter.

1 *What tubular processes tend to make urine more dilute than plasma? More concentrated? What hormone regulates urine osmolarity?*
2 *Describe the force that removes water from the urine in the descending limb of the loop of Henle.*
3 *Describe the role of the vasa recta in the formation of concentrated urine.*

INTEGRATED FUNCTION OF THE KIDNEY
Concept of Renal Clearance

The concentrations of filterable solutes that appear in primary urine are the same as their plasma concentrations. However, excretion rates of different solutes may differ because either absorption or secretion of filtered substances may occur. The sequence of processes in which the kidney deals with a given substance is called **renal handling.** Handling of some representative substances is summarized in Table 19-3.

One way to quantify the handling of a particular substance by the kidney is to determine the **renal clearance** of that substance. The clearance is the number of liters of plasma that are completely cleared of the substance by the kidney per a unit of

TABLE 19-3   Renal Handling of Some Representative Substances

Substance	Plasma concentration	24-hour filtered load	Actively reabsorbed/ secreted	Typical percent of filtered load excreted	24-hour clearance (L)
Glucose	100 mg%	174 gm	R	0	0
Na$^+$	142 mEq/L†	24.7 Eq	R	1	1.7
Urea	30 mg%	47.5 gm	‡	60	105
Phosphate	2 mEq/L	348 mEq	R	20	35
K$^+$	5 mEq/L	518 mEq	R+S	10	17
H$^+$	10^{-4} mEq/L	0.1 mEq*	S	800*	1394
Bicarbonate	24 mEq/L	4.2 Eq	‡	0	0
Cl$^-$	103 mEq/L	18 Eq	R	1	1.7
Creatinine	1 mg%	1.4 gm	Neither	100	171

*Essentially all H$^+$ loss is the result of tubular secretion.
†Milliequivalents/L (mEq/L) is equal to the product of the concentration in millimoles/L times the valence of an ion.
‡See text.

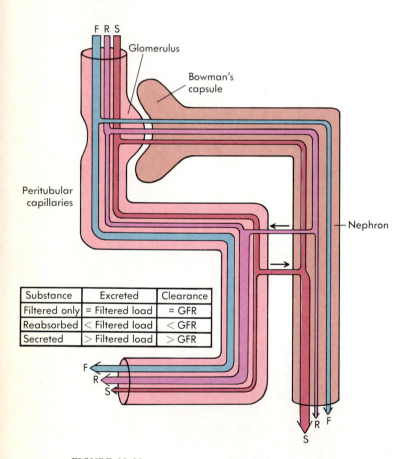

FIGURE 19-22
The three ways in which the kidney may handle materials appearing in the glomerular filtrate. Some substances *(aqua line)* are filtered into Bowman's capsule, but then are neither reabsorbed nor secreted. Others are filtered and reabsorbed into the peritubular capillaries *(purple line)*. Finally, some materials are both filtered and actively secreted from the peritubular capillaries into the renal tubule *(red line)*. The rate of loss of a substance in the urine can thus be equal to, less than, or greater than the filtered load *(inset)*. The thicknesses of the arrows indicate the amount of each of the substances flowing in the plasma and urine.

time. The idea of completely clearing some liters of plasma of one of its solutes has an imaginary aspect because very few substances are completely removed from the plasma during a single pass through the kidney.

The clearance (C_X) of a substance (X) is computed as follows. Plasma concentration of X (P_X) is measured in weight/volume or milliequivalents/volume, depending on whether X is an organic molecule or an ion. For example, glucose concentration is expressed as mg/ml, whereas Na$^+$ concentration is given as mEq/ml (see Chapter 2, p. 24, for the difference between milliequivalent and millimolar). The concentration of X in the urine (U_X) is measured using the same units, and the rate of urine flow is V (in units of volume/time). The amount of X excreted by the kidney is (U_X)(V), and the units will be either weight/time or mEq/time (mg/min for glucose; mEq/min for Na$^+$).

$$C_X = \frac{(U_X)(V)}{P_X}$$

Because U_X and P_X have the same units, these units cancel and clearance has units of volume/time.

If the clearance of a solute equals the GFR, the amount excreted equals the filtered load and the solute did not undergo net secretion or reabsorption during its passage down the renal tubule (Figure 19-22). If the clearance is less than the GFR, net reabsorption has occurred because more of the solute was present in the volume of plasma filtered than appears in the urine (see Figure 19-22, *inset*, line 2). Any solute that is fully reabsorbed by the kidney has a clearance of zero. Solutes that have clearances greater than the GFR have undergone net secretion.

The flow rate of final urine is much less than the GFR because much filtered fluid is reabsorbed. A substance that is filtered but not secreted or reabsorbed becomes concentrated in the urine as fluid is reabsorbed. Because the rate of filtration of such a substance is proportional to the GFR, its clearance is an accurate measure of the GFR. For example, substance F in Figure 19-22 was filtered into Bowman's capsule but not subsequently reabsorbed or secreted. What is the clearance of such a substance? The amount of F excreted in the urine must equal the original filtered load:

$$\text{Filtered load of F} = (GFR)(P_F)$$
$$= (V)(U_F)$$

The clearance of F must be equal to the volume of fluid filtered along with it. Solving for the glomerular filtration rate gives:

$$GFR = \frac{(U_F)(V)}{P_F}$$
$$= C_F$$

Inulin, a plant polysaccharide, is one example of a solute that is filtered but neither secreted nor reabsorbed. The inulin clearance is determined by injecting inulin into a subject, waiting for the injected inulin to be distributed throughout the blood, and then measuring the plasma and urine concentrations of inulin and the rate of urine production. It is frequently more practical to use the **creatinine** clearance as a measure of GFR. Creatinine is naturally produced at a low rate as a metabolite of muscle creatine. It is neither reabsorbed nor metabolized by the kidney and is secreted at a negligible rate. Measurement of creatinine clearance only requires samples of blood and urine and measurement of the rate of urine production and gives results nearly the same as those obtained by measuring inulin clearance.

For those substances which are filtered and reabsorbed (*substance R* in Figure 19-22), some or all of the substance filtered from glomerular blood is returned to the peritubular blood, making the clearance less than the GFR. For substances such as glucose that are essentially totally reabsorbed, the clearance is zero. For solutes that are filtered and secreted (*substance S* in Figure 19-22), some solute enters the urine at the filtration step, an additional amount is transferred from peritubular blood to urine, and the clearance is greater than the GFR.

If the kidney secretes a solute so efficiently that the solute is completely eliminated by the time the blood leaves the kidney, its clearance will be equal to the total **renal plasma flow (RPF)** because the amount of plasma cleared of the solute is all the plasma that entered the kidney. This includes both the filtered fraction entering Bowman's capsule and the unfiltered plasma in the peritubular capillaries. One example of such a solute is **paraaminohippuric acid, or PAH.** PAH is produced in the body in

small amounts as an end product of metabolism of aromatic amino acids. About 90% of the PAH entering a healthy kidney is removed. The rate of PAH excreted in the urine ($U_{PAH} \times V$) equals the plasma concentration of PAH (P_{PAH}) times the RPF; thus the clearance of PAH approximates the RPF. The total **renal blood flow (RBF)** can be calculated from the PAH clearance by dividing it by the fraction of whole blood that is plasma. If the hematocrit is 45%, 0.55 of the blood is plasma, and the renal blood flow RBF is

$$RBF = \frac{C_{PAH}}{0.55}$$

If both the RPF and the GFR are known, the filtered fraction (FF) can be computed:

$$FF = \frac{GFR}{RPF}$$

Clearance measurements tell what the kidney does, but they do not tell how the kidney does it. They do not provide information about the events happening in different regions of the nephron. For example, reabsorption of a filtered substance from one part of the nephron might be accompanied by secretion into another. If reabsorption and secretion happened to balance exactly, the clearance of the substance would equal its filtration rate. It would be a mistake to conclude from this that the substance was filtered but neither secreted nor reabsorbed.

---

1 What are the units of renal clearance?
2 What is the physiological basis for using PAH for measurement of RBF? For using inulin or creatinine for measurement of GFR?
3 What are the effects of tubular secretion and tubular reabsorption on the clearance of a substance? How do the clearances of these substances compare with the GFR as measured by inulin clearance?

How Na$^+$ and K$^+$ are Handled by the Kidney

About 90% of the filtered load of Na$^+$ is reabsorbed in the proximal tubule and thick ascending limb of the loop of Henle. This **obligatory Na$^+$ reabsorption** is not under hormonal control and occurs at about the same rate no matter what the state of Na$^+$ balance in the body. Reabsorption of Na$^+$ in the distal tubule and collecting duct is controlled by the adrenal cortical hormone, aldosterone. In the absence of aldosterone, the distal tubule reabsorbs about 8% of the filtered load, leaving 2% to be excreted. This level of Na$^+$ excretion corresponds to a daily loss of an amount of Na$^+$ equal to the total Na$^+$ content of the body. This condition is typical of Addison's disease, in which inadequate amounts of adrenal cortical hormones are secreted. In the presence of typical plasma levels of aldosterone, Na$^+$ excretion is much less than 1% of the filtered load.

Aldosterone, like other steroid hormones (see

Mechanisms of Action of Diuretic Drugs

Drugs that increase the rate of urine production over the normal rate of 2 to 3 ml/min are called diuretics. Diuretics are used to reduce extracellular fluid volume—an important therapy for hypertension and congestive heart failure. Because of the large market for diuretics to control hypertension, pharmaceutical firms have sponsored much research into their mechanisms. Diuresis can be the result of interference with several different specific renal mechanisms.

A simple factor that can cause diuresis is the filtration of large amounts of a substance that cannot be reabsorbed into the glomerular fluid (Figure 19-C, *1*). One such substance is mannitol. When the proximal tubule reabsorbs Na^+, the tubular osmolarity decreases. Water begins to follow, but this increases the concentration of mannitol. Because mannitol cannot be reabsorbed, it limits water reabsorption. The failure of water to follow Na^+ results in a lower tubular Na^+ concentration and a larger concentration gradient for Na^+. As a result of the increased gradient, active Na^+ transport is decreased. The decreased Na^+ reabsorption further inhibits water reabsorption.

Substances that the kidney handles as it handles mannitol are called **osmotic diuretics.** Not all osmotic diuretics are drugs; some metabolites can have the same effect. One reason the proximal tubule must be permeable to urea is because otherwise urea would act as an osmotic diuretic. Osmotic diuresis occurs in diseases such as diabetes mellitus. Here the amount of filtered glucose often exceeds its transport maximum, and the excess glucose acts as an osmotic diuretic.

Some drugs inhibit Na^+ transport across the tubule, with consequences similar to those of osmotic diuretics. This is the mechanism of action of drugs containing mercury **(mercurials),** which were used before more modern diuretics were developed. **Amiloride** blocks Na^+ reabsorption and acidification by inhibiting gradient-mediated exchange of Na^+ for H^+ (Figure 19-C, *2*). The unreabsorbed Na^+ acts as an osmotic diuretic. Inhibitors of carbonic anhydrase such as **acetazolamide (Diamox)** selectively block bicarbonate reabsorption in the proximal tubule. This failure to reabsorb bicarbonate may have the side effect of increasing the plasma H^+ concentration.

Loop diuretics inhibit active transport of Na^+ or Cl^- in the thick ascending limb of the loop of Henle (Figure 19-C, *3*). **Furosemide** and **bumetanide** are examples of loop diuretics. Loop diuretics also increase the fraction of the filtered Na^+ load reaching the distal tubule, resulting in the loss of both K^+ and H^+ into the urine. Because loop diuretics tend to cause depletion of K^+ as well as of Na^+, patients who take them frequently are given K^+ supplements. Loop diuretics disrupt the kidney's concentrating mechanism.

Antagonists of aldosterone, such as spironolactone (Figure 19-C, 4), allow Na$^+$ that would otherwise have been reabsorbed to pass into the collecting duct and oppose the two-solute mechanism, which works more effectively if urea rather than Na$^+$ is the major solute in the urine.

Inhibitors of antidiuretic hormone secretion include alcohol, caffeine, and water (Figure 19-C, 5). However, because antidiuretic hormone acts on the collecting region of the distal tubule and collecting duct, the resulting diuresis does not usually affect plasma K$^+$ or H$^+$ levels.

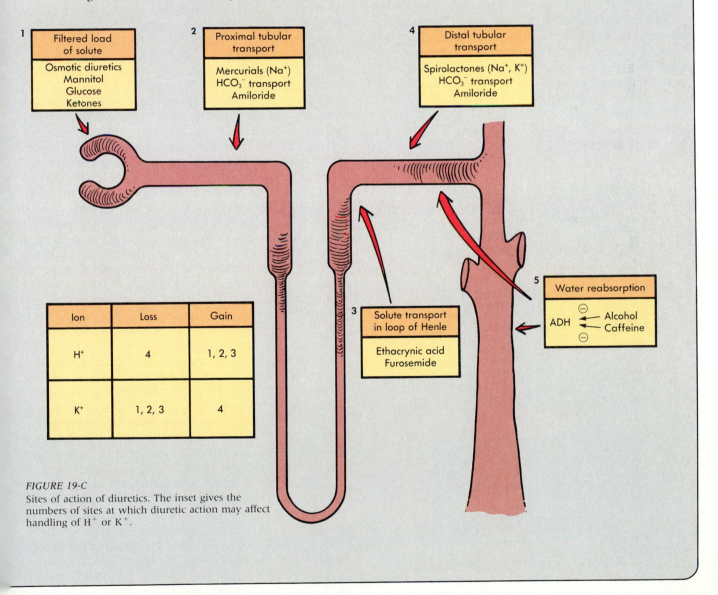

1

Filtered load of solute

Osmotic diuretics
Mannitol
Glucose
Ketones

2

Proximal tubular transport

Mercurials (Na$^+$)
HCO$_3^-$ transport
Amiloride

4

Distal tubular transport

Spirolactones (Na$^+$, K$^+$)
HCO$_3^-$ transport
Amiloride

5

Water reabsorption

ADH ⊖ ← Alcohol
⊖ ← Caffeine

3

Solute transport in loop of Henle

Ethacrynic acid
Furosemide

Ion	Loss	Gain
H$^+$	4	1, 2, 3
K$^+$	1, 2, 3	4

FIGURE 19-C
Sites of action of diuretics. The inset gives the numbers of sites at which diuretic action may affect handling of H$^+$ or K$^+$.

Chapter 5), acts by inducing the transcription of messenger RNAs that code for specific proteins involved in Na$^+$ transport (Figure 19-23). These include Na$^+$ channels that allow Na$^+$ to enter the transporting cells from the urine and Na$^+$-K$^+$– pump proteins (see Chapter 6). Because new proteins must be synthesized and transported to the cell membrane, the response of distal tubular cells to changes in levels of aldosterone requires several hours to be completed.

Potassium handling by the nephron provides an interesting example of interacting filtration, reabsorption, and secretion. Potassium ions are freely filtered. Between 85% and 90% of the filtered K$^+$ is obligatorily reabsorbed by the proximal tubule and the loop of Henle. Depending on the state of K$^+$ balance in the body, the rate of K$^+$ excretion in final urine may be as little as 1% of the filtration rate or as great as 200% of the filtration rate. The distal tubule is able to either reabsorb or secrete K$^+$, depending on the need. This ability is believed to result from the presence of two cell types in the final part of the distal tubule, one of which is specialized for K$^+$ secretion, the other for K$^+$ reabsorption. It is believed that K$^+$ reabsorption goes on at a relatively constant rate, whereas K$^+$ secretion is altered to meet the needs of K$^+$ homeostasis. Secretion of K$^+$

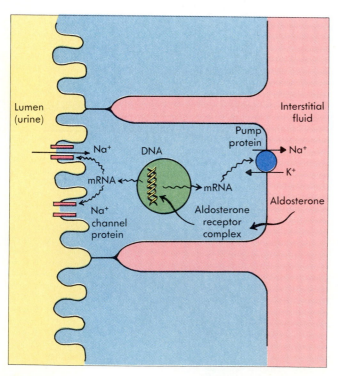

FIGURE 19-23
Mechanism of action of aldosterone. The hormone molecules enter the cytoplasm and bind to receptors; the hormone-receptor complexes enter the nucleus, where they affect the transcription of specific genes that code for proteins involved in Na$^+$ transport, including apical Na$^+$ channel proteins and Na$^+$-K$^+$ pump proteins.

is stimulated by aldosterone, a hormone also involved in Na$^+$ homeostasis. The feedback loops by which aldosterone regulates both Na$^+$ and K$^+$ balance are described in Chapter 20.

Renal Contributions to Acid-Base Regulation

The kidney has a major role in regulating the content of "fixed" acids and bases in the body. Fixed acids are those which cannot be eliminated as gases; recall that the respiratory system has the major responsibility for regulating the level of the volatile acid, CO$_2$. For people who consume a typical diet high in animal protein, catabolism of amino acids and other substances containing phosphorus and sulfur results in production of fixed acids (HCl, H$_3$PO$_4$, H$_2$SO$_4$) that must be eliminated by the kidney. Catabolism of protein results in the formation of HCO$_3^-$ and NH$_4^+$ (ammonium ion; see Chapter 23, p. 612). Some of the HCO$_3^-$ and NH$_4^+$ is converted to urea (CO[NH$_3$]$_2$). Retention of the remaining HCO$_3^-$ would help counter the effect of fixed-acid production, but the remaining NH$_4^+$ must be eliminated from the body.

Acidification of the tubular urine simultaneously accomplishes the goals of eliminating fixed acids, recovering filtered HCO$_3^-$, and eliminating NH$_4^+$. The parts of the tubule where acidification takes place are the proximal tubule, the cortical collecting tubule, and the collecting duct. Acidification of urine in the proximal tubule results in the conversion of filtered HCO$_3^-$ to CO$_2$. The carbonic anhydrase enzyme located on the apical surface of proximal tubular cells is essential for this process. Because CO$_2$ is a nonpolar gas that freely diffuses through tissues, almost all of the resulting CO$_2$ returns to the blood, where it combines with water to form carbonic acid, and then dissociates into HCO$_3^-$ and H$^+$ according to the Law of Mass Action (see Chapter 3, p. 50).

Under normal conditions, almost all filtered HCO$_3^-$ is recovered in this way. Bicarbonate recovery results in no net loss or gain of acid for the body because the H$^+$ and the HCO$_3^-$ involved balance out. After the HCO$_3^-$ has been recovered, the bulk of additional H$^+$ that is secreted must be accepted by other buffers in the urine. For example, the dibasic form of phosphate (HPO$_4^=$) can accept H$^+$, becoming the monobasic form (H$_2$PO$_4^-$) (Figure 19-24).

The gradient of pH across the wall of the tubule results in **pH trapping** of NH$_4^+$ in the urine in the tubule (see Figure 19-24). The process of NH$_4^+$ excretion actually starts in the liver, where the amine (NH$_3$) groups of amino acids are stripped off before conversion of the carbon skeletons to glucose or pyruvate. In a process called **transamination**, the NH$_3$ groups are transferred to the amino acid, glutamate, which becomes glutamine in the process. Glutamine passes into the blood and is taken up by tubular cells, where it is **deaminated**. For each

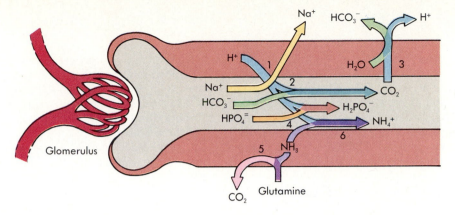

FIGURE 19-24
Interrelationships between H^+ secretion, HCO_3^- reabsorption, and ammonia trapping. H^+ is secreted, largely in exchange for Na^+ *(1)*. The secreted H^+ converts the filtered HCO_3^- to CO_2 *(2)*, which is passively reabsorbed *(3)*, returning both the filtered HCO_3^- and the secreted H^+ to the body. Additional secreted H^+ is accepted by other urinary buffers, mainly phosphate *(4)*. Glutamine is deaminated in the renal tubular cells *(5)*, releasing NH_3, which diffuses into the tubule where it is converted to NH_4^+ *(6)*. The charged NH_4^+ is trapped in the urine and is excreted.

molecule of glutamine, two NH_3 molecules are released within the cells. The carbon skeleton of glutamine is metabolized by the renal cells into CO_2.

Like CO_2, NH_3 is a nonpolar, highly diffusible gas, but it readily reacts with free H^+ to form NH_4^+. The relative concentrations of H^+ and NH_4^+ are determined by the Law of Mass Action. On the apical side of the tubular cells, the concentration of H^+ may be as much as two orders of magnitude greater than that in the interstitial fluid. The equilibrium concentration of NH_4^+ in the urine is high; NH_3 readily diffuses into the urine as NH_3, but once

there it is almost all converted to polar NH_4^+, which is trapped in the urine because it cannot cross the tubular wall.

1 Describe the differences between the handling of Na^+ and K^+ by the kidney. What is meant by obligatory reabsorption?
2 What is the effect of aldosterone on the clearances of Na^+ and K^+?
3 What is accomplished by tubular acidification of urine? Describe the mechanism of pH trapping of NH_3 in urine.

SUMMARY

1. The excretory system consists of the **kidneys,** the **ureters,** which connect the kidneys to the urinary bladder, and the **urethra,** which connects the bladder to the outside of the body. **Urination** is the result of reflexive contraction of the bladder smooth muscle accompanied by opening of the external sphincter (skeletal muscle). The reflex is modulated by pathways from the brain to give voluntary control.

2. **Nephrons,** or renal tubules, are the functional units of the kidney. Cortical nephrons carry out the same functions as juxtamedullary nephrons, except that they do not assist in the formation of the medullary osmotic gradient.

3. About 20% of the renal plasma flow is converted to primary urine by **ultrafiltration** across glomerular capillaries. The primary urine is similar to plasma, except that it contains little of the larger plasma proteins.

4. The functions of the proximal tubule are as follows:
   a. Reabsorption of most of the filtered Na^+ and water.
   b. Passive reabsorption of other small solutes in proportion to their abundance in the filtrate.
   c. Active reabsorption of some essential solutes such as glucose (typically by a Na^+ gradient-coupled process).
   d. Active secretion of some organic metabolic end products.
   e. Active secretion of H^+, which rids the body of its daily production of fixed acid and recovers filtered bicarbonate by conversion to CO_2, which diffuses back into the peritubular capillaries.

5. The **loop of Henle** concentrates descending urine using salt removed from the ascending limb. The result is an osmotic gradient between the cortex and medulla of the kidney. The interstitial fluid in the medulla is further concentrated by entry of urea from the concentrated urine in the collecting ducts. The urine leaving the ascending limb of the loop of Henle is diluted as a result of the reabsorption of salt from it without osmotic movement of water.

6. The distal tubule recovers filtered Na^+ by active transport, diluting the urine and causing other solutes, such as urea, to make up a larger percentage of the total solute content. The rate of Na^+ reabsorption (and K^+ secretion) is increased by the hormone, **aldosterone.**

7. In the presence of ADH, the walls of the collecting duct are permeable to water and the urine in the collecting duct is concentrated by osmotic water movement into the medullary interstitial spaces. Some urea is reabsorbed in this process, contributing to osmotic movement of water and to the solute content of the medullary interstitial fluid. Decreasing the ADH concentration increases the amount of water that is not reabsorbed, making the final urine more dilute.

8. The loop shape of the **vasa recta** mirrors that of the loops of Henle. Its function is to preserve the medullary osmotic gradient by **counter current exchange.** Plasma descending the vasa recta equilibrates with the concentrated medullary interstitial fluid. As the plasma ascends the loop, it equilibrates with the less and less concentrated interstitial fluid and leaves the kidney at the same concentration as normal plasma. In this way, the fluid and solute reabsorbed in the medullary portions of the nephrons are returned to the general circulation and the osmotic gradient is not disrupted.

9. The excretion rate of any component of plasma is the sum of filtration and secretion minus reabsorption. The net of these processes can be expressed as the solute's **clearance,** defined as the number of liters of plasma cleared of the substance per a unit of time. The clearance of substances filtered but not secreted or reabsorbed (for example, inulin and creatinine) is equal to the **glomerular filtration rate** (GFR). The clearance of solutes that are filtered and reabsorbed (for example, Na^+, K^+, Cl^-, glucose, and HCO_3^-) is much less than the GFR. The clearance of secreted substances (for example, PAH and H^+) is greater than the GFR.

1. Juxtamedullary and cortical nephrons differ anatomically and in their physiological roles. Describe the anatomical difference. Explain the functional difference.

2. Describe the blood supply to the kidney. What is the significance of the presence of two capillary beds in series?

3. In response to mild sympathetic activation, the total RBF decreases, but the GFR is almost unchanged. What protects the GFR from changes in perfusion pressure?

4. In what region of the nephron is the tubular fluid always hyperosmotic to plasma? In what region is it always isosmotic?

5. What is Na^+ gradient-mediated cotransport? Give some examples of substances reabsorbed by this process.

6. Account for the fact that inhibition of ADH secretion by alcohol ingestion results in the production of a large volume of dilute urine.

7. In a diabetic the plasma glucose concentration is 10 mg/ml; the inulin clearance is 125 ml/min; and the urine flow rate is 10 ml/min. What is the filtered load of glucose? Assuming a Tm for glucose of 400 mg/min, what is the rate of glucose reabsorption? What is the rate of glucose excretion? What is the concentration of glucose in the urine?

8. Describe the classical and two-solute hypotheses for generation of the medullary concentration gradient. What evidence argues against the classical hypothesis? What is the role of urea in the two-solute hypothesis?

9. What factors can disrupt the ability to produce concentrated urine?

10. What reabsorptive process is driven by tubular acidification? Why does ammonium excretion increase when H^+ secretion is stimulated?

11. A low-protein diet impairs urine concentrating ability. Explain. (Hint—the answer involves an end product of protein metabolism.)

● SUGGESTED READING

BAYLIS, C., and R.C. BLANTZ: Glomerular Hemodynamics, *News in Physiological Sciences*, volume 1, 1986, p. 86. Discusses the factors determining glomerular filtration rate and how these factors are hormonally regulated.

BEEUWKES, R.: Renal Countercurrent Mechanisms: How to Get Something for (Almost) Nothing. In TAYLOR, C.R., K. JOHANSEN, and L. BOLIS, editors: *A Companion to Animal Physiology*, Cambridge University Press, Cambridge, Mass., 1982. Describes the countercurrent multiplier and countercurrent exchange, including the role of urea.

SCHAFER, J.A.: Fluid Absorption in the Kidney Proximal Tubule. *News in Physiological Sciences*, volume 2, 1987, p. 22. Discusses the standing-gradient osmotic-flow hypothesis for proximal tubule salt and water absorption.

SCHULTZ, S.G., and R.L. HUDSON: How Do Sodium-Absorbing Cells Do Their Job and Survive? *News in Physiological Sciences*, volume 1, 1986, p. 185. Also discusses the standing-gradient osmotic-flow hypothesis for proximal tubule salt and water absorption.

SKOTT, O., and J.P. BRIGGS: Direct Demonstration of Macula-Densa Mediated Renin Secretion, *Science*, September 1987, p. 1618. Discusses the renin-angiotensin-aldosterone system for blood pressure regulation.

SILBERNAGL, S.: Tubular Reabsorption of Amino Acids in the Kidney, *News In Physiological Sciences*, volume 1, 1986, p. 167. Describes sodium-dependent cotransport processes in the proximal and distal tubules.

CHAPTER 20

Control of Body Fluid, Electrolyte, and Acid-Base Balance

On completing this chapter you will be able to:

- Describe how total body water is distributed between the intracellular, interstitial, and vascular compartments.

- Understand why cell volume is determined by cytoplasmic protein and extracellular fluid by total body Na^+.

- List the routes for daily water turnover.

- Understand the effects of atrial natriuretic hormone on glomerular filtration rate and Na^+ reabsorption.

- Describe antidiuretic hormone secretion response to plasma osmolarity and atrial stretch.

- Describe the factors affecting aldosterone secretion by the adrenal cortex, and explain how aldosterone increases the recovery of filtered Na^+ in the distal tubule.

- Understand the mechanisms for the reabsorption of filtered K^+ and how plasma K^+ directly stimulates K^+ secretion.

- Describe how plasma Ca^{++} and phosphate are regulated by calcitonin, parathyroid hormone, and 1,25 DOHCC.

- Appreciate how the extracellular fluid pH is buffered by the bicarbonate buffer system, and how the state of this buffer system is controlled by the combined action of lungs and kidneys.

- Describe how H^+ secretion first recovers filtered HCO_3^-, and then results in the formation of "new" HCO_3^- from CO_2, increasing plasma $[HCO_3^-]$ and pH.

*T*he control systems that regulate extracellular fluid provide a constant chemical environment for the cells. It has been said that we are what we eat; in the case of body fluid composition we are also what we drink, and what we do and do not excrete. If the body fluids are to remain in a steady state, entry of salts and water from the diet must exactly match evaporative losses of fluid from the body surface; obligatory losses of salts and water with feces, tears, and other secretions; and excretion of salts and water by the kidneys. The sheer number of routes of gain and loss make regulation of body fluids a web of interacting processes.

Where the regulatory problem is complex, complex regulatory systems evolve. For example, three known interlocking endocrine systems regulate the volume and Na^+ content of extracellular fluid. One of these, the atrial natriuretic peptide system, was only recently discovered, and its full complexity remains to be worked out. Other endocrine systems involved in Na^+ and volume regulation may yet be discovered. Similarly complex problems of matching intakes with losses are faced by the K^+, Ca^{++}, phosphate, and acid-base regulatory systems.

The effects of failure of body fluid homeostasis vary depending on the particular regulatory system and substance involved. Changes in the concentrations of K^+, H^+, or Ca^{++} usually have immediate effects on the excitability of muscles and nerves. In contrast, increases in the total body Na^+ content and extracellular fluid volume may progress for some time without outward symptoms, until the resulting hypertension has taken its toll on the heart and blood vessels.

THE FLUID COMPARTMENTS OF THE BODY
The Intracellular and Extracellular Compartments

In order to understand the topics treated in this chapter, it is necessary to visualize the fluid compartments of the body (see Chapter 1, p. 11 and Chapter 14, p. 368). Total body water averages 60% of an individual's weight in males and 50% in females. The sex difference is due to a greater average contribution of fat to body weight in women than men—fat cells contain less water than most other cell types. Body water exists in two main compartments: (1) the **intracellular compartment,** consisting of fluid inside the cells; and (2) the **extracellular compartment,** containing the fluid outside of the cells. The intracellular compartment contains two thirds of the total body water and includes water in red blood cells; the extracellular compartment contains the remaining one third.

Extracellular fluid is located in many separate spaces of the body. The two main subcompartments are interstitial fluid and blood plasma; other subcompartments are cerebrospinal fluid, the aqueous humor of the eyes, the fluid within the lumen of the gastrointestinal tract, urine in the excretory tract, etc. Neglecting these smaller subcompartments, interstitial fluid makes up approximately 75% of the extracellular fluid; the remaining 25% is blood plasma. Figure 20-1 shows a breakdown of the major body fluid compartments for a 75 kg man with total body water of 45 liters. About 30 liters of total body water is intracellular, 11.25 liters is interstitial, and 3.75 liters is plasma. As a rule of thumb, the intracellular compartment is about twice the volume of the extracellular compartment and two thirds of the total body fluid volume.

The dominant cation of extracellular fluid is Na$^+$ (145 mEq/L), and the most abundant extracellular anions are Cl$^-$ (104 mEq/L) and HCO$_3^-$ (24 mEq/L) (Figure 20-2). The total osmolarity (osmotic pressure—abbreviated P$_{osm}$) of the extracellular fluid (both plasma and interstitial fluid) is normally 300 mOsm (see Figure 20-1). The important difference between the plasma and interstitial fluid is the presence of a higher concentration of proteins in the plasma. The plasma proteins are the only plasma solutes that cannot cross the capillary wall; their restriction to the plasma is responsible for the osmotic gradient across capillary walls which largely counteracts the hydrostatic pressure of capillary blood. The opposing effects of plasma hydrostatic pressure and plasma protein osmotic pressure determine the distribution of fluid between the intracellular and extracellular compartments (Chapter 14).

The major intracellular cation is K$^+$. Chloride is not a major intracellular anion; rather, electroneutrality is maintained by negative charges on protein molecules (see Chapter 6, p. 124). When the two compartments are at equilibrium, the total osmolarity of the intracellular compartment is equal to that of the extracellular fluid. The cell membrane is permeable to K$^+$, but not to protein. Since the steady-state distribution of water between the intracellular and extracellular compartments is determined by the distribution of impermeant solutes, the cytoplasmic proteins largely determine the volume of the intracellular compartment. The system is at equilibrium when the osmolarity of the intracellular proteins balances that of extracellular Na$^+$ (Figure 20-3).

Fluid and Salt Balance

A healthy person maintains a precise balance between water intake and water loss. A person may drink 800 to 1500 ml of water daily and take in an

---

FIGURE 20-1

The fluid compartments of the body. Total body water is divided into intracellular (two-thirds) and extracellular (one-third) compartments. Three-fourths of the extracellular compartment is interstitial fluid and one-fourth is plasma. The osmolarity (P$_{osm}$) of the intracellular and extracellular compartments is always equal when the two systems are at equilibrium. The volume and total solute values given are for a 75 kg man. In this and similar figures, the area of each box indicates its volume.

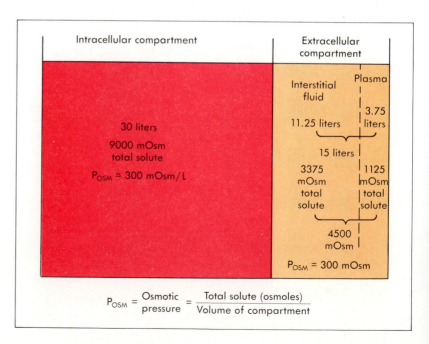

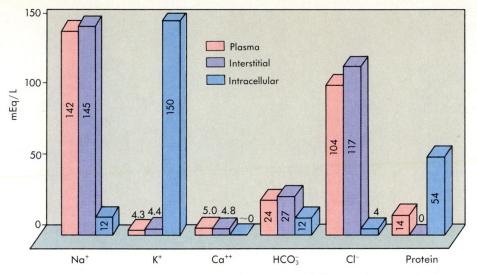

FIGURE 20-2
Electrolyte and protein concentrations of the body fluid compartments.

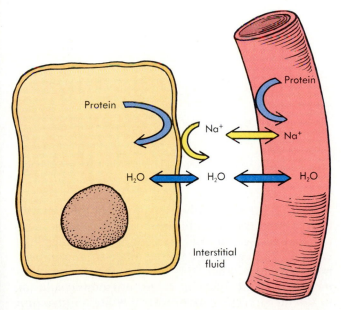

FIGURE 20-3
The permeability properties of the plasma—interstitial fluid barrier and the interstitial fluid—cytoplasm barrier determine which solutes may equilibrate throughout the system and which must remain confined in one or more compartments. Plasma proteins and cytoplasmic proteins do not freely enter the interstitial fluid, so its protein content is low. Na$^+$ can permeate the capillary wall but is effectively excluded from the cytoplasmic water. Water can freely equilibrate across all three barriers, so at equilibrium the osmolarity of fluids must be equal throughout the system.

additional 500 to 700 ml of water in food (Figure 20-4). Oxidative energy metabolism produces 200 to 300 ml of water per day. The total daily water gain is therefore 1.5 to 2.5 liters. This water gain is balanced by **bulk fluid losses** of 800 to 1500 ml as urine and 100 to 150 ml in feces. In addition, 600 to 900 ml of water is lost by evaporation from the lungs and the skin surface; water loss by these routes is called **insensible water loss.** Sweating is responsible for additional evaporative loss.

Depletion of body water **(dehydration)** occurs when losses consistently exceed gains. Some common causes of dehydration are excessive sweating and insensible loss in a hot environment, severe vomiting or diarrhea, and kidney diseases that result in production of large quantities of dilute urine. Excessive ingestion of alcohol causes dehydration because alcohol inhibits secretion of antidiuretic hormone, causing production of a large volume of dilute urine.

Salt gains must balance losses to maintain body fluid homeostasis. In addition to the direct consumption of table salt, many foods contain "hidden" salt in the form of preservatives, such as sodium bisulfide, or additives. Some beverages we drink contain a significant amount of salt. Salt is lost in the urine and in various body fluids, such as sweat, saliva, tears, and secretions of the gastrointestinal tract. Abnormal losses of body fluids occur in vom-

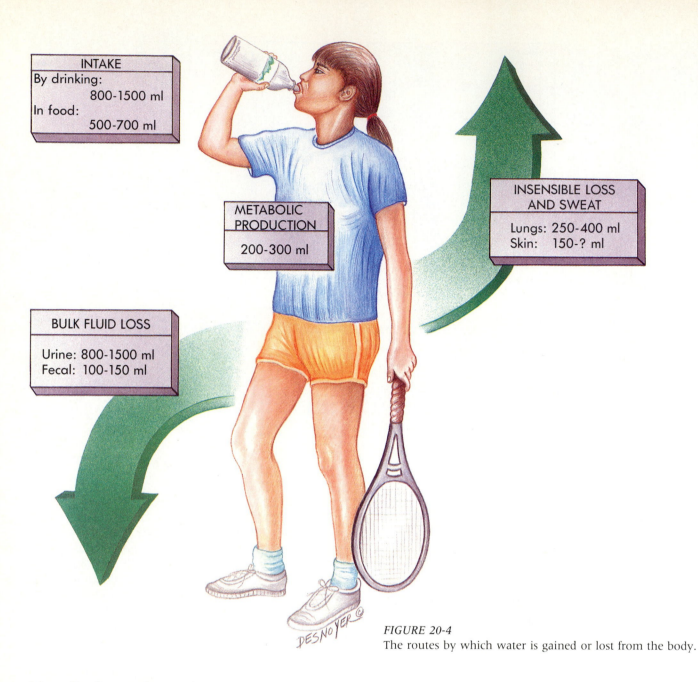

INTAKE
By drinking:
        800-1500 ml
In food:
        500-700 ml

METABOLIC PRODUCTION
200-300 ml

INSENSIBLE LOSS AND SWEAT
Lungs: 250-400 ml
Skin: 150-? ml

BULK FLUID LOSS
Urine: 800-1500 ml
Fecal: 100-150 ml

FIGURE 20-4
The routes by which water is gained or lost from the body.

iting, diarrhea, and excessive sweating and can result in large salt losses.

Challenges to Salt and Water Homeostasis

The problems of extracellular fluid homeostasis can be reduced to three simple homeostatic challenges: (1) gain or loss of isotonic NaCl solution, (2) gain or loss of pure water, and (3) gain or loss of pure NaCl. To illustrate the changes in the body fluid compartment for an individual with 45 liters total body water (30 liters intracellular and 15 liters extracellular), the gains and losses discussed have been exaggerated.

Figure 20-5 shows the effects of gain or loss of 4.5 liters of isotonic saline on the fluid volumes of the 75 kg man of Figure 20-1. Since cells are effectively impermeable to Na^+, additions or losses of isotonic NaCl affect only the volume of the extracellular compartment —the intracellular volume does not change. Seventy-five percent of the extracellular volume change occurs in the interstitial compartment and the other 25% in the plasma compartment. The total solute in 1 liter of isotonic solution is 300 mOsm. The gain or loss of N liters of isotonic saline increases or decreases the total extracellular solute by 300 × N mOsm. A condition of abnormally increased plasma volume is called **hypervolemia** (hyper = high; vol = volume; emia = blood); the corresponding term for abnormally low blood volume is **hypovolemia.**

In reality, gain or loss of isotonic saline could occur in a variety of ways—it is not necessary to consume isotonic solution. Consumption of salt and water in the proportions that they occur in extracel-

KIDNEY AND BODY FLUID HOMEOSTASIS

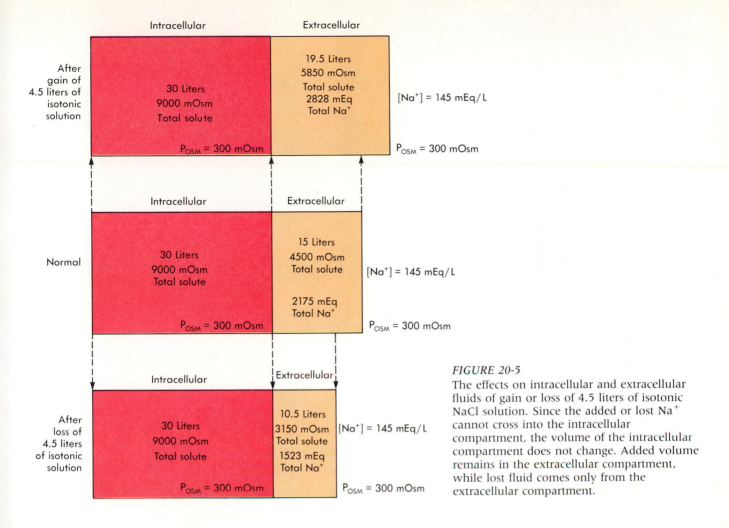

FIGURE 20-5
The effects on intracellular and extracellular fluids of gain or loss of 4.5 liters of isotonic NaCl solution. Since the added or lost Na^+ cannot cross into the intracellular compartment, the volume of the intracellular compartment does not change. Added volume remains in the extracellular compartment, while lost fluid comes only from the extracellular compartment.

lular fluid will have the same effect. For example, a teaspoon of salt (about 5 grams) in a liter of water produces a solution approximately isotonic to plasma. Some salad dressings have a sodium content as high as 150 mg per serving, so the consumption of a 16 ounce bottle involves the gain of 2 to 3 grams of salt. Production of isotonic urine would be a simple example of isotonic fluid loss, but the same effect could result from loss of some pure water by evaporation and simultaneous production of hypertonic urine.

The second simple homeostatic challenge is gain or loss of pure water. Typically, pure water gain is the result of drinking water or other dilute fluids. Pure water loss is caused by production of dilute urine or insensible evaporative loss. For example, dilute urine can be thought of as consisting of some volume of isotonic urine to which an additional volume of pure water has been added. The volume of isotonic solution represents a loss of isotonic solution; the volume of pure water represents a net loss of pure water.

If there is a net loss of pure water from the extracellular compartment, the extracellular osmolarity increases. The resulting osmotic gradient between the intracellular and extracellular fluid causes water to move from cells to extracellular fluid until osmotic equilibrium is restored. The reverse occurs when there is a net gain of water. The volume of water gained or lost from each compartment is proportional to the original volume of that compartment, so two thirds of the change will occur in the intracellular compartment and one third in the extracellular compartment (Figure 20-6). For example, starting with a normal osmolarity (P_{osm}) of 300 mOsm, each liter of water gained or lost will change the osmolarity of both the intracellular and the extracellular fluids by 7 to 8 mOsm. A fundamental difference between the challenge of gain or loss of pure water and the challenge of gain or loss of isotonic solution is that the water challenge affects both the intracellular and the extracellular volumes, while the isotonic challenge affects only the extracellular volume.

The third simple challenge is gain or loss of pure NaCl. For example, if a 75 kg man eats 50 grams of salt (approximately 1.7 Osm) without drinking any water, his extracellular fluid osmolarity (P_{osm}) will initially increase to about 413 mOsm. In this case the elevated osmolarity is the result of

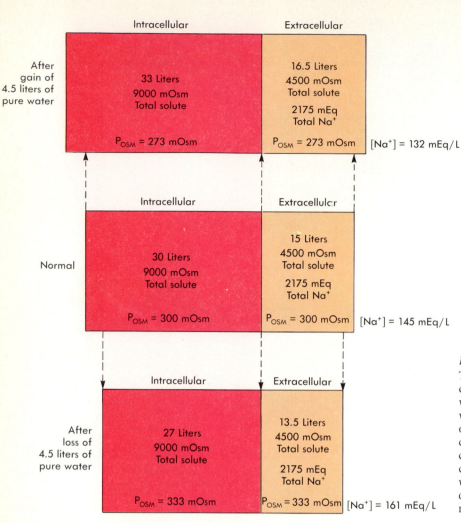

FIGURE 20-6
The effects on the volume and solute concentration of gain or loss of 4.5 liters of pure water. In the case of water gain, the additional water equilibrates between the two compartments on the basis of the total solute in each compartment, so the final volume of both compartments is increased and the final solute concentrations are decreased. In the case of pure water loss, the volumes of both compartments are decreased and the solute concentrations of both rise.

an increase in the Na$^+$ concentration of the extracellular compartment (**hypernatremia**). The increased osmolarity of the extracellular fluid will cause net movement of water from the intracellular compartment to the extracellular compartment until the osmolarities of the intracellular fluid and the extracellular fluid are equal (Figure 20-7). The final osmolarity of intracellular and extracellular fluid will be 300 mOsm + (1.7 Osm/45 liters) = 338 mOsm. The volume of the extracellular compartment will increase by about 3.4 liters to a total of 18.4 liters, while the intracellular compartment decreases by the same amount. Loss of the same amount of pure salt would reduce the extracellular fluid volume by 4.3 liters. This example illustrates the importance of the total body content of Na$^+$ in determining the volume of the extracellular compartment.

An increase in plasma volume increases the venous pressure and cardiac output. If the total peripheral resistance did not change, the mean arterial blood pressure would increase. The effect of salt ingestion on venous and arterial blood pressure is the reason that physicians advise individuals with high blood pressure or heart disease to limit their salt intake. Probably very few people would voluntarily

consume as much as 50 grams of salt at one time. However, in individuals predisposed to hypertension, a defect in Na$^+$ homeostasis could allow as much as 50 extra grams of Na$^+$ to accumulate in the body over time, especially if the regulatory systems were stressed by extra dietary salt. Similarly, loss of pure salt does not occur as such, but net loss occurs when salt consumption fails to replace salt losses. One form of loss of pure salt is production of urine more concentrated in Na$^+$ than plasma because concentrated urine can be thought of as consisting of a volume of isotonic solution to which enough pure salt has been added to bring the urine to its final hypertonic concentration.

1 *What impermeant solutes determine distribution of fluid between the intracellular and extracellular compartments?*
2 *What impermeant solute exercises a key role in determining the distribution of fluid between the vascular and interstitial subcompartments?*
3 *Why is regulation of total body Na$^+$ essential to long-term regulation of blood pressure?*

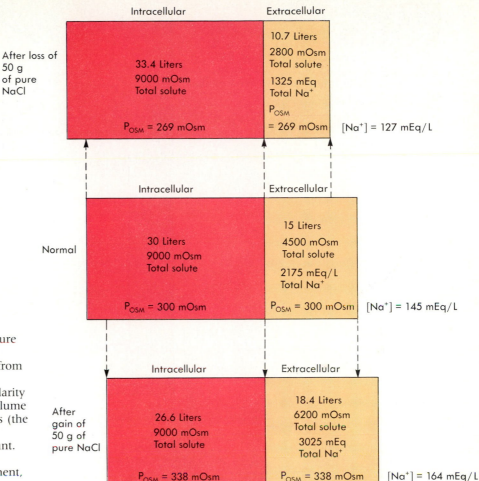

FIGURE 20-7

The effect of gain or loss of 50 grams of pure NaCl. The added salt is confined to the extracellular compartment. Water moves from the intracellular compartment to the extracellular compartment until the osmolarity of the two compartments is equal. The volume of the intracellular compartment decreases (the cells shrink) and that of the extracellular compartment increases by the same amount. When salt is lost, water moves from the extracellular to the intracellular compartment, and the cells swell.

REGULATION OF WATER AND NA$^+$ BALANCE

Neural and Endocrine Effects on Glomerular Filtration Rate

Since primary urine is isosmotic to plasma, increases or decreases in glomerular filtration rate (GFR) would amount to an increase or decrease in loss of isotonic fluid. Mild sympathetic activation, such as occurs when a person stands up or begins to exercise, constricts renal arterioles, reducing the rate of blood flow to the kidney (see Chapter 15, p. 386). However, the glomerulotubular balance described in Chapter 19 (p. 483) maintains GFR constant in the face of such changes in renal perfusion. Only when the sympathetic nervous system is intensely activated, as in severe hemorrhage, does the renal blood flow rate fall low enough to significantly decrease GFR.

The kidneys normally respond to an increase in total body Na$^+$ and extracellular fluid volume with natriuresis, an increase in the rate of urinary loss of Na$^+$ and water. In 1980, muscle cells of the heart atria were discovered to contain a factor that significantly increased natriuresis. This 28 amino acid factor can now be called **atrial natriuretic hor-**

mone (ANH). A significant positive correlation between right atrial pressure and ANH release has been demonstrated in human subjects. Increasing Na$^+$ intake in human volunteers also resulted in significant alterations in plasma ANH. One mechanism by which ANH increases Na$^+$ and water loss is believed to be by increasing the GFR (Figure 20-8), either by an effect on the renal arterioles, by changing the leakiness of the glomerular capillary wall, or both. Other effects of ANH will be described in subsequent sections.

Control of Renal Water Reabsorption by the Antidiuretic Hormone and Atrial Natriuretic Hormone Systems

The hypothalamus contains **osmoreceptor** neurons that are stimulated by increases in plasma osmolarity. The result of increased activity of the hypothalamic osmoreceptors is an increased release of antidiuretic hormone (ADH) from the posterior pituitary (Chapter 5, p. 99). ADH increases the water permeability of the collecting duct (see Chapter 19, p. 495). The greater the plasma levels of ADH, the more water can be recovered in response to the osmotic gradient provided by the medullary intersti-

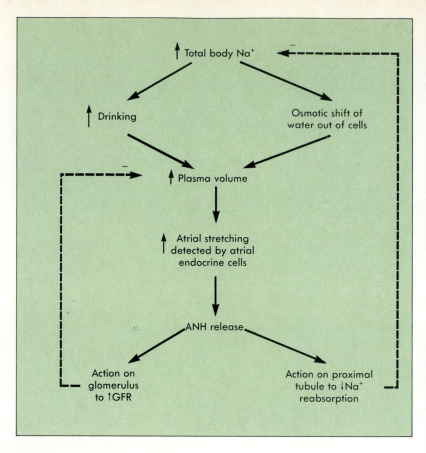

FIGURE 20-8
The sequence of events in the response of the ANH system to an increase in total body Na$^+$. The feedback loop is completed when the decreased reabsorption of Na$^+$ and volume triggered by the hormone restores extracellular fluid volume to normal.

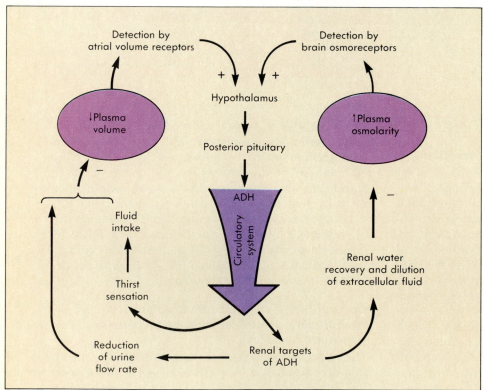

FIGURE 20-9
The response of the ADH system to a decrease in plasma volume or to an increase in plasma osmolarity is an increase in ADH secretion. The negative feedback pathways show the corrective effect of increased renal water reabsorption and increased water consumption on plasma volume and osmolarity.

tial fluid. In the presence of high levels of ADH, the kidney excretes less water and the urine is more concentrated.

For example, if extracellular fluid osmolarity were elevated (a pure salt challenge as in Figure 20-

7), water would move from the cellular compartment into the extracellular compartment. The loss of 250 ml of urine that had an osmolarity of 1200 mOsm would decrease total body water by 250 ml, but the excretion of the solute would cause plasma osmolarity to decrease toward normal, and 500 ml of water would shift back into the intracellular compartment. This water shift would make the intracellular volume and extracellular fluid osmolarity more nearly normal at the expense of the extracellular volume. The feedback loop that stimulates ADH secretion during increases in plasma osmolarity is shown in Figure 20-9.

Water loads that dilute the plasma decrease ADH secretion profoundly, even though plasma volume is also increased. When ADH levels fall, urine passing through the collecting duct gives up relatively little of its water, and a large volume of dilute urine is produced. This has the same effect as loss of a small amount of isotonic solution plus loss of a large volume of pure water. The net effect is now an intracellular-to-extracellular water shift.

Isotonic decreases in plasma volume are sensed by stretch receptors in the atria. When plasma volume is normal, the stretching of the atria during the cardiac cycle has an inhibitory effect on ADH secretion; when plasma volume is decreased, the inhibition is removed. After hemorrhage, for example, total body water and venous pressure are decreased but the plasma osmolarity is normal. Since the osmolarity remains normal, ADH secretion is controlled entirely by the atrial volume receptors, which promote water retention (see Figure 20-9). In dehydration, ADH secretion is stimulated both by the decrease in venous pressure and by the increase in osmolarity. The net secretion of ADH produced by the combined volume receptor reflex and osmoreceptor reflex is greater than that which would be caused by either factor acting alone.

In cases in which both the blood volume and the blood osmolarity deviate from normal, ADH secretion is usually dominated by input from the hypothalamic osmoreceptors. It is adaptive for osmolarity regulation to receive a higher priority than volume regulation, because the volumes of cells and the activity of nerves and muscles depend from minute to minute on the correct osmolarity, while changes in volume can be tolerated and may be compensated for by the cardiovascular reflexes.

The ANH system opposes the ADH system at almost every step (see Figure 20-8). Increased stretch of the atria stimulates ANH secretion, while it inhibits ADH secretion. ANH is a potent inhibitor of ADH release, and there is evidence that it may oppose the ADH effect in the collecting duct by activating an antagonistic second messenger system.

Control of Sodium Loss and Vascular Tone by the Renin-Angiotensin-Aldosterone and Atrial Natriuretic Hormone Systems

The endocrine cells of the juxtaglomerular apparatus are sensitive to three factors: (1) the filtered load of Na^+, (2) the blood pressure in afferent arterioles, and (3) the inputs to the juxtaglomerular cells from

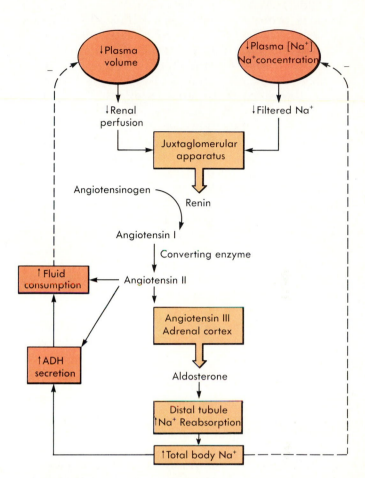

FIGURE 20-10

The response of the aldosterone system to a decrease in plasma volume or in plasma Na^+ concentration is an increase in renin secretion, which leads ultimately to an increase in aldosterone secretion. The increased recovery of filtered Na^+, along with continued Na^+ intake from diet, increases total body Na^+. The dashed lines show the corrective effect of aldosterone secretion. If the initial problem were a volume decrease, the osmoreceptors of the ADH system will ensure that increases in total body Na^+ result in restoration of volume.

sympathetic nerves. The first factor is an indicator of plasma Na^+ concentration; the second and third factors are indicators of plasma volume. If the Na^+ concentration or the blood pressure to the kidney drops or if sympathetic activity rises, the juxtaglomerular cells increase their secretion of **renin** (Figure 20-10).

Renin is a protease that catalyzes the first step of a regulatory cascade. Each renin cleaves a 10 amino acid fragment, **angiotensin I,** from each of a number of molecules of a blood protein **angiotensino-gen.** Angiotensin I is converted to the 8 amino acid peptide **angiotensin II** by a converting enzyme that resides primarily in the endothelial cells of blood vessels. Angiotensin II itself is a potent vasoconstrictor and also stimulates ADH. In the adrenal cortex it is converted to the 6 peptide **angiotensin III,** which stimulates secretion of the steroid hormone **aldosterone** by the adrenal cortex.

Aldosterone increases the reabsorption of Na^+ by the distal tubule. Because so much of the filtered load of Na^+ is reabsorbed by more proximal parts of the nephron, aldosterone can affect only the potential reabsorption of the last 3% of the filtered load. Nevertheless, this 3% (equivalent to about 30 to 50 grams of NaCl per day) constitutes an essential contribution to overall Na^+ balance of the body. Aldosterone also stimulates Na^+ reabsorption at several other sites, including the colon of the intestine and the reabsorptive portion of sweat and salivary glands. At all of these sites, aldosterone decreases the rate of Na^+ loss. If the rate of Na^+ intake remains unchanged, increasing plasma levels of aldosterone will increase the total body Na^+ content. Because the extracellular fluid volume is closely linked to total body Na^+, the renin-angiotensin-aldosterone system (RAA) exercises a strong influence on extracellular fluid volume and thus on blood pressure. Furthermore, angiotensins are potent vasoconstrictors, so that this effect may also contribute to regulation of arterial blood pressure in case of a decrease in plasma volume.

Under normal conditions, the effect of increases or decreases in plasma aldosterone is to change the extracellular fluid volume rather than extracellular fluid Na^+ concentration. The reason for this lies in the relative response times of the ADH and aldosterone systems. Since aldosterone takes effect only after several hours and controls such a small part of the filtered load of Na^+, it takes several days for aldosterone to affect the plasma Na^+ concentration. ADH acts on 20% of the filtered load of water, and its effect is rapid. Therefore water reabsorption keeps up with Na^+ reabsorption, and aldosterone-induced Na^+ retention is equivalent to gain of isotonic saline.

Disorders of the RAA system cause the mean arterial pressure to depart from its setpoint. For example, renin secretion is inappropriately high in some adults. The effects of the regulatory cascade cause a chronic **high-renin hypertension.** In several adrenal disorders there is hypersecretion of aldosterone, which results in a characteristic Na^+ retention and hypertension. In Addison's disease, a general failure of secretion of adrenal steroids, one symptom is negative Na^+ balance and hypotension.

The ANH system (see Figure 20-8) acts as a brake on the RAA system as well as the ADH system. ANH inhibits reabsorption of filtered Na^+ in the distal parts of the nephron, antagonizing the stimulatory effects of aldosterone. The feedback loop is completed when the decreased reabsorption of Na^+ and volume triggered by the hormone restores extracellular fluid volume to normal. Another aspect of ANH-RAA antagonism is the fact that angiotensins are vasoconstrictors, whereas ANH is a vasodilator. Although the natriuretic effect of ANH is fully confirmed in human volunteers, much of what is known about the mechanisms of action of this newly discovered hormone is based on animal experiments.

Thirst and Salt Appetite

The reserve capacity of the kidneys to clear water from the body is so great and the ADH response so rapid that excessive voluntary fluid intake cannot normally cause more than a transient decrease in plasma osmolarity. On the other hand, thirst is the primary defense against decreases in plasma volume. The kidneys can, at best, only slow the loss of water through one of the several routes by which it leaves the body. The ADH, RAA, and ANH systems all seem to be involved in control of thirst. The stimuli that elicit thirst (hypovolemia, increased plasma osmolarity) are the same as those that increase ADH secretion. This fact and the close association of ADH neurons to thirst centers of the hypothalamus suggest that ADH secretion and thirst share a common pathway. Angiotensin II is the most powerful thirst-causing substance known, so the RAA system affects ultimately not only renal performance but also the rate of fluid intake (Figure 20-10). ANH appears to antagonize the thirst response as well as secretion of ADH by the hypothalamus.

Before technology made table salt abundant and cheap, salt was a scarce and valuable commodity for most people. Human beings have a powerful drive to consume salt; when salt is readily available most people voluntarily consume much more of it than the minimum needed to maintain Na^+ balance. Therefore body salt content almost never falls so low that it is regulated by feedback in human beings, although such regulation is important in other animals.

1 *What process prevents urine production from slowing when sympathetic activation decreases renal blood flow? What effect does ANH have on glomerular filtration rate?*
2 *What two types of sensory input are integrated to regulate ADH secretion? Which of the two inputs dominates?*
3 *What effectors are affected by the regulatory cascade set in motion by renin?*

POTASSIUM HOMEOSTASIS
Location of K^+ in the Body

About 98% of total body K^+ is in the intracellular compartment. Each cell is largely responsible for its own K^+ homeostasis, and the cytoplasmic concentration of K^+ differs between cell types, although K^+ is the major cytoplasmic cation in all cells. Assuming that the plasma K^+ concentration remains within normal limits, the total body K^+ is regulated by the rate of K^+ pumping of cells. The regulatory system for plasma K^+ acts on less than 2% of the total body K^+. Why is its job so important?

The resting potential of cells depends mainly on the K^+ concentration gradient across their membranes. The resting membrane potential is largely determined by the ratio of the intracellular K^+ concentration (a large value) to the extracellular K^+ concentration (a small value). In this ratio, the small number exercises a kind of leverage. A change of 1 mEq/L in the intracellular concentration would have a trivial effect on cell membrane potentials. In contrast, because the extracellular K^+ concentration is a relatively small number, increasing it or decreasing it by even as little as 1 mEq/L would have a significant effect on the resting potentials of all cells in the body. Precise regulation of the relatively small amount of K^+ that is in the extracellular fluid is an important homeostatic function of the kidney.

Regulation of K^+ Excretion by Aldosterone

The kidneys regulate plasma K^+ by a combination of K^+ reabsorption and secretion. About 65% of the filtered load of K^+ is reabsorbed in the proximal tubule as a consequence of fluid reabsorption. An additional 20% to 30% is reabsorbed in the ascending loop of Henle as part of the Na^+, K^+, 2 Cl^- cotransport system. The distal tubule receives the remaining 10% of the filtered load of K^+, and it is in the distal tubule that all K^+ regulation occurs.

The rate of loss of filtered K^+ in the urine depends on the relative rates of opposing processes of secretion and absorption that go on in the distal tubule. Active reabsorption is carried out by cells that possess an active K^+ pump in their lumen-facing membranes. The K^+ that is actively transported into reabsorptive cells diffuses out of them into the cortical interstitial fluid. The cells that reabsorb K^+ have the ability to recover all of the K^+ that is delivered to the distal tubule. In **hypokalemia** (low plasma K^+), reabsorption predominates, and nearly all of the filtered load of K^+ is recovered.

Active K^+ secretion in the distal tubule is driven by the basolateral Na^+-K^+ pump (see Chapter 19, p. 488). In secretory cells there is a passive K^+ permeability pathway in the luminal membrane, so K^+ transported into the cells leaks out into the tubule lumen. The activity of the basolateral pump and the permeability of the passive pathway are

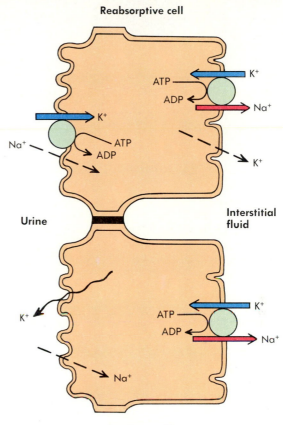

FIGURE 20-11
The distal tubular mechanisms for the renal handling of K^+. In the reabsorptive cell the apical K^+ pump drives K^+ into the cell and it leaks out into the extracellular fluid across the basolateral membrane. These cells also reabsorb Na^+. Na^+ leaks into the cell and is pumped out across the basolateral membrane in exchange for K^+. The amount of K^+ that passes through the basolateral membrane is the sum of that transported in across the apical membrane and that transported in by the Na^+-K^+ pump across the basolateral membrane. In the secretory cell, aldosterone induces a large leak of K^+ across the apical membrane and stimulates the Na^+-K^+ pump. The K^+ transported into the cell across the basolateral membrane leaks into the urine. At the same time, Na^- is leaking into the cell and being transported out by the Na^+-K^+ pump, so both cell types also reabsorb Na^+.

both increased by aldosterone (Figure 20-11). When the plasma K^+ concentration rises above normal **(hyperkalemia)**, the secretion of aldosterone by the adrenal cortex increases (Figure 20-12). This is a direct effect of K^+ on the adrenal cortex and does not involve the renin-angiotensin pathway. In hyperkalemia, the reabsorptive process is partly canceled by the aldosterone-stimulated secretion, and some K^+ is lost in the urine.

In patients with adrenal gland tumors that secrete aldosterone at abnormally high rates, plasma Na^+ levels rise above normal and plasma K^+ falls

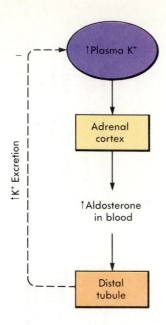

FIGURE 20-12

The direct response of the adrenal cortex to hyperkalemia. The effect of the increased aldosterone secretion is to increase K^+ secretion by the secretory cells of the distal tubule and thus to increase K^+ clearance by the kidney. The corrective effect is shown by the dashed line.

below normal. The opposite condition results when the aldosterone-secreting cells of the adrenal cortex are damaged, as in Addison's disease. In this case, hypovolemia and low total body Na^+ are accompanied by hyperkalemia. Increased plasma K^+ can cause fatal cardiac arrhythmias and disorders of central nervous function and is the main reason that loss of adrenal cortical function is life threatening.

Interaction Between Na^+ and K^+ Regulation and Acid-Base Balance

Every Na^+ reabsorbed from the urine must be matched either by absorption of an anion or by secretion of another cation. In the proximal tubule, Cl^- and other anions can follow the reabsorbed Na^+, but in the distal tubule the passive permeability of the tubule wall to anions is low. Reabsorption of Na^+ must be matched by secretion of K^+ and H^+. Consequently, H^+ and K^+ behave as if they competed with one another for secretion (Figure 20-13, A; note, however, that H^+ and K^+ actually are secreted by different pathways). Therefore the plasma K^+ concentration and H^+ concentration tend to vary in opposite directions.

A change in plasma H^+ concentration can cause a change in plasma K^+ concentration. Acidosis

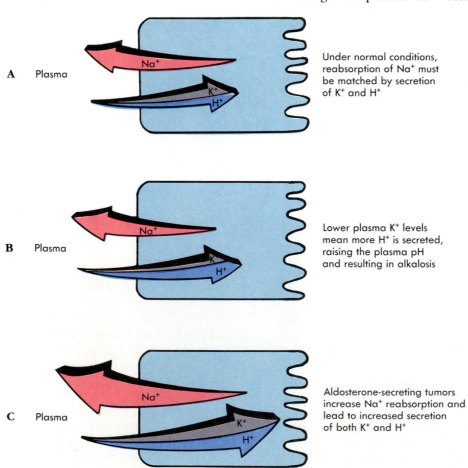

FIGURE 20-13

A summary of the interactions between the plasma concentrations of Na^+, K^+, and H^+.

KIDNEY AND BODY FLUID HOMEOSTASIS

leads to increased H^+ secretion. This is associated with a decrease in K^+ secretion, resulting in an increase in plasma K^+ concentration. Alkalosis, on the other hand, is often associated with a decrease in plasma K^+ concentration. When the plasma K^+ concentration decreases, secretion of H^+ increases and the plasma pH becomes less acidic, causing alkalosis (Figure 20-13, B). Conversely, acidosis can be caused by an increase in plasma K^+ concentration.

A reciprocal relationship between H^+ and K^+ is seen in several pathological situations. In Addison's disease less Na^+ is reabsorbed so less H^+ can be secreted, causing acidosis. Aldosterone-secreting adrenal tumors have the opposite effect, resulting in alkalosis as well as reduced plasma K^+ (Figure 20-13, C). Osmotic diuretics increase the amount of Na^+ reaching the distal tubule, and the increase in Na^+ reabsorption causes increased secretion of K^+ and H^+, which tends to decrease plasma K^+ concentration and produce alkalosis. Finally, when renal function is impaired by infections or decreased renal blood flow, both H^+ and K^+ secretion are reduced. This causes an increase in plasma K^+ concentration and acidosis.

> 1 Why is the regulation of K^+ concentration so important?
> 2 What factors affect the relative rates of H^+ and K^+ secretion by the distal tubule?
> 3 Why do secretion of both K^+ and H^+ diminish in Addison's disease?

CALCIUM AND PHOSPHATE HOMEOSTASIS
Transfer of Ca^{++} and Phosphate Between Bone, Intestine, and Kidney

Calcium's chemical activity in solution is significantly less than would be expected from its concentration, because calcium salts do not dissociate completely. The value that is physiologically significant is the concentration of "free" ionic Ca^{++} rather than the total plasma Ca^{++}. Plasma normally contains 2.5 mM Ca^{++}, of which about 46% is protein bound and about 6% is bound by phosphate ($HPO_4^=$) and other small anions. This leaves about half of the total Ca^{++} as free Ca^{++}. The free and bound Ca^{++} exist in chemical equilibrium with each other according to the Law of Mass Action. Increases in the concentration of any agent that binds Ca^{++}—plasma protein or phosphate, for example—will decrease the free plasma Ca^{++}. Also, small changes in plasma pH have significant effects on the ratio of bound to free Ca^{++}: acidosis increases the concentration of the ionized form, and alkalosis decreases it.

The plasma free Ca^{++} must be closely regulated because of its importance for membrane excitability. Elevated extracellular Ca^{++} (hypercalcemia) reduces

excitability, causing lethargy, fatigue, and memory loss. Depressed plasma Ca^{++} causes muscle cramps, convulsions, and other symptoms of increased neuromuscular excitability. Because plasma phosphate affects the ratio of free to bound Ca^{++} and because Ca^{++} is deposited in bone as a phosphate salt, Ca^{++} homeostasis and phosphate homeostasis must be closely linked.

Most of the Ca^{++} in the body is contained in bone. Although bone seems substantial and permanent, it undergoes continuous remodeling. Deposition of bone is carried out by bone cells called **osteoblasts;** bone breakdown is carried out by a second population of cells called **osteoclasts.** When deposition goes on at a greater rate than breakdown, the mass of bone increases and there is a net transfer of Ca^{++} and phosphate from the plasma to bone. When breakdown predominates, both Ca^{++} and phosphate are released into the plasma.

Bone serves as a reserve to and from which Ca^{++} and phosphate can be transferred to regulate plasma Ca^{++} and phosphate. The total amount of Ca^{++} in bone is so much greater than that in plasma that breakdown of some bone to maintain plasma Ca^{++} does not significantly weaken the skeleton. However, over time, Ca^{++} uptake from the diet must equal Ca^{++} loss in urine, or the strength of bone may be decreased. Active intestinal uptake of Ca^{++} from the diet ranges from 20% to 70% of the available Ca^{++}; the efficiency of the uptake is regulated in response to changes in plasma Ca^{++}. The kidneys are the major route of Ca^{++} and phosphate loss. Typically the renal clearances of both Ca^{++} and phosphate are less than the glomerular filtration rate because of tubular reabsorption. The rates of reabsorption of both Ca^{++} and phosphate are hormonally regulated, as well.

Endocrine Regulation of Total Body Calcium and Phosphate

The hormones involved in Ca^{++} homeostasis are: (1) **calcitonin**, a protein produced and released by the parafollicular cells of the thyroid gland; (2) **parathyroid hormone** or **parathormone (PTH),** a protein secreted by the parathyroid glands near the thyroid; and (3) **1,25 dihydroxycholecalciferol (1,25 DOHCC),** a steroid (see Chapter 3, p. 40).

Of the three hormones, calcitonin is perhaps the least significant in humans. Calcitonin stimulates bone deposition and decreases plasma Ca^{++} (Figure 20-14). Decreases in calcitonin from its basal level allow for Ca^{++} release from bone, but it is possible for plasma Ca^{++} to be regulated adequately in the complete absence of the thyroid gland.

Secretion of PTH increases when the free Ca^{++} concentration of plasma flowing through the parathyroid glands decreases. Parathyroid hormone has three distinct effects (see Figure 20-14), two of which are on the kidney: (1) it causes breakdown of bone;

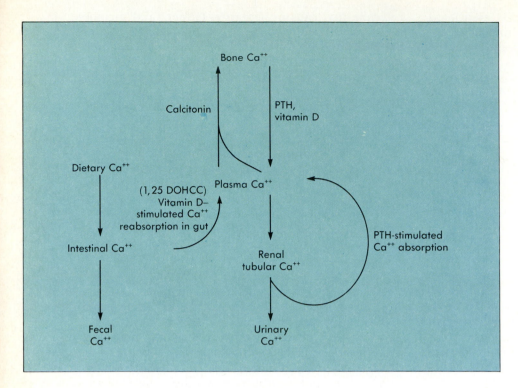

FIGURE 20-14

Routes of gain and loss of Ca^{++} from the body. Some pathways are stimulated by hormones as indicated.

(2) it increases renal reabsorption of Ca^{++}; and (3) it decreases renal reabsorption of phosphate, increasing the clearance of phosphate. The increase in phosphate clearance is an important part of the picture. If the phosphate released in bone mobilization remained in the plasma, it would bind Ca^{++} and oppose restoration of normal free Ca^{++} levels.

1,25 DOHCC is a derivative of **vitamin D_3** from meat and fish and **vitamin D_2** from plant sources. In the body, vitamin D_3 may be synthesized from cholesterol. A key event in the process is the opening of one of the rings of the steroid structure, which occurs in the skin as an effect of ultraviolet light. Irradiation of milk with ultraviolet light is the method used to increase its content of vitamin D_3. Both vitamin D_2 and D_3 are of identical potency in humans and are collectively referred to as cholecalciferol.

The liver adds a hydroxyl group to the 25th carbon of cholecalciferol, while the kidney performs a second hydroxylation, yielding 1,25 DOHCC. Both increased PTH and decreased plasma phosphate stimulate formation of 1,25 DOHCC, so either low plasma Ca^{++} or low plasma phosphate will increase levels of 1,25 DOHCC.

In the intestine, 1,25 DOHCC stimulates Ca^{++} and phosphate absorption. It must be present for normal absorption of Ca^{++} and phosphate from the diet. In its absence, children develop **rickets,** a condition of impaired bone growth. In adults the corresponding condition is called **osteomalacia** ("bad bone"). Deficiency of the hormone became more common when industrialization forced people to spend the daylight hours indoors, diminishing the body's intrinsic production of 1,25 DOHCC. Given that the intestinal effect of 1,25 DOHCC is necessary for normal bone growth and maintenance, it may be surprising to find that 1,25 DOHCC, like PTH, causes bone breakdown. Both the intestinal effect and the bone effect of 1,25 DOHCC tend to increase both plasma Ca^{++} and plasma phosphate.

To understand how PTH and 1,25 DOHCC work together to regulate both plasma Ca^{++} and plasma phosphate, consider the hormonal responses to decreases of either Ca^{++} alone (**hypocalcemia**) or phosphate alone (**hypophosphatemia**) (Figure 20-15). If plasma Ca^{++} decreases, PTH levels increase, causing an increase in 1,25 DOHCC levels. The 1,25 DOHCC and PTH release Ca^{++} and phosphate from bone, 1,25 DOHCC increases uptake of Ca^{++} and phosphate from the diet, and the effect of PTH on the kidney causes it to retain the Ca^{++} and clear the extra phosphate.

If plasma phosphate falls, 1,25 DOHCC levels will increase because of the effect of phosphate on 1,25 DOHCC activation in the kidney. But PTH levels will decrease because lower levels of plasma phosphate automatically increase levels of plasma free Ca^{++} according to the Law of Mass Action as described above. In hypophosphatemia, as in hypocalcemia, 1,25 DOHCC releases both Ca^{++} and phosphate from bone, but the kidney, responding to decreased PTH, increases its clearance of Ca^{++} and decreases its clearance of phosphate. Thus in hypophosphatemia the phosphate mobilized from bone is retained by the kidney and the calcium is

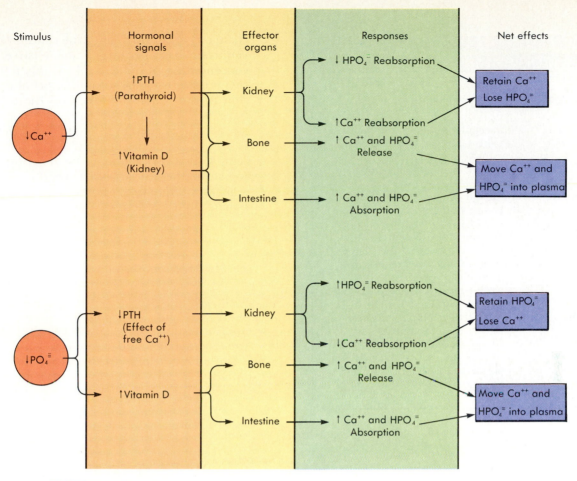

Stimulus	Hormonal signals	Effector organs	Responses	Net effects

FIGURE 20-15
Comparison of the sequence of events in the homeostatic responses to hypocalcemia and hypophosphatemia.

cleared. The key difference between the responses to decreased plasma Ca^{++} and decreased plasma phosphate is that in hypocalcemia both PTH and 1,25 DOHCC levels rise, while in hypophosphatemia 1,25 DOHCC levels rise but PTH levels fall (see Figure 20-15).

> 1 Why is it necessary to distinguish plasma free Ca^{++} from total plasma Ca^{++}? What factors affect the ratio of free Ca^{++} to bound Ca^{++} in the plasma?
> 2 Why must Ca^{++} regulation and phosphate regulation be closely linked?
> 3 What is the one essential difference between the hormonal response to hypocalcemia and the hormonal response to hypophosphatemia?

ACID-BASE HOMEOSTASIS
The Importance of pH Regulation

Small changes in the relative abundance of H^+ and OH^- have widespread effects in the body. Perhaps the most general effect of pH is on enzyme function. Typically enzymes are able to perform their catalytic functions optimally only when the pH is within a narrow range that is characteristic for each individual enzyme. This range is referred to as its **pH optimum.** For most enzymes that operate within the physiological interior, the pH optimum is close to the normal pH of cytoplasm or extracellular fluid, depending on the location of the enzyme. Changes in pH also affect the excitability of nerve and muscle cells: decreases in pH reduce excitability, while increases increase it.

The pH of arterial plasma is regulated at 7.4. The pH of cytoplasm is usually slightly lower than that of plasma. The difference is due largely to the electrical potential across cell membranes; in many cells the H^+ and OH^- ions apparently distribute themselves passively across the cell membrane according to the inside-negative membrane potential. Some cells appear to be able to regulate their internal pH by active transport of H^+ or OH^-.

Buffer Systems

To appreciate the remainder of this chapter it is necessary to fully understand buffer systems. An appendix has therefore been provided, summarizing the chemistry of acids and bases with emphasis on buffer chemistry. The major buffer systems of the

TABLE 20-1 Buffer Systems of the Body

Buffer system	Description/function
Hemoglobin (protein)	Intracellular proteins and the major plasma protein, hemoglobin, act as buffers and compose about 75% of the total buffer capacity of the body
Bicarbonate	Although components of the bicarbonate buffer system are present at relatively low concentrations and the buffer system has a pK_a of 6.1, the fact that the components of this system are regulated causes it to play an important role in controlling the pH of the extracellular fluid
Phosphate	The concentration of phosphate in the extracellular fluid is low compared with the other buffer systems, but phosphate is an important intracellular buffer and is the major buffer in the urine

body are summarized in Table 20-1. For present purposes it is useful to keep in mind the following facts:

1. Acids are solutes that dissociate into H^+ and a **conjugate base.** The ratio between the undissociated acid and its products is determined by the law of mass action according to an equilibrium constant (K_a) that is characteristic for each acid. It is frequently convenient for computations to use the negative logarithm of the K_a, the pK_a. **Strong acids** dissociate relatively completely and thus have large K_a values. In the case of **weak acids,** a substantial concentration of undissociated acid is maintained at equilibrium, so the K_a is low.

2. Weak acids and their conjugate bases can serve as buffer systems. If H^+ is added to the systems, some of the added H^+ will react with conjugate base and be converted to the undissociated acid form, while if H^+ is removed from the system, some undissociated acid will dissociate and partly restore the H^+ concentration. Weak acids are most effective as buffers when the pH of the system is close to the pK_a of the acid, because under these conditions the concentrations of the undissociated acid and its conjugate base are nearly equal. Buffer effectiveness also depends on the total concentration of buffer present, referred to as the **buffer capacity.**

3. As strong acid is added to a buffer system in which the pH is near the pK_a, the resulting pH

change is quite small until the concentration of conjugate base becomes small relative to the concentration of buffer acid. At this point further additions of acid cause the pH of the solution to drop more rapidly, and the conjugate base is said to be titrated. The amount of strong acid needed to titrate the conjugate base is a measure of the buffer capacity of the system. A titration could have been performed using strong base with similar result.

4. A major buffer system in the body is the bicarbonate buffer system (see Table 20-1):

$$CO_2 + H_2O \rightleftharpoons H_2CO_3 \rightleftharpoons H^+ + HCO_3^-$$

The first reaction, facilitated by the enzyme carbonic anhydrase, is the combination of water and carbon dioxide to form carbonic acid. Carbonic acid is a weak acid and dissociates to H^+ and HCO_3^-. The equilibrium constants for the two reactions can be summed together and treated as a single constant. The summed pK_a of the bicarbonate buffer system is 6.1 at body temperature. The Henderson-Hasselbalch equation applies the Law of Mass Action to the bicarbonate buffer system (see the Appendix):

$$pH = 6.1 + \log \frac{[HCO_3^-]}{0.03\ P_{CO_2}}$$

Multiplying the plasma P_{CO_2} by 0.03 in the denominator converts it to carbonic acid concentration.

Respiratory and Renal Contributions to Regulation of Plasma pH

The bulk of the buffer capacity of the whole body is due to protein (Table 20-1). Most of the protein is intracellular (including hemoglobin in red blood cells), but it is difficult to assess the state of intracellular buffer systems, so most of this section has to do with regulation of the extracellular pH.

The normal values for components of the bicarbonate buffer system of arterial plasma are: (1) pH = 7.4; (2) P_{CO_2} = 40 mm Hg; and (3) $[HCO_3^-]$ = 24 mEq/L. Acidosis and alkalosis can arise in two fundamentally different ways. One way is by an excess or deficit of CO_2, the other is by an excess or deficit of fixed acid. **Respiratory acidosis** is acidosis due to excessive plasma P_{CO_2}. Carbon dioxide is referred to as a volatile acid to distinguish it from fixed acids that cannot be eliminated by the respiratory system. **Respiratory alkalosis** arises from a subnormal plasma P_{CO_2}. Respiratory acidosis can be caused by a failure of respiratory function (Table 20-2), either at the respiratory control system level or at the lung level. Respiratory alkalosis can be caused by deliberate hyperventilation, anxiety, or rapid ascent to high altitudes (see Table 20-2).

TABLE 20-2 — Typical Causes of Acid-Base Abnormalities

Condition	Typical causes
Metabolic acidosis	Diarrhea resulting in excess loss of bicarbonate from the intestine
	Renal failure resulting in an inability to secrete H^+ ions
	Untreated diabetes resulting in excess keto acid formation
	Hyperkalemia
Metabolic alkalosis	Vomiting resulting in the loss of gastric acid
	Abuse of antacids
	Aldosterone-secreting tumors
Respiratory acidosis	Lung diseases such as chronic bronchitis and emphysema that decrease alveolar ventilation
	Injury to respiratory control centers; barbiturate overdose; hypoventilation*
Respiratory alkalosis	Hyperventilation*; rapid ascent to high altitude

*NOTE: In patients who are artificially ventilated, failure to monitor blood gases can result in either respiratory acidosis or alkalosis.

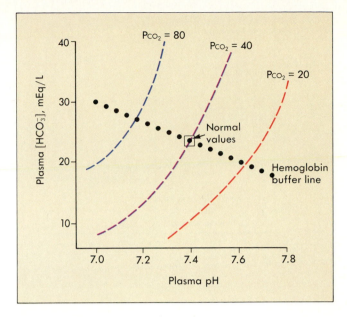

FIGURE 20-16

A Davenport diagram predicts the behavior of the bicarbonate buffer system in response to added fixed acid or base and to changes in P_{CO_2}. The squared point corresponds to the normal values of plasma pH, bicarbonate concentration, and P_{CO_2}. When any two of these values are normal, the Law of Mass Action dictates that the third value will be normal as well. A change in the P_{CO_2} changes the plasma bicarbonate and plasma pH values, tracing out the dotted line (the buffer line). The dashed lines are titration curves for the plasma system at the three different values of P_{CO_2} indicated.

Metabolic acidosis is the traditional term for an excess of any fixed acid, whether of metabolic origin or not. The term **metabolic alkalosis** is used to describe an excess of fixed base. Both metabolic acidosis and alkalosis can be caused by the excessive dietary intake of acid or base (see Table 20-2). Metabolic acidosis and alkalosis can also arise from a wide variety of disease conditions and intoxicants. For example, in renal failure the kidney becomes unable to secrete H^+, resulting in acidosis.

The Davenport diagram (Figure 20-16) is a useful graphic representation of the Henderson-Hasselbalch equation. The vertical axis is the arterial plasma HCO_3^- concentration; the horizontal axis is the arterial pH. The circled point indicates the normal state of the bicarbonate buffer system. It is the point on the graph specified by the normal value of plasma pH (7.4) and the normal plasma HCO_3^- concentration (24 mEq/L). If P_{CO_2} is held constant at 40 mm Hg, addition of fixed acid or base will cause the pH and $[HCO_3^-]$ values to trace out the central dashed line. This is a titration curve for the bicarbonate buffer system at this P_{CO_2}.

Sometimes such titration curves are referred to as CO_2 isobars (isobar means "constant pressure"). A decrease in P_{CO_2} to 20 mm Hg (right-hand curve) results in an increase in the plasma pH to a value slightly above 7.6 and a decrease in bicarbonate to about 20 mEq/L (respiratory alkalosis). An increase in P_{CO_2} to 80 mm Hg (left-hand curve) results in a plasma pH less than 7.2 and a plasma HCO_3^- concentration of about 28 mEq/L.

The pK_a of the bicarbonate buffer system is 6.1. At pH values around the physiological range, the bicarbonate buffer system is too far from its pK_a to be a good buffer. At the lowest pH values shown in Figure 20-16, the titration curves flatten slightly because as the pH decreases, the system is approaching the range in which bicarbonate is a good buffer. During additions of fixed acid to the plasma, the P_{CO_2} remains constant unless there is a change in ventilation. The CO_2 that is formed during the titration of the bicarbonate buffer is lost to the atmosphere because the arterial blood is in equilibrium with alveolar gas. Thus the blood buffer system is referred to as an "open" system because formation of CO_2 in the buffer reaction does not increase the P_{CO_2}.

If the plasma P_{CO_2} is increased, the law of mass action dictates that the pH will decrease and the HCO_3^- concentration will increase. Similarly, if the P_{CO_2} is decreased, the pH rises and the HCO_3^- concentration decreases (see Figure 20-16). In response to changes of P_{CO_2}, the pH and $[HCO_3^-]$ of the system will trace out the dotted line, which is called the **buffer line.** The slant of the buffer line is due to the presence of other buffers (hemoglobin, protein,

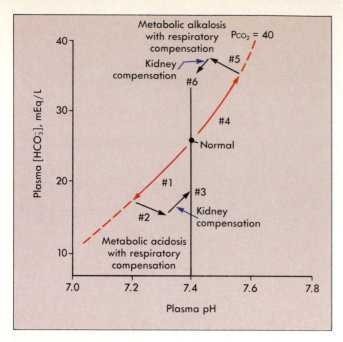

FIGURE 20-17

The effect on plasma pH of challenges to the buffer system and the regulatory systems. *Arrow 1* shows the change in the buffer system induced by injection of acid, followed by rapid but incomplete respiratory compensation *(arrow 2),* and then by slower renal compensation *(arrow 3).* The effect of injection of alkali *(arrow 4)* is similarly followed by rapid, incomplete respiratory compensation *(arrow 5)* and slower renal compensation *(arrow 6).*

etc.) in blood. Each point on the buffer line lies on a titration curve for blood at that P_{CO_2}. Figure 20-16 shows just three of the many such curves possible, each corresponding to a different P_{CO_2}.

Most of the organic and inorganic acids produced by cellular metabolism are almost completely dissociated at the pH of body fluids (that is, they are strong acids). The rate of acid production in individuals who consume a typical high-protein diet is about 1 mM/kg body weight/day. The bicarbonate buffer system minimizes the immediate impact of this acid on the pH of extracellular fluid. The daily production of metabolic acid would produce only a small reduction of bicarbonate stores, but this small

reduction must be opposed by renal secretion of acid to prevent weakening of the bicarbonate reserve that protects the pH against the effects of abnormal acid loads.

The respiratory system and kidneys are both involved in minimizing the change in arterial pH that might be caused by the normal metabolic acid production (lactic acid, for example) and by abnormal loads or losses of fixed acid or base. This cooperation is illustrated by the normal response to a load of fixed acid, such as might be caused experimentally by injection of a strong acid such as HCl. Arrow 1 in Figure 20-17 shows the immediate effect of injection of the acid; the injected acid is buffered by the bicarbonate, and both the pH and the plasma [HCO_3^-] decrease along the titration curve. The amount of CO_2 produced by the buffer titration is relatively small compared with the body's own production of metabolic CO_2 and is easily lost by way of the lungs, so that P_{CO_2} remains constant.

Within a few seconds the decrease in arterial plasma pH is sensed by the peripheral chemoreceptors, and the respiratory response described in Chapter 17 (p. 459) begins. Ventilation increases and arterial P_{CO_2} decreases. When P_{CO_2} decreases, the law of mass action mandates that both the H^+ concentration and the HCO_3^- concentration decrease. The decrease in H^+ concentration carries the arterial pH back in the direction of the normal value, but at the expense of a further decrease in [HCO_3^-] *(arrow 2* in Figure 20-17).

The response of the respiratory system to acute changes in plasma pH is rapid but usually cannot be complete (Chapter 18). The injected lactic acid cannot easily penetrate the blood-brain barrier, so the pH of the CSF is not decreased by the injection and is in fact increased by the respiratory response to peripheral acidification. At some point in the response, the increase in CSF pH cancels out the respiratory response to the peripheral acid. At this point the state of the buffer system can be described as that of **metabolic acidosis with respiratory compensation** (Figure 20-17; Table 20-3).

TABLE 20-3   Renal and Respiratory Responses to Acid-Base Abnormalities

Term	Compensation (before or after)	[HCO_3^-]	P_{CO_2}	pH
Metabolic acidosis	Before	Decreased	Normal	Decreased
	After	Decreased or normal	Decreased or normal	Decreased or normal
Metabolic alkalosis	Before	Increased	Normal	Increased
	After	Increased or normal	Increased or normal	Increased or normal
Respiratory acidosis	Before	Increased	Increased	Decreased
	After	Increased	Increased	Less decreased
Respiratory alkalosis	Before	Decreased	Decreased	Increased
	After	Decreased	Decreased	Increased or normal

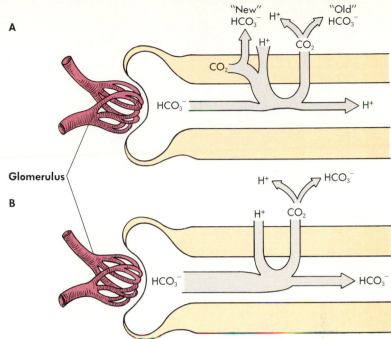

A

FIGURE 20-18
Tubular responses to acidosis and alkalosis. The proximal tubule is shown, but similar processes go on in the distal tubule.

A The tubular response to acidosis. Secretion of H^+ increases. If the acidosis is due to fixed acid, the filtered load of HCO_3^- decreases; if it is due to a respiratory problem, it increases. In either case, the secretion of H^+ is more than enough to recover all the filtered HCO_3^-. Extra secretion of H^+ has the effect of adding "new" HCO_3^- to the blood in addition to the recovery of filtered HCO_3^-.

B The tubular response to alkalosis. In respiratory alkalosis the filtered load of HCO_3^- decreases; in metabolic alkalosis it increases. In either case, the secretion of H^+ is less than normal, and some of the filtered HCO_3^- is not recovered. The unreabsorbed HCO_3^- constitutes a net loss of alkali from the body, while all of the secreted H^+ is reabsorbed as CO_2.

Metabolic acidosis can become chronic in conditions in which there is an increase in the production of acidic end products. If acidosis continues for several days or longer, there is time for renal compensation to become effective. Normally the kidney recovers all of the filtered bicarbonate as a result of its acidification of the urine (Chapter 19, p. 501). To solve the deficiency of bicarbonate in chronic metabolic acidosis, the kidney must not only save all of the filtered bicarbonate, it must also make some "new" bicarbonate to replace what was titrated away by the acid (Figure 20-18, *A*).

The "new" bicarbonate is new only in the sense that it does not represent recovery of filtered bicarbonate. Under acidotic conditions the proximal tubular secretion of H^+ exceeds the amount necessary to recover all of the filtered bicarbonate. After all the filtered bicarbonate is recovered, the proximal tubular cells continue to secrete H^+ from their cytoplasm into the tubular lumen. The proximal tubular cells contain the enzyme carbonic anhydrase, which catalyzes conversion of CO_2 to H^+ and HCO_3^-. The H^+ is secreted into the urine and accepted by other urinary buffers such as phosphate, leaving HCO_3^- behind in the cells. The "new" HCO_3^- resulting from this process is returned to the blood to replace the HCO_3^- that reacted with the lactic acid.

As the renal generation of "new" bicarbonate and loss of H^+ proceeds, the plasma HCO_3^- concentration and arterial pH increase toward normal (*arrow 3* in Figure 20-17). When this process is complete the respiratory system and kidneys have completely compensated for the acidosis, in that the plasma pH is again normal. But the Pco_2 and the plasma $[HCO_3^-]$ are both displaced from their normal values by the compensation.

The coordinated response to an alkaline load can be traced in the same way. In this case, the immediate response to injection of base (metabolic alkalosis) would be conversion of some H_2CO_3 to HCO_3^-. The plasma pH would increase along the titration curve (*arrow 4* in Figure 20-17). In response to the increase in plasma pH, respiration would be depressed, increasing the plasma Pco_2. The result of the respiratory compensation would be a further increase in HCO_3^- concentration and restoration of the arterial pH almost to normal (*arrow 5* in Figure 20-17). At this point the state of the system would be described as **metabolic alkalosis with respiratory compensation** (see Table 20-3).

In the absence of kidney disease, chronic metabolic alkalosis is frequently caused by habitual ingestion of antacid medication. In chronic metabolic alkalosis the task of the kidney is to dispose of excess HCO_3^- and conserve H^+. As HCO_3^- filtration is increased and H^+ secretion is simultaneously decreased, part of the filtered load of HCO_3^- is not reabsorbed, and HCO_3^- begins to spill into the final urine (Figure 20-18, *B*). For every HCO_3^- lost in the final urine, there is a net gain of one H^+ in the plasma, which will return the plasma pH to normal (*arrow 6* in Figure 20-17).

Renal Compensatory Responses in Respiratory Disease

Renal or pulmonary disease places a burden of compensation on the system that remains intact. For example, in acute lung disease with impaired gas ex-

Alcoholics and Antifreeze

Acid-base disorders are usually not the direct result of consumption of acid or base, but there are some drugs and toxins that are metabolized to acids by the body, resulting in an acid-base problem. A family of toxicological problems may be seen in alcoholics who resort to drinking substances that are regarded as substitutes for beverage alcohol, such as antifreeze and windshield deicer. The major ingredient of antifreeze is ethylene glycol, which has a sweet flavor. The flavor of ethylene glycol appeals to animals and small children as well, which is the reason why household supplies of it should be kept carefully capped.

Ethylene glycol itself is relatively harmless, but it is metabolized by the liver to extremely toxic ketoaldehydes and ultimately to oxalic acid. Many of the consequences of ethylene glycol intoxication are due to the resulting metabolic acidosis. As with diabetic ketoacidosis, the anion gap increases at the expense of plasma [HCO_3^-].

The diagnosis of ethylene glycol intoxication is difficult because blood and urine of patients with suspected drug overdose are not routinely assayed for ethylene glycol. The intoxication without indications of ethanol ingestion, the metabolic acidosis, and the presence of crystals of oxalate in urine are important clues. The treatment consists of three basic approaches: removing the ethylene glycol from the blood, correcting the metabolic acidosis, and preventing further metabolism of the ingested ethylene glycol to oxalic acid. The last of these is frequently accomplished by giving ethanol, since ethanol competes with the ethylene glycol for metabolism by the liver, and the metabolites of ethanol are somewhat less toxic.

---

change, the increased P_{CO_2} would decrease the plasma pH and increase the HCO_3^- concentration (*arrow 1* in Figure 20-19). The compensatory renal response would increase the HCO_3^- concentration further, but with time would probably be capable of restoring the arterial pH to normal *(arrow 2)*. At this point, the existence of a **respiratory acidosis with renal compensation** could not be diagnosed from the arterial pH, but could be recognized readily from the elevated plasma [HCO_3^-] and P_{CO_2}. Furthermore, there could be no doubt that the acid-base disturbance was of respiratory and not metabolic origin, because in metabolic acidosis the [HCO_3^-] is depressed by the fixed acid and even further depressed by the respiratory response that reduces arterial P_{CO_2}.

Respiratory alkalosis occurs as an initial reaction to high altitude (Chapter 18), as a consequence of intoxication with poisons that stimulate the respiratory center, and in some neurotic anxiety states. Acute respiratory alkalosis is characterized by increased arterial pH and decreased plasma [HCO_3^-] (*arrow 3* in Figure 20-19). The compensatory renal

response restores arterial pH at the expense of a further decrease in [HCO_3^-] *(arrow 4).*

Renal Compensatory Responses in Metabolic Acidosis and Alkalosis

Untreated diabetes mellitus presents an illustrative example of metabolic acidosis. In diabetes mellitus, the lack of insulin, the hormone necessary for glucose entry into most cells, inhibits glucose metabolism and causes a high rate of fat catabolism. The end products of fat catabolism, the ketone bodies acetone, β-hydroxybutyric acid, and acetoacetic acid are released into the plasma in large quantities, causing **ketoacidosis.** The production of metabolic acid may exceed the rate at which H^+ can be secreted by the kidney, placing a large burden of compensation on the respiratory system.

In metabolic acidosis, blood chemistry may reveal the presence of unusual organic anions. The cation and anion composition of blood plasma can be shown in a type of bar graph called a **Gamblegram** (Figure 20-20). The Gamblegram is a rearrangement of the bar graph shown in Figure 20-2

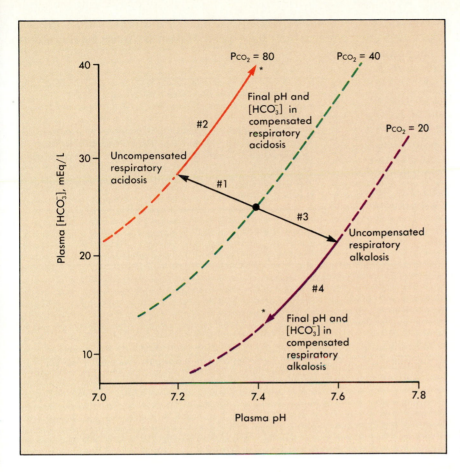

FIGURE 20-19
Respiratory acidosis *(arrow 1)* followed by slow but complete renal compensation *(arrow 2)*. Similarly, respiratory alkalosis *(arrow 3)* is followed by complete renal compensation *(arrow 4)*.

that makes it clear that the total concentration of positive charges in the plasma exactly equals the total concentration of negative charges.

In standard clinical analyses of blood plasma, only the Na^+, K^+, Cl^-, HCO_3^- concentrations and pH are determined. The sum of $[Na^+] + [K^+]$ accounts for the bulk of plasma cation (see Figure 20-20, *A*), but the sum of $[Cl^-] + [HCO_3^-]$ leaves a substantial amount of plasma anion unaccounted for: lactate, phosphate, ketone bodies, and protein. The difference between the sum of $[Na^+] + [K^+]$ and $[Cl^-] + [HCO_3^-]$ is an approximate measure of the other plasma anions and is traditionally called the **anion gap**. This is a deceptive name, since it implies that some anions are missing, whereas they must be there to satisfy the principle of electroneutrality, but are simply not measured in standard assays.

FIGURE 20-20
A A Gamblegram of plasma ionic constituents showing typical normal values.
B A Gamblegram of an untreated diabetic in ketoacidosis. Notice that the contributions of K^+, Ca^{++}, and Mg^{++} to the total cations have increased as a result of the acidosis. The contributions of Cl^- and HCO_3^- to the total anions have decreased to "make room" for the conjugate base ions of the ketoacids, increasing the anion gap.

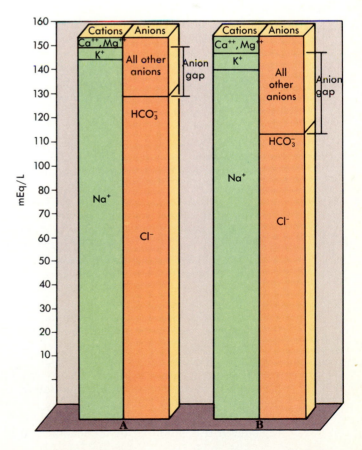

Control of Body Fluid, Electrolyte, and Acid-Base Balance   **525**

Normally the anion gap is about 26 mEq/L. In metabolic acidosis the conjugate base molecules of the acids that are present in abnormal abundance have replaced the bicarbonate ions that were titrated to CO_2 when the acids were added. Thus in metabolic acidosis the total anion concentration may well be normal, but the contribution of HCO_3^- to the total anion is decreased in favor of anions that are not measured in the standard assay. Thus the anion gap increases (see Figure 20-20, *B*), and this increase indicates the presence of unmeasured anions. In the case of diabetic ketoacidosis, the unknown anions are acetoacetate and β-hydroxy-butyrate. The HCO_3^- has been titrated out by the acids, which have partly taken its place in the anion bar.

1 *What physiological processes make it possible for bicarbonate to be an effective buffer for extracellular fluid?*
2 *What is alkalosis? Acidosis? Is it possible for a patient to have a plasma pH that is within normal limits and still be acidotic or alkalotic? Is it possible for a patient to be acidotic and alkalotic at the same time?*
3 *What is meant by "anion gap"? Is the gap caused by a lack of anions?*

SUMMARY

1. **Total body water** is distributed between the **intra-cellular compartment** (two thirds of the total) and the **interstitial** and **vascular** subcompartments of the **extracellular fluid.** The interstitial fluid comprises about 80% of the extracellular fluid.
2. The effective **osmotic pressures** of the intracellular and extracellular fluids are equal. The distribution of fluid between the two compartments is determined by the impermeant solute content of each compartment. Thus cell volume is determined by cytoplasmic protein and extracellular fluid by total body Na^+.
3. Total water turnover is 1.5 to 2.5 liters/day. Roughly one-half of the volume lost is in the form of urine, about one-tenth as fecal water, and the rest by evaporation from skin and respiratory surfaces and secretions such as sweat and tears. Of these routes of loss, only the urine composition can effectively be regulated to balance water loss with gains from drinking and metabolism.
4. Secretion of the peptide hormone **atrial natriuretic hormone** (ANH) is stimulated by atrial stretch. This hormone increases **glomerular filtration rate** and antagonizes the Na^+-conserving tubular mechanisms. An increase in plasma levels of the hormone tends to reduce extracellular fluid volume.
5. The posterior pituitary secretes the peptide **antidiuretic hormone** (ADH) in response to increased osmolarity of plasma. Secretion of the hormone may be inhibited by increased atrial stretch. ADH increases water reabsorption across the wall of the collecting duct, stimulating water conservation and resulting in the production of a hyperosmotic urine. ADH could be said to be the hormone of water conservation.
6. The steroid hormone **aldosterone** is secreted by the **adrenal cortex** in response to a regulatory cascade that begins with secretion of **renin** from the **juxtaglomerular apparatus.** Renin catalyzes the conversion of plasma angiotensinogen to **angiotensin I.** Angiotensin II is produced by an endothelial converting enzyme. **Angiotensin III** is produced by the adrenal cortex and stimulates its secretion of **aldosterone.** Secretion of renin increases in response to decreases in glomerular filtration or the Na^+ concentration of the filtrate. Aldosterone increases the recovery of filtered Na^+ in the distal tubule. Under normal conditions increases in total body Na^+ are matched so rapidly by water retention that the main effect of the hormone is to increase extracellular fluid volume.
7. All filtered K^+ is reabsorbed in the proximal tubule and loop of Henle. Thus the distal tubule's secretion of K^+ is the regulatory step in renal K^+ handling. Plasma K^+ directly stimulates aldosterone release; aldosterone stimulates distal tubular K^+ secretion. If Na^+ reabsorption is constant, K^+ secretion and H^+ secretion behave as if they were competitive; both are stimulated by aldosterone.
8. Plasma Ca^{++} and phosphate are regulated by three hormones: **calcitonin** from the thyroid, **parathyroid hormone** (PTH) from the parathyroid, and **1,25 DOHCC** activated by the kidney. Calcitonin stimulates bone deposition, while 1,25 DOHCC and PTH together mobilize Ca^{++} and phosphate from bone, increase intestinal absorption of Ca^{++} and phosphate, and stimulate Ca^{++} reabsorption and phosphate clearance by the kidney.
9. Extracellular fluid pH is buffered by the **bicarbonate buffer system,** whose state is controlled by the combined action of lungs and kidneys. **Respiratory compensation** for metabolic acidosis/alkalosis involves increases or decreases in arterial P_{CO_2}, with corresponding effects on plasma $[HCO_3^-]$ and pH. **Renal compensation** for acidosis involves an increase in tubular H^+ secretion. The H^+ secretion first recovers all of the filtered HCO_3^-, then results in the formation of "new" HCO_3^- from CO_2, increasing plasma $[HCO_3^-]$ and pH. Renal compensation for alkalosis involves decreased secretion of H^+, so that some HCO_3^- is not reabsorbed and constitutes a net loss of alkali in the urine.

1. A 40-year-old female weighs 48 kg. Plasma Na^+ is 145 mEq/L. Assume that total body water is 50% of total body weight. What is the intracellular volume? What is the total intracellular solute? The intracellular osmolarity? The total body Na^+? If this individual were given 3 liters of isotonic saline intravenously, what would be the new values for each of these variables?

2. In congestive heart failure, plasma levels of ANH increase. Explain why.

3. Explain why dietary Na^+ consumption usually does not lead to hypernatremia but may well contribute to hypertension.

4. Salt substitutes sold under such names as "light salt" usually consist in part of KCl, which has a salty taste that is not as pleasant as that of NaCl. Why would consumption of KCl as a substitute for NaCl be less likely to increase blood pressure?

5. Atherosclerosis of the renal artery restricting blood flow to one kidney results in hypertension. Explain why. Hint: The hypertension is not due to failure of urine formation by the flow-impaired kidney, since hypertension does not develop as a matter of course if one kidney is entirely removed.

6. Secretion of both ADH and ANH is affected by atrial stretching. Why is this part of the circulation monitored so closely by the systems that regulate extracellular volume?

7. A constituent of natural licorice, glycyrrhizic acid, causes hypokalemia, alkalosis, and Na^+ retention. What single action of the drug on the kidney could account for all these symptoms?

8. Hyperparathyroidism is characterized by lethargy, but frequently the disease is discovered when minor accidents result in broken bones. Explain the lethargy and bone fragility. What differences in plasma phosphate concentration would you expect in this condition?

9. From a strictly theoretical point of view, the bicarbonate buffer system is not a good one for extracellular fluid. Why not? What physiological processes make it possible for bicarbonate to be an effective buffer for extracellular fluid?

10. Respiratory compensation for acute metabolic alkalosis or acidosis is usually not complete. Explain.

11. An individual has a plasma $[HCO_3^-]$ of 12 mEq/L and a P_{CO_2} of 40 mm Hg. What is the plasma pH? What would P_{CO_2} need to be to restore the plasma pH to normal if plasma $[HCO_3^-]$ remained at 12 mEq/L?

12. An individual has a plasma pH of 7.1, a plasma $[HCO_3^-]$ of 24 mEq/L, and a P_{CO_2} of 80 mm Hg. Can a single acid-base abnormality explain all of these figures? How would you describe the acid-base problem in this case?

● SUGGESTED READING

BEAUCHAMP, G.K.: The Human Preference for Excess Salt, *American Scientist*, volume 75, p. 27, 1987. Discusses salt appetite and the possible relationship of salt consumption to hypertension.

BLESSING, W.W.: Central Neurotransmitter Pathways for Baroreceptor-Initiated Secretion of Vasopressin, *News in Physiological Sciences*, volume 1, p. 90, 1986. Discusses the way ADH secretion is regulated by arterial blood pressure and how this interacts with plasma osmolarity and central venous pressure to determine ADH release.

CANTIN, M. and J. GENEST: The Heart as an Endocrine Gland, *Scientific American*, volume 254, p. 76, 1986. Discusses the role of ANH in the regulation of plasma volume and osmolarity.

COGEN, M.G.: Atrial Natriuretic Factor Ameliorates Chronic Metabolic Alkalosis by Increasing Glomerular Filtration, *Science*, September 27, 1985, p. 1405. Describes the action of ANF on the kidney and the factors regulating its release.

COHEN, J.J., and J.P. KASSIRER: *Acid-Base*, Little, Brown & Co., Boston, 1982. Short summary of acid-base physiology.

DIBONA, G.F.: Neural Regulation of Renal Tubular Sodium Reabsorption and Renin Secretion, *Federation Proceedings*, volume 44, p. 2816, 1985. Review of the renin-angiotensin-aldosterone system, including its possible relevance to hypertension.

KNOX, F.G., and J.P. GRANGER: Control of Sodium Excretion: the Kidney Produces under Pressure, *News in Physiological Sciences*, volume 2, p. 26, 1987.

LOTE, C.J. (EDITOR): *Advances in Renal Physiology*, Alan R. Liss Inc, New York, 1986. Collection of advanced papers.

SCHNERMANN, J., and J. BRIGGS: Role of the Renin Angiotensin System in Tubuloglomerular Feedback, *Federation Proceedings*, volume 45, p. 1426, 1986. Discusses the reasons that the kidney is largely self-regulating.

Acids, Bases, and Buffer Systems

The Law of Mass Action and Acid-Base Equilibrium

A reversible chemical reaction between two reactants (A and B) to form products (C and D) is described by the equation:

$$A + B \underset{k_2}{\overset{k_1}{\rightleftharpoons}} C + D$$

1

where k_1 and k_2 are called *unidirectional rate constants* and describe the probability for the reaction to occur in either direction. The rates for the left-to-right (forward) and right-to-left reactions are $k_1[A][B]$ and $k_2[C][D]$, respectively (the brackets denote concentrations). The larger the concentrations on one side of the reaction, the greater the unidirectional rate and the more the reaction is driven toward the opposite side. At equilibrium all concentrations must be constant, and this means that the unidirectional reaction rates are equal $(k_1[A][B] = k_2[C][D])$. Solving for the ratio of k_1/k_2 gives the Law of Mass Action:

$$K_{eq} = \frac{k_1}{k_2} = \frac{[C][D]}{[A][B]}$$

2

We call K_{eq} the *equilibrium constant* for the reaction (1). A large K_{eq} means that A and B react easily to form large amounts of C and D (using up A and B). A small K_{eq} means that A and B do not readily react and, at equilibrium, the concentrations of C and D are low compared with those of A and B. Acids are defined as compounds that dissociate in water to yield H^+ ion, whereas bases combine with H^+ ions. If acids are designated by the form HA, reaction (1) becomes:

$$HA \rightleftharpoons H^+ + A^-$$

3

The product A^- is called the *conjugate base* of the acid HA because it can combine (or conjugate) with H^+ to regenerate HA. For a general acid dissociation, the law of mass action (equation 2) is:

$$K_a = \frac{[H^+][A^-]}{[HA]}$$

4

Solving this equation for $[H^+]$ gives:

$$[H^+] = K_a \frac{[HA]}{[A^-]}$$

5

In equations 4 and 5 the equilibrium constant for an acid dissociation is designated as K_a to distinguish it from equilibrium constants of other types of chemical reactions. Strong acids have large values of K_a, meaning that HA readily dissociates to H^+ and A^- and the concentration of undissociated HA is very small. Weak acids partly dissociate (some undissociated HA remains).

The fact that acids and bases are dissolved in an abundance of water cannot be neglected. Water exists in equilibrium with its ions H^+ and OH^-. Such a small fraction of the water is in the undissociated form that the concentration of water can be treated as a constant. If that is so, then the quantity $[H^+][OH^-]$ must also be constant. The latter constant is called K_w and is 10^{-14} moles/L at 25° C.

The pH Scale

The degree to which a solution is acidic is a measure of its $[H^+]$ (see Chapter 2, p. 27). In an acidic solution with $[H^+] = 10^{-2}$ M, the K_w dictates that the $[OH^-]$ be 10^{-12} M. The K_w also dictates that an aqueous solution can have an $[H^+]$ no greater than 1 M and no less than 10^{-14} M. Thus the pH scale, a convenient means of indicating the $[H^+]$, arises from K_w. When $[H^+] = [OH^-]$ at 25° C, the concentration of both ions is 10^{-7} M (pH = 7.0). The most acidic solutions have a pH of 0 ($[H^+] = 1$ M); the least acidic solutions have a pH of 14 ($[H^+] = 10^{-14}$). Since the pH scale is a logarithmic scale, its advantage is that it covers the enormous range of variation of $[H^+]$ in a small range of numbers. It is important to remember that a change in pH of one unit corresponds to a ten-fold change in $[H^+]$.

The pK_a is a Measure of the Strength of an Acid

Just as it is easy to handle a wide range of $[H^+]$ using the pH scale, so also is it easier to express and calculate with K_a values that are on a similar scale. The *pK_a* of an acid is the negative logarithm of its K_a. Strong acids thus have low pK_a values, and

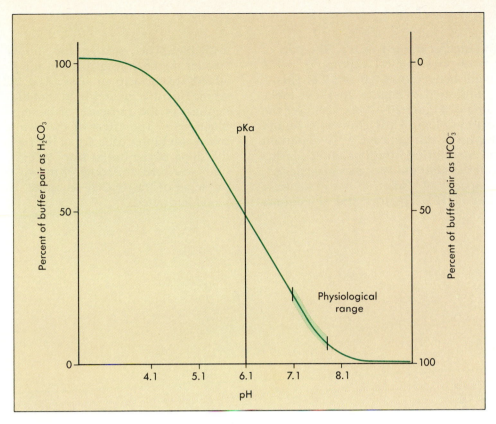

FIGURE 20-A
A titration curve for the HCO_3^- buffer system of plasma. At the pKa for the system the concentrations of HCO_3^- and H_2CO_3 are equal. The physiological range occupies a small segment of the curve more than 1 pH unit from the pKa.

weak acids have higher ones. If negative logarithms are taken of both sides of equation 5, the result is in terms of pH and pK_a:

$$pH = pK_a + \log \frac{[A^-]}{[HA]}$$

6

This is the Henderson-Hasselbalch equation. The equation has three variables: the concentration of H^+, the concentration of the conjugate base A^-, and the concentration of the undissociated acid HA. If any two of the variables are known, the third can be calculated. If the acid in the system is a weak one, the equilibrium of this acid with its conjugate base constitutes a buffer system. The undissociated acid acts as a reserve of H^+; the conjugate base acts as a sink for extra H^+.

Titration Curves are Determinations of the Buffering Ability of Buffer Solutions

If a solution of weak acid is titrated with strong base while the pH is measured, a *titration curve* (Figure 20-A) is generated. At first the pH rises rapidly, but when the pH approaches to within one unit of the pK_a of the weak acid, the curve flattens so that the pH changes relatively slowly as more base is added. The change in pH per amount of base added is least at the point where the pH of the system is equal to the pK_a of the acid. At this point the concentrations of conjugate base and undissociated acid are equal. As still more base is added, the pH begins to rise rapidly again, especially after the pH is more than one unit above the pK_a. Ultimately, essentially all of the weak acid is converted to conjugate base, and the buffering capacity of the buffer is exhausted.

The two reactions of the bicarbonate/carbonic acid/CO_2 buffer system (hereafter called the *bicarbonate buffer system*) are:

$$CO_2 + H_2O \rightleftharpoons H_2CO_3 \rightleftharpoons H^+ + HCO_3^-$$

7

The equilibrium constants for the two reactions can be summed together and treated as a single constant. The summed pK_a of the bicarbonate buffer system is 6.1 at body temperature. The Henderson-Hasselbalch equation written for this system is:

$$pH = 6.1 + \log \frac{[HCO_3^-]}{0.03 \, P_{CO_2}}$$

8

Multiplying the plasma P_{CO_2} by 0.03 in the denominator converts it to CO_2 concentration.

Figure 20-A shows a titration curve for the HCO_3^- buffer system of blood. The concentrations of HCO_3^-

and H_2CO_3 are equal at the pH of 6.1. The part of the curve that is of physiological importance is the very short segment between pH 7.0 and 7.6. This is the buffer line indicated in Figure 20-17; it looks like a straight line in Figure 20-17 because over the small range of physiological pH shown in Figure 20-17 the curvature is insignificant. It is important to keep in mind that during titrations of the plasma bicarbonate buffer system the P_{CO_2} stays constant.

The buffer capacity of a solution is a measure of the total number of H^+ acceptors and H^+ donors available. The greater the buffer capacity, the smaller the change in pH for any given quantity of added strong acid or base. The buffer capacity of whole blood is greater than that of the plasma alone because of the presence of hemoglobin in the red blood cells. The effect of the hemoglobin is to increase the buffer capacity of the blood and to reduce the slope of the buffer line. In effect, the hemoglobin substitutes for some of the HCO_3^- acceptor, and for some of the H_2CO_3 as an H^+ donor, when the blood is titrated.

KIDNEY AND BODY FLUID HOMEOSTASIS

THE GASTROINTESTINAL TRACT

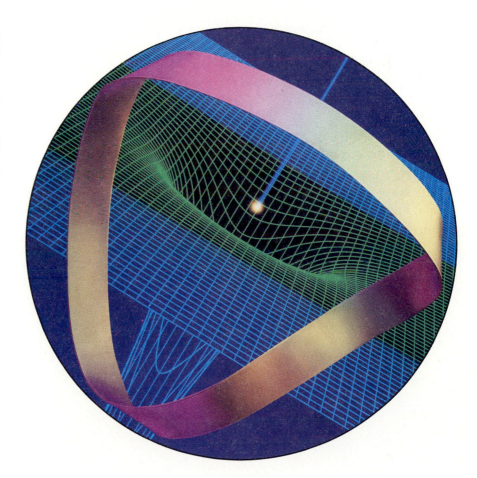

CHAPTER 21

Gastrointestinal Organization, Motility, and Secretion

On completing this chapter you will be able to:

- Describe the anatomy of the gastrointestinal tract and assign general physiological functions to each of its components.
- Identify the location and role of secretory and absorptive epithelial cells.
- Describe the blood supply to the gastrointestinal tract, identifying systemic blood vessels and lymph lacteals.
- Appreciate how muscular contractions mix and move chyme from one region of the gastrointestinal tract to another.
- Distinguish between segmentation and peristalsis.
- Understand the function of mucus.
- Describe the composition of secretions in the stomach and small intestine and the nature of bile secreted by the liver.
- Understand the transport mechanisms responsible for acid and bicarbonate secretion.
- Describe the circulation and physiological role of bile salts.
- Understand how intestinal motility mixes the chyme with digestive secretions, promotes absorption of nutrients, and propels the chyme from one region of the intestine to another.

*T*he gastrointestinal (GI) system consists of a series of internal spaces that are, physiologically speaking, outside the body. The smooth muscle of the gastrointestinal tract mixes the food in these spaces and moves it from one part of the tube system to another. The conditions created within these spaces by warmth, secretion, and mixing result in digestion, the process in which highly organized food structures containing complex macromolecules are reduced to simple molecular forms that can be absorbed by the intestine.

Much human food consists of relatively intact parts of other organisms. A salad of raw vegetables, for example, consists of large fragments of plant parts, containing cells with intact cell walls. The proteins and other large molecules that had been synthesized by the plant cells are potential nutrients locked within the cells. The first task of the gastrointestinal system is to break up the food structure to make these potential nutrients available.

After potential nutrients have been released from food, further steps of digestion chemically convert them into simpler molecules that can be absorbed by the intestine. For example, dietary proteins are, as proteins, nutritionally useless. They cannot be absorbed efficiently from the intestine, and even if they were absorbed they could not be incorporated directly into human cells. Their usefulness depends on the ability of the GI tract to convert them to the common currency of free amino acids.

The task of the GI tract is both helped and hindered by gastronomy—cooking and other cultural techniques of food preparation assist in the disruption of the food structure, but they also may result in loss of some valuable nutrients.

ORGANIZATION OF THE GASTROINTESTINAL SYSTEM
The Anatomy of the Mouth and Esophagus

Figure 21-1, A illustrates the location of the major components of the digestive system in the body. Individual elements of the digestive system can be visualized by having a person drink a suspension of barium (a contrast medium that absorbs X-rays), and then taking X-rays at various times thereafter (Figure 21-1, B). Table 21-1 lists the subdivisions of each of the components in Figure 21-1, A.

In the *mouth*, mastication (chewing) reduces the

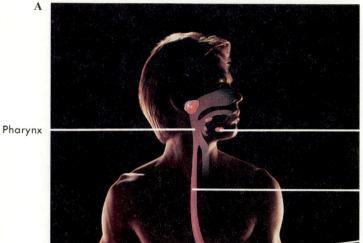

FIGURE 21-1

A The digestive system depicted in place in the body. All organs except the mouth and esophagus are located in the abdomen beneath the diaphragm.

B An X-ray of the stomach and small intestine. A subject was given a barium milk shake (barium is opaque to X-rays) to drink shortly before the X-ray was taken.

Pharynx

Gall bladder

Large intestine

Rectum

Salivary glands

Esophagus

Liver

Stomach

Small intestine

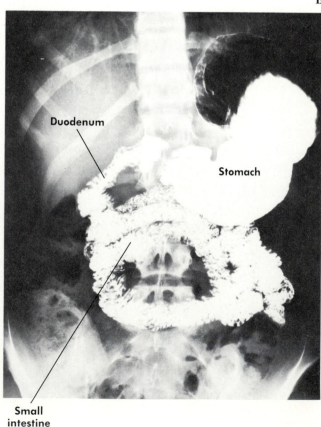

Duodenum

Stomach

Small intestine

THE GASTROINTESTINAL TRACT

TABLE 21-1   Subdivisions of the Gastrointestinal Tract

Major component	Subdivision	Functional role
Mouth		Gross mechanical breakdown; lubrication with saliva
Pharynx		Swallowing; junction with the esophagus is termed the upper esophageal sphincter
Esophagus		Movement of food from the pharynx to the stomach; the junction of the esophagus with the stomach is the lower esophageal sphincter
Stomach	Antrum	Initial storage of food
	Body	Storage/secretion
	Fundus	Vigorous mixing of food with secretions to form a semisolid chyme; the pyloric sphincter (or pylorus) separates the stomach from the duodenum
Small intestine	Duodenum	Segment receiving liver and pancreatic secretions; important site for the regulation and overall coordination of GI function
	Jejunum	Absorption of the majority of the end products of digestion
	Ileum	Fluid reabsorption; junction with the large intestine is termed the ileocecal sphincter
Large intestine	Ascending colon	Fluid reabsorption
	Transverse colon	Fluid reabsorption
	Descending colon	Fluid reabsorption
	Sigmoid colon	Storage of feces
	Rectum	Storage and elimination of feces
	Anus	Most distal opening of the GI tract; it is regulated by the internal and external anal sphincters

size of the food particles and mixes them with secretions from the three sets of salivary glands (Figure 21-2) that open into the mouth. The enzyme **α-amylase** (sometimes called ptyalin) present in the saliva initiates digestion of starches. The **esophagus** connects the **pharynx** with the **stomach** (see Figure 21-1, *A*). The upper portion of the esophagus is formed from striated muscle, the middle portion contains a mixture of striated and smooth muscle, and the lower part contains only smooth muscle. No digestion occurs in the esophagus beyond that which begins in the mouth.

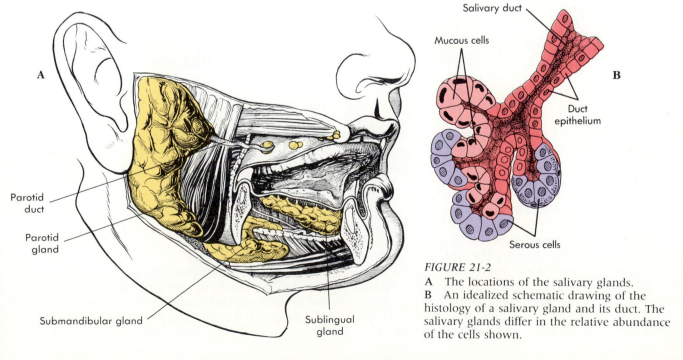

FIGURE 21-2
A   The locations of the salivary glands.
B   An idealized schematic drawing of the histology of a salivary gland and its duct. The salivary glands differ in the relative abundance of the cells shown.

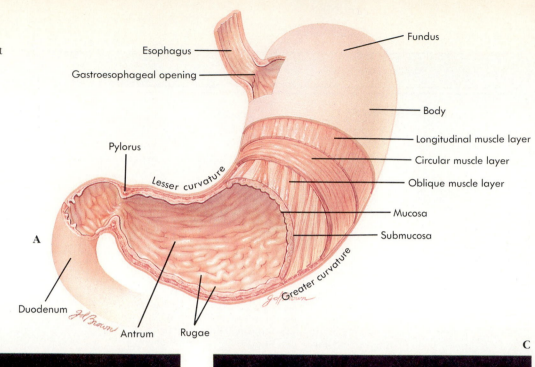

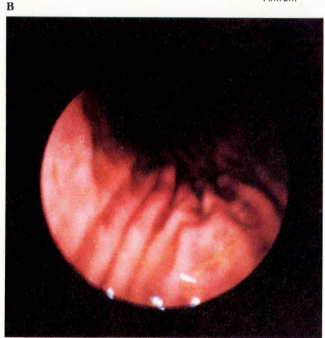

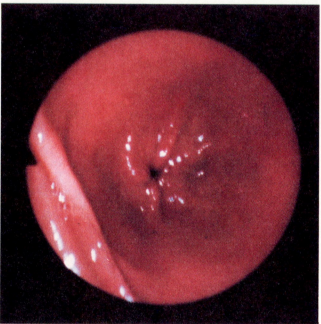

The Anatomy of the Stomach

The stomach (Figure 21-3, *A*) is surrounded by several layers of smooth muscle and is divided into two regions that serve different functions. The **body** and **fundus** are specialized for storage (see Table 21-1). Relaxation of the smooth muscle in the body and fundus allows the stomach to expand when a quantity of food enters it. The **antrum** is specialized for mixing food with gastric secretions to form a semisolid material called **chyme.** The digestion of protein begins in the stomach, but the mixing and storage functions of the stomach are more significant than the protein digestion that occurs there.

The stomach contains numerous folds, termed **rugae,** that can be easily seen without the aid of a microscope (Figure 21-3, *B*). The antrum of the stomach and the small intestine are separated by a region of smooth muscle with a relatively high intrinsic tone, the **pylorus** (Figure 21-3, *C*; Table 21-1). The pylorus is normally partially contracted and acts as a sphincter. The pylorus allows water and other fluids to exit the stomach, but it retards movement of any semisolid gastric contents.

Intestinal Anatomy

The **small intestine** (see Figure 21-1, *A*), also surrounded by layers of smooth muscle, is the major site for both digestion and absorption of food. It is 6 to 7 meters long, has an average diameter of 4 cm, and is divided into three parts on the basis of their

TABLE 21-2 Layers of the Gastrointestinal Tract

Layer	Sublayer	Type of cell present
Serosa		Connective tissue
Muscularis	Longitudinal	Longitudinally oriented muscle
	Myenteric plexus	Enteric nerve fibers; autonomic fibers
	Circular	Circularly oriented muscle
	Submucosal plexus	Enteric nerve fibers; autonomic fibers
Submucosa		Many glands
Mucosa	Muscularis mucosa	Smooth muscle
	Lamina propria	Lymph nodules
	Epithelium	Absorptive cells; glands; receptor cells

the overall control of the GI tract (see Chapter 22, p. 581). The next 250 cm of the small intestine is termed the **jejunum**, while the remaining 400 cm is called the **ileum**. The surface area for digestion and absorption in all regions of the small intestine is much greater than the length of the region indicates because of the structural modifications that increase surface area.

The large intestine (see Figure 21-1, A; Table 21-1) consists of the **colon** and **rectum**. The colon is subdivided anatomically into the **ascending, transverse, descending,** and **sigmoid** regions and is one third the length of the small intestine. It stores feces and absorbs water. The **anus,** the opening of the most distal portion of the large intestine, or rectum, is regulated by two sphincters: the **internal anal sphincter** is a thickened region of the smooth muscle, and the **external anal sphincter** is made up of striated muscle.

Histology of the Gastrointestinal Tract

The wall of the GI tract is composed of four types of tissues: connective, smooth muscle, neural, and epithelial. These are organized into concentric layers that vary in thickness in different regions: the **serosa, muscularis externa, submucosa,** and **mucosa** (Table 21-2). A cross section of the small intestine is illustrated in Figure 21-4. While the relative thickness of each layer and the size of the lumen vary, the general arrangement of the layers is the same throughout the GI tract.

The serosa is the external covering of the GI

function (see Table 21-1). The **duodenum** is the 20 cm long segment just distal to the pylorus. The secretions of the **liver** and **pancreas** enter the small intestine at the level of the duodenum, which also contains several hormone-secreting cells involved in

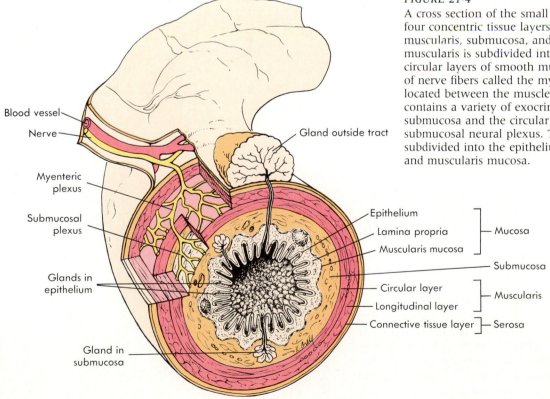

Blood vessel
Nerve
Myenteric plexus
Submucosal plexus
Glands in epithelium
Gland in submucosa
Gland outside tract

Epithelium
Lamina propria — **Mucosa**
Muscularis mucosa
Submucosa
Circular layer — **Muscularis**
Longitudinal layer
Connective tissue layer — **Serosa**

FIGURE 21-4

A cross section of the small intestine. There are four concentric tissue layers: the serosa, muscularis, submucosa, and mucosa. The muscularis is subdivided into longitudinal and circular layers of smooth muscle fibers. A network of nerve fibers called the myenteric plexus is located between the muscle layers. The submucosa contains a variety of exocrine glands. Between the submucosa and the circular muscle is the submucosal neural plexus. The mucosa is subdivided into the epithelium, lamina propria, and muscularis mucosa.

A

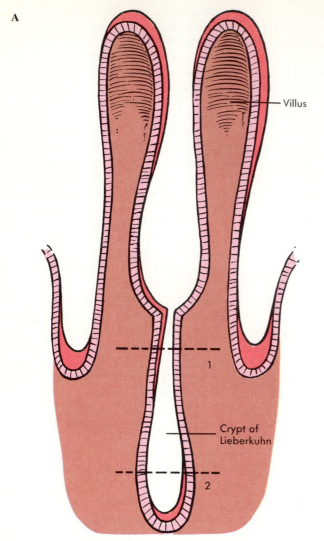

Villus

1

Crypt of
Lieberkuhn

2

FIGURE 21-5
A Structure of a crypt of Lieberkühn.
B and **C** Transverse sections showing cells surrounding the crypt lumen in the upper (*line 1 in* **A**) and lower (*line 2*) portion of the crypt.

B

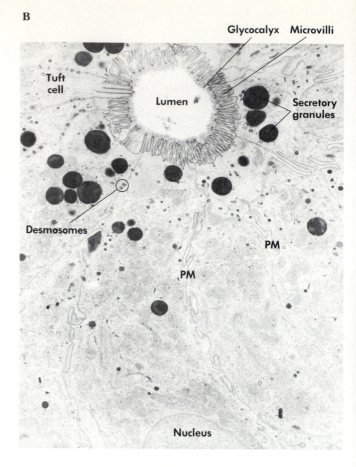

Tuft
cell

Glycocalyx   Microvilli

Lumen

Secretory
granules

Desmosomes

PM

PM

Nucleus

C

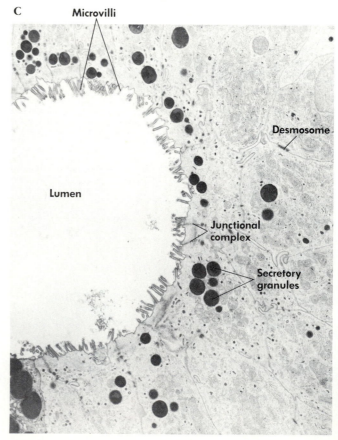

Microvilli

Desmosome

Lumen

Junctional
complex

Secretory
granules

tract and is composed of epithelial cells and connective tissue. The muscularis is subdivided into layers of longitudinal and circular smooth muscle fibers. A network of autonomic axons and ganglia (neuron cell bodies and synaptic connections) called the **myenteric plexus (Auerbach's plexus)** is located between the muscle layers of the muscularis. Nerve fibers from the plexus pass among the muscle fibers, making diffuse synaptic connections with them (see Figure 21-4; Table 21-2).

The submucosa lies beneath the muscularis externa and is made up of connective tissue, blood and lymphatic vessels, and a variety of exocrine glands. Recall, also, that exocrine glands secrete their products via ducts, in contrast to endocrine glands that secrete hormones directly into the blood (see Chapter 5, p. 96). Surrounding the blind ends of exocrine glands are **acini,** the secretory cells. Ducts connect the secretory portion of the glands with the lumen of the GI tract.

Between the submucosa and the circular muscle layer of the muscularis is a second network of autonomic nerve fibers (see Figure 21-4; Table 21-2), the **submucosal plexus** (or **Meissner's plexus**). The two autonomic plexuses make up the **enteric nervous system,** which is responsible for coordination of muscular and glandular functions. Reflexes that control the GI tract may be of the conventional (or "long") variety, in which afferents carry information to the central nervous system and responses are made by way of autonomic nerves. The vagus nerve is the primary pathway for such long reflexes. Some reflexive pathways involve only the plexuses—these are called **enteric reflexes.**

The innermost layer of the wall, the mucosa, contains three subdivisions: (1) the epithelium, (2) the **lamina propria,** the basement membrane the epithelial cells rest on, and (3) a layer of circular muscles called the **muscularis mucosa** (see Figure 21-4; Table 21-2). The epithelium is composed of an enormous number of small, fingerlike projections into the lumen. The projections, which are a millimeter or so long, are called **villi** (Figure 21-5, *A*). At the feet of the villi are the openings of tubelike **crypts of Lieberkühn** (Figure 21-5, *B* and *C*).

The lumen surface appears furry because of the villi. Inside each villus there are blood capillaries and lymphatic capillaries called **lacteals** (Figure 21-6). The lacteals and blood capillaries provide the two routes by which nutrients move from the vicinity of the intestine into the general circulation. Beneath the epithelial cells lies the lamina propria, which is made up of connective tissue, elastin, and collagen fibers and contains blood vessels, some glands, and lymph nodes. The muscularis mucosa is a thin layer of smooth muscle that thrusts the epithelium into permanent, visible ridges called the **plicae circularis (Kerckring's folds;** Figure 21-7).

> 1 What are the general functions of the GI tract?
> 2 What are the four layers of the wall of the tube system?
> 3 What are the major anatomical components of the GI tract?

Surface Area of the Gastrointestinal Tract

The small intestine has the largest internal surface area in relation to its external surface of any region in the GI tract. The plicae circularis and villi combine to increase its luminal area 10 to 30 times over that of a smooth cylinder of the same external diameter. The large intestine is 50% larger in diameter and 33% of the length of the small intestine but, because it lacks villi, the large intestine accounts for only 3% to 4% of the total GI surface area.

Additional enlargement of the surface area is attributable to membrane specializations of the epithelial cells. The luminal membranes of the epithelial cells of the stomach and intestines contain minute projections from the surfaces of the villi

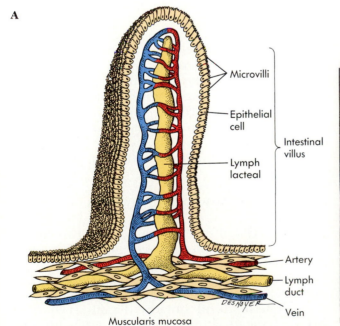

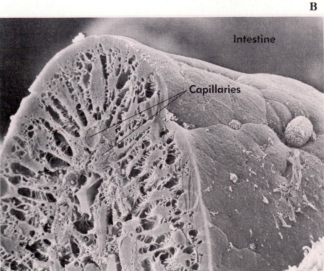

FIGURE 21-6
A   Diagram of the organization of intestinal villi.
B   A scanning electron micrograph showing the cross section of a single villus.

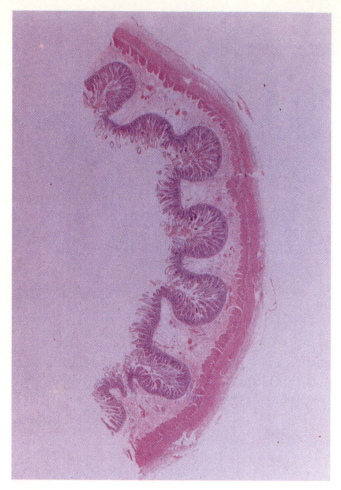

FIGURE 21-7
Photomicrograph of a section of intestine showing villi and plicae circularis.

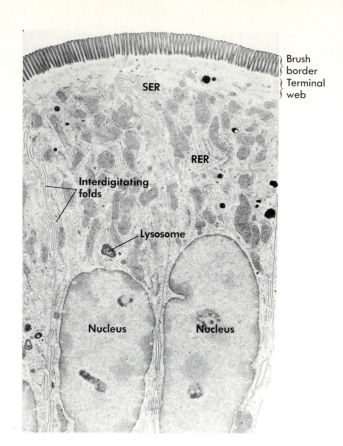

FIGURE 21-8
An electron micrograph of two adjacent epithelial cells showing the details of the microvilli or brush border.

called **microvilli,** each about 1 μm long and 0.1 μm in diameter (Figure 21-8). These microvilli were named the **brush border** because of their appearance under the light microscope. There are about 2 billion microvilli per square centimeter of intestinal epithelium, and they further increase the surface area of the small intestine by a factor of 20 to 40. When the increase in surface area attributable to the villi and plicae circularis folds is added to that attributable to the microvilli, the total absorptive area of the small intestine is about 200 square meters.

The Life Cycle of Cells of the Mucosal Epithelium

In the small intestine, new cells are produced by mitosis in the crypts of Lieberkühn (see Figure 21-5). Immature cells lack digestive or absorptive abilities. Over a period of 2 to 5 days, they migrate up the sides of the villi. During their migration, digestive enzymes and the transport proteins responsible for absorption of organic materials and ions are synthesized and inserted into the cell membranes. Mi-

gration of the new cells pushes the older cells to the tips of the villi, where they slough off into the intestinal lumen. The enzymes and mucus present in secretory cells are released when they are sloughed, and this supplements the release from intact cells.

The entire lining of the small intestine is replaced in 5 to 6 days. Continuous replacement counteracts the damage normally caused by abrasion and digestive enzymes and allows the system to recover rapidly from acute intestinal infections. A similar replacement process occurs in the stomach, where the life cycle of the cells is even shorter than in the intestine.

Secretory and Absorptive Functions of the Mucosa

Epithelial cells of the GI tract perform two general functions: absorption and secretion (Table 21-3). Absorptive epithelial cells contain numerous transport systems and digestive enzymes in their **apical** (lumen-facing) surfaces and have Na⁺-K⁺ pumps on their **basolateral** surfaces (Figure 21-9). They transport sugars, amino acids, and water from the lumen to the interstitial fluid and then to the capillary blood. Adjacent epithelial cells have tight junctions at their apical surfaces that allow small solutes

TABLE 21-3   *The Functions of Intestinal Epithelial Cells*

Function	Cell type	Characteristics	Physiological role
Absorption	Columnar	Microvilli	Increase surface area
		Membrane enzymes	Digestion
		Facilitated diffusion and cotransport	Absorb salt; absorb products of digestion
		Basolateral Na^+-K^+ pump	Absorb water and salt
		Tight junctions	Seal lumen
Exocrine secretion	Goblet	Mucin-filled vesicles	Lubrication; protection
	Crypt and gland	Located in mucosa and submucosa; connected via ducts	Fluid secretion; secretion of enzymes
Endocrine secretion	Paracrine	Sensory microvilli	Sensory receptors
		Vesicles at base	Hormone secretion

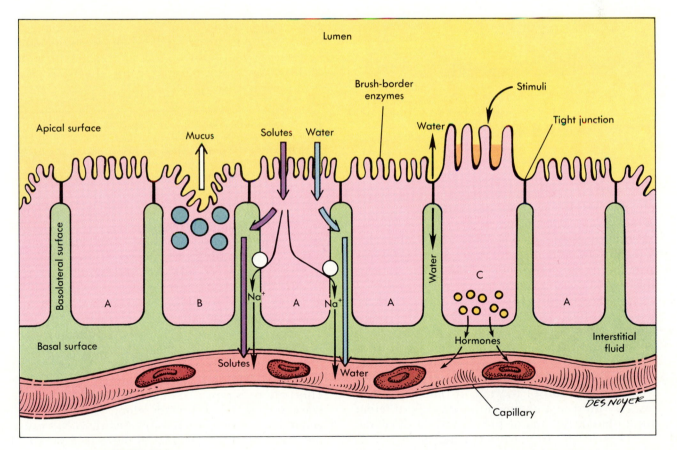

FIGURE 21-9

Types of epithelial cells in the GI tract. In the intestine, absorptive cells **(A)** have transport systems and brush-border digestive enzymes in their lumen-facing (apical) surface and Na^+ pumps on their basolateral surfaces. They transport sugars and amino acids *(purple arrow)* and water *(blue arrow)* from the lumen to the interstitial fluid. The solutes then enter the capillary blood. Adjacent epithelial cells have tight junctions that prevent large molecules from reaching the interstitial space but that can allow water and small ions to move in response to osmotic forces. Goblet cells **(B)** secrete mucins. Endocrine cells **(C)** act as chemoreceptors or mechanoreceptors and release hormones from their basal surfaces.

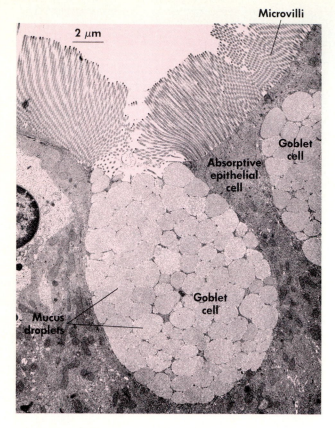

FIGURE 21-10
An electron micrograph of the apical portion of a goblet cell, full of vesicles of mucin ready to be secreted.

with food intake. Mucus also serves to clump indigestible materials together in the large intestine.

Most exocrine cells in the GI tract are located in glands in the mucosa and submucosa and are connected to the lumen via ducts (see Figure 21-4). In the stomach, however, the openings of the gastric glands appear as pits on the surface (Figure 21-11).

The GI tract has smaller numbers of endocrine or endocrine-like cells called **enteroendocrine** or **argentaffin** (silver-staining) cells (see Figure 21-9), concentrated particularly in the stomach and in the small intestine. These enteroendocrine cells act as chemoreceptors or mechanoreceptors. Their apical surface contacts the lumen, where sensory microvilli monitor the chemical and physical properties of the chyme. Many enteroendocrine cells secrete substances into the blood that satisfy the definition of a hormone. However, some enteroendocrine cells release substances that act locally and do not enter the general circulation (paracrine agents; Chapter 5, p. 97).

In some instances it is difficult to draw a distinction between a hormone and a paracrine agent because it is hard to determine how widespread the agent's actions are and where the receptors are located. Two well-known paracrine agents, histamine and prostaglandin, have opposing effects on gastric acid secretion (see p. 552).

> 1 What modifications increase membrane surface area in the stomach and small intestine?
> 2 What are the functional categories of epithelial cells of the GI tract?
> 3 How do the secretions of the GI tract differ in the stomach and small intestine?

Gastrointestinal Motility

Motility in the GI tract falls into three major categories: (1) **peristalsis,** (2) **segmentation,** and (3) mixing motions of villi and microvilli. Peristalsis involves waves of contraction and relaxation that travel for variable distances, depending on the location and the stage of digestion (Figure 21-12, *A*). The first phase of peristalsis is contraction of the longitudinal muscle layer and relaxation of circular muscle. In the second phase, the circular muscle layer contracts and the longitudinal layer relaxes. This pattern is repeated in an adjacent region, as the wave of peristalsis spreads along the GI tract. In general, the exact pattern of peristaltic waves causes the contents of the GI tract to move in a proximal-to-distal direction.

The total distance the wave travels is determined in part by the effectiveness of the electrical connections between muscle cells in the region of the GI system under consideration. The wave of contraction creates a pressure gradient that can move part of the contents along the GI tract. Peristalsis occurs only if the enteric nervous system is

to pass but prevent the movement of large molecules (proteins, for example) from the intestinal lumen to the interstitial space.

The most abundant type of secretory epithelial cell, the **goblet cells** (Figure 21-10; Table 21-3), are distributed throughout the GI tract. The lumen-facing region of mucus-secreting goblet cells has large vesicles filled with **glycoprotein mucins.** When mucin-filled vesicles undergo exocytosis (see Chapter 4, p. 66), the released mucin combines with water to form **mucus.** In the intestines, the absorptive cells are the most numerous, whereas the lining of the stomach is largely composed of secretory cells.

The composition of the mucins and the pH of the watery fluid that accompanies them is different in the different regions of the GI tract. In general, the fluid is similar to plasma but higher in HCO_3^-, so that the mucus layer that lines the lumen is slightly alkaline. The protection provided by alkaline mucus is most important to the cells in the stomach and the initial part of the small intestine, especially the duodenum, because the acidified chyme from the stomach enters this region and its pH is only gradually neutralized by pancreatic secretions. The rate of mucus and fluid secretion increases

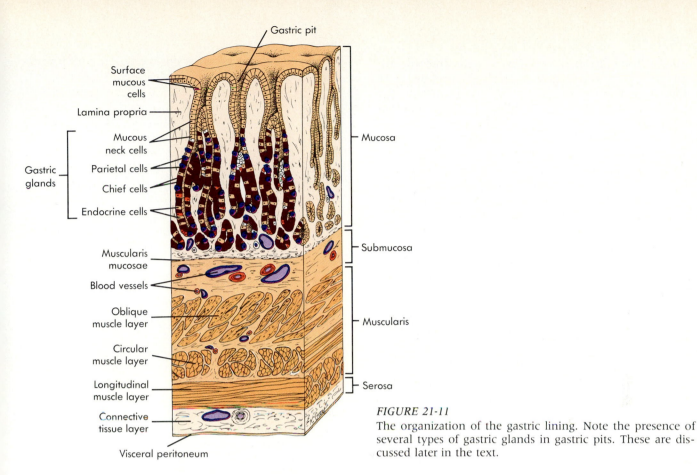

Gastric pit

Surface
mucous
cells

Lamina propria

Mucous
neck cells

Gastric
glands

Parietal cells

Chief cells

Endocrine cells

Muscularis
mucosae

Blood vessels

Oblique
muscle layer

Circular
muscle layer

Longitudinal
muscle layer

Connective
tissue layer

Visceral peritoneum

Mucosa

Submucosa

Muscularis

Serosa

FIGURE 21-11
The organization of the gastric lining. Note the presence of several types of gastric glands in gastric pits. These are discussed later in the text.

A Peristalsis

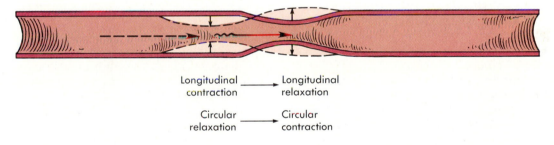

Longitudinal
contraction ——→ Longitudinal
relaxation

Circular
relaxation ——→ Circular
contraction

B Segmentation

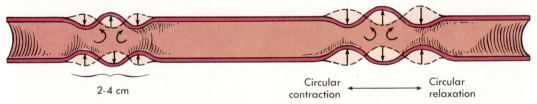

2-4 cm

Circular
contraction ←——→ Circular
relaxation

FIGURE 21-12
Forms of gastrointestinal motility.
A Peristalsis involves coordinated waves of contraction and relaxation that travel for variable distances and cause propulsion.
B Segmentation involves alternating contraction and relaxation of the circular muscle layer that mix the chyme.

intact. Propagation of the wave is assisted by distension of the tract wall by chyme. Although peristalsis occurs throughout the tract, it is strongest in the swallowing pattern of the esophagus (so one can swallow while standing on their head), moderately strong in the stomach, and relatively weak in the intestines.

Segmentation (Figure 21-12, *B*) occurs only in the intestines, where it dominates peristaltic movements. Segmentation involves an alternating contraction and relaxation of the circular muscle layer in a restricted region of the intestine, with the segmentation site moving from one place to another without displacing the intestinal contents very far in either direction. As is true for the peristaltic wave of contraction, the coordination of segmentation is mediated by the enteric nervous system and also is assisted by the mechanical stimulation that results from distension. The result of segmentation is to mix and circulate the chyme by periodically dividing the intestine into segments.

Gastrointestinal Smooth Muscle

The smooth muscle cells of the GI tract are typically of the single-unit type (see Chapter 12, p. 310). GI smooth muscle cells are 50 to 100 microns long and 2 to 5 microns wide. Smooth muscle action potentials have a smaller amplitude and last longer (10 to 20 milliseconds) than those in neurons but are briefer than cardiac action potentials. As in cardiac muscle, an inward current of Ca^{++} is an important component of the action potential in smooth muscle. The conduction velocity of action potentials along smooth muscle fibers is low, because activation of the Ca^{++} channels is slow.

Smooth muscle cells are usually arranged in sheets, with the sheets oriented longitudinally or circularly (see Figures 21-3, *A*, and 21-4). A dual orientation is needed to generate peristaltic and segmentation movements. Smooth muscle cells can be excited by action potentials from other cells or can exhibit an intrinsic pattern of periodic depolarizations called **pacemaker activity** (see Chapter 12, p. 310). In most parts of the GI tract, pacemaker activity takes the form of **slow waves** of depolarization followed by repolarization (Figure 21-13, *A*). Each slow wave lasts 3 to 20 seconds, depending on the location. Slow waves comprise the **basic electrical rhythm** that controls the frequency and progression of peristaltic waves. Groups of cells in the longitudinal muscle layer usually act as the slow wave pacemakers.

Slow wave depolarizations maintain a level of tonic tension in smooth muscle, but these depolarizations by themselves are not large enough to cause appreciable contraction. Under certain conditions, however, a slow wave may exceed the threshold for action potential generation, in which case one or more action potentials occur during the wave. These

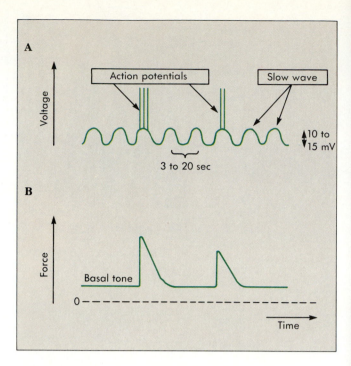

FIGURE 21-13

A Smooth muscle cells exhibit pacemaker activity called slow waves that arise in the longitudinal muscle layer. The membrane voltage varies by 10 to 15 mV, and the slow waves last 3 to 21 seconds. Bursts of action potentials generated at the top of the slow wave can initiate smooth muscle contraction. The force depends on the number of action potentials.
B Most GI muscle exhibits resting tension, or basal tone.

action potentials initiate smooth muscle contraction (Figure 21-13, *B*). Smooth muscle contracts slowly, and the force generated depends on the number of action potentials occurring at the peak of the slow wave (compare *1* and *2* in Figure 21-13, *B*).

In the interval between slow wave–initiated bursts of action potentials, smooth muscle tension is usually above zero, termed **basal tone** (see Figure 21-13, *B*). Most extrinsic control of motility occurs by modifying the basal tone of the GI smooth muscle. Stretch elicits two responses (Figure 21-14): (1) in **stress activation** a brief vigorous stretch can increase the force of contraction and is one means of initiating peristalsis and segmentation; (2) in **stress relaxation** the stomach and colon can adjust their volume to accommodate an increased load with little increase in internal pressure.

Individual villi are capable of motion due to the presence of strands of smooth muscle within the submucosa. The motion of villi is greatly increased when food is present; the increase in motion is attributed to an as yet unidentified hormone believed to be released by endocrine cells in the GI tract.

The structure of microvilli suggests that they may also be capable of motion. Each microvillus contains 20 to 30 actin microfilaments (Figure 21-

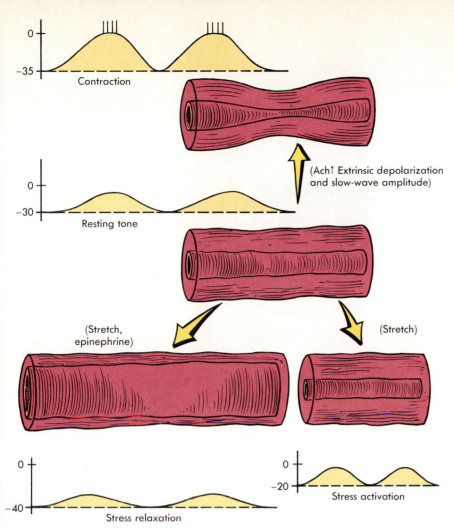

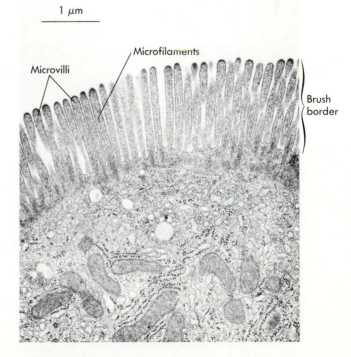

FIGURE 21-14
Factors that regulate intestinal contractility and the underlying electrical events. Extrinsic control is exerted by the parasympathetic branch of the autonomic nervous system. Intrinsic responses include stress activation and stress relaxation.

15) that extend some distance into the apical cytoplasm of the epithelial cells, a region referred to as the **terminal web** (see Figure 21-8). Actin is involved in muscle contraction (see Chapter 12, p. 289) and may make movement and extension of the microvilli possible. The movement of villi and microvilli helps reduce the effect of the **unstirred layer** of fluid immediately adjacent to the apical surface of the cells. Within the unstirred layer, solute movement is slow because it occurs primarily by diffusion rather than bulk flow.

Neural and Hormonal Control of Motility

Both segmentation and peristalsis are initiated by the intrinsic depolarizations of smooth muscle cells, but the coordination of the circular and longitudinal muscle layers requires the enteric nervous system to act as a "visceral brain" for the GI tract (Figure 21-16). The mucosal layer contains the endings of sensory cells specialized for chemoreception and mechanoreception. The cell bodies of these receptor cells are in the submucosal plexus. Enteric interneurons transmit information between the submucosal plexus and the myenteric plexus. Effector neurons in the myenteric plexus respond to the sensory signals by initiating local contractile or secretory re-

FIGURE 21-15
High-magnification electron micrograph showing the microfilament bundles of intestinal microvilli.

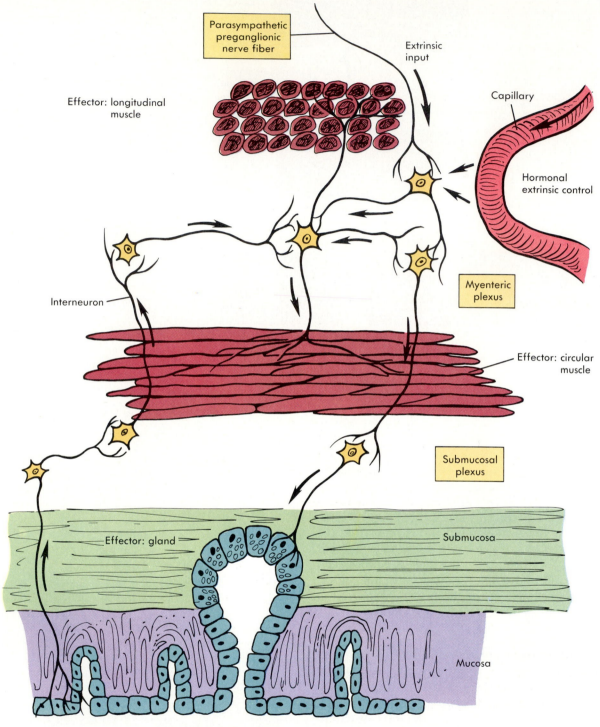

FIGURE 21-16

The organization of the enteric nervous system. Neurons in the myenteric plexus primarily control muscle contraction, whereas the submucosal plexus controls secretion. However, the two networks are extensively interconnected. Extrinsic parasympathetic input is exerted on the myenteric plexus; hormones can also exert extrinsic control.

Gland	Cell type	Secretion	Functional role
Salivary	Duct	Fluid ($NaHCO_3$)	Suspension; lubrication; starch digestion
	Mucus	Mucus	
	Serous	Amylase	
Gastric	Parietal	Acid (HCl)	Solubilization; destruction; vitamin B_{12} absorption
		Intrinsic factor	
	Chief	Pepsin	Protein digestion
	Surface mucus	Thick mucus	Protection
	Mucus neck	Thin mucus	Lubrication
Pancreas	Acinar	Proteases	Protein digestion; fat digestion; starch digestion
		Lipases	
		Amylase (as zymogen granules)	
	Duct	Fluid ($NaHCO_3$)	Buffer acid from stomach
Liver		Fluid ($NaHCO_3$)	Buffer acid; emulsification of fats; elimination of waste materials
		Bile acids	
		Cholesterol	
		Bile pigments	
		Varied organic compounds	
Intestine	Most cells	Fluid ($NaHCO_3$)	Buffer acid; lubrication
	Goblet	Thin mucus	

sponses. Some myenteric motor neurons innervate the longitudinal and circular muscle layers of the muscularis externa and the muscularis mucosa, and other myenteric neurons synapse with the neurons in the submucosal plexus that control secretory cells.

The enteric nervous system is also influenced by inputs from the autonomic nervous system (see Figure 21-16). Postganglionic parasympathetic fibers enter the myenteric plexus and activate motor neurons or interneurons to influence contraction or secretion. Autonomic control of the circulatory supply to the GI system is coordinated within the central nervous system to match the inputs on muscles and glands, so that when digestion and absorption are promoted, an appropriately large blood supply to the stomach and intestines provides for uptake of the nutrients. In addition, the enteric nervous system possesses receptors for a number of hormones that can modify its functioning (see Figure 21-16).

---

1 What is the electrical mechanism of the rhythmic contraction of GI smooth muscle?
2 What structures make up the "visceral brain"?
3 The GI system is rather unique among systems of the body in having its function regulated by both intrinsic and extrinsic reflexes. What is the distinction between the two types of circuits, and what types of changes can they effect?

THE ROLE OF THE MOUTH AND ESOPHAGUS
Function and Control of Salivary Secretion

Saliva has three components: (1) an electrolyte solution that moistens food, (2) mucus, and (3) the enzyme α-amylase. α-Amylase initiates starch digestion. Although the similar pancreatic enzymes can digest the entire carbohydrate load, the α-amylase present in the saliva alters the taste of certain foods by breaking down starches to sugars. The digestive action of the salivary amylase assists in cleansing the teeth by reducing starches to more soluble compounds.

Of the three pairs of salivary glands (see Figure 21-2), the submandibular glands produce 70% of the normal daily secretion of saliva, the parotid glands another 25%, and the sublingual glands the remaining 5%. The rate of flow of saliva into the mouth varies from 0.1 ml/min at rest to 4 ml/min during maximum secretion. There are three types of cells in salivary glands: (1) **mucous cells** secrete mucins; (2) **serous cells** secrete the primary salivary fluid into the **acini;** and (3) **duct cells** line salivary ducts and modify the primary fluid (Table 21-4; Figure 21-17). Submandibular glands have both serous and mucus cells, parotid glands contain only serous cells, and sublingual glands mainly have mucus cells. As a result, the composition of saliva in the mouth depends on the relative secretion from each of the three different pairs of glands.

Salivary glands are innervated by the autonomic

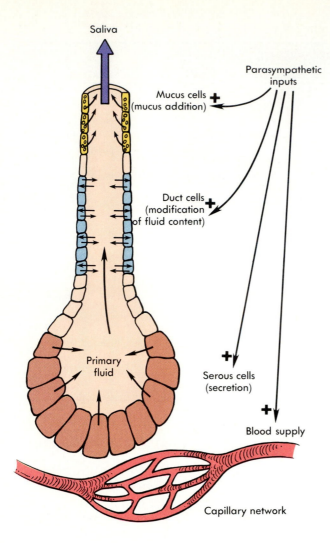

FIGURE 21-17
Summary of parasympathetic control of salivary secretion.

nervous system. Although the two branches typically have contrasting actions on other parts of the GI tract, parasympathetic and sympathetic stimulation both promote salivary secretion. The sympathetic effect is to temporarily increase secretion by causing contraction of smooth muscle associated with the glands, whereas the parasympathetic effect produces a longer lasting increase in the blood flow to the gland that is important for promoting secretion (see Figure 21-17). Salivary secretion is enhanced by the physical presence of food in the mouth (an intrinsic reflex loop) and by visual, olfactory, or mental stimuli (extrinsic controls).

Mechanism of Salivary Secretion
The mechanism of secretion in salivary glands has elements that are common to all electrolyte-secreting glands in the GI system. Secretion is produced in two stages (see Figure 21-17). In the first stage, a primary fluid similar to plasma is secreted into the acini by a combination of active ion trans-

port and plasma filtration. Serous cells also secrete α-amylase. In the second stage, the primary fluid is modified as it flows to the lumen. Salivary duct cells modify the primary serous cell secretion in two ways: (1) they actively transport Na^+ out of the lumen; and (2) they secrete K^+ into it. Active Na^+ transport decreases the osmolarity of the luminal fluid and, if the duct cells are water permeable, the osmotic gradient created by Na^+ will drive water absorption from the lumen. Unstimulated duct cells are not very permeable to water, but parasympathetic stimulation increases the water permeability of duct cells at the same time that it stimulates serous cell secretion.

The composition of saliva reaching the mouth depends on the rate of secretion. At high rates of serous cell secretion, duct cells do not have much time to modify bicarbonate content, and saliva contains a lot of HCO_3^-, whereas at low flow rates, the pH is neutral. Salivary osmolarity is similar to that of plasma at high flow rates because duct cell water permeability is high enough that when Na^+ is absorbed, water follows. At low flow rates, most of the luminal Na^+ and Cl^- is absorbed by duct cells, but because water permeability of duct cells is low at low flow rates, water cannot follow the salt, and the osmolarity of saliva is only about 25% that of plasma.

Swallowing
Swallowing is a stereotyped motor pattern that involves coordination of striated muscle and smooth muscle. There are three phases of swallowing, corresponding to the anatomical location of the material being swallowed. After a voluntary **oral phase** (Figure 21-18, *A*), in which the readiness of the material to be swallowed is assessed and the material is pushed to the back of the mouth by the tongue, the **pharyngeal phase** and **esophageal phase** follow reflexively.

Both of these phases are controlled by a **swallowing center** in the brainstem. In the pharyngeal phase (lasting about 1 second), the entry of a **bolus** of food into the pharynx stimulates receptors that initiate an orderly sequence of contractions of pharyngeal muscles. The pharyngeal contractions generate a pressure that propels the bolus of food toward the esophagus (Figure 21-18, *B*). As the bolus approaches the esophagus, the **upper esophageal sphincter** relaxes and the epiglottis covers the opening of the trachea (Figure 21-18, *C* and *D*). Once the bolus enters the esophagus, the upper esophageal sphincter constricts to prevent reflux of food into the pharynx (Figure 21-18, *E*).

In the esophageal phase of swallowing, a primary peristaltic wave of contraction sweeps along the entire esophagus in the direction of the stomach, propelling the bolus of food down the esophagus in 5 to 10 seconds (Figure 21-18, *F*). As the bo-

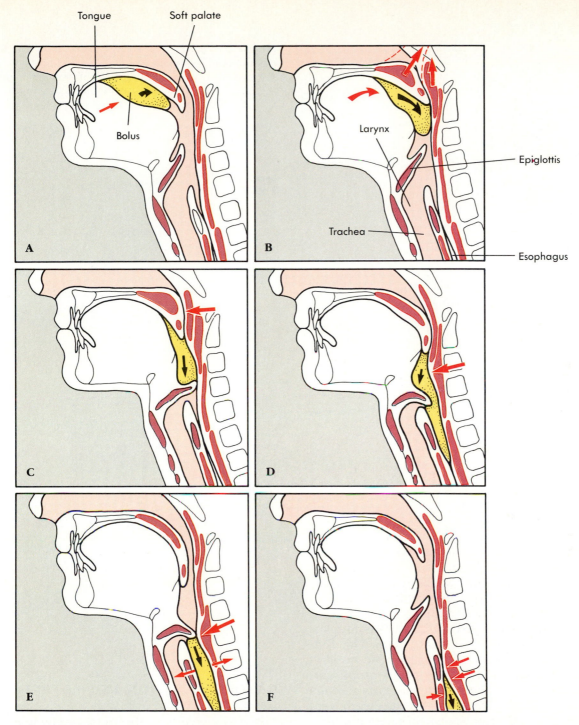

FIGURE 21-18

Swallowing.

A During the voluntary phase a bolus of food *(yellow)* is pushed toward the pharynx.

B-E During the pharyngeal phase the soft palate closes off the entrance to the nasopharynx, and the epiglottis closes off the entrance to the trachea with contractions forcing the bolus of food into the esophagus.

F In the esophageal phase the bolus of food is moved by successive contractions toward the stomach.

lus nears the stomach, the **lower esophageal sphincter (cardiac sphincter)** relaxes. After the bolus enters the stomach, the lower esophageal sphincter contracts, preventing reflux. The esophageal phase also involves a **receptive relaxation** of the stomach. The primary peristaltic wave usually empties the esophagus, but any remaining food triggers secondary peristaltic waves.

The pharyngeal phase is depressed by anesthetics and is affected by brainstem damage. Caffeine

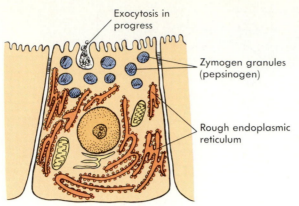

A
Gastric chief cell

Exocytosis in progress

Zymogen granules (pepsinogen)

Rough endoplasmic reticulum

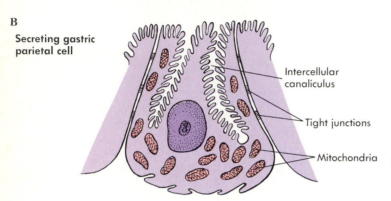

B
Secreting gastric parietal cell

Intercellular canaliculus

Tight junctions

Mitochondria

FIGURE 21-19
Gastric gland cell types.
A   Chief cell, showing rough endoplasmic reticulum.
B   Parietal cell, showing large numbers of mitochondria.
C   Gastric G cell, showing secretory granules filled with the hormone gastrin.
D   Mucus neck cells, showing large mucus granules.

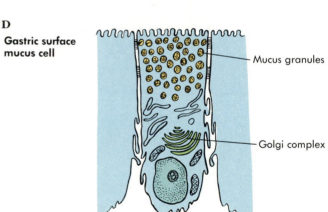

C
Gastric G cell

Secretory granules (gastrin)

Exocytosis in progress

D
Gastric surface mucus cell

Mucus granules

Golgi complex

and alcohol inhibit the contraction of the smooth muscle in the lower esophageal sphincter, and if the strength of contraction drops below a certain level, **gastric reflux** may occur. The resulting irritation of the esophageal lining by HCl causes heartburn.

If the lower esophageal sphincter fails to relax, food may be unable to enter the stomach. This condition is called **achalasia**; it is usually the result of damage to the myenteric plexus of the esophagus. The esophagus enters the abdominal cavity through an opening in the diaphragm called the **esophageal hiatus.** If the lower esophageal sphincter or a portion of the stomach protrudes into the thoracic cavity (a condition called a **hiatal hernia**), gastric reflux and heartburn often result. It is not unusual for a woman to experience this condition late in pregnancy due to the pressure of the uterus on the abdominal contents.

> 1 *What basic mechanisms are involved in production of saliva? Why is the composition of saliva more like that of plasma when the flow rate is high?*
> 2 *What are the functions of salivary amylase?*
> 3 *What structures are involved in swallowing? Can the process of swallowing be voluntarily halted once food enters the esophagus? Why or why not?*

GASTRIC SECRETION AND MOTILITY
Gastric Secretion

The acid secreted by the stomach has four primary roles: (1) it facilitates the reduction of food particles into chyme; (2) it denatures proteins and nucleic acids (see Chapter 3, p. 48); (3) it transforms the inactive form of gastric proteolytic enzymes, termed pepsinogens, to pepsins (which then can activate more pepsinogens); and (4) it destroys ingested bacteria.

Each square centimeter of gastric mucosa contains about 20,000 depressions called **gastric pits** (see Figure 21-11). Each pit receives the secretion of 3 to 7 gastric glands. The glands include several types of cells (see Table 21-3). **Chief** (peptic) cells are located at the base of the glands and secrete an inactive precursor of pepsin called pepsinogen (Figure 21-19,

FIGURE 21-20
Diagram illustrating the processes thought to be involved in
the secretion of HCl in the stomach.

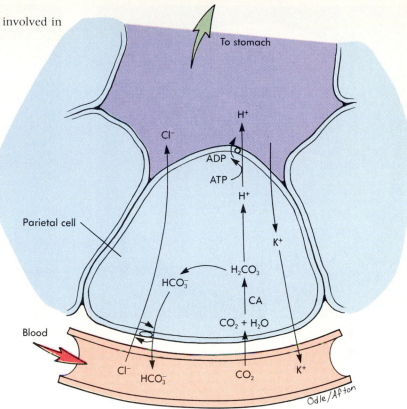

A). **Parietal** (oxyntic) cells are distributed along the length of the gland (Figure 21-19, *B*). They secrete HCl and an **intrinsic factor** necessary for vitamin B_{12} absorption. Gastric glands in the antrum lack parietal cells and cannot secrete HCl, but by the time chyme reaches the antrum it has already been acidified. Glands in the lower part of the stomach also contain argentaffin cells called **G cells** that secrete the hormone **gastrin** (Figure 21-19, *C*). The function of this hormone is described in Chapter 22.

Surface mucus cells (Figure 21-19, *D*) secrete the thick mucin that protects the epithelium of the stomach and duodenum from the harsh, acidic conditions in the lumen. Mucin-secreting cells are stimulated by mechanical or chemical irritation and by parasympathetic inputs. The protective mucus barrier can be damaged by bacterial or viral inflammation, by certain foods, and by drugs such as aspirin. Since aspirin also increases acid secretion by inhibiting prostaglandin synthesis, it can be particularly harmful when taken in large amounts or consumed by sensitive individuals.

The Cellular Mechanism of Acid Secretion

Figure 21-20 shows the cellular mechanisms presently thought to be involved in HCl secretion by parietal cells. The H^+ that is secreted results from the combination of H_2O with CO_2 produced in the metabolic activity of the cell:

$$(CO_2 + H_2O \rightleftharpoons H_2CO_3 \rightleftharpoons HCO_3^- + H^+)$$

A parietal cell has an ATP-dependent active transport system on its luminal surface that accumulates K^+ in exchange for the secreted H^+. The HCO_3^- that is left behind exits the basolateral side of the cell and enters the circulation.

The increase in H^+ secretion that accompanies a meal results in an increase in plasma $[HCO_3^-]$ called the **alkaline tide**. Some of the HCO_3^- movement across the basolateral membrane is in exchange for Cl^- entry, and some Cl^- entry appears to be coupled to Na^+ entry into the cell, although a coupling mechanism has not been characterized. Cl^- passes across the lumen side of the cell by facilitated transport, which may or may not be linked to the K^+ efflux that accompanies elevated HCl secretion. The net result of the transport systems in the parietal cells is secretion of HCl.

Since the plasma pH is 7.4 and the pH of the secreted fluid is 0.84, H^+ ions are secreted against a concentration gradient of about 2 million to 1. This requires large amounts of ATP and accounts for nearly all the O_2 consumed by the stomach. The energy demands of this task are attested to by the large number of mitochondria present in the cytoplasm of this type of cell (see Figure 21-19, *B*). In addition to their special transport systems, parietal

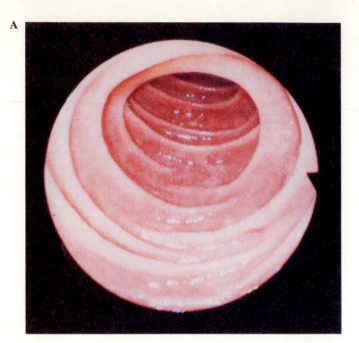
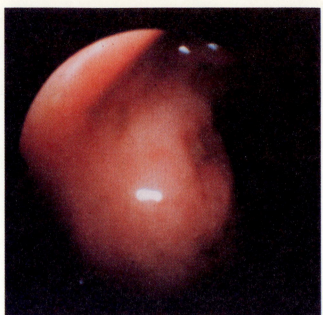

FIGURE 21-21
The normal duodenum (A) and a duodenal ulcer (B) as seen through the endoscope.

cells have an Na$^+$-K$^+$ pump on the basolateral surface that maintains intracellular concentration gradients for these two ions.

Any factor that increases acid secretion above normal levels can overcome the protective mucus barrier and damage the lining of the stomach, causing a **peptic ulcer.** More commonly, excessive gastric acid secretion ulcerates the duodenum (Figure 21-21) or esophagus, because these parts of the tube system are not as well protected against low pH as the stomach lining. There are several possible underlying causes of excessive acid secretion. Anxiety can cause increased parasympathetic activity, which indirectly brings about HCl secretion, and alcohol directly stimulates parietal cells. Prostaglandins normally present in the blood inhibit acid secretion. Since aspirin inhibits prostaglandin synthesis, it can increase acid secretion. Excessive HCl secretion can be controlled by drugs. **Histamine** has the most important role in the potentiation of parietal cell acid secretion, and these histamine receptors are blocked by the drug **cimetidine.**

Gastric Motility
Three functions of the stomach are related to its motility: (1) it stores meals of varying size, (2) mixes food with gastric secretions to form chyme, and (3) releases chyme into the small intestine at a rate the remainder of the GI tract can handle. In the first few hours after a meal, slow, mild contractions of the body of the stomach force the gastric contents toward the antrum. Pyloric and antral motility is highest during and just after meals. It decreases

with time, but the rate of decrease depends on the quantity and characteristics of foods that have been eaten.

Vigorous mixing of food with HCl, mucus, and enzymes to form chyme occurs in the lower portion of the body and the antrum. Mixing in the stomach can be compared with the operation of a food blender: if a lot of food is put in, the rotating blade in the lower part of the container will only act on the food in the lower region, leaving the material in the upper region as unprocessed lumps. If the food processor had an outlet that would allow the liquified portion in the lower region to be drawn off, so as to make room for the food in the upper regions to move down, it would operate more efficiently, and that is what happens when material moves from the pyloric region of the stomach into the intestine.

The motility of the body and antrum are controlled by a pacemaker region located near the middle of the body. This pacemaker initiates slow waves in the longitudinal muscle (about three per second) that spread into the circular muscle layer and sweep into the antrum toward the pylorus. The result is a wave of peristalsis, or **propulsion** (Figure 21-22, *A*). As this peristaltic wave nears the pylorus, it overtakes and passes the contents that are being propelled along. When the wave reaches the pylorus, the antrum and pylorus contract, closing the pyloric sphincter. A few milliliters of material may escape through the sphincter, but most of the gastric contents are forced back towards the antrum (**retropulsion:** Figure 21-22, *B*).

When food first enters the stomach, it is a dense

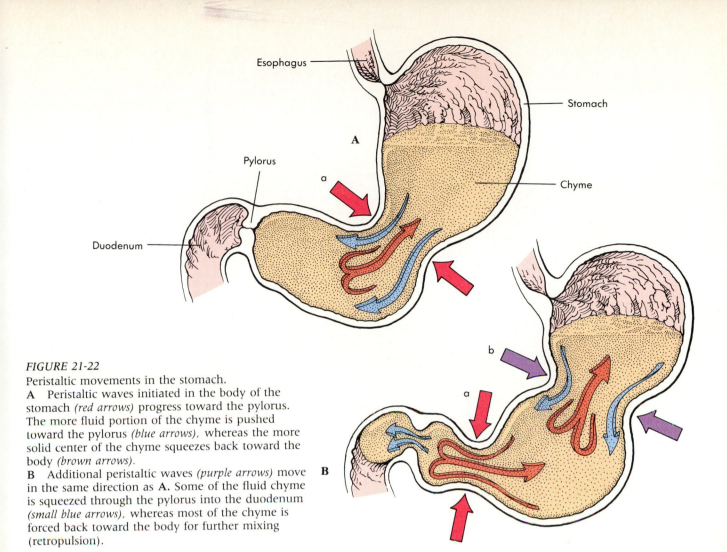

FIGURE 21-22
Peristaltic movements in the stomach.
A Peristaltic waves initiated in the body of the stomach *(red arrows)* progress toward the pylorus. The more fluid portion of the chyme is pushed toward the pylorus *(blue arrows)*, whereas the more solid center of the chyme squeezes back toward the body *(brown arrows)*.
B Additional peristaltic waves *(purple arrows)* move in the same direction as **A**. Some of the fluid chyme is squeezed through the pylorus into the duodenum *(small blue arrows)*, whereas most of the chyme is forced back toward the body for further mixing (retropulsion).

Labels in figure: Esophagus, Stomach, Pylorus, Chyme, Duodenum, A, B, a, b

mass. Propulsion is weak, so little material enters the duodenum. The contents become more fluid over the next few hours because secretions are added, and both propulsion and retropulsion increase. Gastric emptying gradually accelerates, but complete emptying of the gastric contents after a meal usually requires several hours because gastric motility is inhibited by signals that arise in the intestine. The mechanisms of the negative feedback exerted by the intestine on stomach emptying are described in Chapter 22.

> 1 What cellular mechanisms contribute to gastric acid secretion?
> 2 What aspects of gastric motility are involved in vigorous mixing of food with gastric secretions?
> 3 What is the function of the pylorus?

SECRETION AND MOTILITY IN THE INTESTINE
Alkaline Secretions of the Intestine and Pancreas

A thick mucus is needed to protect the duodenum from the enzymes and acid in the chyme that is just leaving the stomach, whereas a thinner mucus is adequate in the remainder of the small intestine. Along with the mucus, intestinal epithelial cells secrete an electrolyte solution that contains HCO_3^-. The majority of intestinal secretion is carried out by cells of the crypts of Lieberkühn (see Figure 21-5). In the duodenum, a significant amount of fluid is secreted by submucosal glands that empty into the crypts. Intestinal fluid secretion is increased by mechanical irritation and by factors that stimulate pancreatic secretion. It also can be stimulated by bacterial toxins, such as those produced by the organism that causes cholera. Excess intestinal fluid secretion results in severe diarrhea, dehydration, and the loss of HCO_3^-.

The pancreas incorporates two quite separate functions—an exocrine function and an endocrine function. These functions are served by distinct parts of the pancreas. Its exocrine portion secretes a solution of $NaHCO_3$ and digestive enzymes into the lumen of the duodenum by way of the pancreatic duct (Figure 21-23). The exocrine secretion is produced by two main classes of cells (see Table 21-3): **acinar cells** that secrete enzymes, and (2) **duct cells** that secrete the primary electrolyte solution.

Pancreatic secretion of fluid averages 2 liters/

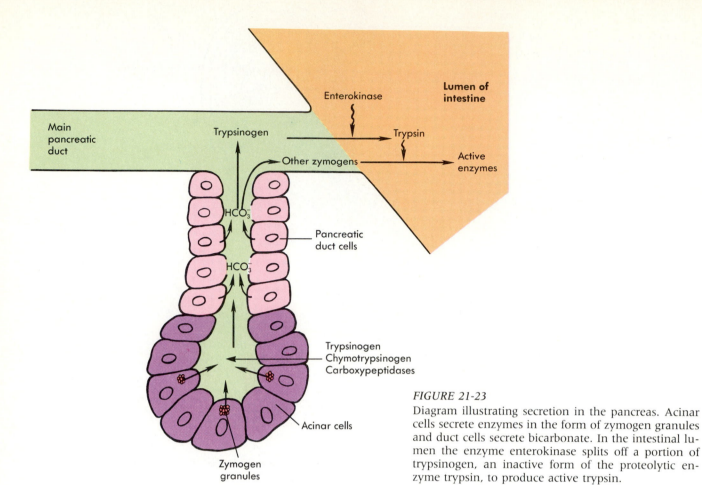

FIGURE 21-23
Diagram illustrating secretion in the pancreas. Acinar cells secrete enzymes in the form of zymogen granules and duct cells secrete bicarbonate. In the intestinal lumen the enzyme enterokinase splits off a portion of trypsinogen, an inactive form of the proteolytic enzyme trypsin, to produce active trypsin.

day, with most secretion occurring after meals. The cellular mechanism that forms the primary pancreatic fluid is similar to that of gastric acid secretion, but opposite in its direction (Figure 21-24). Within the cells, CO_2 and H_2O react to form H_2CO_3, which dissociates to H^+ and HCO_3^-. Na^+ leaks down its concentration gradient, driving the coupled movement of H^+ ions out the basal surface of the duct cell. The HCO_3^- moves into the pancreatic ducts in exchange for Cl^-, and a basolateral Na^+-K^+ pump maintains the intracellular Na^+ concentration of the duct cells at a normal low level.

Pancreatic duct cells are freely water permeable, so the pancreatic fluid is isotonic to plasma. As in the salivary glands, the primary fluid is modified by transport processes in duct cells, and the composition of fluid leaving the pancreas depends on flow rate. At high flow rates it resembles the primary fluid and contains 20 mEq/L $NaHCO_3$, 30 mEq/L NaCl, and a small amount of K^+. At lower rates the Cl^- concentration increases, the HCO_3^- concentration decreases, while the Na^+ concentration remains nearly constant.

Pancreatic Enzymes

The acinar cells located at the base of the pancreatic ducts secrete digestive, or hydrolytic, enzymes in

the form of **zymogen granules** (see Figure 21-23). Zymogens are inactive precursors of digestive enzymes that are subsequently activated in the intestine. Intracellular production and storage of the enzymes in an inactive form prevents the enzymes from digesting the secretory cells themselves. The membrane-bound vesicles containing the zymogens are released by exocytosis.

The five categories of enzymes released from the pancreas are: (1) **proteases** (peptidases) specific for particular kinds of peptide bonds, (2) enzymes that hydrolyze RNA and DNA to their component nucleotides, (3) **phospholipases** that reduce phospholipids to fatty acids and a polar compound, (4) **lipases** that degrade triglycerides into fatty acids and glycerol, and (5) an α-amylase similar to that in saliva.

How are inactive pancreatic zymogens converted to their active forms? It was noted earlier that pepsins in the stomach are activated by acid. There is very little HCl in the intestine, so a different mechanism is required from that which activates pepsins in the stomach. Intestinal epithelial cells have a membrane-bound enzyme called **enterokinase** (see Figure 21-23). Enterokinase converts the inactive proteolytic enzyme **trypsinogen** to **trypsin,** which then activates other pancreatic zymogens by removing

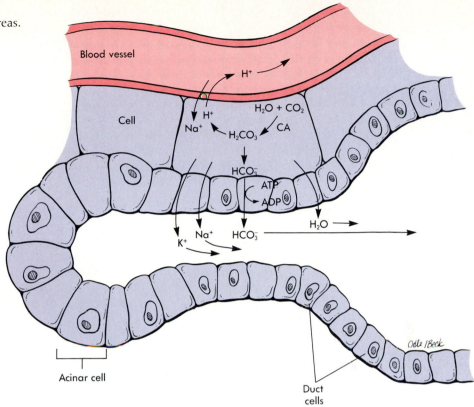

FIGURE 21-24
Bicarbonate production in the pancreas.

specific small peptides from them. Trypsin is also capable of activating trypsinogen, so once a small amount of trypsin is produced, the process of zymogen activation is rapid.

If the pancreatic duct is blocked, the normally low concentration of endogenous trypsin in the pancreas can rise high enough to overcome the protective effects of the **trypsin inhibitor** that is also present in the pancreas. If this occurs, the pancreas begins to digest itself, a disease called **pancreatolysis.**

The Liver and Bile
Fat digestion requires **bile,** a mixture of substances synthesized by the liver. Bile has six components: (1) **bile acids;** (2) the phospholipid **lecithin;** (3) **cholesterol;** (4) the end product of hemoglobin degradation, **bilirubin,** also referred to as a **bile pigment;** (5) toxic chemicals that have been converted into a harmless form, or **detoxified,** by the liver; and (6) an NaHCO₃ solution similar to that produced by the pancreas (see Table 21-4).

Bile acids are synthesized from cholesterol at a rate of about 0.5 grams/day and comprise about half the bile. Before they are secreted, bile acids are combined, or **conjugated,** with the amino acids glycine or taurine to make them more water soluble. Conjugated bile acids are called **bile salts.** The total bile acid and bile salt content of the body is only 3 to 4 grams. Bile salts are reused, and the entire bile acid content of the body recirculates about 10 times a

day (Figure 21-25). The 0.5 gram/day of bile acid that is lost in the intestines (open arrow) is about equal to the normal production rate by the liver. Recycling of secreted bile salts occurs in stages. Bacteria in the intestine deconjugate bile salts into bile acids, and these bile acids are absorbed in the small intestine. They travel to the liver in the hepatic portal circulation. The liver converts the reabsorbed bile acids into bile salts and secretes them along with newly synthesized bile acids.

Between meals, bile is stored in the gallbladder (see Figure 21-25). The **sphincter of Oddi,** located where the **common bile duct** and the **pancreatic duct** enter the duodenum, is normally closed, so bile that is continuously secreted by the liver is forced through the **cystic duct** into the gallbladder. The gallbladder actively absorbs Na⁺ and water and concentrates the bile five- to twentyfold. Since its capacity is about 30 to 40 ml, the gallbladder can store the equivalent of 200 to 800 ml of bile. Upon appropriate stimulation, the sphincter of Oddi relaxes and the smooth muscle of the gallbladder contracts, releasing concentrated bile.

Excessive secretion of cholesterol by the liver or overconcentration of the bile in the gallbladder can result in supersaturation of bile with cholesterol. If this condition is chronic, the cholesterol that precipitates from solution may form sandlike or pebble-like aggregates called **bile duct stones** or **gallstones,** depending on their location (Figure 21-26, *A*). These stones cause severe discomfort and impair

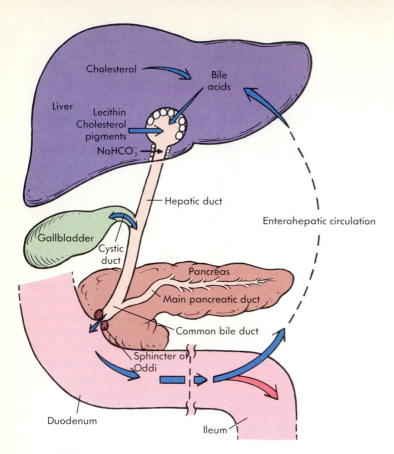

FIGURE 21-25
Bile secretion into the duodenum. Bile is manufactured in the liver and stored in the gallbladder. The sphincter of Oddi regulates the passage of bile and pancreatic secretion into the duodenum. The majority of the bile salts are transformed to bile acids in the ileum and recirculated to the liver.

digestion if they migrate into the common bile duct and block it. A new technique, lithotripsy, has been used successfully to disrupt gallstones that, in the past, have required surgery for removal (Figure 21-26, *B*).

> 1 *What categories of enzymes are secreted by the pancreas?*
> 2 *Why are pancreatic enzymes stored as zymogens? How are they activated in the intestine?*
> 3 *What are the major secretions of the liver? What is the role of the gallbladder? What are the possible fates of bile acids that reach the intestine in bile?*

Intestinal Motility

Intestinal motility has three functions: (1) mixing the chyme with the digestive enzymes, fluids, and bile secreted by the pancreas and gallbladder; (2) circulation of chyme within the lumen to facilitate contact with the absorptive epithelial cells; and (3) propulsion of the chyme toward successive regions of the GI tract. When the rate of segmentation is high, chyme moves more slowly and the mixing promotes digestion and absorption.

Peristaltic contractions are superimposed on segmentation, and their relative importance varies along the small intestine. In the duodenum and jejunum, peristaltic waves are infrequent and they propagate for only about 10 cm, while segmentation activity is high. This limits chyme movement to about 1 cm/min (0.5 meter/hour). In the later portions of the jejunum and in the ileum, peristaltic waves increase in amplitude and duration and begin to dominate segmentation, increasing the propulsion of material toward the large intestine. Over-

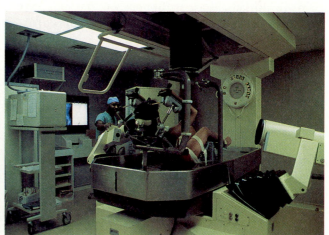

FIGURE 21-26
A X-ray of a patient who had both gallstones and bile duct stones.
B A new technique called lithotripsy, illustrated here, involves immersing a patient with gallstones in water and focusing high-frequency sound waves on the gallstones. These sound waves often disrupt gallstones, and the patient is able to avoid surgery.

all, material travels the 7 meters from the pylorus to the large intestine in about 5 hours.

The large intestine has little longitudinal smooth muscle, and what it has is concentrated in three bands called the **teniae coli** (Figure 21-27). Since circular muscle activity is high, motility in the large intestine is characterized by long-lasting segmentations known as **haustra.** The haustra slowly move the 1 to 2 liters of chyme that enters the large intestine each day back and forth, allowing ample time for absorption of most of the remaining fluid from the chyme. Haustra are more frequent in the descending and sigmoid colon than in the ascending and transverse colon, and the rate of movement of the contents of the colon is about 5 cm/hour.

Most propulsion in the large intestine takes place a few times each day during **mass movements** (Figure 21-28). In a mass movement, segmentation ceases and a strong peristaltic wave sweeps along the colon, moving some material into the rectum. Mass movements can transport chyme all the way from the cecum to the sigmoid colon and often occur following a meal.

The end of the ileum and the colon are separated by the **ileocecal sphincter,** which governs the rate of emptying of the small intestine (see Figure 21-27, A). The ileocecal sphincter is closed most of the time, preventing reflux of material from the large intestine into the ileum. The ileocecal sphincter relaxes when the end of the ileum is distended by chyme, and it contracts more strongly when the colon contracts. Factors that increase gastric motility tend to relax the ileocecal sphincter and permit an increase in the rate of entry of material into the colon.

Defecation

The anal canal has two sphincters (see Figure 21-27, A): the **internal anal sphincter** is composed of smooth muscle and is under autonomic control, whereas the **external anal sphincter** is composed of striated muscle and is under both conscious and

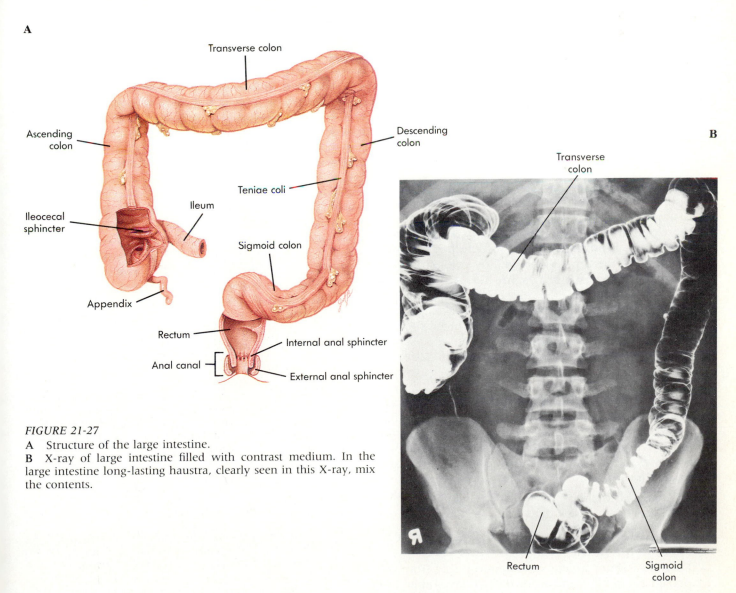

FIGURE 21-27
A   Structure of the large intestine.
B   X-ray of large intestine filled with contrast medium. In the large intestine long-lasting haustra, clearly seen in this X-ray, mix the contents.

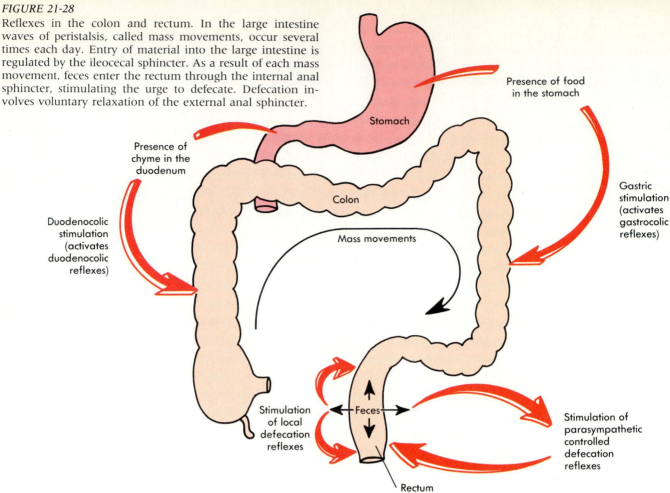

FIGURE 21-28
Reflexes in the colon and rectum. In the large intestine waves of peristalsis, called mass movements, occur several times each day. Entry of material into the large intestine is regulated by the ileocecal sphincter. As a result of each mass movement, feces enter the rectum through the internal anal sphincter, stimulating the urge to defecate. Defecation involves voluntary relaxation of the external anal sphincter.

Presence of food in the stomach

Stomach

Presence of chyme in the duodenum

Colon

Gastric stimulation (activates gastrocolic reflexes)

Duodenocolic stimulation (activates duodenocolic reflexes)

Mass movements

Stimulation of local defecation reflexes

Feces

Stimulation of parasympathetic controlled defecation reflexes

Rectum

autonomic control. The internal anal sphincter is normally contracted. As the rectum is increasingly distended, the internal anal sphincter relaxes and the external anal sphincter contracts. Distension of the rectum produces an urge to defecate. If conditions allow, the urge is followed by voluntary relaxation of the external anal sphincter. A series of reflex contractions in the descending and sigmoid colon and rectum then permits the evacuation of a relatively large amount of material, including that stored in the rectum. These contractions are assisted by an increase in pressure in the abdominal cavity brought about by the contraction of skeletal muscles.

If conditions do not allow defecation, the contraction of the external sphincter is sufficient to halt the rectal contractions, and the urge to defecate subsides. This sequence of stimulation followed by voluntary postponement of defecation can be repeated many times because the rectum can accommodate a fairly large quantity of material through stress relaxation.

THE SCALE OF DAILY FLUID AND SOLUTE Movements in the Gastrointestinal Tract

Food and water intake varies enormously from one person to another and from day to day. Table 21-5

summarizes the typical daily intake of an adult in the United States. Solid food (upper portion of the table) includes 300 to 800 grams of carbohydrate, 75 grams of protein, and 60 to 100 grams of fat (mostly as neutral fats, or triglycerides). Fats have twice the energy content of carbohydrates and proteins per

TABLE 21-5	Daily Gastrointestinal Inputs	
Material	**Source**	**Amount**
Carbohydrate	Food	300 to 800 g
Protein	Food	70 to 80 g
	Secreted enzymes	20 g
Fat	Food	60 to 100 g
Salt (NaCl)	Food	20 to 30 g
	Secretions	30 g
Water	Pure water and water in foods	1 to 1.5 L
Secretions	Salivary glands (NaHCO$_3$)	1.5 L
	Stomach (HCl)	3.0 L
	Pancreas (NaHCO$_3$)	2.0 L
	Liver (bile)	0.5 L

Effect of Gastrointestinal Tract Disorders on Body Fluid Homeostasis

Both the composition and volume of body fluid are threatened by disorders in which large amounts of the secretions of the GI tract are lost. Vomiting is a response to irritation of the stomach and intestine by chemical agents, bacterial or viral infection, or overdistension. For healthy people, vomiting has no lasting consequences, but in some illnesses and eating disorders, vomiting occurs repeatedly and constitutes a significant loss of fluid and acid from the body. Because H^+ and K^+ act as if they compete for a common carrier (see Chapter 20, p. 576), the alkalosis resulting from vomiting causes an increase in K^+ secretion by the distal tubule, so that hypokalemia results. The hypokalemia caused by chronic vomiting may result in death from cardiac arrhythmia.

Diarrhea is a symptom of irritation or infection of the intestine. It may result from increased motility or from interference with the normal control of secretion and reabsorption of fluid and solutes, or both. For the majority of dietary fluid and the secretions of the stomach, pancreas, liver, and initial portion of the small intestine to be absorbed, the chyme must pass through the small intestine and large intestine slowly. A large number of bacteria and viruses can infect the intestinal mucosa, releasing substances that inhibit segmentation and enhance peristalsis. The increase in motility moves the chyme through the large intestine before water and solutes can be absorbed, resulting in diarrhea.

The most serious varieties of diarrheal disease are the result of bacterial toxins that greatly increase intestinal secretion. Cholera toxin is one example. Cholera toxin is believed to bind to a receptor that normally recognizes a peptide hormone that regulates secretion in the intestine. Binding of the toxin with the receptor increases intracellular levels of cyclic AMP. The second messenger stimulates secretion and inhibits solute absorption. The secretion contains quantities of Na^+, K^+, and HCO_3^-. The loss of this secretion, together with dietary fluid and solutes that would otherwise have been absorbed, can rapidly result in a condition of dehydration and acidosis if not treated.

These symptoms, and not the infection itself, are the aspect of the disease that threatens the patient's life. The course of the disease is particularly rapid in infants and small children and is a major cause of infant mortality in developing nations. Death can usually be prevented if the patient is rehydrated orally with a simple solution that contains Na^+, K^+, HCO_3^-, and glucose; only in severe cases is intravenous rehydration or other medication necessary. Mothers in third-world countries can be supplied with this solution very cheaply, and they can be taught to use this treatment even if trained medical assistance is unavailable. As a sad postscript, it is very likely that diarrhea-producing toxins have been purified for possible use in chemical warfare, since their effects are both immediate and very disorganizing.

gram, so 30% to 40% of the calories come from fats, half of the calories come from carbohydrates, and the remaining 10% to 20% come from protein. Of the carbohydrates, typically 60% is starch, 30% sucrose (table sugar), and 10% lactose (milk sugar). The GI tract secretes about 20 grams of protein in the form of digestive enzymes each day. These proteins are broken down along with the ingested proteins, and the amino acids are recycled.

It takes about 4 hours to digest an average meal, and little useful material escapes. Indeed, the GI system can theoretically absorb as much as 20 pounds of sucrose per day! Fat absorption is more limited but still exceeds the body's requirement for lipids. The digestive system is so efficient that feces consist mainly of indigestible materials such as plant cellulose, inorganic matter, cellular debris shed from the intestinal lining, and bacterial debris. A normal diet can include daily ingestion of 20 to 30 grams of NaCl (see Table 21-5), and in addition to this, 30 to 40 grams of NaCl is secreted into the mouth and stomach along with watery secretions. Most of this ingested and secreted NaCl is absorbed. Other electrolytes consumed in gram amounts include K^+, Mg^{++}, Ca^{++}, and phosphate. The K^+ content of food and the K^+ secretion are 10 to 15 times lower than the values for Na^+. In contrast to the high degree of Na^+ conservation, about 25% of the K^+ that enters the intestine is normally excreted. A number of vital trace elements, including Ca^{++}, Fe^{++}, Zn^{++}, Cu^{++}, I^-, and F^-, are also ingested and are absorbed to varying extents.

Inputs of fluid into the GI tract are shown in the bottom two sections of Table 21-5. An average of 1.0 to 1.5 liters of water are consumed each day. GI secretions average 7 liters per day, including 1.5 liters of saliva, 3 liters of gastric HCl, 2 liters of pancreatic fluid (mostly $NaHCO_3$ solutions secreted to buffer the HCl), and 0.5 liters of bile. Most of the total fluid load of 8.5 liters/day is absorbed in the small intestine and some of the remaining water is absorbed in the large intestine, so only around 100 ml is present in the feces. In the stomach and small intestine, secretion and absorption of fluid roughly balance. In the colon the rate of absorption exceeds secretion, and the volume of the intestinal contents decreases.

1 What are the functions of intestinal motility?
2 What is a mass movement?
3 In what part of the intestine are haustra found?

SUMMARY

1. The organization of the GI system allows sequential processing of the food. In the mouth, food particles are reduced in size and mixed with saliva. In the stomach, the food is acidified, protein digestion begins, and food is stored until the chyme produced by mixing food with secretions is delivered to the small intestine. In the small intestine, the contents are neutralized by bicarbonate secretions, and digestion and absorption of proteins, carbohydrates, and lipids occur. The role of the large intestine is to store unabsorbed material.

2. The wall of the GI tract includes structures that perform many functions. The epithelium has both **secretory** and **absorptive** cells. Smooth muscles in the gastrointestinal tract wall are responsible for movement of the contents. Blood vessels and **lymph lacteals** are available for supplying the tissues' needs and for absorbing nutrients.

3. The muscular contractions that are responsible for mixing and moving the food from one region to another are the result of smooth muscle membrane depolarizations and reflex pathways of the enteric neural plexuses. The patterns include **segmentation** and **peristaltic waves.**

4. The secretions of the GI tract perform several functions. **Mucus** lubricates and protects all regions of the GI tract. It is present in saliva, along with fluid and α-amylase. In the stomach HCl and **pepsin** are secreted. In the small intestine, HCO_3^- and enzymes for the digestion of carbohydrates, proteins, and lipids are added from the pancreas. The **bile** secreted by the liver enters the duodenum along with the pancreatic secretions.

5. Acid and bicarbonate secretion is the result of active transport and facilitated transport processes. Blood leaving the stomach is more alkaline than blood in the rest of the circulation, especially after a meal. Blood leaving the intestine and pancreas, the sites of bicarbonate secretion, is more acidic than arterial blood. Over time, the secretion of acid and base tend to cancel one another.

6. **Bile salts** produced in the liver and released with bile into the duodenum are essential for fat digestion and absorption because of their **emulsifying** effect. After they have served their function in the proximal part of the small intestine, they are acted on by bacteria and reabsorbed in the form of **bile acids.** The bile acids return to the liver in the form of bile salts.

7. Intestinal motility mixes the chyme with digestive secretions, promotes absorption of nutrients, and propels the chyme from one region of the intestine to another. Segmentation (mixing) predominates in the early portions of the small intestine; peristalsis in the later portions; **mass movements** are the major propulsive activity of the large intestine.

● **STUDY QUESTIONS**

1. Describe the roles of the mouth, esophagus, stomach, small intestine, and large intestine in the sequential processing of food.

2. What are the two steps in saliva formation?

3. The tonic tension of smooth muscles in the GI tract can be increased or decreased. How are these changes accomplished, and how does the altered tension contribute to GI system function?

4. Explain why retropulsion dominates propulsion and only a small amount of chyme enters the small intestine with each peristaltic wave.

5. Which cells in the stomach show the greatest evidence of oxidative metabolism? What is the function of these cells?

6. List the categories of enzymes secreted by the pancreas.

7. What is enterokinase, and what role does it play in digestion?

8. What voluntary and involuntary processes contribute to defecation?

● **SUGGESTED READING**

FURNESS, J.B., and M. COSTA: *The Enteric Nervous System,* Churchill-Livingstone, Edinburgh, 1987. Complete description of neural control of gastrointestinal function.

HARRIS, A.R.: Why do more Women get Ulcers? *Chatelaire,* volume 61, March 1988, p. 32. Discusses the link between ulcers in women and risk factors such as smoking, alcohol, and stress.

JOHNSON, L.R.: *Gastrointestinal Physiology,* 3rd edition. The C.V. Mosby Company, St. Louis, 1985. A comprehensive textbook.

MARANTO, S.: Bugged by an Ulcer? You Could have a Bug, *Discover,* volume 8, May 1987, p. 10. Considers whether some ulcers are not primarily related to stress, but are caused by specific infections.

CHAPTER 22

Digestion and Absorption

On completing this chapter you will be able to:

- Understand how water, amino acids, and sugars are absorbed in the gastrointestinal tract.
- Appreciate the role of the transmembrane Na^+ gradient in carrier-mediated cotransport of most amino acids and sugars into the intestinal epithelial cells.
- Describe carbohydrate digestion in the mouth, stomach, and small intestine, noting the role of brush-border enzymes in the small intestine.
- Understand the steps in lipid digestion, noting the role of the bile in emulsifying fats.
- Describe how lipids are absorbed and resynthesized in the epithelial cells and transported in the lymph.
- Describe the processes involved in iron absorption.
- Understand the mechanisms for absorption of water-soluble and fat-soluble vitamins, noting the specialized transport mechanisms for vitamins.
- Characterize the cephalic, gastric, and intestinal phases of digestion.
- Compare long-loop and short-loop reflexes in the control of digestion.
- Identify the gastrointestinal tract hormones—gastrin, motilin, gastric inhibitory peptide, secretin, and cholecystokinin—and describe how each regulates gastrointestinal tract function.

*A*bsorption of water, electrolytes, and nutrients is carried out almost entirely in the small intestine. The maximum absorptive capacity of the intestine is much greater than the normal intake of most nutrients. Even when the amount of food a person consumes greatly exceeds his or her nutritional needs, the intestine still absorbs nearly 100% of the fat, protein, and carbohydrate ingested. Caloric intake must be therefore regulated at the level of ingestion because it is not regulated at the step of absorption. In contrast, a few substances, such as calcium and iron, are usually not completely absorbed by the intestine, so adjustments in the rate of absorption of these substances from the gastrointestinal (GI) tract are important in regulating blood levels.

The GI system is intrinsically regulated to an extent unmatched among organ systems. Its single-unit smooth muscle is capable of spontaneous rhythmic contraction even in the absence of nervous or hormonal inputs. The "visceral brain" of the GI tract, consisting of the two plexuses with their web of interconnections, contains about 10^8 neurons, a number similar to that in the spinal cord. Although the plexuses are technically parasympathetic ganglia, most of the neurons are neither adrenergic nor cholinergic, but instead use small peptides as neurotransmitters and neuromodulators. The GI tract also has its own endocrine systems, which help coordinate motility and secretion in different parts of the tube system. The underlying kinship between GI endocrine cells, the neurons of the plexuses, and the central nervous system is seen in their chemical signals. Peptides that were first known as gut hormones are now known to also serve as neurotransmitters in the plexuses and in the central nervous system.

SOLUTE AND FLUID ABSORPTION INTO BLOOD AND LYMPH
Pathways for Absorption from the Gastrointestinal Tract

The structures that lie between the lumen of the intestine and the blood (Figure 22-1) are (1) the layer of mucus covering the microvilli, (2) the mucosal epithelium, (3) the interstitial space, and (4) the capillary walls. The pathways taken through these barriers differ for different substances. The forms in which nutrient molecules can be absorbed into the body are usually quite different from the more complex molecules provided by the diet. The digestion of lipids, carbohydrates, and proteins has one factor in common: they all involve hydrolysis (see Chapter 3, p. 45). In hydrolytic reactions, a bond that links subunits of the food molecule is split, and a hydrogen and hydroxyl ion from a water molecule are added to the two ends of the subdivided molecule. Digestive enzymes catalyze the hydrolysis of specific categories of bonds; their action is optimized by secretions that adjust the pH and osmotic strength of the contents of the GI tract.

After food is reduced to molecules that can be taken into the body, the process of absorption begins. Most sugars and amino acids are transported into epithelial cells of the small intestine by carrier-mediated cotransport systems that are driven by Na^+ moving down its concentration gradient into the cells. The energy cost of this transport system is paid by the ATP expended by the Na^+-K^+ ATPase to pump out the Na^+ ions entering in this cotransport. Large, insoluble fat droplets must be broken down into smaller ones before they can be fully digested in the small intestine. After digestion, rather than moving from the epithelial cells into the plasma along with sugars and amino acids, lipids follow another pathway. They move first into the lacteals and then enter the systemic circulation along with the lymph.

Water and Electrolyte Movement across the Intestinal Wall

Of the 8.5 liters of water that enters the GI system each day, normally about 7.5 liters is absorbed by the small intestine and about 0.9 liter is absorbed by the large intestine, leaving approximately 100 milliliters in the feces. Most water absorption in the GI tract occurs by the same mechanism found in the proximal tubules of the kidney. That is, water reabsorption is secondary to active Na^+ transport. A Na^+ pump located on the basolateral surface of the intestinal epithelial cells actively transports Na^+ from the epithelial-cell cytoplasm to the interstitial fluid, reducing the intracellular Na^+ concentration (Figure 22-2). As a result, Na^+ enters the epithelial cells

FIGURE 22-1
To enter the blood or lymph, materials absorbed from the intestinal lumen must traverse an external mucous layer, the epithelial cell membrane, the epithelial cell cytoplasm (which includes the core of the microvilli), the basolateral membrane, the interstitial fluid, and the membranes and cytoplasm of the capillary or lymph endothelial cells. The route through the mucosal barrier and into the blood is different for different chemical classes of nutrients.

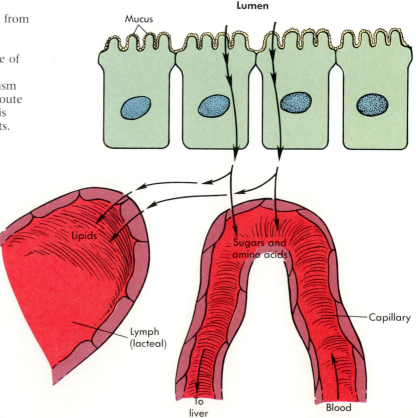

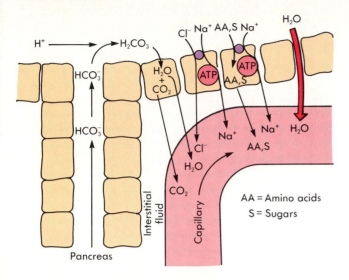

Lumen of intestine

AA = Amino acids
S = Sugars

FIGURE 22-2
In the small intestine, reaction of H^+ with HCO_3^- produces H_2CO_3, which enters the bloodstream as CO_2 and H_2O. Na^+ absorption involves passive Cl^- movement along with the Na^+, which is actively pumped out the basolateral surface of the cells. Water follows passively. Sugars and amino acids are moved from the lumen into the blood by coupled transport systems, and their absorption also drives water absorption.

from the lumen of the GI tract by diffusion, and the passive movement of Cl^- along with the Na^+ maintains neutrality. The net effect is the movement of NaCl, first to the interstitial fluid and then to the capillaries. Salt transport creates an osmotic gradient between the lumen and interstitial fluid, and this Na^+-transport–dependent osmotic gradient is the main driving force for water absorption into the capillaries (see Figure 22-2).

Amino acids and sugars are taken up by intestinal epithelial cells by specific Na^+-dependent cotransport systems (see Figure 22-2). This is a sec-

ond mechanism that creates an osmotic gradient for water movement because the net effect is NaCl uptake. In addition, the absorption of sugars and amino acids decreases the osmolarity of the chyme, creating an osmotic driving force for water absorption. An interesting but minor factor contributing to water absorption is the reaction of the H^+ released in the stomach with the HCO_3^- secreted by the pancreas and secretory cells of the intestine. These two ions combine to form H_2CO_3, which is broken down in the epithelial cells to produce CO_2 and H_2O. Thus two particles that increase the osmotic activity of the chyme react to produce a molecule of H_2O and the gas CO_2, both of which move into the blood (see Figure 22-2).

Chyme entering the duodenum from the stomach is normally hypertonic. The water permeability of the duodenum is high (Table 22-1), and water flows osmotically into the chyme, making it isotonic. Following duodenal equilibration, water, electrolytes, and digestive end products are reabsorbed as the chyme passes through the rest of the GI tract. Because the small intestine is permeable to water, the chyme remains isotonic in the jejunum and most of the ileum as solute and water reabsorption keep pace with each other (see Table 22-1).

The amount of water absorbed in the GI system depends on the speed with which chyme is moved through the intestine. Rapid passage of material through the small intestine does not allow sufficient time for the transport and absorption of solutes, and water is also not absorbed. Reduced absorption of fluids and electrolytes can occur if motility increases because of irritation of the epithelium. Conversely, the longer material remains in the large intestine, the greater the amount of water reabsorbed.

Ions that cannot be absorbed, such as Mg^{++} (an ingredient in several laxatives and antacids), can be ingested, and the presence of such ions keeps the osmolarity of the intestinal contents high and prevents water from being absorbed out of the lumen. The diarrhea resulting from this failure of water reabsorption is produced by a mechanism that is similar to the action of osmotic diuretics.

TABLE 22-1   Water and Salt Reabsorption

Region	Salt transport (lumen to blood)	Water permeability	Secretion	Tonicity of content
Duodenum	Moderate	High	High	Hypertonic
Jejunum	High	High	High	Isotonic
Ileum	Moderate	Moderate	Low	Isotonic
Large intestine	Moderate	Low	Low	Hypertonic

TABLE 22-2 *Functions of the Liver*

Physiological process	Product(s)	Action
Digestion	Bile salts; bile acids; lecithin	Emulsifies fats so they can be absorbed.
	Cholesterol	Recirculates bile salts via the entero hepatic circulation.
Detoxification	Urea; detoxified drugs; inactivated hormones	Adds polar groups to drugs, some hormones, and several metabolites so that they can be excreted by the kidney. Converts ammonia to the much less toxic urea.
Biosynthesis	Plasma proteins	Synthesizes all plasma proteins except the immunoglobulins; include clotting factors and complement (discussed in Chapter 24).
	Lipoproteins	Makes protein carriers for cholesterol and the triglycerides.
Energy metabolism	Glucose; fatty acids; ketones	Stores glucose as glycogen; converts amino acids to glucose; synthesizes and degrades lipids.
Other functions	Iron; heme; transferrin	Destroys aging or damaged red blood cells.

The Liver and Hepatic Portal Circulation

The **liver** plays important roles in digestion, energy metabolism, biosynthesis, and detoxification (Table 22-2). The role of the liver in energy metabolism is described in Chapter 23. Its digestive function is the secretion of **bile,** which is essential for lipid digestion in the intestine. Bile is secreted continuously and passes from the liver into the gallbladder where it is stored and concentrated (see Chapter 21, p. 555). During digestion, contractions of the gallbladder expel bile into the duodenum.

Blood that has passed through the capillaries of the stomach, large and small intestines, spleen, and pancreas is carried by the **hepatic portal vein** to the liver (Figure 22-3). This blood passes through a second set of capillaries in the liver before returning

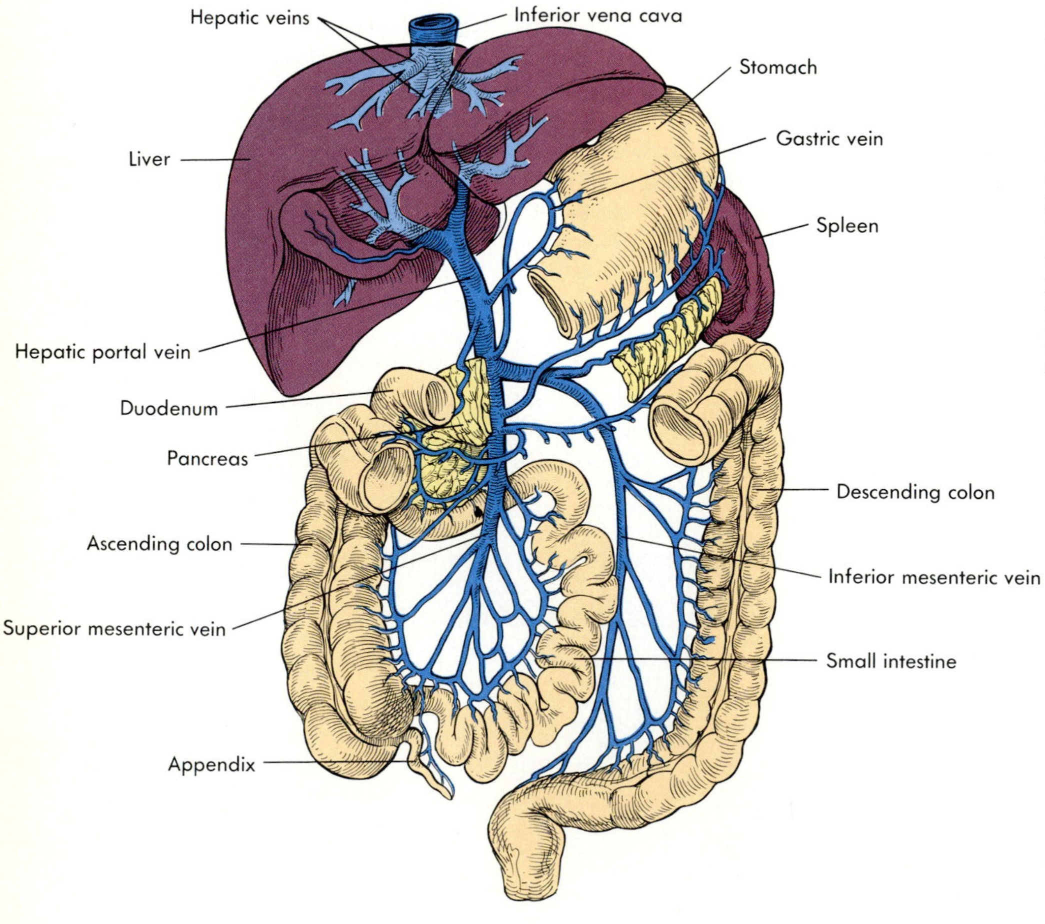

FIGURE 22-3
The hepatic portal system. Most of the blood supply of the liver first passes through the capillaries of the GI system and then enters the hepatic portal vein. A smaller portion of the liver blood supply is arterial blood, which enters through the hepatic artery and supplies the liver cells with oxygen. The blood in the hepatic vein and a small portion of the blood from the GI tract join the blood that flows into the heart from other organs of the body.

to the heart. About 75% of the blood that enters the liver comes from the GI system by way of the hepatic portal vein, and the other 25% (bringing vital oxygen to the liver cells) comes from the systemic circulation by way of the hepatic artery. Blood from both origins leaves the liver in the **hepatic vein.**

Liver cells, or **hepatocytes,** are arranged in a series of sheets only one or two cells thick, separated by the liver capillary network (Figure 22-4). The **lobules,** or functional units of the liver, are sheets of hepatocytes organized around a central vein. Large capillaries, or **sinusoids,** run between the sheets of hepatocytes. The sinusoids receive blood enriched by absorbed nutrients from a branch of the hepatic portal vein. This blood has a low O_2 content because it has already delivered O_2 to the GI tract. However, the sinusoids also receive blood with a high O_2 content from the hepatic artery, and this mixes with the portal blood. The hepatic artery blood supply to the liver meets the liver's needs for gas exchange and also allows the liver to regulate the amounts of sugars and amino acids in the general circulation (see Chapter 23, p. 593). The bile secreted by the hepatocytes does not move into the blood but into specialized ducts that form a system of canals, called **bile canaliculi,** located within each lobule of the liver. These eventually converge to form the bile duct.

The liver plays an important part in the turnover of numerous substances found in the blood, including drugs, hormones, and metabolic end products. The liver converts some substances to more water-soluble forms that can be eliminated by the kidney. Generally this is accomplished by **conjugation**—adding to the substance a polar group such as glucuronic acid, taurine, or glycine. An example of a substance handled in this way is bilirubin, the product of toxic porphyrins released in the degradation of hemoglobin. Conjugated substances may pass into the bile or reenter the blood and be excreted by the kidneys via a secretory pathway specific for organic anions (see Chapter 19, p. 488). The liver also converts some substances to less toxic forms. For example, ammonia is converted to the less toxic urea. Enzymes in the liver degrade some hormones and convert certain drugs and toxins to inactive forms, whereas other drugs (and some toxins) are actually made more toxic by reactions in the liver. The ability of the liver to detoxify drugs decreases with age, so drug dosages appropriate for young adults may produce excessively high plasma levels, and thus toxic side effects, in the elderly.

Another role of the liver is to synthesize all plasma proteins except the γ-globulins. The proteins secreted by the liver include albumin (important in maintaining the plasma osmotic pressure), the protein carriers for plasma transport of cholesterol and triglycerides, clotting factors, angiotensinogen, and the complement proteins involved in the immune response (see Chapter 24, p. 632).

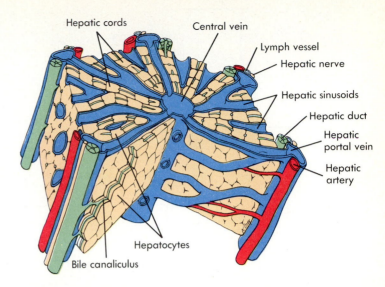

FIGURE 22-4
The liver is composed of lobules, which consist of stacks of sheets of cells through which blood drains in large spaces, called sinusoids, into a central vein. In the portion of a lobule shown here, the mixing of blood from the hepatic artery and hepatic portal vein is illustrated, as are the canaliculi through which bile drains in the opposite direction to the bile duct.

1 What is the role of the hepatic portal system?
2 What role does the blood that enters the liver from the hepatic artery play in hepatic function?
3 What forces determine the distribution of water across the intestinal epithelium?

CARBOHYDRATE DIGESTION AND ABSORPTION
Dietary Sources of Carbohydrate
Dietary carbohydrates fall into three major categories: **monosaccharides, disaccharides,** and **polysaccharides** (see Chapter 3 and Figures 3-5 and 3-6). Monosaccharides include **glucose** (a six-carbon sugar) and **fructose** (a five-carbon sugar). Dietary disaccharides include **sucrose** (table sugar), a disaccharide of glucose and fructose that constitutes 30% of the carbohydrate intake; and **lactose** (milk sugar), a disaccharide of glucose and **galactose** that accounts for 6% of the dietary carbohydrate. Ribose and deoxyribose (present in RNA and DNA) and the disaccharide **maltose** are in the diet in small amounts. **Starch** is the major dietary polysaccharide. Plant starch (**amylopectin**) and animal starch (**glycogen;** see Figure 3-7) consist of branching chains of glucose and together constitute over 50% of the total daily carbohydrate intake.

Although it is a polysaccharide of glucose, **cellulose,** the main structural element in plants, cannot be digested because cellulose contains chemical bonds that are resistant to attack by the enzymes in the GI system. In some animals, such as cattle, particular microbes that live in the gut provide enzymes that

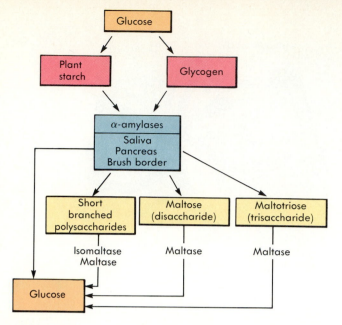

FIGURE 22-5

The two dietary sources of starch are both composed of glucose molecules and can be hydrolyzed by α-amylases. The four major products of starch digestion by the amylases are listed, along with the brush-border enzymes, which complete their digestion to glucose.

can digest cellulose. The cellulose in whole grains, vegetables, and fruits is partially degraded by bacteria in the human large intestine, but this does not release enough glucose to be a significant factor in nutrition. Undigested cellulose contributes to the volume of material in the alimentary canal and stimulates motility of the large intestine. Cellulose is therefore the major component of dietary **fiber**.

Starch Digestion by Amylases and Brush-Border Enzymes

Starch digestion begins in the mouth, where the enzyme, **α-amylase**, is secreted by the salivary glands (Figure 22-5). This enzyme attacks the linkages between adjacent glucose molecules in the straight-chain regions of polysaccharides. Salivary amylase is inactivated by stomach acid, but the pancreas secretes another, very effective, α-amylase (Table 22-3). Digestion of bonds in the starches that cannot be attacked by the amylases is accomplished by **brush-border enzymes**, which are bound to the mucosal membranes of intestinal cells (see Figure 22-5).

Monosaccharide Absorption

Although starch digestion produces only glucose, the dietary carbohydrates, sucrose and lactose, are digested to glucose, galactose, and fructose. All three of these monosaccharides are then taken up by the epithelial cells. There is an Na^+-dependent cotransport system for glucose and galactose in intestinal epithelial cell membranes (Figure 22-6) that is similar to that found in the proximal tubular cells of the kidney (see Chapter 19, p. 486). Because glucose and galactose compete for the same carriers, their carriers can be saturated by a lot of either sugar or by some of both. The energy that indirectly supports this transport is expended by the Na^+-K^+ pump, which bails out Na^+ that enters with the sugar molecules. This expenditure of energy ensures that the Na^+ gradient driving rapid absorption is maintained.

In addition to the Na^+-dependent system, there are Na^+-independent carrier systems for glucose and fructose. Monosaccharides in the cytoplasm of the intestinal epithelial cells reach the interstitial fluid by facilitated diffusion across the basolateral membranes (see Figure 22-6). The monosaccharides subsequently diffuse into the blood.

> 1 What is the difference between a monosaccharide, a disaccharide, and a polysaccharide? In which of these forms is carbohydrate absorbed?
> 2 What types of enzymes are involved in carbohydrate digestion? Explain why cellulose cannot be digested.
> 3 What are the mechanism of intestinal sugar absorption? What driving force results in absorption of glucose and galactose?

TABLE 22-3	Pancreatic Enzymes	
Enzyme	**Substrate**	**Products**
Amylase	Plant starch, glycogen	Maltose, short chains of glucose molecules
Trypsin, chymotrypsin	Internal peptide bonds	Free amino acids, small peptides
Carboxypeptidases	Peptide bonds at carboxy end of peptides	Free amino acids
Aminopeptidases	Peptide bonds at amino end of peptides	Free amino acids
Elastase, collagenase	Internal peptide bonds of collagen	Small peptides
Lipase	Triglycerides	Free fatty acids, 2-monoglycerides, glycerol
Phospholipase	Phospholipid	Lysophosphatides, free fatty acids
Hydrolase	Cholesterol esters	Cholesterol, free fatty acids
RNAase	RNA	Short chains of ribonucleic acid
DNAase	DNA	Short chains of deoxyribonucleic acid

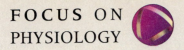
Lactose Intolerance

Normally, carbohydrate digestion and absorption is complete midway through the jejunum, and little monosaccharide reaches the ileum. The ileum has a much lower capacity for absorbing monosaccharides and can be overloaded if the small intestine is diseased. The pancreas nearly always secretes amylase vastly in excess of needs. Most disorders of digestion involve deficiencies of the carbohydrate-specific enzymes present on the membranes of the intestinal epithelial cells. One such condition is called **lactose intolerance.**

Lactose is the major carbohydrate component of milk, and it is degraded by the brush-border enzyme, **lactase,** to glucose and galactose. This enzyme is normally present in infants, but it disappears from the intestine after the first few years of life in most of the world's population. However, most white Europeans and Americans continue to express the enzyme, although a significant fraction (10%) either loses lactase as adults or congenitally lacks it. When lactase is absent, lactose cannot be digested or absorbed and instead remains in the lumen, where it is osmotically active, like the Mg^{++} present in laxatives. The result is diarrhea and fluid loss. In addition, bacteria in the large intestine metabolize lactose to produce large quantities of CO_2 gas, which causes distension and pain. A temporary intolerance of lactose can occur even in children who possess lactase if an intestinal infection causes the cells lining the intestine to be shed more rapidly than usual. Because the immature cells do not have as many functioning enzymes, digestion of lactose is less efficient. Individuals with lactose intolerance can consume milk products such as yogurt and cheese perfectly well because the action of the bacteria used to make these products breaks down the lactose in them before they are consumed. Milk containing cultures of lactose-digesting bacteria (acidophilus milk) is another solution to lactose intolerance.

FIGURE 22-6
Two transport systems for sugars are shown: galactose and glucose share one Na^+-dependent cotransport system, and fructose uses a second Na^+-independent transport system. The sugars are probably moved out of the cells by specific carriers *(black dots)*. Movement into the blood is by diffusion.

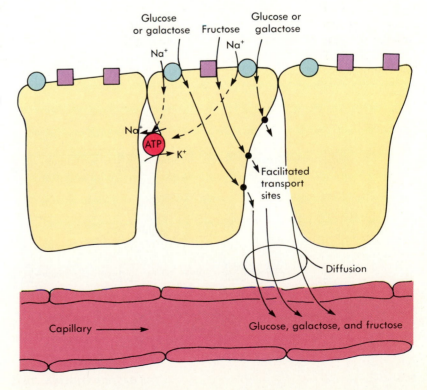

PROTEIN DIGESTION AND ABSORPTION
Enzymes of Protein Digestion

There are more types of amino acids in proteins than there are simple sugars in carbohydrates (see Chapter 3, Figures 3-8, 3-10, and 3-11). Proteins are formed in a dehydration reaction that unites the **carboxy terminal** of one amino acid with the **amino terminal** of another (see Figure 3-10). Enzymes (**peptidases**) that digest proteins recognize particular amino acids or classes of amino acids and attack the **peptide bonds** between them in a process termed hydrolysis (Figure 22-7, A). Peptidases fall into two classes, those which break bonds in the center of polypeptide chains (**endopeptidases**), and those which chop away at polypeptide chains from their free ends (**exopeptidases**). Of the latter, those attacking the free carboxy end are termed **carboxypeptidases**; those attacking the amino end are called **aminopeptidases** (Figure 22-7, B).

The pepsins secreted in the stomach are endopeptidases that are only active when the pH is below 3.0. A low pH breaks some internal bonds that determine the three-dimensional structures of ingested proteins (a process called **denaturation**). Denaturation exposes peptide bonds that are normally deep inside the protein to the action of the digestive endopeptidases. Because a particular endopeptidase is specific for only some of the peptide bonds in a protein, the action of the gastric endopeptidases primarily cuts ingested proteins into polypeptides. Gastric pepsins are inactivated when the acidity of the chyme entering the duodenum is neutralized by pancreatic and duodenal fluid secretions. Gastric peptidases begin the job of protein digestion, but 85% of the peptide bonds must be hydrolyzed in the small intestine by pancreatic and intestinal brush-border proteolytic enzymes.

About 50% of the protein intake is digested by enzymes secreted by the pancreas. Both classes of peptidases (endopeptidases and exopeptidases) are present in the pancreatic secretions, act as free enzymes, and are mixed with the chyme. Examples of endopeptidases secreted by the pancreas are **trypsin** and **chymotrypsin** (see Table 22-2). Brush-border peptidase activity accounts for the remaining 35% of protein digestion, which occurs in the small intestine. The brush border of intestinal epithelial cell membranes has several types of endopeptidases and exopeptidases.

Mechanisms of Amino Acid Absorption

The products of GI protein digestion are amino acids and a few small peptides. Many amino acids can be classified as neutral, basic, or acidic (see Figure 3-8). Others have special structural features that dominate their chemical reactions. There are transport systems that recognize classes of amino acids, such as the acidic or basic groups, as well as a transport system that can accept small peptides. Peptides transported into the intestinal epithelial cells are degraded into their constituent amino acids by enzymes in the epithelial cell cytoplasm.

Most amino acids are known to be driven from the intestinal lumen into the cytoplasm of intestinal

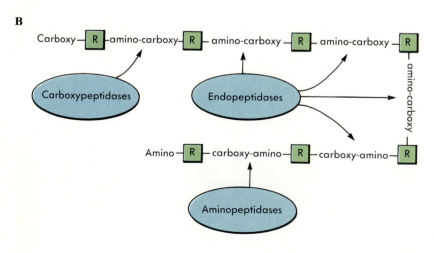

FIGURE 22-7

A A dipeptide (its amino acids are labeled *R*) is hydrolyzed to yield two amino acids. The peptide bond and the amino and carboxy terminals of the dipeptide are indicated, and the sites where the H+ and OH- are added when the peptide bond is broken are shown on the right.

B A polypeptide (its amino acids are labeled *R*) is digested by three classes of peptidases: the carboxypeptidases act from the carboxy end, the aminopeptidases act from the amino end, and the endopeptidases are a diverse family of enzymes that can attack specific classes of peptide linkages within the chain.

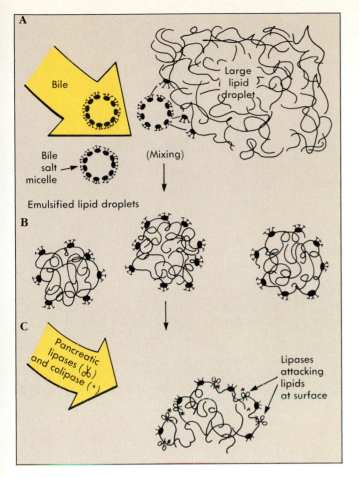

FIGURE 22-8

The steps in lipid solubilization and digestion.

A Bile salts enter the lumen of the GI tract in the form of spherical micelles, with polar groups extending toward the surrounding water and hydrophobic regions oriented toward the center of the micelle. Mixing of intestinal contents brings the bile salt micelles into contact with large aggregates of lipids.

B Some of the lipids are separated from the large masses when their surfaces are covered with bile salts. This process, called emulsification, increases the surface area of the lipids.

C Pancreatic lipases, with the help of colipase, attack the lipid bonds exposed on the surface of the emulsified lipid droplet.

epithelial cells by Na^+ moving down its concentration gradient (see Figure 22-2). Amino acids probably move out the other side of the epithelial cell into the interstitial fluid by an Na^+-independent mechanism. Once in the interstitial fluid, the amino acids enter the plasma and travel along with the products of carbohydrate digestion through the hepatic portal vein to the liver.

> 1 How are the conditions for protein digestion different in the stomach and small intestine?
> 2 What are the most common end products of protein digestion in the stomach and small intestine?
> 3 How is the mechanism for amino acid absorption similar to the mechanism for sugar absorption?

FAT DIGESTION AND ABSORPTION
Fat Emulsification in the Intestine

Most dietary fat consists of **triglycerides** (neutral fats), **phospholipids, cholesterol** or **cholesterol esters** (cholesterol joined to a single fatty acid chain), and some **fatty acids** (see Chapter 3, p. 40). Neutral fats are insoluble in water and form large droplets that coalesce and separate completely from the water phase. Fatty acids and phospholipids are slightly more water soluble, but they still tend to aggregate into relatively large sheets, or bilayers, with the polar groups facing outward and the hydrocarbon chains facing inward. The fats tend to float on top of the other ingested material, and the churning of the stomach reduces the size of the lipid droplets only slightly.

The first problem that must be overcome to digest lipids is to break up large lipid droplets into smaller ones. This increases the surface area for attack by enzymes that are in the aqueous phase. The bile secretions contain a class of molecules that reduce the size of the lipid aggregations by acting very much the way detergents act to "cut" grease. The process, called **emulsification,** is carried out by bile salts derived from the steroid molecule, cholesterol.

Bile salts have polar groups attached to one side, causing them to be **amphipathic** (see Chapter 3, p. 40)—able to associate with polar molecules such as water and with nonpolar molecules like lipids. The bile salts associate with each other and enter the intestine in the form of **micelles** (Figure 22-8, *A*). Bile salts reduce the surface tension of fat droplets and make them more susceptible to disruption into smaller droplets by segmentation of the intestine. The smaller droplets are coated with bile salts and prevented from recoalescing (Figure 22-8, *B*). The result of mixing bile salts with the intestinal contents is the formation of **emulsified droplets** about 1 μm in diameter. This greatly increases the surface area available for the actions of digestive enzymes and therefore reduces the time required for lipid digestion.

Fat Digestion by Pancreatic Lipases

The digestion of fat **(lipolysis)** is accomplished by lipid-digesting enzymes called **lipases,** which are secreted by the pancreas (see Table 22-3; Figure 22-8, *C*). Unlike the proteases, the lipases are secreted in their active forms. Pancreatic lipase action on triglycerides produces **free fatty acids, 2-monoglycerides,** and small amounts of **glycerol** (Figure 22-9). The reason that fairly large amounts of 2-monoglycerides are produced is that lipase tends to attack the fatty acid chains on the end carbons of the glycerol backbone, leaving the one on the center carbon intact.

In addition to lipase, the pancreas secretes an enzyme called **hydrolase** (see Table 22-3), which

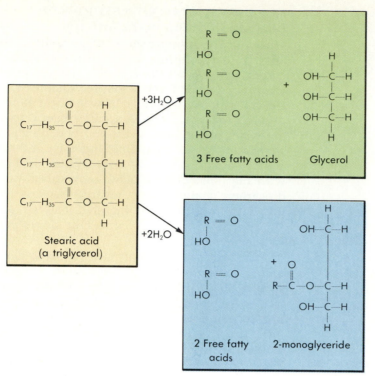

FIGURE 22-9

A triglyceride (for example, stearic acid) is hydrolyzed to glycerol and three fatty acids (long chain represented by R) by the addition of three molecules of H_2O. Alternatively, and more commonly, the triglyceride is hydrolyzed with two H_2O molecules to yield two free fatty acids and a 2-monoglyceride.

breaks down cholesterol esters into fatty acids and cholesterol. Another enzyme that digests lipids is **phospholipase** (see Table 22-3), whose end products are **lysophosphatides** (phospholipids from which the second fatty acid has been cleaved) and fatty acids. Pancreatic lipases act only at the boundary between a lipid droplet and water (see Figure 22-8, C). Phospholipases also require bile salts for their activity. The lipid-digesting enzymes reduce ingested lipids to fatty acids, cholesterol, lysophosphatides, 2-monoglycerides, and glycerol.

By themselves, pancreatic lipases are able to attack lipids in emulsified droplets, but the action of lipase is greatly enhanced by a pancreatic polypeptide known as **colipase.** Colipase has three effects: (1) it increases the ability of pancreatic lipase to adhere to lipid droplets, (2) it prevents bile salts from inhibiting lipase activity, and (3) it changes the pH required for maximum lipase activity so that this pH optimum more nearly corresponds to the actual conditions in the small intestine.

Micelle Formation and Fat Absorption
The end products of lipase and phospholipase activity (free fatty acids, cholesterol, lysophosphatides, 2-monoglycerides, and glycerol) combine with the bile phospholipid lecithin, bile salts, and cholesterol to form structures 50 to 100 nm in diameter called **mixed micelles** (Figure 22-10). In the mixed mi-

celles, the hydrophobic molecules and the parts of amphipathic molecules that are hydrophobic are located in the center of the sphere, whereas the surface of the sphere is studded with the polar projections, consisting primarily of the charged groups on the bile salts. These mixed micelles move around within the chyme and eventually come into contact with the microvilli of the intestinal epithelial cells (see Figure 22-10).

Mixed micelles do not pass as intact structures from the lumen of the intestine into the cytoplasm of epithelial cells. Instead, free fatty acids, 2-monoglycerides, and glycerol diffuse as individual molecules through the lipid bilayer structure of the cell membranes. The products of lipolysis are in much higher concentration in the micelles than in the aqueous phase, but an equilibrium exists between the two phases. When the concentration of lipid in the aqueous phase becomes very high, as it is near the site of digestion, the micelles accumulate dissolved lipids. As mixing of the intestinal contents carries these loaded micelles into the vicinity of the microvilli, the local concentration of fatty acids and 2-monoglycerides in the aqueous phase is much lower because they are dissolving in and disappearing across the lipid bilayer of the epithelial cell membranes. A new equilibrium is established, under these circumstances, and the products of lipolysis move out of the mixed micelle down their con-

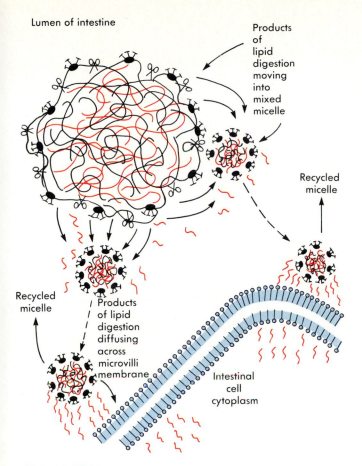

Lumen of intestine

Products of lipid digestion moving into mixed micelle

Recycled micelle

Recycled micelle

Products of lipid digestion diffusing across microvilli membrane

Intestinal cell cytoplasm

FIGURE 22-10
Absorption of lipids is facilitated by the accumulation of fatty acids, glycerol, and other products of fat digestion in mixed micelles. This occurs near the site of lipase activity. The mixed micelles give up their by-products of lipid digestion in the vicinity of the microvilli, where they diffuse across the lipid bilayer membrane. The micelles may be carried back to the site of lipid digestion and then pick up products of lipid digestion.

centration gradient into the aqueous environment, where they are free to diffuse into the epithelial cells. Most bile salts do not enter the cells, and bile salts can continue the process of micelle formation when the bile salts are removed from the intestine, reprocessed, and returned to the site where lipids are being digested.

When bile salts reach the ileum, they are actively transported into the epithelial cells and returned to the liver in the portal circulation (see Chapter 21). Their uptake by the liver stimulates bile salt release, an action that is effective in delivering them to the duodenum as long as the sphincter of Oddi is relaxed (see actions of cholecystokinin, below). This rapid cycling of the bile salts allows the limited supply of bile salts to be used repeatedly in digestion of a meal high in lipids.

Chylomicron Formation in Intestinal Epithelial Cells

Once inside epithelial cells, fatty acids, 2-monoglycerides, cholesterol, and lysolecithin do not simply cross the basolateral membrane and directly enter the interstitial fluid. Triglycerides are resynthesized from fatty acids and 2-monoglycerides by enzymes located in the smooth endoplasmic reticulum (Figure 22-11). Other enzymes in the smooth endoplasmic reticulum reesterify cholesterol or synthesize phospholipids. Intracellular triglycerides, phospholipids, and cholesterol esters are then packaged into tiny vesicles in the Golgi complex (see Figure 22-11). Finally, a lipoprotein synthesized by the intestinal cells that acts in the same way as bile salts is incorporated into the surface of the vesicles and surrounds a lipid droplet with polar groups. The droplets formed within the epithelial cell are called **chylomicrons.**

The chylomicrons are transported to the basolateral surface of the epithelial cell, fuse with the basolateral cell membrane, and are released by exocytosis into the extracellular fluid. Chylomicrons do not enter the capillaries because they are too large to pass through the pores between the capillary endothelial cells. Instead they move into the lacteals, the lymphatic vessels that extend into each of the villi (see Figures 22-1 and 22-11). The intercellular junctions of the lacteals are large enough to allow the entry of chylomicrons. The chylomicrons are carried in the lymph to the systemic venous circulation where the main lymphatic duct (the thoracic duct) empties into the venous system at the junction of the left subclavian and internal jugular veins (see Chapter 14, p. 370). During digestion of a meal rich in fat, the lacteals appear white because they are filled with chylomicrons. This appearance gave them their name, which means "milky."

> 1 What features of the structure of bile acids are responsible for their ability to emulsify fats?
> 2 Why is the action of lipases so dependent on the increase in surface area that results from emulsification?
> 3 How are digestive products of lipids absorbed into the blood?

IRON AND VITAMIN METABOLISM
Iron Absorption

The diet contains materials other than carbohydrates, protein, and fats. On the positive side, there are the vitamins and trace elements necessary for health, and on the negative side, there are substances that are potentially toxic. Regulation of the absorption and metabolism of Na^+, K^+, Ca^{++}, and phosphate were discussed in relation to kidney function (see Chapter 20, pp. 512-518). Iron is another mineral that is present in relatively large

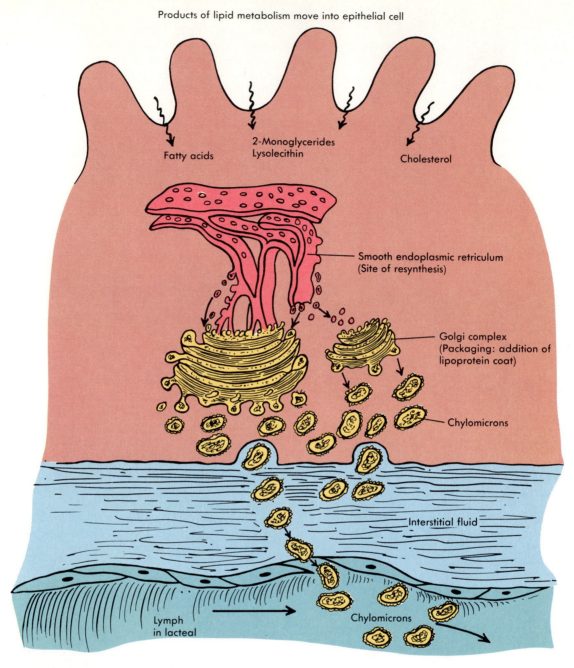

Lumen

Products of lipid metabolism move into epithelial cell

Fatty acids

2-Monoglycerides
Lysolecithin

Cholesterol

Smooth endoplasmic reticulum
(Site of resynthesis)

Golgi complex
(Packaging: addition of
lipoprotein coat)

Chylomicrons

Interstitial fluid

Lymph
in lacteal

Chylomicrons

FIGURE 22-11
Once the products of lipid metabolism move into the epithelial cell, they enter the smooth endoplasmic reticulum and are resynthesized into larger molecules (cholesterol esters and triglycerides). They are then packaged in the Golgi complex and rendered hydrophilic by the addition of β-lipoprotein. The resulting droplets of lipids, called chylomicrons, leave the cell by exocytosis, travel through the interstitial fluid, and enter the lymph lacteals.

quantities in the body. Most of the iron is very efficiently recycled (Figure 22-12).

Red blood cells have an average life span of 120 days. About 10% die in the circulation. The rest are removed by phagocytes in the liver and spleen when they near the end of their effective life span. The products of red blood cell disintegration and the digestion products of the cells trapped by the phagocytes are broken down by spleen and liver cells. The iron from the hemoglobin is either returned to the plasma, where it travels bound to **transferrin,** an iron-carrying protein, or it is stored in an accessible form. The iron stored in the liver and spleen can be released in the bile and absorbed

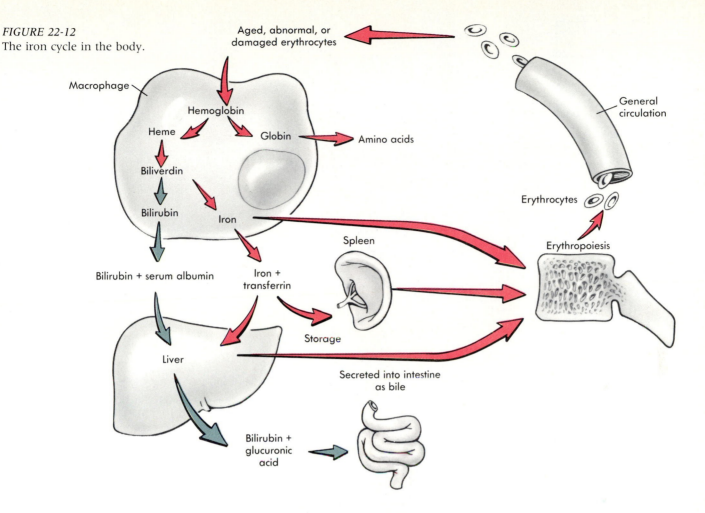

FIGURE 22-12
The iron cycle in the body.

Aged, abnormal, or damaged erythrocytes

Macrophage

Hemoglobin

Heme → Globin → Amino acids

Biliverdin

Bilirubin → Iron

Bilirubin + serum albumin

Iron + transferrin

Spleen

Storage

Liver

Secreted into intestine as bile

Bilirubin + glucuronic acid

General circulation

Erythrocytes

Erythropoiesis

in the intestine when blood levels of iron drop.

Transferrin-bound iron in the plasma travels to the site of red blood cell formation, the bone marrow. There it can either be used immediately in the synthesis of more hemoglobin in the developing red blood cells or it can be stored. Because the iron present in the body is reused so efficiently, little iron needs to be accumulated from the diet unless blood is lost. When blood loss occurs, the bone marrow cells call on their own stored iron and take up the circulating iron at an increased rate. The level of transferrin-bound iron can be replenished in part by release of the iron stored in the liver and spleen, but uptake is still regulated by the absorption mechanisms, which are not increased for several days after blood loss.

The amount of iron absorbed daily is in the range of 0.5 to 1.0 mg for adult men and about twice that for women in their reproductive years, who must compensate for the loss of blood in menstruation. Children and pregnant women need still greater quantities of iron because the increasing volume of the circulatory systems demands a net increase in body iron. Also, individuals who move to a higher altitude or suffer from chronic anoxia respond by increasing the number of red blood cells

present in the circulatory system, thus increasing the O_2 carrying capacity of the blood.

Ionized iron can exist in the Fe^{++} (ferrous) and Fe^{+++} (ferric) oxidation states. Dietary sources of iron include animal muscle myoglobin, vegetables, and fruits. Gastric HCl releases the iron from plant sources and also reduces it to Fe^{++}, which forms insoluble compounds with other food constituents less readily than does Fe^{+++} (Figure 22-13). Transferrin secreted by the epithelial cells of the intestine binds Fe^{++}, allowing it to be absorbed by the epithelial uptake system. The complex of two Fe^{++} per transferrin is recognized by a receptor present in the epithelial cells of the intestine, and the binding stimulates endocytosis, which moves the transferrin carrier and iron into the cell (see Figure 22-13). The entire heme group of myoglobin or hemoglobin can also enter the intestinal epithelial cells, where an enzyme splits the Fe^{++} from heme. Iron from either source has one of two fates in the epithelial cell: (1) some is bound to a carrier molecule and is subsequently actively transported across the basolateral surface into the interstitial fluid, or (2) iron is also bound in insoluble complexes with an intracellular iron-binding protein, **ferritin,** where it is stored for later use. If the supply of iron in the diet drops,

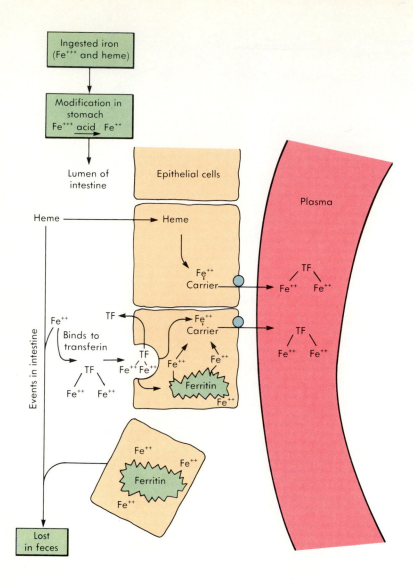

FIGURE 22-13
Steps in the absorption of iron from the intestine and its transport into the blood.

stored iron is released from the ferritin, binds to the carrier, and moves into the blood. When an intestinal cell dies, the iron bound to ferritin is lost in the feces (see Figure 22-13).

When iron is transported out of the epithelial cell, it moves by an unknown mechanism (probably bound to other carriers) through the interstitial fluid and into the plasma, where it binds to transferrin molecules produced by vascular endothelial cells. Transferrin-bound iron is available to the cells that form erythrocytes in the bone marrow (see Figure 22-12). The regulation of iron absorption by transferrin secretion in the stomach and intestine and the further restriction imposed by the limited number of intracellular transferrin molecules prevents excess iron from entering the circulation in adults. However, in young children, the mechanisms regulating iron uptake are geared to admit much more iron. Poisoning can result from inges-

tion of iron supplements intended for adults because excess iron can form insoluble deposits in the brain.

Vitamin Absorption

Vitamins are organic substances essential for proper functioning of the body. They cannot be synthesized in adequate amounts by the body and therefore must be absorbed. In most cases, vitamins are obtained from ingested food. A few vitamins come from bacteria present in the large intestine. Table 22-4 provides a list of vitamins, their sources, their physiological actions, and the consequences of deficiencies.

The vitamins fall into two major groups based on their solubilities. The water-soluble vitamins, which are readily lost during food preparation because of their solubility and heat lability, are absorbed into the intestinal epithelial cells by various

TABLE 22-4   *Major Vitamins*

Vitamin name	Dietary sources	Function	Effects of deficiency
Water-soluble vitamins			
B$_1$ (thiamine)	Whole grains; yeast; nuts; eggs; liver	Coenzyme for carbohydrate metabolism	Beriberi; impaired neural function
B$_2$ (riboflavin)	Legumes; fish; root crops; meat; whole grains	Coenzyme for protein and carbohydrate metabolism	Blurred vision; dermatitis; nausea
Niacin	Yeast; meats; whole grains; nuts	Coenzyme in Krebs' cycle	Pellagra
B$_6$ (pyrodoxine)	Meats; salmon; tomatoes; corn; spinach; yogurt; whole grains	Coenzyme in amino acid and lipid metabolism	Dermatitis; nausea
B$_{12}$ (cyanocobalamin)	Meat; liver; kidney; milk products; eggs	Erythrocyte formation; amino acid metabolism	Pernicious anemia; neural malfunction
Pantothenic acid	Green vegetables; cereals; liver; kidney; yeast	Coenzyme in steroid hormone synthesis and Krebs' cycle	Neuromuscular degeneration; adrenal hormone deficiency
Folic acid	Green leafy vegetables; liver	Enzyme in purine and pyrimidine synthesis and blood cell production	Anemia; abnormally large blood cells
Biotin	Yeast; liver; eggs; kidney; GI bacteria	Coenzyme in fatty acid synthesis; pyruvate conversion to glucose in liver	Mental depression; muscular fatigue; dermatitis
C (ascorbic acid)	Citrus fruit; tomatoes; green vegetables	Promoter of metabolic reactions; wound healing, antibody production, and collagen formation	Scurvy; anemia; connective tissue degeneration
Fat-soluble vitamins			
A (retinol)	Fish liver oils; green and yellow vegetables; cheese	Formation of rhodopsin; regulates bone and tooth formation	Night blindness; epithelial and neural disorders
D (1,25 dihydroxy-calciferol)	Fish liver oils; egg yolk; milk; ultraviolet light on skin	Calcium and phosphorus absorption	Rickets in children; softened or deformed bones
E (tocopherol)	Nuts; wheat germ; vegetable oils	Antioxidant; RNA, DNA, and red blood cell formation	Abnormal membrane function
K	GI bacteria; green leafy vegetables; liver	Coenzyme for prothrombin and clotting factors synthesis	Spontaneous bleeding; delayed clotting

mechanisms. In most cases, a specific recognition molecule has been identified for the water-soluble vitamins, and they may move into the epithelial cells by simple passive transport, facilitated transport, or active transport.

The water-soluble vitamins follow the same pathway traveled by amino acids and sugars into the circulation. The exception is the water-soluble vitamin B$_{12}$, which is simply too large and highly charged to be transported in the usual way. It is absorbed by a pinocytotic uptake system in the intestine (see Chapter 4, p. 66). This uptake system recognizes vitamin B$_{12}$ only when it is bound to **intrinsic factor,** a glycoprotein synthesized and released by parietal cells of the stomach. When vitamin B$_{12}$ enters the epithelial cells, intrinsic factor remains behind in the lumen. At the basolateral surface of the epithelial cells, vitamin B$_{12}$ attaches to a specific plasma protein that serves as a carrier.

Fat-soluble vitamins are absorbed in the same way as by-products of fat digestion, partitioning into micelles and passing into the lymph before entering the general circulation. In general, these vitamins are present in foods containing lipids, and absorption of these vitamins from dietary supplements is facilitated by consuming them with foods that contain lipids. Excess consumption of some vitamins, particularly vitamins A and D, can lead to **hypervitaminosis,** a toxic condition that occurs because these vitamins are normally stored by the liver and potentially can accumulate to high levels. Vitamin E is stored in many different sites, particularly in the lipids of the body, and higher levels of consumption can be tolerated.

> 1 How do the actions of HCl and the presence of transferrin in the lumen both contribute to iron absorption?
>
> 2 When hemorrhage reduces the amount of iron present in the circulatory system, what sources of iron are available in the body?
>
> 3 Why is absorption of fat-soluble vitamin supplements more effective when the vitamins are taken after a meal rather than on an empty stomach?

REGULATION OF THE GASTROINTESTINAL FUNCTION

Gastrointestinal Control Systems

The orderly movement of material through the GI tract and the secretion of fluid and enzymes required for digestion are coordinated by reflex pathways. These pathways are activated by the chemical composition of the food and the fact that its presence is detected by stretch receptors. A large portion of the blood is directed to the digestive tract, particularly when absorption is occurring. Many of the inputs that promote motility, digestion, and absorption are mediated by the parasympathetic branch of the autonomic nervous system along spinal reflex pathways (Table 22-5). The pathways of a second class of reflexes lie entirely within the enteric nervous system.

A third mechanism for regulating the activity of the GI tract involves hormones secreted from cells in the GI epithelium. These cells are sensitive to the presence of food and secretions in the canal and can also be activated by neural inputs, so they play a major role in coordinating activities in the different regions of the system. Table 22-5 summarizes the stimuli that cause the release of the best understood hormones, their effects on GI secretion, GI motility, and the secretion of other hormones, and their roles in GI control.

Cephalic Phase of Digestion

Thinking about food can initiate the first stage of control, which is called the **cephalic phase** of digestion. Added to the purely mental anticipation of eating, the sight and smell of food can also trigger responses in the cephalic phase, which prepare the mouth and stomach for the reception of food. The rate of food intake is limited by the rate at which it can be chewed, mixed with saliva, and reduced to a consistency that is appropriate for swallowing. Meanwhile, a parallel to the increased rate of salivation in the mouth occurs in the stomach, where blood flow and acid, mucus, and pepsin secretion increase. These responses continue and are enhanced as long as food is being chewed and swallowed (Figure 22-14).

The cephalic phase is mediated by reflexes that involve the parasympathetic branch of the autonomic nervous system and by a hormonal reflex pathway. Parasympathetic inputs stimulate the salivary glands and centrally inhibit the sympathetic tone of smooth muscle of arterioles supplying the stomach. Parasympathetic inputs also stimulate the chief cells to release HCl, the parietal cells to release pepsin, and the G cells to release the hormone, **gastrin**. Acetylcholine is not the final transmitter in all of these pathways; several peptide neurotransmitters and neuromodulators are believed to play important roles in the plexuses of the enteric nervous system, so the pharmacology is rather complex. Gastrin moves into the blood and travels through the circulation to the stomach epithelium, where it

---

TABLE 22-5   Some Reflexes of the Gastrointestinal Tract

Reflex name	Stimulus and effect
Reflexes involving the autonomic nervous system	
Receptive relaxation	Swallowing food relaxes the fundus of the stomach.
Enterogastric reflex	Acid and hypertonic solutions in the duodenum inhibit gastric emptying.
Gastrocolic reflex	Distension of the stomach and duodenum increase the motility of the ileum, cecum, and colon.
Gastroileal reflex	Ileocecal sphincter opens when gastric emptying begins.
Reflexes involving only the enteric nervous system	
Myenteric reflex	Local stimulation of the intestine causes contraction of the muscle immediately above the point of stimulation and relaxation of the muscle below the point of stimulation.

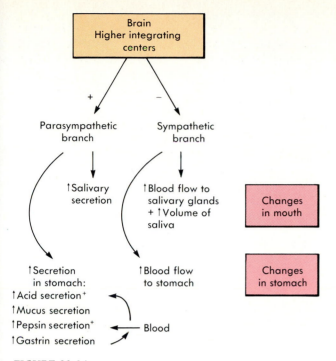

FIGURE 22-14

Activation of the cephalic phase of digestion and the responses it evokes. The consequences of parasympathetic activation are increases in secretion; inhibition of sympathetic activity allows an increase in blood flow to the salivary glands and stomach.

amplifies the parasympathetic stimulation of the parietal cells, which secrete acid, and especially of the chief cells, which secrete pepsinogens.

Gastric Phase of Digestion

When the stomach is empty, its wall is collapsed in folds. As soon as food is swallowed and enters the stomach, the **gastric phase** of digestion is initiated (Figure 22-15). The stimuli that trigger this phase relate to the presence of the food. The two major factors are distension of the stomach wall and the chemical content of the food. Distension of the antrum promotes gastrin release by both local enteric reflexes and by long-loop parasympathetically mediated reflexes.

Even in the presence of favorable stimuli, little gastrin can be released while the pH of the stomach lumen remains near the low values characteristic of the empty stomach. At the beginning of the gastric phase, food arriving in the stomach dilutes and buffers the gastric acid, and the pH of the stomach contents rises. The rise in pH permits gastrin secretion to occur in response to chemical and neural stimuli.

Dietary protein is a powerful stimulant of gastric secretion. The proteins arriving in the stomach

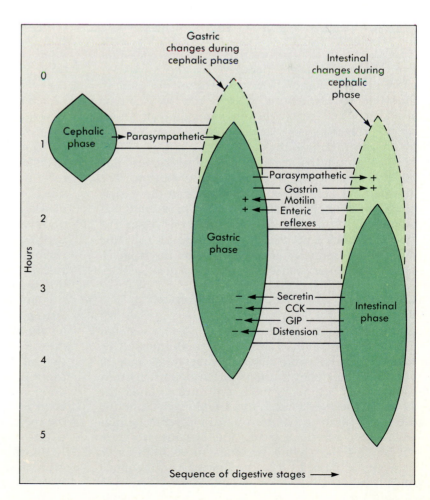

FIGURE 22-15

The temporal overlap between the three phases of digestion and the signals that travel between the different regions in the coordination of digestion.

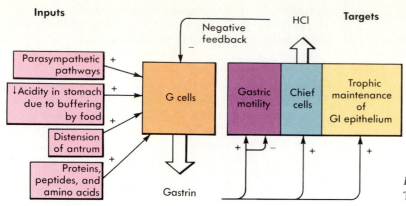

FIGURE 22-16
The control of gastrin secretion and its targets.

are met by the HCl and pepsins that were released in the cephalic phase and that continue to be released in response to cephalic stimuli, which last as long as food intake continues. The acid attacks the secondary and tertiary structure of ingested proteins (see Figure 3-11), opening folds so that the peptide bonds are more accessible to digestion by the gastric pepsins. As digestion proceeds, the amino acid concentration of the gastric contents rises. These free amino acids are **secretagogues**—substances that stimulate gastric secretion by promoting release of gastrin and pepsin (Figure 22-16).

Certain beverages and foods often served as first courses of meals contain secretagogues and therefore promote gastrin release. For example, using clear soups as the first course of a meal provides amino acids that are detected by the stomach's chemoreceptors and speed preparation for protein digestion by promoting gastrin release. Caffeine also acts as a secretagogue and can aggravate an ulcer by promoting acid secretion when food may not be present to absorb the acid.

The stimuli that promote gastrin release provide positive feedback as long as protein is present to be digested. As acid secretion continues in the gastric phase, the buffering capacity of the proteins is exceeded. Gastrin secretion is inhibited when the chyme reaches a pH of 3.0 (see Figure 22-16). This inhibition terminates acid secretion by the stomach as the end of the gastric phase approaches.

Control of Gastric Motility and Stomach Emptying

At the beginning of the gastric phase, the smooth muscle of the stomach relaxes as food enters. This **receptive relaxation** allows a substantial amount of food to be stored in the stomach before vigorous contractions begin. As food continues to enter the stomach, the fundus and upper body store it, while the lower body and antrum mix food with the mucus, buffer solutions, acid, and enzymes to form chyme. As the chyme becomes more fluid, it is ready to be passed on to the first part of the small intestine.

As long as gastrin levels are high, the fundus and body of the stomach are relaxed and serve a storage function. Also, the excitability of smooth muscles in the pyloric region is high, so pyloric contraction occurring during the wave of peristalsis severely limits delivery of chyme to the intestine. Gastrin also increases lower esophageal sphincter tone, preventing stomach contractions from causing a reflux of food into the esophagus. The **gastroileal reflex** (see Table 22-5) relaxes the sphincter located between the ileum and the large intestine; this reflex relaxation is seen when gastric contractions are vigorous, and it promotes the passage of material from the small intestine into the colon in anticipation of the movement of chyme into the initial regions of the small intestine. The stimulation of large intestine motility during the gastric phase is called the **gastrocolic reflex** (see Table 22-5). This is probably mediated by gastrin and by long neural reflexes, and it causes the mass movement that often follows a meal.

During the gastric phase, pancreatic secretion is stimulated so that the intestine is prepared to receive the food that will soon be entering it (see Figure 22-15). Distension of the stomach directly stimulates pancreatic fluid and enzyme secretion via parasympathetic reflexes. Gastrin also stimulates pancreatic enzyme secretion because gastrin is chemically very similar to the hormone, **cholecystokinin (CCK),** which regulates pancreatic secretion in response to food in the intestine.

The stomach also receives signals from the intestine before and just after the chyme begins to enter the intestine. The hormone, **motilin,** increases the strength of gastric contractions and the tone of the pyloric sphincter (Table 22-6; see Figure 22-17). Motilin is released from **argentaffin cells** in the duodenum when the pH of the duodenal contents is greater than 4.5. Motilin secretion is high early in the period following a meal, when little acidic chyme has entered the small intestine, and its role is to stimulate the mixing action of the stomach. Motilin secretion may also stimulate pepsin secretion.

TABLE 22-6   *The Gastrointestinal Hormones*

Hormone	Stimuli	Effects	Role
Gastrin	Amino acids Distension pH > 3 Parasympathetic activity	+Acid secretion +Antral motility +Pepsinogen secretion	Facilitates gastric digestion
Gastric inhibitory peptide	Glucose Fats Hypertonicity	+Intestinal motility −Gastric secretion and motility	Limits gastric emptying; prepares the intestine for substrate
Motilin	pH > 4.5	+Gastric motility +Intestinal motility +Pepsinogen secretion	Facilitates digestion in intestine
Secretin	Hypertonicity pH < 4.5 Parasympathetic activity	+Pancreatic bicarbonate and zymogen secretion −Gastric motility and secretion	Limits gastric emptying; neutralizes the chyme leaving the stomach
Cholecystokinin	Fats Amino acids	+Pancreatic zymogen secretion +Intestinal secretion and motility −Gastric secretion +Gallbladder contraction	Limits gastric emptying; promotes digestion in the intestine

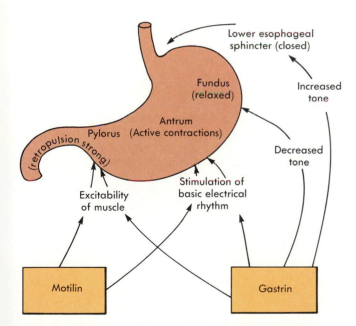

FIGURE 22-17
Control of the stomach by inputs from gastrin and motilin.

Intestinal Phase of Digestion

As soon as food begins to enter the duodenum, the **intestinal phase** of digestion is triggered. Hormones activated and the effects they have are listed in Table 22-6. As was true for the gastric phase, the stimuli for the intestinal phase relate to the stretching of the wall of the intestine and the chemical content of the food. The intestine is the site for digestion of all categories of nutrients, and the quantities of enzymes that are secreted and the time allotted for digestion are determined by the amounts of protein, carbohydrate and fat in the chyme. Both reflexes and hormones are mediators of these effects of meal composition on intestinal activity (see Figure 22-15).

Carbohydrates and proteins are digested more readily than are lipids, so a meal high in lipids requires a much longer time for digestion and absorption. Early in the intestinal phase, when the chyme is buffered by the intestinal mucus and bicarbonate secretions that were stimulated as part of the gastric phase, amino acids arriving in the duodenum stimulate still more gastrin release via reflexes in the enteric nervous system. Thus, during this early period, the feedback from the intestinal receptors may have a positive effect on stomach activities, and HCl secretion in the stomach remains high. This facilitates the preparation of proteins for intestinal digestion. Later in the intestinal phase, the feedback to the stomach becomes mainly inhibitory (see Figure 22-15).

Duodenal mechanoreceptors detect tension in the walls of the duodenum and thus the amount of material entering the intestine from the stomach. These mechanoreceptors inhibit gastric motility and increase pyloric tone by a parasympathetic feedback pathway called the **enterogastric reflex** (see Table 22-5). Surgical removal of the parasympathetic innervation of the GI tract is sometimes used to control ulcers. But because this procedure also elimi-

Appetite, Satiety, and Food Addiction

For body mass to be in a steady state, total caloric expenditure must closely match caloric intake. The closeness of the match is critical because on a time scale of months or years the values of intake and expenditure to be matched are large as compared to the amount of energy stored as fat. Even a slight error would result in significant weight gain if continued over time. From this point of view, most people regulate their body weights quite well. Unfortunately, the value at which weight is regulated is frequently somewhat higher than the value that is optimum for health and self-esteem. Roughly a third of adults in developed nations are overweight; an exact figure is impossible to determine because ideal weights are difficult to define precisely. The equilibrium body weight for any given individual depends on a complex interaction of metabolic regulatory systems that involve appetite, food availability, and the regulation of energy into heat production and storage. Some of the aspects of metabolic regulation are treated in Chapter 23.

Appetite regulation is believed to involve interactions among several regulatory pathways in the brain, some of which receive information from the stomach and intestine. Some of the pathways involve endogenous opiates (enkephalins and endorphins) as modulators (see Chapter 8). These neuromodulators are generally involved in pathways that involve stress and reward, and they stimulate eating. A hypothetical cycle of overeating develops when emotional stress increases opiate levels to a point that overrides the normal mechanisms of satiety. The resulting weight gain increases the stress, thus increasing the drive to eat. Ultimately an addiction to the high opiate levels develops, and the behavior persists because returning to normal eating patterns would involve discomfort similar to the withdrawal syndrome experienced by drug addicts.

The incidence of the eating disorders, **anorexia nervosa** and **bulimia,** has increased dramatically in recent years. Approximately 1% of adolescent females have anorexia and up to 19% of college women may have bulimia. The incidence in males is considerably lower. Anorexia nervosa is best described as self-inflicted starvation; its victims become dangerously thin while denying hunger. Bulimia is at the opposite extreme; people with bulimia alternate between binges of eating and daily fasting, self-induced vomiting (purging), and abuse of laxatives and diuretics to control their weight. These diseases probably represent aspects of a single disorder because some patients show features of both. In the view of some therapists, this disorder could also be described as a food addiction. People with anorexia are successfully denying their addiction; those with bulimia are yielding to it.

Eating disorders are signs of more generalized disturbances of mental functioning, social adjustment, and body image. In addition to the stress such disorders place on the energy balance of the body, they are potentially fatal threats to homeostasis of body fluids. For example, repeated purging may lead to alkalosis and hypokalemia and ultimately to heart arrhythmia. Abuse of diuretics and laxatives causes hypovolemia, electrolyte imbalances, and acid-base disorders.

Because those with anorexia and bulimia characteristically deny that they have a health problem, their conditions may not receive timely medical attention. Repeated purging causes a distinctive pattern of erosion of dental enamel and also an easily recognized enlargement of the salivary glands. Dentists have an important role in diagnosing eating disorders and bringing them to the attention of the patient's physician.

---

nates the enterogastric reflex, it causes rapid dumping of the gastric contents into the small intestine and thus chronic diarrhea.

Osmoreceptors respond to the osmotic strength of the chyme and allow the system to differentiate between stretching caused by water ingestion and that related to food ingestion. If food is present, stretching, elevated osmotic pressure, and nutrient molecules are all detected. These stimuli cause **gastric inhibitory peptide (GIP)** to be released from **K cells** in the mucosa of the intestine. This hormone acts with other signals from the intestine to slow the digestive activity of the stomach and induce the stomach to assume more of a storage role (see Figure 22-15). GIP accomplishes these effects by reducing pepsin and HCl secretion and gastric motility.

The enzymes that function in the intestine require a neutral pH. The acidity of the incoming chyme is one of the most potent stimuli for triggering the changes associated with the intestinal phase of digestion. Chemoreceptors that detect the acid stimulate the release of the hormone, **secretin,** from duodenal cells (Figure 22-18). Secretin is released when the pH of the chyme in the duodenum falls below 4.5. Hypertonic conditions and fat content of

FIGURE 22-18
The control of secretin release and its targets.

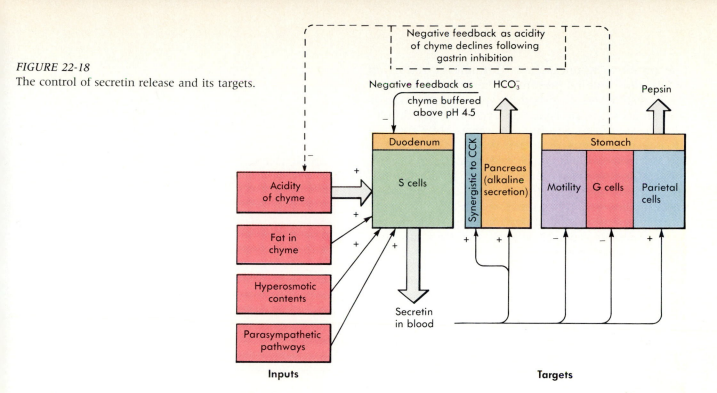

the chyme also stimulate secretin release. Secretin causes the pancreas to secrete a solution high in bicarbonate, which has the effect of buffering the HCl. Secretin also tends to inhibit the motility of the stomach but to stimulate secretion of pepsin.

At the neutral pH characteristic of the intestine, pepsins released in the stomach are not functional, but the rapid buffering of the chyme allows the powerful intestinal enzymes released from the pancreas and present in the brush border to begin digesting the food. Secretin is the single most important inhibitor of gastrin release and therefore of acid secretion. Thus two mechanisms for negative feedback of secretin release exist: (1) reducing the acidity of the incoming chyme through actions on the

stomach acid secretion, and (2) reducing the acidity through the buffering action of the pancreatic bicarbonate secretion (see Figure 22-18).

Fats and fatty acids present in the chyme are detected by duodenal chemoreceptors. These and, to a lesser extent, amino acids, polypeptides, and proteins stimulate the release of CCK (Figure 22-19). This hormone has seven roles: (1) it increases pancreatic enzyme secretion, (2) it relaxes the sphincter of Oddi, (3) it causes contraction of muscle cells of the gallbladder, (4) it inhibits gastric emptying, (5) it potentiates the action of secretin on the pancreatic secretion of enzymes, (6) it promotes secretion and motility in the small intestine, and (7) it is a trophic hormone for the pancreas. The negative feedback

FIGURE 22-19
The control of CCK secretion and its targets.

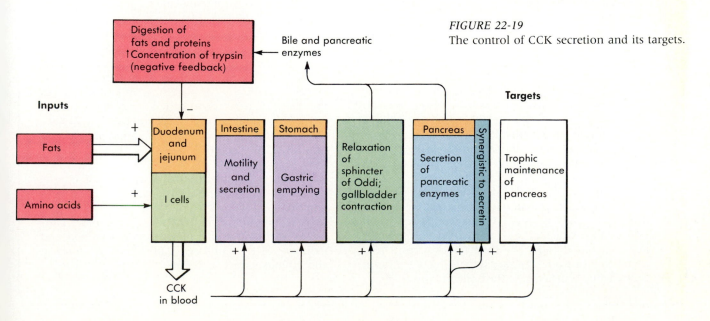

that limits CCK release acts through the reduction of fats and proteins present in the intestine as these are digested and absorbed. A specific inhibition is provided by the presence of trypsin in the duodenum because this inhibits the release of CCK and thus decreases enzyme secretion (see Figure 22-19).

Fat in the duodenum also stimulates the secretion of GIP. Secretin, GIP, and CCK all directly decrease gastric motility (see Table 22-6). The presence of many lipids in a meal will greatly extend the gastric and intestinal phases of digestion as the slow processing of the fats by the intestine suppresses gastric emptying.

1 When peristaltic activity in the stomach is high, the excitability of muscles in the pyloric sphincter region is also high. What is the effect of sphincter excitability on gastric emptying?
2 How does acidic chyme in the duodenum result in increased pancreatic secretion?
3 By what mechanisms will a meal high in fats delay gastric emptying longer than would a meal high in carbohydrates?

SUMMARY

1. The movement of water, amino acids, and sugars into the body from the alimentary canal depends largely on the movement of Na^+. Sodium ions diffuse from the lumen into the epithelial cells in response to a gradient created by active transport of sodium out of the cells into the interstitial fluid. As Na^+ moves out of the lumen, water follows by osmosis. The gradient for Na^+ also drives the carrier-mediated transport of most amino acids and sugars into the epithelial cells. The carriers recognize particular classes of sugars or amino acids.

2. Digestion of food macromolecules is accomplished by enzymes that catalyze hydrolytic reactions. **Amylase** begins carbohydrate digestion in the mouth, and **pepsins** begin protein digestion in the stomach, but most digestion is accomplished in the small intestine by enzymes from the pancreas or attached to the **brush-border** membranes of the intestine.

3. Digestion of lipids requires the emulsifying action of **bile acids** to increase the lipid surface area so that pancreatic **lipases** can hydrolyze the lipids. **Colipase** facilitates the actions of lipases. Bile acids also facilitate absorption by collecting the products of lipid digestion into **mixed micelles.** Some resynthesis of lipids occurs in the epithelial cells, and these lipids are released into the lymph in small droplets called **chylomicrons.**

4. Iron absorption is regulated by the secretion of the carrier molecule, **transferrin,** into the lumen. Most iron in the body is recycled between old red blood cells and the cells in the bone marrow that produce new red blood cells. The liver removes iron from **heme** and the iron is released into the blood, where it is carried by transferrin to the bone marrow.

5. Vitamins fall into two categories depending on their solubilities: **water-soluble** and **fat-soluble.** Most water-soluble vitamins are absorbed by transport mechanisms, but **vitamin B_{12}** must be bound to the carrier, **intrinsic factor,** which is secreted by the stomach, to be taken into intestinal cells. Fat-soluble vitamins partition with the fats into mixed micelles and chylomicrons.

6. Digestion can be divided into three phases, depending on where food stimuli are acting. The **cephalic phase** begins even before food ingestion, when the idea or stimuli arising from the presence of food are detected. The **gastric phase** of digestion begins as soon as food enters the stomach, and the **intestinal phase** begins as soon as food enters the intestine. The three phases therefore have considerable overlap in time.

7. Stimuli arising from the physical and chemical presence of food in the mouth, the stomach, and the intestine regulate the movement of material through the GI tract and the rate of secretion of enzymes. Regulation of digestion involves reflexes that use "long" parasympathetic pathways, "short" reflex loops, and hormonal secretions.

8. The best-understood GI-tract hormones are **gastrin, motilin, GIP, secretin,** and **CCK.** Gastrin stimulates gastric acid and pepsinogen secretion and gastric motility. Motilin stimulates gastric and intestinal motility and pepsinogen release. GIP stimulates intestinal and pancreatic secretion and inhibits gastric secretion and motility. Secretin stimulates pancreatic secretion and inhibits gastric secretion and motility. CCK stimulates gallbladder contraction and inhibits gastric secretion.

1. How are each of the following regions of the GI tract functionally specialized for digestion and reabsorption?

    Mouth   Duodenum
    Esophagus   Jejunum and ileum
    Stomach   Large intestine

2. What enzymes digest carbohydrates? Where does most carbohydrate digestion occur? How are carbohydrates absorbed?

3. How are proteins digested? What are the two sites of protein digestion? How are amino acids absorbed? Can peptides be absorbed?

4. What are the roles of the bile salts in fat digestion and absorption?

5. In what ways is absorption of fat different from absorption of amino acids and carbohydrates?

6. For each of the following GI hormones, state (a) its source, (b) the most important physiological stimulus for its release, (c) the most important inhibitor of its release, and (d) its target organs.

    Gastrin   Cholecystokinin
    Secretin   Gastric inhibitory peptide

7. What events occur during the cephalic phase of digestion? The gastric phase? The intestinal phase?

8. What factors determine the rate of gastric emptying? Will a meal with a high fat content empty more or less rapidly than a meal with a high carbohydrate content? What feedback control is involved?

● SUGGESTED READING

ABRAHAM, S., and LLEWELLYN-JONES: *Eating Disorders—The Facts*, Oxford University Press, Oxford, England, 1984. Covers the physiological and psychological factors leading to obesity, anorexia, and bulimia.

The Digestive System, *Consumer Guide Magazine*, April 1986, p. 224. A general discussion of digestion in the family and medical health guide series.

Fighting Fat, the Seal Way. *Discover*, August 1988, p. 7. An article discussing the fact that seals and sea lions lack apoprotein E, a necessary cofactor in the digestion of fats.

GREEN, M., and H.L. GREEN: *The Role of the Gastrointestinal Tract to Nutrient Delivery*, Academic Press, Inc., Orlando, Fla., 1984. Summary of digestion and absorption of carbohydrates, proteins, and fats.

PODOLSKY, D.M.: Shock-Wave Therapy Beats Surgery: Gallstones May Be Next, *American Health*, July 1983, p. 16. See the description in the following article.

WININCK, M: Control of Appetite. In: *Current Concepts in Nutrition*, John Wiley & Sons, Inc., New York, 1988. Describes appetite regulation by control centers in the brain and receptors in the GI tract. Discusses the fact that the CNS pathways involved in satiety involve endogenous opiates and mediate responses to stress and reward.

What Our Ancestors Ate, *The New York Times Magazine*, June 5, 1988, p. 54. Compares the modern diet with archaeological findings regarding past eating habits.

GROWTH, METABOLISM, REPRODUCTION, AND IMMUNE DEFENSE

CHAPTER 23

Endocrine Control of Organic Metabolism and Growth

On completing this chapter you will be able to:

- Understand how insulin and glucagon regulate blood glucose.
- Distinguish between the absorptive and postabsorptive states.
- Understand the differences between type I diabetes and type II diabetes.
- Describe how increases in blood levels of glucose, bulk aspects of ingested food, and neural and hormonal signals produce satiety.
- Understand the factors involved in long-term regulation of body weight.
- Describe how the relative proportions of fat, muscle, and ossified skeleton vary over the life cycle.
- Describe the processes by which thyroid hormones set the body's metabolic rate and respond to cold.
- Understand the mechanisms involved in the production of fever in response to infections.
- Describe the adaptive responses of the body to periods of starvation.

Organic metabolism is regulated on several time scales. Blood levels of nutrients such as glucose are controlled over a time scale of minutes. Appetite for food is matched with requirements for energy, repair, and growth over a time scale of hours to days. The longest regulatory time scale reflects the differing metabolic requirements of different life stages. The overall metabolic picture is different for young children, adolescents, young adults, mature adults, and older adults. It differs between men and women, and is greatly affected in women by pregnancy and milk production.

In adults maintaining a stable weight, overall organic metabolism is balanced, although most tissue components are continually being degraded and replaced. A slight excess of carbohydrate and fat intake over metabolism to CO_2, water, and energy leaves a small extra amount of fat in fat cells each day. Continued over time, even small imbalances would lead to large changes in the mass of fat. A minor imbalance in favor of carbohydrate and fat breakdown over intake would have the opposite result: a steady erosion of first the fat reserve and ultimately the protein structure of the body.

In many less well-developed societies, periodic famine is a fact of life for most people. It is also true that in most highly developed societies, some people are sometimes unable to meet their daily nutritional needs. In undernutrition, the metabolic state of the body is an exaggeration of the state occurring during the short fasts that intervene between the last meal of the day and breakfast on the next day. Stored nutrients move into the blood and are converted to forms that can support energy metabolism. In undernutrition, metabolism is fundamentally altered, resulting in a decrease in the amount of energy spent by cells on turnover of their constituents, so that the cells become more thrifty in their use of energy. This pattern is the opposite of the one seen during growth in children and adolescents. Undernutrition, particularly when it involves protein intake, is fundamentally incompatible with normal growth of organ systems such as the central nervous system and skeleton, which take on the adult pattern gradually during growth.

CHARACTERISTICS OF THE ABSORPTIVE STATE

Role of Insulin

In the bodies of adults, some periods of the day are dominated by the process of **anabolism** (synthesis), whereas other periods are dominated by the process of **catabolism** (degradation) (Table 23-1; see Chapter 4, pp. 76-82). Anabolic processes include glycogen formation from glucose (**glycogenesis**), triglyceride formation (**lipogenesis**), ketone synthesis (**ketogenesis**), and glucose formation from amino acids (**gluconeogenesis**). Catabolic processes include glycogen breakdown (**glycogenolysis**), lipid breakdown (**lipolysis**), and protein breakdown (**proteolysis**).

Because many processes must go on continuously (for example, the simultaneous loss and replacement of the epithelial cell lining of the GI tract), the difference between net anabolism and net catabolism is usually determined by what the body is doing with nutrients that enter during absorption and what it is doing when absorption is not occurring.

The period during which nutrients are being absorbed from the intestine is called the **absorptive** state of metabolism. During the **postabsorptive state,** nutrients are not entering the body and must be made available from internal stores. The change in disposition of nutrients that occurs at the end of the absorptive phase is the result of changes in levels of hormones and in the pattern of activity of the autonomic nervous system. Storage of nutrients as glycogen and fat dominates the absorptive state. This temporary storage is reversed in the postabsorptive stage, when stored glycogen and fat molecules are broken down and released to supply the body's needs for energy and synthesis.

During the absorptive period, the hormone, **insulin,** regulates the entry of glucose and amino acids into cells. Insulin is released from **beta cells** of the pancreatic **islets of Langerhans** (Table 23-2 and Figure 23-1). Insulin is synthesized as an inactive precursor called **proinsulin** and subsequently packaged into secretory vesicles in the Golgi apparatus. Proinsulin is hydrolyzed to insulin inside these secretory vesicles by a proteolytic enzyme. Insulin is then released by exocytosis, like a neurotransmitter. A normal plasma glucose concentration is in the range of 80 to 110 mg/dl of plasma, and at this level, plasma insulin is about 10 microunits/ml

---

TABLE 23-1   Anabolic and Catabolic Processes

Category	Process	Description
Anabolic	Glycogenesis	Formation of glycogen from glucose
	Lipogenesis	Formation of fatty acids and then triglycerides (esterification)
	Ketogenesis	Synthesis of ketones
	Gluconeogenesis	Synthesis of glucose from amino acids
Catabolic	Glycogenolysis	Breakdown of glycogen to glucose
	Lipolysis	Breakdown of triglycerides to fatty acids and glycerol, and then to acetyl-CoA
	Proteolysis	Breakdown of protein to amino acids

TABLE 23-2   Hormones that Regulate Metabolism

Hormone	Chemical characteristics	Secreted by
Insulin	Protein	Beta cells of pancreatic islets of Langerhans
Glucagon	Protein	Alpha cells of pancreatic islets of Langerhans
Epinephrine	Catecholamine	Adrenal medulla
Cortisol	Steroid (glucocorticoid)	Adrenal cortex
Thyroid hormones (T_3 and T_4)	Iodinated thyronines (tyrosine derivatives)	Thyroid gland
Growth hormone (somatotropin)	Protein	Anterior pituitary
Somatostatin	Small protein (14 amino acids)	Delta cells of pancreatic islets of Langerhans

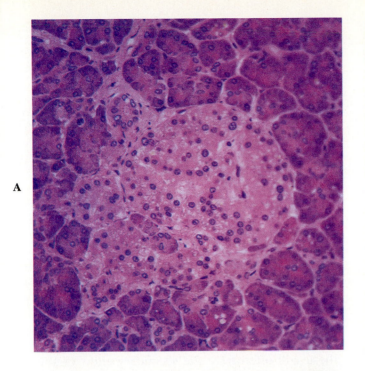

A

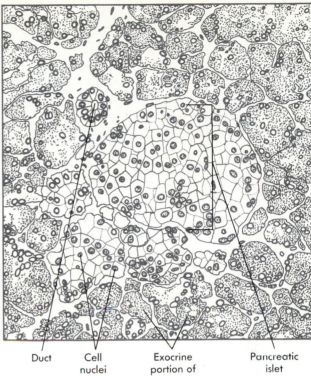

Duct   Cell   Exocrine   Pancreatic
        nuclei   portion of   islet
                pancreas

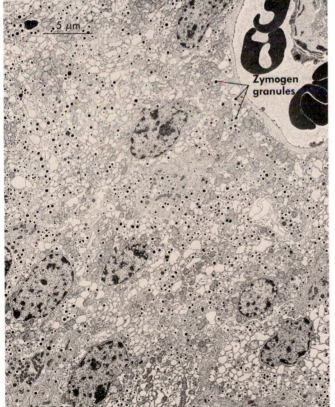

B

FIGURE 23-1

A Islet of Langerhans surrounded by exocrine tissue (acini and ducts) in the pancreas. Beta cells are the most common cell type in the islets and are usually found in the center of the islet. Alpha cells are almost always located near the outside of the islet. Delta cells are the least common type.

B Electron micrograph of islet cells, showing cytoplasm filled with vesicles of hormone. A capillary with red blood cells is seen in the upper right corner.

(μU/ml—an arbitrary measure of concentration).

The most potent stimulus for insulin secretion is elevated glucose levels in the blood (Figure 23-2 and Table 23-3). Insulin secretion is blocked when the plasma glucose level drops below 50 mg/dl, and maximum insulin secretion occurs at a plasma glu-

cose level of 300 mg/dl. The release of insulin from beta cells also depends on how many glucose receptors are present on the beta cells. High-carbohydrate diets increase the density of glucose receptors on beta cells, making this feedback system more sensitive. That is, after adaptation to high levels of glu-

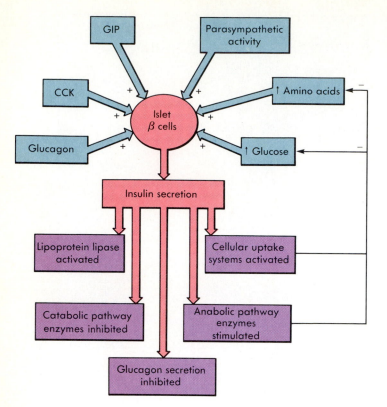

FIGURE 23-2

Inputs to beta cells and effects of insulin, including negative feedback on glucose and amino-acid levels.

Factor	Effect on insulin secretion	Effect on glucagon secretion
Hyperglycemia	Increase	Decrease
Hypoglycemia	Decrease	Increase
Increased plasma amino acids	Increase	Increase
Parasympathetic input to pancreas	Increase	Decrease
Sympathetic input to pancreas	Decrease	Increase
Somatostatin	Decrease	Decrease

cose, a given level of plasma glucose causes a greater rate of insulin secretion. Conversely, low-carbohydrate diets decrease insulin receptor density and thus decrease insulin secretion at a given level of plasma glucose.

In addition to elevated levels of glucose, the beta cells are also stimulated to release insulin in response to elevated levels of amino acids in the blood, to the hormones, gastric inhibitory peptide (GIP), and cholecystokinin (CCK), and to increased parasympathetic activity (see Figure 23-2). Other hormones can also raise plasma glucose levels, and these will often have the secondary effect of stimulating insulin release.

Insulin is called the hormone of abundance because it promotes the use of glucose, amino acids, and fats as they come into the blood during the absorptive period (Table 23-4). Insulin is required for glucose transport into nearly all cells. The primary exceptions are liver cells, brain tissue, erythrocytes, and cells in the renal medulla.

Insulin is an anabolic hormone because it promotes repair of structures composed of proteins and lipids. This stimulation of synthesis is partly due to the fact that increased uptake of glucose, amino acids, and fats leads to high levels of these reactants in the cells, driving reactions in the direction of net synthesis by the Law of Mass Action. Insulin further promotes the storage of extra nutrients as (1) glycogen in muscle and liver cells, and (2) fat in liver and adipose cells by stimulating specific en-

TABLE 23-4 *Major Effects of Insulin during the Absorptive State*

Target tissues	Insulin effects	Metabolic consequences
Muscle	Increases glucose uptake	Increased muscle glycogen
	Increases uptake of some amino acids	Increased protein synthesis
Liver	Stimulates enzymes of glycogen synthesis; inhibits enzymes that break glycogen down	Increased liver glycogen
	Stimulates enzymes of pathways leading to synthesis of fat from fatty acids and keto acids	Ingested amino acids and fatty acids are converted into lipids for shipment to fat cells
Fat	Increases glucose uptake	Plasma glucose is converted to alpha-glycerophosphate and used in fat synthesis
	Stimulates lipoprotein lipase	Plasma lipids are taken up for fat synthesis
	Inhibits hormone-sensitive lipase	Decreased fat breakdown in fat cells

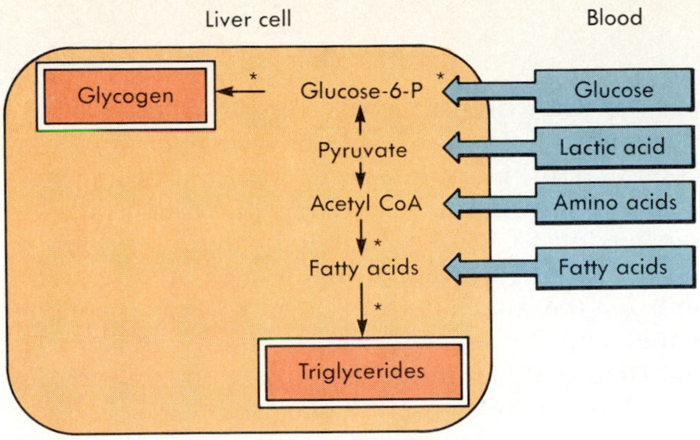

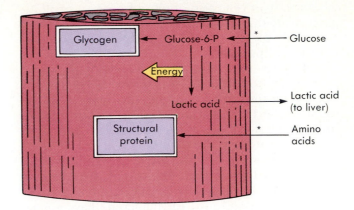

FIGURE 23-3

Reactions in liver cells in the absorptive state. Steps stimulated by insulin are indicated by an asterisk.

FIGURE 23-4

Reactions in white muscle cells in the absorptive state. Steps stimulated by insulin are indicated by an asterisk.

zymes. While promoting substrate storage, insulin exerts a strong inhibition on catabolic pathways. It prevents the breakdown of glycogen and fat reserves by its inhibition of hormone-sensitive enzymes in several key catabolic pathways. Figure 23-2 summarizes the general effects of insulin on the body.

1 *Define anabolism and catabolism. Give examples for the treatment of protein, carbohydrates, and fats by the body.*
2 *How does the Law of Mass Action combine with the actions of insulin to direct metabolic pathways during the absorptive phase?*
3 *In what ways does insulin stimulate anabolism in the absorptive phase?*

Roles of the Liver

The hepatic portal system transfers nutrient-rich blood from the GI tract directly to the liver, where approximately 75% of the glucose and amino acids absorbed from the intestine can be taken up. The role of the liver during the absorptive period is to transform incoming nutrients into more complex forms suitable for storage. Depending on their structure, amino acids can be converted to pyruvate for use in glycogen formation, or they can be converted directly into intermediates in fatty acid synthesis (Figure 23-3).

Liver cells do not need insulin for uptake of amino acids and glucose. However, insulin exerts a powerful effect on carbohydrate metabolism in the liver by its effects on enzymatic pathways within liver cells. In the first step of glycolysis (see Chapter 4, p. 76), glucose is converted to glucose-6-phosphate, a molecule that cannot diffuse back out across the liver cell membrane. This conversion is stimulated by insulin (see Figure 23-3), effectively trapping glucose inside the liver cells. Thus this insulin-sensitive enzymatic reaction favors storage during the absorptive period when insulin levels are high. The synthesis of glycogen from glucose-6-phosphate is simultaneously stimulated by another

rate-limiting insulin-sensitive enzyme, **glycogen synthetase**. Although the liver forms glycogen mainly from glucose, some of the glycogen stored by the liver during the absorptive state is synthesized from lactate carried by the blood to the liver from white (fast glycolytic) muscle cells (Figure 23-4; see also Figure 23-3 and Chapter 12, pp. 301-306).

Transport and Storage of Lipids

Although an average of about 200 grams of glycogen can be stored in the liver, most of the nutrients entering the liver follow metabolic pathways to lipids. Glucose metabolized to acetyl-CoA is used to form the fatty acids of triglycerides. Triglycerides can be stored in the liver as triglycerides, or they can be released into the plasma and travel to **adipose** (fat) cells (Figure 23-5). When excess acetyl-

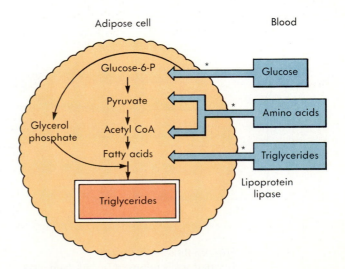

FIGURE 23-5

Reactions in adipose cells in the absorptive state. Steps stimulated by insulin are indicated by an asterisk.

CoA is formed in carbohydrate or amino acid metabolism, liver cells also produce **ketones,** small organic acids that can be metabolized by most tissues (see Table 23-1). In this way amino acids also participate in lipogenesis.

Blood is rich with lipids following a typical meal, both from the chylomicrons that entered the blood from the lymph and from the lipids manufactured in the liver (Figure 23-6). The liver releases lipids in the form of particles called **very low density lipoproteins (VLDLs),** which spend an average of 3 hours in the circulation. When the VLDLs deliver their lighter elements to cells, they become **low-density lipoproteins (LDLs)** and return to the liver (see Figure 23-6). Chylomicrons have a relatively short average lifetime in the plasma—about 8 minutes. Both VLDLs and chylomicrons contain triglycerides that must be broken down before they can enter cells. Two factors are of particular importance in fat uptake. The first, **lipoprotein lipase (capillary lipase),** is present in the endothelial cells of the capillaries that supply adipose tissue (see Figure 23-6) and, to a lesser extent, cardiac and skeletal muscle. Lipoprotein lipase enzymatically liberates free fatty acids from the triglycerides, so the presence of this enzyme determines which tissues will receive the lipids. Insulin is one of the hormones that stimulates lipoprotein lipase activity (see Figure 23-5 and Table 23-4).

The second factor that is important for lipid uptake is the **high-density lipoproteins (HDLs)** which are synthesized mainly by the liver (and, to a lesser extent by the intestine). HDL particles circulating in the plasma facilitate uptake of lipids by transferring to VLDLs and chylomicrons a class of molecules called **apoproteins,** which are important both for the attachment of HDL particles to membranes and for the activation of lipoprotein lipase. The HDL particles speed the uptake of triglycerides and assist in collecting the cholesterol that is liberated from cell membranes into the plasma. A high ratio of HDL particles to LDL particles slows the development of **atherosclerosis.** Factors that have been identified as promoting the amount of HDLs in the blood include exercise and consumption of fish oils.

After fatty acids enter adipose cells, they are esterified on a glycerol backbone (see Chapter 3, p. 42) to form triglycerides (see Figure 23-5). This process is more complex than it appears, because intracellular synthesis (esterification) of triglycerides requires an intermediate phosphorylated form of the glycerol molecule called **glycerol phosphate.** Adipose cells do not take up glycerol along with the fatty acids, but instead must synthesize glycerol phosphate from glucose. This imposes two restrictions on fat synthesis and storage in the adipose tissue: (1) the blood must have a sufficient supply of glucose so that it can be absorbed to form glycerol phosphate, and (2) insulin must be present to facil-

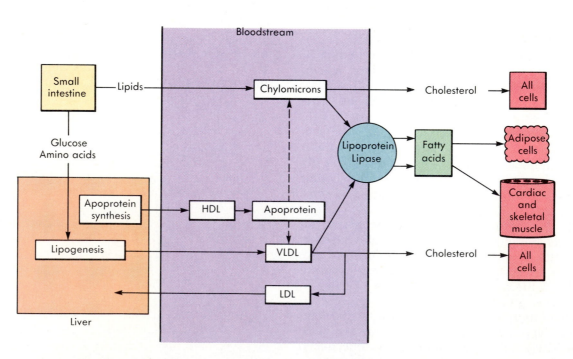

FIGURE 23-6
The pathways followed by fats during the absorptive state. Apoprotein is synthesized by the liver and delivered to the very low-density lipoproteins by the high-density lipoproteins. The apoproteins facilitate uptake of lipids. Low-density lipoproteins and chylomicron remnants are recycled.

itate glucose entry (see Figure 23-5). Thus, although the presence of lipoprotein lipase allows fatty acids to enter adipose cells in a hormone-independent step, the storage of triglycerides requires the presence of insulin.

Protein Anabolism

High levels of amino acids in the blood promote protein synthesis. This occurs all over the body, but muscles represent about half of the body weight and therefore represent the major concentrations of protein in the body. Unlike glycogen and fat, protein is not principally used for energy storage, but constitutes the structural and functional elements of cells. Increases in muscle mass require muscle-building exercise in addition to adequate protein intake. Nevertheless, there is a normal rate of turnover of muscle protein, and synthesis increases in the absorptive period when amino acids are in abundance (see Figure 23-4).

Protein synthesis requires not only that many amino acids be present, but that all the **essential amino acids** (those the body cannot synthesize; see Chapter 3, p. 47) be present in the proper proportions at the same time. Otherwise, synthesis of some proteins will not occur. Amino acids are not stored as such and thus cannot be saved for future use. Instead, excess amino acids enter metabolic pathways that lead to fatty-acid synthesis. The requirement for a balance of essential amino acids means that dietary protein sources must either be **complete proteins,** in that they contain all the essential amino acids (an example is meat), or, alternatively, be **complementary incomplete proteins** consumed in the same meal. Vegetarian diets typically are built around the combination of one of the whole grains with legumes or nuts and dairy products, and such diets provide all of the essential amino acids.

Somatostatin

Somatostatin (see Table 23-2) is secreted by pancreatic **delta cells** (see Figure 23-1) and by similar endocrine cells in the intestinal mucosa. Somatostatin decreases GI motility and secretion, pancreatic secretion, and the release of insulin and glucagon (discussed below). The release of somatostatin is increased by high plasma levels of both glucose and amino acids. Plasma somatostatin levels normally rise only slightly during the absorptive state, but they increase more dramatically if consumption of large amounts of food causes a rapid rise in blood levels of sugars and amino acids. Under these circumstances, somatostatin is capable of slowing all aspects of digestion and absorption (Figure 23-7). It thus prolongs the absorptive state and prevents nutrients from entering the body more rapidly than processing and storage mechanisms can handle.

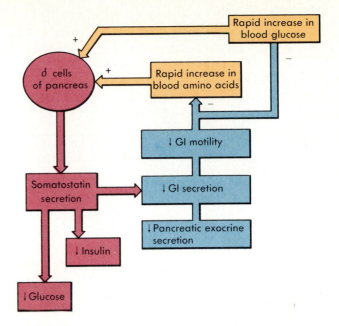

FIGURE 23-7
Inputs to delta cells and effects of somatostatin, including negative feedback, which reduces entry of glucose and amino acids into the circulation.

---

1 How is the number of glucose receptors on the beta cells of the pancreas adjusted by diet?
2 Why does protein anabolism depend on consumption of proteins containing all the essential amino acids in the same meal?
3 What is the role of apoproteins in lipid absorption, and why is it good to have a high proportion of the HDL particles to LDL particles in the blood?

CHARACTERISTICS OF THE POSTABSORPTIVE STATE
Onset of the Postabsorptive State

In the typical pattern of three meals a day, the full effects of the postabsorptive state are rarely encountered during the day, so the only period in which the body enters the postabsorptive state to any significant extent is during the night, before "breaking the fast" at breakfast. In the first portion of the 8 to 12 hours between the evening meal and breakfast, insulin promotes glycogen and fat storage in the liver and adipose tissue and net synthesis in all body cells. As the levels of glucose and amino acids begin to drop, the stimulation of beta cells diminishes. This automatically produces a drop in insulin blood levels, which correlates with the beginning of the postabsorptive state (see Table 23-3).

By itself, the drop in insulin levels adapts the body for a period in which nutrients must be supplied from the body's stores. Withdrawal of insulin removes the facilitation of insulin-dependent pathways and entry steps and also removes the inhibition exerted by insulin on catabolic pathways in carbohydrate and lipid metabolism. The rate of gly-

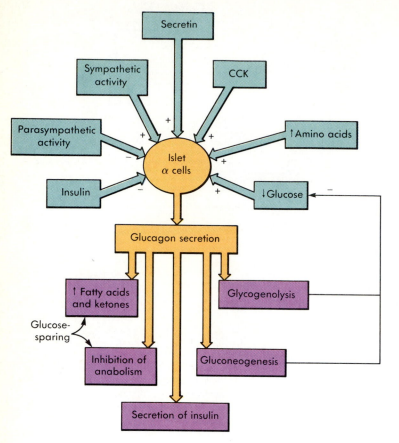

FIGURE 23-8

Inputs to alpha cells and effects of glucagon, including negative feedback, which increases plasma glucose levels.

cogenolysis in the liver and of lipolysis in both liver and adipose tissue increases as soon as insulin levels begin to drop (see Table 23-1). Liver glycogen stores by themselves are usually sufficient to supply the body's need for glucose in the period between the end of the absorptive period following dinner

and breakfast. However, the postabsorptive state elicits its own hormones, which oppose the actions of insulin and participate in the transformation of glycogen, fats, and protein into molecules that can be metabolized for energy. The extent to which any particular hormone is active in the postabsorptive phase is determined by how long the postabsorptive period lasts and by other factors such as exercise or emotional stress.

Role Of Glucagon

Glucagon (see Table 23-2), the major hormone of the postabsorptive state, is released from the alpha cells of the islets of Langerhans (see Figure 23-1). Glucagon is a catabolic hormone (Figure 23-8 and Table 23-5) that promotes glycogenolysis in the liver. At the same time that the liver is stimulated to release glucose derived from stored glycogen, it is also stimulated to convert other molecules (chiefly pyruvate, lactic acid, amino acids, and glycerol) into glucose (gluconeogenesis, see Table 23-1). The major source of reactants for gluconeogenesis is the muscles, which have both a store of glycogen and a large amount of contractile protein. Muscle cell reactions in the postabsorptive state are shown in Figure 23-9, *B*. (Note that in this and the corresponding illustrations for liver and adipose cells, the reactions of the absorptive state have been duplicated as part *A* for ease of comparison.)

Glucagon also stimulates lipolysis in adipose tissue by stimulating the hormone-sensitive lipase (Figure 23-10, *B*; see Table 23-1). The fatty acids that are released in lipolysis travel in the blood bound to serum albumen until they are taken up by cells. Through its alteration of metabolic pathways, glucagon increases the levels of glucose, fatty acids, and glycerol in the blood. When insulin levels are low, glucose does not easily enter most cells. Cells preferentially metabolize fatty acids, reserving glucose for those cells which do not have insulin-

TABLE 23-5	*Effects of Glucagon and Catecholamines during the Postabsorptive State*	
Target tissue	*Hormone effects*	*Metabolic consequences*
Liver	Stimulates enzymes of glycogen breakdown; inhibits enzymes of glycogen synthesis	Conversion of glycogen to glucose, which is released into blood
	Stimulates enzymes of gluconeogenesis	Conversion of lactate made by muscle into glucose for use by CNS
	Stimulates enzymes of fatty acid oxidation; inhibits enzymes of lipogenesis	Conversion of fatty acids released by fat into ketone bodies, which can be used for energy by most cells, sparing glucose for CNS
Fat	Stimulates hormone-sensitive lipase	Increased lipolysis in fat, with release of fatty acids and glycerol for conversion to ketones by liver

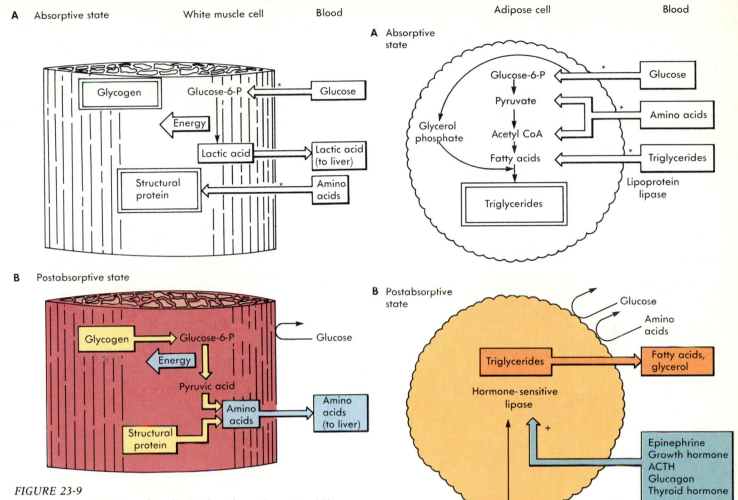

A Absorptive state   White muscle cell   Blood

B Postabsorptive state

A Absorptive state

B Postabsorptive state   Adipose cell   Blood

FIGURE 23-9
Reactions in white muscle cells in the absorptive state (**A**) and the postabsorptive state (**B**). Steps stimulated by insulin are indicated by an asterisk. The absence of insulin in the postabsorptive state inhibits glucose uptake.

FIGURE 23-10
Reactions in adipose cells in the absorptive state (**A**) and the postabsorptive state (**B**). Steps stimulated by insulin are indicated by an asterisk. The absence of insulin blocks the uptake of glucose and amino acids during the postabsorptive state.

dependent uptake mechanisms. Neurons are the major cell type in this category.

Glucagon also promotes the use of fatty acids as an alternative energy source. The metabolism of fatty acids for energy increases the levels of acetyl-CoA in the liver, and this favors release of **ketones (ketone bodies)** from liver cells (Figure 23-11, *B*). These ketones, which include acetone, acetoacetic acid, and β-hydroxyacetoacetic acid (β-hydroxybutyric acid), serve as alternative energy sources. This promotion of fatty acids and ketones as energy sources is called **glucose sparing.** In addition to its stimulation of catabolic and gluconeogenic pathways, glucagon inhibits the enzymes stimulated by insulin in glycogenic and lipogenic metabolic pathways (see Figure 23-8 and Table 23-5).

Control of Glucagon Secretion
Glucagon secretion is inhibited by the blood levels of glucose typical of the absorptive state, as well as

by the high level of parasympathetic activity that exists during the early phases of digestion. Through interactions within the pancreas, the glucagon-producing alpha cells are inhibited when the insulin-producing beta cells are stimulated and vice versa (see Table 23-3). In the postabsorptive state, plasma glucose levels, insulin secretion, and parasympathetic activity decrease and glucagon secretion increases. Thus intervention by glucagon prevents a further fall in plasma glucose.

In certain circumstances, glucagon plays an important role in the absorptive state. Plasma amino acids and the hormones, CCK and secretin, stimulate glucagon release at the same time that they are

A Absorptive state

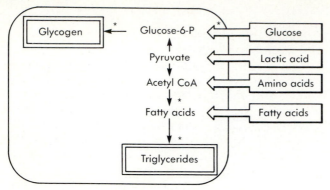

Liver cell   Blood

B Postabsorptive state

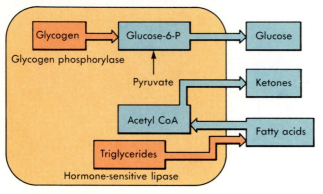

FIGURE 23-11

Reactions in liver cells in the absorptive state (**A**) and the postabsorptive state (**B**). Steps stimulated by insulin are indicated by an asterisk.

stimulating insulin secretion (Figure 23-12; see Table 23-3). When a carbohydrate-rich meal is consumed, the elevation of blood glucose overrides amino acid, CCK, and secretin stimulation and keeps glucagon levels low. If a protein or fat-rich meal is consumed, little glucose is entering the bloodstream, and the stimulatory effects on glucagon secretion dominate. If glucagon were not secreted along with insulin to mobilize glucose stores there would be an uncontrolled drop in plasma glucose (see Figure 23-12).

Glucagon secretion also increases whenever the sympathetic branch of the autonomic nervous system is activated (see Figure 23-8 and Table 23-3). This occurs during exercise and in stressful situations. In exercise and stress, blood glucose is elevated in anticipation of increased tissue demands. Glucagon release does not inhibit insulin secretion, but rather promotes it. Consequently, in exercise or stress the glucose released in response to glucagon intervention can be used by the body because insulin is also present.

> 1 Why is the decrease in insulin the single most important aspect of the hormonal pattern of the postabsorptive state?
> 2 What is the role of glucagon in the postabsorptive state? Under what circumstances is its secretion needed in the absorptive state?
> 3 What effect does insulin secretion have on glucagon secretion? Does glucagon have the same effect on insulin secretion?

FIGURE 23-12

Comparison of the hormonal responses to a meal with a typical mixture of carbohydrates and proteins (**A**) with the responses to a meal with high proteins in the absence of carbohydrates (**B**). When no glucose is coming from the intestine, insulin still increases the transport of glucose into cells, and preservation of the normal level of blood glucose is accomplished by glycogenolysis, which is stimulated by glucagon.

A Both carbohydrates and proteins consumed:

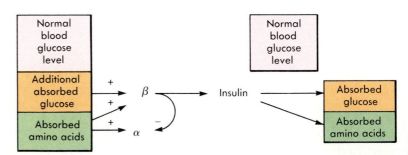

B High protein meal consumed:

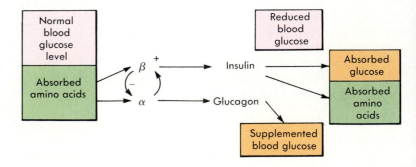

Conversion of Protein to Glucose

Turnover of protein can take place in all tissues, but most protein in the body is in skeletal muscles, so the protein turnover in muscle is the most significant. In the postabsorptive state, movement of amino acids out of the muscles exceeds the rate of amino-acid entry (see Figure 23-9, B). The amino acids that are present in muscle protein do not enter the blood in the same proportions at which they are found in the muscle. This is because amino acids are not all treated the same way when proteins are broken down. The essential amino acids are likely to be reused in protein synthesis, whereas nonessential amino acids tend to be released into the blood to be metabolized for energy.

Some nonessential amino acids are synthesized from carbohydrates in the muscle and released to supply the energy needs of other tissues. For instance, pyruvate is not able to leave the muscle cells, but it can be aminated (with amino groups taken from other amino acids) to form alanine or glutamate. Alanine increases in concentration in the blood during the postabsorptive state (see Figure 23-9, B).

Adrenal Hormones

The hormone, **cortisol**, a steroid hormone produced by the adrenal cortex (see Table 23-2), must be present for glucagon to stimulate catabolic pathways. Although cortisol is not typically elevated in the postabsorptive state, it is needed for the responses made by the body in this state. Cortisol is one of a family of steroids called **glucocorticoids** because of their importance for maintaining blood glucose levels.

Epinephrine (see Table 23-2) may also contribute to the postabsorptive state (Figure 23-13), especially if stress occurs during this period. The degree of stress associated with the postabsorptive state is determined in part by how fast plasma glucose levels drop. When the plasma glucose concentration falls rapidly, the sympathetic nervous system is activated and epinephrine is released from the adrenal medulla. Nervousness and irritability associated

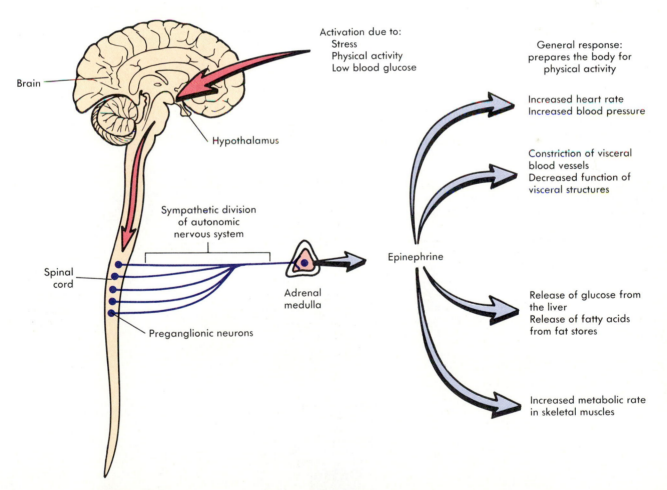

Activation due to:
Stress
Physical activity
Low blood glucose

General response:
prepares the body for
physical activity

Increased heart rate
Increased blood pressure

Constriction of visceral
blood vessels
Decreased function of
visceral structures

Epinephrine

Release of glucose from
the liver
Release of fatty acids
from fat stores

Increased metabolic rate
in skeletal muscles

Brain

Hypothalamus

Sympathetic division
of autonomic
nervous system

Spinal
cord

Adrenal
medulla

Preganglionic neurons

FIGURE 23-13
Regulation of the secretion of epinephrine from the adrenal medulla. Inputs that cause increased secretion of epinephrine include stress, physical activity, and low blood glucose. Epinephrine elevates glucose, blocks glucose use in cells that have insulin-dependent uptake systems, induces glucose sparing, and promotes lipid availability.

with hunger can result from these elevated levels of epinephrine. The liver and adipose tissue are innervated by sympathetic endings, and norepinephrine released from those endings, as well as blood-borne epinephrine, stimulates liver gluconeogenesis and glycogenolysis and hormone-sensitive-lipase activation in adipose cells. Epinephrine also stimulates glucagon secretion and inhibits insulin secretion.

Diabetes Mellitus: Cellular Famine in the Midst of Plenty

Diabetes mellitus is a condition that results when inadequate uptake of glucose from the blood by the body's cells causes high levels of glucose in the blood **(hyperglycemia).** The high levels of glucose saturate the glucose transport system in the kidney so that glucose "spills over" into the urine **(glycosuria),** making it sweet, a property that could be diagnosed by physicians in ancient times.

Abnormally high levels of glucose in the blood and the nutritional deficits suffered by cells unable to use the glucose produce various abnormalities, including blindness, atherosclerosis, and poor circulation. The last leaves the patient susceptible to complications of infections such as gangrene. In the urine, there are not only high levels of glucose, but also of urea (from increased protein catabolism) and ketones (from increased fat oxidation). All of these act as osmotic diuretics (Figure 23-14), causing the increased urine production characteristic of untreated diabetes mellitus.

There are two types of diabetes mellitus: **Type I diabetes,** which affects about 20% of all patients, results from inadequate secretion of insulin from the beta cells. Type I diabetes is also called **insulin-dependent** diabetes because it can be treated by insulin replacement therapy. If untreated, the failure to use glucose and inhibit the catabolic pathways that contribute to gluconeogenesis by the liver result in several metabolic derangements (see Figure 23-14). Most threatening is the production of acidic ketones from fatty acids in the liver. Ketone

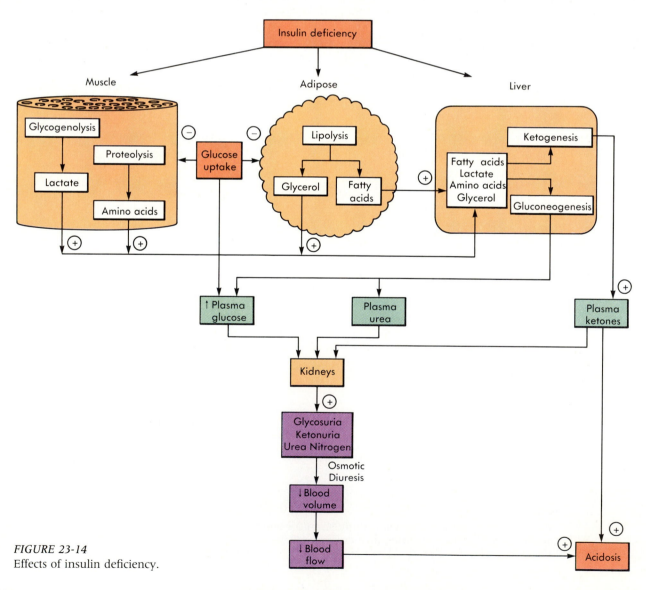

FIGURE 23-14
Effects of insulin deficiency.

production in a nondiabetic, fasting individual cannot reach levels that result in **metabolic acidosis** because elevated ketone levels elicit secretion of insulin, which opposes ketone production. In diabetics, this negative feedback pathway does not work, and the resulting **ketoacidosis** may be so severe that it causes **diabetic coma** or death.

Diabetics who take insulin without also eating carbohydrate are in danger of a rapid drop in blood glucose. Decreases of plasma glucose to the range of 20 to 30 mg/dl may cause **insulin shock,** which can be fatal because of its effects on the brain. Rapid administration of glucose reverses insulin shock. Newer methods of supplying insulin (such as minipumps) have greatly improved the patient's ability to match the supply of insulin to the body's needs.

The remaining 80% of diabetic patients have **insulin-independent,** or **type II, diabetes,** in which release of insulin appears to be in the normal range, but either insulin receptors do not respond normally to insulin, or the density of insulin receptors is abnormally low. Although in principle insulin injection might be used to overcome the low receptor density, insulin may lead to further reduction of receptor number by the target cells **(down regulation);** thus most patients with type II diabetes are treated by controlling their diet and by increasing exercise. Beneficial effects of exercise include stimulation of glucose uptake by the muscle cells and the production of more insulin receptors.

In many cases, type II diabetes is associated with obesity, and the scarcity of insulin receptors may be a response in these sensitive individuals to the elevated levels of insulin secreted in response to consumption of excessive quantities of food. Reduced caloric intake can result in **up regulation,** an increase in the number of insulin receptors in the membranes of target cells. The change can be seen in a few days, even before significant weight loss has occurred.

1 What are the roles of cortisol and epinephrine in the postabsorptive state?
2 Compare the regulation of insulin receptors on target cells in type II diabetics with the regulation of glucose receptors on target cells in normal individuals.
3 Why is the diabetic response to elevated glucose maladaptive?

REGULATION OF FOOD INTAKE
Short-Term Regulation of Food Intake and Satiety Signals

Food is not consumed continually, but is taken in meals with intervening periods of nonconsumption. Proteins, carbohydrates, and fats also differ in their energy content per gram (Table 23-6). Factors that influence total intake are how long meals last, the quantity of food consumed in each meal, the nutrient content of a meal, and how frequently meals occur (Figure 23-15). What signals initiate and terminate food consumption? Hunger does not correspond particularly well with the nutritional state of the body as measured by the circulating levels of glucose or stored fats. The desire to eat is determined by many factors associated with social patterns of food intake or the aromatic and visual appeal of food. Yet there appear to be some signals that regulate the amount of food consumed and the total caloric intake.

Changes in metabolic rate occur even before nutrients begin to be consumed. These are caused by hormonal stimulation in the cephalic stage of digestion. Estimates of the richness of food can be made on the basis of past experience, and this influences the level of circulating hormones before ingestion. Thus the association cortex is involved, along with areas of the brain involved with reward systems.

Both insulin and glucagon levels are elevated in

TABLE 23-6	Caloric Content of the Three Major Categories of Nutrients	

Category	Kcal/gram*	Kcal/1 O₂ consumed
Carbohydrate	4.2	5.0
Protein	4.3	4.5
Fat	9.4	4.7

*A kilocalorie (kcal) is the amount of heat required to raise the temperature of 1000 grams of water by 1° C. The caloric content is measured by measuring the heat liberated by complete combustion of a known weight of the nutrient, using an instrument called a bomb calorimeter.

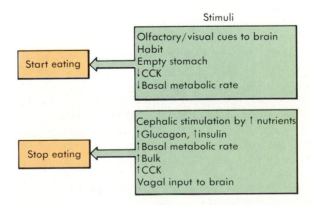

FIGURE 23-15
Regulation of food consumption is determined by the amount of food consumed in each meal and the length of the interval between meals. Food consumption is regulated by signals. Some of the best-studied stimuli for the onset and termination of feeding are listed on the right.

the cephalic stage of digestion before food absorption; thus stored nutrients are mobilized, and their uptake into cells is stimulated. A neural connection between sugar receptors in the mouth and the hypothalamus is one pathway that promotes these hormonal changes. Early signals related to incoming food lead to metabolic changes that provide a feeling of **satiety**, of having consumed enough, even before significant absorption takes place (see Figure 23-15).

Termination of a meal is also influenced by signals arising from the GI tract that indicate both bulk and nutrient levels of ingested food (see Figure 23-15). In the stomach, stretch receptors detect the presence of food, and the hunger pangs resulting from contraction of the empty stomach are alleviated by the bulk aspects of food. In experiments in which nutrient-poor bulk is ingested, meals are terminated on the basis of bulk before adequate caloric intake occurs. However, this control is overridden within days by mechanisms that respond to the body's overall nutritional state so that such experimental subjects will increase intake of the nutrient-poor food over time.

In spacing meals, signals of satiety that arise from the duodenum and from nutrients flowing to the liver in the hepatic portal system are more important than are gastric satiety signals. Entry of chyme into the duodenum increases the level of CCK in the blood, and many regions of the brain have CCK receptors. The intermeal interval is affected by how long CCK levels are elevated in the blood. Finally, the elevation of temperature that accompanies digestion continues all during the absorptive phase, and the tendency to eat again is correlated with the drop in metabolic rate as the body approaches the postabsorptive state (see Figure 23-15).

Two centers in the hypothalamus are involved in regulation of feeding and satiety. Animal studies in which specific hypothalamic regions were destroyed indicated the presence of a **feeding center** in the lateral hypothalamus and a **satiety center** in the ventromedial hypothalamus. The hypothalamus is not the only part of the brain involved in food intake because many of the relevant signals project to higher brain regions. Recent research has greatly increased knowledge of the complexity of the pathways and the various signals that contribute to regulation of food intake. However, a simple neural control mechanism for feeding that satisfies all the experimental observations is not yet available.

Long-Term Control of Food Intake by Nutritional State

Although the level of glucose in the blood is not well correlated with the daily pattern of eating, long-term carbohydrate deprivation does produce hunger. The **glucostatic** hypothesis of long-term regulation of eating is based on feedback relating to the availability of glucose to body cells. However, this hypothesis has been partly supplanted by the **lipostatic** hypothesis, which asserts that feeding frequency and duration are regulated by a signal that arises from the fat deposits in the body.

The general nutritional state of the body is related to the intake of all categories of food. Carbohydrates, fats, and amino acids all contribute to the anabolic reactions of the absorptive state, and the heat that is produced in these reactions is another signal related to total nutrient intake. There is some evidence that heat from nutrient anabolism may serve as a satiety signal. As body fat accumulates, metabolic rate increases, with an accompanying elevation of heat production. If heat production during the anabolic phase is a factor that tends to limit consumption, then the elevation of body temperature associated with accumulation of excess body fat would be an automatic mechanism that tends to return the body to a leaner weight. This is the **thermostatic** hypothesis of body weight regulation.

CHANGES IN METABOLISM OVER THE LIFE CYCLE
Growth and Growth Hormone

Fetal and early postnatal development does not require **growth hormone** (also called **somatotropin** or **human growth hormone [HGH]**). The principal effect of HGH is seen in children. The rapid metabolic rate of children and adolescents is largely attributable to the stimulation of net anabolism by HGH in these ages. HGH is needed for cartilage formation (**chondrogenesis**) and for the subsequent calcification (**ossification**) of bone by osteoblasts. These processes are promoted by HGH's stimulation of mitosis and protein synthesis in cells in the **epiphyseal plates**, the growing regions of long bones (Figure 23-16). Protein synthesis is seen in soft tissues such as muscle as well as at the epiphyseal plates, and protein synthesis is favored over all other metabolic pathways by HGH. Rapidly growing children and adolescents, who typically have little body fat, exemplify the effects of this hormone on metabolism. Body fat is mobilized as an energy source by HGH at the same time that amino acids, another potential energy source, are being incorporated into protein.

HGH is secreted from the anterior pituitary in an oscillatory pattern that usually peaks during the early hours of the sleep cycle (Figure 23-17). The release is stimulated by **growth hormone-releasing factor (GRH)** from the hypothalamus and inhibited by **somatostatin (growth hormone-inhibiting hormone)** from the hypothalamus.

HGH exerts its effects through induction of **somatomedins**, a family of hormones secreted into the blood by the liver (and possibly other tissues). The somatomedins are similar to insulin in molecular structure, and they are the agents that act on target tissues to promote mitosis and protein synthesis

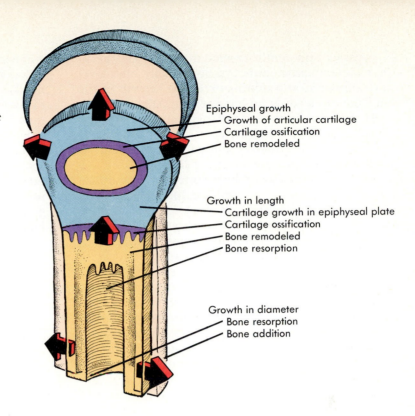

FIGURE 23-16
Growth of long bones occurs at the bone surface and at the epiphyseal plates. Surface growth increases bone thickness and can occur at any stage of growth. The surface is a major site of bone remodeling in response to mechanical stress. Lengthening occurs at the epiphyseal plates and continues only until the plates close. Epiphyseal closure is normally complete by age 20 or sooner.

Epiphyseal growth
— Growth of articular cartilage
— Cartilage ossification
— Bone remodeled

Growth in length
— Cartilage growth in epiphyseal plate
— Cartilage ossification
— Bone remodeled
— Bone resorption

Growth in diameter
— Bone resorption
— Bone addition

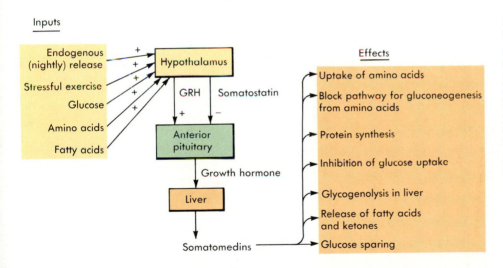

Inputs

Endogenous (nightly) release
Stressful exercise
Glucose
Amino acids
Fatty acids

Hypothalamus

GRH   Somatostatin

Anterior pituitary

Growth hormone

Liver

Somatomedins

Effects

Uptake of amino acids

Block pathway for gluconeogenesis from amino acids

Protein synthesis

Inhibition of glucose uptake

Glycogenolysis in liver

Release of fatty acids and ketones

Glucose sparing

FIGURE 23-17
Control of the secretion of growth hormone and its stimulation of the somatomedins that mediate its effects. The biological rhythm that controls the release of growth hormone produces a nightly peak of secretion; other inputs act to increase the overall amount of growth hormone that is released.

(see Figure 23-17). The induction of new enzymes and an increase in ribosomes, which are needed to synthesize protein in the target cells, requires some time, so the effects of HGH occur more slowly than those of hormones which act by changing the effectiveness of enzymes already present in the cell. Somatomedins promote protein synthesis by increasing the rate of uptake of amino acids into muscle (an insulin-like effect), and they increase the availability of amino acids to the muscle cells by opposing the use of amino acids in gluconeogenesis.

HGH opposes fat deposition by decreasing glucose uptake and utilization in adipose tissue. Conversely, HGH increases synthesis of the capillary lipases needed for dietary triglyceride uptake into adipose tissue and promotes the breakdown of the triglycerides allowing fatty acids to be released. In this way, HGH maintains blood glucose levels, decreases storage of fat, and promotes the use of fatty acids for energy. This effect of HGH on fat metabolism is seen only if the plasma insulin levels are low because insulin normally inhibits the catabolism of fat through its inhibition of intracellular hormone-sensitive lipases.

Deficiencies of HGH or a lack of HGH receptors in early childhood results in **dwarfism.** The stature of pygmies apparently results from lack of one of the major somatomedins. An abnormally high level of HGH secretion produces excessive growth of the long bones, leading to **gigantism.** Administering HGH during childhood and puberty can likewise result in additional growth of long bones. This treatment, which allows normal development in children with deficient pituitary function, can also increase the rate of growth in normal children. Biotechnology has now reduced the cost of producing HGH. Some parents ask for the use of this hormone in children with normal growth potential in an attempt to promote growth and increase heights.

In adults, HGH secretion increases in response to the stress experienced during sustained exercise and in starvation. In both cases, HGH acts to conserve the protein stores of the body. Because the epiphyseal plates are closed, HGH ceases to have an effect on bone growth in adults, but there is still a potential effect on bone width. Thus excessive secretion of HGH, which can result from tumors in the anterior pituitary, produces **acromegaly.** This condition is characterized by thickening of the bones of the fingers and toes, some muscle hypertrophy, disfiguring growth of the flat bones of the back and face, and coarse skin. Deficiencies of HGH in adults have little noticeable effect because many forms of growth such as exercise-induced muscle hypertrophy, most healing processes, and the normal turnover of somatic cells, do not require HGH.

Effect of Aging on Body Composition

The relative contributions of the skeleton, muscle mass, and adipose tissue to total body weight change over the life cycle. Children tend to be plump in early childhood and then become thin in the period dominated by HGH when rapid long-bone growth is occurring. In adulthood, a gradual shift in the proportions of weight is attributable to changes in body protein, fat, and skeleton, even when body weight is maintained at a constant level.

Almost all people lose muscle tissue as they get older. This loss is attributable both to a reduction of muscle mass through loss of filament proteins and to a reduction in the number of muscle cells. Loss of muscle mass is much less evident in individuals involved in daily physical labor and is largely a function of disuse. As is apparent from the data presented in Figure 23-18, the percentage of total body weight represented by fat more than doubles for men ranging in age from 20 to 55 years. Taken together with the loss of muscle mass, it could be argued that the ideal weight of physically inactive individuals decreases with increasing age rather than remaining constant.

Because muscle has a much higher metabolic rate than adipose tissue, the shift in weight from muscle mass to fat reduces the overall metabolic rate of the body. Ideally, the changes in body composition should automatically result in reduced food consumption. Unfortunately, eating habits formed over a lifetime are difficult to change, and lack of exercise exacerbates the problem just at a time in life when weight gain and the threat of atherosclerosis become significant. Because the beneficial effects of exercise are well known, it is wise to develop and maintain a pattern of exercise that can be continued throughout adulthood.

> 1 How do the relative proportions of muscle and fat affect the overall metabolic rate?
> 2 In what ways does growth hormone (the somatomedins) affect metabolism?

METABOLIC RESPONSES TO THREATENING SITUATIONS
Thyroid Hormones and the Basal Metabolic Rate

A standard way of comparing the metabolic rate of different individuals is to measure their **basal metabolic rate.** This is the rate at which the body

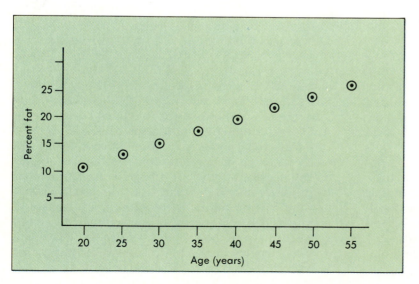

FIGURE 23-18
Average increase in fats as a proportion of total body weight with increasing age in men.

The loss of muscle mass with advancing age is accompanied by net decalcification of the skeleton, called **osteoporosis.** In osteoporosis, the loss of calcium is accompanied by loss of the cells that deposit the calcium salts and is therefore irreversible. Osteoporosis is a real threat to women in their postmenopausal years because, in women, maintenance of bone mass is to some extent estrogen dependent, and the level of estrogen drops with the cessation of menstruation. The problem of osteoporosis develops more slowly in men, who typically have greater bone mass to begin with and therefore do not experience the weakening that results in easy fractures until their 70s and 80s.

Concern about the increased likelihood of bone fractures that accompanies osteoporosis has led to recommendations for increased intake of calcium. This has fostered the addition of calcium to all types of foods. However, much of the added calcium is not in a form that the body can use; even the calcium in spinach cannot be absorbed. Dairy products remain the best and safest sources of calcium, but it is now recognized that, although ingestion of adequate amounts of calcium is necessary, estrogen supplements are also beneficial in reducing the destruction of bone in postmenopausal women. Another substance that is receiving increasing attention as a protector of bone structure is fluoride, which protects the other major calcium-based structures in the body, the teeth.

Recent studies of the development of osteoporosis compared groups of women who were matched with respect to known or suspected risk factors for development of osteoporosis (that is, low calcium intake, cigarette smoking, moderate alcohol consumption, and skeletal type) and who differed only in their amount of daily exercise. The conclusion of the study was that a lifelong pattern of regular exercise confers significant protection against the ravages of osteoporosis. The exercise must be of the weight-bearing sort, such as jogging, because some otherwise beneficial exercises, such as swimming, consume energy but do not seem to help maintain the skeleton. It appears that, with advancing age, bone follows the "use it or lose it" pattern that applies to many aspects of human physiology, including mental and sexual function.

---

breaks down stored energy when the person is in an alert but resting condition and has fasted for 12 hours. The direct measurement of the basal metabolic rate is made with the person in a small chamber that has sensors in the walls that can measure heat liberated from the body. An indirect method is based on the measurement of the rate of O_2 uptake in respiration because the rate of release of energy from food is quantitatively related to O_2 consumption (see Table 23-6). Typically, a mixture of fuels is oxidized during the postabsorptive state, so that about 4.8 kcal of energy are released for every liter of O_2 consumed in respiration. The basal metabolic rate of adults is about 20 kcal/kg body weight/day, or roughly 1 kcal/min for a typical adult. The total metabolic rate of an individual depends on sex, physical activity, and genetic factors. Average rates are about 39 kcal/kg/day for men and about 34 kcal/kg/day for women.

Many hormones affect the overall rate of conversion of energy sources to CO_2, water, ATP, and heat. These include metabolic hormones that respond to short-term changes in the nutritional state (see Table 23-2); for example, the effect of insulin may increase the basal metabolic rate by as much as 20% during the absorptive state. Sex hormones have effects on metabolism (as suggested by the sex difference in metabolic rate) but are not involved in regulation of metabolic rate. The **thyroid hormones** (see Table 23-2) are of particular importance in regulating the basal metabolic rate and in long-term adaptation to stress, including prolonged exposure to cold. The major form of thyroid hormone secreted by the thyroid gland is **thyroxine (T_4).** Much of the secreted T_4 is converted by the liver and kidney to a more active form, **triiodothyronine (T_3),** by removal of one iodine atom (I). Most of the metabolic effect of the thyroid hormones is actually caused by T_3 rather than by the secreted form of T_4.

The thyroid contains spherical follicles into which a large storage protein, thyroglobulin, is secreted (Figure 23-19). Thyroid hormones are formed from the amino acid, tyrosine, in a two-stage process (Figure 23-20). First, tyrosine residues on thyroglobulin are singly or doubly iodinated to form monoiodotyrosine (MIT) or diiodotyrosine (DIT). Next, an MIT and a DIT are joined to form T_3, or two DIT are joined to form T_4. Then the hormone molecule is cleaved from the thyroglobulin and secreted into the blood (Figure 23-21). About 80% of the total hormone secreted by the thyroid is T_4. The requirement for iodine to synthesize the hormone is met by active accumulation of iodide (I^-) by the thyroid from the blood.

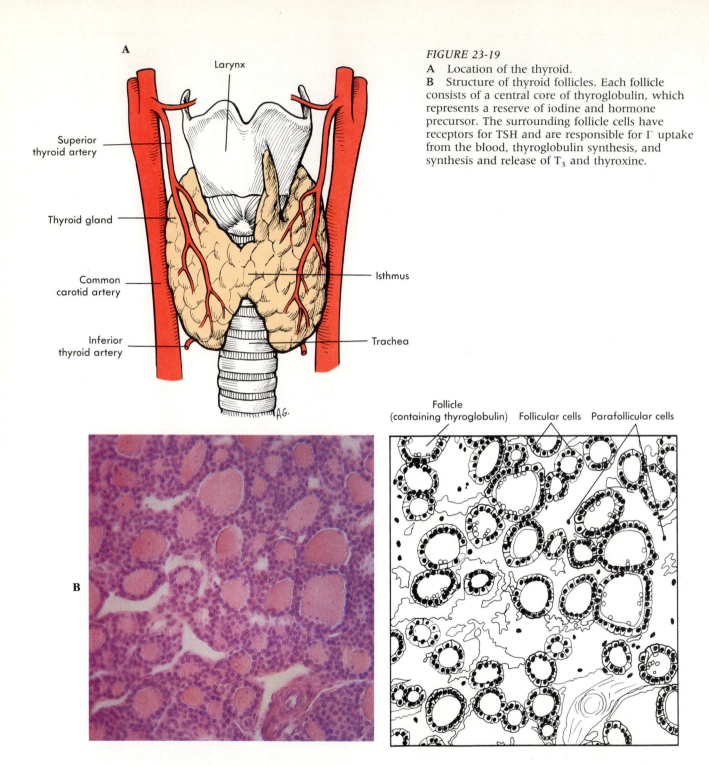

A Location of the thyroid.

FIGURE 23-19
A Location of the thyroid.
B Structure of thyroid follicles. Each follicle consists of a central core of thyroglobulin, which represents a reserve of iodine and hormone precursor. The surrounding follicle cells have receptors for TSH and are responsible for I⁻ uptake from the blood, thyroglobulin synthesis, and synthesis and release of T_3 and thyroxine.

Follicle (containing thyroglobulin)   Follicular cells   Parafollicular cells

The receptors for thyroid hormones are in the nuclei of cells. The affinity of the receptors for T_3 is much higher than for T_4. The major effect of thyroid hormones is to alter enzyme systems that affect metabolic rate. The thyroid hormones increase metabolic rate through induction of specific proteins, such as receptors that are incorporated in the cell membrane, the Na^+-K^+ ATPase, and enzymes involved in specific anabolic pathways. The result of these actions is an increase in O_2 consumption (basal metabolic rate) and protein synthesis within

cells. Thyroid hormone effects on lipid and carbohydrate metabolism include increasing glucose uptake, glycogenolysis, and fatty acid use.

Thyroid hormones potentiate the action of many other hormones. Thyroid hormone secretion is required for normal growth and development, especially that of the central nervous system. Absence of thyroid hormones in early infancy causes severe mental retardation. In adults, changes in the level of thyroid hormone secretion affect the central nervous system, with normal levels enhancing alertness,

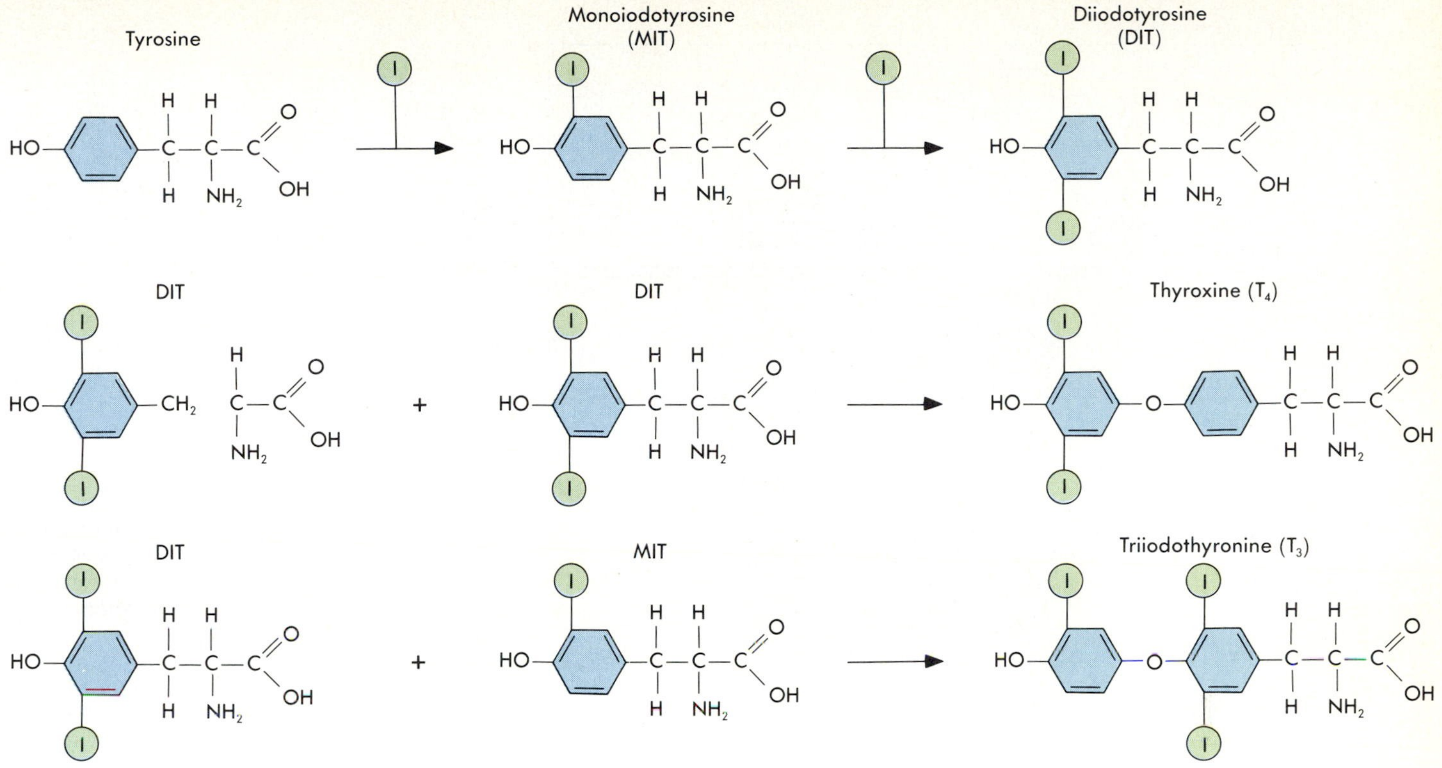

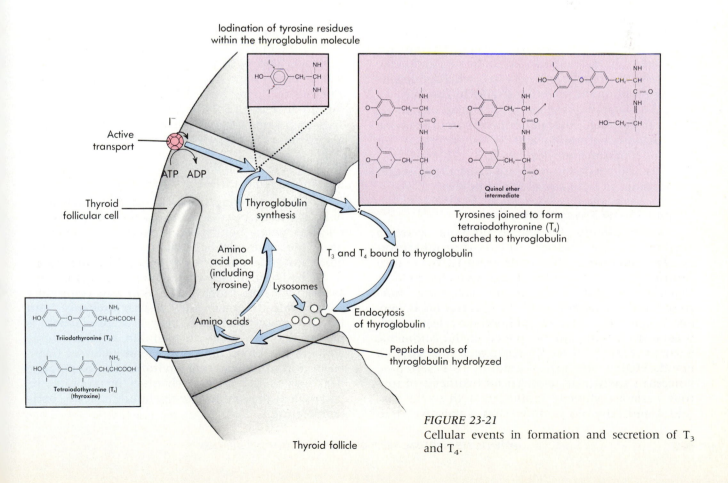

FIGURE 23-20

The two steps in synthesis of T_3 and thyroxine from tyrosine in the thyroid follicular cells. The first step is formation of either monoiodotyrosine (MIT) or further iodination to form diiodotyrosine (DIT). In the second step, either the diiodinated benzene ring from one molecule of DIT is joined to a second DIT to form thyroxine, or the monoiodinated iodinated ring from an MIT is joined to a DIT to form T_3. Formation of additional T_3 in target tissues is the result of removal of an I from T_4 (pathway not shown).

FIGURE 23-21

Cellular events in formation and secretion of T_3 and T_4.

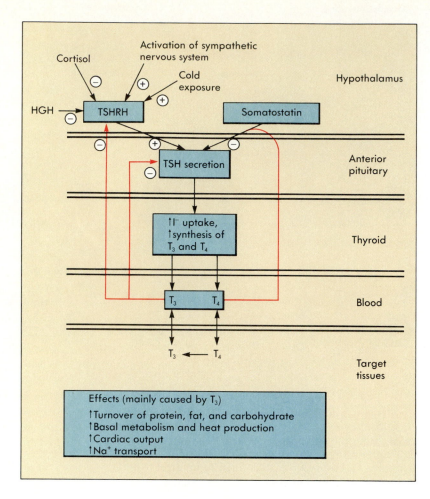

FIGURE 23-22
Control of secretion of the thyroid gland hormones, T_3 and T_4. The hypothalamus has both a releasing factor, thyrotropin-releasing hormone (TSHRH), and an inhibiting hormone, somatostatin (thyroid-inhibiting hormone), which regulate the release of thyroid-stimulating hormone (TSH) by the anterior pituitary. Some of the factors that affect TSHRH secretion are shown. See Figure 23-17 for additional factors that affect somatostatin secretion. Blood levels of T_3 and T_4 are affected by rates of secretion from the thyroid and by rates of conversion of T_3 to T_4 in target tissues. Feedback control of thyroid function is mediated in part by effects of plasma levels of T_3 and T_4 on pituitary TSH secretion and on hypothalamic secretion of TSHRH and somatostatin (feedback loops shown in red).

memory, and overall emotional tone. Excesses of thyroid hormone secretion **(hyperthyroidism)** pathologically increase the metabolic rate and make the patients irritable and nervous, whereas deficits of thyroid hormone secretion **(hypothyroidism)** reduce the metabolic rate and physical and mental activity. In children, thyroid hormone stimulates bone growth and maturation and is also necessary for normal reproductive development. Thyroid hormone is also needed to maintain many aspects of renal function in both adults and children.

Regulation of Thyroid Function

The major stimulus for thyroid hormone secretion (Figure 23-22) is anterior pituitary **thyroid-stimulating hormone (TSH)**. Secretion of TSH is in turn controlled by the hypothalamus, which produces **thyroid-stimulating hormone releasing hormone (TSHRH)**. Normally, TSH secretion is very nearly constant, as is thyroid hormone release. Activation of the sympathetic branch of the autonomic nervous system directly stimulates the thyroid gland. HGH and cortisol inhibit TSH secretion, whereas I⁻, which is required for synthesis of thyroid hormone, directly inhibits growth of the thyroid gland. Thyroid hormone has a negative feed-

back effect on TSHRH release and TSH secretion.

An example of the effect of interrupting this feedback loop is seen in dietary iodine deficiency. When there is not enough I⁻ to make normal amounts of T_4, the plasma levels of T_4 decline, causing an increase in TSH secretion. The additional TSH stimulates thyroid enlargement. The enlarged gland causes a characteristic swelling of the neck called a **goiter**.

Metabolic Adaptation to Cold

Body heat is produced as a consequence of metabolic reactions because some of the energy that is released when bonds are broken is liberated as heat. The more metabolically active a tissue is, the more heat it produces. Major heat-producing organs in the body are muscle and the visceral organs, such as the liver and kidney. The skin and respiratory surfaces are the major sites of heat exchange with the environment. When body heat production is greater than necessary to maintain a normal body temperature, blood flow to the skin increases transfer of heat to the external environment. Evaporative heat loss is the only mechanism for cooling the body when the ambient temperature is above body temperature.

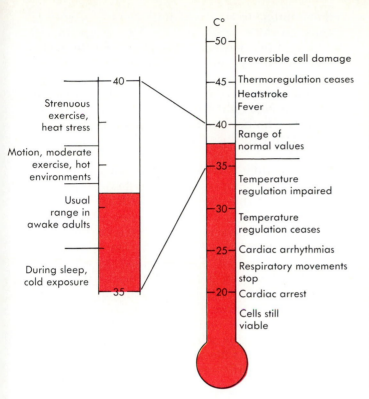

FIGURE 23-23
The normal range of core body temperature, and effects of departures from this range.

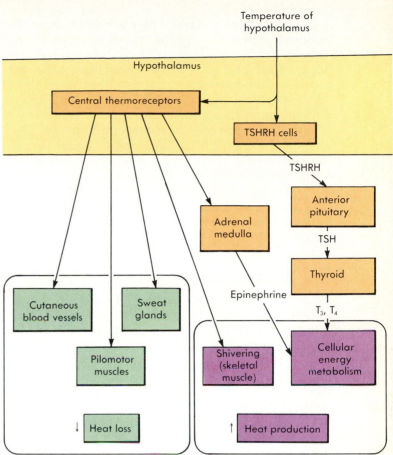

FIGURE 23-24
Pathways in central control of core body temperature by hypothalamic centers. The responses of the effectors to departures from the thermostatic set point are given in Table 23-7.

The usual range (set point; see Chapter 5, p. 86) of values of the **core body temperature**—the temperature of the heart, brain, thoracic and abdominal organs, and blood in major central vessels—is between 36° C and 37.6° C in adults (Figure 23-23). Departures from this narrow range have life-threatening implications, especially when body temperature rises above 45° C. Body temperature typically falls to the low end of the normal range in sleep and cold exposure. It rises to the high end of the normal range in exercise and heat stress. Body temperature may rise as high as 43° C when the normal set point is elevated in fever. Control of body temperature involves two basic types of responses: (1) modulation of heat loss, and (2) modulation of heat production (Figure 23-24 and Table 23-7).

When the external temperature is well below body temperature, blood is shunted toward internal organs, and behaviors that decrease the area of skin available for heat exchange, such as folding the arms or hunching the shoulders, are elicited by the sensation of cold. **Shivering thermogenesis** results from activation of the sympathetic nervous system. Shivering increases the body's heat production by activating a special pattern of repetitive, asynchronous contractions of skeletal muscle fibers, which requires synthesis and hydrolysis of ATP with its accompanying heat production. Although periph-

	TABLE 23-7	Summary of Centrally Mediated Thermoregulatory Responses

	Hypothalamic temperature	
Effectors	Above the set point	Below the set point
Cutaneous arterioles	Dilated	Constricted
Sweat glands	Secreting sweat	No sweat
Pilomotor muscles	Relaxed	Contracted ("goose bumps")
Skeletal muscles		Shivering thermogenesis
Adrenal medulla		Epinephrine secretion increased
Thyroid		Thyroid hormone secretion increased

eral temperature receptors play a role in these responses, their input is coordinated with the receptors in the hypothalamus that are sensitive to the temperature of the blood, as evidenced by the fact that localized heating or cooling of specific regions of the hypothalamus elicits changes in blood distribution, sweating, or shivering.

The increased basal metabolic rate and heat production that result from long-term exposure to cold (**nonshivering thermogenesis**) are controlled by responses of the thyroid gland acting in conjunction with the sympathetic branch of the autonomic nervous system and epinephrine released from the adrenal medulla. Exposure to cold in many types of mammals stimulates TSHRH secretion via a neural input to the hypothalamus and this increases TSH and thyroid hormone release. In man, the thyroid is necessary for adaptive changes made in response to cold, but the circulating levels of the hormone do not become significantly elevated, which may simply reflect increased rates of uptake. Although thyroid hormone actions can increase the metabolic rate of cells, a primary contribution of thyroid hormones to nonshivering thermogenesis is to stimulate synthesis of increased numbers of receptors for epinephrine on the target tissue cells (especially of the liver, pancreas, and muscle). These tissues then increase their metabolic rate under the control of epinephrine.

T_4 and T_3 may also increase the basal metabolism of cells by increasing the permeability of the cell membrane to Na^+, which imposes increased demands on ATP production to drive the Na^+-K^+ pump, and by increasing the rates of **futile cycles**, in which storage molecules are repeatedly synthesized and degraded in energetically wasteful but heat-producing reactions. The result is that the basal metabolic rate of the body is elevated, and individuals who have adapted in this way are able to survive cold exposure that might kill unadapted individuals.

Human infants have a very limited capacity to regulate body temperature by either sweating or shivering, but they do have a mechanism for producing heat that is lost after the first year or so of life (although this mechanism is present throughout life in many mammals). **Brown fat** is a special, metabolically active tissue located in the neck, chest, and between the scapular bones in the human infant. The location of brown fat allows infants to channel heat to the thorax and to blood in the arteries that serve the brain. The color of brown fat results from the many mitochondria present in the fat cells, which reflect the high metabolic capabilities of this tissue as compared with white adipose cells and with numerous blood vessels. Activation of brown fat by sympathetic inputs elicits heat production through pathways that allow the process of oxidative metabolism to be largely uncoupled from phosphorylation. The result is that stored lipids are catabolized, but very little useful work (ATP production) is performed, so the sole purpose served by these fat deposits is to rapidly generate heat.

Body Temperature Regulation and Fever

The centers for regulation of body temperature in the hypothalamus work in a push-pull fashion to increase or decrease body heat retention. Body temperature is regulated around a normal body temperature set point by mechanisms discussed above. **Fever**, the elevation of body temperature seen in response to infection, results from a resetting of the hypothalamic set point to a higher temperature. The way this change is achieved and the benefit it confers in the body's fight against infection have been appreciated only within the last decade.

Fever-producing agents that enter the body from external sources are called **exogenous pyrogens** (Figure 23-25). Bacterial toxins and bacterial wall fragments can serve as exogenous pyrogens. In addition, when macrophages interact with bacteria or other fever-producing agents, they release a hormone, **interleukin-1**, that acts as an **endogenous pyrogen** (in addition to various nonspecific effects this hormone has on target cells; see Chapter 24). The interaction of interleukin-1 with the hypothalamus indirectly resets the set point for body temperature at a higher level because interleukin-1 cannot itself cross the blood-brain barrier; the resetting is thought to occur through the induction of pros-

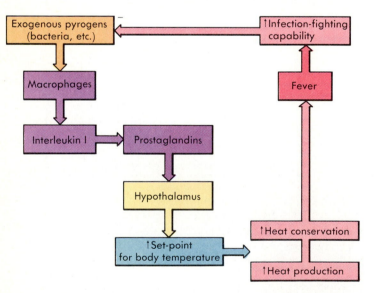

FIGURE 23-25
Pathway for production of fever by pyrogen resetting of the hypothalamus body temperature set-point. The elevation of body temperature is achieved in a matter of hours by changes in the regulation of body heat loss, and additional heat may also be generated by shivering if that is needed. The fever has been shown to assist the disease-fighting mechanisms of the body.

Control of Blood Cholesterol Levels by the Liver

Cholesterol is the basic molecule from which steroid hormones and bile salts are synthesized. Cholesterol enters the bloodstream from two sources: (1) intake in the diet, and (2) synthesis in the liver. The dietary cholesterol enters the blood in chylomicrons, which move relatively rapidly from the circulation into the liver. Cholesterol obtained from the diet reduces but does not abolish the liver's synthesis of cholesterol. Sources of lipids and cholesterol are easily obtained in the typical American diet, and the problem is not too little cholesterol, but too much. Dietary strategies include reduction of consumption of animal fats, which are high in cholesterol, and consumption of foods that reduce cholesterol absorption from the intestine or that block bile salt recycling. Oat bran has been shown to reduce blood cholesterol levels not just because it introduces bulk, but also because it has some special affinity for the cholesterol.

Elevated levels of cholesterol in the blood are known to cause development of **atherosclerosis** (formation of deposits on arterial walls that ultimately lead to blockage of the arterial lumen). Information on why blood cholesterol levels are not always maintained in a healthy range has only recently been obtained. One contribution to the story resulted from the lucky observation of obesity in an individual rabbit, which led to the development of a strain of rabbits with a deficiency in cholesterol uptake. This model system allowed the path followed by cholesterol from blood into tissues to be studied. Another contribution was the discovery of compounds in soil molds that inhibit a key enzyme in the synthesis of cholesterol.

Cholesterol from the liver follows the same path as triglycerides, entering the blood in VLDL particles. However, when triglycerides enter cells under the control of lipoprotein lipase, most of the cholesterol remains in the remnants of VLDL particles, which become LDLs after VLDL particles have unloaded their lighter elements. The LDLs normally return to the liver, where they are taken out of the plasma by two uptake systems. The most important uptake mechanism for LDL is controlled by LDL receptors synthesized by the liver. Excessive intake of cholesterol can saturate the LDL receptors, and the liver synthesizes fewer receptors if the blood level of LDL is high (an example of down regulation). Therefore LDL uptake is even less effective in the face of plenty of cholesterol. Intake of saturated triglycerides, such as those obtained from animal sources, worsens the problem because saturated triglycerides suppress the LDL-receptor uptake system of the liver while stimulating synthesis of additional cholesterol. The unavoidable conclusion is that diet plays a very significant role in determining cholesterol levels in people with normal lipid metabolism. Dangerously high LDL plasma levels found in a small portion of the population are now known to result from a genetic defect in the ability to synthesize the LDL receptor.

Aging increases the levels of LDL in blood. It is not known whether this increase, in the face of no dietary changes, is attributable to a decrease in the effectiveness of LDL uptake, greater cholesterol synthesis by the liver, or a combination of the two factors. Knowledge of the pathway followed by cholesterol in the body has allowed development of treatments based on the stimulation of LDL-receptor synthesis (thanks to the rabbits) and inhibition of cholesterol synthesis by the liver (the enzyme inhibitors of the soil molds).

---

taglandin release within the brain. The prostaglandins released in the brain are therefore also called endogenous pyrogens. It is common practice to block the development of fever by administering **antipyretic** (fever-opposing) agents such as aspirin and acetaminophen, both of which act by blocking prostaglandin synthesis.

After the hypothalamus has been reset for a higher temperature by pyrogens, that temperature is regulated just as closely around the new set point as the normal body temperature is in a healthy individual. The difference is that the sum of heat-loss and heat-gain mechanisms add up to a higher temperature. Depending on the external conditions, an individual may not only conserve body heat but even begin to shiver to increase body heat production. This response is accompanied by the sensation of being chilled, even as body temperature is rising.

Fever is beneficial because it enhances the immune system's defenses, with all aspects of the engulfment and killing of invading organisms becoming more effective. Experiments with various warm-blooded and cold-blooded animals (the latter can increase body temperature only through behavioral adjustments) indicate that the ability to survive infections is greater if elevation of body temperature is not blocked. The conclusion that fevers are an adaptive response is unavoidable, and many physicians now recommend that antipyretic drugs not be routinely used to eliminate mild and moderate fevers.

Some of the elevation of body temperature that occurs during exercise is also linked to release of interleukin-1 and its temporary elevation of the hypothalamic set point. Thus exercise could have some immediate disease-fighting benefits in addition to maintaining aerobic capacity and a healthful body weight.

Starvation

During prolonged fasting or the low nutrient intake levels that produce starvation, the body enters a new state that is somewhat different from the postabsorptive state. One of the changes seen in fasting and starvation is a progressive decline in basal metabolic rate. This change increases the chances of survival by slowing the rate at which the body's energy stores are consumed. The stores of nutrients that form the basis of survival consist mostly of fats and muscle protein. Fats contain about twice as much energy per gram as do carbohydrates or protein. The 10 kg of fat available in a typical individual is sufficient for survival of a 30- to 40-day fast. A smaller amount of fat is also available from the liver. Fats are used for energy during starvation just as in the postabsorptive state because insulin secretion is depressed and glucagon levels are elevated. The proteins present in the muscles and bones can be metabolized to extend the survival period up to two months (Table 23-8).

The stress of fasting and starvation activates the sympathetic nervous system. The resulting secretion of epinephrine by the adrenal medulla promotes glucose release from the liver. Epinephrine also inhibits insulin release, so uptake of glucose by insulin-dependent uptake systems is severely restricted, and potentiates glucagon release. Epinephrine stimulates the enzymes that mobilize fat stores and promote gluconeogenesis, increases lactic acid uptake by the liver, and promotes glycogenolysis to elevate plasma glucose levels. Epinephrine also spares glucose by stimulating lipolysis in adipose tissue and the liver and the uptake of fatty acids into muscle. The overall result of epinephrine is to produce a metabolic pattern similar to, but more exaggerated than, the postabsorptive state.

Many of the actions of epinephrine require that cortisol also be present. Epinephrine stimulates cortisol secretion by acting on the anterior pituitary cells that secrete **adrenocorticotropic hormone (ACTH)**. Control of glucocorticoid secretion (Figure 23-26) is also exerted on the hypothalamus, which releases three factors that initiate ACTH secretion: **corticotropin-releasing hormone (CRH), vasopressin,** and **CCK.** Stressful situations activate antidiuretic responses through peripheral actions of vasopressin.

Cortisol accentuates the effects of glucagon and norepinephrine. Like epinephrine secretion, cortisol secretion is increased whenever plasma glucose levels fall rapidly. Other stimuli causing cortisol release include tissue damage and infections. Cortisol is strictly a catabolic hormone that antagonizes the anabolic effects of insulin and mobilizes stored nutrients, and so its presence during starvation is required for the body to significantly tap the muscle proteins to provide amino acids for energy, lipogenesis, and gluconeogenesis (Figure 23-27). In the liver, cortisol increases the rate of conversion of fatty acids to ketone bodies (see Figure 23-27).

Muscle protein could maintain total body metabolism for several weeks, but continued protein breakdown ultimately leads to serious protein deficiencies. HGH plays an important role in prolonged fasting because it prevents the extensive protein breakdown that would occur if only epinephrine and cortisol levels increased.

The liver's role in conversion of amino acids to glucose is crucial during starvation because the blood level of glucose must be maintained for normal brain function to continue, and fats cannot be converted to glucose. Most cells use fats for energy, but the brain, which is an obligate consumer of glucose when food intake levels are normal, is able to

TABLE 23-8	Substrate Use and Storage				
Substance	Daily turnover (grams)	Storage form	Organ	Amount stored (grams)	Functional reserve
Glucose	250	Glycogen	Liver	200	<1 day
			Muscle	200	
Amino acids	150	Protein	Muscle	6000	1 to 2 weeks
Fatty	100	Triglycerides	Adipose tissue	10,000	4 to 6 weeks
			Liver	400	1 to 2 days

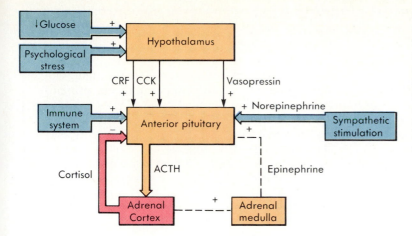

FIGURE 23-26
Control of glucocorticoids (cortisol) by releasing factors from the hypothalamus; these factors are stimulated by a drop in plasma glucose levels and stressful stimuli of psychological origin. The immune system and the sympathetic nervous system and epinephrine can act at the level of the anterior pituitary to stimulate cortisol release, and cortisol stimulates synthesis of epinephrine. Negative feedback is provided by circulating levels of glucocorticoids on the anterior pituitary.

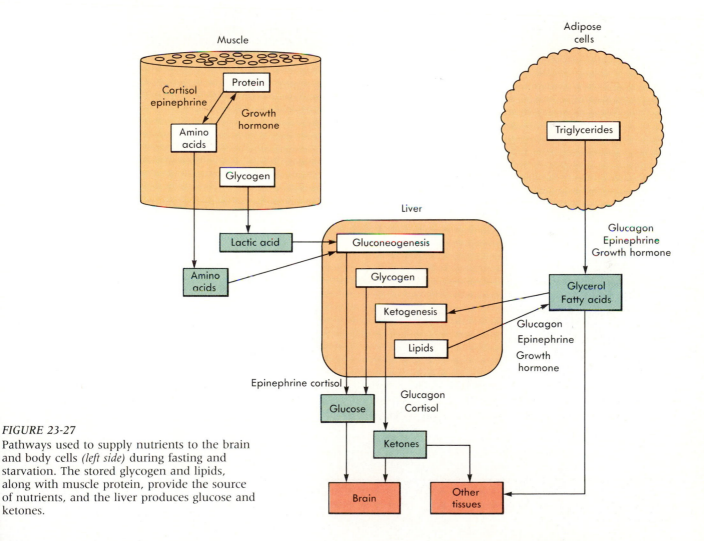

FIGURE 23-27
Pathways used to supply nutrients to the brain and body cells (left side) during fasting and starvation. The stored glycogen and lipids, along with muscle protein, provide the source of nutrients, and the liver produces glucose and ketones.

adjust its enzymes to allow it to use ketones for up to half of its energy needs after a week of fasting (see Figure 23-27). The two parts of the body that are protected from net catabolism the longest during starvation are the heart and brain.

1 There is more T_4 than T_3 in the blood, yet the T_3 is more potent. Why?
2 HGH is released in starvation. What role does it play?
3 By what sequence of events can a bacterial infection induce a fever? How is a fever beneficial?

SUMMARY

1. Hormones secreted by the pancreas regulate blood glucose; **insulin** promotes uptake of glucose when the supply in the blood is high; **glucagon** increases the level of glucose in the blood by promoting **glycogenolysis, gluconeogenesis,** and **glucose sparing.**

2. The **postabsorptive state** is defined by the drop in glucose and therefore in the insulin concentration in the blood. This drop in insulin removes catabolic pathway inhibition, and glucagon secretion stimulates catabolic pathways.

3. Diabetes produces the postabsorptive state in the presence of excess glucose because either insulin secretion is insufficient or absent **(type I diabetes),** or the body does not have enough insulin receptors to allow insulin in the blood to have its normal effects on uptake and synthesis **(type II diabetes).**

4. Stimulation of catabolic pathways begins even before food is ingested or absorbed. The resulting increase in blood levels of glucose, bulk aspects of the ingested food, and neural and hormonal signals (CCK) all produce **satiety.**

5. Long-term regulation of body weight requires that intake be matched to nutrient use. The amount of glucose available to body cells, signals arising from fat cells, and heat production associated with metabolism of fats, carbohydrates, and proteins may all contribute to regulation of intake.

6. The relative proportions of fat, muscle, and ossified skeleton that make up the body weight vary over the life cycle, with muscle and skeletal weight peaking around age 35, and fat increasing as a proportion of body weight as the muscle and skeletal weight decline. Substitution of less metabolically active fats for muscle decreases overall metabolic rate.

7. **Thyroid hormones** set the body's metabolic rate, which can be increased in response to cold. Thyroid hormones and activation of the sympathetic nervous system promote adaptation to cold.

8. **Interleukin-1** is released from macrophages in response to infections; it stimulates **prostaglandin** release within the brain, and the hypothalamic thermostat is reset to produce a fever. The elevated temperature promotes the body's disease-fighting mechanisms and is therefore beneficial.

9. In starvation, epinephrine promotes the postabsorptive pattern of energy use, and HGH protects muscle protein by mobilizing fatty acids and promoting their use as fuel.

STUDY QUESTIONS

1. Where are lipids stored in the body? Carbohydrates? Where is most of the protein in the body located?

2. What molecules supply most of the energy for body cells during the absorptive state? During the postabsorptive state?

3. Carbohydrates and amino acids can be transformed into lipids by liver cells. What happens to the lipids then?

4. What are the effects of insulin on carbohydrate, protein, and lipid metabolism in (a) muscle, (b) adipose cells, and (c) liver cells?

5. What factors increase insulin secretion?

6. What hormones oppose the effects of insulin? What signals lead to elevation of both insulin and glucagon levels after a protein-rich meal?

7. What factors are thought to contribute to the sensation of satiety? Where are the feeding and satiety centers located in the brain?

8. Why is a low ratio of HDL to LDL associated with elevated risk of developing atherosclerosis?

9. Why are women more likely to suffer from osteoporosis than are men? What factors are known to retard the progress of osteoporosis?

10. Describe the major effects of (a) HGH and (b) epinephrine.

11. What hormones promote cold adaptation, and how do they interact to effect changes? How is the response of an infant different from that of an adult? What metabolic responses can be made to chronic cold exposure, and what hormones are involved?

12. A starving person uses up the body's supply of nutrients in what order? What role is played by (a) epinephrine, (b) cortisol, and (c) HGH during starvation?

SUGGESTED READING

ALTSCHUL, A.M.: *Weight Control, A Guide For Counselors and Therapists,* Praeger Publishers, New York, 1987. Practical manual for an integrated program of weight reduction and control.

BATES, S.R., and E.C. GANGLOFF: *Atherogenesis and Aging,* Springer-Verlag New York, Inc., New York, 1987. Discusses the changes in the handling of dietary fat during aging and the implications for the formation of plaques in blood vessels.

BEGLEY, S.: The Deadliest Kind of Cold, *Newsweek,* March 9, 1987, p. 63. Discusses the causes of hypothermia and how it can be treated.

BENZALA, D.: The Lifestyle Disease Only You Can Prevent, *Consumer's Digest,* June 1986, p. 49. Discusses diabetes and its treatment by diet modification and exercise. Also describes the recent use of insulin pumps in treating diabetes.

COHEN, L.A.: Diet and Cancer, *Scientific American,* November 1987, p. 42. Summary of evidence for an association between high-fat, low-fiber diets and certain cancers.

Raises the question of whether the human GI tract evolved to handle a very different set of foodstuffs than those currently consumed.

Cuddle That Rat, *Discover,* July 1988, p. 6. An article discussing the physiological and psychological responses to stress.

DAWBER, T.R.: *The Framingham Study, The Epidemiology of Atherosclerotic Disease,* Harvard University Press, Cambridge, Mass. 1980. Landmark investigation of the epidemiology of atherosclerosis in the inhabitants of a small Massachusetts town linking elevated lipid levels, cigarette smoking, obesity, and low physical activity to arteriosclerosis.

FRISCH, R.E.: Fatness and Fertility, *Scientific American,* March 1988, p. 88. Argues that a woman must store a threshold minimum of fat to reproduce and how gaining or losing just a few pounds can dramatically affect fertility.

HALL, T.: Cravings, *New York Times Magazine,* September 27, 1987, p. 22. Discusses whether the body really knows what it needs in the way of nutrition.

CHAPTER 24

The Immune System

On completing this chapter you will be able to:

- Describe the cellular components of the immune system, noting the subclasses of leukocytes and lymphocytes.
- Describe the structure of antibodies (the agents of humoral immunity) and explain how the extreme variability of the antibodies is generated by somatic rearrangement.
- Describe agglutination, opsonization, and activation of the complement system.
- State the physiological roles of the antibody classes: IgG, IgM, IgE, IgA, and IgD.
- Appreciate the role of histocompatibility antigens in cell-mediated immunity.
- Describe clonal differentiation of T cells in the thymus and compare the properties of T-cell receptors to antibodies.
- Understand how killer and cytotoxic T cells recognize and attack cells.
- Appreciate the role of helper T cells in the immune response.
- Describe the process of inflammation, noting the role of bradykinin and histamine.
- Understand how the complement cascade is activated by antibody binding and by the presence of bacterial cell wall components.
- Describe the ABO blood system and be able to predict the occurrence of transfusion reactions.
- Identify how autoimmune diseases can result when the mechanisms that normally suppress immune responses to self antigens fail.

*I*mmunology is the science of the body's resistance to invasion by other organisms. The immune system is able to distinguish friend from foe and to "remember" old enemies. The molecular basis of recognition and memory is similar in principle to the binding of hormones or neurotransmitters with their receptors, but involves an entirely different class of specific proteins.

Without the immune system, the body would rapidly be colonized by bacteria that are not a threat to the health of people with intact immune systems. Without the vigilance of the immune system, the continual transformation of small numbers of normal cells into precancerous cells would result in many more cases of cancer. Some of the consequences of loss of the immune system are seen in acquired immunodeficiency syndrome (AIDS), in which a viral infection weakens some elements of the immune system. Death from cancer or infection is a likely outcome of infection with the AIDS virus.

Misdirections of the immune system against the tissues of its host are responsible for numerous diseases. Inappropriately directed responses are also responsible for allergies. Positive-feedback elements of the system that normally accelerate the initial stages of an immune response can, under some circumstances, generate responses that are out of proportion to the threat and that in themselves may threaten the life of the host.

The techniques of molecular biology have contributed to understanding of the immune system, and the immune system has also provided biologists with some potent tools for molecular recognition. Immunology is a rapidly advancing science; this single chapter can only briefly outline recent progress in this area.

DIVERSITY IN THE IMMUNE SYSTEM
Specific and Nonspecific Immune Responses

The immune system consists of a family of cells derived from the same stem-cell population as red blood cells and platelets (Figure 24-1), together with various chemical factors that coordinate the activities of the immune system with those of other tissues and direct an attack against foreign substances and invading cells. The immune system responds to injuries and invasions of all types with a **nonspecific immune response,** consisting of inflammation and phagocytosis of foreign substances by macrophages (discussed later). **Specific immunity** refers to the process whereby an individual's immune system distinguishes substances that normally circulate in the blood and an individual's nor-

FIGURE 24-1
Classes of blood cells.

Blood cell	Function
Erythrocyte	 O₂ and CO₂ transport
Neutrophil	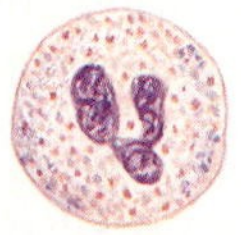 Immune defenses (Phagocytosis)
Eosinophil	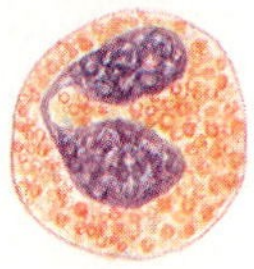 Defense against parasites
Basophil	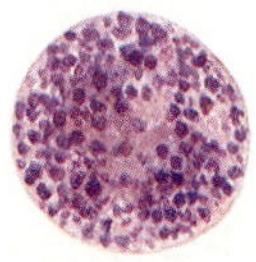 Inflammatory response
Monocyte	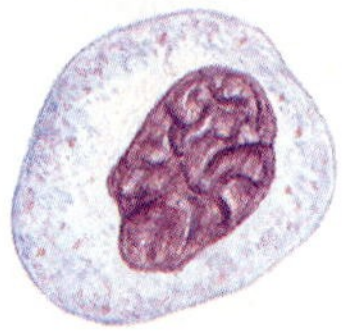 Immune defenses (precursor of tissue macrophage)
B-Lymphocyte	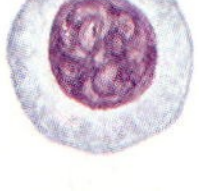 Antibody production (precursor of plasma cells)
T-Lymphocyte	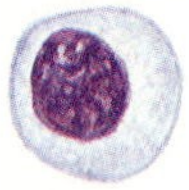 Cellular immune response
Platelets	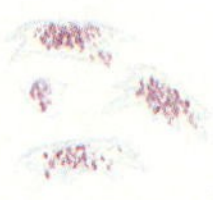 Blood clotting

mally functioning cells (self) from substances that are not normally in the body or cells that are abnormal in their function (nonself). Specific immunity results in a particularly vigorous response against invaders or substances that have entered the body previously and triggered processes of chemical recognition by the immune system.

Substances capable of being recognized by the immune system are termed **antigens.** To be an antigen, a molecule must be of a certain minimum size and must possess certain chemical features. Small molecules (**haptens**) may serve as antigens if they become chemically bound to larger molecules. Antigenic substances which normally circulate in their blood are called **self antigens** and those which are not normally there are called **nonself antigens**). Generally, immune responses are mounted only against nonself antigens, but the distinction occasionally fails, causing an **autoimmune disease.**

Cellular Elements of the Immune System

The **white blood cells,** or **leukocytes,** constitute the circulating agents of the immune system. The major class of leukocytes is termed **granulocytes** because their cytoplasm contains numerous small vesicles. There are three types of granulocytes, named on the basis of their affinity for different stains used in microscopy. In order of abundance, they are **neutrophils, eosinophils,** and **basophils** (see Figure 24-1). Neutrophils make up 50% to 75% of the total leukocyte population. They are **phagocytes;** they can engulf and digest bacteria and other foreign materials. Eosinophils, which make up 1% to 6% of the leukocytes, are especially effective against parasite infections. Basophils, less than 1% of the leukocyte population, release **histamine** and other local chemical mediators that are responsible for some aspects of a complex immune response called **inflammation,** discussed later in this chapter. Histamine release is also triggered by specific immune responses. **Mast cells** play the same role as basophils, except that they reside permanently in the tissues rather than circulating in the blood.

Monocytes compose 2% to 10% of the leukocyte population. They leave the bloodstream and enter interstitial spaces, where they transform into **macrophages,** cells that are specialized for phagocytosis and that also provide important assistance in initiating and controlling specific immune responses. Some macrophages become permanent residents in lymph nodes or internal organs, whereas other "wandering" macrophages circulate through the interstitial spaces.

The specific immune responses are carried out by circulating cells called **lymphocytes.** Lymphocytes make up 20% to 45% of the leukocyte population and are divisible into two general types: **B cells** and **T cells.** These two cell types have distinct functions. The life history and function of lympho-cytes involves several **lymphoid tissues,** including lymph nodes, bone marrow, spleen, and thymus.

> *1 What is the fundamental difference between specific and nonspecific immunity?*
> *2 What is meant by the concept of memory in the immune system?*
> *3 What is an antigen?*

B Lymphocytes and Antibody-Mediated Immunity

To understand the immune system, we must have answers to two questions: How does the immune system recognize specific antigens? How does it distinguish self antigens from nonself antigens? An important step in understanding immunity was made with the discovery that individuals who had recovered from certain diseases had substances in their blood plasma that could make another susceptible individual temporarily resistant to the disease. The specific disease resistance that is conferred by substances present in the plasma of immune individuals is referred to as **humoral immunity.**

The substances that confer humoral immunity are **antibodies.** The term was coined by the German scientist E.A. von Behring late in the last century before anything about the structure or origin of antibodies was known. Antibodies compose a family of plasma proteins called **immunoglobulins,** or **gamma globulins.** Humoral immune responses involve the secretion of specific antibodies into the plasma by **plasma cells,** specialized cells derived from B lymphocytes (Figure 24-2). A preparation of

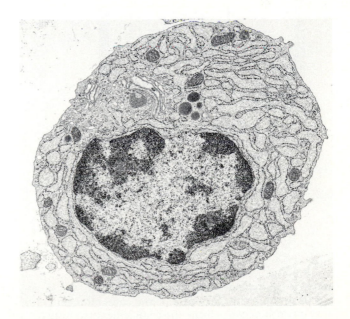

FIGURE 24-2
The structure of plasma cells. The cytoplasm is filled with rough endoplasmic reticulum for the synthesis of antibody protein.

TABLE 24-1   *Comparison of Active and Passive Immunity*

		Examples	
Type	**Characteristics**	**Natural**	**Artificial**
Active immunity	The person produces his or her own immune agents, either antibodies or T cells	Infection by bacteria or viruses	Vaccination with weakened virus or bacterial antigens
Passive immunity	The person is given antibodies produced by an actively immune person (or animal)	Transfer of IgG from the mother to the child across the placenta and IgA in breast milk	Immunoglobulin administration; monoclonal antibody therapy

the serum (the fluid portion of plasma left after blood clots) of an immunized individual or animal containing these substances is called an **antiserum.**

The acquisition of immunity following an initial exposure to a source of non-self antigens is dramatically illustrated by measles, mumps, and chicken pox. If a child contracts any of these diseases, the immune system memory mechanisms render the

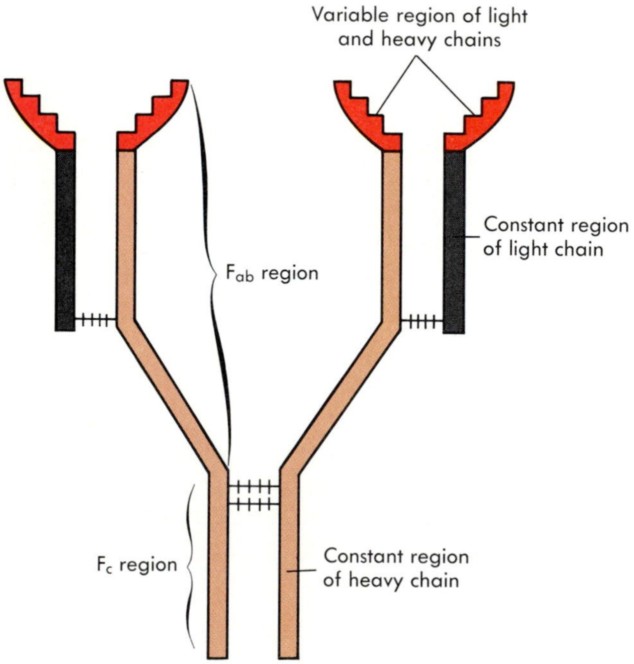

FIGURE 24-3
Structure of an antibody. Antibodies consist of two heavy and two light chains joined by disulfide bonds. The variable region *(red)* binds to the antigen. The constant region of the antibody *(black)* can activate the complement pathway or attach the antibody to the membrane of cells such as macrophages, basophils, or mast cells. The disruption of the disulfide bonds results in the formation of two F*ab* (for antigen-binding) fragments and an F*c* (from crystalizable) fragment.

child **permanently immune** to any reoccurrence of the disease. Permanent immunity can also be produced by injecting a weakened bacterium or virus in **vaccination.** The production of specific antibodies in response to the presence of an antigen is referred to as **active immunization** (Table 24-1).

In contrast to active immunity, the protection given a susceptible individual by transfer of immunoglobulins produced by another individual is called **passive immunization** because it is only temporary, lasting for a few days or weeks (see Table 24-1). Passive immunization is accomplished by injecting an antiserum or purified immunoglobulins. The transfer of some types of antibodies from mother to child across the placenta during prenatal development and in breast milk after birth is an example of naturally occurring passive immunization.

The basic functional unit of all types of antibodies is a Y-shaped protein comprising a pair of light and a pair of heavy amino acid chains joined by disulfide bonds (Figure 24-3). At the forked, or **Fab,** end of the molecule, the heavy and light chains form two identical binding sites for a single, specific antigen (Fab stands for antigen-binding fragment). The amino acid sequences in the Fab end must be different for each different antigen; thus this portion of the molecule is termed the **variable region.**

Within the variable region, three short **hypervariable segments** form a cleft that acts as a binding site for the antigen. The portion of the antigen molecule recognized and bound by the Fab region of an antibody is called the **determinant** portion. The fit between the cleft and the determinant portion of the antigen is similar to the way a lock fits a key. Determinant portions must meet certain requirements of size and chemical reactivity; thus molecules that are too small or too chemically unreactive to interact with a Fab binding site do not generate antibody-mediated immune responses. Conversely, a large antigen molecule may possess more than one determinant site, so there may well

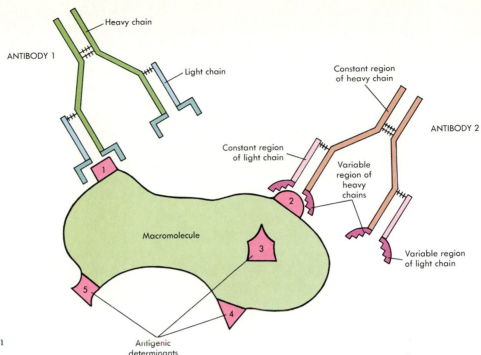

ANTIBODY 1

Heavy chain

Light chain

Constant region of heavy chain

ANTIBODY 2

Constant region of light chain

Variable region of heavy chains

Macromolecule

Variable region of light chain

Antigenic determinants

FIGURE 24-4
Antibodies may recognize different regions (determinants) of an antigen molecule.

be several antibodies specific for a single antigen, each recognizing a different determinant portion of the antigen (Figure 24-4).

The studies of Austrian scientist Karl Landsteiner in the first decades of this century demonstrated that the immune system is capable of making antibodies to a very large number of different antigens. Presently, estimates of the number of potential unique antibodies range from 10^6 to as high as 10^{20}. The extreme variability in the amino acid sequences of the hypervariable segments is largely responsible for this very large number of possible antibodies. In the chromosomes of each person, there are only a few hundred genes that code for immune proteins, not nearly enough to explain the large number of different amino acid sequences seen in the Fab region. What is the source of this prodigious variability?

Two hypotheses have been proposed to explain the great diversity of antibodies. In one hypothesis, the **instructional theory**, it was proposed that the structure of the antigen determinant acted as a template for the structure of the corresponding antibody. This appealing hypothesis has been proven incorrect. In the currently accepted hypothesis, the **clonal selection theory**, it has been proposed that lymphocytes differentiate randomly during development into many **clones**, each of which consists of a small number of identical cells that can synthesize

one specific antibody. Clones that happen to make antibodies to self antigens are eliminated during subsequent development. The remaining clones remain quiescent unless their corresponding antigens enter the bloodstream. The first appearance in the body of a nonself antigen causes the clones that can manufacture antibodies specific for that antigen to proliferate and begin to synthesize antibodies.

The genetic sources of the variability of clones have been worked out relatively recently. The major source of variability is a process called **somatic rearrangement**. Immune receptor genes are not continuous sequences of nucleotides but must be assembled by joining several separate segments of DNA that are initially in different locations on the chromosome. Assembly of these fragments of genetic information is a key event in the process of clonal formation, which takes place after immune cells have differentiated from other fetal somatic cells. Within the founder cell of each clone, the gene specifying the single variable region of the antibody that the clone will make is assembled from a library of possible sequences. There are about 16,000 possible heavy chains and about 1200 possible light chains. Thus, in principle, there are about 200 million (that is, $16,000 \times 1200$) possible combinations of different light and heavy chains (Figure 24-5). Additional variability is introduced by **somatic mutations,** errors in base pairing that occur during

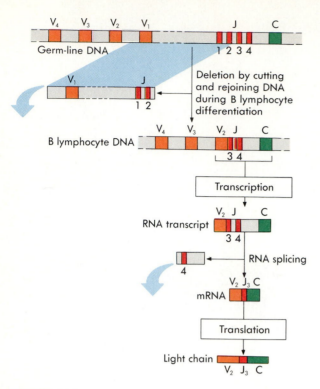

FIGURE 24-5
The gene for the variable region of a heavy chain is assembled from segments chosen at random from a large library of DNA segments. The choice is different in each clone, and a large number of different variable regions is possible.

chromosomal replication in mitosis (see Chapter 3, p. 55), and by **frame-shift errors**, which occur when the segments of DNA are initially assembled to form the antibody gene. The combination of these three sources of variation results in the large number of Fab varieties. Each B-cell clone can produce a single type of Fab.

The sequence of events that is believed to occur in development of the B-lymphocyte clones is as follows: First, the precursors of B lymphocytes are generated by mitosis of stem cells in the bone marrow (Figure 24-6). They may then migrate to some embryonic lymphoid tissue, where clonal differentiation takes place. In birds, clonal differentiation occurs in a lymphoid structure called the **bursa of Fabricius,** which has no homologue in mammals. The name *B cells* is short for **bursa cells.** The sites of clonal differentiation in mammals are not known but are believed to be the spleen, liver, and bone marrow. After clones differentiate, most clones that make antibodies to substances circulating in the fetal plasma at that time are eliminated. The remaining clones disperse to the blood and lymphoid tissues. The fate of individual clones in postnatal life depends on the antigens to which the body is exposed.

The first response to a new antigen is called a **primary immune response.** The memory effect arises from the fact that during a primary immune response, the responding clones undergo mitosis (Figure 24-7). Some of the daughter cells become

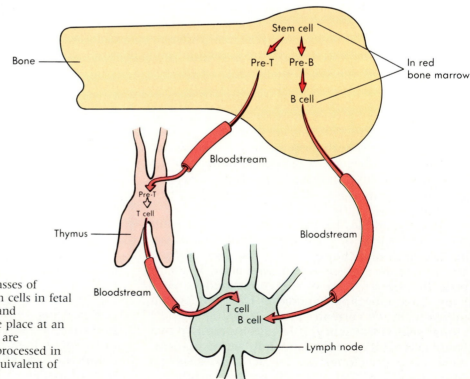

FIGURE 24-6
Differentiation of the two different classes of lymphocytes, B and T cells, from stem cells in fetal bone marrow. Clonal differentiation and elimination of self-specific clones take place at an intermediate site in the body. (T cells are processed in the thymus. B cells are processed in the bone marrow, the mammalian equivalent of the bursa of Fabricius for B cells.)

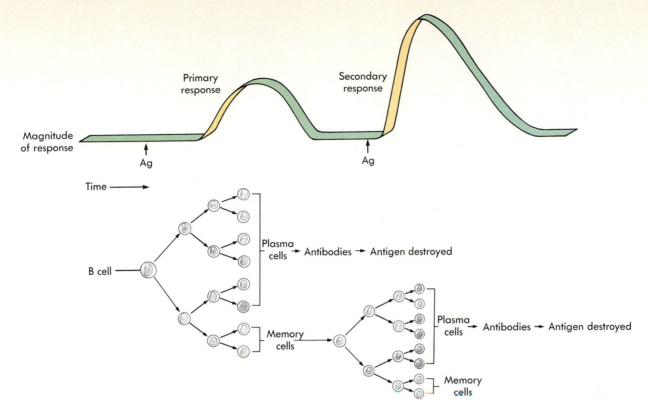

FIGURE 24-7
A primary immune response also lays the foundation for the more rapid and vigorous secondary immune responses to subsequent exposures to the same antigen. The responses of only one clone are shown here; each antigen typically elicits responses from more than one clone because antigens typically have more than one determinant.

plasma cells that secrete antibodies; others become **memory cells** that can quickly give rise to a **secondary immune response** if the antigen is experienced again. The primary immune response requires about 20 days to reach peak production of antibody. The primary immune response is usually too slow and too weak to prevent development of the disease. In a secondary response, the time required for maximum production of antibody is 5 to 14 days, and the total amount of antibody produced is much greater because the number of cells in responding clones is greater from the beginning. For many diseases, the secondary response is usually sufficient to prevent development of the disease. In other cases, the secondary response simply makes subsequent episodes of a particular disease less severe.

> *1 What is meant by the term clonal selection?*
> *2 What is a hapten?*
> *3 What is the difference between a primary and a secondary immune response?*

Classes of Antibodies

The crystallizable fragment (Fc portion) of the antibody molecule is the stem of the Y (see Figure 24-3). In contrast to the very large number of possible Fab regions, there are only five Fc structures, desig-

nated IgG, IgM, IgE, IgA, and IgD (Table 24-2 and Figure 24-8; *Ig* stands for *immunoglobulin*, so IgG could also be referred to as immunoglobulin G). When the Fab region has bound to an antigen, the

TABLE 24-2	Classes of Immunoglobulins	
Class	**Property**	**Functional role**
IgG	Stimulates phagocytosis and complement reactions (can cross placenta)	Identifies microorganisms for engulfment or lysis and provides passive immunity in newborn infants
IgE	Binds to mast cells and basophils Stimulates histamine release	Inhibits parasite invasion; involved in allergic reactions
IgD	Unknown	Unknown
IgA	Present in saliva, tears, breast milk, and intestinal secretions	Agglutinates infectious agents in secretions outside the body
IgM	Stimulates phagocytosis and complement reactions	Identifies microorganisms for engulfment or lysis

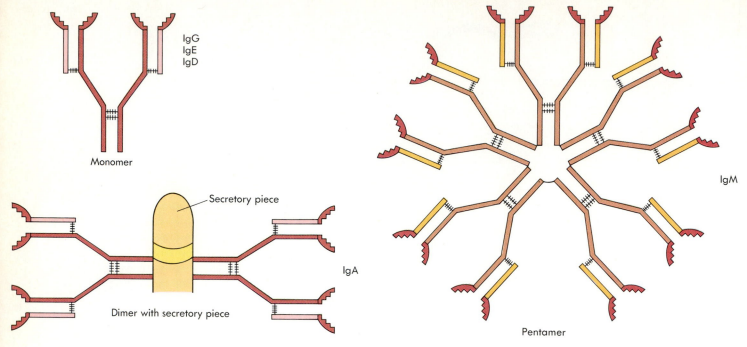

FIGURE 24-8
The basic antibody unit is a monomer. IgA is a dimer, and IgM a pentamer.

Fc region initiates reactions that can destroy the invading organisms. The structure of the Fc region determines the types of defensive responses that can be elicited by antibody binding.

A clone may use its single type of Fab to make more than one class of antibody by attaching the Fab to more than one type of Fc structure. This is seen in the differing mix of antibody classes made in the primary and secondary immune responses to the same antigen. The major class of immunoglobulin produced by virginal clones of B lymphocytes in a primary response is IgM; production of IgG begins more slowly (Table 24-3). The much more vigorous and more rapid production of antibody in a secondary immune response is caused by a larger, more rapid production of IgG; production of IgM remains at almost the same level as in the primary response (see Table 24-3).

The effects of antibodies on antigen-bearing invaders fall into three general categories. Because the basic antibody structure provides two identical binding sites, each antibody unit can join two antigens together. For example, if the two antigens are on two different bacterial cells, the two foreign cells are linked together like chain-gang prisoners, and their ability to disperse in the body is reduced. This process is termed **agglutination,** and large clumps of antibody and antigen can be formed in this way. The presence of bound antibodies on the surface of an invading cell tags it as an invader so that it is more readily identified by macrophages; this tagging is termed **opsonization.** Finally, antibody binding activates a collection of plasma proteins called the **complement system,** a cascade of reactions involving positive and negative feedback (see Chapter 5, p. 86). Some elements of the complement

TABLE 24-3	Comparison of Virgin and Memory B Cells	
Characteristics	**Virgin B cells**	**Memory B cells**
Response to antigen	Differentiate to plasma cells and/or transform into memory cells (slow response)	Begin to produce antigen after short delay
Antibody classes produced	IgM first; then IgG	Almost all IgG
Lifespan	Virgin clones may last for life; plasma cells persist a few weeks after primary response	Memory cells survive for months to years

Using Monoclonal Antibodies in Physiology and Medicine

Since the discovery of antibodies, scientists and physicians have recognized that antibodies can be used to recognize specific molecules with great precision. Antibodies can be raised in the laboratory by injecting animals with antigen and collecting blood plasma after the animals have mobilized a secondary immune response. Because every antigen has at least several potential determinants, the population of antibodies raised in this way is a mixture of antibodies to different determinants. Such a mix of antibodies is said to be **polyclonal.**

Polyclonal antibodies not only recognize different determinants on the same antigen molecule, but also have a relatively high probability of including some antigen-binding sites that cross-react with other molecules. The precision of the recognition and targeting process

could be refined if there were some way to ensure that the antibodies raised to an antigen were **monoclonal;** that is, derived from a single B-cell clone and specific for a single determinant. This could be done if a single plasma cell of the appropriate clone were able to multiply indefinitely in cell culture, but plasma cells do not multiply well or live long either in culture or in the body.

However, in 1984, British scientist C. Milstein and Swiss scientist G. Kohler received a Nobel Prize for their discovery of a method for making large quantities of monoclonal antibodies. They discovered a way to make plasma cells multiply in culture by hybridizing them with myeloma cells, a variety of cancer cell that divides ceaselessly in culture when provided with the proper nutrients. These hybrids (called **hybridoma cells**) are obtained by **cell fusion;**

the plasma cells are cultured together with myeloma cells and some of the cells fuse, resulting in cells that multiply like the cancer cells but that also continue to make antibodies (Figure 24-A). The cells that fail to fuse can be eliminated; the cancer cells die if deprived of a necessary nutrient, and the unfused plasma cells do not survive long in culture. Single hybridoma cells can be isolated and allowed to multiply into a population of genetically identical cells that constitute a clone and can be cultured indefinitely. The antibody produced by each clone established in this way is screened for its specificity.

Monoclonal antibodies can be coupled (**conjugated**) to fluorescent molecules that make the sites at which they bind visible under the light microscope or to molecules containing a heavy metal such as gold that makes the location of the antigen visible with the electron microscope. Monoclonal antibodies to specific neurotransmitters can be used to trace neural pathways in the nervous system. Monoclonal antibodies can also be used to purify large amounts of their antigens by a process called **affinity chromatography,** in which tissue extracts are passed through a bed of particles to which the antibodies have been attached. The antibodies retain the desired antigen, allowing other substances to pass. Interferon was first commercially purified with this technique.

There are potential uses for monoclonal antibodies in both medical diagnosis and therapy. Tumor cells can be identified with monoclonal antibodies specific for a particular characteristic antigen. The location of the tumor in the body can then be determined by conjugating the antibody to a radioactive label that can be detected by scanning the body. Monoclonal antibodies might direct the patient's immune system to a tumor, or they could be made to deliver toxic drugs specifically to tumor cells but not to normal cells, serving as a type of "magic bullet."

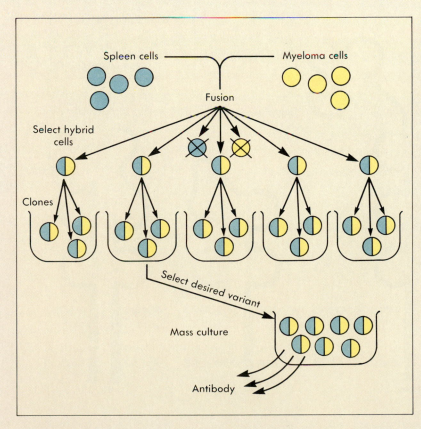

FIGURE 24-A
Production of hybridoma clones by cell fusion followed by selection of antibody-secreting clones and further selection and mass culture of a clone that secretes the desired antibody.

system destroy invading cells, and others amplify the nonspecific immune responses to infection, which are collectively termed **inflammation.**

The major antibody class present during a primary immune response is IgM, much of which is located on the surface of B cells. These immunoglobulins can activate the complement system and cause particulate antigens to agglutinate, but they do not activate macrophages. With continued exposure to antigen, or upon reexposure, large amounts of IgG are secreted, and this is usually the most abundant type of immunoglobulin in the blood. The Fc region of IgG activates the complement system, but it also enhances the activity of macrophages.

The IgA class of immunoglobulins is found in exocrine secretions such as breast milk, respiratory and GI-tract mucus, saliva, and tears. It plays a major role in defending against entry of foreign antigens across body surfaces. The presence of IgA in breast milk may account for the increased resistance of breast-fed infants to GI disease. IgA inhibits bacterial adherence, opposing formation of bacterial colonies on the body surface, but it is unable to damage bacteria because the proteins of the comple-

ment system are not present in exocrine secretions.

The roles of IgE and IgD are less clear. IgE causes mast cells and basophils to release histamine, stimulating some pathways of the inflammatory response. IgE appears to have a significant role in allergic reactions and contributes to defense against parasitic infections. IgD is located on the surface of B cells, but its function is unknown.

Histocompatibility Antigens: The Identification Tags of Cells

All nucleated cells are marked with a combination of **histocompatibility antigens** on their surfaces. Two families of histocompatibility antigens, **MHC-1** and **MHC-2,** are present in humans. MHC stands for **major histocompatibility complex.** As is explained below, the two families of MHC proteins play different functional roles. MHC-1 antigens are present on the surfaces of all body cells (except erythrocytes and platelets); MHC-2 antigens are present only on the surfaces of B cells, macrophages, and some types of T cells. The structure of MHC antigens is shown in Figure 24-9. The MHC antigens are coded for by the most highly variable human

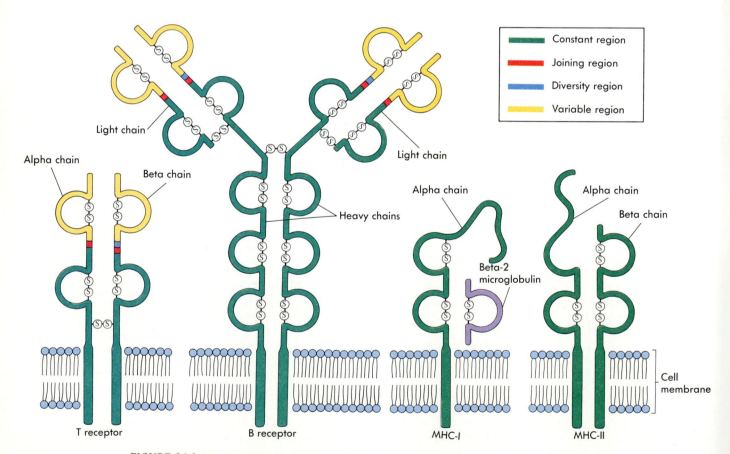

FIGURE 24-9

Similarity of structures of MHC antigens, T-cell receptors, and antibodies (B receptors). All four classes of molecules have similar molecular structures and share similar amino acid sequences. The loops are formed by connecting disulfide bonds.

GROWTH, METABOLISM, REPRODUCTION, AND IMMUNE DEFENSE

genes known. Each gene may be represented by 1 of at least 50 variant forms called **alleles**. The combination of MHC genes is as unique to each individual as a fingerprint, and only identical twins share the same combination of MHC antigens. The more closely related individuals are to each other, the more likely they are to possess some common MHC antigens. This is the reason that tissue transplants are more likely to succeed if the donor and recipient are close relatives.

T Lymphocytes and Cell-Mediated Immunity

T lymphocytes do not secrete immunoglobulins, but they have **antigen-specific receptors** on their membrane surfaces that recognize antigens by a similar mechanism. The diversity of these receptors is similar to that of antibodies and is generated by similar mechanisms. The structure of the T-cell receptor (see Figure 24-9) can be likened to one arm of an antibody. It consists of an **alpha** and a **beta** chain.

The T lymphocytes derive their name from the thymus. Precursors of T cells seed into this organ from the liver, spleen, and bone marrow (see Figure 24-6) and undergo differentiation into a large number of antigen-specific clones whose diversity is similar to that of B cells. Self-directed clones are eliminated. The clones then migrate to blood and other lymphoid tissues. The thymus is large in early life, but it begins to atrophy during the second decade of life and has almost vanished by adulthood. In addition to its role as a site of differentiation of T cells during early life, the thymus secretes several hormones that are important for formation of new T cells. The atrophy of the thymus after puberty may contribute to a decrease in the rate of T-cell formation in the course of aging.

---

1 What is an MHC antigen?
2 Describe the structural similarities and differences between antibody and the T-cell receptor.
3 Why is the T-cell receptor incapable of binding free antigen to which it is specific?

Killer T Cells and Tissue Rejection Reactions

Cytotoxic, or **killer**, **T cells** are the basis of what is termed **cell-mediated immunity** (Table 24-4). When tissue from one individual is transplanted into another individual, the recipient's immune system quickly reacts to destroy the foreign tissue in what is termed a **transplant rejection**. This is the classic example of cell-mediated immunity. However, this must be an incidental role of the cell-mediated immune system because it is unlikely that natural selection has shaped the immune system to resist organ transplants. Instead, it appears that the main role of killer T cells is recognizing self cells with which something has gone wrong. In addition

TABLE 24-4	Subclasses of T Lymphocytes
Designation	**Functional role**
Cytotoxic (killer) T cells	Recognition/destruction of cells that have been infected with a virus or have been transformed in some way (for example, cancer cells).
Natural killer cells	Resemble cytotoxic T cells, but their activity does not require previous exposure to antigen.
Helper T cells	Regulators of the immune response; they increase the numbers of cytotoxic T cells, activate B cells, and secrete various lymphokines.
Inducer T cells	Activated by helper T cells; they stimulate production of new T cells.
Suppressor T cells	Inhibit both cell-mediated (T-cell activity) and humoral (B-cell mediated) immunity; limit what might otherwise be an uncontrolled response.

to nonself cells, specific T-cell clones recognize cells that have expressed novel antigens as a result of viral infection or that have become abnormal. To bind to the novel antigen, the T cell must also recognize the histocompatibility antigen (Figure 24-10); free antigen molecules cannot be recognized by T cells.

As cells of the fetus differentiate into the specialized cell types characteristic of particular tissues, some of their genes are turned off because the proteins they code for are no longer needed. Cancer is regarded as a consequence of inappropriate activation of genes that played a role in early development but that should have remained turned off afterwards. The process by which genes are turned on at an inappropriate stage of development is termed **transformation**. Expression of these genes in a fully differentiated cell may cause the cell to multiply abnormally and to engage in inappropriate metabolic activities. The result is a population of descendants of the original transformed cell that constitute a **cancer**. Transformed cells express unusual surface antigens that can be recognized by both T-cell receptors and antibodies; the specific immune

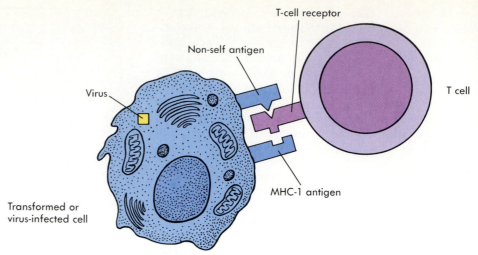

FIGURE 24-10
Recognition of a virus-infected or a
transformed cell by a killer T cell
requires the simultaneous recognition
of the MHC-1 antigen and a nonself
antigen.

system thus constitutes a line of defense against
cancer, and the establishment of a population of
cancer cells in the body represents a failure of the
immune system.

Viruses are small packets of genetic material enclosed in a protein coat that are able to use the intracellular environment to replicate their genes. Viruses take over the protein-synthesizing apparatus
of a cell and force it to synthesize viral proteins. After a time, an infected cell becomes filled with viruses, ruptures, and releases the viruses to infect
other cells. Viral infections can be fought in several
ways. First, infected cells produce a substance called
interferon that inhibits the replication of some vi-

ral genes. Interferon can be passed from infected
cells to uninfected cells as a form of nonspecific immunity. Second, antigens of the protein coat of free
viruses can be recognized by antibodies. Finally, virus-infected cells, like transformed cells, express
novel surface antigens and can be attacked by killer
T cells to prevent the infection from spreading (see
Figure 24-10).

Binding of a killer T cell to the target cell is followed by secretion of toxins from the T cell that
cause the destruction of the target cell. One such
toxin is a protein called **perforin.** Several perforin
molecules join into a ring (Figure 24-11) to form a
pore in the membrane of the target cell. Interaction

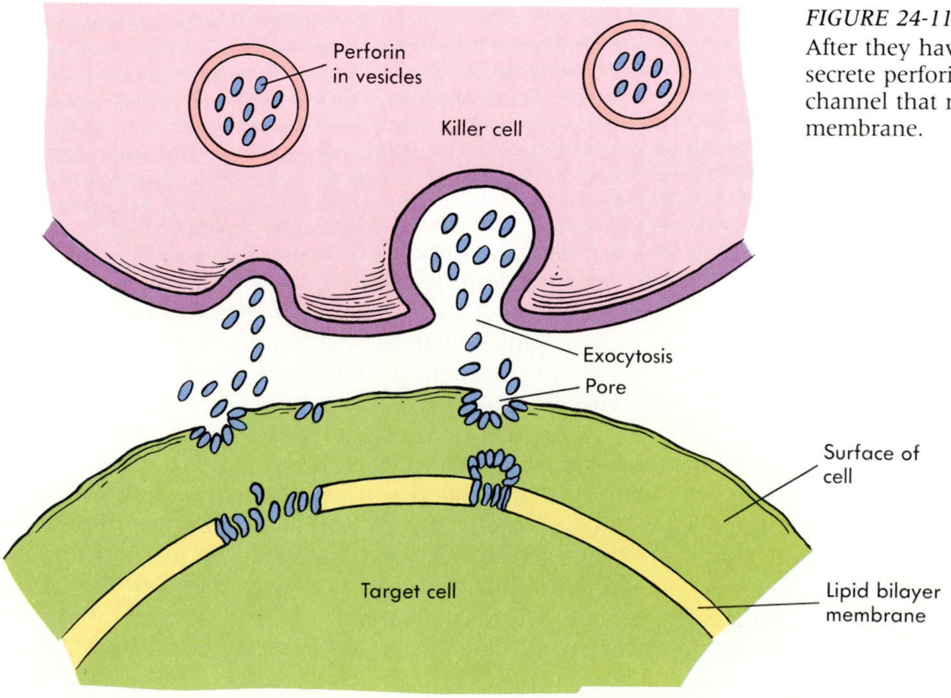

FIGURE 24-11
After they have bound to target cells, killer T cells
secrete perforin, which assembles itself into a large
channel that makes holes in the target cell's
membrane.

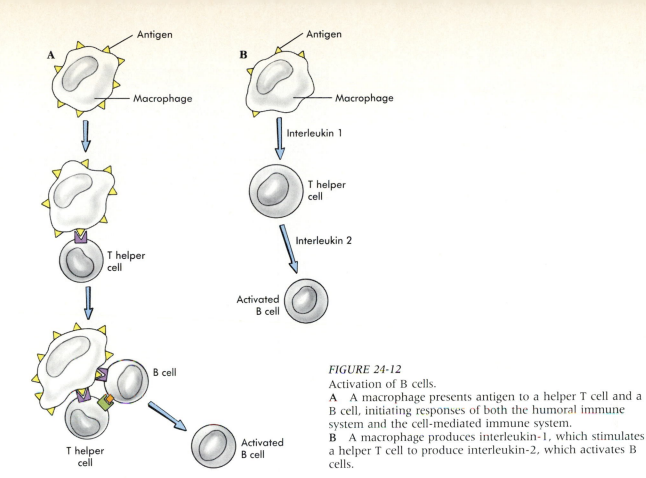

FIGURE 24-12
Activation of B cells.
A A macrophage presents antigen to a helper T cell and a B cell, initiating responses of both the humoral immune system and the cell-mediated immune system.
B A macrophage produces interleukin-1, which stimulates a helper T cell to produce interleukin-2, which activates B cells.

between the perforin molecules is Ca^{++} dependent. The result of pore formation is that the target cell suffers an influx of water and bursts, or **lyses.**

Natural killer (NK) cells resemble cytotoxic T cells except that NK cells are not antigen specific, and their activity does not require any previous exposure to the antigen (see Table 24-4). There seem to be several subsets of NK cells, each of which recognizes certain categories of nonself antigens. These cells may be involved in resistance to some types of cancer. Interferon stimulates the activity of NK cells and thus may be useful in cancer therapy.

COORDINATION IN THE IMMUNE SYSTEM
Control of Immune Responses by Helper and Suppressor T Cells

Proliferation of both B and T cells is controlled by **helper T cells,** which have a critical role in initiating immune responses and in coordinating the humoral and cell-mediated systems (see Table 24-4). The triggering event in a specific immune response is the arrival in lymphoid tissues of macrophages that have ingested antigens. When macrophages ingest foreign antigens, the ingested antigens are modified and displayed on the macrophages' cell surfaces in a process called **antigen presentation** (Figure 24-12, *A*). Presentation greatly enhances the ability of the antigens to elicit a response from the corresponding clones of B and T cells.

Macrophages are continuously moving to sites of infection by way of the bloodstream and then returning to lymph nodes along the lymphatic veins. When passing through lymph nodes, the macrophages come into contact with helper T cells. The T cells that belong to the corresponding clones recognize the combination of nonself antigen and MHC-2 antigen (Figures 24-13 and 24-14) and become activated. The activated helper T cells then stimulate the production of cytotoxic T cells. The helper T cells also activate the corresponding B-cell clones. The process appears to work as follows. Activated T cells can bind to those B cells which are displaying the appropriate antibody on their surfaces. This binding involves an interaction between four molecules: (1) the foreign antigens, (2) the B cell's MHC-2 antigens, (3) the B cell's antibodies, and (4) the T cell's receptor (Figure 24-15). This binding activates the B cells to proliferate and gives rise to memory cells and plasma cells.

Helper T cells trigger an immune response to the antigens delivered by the macrophages in several ways. In response to the hormone, **interleukin-1,** which is released by macrophages (see Figure 24-12, *B*), helper T cells increase the number of corresponding killer T cells by (1) secreting **interleukin-2,** which stimulates T-cell multiplication, and (2) by activating **inducer T cells,** which stimulate differentiation of new T cells (see Table 24-4). Activated helper T cells also secrete a family of substances called **lymphokines,** which stimulate the

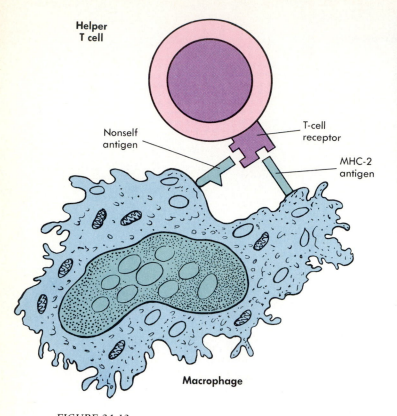

Helper
T cell

Nonself
antigen

T-cell
receptor

MHC-2
antigen

Macrophage

FIGURE 24-13

Presentation of antigen to a helper T cell by a macrophage is mediated by simultaneous recognition of the macrophage's MHC-2 antigen and the nonself antigen.

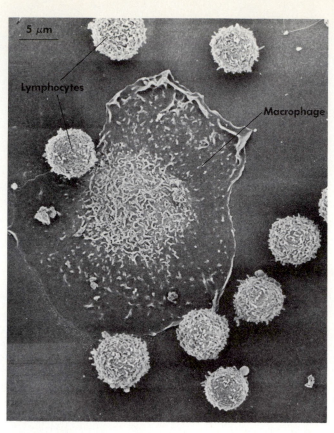

5 μm

Lymphocytes

Macrophage

FIGURE 24-14

Scanning electron micrograph showing a macrophage (large central cell) contacting three T lymphocytes.

FIGURE 24-15

The helper T cell activates a cell of the appropriate B cell clone. The T cell recognizes the B cell by its MHC-2 antigen; the B cell recognizes the antigen with its antibody.

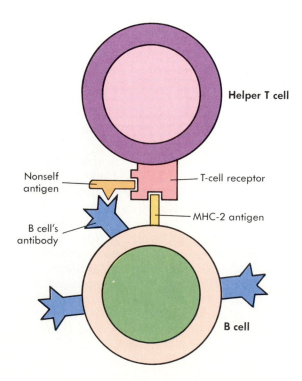

Helper T cell

Nonself
antigen

T-cell receptor

B cell's
antibody

MHC-2 antigen

B cell

activities of macrophages, converting them to "angry" macrophages. One of these lymphokines is **macrophage migration-inhibitory (MMI) factor,** which causes macrophages to congregate at the site of the infection. The assistance of helper T cells is needed to stimulate the proliferation of specific B-cell clones.

At the same time that helper T cells are stimulating an immune response, they are also activating **suppressor T cells,** which are capable of inhibiting both the cell-mediated and the humoral components of the response (see Table 24-4). This action might at first appear to be self-defeating, but the result is to limit what would otherwise be an uncontrolled positive-feedback response. If suppressor T cells are defective, immune reactions may be triggered inappropriately or may result in excessive destruction of host tissues.

> 1 What are the functional differences between killer T cells and helper T cells? How are these differences reflected in the differences in their MHC receptors?
> 2 What is the importance of antigen presentation in the specific immune response?
> 3 What is the importance of suppressor T cells?

Inflammation

When a tissue is damaged, injured, or infected, the surrounding area displays the redness, swelling, warmth, and pain characteristic of inflammation (Table 24-5). Three major processes contribute to this response: (1) blood flow to the inflamed area increases (causing redness); (2) capillary permeability increases, allowing humoral agents of the immune system to get to the affected area (swelling is a consequence of the increased capillary permeability); and (3) leukocytes migrate from the blood vessels into the affected tissue, attracted by substances released in the inflammatory response. This last process is called **chemotaxis.**

TABLE 24-5	Mediators of Inflammation	
Substance(s)	**Effect(s)**
Bradykinin (histamine)	Vasodilation; increased capillary permeability; activation of pain fibers
Complement	Bacterial cell lysis; attracts leukocytes; stimulates phagocytes
Prostaglandins	Potentiates bradykinin; increases set point for body temperature

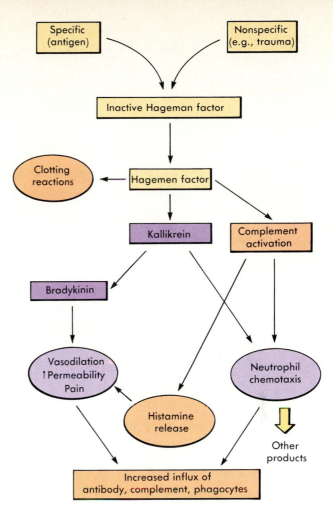

FIGURE 24-16
The reactions of the kallikrein-kinin system.

The inflammatory response is nonspecific in that many elements of it are elicited by substances released from damaged tissues, no matter what the cause of the damage. The responses are augmented if the injury results in introduction of antigens that are carried to the lymph nodes by macrophages and recognized by the specific immune system.

Inflammation is partly mediated by two cascade systems of proteolytic enzymes present in body fluids. Such systems involved in the formation and dissolution of blood clots were described in Chapter 14, p. 349. The two cascade systems that are important in inflammation are the **kallikrein-kinin system** and the **complement system.** A third class of substances, prostaglandins, appears to be important regulators of the intensity of inflammation.

Kallikrein-Kinin System in Inflammation

The kallikrein-kinin system is activated by damaged tissue surfaces. The end products of the cascade are **kinins,** the most important of which is **bradykinin** (Figure 24-16). Bradykinin causes vasodilation, which causes the redness and warmth of inflamed areas. It also causes increased permeability of capillaries, resulting in a local edema and in-

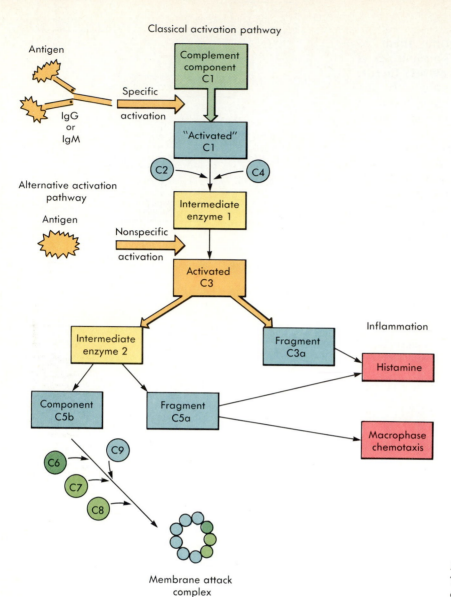

Classical activation pathway

Antigen

IgG or IgM

Specific activation

Complement component C1

"Activated" C1

C2 → ← C4

Alternative activation pathway

Antigen

Nonspecific activation

Intermediate enzyme 1

Activated C3

Intermediate enzyme 2

Fragment C3a

Inflammation

Histamine

Component C5b

Fragment C5a

Macrophase chemotaxis

C6 → C9
C7 → C8 →

Membrane attack complex

FIGURE 24-17
The pathways of activation of the complement system.

creased lymph flow from the damaged area. Bradykinin activates neural afferents, whose activity is felt as the burning pain characteristic of skin abrasions and burns. Histamine released from mast cells and basophils has similar effects to those of bradykinin.

Complement System

The complement system organizes an attack on bacterial invaders. It is composed of nine plasma proteins designated C1 through C9 and several fragments indicated by lower-case letters. The complement system can be activated by elements of the specific immune system (the **classical pathway**) or by elements of the nonspecific immune system (the **alternative pathway**) (Figure 24-17). The classical pathway starts with the binding of an antibody (IgG or IgM) to an antigen molecule on the surface of a bacterial cell. The alternative pathway may be initiated by molecules typical of bacterial cell walls

and does not require antibody binding. The point at which these two pathways converge is at the formation of complement fragment C3a to its active form. Active C3a then activates the rest of the components of the cascade, which join together to form a **membrane attack complex** (Figure 24-18) that attaches itself to the bacterial cell wall. This complex forms a hole, which leads to lysis of the cell in a process very similar to that by which target cells are lysed by perforin secreted by T cells.

In addition to their importance in the formation of membrane attack complexes, the components of the complement cascade have effects that stimulate inflammation and direct phagocytes to attack invading cells. Some of the complement proteins have vascular effects similar to those of kinins. Others cause the release of histamine from mast cells. Still others cause chemotaxis of wandering neutrophils to the inflamed area and stimulate phagocytosis (Figure 24-19).

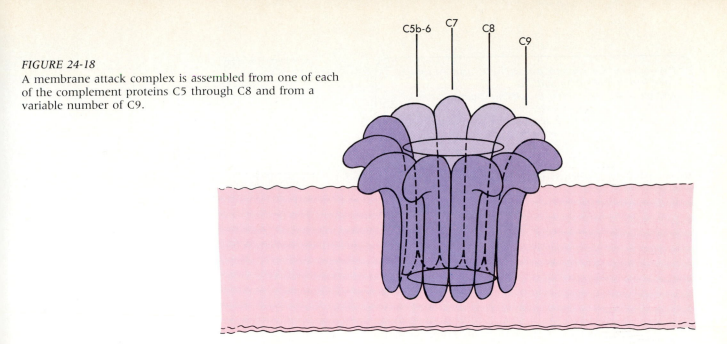

FIGURE 24-18
A membrane attack complex is assembled from one of each of the complement proteins C5 through C8 and from a variable number of C9.

C5b-6 C7 C8 C9

Cell multiplication

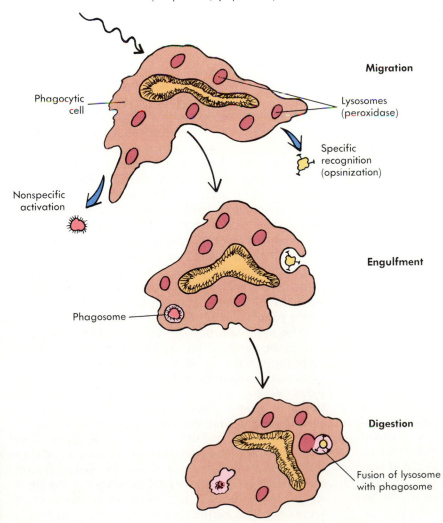

Chemotaxis
(complement/lymphokines)

Phagocytic
cell

Lysosomes
(peroxidase)

Migration

Specific
recognition
(opsinization)

Nonspecific
activation

Phagosome

Engulfment

Digestion

Fusion of lysosome
with phagosome

FIGURE 24-19
Phagocytosis is stimulated by specific and nonspecific factors.

Roles of Prostaglandins in Inflammation and Fever

Prostaglandins are derivatives of **arachidonic acid.** Macrophages are the greatest source of prostaglandins in the immune system. The major prostaglandin produced by macrophages is PGE_2. This prostaglandin greatly potentiates both the ability of bradykinin to stimulate pain afferents and the vascular changes of inflammation. Several drugs commonly used against pain and inflammation owe their effectiveness to their action on some step in the metabolic pathways that lead from arachidonic acid to prostaglandins (Figure 24-20). Aspirin, a salicylate derived from the inner bark of willow trees, has been used medicinally by various cultures since prehistoric times. The antiinflammatory effect of high concentrations of glucocorticoid steroids was discovered relatively recently; preparations of such drugs are used to suppress allergic reactions and skin irritations.

PGE_2 is the prostaglandin that is thought to increase the set point of the thermoregulatory reflexes, thus causing fever (see Chapter 23). The immediate stimulus for production and release of prostaglandins in the brain is interleukin-1 from macrophages. This hormone is responsible for several nonspecific responses to infection. One of the most interesting responses to interleukin-1 is reduction of the plasma concentrations of iron and zinc. The decreased rate of release of iron bound to transferrin by the liver and the binding of free iron by **lactoferrin** released from neutrophils reduce the iron available for uptake by bacteria and thus significantly inhibit bacterial multiplication. The effect of reduction of plasma levels of zinc may be similar.

Transfusion Reactions: One Form of Tissue Graft Rejection

Blood was the first tissue to be successfully transplanted. Red blood cells do not carry MHC antigens, but they do possess other surface antigens that can lead to **transfusion reactions** from the recipient's immune system. At least 15 different families of blood-group antigens have been recognized. In most cases, an individual whose own red blood cells do not possess the antigen in question can be successfully transfused at least once with blood that includes it. After the initial transfusion has caused a primary immune response to the antigen, subsequent transfusions incite an immune response that

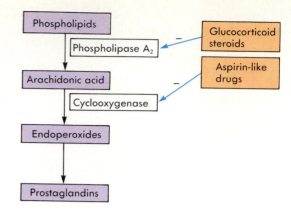

FIGURE 24-20
An abbreviated pathway of prostaglandin synthesis showing the steps at which antiinflammatory steroids and aspirin-like drugs act.

destroys the transfused cells and makes the recipient ill.

One example of such an antigen is the **Rh system.** Those who possess the antigen are termed **Rh-positive; Rh-negative** individuals lack the antigen. This antigen is the cause of Rh disease, an instance in which the placenta fails to protect the fetus from an immune attack by its mother's body. The antibodies to the Rh antigen can cross the placenta. Fetal red blood cells normally do not enter the mother's circulation, but they may do so late in gestation or at birth when placental blood vessels are disrupted. If the mother is Rh-negative but the fetus is RH-positive (having inherited the gene for the antigen from its father),the fetal red blood cells trigger a primary immune response in the mother. In a subsequent pregnancy, the anti-Rh antibodies crossing the placenta will attack the fetal blood cells, causing **erythroblastosis fetalis,** or **Rh disease,** in the infant that is RH-positive (Figure 24-21). The disease can be prevented by flooding the maternal bloodstream with anti-Rh antibodies just after each child is delivered. The injected anti-Rh destroys any fetal cells that may have gotten into the mother's blood before the antigen they bear can initiate a primary response.

The most familiar family of red blood cell antigens is the basis for the **ABO blood group system.** Each individual possesses two homologous genes for antigens of this family, and there are three different alleles. The determinants of these antigens are small, branched chains of sugar molecules (oligosaccharides) that are attached to either lipid or protein molecules of the cell membrane of red blood cells. The three determinants coded for by the three alleles are termed A, B, and H. The A and B antigens are obtained by attaching either a single n-acetyl galactosamine (A) or a galactose (B) to a precursor oligosaccharide (H). People with different al-

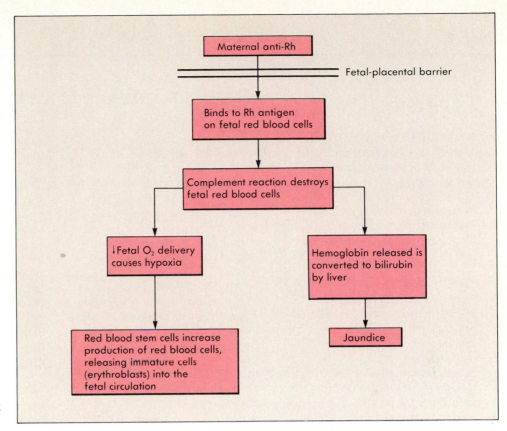

FIGURE 24-21
Steps in the development
of Rh disease.

leles make some of each antigen; for example, those with the blood type designated AB make both antigen A and antigen B. People with the O blood type do not have either the A or the B allele and produce only the precursor glycoprotein H.

The unusual feature of the ABO system is that immunity to non-self antigen is almost always developed even if there is no exposure to it by transfusion. Thus those who have type A blood possess antibodies to the B antigen even if they were never transfused with incompatible blood, and vice versa. Those who lack both the A and B alleles have type O blood and produce antibodies to both antigens. Individuals with type AB blood do not produce antibodies to either antigen. The precursor oligosaccharide is a self antigen in everyone, except in a few rare people in whom the gene for the H oligosaccharide has been lost. How the anti-A and anti-B antibodies arise in the absence of exposure to the antigen is still not understood. These might be antibodies which develop in response to antigens to which almost everyone is exposed, such as a common microorganism, and which only coincidentally cross-react with the A and B antigens.

Incompatibilities among the different types of the ABO system are shown in Table 24-6. The severity of a transfusion reaction is affected by the dilution of the donated plasma in the larger volume of the recipient's blood. For example, if a person with type A blood receives type B blood, two antibody-antigen reactions occur: (1) the donor's anti-A reacts with the A antigen of the recipient's red blood cells, and (2) the recipient's anti-B reacts with the B antigen of the donated red blood cells. The first reaction is of little consequence because the donated antibodies are greatly diluted by the recipient's plasma. The second reaction is severe because the recipient's plasma is diluted only slightly by the transfused plasma. The second reaction destroys the donated cells, releasing hemoglobin and cell fragments into the plasma that may interfere with renal filtration. It also sets off a widespread complement reaction that makes the recipient quite ill. People with type AB blood can receive either types A or B

TABLE 24-6	Summary of Transfusion Reactions			

| Recipient blood group | Blood group of donor | | | |
	Group A	Group B	Group AB	Group O
A	No*	Yes†	Yes	No
B	Yes	No	Yes	No
AB	No	No	No	No
O	Yes	Yes	Yes	No

*No: No transfusion reaction will occur.
†Yes: Transfusion reaction will occur.

Acquired Immunodeficiency Syndrome

In 1981, the Centers for Disease Control in Atlanta recognized the first cases of a new sexually transmitted disease that came to be called **acquired immunodeficiency syndrome (AIDS)**. The illness and death associated with this disease are largely caused by the body's loss of its ability to mount an effective specific immune defense. The syndrome is characterized by forms of cancer and infectious disease that are seldom seen in the population at large, including **pneumocystosis**, which is caused by the protozoan *Pneumocystis carinii,* and a cancer called **Kaposi's sarcoma.** Some individuals fall prey to infections by organisms that are always present in the body but are ordinarily not able to proliferate and cause disease. The progress of the disease varies between individuals. About 30% to 40% of those who have been infected for 3 years will have developed mild symptoms (termed the **AIDS-related complex**), including fever, night sweats, fatigue, weight loss, and damage to lymph nodes. At least 15% to 20% of those who have been infected for 3 years will have developed the full syndrome, which typically is fatal within 2 years.

In developed nations, the groups presently at highest risk for the disease are homosexual and bisexual men, intravenous drug users, people who for medical reasons must use large quantities of blood products (such as hemophiliacs), and the sexual partners of people in these risk groups. In Africa, where the disease is thought to have originated, the pattern of infection differs in that heterosexual contact is the major mode of transmission. In some African nations, the disease has become very widespread. The mode of contact is not intrinsically important; the disease is spread by contact with the semen or vaginal secretions of infected sex partners, by blood transfused from infected blood donors, or by contaminated needles and syringes passed between intravenous drug users.

The **human immunodeficiency virus** (HIV) that causes AIDS was identified independently by F. Barre-Sinoussi, J.-C. Chermann, and L. Montagnier of the Pasteur Institute in Paris and by R.C. Gallo and his associates at the U.S. National Cancer Institute. It is a **retrovirus,** whose genetic material consists of RNA rather than DNA (Figure 24-B). When retroviruses infect a cell, the nucleotide sequence of the viral RNA is transcribed into DNA by **reverse transcriptase** carried by the virus. This reverses the usual sequence in cells, in which the genetic information normally resides in chromosomal DNA and is transcribed into RNA to direct protein synthesis (see Chapter 3, p. 56). The new viral DNA may then be transcribed into viral proteins by the host cell. In some cases the viral DNA joins itself to the host's DNA and is treated as if it were part of the original genetic material of the cell. In such cases, the viral DNA may result in a **latent infection** that can persist for many years before the viral genes begin to be expressed and symptoms appear (Figure 24-C).

The AIDS virus may infect

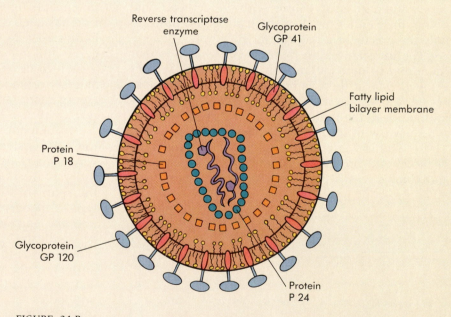

FIGURE 24-B
The structure of the HIV virus. The virus is surrounded by a protein coat that is studded with receptors for corresponding molecules on the lymphocyte cell membrane. The interior of the coat contains viral RNA and reverse transcriptase.

Labels: Reverse transcriptase enzyme; Glycoprotein GP 41; Fatty lipid bilayer membrane; Protein P 18; Glycoprotein GP 120; Protein P 24

several types of cells, including epithelial cells, neurons, and macrophages, but the immunosuppression characteristic of AIDS results from infection of helper T cells by the virus. The virus recognizes helper T lymphocytes by binding to a T-cell-specific cell-surface receptor that stimulates the lymphocytes to ingest the receptor and the bound virus (this process, termed endocytosis, is described in Chapter 4, p. 66). The infection remains latent until the infected lymphocyte is stimulated by a secondary infection. A control protein produced by the lymphocytes on activation stimulates production of lymphokines; this control protein also activates synthesis of viral proteins. A furious burst of viral reproduction in the cell follows, killing the cell and releasing viruses to infect other cells. Ultimately helper T cells almost disappear from the lymphocyte population. Because helper T cells are the cornerstone of the specific immune response, the immune system becomes less able to direct a response either against the AIDS virus itself or against other infections or cancer cells.

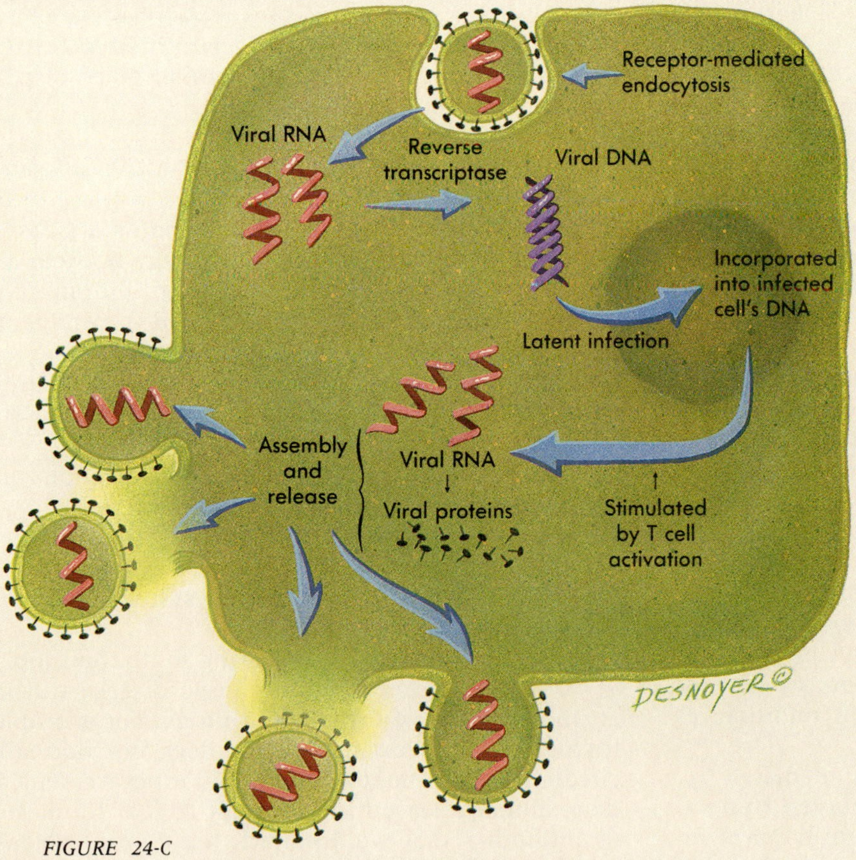

Diagnosis of AIDS carriers and rejection of contaminated donor blood depends, ironically, on the presence of antiviral antibodies in the blood. In the early stages of infection, antibodies may not be present in quantity. The latent period between initial infection and development of full-blown symptoms is typically more than 1 year and may be as long as 5 to 8 years. The disease is sufficiently new that the incidence of infection in the U.S. population is impossible to know with certainty; some estimates suggest that as many as several million Americans may carry latent infections. Most of them do not know that they are infected.

How big a problem will AIDS be by the end of the century? Will AIDS spread throughout the American population? The U.S. Public Health Service estimates that by 1992 there will have been 270,000 cases of AIDS and 179,000 deaths. Further transmission of the disease will be reduced, but not stopped, by public health measures that encourage the use of condoms and discourage the use of intravenous drugs. Successful development of a vaccine is a challenge because the antigens of the AIDS viral protein coat, like those of many other viruses, undergo very rapid evolution. Drugs such as **azidothymidine (AZT)** that interfere with viral multiplication may be effective in blunting the onset of the disease, but do not presently protect people against infection.

FIGURE 24-C
The sequence of events that leads from infection of a T cell to a latent infection followed at some later date by activation of viral replication and destruction of the infected cell, releasing the new viruses.

blood with only a minor reaction because the donor's antibodies are greatly diluted. Similarly, although type O blood contains both anti-A and anti-B, people with types A, B, or AB blood can receive type O blood because the dilution effect protects the recipient's cells from a severe reaction.

It should be emphasized that the ABO and Rh types are only two of numerous blood type factors. Before a blood transfusion is performed, the compatibility of donor and recipient blood can be tested by mixing some of the prospective donor blood with plasma from the recipient. Incompatibility is indicated by agglutination of the donor's cells. As Table 24-6 shows, blood can, in principle, be transfused between some individuals whose ABO types are not the same without provoking a transfusion reaction. However, such cross-type transfusions are now discouraged except in emergency situations.

Allergy and Hypersensitivity

Allergies are disorders of immune function in which repeated exposures to an antigen elicit immune responses severe enough to cause discomfort, illness, or even death. The immune system has become **hypersensitive** to these antigens, which are called **allergens.** Allergens are frequently common components of the enviroment, such as plant pollen, small organisms, or household chemicals. Many allergens are harmless except for their ability to stimulate hypersensitivity in susceptible individuals.

Immediate hypersensitivity is characterized by responses that occur within a few seconds or minutes after exposure to the allergen. Hay fever, asthma, food allergies, and hives are examples. In these allergies, the allergen is recognized by IgE antibodies, resulting in a release of histamine, leukotrienes, and other inflammation mediators at the site of response. In the case of inhaled allergens, the inflammation mediators cause a powerful constriction of small airways in the lung, making breathing difficult. In severe immediate responses, inflammation mediators may diffuse into the blood in sufficient amounts to cause **anaphylaxis,** a condition of widespread vasodilation, decreased blood pressure, and even cardiac failure. Any situation in which a large number of immune complexes are formed in the blood or tissues can cause a similar condition of widespread inflammation.

Some cases of hypersensitivity are mediated by T cells. These responses are called **delayed hypersensitivity** because they require hours or days to reach full intensity. In delayed hypersensitivity, the T-cell response is so vigorous and broadly directed that otherwise healthy tissues are damaged. The rash of measles and the irritation caused by the toxins of poison ivy and poison oak are examples of delayed hypersensitivity.

The severity of allergic responses is affected by health and emotional state, but susceptibility to particular allergies appears to be determined largely by heredity and probably involves inheritance of particular MHC-complex genes.

Autoimmune Diseases: Misdirected Responses of the Immune System

Several diseases either are recognized as attacks of the immune system on some part of the body or are thought to be such (Table 24-7). An autoimmune disease could, in principle, arise from the accidents listed below.

1. Releasing into the blood an endogenous antigen that was not present in fetal life when self-directed clones were eliminated.
2. An antibody induced against an exogenous antigen cross-reacting with a self antigen.
3. An immune response to an exogenous antigen damaging surrounding tissue.
4. The MHC self-identification system failing to protect healthy cells.
5. The suppressor T cells failing to do their job.

Different forms of autoimmune disease can be attributed to all of these possibilities. For example, thyroglobulin (see Chapter 23) does not normally circulate in the blood, but an initiating accident or disease that released some might trigger formation of antibodies that would attack the thyroid, releasing more thyroglobulin and starting a positive-feedback cycle that would ultimately result in destruction of the thyroid and thus in hypothyroidism (thyrotoxicosis). A similar process may cause Type I (juvenile-onset) diabetes (see Chapter 23).

In rheumatic fever, an antibody raised against an antigen of streptococcal bacteria cross-reacts with a protein present in the heart valves, damaging

TABLE 24-7	Autoimmune Diseases
Disease	*Affected organ systems*
Juvenile-onset diabetes	Pancreatic beta cells
Rheumatoid arthritis	Joints
Ankylosing spondylitis	Spine
Multiple sclerosis	Myelin in the central nervous system
Thyrotoxicosis	Thyroglobulin
Rheumatic fever	Valves in the heart
Myasthenia gravis	Acetylcholine receptors at the neuromuscular junction
Ulcerative colitis	Intestine
Male infertility (some)	Spermatozoa
Systemic lupus erythematosus	Most organs
Amyotrophic lateral sclerosis	Motor neurons in the spinal cord

them. Some acute bacterial diseases result in the formation of immune complexes of antigens and antibodies that lodge in sites such as the glomeruli of the kidney. The inflammation triggered by the presence of these complexes results in glomerulonephritis. Numerous autoimmune diseases can be associated with one or another allele of the MHC system. In an extreme example, 90% of people who develop a joint disease called ankylosing spondylitis possess the B27 antigen of the MHC complex, whereas it occurs in only 8% of those without the antigen. Rheumatoid arthritis, multiple sclerosis, and myasthenia gravis are examples of immune attacks directed against the joints, the myelin coats of nerves, and the acetylcholine receptors, respectively. The incidence of particular MHC antigens in patients who have these diseases is approximately double the incidence in the population at large. These relationships suggest that a failure of the self-recognition property, which should be mediated by the MHC system, is at least a contributing factor in these diseases.

Development of autoimmune disease may require two defects: (1) the existence of self-directed B-cell clones, and (2) an ineffective suppressor T-cell system. For example, in **systemic lupus erythematosus,** antibodies are generated by B cells against almost all tissues of the body. Sufferers of this disease possess a genetic defect in the generation of a particular subclass of suppressor T cells, which appears to put them at risk for the disease.

1 What factors result in a transfusion reaction? Why is this harmful to the recipient?
2 What is an autoimmune disease?
3 How can the existence of autoimmune diseases be reconciled with the theory that self-directed clones are eliminated from the immune system in early development?

SUMMARY

1. The immune system consists of the following cell types
   A. Leukocytes
      a. Granulocytes—neutrophils, eosinophils, and basophils
      b. Lymphocytes
         1. B cells and plasma cells
         2. Numerous T cell types, including helper, killer, suppressor, and inducer T cells
   B. Humoral components
      a. Antibodies secreted by plasma cells
      b. Complement proteins
   C. Macrophages
   D. Mast cells
2. **Antibodies,** the agents of humoral immunity, are composed of basic units made up of two heavy protein chains linked together with two light chains to form a **Y,** whose arms, the **Fab regions,** contain highly variable amino acid sequences. The effect of the variability is to form a large variety of binding sites that can recognize various **antigen determinants.** The extreme variability of the Fab regions is obtained by a process of somatic rearrangement in which DNA segments coding for this region are joined together randomly during the process of clonal differentiation of B cells.
3. Antibody binding may affect the target antigen by agglutination or opsonization or by activation of the complement system.
4. Antibodies are classed as IgG, IgM, IgE, IgA, or IgD according to the function of the **Fc portion** of the molecule (the stem of the **Y**).
5. **Histocompatibility antigens** are identity tags for the cell-mediated immune system. They are unique to each individual, arising from the most polymorphic gene loci known. **MHC-1** antigen is found on all nucleated body cells and directs killer T lymphocytes to their targets; **MHC-2** antigen is found on B cells, macrophages, and some types of T cells, and it channels interactions between T cells and B cells that occur in the course of a specific immune response.
6. T cells undergo clonal differentiation in the thymus. During this process, various clones arise, each with a different receptor specificity. The **T-cell receptor** resembles antibody in some respects, but it can bind a foreign antigen only if the histocompatibility antigen is also present.
7. **Killer,** or **cytotoxic,** T cells recognize and attack cells that are simultaneously displaying MHC-1 antigen and nonself antigen. **Natural killer** cells are not specific but are directed to major classes of nonself antigens.
8. **Helper T cells** are activated by antigen presented by macrophages, stimulating inducer T cells to deliver the antigen to activate B-cell clones to proliferate and give rise to memory cells and plasma cells. At the same time, helper T cells stimulate the activity of **suppressor T cells,** which limit the duration and scope of the immune response.
9. **Inflammation** is characterized by vascular dilation and nervous activity that results in local redness, swelling, warmth, and pain. All these characteristics are attributable to bradykinin released by damaged tissues and to **histamine** released by mast cells and basophils at a site of infection. The responses are nonspecific but are greatly augmented by the specific immune system.
10. A cascade of **complement proteins** is activated by antibody binding and by the presence of bacterial cell wall components. One effect of complement activation is the formation of membrane attack complexes that form holes in bacterial cell membranes, leading to lysis. Some complement proteins have kininlike effects.
11. Red blood cells carry blood type antigens that may be recognized by the immune system. Transfusion reactions, in which the transfused cells are destroyed by the recipient's antibodies, may result if the donor's and recipient's ABO **blood groups** do not match or if the recipient has developed antibodies to another blood type antigen as a result of previous exposure. Rh disease occurs in fetuses with the Rh antigen whose mothers do not carry the Rh antigen and who developed antibodies to it in a previous pregnancy.
12. **Autoimmune** diseases result when a disease causes the release of self antigens that did not previously circulate in the blood, when antibodies to nonself antigens also recognize self antigens, or when the mechanisms that normally suppress immune responses to self antigens fail.

1. What events give rise to swelling and pain at the site of an infection?

2. Cells of the immune system fall into classes that overlap. Which cell types are phagocytes? Which are granulocytes? Which are leukocytes? Which are lymphocytes? What is the functional similarity between basophils and mast cells?

3. What are the advantages of having a cascade of complement reactions that is activated by antibody binding?

4. Why is it that organ transplants are more likely to succeed if the donors are closely related? Why might even closely related donors not be acceptably matched to a recipient?

5. What are the mechanisms by which antibody binding results in damage to infecting organisms?

6. What do the killer T cells and the complement system have in common?

7. People with juvenile-onset diabetes who have been treated with insulin may develop antibodies to insulin; those without diabetes do not make such antibodies. Explain.

8. What is the role of the immune system in preventing cancer? Could aging in the immune system contribute to the fact that cancer is much more frequent in older people?

9. Trace the events of a primary immune response to a local infection. How would a secondary response be different?

10. What failures of immune differentiation and control can cause autoimmune disease?

11. What is the role of suppressor T cells in an immune response?

12. In most cases, transfusion reactions are the result of the recipient's antibodies attacking the transfused cells. In what situations is an attack of donor antibodies on the recipient's cells possible? Such reactions are usually not a problem—can you figure out why?

● SUGGESTED READING

ADA, G.L., and G. NOSSAL: The Clonal Selection Theory, *Scientific American*, August 1987, p. 62. Lucid description of an important concept in immunology.

BOWER, B.: Taking Care of Immunity, *Science News*, September 12, 1987, p. 168. Discusses how stress involved in caring for relatives with chronic disease may undermine the caregiver's own immune response.

COHEN, I.R.: The Self, the World and Autoimmunity, *Scientific American*, April 1988, p. 52. Shows how the self-recognizing and self-attacking properties of the immune system may be exploited to police those aberrant cells responsible for autoimmune diseases such as multiple sclerosis, myasthenia gravis, and diabetes.

DINARELLO, C.A.: Biology of Interleukin-1, *FASEB Journal*, volume 2, 1988, p. 108. Discusses humoral mediators of inflammation and the regulation of the immune response.

ESSEX, M., and P.J. KENKI: The Origins of the AIDS Virus, *Scientific American*, October 1987, p. 64. A discussion of how AIDS-related HIV viruses interact with human beings and monkeys and how some HIV viruses seem to have evolved toward disease-free coexistence with their hosts.

GALLO, R.C., and L. MONTAGNIER: AIDS in 1988, *Scientific American*, October 1987, p. 40. In their first collaboration, the two investigators who established the cause of AIDS describe how HIV was isolated and linked to AIDS, the current status of AIDS research, and the prospects for an AIDS therapy.

HASELTINE, W.A., and F. WONG-STAAL: The Molecular Biology of the AIDS Virus, *Scientific American*, October 1988, p. 52. Shows how just three viral genes direct the machinery of a cell infected with the AIDS virus.

KOLATA, G.: Immune Boosters, *Discover*, September 1987, p. 58. Discusses various drugs that are used to augment the response of the immune system.

LEWIS, G.P.: *Mediators of Inflammation*, John Wright & Sons, Ltd., Bristol, England, 1986. Comprehensive description of inflammation, including the complement system.

METZGER, H., and J-P. KINET: How Antibodies Work: Focus on Fc Receptors, *FASEB Journal*, volume 2, 1988, p. 3. Summary of antibody production and antibody-antigen interactions.

WEBER, J.N., and R.A. WEISS: HIV Infection: The Cellular Picture, *Scientific American*, October 1987, p. 100. Describes the binding of the AIDS virus to the CD4 antigen on helper T cells.

YOUNG, J.D., and Z.A. COHN: How Killer Cells Kill, *Scientific American*, January 1988, p. 36. Discusses the way killer T cells recognize and destroy their target cells.

CHAPTER 25

Reproduction and Its Endocrine Control

On completing this chapter you will be able to:

- Describe the anatomy of male and female reproductive systems.

- Describe how gonadal sex is determined by the sex chromosome complement, and how the presence of male or female gonads determines the course of development of male or female internal accessory structures and genital sex.

- Understand the genetic and endocrine causes of representative disorders of sexual development.

- Trace the sequence of events in formation of male and female gametes, and know the similarities and differences between this process as it occurs in male and female gonads.

- Describe the changes characteristic of male and female puberty and their endocrine basis.

- Describe the endocrine feedback loops that control gonadal function in men and women.

- Trace the events and changes of the menstrual cycle and explain the endocrine roles of the hypothalamus, anterior pituitary, ovarian follicle, and corpus luteum in control of the cycle.

*F*rom the moment of conception to maturity as a man or woman, sexual development is a road with many branches. The first branch occurs at conception—one of the father's contributions to his children are genes that determine their sexes. A specific signal from the genetic makeup causes the reproductive organs of male babies to develop along the male pattern before birth. Normally this signal is not present in female babies, and in its absence female reproductive organs develop automatically.

Starting as soon as 6 months after birth, the child begins to develop a sense of identity as a girl or boy. The development of an appropriate sexual identity is highly dependent upon interactions with the adults who care for the child. Ten to fourteen years after birth, hormones secreted by the brain cause reproductive capacity to develop, and the physical characteristics that distinguish mature men and women begin to emerge. At each turning point in development, the body has the potential to take either the female or male direction.

The ultimate consequence of sexual development is production of reproductive cells that will pass the genes to the next generation. Each individual produces an enormous number of reproductive cells. Of the stock of several million reproductive cells each female possesses at birth, only several hundred will ever have a chance to be fertilized—the rest will degenerate at various stages. In males, reproductive cells are formed continuously after sexual maturity is reached, and each ejaculation releases perhaps a hundred million. Each reproductive cell produced by an individual carries a unique assortment of his or her genes. As a result, each child possesses one of a very large number of different possible combinations of the genes of the parents.

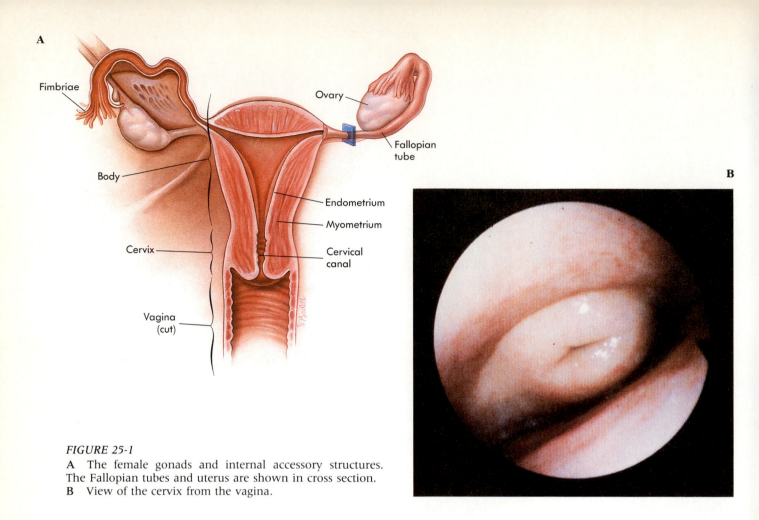

FIGURE 25-1

A The female gonads and internal accessory structures. The Fallopian tubes and uterus are shown in cross section.
B View of the cervix from the vagina.

TABLE 25-1	The Male and Female Reproductive Systems	

	Male	Female
Gonads	Testes	Ovaries
Gonadal steroids	Testosterone	Estrogen (mainly estradiol), progesterone
Accessory structures	Penis	Fallopian tubes
	Scrotum	Uterus
	Duct system composed of	Vagina
	Rete testis	Labia majora
	Efferent ductules	Labia minora
	Epididymis (2)	Clitoris
	Vas deferens (2)	
	Ejaculatory duct	
	Penile urethra	
	Glands include	
	Prostate	
	Seminal vesicles (2)	
	Cowper's glands (2)	
Secondary sex characteristics	Male pattern of body, pubic, and facial hair; larger larynx (deeper voice); greater bone mass and growth of long bones; narrow pelvis; greater ratio of muscle mass to total body weight; higher hematocrit level; male behavior (?)	Female pattern of body and pubic hair; wider pelvis; greater ratio of fat mass to total body weight; female pattern of fat deposition; female behavior (?)

THE ANATOMY OF MALE AND FEMALE REPRODUCTIVE ORGANS
Female Reproductive System

In both sexes, the reproductive systems, or **genitalia,** consist of the primary reproductive organs (**gonads**) that produce reproductive cells and secrete sex hormones and **accessory structures** that support the function of the gonads. In females, the gonads are **ovaries** (Figure 25-1; Table 25-1) located within the abdomen. Each **ovary** weighs about 15 grams and is attached to the pelvic wall and to the uterus by a fine network of ligaments. A 10 to 12 cm long **Fallopian tube,** or **oviduct,** associated with each ovary provides a route for the female reproductive cells (**ova;** one is an **ovum**) to travel to the uterus. Although the Fallopian tubes are attached to the uterus, they are not connected to the ovaries. Instead, fingerlike processes of the Fallopian tubes called **fimbriae** partially surround each ovary.

The **uterus** (see Figure 25-1, *A*) is a highly muscular organ, located between the bladder and rectum and connected to the abdominal wall by ligaments. The uterus encloses and supports the developing infant, called the **fetus,** and expels the mature fetus in the process of **labor.** The wall of the uterus consists of a lining, the **endometrium,** and a coat of smooth muscle, the **myometrium.** The interior of the uterus is normally quite small and enlarges only during pregnancy. The narrow passage at the lower end of the uterus is the **cervical canal.** The **cervix** of the uterus (see Figure 25-1, *B*) leads to a muscular canal, the **vagina,** which in turn leads to the body exterior. The vaginal opening lies posterior to the urethra and anterior to the rectum

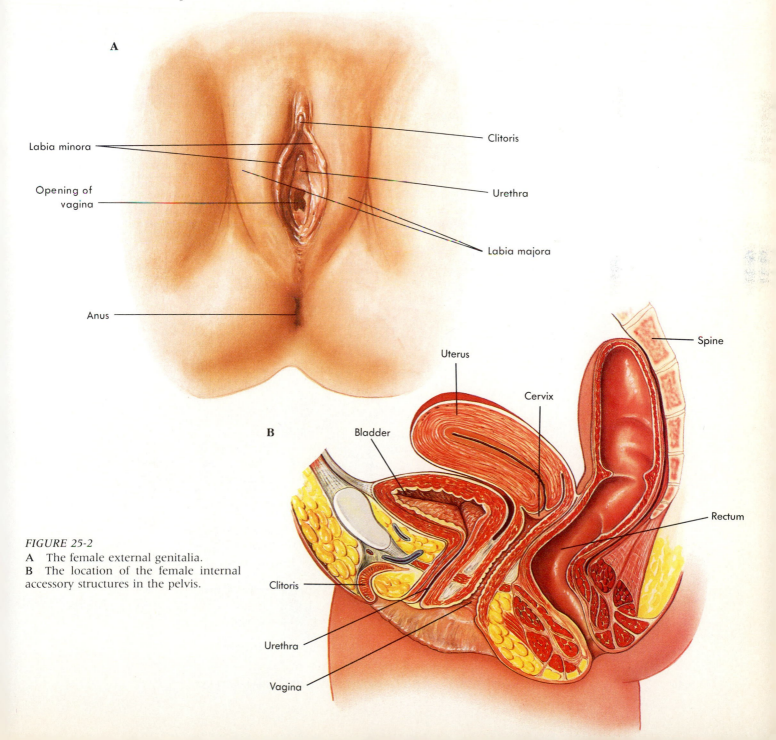

FIGURE 25-2
A   The female external genitalia.
B   The location of the female internal accessory structures in the pelvis.

(Figure 25-2). Although the vagina is open at one end, the interior of the uterus is protected from the entry of foreign materials because the cervix is narrow and usually filled with a thick mucus.

The structures that surround the opening of the vagina are collectively referred to as the **external genitalia.** These include the **clitoris** (see Figure 25-2), a structure that serves as a sensory focus in the sexual response, and two pairs of skin folds, the **labia majora** and **labia minora,** which surround and protect the openings of the vagina and urethra. The Fallopian tubes, uterus, vagina, clitoris, labia minora, and labia majora are the female **accessory organs** (see Table 25-1).

1 What are the female gonads and where are they located?
2 What are the female accessory organs?

Male Reproductive System

The male gonads, the **testes** (Figure 25-3) produce male reproductive cells (**spermatozoa** or **sperm cells**). In the fetus, the testes are within the abdomen. Before birth they normally descend into a sac (the **scrotum**) located outside the abdomen, where the temperature is slightly below that of the body core. If descent of the testes fails to occur, the testes remain at core body temperature and spermatozoa cannot develop normally. Within the testes, production of spermatozoa goes on in **seminiferous tu-**

bules (Figure 25-4), which make up 80% of the testicular mass. Seminiferous tubules possess cells called spermatogonia that give rise to spermatozoa, whereas **Sertoli cells** nurture the developing spermatozoa. Cells that are in intermediate stages in formation of spermatozoa (**spermatocytes, spermatids**) are also present in the seminiferous tubules. **Leydig cells (interstitial cells)** located between the seminiferous tubules secrete the male sex hormone **testosterone.**

The male accessory structures (see Table 25-1 and Figures 25-3 and 25-4, *A*) include a system of ducts and glands, together with the **penis,** which introduces spermatozoa into the vagina of the female and is the focus of sensation in the male sexual response. Starting from the seminiferous tubules, the ducts are: the **rete testis, efferent ductules** (both of these structures are within the testes as shown in Figure 25-4), the **epididymis** (one is located next to each testis), the **vas deferens,** leading from each testis to the **ejaculatory duct,** which leads to the **urethra.** Of these, all but the last two are paired.

The glands of the male system (see Figure 25-3) include the two **seminal vesicles** and the **prostate gland,** which empty into the vas deferens and ejaculatory duct, and **Cowper's glands** (also called **bulbourethral glands**), which empty into the urethra. The combination of the secretions of all these glands with the spermatozoa and fluid leaving the vas deferens forms a mixture called **semen.** The spermato-

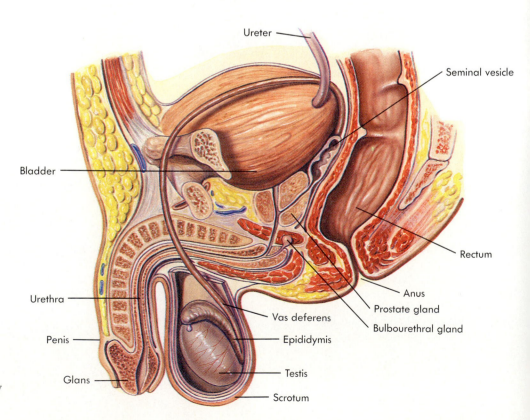

FIGURE 25-3
The testes and male accessory reproductive structures.

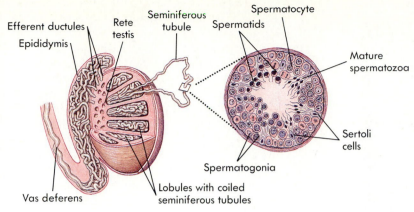

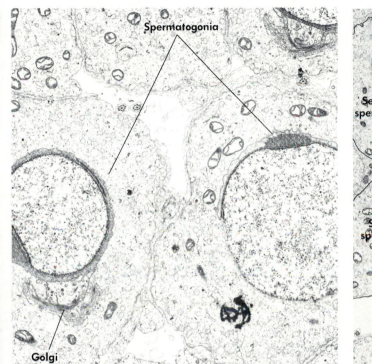

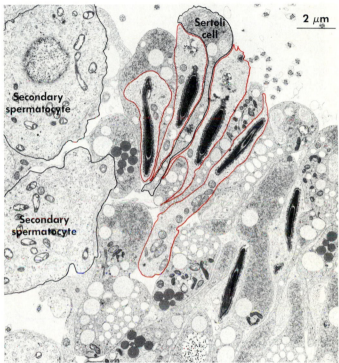

FIGURE 25-4

A *(Left)* Testis with epididymis and vas deferens, showing arrangement of the rete testis and seminiferous tubules.

(Right) Cross section of a single seminiferous tubule showing spermatogonia, Sertoli cells, and cells in various stages of meiosis and spermiogenesis.

B Electron micrograph of portion of seminiferous tubule showing two spermatogonia.

C Electron micrograph of seminiferous tubule showing four spermatids surrounded by Sertoli cells. The four spermatids outlined probably arose from a single spermatogonium. Secondary spermatocytes are adjacent.

zoa themselves make up less than one-tenth of the total volume of semen. The bulk of the volume of semen is contributed by the seminal vesicles, with smaller contributions from the prostate and bulbourethral glands. The constituents of semen are mixed and expelled through the urethra by contractions of smooth muscle of the duct system during **ejaculation.**

Semen contains several constituents that contribute to the viability of spermatozoa deposited in the vagina. These include fructose, a sugar that supplies energy for motility; bicarbonate that buffers vaginal acidity; prostaglandins that stimulate contractions of the female reproductive tract to help transport spermatozoa; and factors that promote clotting of the semen to keep it in the vagina.

The penis and the clitoris both develop from the same embryonic structure. The penis contains three vascular sinuses (Figure 25-5): the two **corpora cavernosa** and the **corpus spongiosum** through which the urethra passes. Together these are referred to as the **erectile tissues** because they become filled with blood during erection of the penis. Similar but smaller structures are responsible for clitoral erection in females. Much of the sensory input from the penis comes from a dense population of receptors in the **glans** at the tip of the penis (see Figure 25-3); the clitoral glans has an even denser population of similar receptors.

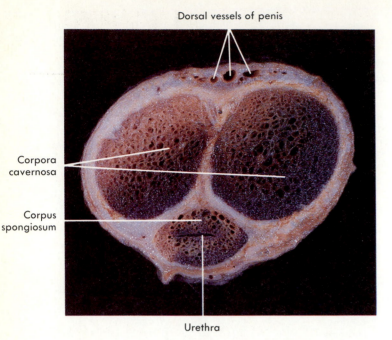

FIGURE 25-5

Cross section of penis showing erectile tissue (corpora cavernosa and corpus spongiosum), urethra, and dorsal vein and arteries.

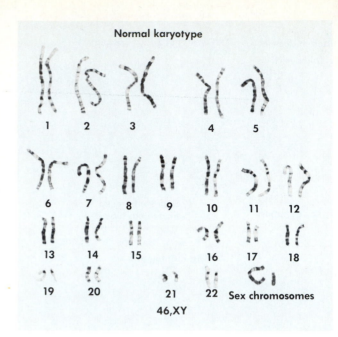

FIGURE 25-6

Chromosomes of a cell as they appear during metaphase of mitosis. The chromosomes are cut out of a photomicrograph of the cell nucleus and arranged in homologous pairs to form a karyotype. The sex chromosome complement is XY.

1 Name and give the functions of the three major categories of testicular cells.
2 What and where are the erectile tissues in the male system? Where are their counterparts in the female system?
3 What are the components of the male gland and duct system?

THE GENETIC BASIS OF REPRODUCTION AND SEX DETERMINATION
Meiosis—Formation of Haploid Gametes

Human cell nuclei contain a total of 46 chromosomes, including 22 **homologous pairs** of **autosomal** (or **somatic**) **chromosomes** and a homologous pair of **sex chromosomes** (Figure 25-6). One chromosome of each homologous pair is contributed by each parent. Except for the sex chromosomes, the members of a pair of homologous chromosomes are of identical size and appearance under the microscope and carry the same genes. Each pair of sex chromosomes is normally composed of either an **X chromosome** and a **Y chromosome** (the normal situation in males) or two X chromosomes (the normal situation in females).

The total of 46 chromosomes is twice the number of chromosomes contributed by each parent (N) and is referred to as the **diploid** or **2N** complement. In mitotic cell division that occurs in growth (see Chapter 3), each chromosome replicates itself, and each daughter cell receives a set of chromosomes

that, except in occasional accidents, is identical to that of the mother cell.

Meiosis is the form of cell division that produces **gametes** (reproductive cells) in the ovary and testis. Meiosis results in cells that have the **haploid** chromosomal complement (N). Meiosis involves two successive cell divisions, called the **first** and **second meiotic divisions.** At the completion of meiosis, each of the four resulting gametes has one representative of each of the homologous pairs of autosomal chromosomes, and one of the sex chromosomes. Combination of the nuclei of two gametes in fertilization restores the diploid complement of chromosomes in the fertilized ovum (a **zygote**).

On the chromosomal level, the events of meiosis are similar in the ovary and testis. To simplify the explanation, the process of **spermatogenesis** in the testis will be followed first (Figure 25-7), and then the important differences between spermatogenesis and oogenesis will be noted.

In the testes of sexually mature males, spermatogonia continuously undergo mitotic divisions. Of the two daughter cells produced by each mitotic division, one will remain a spermatogonium, and the other will begin a cell line that will give rise to 16 spermatozoa. In this way, the population of spermatogonia is maintained, and at the same time production of spermatozoa can go on continuously throughout mature life.

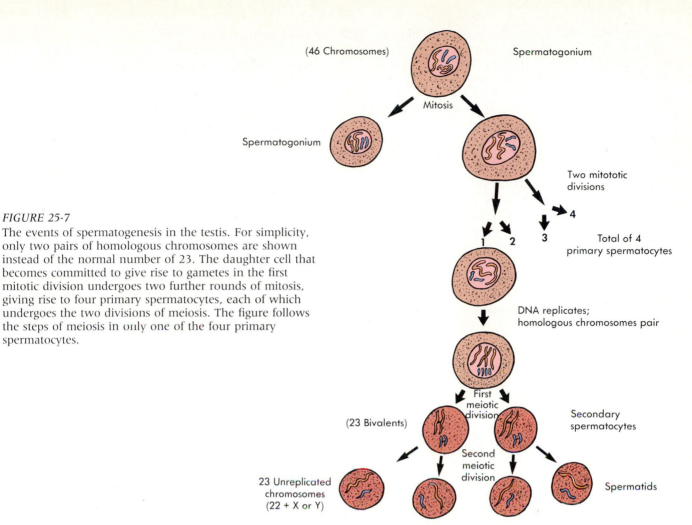

(46 Chromosomes) Spermatogonium

Mitosis

Spermatogonium

Two mitototic
divisions

4

1   2   3

Total of 4
primary spermatocytes

DNA replicates;
homologous chromosomes pair

First
meiotic
division

(23 Bivalents)

Secondary
spermatocytes

Second
meiotic
division

23 Unreplicated
chromosomes
(22 + X or Y)

Spermatids

FIGURE 25-7

The events of spermatogenesis in the testis. For simplicity, only two pairs of homologous chromosomes are shown instead of the normal number of 23. The daughter cell that becomes committed to give rise to gametes in the first mitotic division undergoes two further rounds of mitosis, giving rise to four primary spermatocytes, each of which undergoes the two divisions of meiosis. The figure follows the steps of meiosis in only one of the four primary spermatocytes.

The cell line that is ultimately to give rise to spermatozoa first goes through two further cycles of mitosis (see Figure 25-7), resulting in four **primary spermatocytes,** each of which will undergo meiosis. The primary spermatocytes, like spermatogonia, are diploid cells. Prior to the **first meiotic division,** the chromosomes of each primary spermatocyte replicate. Each original chromosome and its copy remain attached to one another, forming a **bivalent.** At this point each primary spermatocyte contains 23 homologous pairs of bivalents. As primary spermatocytes undergo the first meiotic division, each daughter cell (a **secondary spermatocyte**) receives one bivalent from each of the homologous pairs. Thus each of the eight secondary spermatocytes derived from a single spermatogonium possesses two copies of one of each of the original 23 pairs of homologous chromosomes.

Particular genes are located at specific sites on specific chromosomes. The site of a gene on a chromosome may be occupied by one of several alternative forms of the gene called **alleles.** Different alleles are largely responsible for different traits. For example, the characteristics of body and head hair are coded for by a family of genes. The presence and interaction of different alleles of these genes determines whether one's hair will be curly or straight, brown, black, or blond, etc. Even though homologous chromosomes look identical under the microscope and carry the same genes, they are not identical because they carry different alleles for many of those genes. Consequently, the two secondary spermatocytes resulting from the first meiotic division are not genetically identical.

The first meiotic division is a major source of the immense genetic variability among the reproductive cells produced by a single individual. Because bivalents are distributed randomly between the two daughter cells, the chromosomes derived from one's parents may be assorted into a large number of new combinations. The primary spermatocytes of an individual can give rise to more than 4 million different combinations of bivalents in the secondary spermatocytes. Still more variation in the possible genetic makeup of secondary spermatocytes is introduced by a process called **crossing** over, in which homologous bivalents trade pieces of each other while they are aligned during the first meiotic division (see Figure 25-7).

Each of the two secondary spermatocytes resulting from the first meiotic division of a primary spermatocyte undergoes a **second meiotic division** (see Figure 25-7). In the second meiotic division, each bivalent splits, and each of the two daughter cells **(spermatids)** receives one of the replicates. Thus each spermatid has one copy of one chromosome from each of the pairs of homologous chromosomes that were present in the secondary spermatocyte. Because there were two cycles of meiotic division, each of the four primary spermatocytes was the ancestor of two secondary spermatocytes and four spermatids. Thus the cell line established by the original mitotic division of a single spermatogonium terminates in 16 spermatids. Each spermatid has 23 chromosomes, the haploid number.

Spermiogenesis

Although the spermatids resulting from spermatogenesis are haploid, they cannot function as reproductive cells because they lack the cellular equipment needed to reach and fertilize an ovum. In a process called **spermiogenesis** or **spermiation**, spermatids acquire the specialized structures of spermatozoa (Figure 25-8, A and B).

In spermiogenesis, most of the cytoplasm is discarded, leaving a sperm head 2 to 3 microns in diameter and about 5 microns long, which consists of a nucleus surrounded closely by plasma membrane. Each spermatid develops a **flagellum** (or "tail") for locomotion, a **midpiece** containing mitochondria to provide the ATP to drive the flagellum, and an **acrosome** on the head (see Figure 25-8, A and C). The acrosome is located beneath the plasma membrane and contains enzymes needed for penetration of an ovum, including hyaluronidase, protease, and corona-penetrating enzyme.

Unlike most cells which depend on glucose, spermatozoa depend on fructose (secreted by the seminal vesicles) for energy. The average velocity of locomotion of spermatozoa is about 50 microns/second (4 mm/min). Ejaculated spermatozoa cannot fertilize an ovum until they have been **capacitated** by unknown chemical factors in the female repro-

A

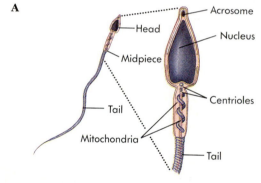

FIGURE 25-8
A   The structure of a spermatozoan.
B   Light micrograph of human spermatozoa after ejaculation. The heads are glowing because they have been labeled with a fluorescent antibody.
C   Electron micrograph of sectioned rat spermatozoa. Parts of heads, midpieces with mitochondria, and flagella are visible.

B

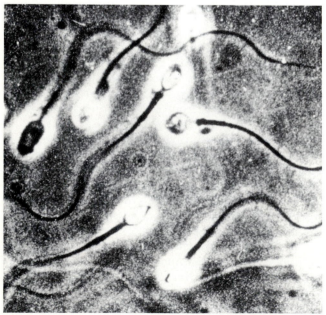

C

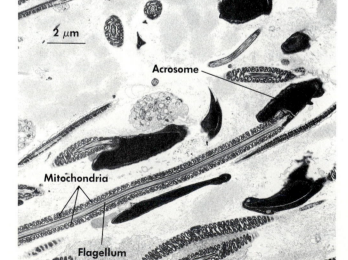

TABLE 25-2   *Comparison of Oogenesis and Spermatogenesis*

	Oogenesis	*Spermatogenesis*
Stem cells	Oogonia	Spermatogonia
Timing of meiosis	Arrested at first meiotic metaphase before birth; resumed at ovulation after puberty; second meiotic division occurs after fertilization; process ceases at menopause	Begins at puberty and continues until death
Fates of daughter cells	Only primary oocytes remain at birth; each primary oocyte gives rise to one ovum; the other daughter cells (polar bodies) degenerate	Each spermatogonium gives rise to 16 spermatids that all become spermatozoa
Cytoplasmic division	Each oocyte receives most of the cytoplasm; polar bodies get chromosomes but almost no cytoplasm	Cytoplasm is divided equally between daughter cells; almost all cytoplasm is shed during spermiogenesis

ductive tract. The need for spermatozoa to be capacitated was a problem in the development of in vitro fertilization (see Essay, p. 672).

Oogenesis and Follicle Development

Oogenesis and spermatogenesis differ in the timing of the processes in the life cycle, in the fates of daughter cells, and in the differentiation of the cytoplasmic parts of the cells (Table 25-2). Spermatogenesis begins at puberty and continues until old age. In the ovaries all oogonia are converted to primary oocytes before birth. The first meiotic division of primary oocytes begins during embryonic development but is arrested at metaphase (Figure 25-9; see Chapter 3 for stages of cell division). About the time of birth each ovary contains about 2 million **primordial follicles,** each consisting of a primary oocyte surrounded by a single layer of supporting **granulosa cells.** Each primary oocyte contains 46 homologous pairs of bivalents.

Throughout life, follicles are lost by a degenerative process called **atresia.** About 400,000 follicles remain at the time of puberty. After puberty, cycles of **ovulation** begin. In each cycle, several follicles begin to mature, becoming **primary follicles,** but usually only one follicle completes maturation and releases its oocyte. By the time of **menopause,** the cessation of ovarian function that typically occurs in

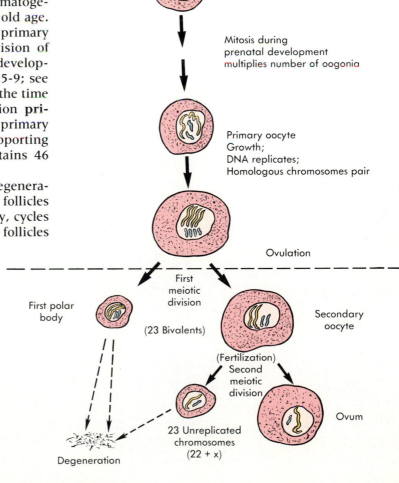

FIGURE 25-9
The events of oogenesis in the ovary. Because of the unequal divisions of cytoplasm in ovarian meiosis, each primary oocyte gives rise to a single ovum.

Oogonium
(46 chromosomes)

Mitosis during
prenatal development
multiplies number of oogonia

Primary oocyte
Growth;
DNA replicates;
Homologous chromosomes pair

Ovulation

First
meiotic
division

First polar
body

(23 Bivalents)

Secondary
oocyte

(Fertilization)
Second
meiotic
division

23 Unreplicated
chromosomes
(22 + x)

Ovum

Degeneration

the 5th decade of life, the stock of oocytes is essentially exhausted. In a woman's reproductive life, only a few hundred of the original number of follicles are actually ovulated; the rest degenerate.

The first meiotic division of the primary oocyte (see Figure 25-9) is resumed about the time of ovulation. The division of cytoplasm is not equal. One of the two daughter cells retains almost all of the cytoplasm, becoming a **secondary oocyte.** The other daughter cell, a tiny **first polar body,** has no future but may divide before degenerating. If the secondary oocyte is penetrated by a spermatozoan, it undergoes the second meiotic division (see Figure 25-9). As in the first meiotic division, the division of chromosomes is equal but that of the cytoplasm is unequal, resulting in a tiny **second polar body** and a relatively large **ovum.** After fertilization, the spermatozoan's nucleus fuses with that of the ovum, restoring the diploid (2N) chromosome number. Successive cycles of mitosis begin the development of an embryo. The embryo will undergo a substantial amount of development and cell division before it can begin to receive significant nutrition from the mother's body. The unequal division of cytoplasm in ovarian meiosis is therefore an adaptation that

provides the ovum with a larger store of nutrients and cellular organelles to distribute to the cells of the early embryo.

> 1 *What is meant by the diploid chromosomal complement? What are homologous chromosomes? What is a bivalent?*
> 2 *What are the differences between spermatogenesis and oogenesis?*
> 3 *Distinguish between spermatogenesis and spermiogenesis.*

Determination of Genetic Sex

In spermatogenesis, the reduction of the chromosomes to the haploid number during the first meiotic division results in spermatozoa that have either an X or a Y chromosome. The autosomal chromosomes and the X chromosome carry genes that specify the basic structural and functional aspects of the body that are common to both sexes—they do not normally carry genes that determine the course of sexual development. The Y chromosome is much smaller than the X chromosome (see Figure 25-6) and carries only a few genes, but among them are a

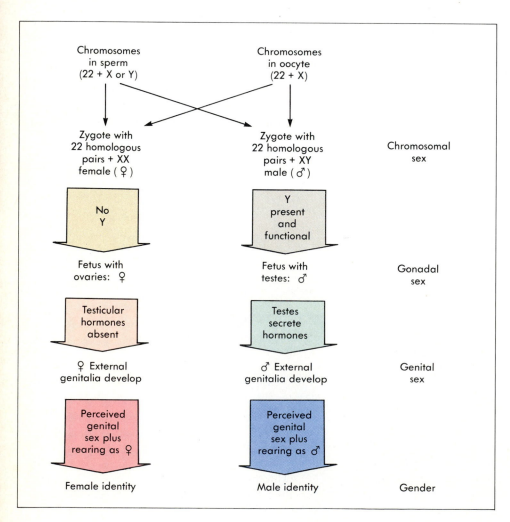

FIGURE 25-10
Inheritance of sex chromosomes begins a sequence in which prenatal development differentiates appropriate gonads and genitalia. Rearing helps the child acquire an appropriate gender identity and sexual orientation. Disorders of sexual development can arise at any step of this sequence.

gene or genes that cause the gonads of a fetus to become testes.

In oogenesis all of the gametes carry an X chromosome because the oogonia, like all the rest of the cells of a woman, do not have any Y chromosomes. Spermatozoa carry either an X or a Y chromosome. The genetic sex of the zygote is determined by the spermatozoan. If it carries a Y chromosome, the sex chromosome complement of the zygote will be XY (see Figure 25-6); if it carries an X, the zygote's complement will be XX. Possession of a Y chromosome confers **male genetic sex;** absence of a Y chromosome confers **female genetic sex.**

Embryonic Development of Reproductive Organs

The development of reproductive organs before birth can be divided into two steps (Figure 25-10). In the first step, the genetic sex of the fetus, determined by the sex chromosome transmitted by the spermatozoan at fertilization, causes the development of the corresponding gonads, establishing the **gonadal sex.** The second step is the formation of appropriate male or female accessory organs, including both internal and external genitalia. Development of male structures is the result of the presence of testes. The development of appropriate external genitals establishes the **genital sex** of the fetus, the basis for the assignment of a child as male or female at birth.

An embryo has **indifferent gonads** that can become either testes or ovaries and a **primitive urogenital tract** that can give rise to either male or female accessory organs. Indifferent gonads contain three types of cells: (1) cells that will give rise to gametes (oogonia or spermatogonia), (2) precursors of cells that will nourish the developing gametes (granulosa cells in the ovary; Sertoli cells in the testis), and (3) precursors of cells that will produce sex hormones (thecal cells in the ovary; Leydig cells in the testis).

The primitive urogenital tract consists of two duct systems: the **Wolffian ducts,** which can give rise to the male gland and duct system; and the **Müllerian ducts** (Figures 25-11), which can develop into female internal accessory structures. The embryo also has an external **genital tubercle, urogenital groove,** and **genital swellings** (Figure 25-12) that can develop into male or female external genitalia. Thus the gonads and accessory reproductive structures of the early embryo are **bipotential,** but unless a signal from the Y chromosome is received to cause the realization of the male potential, the female potential is automatically expressed in development.

Recent studies suggest that a key gene carried by the Y chromosome codes for a protein of the type called **finger proteins.** The fingerlike structure of these proteins can interact directly with the DNA of specific genes to control transcription. Presumably the finger protein affects the activity of a gene or genes whose products switch the indifferent gonads from their built-in female pattern of development to the male pattern.

In the absence of a Y chromosome, the indifferent gonads differentiate into ovaries at 9 to 10 weeks after conception (see Figure 25-11), establishing female gonadal sex. At about 18 to 20 weeks the Müllerian ducts develop into Fallopian tubes, uterus, and the innermost one-fourth of the vagina. The Wolffian ducts spontaneously degenerate at 10 to 11 weeks, so that male accessory sex organs do not appear. The external genital tubercle differentiates into the outer three-fourths of the female vagina, the labia majora and minora and clitoris, establishing **female genital sex.**

The presence of the Y chromosome with its control gene prevents the sequence of events just described. The control gene causes the indifferent gonads to differentiate into testes at 6 to 8 weeks, 2 to 3 weeks before the ovaries would otherwise develop. The presence of testes establishes **male gonadal sex.** Shortly after the testes differentiate, the Leydig cells begin to secrete testosterone. Early production of testosterone stimulates the Wolffian ducts to differentiate into the male gland and duct system: epididymis, vas deferens, and seminal vesicle (see Figures 25-3 and 25-11). The Sertoli cells of the testis differentiate between 9 and 10 weeks and secrete a protein called **Müllerian regression factor** (MRF) that suppresses development of the structures derived from the Müllerian ducts. Testosterone also inhibits differentiation of the external genitalia into female structures and stimulates the primitive urogenital tract to develop into the prostate gland and male external genitalia (see Figures 25-11 and 25-12), establishing **male genital sex.**

> 1 What is the developmental basis for the sometimes-heard statement that the basic human body plan is female?
> 2 Name the factors that determine a person's
>   a. Genetic sex
>   b. Gonadal sex
>   c. Genital sex

HORMONAL CONTROL OF SEXUAL MATURATION
Role of the Hypothalamus and Anterior Pituitary in Puberty

The gonads and accessory sexual structures grow slowly, do not produce gametes, and secrete very little steroid hormone during the 10 to 12 years between infancy and the onset of sexual maturation, or **puberty.** The first detectable event of puberty is an increase in the secretion of the gonadotropins **luteinizing hormone (LH)** and **follicle-stimulating hormone (FSH)** by the anterior pituitary.

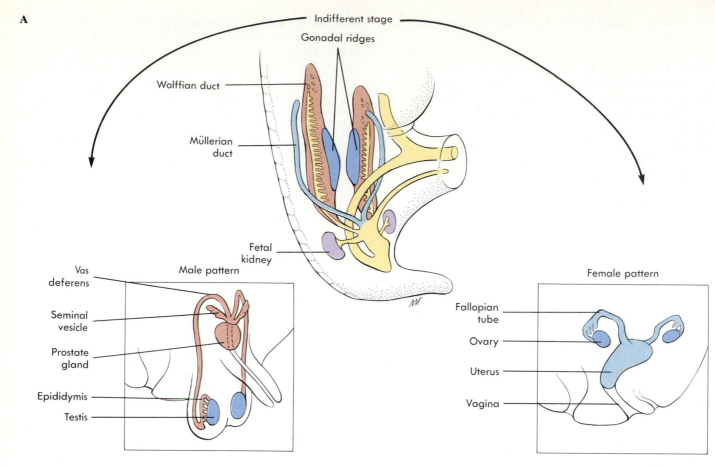

A

Indifferent stage

Gonadal ridges

Wolffian duct

Müllerian duct

Fetal kidney

Male pattern

Vas deferens

Seminal vesicle

Prostate gland

Epididymis

Testis

Female pattern

Fallopian tube

Ovary

Uterus

Vagina

B

Sequence of differentiation

Y chromosome — Indifferent gonads — No Y chromosome

Testes
6 to 8 weeks

Ovaries
9 to 10 weeks

Müllerian inhibitory factor — (−) Müllerian ducts

Degenerate
9 to 10 weeks

Testosterone — (+) Wolffian ducts

Male accessory structures
6 to 8 weeks

Degenerate
10 to 11 weeks

Female accessory structures
18 to 20 weeks

5 α reductase — Dihydro-testosterone — (+) Genital tubercle and genital swelling

Male external genitalia
10 to 39 weeks

Female external genitalia
10 to 39 weeks

FIGURE 25-11
A The early embryo *(center)* is sexually bipotential. The embryonic gonads can become either ovaries *(right)* or testes *(left)*, depending on the presence or absence of a Y chromosome. In the case of male gonadal sex, secretion of testosterone causes the Wolffian ducts *(brown)* to develop into the male internal accessory structures *(left)*; Müllerian inhibiting factor secreted by the testes causes the Müllerian ducts *(blue)* to deteriorate. Testosterone also causes the external genitalia to assume the male form. In the absence of testes, the Wolffian ducts degenerate, the Müllerian ducts develop into the female internal accessory structures, and the external genitalia assume the female form.
B The sequence and timing of the events of sexual differentiation.

FIGURE 25-12

The anatomical differentiation of external genitalia. The alternative fates of the bipotential embryonic structures are color-coded.

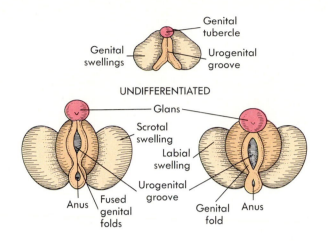

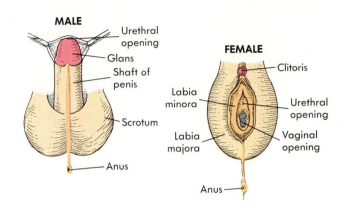

These hormones were named for their effects in women, but they also are involved in control of reproduction in men (Table 25-3).

As described in Chapter 5, secretion of anterior pituitary hormones is regulated by secretion of releasing hormones and release-inhibiting hormones by the hypothalamus. The secretions of LH and FSH are believed to be controlled by a single **gonadotropin-releasing hormone (GnRH)**. The effect of gonadotropin is to stimulate growth of the gonads to their adult size, to stimulate gametogenesis, and to initiate secretion of sex steroids. The gonadal steroid secretion initiates sex-specific changes in body growth and development of **secondary sexual characteristics,** the sex-specific body features that develop after puberty (see Table 25-1).

In women, release of gonadotropin occurs on a cyclical basis, driving the female reproductive, or **menstrual, cycle.** In men, the average release is relatively steady over periods of time as long as several days, but there are a number of pulses of gonadotropin release during each day.

The factors that affect the brain's timing of puberty are poorly understood. In girls the age at first menstruation (**menarche**) is an easily dated indicator of the onset of puberty. Some evidence suggests that a threshold body weight must be reached for puberty to occur, so that nutritional factors probably have some part in initiating puberty. The mean age of menarche has decreased continuously since the beginning of the last century in populations of European origin (Figure 25-13). The causes of this trend are not clear, but changes in nutrition and other cultural influences are likely to be involved.

Gonadal Steroids and Sexual Maturation

The major steroid secreted in males is testosterone, which masculinizes the secondary sexual characteristics at puberty and maintains the male accessory structures (Table 25-4). Testosterone causes enlargement of the larynx and growth of the penis and seminal vesicles. Testosterone is mainly responsible for initiating the growth spurt of male puberty, and it ultimately brings growth in stature to an end by stimulating closure of epiphyses at the ends of long bones. Testosterone stimulates growth of muscles and is probably responsible for the higher hemat-

TABLE 25-3	Effects of Gonadotropins	
	Males	*Females*
Follicle-stimulating hormone (FSH)	Stimulates spermatogenesis and spermiogenesis by stimulating Sertoli cells	Stimulates follicle maturation; is required for follicles to progress to the secondary stage and begin to secrete estradiol
Luteinizing hormone (LH)	Increases testosterone secretion by stimulating Leydig cells; LH is required for spermatogenesis because spermatogenesis requires testosterone	A surge of LH causes the mature Graafian follicle to ovulate and transform into a corpus luteum (luteinization)

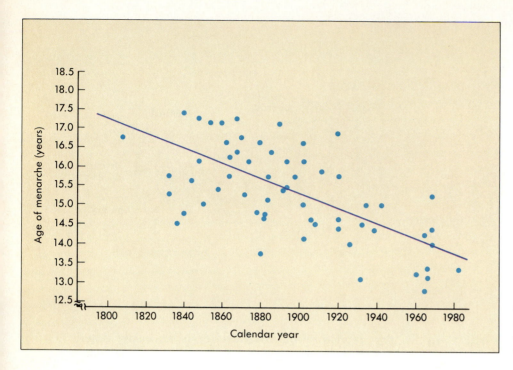

FIGURE 25-13

Historical decline in the mean age of menarche in a number of populations of European origin.

ocrit level of men. It is required for spermatogenesis and plays a major role in establishment and maintenance of sexual interest, or **libido.**

Some target tissues must convert testosterone to **estradiol, dihydrotestosterone,** or **5 α-andro-**

TABLE 25-4	Effects of Testosterone and Its Metabolites

Prenatal

1. Inhibits development of Müllerian structures (female internal accessory organs)
2. Masculinizes external genitalia
3. Establishes male pattern of steady release of gonadotropin after puberty

At Puberty

1. Stimulates protein synthesis, increasing growth and mature muscle mass
2. Increases basal metabolic rate
3. Stimulates long bone growth (and also closure of epiphyses)
4. Stimulates development of male accessory sexual organs and maintains them in functional state
5. Establishes male secondary sexual characteristics
6. Necessary for spermatogenesis
7. Stimulates formation of red blood cells
8. Stimulates libido (and may have other behavioral effects)

stenediol before it can bind with their receptors. These conversions are catalyzed by enzymes the target tissues possess: aromatase in the case of those that require estradiol; 5 α-reductase in the case of dihydrotestosterone; and both 3 α-reductase and 5 α-reductase in the case of 5 α-androstenediol (Figure 25-14).

Dihydrotestosterone stimulates the growth of the scrotum and prostate and maintains prostatic secretion. It is the form of testosterone necessary for masculinizing the external genitalia in fetal development (see Figure 25-10). An inborn deficiency of reductase causes a disorder of sexual development in which the Wolffian structures (which are activated by testosterone) develop normally but the external genitalia fail to masculinize completely (Table 25-5). Both dihydrotestosterone and 5 α-androstenediol stimulate the growth of facial, body, and pubic hair and are a contributing factor in male pattern baldness.

In females there are two major reproductive steroids, **estradiol** (an **estrogen**) and progesterone (a **progestin;** see Figure 25-14). Estradiol causes almost all female-specific changes that occur after puberty (Table 25-6). Both the mammary glands and the adipose tissue of the breasts are stimulated to increase in size by estradiol. It favors the formation of characteristic subcutaneous fat deposits of women, with the result that in women the total adipose mass is twice the male average, while the muscle and skeletal mass are only two thirds the male average. The broader pelvis of the female skeleton and a shorter period of growth of long bones are also effects of estradiol. After puberty, estradiol

FIGURE 25-14

Pathways of steroid synthesis and metabolism. In the testis, testosterone is synthesized from cholesterol by a pathway that includes progesterone. Multiple arrows indicate places where several intermediate steps in the pathway are not shown in the figure. In some of its target tissues, testosterone may be metabolized to estradiol, dihydrotestosterone, or 5 α-androstenediol. (The ovary makes estradiol by a similar route.) The individual carbons of the steroid backbone are numbered so that reactions involving them can be more easily identified—so, for example, the enzyme 5 α-reductase breaks a double bond joining carbons 4 and 5.

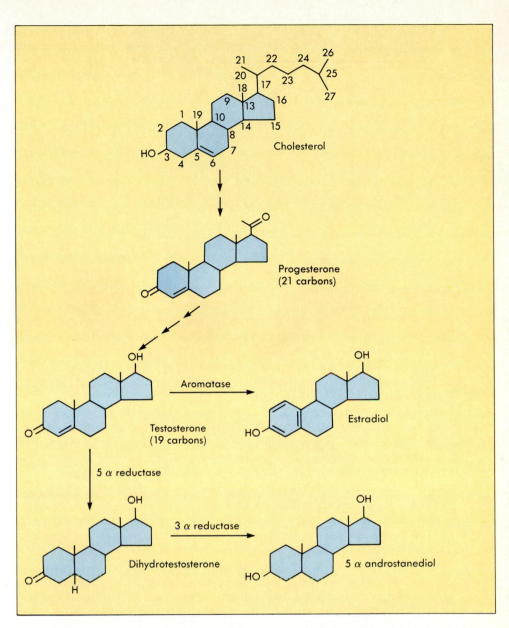

TABLE 25-5	Some Endocrine Disorders That Affect Genital Sex

Name	Defect	Effects
Adrenogenital syndrome	Adrenals produce excess of DHEA	Masculinized female external genitals
Adrogen-insensitivity (testicular feminization) syndrome	Defective androgen receptor	XY genotype has undescended testes, female body plan, and external genitalia; no internal accessory structures
5 α-reductase deficiency	Testosterone not converted to dihydrotestosterone	XY genotype has testes and male internal accessory organs but incompletely masculinized external genitalia
Hypopituitary hypogonadism	Secretion of FSH and/or LH inadequate	Weak puberty, reduced fertility (both sexes); decreased libido in males

maintains the normal function and structure of the accessory sex organs.

Shortly before sexual maturation, the adrenals of adolescents of both sexes undergo an increase in growth and activity. The adrenal cortex produces small amounts of androgen as a by-product of synthesis of cortisol and aldosterone. The major androgens produced by the adrenal are **dehydroepiandrosterone (DHEA)** and androstenedione. Although these are weaker androgens than testosterone, they are important in female maturation because in women the adrenal is a major source of the relatively small amount of androgen produced. Adrenal androgen is mainly responsible for the growth of pubic and underarm hair in women, and its metabolic effects may contribute to the growth spurt that occurs at the time of female puberty. The sexual interest of women also appears to depend on adequate levels of androgen, rather than the "female" sex hormones estradiol and progesterone. Of the androgen produced in women, about half is from the adrenals, the rest being produced by theca cells. At menopause, secretion of steroids by the ovaries drops, but androgen continues to be produced by the adrenals, maintaining sexual interest.

1 What causes the gonads to begin to function at puberty?
2 What are the male and female secondary sexual characteristics? What causes them to develop in males? In females?
3 What are the sources and roles of androgen in women?

Sex Hormones and Development of the Nervous System

There are anatomical and functional differences between the nervous systems of males and females. In animals, exposure of the brain and anterior pituitary to testosterone before birth appears to establish the male pattern of steady release of gonadotropin after puberty, and testosterone maintains the function of lumbar spinal neurons that are involved in reflexive control of erection and ejaculation. Interestingly, brain cells must convert testosterone to estradiol (see Figure 25-14) before the hormonal message can be transduced. There are slight differences in the structures of human male and female brains; the role of similar differences in male and female brains of other mammals is being investigated currently.

Differences in the behavior of male and female animals are characteristic in mammals. In experimental animals it is clear that prenatal exposure to testosterone is important for development of normal male sexual behavior after maturity in males, as well as for the establishment of other behaviors typical of males, and that prenatal testosterone can stimulate the expression of male behavior in genetic females. The evidence for any prenatal hormone effects on human behavior is much less conclusive, although the regions of the brain involved in such behavior are fundamentally the same in humans as in other mammals. In all mammals, receptors for steroid hormones are widespread throughout the brain.

Some children are born with external genitals that are intermediate in structure between the male and female forms, making it difficult to assign them to their correct gonadal sex. Studies of children who have been raised with a gender at odds with their gonadal sex show clearly that gender identification is much more dependent on early social conditioning than on gonadal sex or genetic sex.

Sexual orientation—one's preference for partners of the same or the opposite gender—does not seem to be reflected in the hormonal picture after sexual maturity. Whether there is a prenatal determination of sexual orientation is not known. Some people are **gender-dysphoric**—intensely uncomfortable with their genital sex and the gender that should have accompanied it. These individuals can sometimes be successfully treated by sex-change operations and subsequent treatment with the appropriate sex hormone. For male-to-female transsexuals it is relatively easy to construct a functional vagina. Penises constructed by plastic surgeons are much less satisfactory. Present evidence suggests that gender dysphoria should be considered a psychological rather than an endocrine problem.

Disorders of Sexual Differentiation

The homologous sex chromosomes may fail to separate properly in meiosis, resulting in spermatozoa or oocytes that carry either no sex chromosome or two sex chromosomes. If these abnormal gametes are involved in production of a zygote, chromosomal complements of XO (the O means that one of the gametes contributed no sex chromosome), XXY, XYY, or XXX may result (Table 25-7).

TABLE 25-7   Disorders of Genetic Sex

Name	Sex chromosomes	Body plan	Effects
Turner's syndrome	XO	Female	Infertility, no puberty
Klinefelter's syndrome	XXY	Male	Underdeveloped testes, infertility, weak puberty
Supermale	XYY	Male	Effects (if any) are not certain
Superfemale	XXX	Female	Reduced fertility

The first possibility (XO), called **Turner's syndrome,** results in female gonadal sex and the female body plan, as would be expected from the principle that the body develops as female in the absence of a Y chromosome. However, two X chromosomes are apparently needed for normal ovarian function. The ovaries of females with Turner's syndrome do not develop beyond a primitive stage, and do not assume either a reproductive or an endocrine function at puberty. These women are thus infertile and must be given estrogen to initiate a feminizing puberty and maintain feminine secondary sexual characteristics.

The second possibility (XXY) results in **Klinefelter's syndrome.** The gonads and body plan are male, but the function of the gonads is impaired. Males with this symptom undergo what is at best a weakly masculinizing puberty. They are typically tall, because the growth of long bones continues for longer than it would if testosterone were present. The third possibility, **supermale** (XYY), results in men who appear normal.

The fourth possibility (XXX) has been called the **superfemale** complement; the women who have this complement are essentially normal, but their fertility is reduced. All possible sex chromosomal abnormalities are accompanied to a greater or lesser extent by mental retardation or problems of personality development, which suggests that the products of genes on the sex chromosomes are important for other physiological functions as well as for sexual differentiation.

Excessive androgen production by the adrenals can masculinize the external genitalia of a fetus whose genetic and gonadal sex is female; this is the **adrenogenital syndrome** (see Table 25-5). Typically this occurs as a result of a genetic defect in the fetus or its mother that blocks a step in the pathways of steroid metabolism leading to cortisol and aldosterone. The lack of cortisol causes an increase in secretion of ACTH, which powerfully stimulates steroid metabolism in the adrenal. Excessive production of androgen results when steroid molecules that normally are precursors for cortisol and aldosterone accumulate upstream from the blocked steps

in the pathway. These precursors then enter the alternative pathway leading to DHEA and androstenedione, raising their blood levels high enough to masculinize the developing external genitals. The degree of masculinization is variable but may be so complete that an error of sex assignment is made at birth. This error may become apparent only when a feminizing puberty occurs.

In some genetic males, there is a defect in the gene for 5 α-reductase enzyme. Some parts of the primitive urogenital groove depend on dihydrotestosterone rather than testosterone, so the lack of this enzyme results in failure of the external genitalia to masculinize before birth. The urogenital groove may fail to close, and the penis remains small.

An extreme example of failure to masculinize is provided by the **testicular feminization syndrome,** in which the genetic and gonadal sex are male, but androgen receptors are defective. The Müllerian structures degenerate because MRF is secreted, but the Wolffian structures also fail to develop because their receptors cannot respond to the testicular androgen. Thus there is neither a female nor a male internal accessory system, and the testes remain in the abdomen. The insensitivity of the genital tubercle to androgen causes the external genitalia to take the female pattern even though testosterone is present. Consequently those who have this condition appear female at birth. They undergo a feminizing puberty under the influence of the estrogen produced by the testes in the course of their much larger production of androgen. The women who have this condition are genetic and gonadal males but their gender and genital sex are female. The condition typically comes to medical attention because menstruation does not begin at puberty, or because of a complaint of infertility.

---

1 What factors are believed to be involved in the development of gender identity?
2 What effects result from having too many or too few X chromosomes? Contrast these effects with those of having too many or too few Y chromosomes.

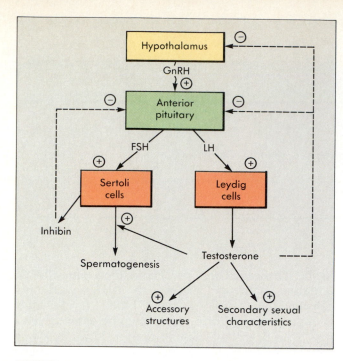

FIGURE 25-15
Negative feedback effects of testosterone and inhibin from the testes regulate the anterior pituitary's secretion of the gonadotropins LH and FSH. Testosterone is necessary for spermatogenesis; it also stimulates the growth and maintains the function of accessory structures and secondary sexual characteristics.

ENDOCRINE REGULATION OF MALE REPRODUCTIVE FUNCTION
Feedback Regulation of Spermatogenesis and Testosterone Secretion

The male hypothalamus secretes GnRH in bursts throughout the day, but the total amount of GnRH secreted remains essentially constant from one day to another. One effect of GnRH is to increase secre-tion of the gonadotropins FSH and LH by the pituitary.

Pituitary FSH controls spermatogenesis by stimulating Sertoli cells (Figure 25-15; see Table 25-3), and the Sertoli cells exert negative feedback control on the pituitary through their secretion of the hormone **inhibin**. Inhibin secretion completes a feedback loop that allows spermatogenesis to be regulated independently of the endocrine function of the testis.

Production and secretion of testosterone is carried out by the Leydig (interstitial) cells. Testosterone exerts feedback control on LH secretion at several levels (see Figure 25-15). Grossly elevated testosterone levels inhibit secretion of both LH and FSH by inhibiting secretion of GnRH. This is the reason that abuse of androgens in body-building can cause infertility. A more subtle effect occurs at the level of the anterior pituitary, where elevated testosterone levels specifically inhibit secretion of LH. This completes a feedback loop by which the endocrine function of the testis can be regulated somewhat independently of spermatogenesis.

Changes in Testicular Function with Age
Levels of testosterone rise after puberty and stabilize at their maximum value by about age 20. Thereafter testosterone levels remain relatively steady until about the 4th decade of life, when a trend toward decreasing levels becomes apparent (Figure 25-16). The degree of decrease with age is highly variable from one individual to another.

Coincidental with the decrease in testosterone secretion is an increase in gonadotropin levels (see Figure 25-16). This pattern shows that the decrease in testosterone secretion is not due to a change in gonadotropin secretion with age. The brains of

FIGURE 25-16
Rates of urinary excretion of androgens and gonadotropin over the male life span. These rates are rough indications of the plasma levels of the corresponding hormones. Most of the androgen in male plasma is testosterone. As plasma testosterone levels fall in later life, the decrease in negative feedback causes gonadotropin levels to rise. (From data of Pedersen-Bjergaard and Tonneson, *Acta Med. Scand.* 131, supplement 213, p. 284, 1948.)

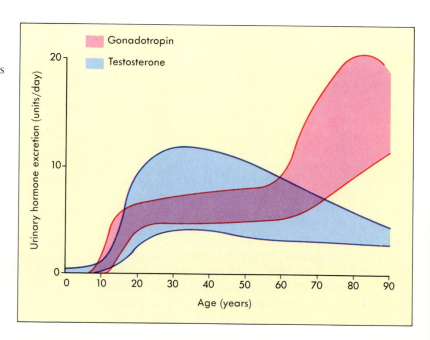

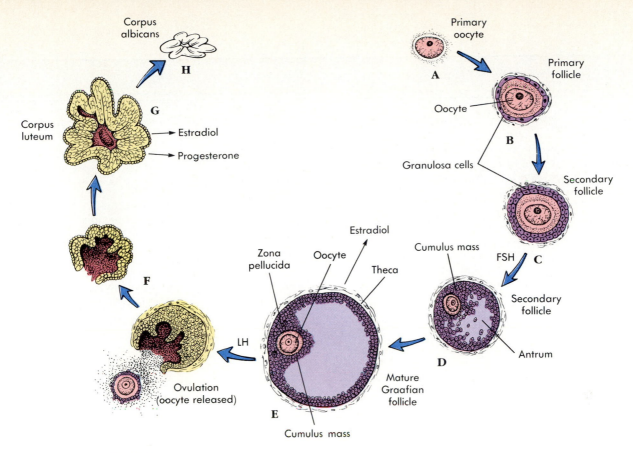

FIGURE 25-17
The life cycle of an ovarian follicle. A single primary oocyte (A) serves as the basis for a primary follicle (B). Several primary follicles advance to the secondary follicle stage (C) over several menstrual cycles. Usually only one follicle per cycle continues to mature (D) under the influence of FSH and reaches the Graafian stage (E). A surge of LH at midcycle causes the Graafian follicle to ovulate (F). After ovulation, the remnant of the follicle becomes a corpus luteum (G). If pregnancy does not occur, the corpus luteum degerates after about 12 to 14 days, leaving a scarlike corpus albicans (H).

older men are just as competent to release GnRH, and their anterior pituitaries just as able to release LH and FSH, as those of young men. The hypothalamus and anterior pituitary respond to the decrease in testosterone secretion as would be expected from the negative feedback loops shown in Figure 25-15. Thus it is the testes themselves that must be affected by age rather than their hormonal control system.

Target tissues up-regulate their testosterone receptors as testosterone levels fall, somewhat cushioning the physiological effects of the change in testosterone level. In spite of the decrease, testosterone levels are adequate to support spermatogenesis throughout the male lifespan after puberty, to maintain the function of accessory structures, and to maintain some level of sexual interest.

> 1 Describe the functions of the following hormones in men:
> a. FSH   c. Testosterone
> b. LH   d. Inhibin
> 2 Describe the two hormonal feedback loops that regulate gonadal function in men.

ENDOCRINE REGULATION OF FEMALE REPRODUCTIVE FUNCTION
Female Reproductive Cycle

The ovaries of sexually mature women undergo regular cycles (Figure 25-17) in which maturation and ovulation of a follicle are followed by a period during which hormones secreted by the remnant of the ovulated follicle make the uterus receptive to implantation of an embryo. The maturing follicle also functions as an endocrine organ, secreting estradiol. In ovulation, the mature follicle ruptures, releasing an ovum (actually a secondary oocyte) into the body cavity, from which it is swept by the fimbriae into the adjacent Fallopian tube, where it may be fertilized.

After ovulation the remnant of the follicle becomes a **corpus luteum,** which maintains the receptivity of the uterus to a pregnancy by secreting estradiol and progesterone. If pregnancy does not occur, the uterine receptive period is terminated approximately 2 weeks after ovulation by the degeneration of the corpus luteum. The resulting decrease in estradiol and progesterone causes the uterine endometrium to be shed. The ensuing flow of blood

TABLE 25-8   *Phases of the Menstrual Cycle*

Days	Ovary	Uterus	Hormonal picture
1 to 4	*Follicular phase begins:* corpus luteum degenerates; several follicles begin to mature	*Menstrual phase:* outer layer of endometrium sheds after estradiol and progesterone levels fall	Estradiol, progesterone, FSH, and LH levels all low
5 to 13	*Follicular phase continues:* FSH stimulates follicle maturation	*Proliferative phase:* endometrium regrows	Estradiol: rising Progesterone: low FSH and LH: low
14	*Ovulatory phase:* FSH surge causes ovulation	Cervical mucus becomes thin and watery	LH and FSH: sharp rise Estradiol: falls after follicle ovulates
15 to 28	*Luteal phase:* corpus luteum secretes estradiol and progesterone	*Secretory phase:* endometrium secretes uterine milk	Estradiol and progesterone: high FSH and LH: low

and loss of tissue is **menstruation.** Because the menstrual phase of the cycle is the easiest phase to detect, the human female reproductive cycle is called a **menstrual cycle.** In many mammals, the increase in estrogen levels near the time of ovulation is accompanied by a dramatic increase in sexual interest called **estrus;** this is the origin of the term estrogen.

In most women the cycle lasts about 28 days; the normal range of values is 21 to 35 days. The first menstrual day is numbered as day 1. Ovulation, which occurs about day 14 of a typical 28 day cycle, splits the ovarian cycle into two parts, the **follicular** and **luteal phases** (Table 25-8). The follicular phase lasts from the first day of menstruation to ovulation, and the luteal phase lasts from ovulation to the first menstrual day of the next cycle. The variability in the length of the cycle makes it difficult for human couples to determine the most fertile part of the cycle, but it is possible to predict the time of ovulation by recording body temperature and features of cervical mucus and the cervix itself. Body temperature, typically measured upon awakening, increases slightly as a reflection of the LH surge and remains elevated for the rest of the cycle.

The **cervical mucus,** a secretion that protects the mouth of the uterus, becomes thin and fluid in the first half of the cycle under the influence of es-

FIGURE 25-18
A photomicrograph of ovarian tissue showing a primary follicle *(right)* and several primordial follicles *(left).*

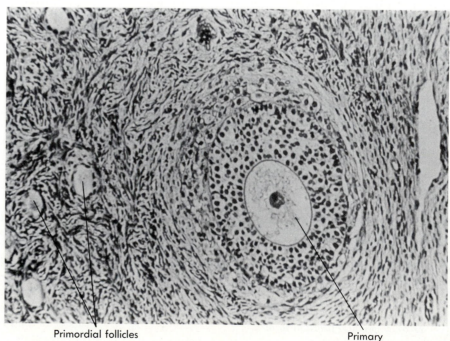

Primordial follicles

Primary follicle

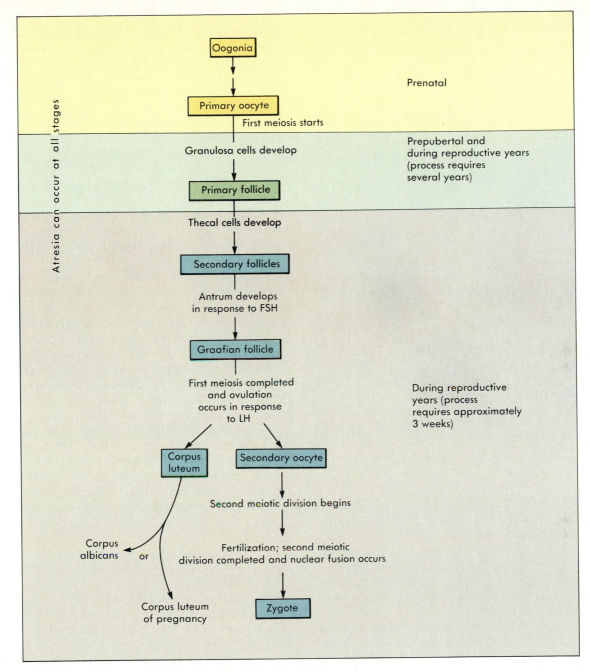

FIGURE 25-19
Flow-chart of oogenesis and follicle development.

trogen and is most easily penetrated by sperm at the time of ovulation. After ovulation, the mucus becomes thick and relatively impenetrable to sperm. The opening of the cervical canal changes visibly near the time of ovulation: it appears closed during most of the proliferative and secretory portions of the cycle (as observed using a speculum, see Figure 25-1, *B*), but the smooth muscle allows the canal to remain open around the time of ovulation. With experience, a woman can detect the gaping of the cervical canal by finger contact with the cervix.

Maturation and Ovulation of Ovarian Follicles

The initial stage of maturation of a follicle (Figures 25-18 and 25-19; see Figure 25-17) involves division of the granulosa cells that surround the ovum and the secretion by granulosa cells of a viscous substance that forms the **zona pellucida,** a layer between the ovum and granulosa cells. However, the ovum and granulosa cells remain in contact with one another via thin strands of cytoplasm. As the follicle continues to develop by granulosa cell proliferation, an outer layer of thecal cells is added by dif-

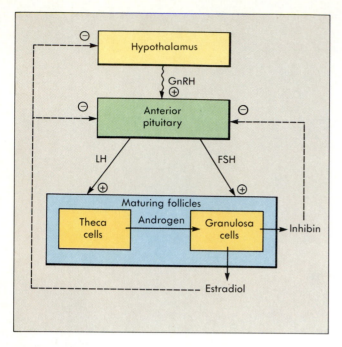

FIGURE 25-20
Negative feedback loops by which estradiol and inhibin secreted by the ovaries regulate gonadotropin secretion by the anterior pituitary (compare with the male pattern shown in Figure 25-15). Negative feedback is the dominant factor that regulates gonadotropin secretion during the luteal phase of the menstrual cycle.

ferentiation of surrounding interstitial cells. The result is that the ovum of a partially developed follicle is surrounded by concentric structures: (1) the zona pellucida, (2) a layer of granulosa cells, and (3) thecal cells (see Figure 25-17). A follicle that has reached this stage is called a **secondary follicle.** This initial stage of follicle maturation does not depend on FSH.

As a follicle continues its maturation into the secondary stage, it becomes an active endocrine tissue. Thecal cells supply the granulosa cells with androgen, which is converted to estradiol and secreted into the blood (Figure 25-20; see Figure 25-14). The follicle also secretes the hormone inhibin. Just as in the male reproductive system, inhibin provides information to the pituitary about the rate of gamete formation, allowing for regulation of FSH secretion independent of LH secretion.

The continued maturation of a secondary follicle requires the presence of FSH (see Figures 25-17 and 25-19). Some proliferation of granulosa and thecal cells occurs, but the major feature of this stage is development of a fluid-filled cavity called the **antrum.** The antrum is formed by secretion of fluid by the granulosa cells into the center of the follicle. A thin ring of granulosa cells surrounds the margin of the follicle just inside the thecal cells, while other gran-

ulosa cells form a mound (the **cumulus mass** or **cumulus oophorus**) that supports the ovum. Because continued follicle maturation can only take place when FSH levels are adequate, follicles can progress into the secondary stage only after gonadotropin secretion rises at puberty.

In the final stage of follicular maturation, the antrum becomes so large that the ovum and its associated granulosa cells effectively float within the antral fluid, remaining connected to the outer granulosa and thecal cells by a thin strand of granulosa cells. This stage of follicular development is called a **Graafian follicle** (Figure 25-21, *A* and *B;* see Figure 25-17). A Graafian follicle is about 500 microns in diameter and visibly bulges from the surface of the ovary (see Figure 25-21, *B*).

During each menstrual cycle several secondary follicles begin to form antra, but normally only a single follicle in one ovary will become a Graafian follicle and be ovulated. The other partially developed follicles undergo the process of atresia. Follicles appear to be able to inhibit each other's maturation. Maturation of a single Graafian follicle can be thought of as the outcome of a race in which the winner took the lead at some point and inhibited the further maturation of its competitors. However, in a few percent of the menstrual cycles the race is a tie, and more than one Graafian follicle is matured and ovulated. Multiple ovulation occurs routinely in many animals and is the most common cause of multiple births in women. The resulting children are called **fraternal twins.** They are not genetically identical to one another and may be of different sexes.

Feedback Control of the Endocrine Function of the Ovaries

The ovaries are the main source of estrogen in women who are not pregnant. In response to LH secreted by the pituitary (see Figures 25-19 and 25-20), thecal cells of maturing follicles synthesize androgen, which enters nearby granulosa cells where it is converted to estrogen by enzyme systems that depend on the presence of FSH. The estrogen (mainly estradiol) is secreted into the blood and the follicle.

Both FSH and LH are needed for ovarian cycles. In the absence of FSH, LH cannot elicit estrogen production because the granulosa cells cannot convert androgens to estrogen. As explained earlier, pituitary FSH causes the granulosa cells to proliferate as the follicle matures, so that FSH is an estrogen-stimulating tropic hormone as well as a regulator of granulosa cell androgen-to-estrogen conversion and secretion. Furthermore, thecal cells have a high affinity for LH only when FSH is present.

Generally, estradiol exerts negative feedback on both the hypothalamic release of GnRH and on the

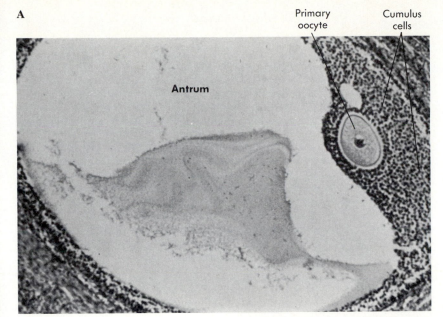

A

Antrum

Primary oocyte

Cumulus cells

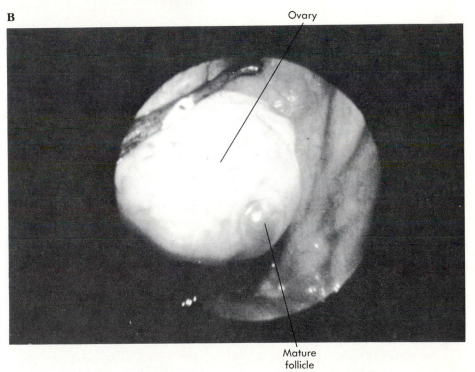

B

Ovary

Mature follicle

FIGURE 25-21

A   A photomicrograph of a section through a Graafian follicle, showing the fluid-filled antrum and the primary oocyte surrounded by cumulus mass cells.

B   A photograph of an ovary showing a Graafian follicle protruding from its surface. This photograph was made through a laparoscope, a slender tube carrying surgical instruments and a fiber-optics viewing system. The laparoscope is inserted through a small incision in the body wall for diagnosis and surgery.

pituitary release of FSH and LH (see Figure 25-20). In addition to the direct feedback to the pituitary and hypothalamus from estradiol, granulosa cells secrete inhibin, which specifically inhibits the secretion of FSH from the anterior pituitary. But LH secretion does not always behave as if it were regulated by negative feedback—the surge of LH that causes ovulation occurs even though estradiol levels are rising (Figure 25-22). The LH surge can be explained by the fact that the anterior pituitary responds to changes in plasma levels of estradiol.

When plasma levels of estradiol are steady, and especially when they are steady at a moderate level, the negative feedback control of gonadotropin secretion predominates. When estradiol levels rise rapidly to a high level, as they do during the follicular phase (see Figure 25-22), the negative feedback loop is ineffective—the pituitary is actually stimulated to release more gonadotropin by the rising estradiol levels.

Three factors are believed to be responsible for positive feedback on estradiol secretion during the

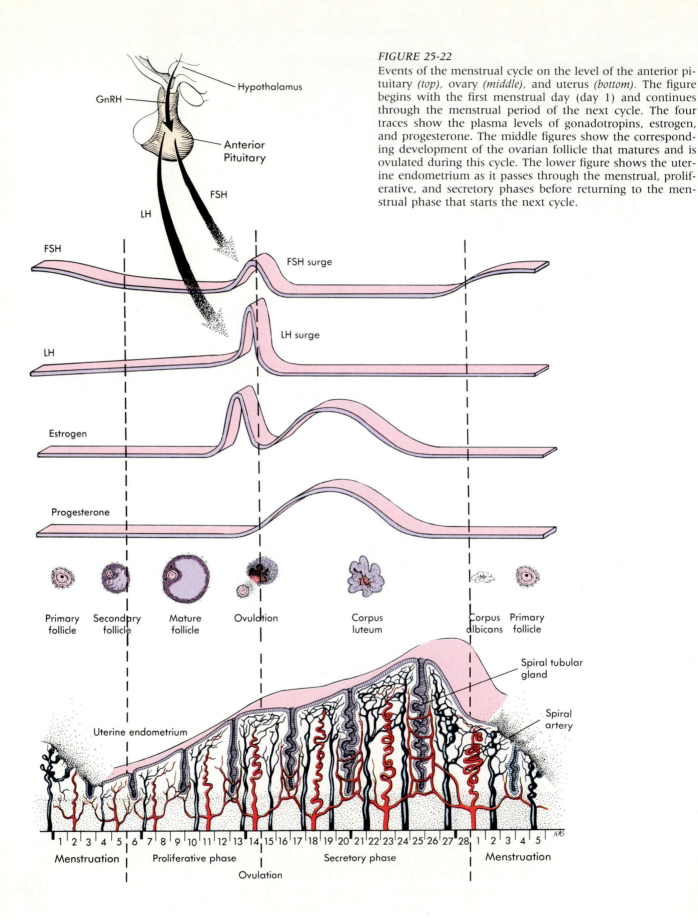

FIGURE 25-22

Events of the menstrual cycle on the level of the anterior pituitary *(top)*, ovary *(middle)*, and uterus *(bottom)*. The figure begins with the first menstrual day (day 1) and continues through the menstrual period of the next cycle. The four traces show the plasma levels of gonadotropins, estrogen, and progesterone. The middle figures show the corresponding development of the ovarian follicle that matures and is ovulated during this cycle. The lower figure shows the uterine endometrium as it passes through the menstrual, proliferative, and secretory phases before returning to the menstrual phase that starts the next cycle.

Hypothalamus

GnRH

Anterior Pituitary

FSH

LH

FSH

FSH surge

LH

LH surge

Estrogen

Progesterone

Primary follicle

Secondary follicle

Mature follicle

Ovulation

Corpus luteum

Corpus albicans

Primary follicle

Spiral tubular gland

Spiral artery

Uterine endometrium

1 2 3 4 5 6 7 8 9 10 11 12 13 14 15 16 17 18 19 20 21 22 23 24 25 26 27 28 1 2 3 4 5

Menstruation | Proliferative phase | Secretory phase | Menstruation

Ovulation

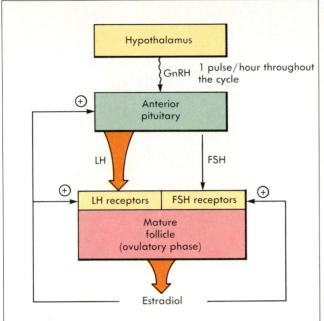

FIGURE 25-23
Pathways that provide positive feedback to LH secretion during the follicular stage of the menstrual cycle. The positive feedback culminates in a surge of LH that causes a Graafian follicle to ovulate (see Figure 25-22).

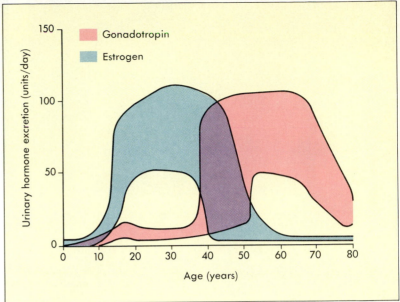

FIGURE 25-24
Normal ranges of rates of urinary excretion of gonadotropin and estrogens in a population of women of various ages. These rates are rough indicators of the plasma levels of the corresponding hormone. During the reproductive years, much of the variation in the range of normal values is due to changes over the menstrual cycle. (From data of Pedersen-Bjergaard and Tonneson, *Acta Endocrinol.*, volume 1, p. 38, 1948.)

follicular phase (Figure 25-23): (1) estradiol stimulates granulosa cell division and promotes further estradiol secretion; (2) estradiol increases the ability of granulosa cells to bind FSH; and (3) estradiol and FSH combine to cause the appearance of LH receptors on granulosa cells, and the binding of LH to granulosa cells further stimulates estradiol production. The result is that the plasma estradiol levels increase rapidly, while at the same time granulosa cells become exquisitely sensitive to pituitary LH release. Granulosa cell LH receptors are also responsible for the increase in antral size of a Graafian follicle and the release of hydrolytic enzymes at ovulation.

This positive feedback results in the increase in LH (**LH surge**) and the smaller increase in FSH that occur about the middle of the cycle just before ovulation (see Figure 25-22). The effect on LH secretion is greater than that on FSH secretion because the positive feedback effects of estrogen on the pituitary primarily influence LH secretion, and inhibin continuously inhibits FSH secretion. The function, if any, of the small increase in FSH secretion that occurs along with the LH surge is unknown.

Recent studies indicate that in primates the entire cycle of positive feedback that leads to ovulation can be carried out by the ovaries and pituitary. If the pathway between the hypothalamus and the pituitary is experimentally interrupted and pulses of

GnRH are given at the rate of about one every hour, ovulation occurs normally. The cycle cannot be carried out if GnRH is not present, or if the timing of the pulses is abnormal. These results suggest that the hypothalamus does not play a direct role in the timing of LH and FSH increases that trigger ovulation.

During the luteal part of the cycle, the presence of progesterone modifies the responsiveness of the anterior pituitary to high levels of estradiol, preventing the positive feedback that could cause another surge of LH. During the luteal phase, the negative feedback system dominates, and gonadotropin secretion is regulated at a stable, relatively low level (see Figure 25-22).

These cycles of estrogen and progesterone secretion by the ovaries continue until a gradual decrease in cycle regularity leads to menopause. By the time of menopause, the supply of follicles seems exhausted, and the ovaries become unable to secrete steroids in response to gonadotropin. This loss of ovarian function is complete, in contrast to the slow decrease in testicular function with age. Evidence for a change in the ovaries themselves, rather than their control system, is seen in a rise in gonadotropin levels after menopause (Figure 25-24). Thus, as with the testis, the effect of aging is on the ovary, ultimately making it incompetent to perform its

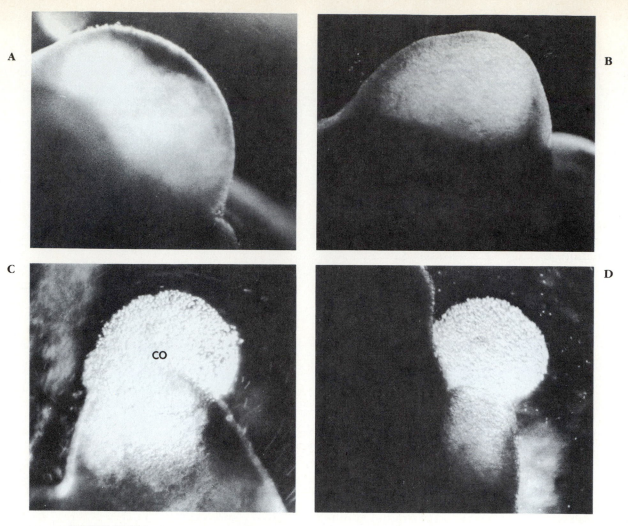

A

B

C

D

CO

FIGURE 25-25
This series of photographs shows the events of ovulation in a hampster.
A and B The follicle is swelling while its surface is being weakened by enzymatic action.
C and D The fluid and cells are emerging as the follicle begins to collapse.

functions while the negative feedback loops continue to function.

The decline of estrogen at menopause results in some reduction of the feminization that occurred at puberty. The sex-specific fat deposits and the mass of the accessory reproductive structures diminish, and decrease in bone calcification may increase the risk of osteoporosis (see Chapter 23).

1 The menstrual cycle of the uterus can be divided into three phases. What events and processes are characteristic of each phase? How do the uterine phases correspond to what is going on in the ovaries?

2 What are the roles of the granulosa cells and the thecal cells in estradiol production in the ovary?

3 Under what conditions does estradiol inhibit gonadotropin secretion? Under what conditions does it stimulate it? What is the function of positive feedback of estradiol secretion on gonadotropin secretion in the menstrual cycle?

Control of Ovulation and Luteinization by LH

The approach of the ovulatory phase of the menstrual cycle is accompanied by a rise in plasma levels of estradiol (see Figure 25-22). Ovulation of the mature Graafian follicle (Figure 25-25) is triggered by the LH surge that occurs in midcycle. Within 48 hours of the LH surge, a Graafian follicle undergoes four changes. First, the primary oocyte completes the first meiotic division and becomes a secondary oocyte, giving off a first polar body in the process (see Figure 25-9). The other changes that occur in response to the LH surge are that the granulosa cells begin to secrete progesterone in addition to estradiol, there is a sudden increase in antral fluid so that the follicle swells, and hydrolytic enzymes are secreted at the point where the Graafian follicle bulges from the ovary (see Figure 25-21, *B*). These enzymes break down the surface of the ovary, allowing the release of the oocyte (now conventionally called an ovum, although the second meiotic division has not yet occurred) into the peritoneal cavity near the opening of the Fallopian tube. Following ovulation, the ovum is surrounded by the zona pellucida and the granulosa cells of the cumulus oophorus.

The ovum is swept into the Fallopian tube by the cilia of the fimbriae (Figure 25-26). The ovum is carried toward the uterus by continued ciliary activity assisted by contractions of the smooth muscle cells of the Fallopian tube. An ovum can be fertilized for only about 10 to 12 hours after ovulation. Spermatozoa can survive in the female reproductive tract for perhaps 3 days, so intercourse (or artificial insemination) is likely to cause pregnancy only during a relatively brief window of time in the menstrual cycle. Fertilization normally occurs at the ovarian end of the Fallopian tube, so spermatozoa deposited into the vagina in sexual intercourse must travel through the uterus and much of the length of the Fallopian tube to meet the ovum.

Some spermatozoa may appear in the Fallopian tubes within minutes after being deposited in the vagina. The spermatozoa could not have made the trip so rapidly if they depended only on their own motility. This suggests that the Fallopian tubes perform a rather tricky task of assisting the movement of both spermatozoa and ovum toward one another. However, even with assistance, so many spermatozoa are lost along the way that only a relatively small fraction of the spermatozoa ejaculated succeed in getting to the vicinity of the oocyte. Consequently, men whose semen contains fewer than 20 million spermatozoa per milliliter are regarded as infertile.

Fertilization

To achieve fertilization (Figure 25-27), a spermatozoan must recognize its target and then penetrate the coat of granulosa cells, the zona pellucida, and the oocyte's cell membrane. This task seems all the more forbidding when the enormous difference in size between oocytes and spermatozoa is considered. The recognition that occurs when the sperm head touches the outlying granulosa cells (Figure 25-28) can be compared to receptor binding. The acrosome of the sperm head contains enzymes that when released allow the head to dissolve its way through the layers surrounding the oocyte. Binding

FIGURE 25-26
Electron micrograph of a section of Fallopian tube showing the coat of cilia on its surface. Many cilia are seen in cross section *(center)*; in a few cases the plane of the section passes for some distance along the length of the cilium *(upper right, bottom)*.

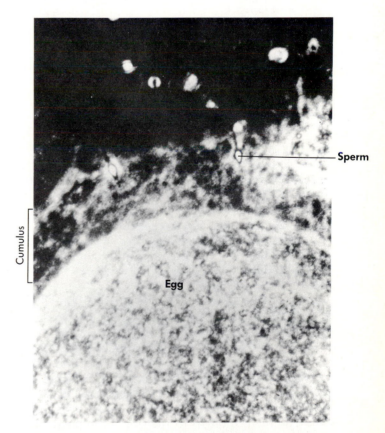

FIGURE 25-27
Two spermatozoa are penetrating the corona radiata of an ovum. As in Figure 25-8, *B,* the heads of the spermatozoa are labeled with fluorescent antibody to make them more visible under the microscope.

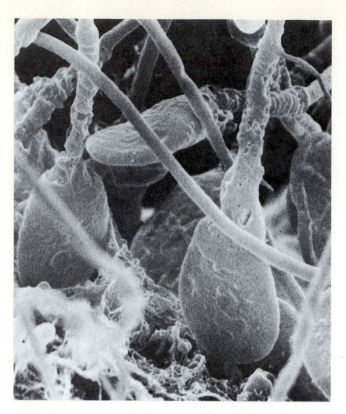

FIGURE 25-28
A scanning electron micrograph of spermatozoa attaching themselves to the surface of an ovum. This picture was made under laboratory conditions; the density of spermatozoa is much greater than would normally occur in the Fallopian tube.

of recognition receptors causes the acrosome to release its enzymes; this **acrosomal reaction** does not readily occur in spermatozoa that have not been capacitated. When the sperm head comes into contact with the oocyte (see Figure 25-27), the cell membranes of the two cells fuse and the sperm nucleus enters the oocyte cytoplasm.

Within seconds after the entry of the sperm nucleus, the oocyte's cell membrane undergoes a **cortical reaction.** Vesicles that lie just under the cell membrane fuse with it, forming a **fertilization membrane** (Figure 25-29). This reaction involves entry of Ca^{++}, and the process may be similar to that involved in release of synaptic vesicles.

Simultaneous fertilization of an oocyte by more than one spermatozoan results in a condition, called **polyspermy,** that prevents normal development of the zygote. The probability of polyspermy is usually small because relatively few spermatozoa successfully complete the trip to the upper Fallopian tube, and because the formation of the fertilization membrane makes the oocyte unreceptive to additional sperm heads.

After the head of the spermatozoan enters the cytoplasm of the ovum, the second meiotic division

occurs, and the second polar body is formed (see Figure 25-29). The fertilized ovum, or zygote, continues to move slowly down the Fallopian tube. Cycles of mitosis result in a solid ball of cells called a **morula.** The mass of the morula remains constant, so that with each cycle of cell division, each of the cells becomes smaller.

Formation of the Corpus Luteum

While the zygote is traversing the Fallopian tube, the granulosa cells remaining in the ruptured follicle proliferate and fill the vacant antral cavity, transforming it into a corpus luteum (see Figure 25-17). The formation of a corpus luteum (**luteinization)** requires LH. Although relatively large amounts of LH are needed to form the corpus luteum, lower plasma LH levels will maintain its function. The portion of the menstrual cycle in which the corpus luteum is functional is called the luteal phase of the cycle (see Figure 25-22).

The granulosa cells of the corpus luteum continue to secrete large amounts of estrogen and progesterone, so that after luteinization blood levels of estrogen remain high and blood levels of progesterone rise (see Figure 25-22). Luteal progesterone maintains the uterus in a condition capable of receiving and nurturing a fertilized ovum. If fertilization does not occur, the corpus luteum attains its maximum development in 10 to 14 days and then degenerates into a structure called a **corpus albicans** (see Figures 25-17 and 25-22). A prostaglandin secreted by the aging corpus luteum is responsible, so that the corpus luteum literally self-destructs.

Endometrium of the Uterus and Its Role in the Menstrual Cycle

At the start of the menstrual cycle the endometrium (the lining of the uterus) is about 1 mm thick. As estrogen levels rise during the follicular phase of the ovaries (see Figure 25-23), the cells of the outer layers of the endometrium multiply, increasing endometrium thickness to 2 to 3 mm. More blood vessels (**spiral arteries)** are added, and exocrine glands (**spiral glands)** develop (see Figure 25-22 and Table 25-8). This is the **proliferative phase** of endometrial development (see Table 25-8).

The uterine endometrium is the principal site of progesterone action in the luteal phase of the menstrual cycle, but the endometrium can respond to progesterone only if it has been primed by estrogen during the follicular phase. After ovulation, the estrogen and progesterone secreted by the corpus luteum convert the endometrium into a structure specialized to receive an embryo. The combined effect of estrogen and progesterone further thickens the lining of the endometrium but, more importantly, causes the endometrium to become secretory. The inner surface of the endometrium is thrust into folds, while its spiral glands become convoluted

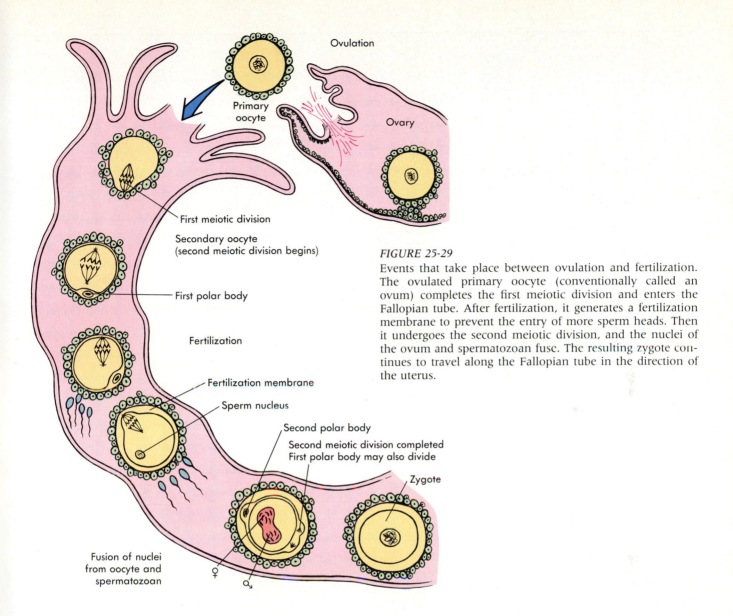

Ovulation

Primary oocyte

Ovary

First meiotic division

Secondary oocyte
(second meiotic division begins)

First polar body

Fertilization

Fertilization membrane

Sperm nucleus

Second polar body
Second meiotic division completed
First polar body may also divide

Zygote

Fusion of nuclei
from oocyte and
spermatozoan

♀

♂

FIGURE 25-29
Events that take place between ovulation and fertilization. The ovulated primary oocyte (conventionally called an ovum) completes the first meiotic division and enters the Fallopian tube. After fertilization, it generates a fertilization membrane to prevent the entry of more sperm heads. Then it undergoes the second meiotic division, and the nuclei of the ovum and spermatozoan fuse. The resulting zygote continues to travel along the Fallopian tube in the direction of the uterus.

and secrete a carbohydrate-rich fluid ("**uterine milk**") that can nourish the developing embryo until it is able to implant in the uterine wall. This is the **secretory phase** of the uterine cycle (see Table 25-8); it corresponds to the luteal phase of the ovarian cycle (see Figure 25-22). The thickness, metabolic rate, and glycogen content of the endometrium are greatest when estrogen and progesterone levels peak at days 22 to 23 of the menstrual cycle, about 7 to 8 days after ovulation.

The uterine cycle also affects the composition of the cervical mucus. During the proliferative phase the cervical mucus changes from a thick, inelastic secretion to a thin, watery material. After luteinization, the mucus becomes as thick as it was at the beginning of the cycle. Women who prefer not to use other methods of birth control can learn to examine their own cervical mucus to avoid sexual intercourse during the part of the reproductive cycle

in which a pregnancy is most likely to occur.

The outer layers of the endometrium degenerate when the estrogen and progesterone levels fall at the end of the menstrual cycle (see Figure 25-22). Much of this degeneration is caused by the fact that estrogen and progesterone inhibit the synthesis of uterine prostaglandins that act as local vasoconstrictors. When hormone levels decrease, prostaglandin synthesis increases and causes vasoconstriction of the spiral arteries. This impairs blood flow, leading to local tissue breakdown. Lysosomal enzymes also disrupt the extracellular matrix that normally holds the endometrial cells together. The result is that the outer portion of the endometrial lining is sloughed off into the uterine cavity. The menstrual flow usually lasts from 4 to 5 days and involves a blood loss of about 50 ml. The period of menstruation corresponds to the first few days of the follicular phase of the ovary (see Figure 25-22 and Table 25-8).

In Vitro Fertilization—A Last Resort for Infertile Couples

Many cases of infertility involve physical damage to some part of the reproductive system. Nearly one third of female infertility cases involve the Fallopian tubes. Scarring or blockage of the Fallopian tubes can occur as a result of pelvic inflammatory disease or sexually transmitted disease, cancer, or endometriosis (an abnormal growth of uterine endometrium in sites outside the uterus). Other causes of female infertility include disorders of the uterus or cervix that may block sperm transport through these sites or prevent the entry of a zygote into the uterus. In the male system, blocked seminal ducts sometimes occur. Varicose veins of the scrotum can reduce fertility by increasing the temperature of the testes. Structural defects in both the male and female systems can frequently be treated by surgery.   In some women, the mucous secretions of the cervix never become penetrable to spermatozoa. In others, the vaginal secretions are too acidic or contain antibodies to spermatozoa. In these cases, artificial insemination, which deposits a fresh sample of the husband's semen directly into the uterus, can bypass the problem.

If other measures fail, a last resort for infertile couples is in vitro fertilization (Figure 25-A). In this procedure, gonadotropin treatments are used to induce the maturation of several follicles in a single cycle. The maturation of the follicles is followed with ultrasonography. When several follicles are judged to be close to maturity, a small incision is made in the woman's abdomen for the insertion of a laparoscope, a fiber-optic device that allows the physician to see the ovaries, and also to puncture mature follicles and to suck up the oocytes that are released. Several oocytes are collected in this way and fertilized with spermatozoa donated by the husband. The spermatozoa must be capacitated by washing them in a culture medium that simulates a trip through the uterus, cervix, and Fallopian tubes. Development of this step was critical for the success of the method, because spermatozoa that do not go through the process do not successfully fertilize ova. When the ova have demonstrated that they are successfully fertilized by beginning cell division, they are introduced into the woman's uterus. One or more of them may succeed in implanting, resulting in a pregnancy that proceeds from that point just like a natural pregnancy. The children conceived in this procedure have received the unfortunate name of "test-tube babies."

The method still has some imperfections. The procedure has about one chance in three of resulting in a pregnancy, so in many cases it must be repeated, but when it does succeed it frequently results in multiple births. Some observations suggest that the hormone treatment used to induce multiple ovulation also makes the uterine environment less favorable for successful implantation. This problem could be overcome by freezing some embryos and introducing them during later

---

1 *What are the events of ovulation? What anatomical path is followed by the ovum after ovulation? Where does fertilization occur? What are the events of fertilization?*
2 *What is the function of the corpus luteum? From what structure is it derived?*
3 *What changes are seen in the endometrium over the menstrual cycle? What are the effects of estradiol and progesterone on the endometrium? What processes are responsible for the onset of menstruation?*

Hormonal Birth Control and Fertility Enhancement

The first hormonal birth control agents were developed on the basis of observation that steady levels of estrogen in the presence of progesterone inhibited gonadotropin secretion and thus interfered with follicle maturation and ovulation. In most cases the regimen followed in hormonal birth control simulates the normal menstrual cycle; estrogen alone or in combination with progesterone is taken for 20 days, followed by a 5 day period in which both steroids are withdrawn and menstruation occurs. If a periodic menstruation is not allowed, the endometrium becomes overdeveloped and small episodes of bleeding occur.

No drug that interferes with a process as complex as reproduction can be expected to be entirely free of side-effects. As the potential hazards of large doses of estrogen and progesterone were recognized, formulations with lower doses were developed. In many cases the low-dose formulations do not reliably prevent ovulation but are still effective contraceptives (Table 25-9). This is because other processes that contribute to fertility, such as the

unstimulated cycles. The success of the method could also be improved if there were some way to determine which of the embryos was most likely to survive in the uterus and discarding the rest. Ethical problems arise from discarding or conducting experiments on embryos that, in the views of many, are individuals with the same rights to live and be cared for as full-term infants.

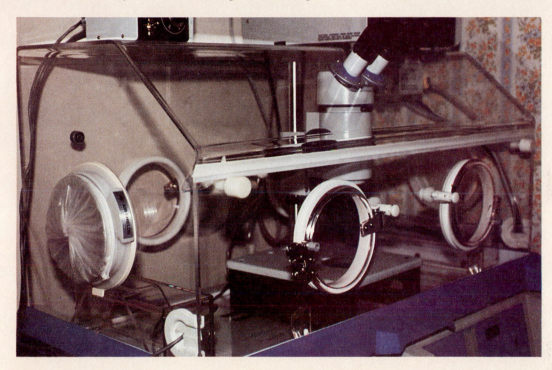

FIGURE 25-A
Originally a research tool, the apparatus for in vitro fertilization is now manufactured commercially.

transport of gametes in the Fallopian tubes, the hospitability of the uterine environment, the penetrability of cervical mucus to spermatozoa, and the pH of vaginal fluid, are affected by the presence of the exogenous steroids.

With better understanding of the hormonal control of reproduction, it may be possible to replace steroid hormones with nonsteroid hormones. For example, inhibin has considerable promise as a contraceptive for both men and women because in principle it should suppress formation of gametes without affecting the endocrine function of the gonads. Prostaglandin-based contraceptives that interfere with implantation have been developed, but they pose ethical problems for some people because their action could be regarded as termination of pregnancy.

Infertility is a medical problem for one of every seven married couples in the United States. Female infertility may be overcome in some cases by treatment with gonadotropins to stimulate follicle development and ovulation. These drugs are sometimes able to deliver too much of a good thing, because they may result in multiple ovulation and the conception of as many as six or seven fetuses simultaneously. Multiple births make exciting news stories but represent injudicious medicine. Multiple birth is hazardous to both fetuses and the mother, and such pregnancies require a disproportionate amount of medical attention and expense. Using ultrasonography and other methods, physicians can estimate the number of follicles that will ovulate in a given cycle in a patient being treated with gonadotropin; conception can be avoided if it appears that multiple ovulations will occur.

Method	Theoretical effectiveness*	Use effectiveness
Tubal ligation	0.004	0.006
Vasectomy	0.15	0.15
Estrogen/progesterone pill	0.34	4 to 10
Progestin-only pill	1 to 15	5 to 10
Condom with spermicide	Less than 1	5
IUD	1 to 3	0.5 to 4
Condom	3	10
Diaphragm with spermicide	3	2 to 20
Cervical cap	3	2 to 20
Spermicide alone	3	2 to 30
Vaginal sponge	9 to 11	13 to 16
Coitus interruptus	0 to 9	20 to 25
Rhythm	2.5 to 13	21 to 40
Unprotected intercourse	90	90
Abstinence	0	?

Data from Vessey et al.: Efficacy of different contraceptive methods, *Lancet*, April 10, 1982; and Sloane, *Biology Of Women* (2nd edition), John Wiley & Sons, Inc., New York, 1985.

*The theoretical effectiveness is an estimate of the failure rate based on the intrinsic characteristics of the method; the use rate reflects additional failures resulting from variations in anatomy, physiology, behavior, and motivation in the user population. Use failure rates may vary greatly in different populations.

Nonhormonal Methods of Contraception

A number of barrier contraceptive methods, the **condom,** the **diaphragm,** the **cervical cap,** and the **vaginal sponge,** act by blocking access of spermatozoa to the cervix. The physical barriers may be combined with a chemical barrier of **spermicide.** A relatively permanent block to gamete transport is attained by **tubal ligation,** a surgical procedure in which the Fallopian tubes are ligated and severed, closing the connection between the abdominal cavity and the uterus. The analogous operation in men is called a **vasectomy;** the two vasa deferentia are exposed with small incisions, ligated, and cut.

The **intrauterine device,** or **IUD,** is not a barrier method of contraception. The mechanisms of action of IUDs are not well understood. The presence of the device in the uterus apparently alters the uterine environment sufficiently to interfere with the process by which an embryo implants in the uterine wall. Almost all of the IUDs marketed so far have generated unacceptable side-effects in a significant fraction of their users.

The practical effectiveness of different means of contraception (see Table 25-9) depends on intrinsic physiological and anatomical factors, which may vary from individual to individual. The effectiveness of methods that require attention and action on the part of the user, such as abstinence, the rhythm method, and the barrier methods that are applied before sexual intercourse, obviously depends on conscientious application. The low but finite failure rate of the surgical methods is due to the occasional failure of the vas deferens and the Fallopian tubes to close fully after they are ligated.

> 1 Describe the currently available methods of hormonal birth control. What alternatives to hormonal contraception are available?
> 2 Explain the physiological mechanism for the action of estrogen/progesterone birth control pills.

SUMMARY

1. The primary reproductive organs (ovaries and testes) arise from the bipotential indifferent gonads in fetal development. Ovaries develop automatically unless a protein coded for on the Y chromosome is present.
2. The female internal accessory organs (Fallopian tubes, uterus, and inner vagina) arise from the embryonic Müllerian duct system automatically unless MRF, secreted by the testes, is present. The male internal gland and duct system (rete testis, efferent ducts, epididymis, vas deferens, seminal vesicle, prostate gland, and ejaculatory duct) arise from the embryonic Wolffian duct system in response to testosterone secreted by the fetal testis.
3. The female external genitalia (labia majora, labia minora, clitoris, and vagina) develop from the embryonic urogenital groove and genital tubercle automatically unless dihydrotestosterone, a derivative of testicular testosterone, is present. In the presence of dihydrotestosterone, the urogenital groove closes completely, forming the penile urethra, and the genital swellings fuse to form the scrotum into which the testes descend. The genital tubercle becomes the penis.
4. The stem cells for gamete formation (spermatogonia in the testis and oogonia in the ovary) undergo mitosis, producing cells that become committed to gamete formation (primary oocytes or primary spermatocytes). The latter undergo meiosis which reduces the chromosomal complement to the haploid number (23 chromosomes). Each primary spermatocyte gives rise to four spermatids, which subsequently differentiate into spermatozoa. In oogenesis, the end products are one ovum and polar bodies.
5. Puberty is initiated by an increase in the pituitary's secretion of gonadotropin, which causes the gonads to secrete sex steroids and produce gametes. The testes secrete the steroid testosterone; the ovaries secrete estrogen (mainly estradiol) and progesterone. Testosterone secretion is relatively steady; secretions of estrogen and progesterone follow an ovarian cycle of maturation of follicles followed by ovulation and conversion of the follicle to a corpus luteum.
6. At puberty, the sex hormones stimulate sex-specific developmental changes, including effects on fat deposition, muscle, and bone growth; growth of the accessory reproductive structures to their adult size; and effects on behavior and the intensity of sexual interest.
7. Sertoli cells of the testis support spermatogenesis and spermiogenesis; Leydig cells secrete testosterone. Spermatogenesis is controlled by a feedback loop in which inhibin secreted by Sertoli cells inhibits FSH secretion. Testosterone secretion is controlled by feedback loops in which high testosterone levels inhibit secretion of GnRH by the hypothalamus and the release of LH by the pituitary.
8. The ovarian cycle consists of follicular and luteal phases. Each maturing follicle includes a primary oocyte, a layer of granulosa cells, and a theca. Thecal cells synthesize androgen, which is converted to estradiol by granulosa cells. Ultimately one follicle becomes a Graafian follicle. The secretion of estrogen rises rapidly as the Graafian follicle grows, triggering a surge of LH secretion which results in ovulation. After ovulation, FSH transforms the remnant of the follicle into a corpus luteum which secretes both estradiol and progesterone during the subsequent luteal phase of the cycle, which is ended by breakdown of the corpus luteum.
9. The uterine cycle begins with menstruation, loss of the lining of the uterine endometrium. During the follicular phase of the ovary, estradiol stimulates regrowth of the endometrium (the proliferative phase). After ovulation, estradiol and progesterone secreted by the corpus luteum cause the endometrium to secrete uterine milk (the secretory phase), creating an environment favorable for implantation of an embryo.

1. List the primary reproductive organs and the accessory reproductive structures for males and females.

2. Describe the factors that determine whether internal sexual organs will differentiate into male or female structures. What structures arise from the Müllerian duct? The Wolffian duct?

3. What factors normally determine whether the external genitalia develop as male or female? What are some ways in which disorders of development can result in a person whose genital sex is at odds with his/her gonadal and genetic sex?

4. Describe the function of each of the following:
   Graafian follicle   Granulosa cell
   Thecal cell   Corpus luteum

5. Trace the changes in LH, FSH, estradiol, and progesterone over the menstrual cycle. What is the physiological role of each of the four hormones?

6. What is the major physiological role of each of the following structures?
   Sertoli cells   Vas deferens
   Leydig cells   Seminal vesicle

7. What is the function of the midpiece of a spermatozoan? The acrosome? What is the importance of capacitation for spermatozoa? What is the function of the cortical reaction of oocytes?

8. Why is it that a Graafian follicle can trigger ovulation by secreting estrogen, while birth control pills containing estrogen can prevent ovulation?

9. What are the sites of secretion and physiological effects of each of the following hormones? For the ones that are abbreviated, supply full names.
   GnRH   Estradiol
   FSH   Progesterone
   LH   Inhibin
   Testosterone

10. What consequences would you expect for male children who have the genetic defect that causes adrenogenital syndrome in female children?

● SUGGESTED READING

BROMWICH, P., and T. PARSONS: *Contraception—The Facts,* Oxford University Press, Oxford, England 1984. Straightforward presentation of contraceptive methods.

EDWARDS, D.D.: Keeping Sex Under Control, *Science News,* volume 133, January 9, 1988, p. 88. Reports on discovery of FSH antagonists in women treated with a GnRH analog. These substances have been given the name sex "antihormones." A more detailed paper by Dahl et al. appeared in *Science,* volume 239, January 1, 1988, p. 72.

FRISCH, R.E.: Population, Food Intake and Fertility: Historical Evidence for a Direct Effect of Food Intake on Reproductive Ability, *Science,* volume 199, 1978, pp. 22-30. Presents evidence for nutritional effects on timing of puberty and fertility in women.

MONEY, J.: *Venuses Penuses,* Prometheus Books, Buffalo, N.Y., 1986. A collection of the writings of one of the foremost authorities on the physiology and psychology of sexual differentiation.

REINISCH, J.M., L.A., ROSENBLUM, and S.A. SANDERS: *Masculinity/Femininity, Basic Perspectives,* Oxford University Press, New York, 1987. Presents discussion of possible innate differences between men and women.

WASSARMAN, P.M.: Fertilization in Mammals, *Scientific American,* December 1988, pp. 78-84. Presents recent studies of the mechanism of recognition of oocytes by spermatozoa.

CHAPTER 26

Sexual Response, Pregnancy, Birth, and Lactation

On completing this chapter you will be able to:

- Describe the overall pattern and important events of the male and female sexual responses.
- Describe the early development of the human zygote, its implantation in the uterus, and differentiation of the embryo and chorion.
- Understand the endocrine functions of the corpus luteum and placenta.
- Understand the architecture of the fetal circulation, how it functions in oxygen transport between maternal blood and fetal tissues, and the changes in the circulation that must occur at birth.
- Know the timing of key events of embryonic and fetal development and describe the changes in maternal physiology caused by pregnancy.
- Describe the stages of labor and the endocrine changes that are hypothesized to initiate labor and increase the strength of uterine contractions as labor progresses.
- Describe the control of breast development and milk production by prolactin, placental lactogen, progesterone, and estrogen.
- Trace the hormonal reflex that results in oxytocin secretion and milk letdown when an infant suckles.

*T*he preceding chapter described the structure, development, and control of the reproductive system—the equipment of reproduction. It remains to be shown in this chapter how the reproductive systems of men and women bring together gametes, how the mother's system supports the developing child until it can live outside her body, and how it continues the child's nourishment after birth.

Prenatal development is dependent on the nutritive and endocrine functions of the placenta, an organ that links the developing infant to its mother's circulatory system and maintains a favorable environment for the fetus. After about 39 weeks of development, strong contractions of the muscular wall of the uterus expel the infant into life as a physically separate individual. The events that trigger the onset of labor are not presently understood.

To survive, the newborn infant must rapidly make dramatic adjustments in its circulatory system, respiratory system, and metabolism. After birth the child and mother continue their interdependent relationship. Through breastfeeding the mother can supply both nutrients and immunity. The child's suckling in turn has a potent effect on the mother's endocrine system, providing for continued milk production and suppressing ovulation.

TABLE 26-1   Phases of the Sexual Response

	Characteristic changes		
	---	---	---
Phase	**Male**	**Female**	**Both sexes**
1. Excitement	Erection	Vagina lubricates; clitoris swells	Blood pressure and skeletal muscle tone increase; blood flow to pelvic area increases; hyperventilation
2. Plateau	Scrotum contracts, elevating testes	Orgasmic platform forms; uterus is elevated, causing vaginal tenting	Sex flush appears; hyperventilation; heart rate 100 to 160 beats/minute
3. Orgasmic	Ejaculation	Orgasmic platform contracts rhythmically	Heart rate 110 to 180 beats/minute; blood pressure and respiration maximum
4. Resolution	Erection subsides; refractory period	Uterus returns to normal position	Respiration and blood pressure return to normal; pelvic vasodilation reversed; skeletal muscle tone decreases

SEXUAL RESPONSE
The Sexual Response—A Whole-Body Response

Beginning in the late 1950s William Masters and Virginia Johnson began to measure physiological changes in the bodies of men and women during the sexual response. Until then, physiologists and physicians had virtually no information about the physiological changes accompanying sexual function. Masters and Johnson demonstrated that the emotional and genital components of the response are accompanied by changes in blood pressure, respiration, and regional blood flow. These changes involve both branches of the autonomic nervous system.

Considering both the genital and the extragenital changes, the sexual response pattern of both men and women can be divided into four phases (Table 26-1): an initial **excitement phase** during which the genital and extragenital changes occur, a **plateau phase** during which the changes are sus-

tained, an **orgasmic phase** during which both the emotional and physiological responses are most intense, and a **resolution phase** in which excitement subsides and the physiological parameters return to their baseline values. In men there is typically a refractory period of varying length between orgasms during which at least some resolution occurs. In women the intensity and number of orgasmic peaks are variable, and women may return repeatedly to the orgasmic phase without having to pass through the resolution phase (Figure 26-1).

In both sexes there are elevations of blood pressure and skeletal muscle tone during the excitement phase, and even greater increases during the orgasmic phase. As the orgasmic phase approaches, both men and women may display a rashlike vasodilation, the **sex flush,** over the face and chest. Specific vasodilation in the genital area is driven by parasympathetic fibers that run to the blood vessels of the gentilia. These fibers constitute an exception to

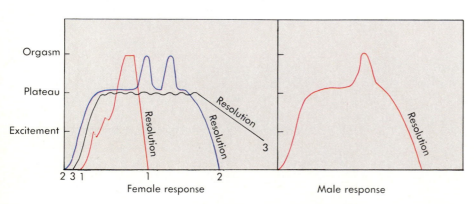

Female response

Male response

FIGURE 26-1

Comparison of the phases of sexual response in males and females. The female response may follow one of three basic patterns, including a shorter or longer plateau phase and from none to several orgasmic phases. The male pattern is typically more uniform.

the rule that blood vessels receive only sympathetic input. As described in the preceding chapter, the male and female external genitalia arise from the same embryonic structures, and in both cases vasodilation is important for the genital component of sexual response.

One might expect that, since the structure and functional roles of male and female genitalia are different, the subjective feelings of men and women during the sexual response would be different also. However, in one study in which male and female subjects described their feelings during sexual response, readers found it impossible to reliably distinguish the male and female responses if all direct references to the gender of the subjects were deleted from the accounts.

The Male Genital Responses

The penis must become stiff in order to penetrate the female vagina; this process is called **erection.** The penis has an extensive network of venous sinuses, the corpus spongiosum and corpora cavernosa (see Figure 25-2); entry of blood into them is under neural control. In the absence of sexual stimulation, the arterioles leading to the sinuses are constricted and the resistance of the veins draining the sinuses is low. Consequently there is little blood in the venous sinuses, and the penis is flaccid. During the excitement phase, visual, olfactory, tactile, or imaginary stimuli increase parasympathetic activity and decrease sympathetic activity. As a result of these signals, the arterioles leading to the corpus spongiosum and corpora cavernosa dilate, and blood flows into the sinuses under higher pressure. As the venous sinuses fill with blood, they compress the veins draining the erectile tissue and increase their flow resistance. As a result of these changes the pressure of blood in the venous sinuses approaches the mean arterial pressure. Failure to achieve erection in response to appropriate stimulation (**impotence**) can be caused by worry or anxiety and by drugs that affect the autonomic nervous system.

Semen is expelled through the penis during **ejaculation,** the major genital event of the male orgasmic phase. Ejaculation is a spinal reflex mediated by sympathetic nerves that stimulate contraction of the smooth muscles of the tubular system, of the accessory glands, and of muscles at the base of the penis. Ejaculation is normally a two-stage process. In the first stage (**emission**) the smooth muscle lining the internal gland and duct system contracts, forcing semen from the epididymis into the urethra. In the second stage (**expulsion**) the urethral muscle and skeletal muscle at the base of the penis contract rhythmically, expelling the semen from the penile urethra. These contractions are accompanied by sensations of intense pleasure. The sphincter at the base of the bladder constricts during ejaculation, thus semen cannot enter the bladder.

The Female Genital Responses

The vasodilation that occurs during the excitement phase causes erection of the clitoris and an increase in the secretion of vaginal fluid across the vaginal wall (**vaginal lubrication**). This fluid helps to neutralize the normally acidic pH of the vagina and therefore is important for survival of spermatozoa in this environment.

Like the penis, the clitoris (see Figure 25-2) is the sensory focus for stimulation during sexual activity. Stimulation of the clitoris, either by direct contact or contact with the labia, is almost always necessary for orgasm. It has been characterized as a sex organ that is not also a reproductive organ, since unlike the penis it has no direct part in bringing about the meeting of gametes.

The **orgasmic platform,** a constriction of the outer one third of the vagina, forms near the end of the plateau phase. The orgasmic phase is characterized by rhythmic contractions of the orgasmic platform and the uterus. During the resolution phase the uterus returns to its usual position in the pelvis and the blood flow to the pelvic area diminishes to its normal level.

---

1 What are the stages in the female and male sexual response cycles? What changes characterize each phase?

2 What autonomic inputs are responsible for erection and ejaculation?

3 In what ways are the penis and the clitoris functionally similar and in what important way do they differ in function?

PREGNANCY
From Fertilization to Implantation

After fertilization, about 4 days are required for the zygote, now a **morula** (Figure 26-2), to reach the uterus. While the morula is traveling in the Fallopian tube, the postovulatory follicle becomes a corpus luteum and the mother's menstrual cycle enters the luteal phase. Estradiol secreted by the corpus luteum sustains the thick uterine endometrium, and the progesterone induces secretion of **uterine milk.** By the time the morula arrives in the uterus, it usually consists of 48 cells. The morula floats freely in the lumen of the uterus and is nourished by uterine secretions as it undergoes further development to an embryonic form called a **blastocyst** (Figure 26-3).

The blastocyst consists of a hollow sphere of cells, the **trophoblast** or **trophoderm,** which encloses an **inner cell mass** or **embryoblast.** The fates of these two parts of the blastocyst are very different. The inner cell mass will become the **embryo** and ultimately develops into a baby's body; the trophoblast will become the **chorion,** which forms an outer covering of the embryo. The chorion gives rise to the fetal component of the **placenta,** the organ that mediates transfer of nutrients, gases,

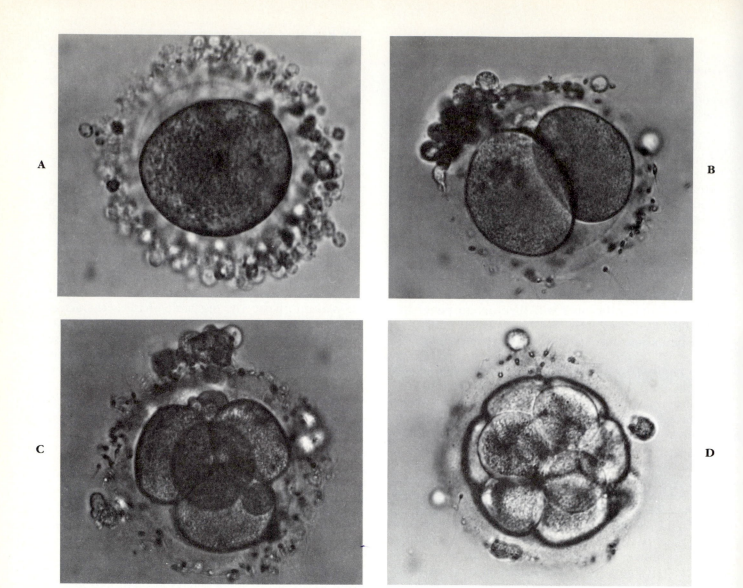

FIGURE 26-2
Early stages of human development.
A   The zygote within its fertilization membrane and still surrounded by the corona radiata
of follicular cells (see Chapter 25).
B   The first mitotic division after fertilization.
C   The eight-cell stage after three rounds of mitosis.
D   Morula.

and wastes between maternal and fetal blood and secretes hormones that support the pregnancy.

About the seventh or eighth day after ovulation, the blastocyst lodges against the wall of the uterus. Secretions of the trophoblast dissolve the endometrium, and the blastocyst literally eats its way into the endometrium since it is nourished by the products of the enzymatic actions (Figure 26-4, *A*). This process is called **implantation**. Finally the blastocyst is completely embedded in the endometrium, surrounded by its chorion (Figure 26-4, *B*).

Since the beginning of the luteal phase, progesterone secreted by the corpus luteum has maintained the uterus in a state not unlike that of pregnancy, even though a fetus is not yet present. True pregnancy begins with implantation. Some prostaglandin-based contraceptives now reaching the public act by interfering with implantation. Even without contraception, about half of all zygotes that reach this stage are believed to die after failing to implant. In such cases the mother never becomes aware that fertilization has occurred.

The situation of the embryo that succeeds in implanting is still far from secure. It has already completed a journey that in proportion to its size, was immense, but it is still tiny in comparison to the

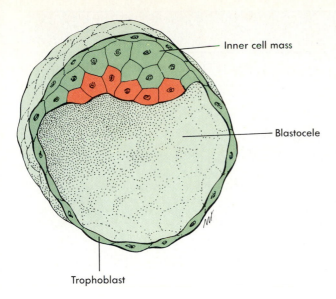

FIGURE 26-3
Preimplantation blastocyst.

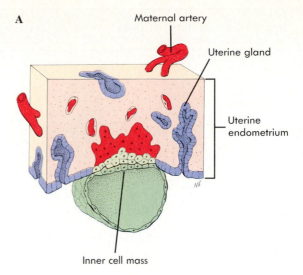

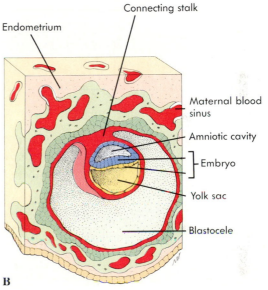

FIGURE 26-4
A The beginning of implantation. The implanting blasto-cyst is shown in green.
B After implantation. The inner cell mass is differentiating into an embryo, amnion, and yolk sac. The chorion (shown in green) is stimulating the growth of surrounding maternal blood vessels.

size of its mother, consisting of only several hundred cells. From the point of view of its mother's immune system, the embryo is an invader. Finally, the embryo is dependent for its survival on continued secretion of estrogen and progesterone by the corpus luteum. Without some signal that pregnancy has begun, the corpus luteum will self-destruct in only a few days. The resulting loss of the uterine endometrium would terminate the pregnancy. Even normal spontaneous contractions of the uterus would probably be enough to dislodge the embryo. The chorion and placenta provide solutions to all of these problems. The evolution of the mammalian trophoblast and placenta was an important adaptation for internal fetal development.

Endocrine Functions of the Trophoblast and Placenta

The chorion, and later the placenta, secrete a number of hormones that affect the physiology of the mother (Table 26-2). Immediately after implantation the trophoblast begins to secrete **human chorionic gonadotropin (HCG)**, which prevents the self-destruction of the corpus luteum and causes it to maintain the secretion of estrogen and progesterone needed to prevent shedding of the uterine lining. The appearance of HCG in the blood and urine of the mother (Figure 26-5) is the first indicator that pregnancy has begun. An assay for HCG is the basis of early pregnancy tests, both those used by physicians and those commercially available for home use (Figure 26-6). The secretion of chorionic gonadotropin by the trophoblast (and later by the placenta) continues during the first 3 months (**first trimester**) of pregnancy.

By the beginning of the second trimester, the placenta reaches full development and assumes responsibility for secretion of estrogen, progesterone, and HCG. Although the placenta can synthesize progesterone from cholesterol, it is unable to perform the complete chemical synthesis of estrogens from cholesterol. Both maternal and fetal adrenals can convert progestins to dehydroepiandrostened-ione (DHEA) (see Chapter 25). The placenta then completes the synthesis of estrogens from DHEA. The fetal adrenal also depends on the placenta for supplying progesterone for conversion to cortisol.

TABLE 26-2   Development of Organ Systems during Trimester

Organ system	First trimester	Second trimester	Third trimester
		Months of gestation	
Nervous	Major divisions established Eyes and ears begin to develop Nerve cells in brain	Limbs innervated Vestibular system formed Olfactory system formed Some reflexes appear Spontaneous body movements	Eyes develop
Circulatory	Tubular heart formed Heartbeat begins Hematopoiesis begins	Heart has four chambers Liver makes blood cells	Heartbeat audible Bone makes blood cells
Respiratory	Primary bronchi, trachea	Secondary, tertiary bronchi Tracheal cartilage	
Musculoskeletal	Body segments form Limb buds form Muscle precursors enter limb buds Fingers and toes form Cartilage skeleton	Bone formation starts Muscles assume adult form	
Endocrine	Thyroid forms Pituitary forms	Thyroid in adult position Adrenals form	
Integumentary		Sensory receptors appear in skin	
Urinary		Kidney differentiates	Kidney is functional
Reproductive	Wolffian, Müllerian systems form	Gonads differentiate Internal accessory systems differentiate	External genitals differentiate
Digestive	Stomach and intestine recognizable Major divisions of intestine form	Liver, pancreas appear Tooth buds form Swallows amniotic fluid	
Immune	Thymus appears	Spleen appears	

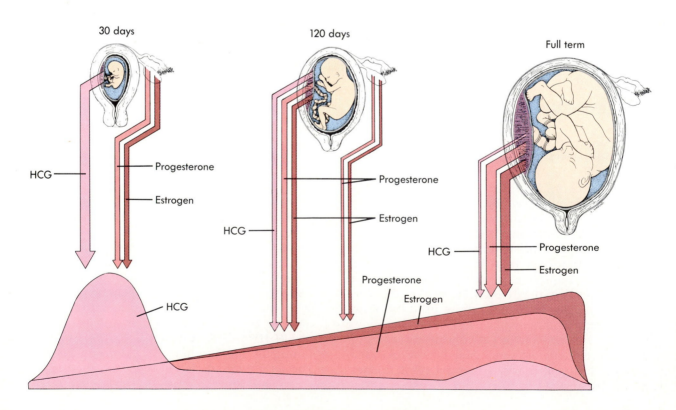

A Urine of pregnant woman

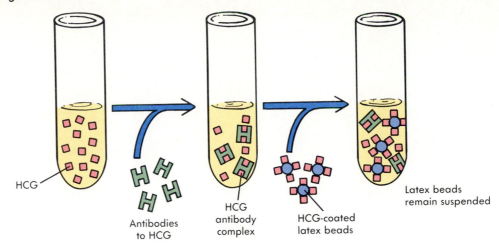

HCG

Antibodies
to HCG

HCG
antibody
complex

HCG-coated
latex beads

Latex beads
remain suspended

B Urine of nonpregnant woman

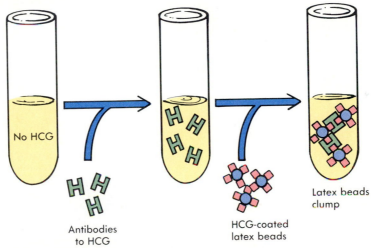

No HCG

Antibodies
to HCG

HCG-coated
latex beads

Latex beads
clump

FIGURE 26-6

A   The presence of HCG in a sample of blood or urine of a pregnant woman is detected by an assay that uses antibodies to HCG. When the solution containing antibodies is added to the sample, the HCG binds to the antibodies, saturating the antibody binding sites. When latex beads with HCG attached are added subsequently, the beads are not agglutinated because all of the antibody binding sites are already occupied.

B   The blood or urine sample from a nonpregnant woman contains no HCG, so the antibodies from the test kit are still able to bind the HCG on the latex beads when they are added, causing visible clumping that indicates a negative test.

FIGURE 26-5

Changes in the concentrations of some reproductive hormones during pregnancy. During the first trimester *(left)*, HCG secreted by the chorion and placenta maintains secretion of estrogen and progesterone by the corpus luteum in the ovary. During the second trimester *(middle)*, the placenta begins to take over secretion of estrogen and progesterone from the corpus luteum, and secretion of HCG decreases. By the last trimester *(right)*, the placenta is the major source of estrogen and progesterone. Levels of estrogen and progesterone drop shortly before term.

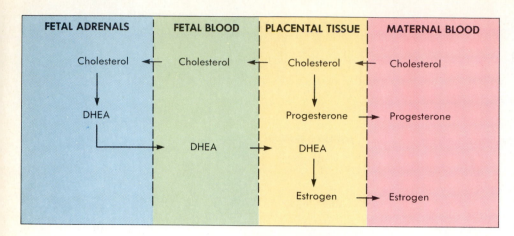

FIGURE 26-7
Metabolism of cholesterol to progesterone and estrogen in the placenta requires the cooperation of the adrenals of mother and fetus. The contribution of the adrenals is conversion of progestins to DHEA, which the placenta can convert to estrogens.

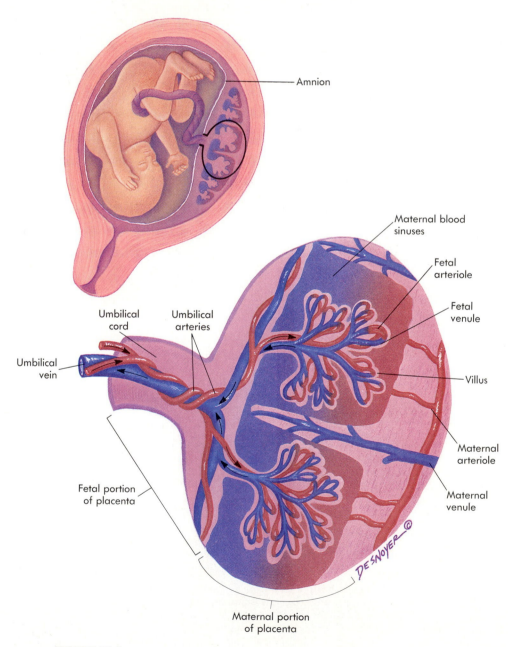

FIGURE 26-8
The structure of the interface between maternal and fetal circulations in the placenta.

These pathways of steroid metabolism are an excellent example of interdependence within the maternal-fetal-placental unit (Figure 26-7).

When the placenta begins to secrete sufficient estrogen and progesterone to maintain the pregnancy, placental secretion of HCG decreases and the corpus luteum regresses. During the second and third trimesters the plasma levels of estrogen and progesterone rise continuously, reaching a peak at the time of labor (see Figure 26-5).

In addition to HCG, estrogen, and progesterone, the placenta secretes a hormone called **human placental lactogen (HPL)** or **chorionic somatomammotropin.** Levels of HPL rise throughout the first and second trimesters and reach a peak in the last month of pregnancy. Placental lactogen is structurally similar to both growth hormone (Chapter 23) and prolactin (a hormone involved in milk production, discussed later in this chapter), and HPL has similar effects on the mother. Placental lactogen stimulates breast development in preparation for postnatal **lactation** (milk production), supports fetal bone growth, and alters the way the mother uses metabolic substrates by substituting lipids for glucose as the cellular energy source. Lipid utilization by the mother makes more glucose available to the fetus. The placenta also secretes antidiuretic hormone, aldosterone, and renin. All these hormones act to increase the maternal extracellular fluid volume. The increased blood volume enhances the ability of the mother's circulatory system to meet the demands placed on it by the fetus and placenta.

One of the most recently recognized hormones of pregnancy is **relaxin.** Like estrogen and progesterone, relaxin is initially secreted by the corpus luteum, but later in pregnancy most relaxin probably comes from the uterine muscle and placenta. Relaxin softens the connective tissue between the bones of the pelvic girdle, so that the pelvic aperture is enlarged and the infant passes through more easily at the time of delivery. Relaxin also softens the cervix and may have a role in precipitation of labor at the end of the gestation period. The gene for relaxin has been cloned, and relaxin is now under production for use as a drug.

The Structure and Functions of the Placenta

As the embryo continues to grow, it can no longer depend on diffusion to supply its needs for nutrients and oxygen and to carry away carbon dioxide and wastes. This problem is partly solved as the fetus develops a circulatory system, and the chorion gives rise to the fetal part of the placenta (termed **placentation**). The fetal blood must be brought in close proximity to the maternal blood for exchange to take place, but actual mixing of the two blood supplies would result in an immune attack being directed at the fetus by the mother. Thus the fetal and maternal bloods cannot be allowed to mix, but rather the structure of the placenta is such that placental villi containing fetal capillaries are immersed in **maternal blood sinuses** in the endometrium (Figures 26-8 and 26-9). Diffusional exchange of gases and some nutrients occurs, and other nutrients are actively transported across the capillary walls. These walls constitute the **fetal-maternal barrier** that usually protects the placenta and fetus from a maternal immune response. Moreover, some

FIGURE 26-9
Photomicrograph of a section of placenta showing fetal capillaries in a maternal blood sinus.

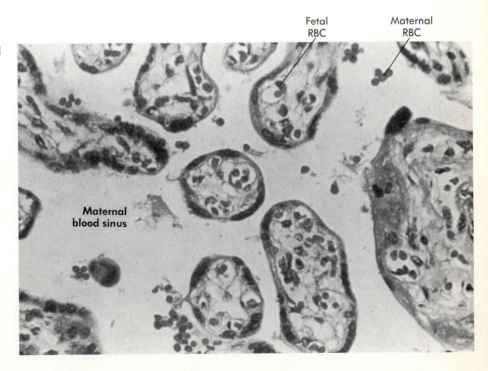

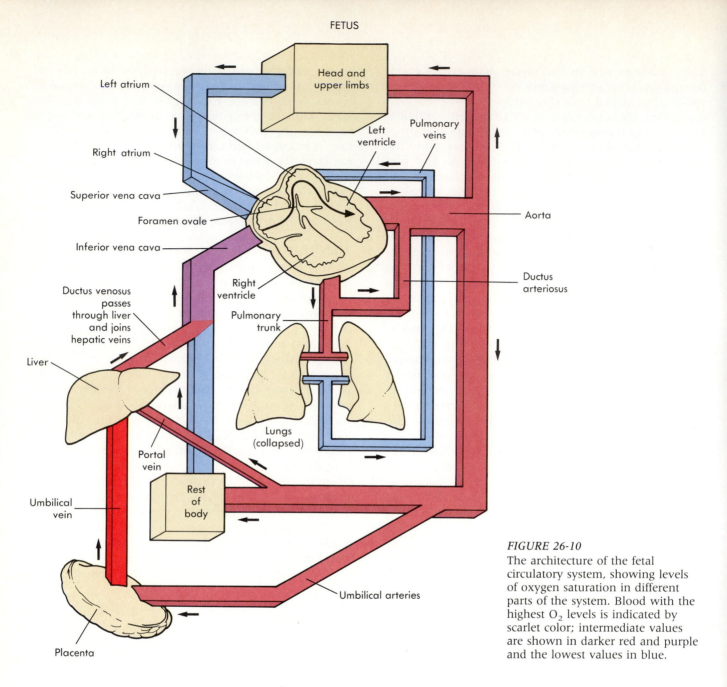

FETUS

Left atrium

Right atrium

Superior vena cava

Foramen ovale

Inferior vena cava

Ductus venosus
passes
through liver
and joins
hepatic veins

Liver

Portal
vein

Umbilical
vein

Placenta

Head and
upper limbs

Left
ventricle

Pulmonary
veins

Right
ventricle

Pulmonary
trunk

Lungs
(collapsed)

Rest
of
body

Aorta

Ductus
arteriosus

Umbilical arteries

FIGURE 26-10
The architecture of the fetal
circulatory system, showing levels
of oxygen saturation in different
parts of the system. Blood with the
highest O_2 levels is indicated by
scarlet color; intermediate values
are shown in darker red and purple
and the lowest values in blue.

maternal antibodies are transferred to the fetus, protecting it from those diseases to which the mother is immune.

The remarkable ability of the placenta to organize maternal support for the needs of the fetus is illustrated by **ectopic** pregnancies, in which the blastocyst implants at a site other than the uterus. Possible sites are the abdominal cavity and the Fallopian tube. Even in ectopic sites the placenta can protect the fetus from immune attack and stimulate sufficient new maternal blood vessels to support fetal development. Embryos that implant in a Fallopian tube, the most common ectopic site, will eventually rupture the tube, causing abdominal bleeding that threatens the life of the mother. These pregnancies must be terminated. Implantation at other ectopic sites in the body cavity is not necessarily life threatening for mother or child and may go unde-

tected until the pregnancy is quite advanced. In these cases it is frequently possible to deliver the baby by Caesarean section.

1 What maternal hormones are important for maintaining the pregnancy? What hormones does the fetus produce, and what is the function of each?
2 What is a blastocyst? What part of the blastocyst gives rise to the fetal portion of the placenta? What part gives rise to the embryo?
3 What is meant by the terms luteal phase and placental phase of gestation? Do they overlap in time? What are the endocrine functions of the corpus luteum and the placenta?

The Fetal-Placental Circulation

The fetal circulatory system is adapted to include the placental circulation (Figure 26-10). Two **um-**

bilical arteries branch from the fetal descending aorta and pass into the umbilical cord that connects the fetus with the placenta. A single **umbilical vein** returns blood from the placenta. The umbilical vein joins the venous circulation of the fetal liver, which is connected to the inferior vena cava by the **ductus venosus.** This shunt allows blood returning from the placenta to pass directly to the vena cava by a low-resistance route. The blood that enters the fetal vena cava is a mixture of relatively well-oxygenated blood from the umbilical vein and less well-oxygenated blood flowing in the hepatic portal vein. As this blood flows from the ductus venosus into the vena cava, there is a further dilution with fetal venous blood.

The lungs of the fetus are collapsed and present a high resistance to blood flow. Since they serve no respiratory function before birth, it is unnecessary that all of the cardiac output pass through them. Venous blood returning to the right heart from the body mixes with placental blood and enters the right heart. Most of this blood is not pumped to the lungs, but instead takes one of two shunts to the systemic circulation. One shunt pathway is provided by an opening, the **foramen ovale,** in the septum between the atria. The other is the **ductus arteriosus,** a connection between the aorta and the pulmonary artery. This arrangement of the placental circulation has some important consequences. The placental blood equilibrates with partly deoxygenated maternal blood, returning to the fetus only about 80% saturated with oxygen. The placental blood mixes with fetal venous blood that is about 27% saturated, so that the arterial blood is only about 58% saturated with oxygen.

Clearly oxygen delivery is a physiological problem for the fetus, and its tissues must be adapted to function at lower oxygen partial pressures than those that will be typical after birth. The situation would be even less favorable if **fetal hemoglobin** were not adapted for effective O_2 transport at low O_2 partial pressures. Fetal hemoglobin is a tetramer like adult hemoglobin (see Chapter 17) except that it contains two gamma subunits in place of the two beta subunits of adult hemoglobin. The substitution shifts the O_2 dissociation curve leftward compared to that of adult hemoglobin (Figure 26-11). The shift allows fetal hemoglobin to load from partly deoxygenated maternal blood and increases its oxygen carrying capacity by 20% to 30% at the P_{O_2} of placental venous blood.

At birth, the fetal circulatory and respiratory systems must undergo dramatic changes to assume their function in the extrauterine environment (Figure 26-12). When the umbilical circulation is interrupted, the ductus venosus and umbilical arteries close off and eventually wither. The ductus arteriosus constricts and later becomes permanently closed. The inflation of the lungs with the first breath lowers the resistance of the pulmonary loop,

and all of the blood pressures in this loop fall below their respective values in the systemic loop. The foramen ovale is provided with a flap that permits blood to flow from the right atrium to the left atrium, but not in the reverse direction. The development of a pressure gradient between the right and left atria causes the flap to be pushed shut. Eventually the flap fuses to the rest of the interatrial septum, and permanent separation of the pulmonary and systemic loops is accomplished.

In rare cases there are shunts between the pulmonary and systemic loops after birth. For example, the ductus arteriosus may fail to close (**patent ductus**). Any such connection between the two loops after birth results in blood leakage from the systemic loop into the pulmonary loop, because blood pressure in the pulmonary loop is lower than in the systemic loop.

1 What is a vascular shunt? What are the three vascular shunt pathways of the fetal circulation? What happens to these shunts at birth?
2 Where in the fetal circulation is the blood most oxygenated? Least oxygenated? Why is blood in the fetal dorsal aorta less saturated with oxygen than blood in the umbilical vein?
3 Why is the fetal pulmonary loop a high-resistance pathway? What effect does this have on blood flow through the loop?

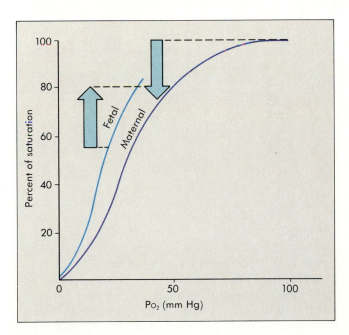

FIGURE 26-11
The oxygen dissociation curves of maternal and fetal hemoglobin. The leftward position of the curve for fetal hemoglobin indicates a higher oxygen affinity. As maternal blood enters the placental blood sinuses, the P_{O_2} drops and O_2 is unloaded from maternal hemoglobin (*indicated by arrow pointing down*). The oxygen given up by the maternal hemoglobin is accepted by the fetal hemoglobin (*arrow pointing up*). Even with the leftward shift of its dissociation curve, the fetal hemoglobin is still not saturated when it leaves the placenta.

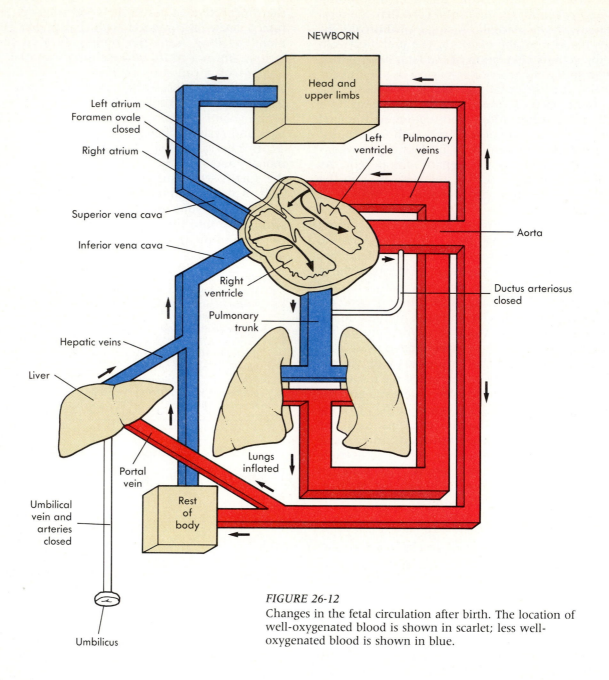

Left atrium
Foramen ovale closed
Right atrium

Left ventricle
Pulmonary veins

Superior vena cava

Inferior vena cava

Aorta

Right ventricle

Ductus arteriosus closed

Pulmonary trunk

Hepatic veins

Liver

Lungs inflated

Portal vein

Rest of body

Umbilical vein and arteries closed

Umbilicus

FIGURE 26-12
Changes in the fetal circulation after birth. The location of well-oxygenated blood is shown in scarlet; less well-oxygenated blood is shown in blue.

From Implantation to Birth—The Developmental Timetable

The human **gestation period,** or duration of prenatal development, is 39 weeks, counting from the last menstrual period. Conventionally the gestation period is divided into three 3-month parts—the first, second, and third **trimesters.** Most of the organ systems begin to differentiate within the **embryonic phase** of development—the first 8 weeks of gestation (Figure 26-13, *A* and *B*). This is the part of gestation in which sensitivity to drugs and other chemicals is greatest. The subsequent 31 weeks of development are the **fetal phase.**

By about a week after implantation, the beginnings of the heart and nervous system are recognizable. By 2 to 3 weeks the heart has begun to beat, but unless she has had a pregnancy test, the mother may not yet suspect that she is pregnant. By 3 to 4 weeks, most of the major organ systems have begun to differentiate (see Table 26-2): the embryonic thyroid, parathyroids, liver, and pancreas are present; the genital tubercle is recognizable; and limbs, eyes, and nose have begun to form. At this point the body of the embryo is about 0.6 cm long (see Figure 26-13, *A*). It is enclosed in a double-walled chamber attached to the uterine endometrium by the placenta. The outer wall of the chamber is the chorion, which is continuous with the placenta; the inner wall is

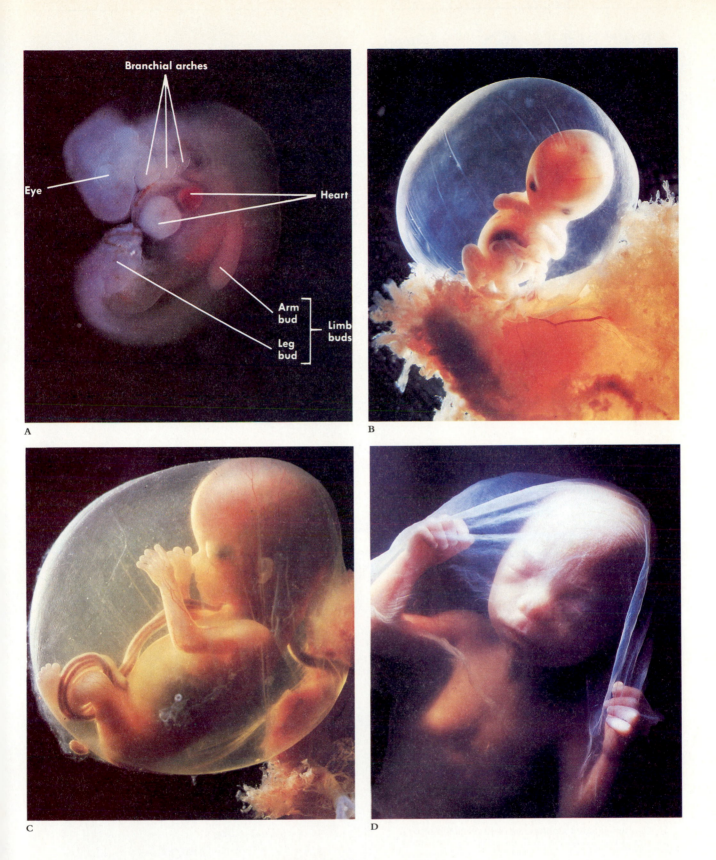

FIGURE 26-13
Human embryos and fetuses.
A   At 35 days.
B   At 49 days.
C   At the end of the first trimester.
D   At 4 months.

Parents whose age or genetic
makeup puts them at risk for
conceiving a child with a genetic
disease may elect to undergo a
process in which a small sample of
fetal cells is removed and cultured
for diagnosis. The most common
method is called *amniocentesis*. The
fetus is enclosed in an amniotic sac
composed of fetal cells. A needle is
passed through the mother's
abdomen and into this sac, using
ultrasonography (Figure 26-A, *1*) to
direct the needle away from the
fetus. A sample of amniotic fluid

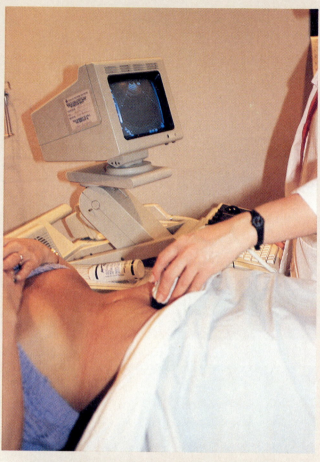

1

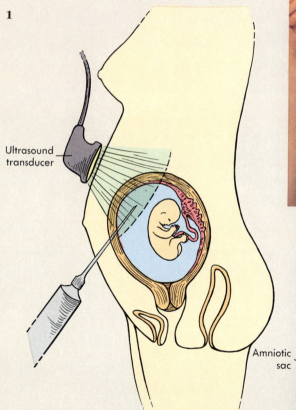

Ultrasound
transducer

Amniotic
sac

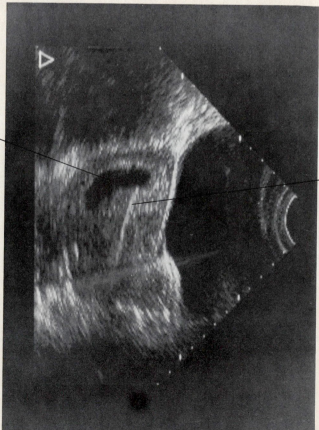

Needle

FIGURE 26-A

1   Ultrasonogram of amniotic sac with needle in place
for amniocentesis. As shown in the inset, the ultrasound
imager reconstructs an image of a two-dimensional
section through the uterus. In this case the section
passed through the uterus along a plane that included
the amniotic sac and part of the placenta, but not the
fetus itself. Passing the needle into the amnion along this
plane ensures that it will miss the fetus.

2   Ultrasonogram showing a 4-week-old embryo in its
amniotic sac.

3   In this ultrasonogram of a 6-week embryo the plane
of the image passes through the head and trunk. The
hollow cerebral ventricles are visible.

usually contains viable cells that may be cultured. The chromosomes of the dividing cells can be examined microscopically for abnormalities, such as the extra chromosome number 21 that results in Down's syndrome, a form of mental retardation whose incidence is higher in children of older mothers. The fetal cells can also be analyzed for marker substances that indicate genetic abnormalities, such as the abnormal enzyme responsible for Tay-Sachs disease, in which myelination of neurons is impaired. As part of this process, the genetic sex of the fetus is determined. Parents who wish to enjoy the suspense of not knowing the child's sex until birth may ask not to be given this information.

Amniocentesis has several disadvantages: it cannot be carried out until the fourteenth week of pregnancy, when the amnion is sufficiently large to allow fluid to be collected. Two weeks are needed to culture the cells, so that the pregnancy is relatively advanced by the time a diagnosis is available. A newer technique, chorionic villus biopsy, involves inserting a catheter through the cervix to collect a small piece of the placenta. This method can provide diagnostic information as early as the tenth week of gestation. Both methods pose a slight but finite risk to the fetus. Both give the parents information that may cause them to terminate the pregnancy, or allow them to prepare to deal with the additional problems posed by a disease if they elect to continue the pregnancy. Ultimately prenatal diagnoses like these may give physicians the opportunity to provide therapy to fetuses with genetic and developmental diseases.

The images provided by ultrasonography (Figure 26-A, *2-3*) allow the progress of many aspects of fetal development to be followed. Surgery on fetuses in the uteri of laboratory animals is a fairly common experimental technique. In the future, it might be possible to perform corrective surgery on developmentally impaired human fetuses. At present the risks of such therapy appear to outweigh the benefits.

2

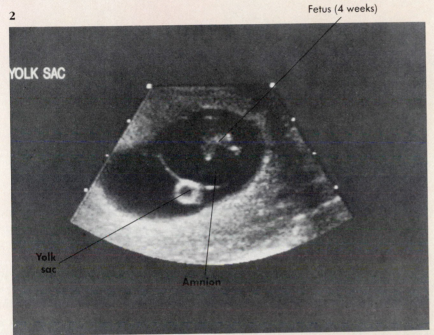

Fetus (4 weeks)

YOLK SAC

Yolk sac

Amnion

3

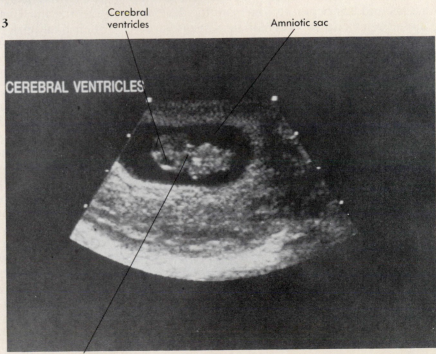

Cerebral ventricles

Amniotic sac

CEREBRAL VENTRICLES

Fetus (6 weeks)

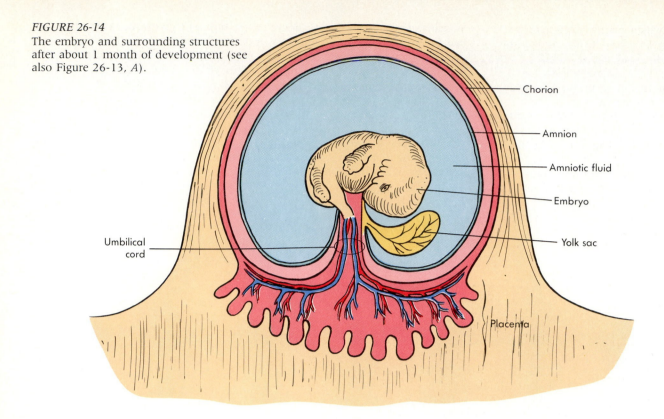

FIGURE 26-14
The embryo and surrounding structures after about 1 month of development (see also Figure 26-13, A).

Chorion

Amnion

Amniotic fluid

Embryo

Yolk sac

Umbilical cord

Placenta

the **amnion** (Figure 26-14). Within the amnion, the embryo, attached to the placenta by its umbilical cord, floats in a miniature sea of **amniotic fluid**. A **yolk sac** attached to the abdomen of the embryo contains stored nutrients and is one site of formation of blood cells (see Figure 26-14). By the sixth week of development, the embryonic central nervous system, at first a simple tube of neural precursor cells, shows subdivisions characteristic of the adult CNS and has begun to innervate the developing appendages and to differentiate sensory structures such as eyes and ears (see Table 26-2).

By the end of the first trimester, the embryo, now a **fetus,** is 7 to 8 cm long (Figure 26-13, C and D). Its heartbeat can be detected. Its face is recognizably human, its appendages are complete, and its genital sex can be determined. The development of each individual organ system can be thought of as a thread that is woven together with other threads to generate an organism that, as development goes forward, is increasingly capable of independent life (see Table 26-2). By the end of the second trimester, the fetus is 25 to 37 cm long, weighs 550 to 700 grams, and differentiation of all of its organ systems is well advanced. It is approaching the stage at which, with medical assistance, it might survive if born prematurely. During the third trimester, the fetus grows rapidly and also accumulates a store of fat important for survival in the first few days after delivery. Some organs, such as the heart and kidney, are already functioning. At about the eighth month

the lung begins to prepare for function by secreting **pulmonary surfactant.** The presence of surfactant greatly increases the chance of survival if birth were to occur.

At birth the baby typically weighs between 3 to 3.5 kg and is about 50 cm long. Birth is a milestone in the developmental process but does not represent the end of it, since many organ systems, including the central nervous system, the reproductive system, and the muscular-skeletal system, do not arrive at their final form until puberty or later.

Effects of Pregnancy on the Mother
Pregnancy has dramatic effects on the physiology of the mother. Most of these are apparently due to the hormones produced by the placenta (Table 26-3). Typically women gain weight during pregnancy, especially during the last two trimesters (Figure 26-15). Only about 30% of the weight gained is accounted for by the fetus and placenta; the rest is due to changes in the mother's body that support the pregnancy and subsequent lactation (Table 26-4). By the end of pregnancy the mother's blood volume typically has increased by about 30%, the result of the placental steroids and increased secretion of aldosterone by the mother's adrenals. The increase in blood volume causes her cardiac output to increase by a similar percentage. Several athletes have turned in excellent Olympic performances while pregnant, suggesting that adaptations of the cardiovascular system and metabolism induced by

TABLE 26-3   The Hormones of Pregnancy, Labor, and Lactation

Hormone	Secreted by	Effects
Estrogen	Corpus luteum, placenta	Maintains endometrium; contributes to breast development
Progesterone	Corpus luteum, placenta	Maintains endometrium; inhibits contraction of myometrium; stimulates breast development
Placental somatomammo-tropin (fetal lactogen)	Placenta	Increases fat and carbohydrate catabolism in mother, making stored nutrients available to fetus; contributes to breast development
Oxytocin	Posterior pituitary	Stimulates uterine contraction; causes milk letdown
Relaxin	Corpus luteum, placenta, uterus	Softens cervix; loosens ligaments of pelvis; inhibits uterine contractions; potentiates effect of oxytocin
Prolactin	Anterior pituitary	Maintains milk production after delivery

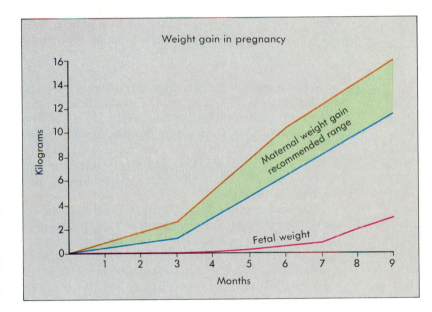

FIGURE 26-15
Maternal weight gain during pregnancy (in kilograms). The portion of the weight gain due to the growth of the fetus itself is shown in the lower curve.

the placenta do not impair performance and may even improve endurance and maximum power output.

The quality of the mother's nutrition during the first trimester is important because during this time her intake of protein, vitamins such as folic acid, and minerals such as calcium, phosphorus, and iron can exceed the immediate demands of the embryo. The extra nutrients and minerals stored during the first trimester are drawn on during the second and third trimesters, when fetal growth may impose greater demands than can be met by intake. This fact underscores the value of beginning a program of good prenatal nutrition as early as possible, perhaps even before pregnancy.

TABLE 26-4   Typical Components of Maternal Weight Gain during Pregnancy

Component	Weight (kg)
Baby	3.2
Placenta	0.5
Amniotic fluid	0.9
Breasts	0.9
Blood	1.4
Uterus	0.9
Other	8.1
TOTAL	15.9

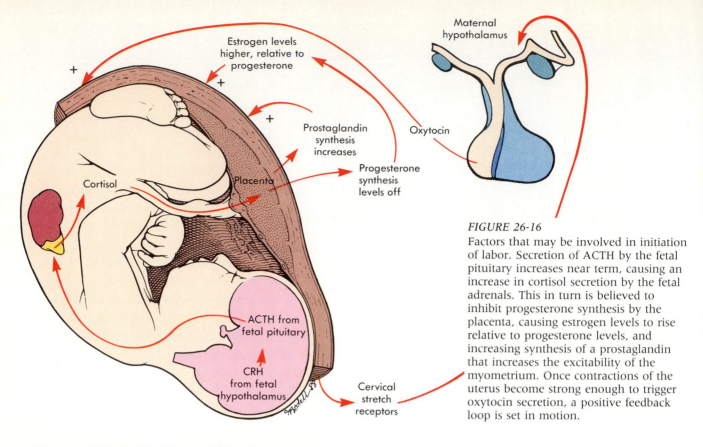

Estrogen levels
higher, relative to
progesterone

Prostaglandin
synthesis
increases

Progesterone
synthesis
levels off

Cortisol

Placenta

Oxytocin

Maternal
hypothalamus

ACTH from
fetal pituitary

CRH
from fetal
hypothalamus

Cervical
stretch
receptors

FIGURE 26-16
Factors that may be involved in initiation
of labor. Secretion of ACTH by the fetal
pituitary increases near term, causing an
increase in cortisol secretion by the fetal
adrenals. This in turn is believed to
inhibit progesterone synthesis by the
placenta, causing estrogen levels to rise
relative to progesterone levels, and
increasing synthesis of a prostaglandin
that increases the excitability of the
myometrium. Once contractions of the
uterus become strong enough to trigger
oxytocin secretion, a positive feedback
loop is set in motion.

> 1 *What major events are characteristic of the embryonic period?*
> 2 *What are some of the effects of pregnancy on the mother's physiology?*

BIRTH AND LACTATION
What Determines the Length of Gestation?

During most of the normal gestation period, the uterine smooth muscle is relatively quiet, but as the fetus grows, the muscle is stretched. Stretching increases the contractility of the muscle (**stress-activation**), and some uncoordinated contractions may occur up to 1 month before labor begins (**Braxton Hicks contractions**). It is thought that these contractions may represent "training exercises" for the uterine muscle. At the onset of labor, the pressure of the head of the fetus against the cervix initiates a hormonal reflex (Figure 26-16) that increases secretion of the peptide hormone **oxytocin** from the posterior pituitary. Oxytocin greatly potentiates uterine contractions, setting up a positive feedback loop of strong, coordinated waves of contraction that continues until delivery is complete.

What factors determine when labor begins? The potential of oxytocin to stimulate labor increases as the end of the gestation period approaches. Early in pregnancy there are no uterine oxytocin receptors. Oxytocin receptors appear in increasing numbers as pregnancy progresses. Binding of oxytocin to its receptors has two effects: (1) it causes uterine smooth muscle contraction, and (2) it stimulates the uterus

to make a prostaglandin that also activates uterine smooth muscle. The afferents for this neuroendocrine reflex are stretch receptors in the cervix. While labor depends on oxytocin secretion, uterine prostaglandin production is also necessary. In the absence of uterine prostaglandins, the cervix does not dilate properly, and therefore labor does not progress normally.

An understanding of the factors that normally prevent stretching from prematurely activating uterine contraction would have medical applications, because premature babies require much medical attention and have a much poorer chance of survival than do full-term infants. Factors involved in the timing of labor are poorly understood and may differ among the different species of animals that have been used to investigate the question. It appears probable that some hormonal signal from the fetus is needed to set the stage for normal labor. Attention has focused on the secretion of cortisol by the fetal adrenal cortex. Cortisol secretion is needed to prepare the infant for the rigors of postnatal life. Cortisol stimulates secretion of lung surfactant and increases the ability to mobilize metabolic stores between meals and to thermoregulate. In some animals there is a dramatic increase in cortisol secretion by the fetal cortex several weeks before birth, and some investigators believe this cortisol could serve as a signal to the maternal endocrine system that labor should begin (see Figure 26-16).

Oxytocin levels in maternal plasma do not suddenly increase just before labor begins, but are rela-

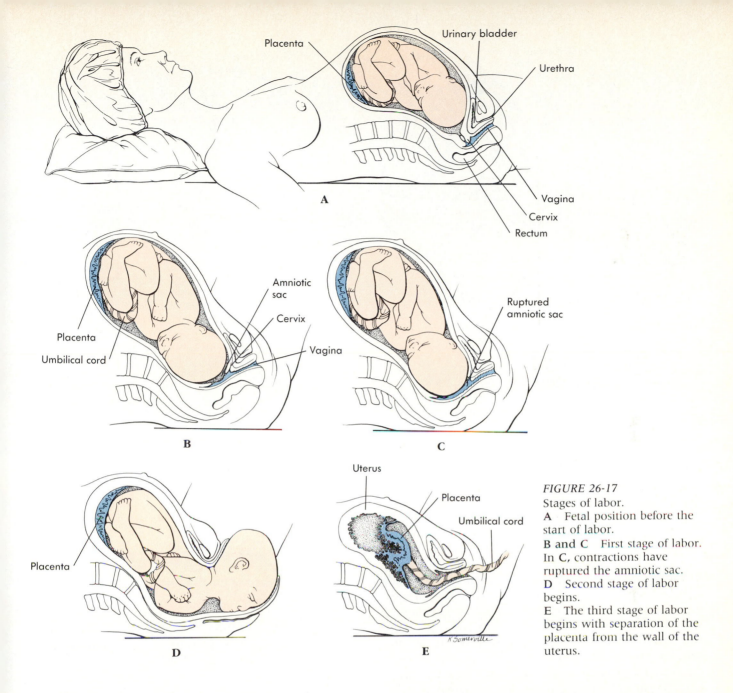

Placenta

Urinary bladder

Urethra

Vagina

Cervix

Rectum

A

Amniotic sac

Cervix

Placenta

Umbilical cord

Vagina

B

Ruptured amniotic sac

C

Placenta

D

Uterus

Placenta

Umbilical cord

E

FIGURE 26-17
Stages of labor.
A   Fetal position before the start of labor.
B and C   First stage of labor. In **C**, contractions have ruptured the amniotic sac.
D   Second stage of labor begins.
E   The third stage of labor begins with separation of the placenta from the wall of the uterus.

tively high throughout the later stages of pregnancy. Since levels of oxytocin are relatively high, and oxytocin receptors are present, it is not clear what prevents premature labor. One inhibitory factor is that progesterone levels are high during pregnancy, and progesterone directly inhibits uterine smooth muscle, while simultaneously blocking its response to oxytocin and prostaglandin. Used as a drug, progesterone is frequently effective in preventing threatened premature labor. However, maternal progesterone levels do not fall abruptly prior to labor. Another possible factor in the timing of labor is relaxin, which promotes an increase in the number of oxytocin receptors on uterine muscle. The amounts of relaxin in the maternal blood peak just prior to delivery and then drop off rapidly, so that the relaxing effect it exerts on uterine muscle is removed.

Whatever factors are involved, the protection from premature labor is not absolute, because labor may be **induced** medically by administration of oxytocin. Once labor begins, secretion of oxytocin is driven to high levels by the positive feedback loop (see Figure 26-16) that continues until the fetus is delivered.

The Stages of Labor
During the last trimester the fetus usually takes a head-downward position so that its head is against the cervix. Normally, it will be delivered head-first. Any other position predisposes it to **breech birth**, in which some other part of its body enters the cervix first. The approach of labor is indicated by **effacement**, or thinning of the cervix in response to relaxin, and promoted by uterine contractions.

There are three stages of labor (Figure 26-17). Stage one is marked by the onset of regular uterine

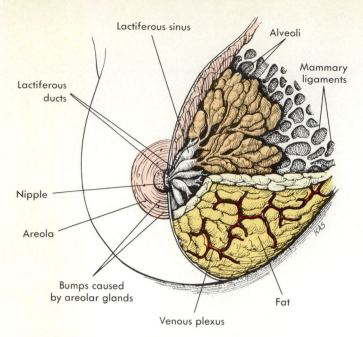

Lactiferous sinus

Alveoli

Mammary ligaments

Lactiferous ducts

Nipple

Areola

Bumps caused by areolar glands

Venous plexus

Fat

FIGURE 26-18

The structure of the breast. The mass of the breast is largely fat; the shape is maintained by a network of supporting ligaments. The mammary glands consist of alveoli that empty into a system of sinuses emptied to the nipple through ducts.

contractions that open or **dilate** the cervix to 7 to 10 cm from its prelabor diameter of 0 to 2 cm. During this stage, uterine contractions are occurring in waves in which the intervals between contractions are about 3 minutes. The duration of stage one is very variable.

Stage two begins when the baby begins to move through the cervix and ends when it is delivered. Stage two may last as long as several hours in new mothers, but may be as brief as a few minutes in women who have previously delivered. In stage three, the placenta ("afterbirth") is delivered; this stage lasts only a few minutes.

> *1 What factors are hypothesized to be responsible for the onset of labor? What is the role of oxytocin in labor?*
> *2 What are the three stages of labor?*

Prolactin And Lactation

The evolution of mammary glands from sweat glands was the defining step in mammalian evolution. The form of the breasts of nonlactating women is determined almost entirely by the distribution of adipose tissue, a secondary sexual feature that does not contribute to milk production. During lactation, milk production goes on in the **alveoli** of mammary glands (Figure 26-18). Milk from the alveoli is secreted into a series of **lactiferous sinuses,** surrounded by smooth muscle, that eventually lead by way of ducts to the nipple.

At puberty, breast development is stimulated mainly by estradiol (Figure 26-19), although insu-

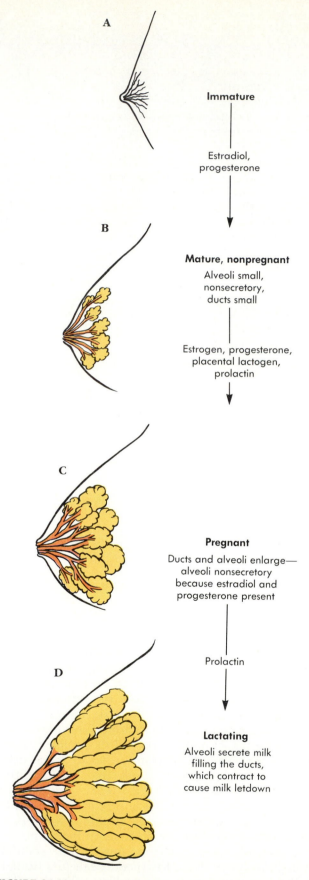

A

Immature

Estradiol, progesterone

B

Mature, nonpregnant
Alveoli small, nonsecretory, ducts small

Estrogen, progesterone, placental lactogen, prolactin

C

Pregnant
Ducts and alveoli enlarge—alveoli nonsecretory because estradiol and progesterone present

Prolactin

D

Lactating
Alveoli secrete milk filling the ducts, which contract to cause milk letdown

FIGURE 26-19
Hormonal influences on the breast at puberty, during pregnancy, and during lactation.

lin, thyroid hormones, cortisol, and growth hormone have permissive effects. Progesterone results in proliferation of the sinuses and ducts and in generalized breast enlargement (so that the size of the breast varies during the menstrual cycle).

During pregnancy, estrogen and progesterone levels are elevated well beyond those of the luteal phase of the menstrual cycle (see Figure 26-5), and therefore the alveoli and alveolar ducts proliferate considerably. Two other hormones, prolactin and placental lactogen, are also important for preparing the breasts for milk production.

Prolactin, an anterior pituitary hormone, stimulates alveolar development and milk production. How the hypothalamus controls release of prolactin is not completely understood. Current evidence suggests that prolactin release may be inhibited by the neurotransmitter dopamine and stimulated by a small polypeptide from the posterior pituitary. This stimulation by a blood-borne factor is an exception to the general rule that hypothalamic releasing factors reach the anterior pituitary by way of the hypothalamic-anterior pituitary portal circulation (see Chapter 5).

Prolactin begins to be secreted in small amounts at puberty, but prolactin secretion is greatly increased during pregnancy. Further increases in prolactin secretion occur during lactation. Prolactin stimulates growth and secretory activity of the mammary alveoli. Placental lactogen acts like pituitary prolactin, augmenting alveolar growth. Although prolactin and placental lactogen reach high levels in the plasma toward the end of pregnancy, milk production does not begin because of the inhibitory effects of the high levels of estrogen and progesterone that exist in the later stages of pregnancy. In effect, the breasts are primed for milk production during pregnancy but not switched on. Levels of estrogen and progesterone fall rapidly after delivery, because the placenta is no longer present and follicles are not maturing in the ovaries. The drop in estrogen and progesterone allows milk production to begin.

Milk production is sustained by continued prolactin secretion as long as suckling continues. According to a current hypothesis, when an infant nurses, sensory signals from the nipples (Figure 26-20) cause the hypothalamus to increase the secretion of prolactin releasing factor from the posterior pituitary, maintaining plasma prolactin levels and milk production. Prolactin release could also be sustained by decreases in dopamine release or by other controlling factors as yet undiscovered. Increases in the frequency or duration of suckling stimulate prolactin release, with a time lag of about a day, so that rate of milk production increases to match the needs of the growing baby. If a mother does not nurse her infant, prolactin levels fall and milk production stops after a few days.

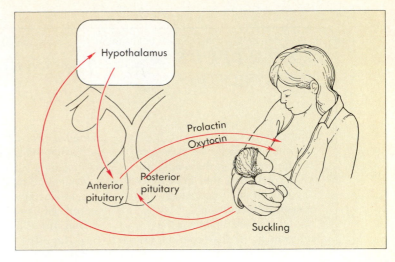

FIGURE 26-20
The hormonal reflexes caused by suckling. Both prolactin release from the anterior pituitary and oxytocin release from the posterior pituitary are stimulated by afferents from the nipple. Prolactin sustains milk production, and oxytocin causes milk letdown.

Oxytocin and Milk Letdown

Milk production and the movement of milk from the alveoli into the ducts, called **milk letdown,** are distinct processes. Milk letdown involves a reflex in which stimulation of the nipple by an infant's suckling causes oxytocin release from the posterior pituitary (see Figure 26-20). Oxytocin causes contraction of the smooth muscle cells around the alveoli and delivers milk to the nipples. The reflex becomes conditioned after a short time, so that the sound or sight of the infant can cause milk letdown to start even before suckling begins. Oxytocin released in response to suckling also stimulates uterine contraction. This side-effect of breast-feeding is important after delivery because it speeds the recovery of the stretched uterus to its normal size.

Milk production does not begin for a day or so after birth. During this time the breasts secrete **colostrum,** a yellow watery fluid that contains large quantities of maternal antibodies. Antibody secretion continues after milk production begins, and by the fourth month of lactation it may amount to as much as 0.5 gram/day. This is an example of **passive immunization** (see Chapter 24).

The Contraceptive Effect of Lactation

Pregnancy and lactation both place significant burdens on the mother's metabolism and nutrient balance. For most of human evolution, children were breast-fed for several years after birth, and this is still the practice in some societies. There must have been significant selective pressure on the evolving human endocrine system to avoid pregnancy while mothers were still nursing.

Recent evidence suggests that lactation, while not a reliable means of contraception under the con-

The Birth Experience Contributes to the Survival of the Newborn

During birth, infants are intermittently deprived of oxygen by compression of the umbilical cord. After birth, they must quickly adapt to an environment in which they must breathe air, thermoregulate, and be fed at intervals rather than continuously. Recent evidence suggests that the stress of normal vaginal birth stimulates secretion of catecholamine hormones (primarily norepinephrine) that improve infant survival during vaginal delivery and help initiate physiological adaptations to extrauterine life. The first evidence for an important role of catecholamines during birth was the finding that the catecholamine levels in the blood of healthy infants at birth were more than five times higher than the highest levels achieved by adults during exercise. Removal of the adrenals from new born animals reduced their ability to survive episodes of hypoxia like those caused by uterine contractions during labor.

In fetal animals, injection of epinephrine stops secretion of fluid into the lungs and causes fluid to be reabsorbed. This effect would contribute to clearing fluid from the lungs so that air breathing can be effective. Catecholamines also stimulate secretion of pulmonary surfactant, which reduces the work of breathing (Chapter 16).

Mobilization of nutrient stores and production of heat are stimulated by catecholamine in infants just as in adults (Chapter 23). This effect may help the infant survive the first day of postnatal life, since milk production typically does not start until the second day. Cooling after birth is a significant threat, since it is estimated that a wet newborn can lose heat at a rate as great as 200 cal/kg/min. In contrast the maximum rate of heat production in a resting adult is only about 90 cal/kg/min. Newborn infants possess a special heat-producing organ: a deposit of **brown fat** (Chapter 23) that forms a mantle beneath the skin of the upper thorax, accounting for 2% to 6% of body weight. Brown fat is brown because of its abundant mitochondria. Under the influence of catecholamines, brown fat begins to release large amounts of heat. This is possible because, in the mitochondria of brown fat, oxidative metabolism is not coupled to phosphorylation, so all of the potential energy of the lipid appears as heat. The brown fat disappears a few weeks after birth.

Secretion of epinephrine and norepinephrine from the adrenal medulla in children and adults in response to a stressful situation is stimulated by activity in sympathetic preganglionic nerves leading to the adrenal glands. However, in infants the autonomic nervous system is not fully developed at the time of birth, and the adrenals are not yet connected to the nervous system; secretion of adrenal catecholamines in infants must be stimulated by an indirect mechanism. The nature of the indirect pathway is not clear, but two factors that are involved are hypoxia and pressure on the head. These stimuli are experienced by infants during vaginal delivery, but not during surgical delivery **(Caesarean section)**. The use of surgical delivery as a means of circumventing the hazards of vaginal delivery has increased in developed countries, as improvements in surgical practice have reduced the risk of the surgery to the mother. Ironically, the use of surgical delivery would seem to deprive infants of a hormonal response that may contribute significantly to their survival in the hours immediately following birth.

ditions of child-rearing that prevail in developed countries, can have a significant effect on birth-spacing in developing countries. The contraceptive effect is due to reflexive inhibition of gonadotropin secretion by the stimulus of suckling. The effect is maximum if the child's diet is not supplemented and if the mother breast-feeds her child at least 6 times daily for a total of at least 1 hour per day of feeding. Under these conditions ovulation does not occur, so that lactation is at least as effective as any artificial means of contraception. More than 80% of third-world couples use no artificial means of birth control, so lactation is the most widely used contraceptive in this population. Extensive marketing of artificial milk formulations in third-world countries can be expected to greatly diminish the natural birth-spacing effect of lactation.

> 1 What factors are responsible for priming of the breasts for milk production during gestation?
> 2 What are the roles of prolactin and oxytocin in lactation?
> 3 What is the effect of prolactin secretion on gonadotropin secretion and ovulatory cycling?

SUMMARY

1. The **sexual response** is a whole-body response involving both branches of the autonomic nervous system as well as the central nervous system. The response can be divided into **excitement, plateau, orgasmic,** and **resolution** stages.

2. After fertilization in the upper Fallopian tube, the zygote divides to form first a **morula** of undifferentiated cells and then a **blastocyst** consisting of a **trophoblast** and **embryoblast.** After implantation the trophoblast becomes the **chorion,** which gives rise to the fetal component of the **placenta.** The embryoblast becomes an **embryo,** which after the eighth week of development is termed a **fetus.**

3. Implantation is followed quickly by secretion of **human chorionic gonadotropin,** which sustains the corpus luteum for about the first 4 weeks of pregnancy.

4. The placenta exchanges nutrients and gases with maternal blood and protects the fetus from attack by the mother's immune system. The placenta also secretes estrogen and progesterone to maintain a hospitable uterine environment and **placental lactogen** to prepare the breasts for lactation and to mobilize nutrients from maternal stores.

5. The fetal circulatory system has three shunts: the **ductus venosus** that allows blood to reach the vena cava from the umbilical vein without passing through the hepatic circulation; the **ductus arteriosus,** a route between the pulmonary arteries and the aorta; and the **foramen ovale** between the right and left atria. The last two shunts allow blood to bypass the pulmonary loop while the lungs are collapsed and nonfunctional. The shunts normally close shortly after birth. The fetal blood must load and transport adequate oxygen at a lower partial pressure than adult blood, so the oxygen dissociation curve of fetal hemoglobin is left-shifted compared to adult hemoglobin.

6. Development of organ systems is initiated during the embryonic period—the first 8 weeks of development. During this period the development process is most sensitive to drugs and toxins. By the end of the second trimester, organ systems are almost completely differentiated. The third trimester is a period of growth and nutrient storage that prepares the fetus for independent life.

7. **Labor** is normally initiated after 39 weeks of gestation and consists of three stages: in stage one uterine contractions dilate the cervix; in stage two the baby passes through the cervix; and in stage three the placenta is delivered. Initiation of labor is not fully understood. The progress of labor is assisted by the hormone **oxytocin,** which stimulates uterine contraction. As term approaches, there is an increase in the number of oxytocin receptors. The first contractions of labor initiate a positive feedback reflex that increases oxytocin secretion from the posterior pituitary. Oxytocin also stimulates secretion of a prostaglandin that is important in dilation of the cervix.

8. The breasts are primed for **milk production** by high levels of estrogen and progesterone secreted by the placenta during pregnancy. Milk production is hypothesized to be initiated after delivery by release of a small polypeptide from the posterior pituitary that stimulates secretion of prolactin from the anterior pituitary.

9. **Milk letdown** is a hormonal reflex in which stimulation of the nipples causes oxytocin secretion, which in turn causes the smooth muscle of the mammary ducts to contract.

10. Lactation can suppress ovulation for some time after pregnancy, especially if the infant is fed at frequent intervals during the day and receives no other food.

STUDY QUESTIONS

1. What factors stimulate oxytocin secretion? What is the role of oxytocin in labor? In lactation? Are the target tissues involved similar in any way?

2. How is prolactin secretion controlled? What is the role of prolactin in lactation?

3. What is an ectopic pregnancy? What are the possible outcomes of such a pregnancy?

4. Name six placental hormones and describe their effects on maternal physiology.

5. Why is it important to augment the mother's nutrition early in pregnancy?

6. If premature delivery is anticipated, the fetus may be treated with cortisol. What is the benefit of this?

7. During the first 8 weeks of development, the development of a child is most threatened by drugs and toxins. What developmental processes are going on during this time?

8. If the ductus arteriosus fails to close after birth, what effect would this probably have on:
   a. Blood flow through the systemic loop
   b. The oxygenation of systemic arterial blood
   c. The pressure of blood in the pulmonary loop

9. What are the roles of progesterone in pregnancy? Progesterone is sometimes used medically to prevent a threatened premature delivery. What is the reasoning behind this use?

10. How do the structure and oxygen affinity of fetal hemoglobin compare to those of adult hemoglobin? What is the adaptive value of the difference in oxygen affinity?

SUGGESTED READING

BEN-JONATHAN, N., J.F., HYDE, and I. MURAI: Suckling-Induced Rise in Prolactin: Mediation by Prolactin-Releasing Factor from Posterior Pituitary, *News In Physiological Sciences*, volume 3, 1988, pp. 172-175. Short review of evidence for control of prolactin secretion by dopamine and an unknown peptide from posterior pituitary.

LAGERCRANTZ, H., and T.A. SLOTKIN: The "Stress" of Being Born, *Scientific American*, April 1986, pp. 100-107. Gives more information on the adaptive effects of catecholamine secretion during birth.

SHORT, R.V.: Breast Feeding, *Scientific American*, April 1984, pp. 35-41. Describes some recent hypotheses about the control of prolactin secretion and the inhibition of ovulation in lactating women.

SLOANE, E.: *Biology Of Women*, 2nd edition, John Wiley & Sons, New York, 1985. A general text with excellent treatment of sexual physiology, contraception, and pregnancy.

APPENDIX *I*

Directional Terms for Humans

Term	Definition
Inferior	A structure lower than another
Superior	A structure higher than another
Anterior	The front of the body (ventral)
Ventral	Towards the belly (anterior)
Posterior	The back of the body (dorsal)
Dorsal	Toward the back (posterior)
Cephalic	Closer to the head
Caudal	Closer to the tail
Proximal	Closer than another structure to the point of attachment to the trunk
Distal	Further than another structure from the point of attachment to the trunk
Lateral	Away from the midline of the body
Medial	Toward the midline of the body
Ipsilateral	On the same side of the body
Contralateral	On the opposite side of the body
Superficial	Toward or on the surface
Deep	Away from the surface (internal)

APPENDIX II

Physical Constants

Constant	Symbol	Value
Atomic mass unit	amu	1.6606×10^{-27} kilograms
Avogadro's number	N	6.02214×10^{23}/moles
Boltzman constant	k	1.3807×10^{-23} joules/°K
Electron charge	e	1.60218×10^{-19} coulomb
Faraday constant	F	96,845 coulombs/mole
Gas constant	R	8.3144 joules/°K/mole

APPENDIX III

Units and Conversion Factors

Property	SI unit	Conversion factors
Length	Meter	1 meter = 1.0936 yards
		1 centimeter = 0.3937 inch
		1 inch = 2.54 centimeters
		1 mile = 1.6093 kilometers
		1 kilometer = 0.62137 mile
		1 angstrom = 0.1 nanometer
Mass	Gram	1 kilogram = 2.2046 pounds
		1 pound = 453.59 grams
Pressure	Pascal	1 pascal = 1 kilogram/m/sec^2
		1 atmosphere = 106 pascals
		1 atmosphere = 760 mm Hg (torr)
		1 millibar = 100 pascals
Energy	Joule	1 joule = 0.239 calorie
		1 calorie = 4.184 joules
		1 British Thermal Unit (BTU) = 1055.06 joules = 252.2 calories
Volume	Liter	1 liter = 1.0567 quarts
		1 quart = 0.94633 liter
Temperature	Kelvin	0 °K = −273.15 °C = −459.67 °F
		°C = 5/9 (°F − 32)
		°F = 9/5 (°C + 32)

Physiological Reference Values

Parameter	Normal range	Units
Cellular elements of the blood		
Hematocrit	40 to 54	
Hemoglobin	14 to 18	gm/dl
Red blood cells	4.6 to 6.2	Million/mm^3
White blood cells	5000 to 10,000	Cells/mm^3
Blood plasma		
Albumin	3.2 to 5.6	gm/dl
Bicarbonate	21 to 27	mEq/L
Calcium (total)	4.3 to 5.3	mEq/L
Calcium (ionized)	2.1 to 2.6	mEq/L
Chloride	95 to 103	mEq/L
Cholesterol	150 to 250	mg/dl
Immunoglobulins	2.3 to 3.5	gm/dl
Glucose	65 to 100	mg/dl
Iron	50 to 150	μg/dl
Lactic acid	5 to 20	mg/dl
Magnesium	1.5 to 2.5	mEq/L
Osmolarity	280 to 290	mOsm/L
pH	7.35 to 7.45	
Phosphate	1.8 to 2.6	mEq/L
Potassium	4.0 to 4.8	mEq/L
Sodium	136 to 142	mEq/L
Sulfate	0.2 to 1.3	mEq/L
Urea	8 to 20	mg/dl
Uric acid	2.1 to 7.6	mg/dl
The cardiovascular system		
Cardiac output	4.0 to 8.0	L/min
A-V O$_2$ difference	3.5 to 4.7	ml/dl
Systemic arterial		
Central venous pressure	2 to 7	mm Hg
Diastolic pressure	60 to 90	mm Hg
Mean pressure	70 to 105	mm Hg
Systolic pressure	90 to 140	mm Hg
Pulmonary arterial		
Diastolic pressure	4 to 13	mm Hg
Mean pressure	9 to 19	mm Hg
Systolic pressure	17 to 32	mm Hg
Left ventricle		
Diastolic pressure	5 to 12	mm Hg
Systolic pressure	90 to 140	mm Hg

Parameter	Average value	Units
The lung volumes		
Static lung volumes and capacities		
Alveolar ventilation	5000	ml/min
Anatomical dead space	150	ml
Expiratory reserve volume (ERV)	1200	ml
Functional residual capacity (FRC)	2500	ml
Inspiratory capacity (IC)	3500	ml
Minute volume	7500	ml/min
Residual volume (RV)	1200	ml
Tidal volume (TV)	500	ml
Total lung capacity (TLC)	6000	ml
Vital capacity (VC)	5000	ml
Dynamic lung volumes		
Forced expiratory volume (FEV$_{1.0}$)	4000	ml
Forced vital capacity (FVC)	5000	ml

Parameter	Normal range	Units
The blood gases		
Arterial blood gases		
Carbon dioxide content	47 to 51	ml/dl
Carbon dioxide partial pressure (P$_{CO_2}$)	35 to 45	mm Hg
Oxygen content	16.5 to 20.0	ml/dl
Oxygen partial pressure (P$_{O_2}$)	93 to 97	mm Hg
Oxygen saturation	96 to 98	Percent
Venous blood gases		
Carbon dioxide content	51 to 55	ml/dl
Carbon dioxide partial pressure (P$_{CO_2}$)	41 to 51	mm Hg
Oxygen partial pressure (P$_{CO_2}$)	35 to 45	mm Hg
Oxygen saturation	70 to 80	Percent
The kidney and urine		
Ammonia (normal)	20 to 70	mEq/24 hr
Ammonia (acid load)	70 to 160	mEq/24 hr
Calcium	100 to 250	mEq/24 hr
Chloride	110 to 250	mEq/24 hr
Creatinine clearance	100 to 135	ml/min
Osmolarity (normal)	500 to 800	mOsm/L
Osmolarity (extremes)	40 to 1400	mOsm/L
pH (normal)	6.0 to 6.6	
pH (extremes)	4.6 to 8.0	
Phosphate	0.9 to 1.3	gm/24 hr
Potassium	25 to 100	mEq/24 hr
Sodium	130 to 260	mEq/24 hr
Urea	6 to 17	gm/24 hr
Uric acid	250 to 750	mg/24 hr
Volume (normal)	600 to 1600	ml/24 hr

Minerals and Vitamins

Mineral	Minimal daily requirement	Functional role
Important minerals in the body		
Calcium	0.8 to 1.2 gm	Bone and teeth formation; blood clotting; muscle contraction; nerve activity
Chlorine	1.7 to 5.1 gm	Blood acid-base balance; acid production (stomach)
Chromium	0.05 to 0.20 mg	Associated with enzymes in glucose metabolism
Cobalt	Unknown	Component of vitamin B_{12}; erythrocyte production
Copper	2.0 to 3.0 mg	Hemoglobin and melanin production; component of electron transport chain
Fluorine	1.5 to 4.0 mg	Extra strength to teeth
Iodine	100 to 150 μg	Thyroid hormone production; maintenance of normal metabolic rate
Iron	10 to 18 mg	Component of hemoglobin; ATP production in the electron transport chain
Magnesium	300 to 350 mg	Coenzyme constituent; bone formation; muscle and nerve function
Manganese	2.5 to 5.0 mg	Hemoglobin synthesis; growth; activation of several enzymes
Molybdenum	0.15 to 0.5 mg	Enzyme component
Phosphorus	800 to 1200 mg	Bone and teeth formation; important in energy (ATP) transfer; component of nucleic acids
Potassium	1.8 to 5.6 gm	Muscle and nerve function; heart rate
Selenium	0.05 to 0.2 mg	Component of many enzymes
Sodium	1.1 to 3.3 gm	Osmotic pressure regulation; nerve and muscle function
Sulfur	Unknown	Component of hormones, vitamins, and proteins
Zinc	10 to 15 mg	Component of several enzymes; carbon dioxide transport and metabolism; requirement for protein metabolism

Vitamin	Fat or water soluble	Minimal daily requirement	Function
The principal vitamins			
A$_1$ (retinol)	F	1.7 mg	Rhodopsin synthesis; bone growth; health of epithelial cells
B$_1$ (thiamine)	W	1.6 mg	Carbohydrate and amino acid metabolism; growth requirement
B$_2$ (riboflavin)	W	1.8 mg	Component of FAD; involved in Krebs cycle
Pantothenic acid	W	Unknown	Constituent of coenzyme A; glucose production; steroid hormone synthesis
B$_3$ (nicotinic acid)	W	20 mg	Component of NAD; glycolysis and Krebs cycle
B$_6$ (pyridoxine)	W	Unknown	Involved in amino acid metabolism
Folic acid	W	0.25 mg	Hematopoiesis; nucleic acid synthesis
B$_{12}$ (cyanocobalamin)	W	1.2 mg	Erythrocyte production; amino acid metabolism
C (ascorbic acid)	W	80 mg	Collagen synthesis; protein metabolism
D (cholecalciferol)	F	11 mg	Promotes calcium and phosphorus use; growth; bone and teeth formation
E (α-tocopherol)	F	Unknown	Prevents catabolism of some fatty acids
H (biotin)	W	Unknown	Fatty acid and purine synthesis; entry of pyruvate into Krebs cycle
K	F	100 µg	Required to make clotting factors

APPENDIX VI

Answers to Study Questions

CHAPTER 1

1. "Animals eat because they need nourishment." "The heart beats faster during exercise because the exercising muscles need more blood." "Body temperature increases during infection because the higher temperature inhibits reproduction of microorganisms." If some life principle made organisms different from inanimate objects, then all questions could be settled with the statement, "Organisms are the way they are because the life principle causes them to be like that."
2. The internal environment of multicellular animals can be regulated; the presence of many cells allows for functional specialization of different cell types.
3. Movement of solutes by diffusion occurs at rates high enough to be physiologically significant if the distance is of the order of microns to tens of microns; if the distance is greater, diffusion rates are much too low to be of physiological importance.
4. Environmental stress is the sum of processes that tend to bring the organism into chemical and physical equilibrium with the environment. Depending on the environment, these processes might include heating, cooling, drying, dilution, and so forth.
5. Natural selection is the term for the process by which environmental stress, through its differential effects on survival and reproduction of genetically different individuals in a population, results in predominance of genes that favor reproduction and survival.
6. See p. 14 and Figure 1-5.

CHAPTER 2

1. See the Glossary.
2. Examples of the bond types are given in order of stability. Examples of covalent bonding: the bond between oxygen atoms to form O_2; bonds between C and H in carbohydrates.
   Examples of ionic bonding: between Na^+ and Cl^- to form NaCl; between Na^+ and HCO_3^- to form $NaHCO_3$.
   Examples of hydrogen bonding: between the $-H$ and $-O-$ atoms of adjacent water molecules; between -H and -OH or -NH groups of proteins.
   An example of van der Waals interactions: attraction between adjacent nonpolar side groups of proteins.
3. The number of unpaired electrons.
4. Polar molecules have unequal distributions of charge. The bond angle between the H atoms of water is determined by the positions of the orbitals of the unpaired electrons of the O atom. See Table 2-2.
5. When each molecule of the solute dissociates into more than one particle in solution, since osmotic pressure is determined by the concentration of particles rather than by the molar concentration.
6. They constitute one of the forces between different regions of protein molecules that cause the amino acid chains of proteins to assume their normal folded shape.
7. 10^{-7} M or 0.1 μM
8.

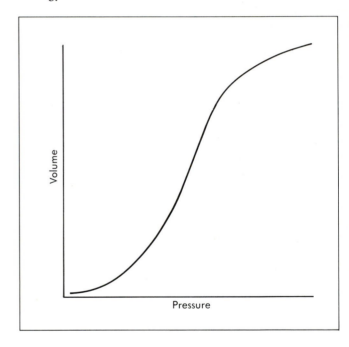

9. Assembling a macromolecule requires energy to bring the constituents together and to form chemical bonds; the Second Law of Thermodynamics says that the increased order (decreased entropy) of the macromolecule as compared to its free constituent atoms must be more than matched by increases in entropy elsewhere in the universe.
10. See Figure 2-17.

CHAPTER 3

1. See the Overview and relevant subheads in the chapter.
2. All enzyme-mediated reactions involve binding of the reactants to an active site of the enzyme, which reduces the barrier of activation energy that must be overcome to allow the reaction to occur. The reaction rate depends on the concentration of enzyme and of reactants and products; when the

concentrations of reactants are such that the active sites of all enzyme molecules are continuously occupied, the enzyme is said to be saturated and the reaction rate is maximal.

3. DNA replication—S period
   chromosome condensation—prophase of mitosis
   chromosome separation—anaphase of mitosis

4. See "DNA Replication" under **Nucleic Acids.**

5. See "DNA Transcription into RNA" under **Protein Synthesis.**

6. In posttranscriptional processing, intron portions of the primary transcript are removed and the remaining exons may be spliced together in different combinations to form several different mRNA sequences.

7. The genetic code is *redundant* in that a given amino acid may be specified by more than one base triplet, or codon.

CHAPTER 4

1. See relevant subheadings in the chapter.

2. See Table 4-2.

3. The cells of tight epithelia are joined by one or more continuous bands of tight junctions that prevent larger solutes from passing between the cells. In leaky epithelia the "tight" junctions are discontinuous, allowing large solutes and water to pass through the epithelium by paracellular pathways.

4. In the absence of a nucleus, the cell can neither reproduce by mitosis nor direct the synthesis of new messenger RNA for protein synthesis. Without mitochondria, the red cells must rely on anaerobic glycolysis for energy production.

5. The two major pathways of energy metabolism are glycolysis, and the Krebs cycle. Glycolysis starts with carbohydrate and mediates substrate-level phosphorylation. In cells with mitochondria, glycolysis is a source of pyruvate for the Krebs cycle. The Krebs cycle also is the major pathway for energy production from catabolism of protein and fat. Glycolysis can be carried out without oxygen and can respond rapidly to changes in demand, appropriate characteristics for some skeletal muscle cells. Glycolysis is far less efficient than the Krebs cycle, which is used by almost all cells that must sustain a steady, high rate of energy production. Organs that depend primarily on the Krebs cycle include the heart, brain, and skeletal muscle cells that fatigue slowly.

6. Aerobic energy metabolism is roughly 10 to 20 times more efficient than anaerobic glycolysis at converting the energy of substrates into ATP.

7. As electrons move from NADH and FADH along the series of iron-containing molecules called cytochromes, a small amount of energy is released at each step. Some of this energy is used to drive the transport of H^+ across the mitochondrion to the outside. As the H^+ moves out, energy is transferred to protein complexes that can phosphorylate ADP.

8. The amino acids produced by protein degradation and the fatty acids and glycerol produced by fat metabolism are converted to acetyl-CoA, which enters the Krebs cycle. The energy yield of protein metabolism is very similar to that of carbohydrate metabolism, but fats yield 2.3 times as much energy per gram as glucose.

CHAPTER 5

1. A simple feedback loop comprises one or more *sensors*, an *integrator*, and one or more *effectors*.

2. The gain affects how far the regulated variable will depart from the set point in response to perturbing factors. Given an error signal of constant magnitude, the higher the gain, the greater the effectors' response and the sooner the regulated variable returns to the set point.

3. Time delays in the feedback loop tend to cause the effectors' response to lag behind the changes in the error signal. If the error signal is continuously changing, the effectors' response never quite catches up with the changes; the result is termed a phase lag.

4. At the intracellular level, the local or tissue level, and the extrinsic level of nervous or endocrine responses that regulate organ systems.

5. Some examples are given in the section "Physiological Examples" under **Homeostatic Control Theory;** others are given under **Endocrine Reflexes.** Perhaps you can think of still others.

6. See "Hormones as Chemical Messengers" under **Endocrine Reflexes.**

7. Whether the target organ possesses receptors for the hormone.

8. Generally, hormones may affect the activity of enzyme systems already present in the cell, or they may affect the rate of synthesis of enzymes by affecting the relevant genes. For specifics see "Hormone Functional Classes" under **Endocrine Reflexes.**

9. Prostaglandins are synthesized from cell membrane lipids and typically act at the local level.

10. See "Hormone Functional Classes" under **Endocrine Reflexes.**

11. The posterior pituitary contains secretory nerve terminals that release small peptide neurohormones. The anterior pituitary cells are not neurons; their hormones are proteins.

12. In many cases hormone secretion does not occur at a steady rate, but rather occurs in pulses throughout the day or at particular times of day, or in response to particular temporary challenges such as exercise, eating, or stress or excitement.

CHAPTER 6

1. In the fluid mosaic model, the lipids of the cell membrane are fluid enough to allow membrane proteins to be mobile within the plane of the membrane but not to rotate through the plane of the membrane or to migrate from one side to the other. Thus proteins that initially are inserted into one or the other of the membrane surfaces usually remain there, and proteins that span the membrane maintain a consistent orientation with respect to the two membrane surfaces.

2. Some factors that might be involved, depending on the solute:
   a. The lipid solubility of the solute
   b. The size of the solute
   c. For ionic solutes, whether the solute could exist in an uncharged (and thus more lipid-soluble) form
   d. The existence in the membrane of proteins that formed specific permeability pathways or acted as transport proteins for that solute

3. a. For a nonelectrolyte, 300 mOsm = 300 mM

b. For a solute that dissociated into two univalent ions (for example, NaCl), 150 mM = 300 mOsm (neglecting the isotonic coefficient)

c. For a solute that dissociated into one divalent and two univalent ions (for example, $CaCl_2$), 100 mM = 300 mOsm (again neglecting the isotonic coefficient)

4. The answer depends on which solutes the cell membrane is permeable to. There cannot be an osmotic gradient across the cell membrane, but only impermeant (or effectively impermeant) solutes can contribute to osmotic movement of water across the membrane.

5. The Na^+-K^+ pump makes the membrane effectively impermeable to Na^+, allowing extracellular Na^+ to balance the osmotic effect of intracellular proteins.

6. Two facts explain the inside-negativity of the resting potential: the membrane is mainly permeable to K^+ and the K^+ gradient is outwardly directed. See "Determinants of the Resting Potential of a Cell" under **The Membrane Potential.**

7. Increasing the extracellular K^+ decreases the magnitude of the transmembrane K^+ gradient and shifts the membrane potential nearer to the Na^+ equilibrium potential. The result is depolarization.

8. Mediated transport systems involve membrane transport proteins ("carriers") that are specific for particular molecular types and that can become saturated as the concentrations of the transported molecules are increased.

CHAPTER 7

1. These venoms prolong the open time of Na^+ channels, increasing the duration of the spike portion of the action potential. Since the channels remain open longer, the peak of the action potential has more time to approach the Na^+ equilibrium potential, increasing its amplitude slightly. Since the spike part of the action potential lasts longer, the absolute refractory period, which corresponds to this part of the action potential, also is longer.

2. See "Spread of Excitation" under **Propagation of Action Potentials.**

3. See Figures 7-13, 7-16, and 7-17.

4. The height or amplitude of the potential is determined by the number of Na^+ channels that open (in other words, the magnitude of the inward Na^+ current relative to the constant outward K^+ leak) and the length of time the channels remain open. A stimulus that exceeds the threshold level of depolarization sets in motion a positive feedback cycle that causes all the Na^+ channels to open. Thus action potentials are not graded to the stimulus magnitude.

5. Inactivation of Na^+ channels and opening of the voltage-sensitive K^+ channels.

6. Inactivated channels are latched in the closed state and cannot open again until the membrane repolarizes. Removing inactivation leaves the channels in the closed state; in this state they can open again in response to depolarization.

7. See "Effect of Diameter and Myelination on Propagation Velocity" under **Propagation of Action Potentials.**

8. Within a nerve trunk, each axon individually follows the all-or-none law; the compound action potential is the sum of electrical currents caused by the population of axons in the trunk that are excited. The threshold stimulus intensity for different classes of axons varies. Some are activated at low stimulus intensity; additional classes can be activated at higher stimulus intensities.

9. A reduction in the density of active channels raises the threshold - the initial depolarization must be much greater to get the positive feedback cycle started. If enough channels are blocked, propagation of an action potential becomes impossible and sensations cannot be transmitted from the anesthetized area to the central nervous system.

CHAPTER 8

1. See "Input and Output Segments of Neurons" under **Neuron Structure and Information Transfer.**

2. Typically telephones are not labeled lines because each one can receive calls from any other telephone. In contrast, an intercom system connecting your telephone with a second set at a known location would be a labelled line.

3. The receptive field of a cutaneous receptor could be mapped by recording from the receptor's axon while stimulating the skin with single touches of a mechanical stimulator such as a bristle or fine needle. Summated responses could be elicited by rapidly repeated touches in the same spot or by simultaneously touching more than one spot in the receptive field.

4. Temporal summation refers to repeated stimulation of the same restricted part of the input segment of a neuron. Spatial summation refers to simultaneous stimulation at different spots on the cell's input segment.

5. Depolarization is the result of an increase in Na^+ permeability, although some channels that mediate depolarizing receptor potentials admit other ions as well. Hyperpolarization typically is the result of increased K^+ permeability. Photoreceptors are an exception to this rule; their hyperpolarization in response to light is the result of a decrease in Na^+ permeability (some Na^+ channels are constantly open in the dark and are closed by light).

6. See **Generation of the Neural Code.** The key to the process is an initial graded response (the receptor potential) that gives rise to a proportionate frequency of action potentials in the output segment of the receptor.

7. Phasic receptors give information about the onset of a stimulus and the rate of change of its intensity; tonic receptors give information about steady states of stimulus intensity. Frequently, different responses are appropriate for "new" as compared to "old" stimuli. This could be achieved, for example, by neuronal wiring in which one set of effectors was excited by input from phasic receptors and inhibited by input from tonic receptors.

8. Conventionally, neurotransmitters mediate the millisecond-to-millisecond changes in permeability that are responsible for chemical transmission across synapses, whereas neuromodulators are responsible for longer-lasting changes in synaptic efficacy. By definition neurohormones must be delivered to their targets by blood flow. A number of substances can act as neurotransmitters, neuro-

modulators, or neurohormones; there is no reason why the same substance could not serve all three functions in different locations in the body.

CHAPTER 9

1. Typically areas with fine two-point discrimination have a disproportionately large representation in the primary somatosensory cortex (see Figure 9-17). The receptors of such areas typically have small receptive fields and the receptive fields overlap extensively, so the receptor density is high.
2. See "The Reticular Activating System" under **The Somatosensory System** and also **Sleep and the Wakeful State.**
3. Once laid down in long-term memory, memories usually are not completely abolished by damage to any discrete part of the brain.
4. Such deficits might be seen in people who have sustained damage to the language areas of the dominant hemisphere, particularly the Broca area or connections between it and other language areas. Musical skills typically are localized in the non-dominant hemisphere, particularly so in individuals who have not received extensive music training.
5. Mixed nerves include both afferent and efferent axons. This obtains for all spinal nerves and some but not all cranial nerves.
6. See Figure 9-14.
7. White matter contains myelinated axons; gray matter contains neuronal cell bodies. In the brain the outermost layers are gray and the innermost layers are white; in the spinal cord the reverse is true.
8. The dorsal column pathway (see Figure 9-16) carries modality-specific fine touch, proprioception, and sharp pain information to the primary somatosensory cortex by way of the thalamus; the non-specific pathway (see Figure 9-19) projects mainly to parts of the limbic system and carries information about sensations such as heat, cold, and dull or burning pain.
9. See **The Somatosensory System.**

CHAPTER 10

1. Sudden loud noises can damage the ear during the several tens of milliseconds of delay of the reflex loop.
2. See "Frequency Localization in the Cochlea" under **The Auditory System.**
3. Lateral inhibition between afferents from the cochlea is important in resolving pure tones of similar frequencies.
4. The polarity of the receptor potential of hair cells is determined by the direction of bending of the stereocilia. Rotations of the head cause the endolymph of the ampulla to move relative to the cupula; which way the stereocilia are bent depends on the direction of rotation of the head.
5. When the head begins to rotate, inertia causes the endolymph to lag behind the wall of the ampulla, bending the cupula. Stimulation of the hair cells drives compensatory eye movements (nystagmus). After a time, friction between the ampulla and the endolymph accelerates the endolymph so that both endolymph and ampulla are rotating at

the same velocity, and the cupula is no longer bent. When the head stops rotating, the endolymph's momentum causes it to surge ahead, bending the cupula in the opposite direction. The resulting nystagmus is thus in the opposite direction.
6. See Figure 10-25.
7. See Figure 10-24.
8. As one reasonable hypothesis, a simple cell might receive input from some on-center, off-surround ganglion cells whose receptive fields form a row on the retinal surface (see Figure 10-29). When a stimulus bar or border lies along the excitatory centers of the receptive fields, the simple cell is maximally excited. If the bar lies immediately on either side of the row of on-centers, it stimulates the off-surrounds of the ganglion cells, and the simple cell is maximally inhibited. The connections needed for a complex cell are clearly much more complicated. A complex cell would have to receive input from several parallel rows of ganglion cells, with neural connections between the rows to mediate the motion sensitivity. Currently such connections are thought to be made by amacrine cells, but the details of the circuitry are not clear.
9. Taste receptors are modified epithelial cells, whereas olfactory receptors are neurons. In both cases discrimination is not the product of receptors labeled lines, because individual receptors can respond to more than one flavorant or odorant. Thus to perceive a given odor or flavor, the brain must integrate input from a large number of receptors with different response spectra.
10. The number of distinct odors is so large that if there were primary odors, the number of primary odors would also have to be quite large. In practice almost all odor sensations arise from complex mixtures of odorant molecules.

CHAPTER 11

1. Proximal muscles of the trunk, which are involved in posture and gross movement, are innervated by motor neurons whose cell bodies are in the medial regions of the ventral horn of the spinal cord. Distal muscles located in the appendages and involved in manipulation are innervated by neurons whose cell bodies are in lateral regions of the ventral horn. The direct corticospinal tract or pyramidal tract forms direct connections between the motor cortex and the spinal motor neurons involved in finely controlled movements of arms, hands, and fingers.
2. See "Structure of the Muscle Spindle," "Muscle Spindles as Length and Length Change Detectors," and "Control of the Stretch Reflex by Gamma Motor Neurons" under **Spinal Reflexes and Postural Stability.**
3. As described in the section "Coactivation of Alpha and Gamma Motor Neurons" under **Spinal Reflexes and Postural Stability,** stretch reflexes are activated during voluntary movements if actual muscle length deviates from a new set-point length set by gamma activation. The additional muscle fibers recruited by the reflex increase the gain of the feedback loop. In alpha-gamma coactivation, it is the activation of gamma motor neu-

rons that resets the length set point of the reflex. If the reflex set point were not reset, the reflex would oppose the voluntary length change.

4. See "Golgi Tendon Organs: Monitors of Muscle Tension" under **Spinal Reflexes and Postural Stability.**

5. Muscle groups that move a joint in one direction are opposed by antagonist muscles that move the joint in the opposite direction. Typically the spinal circuits are arranged so that when the motor neurons serving agonists are activated, those serving antagonists are inhibited; this is reciprocal innervation. This innervation is important both for reflexes that mediate limb extension (stretch reflex and crossed extension) and those that mediate rapid flexion (withdrawal reflex), because it prevents the tone of antagonist muscles from acting as a drag on agonists when the reflexes are activated.

6. Facilitation is a use-related increase in the responsiveness of a motor neuron to excitatory synaptic input from afferents. It frequently is the result of spatial summation of increased excitatory input to the motor neurons from central interneurons. The sensitivity of spinal stretch and postural reflex pathways is modulated by input from the vestibular system of the brainstem along the vestibulospinal pathway.

7. See "The Cerebellum and Coordination of Rapid Movements" under **Brainstem Motor Areas.**

8. The caudate nucleus, putamen, and globus pallidus. The evidence that these subcortical nuclei possess motor programs is twofold: in primates, damage to the basal ganglia causes execution of inappropriate motor programs; in nonprimates, a fairly wide range of activities can be carried out by decorticate animals.

9. See "Motor Programs in the Central Nervous System" under **Motor Control Centers of the Brain.**

10. The cholinergic receptor types are nicotinic and muscarinic; the adrenergic receptor types are alpha$_1$, alpha$_2$, beta$_1$, and beta$_2$. See Tables 11-5 and 11-6 for examples of their locations, and Table 11-4 for examples of their effects.

11. See Summary items 1, 2, and 3.

12. Two important examples are vascular smooth muscle, which in most parts of the body receives only sympathetic (vasoconstrictor) input, and cardiac ventricular muscle fibers, which receive only sympathetic (excitatory) input. See Table 11-4. The effect of autonomic input is determined not only by which receptor type(s) the effector possesses, but also by the nature of the second message(s) set in motion by the binding of transmitter to receptor in that particular tissue. Thus activation of the same receptor type may have either excitatory or inhibitory effects, depending on the tissue involved.

CHAPTER 12

1. The events can be divided into three groups: presynaptic events, postsynaptic events, and events that involve the contractile machinery. The presynaptic events culminate in transmitter release; the postsynaptic events begin with binding of ACh to the nicotinic receptors of the postsynaptic membrane and lead to initiation of an action potential in the muscle cell. The muscle action potential causes Ca^{++} release from the SR; the released Ca^{++} activates the contractile machinery. The details of each step in the sequence are found under appropriate headings in the text.

2. *Transverse tubules* conduct action potentials into the interior of muscle cells and mediate release of *inositol triphosphate*, causing the terminal cisternae of the *sarcoplasmic reticulum* to release Ca^{++} that it had sequestered during rest. The Ca^{++} diffuses to regulatory subunits of *troponin*, causing it to undergo a conformational change that moves *tropomyosin* away from actin-myosin binding sites on the thin filaments. Myosin heads bind to actin and contraction begins.

3. In the absence of ATP, myosin heads are irreversibly bound to actin in the "uncocked" state and are unable to undergo power strokes—the resulting condition of stiffness is rigor mortis. In the absence of Ca^{++}, myosin heads are "cocked," but tropomyosin prevents them from binding to actin; this is the condition characteristic of resting muscle.

4. In an isotonic contraction the muscle shortens against a constant load; in an isometric contraction the muscle is prevented from contracting. A person lifting a load is performing what is approximately an isotonic contraction. The contraction is not perfectly isotonic because the effect of leverage changes the load on individual muscles as the elements of a joint move relative to one another. Isometric contractions occur when a person attempts to lift a load that is too heavy or when a load is held suspended in one position.

5. In the flexors of the left arm, which has a trivial load, relatively few motor neurons would be active at any instant and most of the active ones would serve slow muscle units. In the flexors of the right arm, many slow motor units would be active and some would show rapid bursts of action potentials. Some motor units of fast muscle fibers probably will have been recruited also, but since 1 kilogram is considerably less than the maximum load that most people can support with one hand, many fast units probably will not be recruited at all; others will be active for only brief periods.

6. See "Skeletal Muscle Fiber Types and Energy Metabolism" under **Muscle Energetics and Metabolism.**

7. See "Muscle Fatigue" under **Muscle Energetics and Metabolism.**

8. Important differences include:
   Contractile machinery: Compare Figure 12-9 to Figure 12-30.
   Organization of myofibrils within the cell: Compare Figures 12-2 and 12-27.
   Contractile properties: See Figures 12-15 and 12-29.
   Mechanism of excitation-contraction coupling: See Table 12-5.
   Innervation: See Figure 12-31.

9. The contractile machinery of smooth muscle must frequently resist steady forces. For example, the smooth muscle of arteries must constantly resist blood pressure that otherwise would distend the

arterial system. Latch bridges are an economical way of maintaining such resistance to stretch over long periods of time.

10. Single-unit smooth muscles are those in which activation of a few fibers causes the entire population of fibers to contract, as a result of electrical coupling between the fibers. Cardiac fibers contract synchronously during a heartbeat because they are coupled by gap junctions. For the heart, as for single-unit smooth muscle, the stimulus to contract originates in a small population of pacemaker fibers. Each fiber is innervated in multiunit smooth muscle.

CHAPTER 13

1. Since all cardiac fibers are activated during each heartbeat, the only means available to increase the total force is an increase in the force produced by each myocardial fiber. Such increases may result from increased ventricular filling, increased adrenergic input, or both. See "Autoregulation of Stroke Volume" and "Determinants of Cardiac Output" under **The Heart As a Pump.** In skeletal muscle more than enough Ca^{++} is released during a twitch to saturate all of the troponin molecules and thus activate all of the myosin heads. This is believed not to be the case in cardiac muscle, so that in cardiac muscle, unlike skeletal muscle, intracellular Ca^{++} levels can modulate force generation.

2. See Figure 13-19.

3. See Figure 13-19.

4. During ventricular contraction, the atria are repolarizing; the electrical signal of atrial repolarization is not normally seen in the ECG because it is overwhelmed by the much larger signal of ventricular depolarization. During this phase the pressure in the left ventricle is greater than atrial pressure but less than aortic pressure.

5. The plateau reflects the duration of the slow inward current of Na^+ and Ca^{++}; another factor that contributes to the length of the plateau is the the slow onset of the outward K^+ current, compared to nerve and skeletal muscle.

6. See Figures 13-16 and 13-19.

7. The fibers with the most rapid rate of diastolic depolarization have the highest intrinsic rate and drive the rest of the heart at their pace.

8. See "Irregularities of the Cardiac Rhythm" under **The Heart As an Excitable Tissue.**

9. See Figures 13-7 and 13-17. Reentry is described in "Irregularities of the Cardiac Rhythm."

10. See "Excitation-Contraction Coupling in Cardiac Muscle Fibers" under **The Heart As an Excitable Tissue,** and Figure 13-12. The effect of adrenergic input on myocardial cells is an increase in the force of contraction and a decrease in the duration of the active state, making systoles both more forceful and briefer.

11. See "Autoregulation of Stroke Volume" under **The Heart As a Pump.**

12. See "Determinants of Cardiac Output" under **The Heart As a Pump,** and Figure 13-23.

CHAPTER 14

1. As in other regulatory cascades, multiple steps provide for amplification of the signal and allow several processes to be initiated or controlled by a single event. Since each step in the cascade typically is regulated by allosteric effectors, multiple steps also allow for finer control than a single step. Review the discussion of regulatory cascades in Chapter 5. In the early stages of clot formation, platelets that have been made sticky by contact with an abnormal surface release ADP, a signal that causes nearby platelets also to become sticky. As fibrin is formed, more and more platelets are stimulated to release ADP. Negative feedback is provided by formation of the clot-dissolving substance plasmin as one product of the clotting factor cascade.

2. See Figure 14-11. See Figure 14-16.

3. Because there are many more capillaries than arterioles, the net cross-sectional area of all capillaries taken together is greater than that of all arterioles taken together. Thus the total resistance of capillaries is much less than the total resistance of arterioles.

4. An increase in tissue colloid osmotic pressure would decrease the transcapillary osmotic gradient. The resulting increase in interstitial fluid volume would tend to cause edema. An increase in venous pressure would increase the capillary hydrostatic pressure and cause an increase in interstitial fluid volume. See Table 14-4.

5. The elasticity of the walls of the arterial tree normally decreases with age. The resulting decrease in vessel compliance (see Figure 14-10) causes systolic pressure to increase.

6. A decrease in venous compliance would decrease the venous volume and increase the venous pressure. Venous return would increase, but total peripheral resistance would hardly be affected because the contribution of venous resistance to total peripheral resistance is very small in any case. Arteriolar dilation decreases total peripheral resistance, making blood flow from arteries to veins easier and increasing venous return.

7. The outcome depends on two factors, the number of branches and their dimensions. Generally the diameter of branch vessels is less than that of the stem. Because a vessel's resistance is very sensitive to its diameter, each branch vessel has a much higher resistance than the stem. The decrease in diameter with branching tends to increase the net resistance, but this effect may be overcome if the number of branch vessels is very large.

8. See Figures 14-4 and 14-6.

9. The total volume of blood in the tube corresponds to $29 + 21$ mm $= 50$ mm. The hematocrit is $(21$ mm$/50$ mm$) \times 100 = 42\%$.

10. a. Vasoconstriction: decrease in blood flow; the dominant effect in most elements of the peripheral circulation

   b. Vasodilation: increase in blood flow; important in skeletal muscle in the presence of moderate levels of epinephrine

   c. Vasodilation: increase in blood flow; occurs in skeletal muscle before exercise

   d. Vasodilation: increase in blood flow; responsible for increased blood flow to the genital tissues during sexual response

11. The ratio of flow resistances is $1:2^4$, reflecting that resistance is proportional to the fourth power of the radius. This means that flow through the smaller vessel is 1/16 of that through the larger vessel. The total flow through both vessels is 10 ml/min; the flow through the smaller vessel is 10 ml/min $\times$ 1/16 = 0.625 ml/min, and the flow through the larger vessel is 9.375 ml/min.

CHAPTER 15

1. Effects on venous return of:
   transfusion: increase
   arteriolar constriction: decrease
   decreased venous compliance: decrease
   (See Figure 15-8)
2. Arteriolar dilation in exercising muscles increases venous return. The Frank-Starling :aw results in an increase in cardiac output.
3. Assuming there is no reflexive compensation, increases in myocardial contractility increase mean arterial pressure. Cardiac output increases, but central venous pressure decreases (see Figure 15-9).
4. See Figure 15-9.
5. In response to sudden transition to upright posture:
   a. baroreceptor firing rate decreases
   b. parasympathetic input to the SA node decreases and sympathetic input increases, resulting in an increased heart rate
   c. capillary hydrostatic pressure throughout the lower body increases
   d. increased sympathetic input to veins decreases venous compliance
   e. other reflexive responses: increased sympathetic activity constricts arterioles and increases myocardial contractility
6. The decrease in blood volume would decrease mean arterial pressure, except that arterial pressure is protected by the baroreceptor reflex. The reflexive increase in arteriolar constriction decreases the mean capillary hydrostatic pressure, shifting the balance of capillary filtration to favor net movement of fluid from interstitial spaces into the capillaries. The resulting transfer of fluid from interstitial spaces to the plasma replaces the lost plasma volume within a short time, but the lost erythrocytes cannot be replaced so rapidly. Thus the hematocrit decreases.
7. Before compensation for blood loss:
   cardiac output: decreased
   central venous pressure: decreased
   mean arterial pressure: decreased
   total peripheral resistance: unchanged
8. After reflexive compensation for blood loss:
   total peripheral resistance: increased
   central venous pressure: decreased
   cerebral blood flow: unchanged, because the reflex regulates mean carotid arterial pressure
   renal blood flow: reduced by reflexive arteriolar constriction
9. A heat load causes reflexive cutaneous vasodilation. The decrease in peripheral resistance caused by dilated skin arterioles is added to the decrease caused by dilated arterioles in exercising skeletal muscle. Under these conditions, it may not be possible to maintain mean arterial pressure without constriction of cutaneous arterioles, but such constriction would increase the heat load. Generally the blood flow through cutaneous vessels is managed reflexively to meet the needs of thermoregulation, as long as mean arterial pressure is normal. However, if mean arterial pressure drops significantly, the cardiovascular center causes cutaneous vasoconstriction, even if there is also a heat load.

CHAPTER 16

1. See Figure 16-2 and Table 16-1.
2. See Figure 16-5.
3. Normally, anatomical dead space is the volume of the airway structures, including nasal passages, trachea, bronchi, and bronchioles down to the level of respiratory bronchioles. For the diver breathing through a snorkel, the total dead space is 300 ml. The dead space ventilation at 12 breaths/min is 12 breaths/min $\times$ 300 ml/breath = 3600 ml/min. To maintain an alveolar minute volume of 3000 ml/min, the total ventilation must be 3600 ml/min + 3000 ml/min = 6600 ml/min. The tidal volume required at 12 breaths/min is (6600 ml/min) / (12 breaths/min) = 550 ml/breath.
4. See Table 16-3.
5. The "why" can pose at least three different questions. First, "What physical properties of the lung-chest wall cause a negative intrapleural pressure?" The answer is that the negative pressure is the result of the opposing outward spring of the chest wall and inward recoil of the lungs. Second, "What developmental processes led to this situation?" Since the volume of intrapleural fluid remains quite small, growth of the thorax relative to that of the lung during early development must increase the intrapulmonary volume and stretch the lung wall. Third, "What advantage could there be in having the lungs always partly inflated?" To answer this teleological question, the alternative must be considered. If the lungs were fully deflated at the end of a breath, the alveolar surfaces would become attached to one another by surface tension. Reinflating the lung for another breath would require considerable energy. Furthermore, gas exchange would come to a halt during the part of the respiratory cycle when the alveoli were completely deflated.
6. See "The Lung-Chest Wall System" under **Structure and Function in the Respiratory System.**
7. See Figure 16-17 and accompanying text. Some of the energy expended in a respiratory cycle is expended against the elasticity of the lung-chest wall system; the rest is expended against the flow resistance of the airway. The intrapleural pressure change reflects both the elastic and flow-resistive components; the alveolar pressure reflects only the flow-resistive component.
8. See Figure 16-15.
9. In the absence of pulmonary surfactant the alveolar surface tension is greater; therefore lung recoil is greater, and the functional residual capacity is decreased.
10. If lung compliance is decreased, the flow-

resistive component of the work of inflating the lungs increases and ventilation requires extra effort.

11. See Figure 16-22.
12. If airway resistance increases, maintenance of normal alveolar ventilation requires a greater pressure gradient between the atmosphere and the alveoli. Thus both the alveolar pressure and the intrapleural pressure must be more negative during inspiration. The resulting increase in the flow-resistive component of respiratory work increases the total work of breathing. Other consequences of increased airway resistance are noted in "Pulmonary Resistance and Compliance in Disease" under the heading **Structure and Function in the Respiratory System.**

CHAPTER 17

1. The effect of dead space (see Chapter 16) makes exchange of alveolar air with atmospheric air less than perfect (see "The Composition of Alveolar Gas" under **Gas Exchange in Alveoli**). Increases in the rate of oxidative metabolism, if not matched by changes in ventilation, would increase the P_{CO_2} and decrease the P_{O_2} of alveolar gas. Since the blood flowing through pulmonary capillaries equilibrates with alveolar gas, the P_{CO_2} of systemic arterial blood must increase and the P_{O_2} must decrease.
2. The relationship is described by the alveolar gas equation (see p. 431). This equation can be solved for either alveolar P_{CO_2} or alveolar P_{O_2} as the dependent variable.
3. The hemoglobin content. Usually the hematocrit is a good indicator of hemoglobin content. This assumes that essentially all of the hemoglobin is functional—if some of it is nonfunctional (for example, converted to carboxyhemoglobin by carbon monoxide), the oxygen-carrying capacity would be reduced proportionately.
4. If the oxygen-carrying capacity of blood were halved, as for example in a severe case of anemia, the P_{O_2} of systemic arterial blood would remain normal, but the O_2 content of systemic arterial blood would be halved. This would result initially in a decrease in O_2 delivery and a decrease in tissue P_{O_2}. A new steady state would be established when tissue P_{O_2} reached a value sufficiently low to extract O_2 from the blood at the normal rate. The P_{O_2} of mixed venous blood would reflect this new lower tissue P_{O_2}, and the arteriovenous P_{O_2} difference would be increased.
5. The equilibrium concentrations of myoglobin and oxymyoglobin for different P_{O_2} values are determined by the Law of Mass Action. The shape of the resulting function (the dissociation curve) is hyperbolic. The S-shaped dissociation curve of hemoglobin is the result of cooperation between the four subunits of the hemoglobin molecule—an allosteric effect that modifies the rate constant for binding of O_2 in response to the number of O_2 molecules already bound to the hemoglobin molecule.
The O_2 affinity of myoglobin is greater than that of hemoglobin, especially in the range of P_{O_2} values characteristic of systemic capillaries. Thus O_2 can

be efficiently transferred from hemoglobin to myoglobin for delivery to mitochondria.
6. Doubling of alveolar ventilation, without a proportional increase in oxidative metabolic rate, would decrease the P_{CO_2} of alveolar air and thus the P_{CO_2} of systemic arterial blood. Such a decrease would shift the plasma bicarbonate buffer equilibrium to the left, decreasing both plasma $[HCO_3^-]$ and plasma $[H^+]$. Thus plasma pH would increase. This is an example of respiratory alkalosis.
7. Small differences between the P_{O_2} of alveolar gas and arterial blood may arise as the result of ventilation-perfusion mismatch in the lung. This may cause physiological shunt, which is a net overperfusion of alveoli, or physiological dead space, a net underperfusion (see Chapter 16). In emphysema and other fibrotic lung diseases, the diffusional path for gas exchange between alveoli and pulmonary capillaries is increased, and pulmonary blood may not fully equilibrate with alveolar gas (see Figure 17-3). Such diffusional limitations apply both to O_2 and CO_2.
8. A: See Figure 17-11. B: See Figures 16-6 and 17-4, A.
9. See Figure 17-7.

CHAPTER 18

1. See Figure 18-2.
2. See Figure 18-8. Peripheral chemoreceptors respond primarily to P_{O_2}, but also to pH and indirectly to P_{CO_2}. Under normal conditions they contribute relatively little to respiratory drive. Halving the atmospheric pressure would halve the P_{O_2} of inspired air. Alveolar P_{O_2} would decrease; the input of peripheral chemoreceptors becomes important when arterial P_{O_2} falls below 60 to 70 mm Hg or when the pH of plasma is decreased.
3. See Figure 18-5. The exact location of the chemosensitive cells is not known. The central chemoreceptors respond to changes in the pH of cerebrospinal fluid; thus they respond indirectly to the P_{CO_2} of arterial blood. They make the largest single contribution to respiratory drive under normal conditions.
4. See "Arterial P_{CO_2}—the Central Chemoreceptors" under **The Chemical Stimuli for Breathing.**
5. Compare Figures 18-6 and 18-9.
6. See Figure 18-10 and the left side of Figure 18-14. Fourteen thousand feet corresponds to about 4300 meters, so at this altitude arterial hemoglobin is considerably less than fully saturated (which would tend to increase respiratory drive by the peripheral receptors), but plasma P_{CO_2} is also decreased (which depresses the central component of respiratory drive in unacclimated people). Because of the conflicting central and peripheral signals, full ventilatory compensation for the decreased P_{O_2} of inspired air is not possible until altitude acclimation has occurred. The reduced O_2 delivery to working muscles results in more rapid fatigue.
7. This person is hypoxic, but plasma pH and P_{CO_2} are normal. See Figure 18-11 and curve B in Figure 18-13. Ventilation is driven to three to four times the normal value by the peripheral chemoreceptors; however, the input from these

receptors also makes the respiratory control center more sensitive to any change in P_{CO_2}.

8. See Figure 18-12, which predicts a ventilatory increase of fourfold to fivefold. Since plasma P_{CO_2} is normal, the ventilatory response probably is driven by the peripheral chemoreceptors, especially if the change in plasma pH is very recent.

9. Because arterial pH is normal, the increased P_{CO_2} cannot affect peripheral chemoreceptors; therefore any effect must be on central chemoreceptors. If the increased arterial P_{CO_2} is recent, ventilation will be increased by severalfold (see Figure 18-6 and curve A of Figure 18-13).

10. Vagotomy abolishes inhibitory input from lung and chest wall stretch receptors that normally mounts during inspiration (see Figure 18-4). The afferents from aortic arch chemoreceptors also are severed, abolishing part of the input from central chemoreceptors. The immediate effect of cutting stretch receptor afferents is an increase in tidal volume without a change in respiratory rate. The resulting increase in alveolar ventilation causes a transient respiratory alkalosis. The feedback loop that runs from central chemoreceptors to the respiratory center is still intact, and the effect of the alkalosis on this loop decreases the respiratory rate to a level that restores plasma P_{CO_2} and pH to their set-point values. A few minutes after vagotomy the tidal volume is still increased, but the respiratory rate is reflexively decreased, and alveolar ventilation is close to its value before vagotomy.

CHAPTER 19

1. For the anatomical comparison, see Figure 19-4. The main known functional difference between the two is that juxtamedullary nephrons are involved in formation of concentrated urine, whereas cortical nephrons are not.

2. The anatomy of the renal circulation is shown in Figure 19-6. As shown in Figure 19-12, the net pressure gradient in glomerular capillaries favors filtration resulting in urine formation. In the peritubular capillaries the net pressure gradient is reversed, allowing fluid reabsorbed from primary urine to reenter the circulation.

3. Differential regulation of the tone of afferent and efferent arterioles allows the GFR to be maintained despite changes in RBF. In this case constriction of the afferent arterioles by sympathetic input is matched by constriction of the efferent arterioles, keeping the net pressure gradient between glomerular capillaries and the Bowman capsule constant. This form of autoregulation is called glomerulotubular balance and is mediated by the juxtaglomerular apparatus.

4. See Figure 19-21.

5. In Na^+ gradient-mediated cotransport, the energy stored in the transmembrane Na^+ gradient is used to drive simultaneous "downhill" movement of Na^+, which is coupled with "uphill" movement of solute into the cell. Both monosaccharides and amino acids may be transported by this process. See Chapter 6.

6. Inhibition of ADH secretion results in a decrease in the water permeability of the wall of the med-

ullary collecting ducts, reducing the rate of water reabsorption from collecting duct urine. The consequence is that the composition of final urine becomes more like that of distal tubular urine (see Figure 19-21), and the additional unreabsorbed water adds to urine volume.

7. The filtered load of glucose is the plasma concentration (10 mg/ml) multiplied by the GFR (125 ml/min), which equals 1250 mg/min. The filtered load exceeds the Tm of 400 mg/min, so the reabsorption rate is 400 mg/min and the excretion rate is the filtered load (1250 mg/min) minus the reabsorption rate (400 mg/min), which equals 850 mg/min. The urine concentration of glucose is the excretion rate divided by the urine flow rate: (850 mg/min)/(10 ml/min) = 85 mg/ml.

8. See "Loops of Henle and the Collecting Duct—Formation of Concentrated Urine" under **Mechanisms of Urine Formation and Modification.**

9. Some possible factors:
   a. lack of ADH (diabetes insipidus)
   b. inhibition of NaCl transport in the thick ascending limb of the loop of Henle; this is an effect of "loop" diuretics
   c. see the answer to question 11

10. Tubular acidification drives reabsorption of filtered bicarbonate. Since the chemical equilibrium between NH_3^+, H, and NH_4^+ determines the concentration of NH_4^+ in the urine, the more acid the urine, the more ammonium is excreted.

11. In the two-solute hypothesis, a concentration gradient of urea provides part of the driving force for water recovery from the collecting duct. When a diet contains adequate calories but no excess protein, little urea is produced. Reduction of the concentrating ability of the kidney by such diets is evidence in favor of the two-solute hypothesis.

CHAPTER 20

1. Total body water constitutes half of body weight, or 24 liters (1 kg = 1 L). ECF constitutes one third of total body water, or 8 liters, so the remainder (the ICF volume) is 16 liters. If intracellular solute concentration is 300 mOsm/L (see Figure 20-1), the total intracellular solute is 16 L × 300 mOsm/L = 4.8 Osm. Assuming that all of the total body Na^+ is in the ECF, the total body Na^+ is 8 L × 145 mEq/L = 1.16 Eq. After administration of 3 liters of isotonic saline the ECF volume is 11 liters. The ICF volume does not change because Na^+ is effectively impermeant to cells. The ECF osmolarity does not change because the saline is isotonic. Isotonic saline (300 mOsm/L) contains 150 mEq/L of Na^+. The body Na^+ increases by 3 L × 150 mEq/L = 0.45 Eq to a total of 1.16 Eq + 0.45 Eq = 1.61 Eq.

2. In congestive heart failure, right atrial pressure (that is, central venous pressure) increases (see Figure 15-9), stretching the atria. This stimulates release of ANH.

3. The ECF Na^+ concentration is stabilized by osmotic transfer of fluid from the ICF and by potent thirst induction and renal fluid conservation induced by the ADH system. All of these tend to make increased ECF volume the major consequence of increased total body Na^+. Since arterial

blood pressure is also strongly regulated by the cardiovascular reflexes, volume increases are most likely to cause hypertension for individuals whose ability to regulate blood pressure is impaired in some respect.

4. An increase in total body K^+ would affect mainly ICF volume, since ICF contains almost all the total body K^+.

5. Reduction of blood flow to the kidney is a potent stimulus for renin secretion. The affected kidney would release large amounts of renin, raising aldosterone levels inappropriately. The excess aldosterone may well cause an increase in total body Na^+ and ECF sufficient to overwhelm reflexive compensation and increase arterial blood pressure.

6. As shown in Figure 16-8, the right atrial pressure (that is, central venous pressure) is affected directly by changes in plasma volume. Furthermore, the reflexive responses that regulate mean arterial blood pressure in response to changes in plasma volume exaggerate the changes in right atrial pressure. Thus atrial pressure is a far more accurate indicator of plasma volume than mean arterial pressure.

7. Glycyrrhizic acid is an aldosterone agonist.

8. Hyperparathyroidism causes bone breakdown to dominate bone formation, reducing the total mass of bone and increasing both ECF levels of Ca^{++} and phosphate. The increased Ca^{++} depresses excitability of nerves and muscles (see "Extracellular Ca^{++} and Excitability" under **Refractory Period and Firing Frequency,** Chapter 7).

9. See "Buffer Systems" under **Acid-Base Homeostasis.**

10. See "Respiratory and Renal Contributions to Regulation of Plasma pH" under **Acid-Base Homeostasis.**

11. Plasma pH would be about 7.1, and a P_{CO_2} of about 20 mmHg would be needed for full respiratory compensation, as estimated from Figure 20-16.

12. The P_{CO_2} of 80 indicates respiratory acidosis. However, the pH and bicarbonate concentrations are somewhat lower than expected for pure respiratory acidosis, especially if sufficient time had elapsed for renal compensation; this suggests a combined respiratory and metabolic acidosis (see Figure 20-17).

CHAPTER 21

1. See Table 21-1. In answering this question, consider the following:
   physical changes
   food transport within the tube system
   storage
   secretion
   digestion
   absorption

2. See "Function and Control of Salivary Secretion" under **The Role of the Mouth and Esophagus.**

3. A complete answer to this question should take into account the intracellular mechanisms of control of contraction in smooth muscle (see Chapter 12 and "Gastrointestinal Smooth Muscle" in this chapter), the patterns of motility in the gastrointestinal tract (see "Gastrointestinal Motility"), and the effects of autonomic innervation (see "Neural and Hormonal Control of Motility").

4. See "Gastric Secretion and Motility."

5. Parietal cells are densely populated with mitochondria (Figure 21-19), reflecting the large amounts of energy needed to transport H^+ into the gastric lumen against a pH gradient of several orders of magnitude.

6. See "Pancreatic Enzymes" under **Secretion and Motility in the Intestine,** and Table 21-4.

7. Enterokinase is a protease secreted by intestinal cells that activates trypsinogen; the trypsin formed activates the rest of the pancreatic enzymes.

8. See "Defecation" under **Secretion and Motility in the Intestine.**

CHAPTER 22

1. In answering this question, you might consider the following aspects:
   a. the role in digestion of each of the anatomical regions in the list (as described in Chapter 21)
   b. the particular anatomical specializations of each region, such as crypts, gastric glands, salivary glands, and so forth
   c. the chemical environment of each region resulting from secretions of digestive glands and the gastrointestinal mucosa
   d. cellular specializations such as microvilli and membrane transport systems

2. See **Carbohydrate Digestion and Absorption.**

3. See **Protein Digestion and Absorption.**

4. a. Emulsification of lipids by bile salts greatly accelerates lipase and phospholipase activity.
   b. Bile salts stabilize free fatty acids, monoglycerides, and diglycerides in the form of micelles, the form in which digested lipid is carried to the mucosal surface of intestinal cells.

5. Three major differences are:
   a. Lipid absorption into intestinal cells is a passive rather than an active process.
   b. Unlike monosaccharides and amino acids, the products of fat digestion (free fatty acids, monoglyceride, and glycerol) undergo reassembly (into triglyceride) in intestinal cells.
   c. The reassembled lipids traverse the intestinal cells in vesicles rather than as solutes. They enter the blood via the lymphatic system rather than passing directly into intestinal capillaries. Consequently, they enter the general circulation rather than passing directly to the liver via the enterohepatic portal circulation. See Figures 22-8 through 22-11).

6. See Table 22-6.

7. See Figures 22-14 and 22-15.

8. Gastric emptying is affected by reflexes and hormones, both of which are responsive to the volume and acidity of chyme and to the presence of specific nutrients. See Figures 22-16 through 22-19 and Tables 22-4 and 22-5.

CHAPTER 23

1. Lipids: adipose tissue
   Carbohydrates: liver and skeletal muscle
   Protein: muscle typically is the site of most intracellular protein

2. During the absorptive state, most energy comes from glucose and amino acids absorbed from the intestine. During the postabsorptive state, the drop in insulin levels and the rise in glucagon and epinephrine levels causes the body to shift to stored glycogen and fat as energy sources.

3. See "Transport and Storage of Lipids" under **Characteristics of the Absorptive State.**

4. a. In muscle, insulin promotes storage of glucose as glycogen and favors protein synthesis because it allows an influx of amino acids. Some of the glucose is metabolized to lactate. The lactate enters the blood, and some of it undergoes gluconeogenesis in the liver. See Figure 23-9.

   b. In adipose cells, insulin favors fat deposition by stimulating glucose entry. After entry, the glucose is converted into the glycerol backbone of triglyceride. See Figure 23-11.

   c. Uptake of glucose and amino acids by the liver is insulin independent. However, insulin stimulates the enzymes of glycogen synthesis, indirectly favoring glucose uptake and conversion of absorbed glucose to liver glycogen. See Figure 23-10.

5. See Table 23-3.

6. Glucagon, epinephrine, and cortisol. Generally insulin secretion and glucagon secretion are reciprocal, but secretion of both hormones is stimulated by an increase in plasma amino acid levels.

7. See "Short-Term Regulation of Food Intake and Satiety Signals" under **Regulation of Food Intake.**

8. See "Transport and Storage of Lipids" under **Characteristics of the Absorptive State,** and the essay "Control of Blood Cholesterol Levels by the Liver."

9. Both testosterone and estrogen stimulate calcification and favor maintenance of bone mass. Because menopause, the cessation of female reproductive cycles, causes levels of estrogen to decrease, postmenopausal women are more likely to develop osteoporosis than men of the same age. Replacement of estrogen can prevent osteoporosis. Dietary calcium intake must be adequate, but it is not clear whether excessive calcium intake has any beneficial effect. Exercise that loads bone tends to protect bone mass, as does ingestion of small amounts of fluoride.

10. For HGH, see "Growth and Growth Hormone" under **Changes in Metabolism Over the Life Cycle.** For epinephrine, see Table 23-5.

11. See "Metabolic Adaptation to Cold" under **Metabolic Responses to Threatening Situations.**

12. See "Starvation" under **Metabolic Responses to Threatening Situations.**

CHAPTER 24

1. See "Inflammation" under **Coordination in the Immune System.**

2. The two major types of phagocytes are macrophages (transformed monocytes) and neutrophils (one class of granulocytes).
   The three classes of granulocytes are neutrophils, basophils, and eosinophils.
   "Leukocyte" is the general name for white blood cells, including granulocytes, monocytes, and lymphocytes.
   Lymphocytes are the class of leukocytes that are responsible for specific immune responses; these include two general types, B cells and T cells.
   Both basophils and mast cells secrete histamine, a mediator of inflammation.

3. The complement cascade, like other regulatory cascades (see Chapter 5), *amplifies the initial signal* set in motion by binding of antibody to antigen, and *branches the signal* to activate several different elements of immune response.

4. The more closely the donor and the recipient are related, the more likely they are to share similar major histocompatibility antigens. However, there are a number of different MHC antigens, and only identical twins inherit an identical complement of MHC antigens.

5. See "Complement System" under **Coordination in the Immune System.**

6. Both kill target cells by forming large protein pores in the target cells' membranes. However, the proteins involved in pore formation are not the same for the two systems.

7. In juvenile-onset diabetes, the pancreas may perhaps never secrete insulin. According to the clonal selection hypothesis, if insulin is not present in the blood at the time of clonal selection of B cells, clones that would make antiinsulin are not eliminated. Consequently, these clones would survive and generate a primary immune response when insulin is given. On the other hand, the pancreas could be normal initially. If for some reason clones that could direct an immune response against pancreatic cells were not eliminated, this error could result in an autoimmune attack on the pancreas. This attack would stop insulin secretion by damaging β-cells and also probably would involve formation of antibodies to insulin as well as other proteins synthesized by pancreatic cells.

8. The immune system usually succeeds in eliminating cells that express novel antigens, including cancer cells. However, the vigilance of the immune system declines with age.

9. See "Recognition and Memory in Specific Immune Responses" under **Diversity in the Immune System,** and Figure 24-7.

10. See "Autoimmune Diseases: Misdirected Responses of the Immune System" under **Coordination in the Immune System.**

11. Suppressor T cells moderate the vigor of the cell-mediated specific immune response. They form part of a negative feedback system that usually prevents excessive damage of the host tissues in an immune response. See "Control of Immune Responses by Helper and Suppressor T Cells" under **Coordination in the Immune System.**

12. Usually the volume of donated plasma is small in comparison to the recipient's plasma volume, so the donated antibodies are diluted in a much larger volume of the recipient's plasma. Thus, the donor's antibodies are thus not concentrated enough to trigger a significant complement reaction or to agglutinate many of the recipient's red blood cells. The reverse would be true for the recipient's antibodies to donated cells.

CHAPTER 25

1. See Table 25-1 and Summary items 1 and 2.
2. See Figures 25-10 and 25-11.
3. For factors that control differentiation of external genitalia, see Figures 25-10 and 25-11. For disorders, see "Disorders of Sexual Differentiation" under **Hormonal Control of Sexual Maturation.**
4. A Graafian follicle is the final stage of follicular development; it contains and nourishes a maturing oocyte and secretes estrogen. The thecal cells of the follicle synthesize androgen and export it to granulosa cells, which convert it into estrogen, the hormone mainly responsible for female secondary sex characteristics. After ovulation, the remnant of follicle transforms into a corpus luteum, which secretes estrogen and progesterone. The latter hormones maintain the secretory state of the uterine endometrium until the end of the cycle.
5. See Figure 25-22. LH (lutenizing hormone) and FSH (follicle-stimulating hormone) are trophic hormones responsible for maintaining ovarian and testicular function; in females, cycles of FSH and LH release drive maturation and ovulation of follicles and their subsequent transformation into corpora lutea. Estrogen maintains female secondary characteristics; its rise during the follicular stage of the cycle causes the uterine endometrium to proliferate. Progesterone converts the endometrium to the secretory phase in preparation for implantation of the zygote that will result if the ovulated ovum is fertilized.
6. Sertoli cells: the testicular cells that nourish maturing spermatozoa
   Leydig cells: the testicular cells that secrete testosterone.
   Vas deferens: the two vasa deferentia connect the testes to the rest of the male gland and duct system (see Figure 25-3)
   The two seminal vesicles are glands that together with the prostate and Cowper glands secrete the fluid component of semen.
7. See "Spermiogenesis" under **The Genetic Basis of Reproduction and Sex Determination** and "Fertilization" under **Endocrine Regulation of Female Reproductive Function.**
8. See "Hormonal Birth Control and Fertility Enhancement" under **Endocrine Regulation of Female Reproductive Function.**
9. GnRH is gonadotropin-releasing hormone, a hypothalamic hormone that controls release of the anterior pituitary hormones LH and FSH. The functions of LH and FSH are given in Table 25-3. The functions of the gonadal steroid hormones testosterone, estradiol, and progesterone are given in Tables 25-4 and 25-6.
10. The male external genitalia would in any case be masculinized by testicular androgen during early development, but if excessive secretion of adrenal androgen continues after birth, a precocious (early) puberty results. Depending on the location of the defect in the pathways of steroid metabolism in the adrenals, excessive androgen secretion may be accompanied by excessive cortisol secretion, which diminishes immune response, or by excessive aldosterone secretion, which causes hypertension.

CHAPTER 26

1. Oxytocin secretion is stimulated by cervical stretch during labor and by suckling after labor. During labor its role is to stimulate uterine contractions. During lactation it stimulates milk letdown. Both of these processes involve increasing the activity of smooth muscle.
2. Prolactin maintains milk production while suckling continues. The factors that control its release are discussed in "Prolactin and Lactation" under **Birth and Lactation.**
3. Ectopic means "in the wrong place." The outcome depends on the site of implantation. Blastocysts that implant in the fallopian tube do not succeed; when they rupture the fallopian tube, a surgical emergency results. Implantation at other sites within the abdominal cavity can result in a pregnancy that sometimes can be carried to term.
4. See "Endocrine Functions of the Trophoblast and Placenta" under **Pregnancy.**
5. See "Effects of Pregnancy on the Mother" under **Pregnancy.** During the early part of pregnancy the mother can accumulate stores of some nutrients in preparation for the much higher demands of the fetus during the later stages of pregnancy. Folic acid, a vitamin, is an excellent example. The importance of this vitamin in development is such that it is recommended that women who anticipate becoming pregnant to begin taking folic acid supplements.
6. Cortisol stimulates the fetal lung to begin secreting pulmonary surfactant. The importance of this secretion for breathing is explained in Chapter 16 (p. 417). Lack of pulmonary surfactant causes a form of respiratory distress called hyaline membrane disease.
7. Because the events that determine the fates of entire organ systems occur during the embryonic phase of development, damage during this time typically has a widespread impact. See Table 26-2.
8. The effects of patent ductus arteriosus:
   After birth the resistance of the pulmonary loop falls below that of the systemic loop, therefore some of the output of the left heart can pass through the ductus and reenter the pulmonary circulation, diminishing the effective cardiac reserve available for the systemic circulation. This left-to-right shunt raises the pressure of the pulmonary loop. Since some of the blood makes more than one pass through the pulmonary loop for each pass through the systemic loop, the diverted blood actually has an additional opportunity to become oxygenated, and the oxygen saturation of systemic arterial blood is normal.
9. Progesterone maintains the uterine endometrium during pregnancy and inhibits contraction of uterine smooth muscle, a factor that tends to prevent premature labor. Presumably, in women prone to premature labor the amount of progesterone secreted by the corpus luteum is inadequate to prevent the onset of stretch-induced contractions; this can be corrected by supplying additional progesterone.
10. See "The Fetal-Placental Circulation" under **Pregnancy.**

Glossary

A

a band The region of a muscle sarcomere that contains myosin thick filaments. p. 286

abscissa The horizontal, or x-axis, of a graph. p. 28

absolute refractory period The interval following an action potential during which a second action potential cannot be generated. p. 152

absorptive state The period after a meal when excess substrates are stored for future use. p. 590

acceleration The rate of change of velocity with time. p. 28

accommodation The change in the radius of the lens that allows the eye to focus on objects at different distances. p. 238

acetylcholine (ACh) A chemical transmitter released at the junction of a nerve and a skeletal muscle, in the autonomic ganglia, and from postganglionic parasympathetic fibers. p. 182, 277

acetylcholinesterase An enzyme that degrades acetylcholine. p. 277

acetyl-CoA A coenzyme that enables the end products of glycolysis, amino acids, and fatty acids to enter the Krebs cycle. p. 77

achalasia A condition in which food cannot enter the stomach from the esophagus. p. 550

acid A compound that dissociates in solution to produce H^+ ions. p. 26

acid-base balance Maintenance of the plasma pH at a normal value of 7.4. p. 519

acidity The H^+ content of a solution The lower the pH, the more acid the solution. p. 26

acidosis A condition in which the plasma pH is less than 7.4. p. 526

acini The blind ends of secretory (exocrine) glands. p. 406, 539

acromegaly A clinical condition in adults with elevated levels of growth hormone. p. 604

acrosome The apical portion of male sperm; it permits penetration of the ovum. p. 650

F- and G-actin The muscle protein in thin filaments. F-actin filaments are composed of G-actin subunits. Actin and myosin interact when Ca^{++} is present. p. 289

action potential An all-or-none membrane depolarization that is propagated along nerve fibers without any decrease in amplitude. p. 140

activation energy The energy "hill" that must be surmounted for a chemical reaction to proceed in a given direction. p. 50

active force In a muscle fiber, the component of total tension generated by the interaction between the thick and thin filaments. p. 297

active site The region of an enzyme that specifically binds a substrate, reducing the overall activation energy for a reaction. p. 50

active state The state of a muscle fiber when Ca^{++} and ATP are present and the fiber is generating force. p. 291

active transport Transport of a substance against its concentration gradient by direct use of cellular energy (ATP). p. 125

actomyosin The low-energy complex formed between actin and myosin in the absence of ATP; it is also known as a "rigor complex." p. 291

adaptation In sensory receptors, the decrease in amplitude of the generator potential despite the continued presence of a stimulus. p. 169

adenohypophysis See anterior pituitary.

adenosine A nucleotide that vasodilates the coronary arteries. p. 189

adenosine triphosphate (ATP) The "energy currency" of the cell. It is synthesized from adenosine diphosphate (ADP) in glycolysis, the Krebs cycle, and through oxidative phosphorylation. The breakdown of ATP by hydrolysis is coupled with reactions that require energy, rendering the overall reaction favorable. p. 70

adenyl cyclase An enzyme that converts ATP to cyclic AMP. It often is coupled with membrane receptors through the activation of intermediate G-proteins. p. 103

adipose tissue Fat cells; the site of lipid storage and a major factor in endocrine regulation of organic metabolism. p. 593

adrenal cortex That portion of the adrenal gland that secretes the corticosteroid hormones. p. 113, 599

adrenal medulla That portion of the adrenal that secretes epinephrine. It is

an extension of the autonomic nervous system. p. 259

adrenergic receptors Receptors for the neurotransmitter norepinephrine. The four subclasses are designated α_1, α_2, β_1, and β_2. p. 277

adrenocorticotropic hormone (ACTH) A hormone secreted by the anterior pituitary that controls the release of the adrenal corticosteroids. p. 612

aversion learning A form of conditioning specifically involving taste. p. 214

aerobic metabolism The complete degradation of glucose to carbon dioxide and water by the combined action of glycolysis, the Krebs cycle, and oxidative phosphorylation. p. 75

afferent In the nervous system, a nerve fiber leading from a receptor to the spinal cord or brainstem. p. 195

afferent arterioles The renal arterioles proximal to the glomerular capillaries. p. 480

affinity A measure of the relative binding of various substrates to an enzyme, or ligands to a receptor. p. 52

afterload The load against which an activated muscle must try to shorten Greater afterloads result in lower velocities. p. 299, 339

afterpotential The membrane potential following an action potential. p. 152

agglutination The clumping together of cells. p. 638

agonist Any compound, such as a drug, that can activate a receptor; to be distinguished from compounds that activate receptors under normal conditions. p. 277

agonist muscle A muscle that moves a particular joint in the same direction as another muscle. p. 286

airways The bronchi and bronchioles of the lungs. p. 402

albumin One of the plasma proteins, and the major determinant of plasma osmotic pressure. p. 348

aldosterone A hormone secreted by the adrenal that controls Na^+ reabsorption in the distal tubules of the kidneys. p. 381

alkaline tide A term referring to the fact that plasma leaving the stomach has an excess of HCO_3^- as a result of acid secretion. p. 551

alkalosis A condition in which the plasma pH is greater than 7.4. p. 526

allosteric effect The modification of enzyme activity by the binding of a modulator to an allosteric site separate from the active site. p. 52

allosteric site See allosteric effect.

alpha helix One of the secondary structures of proteins. p. 48

alpha motor neuron A spinal cord neuron innervating skeletal muscle fibers. p. 263

alpha receptors See adrenergic receptors.

alpha rhythm One of the inherent electrical rhythms of the brain seen in the electroencephalogram of a relaxed, awake individual with eyes closed. The frequency is about 10 per second. p. 217

alveolar ducts In the lungs, the smallest airways leading to the alveoli. p. 403

alveolar gas equation A formula for calculating the alveolar partial pressure of oxygen if the inspired partial pressure of oxygen, the alveolar partial pressure of carbon dioxide, and the alveolar ventilation are known. p. 431

alveolar ventilation The volume of air moving into and out of the alveoli during each respiratory cycle. p. 410

alveoli (alveolar sacs) Thin, air-filled sacs in the lungs across whose walls oxygen and carbon dioxide are exchanged. p. 402, 698

amacrine cell A cell in the retina that connects adjacent photoreceptors and is primarily sensitive to the movement of an object. p. 239

amiloride A drug that inhibits active Na^+ transport in the kidneys. p. 498

amino acids The building blocks of proteins; there are 20 amino acids. p. 47

aminopeptidase An enzyme that degrades proteins one amino acid at a time, beginning at the free amino end. p. 570

ammonia (NH_3) A lipid-soluble compound. In the kidneys, conversion of ammonia to ammonium ions permits the secretion of large amounts of H^+ ions. p. 500

amnesia Loss of memory. p. 293

ampere (A) The unit of electrical current. p. 34

amphipathic molecules Molecules such as phospholipids that have polar and nonpolar regions. They aggregate as micelles or bilayers. p. 40

ampulla The region of the semicircular canals that contains vestibular hair cells. p. 233

amygdala A component of the limbic system in the brain. p. 213

amylase An enzyme that degrades starches to monosaccharides and disaccharides. p. 535, 568

amylopectin Plant starch. p. 567

anabolism Processes that build body tissues. p. 75, 590

anaerobic metabolism Metabolic processes that occur without oxygen. p. 75

analgesia Loss of the sensation of pain. p. 208

anastomoses Direct connections between arteries and veins. p. 356

anatomic dead space The region of the lung that cannot exchange gas. p. 411

androgens The male sex steroid hormones, principally testosterone. p. 656

anemia A condition in which the number of red blood cells is low. p. 434

anesthesia Loss of all sensation. p. 208

angiotensin I The hormone produced in response to renin release by the kidneys. p. 381, 513

angiotensin II A vasoconstrictor produced from angiotensin I that stimulates secretion of ADH and thirst. p. 514

angiotensin III A substance produced from angiotensin II that stimulates secretion of aldosterone from the adrenal gland. p. 514

angiotensinogen A plasma protein that is cleaved by a converting enzyme in the presence of renin to produce angiotensin I. p. 381, 514

angstrom(Å) An older unit of distance equal to 10^{-10} meters. p. 25

anion A negatively charged ion. p. 21

anion gap A measure of plasma anions other than Cl^- and HCO_3^-. p. 525

anorexia nervosa A condition in which an individual abnormally restricts food intake. p. 582

antagonist In receptor activation, a compound (drug) that will block a receptor. p. 95, 277

antagonist muscle A muscle that opposes the movement produced by activation of another muscle at a joint. p. 286

anterior chamber In the eye, the space in front of the lens. p. 236

anterior pituitary A multifunctional endocrine gland that releases adrenocorticotropic hormone (ACTH), thyroid-stimulating hormone (TSH), follicle-stimulating hormone (FSH), luteinizing hormone (LH), prolactin, and growth hormone (GH). p. 99, 111

anterolateral system One of the major ascending somatosensory tracts; it mediates pain, temperature, and "crude" touch. p. 212

antibody An immunoglobulin molecule. Released by lymphocytes, it specifically recognizes "foreign" macro-

molecules and activates other components of the immune system. p. 68, 619

anticodon The region of the transfer RNA molecule that contains base pairs complementary to base pairs (codons) of messenger RNA. p. 58

antidiuretic hormone (ADH) A hormone released from the posterior pituitary that controls water permeability in the kidneys. Its release is influenced by plasma osmolarity and venous pressure (vasopressin or vasopressin-ADH). p. 380

antigen Any macromolecule that produces a specific immune response. p. 619

antigravity muscles Postural muscles. p. 273

antiport An active transport process in which two substances move in opposite directions across a cell membrane by means of a carrier. p. 131

antipyretic A substance that reduces fever. p. 611

antrum The region of the stomach closest to the duodenum. p. 536, 664

anus The most distal sphincter in the gastrointestinal tract. p. 537

aorta The large blood vessel that receives the output of the left ventricle. p. 321

aortic baroreceptors Stretch-sensitive receptors in the wall of the aorta that monitor arterial blood pressure. p. 386

aortic arch The proximal region of the main artery leading from the heart. p. 386

aortic bodies Oxygen- and pH-sensitive receptors near the aortic arch. p. 454

aortic valve The valve between the left ventricle and the aorta. p. 322

aperture The area of the pupil of the eye. p. 237

apex The portion of the cochlea farthest from the oval window; it is activated primarily by low-frequency sounds. p. 325

aphasia A deficit in expressing or understanding language. p. 215

apneustic center A brainstem region whose activity inhibits respiration. p. 451

apoproteins Proteins needed to activate lipoprotein lipase and to permit chylomicrons and very low density-lipoproteins (VLDLs) to enter the cells lining the intestine. p. 594

aqueous humor The thin, watery fluid in the anterior and posterior chambers of the eye. p. 236

arachidonic acid A phospholipid whose synthesis is activated by many hormones and neurotransmitters. p. 350

argentaffin cells Gastrin-secreting cells in the stomach. p. 542

arrhythmias Irregularities in the heart rate. p. 327

arterial pressure pulse The periodic variation in pressure caused by the heart's pumping action. p. 335

arteries Blood vessels that branch out from the aorta and lead into arterioles. p. 322

arterioles Vessels of the circulation that have the highest flow resistance and therefore control blood distribution. p. 354

ascorbic acid Vitamin C. p. 577

association cortex A term denoting the areas of the brain that are not primarily sensory or motor. p. 214

association neuron An interneuron in the spinal cord or brain. p. 94, 196

associative learning A "higher" form of learning that requires correlations between two different stimuli. p. 213

asthma A disease characterized by the plugging of airways. p. 420

ataxia Unsteadiness of gait or motor coordination. p. 272

atherosclerosis The formation of fatty deposits on the interior surface of arteries; these deposits eventually can obstruct blood flow. p. 594

atmospheric pressure The hydrostatic pressure exerted on objects on the earth's surface by the air. At sea level it normally is 760 mm Hg. p. 429

atom The basic constituent of all matter. p. 18

atomic number The number of protons in the nucleus; it defines an element. p. 18

atomic weight The sum of the protons and neutrons in the nucleus. Elements with the same atomic number but different atomic weights are called isotopes. p. 18

ATPase Any enzyme that hydrolyzes ATP to ADP. p. 125

atresia The degeneration of unfertilized follicles. p. 651

atria The chambers of the heart that receive blood from the systemic and pulmonary veins. p. 333

atrial activation (atrial systole) The phase of the cardiac cycle during which the atria are contracting. p. 335

atrial receptors Stretch receptors in the wall of the right atrium that measure venous blood pressure. p. 380, 511

atrial septum The tissue separating the left and right atria. p. 322

atriopeptin (atrial natriuretic hormone [ANH]) A hormone released in response to increased atrial pressure (blood volume) that increases glomerular filtration in the kidneys. p. 511

atrioventricular (A-V) node A group of cells between the atria and ventricles that relay electrical activity to the ventricles. The delay provided by these cells gives the atria time to contract before the ventricles are activated. p. 322

atropine A drug that blocks parasympathetic postganglionic muscarinic receptors for acetylcholine. p. 279

auditory canal The canal of the outer ear. p. 224

auditory cortex The region of the cerebral cortex that receives direct sensory input from the auditory system. It is located in the temporal lobe. p. 230

auditory system One of the "special senses." It is composed of the ear and the neural pathways leading to the auditory cortex. p. 224

Auerbach plexus A component of the enteric (intrinsic) nervous system of the gut between the circular and longitudinal muscle layers. It is also known as the myenteric plexus. p. 538

autoimmune diseases Diseases in which normal regulation of the immune system breaks down, and tissues in the body are attacked by cells of the immune system. p. 619

automaticity A term denoting the heart's ability to generate its own intrinsic electrical rhythm. See **pacemaker**. p.325

autonomic ganglia Groups of neurons in the autonomic nervous system that are situated outside the central nervous system. p. 197

autonomic nervous system (ANS) The component of the central nervous system that controls visceral smooth muscle and glands. p. 195

autonomic reflex An "involuntary" reflex involving visceral receptors and smooth muscle or glands. p. 95

autoregulation The ability of a tissue or organ to internally regulate its function internally without extrinsic neural or hormonal input. p. 93, 377

autotransfusion The effective increase in blood pressure caused by a sympathetically mediated decrease in venous compliance. p. 388

A-V block The failure of action potentials to pass through the AV node from the atria to the ventricles. p. 327

A-V delay The time required for action potentials to pass through the AV node from the atria to the ventricles. p. 326

average flow velocity In the circulation, the linear velocity of the movement of an element of blood past a given point. p. 359

Avogadro's number The number of molecules of a compound contained in 1 mole It is a universal constant equal to 6.023×10^{23}. p. 24, 429

axon The component of a neuron that carries efferent signals. It usually is the longest component of the neuron. p. 147

axon hillock The region of the nerve cell body from which the axon arises; it is the site of origin of action potentials. p. 155

axoplasmic transport An ATP-dependent process involving neurofilaments and neurotubules that moves materials away from (anterograde) or toward (retrograde) the soma of a neuron. p. 177

B

B lymphocytes A subclass of lymphocytes that can be activated to differentiate into antibodysecreting plasma cells that mediate humoral immune responses. p. 619

baroreceptors Sensory receptors that detect pressure. They are present in the aorta and the carotid arteries. p. 95, 378

basal body A submembranous structure in auditory and vestibular hair cells that determines their polarity. p. 229

basal ganglia A group of subcortical nuclei that are mainly involved in controlling refined muscle movements. p. 269

basal tone Resting tone; a condition in which a tissue has some intrinsic level of activity in the absence of neural or hormonal control. The term applies to smooth muscle tension, spontaneous firing of nerve fibers, and so forth. p. 277

base A compound that will combine with a H^+ ion. In the cochlea, the region nearest the oval window, where high-frequency sounds are localized. p. 325

basic electrical rhythm Periodic waves of depolarization in visceral smooth muscle; this pulsing serves as the pacemaker for gastrointestinal motility. p. 310, 544

basilar membrane In the inner ear (cochlea), the structure that localizes sound frequencies to spatially segregated areas. p. 227

basolateral The surface of renal cells that faces adjacent cells. p. 479, 540

basophils A class of white bood cells that secrete histamine. p. 619

behavioral homeostasis The form of homeostasis that involves modification of the body's external environment. p. 14

beta cells Insulin-secreting cells in the pancreas. p. 590

beta receptors See adrenergic receptors.

beta waves One of the intrinsic electrical rhythms of the brain seen in the electroencephalogram of an alert subject with eyes open. p. 217

bicarbonate (HCO_3^-) The base formed when carbon dioxide dissolves in plasma. p. 27

bile A fluid produced by the liver that contains lecithin (a phospholipid), bile salts, and bile pigments. It is required for lipid digestion. p. 555

bile pigment Bilirubin. p. 555

bile salts One of the ingredients of bile that aids fat digestion. p. 555

bilirubin One of the bile pigments derived from hemoglobin. p. 555

bipolar cells In the retina, the cells that relay activity from the rod and cone photoreceptors to the ganglion cells. p. 239

bipolar neuron A class of nerve cells that are characterized by two branching processes at opposite ends of the cell body. p. 166

bladder The organ in which urine is stored. See **micturition**. p. 476

blastocyst A hollow sphere of cells in the early stages of embryogenesis. See **trophoblast** and **embryoblast**. p. 682

bleaching Splitting of the visual pigment into opsin and retinal. p. 241

blood-brain barrier A term referring to the fact that the cerebrospinal fluid is functionally isolated from the plasma. p. 369

blood The circulating contents of the cardiovascular system. Blood has a fluid component (plasma) and a cellular component (the red and white blood cells). p. 359

blood clot A structure formed from platelets and fibrin that seals ruptures in the walls of blood vessels. p. 349

blood doping A process in which an individual's blood is removed and the erythrocytes later retransfused in an effort to increase oxygen-carrying capacity. p. 366

blood pressure The pressure of the blood; used alone, the term usually refers to the average arterial pressure. p. 335

blood reservoirs The veins; they contain about 80 of the circulating volume. p. 357

blood type A classification of individuals based on the antigen present on the surface of their red blood cells. The blood types are A, B, AB, and O. p. 638

blue-sensitive cones Photoreceptors with maximum sensitivity at 420 namometers. p. 240

body fluid compartments A term denoting the division of the total body water into intracellular, interstitial, and plasma components. p. 506

body of the stomach The middle region of the stomach. p. 536

Bohr effect The shift in the oxyhemoglobin dissociation curve produced by changes in pH, CO_2, or temperature. p. 436

bolus The amount of food normally swallowed. p. 548

botulinum toxin A poison released by a bacterium sometimes found in improperly canned food; it can block neuromuscular transmission. p. 180

Bowman's capsule The proximal part of the nephron into which plasma is filtered. p. 478

Boyle's Law One of the Gas Laws. p. 415

bradycardia An abnormally low heart rate. p. 327

bradykinin A potent local vasodilator released during an inflammatory response. p. 189, 377, 631

brainstem The most primitive region of the brain; it includes the medulla, pons, and midbrain and contains many control centers for the autonomic nervous system. p. 197

Broca's area One of the two "language areas" of the cortex; it is responsible for translating concepts and commands into a verbal format. p. 214

bronchi Large airways in the lungs that contain cartilage. p. 402

bronchioles The small airways of the lungs that are surrounded by smooth muscle. p. 402

brown fat Specialized adipose tissue in infants that is important in thermoregulation. p. 610

brownian motion The random motion of large particles or molecules in a solution, caused by the thermal energy of water molecules. p. 120

brush border The luminal, microvillous surface of intestinal epithelial cells. p. 540

buffer A mixture of an acid and its conjugate base that can minimize the pH changes caused by addition of strong acids or bases. p. 27, 520

buffer line The line on a Davenport diagram that describes hemoglobin buffering. p. 521

bulimia An eating disorder in which an individual cycles between food binges and exaggerated dieting or voluntarily induced vomiting. p. 582

bundle branches The component of the heart's conducting system that relays impulses from the A-V node and bundle of His to ventricles. p. 325

bundle of His The component of the conducting system that is activated immediately after the A-V node. See **bundle branches**. p. 325

bursa of Fabricius A lymphoid structure in birds where lymphocyte differentiation occurs. p. 622

C

C fibers Unmyelinated nerve fibers usually less than 1 μm in diameter. p. 207

calcitonin Thyrocalcitonin; one of the hormones involved in regulating Ca^{++} and phosphate balance. p. 517

calcium (Ca^{++}) A divalent cation; it is involved in muscle contraction and synaptic transmission and may be an intracellular messenger. p. 104

calcium channel A voltage-dependent "pore" in the membrane that is responsible for the plateau of the cardiac action potential, for neurotransmission, and for stimulus-secretion coupling. p. 179

caldesmon A protein that can inhibit crossbridge formation and thus muscle contraction by binding to the actin-tropomyosin complex. p. 312

calmodulin A calcium-binding protein; it mediates smooth muscle contraction and may be a general mechanism for intracellular regulation. p. 105

calorie (cal) A unit of energy; 1 kcal = 1000 cal. p. 605

capacitance The electrical "inertia" of cells that prevents their membrane potential from changing instantaneously. In the cardiovascular system, the change in volume produced by a given change in pressure. p. 358

capacitation In reproduction, the ability of sperm to fertilize an ovum. p. 650

capillary The smallest blood vessel; it is surrounded by a single layer of epithelial cells. p. 321

capillary filtration The movement of water from the interior of a capillary into the interstitial space. p. 368

carbamino compounds (carbamino-hemoglobin) The compounds formed when CO_2 reacts with hemoglobin. p. 436

carbohydrate Organic molecules composed of carbon, hydrogen, and oxygen in the ratio 1:2:1. p. 45

carbon An element; carbon has a valence of +4 and can combine with itself and with many other atoms to form complex organic compounds. p. 20

carbon dioxide (CO_2) One of the respiratory gases; CO_2 is produced by the cells while they consume oxygen. p. 21, 439

carbon monoxide (CO) A gaseous by-product of hydrocarbons that competes with oxygen for hemoglobin and thus inhibits oxygen transport. p. 438

carbonic acid (H_2CO_3) The acid produced when CO_2 reacts with water. p. 439

carbonic anhydrase An enzyme that

accelerates the reaction of CO_2 with water. p. 439, 520

carboxypeptidase An enzyme that degrades proteins one amino acid at a time, beginning at the C-terminal end. p. 570

cardiac cycle The stereotyped pattern of electrical and mechanical activity that occurs during each heartbeat. p. 335

cardiac failure The inability of the heart to pump an adequate amount of blood to supply tissue needs. p. 386

cardiac function curve The relationship between cardiac output and venous pressure. p. 341, 384

cardiac glycosides Drugs that increase the heart's contractile strength. p. 340

cardiac muscle The form of striated muscle found in the heart. p. 306

cardiac output The amount of blood pumped per minute by each ventricle of the heart. p. 340

cardiac reserve The amount that the cardiac output can increase above the normal resting level as a result of changes in venous return. p. 384

carotid baroreceptor A stretch receptor that monitors arterial blood pressure. p. 386

carotid bodies O_2- and pH-sensitive sensory chemoreceptors. p. 454

carotid sinus The region of the carotid artery that contains the carotid baroreceptors. p. 378

carriers Lipid-soluble molecules that bind ions or polar molecules and transport them across membranes. p. 119

catabolism The breaking down of tissues for energy production. p. 75, 590

catalyst A substance that accelerates a chemical reaction but is not consumed in the process. See enzymes. p. 50

catecholamines A generic term for norepinephrine and epipherine. p. 260

caudate nucleus One of the basal ganglia of the brain. p. 269

cecum The most proximal end of the large intestine, separated from the small intestine by the ileocecal sphincter. p. 534

cell The basic element of all living organisms. p. 166

cell membrane The structure composed of lipids and proteins that separates the cytoplasm of a cell from the surroundings. p. 118

cellular immunity A component of the immune response; it is mediated by T lymphocytes. p. 627

cellulose Plant starch; it is a rigid polymer of glucose. p. 567

central chemoreceptors Regions of the medulla (brainstem) that are functionally sensitive to the partial pressure of CO_2 in the arterial blood and that regulate respiration. p. 451

central nervous system (CNS) The brain and spinal cord. p. 195

central venous pressure The blood pressure in the right atrium. p. 339

cephalic phase In digestion, a term referring to the fact that olfactory, taste, and visual stimuli produce gastrointestinal activity. p. 578

cerebellum A region of the brain behind the pons that coordinates and refines motor movements. The cerebellum is important for postural stability. p. 197

cerebral cortex The outer layer of the cerebrum; the most sophisticated and phylogenetically recent area of the brain. p. 198

cerebral dominance The specialization of one hemisphere for language functions. p. 214

cerebrum The most phylogenetically recent area of the central nervous system, composed of the cortex and the corpus striatum (basal ganglia). p. 197

cerebrospinal fluid The extracellular fluid in the central nervous system. p. 452

cervix The distal opening of the uterus facing the vaginal canal. p. 645

channel A permanent or semipermanent "pore" in a membrane through which ions or polar molecules can move at a high rate. p. 119

Charles Law One of the Gas Laws; at constant pressure, the volume of a gas is proportional to temperature. p. 429

chemical bonding The process by which atoms combine to form compounds. The term also refers to the forces that cause molecules to interact with one another. See covalent bonds, hydrogen bonds, ionic bonds, and Van der Waals forces. p. 19

chemical senses Olfaction and taste. p. 250

chemical synapses Synapses at which a molecule (neurotransmitter) released from a nerve cell causes permeability changes in another nerve or muscle cell. p. 175

chemiosmotic hypothesis The process by which energy is transferred from a reduced coenzyme to form ATP in the mitochondria of a cell. p. 80

chemoreceptors Sensory receptors sensitive to chemical compounds. p. 169, 250

chief cells The cells in the stomach that secrete pepsinogens. p. 550

chloride (Cl^-) The principal extracellular anion. p. 131

cholecystokinin (CCK) One of the gastrointestinal hormones; it causes the gallbladder to contract and inhibits secretion of gastric acid. p. 580

cholesterol A steroid used to synthesize the steroid hormones. p. 394, 555

cholinergic A neuron that releases acetylcholine as its neurotransmitter. p. 277

chondrogenesis The initial stage of bone formation. p. 517

chorea The uncoordinated limb movements seen in Huntington's disease. p. 274

chromaffin cells Modified postganglionic neurons in the adrenal medulla that release norepinephrine. p. 259

chromophore Light-sensitive pigment in the photoreceptors. p. 241

chromosome The form of the genetic material (DNA) in the nucleus of a cell. p. 53

chronotropic Affecting the heart rate. p. 327

chylomicron A mixed micelle of fatty acids, phospholipids, cholesterol, and protein that can be absorbed by the epithelial cells of the small intestine during fat digestion. p. 573

chyme The semisolid contents of the gastointestinal tract. p. 536

chymotrypsin An enzyme that degrades amino acids at internal peptide bonds. p. 570

cilia Tiny hair-like projections on the surface of a cell. p. 64

ciliary body The source of aqueous humor in the anterior chamber of the eye. p. 235

cimetidine A drug that inhibits gastric secretion by blocking histamine receptors. p. 552

circular muscle One of the muscle layers in the gastrointestinal tract. p. 537

cis-retinal The form of retinal before it is activated by photons of light. p. 241

clearance In the kidneys, the minimum amount of plasma that would contain the amount of some substance that is excreted. p. 495

clonal selection In immunity, the proliferative response of lymphocytes that recognize a specific foreign antigen. p. 621

coactivation The activation of both alpha and gamma (spindle) motor neurons are activated during normal muscle movements. p. 265

cochlea The inner ear; it contains the auditory receptors. p. 224

cochlear duct An alternative term for the scala media. p. 227

cochlear partition An alternative term for the basilar membrane. p. 227

codon The sequence of three nucleotides that "codes" for each of the 20 unique amino acids in DNA and RNA. p. 57

coenzymes Nonprotein compounds that are not enzymes themselves but that transfer small molecules or H$^+$ between metabolic pathways. They are catalysts, because they are not used up in the process. p. 75

cold vasodilation A condition in which prolonged cooling leads to a breakdown in the reflex vasoconstriction of peripheral blood vessels, causing "rosy cheeks." p. 392

colipase A substance that increases the effectiveness of pancreatic lipase in fat digestion. p. 572

collagen The connective tissue that makes up tendons. p. 286

collateral ventilation In the lungs a term referring to the fact that adjacent alveoli are connected by small pores. p. 422

collecting duct In the kidneys, the most distal portion of the nephronIts water permeability is regulated by ADH. p. 478

colloid osmotic pressure The osmotic pressure of the plasma. It is higher than expected because the plasma proteins are charged. p. 368

colon The part of the large bowel between the cecum and the rectum. p. 537

competitive inhibition Blocking of enzymes or receptors caused by the binding of a molecule at the site normally occupied by a substrate or ligand. p. 52

complement A collection of plasma proteins that are activated during an immune response and that mediate lysis of "foreign" cells. p. 55, 624

complete protein A protein that contains all essential amino acids. p. 595

complex cells Neurons in the visual cortex that are sensitive to the movements of lines and edges. p. 246

compliance The ratio of the change in length (volume) of an object to the applied force (pressure). It is a measure of "stiffness." p. 33, 357, 412

compound action potential The extracellular signal recorded from a nerve trunk. It reflects the number of active fibers in a particular size category. p. 159

compressibility The ratio of the change in volume of a gas to the change in pressure. Liquids are incompressible. p. 428

compression The phase of a sound wave in which pressure increases. p. 224

concentration The amount of solute in a solution; it is usually expressed in moles. p. 24

condensation A chemical reaction involving removal of water. p. 45

conducting fibers In the heart, the bundle of His, right and left bundle branches, and Purkinje fibers. p. 325

conductance The reciprocal of resistance; it is equivalent to permeability. p. 34

conduction aphasia A condition in which an individual cannot repeat what has been said because of a lesion between the language areas of the brain and the auditory cortex. p. 215

cones Color-sensitive photoreceptors in the retina. p. 240

congestive failure In the heart, a condition in which the central venous pressure (and thus end-diastolic volume) are high. p. 340

conjugate base The base produced by the dissociation of an acid. p. 26, 459

connective tissue One of the four basic tissue types; it makes up the tendons and a large part of the body's "loose" tissue. p. 7

connexon The structure in a gap junction that forms a "pore" between adjacent cells. p. 73

constriction The narrowing of a blood vessel or airway in the lungs. p. 277

continuity equation In the absence of sources or sinks, the volume flow of fluids or gases, expressed as the product of the linear velocity and crosssectional area, remains constant. p. 360

contractility In muscle, the ability of the crossbridges to generate force. It is related to internal Ca^{++} levels and myosin ATPase activity. p. 292, 324

contraction The generation of force in a muscle. p. 291

contralateral The opposite side of the body. p. 209, 261

controller The element in a negative feedback loop that evaluates the error signal and regulates the effector(s). p. 86

convergence The arrival of input from many different neurons at a single target neuron. p. 184, 211

Cori cycle The cycle in which lactic acid produced in muscle by glycolysis is reconverted to glucose by the liver and returned to the circulation. p. 303

cornea The transparent curved surface in the front of the eye. p. 235

coronaries The arteries that supply blood to the heart. p. 322

corpus callosum The array of nerve fibers that connects the two hemispheres of the brain. p. 198

corpus luteum In the menstrual cycle, the structure that is formed from a follicle after ovulation. It secretes progesterone and estrogen. p. 661

corpus striatum The area of the central nervous system "beneath" the cortex but "above" the thalamus and hypothalamus. p. 198

corresponding points The points on the retina of each eye upon which an image is formed by light rays from a single point in the visual field. p. 246

cortex See cerebral cortex or renal cortex. p. 476

cortical column A term referring to the fact that cortical neurons at different depths share similar receptive fields. p. 201

cortical nephron The nephrons in the kidney whose loops of Henle do not penetrate very deeply into the renal medulla. p. 479

cortical reaction In fertilization, the initial response to sperm penetration. p. 670

corticorubrospinal tract A descending motor tract that arises from the red nucleus. p. 273

corticospinal tract A descending motor tract that arises from the cortex. p. 273

cortisol The major steroid secreted by the adrenal cortex. p. 599

cotransport Transport of a substance against its concentration gradient, which occurs because the substance is coupled to the "downhill" movement of another substance, usually Na$^+$. p. 127

countercurrent exchanger In the kidneys, a term referring to the fact that the blood vessels of the vasa recta act to conserve the osmotic gradient in the medulla. p. 493

countercurrent multiplication In the kidneys, the processes that increase the osmolarity of the renal medulla and allow production of hypertonic urine. p. 492

countertransport See antiport.

covalent bond A chemical bond formed when electrons are shared among the constituent atoms. Covalent bonds can be polar or nonpolar. p. 19

Cowper's glands In the male reproductive system glands that secrete fluids into the urethra. p. 646

cranial nerves Peripheral nerves that arise from neurons in the brainstem. p. 197

creatine phosphate A high-energy compound that "buffers" ATP in muscle. p. 301

cretinism Mental retardation caused by thyroid hormone deficiency during early development. p. 606

crista The portion of the vestibular system that contains hair cell receptors. p. 233

cristae of mitochondria Foldings of the inner mitochondrial membrane. p. 70

critical period The period during the development of the visual system when visual signals seem to govern formation of neural connections. p. 249

crossbridge The "heads" of the myosin molecules that interact with the actin thin filaments and generate force. p. 286

crossbridge cycling The rate of formation and hydrolysis of the actinmyosin interactions that generate force in a muscle fiber. p. 290, 312

crossed extension The reflex in which a noxious stimulus to one limb leads to oppositely directed movements of limbs on the two sides of the body. p. 267

crypt cells In the intestine, cells located at the base of the villous folds. They secrete water and ions. p. 541

cupula A gelatinous structure that occludes the semicircular canals. When displaced by rotations, it activates the hair cells. p. 233

curare A potent blocker of nicotinic cholinergic receptors such as are found at the neuromuscular junction. p. 278

current The flow of charge; current = voltage/resistance (Ohm Law). p. 34, 167

current of injury The current that flows between a damaged and a normally polarized region of heart muscle. p. 337

cutaneous Pertaining to the skin. p. 205

cyclic AMP One of the cyclic nucleotides; a major intracellular "second messenger." p. 244

cycling rate See **crossbridge cycling.**

cytochromes Components of the mitochondrial electron transport chain. p. 78

cytoplasm All the material inside a cell other than the nucleus. p. 64

cytoplasmic receptor In endocrinology, a hormone-binding site in the cytoplasm rather than in either the cell membrane or nucleus. p. 102

cytoskeleton In cells, a network of tubules and filaments that acts as an internal structural "framework." p. 71

cytotoxins A subclass of T lymphocytes that are involved in cell-mediated immunity. p. 629

D

D cells Pancreatic endocrine cells that secrete somatostatin. p. 595

Dalton Law The Avogadro number expressed in terms of partial pressures. p. 429

Davenport diagram A graphical representation of the Henderson-Hasselbalch equation that relates pH, bicarbonate concentration, and CO_2 partial pressure to one another. It is used to evaluate acid-base balance. p. 521

dead space See **anatomical dead space.**

deamination A chemical reaction that involves removal of an NH_2 group. p. 599

decarboxylation A chemical reaction that involves removal of a carboxyl group. p. 77

decerebration Section of the brainstem above the vestibular nucleus. p. 269

decerebrate rigidity Exaggeration of stretch reflexes characteristically seen in a decerebrate animal. p. 271

decibel A logarithmic unit of measure for sound intensity. p. 224

decomposition A condition in which movements are fragmented into their component parts as a result of injury to the basal ganglia. p. 274

decorticate Lacking a cortex. p. 271

decremental conduction The persistence of voltage changes for a short distance along a nerve or muscle fiber in the absence of propagated action potentials. p. 140

decussate To cross. p. 209

defibrillation A procedure for suddenly depolarizing the entire heart in an effort to restore a normal pattern of electrical activation. p. 329

dehydration Loss of body water with an increase in osmolarity. p. 45, 507

dehydrogenation A chemical reaction that involves removal of a hydrogen atom. p. 52, 177

delta waves One of the intrinsic electrical rhythms of the brain seen during the deeper stages of sleep. p. 217

demyelination Loss of the myelin sheath that surrounds myelinated nerves. p. 161

denaturation The breaking of weak chemical bonds with a concomitant inactivation of proteins and nucleic acids. p. 48

dendrite Processes on a nerve cell that receive input from other neurons. p. 166

dense bodies Structures in smooth muscle that are functionally equivalent to the Z lines of striated muscle; sites where actin filaments attach. p. 310

deoxyribonucleic acid (DNA) One of the nucleic acids; it stores genetic information in its nucleotide sequence and can replicate itself. Regions of DNA can be transcribed into messenger RNA and then into specific proteins. p. 53

dependent variable In algebra, the variable whose value changes when other, independent variables are altered. It usually is seen on the left side of an algebraic equation. p. 28

dephosphorylation A chemical reaction that involves removal of a phosphate group. It often is a way second messengers affect enzyme activity. p. 52

depolarization A condition in which the membrane potential becomes internally less negative. p. 141

dermatome The region of the body innervated by one spinal nerve. p. 205

desensitization The process by which permeability changes in a receptor disappear with prolonged exposure to chemical compounds such as neurotransmitters. p. 187

desmosome A "loose" junction that mechanically couples cells. p. 73

detoxification The conversion of drugs and other toxic compounds in the liver into organic anions that can be excreted by the kidneys. p. 555

deuteranopia A condition in which an individual lacks green-sensitive photoreceptors. p. 243

diabetes insipidus A clinical condition in which large amounts of relatively dilute urine are produced because of a lack of ADH regulation. p. 494

diabetes mellitus A clinical condition in which plasma and urine glucose levels are abnormally high, usually as a result of insulin deficiency. p. 488, 600

diaphragm The muscle that separates the thorax and the abdomen. Its contraction is the major factor in resting respiration. p. 402, 675

diastole The phase of the cardiac cycle during which the ventricles are relaxed. It includes rapid and reduced filling. p. 326

diastolic depolarization In pacemaker cells, a term referring to the fact that between action potentials, the membrane potential gradually becomes internally less negative. The slope of diastolic depolarization determines the heart rate. p. 327

diastolic pressure The minimum pressure in the arteries reached just before blood is ejected from the ventricles. p. 338

dichromat An individual who has only two types of color-sensitive cones. p. 243

dicrotic notch The displacement seen in the aortic pressure pulse when the aortic valve closes; the incisura. p. 335

diencephalon The thalamus and hypothalamus. p. 197

diffusion Net movement of substances from areas of high to low concentration as a result of their inherent thermal energy. p. 121, 407

diffusion coefficient A measure of the rapidity with which a substance diffuses; a kind of molecular mobility. p. 122

digestion The degradation of proteins, fats, and carbohydrates into their constituent amino acids, fatty acids, and simple sugars. p. 564

digestive enzymes Enzymes capable of breaking down proteins, carbohydrates, and fats to amino acids, simple sugars, and fatty acids. p. 547

digestive system The organ system composed of the stomach, intestines, and associated glands. p. 534

digestive vesicle The result of fusion between a lysozyme and an endocytotic vesicle; the site of intracellular breakdown of foreign materials. p. 70

dihydroxyphenylalanine (DOPA) One of the major central neurotransmitters. p. 182

dilation An increase in the cross-sectional area of blood vessels or airways. p. 277

disaccharide A compound formed from two monosaccharides; table sugar. p. 45, 567

distal muscles Muscles involved in fine movements of the extremities. p. 273

distal tubule In the kidneys, the region of the nephron where ADH and aldosterone regulate water and Na^+ reabsorption. It is located between the ascending loop of Henle and the collecting duct. p. 478

disulfide bond A chemical bond between cysteine residues in proteins. p. 48

diuresis The production of large amounts of hypotonic urine. p. 498

diuretic A substance that greatly increases urine flow. p. 393

divergence The sending of processes by a neuron to many target neurons. p. 187

DNA See deoxyribonucleic acid.

DNA-dependent RNA polymerase The enzyme responsible for "reading" a sequence of bases on DNA to form complementary RNA. p. 55

Donnan equilibrium A term referring to the fact that impermeable, negatively charged proteins inside a cell tend to produce voltage and osmotic pressure differences across the cell membrane. p. 124

dopamine One of the major central neurotransmitters. p. 182

dorsal column system The ascending somatosensory tract that carries signals from morphologically specialized cutaneous receptors and proprioceptors. It is the most "direct" sensory pathway. p. 209

dorsal horn The region of the spinal gray matter where the primary sensory afferents first synapse. p. 194

dorsal root ganglion The structure outside the central nervous system that contains the cell bodies of the peripheral spinal nerves. p. 167

double bond A covalent bond formed when each atom donates two electrons. p. 19

double Donnan equilibrium A condition in which the activity of the Na^+ pump effectively makes the cell membrane impermeable to Na^+. p. 124

down-regulation A decrease in receptor density caused by the binding of a hormone, neurotransmitter, or neuromodulator to a cell. p. 101, 601

dual innervation In the autonomic nervous system, the reception by many organs of both sympathetic and parasympathetic input. p. 277

duodenum The most proximal region of the small intestine. p. 537

dwarfism Stunted growth caused by a deficiency of growth hormone. p. 603

dye dilution A procedure in which a dye injected into the circulation is used to determine cardiac output. p. 382

dynamic (gamma) motor neuron Motor neurons that innervate the central region of the muscle spindle stretch receptor. p. 263

E

echocardiography A procedure in which high frequency sound waves are used to examine the heart's pumping action. p. 326

ectopic pacemaker In the heart, usually a region that spontaneously discharges, even though it is not the normal or a reserve pacemaker; this often occurs in the ventricular muscle. p. 328

edema Accumulation of fluid in the interstitial compartment. p. 370

effectors A general term referring to striated muscle, visceral smooth muscle, or glands; the "motor" element of feedback systems. p. 52

efferent A term referring to nerve fibers leading from the central nervous system to muscles or glands. p. 195, 646

efferent arterioles In the kidneys, the arterioles located between the glomerular and peritubular capillary beds. p. 480

efficiency In thermodynamics, the ratio of work done to heat consumed. p. 35

Einthoven triangle In cardiology, the geometrical arrangement of leads I, II, and III of an electrocardiogram. p. 334

ejaculation A coordinated series of contractions of the male reproductive system that expels semen. p. 647, 682

elastic recoil The tendency of the lung (or any object) to shrink to lower volumes and of the chest wall to expand. p. 357

elastic restoring forces See elastic recoil; to be contrasted with externally applied forces. p. 29

elastic work The component of the work of breathing needed to overcome the elastic recoil of the lungs; to be contrasted to flow-resistive work. p. 417

electrical field The imaginary field set up by charged particles that determines the force a small test charge would experience. p. 33

electrical potential The nondirectional energy "well" or "hill" set up by charged particles. p. 33

electrical synapses Junctions between cells that involve direct connections between the two cytoplasmic compartments. p. 175

electrically excitable A term referring to cells capable of generating propagated, all-or-nothing action potentials. p. 141

electrically inexcitable A term referring to cells that undergo no voltage changes or only local and generally graded ones. p. 141

electrocardiogram (ECG) The extracellular signal produced by electrical activation of the heart. p. 34, 333

electrochemical gradient A term referring to the fact that movement of charged substances across cell membranes depends on both the concentration gradient and the electrical field. p. 78

electroencephalogram (EEG) The extracellular signal produced by electrical activation of the cerebral cortex. p. 216

electrogenic pump A term referring to the fact that the Na^+-K^+ ATPase (sodium pump) transports more Na^+ ions than K^+ ions. p. 126

electrolyte A solution composed of ions. p. 23

electron The negatively charged particle that orbits the nucleus. p. 18

electron acceptor An atom that tends to acquire extra electrons. p. 18

electron donor An atom that tends to give up electrons. p. 18

electron shell A term referring to the fact that electrons occupy a set of fixed orbits at varying distances from the nucleus. p. 18

embryoblast The portion of the blastocyst that develops into a fetus. p. 682

emphysema A disease characterized by a loss of lung tissue, resulting in an increase in lung compliance and the collapse of airways during expiration. p. 420

emulsification The process by which water-insoluble fats form microscopic micelles in solution. p. 571

end-diastolic volume The maximum volume of a ventricle just before contraction. p. 335

end-systolic volume The minimum volume of a ventricle just after contraction. p. 338

endocrine reflex A reflex that involves glandular secretion rather than contraction or relaxation of muscles. p. 95

endocrine system The collection of ductless glands that secrete chemical messengers, called hormones, directly into the blood. p. 96

endocytosis The process by which a cell takes up extracellular materials by "pinching off" a vesicle from its membrane. p. 66

endogenous pyrogen A fever-causing substance produced by the body. p. 610

endolymph The fluid in the semicircular canals and a large part of the cochlea. Its composition resembles that of intracellular fluid. p. 227

endometrium The internal layer of cells in the uterus. p. 670

endoplasmic reticulum (ER) An intracellular organelle consisting of large, flattened sheets. Rough ER has ribosomes embedded in its surface. p. 69

endorphins A neuroactive peptide (endogenous opiate) that suppresses pain. p. 94

endotoxin A substance released by bacteria that increases capillary permeability. p. 372

endplate potential (EPP) The voltage change produced at the neuromuscular junction as a result of transmitter release. p. 182

enkephalin One of the neuroactive peptides used as an inhibitory presynaptic transmitter in afferent pain fibers in the dorsal horn of the spinal cord. p. 189

enteric nervous system The network of nerve cells and fibers intrinsic to the gastrointestinal tract. p. 539

enterogastric reflex A gastrointestinal reflex in which increased duodenal tension decreases gastric motility and increases pyloric tone. p. 581

enterohepatic circulation The continual recirculation of the bile between the liver and the gastrointestinal tract. p. 555

enterokinase An intestinal enzyme that activates pancreatic trypsinogen. p. 554

enthalpy In thermodynamics, a measure of heat content. p. 34

entropy In thermodynamics, a measure of "randomness." The lower the entropy, the more ordered a system. In all spontaneous processes, the total entropy of the universe increases. p. 35

enzymes The substrate-specific protein catalysts for biochemical reactions. En-zymes reduce activation energy. p. 93, 125

eosinophils White blood cells that protect against parasites. p. 619

epididymis The male reproductive tract between the sites of sperm production (seminiferous tubules) and the vas deferens. p. 646

epimysium The connective tissue surrounding a muscle. p. 286

epinephrine The catecholamine released by the adrenal medulla. Epinepherine preferentially activates β_2-adrenergic receptors. p. 182

epiphyseal plate The growing region of the bones. p. 517

epithelial cells One of the four basic cell types. They line all body cavities, form the skin, and comprise most glands. p. 5

equilibrium A state in which all a system's variables are constant. p. 32, 121

equilibrium constant In the Law of Mass Action, a measure of the degree to which a chemical reaction takes place. It relates equilibrium concentrations of reactants and products to one another. p. 50

equilibrium potential The voltage that would exist across a membrane if it were exclusively permeable to one ion. p. 130

erection Swelling of the penis caused by its engorgement with blood. p. 681

error signal In negative feedback systems, the difference between the actual value of a variable and its ideal "set-point." p. 86

erythrocyte A red blood cell. p. 346

erythropoeitin A hormone that regulates production of red blood cells. p. 349

esophageal hiatus The opening from the esophagus to the stomach. p. 535

esophageal phase The phase of swallowing during which esophageal distension causes secondary waves of peristalsis. p. 548

esophagus The long tube that connects the pharynx with the stomach. p. 535

essential amino acid Amino acids that the body cannot synthesize. p. 47, 595

estrogen A female sex steroid secreted by the ovaries. p. 656

evoked potential Electrical changes recorded from the sensory and motor areas of the cortex that correspond to sensations and movements. p. 271

excitability The ease with which a nerve or muscle cell can be made to generate an action potential. The lower the threshold, the more excitable the cell. p. 293

excitation-contraction coupling The process by which an action potential in a muscle activates the muscle's contractile "machinery." p. 294

excitatory phase The initial stage of the human sexual response. p. 680

excitatory postsynaptic potential (EPSP) The depolarization observed at a synapse when a neurotransmitter increases the Na^+ permeability. p. 181

exocrine gland A gland that secretes substances through ducts; to be contrasted to endocrine gland. p. 96, 538

exocytosis The process by which a cell extrudes some of its contents enclosed within a secretory vesicle. p. 66

exogenous pyrogen A fever-causing substance released by bacteria. p. 610

exons Regions of DNA used as codes for protein synthesis; to be contrasted to noncoding regions that must be removed before transcription to RNA. p. 57

expiration The phase of respiration during which air leaves the lungs. p. 408

expiratory muscles Muscles of the chest wall that normally are not activated during passive expiration but that are recruited in forced expiration. p. 448

expiratory reserve volume (ERV) The difference between the volume of air in the lungs at the end of a passive expiration and the residual volume of air that cannot be expelled by active effort. p. 409

expressive aphasia A condition in which an individual cannot form gramatically correct sentences or phrases because of damage in the Broca area. p. 214

extensor A muscle that serves to extend a joint. p. 261

extracellular fluid (compartment) Fluid outside the cells. p. 11, 506

extrafusal fiber The "normal" fibers of a muscle. See **intrafusal fibers**. p. 263

extrapyramidal An older term referring to the descending motor tracts that do not pass through the medullary pyramids. p. 273

extrasystole A condition in which ventricular contraction is not preceded by atrial activation. It is felt as a "skipped beat" because the next heartbeat usually is delayed. p. 328

external anal sphincter The most distal anal sphincter, composed of both voluntary and smooth muscle. p. 476

extrinsic pathway Formation of a blood clot in the presence of damaged tissue by a process involving factor III rather than activation of Hageman factor. p. 351

extrinsic protein A protein associated with the internal or external surface of a membrane through weak electrostatic forces. p. 118

extrinsic regulation Control of muscles and glands by the nervous or endocrine systems. p. 94

F

facilitated diffusion Carrier-mediated diffusion; the translocation of substances down their concentration gradients via specialized pathways. p. 125

facilitation In synaptic transmission, an increase in the quantal content either during a train of nerve impulses or as a result of some other presynaptic process. p. 268

fallopian tubes The tubes that connect the ovaries with the uterus. p. 645

fascicles Small groups of muscle fibers; an anatomical rather than functional subunit of a whole muscle. p. 286

fast pain, first pain, or "pricking" pain The component of pain that is relayed via the neospinothalamic tract and that can be well localized. p. 212

fast glycolytic muscle fibers One of the three types of muscle fibers characterized by rapid contraction and susceptibility to fatigue. p. 301

fast oxidative muscle fibers Muscle fibers characterized by moderately rapid contraction, abundant mitochondria, and moderate resistance to fatigue. p. 301

fat-soluble vitamins The water-insoluble vitamins (A, D, E, and K). p. 577

fatigue Tiredness. In muscle, the decrease in contractile force with prolonged use. p. 304

fats Triglycerides. p. 40

fatty acids A hydrocarbon chain with a carboxyl group at one end If the hydrocarbon chain has one or more double bonds, the fatty acid is referred to as unsaturated. p. 40, 571

feature extraction The way in which the visual system resolves the outlines and movements of an object. p. 205

feed-forward regulation In control theory, changes that anticipate fluctuations in the controlled (regulated) variable. p. 87

fenestrations Large pores in the walls of capillaries in vascular beds such as the kidneys and some endocrine glands. p. 359, 484

ferritin An iron-binding protein in cells lining the intestine. p. 575

fertilization The union of a sperm and an ovum. p. 670

fetal hemoglobin A form of hemoglobin characterized by an increased ability to bind oxygen at low partial pressure. p. 438, 689

fiber types In muscle, the difference between the speed of contraction and

metabolic pathways used by a fiber. p. 301

fibrillation In the heart, disorganized electrical activity in the atria or ventricles that cannot produce a normal contraction. p. 329

fibrin The fibrous components of a blood clot. p. 350

fibrinogen The plasma protein that polymerizes to form fibrin. p. 351

fibrosis In the lungs, a condition involving a decrease in lung compliance. p. 406

Fick equation The relationship between the net flux of a substance and its concentration gradient; the diffusion equation. p. 382

filtered load In the kidneys, the amount of a substance that is filtered into the glomerulus. It is expressed in milligrams per minute. p. 487

filtration See capillary filtration.

filtration fraction In the kidneys, the fraction of the renal plasma flow that is filtered into the glomerulus. p. 484

finger proteins Proteins coded by the Y chromosome that control transcription of specific genes. p. 653

first heart sound The sound associated with closing of the A-V valves. p. 335

First Law of Thermodynamics A statement of the principle of conservation of energy. p. 34

fixed receptor In endocrinology, a plasma membrane—associated receptor. p. 103

flavin adenine dinucleotide (FAD) One of the electron-carrying coenzymes. p. 75

flexion reflex The withdrawal of a limb from a noxious stimulus. p. 267

flexor A muscle that bends a joint. p. 261

flow Movement of a fluid, either liquid or gas. p. 30, 361

flow autoregulation See autoregulation.

flow resistance The result of internal frictional forces in a liquid. Flow is equal to a pressure gradient divided by the resistance. p. 30, 361

flow-resistive work In the lungs, the component of the work of breathing needed to overcome frictional forces in the airways. p. 417

flower-spray endings The afferent fibers that innervate the peripheral regions of the muscle spindle. p. 261

fluid mosaic model A conceptualization of a membrane in which intrinsic proteins move freely in a two-dimensional sea of lipid. p. 118

flux The rate of movement of materials across an interface. See **unidirectional flux** and **net flux.** p. 121

folic acid Vitamin B_{12}, required for iron absorption in the gastrointestinal tract. Its absence causes pernicious anemia. p. 577

follicle In the ovaries, the structure containing the ovum, granulosa cells, follicular cells, and thecal cells. One follicle matures and is ovulated during each menstrual cycle. p. 207

follicle-stimulating hormone (FSH) An anterior pituitary hormone required for germ cell (sperm and ovum) maturation. p. 653

follicular phase The phase of the menstrual cycle that precedes ovulation. p. 662

fovea The area of the retina that contains the cone photoreceptors. p. 241

Frank-Starling Law The principle that the length-tension relation of cardiac muscle causes cardiac output to be a function of venous return. p. 340

free energy The energy of a system that is available to do "useful" work. p. 35

free nerve endings Sensory nerve terminals not characterized by any overt structural specialization which, nevertheless, are sensitive to particular modalities such as touch, temperature, or pain. p. 208

frontal lobe The most anterior area of the cerebral cortex; it is involved in motor control and emotional behavior. p. 198

fructose A 5-carbon monosaccharide bound to glucose in table sugar. p. 567

functional residual capacity (FRC) the volume of air normally remaining in the lungs after a passive rather than forced expiration. p. 409

functional residual volume (FRV) The difference between the functional residual capacity and the residual volume attatined in a maximum expiratory effort. p. 410

functional syncytium A group of cells that function as a unit because of electrical connections between them. For example the heart and unitary smooth muscle. p. 307

G

G cells Gastrin-secreting cells in the stomach. p. 551

G-protein A membrane-bound protein that serves as an intermediate between hormone or neurotransmitter binding to receptors and activation of adenyl cyclase. p. 103, 181

gain In control theory, a measure of the sensitivity of a negative feedback system. More generally, it is equivalent to amplification. p. 87

ext-hai

galactose A monosaccharide linked to glucose to form lactose (milk sugar). p. 567

gallbladder The organ that stores the bile produced by the liver. p. 555

gamete Reproductive cells; an ovum or sperm. p. 648

γ-aminobutyric acid (GABA) One of the central neurotransmitters. p. 182

gamma motor neurons Motor neurons that activate muscle spindle intrafusal fibers; to be contrasted to alpha motor neurons. p. 263

ganglion A tightly packed group of neurons outside the central nervous system. p. 197

ganglion cells In the retina, the neurons whose axons form the optic nerve. p. 239

gap junction A junction between cells that couples them electrically. p. 73

gas A phase of matter in which molecules do not interact. p. 429

gastric inhibitory peptide One of the intrinsic gastrointestinal hormones. p. 582

gastric juice The acid and enzyme secretions of the stomach. p. 550

gastric motility Contractions of the smooth muscle of the stomach, including segmentation and peristalsis. p. 550

gastric phase In digestion, the phase in which chyme in the stomach initiates a series of local neural and hormonal reflexes. p. 579

gastrin A gastrointestinal hormone that increases gastric secretion. p. 578

gastrocolic reflex A reflex that involves an increase in the large intestine's motility during the gastric phase of gastric secretion. p. 580

gastroileac reflex An increase in activity in the small intestine (ileum) that occurs during gastric emptying. p. 580

gastrointestinal tract The esophagus, stomach, and intestines. p. 534

gated ion channel Water-filled pores in the cell membrane selected for certain ions. Opening and closing of such channels may be regulated by the electrical field, by molecules such as hormones and neurotransmitters, or by physical stimuli. p. 126, 143

generator potential A receptor potential; the change in voltage that occurs when a sensory receptor is activated by a stimulus. p. 140, 171

Gibbs-Donnan equilibrium See Donnan equilibrium.

gigantism An abnormal increase in body size caused by excess growth hormone. p. 604

gland A collection of cells specialized for secretion. Glands may have ducts (exocrine) or may secrete into the blood (endocrine). p. 96

glaucoma A condition in which intraocular pressure increases. p. 236

glia In the nervous system, the non-excitable satellite cells that surround axons. They may or may not make myelin. See **Schwann cells** and **oligodendrocytes.** p. 157

globus pallidus One of the basal ganglia. p. 269

glomerular filtration In the kidneys, the process by which plasma is filtered into the proximal region of the nephrons. p. 484

glomerular filtration rate The rate at which plasma is filtered into the glomerulus. It usually is expressed in milliliters per minute. p. 482

glomerulotubular balance In the kidney, the tendency of the proximal tubule to reabsorb a constant fraction of its filtered load. p. 482

glomerulus In the kidney, the most proximal portion of the nephron, composed of the glomerular capillaries and the Bowman capsule. p. 478

glucagon A hormone secreted by the pancreas that promotes mobilization of glucose from body stores. p. 596

glucocorticoids Steroids secreted by the adrenal cortex that affect organic metabolism; includes cortisol. p. 599

gluconeogenesis The formation of glucose from amino acids. p. 590

glucoprivation Long-term depletion of carbohydrates. p. 612

glucose A monosaccharide; it is degraded in glycolysis to produce ATP. p. 45, 567

glucose-sparing A term denoting the use of fatty acids and ketones in metabolism. p. 597

glutamate One of the central neurotransmitters. p. 182

glycine One of the central neurotransmitters. p. 182

glycogen The storage form of glucose; a highly branched polymer. p. 45, 567

glycogen synthetase The enzyme that converts glucose phosphate to glycogen. p. 593

glycogenolysis The breakdown of glycogen into glucose. p. 590

glycolipid A molecule composed of a lipid and a sugar residue; a component of many cell membranes. p. 41

glycolysis A metabolic pathway in which glucose is degraded to pyruvate. p. 76

glycoprotein The combination of a protein with carbohydrate residues. p. 542

glycosuria The presence of glucose in the urine. p. 600

goblet cell A secretory cell in the intestinal epithelium. p. 542

Goldman equation An equation that relates the membrane potential to the concentration gradients and relative permeabilities of all ions capable of crossing the cell membrane. p. 133

Golgi apparatus The intracellular organelle that "packages" proteins synthesized in the rough endoplasmic reticulum into secretory vesicles. p. 70

Golgi tendon organs Stretch receptors in tendons that measure muscle tension. p. 265

gonadotropic hormone A generic term that includes the anterior pituitary hormones FSH and LH. p. 110, 660

gonadotropin-releasing hormone (GnRH) The hypothalamic releasing factor that controls the secretion of FSH and LH from the anterior pituitary. p. 655

gonads The ovaries and testes. p. 645

graded In receptor physiology, a term denoting that generator and synaptic potentials vary continuously in amplitude and do not have a threshold. p. 140

gram (g) The unit of mass. p. 18

gramicidin An antibiotic that forms a transmembrane channel in lipid bilayers. p. 129

granular neuron A neuron with an extensive set of highly branched dendrites. p. 257

granulosa cells Cells in the ovary that nurture the ovum. p. 482, 651

gray matter The areas in the central nervous system that contain nerve cell bodies. p. 194

gray rami Afferent connectives between the sympathetic chain and the main part of the spinal cord. p. 259

green-sensitive cones Photoreceptors with maximum absorbance at 530 nanometers. p. 240

growth hormone (GH) An anterior pituitary hormone that has widespread effects on growth, development, and organic metabolism. p. 197, 602

guanine One of the five nucleotides in RNA and DNA. p. 53

H

H zone The region of the thick filament that is free of crossbridges. It corresponds to the "tail" regions of the myosin molecules. p. 286

habituation A form of nonassociative learning present in invertebrates that involves reactions to repetition of a single stimulus. p. 213

Hageman factor A plasma protein; its activation is the initial event in the formation of a blood clot. p. 350

hair cells Ciliated sensory receptors in the cochlea and vestibular apparatus. p. 229

Glossary   **731**

hair follicle receptors One of the somatosensory touch receptors. p. 207

Haldane effect The effect of oxygen on the dissociation curve for carbon dioxide. p. 440

hapten A small molecule that can serve as an antigen if it becomes chemically bound to a larger molecule. p. 619

haustra Segmentation movements in the large intestine. p. 557

heart The organ that pumps blood in the circulatory system. p. 321

heart failure The inability of the heart to maintain an adequate cardiac output. p. 386

heart rate The number of times the heart contracts per minute. p. 327

heart sounds Sounds produced during the cardiac cycle by the closing of valves or by vibrations caused by rapid blood flow. p. 335

heat The transfer of energy resulting from differences in temperature. p. 34

heat stroke A condition in which the core body temperature is abnormally high. p. 392

helicotrema The tiny opening at the distal end of the cochlea; it connects the scala vestibuli and the scala tympani. p. 226

helix A three-dimensional, coiled structure that resembles a spring. p. 48

helper cells A class of T lymphocytes that increase the activity of B cells and other T-lymphocyte subpopulations. p. 629

hematocrit The fraction of the blood occupied by red blood cells. p. 348

heme The component of hemoglobin to which oxygen molecules bind. p. 434

hemicolinium A drug that blocks choline uptake into cholinergic nerve terminals. p. 278

hemodynamic Having to do with the flow of blood. p. 359

hemoglobin (Hb) The oxygencarrying protein in the blood. p. 485

hemorrhage Loss of blood volume through bleeding. p. 389

hemostasis The process of blood clot formation. p. 349

Henderson-Hasselbalch equation The Law of Mass Action applied to the reaction of CO_2 with water to form bicarbonate and H^+. It describes the interrelationships between plasma pH, bicarbonate concentration, and the partial pressure of CO_2. p. 453, 520

Henle loop In the kidney, the nephron between the proximal and distal tubules responsible for the concentration of the medullary interstitium. p. 478

heparin A substance produced in the lungs that inhibits blood clotting. p. 354

hepatic circulation The flow of blood in the liver. p. 566

hepatocytes Liver cells arranged in a series of sheets. p. 567

hexamethonium A drug that blocks nicotinic cholinergic receptors. p. 278

hiatal hernia Bulging of the wall of the stomach into the thorax. p. 550

high-density lipoprotein (HDL) A lipoprotein that facilitates lipid uptake into chylomicrons and very-low-density lipoproteins (VLDLs). p. 594

high-renin hypertension Hypertension caused by renal arteriolar atherosclerosis. p. 514

hippocampus An "older" region of the cortex involved in initial processing of memories and in emotional behavior. p. 198

histamine A substance released by white blood cells that increases capillary permeability. p. 552, 619

histocompatibility antigens Markers on the surface of cells that distinguish "self" from "nonself." p. 626

homeostasis The constancy of the"internal environment." p. 10

homologous chromosome One of the 23 pairs of nonsex chromosomes. p. 648

homunculus The often exaggerated representation of the body surface onto the somatosensory or motor cortex. p. 270

horizontal cells Neurons that couple adjacent photoreceptors in the retina. p. 239

hormone A chemical "messenger" secreted by endocrine glands. p. 95

human chorionic gonadotropin (HCG) A placental hormone that maintains the corpus luteum during pregnancy. p. 683

humoral immunity The immune system response in which B lymphocytes secrete specific antibodies. p. 619

Huntington disease A disease of the central nervous system characterized by uncontrolled limb movements. p. 274

hyaline membrane disease See respiratory distress syndrome.

hydration (shell) The "surrounding" of a dissolved ion by water molecules. p. 23, 122

hydraulic filter In the cardiovascular system, the way in which the arteries convert the pulsatile output of the heart into steady flow. p. 358

hydrocarbon A molecule composed only of carbon and hydrogen. p. 40

hydrochloric acid (HCl) A strong acid secreted by the stomach. p. 26, 550

hydrogen The lightest element. p. 18

hydrogen bond A chemical bond created between two molecules or different parts of the same molecule by sharing of an electron from hydrogen with an acceptor. p. 22

hydrogen ion (H^+) A hydrogen atom stripped of its electron. p. 21

hydrogenation A chemical reaction that involves addition of a hydrogen atom. p. 52

hydrolysis The breaking of a chemical bond by the addition of water. p. 45

hydrophilic "Water loving"; refers to the attraction between polar groups of compounds and water molecules. p. 23

hydrophobic "Water hating"; refers to the tendency for nonpolar compounds to interact with each other in such a way that water is excluded. p. 23

hydrostatic force The force exerted by a column of liquid or gas. p. 30

hypercalcemia A condition in which the plasma Ca^{++} is abnormally high. p. 153

hypereffective (hypoeffective) In cardiovascular physiology, hypereffective indicates an increase in cardiac contractility; hypoeffective, a decrease. p. 341

hyperplasia An increase in the number of cells in a tissue. p. 304

hyperpolarization A condition in which the membrane potential of a cell becomes internally more negative. p. 141

hypertension Abnormally high arterial blood pressure. p. 381

hypertonic solution A solution in which a cell will shrink. p. 124

hypertrophy An increase in the size of a cell such as muscle. p. 304

hyperventilation An increase in alveolar ventilation that causes the alveolar P_{CO_2} to fall below 40 mm Hg. p. 410

hypervolemia A condition in which total body water is abnormally high. p. 381, 508

hypocalcemia A condition in which plasma Ca^{++} is abnormally low. p. 154, 518

hypoglycemia Condition in which plasma glucose is abnormally low. p. 590

hypoosmotic A solution with an osmotic pressure lower than that inside cells. p. 24, 123

hypotension Abnormally low arterial blood pressure. p. 388

hypothalamic-releasing factors Peptides released from terminals of the hypothalamic neurons that control secretion of anterior pituitary hormones. p. 99

hypothalamohypophysial portal system The network of blood vessels connecting the hypothalamus and the anterior pituitary gland that carries the releasing factors responsible for pituitary secretion. p. 100

hypothalamus The region of the brain that controls the release of hormones from the pituitary gland. p. 198

hypotonic solution A solution in which cells will swell. p. 124

hypovolemia A condition in which total body water is abnormally low. p. 389

hypoxemia A condition in which the arterial P_{O_2} is abnormally low. p. 454

hypoxia Lack of oxygen. p. 435

I

I band The region of a muscle sarcomere that contains thin filaments but not thick filaments. In the sliding filament model, the region of nonoverlap. p. 286

ileocecal sphincter The sphincter between the small and large intestines. p. 557

ileogastric reflex The decrease in gastric motility caused by distension of the ileum. p. 580

ileum The most distal region of the small intestine; it empties into the cecum. p. 537

immune system The organ system that "recognizes" and attacks "foreign" materials, including bacteria and viruses. p. 617

immunoglobulins Antibodies Protein molecules in which one end is specialized for recognition of "foreign" molecules (antigens) and the other end activates components of the immune response. p. 348, 619

implantation Insertion of a fertilized ovum into the uterine wall. p. 683

incisura See dicrotic notch.

incomplete heart block A condition in which not all atrial contractions lead to ventricular activation because the AV node has a limited ability to pass impulses at high frequency. p. 327

incus One of the bones of the middle ear. p. 224

inert A term referring to elements whose outer electron shell contains its full complement of electrons and which cannot combine with other atoms to form compounds. p. 18

infarct A damaged region of the heart muscle. p. 329

inflammation Swelling; the term also refers to nonspecific immune responses. p. 619

inhibitory postsynaptic potential (IPSP) At a synapse, the change in membrane potential produced by a neurotransmitter that increases the K^+ permeability. p. 182

inner ear The cochlea. p. 224

inner segment The part of rod and cone photoreceptors that contains the nucleus of the cell and synaptic vesicles. p. 240

innervation ratio The number of muscle fibers innervated by a motor neuron. p. 256

inotropic effect Causing an increase in cardiac contractility. p. 330

inorganic Compounds that do not contain carbon. p. 21

insensible water loss Water lost through normal respiration and evaporation. p. 507

insertion One of the connections of a muscle to the skeleton. Contraction pulls the insertion toward the origin. p. 286

inspiration The phase of respiration during which air enters the lungs. p. 408

inspiratory capacity (IC) The volume that can be inhaled after passive expiration. The difference between total lung capacity and functional residual capacity. p. 409

inspiratory muscles The muscles of the chest wall that can be activated to increase the tidal volume. p. 448

inspiratory reserve volume (IRV) The volume of air that can be inhaled beyond the resting tidal volume. p. 409

insufficiency A term denoting the "leakiness" of a heart valve. p. 338

insulin A hormone secreted by the pancreas that is necessary for glucose utilization in most tissues. p. 497

insulin-dependent diabetes Excess blood sugar caused by decreased secretion of insulin. p. 600

insulin-independent diabetes Excess blood sugar caused by a decrease in the density of insulin receptors. p. 601

integrater (integration center) The collection of neurons in the spinal cord between afferent sensory fibers and motor fibers. p. 86

intention tremor Shaking seen when an individual tries to reach for an object. p. 272

intercalated disks Structures in cardiac muscle that electrically couple adjacent cells. p. 307

interleukins Molecules released during the immune response that activate other cells of the immune system and act as endogenous pyrogens. p. 610

intermediate filaments The most stable elements of the cytoskeleton, thought to provide the basic anatomical structure for a cell. p. 71

intermediate zone In the spinal cord, the gray matter between the dorsal and ventral horns; it contains spinal interneurons. p. 194

internal anal sphincter The more proximal anal sphincter, composed of involuntary smooth muscle. p. 476

internal energy In thermodynamics, the energy "within" a system. p. 34

interneuron The neurons in the spinal cord that are not motor neurons. p. 166

interstitial fluid The fluid in the spaces between cells. p. 4, 506

intestinal phase In digestion, the neural and hormonal reflexes initiated by entry of chyme into the duodenum. p. 581

intestine A generic term for the small and large bowel. p. 536

intracellular compartment The fluid inside the cells. p. 506

intrafusal fibers The muscle fibers inside the spindle controlled by spinal gamma motor neurons. These fibers regulate spindle sensitivity. p. 263

intrapleural pressure The pressure in the intrapleural space. p. 402

intrapleural space The fluid-filled space between the lungs and chest wall. p. 402

intrinsic factor A glycoprotein required for vitamin B_{12} absorbtion. p. 551

intrinsic pathway Formation of a blood clot by activation of Hageman factor. p. 351

intrinsic proteins Proteins tightly embedded in cell membranes, often extending entirely across them to form receptors or channels. p. 118

intron A noncoding region of DNA. See exons. p. 57

inulin A plant polysaccharide. p. 495

inulin clearance A measure of the rate of glomerular filtration. p. 495

ion A charged atom. p. 21, 43

ionic bond A chemical bond formed by donation and acceptance of electrons rather than by sharing. p. 21

ionophore A molecule that can carry ions across membranes. p. 129

ipsilateral On the same side of the body. p. 209, 261

iris The pigmented area surrounding the pupil of the eye. p. 236

iron deficiency anemia A decrease in the hemoglobin content of the blood, caused by a lack of iron in the diet. p. 434

ischemia Inadequate blood supply to a tissue. p. 329

islets of Langerhans The regions of the pancreas that contain the cells that secrete insulin and glucagon. p. 99, 590

isometric contraction Generation of force by a muscle under conditions that prevent the muscle's length from changing. p. 297

isosmotic solution A solution with an osmotic pressure equal to that of a cell. p. 24, 123

isotonic contraction Muscle shortening under conditions of constant load. p. 297

isotonic solution A solution in which the volume of a cell will not change. p. 123

isotopes Elements in which the number of protons in the nucleus is constant, but the number of neutrons varies. p. 18

isovolumetric contraction In the cardiac cycle, contraction of the ventricle under conditions of constant volume; an isometric contraction. p. 335

isovolumetric relaxation In the cardiac cycle, relaxation of the ventricle under conditions of constant volume. p. 335

J

jejunum The "middle" region of the small intestine between the duodenum and the ileum. p. 537

joint receptors Sensory receptors in the joint capsules that provide information about limb position and movement. p. 262

junctions See desmosome, gap junction, and tight junction.

juxtaglomerular apparatus A specialized set of cells in the kidneys that release renin when renal arterial pressure decreases. p. 482

juxtamedullary nephrons Nephrons whose glomeruli are near the medulla and whose Henle loops penetrate the inner medulla. p. 479

K

K cells Cells in the intestine that secrete gastric inhibitory peptide (GIP). p. 582

kallikrein The enzyme that generates bradykinin from its precursor. p. 631

Kerckring folds See plicae circulares.

ketoacidosis Acidosis caused by accumulation of ketones. p. 524, 601

ketone bodies Small molecules that are produced during the breakdown of fatty acids. p. 597

kidneys The principal excretory organs. p. 189

kinetic energy The energy an object has as a result of mass and velocity. p. 34

kinocilium The cilium in hair cells that determines directional sensitivity. p. 233

Korotkoff sounds The sounds heard through a stethoscope placed on a peripheral vein; they are caused by the pulsations of the blood through a partially inflated pressure cuff. p. 363

Kreb cycle A metabolic pathway in which 2-carbon acetyl groups are degraded to CO_2 and water. It occurs in the cental region of mitochondria. p. 77

L

labia minora (majora) Skin folds covering the outer opening of the vagina. p. 646

labelled line In sensory physiology, the ordered way in which afferent fibers enter the central nervous system. p. 169

labyrinth The inner ear; the cochlea plus the vestibular apparatus. p. 224

lacteals Lymph vessels in the intestinal villi. p. 539

lactic acid The end result of glucose metabolism in the absence of oxygen.

lactose Disaccharide of glucose and galactose; milk sugar. p. 567

lactose intolerance A condition in which the enzyme needed to degrade lactose is absent. p. 569

lamina propria The component of the gastrointestinal mucosa that contains glands. p. 539

laminar flow Nonturbulent or steady flow. p. 360

language areas The regions of the cerebral cortex that are specialized for the decoding and encoding of verbal content. See Broca's area and Wernicke's area. p. 214

Laplace Law The relationship between the tension in the wall of a sphere or cylinder and the transmural pressure: (P = T/2r). p. 418

large intestine The most distal region of the gastrointestinal tract. p. 537

larynx The region of the airway that contains the vocal cords. p. 402

latency In muscle, the time between electrical activation and contraction. In spinal reflexes the time between a stimulus and a motor response. p. 299

lateral geniculate One of the brainstem relay stations in the visual system. p. 246

lateral inhibition In sensory physiology, the tendency of input to adjacent regions of a sensory surface to be mutually inhibitive. p. 204

lateral reticular facilitory area The region in the brainstem that contains neurons that enhance spinal reflexes. p. 271

lateral vestibulospinal tract One of the descending motor pathways. p. 273

Law of Mass Action The principle that increasing the concentration of a reactant or product proportionately increases the unidirectional reaction rate. p. 50

leaky epithelia Epithelia characterized by a high water permeability. They are found where osmotically driven water flow is important physiologically. p. 74

learning The ability to modify behavior because of previous experience. p. 213

lecithin A phospholipid; a component of membranes and the bile. p. 555

lens The transparent portion of the eye that focuses light rays on the retina. p. 235

lemniscal system The major ascending sensory pathway, which carries proprioceptive information and fine touch. p. 209

length constant The distance over which currents applied to a nerve fiber will dissipate by a given amount, usually a few millimeters. p. 157

leukocytes White blood cells. p. 348, 619

luteinizing hormone (LH) The anterior pituitary hormone needed for secretion of estrogen and testosterone by the ovaries and testes respectively. LH also controls ovulation. p. 111, 660

Leydig cells Cells in the male reproductive system that secrete testosterone. p. 646

Lieberkuhn crypts Depressions in the gastrointestinal tract in which new epithelial cells are formed by division. p. 539

limb leads The electrode connections to the arms and legs that are used to record an electrocardiogram. p. 334

limbic system An older region of the cerebrum that is involved in emotional behavior. p. 200

linear velocity of flow The rate of movement of an element of fluid past a given point; to be contrasted to volume flow and average flow velocity. p. 359

lipases Enzymes that degrade triglycerides into fatty acids. p. 554

lipid bilayer The basic "framework" of cell membranes. Bilayers consist of aggregated phospholipids whose hydrocarbon tails interact with one another. p. 40, 118

lipids Compounds composed largely of hydrocarbons that usually are not water soluble. See fatty acids, phospholipids, steroids, and triglyceride. p. 40

lipogenesis The synthesis of fatty acids. p. 590

lipolysis The breakdown of fatty acids into 2-carbon acetyl groups. p. 571

lipoprotein A combination of proteins with lipid residues. p. 594

lipoprotein lipase An enzyme on endothelial cell membranes that liberates free fatty acids from triglycerides. p. 594

liquid crystal A term denoting the structures (bilayers and multilamellar vesicles) typical of phospholipid and other amphipathic molecules in solution. p. 64

liter (L) The unit of volume. p. 28

liver The organ that acts as a chemical "reprocessing plant." It can interconvert carbohydrates, amino acids, and lipids. p. 537

load See preload or afterload.

lobules Functional subunits of the liver. p. 567

logarithm The power to which the base (usually 10) must be raised to obtain a given number. Logarithms are never negative. p. 173

long-term memory Memories that cannot be electrically disrupted and are presumably the result of some permanent alteration in neuronal function. p. 213

longitudinal muscle One of the muscle layers in the gastrointestinal tract. p. 537

lower esophageal sphincter The physiological sphincter that separates the esophagus from the stomach. p. 549

lumen The space within a tissue or organ; also, the space between membranes of the endoplasmic reticulum. p. 69, 537

lungs The site of blood-atmospheric gas exchange. p. 402

luteal phase In the menstrual cycle, the phase immediately after ovulation and before menstruation. p. 662

lymph nodes Areas along the lymphatic vessels that contain lymphocytes. p. 619

lymphatic duct The lymph vessel that receives fluid from the right shoulder and the right side of the head. p. 372

lymphatic system The set of vessels that transports the lymph. p. 372

lymphocyte A type of white blood cell that is involved in specific immunity. p. 370

lysis The swelling and subsequent disruption of a cell. p. 123, 664

lysosome An intracellular organelle that contains hydrolytic (digestive) enzymes. p. 70

M

macrocolumns One of the functional ways in which sensory areas of the cortex are organized. p. 248

macrophage One of the white blood cells derived from monocytes. Macrophages can engulf and digest bacteria and "foreign" substances. p. 619

macula The portion of the vestibular system (utricle and saccule) that contains linear acceleration-sensitive hair cells. p. 233

macula densa In the kidney, a region of the early distal tubule adjacent to the glomerular efferent arterioles. It plays a major role in intrarenal regulation. p. 482

malleus One of the bones of the middle ear. p. 224

maltose A disaccharide of glucose and galactose. p. 567

mammary gland The milk-producing gland in the female breast. p. 698

MAO inhibitors Drugs that inhibit the enzyme monoamine oxidase; a class of antidepressants. p. 279

mass movement In the gastrointestinal tract, an organized peristaltic wave that sweeps chyme along the small and large intestines several times a day. p. 557

mean blood pressure Average arterial blood pressure; it is equal to the product of cardiac output and total peripheral resistance. p. 338

mean circulatory filling pressure The pressure that would exist everywhere in the circulatory system, if the heart were "off," as a consequence of blood volume. p. 381

mechanoreceptors Sensory receptors sensitive to mechanical force. p. 169, 207

medial reticular inhibitory area A region of the brainstem that decreases spinal reflex tone. p. 271

mediated transport See facilitated diffusion.

medulla (brain) One of the three regions of the brainstem. The site of many of the visceral control centers. p. 197

medulla (kidney) The innermost region of the kidney; it contains the loops of Henle. p. 476

megacaryocytes Bone marrow cells that give rise to platelets. p. 349

meiosis Cell division without duplication of chromosomes. p. 648

Meissner corpuscles A type of touch receptor in the skin. p. 207

Meissner plexus One of the two neural networks intrinsic to the gut. p. 539

membrane The selectively permeable structure of phospholipid molecules and proteins that surrounds all cells. p. 64, 103

membrane permeability A measure of the ease with which different substances cross cell membranes. p. 128

membrane potential The voltage difference across a cell membrane. p. 131

memory Retention of experiences Long term memories are difficult to disrupt, whereas short-term memory can be affected by trauma. p. 213

memory cells A class of B lymphocytes that can quickly react to a second exposure to a particular antigen. p. 623

memory consolidation The process by which transient shortterm memories are converted to relatively stable long-term memories. p. 213

menarche The onset of menstruation. p. 655

menstrual cycle A monthly cycle of ovulation and menstruation in the female. p. 655, 662

menstruation The monthly shedding of the uterine lining. p. 662

mesencephalic locomotor area A region of the brainstem that causes nearly normal walking movements when stimulated. p. 271

mesencephalon The region of the brain composed of the medulla, midbrain, and pons. p. 197

messenger RNA (mRNA) The ribonucleic acid that represents a "copy" of a region of a cell's DNA (genome), used as a template for protein synthesis. p. 53

metabolic acidosis A decrease in plasma pH below 7.4 caused by addition of acid other than carbonic acid. p. 459, 521

metabolic alkalosis An increase in plasma pH above 7.4 caused by the loss of acid other than carbonic acid or the addition of base. p. 521

metabolism Intracellular chemical reactions. p. 75

metarterioles Vessels that connect arterioles directly to venules; they sometimes are referred to as "shunt capillaries" to distinguish them from true capillaries. p. 356

meter (m) The unit of length. p. 28

micelles The structures into which amphipathic molecules aggregate in solution; they can be planar or spherical. p. 40, 571

microcolumns In the cortex, an organizational feature of many sensory areas. p. 248

microelectrode A glass capillary pulled to have a tip less than 1 μm, used to record voltages from cells. p. 131

microfilaments Thin filaments in the cytoplasm that are composed of actin. p. 71

micropuncture A technique for sampling regions of the renal tubule. p. 484

microtubules Hollow tubular structures in the cytoplasm made up of a protein known as tubulin. Both microtubules and microfilaments seem to be involved in axoplasmic flow, cell division, and motility. p. 71

microvilli Microscopic projections on the luminal surface of epithelial cells. They vastly increase the surface area for transport. p. 540

micturition Urination. p. 476

midbrain The brainstem just below the thalamus and above the spinal cord. p. 197

middle ear The region between the eardrum and the cochlea; it normally contains three small bones that transmit sounds. p. 224

milligram % (mg%) A unit of weight used in clinical settings. It equals the number of milligrams of a substance contained in 100 milliliters of water. p. 28, 496

millimeters of mercury (mm Hg) A unit of measure for pressure. p. 28

mineralocorticoids Steroid hormones secreted by the adrenal cortex that affect salt balance; aldosterone is one example. p. 113, 381, 482

miniature endplate potential (mEPP) The endplate potential that results when one synaptic vesicle releases its neurotransmitter. p. 182

mitochondria The intracellular organelles in which the Kreb cycle and oxidative phosphorylation occur. The "powerhouses" of a cell. p. 70

mitochondrial matrix The central area of the mitochondria; it contains soluble mitochondrial enzymes. p. 71

mitral valve The valve between the left atrium and left ventricle. p. 322

mixed receptor A sensory receptor whose output depends on both the absolute magnitude of a stimulus and its rate of change with time. p. 174

mobile receptor hypothesis In endocrinology, a cytoplasmic hormone receptor. p. 103

modality The nature of a sensory stimulus—touch, smell, sound, visual, kinesthetic, hot, cold, pain, and so forth. p. 169

molarity The amount of a compound in moles in 1 liter of a solution. p. 24

mole A weight of a compound in grams equal to the compound's molecular weight; 1 mole contains the same number of molecules (the Avogadro number). p. 24

molecular weight The sum of the atomic weights of the atoms in a compound. p. 24

molecule A combination of atoms via covalent or ionic bonds. p. 19

monoamine oxidase The enzyme that degrades norepinephrine in adrenergic nerve terminals. p. 182

monochromat An individual who lacks all three cone photopigments and thus color vision. p. 243

monocyte A type of white blood cell. p. 618

monosynaptic Having only one synapse. p. 261

morula A solid ball of cells produced by the initial divisions of a fertilized ovum. p. 670

motility See gastric motility.

motor area The region of the frontal lobe of the cortex that initiates voluntary muscle movement. p. 270

motor endplate The neuromuscular junction; the region in which a motor nerve functionally contacts a muscle. p. 177

motor nerve The nerve fiber leading from the spinal cord to a muscle (skeletal or smooth) or a gland. p. 194

motor neuron A neuron in the central nervous system that controls skeletal or smooth muscle contraction or glandular secretion. p. 95, 166, 263

motor programs Characteristic patterns of muscular activation seemingly stored in certain regions of the central nervous system. p. 269

motor tract Groups of fibers in the spinal cord that carry imupulses from motor areas of the brain. p. 273

motor unit The set of muscle fibers innervated by a motor neuron. p. 256

mucin The protein component of mucus. p. 541

mucosa (mucosal) The internal lining (surface) of the gastrointestinal tract. p. 537

mucus A thick secretion that protects epithelial cells such as those found in the gastrointestinal tract. p. 541

Müllerian duct The structure in the embryonic gonad that gives rise to the fallopian tubes, uterus, and proximal vagina in the female. p. 653

multiunit smooth muscle Smooth muscle in which cells are not electrically coupled. p. 310

multivesicular bodies Membrane-bound remnants of empty synaptic vesicles. p. 79

muscarinic receptors A subclass of cholinergic (ACh) receptors present on the parasympathetic effector organs. See **nicotinic receptors**. p. 277

muscle fatigue See **fatigue**.

muscle fiber The basic functional unit of a muscle; a muscle cell. See also **motor unit**. p. 286

muscle hypertrophy An increase in the size of muscle fibers without any change in the number of cells. p. 304

muscle spindle A specialized receptor within muscles that provides information about muscle length and its rate of change. p. 261

muscle tension The force exerted by a muscle between its ends. p. 289

muscle tone The force generated in a muscle in the absence of any conscious contraction. p. 262

muscle unit See motor unit.

muscularis mucosa The innermost smooth muscle layer of the gut. p. 539

musculoskeletal system The skeletal muscles and bones. p. 7

myelin An insulator around some nerve fibers that consists of several layers of cell membrane. p. 157

myenteric plexus One of the intrinsic neural networks in the gut It is located between the circular and the longitudinal muscle layers. p. 538

myesthenia gravis A disease characterized by muscle weakness involving the block of acetylcholine receptors at the neuromuscular junction by antibody. p. 177, 256

myesthenic syndrome Eaton-Lambert syndrome. A disease characterized by muscle weakness, it involves a decrease in the presynaptic release of acetylcholine at the neuromuscular junction. p. 177, 256

myocardium Muscle cells of the heart.

myofiber A heart muscle cell. p. 322

myofibrils The elements of a muscle that contain the thick and thin filaments organized into sacomeres. p. 286

myofilaments Thick (myosin) and thin (actin) filaments. p. 286

myogenicity The spontaneous activity of smooth muscle cells despite the absence of neural or hormonal input. p. 309

myoglobin An oxygen-carrying protein found in muscle. p. 302, 438

myopia Nearsightedness. p. 238

myosin One of the contractile proteins in muscle. p. 286, 312

myosin heavy chain The high-molecular-weight component of myosin that contains the actin binding site and ATPase activity. p. 286

myosin isozymes Variants of myosin characterized by differences in ATPase activity and thus crossbridge cycling rate; fast and slow myosin. p. 292

myosin light chain The lower-molecular-weight component of myosin, located in the crossbridges, which regulates myosin ATPase activity. p. 292

myosin light chain kinase An enzyme that phosphorylates myosin light chains. p. 292

myotatic reflex See stretch reflex.

myxedema Decreased metabolic rate and mental activity in adults caused by a decrease in thyroid hormone. p. 605

N

Na⁺ activation In nerve and muscle tissue, the voltage-gated opening of Na⁺ selective ion channels. p. 150

Na⁺ inactivation In nerve and muscle tissue, a term referring to the fact that voltage-gated Na⁺ channels spontaneously close after briefly opening. p. 150

Na⁺-K⁺ ATPase The sodium "pump." An enzyme that hydrolyzes ATP to transport Na⁺ against its concentration gradient. p. 125

natural killer cells In immunity, cells resembling cytotoxic lymphocytes that may have a role in resistance to cancer. p. 629

near point The minimum distance over which the eye can focus on an object. p. 238

negative feedback The principle that deviation of a variable from its desired value results in actions that correct the deviation. p. 86

neocortex The most recent region of the cortex not including the hippocampus and the limbic system. p. 198

neospinothalamic tract One of the major ascending sensory tracts. It relays fast pain and crude touch and is somatotopically organized. p. 212

nephron The basic functional unit of the kidney; in essence, it is a long tube. See **collecting duct, distal tubule, glomerulus, loop of Henle,** and **proximal tubule.** p. 478

Nernst equation An equation for the equilibrium potential of an ion. p. 130

net flux The difference between two unidirectional fluxes. For any process at equilibrium, the net flux must be zero. p. 121

neural code A sequence of action potentials. p. 170

neuroendocrine A term referring to the control of pituitary hormone release by the central nervous system. p. 96

neurohypophysis The posterior pituitary. p. 99

neuromodulator A substance that acts to alter the effect of a neurotransmitter. Neuromodulators usually activate some type of intracellular second messenger, which, in turn, modifies an ion channel. p. 188, 260

neuromuscular junction The synapse between a motor nerve and a muscle. p. 177, 256

neuron A nerve cell, including the cell body and all its processes. p. 95, 257

neuropeptides Peptides released from nerve fibers. p. 189

neurotransmitter A chemical released by a neuron that activates specific receptors on a second neuron, muscle cell, or gland. p. 68, 181

neutron The neutral particle in the nucleus of an atom that has approximately the same mass as a proton. p. 18

neutrophil A white blood cell that can ingest and destroy bacteria. p. 619

nicotinamide adenine dinucleotide (NAD) A coenzyme that carries H⁺ ions from one metabolic pathway to another. p. 75

nicotine An alkaloid from tobacco that activates nicotinic parasympathetic receptors. p. 277

nicotinic receptors A subclass of cholinergic receptors; They are present at the neuromuscular junction and in autonomic ganglia. p. 277

nociception Pain. p. 267

nodes of Ranvier The "gaps" along a myelinated nerve fiber where Na⁺ channels produce action potentials separated by an internode wrapped in myelin. p. 157

nonassociative learning A primitive form of learning present in invertebrates that does not require correlations of diverse stimuli. p. 213

noncompetitive inhibition A condition in which enzyme activity is decreased or receptors are blocked when molecules bind to sites other than those at which the normal substrates or ligands bind. p. 52

nonpolar A state in which the constituent atoms forming covalent bonds "share" their electrons equally so that there is no time-averaged local charge. p. 20

nonshivering thermogenesis An increase in heat production caused by increases in metabolic rate and muscle tone. p. 610

norepinephrine The neurotransmitter at sympathetic postganglionic terminals and at many central synapses. It is a catecholamine and one of the biogenic amines. p. 260

nuclear bag fibers Muscle spindle fibers with enlarged central regions. p. 263

nuclear chain fibers Uniform muscle spindle fibers. p. 263

nuclear pore The "hole" in the nuclear double membrane. p. 68

nucleic acid A polymer of nucleotides; DNA or RNA. p. 53

nucleolus A structure in the nucleus of some cells that contains the genetic material. p. 53

nucleotides Compounds formed from a sugar (ribose or deoxyribose), a phosphate group, and a purine or pyrimidine base. p. 53

nucleus The center of an atom; it contains protons and neutrons. It is also the intracellular organelle that contains most of a cell's DNA. p. 68

nystagmus Periodic back-and-forth movements of the eye, usually elicited by stimulation of the vestibular system. p. 235

O

occipital lobe The most posterior region of the cortex; it receives the primary visual input. p. 200

off pathway A way in which cells in the retina are organized to process visual signals. The term "off" is used because light decreases the activity of afferent fibers in the optic nerve. p. 244

ohm The unit of measure for electrical resistance. p. 33

Ohm's Law An equation that relates current flow to the applied voltage: $I = V/R$. p. 34

olfaction Smell. p. 250

olfactory lobe The region of the cortex that contains the neurons that respond to smell. p. 198

oligodendrocytes Glial cells in the central nervous system that produce myelin. p. 157

on pathway A way in which cells in the retina are organized to process visual signals. The term "on" is used because light increases the activity of afferent fibers in the optic nerve. p. 244

oocytes (oogonia) Stem cells in the female reproductive system. p. 651

open-loop system A system that is not controlled. p. 87

operating point In the cardiovascular system, the point at which the venous pressure and cardiac function curves intersect. This defines the cardiac output and central venous pressure. p. 384

operon A regulatory unit of DNA that determines what sequences of base pairs will be used as a template for messenger RNA synthesis. p. 56

opiates Pain-killing drugs related to morphine. See **endorphins.** p. 94

opsin A component of the visual photopigment. p. 241

opsonization The presentation of antigens by macrophages. p. 624

optic chiasm The point where the optic nerve partially divides. p. 247

optic disk The region of the retina where afferent fibers from ganglion cells leave the eye to form the optic nerve. p. 241

optic nerve A cranial nerve formed by axons from the retinal ganglion cells. p. 241

oral phase The phase of swallowing that is initiated voluntarily. p. 548

organ A functional unit of the body composed of one or more tissues. p. 5

organ of Corti The structure in the cochlea that contains auditory hair cells. p. 229

organ system A collection of organs with a common physiological function. p. 7

organelles Identifiable structures within a cell. p. 69

organic A term referring to compounds that contain carbon. p. 21

origin A site where a muscle tendon attaches to the skeleton. p. 286

osmolarity A measure of the number of dissolved particles in a solution. Osmolarity is inversely related to the concentration of water. p. 24

osmoreceptor A sensory receptor sensitive to osmolarity. p. 501

osmosis The process of water diffusion down its concentration gradient. p. 123

osmotically active (or inactive) A term referring to an impermeable substance that can or cannot produce a gradient for osmosis. p. 123

osmotic diuretics Substances that increase urine production because they are not reabsorbed by the renal tubules. p. 393

osmotic pressure The pressure between a solution containing a solute and water. p. 30, 123

ossicles The bones of the middle ear. p. 224

ossification Calcification of bone. p. 517

osteoblasts The cells that produce bone. p. 517

osteoclasts The cells that degrade bone. p. 517

otoliths Dense structures in the vestibular system that contain $CaCO_3$; they produce a shearing force of the hair cells in response to linear acceleration. p. 233

ouabain An herb that increases heart muscle contractility. p. 340

outer ear The external auditory canal, pinna, and eardrum. p. 224

outer segment The region of a photoreceptor that contains photopigments in an array of membranes. p. 240

oval window The point on the cochlea where the last of the middle ear bones transmits its vibrations. p. 224

ovary The female gonad; it contains follicles and secretes the hormones estrogen and progesterone. p. 645

ovulation The release of a mature ovum from its follicle. p. 651

oxidation The gaining of an electron. p. 52

oxidative phosphorylation The metabolic process in which H^+ ions carried by coenzymes are transferred to molecular oxygen while ATP is synthesized. p. 75

oxygen One of the respiratory gases essential for oxidative phosphorylation. p. 429

oxygen content The amount of oxygen carried by the blood. It depends on oxygen partial pressure and hemoglobin concentration. p. 434

oxygen debt A term referring to the period after exercise when accumulated lactic acid is metabolized by the liver. p. 303

oxyhemoglobin dissociation curve A graphic representation of the relationship between the partial pressure of oxygen and the relative saturation of hemoglobin with O_2. p. 434

oxyntic glands Glands in the stomach that secrete hydrochloric acid and intrinsic factor. p. 551

oxytocin A posterior pituitary hormone that causes uterine contractions and milk let-down. p. 99, 189, 696

P

P_{50} The partial pressure of oxygen needed for half-maximum saturation of hemoglobin. p. 433

P wave The component of the electrocardiogram that corresponds to the electrical activation of the atrium. p. 333

pacemaker A cell that spontaneously depolarizes and therefore can initiate activity in adjoining cells. Pacemakers are found in the heart and gastrointestinal tract. See **sinoatrial (S-A) node** and **atrioventricular (A-V) node.** p. 310, 544

pacemaker potential Spontaneous depolarization in a pacemaker. p. 310

pacinian corpuscle A cutaneous mechanoreceptor sensitive to vibration. p. 168

paleospinothalamic tract One of the ascending sensory pathways. It carries slow pain and temperature information but does not reach the cortex. p. 211

pancreatic islets See **islets of Langerhans.**

pancreatic juice A general term referring to the secretion of bicarbonate and enzymes by the pancreas. p. 554

pancreatic lipases Enzymes that degrade triglycerides to fatty acids. p. 554

pancreatolysis Self-digestion of the pancreas, caused by a lack of trypsin inhibitor. p. 555

papillary muscles In the heart, the muscles that restrain the A-V valves. p. 325

paraaminohippurate (PAH) An end product of aromatic amino acid metabolism that is totally secreted by the kidneys and thus is a measure of renal plasma flow. p. 497

paracellular route A term referring to the movement of water through the "tight junctions" that connect adjacent epithelial cells in the gastrointestinal tract, kidneys, and some glands. p. 74

paracrine agent (gland) A hormone that has only local effects, and the gland from which it is secreted. p. 93

parasympathetic nervous system One of the two principal divisions of the autonomic nervous system. Characterized by long preganglionic fibers and short postganglionic fibers. Uses ACh as a transmitter at both its synapses. p. 256

parathyroid glands Glands near the thyroid that secrete parathyroid hormone. p. 517

parathyroid hormone A hormone that affects calcium and phosphate balance by altering renal transport. p. 517

paravertebral chain The series of sympathetic ganglia that lie adjacent to, but outside of, the spinal cord. p. 258

parietal cells Cells in the stomach that secrete hydrochloric acid. p. 551

parietal lobe The region of the cortex located between the frontal and occipital lobes that contains the somatosensory areas. p. 198

parietal pleura The tissue lining the thoracic cavity. p. 402

Parkinson's disease A disease of the central nervous system characterized by tremor caused by a deficit in the transmitter dihydroxyphenylalanine (DOPA). p. 272

parotid gland One of the salivary glands. p. 547

partial pressure The pressure that one gas present in a mixture of gases would exert if it were present alone in the same volume. p. 429

passive transport Diffusion of a substance down its concentration gradient via a selective channel or carrier. See **facilitated diffusion.** p. 125

patch-clamp A technique for examining the movement of ions across small areas of a cell membrane. p. 312

pepsin The enzyme that breaks specific internal peptide bonds in a protein. p. 551

pepsinogen The inactive precursor of the enzyme pepsin. p. 551

peptides Short chains of amino acids joined by peptide bonds. p. 47, 260

peptide bond A chemical bond between amino acids that is formed by removal of water. p. 47

peptide hormones Hormones composed of peptides. p. 103

perforin A toxin secreted after an activated T cell binds to its target. p. 628

perfusion pressure Pressure actively generated by the heart. p. 362

perilymph Fluid in the scala media of the ear that resembles intracellular fluid. p. 227

perimysium The connective tissue surrounding a group of muscle fibers referred to as a muscle fascicle. p. 286

peripheral chemoreceptors The oxygen- and pH-sensitive sensory receptors in the cardiovascular system. See **aortic bodies** and **carotid bodies**. p. 454

peripheral nervous system The region of the nervous system that lies outside the brain and spinal cord. p. 195

peristalsis Waves of contraction in the gut that persist for some distance along its length; to be contrasted to the process of segmentation. p. 542

peristaltic rush Coordinated peristalis that occurs two or three times a day. p. 557

peritubular capillaries In the kidneys, the capillaries that surround the proximal and distal convoluted tubules. p. 480

permeability (coefficient) The ability of a substance to pass across an interface such as a cell membrane or epithelial cell. p. 118

permissive In endocrinology, a term referring to the fact that the presence of one hormone is necessary for another to be fully effective. p. 97

pernicious anemia A decrease in hemoglobin content caused by a dietary deficiency of vitamin B_{12}. p. 434

perspiration Sweating. p. 392, 464

perturbation A disturbance. p. 86

pH A measure of the concentration of H^+ ions: pH $= -\log [H^+]$. p. 26

pH trapping The accumulation of NH_4^+ in the renal tubules. p. 500

phagocytosis Endocytosis characterized by the intake of extracellular material, including cellular debris, foreign material, and bacteria. p. 66

pharyngeal phase The phase of swallowing that follows the voluntary oral phase. p. 548

pharynx The throat. p. 402, 535

phase (lag) A time measure of the relative position of sinusoidal signals such as sound waves. p. 327

phasic receptor A rapidly adapting sensory receptor in which the generator potential disappears even though the stimulus persists. p. 173

phosphodiesterase An enzyme that degrades cyclic GMP, often activated by a G-protein. p. 105

phosphoinositides Lipid compounds that often serve as internal second messengers. p. 105

phospholipid An amphipathic lipid molecule composed of two fatty acid chains and a polar group esterified on a glycerol "backbone." p. 40, 350, 554

phosphorylation The addition of a phosphate molecule. p. 52, 312

photon A light wave. p. 235

photopigment The light-sensitive component of the visual photoreceptors. p. 240

photoreceptors Rods and cones in the retina; they contain photopigments. p. 239

photopic vision Vision mediated by cone photoreceptors; color-sensitive vision. p. 241

phrenic nerve The motor nerve fiber that innervates the diaphram. p. 448

physiological shunt In the lungs, a condition in which some regions are not adequately ventilated compared to local blood flow. p. 422

piloerection "Goose bumps." p. 392

pinna The external portion of the ear. p. 224

pinocytosis The uptake of materials into a cell by means of a pinching off of small membrane-bound vesicles. p. 66

pituitary gland An endocrine gland located below the hypothalamus. The anterior portion secretes six hormones (ACTH, TSH, LH, FSH, prolactin, and GH) and is controlled by releasing factors secreted by the hypothalamus. The posterior portion is an extension of the hypothalamus and secretes ADH and oxytocin. p. 198

placenta The organ that develops during pregnancy from maternal and fetal tissue and serves as an interface between the maternal and fetal circulations. p. 682

plasma The fluid component of the blood. p. 348

plasma cell Cells derived from B lymphocytes that secrete antibodies. p. 619

plasma proteins Proteins in the fluid component of the blood. p. 348

plasma membrane See **membrane**.

plasticity In neurophysiology, a term referring to a change in synaptic efficacy. p. 180

plateau In the cardiac action potential, the component during which the membrane potential remains in the vicinity of 0 millivolts; also called phase 2. It results from an increase in Ca^{++} permeability. p. 327, 680

platelets Small cell fragments in the blood that contribute to the formation of blood clots (hemostasis). p. 348

pleated sheet One of the two secondary structures of proteins characterized by an association of adjacent amino acid chains. p. 48

plethysmograph A device for measuring lung volumes; it can measure the residual volume of the lung. p. 409

plicae circulares Visible folds in the lining of the gastrointestinal tract. p. 539

pneumotaxic center The region of the brainstem that can stimulate respiration. p. 451

pneumothorax A condition in which air enters the intrapleural space. p. 413

poise The unit of viscosity. p. 363

Poiseuille Law The equation that describes the flow resistance of a blood vessel in terms of its radius and the blood viscosity. p. 363

polar A molecule that has an unequal charge distibution, or dipole moment. p. 20

polycythemia An abnormally high concentration of red blood cells. p. 348

polymer A long chain made up of identical molecular subunits. p. 45

polypeptide A long chain of amino acids linked by peptide bonds. p. 48

polysaccharides A long chain of monosaccharides that may be unbranched (cellulose) or branched (glycogen and the starches). p. 45, 567

polysynaptic Having more than two synapses. p. 261

pons The region of the brainstem between the medulla and the midbrain. p. 197

portal circulation A term denoting vessels that link two organ systems. p. 100, 566

positive feedback The principle that deviation of a variable from its desired value results in further deviation. p. 90

postabsorptive state A condition in which substrates are not being absorbed from the gastrointestinal tract and must be provided from body stores. p. 590

posterior chamber One of the spaces in the eye outside the lens; it contains a watery fluid, the aqueous humor. p. 236

posterior pituitary The region of the pituitary gland that secretes ADH and oxytocin; it is an extension of the central nervous system. p. 99

postganglionic In the autonomic nervous system, a term referring to the neurons or nerve fibers that lead from the autonomic ganglia to target tissues. p. 257

postsynaptic inhibition In synaptic transmission, a condition in which a terminal on a cell body or dendrite releases a transmitter that increases the K^+ permeability, decreasing the excitability of the postsynaptic neuron. p. 182

postsynaptic potential The voltage change in a neuron produced by the release of a neurotransmitter. See **excitatory postsynoptic potential** and **inhibitory postsynoptic potential**. p. 181

posttranscriptional processing Modification of the messenger RNA produced from DNA templates by removal of bases corresponding to noncoding regions. p. 57

postural reflexes Reflexes that maintain balance. p. 268

potassium (K^+) The major intracellular positively charged ion. p. 132

potassium equilibrium potential (E_K) The voltage that would exist across a membrane if it were exclusively permeable to K^+. p. 130

potential Voltage. p. 182

potential energy The energy an object has because of its position; to be contrasted to kinetic energy and internal energy. p. 34

PR interval The period in the electrocardiogram between the P wave and the ORS complex. p. 333

precapillary sphincter A small group of smooth muscle cells that controls the entry of blood to a capillary. p. 355

precordial leads A set of electrodes placed on the chest to record an electrocardiogram. p. 334

preganglionic neurons In the autonomic nervous system, the neurons or nerve fibers that lead from the central nervous system to the autonomic ganglia. p. 257

preload In muscle, the amount of force that must be exerted on a resting muscle to stretch it to a given initial length. In the heart, the preload corresponds to the venous pressure and determines the end-diastolic volume. p. 299, 339

presbyopia A loss in the normal ability of the lens of the eye to focus on objects at various distances. p. 238

pressure Force per unit area. p. 429

pressure overload A condition such as aortic stenosis in which the ventricles are forced to generate abnormally high pressure to eject blood. p. 338

pressure-volume loop See work loop.

presynaptic facilitation An increase in transmitter release at a synapse. p. 184

presynaptic inhibition In synaptic transmission, a condition in which an axon makes contact with the terminal of another axon (an axoaxonic synapse) and releases a transmitter that decreases the quantal content. p. 183

presystolic murmur An abnormal heart sound that precedes ventricular activation; it is associated with mitral stenosis. p. 338

prevertebral ganglion A sympathetic ganglion adjacent to the aorta. p. 258

primary afferent Nerve fibers whose cell bodies are located in the dorsal root ganglia and that relay sensory information into the spinal cord. p. 195

primary structure The amino acid sequence of a protein. p. 48

primary transcript The immediate result of DNA transcription; it contains exons and introns. Primary messenger RNA is modified to final messenger RNA in the process of posttranscriptional processing. p. 57

priming In endocrinology, the stimulation by one hormone of synthesis of the receptor for another. p. 102

proenzyme An inactive enzyme precursor that, when cleaved by another enzyme, produces an active enzyme. p. 58

progesterone A female sex steroid; it is secreted by the corpus luteum (and placenta) and affects uterine development. p. 656

proinsulin The inactive precursor to the hormone insulin. p. 590

prolactin One of the anterior pituitary hormones; it stimulates milk secretion by the mammary glands. p. 699

proliferative phase In the menstrual cycle, the phase before ovulation during which the endometrium of the uterus is increasing in thickness but is not secretory. p. 670

promoter The region of a gene that acts as a switch to turn on transcription of a sequence of base pairs (an operon). p. 56

propagation The ability of an action potential to travel long distances along a nerve fiber. p. 140

proprioceptors Receptors in joints and muscles that provide information about limb position and movement and muscle length and tension. p. 262

propriospinal neurons Spinal interneurons whose processes extend over several spinal cord segments and thus coordinate groups of muscles. p. 273

propulsion Contractions of the stomach that move chyme toward the pylorus. p. 552

prostaglandins A group of modified fatty acids that act as chemical messengers. p. 92, 377, 631

prostate A gland in the male that secretes fluid into the urethra. p. 646

protanopia A condition in which red-sensitive photopigment is absent. Individuals with this condition cannot distinguish between red and green. p. 243

protease An enzyme that degrades proteins to smaller peptides. p. 554

protein A large molecule composed of one or more polypeptide chains. p. 46, 119

protein kinase A generic term for an enzyme that phosphorylates proteins. p. 53, 105

prothrombin A plasma protein that, when activated by one of the clotting factors, catalyzes the conversion of fibrinogen to fibrin. p. 351

proton The positively charged particle present in the nucleus of an atom The number of protons defines an element. p. 18

proximal muscle Muscle near the trunk that is primarily involved in maintaining postural stability. p. 273

proximal tubule The region of the nephron immediately after the glomerulus. p. 478

ptyalin A salivary enzyme that breaks down starch (amylase). p. 547

pulmonary circulation Blood flow through the lungs. p. 422

pulmonary edema Accumulation of fluid in the alveoli of the lungs. p. 370

pulmonary surfactant See surfactant.

pulse pressure The difference between arterial systolic and diastolic pressure. p. 338

punctate In sensory physiology, a term referring to the fact that primary afferent fibers innervate a limited area of the skin. p. 168

pupil The dark opening in the eye whose diameter is controlled by the iris. p. 235

pure tone The sound made by a single frequency. p. 224

Purkinje fibers The most distal component of the conducting system of the heart; it activates the interior of the ventricles. p. 325

putamen One of the basal ganglia of the brain. p. 198, 269

pylorus The physiological sphincter between the stomach and the duodenum. p. 535

pyrimidines Cytosine, thiamine, and uracil. Three of the nucleotide bases. p. 53

pyruvate The end product of glycolysis. p. 77

Q

QRS complex The component of the electrocardiogram that corresponds to the electrical activation of the ventricles. p. 333

quantal content The number of vesicles released from a synaptic terminal in response to a single presynaptic action potential. p. 180

quantum The amount of neurotransmitter contained in a single synaptic vesicle. p. 180

quaternary structure The subunit structure of proteins. p. 48

random coil One form of the secondary structure of proteins. p. 48

rapid ejection The phase of the cardiac cycle during which the heart is ejecting blood into the aorta and aortic pressure is increasing. p. 335

rapid eye movement (REM) sleep The phase of the sleep cycle characterized by desynchronized brain activity and dreaming; also called paradoxical sleep. p. 212

rapid filling The phase of the cardiac cycle during which the ventriclar volume is increasing but ventricular pressure is still decreasing. p. 335

rarefaction The phase of a sound wave during which pressure decreases. p. 224

readiness potential An electrical potential that can be recorded from the cortex before the onset of a movement. p. 271

receptive aphasia A condition in which an individual can form sentences gramatically but cannot comprehend. p. 215

receptive field The region of a sensory surface that, when stimulated, changes the response of a neuron or nerve fiber. p. 168, 204

receptive relaxation Relaxation of the stomach during the entry of food. p. 549

receptors Membrane-bound molecules with specific binding sites for molecules such as neurotransmitters or hormones. p. 86, 181

receptor potential See generator potential.

reciprocal innervation A term referring to the fact that activation of the motor neurons to one muscle usually inhibits motor neurons of antagonists at that joint. p. 261

recombinant DNA Foreign DNA that has been incorporated into a cell's own DNA. p. 60

recruitment The process by which additional motor units are activated during the contraction of a skeletal muscle. p. 256, 422

rectum The most distal region of the gastrointestinal tract. p. 537

red blood cells The cellular elements of the blood that contain hemoglobin. p. 348

red nucleus A collection of neurons in the brain that are involved in control of refined movements. p. 269

red-sensitive cones Photoreceptors with maximum sensitivity at 560 nanometers. p. 240

reduced ejection The phase of the cardiac cycle during which blood is being ejected from the heart into the aorta, but aortic pressure is decreasing. p. 335

reduced filling The phase of the cardiac cycle during which the ventricular volume and pressure are both increasing. It follows rapid filling. p. 335

reduction A reaction involving the loss of an electron. p. 52

reentry A condition in which electrical activity in the heart is altered so that different regions excite one another in a vicious cycle. p. 329

referred pain Visceral pain associated with the sensory nerve fibers leading from corresponding areas of the body surface. p. 209

reflex A negative feedback system in which a sensory stimulus leads to a motor response that diminishes the magnitude of the eliciting stimulus. p. 95

refractory period The interval after an action potential when a second action potential either cannot be produced (absolute refractory period) or has an increased threshold (relative refractory period). p. 152

Reissner's membrane In the cochlea, a layer of cells separating the scala vestibuli and the scala media. p. 224

relative refractory period A period of time after an action potential when a second action potential can be obtained, but the threshold is increased. p. 152

relative saturation The percent saturation of hemoglobin at a given partial pressure of oxygen. p. 434

relaxin A hormone used to relax the birth canal during labor. p. 687

releasing (or release-inhibiting) hormone A hypothalamically released substance that elicits (or inhibits) secretion of hormones by the anterior pituitary. p. 100

renal calyx The interior potion of the kidney; it empties into the ureter. p. 476

renal cortex The outermost layer of the kidney; it contains glomeruli and the proximal and distal tubules. p. 476

renal medulla A structure located below the cortex that contains the loops of Henle. p. 476

renal pyramids A structure located beneath the medulla that has numerous collecting ducts, giving it a striped appearance. p. 476

renin A hormone released by the juxtaglomerular cells of the kidneys that catalyzes production of angiotensin and secretion of aldosterone. p. 482, 513

repetitive firing The train of action potentials elicited by an applied stimulus. p. 152

replication The process by which DNA generates a copy of itself. p. 55

residual volume The volume of air in the lungs that cannot be expelled. p. 409

resolution phase The phase of the human sexual response after climax. p. 680

resonant frequency The frequency at which external forces applied to harmonic oscillators (objects such as pendulums or springs) cause a progressive increase in the amplitude of the resulting oscillations. p. 228

respiratory acidosis A decrease in plasma pH below 7.4 caused by the addition of carbonic acid; that is, by decreased alveolar ventilation. p. 520

respiratory alkalosis An increase in plasma pH above 7.4 caused by a deficit in carbonic acid; that is, by increased alveolar ventilation. p. 520

respiratory distress syndrome A condition in newborns that involves decreased pulmonary surfactant. p. 48

respiratory drive The combined input to the brainstem centers controlling respiration. p. 451

respiratory pump A term referring to the fact that venous return increases during inspiration as a result of a decrease in intrapleural pressure. p. 388

respiratory quotient The ratio of CO_2 produced to O_2 consumed. p. 431

resting potential The voltage difference across a cell that is not producing an action potential or being otherwise excited. p. 131, 141

resting tone See basal tone.

restriction endonucleases Enzymes that cleave DNA at specified base pair sequences; they are used in genetic engineering to restructure genes. p. 60

reticular formation A region of the brainstem consisting of a diffuse network of neurons. It is thought to be involved in monitoring the flow of information between the brain and spinal cord and in setting a level of cortical "awareness." p. 197

retina The interior surface of the eye; it contains photoreceptors and other visual system neurons. p. 239

retinal A modified form of vitamin A that serves as the immediate receptor for photons of light. p. 241

retrograde amnesia Loss of memory for preceding events. p. 213

retropulsion Contractions of the stomach that move chyme away from the pylorus. p. 552

reverse transcriptase An enzyme capable of synthesizing DNA from RNA. p. 60

reversible reaction A reaction that can proceed in either direction. p. 35

Rh factor Antigens on red blood cells that are different from A, B, and O antigens. If Rh antigens are present, an individual is termed Rh positive (Rh$^+$). p. 634

rhodopsin The rod photopigment (retinal plus opsin). p. 241

ribosomal RNA The RNA component of the ribosome. p. 53

ribosome A large cytoplasmic particle made up of proteins and RNA to which messenger RNA molecules attach during their translation into protein. p. 58

rickets Bone weakness and underdevelopment in children caused by Ca^{++} deficiency. p. 518

right axis deviation A condition in the heart in which certain waves of the electrocardiogram are increased because of an increase in the muscle mass of the right ventricle. p. 334

righting reflex A postural reflex that maintains the position of the head in space; it is mediated by stretch receptors in the neck. p. 269

rigor mortis A condition after death in which the depletion of ATP results in a permanent actin-myosin complex. p. 291

rod One of the photoreceptors; it is not color sensitive. p. 240

rough endoplasmic reticulum An intracellular membrane system characterized by the presence of ribosomes. p. 69

round window The opening in the cochlea that faces the middle ear but is not attached to the ossicles. p. 224

Ruffini endings Sensory receptors in the skin that are sensitive to touch. p. 208

S

saccule A receptor in the vestibular system that is sensitive to linear acceleration and head position. p. 233

safety factor The ratio of the current flowing along a nerve fiber to the minimum amount needed to sustain conduction. p. 159

saliva The fluid secreted by salivary glands. p. 547

salivary glands Glands in the mouth that secrete the enzyme amylase, some electrolytes, and mucus. p. 54

salt A crystalline compound made up of positively and negatively charged ions that fully dissociate in water. p. 21

saltatory conduction The conduction of nerve impulses in myelinated nerve fibers. p. 159

sarcolemma The plasma membrane surrounding a muscle cell. p. 286

sarcomere Subunits of myofibrils that contain the thick and thin filaments. p. 286

sarcoplasmic reticulum The smooth endoplasmic reticulum of a muscle fiber. It releases and resequesters Ca^{++} during contraction. p. 69, 293

satiety center The region of the brainstem where stimulation inhibits feeding. p. 602

saturated Having no double or triple bonds, as in saturated fatty acids. p. 40

saturation A condition in which increasing the concentration of a reactant (or transported substance) fails to increase the reaction (or transport) rate. p. 52

scala media The fluid-filled region of the cochlea that contains the organs of Corti. p. 227

scala tympani The fluid-filled region of the cochlea nearest the round window. p. 227

scala vestibuli The fluid-filled region of the cochlea nearest the oval window. p. 227

Schlemm canal A small duct connecting the anterior and posterior chambers of the eye; if it becomes plugged, glaucoma results. p. 236

Schwann cells The myelin-forming glial cells in the peripheral nervous system. p. 157

scientific notation A system in which numbers are expressed as the product of a number between 1 and 10 and a power of 10. For example, in scientific notation, $1,559,000 = 1\ 559 \times 10^6$ and $0.00034 = 3.4 \times 10^{-4}$. p. 173

sclera The outer layer of the eye; it appears to be white in color. p. 235

scopalamine A drug that blocks muscarinic cholinergic synapses. p. 277

scotopic vision Vision mediated by rods; not color sensitive. p. 241

scrotum The external pouch containing the male testes. p. 646

Second Law of Thermodynamics The principle that heat can flow only from higher to lower temperatures; also, a definition of entropy. p. 35

second heart sound The sound associated with the closing of the aortic and semilunar valves. p. 338

second messenger A molecule formed by the interaction of a ligand with a receptor that subsequently acts to modify intracellular metabolic processes. p. 103

secondary peristalsis Waves of contraction in the esophagus triggered by the presence of residual food. p. 535

secondary structure The α helix or random coil conformation of a polypeptide chain. p. 48

secretin A hormone secreted by duodenal cells that stimulates pancreatic bicarbonate secretion. p. 582

secretory phase In the menstrual cycle, the period between ovulation and menstruation during which the uterine lining develops a secretory characteristic. p. 671

segmental reflexes Reflexes mediated by interneurons in one segment of the spinal cord. p. 209

segmentation A series of more or less stationary contractions of the gut that act to mix its contents. p. 542

semen The combination of sperm and fluids secreted by the accessory glands of the male reproductive system. p. 646

semicircular canals The rotation-sensitive components of the vestibular system. p. 233

semilunar valves Valves between the right ventricle and the pulmonary artery. p. 324

seminal vesicles In the male, glands that secrete fluid into the vas deferens. p. 646

seminiferous tubules In the male, the sites of sperm production. p. 646

sensitization A primitive, nonassociative form of learning in which the repeated presentation of a stimulus enhances an avoidance response. p. 213

sensory receptor See receptors.

sensory tract A bundle of fibers in the spinal cord and brainstem that carry information from one of the special senses or from nonspecific sensory nerve endings. p. 194

series elastic element In muscle physiology, a term referring to the fact that muscles act as if an elastic structure must be stretched before the force generated by the crossbridges can be "felt" externally. p. 297

serotonin One of the central neurotransmitters. p. 182

Sertoli cells In the male, the cells that line the seminiferous tubules and "nurture" sperm maturation. p. 646

serum Plasma devoid of the plasma proteins that are used to form blood clots. p. 348

set point In control theory, the value at which a negative feedback system will regulate a controlled variable. p. 86

sex chromosome Chromosome 24; in males, Y, and females, X. p. 648

sex steroids Estrogen, progesterone, and testosterone. p. 114

shivering thermogenesis Increased heat production caused by consciously sensed contractions of the skeletal muscles. p. 609

shock A condition in which the blood pressure is abnormally low. It can be caused by loss of blood volume, a decrease in the heart's pumping ability, or loss of normal venous tone. p. 390

short-term memory Memory easily disrupted by trauma. p. 212

signal sequence A particular sequence of base pairs that serves to "direct" messenger RNA to ribosomes in the rough endoplasmic reticulum. p. 70

simple cell   In the visual cortex, a cell sensitive to lines and edges. p. 246

single-unit smooth muscle   A condition in which smooth muscle cells are electrically coupled to one another. p. 310

sinoatrial (S-A) node   The usual pacemaker of the heart; it is in the right atrium.

sinus arrest   A condition in which the S-A node ceases to generate periodic action potentials. p. 327

sinusoids   Large capillaries in the liver between sheets of hepatocytes. p. 567

size principle   The principle that in the activation of a muscle, the smaller motor units are recruited first. p. 256, 422

skeletal muscle   A muscle that attaches to the skeleton. p. 286

skeletal muscle pump   A term referring to the fact that muscle contraction combined with unidirectional valves in the veins helps propel blood to the heart. p. 388

slow muscle fibers   Fibers that contain a myosin with low ATPase activity. p. 302

slow Na$^+$/Ca^{++}channel   An ion channel in heart muscle that allows both Na$^+$ and Ca^{++} to pass through it and whose rate of activation is slower than the "fast" Na$^+$ channel of nerve and skeletal muscle fibers. p. 327

slow oxidative muscle fibers   One of the three classes of skeletal muscle characterized by abundant mitochondria, fatigue resistance, and myosin with low ATPase activity. p. 302

slow pain, second pain, or burning pain   The component of pain that is not easily localized and that persists for some time after removal of a noxious stimulus. p. 208

slow waves   Slow changes in the membrane potential of smooth muscle cells in the gut. Action potentials tend to occur at the peaks of these slow waves. p. 217, 544

smooth endoplasmic reticulum   An intracellular structure involved in the synthesis of membranes. See **rough endoplasmic reticulum**. p. 69

smooth muscle   See **unitary smooth muscle** and **multiunit smooth muscle**.

sodium (Na$^+$)   The main extracellular positively charged ion. p. 132

sodium equilibrium potential (E$_{Na}$)   The voltage that would exist across a cell membrane if it were solely permeable to Na$^+$ ions. p. 130

sodium pump   See **Na$^+$-K$^+$ ATPase**.

solubility   The amount (weight) of a solute that can be dissoved in a given amount of solvent. p. 22

solute   Molecules dissolved in a solution. p. 22

solvent   The liquid in which molecules are dissolved, usually water. p. 22

soma   The cell body of a neuron. p. 166

somatic motor system   The collection of neurons that activates the skeletal muscles. p. 197

somatic reflex   A reflex involving the skeletal muscles. p. 95

somatomedin   An Insulinlike hormone secreted by the liver in response to growth hormone. p. 602

somatosensory system   The collection of neurons that receives information from sensory receptors on the skin and from internal proprioceptors. p. 206

somatostatin   A hormone released by the pancreas that acts to prevent an overload of the gastrointestinal system. p. 189, 595

somatotopic organization   In sensory physiology, a term referring to the fact that afferent fibers enter the central nervous system in an orderly way, thus preserving a "map" of the sensory surface in ascending fiber tracts, relay nuclei, and the cortex. p. 209

somatotropin   Growth hormone. p. 97, 602

sound waves   Sinusoidal variations in the pressure in the air. p. 228

spatial summation   A process by which two stimuli close together in time produce a response greater than would have been seen if either were present alone. p. 171

spermatogenesis   The production of sperm. p. 648

spermatozoa   Sperm. The male gamete. p. 646

sphincter   A muscular ring that, when contracted, closes off a blood vessel or a portion of the gastrointestinal tract. p. 555

spike   An action potential. p. 140

spinal cord   The most primitive region of the central nervous system; it contains the "circuitry" for basic reflexes. p. 194

spinal nerve   One of the peripheral nerve trunks that enter the spinal cord; to be contrasted to cranial nerves. p. 197, 269

spinal reflexes   Reflexes that are "hard wired" into the spinal cord. p. 263

spindle   A length-sensitive stretch receptor in muscles. p. 261

spinothalamic tract   The set of sensory nerve fibers that carries kinesthetic information and fine touch. p. 212

spirometer   A device for measuring some lung volumes; it cannot be used to determine the residual volume. p. 408

splanchnic circulation   Blood flow through the abdominal viscera. p. 534

spleen   The largest lymph tissue; it is located between the stomach and diaphragm. p. 370

split brain   A condition in which the fibers normally connecting the two hemispheres of the brain have been severed or damaged. p. 216

S-T segment   The interval in the electrocardiogram that corresponds to the plateau phase of the ventricular action potential. p. 334

staircase phenomenon   In the heart, a term referring to the fact that at high heart rates, the force of contraction increases; also called treppe. p. 330

standing-gradient osmotic flow   A process in the kidney whereby Na$^+$ transport across the basolateral surface creates an osmotic gradient for water. p. 479

stapes   One of the bones of the middle ear. p. 225

starch   A glucose polymer characterized by a degree of branching intermediate between that of glycogen and cellulose. p. 567

Starling Law   See **Frank-Starling Law**.

static gamma motor neurons   Motor neurons innervating the peripheral ends of muscle spindles and thus selectively modifying the muscle's tonic properties. p. 263

static lung volumes   Steady-state lung volumes measured with a spirometer. p. 410

steady state   A condition in which some variables of a system are constant with time. p. 30

stenosis   Narrowing; it usually refers to partial occlusion of a blood vessel. p. 338

stereocilia   The cilia on the auditory and vestibular hair cells. p. 229

steroids   A class of lipids in which polar groups are attached to a skeleton of four carbon rings. p. 40

stimulus (stimulation)   A chemical or physical change in the environment. p. 140

stimulus-gated ion channel   A ion channel that is opened or closed by a physical stimulus such as light, mechanical movement, odors, tastes, and so forth. p. 43

stimulus-secretion coupling   In secretory cells, the process by which a stimulus elicits exocytosis. p. 175

stomach   The component of the gastrointestinal tract that receives food from the esophagus. p. 535

stress activation (or relaxation)   In smooth muscle, a process in which increased tension leads to contraction (or relaxation) of the smooth muscle cells. p. 544

stretch receptors   Sensory receptors sensitive to stretch. p. 261

stretch reflex   The only spinal reflex with a monosynaptic component In it stretching a muscle leads to its reflex contraction. p. 95, 261

striated muscle Muscle characterized by a pattern of light and dark bands that arises from the ordering of its contractile proteins. p. 292

stroke volume The amount of blood ejected from a ventricle in one heartbeat; the difference between end-diastolic and end-systolic volume. p. 335

sublingual gland One of the salivary glands. p. 547

submaxillary gland One of the salivary glands. p. 547

submucosa The connective tissue beneath the gut mucosa. p. 539

substance P One of the neuroactive peptides; it is the transmitter released from primary afferent pain fibers. p. 189

substantia gelatinosa A region of the dorsal horn of the spinal cord where the primary afferent pain fibers synapse. p. 208

substantia nigra A region of the brainstem that has deeply pigmented neurons. p. 212

substrate A molecule that binds to the active site of an enzyme. p. 50

substrate-level phosphorylation Formation of ATP from ADP by transferring a phosphate group from another higher-energy compound and without involving oxygen metabolism. p. 75

subthreshold stimuli Stimuli too low to produce an action potential. p. 143

sulci The grooves in the cortex. p. 198

summation See **spatial summation** and **temporal summation.** The term also refers to the additive contractile effect seen during repetitive stimulation of a skeletal muscle. p. 140

suppressor cell A subclass of T lymphocytes. p. 631

surfactant In the lungs, an amphipathic lipoprotein that reduces the alveolar surface tension and thus the intrapulmonary pressure. p. 418, 634

swallowing center The region of the medulla that coordinates the pharyngeal and esophageal phases of swallowing. p. 548

sweating The secretion of a hypotonic electrolyte solution by glands in the skin As sweat evaporates, it cools the body surface. p. 392

sympathetic nervous system A division of the autonomic nervous system. p. 195

symport Cotransport in which two substances move across the membrane in the same direction by means of a common carrier. p. 127

synapse A junction between neurons or between a nerve and a muscle. p. 175

synaptic cleft The small space between the presynaptic and postsynaptic elements at a synapse. p. 175

synaptic delay The time between release of transmitter from the presynaptic terminal and the beginning of the postsynaptic potential. p. 179

synaptic efficacy Refers to the amount of neurotransmitter released as a result of a single presynaptic action potential. p. 180

synaptic potential See **excitatory postsynaptic potential** and **inhibitory postsynaptic potential.** p. 140

synergist A muscle that moves a joint in the same direction as another muscle. p. 98, 261

systemic circulation The blood flow through all the organs except the lungs. p. 319

systole The period during which the ventricles are contracting. p. 326

systolic murmur A sound that occurs during rapid ejection in cases of backflow of blood into the left atrium caused by a leaky mitrial valve. p. 338

systolic pressure The maximum arterial pressure reached during systole. p. 338

T

T lymphocytes A subclass of lymphocytes. p. 619

T tubule See transverse tubules.

T wave The component of the electrocardiogram that corresponds to the repolarization of the ventricle. p. 333

tachycardia An abnormally rapid heart rate. p. 327

target cell In endocrinology, the cells that contain receptors for a particular hormone. p. 96

taste buds The sensory receptors for taste. p. 250

tectorial membrane The gelatinous structure on top of the auditory hair cells that exerts a shearing force when the basilar membrane moves up or down. p. 229

temporal lobe An area of the cortex; it contains the auditory areas and one of the language centers. p. 198

temporal summation The addition of stimuli close together in time to produce a response greater than that seen if either were present alone. p. 171

tendon The band composed of collagen fibers that attaches a muscle to the skeleton. p. 95, 286

tendon organs Stretch-sensitive receptors in the tendons that provide information concerning muscle tension. p. 265

teniae coli The longitudinal muscle in the large intestine. p. 557

tension Equivalent to force. p. 297, 417

tensor tympani A muscle that inserts on one of the middle ear bones and regulates the transmission of sound. p. 225

terminal cisternae In muscle, the lateral portions of the sarcoplasmic reticulum from which Ca^{++} is released. p. 293

tertiary structure The threedimensional conformation of a polypeptide chain. p. 48

testes The male gonads. p. 646

testosterone The main male sex hormone secreted by the testes. p. 646

tetanus Sustained contraction in a skeletal muscle at high frequencies. p. 154, 297

thecal cells A type of cell found in the ovaries. p. 651

thermodilution A technique that uses a pulse of heat to measure cardiac output. p. 382

thermoregulation The process by which body temperature is regulated. p. 392

theta waves Low-frequency components of the electroencephalogram seen in stage I sleep. p. 217

thick ascending limb The cortical portion of the Henle loop; it transports Cl^-. p. 479

thick filaments The contractile filament in muscle that contains myosin. p. 286

thin ascending limb The medullary, salt-permeable portion of the loop of Henle. p. 479

thin filaments The contractile filament in muscle that contains actin. p. 286

thoracic cavity The interior portion of the body that contains the heart and lungs. p. 402

thoracic duct The major vessel into which the lymphatic veins from most parts of the body empty and return the lymph to the heart. p. 372

threshold The voltage that a nerve or muscle cell needs to be depolarized to to produce an action potential. p. 142

thrombin An enzyme needed for conversion of fibrinogen to fibrin in clotting. p. 351

thrombocytes Platelets. p. 348

thromboplastin Factor III; a clotting factor released by damaged cells that accelerates formation of a blood clot. p. 351

thymine One of the pyrimidine bases in nucleotides. p. 53

thymus An organ of the lymphatic system and the site of T-lymphocyte production. p. 312, 627

thyroglobulin The glycoprotein in thyroid follicles that binds thyroid hormone. p. 605

thyroid gland A gland in the neck that secretes thyroid hormone. p. 99, 605

thyroid hormone Thyroxin (T_3); a hormone that affects organic metabolism. p. 605

tidal volume (TV) The amount of air entering and leaving the lungs during the respiratory cycle. p. 409

tight epithelia Epithelia in which the water permeability of the tight junctions coupling adjacent cells is low; it is found in regions where osmotic water flow is undesirable. p. 74

tight junction A junction between cells that prevents substances from leaking between them; it is seen in most epithelial tissues. p. 73

tissue A collection of cells The four basic classes are nerve, muscle, connective, and epithelial. p. 5

titratable acidity In the kidneys, the amount of H^+ contained in the urine that has been buffered by phosphate or has acted to decrease its pH. p. 500

tone The sound made by a single frequency. p. 277

tonic receptors Sensory receptors that respond to sustained stimuli with an equally sustained generator potential. p. 173

tonotopic A term referring to an orderly mapping of sound frequencies on neurons and/or fibers within the central nervous system. p. 230

torr (t) A unit of pressure. p. 28

total body water The total amount of water the body contains. p. 508

total lung capacity (TLC) Air in the lungs after a maximum inspiration. p. 410

total peripheral resistance (TPR) The equivalent flow resistance of all the systemic organs "seen" by the heart. p. 362

touch dome endings Specialized touch receptors in the skin. p. 208

trachea The main airway that connects the lungs and pharynx. p. 402

transamination The transfer of amide groups from one compound to another. p. 77

transcellular route A term referring to movement of substances across the membranes of epithelial cells rather than between adjacent cells (paracellular). p. 74

transcription The process by which a portion of a DNA molecule is "copied" to form messenger RNA. p. 53

transduction The process by which a sensory receptor converts a stimulus into a series of nerve impulses. p. 169

transfer RNA (tRNA) The species of RNA that binds an amino acid at one end while "reading" triplet nucleotide sequences on messenger RNA at the other end. Each of the 20 amino acids has a specific transfer RNA. p. 53

translation The process by which a messenger RNA molecule guides the synthesis of a protein. p. 58

transferrin An iron-carrying plasma protein. p. 574

transmitters Chemicals released by neurons that alter the membrane permeability of target cells through the activation of specific receptors. p. 68, 181

transmural pressure Pressure across the walls of an organ or blood vessel. p. 418

transport maximum (T_m) In the kidneys, the maximum rate of tubular reabsorption of some substance. p. 487

trans-retinal The form of retinal after activation by light that cannot bind opsin. p. 241

transverse tubules In muscle, the invaginations of the surface membrane that allow an action potential to "penetrate" the interior of a muscle fiber and eventually cause the release of Ca^{++}. p. 292

tremor Shaking seen in a resting individual. p. 272

treppe See staircase phenomenon.

triacylglycerol See triglyceride.

trichromat An individual with normal color vision; that is, having three types of color-sensitive cones. p. 243

tricuspid valve The valve between the right atrium and right ventricle. p. 322

tricyclic antidepressants A class of antidepressant drugs characterized by a chemical structure consisting of three heterocyclic rings. Imipramine is an example. p. 182

triglyceride Neutral fat composed of three fatty acid chains esterified on a glycerol backbone. p. 40, 571

tritanope An individual who lacks blue-sensitive conesand thus cannot distinguish between blue and green. p. 243

trophoblast The component of the blastocyst that develops into the placenta after implantation in the uterine wall. p. 682

tropomyosin The muscle protein that winds along the thin filament and covers the actin-binding sites in the absence of Ca^{++}. p. 290

troponin The muscle protein that binds actin (subunit I), tropomyosin (subunit T) and Ca^{++} (subunit C). When Ca^{++} is present, it "pulls" tropomyosin to uncover binding sites for myosin. p. 290

trypsin An enzyme that degrades proteins to peptides by cleaving internal bonds. p. 554

trypsin inhibitor A substance that inhibits trypsin activation in the pancreas. p. 555

turnover The rate at which elements of a cell are replaced. p. 10

twitch The contraction of a muscle after a single action potential. p. 292

two-solute model The combined role of active salt transport and differential urea permeability in creating the medullary osmotic gradient. p. 491

two-point discrimination The ability to perceive whether a single or two closely spaced stimuli are present on the skin. p. 211

tympanic membrane The eardrum. p. 224

type IA nerve fibers Sensory nerve fibers with diameters of 13 to 20 μm. p. 205

type IB nerve fibers Sensory nerve fibers with diameters of 6 to 12 μm. p. 205

type II nerve fibers Sensory nerve fibers with diameters of 1 to 5 μm, which carry fast pain and touch. p. 205

U

ulcer An erosion of the wall of the gut, usually in the stomach or duodenum. p. 550

ultrafiltration In the kidneys, the process by which the plasma is filtered free of protein into the glomerulus as a result of arterial blood pressure. p. 484

unidirectional flux The rate of movement of a substance across an interface in a particular direction. The difference between two unidirectional fluxes is the net flux. The unidirectional flux is proportional to concentration. p. 121

unitary smooth muscle Smooth muscle in which all the cells are electrically coupled via gap junctions. p. 310

upper esophageal sphincter Smooth muscle between the esophagus and the larynx. p. 548

unsaturated Having one or more double or triple bonds. p. 40

up-regulation An increase in receptor density caused by the presence of a hormone. p. 102, 601

upstroke The rising phase of an action potential. p. 140

uracil One of the pyrimidine bases in nucleotides. p. 75

urea The major waste product produced by the breakdown of proteins; it contains two ammonia molecules linked with carbon dioxide. p. 490

ureters The tubes that connect the kidneys with the bladder. p. 476

urethra The tube that connects the bladder with the exterior. p. 476, 646

uric acid A waste product produced by the breakdown of nucleic acids. p. 490

urine The final fluid excreted by the kidneys. p. 490

utricle A receptor in the vestibular system that is sensitive to linear accelerations and head position. p. 233

V

vagina The muscular canal connecting the female uterus with the exterior of the abdomen. p. 645

vagus nerve One of the cranial nerves. p. 197

valance A measure of an atom's ability to donate or accept electrons. Electron donors have positive valences. p. 18

van der Waals forces The attraction between molecules that occurs as a result of induced dipole-induced dipole interactions; a weak form of chemical bonding. p. 22

varicose veins A condition in which peripheral veins swell because their compliance has increased. p. 387

varicosities Regions in autonomic nerve fibers at which neurotransmitter is released. p. 177, 260

vas deferens The portion of the male reproductive system between the epididymus and the urethra. p. 646

vasa recta The capillary network in the kidneys that surrounds the Henle loops; part of the peritubular capillary network. p. 480

vasoconstriction Narrowing of a blood vessel. p. 376

vasomotion The random blood flow in a capillary bed, caused by the periodic opening and closing of various precapillary sphincters. p. 355

vasopressin Antidiuretic hormone (ADH). p. 99, 189, 380

vein A thin-walled blood vessel that returns blood to the heart. p. 321

vena cava The largest vein; it leads into the right atrium. p. 321

venodilation An increase in the diameter of the veins. p. 356

venous pooling The increase in blood volume in the veins of the lower extremities that occurs with prolonged standing. p. 386

venous return The return of blood to the heart, measured in liters per minute. p. 341

ventilation Movement of air into and out of the lungs, measured in liters per minute. p. 408

ventilation-perfusion ratio The ratio between the amount of air and blood exchanged in a certain region of the lung. p. 422

ventral horn The portion of the gray matter of the spinal cord that contains the motor neurons. p. 194

ventral root The portion of a spinal nerve that enters the ventral horn of the spinal cord and contains efferent motor fibers. p. 194

ventricles The two chambers of the heart that expel blood into the systemic or pulmonary circulations. p. 452

ventricular fibrillation A disorganized pattern of electrical activity in the ventricle that cannot cause a coordinated contraction. p. 324

ventricular septum The tissue separating the two ventricles. p. 320

venule A small vein. p. 356

very low density lipoprotein (VLDL) A lipoprotein manufactured in the liver that carries low-molecular-weight lipids to the cells After these are removed, VLDL becomes low-density lipoprotein. p. 594

vesicle A small, membrane-bound sphere. p. 66, 176

vestibular membrane Reissner membrane. p. 227

vestibular nucleus A collection of neurons in the brainstem that receives input from the vestibular hair cell receptors. p. 269

vestibulospinal tract Descending motor fibers that originate in the vestibular nucleus; the vestibulospinal tract is part of the medial motor system. p. 273

villi Large foldings of the intestinal mucosa that increase its surface area. p. 539

viscera Internal organs in the thorax and abdomen. p. 475

visceral pleura The tissue found on the exterior of the lungs. p. 402

viscosity A measure of the internal friction in a liquid. p. 363

visual cortex The area of the cortex that receives the primary visual projection; it is located at the rear of the brain. p. 248

visual purple Rhodopsin. p. 241

vital capacity (VC) The maximum amount of air that can be moved into and out of the lungs. Total lung capacity minus residual volume. p. 409

vitreous humor A thick fluid found in the portion of the eye between the lens and retina. p. 236

volume flow The amount of fluid in liters or milliliters per minute that passes a given point. p. 359

volume overload A condition in which the left ventricle must pump abnormally large amounts of blood because of a leaky mitral valve. p. 338

volume receptors Stretch receptors in the right atrium that measure central venous pressure. p. 380, 511

vols% A clinical unit of measure for gases; the number of milliliters of a gas contained in 100 milliliters of blood. p. 433

volt (V) The unit of measure for electrical potentials. p. 34

voltage The difference in electrical potential between two points. p. 34, 143

voltage-clamp A technique for controlling the voltage of a cell membrane and measuring the movement of ions across it. p. 147

voltage-gated ion channel An ion channel that opens or closes when the electrical field across the membrane changes. p. 125

voluntary nerve A somatic motor nerve. p. 194

W

water The fundamental solvent for all biochemical processes. A strongly polar molecule in which other polar molecules and ions are soluble. p. 22

weak acid An acid that incompletely dissociates in water. p. 520

Wernicke's area One of the language areas of the brain; it encodes and decodes concepts and commands. p. 214

white matter Areas of the central nervous system made up of nerve fibers rather than cell bodies. p. 194

white rami Connections between the sympathetic ganglia and the spinal cord. p. 258

withdrawal reflex A spinal reflex involving removal of a limb from a potentially harmful stimulus. p. 267

wolffian duct The part of the embryonic gonad that develops into the male reproductive ducts; it disappears in the female. p. 653

word deafness A condition in which comprehension is impaired only for verbal communication. p. 215

work Force times distance. p. 34

work loop In cardiovascular physiology, the relationship between ventricular volume and ventricular pressure. p. 338

Z

Z line In muscle, the structure into which the thin filaments insert in a regular hexagonal array. The Z lines define the limits of a sarcomere. p. 286

zona compacta The portion of the uterine lining nearest to the lumen. p. 670

zona pellucida One of the layers of a developing follicle. p. 663

zona spongiosa The portion of the uterine lining rich in blood vessels. p. 670

zonular fibers In the eye, fibers that act to change the curvature of the lens. p. 235

zygote The result of the combination of an egg and sperm. p. 648

Illustration Credits

Chapter 1

1-2: Cynthia Turner Alexander/Terry Cockerham, Synapse Media Production
1-3, 1-4: Barbara Stackhouse
1-6: Cardiovascular Unit, University of Alabama Medical Center

Chapter 2

2-5, C: Michael Godomski/Tom Stack & Associates
2-6: William Ober
2-20, B: Courtesy of O.N. Markand, M.D.
2-20, C: Rebecca S. Montgomery

Chapter 3

3-7, B: Brenda R. Eisenberg, Ph.D., University of Illinois at Chicago
3-12: Dr. Andrew P. Evan, Indiana University
3-13, 3-14, 3-19, 3-22: Ronald J. Ervin
3-23, 3-27: Andrew Grivas
3-24, 3-25: William Ober

Chapter 4

4-1, 4-6, 4-7, A, 4-9: Ronald J. Ervin
4-2, 4-12, B, 4-18: Brenda R. Eisenberg, Ph.D., University of Illinois at Chicago
4-3, 4-4, 4-5, 4-11, 4-16: Dr. Andrew P. Evan, Indiana University
4-7, B: Birgit H. Satir, Release of content of secretory vesicle via the fusion of the vesicle membrane and the cell membrane. The content expands and empties to the outside of the cell.
4-8 A, 4-10, 4-12, A, 4-14, A, 4-17, B: William Ober
4-8, B: T.W. Tillack
4-8, C, 4-10: Charles Flickinger
4-14, B: Ellen Chernoff
4-15: Michael P. Schenk
4-17, A: Carmillo Peracchia
4-17, C: Elliot Hertzberg

Chapter 5

5-1, 5-3, 5-4, 5-6, 5-9, A: Trent Stephens
5-7, 5-17: Barbara Stackhouse
5-9, B, 5-13, 5-19, 5-20, 5-21, 5-23: Andrew Grivas
5-11: Scott Bodell
5-16: Patricia Kane, Indiana University

Medical Center, Radiology Department
5-A: Cynthia Turner Alexander/Terry Cockerham, Synapse Media Production
5-B: William Ober

Chapter 6

6-1, 6-2, 6-10: William Ober
6-6, 6-7: Ronald J. Ervin
6-9: Barbara Stackhouse
6-17: Joan M. Beck

Chapter 7

7-1, A, 7-14, 7-15, 7-21, 7-23, 7-25: Barbara Stackhouse
7-1, B: Alan Peters
7-11: Roger T. Hanlon
7-22, A: Scott Bodell
7-22, B: Brenda Eisenberg
7-24: Dr. Andrew P. Evan, Indiana University

Chapter 8

8-1, A, 8-1, B, 8-24: Scott Bodell
8-3, 8-4, 8-6, A, 8-6, B, 8-8, 8-9, 8-10, 8-13, 8-16, 8-19, 8-26, 8-A: Barbara Stackhouse
8-14: Michael V.L. Bennett
8-15, 8-18: Brenda R. Eisenberg, Ph.D., University of Illinois at Chicago
8-17: Joan M. Beck/Andrew Grivas
8-20: M. Klymkowsky; In Stroud, R.M.: Acetylcholine Receptor Structure, Neuroscience Commentaries 1, 124-138, 1983. Reprinted by permission of the Society for Neuroscience

Chapter 9

9-3, A, 9-5, 9-8: William Ober
9-3, B: Brainslav Vidic
9-4, 9-16, 9-19: Michael P. Schenk
9-6, 9-7, 9-9: Scott Bodell
9-10, A, 9-12: Barbara Stackhouse
9-10, B: A.W. Campbell: Histological Studies on the Localization of Cerebral Function, Cambridge University Press, England, 1905
9-13: G. David Brown
9-15: Janis K. Atlee/Michael P. Schenk
9-18: Terry Cockerham, Synapse Media Production

9-A, 9-B: Patricia Kane, Indiana University Medical Center, Radiology Department

Chapter 10

10-2, 10-3, 10-6, 10-8, 10-11, A, 10-12, A, 10-15, 10-16, 10-22, A, 10-28, A, 10-28, B, 10-32, 10-33, 10-34: Marsha J. Dohrmann
10-4: Dr. Tilney
10-5, 10-9, 10-10, 10-14, 10-24, 10-29, 10-30: Barbara Stackhouse
10-11, B, 10-12, B: Kathy Mitchell Grey
10-13, 10-18, 10-19: G. David Brown
10-17, 10-20, B, C: Kenneth G. Julian, C.R.A., F.O.P.S.
10-20, A: William Ober
10-22, B: E.R. Lewis/F.S. Werblin/Y.Y. Feevi, University of California, Berkeley
10-23: Donna Odle
10-28, C: Patricia Kane, Indiana University Medical Center, Radiology Department
10-A: S. Ishihara: Tests for Colour-Blindness, Kanhara Shuppan Co., Ltd. Tokyo, Japan, 1973. Provided by the Washington University Department of Ophthalmology.

Chapter 11

11-2: Lennart Nilsson
11-3, 11-6, 11-9, 11-10, 11-15, 11-20, 11-23, 11-24, 11-25, 11-26: Barbara Stackhouse
11-4: John Martini
11-11, 11-12: Michael P. Schenk
11-16: William DeMyer, Professor of Child Neurology, Indiana University School of Medicine
11-18: A.W. Campbell: Histological Studies on the Localization of Cerebral Function, Cambridge University Press, England, 1905

Chapter 12

12-1, 12-16, 12-17, 12-18, 12-19, 12-26, 12-27, 12-30, 12-31: Barbara Stackhouse
12-2, 12-4, 12-5, 12-6, A, 12-7, 12-9, 12-11, 12-13: Joan M. Beck
12-3, 12-8, 12-12, 12-21, 12-24, 12-28: Brenda R. Eisenberg, Ph.D., University of Illinois at Chicago
12-23: Kathy Mitchell Grey

Chapter 13

13-1: Cynthia Turner Alexander/Terry Cockerham, Synapse Media Production
13-2, 13-5 (inset), 13-6, B, C, 13-13, 13-15, 13-17, 13-20: Barbara Stackhouse
13-3, 13-6, A, 13-18: Patricia Kane, Indiana University Medical Center, Radiology Department
13-4: David Mascaro & Associates
13-5, 13-19, 13-A: Rusty Jones
13-10: Dr. Andrew P. Evan, Indiana University
13-16, C: Rebecca S. Montgomery

Chapter 14

14-1: William Ober
14-2, 14-4, 14-8, 14-13, 14-15, 14-17, 14-18: Barbara Stackhouse
14-3, A, 14-5, 14-7, B: Dr. Andrew P. Evan, Indiana University
14-3, B, 14-7, A: Brenda R. Eisenberg, Ph.D., University of Illinois at Chicago
14-9, 14-14: Ronald J. Ervin
14-11: Trent Stephens
14-20, 14-B: Joan M. Beck/Donna Odle
14-21: G. David Brown

Chapter 15

15-1, 15-2, 15-3, 15-5, 15-10, 15-11, 15-12, 15-17, 15-A: Barbara Stackhouse
15-B, 1: Dr. Andrew P. Evan, Indiana University
15-B, 2: Rebecca S. Montgomery
15-C: Patricia Kane, Indiana University Medical Center, Radiology Department

Chapter 16

16-1, 16-3, A, 16-5, 16-6, 16-8, 16-10, 16-11, 16-12, A, 16-13, 16-14, 16-22, 16-23: Barbara Stackhouse
16-2: Jody L. Fulks
16-3, B: Glenn Lehman, M.D., Indiana University Medical Center
16-4, 16-20: Dr. Andrew P. Evans, Indiana University
16-7: Brenda R. Eisenberg, Ph.D., University of Illinois at Chicago
16-12, B: Patricia Kane, Indiana University Medical Center, Radiology Department

Chapter 17

17-2, 17-3, 17-4, A, 17-5, 17-11, 17-13: Barbara Stackhouse
17-4, B: Brenda R. Eisenberg, Ph.D., University of Illinois at Chicago

Chapter 18

18-1: Trent Stephens
18-2, 18-3, 18-5, 18-8, 18-14: Barbara Stackhouse

Chapter 19

19-1, A, 19-2, 19-3, A, 19-4, 19-5, A, 19-6, 19-7, 19-12, 19-13, 19-14, 19-17, 19-18, 19-19, 19-21, 19-22, 19-23, 19-24, 19-A, 19-B, 19-C: Barbara Stackhouse
19-1, B: Patricia Kane, Indiana University Medical Center, Radiology Department
19-3, B, 19-5, B, C, 19-8, 19-9, 19-10, 19-11: Dr. Andrew P. Evan, University of Indiana

Chapter 20

20-3, 20-4, 20-11, 20-13, 20-18: Barbara Stackhouse

Chapter 21

21-1, A: Cynthia Turner Alexander/Terry Cockerham, Synapse Media Production
21-1, B, 21-26, A, 21-27, B: Patricia Kane Indiana University Medical Center, Radiology Department
21-2, A: David J. Mascaro & Associates
21-2, B, 21-4, 21-11: Kathy Mitchell Grey
21-3, A, 21-18, 21-22, 21-27, A: G. David Brown
21-3, B, C, 21-21: Glenn Lehman, M.D., Indiana University Medical Center
21-5, A, 21-6, A, 21-9, 21-14, 21-16, 21-17, 21-19, 21-23, 21-25: Barbara Stackhouse
21-5, B, C, 21-6, B, 21-7, 21-8, 21-15: Courtesy of Ralph A. Jersild Jr.
21-10: Brenda R. Eisenberg, Ph.D., University of Illinois at Chicago
21-15, 21-24: Joan M. Beck/Donna Odle
21-26, B: Rebecca S. Montgomery
21-28: Joan M. Beck

Chapter 22

22-1, 22-1, 22-6, 22-11: Barbara Stackhouse
22-3: Karen Waldo
22-4: Kathy Mitchell Grey
22-12: Andrew Grivas

Chapter 23

23-1, A, 23-13, 23-18, B: Trent Stephens
23-1, B: Brenda R. Eisenberg, Ph.D., University of Illinois at Chicago
23-18, A, 23-20: Andrew Grivas

Chapter 24

24-1: William Ober
24-2: Brenda R. Eisenberg, Ph.D., University of Illinois at Chicago
24-3, 24-6, 24-7, 24-12, 24-21: Michael P. Schenk
24-4, 24-8, 24-10, 24-11, 24-13, 24-15, 24-18, 24-19, 24-B, 24-C: Barbara Stackhouse
24-14: Steven L. Goodman/Ralph M. Albrecht, University of Wisconsin, Madison

Chapter 25

25-1, A: Scott Bodell
25-1, B, 25-4, B, C, 25-8, B, C, 25-18, 25-21, 25-27: Carolyn B. Coulam/John A. McIntyre
25-2, A: David J. Mascaro & Associates
25-2, B, 25-3, 25-4, A, 25-8, A: William Ober
25-5: Branislav Vidic
25-6: Nyla Heerema
25-7, 25-9, 25-29: Barbara Stackhouse
25-11, A: Marcia Hartsock
25-17: Kevin Somerville/Kathy Mitchell Grey/Scott Bodell
25-25: Prudence Talbot: J. Exp. Zool. 225:144-148, 1983
25-26: Brenda R. Eisenberg, Ph.D., University of Illinois at Chicago
25-28: Lennart Nilsson
25-A: Rebecca S. Montgomery

Chapter 26

26-2: Lucinda L. Veeck, Jones Institute for Reproductive Medicine, Norfolk, Virginia
26-3, 26-4: Marcia Hartsock
26-5, 26-17, 26-18: Kevin Somerville
26-6, 26-8, 26-14, 26-19, 26-A, 1: Barbara Stackhouse
26-9, 26-A, 2: Carolyn B. Coulam/John A. McIntyre
26-10, 26-12: G. David Brown
26-13: Lennart Nilsson: A Child is Born, New York, Dell Publishing Co., 1966
26-16: Scott Bodell
26-A, 1: Rebecca S. Montgomery

Section Openers

I, II, III, IV, V, VI, VII: FPG Orion Press

Index

A

A band in sarcomere, 286
ABO blood group system, 634-635, 638
Absolute refractory period, 152
Absorption, 562-585
  amino acid, mechanisms of, 570-571
  carbohydrate, 567-568
  fat, 571-573
  iron, 573-576
  monosaccharide, 568
  protein, 570-571
  of solutes and fluids into blood and
    lymph, 564-567
  vitamin, 576-578
Absorptive state of metabolism
  characteristics of, 590-595
  insulin in, 590-593
  lipid transport and storage in, 593-595
  liver in, 593
  protein anabolism in, 595
  somatostatin in, 595
Acceleration, 28-29
  angular, detection of, 233, 235
  linear, detection of, 233
Acclimatization to altitude, 457-459
Accommodation of eye, 237-238
Acetylcholine
  membrane receptors for, 277, *278*
  as transmitter in motor pathways, 260
Achalasia, 550
Acid(s), 26-27
  bile, 555
  fixed, 459
  nucleic, 53-56
  secretion of, gastric, 550-552
  strength of, pKa as measure of, 528-529
  strong, 520
  weak, 520

Acid-base balance
  abnormalities of, causes of, 521*t*
  law of mass action and, 528
  sodium and potassium regulation and,
    interaction between, 516-517
Acid-base homeostasis, 519-526
Acid-base regulation, renal contributions
  to, 500-501
Acidosis
  acute, 459-460
  metabolic, 459-460, 521
    in diabetes mellitus, 601
    renal compensatory responses in,
      524-526
    with respiratory compensation,
      522-523
  respiratory, 520
    with renal compensation, 524
Acinar cells of pancreas, 553, 554
Acini, 406-408, 547
Acquired immunodeficiency syndrome
  (AIDS), *636-637*
Acromegaly, 604
Acrosomal reaction in fertilization,
  670
Acrosome of spermatid, 650
Action potential(s), 140, 142-143
  changes in permeability during,
    150-152
  compound, in nerve trunks, 159-160
  initiation of, in receptors, 171, *172*
  propagation of, 155-161; *see also*
    Propagation of action potentials
Activation energy of reaction, 50
Active immunization, 620
Active transport, 125
Activity, adaptive effects of, 47
Acuity, visual, 244-245
Adaptive effects of activity and training,
  471
Addiction, food, *582*
Adenine in deoxyribonucleic acids, 53
Adenosine diphosphate (ADP) in energy
  metabolism, 75

Adenosine in intrinsic control of vascular
  smooth muscle, 377
Adenosine monophosphate, cyclic, in
  peptide/protein hormone action,
  103
Adenosine triphosphate (ATP), 70
  in energy metabolism, 75
  production of, in oxidative
    phosphorylation, 80
Adenylate cyclase in peptide/protein
  hormone action, 103, 105
Adipose cells in absorptive state of
  metabolism, 593
Adrenal glands
  exercise and, 467
  hormones produced by, 113
Adrenal hormones in postabsorptive state
  of metabolism, 599-600
Adrenal medulla, 259
Adrenergic receptors, 277-278
Adrenergic synapses, transmitter synthesis
  in, 280
Adrenocorticotropic hormone (ACTH)
  source, target and action of, 111
  in starvation, 612
Adrenogenital syndrome, 659
Aerobic training, 471
Afferent arterioles, 480
Afferent fibers, 195
Afferent neurons as labelled lines, 169
Affinity for substrate, 52
Afterload of heart, 339
Afterloaded contractions, latent period in,
  299-301
Age, testicular function changes with,
  660-661
Agglutination, 624
Aging, body composition in, 604
Agonists, 277
  hormones as, 98
Airway(s)
  constriction and dilation of, dual
    autonomic inputs and, 277
  pressure, flow, and resistance in,
    415-417

Bulk fluid losses, 507
Bundle branches, 325
Bundle of His, 325
Bursa cells, 622
Bursa of Fabricius, 622

C

C-terminal in dipeptide, 48
Calcitonin
  in calcium homeostasis, 517
  source, target and action of, 112
Calcium
  excitation-contraction coupling and,
    292-297
  and phosphate homeostasis, 517-519
  total body, endocrine regulation of,
    517-519
Calcium channels, control of
    neurotransmitter release by, 179
Calcium entry, learning and memory and,
    188
Calcium homeostasis, disorders of,
    153-154
Calcium ions, extracellular, excitability
    and, 153-154
Caldesmon in smooth muscle contractility
    control, 312-313
Calmodulin as second messenger, 105
Calsequestrin, 294
Calyes, renal, 476
Canal
  auditory, 224
  of Schlemm, 236
Cancer
  gene transformation and, 628-629
  incidence of, dietary fat and, 43
Capacitance elements, veins as, 358
Capacitation of spermatozoa, 650-651
Capacity(ies)
  definition of, 408-409
  forced vital, 410
  functional residual, 409, 410t
  inspiratory, 409, 410t
  lung, 408-410
  total lung, 410
  vital, 409, 410t
Capillary(ies), 355-356
  discontinuous, 359
  fenestrated, 359
  and interstitial spaces, fluid exchange
    between, 368-370
  lymphatic, 370
  pulmonary, 321
  structure of, 359
  systemic, 321
    carbon dioxide loading in, 440-441
    oxygen unloading in, 436
  and tissues, transfer of fluids and solutes
    between, 368-372
  walls of, exchange of nutrients and
    wastes across, 368
Capillary bed(s), 356
  peritubular, 480
Capillary filtration and reabsorption in
    lung, 422

Capillary lipase in fat uptake, 594
Carbohydrate(s), 43-45
  dietary sources of, 567-568
  digestion and absorption of, 567-568
Carbohydrate loading for exercise, 465
Carbon compounds, 20-21
Carbon dioxide
  effects of, on oxygen unloading,
    436-437
  partial pressure of, arterial, breathing
    and, 451-455
  transport of, blood buffers and, 439-443
Carbon dioxide loading in systemic
    capillaries, 440-441
Carbon dioxide unloading in lung, 443
Carbon dioxide-carbonic acid-bicarbonate
    equilibrium, 439-440
Carbon monoxide poisoning, physiology
    of, 438
Carbonic anhydrase
  carbonic acid formation and, 439
  as catalyst, 50
Carboxy terminal of amino acid, 570
Carboxypeptidases in protein digestion,
    570
Cardiac catheterization, 334
Cardiac cycle, phases of, 335, 336, 337t
Cardiac function curve, 341, 342
Cardiac muscle, 5
  lack of summation in, 307-309
  physiology of, 306-309
  structure of, 306-307
Cardiac muscle cells, functional
    specializations of, 324-325
Cardiac output
  determinants of, 340-342
  effect of, on central venous pressure,
    381-383
  measurement of, 382-383
  vascular determinants of, 384-385
Cardiac reserve, 384-385
Cardiac rhythm
  cellular basis of, 326-328
  irregularities of, 328-329
Cardiac sphincter in swallowing, 549
Cardiopulmonary resuscitation, 421
Cardiovascular centers of pons, 276
Cardiovascular limit, performance in
    sustained exercise and, 390-392
Cardiovascular operating point, 383-384
Cardiovascular reflexes, 376-381
Cardiovascular system, 317-397
  components of, 6t, 376
  in disease, 386-393
  exercise and, 386-393, 468-469
  functions of, 6t
  heart in, 318-344; see also Heart
  location of, 7
  regulation of, 374-397
Carotid chemoreceptors, 454-456
Carotid sinus, 378-379
Carrier-mediated transport, 126-127
Carriers, transport proteins as, 119
Catabolism, 75, 590
Catalysts, 50, 51
Catalytic subunit in peptide/protein
    hormone action, 105
Cataracts, 235-236

Catecholamines as neurohormones, 260
Catheterization, cardiac, 334
Caudate nucleus, 269
Cell(s)
  amacrine, 239
    of retina, 246
  bipolar, 239
  cardiac muscle, functional
    specializations of, 324-325
  chemistry of, 38-61; see also Chemistry
    of cells
  chromaffin, 259
  complex, in detection of shape and
    movement, 246
  ganglion, 239
    pathways from photoreceptors to,
      244
  glial, 5
  horizontal, 239
    of retina, 245-246
  hypercomplex, in detection of shape
    and movement, 246, 248
  junctions between, 72-74
  membranes of, 64-66; see also
    Membrane(s), cell
  muscle, connective tissue boundaries of,
    286
  nucleus of, 68-69
  replacement of, tissue growth and,
    55-56
  resting potential of, determinants of,
    132-133
  sickled, 48, 50
  simple, in detection of shape and
    movement, 246
  single, to organ systems, 4-10
  stem, 5
  structure of, 62-72
  target, in hormonal information
    transfer, 96-97
  volume of, determinants of, 123-125
Cell body(ies), 166
  and axons, segregation of, 194-195
Cell-mediated immunity, T-lymphocytes
    and, 627
Cellular elements of immune system, 619
Cellulose, 567-568
Central chemoreceptors, breathing and,
    451-453
Central motor program for respiratory
    cycle, 448-450
Central nervous system, 195
  motor programs in, 270
  regions of, 197-198
  rhythm of breathing generated by,
    450-451
Central neurons, processing of synpatic
    inputs in, 187
Central olfactory pathways, 252
Central venous pressure, cardiac output
    and, 381-383
Central visual pathways, 246, 247
Cephalic phase of digestion, 578-579
Cerebellum, 198
  coordination of rapid movements and,
    269-270
Cerebral cortex, 198, 200
  organization of, 200-201

Pupil, 235
Pure tone, 224
Purine base, 53
Purkinje fibers, 325
Purple, visual, 241
Putamen, 269
Pylorus, 536
Pyramidal tract, 273
Pyramids, renal, 476
Pyrimidine base, 53
Pyrogens
  endogenous, 610
  exogenous, 610
Pyruvate
  in glycolysis, 77
  and lactate, interconversion of, 77

Q

QRS complex on ECG, 333
Quantal content, 179-180
Quantal transmission, 177

R

Raphe nuclei, 212
Rapid eye movement (REM) sleep, 212,
    217-219
Rapid movements, coordination of,
    cerebellum and, 269
Rarefaction of air in sound waves, 224
Readiness potentials, 271, 273
Receptive aphasia, 215
Receptive field(s)
  concept of, 168
  ganglion cell, 244-245
  sensory, 204
  somatosensory, dermatome organization
      of, 205-207
Receptive relaxation of stomach, 580
  in swallowing, 549
Receptor(s)
  antigen-specific, of T lymphocytes, 627
  auditory, hair cells as, 229-230
  cholinergic, 277
  cold, 208
  cytoplasmic, for steroid hormones, 103
  fixed, in protein/peptide hormone
      action, 103, 104
  hair follicle, 207-208
  in hormonal information transfer, 96-97
  hormone, regulation of, 101-102
  initiation of action potentials in, 171,
      172
  irritant, in airway, 450
  linear, 172-173
  logarithmic, 172-173
  membrane-bound, in protein/peptide
      hormone action, 103, 104
  mixed, 174
  modified epithelial cells as, 174, 175
  muscarinic, 277
  in negative feedback, 86
  neurotransmitter-sensitive, 181

Receptor(s)—cont'd
  nicotinic, 277
  nuclear, in steroid hormone action,
      103
  olfactory, 250-252
  pain, 208-209
  sensory
    physiological properties of, 169-170
    as transducers, 169-170
  temperature, 208
  touch, 207-208
  vestibular, 231-233
  warm, 208
Receptor potentials, 140, 142, 171
  production of, 170-171
Receptor proteins, 119
  stimulus-sensitive, 170-171
Receptor-mediated endocytosis, 66
Reciprocal inhibition, 261-262
Recognition proteins, 119
Recruitment, 422
  definition of, 297
Rectum, 537
Red blood cells, 348
Red nucleus of midbrain, 198
Reduction, 52
Redundancy, 57
Referred pain, 209
Reflex(es)
  autonomic, 95, 96
  baroreceptor, 378-380
    standing upright and, 386-389
  cardiovascular, 376-381
  endocrine, 95, 96-106
  enteric, 539
  enterogastric, 581-582
  gastrocolic, 580
  gastroileal, 580
  of GI tract, 578t
  Hering-Breuer, 450
  knee jerk, 261
  myotatic, 95, 261
  neural, 95
  righting, 269
  spinal
    autonomic, 276
    postural stability and, 261-263
  stretch, 95, 261
  tendon, 95
  withdrawal, crossed extension and,
      267-268
Reflex arcs, 94
Reflux, gastric, 550
Refractory period
  absolute, 152
  relative, 152
Regulation
  down, of insulin receptors in type II
      diabetes, 601
  extrinsic, 90-106; see also Extrinsic
      regulation
  of food intake, 601-602
  hormonal, 99
  of hormone receptors, 101-102
  intracellular, 90-92
  intrinsic, 90
  local, 90, 92-94
  by prostaglandins, 93

Regulation—cont'd
  up, of insulin receptors in type II
      diabetes, 601
Regulators, 52
Regulatory subunit, calmodulin as, 105
Relative refractory period, 152
Relaxation
  receptive, of stomach, 580
  stress, of intestinal contractility, 544,
      545
Relaxin
  in pregnancy, 687
  source, target and action of, 114
Releasing hormones, 100
Release-inhibiting hormones, 100
REM sleep, 212, 217-219
Renal blood flow (RBF), 497
Renal calyces, 476
Renal clearance, 495-497
Renal compensation, respiratory acidosis
      with, 524
Renal corpuscle, 478
Renal handling, 495
Renal pelvis, 476
Renal plasma flow (RPF), 497
Renal pyramids, 476
Renal tubules, 478
Renin, 513
  in extracellular fluid homeostasis
      regulation, 482
Renin-angiotensin-aldosterone, control of
      sodium loss and vascular tone by,
      513-514
Renin-angiotensin-aldosterone system in
      blood pressure regulation, 381
Repetitive firing during prolonged
      depolarization, 152-153
Replication of DNA, 55
Repolarization, membrane, 144-146
Representation, somatotopic, 209
Reproduction, genetic basis of, 648-653
Reproductive cycle, female, 661
Reproductive function
  female, endocrine regulation of, 661
  male, endocrine regulation of, 660-661
Reproductive hormones, 114
Reproductive organs, embryonic
      development of, 653
Reproductive system
  components and functions of, 6t
  female, 645-646
  location of, 9
  male, 646-647
Residual volume, 409-410
Resistance, airway, 403, 417
Resistance vessels, 367
Resolution phase of sexual response, 680
Resonance
  principle of, 228
  sympathetic, 28
Respiratory acidosis, 520
  with renal compensation, 524
Respiratory alkalosis, 520
Respiratory bronchioles, 403
Respiratory centers, 451
Respiratory compensation
  metabolic acidosis with, 522-523
  metabolic alkalosis with, 523

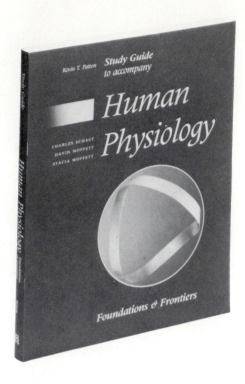

STUDY GUIDE to Accompany Human Physiology: Foundations and Frontiers By Kevin Patton

This valuable STUDY GUIDE can help you be more successful in your physiology course by:
• Making your study time more effective.
• Improving your science comprehension.
• Enhancing your recall of important information.
• Improving your test scores.

Features include:
• Practice tests for each chapter, many written by the author of *Human Physiology: Foundations and Frontiers*, reflect the actual exams you will face in the classroom. Answers for all questions are provided.
• A *Student Survival* guide shows you how to study effectively without spending more time.
• Features to help you master key terms include grouped word lists, pronunciation keys, definitions, and learning word parts.
• Specific learning objectives focus on the important information in each chapter.
• Over 70 illustrations, such as drawings, flow charts, graphs, and other aids, help you visualize and understand important principles.
• A variety of exercises help you learn through graphing, problem-solving, fill-in paragraphs, tables, and more.

To purchase a copy ask your bookstore manager or call toll-free (800) 633-6699. In Missouri, call (800) 325-4177.